HANDBOOK OF APPLIED ANALYSIS

For other titles published in this series, go to
www.springer.com/series/5613

Advances in Mechanics and Mathematics

VOLUME 19

Aims and Scope

Mechanics and mathematics have been complementary partners since Newton's time, and the history of science shows much evidence of the beneficial influence of these disciplines on each other. The discipline of mechanics, for this series, includes relevant physical and biological phenomena such as: electromagnetic, thermal, quantum effects, biomechanics, nanomechanics, multiscale modeling, dynamical systems, optimization and control, and computational methods.

Driven by increasingly elaborate modern technological applications, the symbiotic relationship between mathematics and mechanics is continually growing. The increasingly large number of specialist journals has generated a complementarity gap between the partners, and this gap continues to widen. *Advances in Mechanics and Mathematics* is a series dedicated to the publication of the latest developments in the interaction between mechanics and mathematics and intends to bridge the gap by providing interdisciplinary publications in the form of monographs, graduate texts, edited volumes, and a special annual book consisting of invited survey articles.

HANDBOOK OF APPLIED ANALYSIS

By

NIKOLAOS S. PAPAGEORGIOU
National Technical University, Athens, Greece

SOPHIA TH. KYRITSI-YIALLOUROU
Hellenic Naval Academy, Piraeus, Greece

Springer

N.S. Papageorgiou
National Technical University
Department of Mathematics
157 80 Athens
Zografou Campus
Greece
npapg@math.ntua.gr

S. Th. Kyritsi-Yiallourou
Hellenic Naval Academy
Military Institute of University Education
Leoforos Chatzikyriakou
185 39 Piraeus
Greece
skyrits@math.ntua.gr

Series Editors:

David Y. Gao
Department of Mathematics
Virginia Polytechnic Institute
and State University
Blacksburg, VA 24061
gao@vt.edu

Ray W. Ogden
Department of Mathematics
University of Glasgow
Glasgow, Scotland, UK
rwo@maths.gla.ac.uk

ISBN 978-0-387-78906-4 e-ISBN 978-0-387-78907-1
DOI 10.1007/b120946
Springer Dordrecht Heidelberg London New York

Library of Congress Control Number: 2009927202

Mathematics Subject Classification (2000): 34xx, 35xx, 46xx, 47xx, 49xx, 90xx, 91xx

Printed on acid-free paper

Springer is part of Springer Science+Business Media (www.springer.com)

To the memory of my father

S. Th. K.

The mathematical sciences particularly exhibit order, symmetry and limitation; and these are the greatest forms of the beautiful.

Aristotle
Metaphysica, 3M, 1078b

Series Preface

As any human activity needs goals, mathematical research needs problems.
—David Hilbert

Mechanics is the paradise of mathematical sciences.
—Leonardo da Vinci

Mechanics and mathematics have been complementary partners since Newton's time, and the history of science shows much evidence of the beneficial influence of these disciplines on each other. Driven by increasingly elaborate modern technological applications, the symbiotic relationship between mathematics and mechanics is continually growing. However, the increasingly large number of specialist journals has generated a duality gap between the partners, and this gap is growing wider.

Advances in Mechanics and Mathematics (AMMA) is intended to bridge the gap by providing multidisciplinary publications that fall into the two following complementary categories:

1. An annual book dedicated to the latest developments in mechanics and mathematics;
2. Monographs, advanced textbooks, handbooks, edited volumes, and selected conference proceedings.

The AMMA annual book publishes invited and contributed comprehensive research and survey articles within the broad area of modern mechanics and applied mathematics. The discipline of mechanics, for this series, includes relevant physical and biological phenomena such as: electromagnetic, thermal, and quantum effects, biomechanics, nanomechanics, multiscale modeling, dynamical systems, optimization and control, and computation methods. Especially encouraged are articles on mathematical and computational models and methods based on mechanics and their interactions with other fields. All contributions will be reviewed so as to guarantee the highest possible scientific standards. Each chapter will reflect the most recent achievements in the area. The coverage should be conceptual, concentrating on the methodological thinking that will allow the nonspecialist reader to understand it. Discussion of possible future research directions in the area is welcome. Thus, the annual volumes will provide a continuous documentation of the most recent developments in these active and important interdisciplinary fields. Chapters published in this series could form bases from which possible AMMA monographs or advanced textbooks could be developed.

Volumes published in the second category contain review/research contributions covering various aspects of the topic. Together these will provide an overview of the state-of-the-art in the respective field, extending from an introduction to the subject right up to the frontiers of contemporary research. Certain multidisciplinary topics, such as duality, complementarity, and symmetry in mechanics, mathematics, and physics are of particular interest.

The *Advances in Mechanics and Mathematics* series is directed to all scientists and mathematicians, including advanced students (at the doctoral and postdoctoral levels) at universities and in industry who are interested in mechanics and applied mathematics.

David Y. Gao
Ray W. Ogden

Preface

The aim of this book is to present the basic modern aspects of nonlinear analysis and then to illustrate their use in different applied problems.

Nonlinear analysis was born from the need to deal with nonlinear equations which arise in various problems of science, engineering, and economics and which are often notoriously difficult to solve. On a theoretical level, nonlinear analysis is a remarkable mixture of various areas of mathematics such as topology, measure theory, functional analysis, nonsmooth analysis, and multivalued analysis. On an applied level, nonlinear analysis provides the necessary tools to formulate and study realistic and accurate models describing various phenomena in different areas of physical sciences, engineering and economics. For this reason, the theoretically inclined nonmathematician (physicist, engineer, or economist), needs to have a working knowledge of at least some of the basic aspects of nonlinear analysis. This knowlegde can help him build good models for the phenomena he studies, study them in detail and extract from them important information which is crucial to the design process. As a consequence, nonlinear analysis has acquired an interdisciplinary character and is a prerequisite for many nonmathematicians, who wish to investigate their problems in detail with the greatest possible generality. This leads to a continuously increasing need for books that survey this large area of mathematical analysis and present its applications.

There should be no misunderstanding. The subject is vast, it touches many different areas of mathematics and its applications cover several other fields in science and engineering. In this volume, we make an effort to present the basic theoretical aspects and the main applications of nonlinear analysis. Of course the treatment is not exhaustive; such a project would require several volumes. Nevertheless, we believe that we touch the main parts of the theory and of the applications. Mathematicians and nonmathematicians alike can find in this volume material that covers their interests and can be useful in their research and/or teaching.

Chapter 1 begins with the calculus of smooth and nonsmooth functions. We present the Gâteaux and Fréchet derivatives and develop their calculus in full detail. In the direction of nonsmooth functions, first we deal with convex functions for which we develop a duality theory and a theory of subdifferentiation. Subsequently, we generalize to locally Lipschitz functions (Clarke's theory). We also introduce and study related geometrical concepts (such as tangent and normal cones) for various kinds of sets. Finally we investigate a kind of variational convergence of functions, known as Γ-convergence, which is suitable in the stability (sensitivity) analysis of variational problems.

In Chapter 2, we use the tools of the previous chapter in order to study extremal and optimal control problems. We begin with a detailed study of the notion of lower semicontinuity of functions. Next we examine constrained minimization problems and develop the method of Lagrange multipliers. This leads to mimimax theorems, saddle points and the theory of KKM-multimaps. Section 2.4 deals with some modern aspects of the direct method, which involve the so-called variational principles, central among them being the so-called "Ekeland variational principle". The last two sections deal with the calculus of variations and optimal control. In optimal control, we focus on existence theorems, relaxation and the necessary conditions for optimality (Pontryagin's maximum principle).

Chapter 3 deals with some important families of nonlinear maps and examines their uses in fixed point theory. We start with compact and Fredholm operators which are the natural generalizations of finite rank maps. Subsequently we pass to operators of monotone type and to accretive operators. Monotone operators exhibit remarkable surjectivity properties which play a central role in the existence theory of nonlinear boundary value problems. Accretive operators are closely related to the generation theory of linear and nonlinear semigroups. We investigate this connection. Then we introduce the Brouwer degree (finite-dimensional) and the Leray–Schauder and Browder–Skrypnik degrees (the latter for operators of monotone type) which are infinite-dimensional. Having these degree maps, we can move to the fixed point theory. We deal with metric fixed points, topological fixed points and investigate the interplay between order and fixed point theory.

Chapter 4 presents the main aspects of critical point theory which is the basic tool in the so-called "variational method" in the study of nonlinear boundary value problems. We start with minimax theorems describing the critical values of a C^1-functional. Then we present the Ljusternik–Schnirelmann theory for multiple critical points of nonlinear homogeneous maps. This way we have all the necessary tools to develop the spectral properties of the Laplacian and of the p-Laplacian (under Dirichlet, Neumann and periodic boundary conditions). Then using the Lagrange multipliers method we deal with abstract eigenvalue problems. Finally we present some basic notions and results from bifurcation theory.

Chapter 5 uses the tools developed in Chapters 3 and 4 in order to study nonlinear boundary value problems (involving ordinary differential equations and elliptic partial differential equations). First we illustrate the variational method based on the minimax principles of critical point theory and then we present the method of upper and lower solutions and the degree-theoretic method. Subsequently we consider nonlinear eigenvalue problems, for which we produce constant sign and nodal (sign changing) solutions. Then we prove maximum and comparison principles involving the Laplacian and p-Laplacian differential operators. Finally we deal with periodic Hamiltonian systems. We consider the problem of prescribed minimal period and the problem of a prescribed energy level. For both we prove existence theorems.

In Chapter 6, we deal with the properties of maps which have as values sets (multifunctions or set-valued maps). We introduce and study their continuity (Section 6.1) and measurability (Section 6.2) properties. Then for such multifunctions (continuous or measurable), we investigate whether they admit continuous or measurable selectors (Michael's theorem and Kuratowski–Ryll Nardzewski, and the Yankov–von Neuman–Aumann selection theorems). This leads to the study of the sets of integrable selectors of a multifunction, which in turn permits a detailed set-valued integration. The notion of decomposability (an effective substitute of convexity) plays a central role in this direction. Then we prove fixed point theorems for multifunctions and also study Carathéodory multifunctions. Finally we introduce and study various notions of convergence of sets that arise naturally in applications.

In Chapter 7, we consider applications to problems of mathematical economics. We consider both static and dynamic models. We start with the static model of an exchange economy. Assuming that perfect competition prevails, which is modelled by a continuum (nonatomic measure space) of agents. We prove a "core Walras equivalence theorem" and we also establish the existence of Walras allocations. We then turn our attention to growth models (dynamic models). First we deal with an infinite horizon, discrete-time, multisector growth model and we establish the existence of optimal programs for both discounted and undiscounted models. For the latter, we use the notion of "weak maximality". Then we determine the asymptotic properties of optimal programs via weak and strong turnpike theorems. We then examine uncertain growth models and optimal programs for both nonstationary discounted and stationary undiscounted models. We also characterize them using a price system. Continuous-time discounted models are then considered, and finally we characterize choice behavior consistent with the "Expected Utility Hypothesis".

Chapter 8 deals with deterministic and stochastic games, which provide a substantial amount of generalization of some of the notions considered in the previous chapter. We start with noncooperative n-players games, for which we introduce the notion of "Nash equilibrium". We show the existence of such equilibria. Then we consider cooperative n-players games, for which we define the notion of "core" and show its nonemptiness. We continue with random games with a continuum of players and an infinite-dimensional strategy space. For such games, we prove the existence of "Cournot–Nash equilibria". We also study corresponding Bayesian games. Subsequently, using the formalism of dynamic programming, we consider stochastic, 2-player, zero-sum games. Finally, using approximate subdifferentials for convex function, we produce approximate Nash equilibria for noncooperative games with noncompact strategy sets.

Chapter 9 studies how information can be incorporated as a variable in various decision models (in particular in ones with asymmetric information structure). First we present the mathematical framework, which will allow the analytical treatment of the notion of information. For this purpose, we define two comparable metric topologies, which we study in detail. Then we examine the ex-post view and the ex-ante view, in the modelling of systems with uncertainty. In both cases we establish the continuity of the model in the information variable. Subsequently, we introduce a third mode of convergence of information and study prediction sequences. We also study games with incomplete information and games with a general state space and an unbounded cost function.

The final chapter (Chapter 10) deals with evolution equations and the mathematical tools associated with them. These tools are developed in the first section and central among them is the notion of "evolution triple". We consider semilinear evolutions, which we study using the semigroup method. We then move on to nonlinear evolutions. We consider evolutions driven by subdifferential operators (this class of problems incorporates variational inequalities) and problems with operators of monotone type, defined within the framework of an evolution triple. The first class is treated using nonlinear semigroup theory, while the second requires Galerkin approximations. We conclude with an analogous study of second-order evolutions.

The treatment of all subjects is rigorous and every chapter ends with an extensive survey of the literature.

We hope that both mathematicians and nonmathematicians alike, will find some interesting and useful for their needs in this volume.

Acknowledgments: We would like to thank Professor D. Y. Gao for recommending this book for publication in the *Advances in Mechanics and Mathematics* (AMMA) series. Many thanks are also due to Elizabeth Loew and Jessica Belanger for their patience and professional assistance.

Athens, September 2008

Nikolaos S. Papageorgiou
Sophia Th. Kyritsi–Yiallourou

Contents

1

Smooth and Nonsmooth Calculus

Summary. *In this chapter we provide an overview of the main aspects of the smooth and nonsmooth calculus. We start with the smooth calculus and we introduce the notions of Gâteaux and Fréchet derivatives. We develop all the important calculus rules and we present the implicit and inverse function theorems together with their most important consequences. Then we pass to nonsmooth functions, starting with convex functions. For such functions, we develop the duality theory using the notion of conjugate functions and then we present the subdifferential theory. Subsequently, we extend the subdifferential theory to locally Lipschitz functions (Clarke's theory). Using these theories, we also examine the local geometry of convex and nonconvex sets. Finally we present and study in detail a notion of convergence of functions distinct from the pointwise convergence, known as the Γ-convergence, which is designed to deal with the stability analysis of variational problems.

Introduction

In this chapter we present an overview of the basic aspects of the smooth and nonsmooth calculus. This chapter provides all the basic tools to approach the theoretical and applied subjects in the chapters that follow.

In Section 1.1 we present the basic items from the classical smooth calculus of functions defined in a Banach space. So we introduce the notions of Gâteaux and Fréchet derivatives and then study their properties. We develop mean value theorems and chain rules, we consider partial derivatives for functions defined on product spaces and we conclude with two basic theorems that are important in analysis and differential geometry, namely the implicit function theorem and the inverse function theorem.

In Section 1.2 we deal with convex functions defined in a Banach space. Convex functions provide the link that connects the smooth calculus with the nonsmooth one. Continuous convex functions have remarkable differentiability properties on certain Banach spaces and in $\mathbb{R}^N$ possess a derivative under conditions that are weaker than the usual ones. For lower semicontinuous convex functions that are not differentiable, we can introduce a multivalued substitute of the derivative, known as the (convex) subdifferential of the function. In Section 1.2 we study the basic

N.S. Papageorgiou, S.Th. Kyritsi-Yiallourou, *Handbook of Applied Analysis*,
Advances in Mechanics and Mathematics 19, DOI 10.1007/b120946_1,

properties of the subdifferential and simultaneously we develop the duality theory for convex functions which is closely related to the subdifferential theory.

In Section 1.3 we extend the notion of subdifferential to nonconvex functions. This is achieved with the use of locally Lipschitz functions. For such functions we define a subdifferential operator, known as the generalized subdifferential, and develop the basic rules of the corresponding calculus.

Section 1.4 deals with the local geometry of nonsmooth convex and nonconvex sets. First we deal with convex sets that are not smooth and introduce for them substitutes of the well-known tangent and normal planes of differential geometry. These are the tangent and normal cones. Then with the help of the generalized subdifferential of the distance function (which is Lipschitz continuous) we extend the notions of tangency and normalcy to nonsmooth nonconvex sets.

Finally in Section 1.5, we introduce a convergence of functions, distinct from the pointwise convergence, which is designed in such a way that makes it the right tool in the stability (sensitivity) analysis of variational problems. This convergence is known as the Γ-convergence and in Section 1.5 we present the basic results concerning it. We also introduce the so-called Mosco convergence of functions, which is more suitable when dealing with problems defined in a reflexive Banach space.

1.1 Gâteaux and Fréchet Derivatives

Let X, Y be Banach spaces and $f\colon X \longrightarrow Y$. In what follows by $\mathcal{L}(X,Y)$ we denote the Banach space of bounded linear operators from X into Y. First we introduce a weak form of differentiability.

DEFINITION 1.1.1 We say that f is *Gâteaux differentiable* at $x \in X$, if there exists $L \in \mathcal{L}(X,Y)$ such that

$$\lim_{\lambda \to 0} \frac{f(x+\lambda h) - f(x)}{\lambda} = Lh \qquad \text{for all } h \in X.$$

The operator $L \in \mathcal{L}(X,Y)$ is called the Gâteaux derivative of f at $x \in X$ and is denoted by $f'(x)$. We say that f is Gâteaux differentiable if it is Gâteaux differentiable at every $x \in X$.

REMARK 1.1.2 The Gâteaux derivative is a directional derivative and so it is essentially a one-dimensional concept. Indeed, if for fixed $x, h \in X$ we introduce $\varphi(\lambda) = f(x+\lambda h)$, $\lambda \in \mathbb{R}$, then $(\mathrm{d}/\mathrm{d}\lambda)\varphi(\lambda)\big|_{\lambda=0} = f'(x)h$. Evidently if the Gâteaux derivative exists at x, then it is unique.

EXAMPLE 1.1.3 (a) If $A \in \mathcal{L}(X,Y)$, then $A'(x) = A$ for all $x \in X$.
(b) If $f = (f_k)_{k=1}^m \colon \mathbb{R}^n \longrightarrow \mathbb{R}^m$, then $f'(x) = \Big((\partial f_i/\partial x_j)(x)\Big)_{i=1,j=1}^{m,n} \in \mathbb{R}^{m\times n}$ (the Jacobian matrix).

We can prove a mean value theorem for Gâteaux differentiable functionals $\varphi\colon X \longrightarrow \mathbb{R}$

PROPOSITION 1.1.4 *If $f: X \longrightarrow \mathbb{R}$ is Gâteaux differentiable and $x, h \in X$, then there exists $0 < t < 1$ such that*

$$f(x+h) - f(x) = f'(x+th)h.$$

PROOF: Let $\varphi(\lambda) = f(x + \lambda h), \lambda \in \mathbb{R}$. Then $(\mathrm{d}/\mathrm{d}\lambda)\varphi(\lambda) = f'(x + \lambda h)h$. So invoking the mean value theorem for real functions, we can find $0 < t < 1$ such that $\varphi(1) - \varphi(0) = (\mathrm{d}/\mathrm{d}\lambda)\varphi(\lambda)\big|_{\lambda=t}$, hence $f(x+h) - f(x) = f'(x+th)h$. □

If Y is not equal to $\mathbb{R}$, then the mean value theorem is not valid.

EXAMPLE 1.1.5 Consider the function $f: \mathbb{R} \longrightarrow \mathbb{R}^2$ defined by $f(t) = (\cos t, \sin t)$.

In this case the mean value theorem takes an inequality form. Namely we have the following.

PROPOSITION 1.1.6 *If $f: X \longrightarrow Y$ is Gâteaux differentiable, $x, h \in X$ and $y^* \in Y^*$, then there exists $0 < t < 1$ such that*

$$\langle y^*, f(x+h) - f(x) \rangle_Y = \langle y^*, f'(x+th)h \rangle_Y$$
$$\text{and} \quad \|f(x+h) - f(x)\|_Y \leq \|f'(x+th)\|_{\mathcal{L}} \|h\|_X.$$

PROOF: Consider the function $\varphi: X \longrightarrow \mathbb{R}$ defined by $\varphi(x) = \langle y^*, f(x) \rangle_Y$. Then $\varphi'(x) = \langle y^*, f'(x)h \rangle_Y$. By virtue of Proposition 1.1.4, we can find $0 < t < 1$ such that

$$\varphi(x+h) - \varphi(x) = \langle y^*, f'(x+th)h \rangle_Y.$$

Because $y^* \in Y^*$ was arbitrary, we can choose $y^* \in Y^*$ with $\|y^*\|_{Y^*} = 1$ such that

$$|\langle y^*, f(x+h) - f(x) \rangle_Y| = \|f(x+h) - f(x)\|_Y.$$

So from the first part of the proof, we have

$$\begin{aligned} \|f(x+h) - f(x)\|_Y &= |\langle y^*, f'(x+th)h \rangle_Y| \\ &\leq \|f'(x+th)\|_{\mathcal{L}} \|h\|_X \quad (\text{recall that } \|y^*\|_{Y^*} = 1). \end{aligned}$$

□

Now we introduce the second stronger form of differentiability.

DEFINITION 1.1.7 We say that $f: X \longrightarrow Y$ is *Fréchet differentiable* at $x \in X$, if there exists $L \in \mathcal{L}(X, Y)$ such that

$$f(x+h) - f(x) = Lh + r(x, h)$$

where $(\|r(x,h)\|_Y)/\|h\|_X \longrightarrow 0$ as $\|h\|_X \longrightarrow 0$. The operator $L \in \mathcal{L}(X, Y)$ is called the *Fréchet derivative* of f at x and it is denoted by $f'(x)$. It is clear from the context which derivative we mean when we write $f'(x)$. We say that f is Fréchet differentiable if it is Fréchet differentiable at every $x \in X$.

REMARK 1.1.8 The Fréchet derivative when it exists, is unique. It is clear from the above definition that Fréchet differentiability implies Gâteaux differentiability. The converse is not in general true.

EXAMPLE 1.1.9 Consider the function $f\colon \mathbb{R}^2 \longrightarrow \mathbb{R}$ defined by

$$f(x_1,x_2) = \begin{cases} \frac{x_1^6}{x_1^8+(x_2-x_1^2)^2} & \text{if } (x_1,x_2) \neq (0,0) \\ 0 & \text{if } (x_1,x_2) = (0,0) \end{cases} \qquad \text{for all } x = \begin{pmatrix} x_1 \\ x_2 \end{pmatrix} \in \mathbb{R}^2.$$

Then it is easy to see that f is Gâteaux differentiable at the origin with zero Gâteaux derivative. On the other hand moving along the curve (u,u^2), we see that

$$\frac{|f(u,u^2)|}{\|\left(\begin{smallmatrix} u \\ u^2 \end{smallmatrix}\right)\|_{\mathbb{R}^2}} = \frac{1}{u^2}$$

and so f is not Fréchet differentiable at $x=0$.

The next proposition gives a situation where the two notions coincide.

PROPOSITION 1.1.10 *If $f\colon X \longrightarrow Y$ is Gâteaux differentiable on some open set $U \subseteq X$, $x \in U$ and the map $u \longrightarrow f'(u)$ is continuous at $x \in X$ from X into $\mathcal{L}(X,Y)$ furnished with the operator norm topology, then f is also Fréchet differentiable at $x \in X$ and the two derivatives coincide.*

PROOF: Let $r(x,h) = f(x+h) - f(x) - f'(x)h$ and $y^* \in Y^*$. Using Proposition 1.1.6 we can find $0 < t < 1$ such that

$$\langle y^*, r(x,h)\rangle_Y = \langle y^*, f'(x+th)h - f'(x)h\rangle .$$

As before, we choose $y^* \in Y^*$ with $\|y^*\|=1$ such that

$$\begin{aligned} & \|r(x,h)\|_Y = |\langle y^*, r(x,h)\rangle_Y|, \\ \Rightarrow\ & \|r(x,h)\|_Y \leq \|f'(x+th) - f'(x)\|_{\mathcal{L}}\|h\|_X, \\ \Rightarrow\ & \frac{\|r(x,h)\|_Y}{\|h\|_X} \leq \|f'(x+th) - f'(x)\|_{\mathcal{L}} \longrightarrow 0 \qquad \text{as} \quad \|h\|_X \longrightarrow 0. \end{aligned}$$

This proves the Fréchet differentiability of f at x and that the two derivatives coincide. □

Existence of the Fréchet derivative at $x \in X$ implies continuity of f at $x \in X$.

PROPOSITION 1.1.11 *If f is Fréchet differentiable at $x \in X$, then f is continuous at $x \in X$.*

PROOF: From Definition 1.1.7, we see that there is $\delta > 0$ such that

$$\|f(x+h) - f(x) - f'(x)h\|_Y \leq \|h\|_X, \quad \text{if} \quad \|h\|_X \leq \delta.$$

Therefore we have

$$\|f(x+h) - f(x)\|_Y \leq (1 + \|f'(x)\|_{\mathcal{L}})\|h\|_X, \quad \text{if} \quad \|h\|_X \leq \delta,$$

from which we conclude the continuity of f at $x \in X$. □

REMARK 1.1.12 The result is not true if Fréchet differentiability is replaced by Gâteaux differentiability. The function in Example 1.1.9 is Gâteaux differentiable at the origin but it is not continuous there.

EXAMPLE 1.1.13 Let $X = C([0,b])$ (equipped with the supremum norm) and $g \in C^1(\mathbb{R})$. Consider the map $f(u) = \int_0^b g\big(u(t)\big)dt$ for all $u \in C([0,b])$. Then $f : X \longrightarrow \mathbb{R}$ is Fréchet differentiable and

$$f'(u)h = \int_0^b g'\big(u(t)\big)h(t)dt \qquad \text{for all } h \in C([0,b]).$$

We have the following chain rule for derivatives.

PROPOSITION 1.1.14 *If X, Y, Z are Banach spaces, $f : X \longrightarrow Y$ is Gâteaux differentiable at x and $g : Y \longrightarrow Z$ is Fréchet differentiable at $f(x)$, then $\xi = g \circ f : X \longrightarrow Z$ is Gâteaux differentiable at x and*

$$\xi'(x) = g'\big(f(x)\big)f'(x).$$

Moreover, if f is Fréchet differentiable at $x \in X$, then ξ is Fréchet differentiable at $x \in X$.

PROOF: If $\lambda \neq 0$, then we have

$$\begin{aligned} &\frac{1}{|\lambda|}\|\xi(x+\lambda h) - \xi(x) - \lambda g'\big(f(x)\big)f'(x)h\|_Z \\ \leq &\frac{1}{|\lambda|}\|g\big(f(x+\lambda h)\big) - g\big(f(x)\big) - g'\big(f(x)\big)\big(f(x+\lambda h) - f(x)\big)\|_Z + \\ &+\frac{1}{|\lambda|}\|g'\big(f(x)\big)\big(f(x+\lambda h) - f(x) - \lambda f'(x)h\big)\|_Z. \end{aligned} \tag{1.1}$$

Due to the Gâteaux differentiability of f at x, we have

$$\frac{1}{|\lambda|}\|g'\big(f(x)\big)\big(f(x+\lambda h) - f(x) - \lambda f'(x)h\big)\|_Z \longrightarrow 0 \tag{1.2}$$

$$\text{and} \quad \|f(x+\lambda h) - f(x)\|_Y \longrightarrow 0 \quad \text{as } \lambda \longrightarrow 0.$$

Then for any $\lambda \in \mathbb{R}$ such that $f(x+\lambda h) \neq f(x)$, due to the Fréchet differentiability of g at $f(x)$, we have

$$\begin{aligned} &\frac{1}{|\lambda|}\|g\big(f(x+\lambda h)\big) - g\big(f(x)\big) - g'\big(f(x)\big)\big(f(x+\lambda h) - f(x)\big)\|_Z \\ = &\left(\frac{\|g\big(f(x+\lambda h)\big) - g\big(f(x)\big) - g'\big(f(x)\big)\big(f(x+\lambda h) - f(x)\big)\|_Z}{\|f(x+\lambda h) - f(x)\|_Y}\right) \\ &\times\left(\frac{\|f(x+\lambda h) - f(x)\|_Y}{|\lambda|}\right) \longrightarrow 0 \quad \text{as } \lambda \longrightarrow 0. \end{aligned} \tag{1.3}$$

Returning to (1.1), passing to the limit as $\lambda \longrightarrow 0$, and using (1.2) and (1.3), we conclude that $g \circ f$ is Gâteaux differentiable at $x \in X$.

In a similar fashion we can show that $g \circ f$ is Fréchet differentiable at $x \in X$, if g is Fréchet differentiable at $f(x)$. $\square$

DEFINITION 1.1.15 We say that $f: X \longrightarrow Y$ is *continuously Fréchet differentiable* at $x \in X$, if f is Fréchet differentiable at $x \in X$ and $u \longrightarrow f'(u)$ is continuous at $x \in X$ from X into $\mathcal{L}(X, Y)$ furnished with the operator norm topology. If this is true at every $x \in X$, we simply say that f is continuously Fréchet differentiable and write $f \in C^1(X, Y)$ (or simply $f \in C^1(X)$ if $Y = \mathbb{R}$).

Now suppose $Y = Y_1 \times Y_2$ with Y_1, Y_2 Banach spaces. It is well-known that Y can be given many different equivalent norms which make it a Banach space (e.g., $\|(y_1, y_2)\|_1 = \|y_1\|_{Y_1} + \|y_2\|_{Y_2}$ or $\|(y_1, y_2)\|_\infty = \max\{\|y_1\|_{Y_1}, \|y_2\|_{Y_2}\}$). The space $\mathcal{L}(X, Y_1 \times Y_2)$ is isometrically isomorphic to the product space $\mathcal{L}(X, Y_1) \times \mathcal{L}(X, Y_2)$. Let $p_1: Y \longrightarrow Y_1$ and $p_2: Y \longrightarrow Y_2$ be the corresponding canonical projection operators. We know that $p_1 \in \mathcal{L}(X, Y_1)$ and $p_2 \in \mathcal{L}(X, Y_2)$. If $f: X \longrightarrow Y = Y_1 \times Y_2$, then $f_1 = p_1 \circ f$ and $f_2 = p_2 \circ f$ are the components of f; that is, $f = (f_1, f_2)$. If $i_1: Y_1 \longrightarrow Y$ and $i_2: Y_2 \longrightarrow Y$ are the canonical injections, then

$$f = i_1 \circ f_1 + i_2 \circ f_2.$$

Recall that $i_1 \in \mathcal{L}(Y_1, Y)$ and $i_2 \in \mathcal{L}(Y_2, Y)$.

PROPOSITION 1.1.16 *A function $f: X \longrightarrow Y = Y_1 \times Y_2$ is Fréchet differentiable at $x \in X$ if and only if f_1 and f_2 are Fréchet differentiable at $x \in X$. We also have*

$$f'(x) = \big(f_1'(x), f_2'(x)\big) \quad \textit{with } f_1'(x) = p_1 \circ f'(x),\ f_2'(x) = p_2 \circ f'(x).$$

PROOF: Suppose f is Fréchet differentiable at x. Then from Proposition 1.1.14, we have that f_1 and f_2 are both Fréchet differentiable at $x \in X$, $f_1'(x) = p_1 \circ f'(x)$, $f_2'(x) = p_2 \circ f'(x)$, and $f'(x) = \big(f_1'(x), f_2'(x)\big)$.

Conversely, suppose that f_1 and f_2 are Fréchet differentiable at $x \in X$. Endow $Y = Y_1 \times Y_2$, for example, with the norm $\|(y_1, y_2)\|_Y = \max\{\|y_1\|_{Y_1}, \|y_2\|_{Y_2}\}$ and for all $h \in X$ set

$$\xi(h) = \big(f_1'(x)h, f_2'(x)h\big).$$

Then for every $u \in X$, we have

$$\begin{aligned} \|f(u) - f(x) - \xi(u - x)\|_Y &= \max_{k \in \{1,2\}} \|f_k(u) - f_k'(x)(u - x)\|_{Y_k} \\ &= o(\|u - x\|_X), \end{aligned}$$

$$\text{where} \quad \frac{o(\|u - x\|_X)}{\|u - x\|_X} \longrightarrow 0 \quad \text{as } u \longrightarrow x \text{ in } X.$$

This shows that f is Fréchet differentiable at $x \in X$ and $f'(x) = \big(f_1'(x), f_2'(x)\big)$. ∎

Next suppose that $X = X_1 \times X_2$ with X_1, X_2 Banach spaces and $f: X \longrightarrow Y$, where Y is another Banach space. Let $x = (x_1, x_2) \in X$. The function $u_1 \longrightarrow f(u_1, x_2)$ is called the first partial function associated with f at x and $u_2 \longrightarrow f(x_1, u_2)$ is called the second partial function associated with f at x. We can consider the Fréchet derivatives of these partial functions. These are called partial derivatives of f and are denoted by $f_{u_1}'(x)$ and $f_{u_2}'(x)$. Then in a straightforward manner, we can show the following result.

PROPOSITION 1.1.17 *If $f\colon X = X_1 \times X_2 \longrightarrow Y$ is Fréchet differentiable at $x = (x_1, x_2) \in X$, then the two partial functions associated with f at x are Fréchet differentiable at $x_1 \in X_1$ and at $x_2 \in X_2$, respectively, and we have*

$$f'(x)h = f'_{x_1}(x)h_1 + f'_{x_2}(x)h_2 \quad \textit{for all } h = (h_1, h_2) \in X_1 \times X_2.$$

REMARK 1.1.18 Clearly the result can be extended to an arbitrary finite product $X = X_1 \times \cdots \times X_n$, $n \geq 2$.

The converse of Proposition 1.1.17 is not in general true. Namely the existence of the partial derivatives of $f\colon X = X_1 \times X_2 \longrightarrow Y$ at $x = (x_1, x_2) \in X$ does not imply the differentiability of f at $x \in X$. The next example illustrates this.

EXAMPLE 1.1.19 Let $f\colon \mathbb{R}^2 \longrightarrow \mathbb{R}$ be defined by

$$f(x_1, x_2) = \begin{cases} \frac{x_1 x_2}{x_1^2 + x_2^2} & \text{if } (x_1, x_2) \neq (0,0) \\ 0 & \text{if } (x_1, x_2) = (0,0) \end{cases} \quad \text{for all } \begin{pmatrix} x_1 \\ x_2 \end{pmatrix} \in \mathbb{R}^2.$$

Then $(\partial f / \partial x_1)(0,0) = (\partial f / \partial x_2)(0,0) = 0$, but f is not continuous at $(0,0)$, hence a fortiori it is not Fréchet differentiable there (see Proposition 1.1.11).

PROPOSITION 1.1.20 *If $\{X_k\}_{k=1}^n$, Y are Banach spaces and $f\colon X = X_1 \times \cdots \times X_n \longrightarrow Y$, then $f \in C^1(X, Y)$ if and only if the n-partial derivatives $f'_{x_k}(x)$ exist at every $x \in X$ and $x \longrightarrow f'_{x_k}(x)$ is continuous from X into $\mathcal{L}(X_k, Y)$ with the operator norm topology, for all $1 \leq k \leq n$.*

PROOF: We do the proof for $n = 2$. The general case follows easily by induction.

First we assume that $f \in C^1(X, Y)$. From Proposition 1.1.17 we know that $f'_{x_1}(x), f'_{x_2}(x)$ exist for all $x \in X$ and we have

$$f'_{x_1}(x)h_1 = f'(x)(h_1, 0), \; f'_{x_2}(x)h_2 = f'(x)(0, h_2)$$

for all $x \in X$, $h_1 \in X_1$, $h_2 \in X_2$.

From this follows at once that $x \longrightarrow f'_{x_k}(x)$ is continuous from X into $\mathcal{L}(X_k, Y)$ with the operator norm topology for $k = 1, 2$.

Now suppose that the partial derivatives f'_{x_k} exist on X and are continuous. For $h = (h_1, h_2) \in X = X_1 \times X_2$, let $\xi(x)h = f'_{x_1}h_1 + f'_{x_2}h_2$. Then for all $u = (u_1, u_2) \in X$, we have

$$\begin{aligned} & \|f(u) - f(x) - \xi(x)(u - x)\|_Y \\ \leq \; & \|f(u_1, u_2) - f(x_1, u_2) - f'_{x_1}(x_1, x_2)(u_1 - x_1)\|_Y \\ & + \|f(x_1, u_2) - f(x_1, x_2) - f'_{x_2}(x_1, x_2)(u_2 - x_2)\|_Y. \end{aligned} \tag{1.4}$$

Then given $\varepsilon > 0$, we can find $\delta > 0$ such that $\|u_2 - x_2\|_X \leq \delta$ implies

$$\|f(x_1, u_2) - f(x_1, x_2) - f'_{x_2}(x_1, x_2)(u_2 - x_2)\|_Y \leq \frac{\varepsilon}{2}\|u_2 - x_2\|_{X_2}. \tag{1.5}$$

Also from Proposition 1.1.6, we have

$$\begin{aligned} & \|f(u_1,u_2)-f(x_1,u_2)-f'_{x_1}(x_1,x_2)(u_1-x_1)\|_Y \\ \le \sup_{z_1\in[x_1,u_1]} & \|f'_{x_1}(z_1,u_2)-f'_{x_1}(x_1,x_2)\|_{\mathcal{L}}\|u_1-x_1\|_{X_1} \\ & \big([x_1,u_1]=\{tx_1+(1-t)u_1:0\le t\le 1\}\big). \end{aligned} \tag{1.6}$$

By hypothesis, we can find $\delta_1\le\delta$ such that $\max\{\|z_1-x_1\|_{X_1},\|u_2-x_2\|_{X_2}\}\le\delta$, implies

$$\|f'_{x_1}(z_1,u_2)-f'_{x_1}(x_1,x_2)\|_{\mathcal{L}}\le\frac{\varepsilon}{2}. \tag{1.7}$$

Using (1.5) through (1.7) in (1.4), we see that if $\|u-x\|_X=\max\{\|u_1-x_1\|_{X_1},\|u_2-x_2\|_{X_2}\}\le\delta_1$, then

$$\begin{aligned} \|f(u)-f(x)-\xi(x)(u-x)\|_Y &\le \frac{\varepsilon}{2}\|u_2-x_2\|_{X_2}+\frac{\varepsilon}{2}\|u_1-x_1\|_{X_1} \\ &\le \varepsilon\max\{\|u_2-x_2\|_{X_1},\|u_1-x_1\|_{X_1}\} \\ &= \varepsilon\|u-x\|. \end{aligned}$$

This proves the Fréchet differentiability of f at x and $f'(x)h=\xi(x)h=f'_{x_1}(x)h_1+f'_{x_2}(x)h_2$ for all $h=(h_1,h_2)\in X_1\times X_2$. $\square$

COROLLARY 1.1.21 *If Y is a Banach space and $f\colon\mathbb{R}^n\longrightarrow Y$, then $f\in C^1(\mathbb{R}^n,Y)$ if and only if the n-partial derivatives $(\partial f/\partial x_k)(x)$ are continuous from $\mathbb{R}^n$ into Y for all $k=1,\dots,n$.*

In this last part of this section we prove the *implicit function theorem* and the *inverse function theorem*, which are important in analysis and differential geometry.

The proof of the implicit function theorem, depends on the Banach fixed theorem. Here we simply state the result and postpone its proof until Section 3.4, where the result is formulated and proved in a more general setting.

PROPOSITION 1.1.22 *If X is a Banach space, $C\subseteq X$ is a nonempty closed subset, and $f\colon C\longrightarrow C$ satisfies*

$$\|f(x)-f(z)\|_X\le k\|x-z\|_X \qquad \textit{for all } x,z\in C \textit{ with } k<1, \tag{1.8}$$

then there exists a unique $x\in C$ such that $x=f(x)$ (fixed point of f). Moreover, if Y is another Banach space, $U\subseteq Y$ nonempty open and we have a family of functions $f(u)\colon C\longrightarrow C$, $u\in U$, satisfying (1.8) with $k<1$ independent of $u\in U$, then the unique fixed point $x(u)\in C$ (i.e., $x(u)=f(u)\big(x(u)\big)$ for all $u\in U$) depends continuously on $u\in U$.

Using this proposition we can prove the so-called *implicit function theorem.* This theorem is about equations of the form $f(x,y)=0$. It tells us under what conditions this equation defines a function $y=g(x)$ (we say that $y=g(x)$ is defined *implicitly*) and we would like to compute $\mathrm{d}y/\mathrm{d}x$. Given such a function f, in general we can not solve for y explicitly and so it is important to know that such a function g indeed exists without having to solve for it.

THEOREM 1.1.23 (***Implicit Function Theorem***) *If X, Y, Z are Banach spaces, $U \subseteq X \times Y$ is a nonempty open set, $(x_0, y_0) \in U$, $f \in C^1(U, Z)$, $f'_y(x_0, y_0) \in \mathcal{L}(Y, Z)$ is an isomorphism, and $f(x_0, y_0) = 0$, then there exist open neighborhoods V of x_0 and W of y_0 with $V \times W \subseteq U$ and a function $g \in C^1(V, W)$ such that*

$$f\big(x, g(x)\big) = 0 \qquad \text{for all } x \in V \tag{1.9}$$

$$\text{and} \quad g'(x) = -f'_y\big(x, g(x)\big)^{-1} \circ f'_x\big(x, g(x)\big) \qquad \text{for all } x \in V. \tag{1.10}$$

Moreover, for every $x \in V$, $g(x)$ is the unique solution in W of (1.9).

PROOF: For notational convenience, we set $S_0 = f'_y(x_0, y_0)$. Evidently the equation $f(x, y) = 0$ is equivalent to

$$y = y - S_0^{-1} f(x, y) = F(x, y). \tag{1.11}$$

The significance of this straightforward reformulation is that on F we can apply Proposition 1.1.22. Because $S_0^{-1} \circ S_0 = \mathrm{Id}_Y$, we have

$$F(x, y_1) - F(x, y_2) = S_0^{-1}\Big(S_0(y_1 - y_2) - \big(f(x, y_1) - f(x, y_2)\big)\Big) \quad \big(\text{see (1.11)}\big).$$

Because of the hypotheses on f, we can find $\delta_1 > 0$ and $r > 0$ such that if $\|x - x_0\|_X \leq \delta_1$ and $\|y_1 - y_0\|_Y \leq r$, $\|y_2 - y_0\|_Y \leq r$ (i.e., $\|y_1 - y_2\|_Y \leq 2r$), then

$$\|F(x, y_1) - F(x, y_2)\|_Z \leq \frac{1}{2}\|y_1 - y_2\|_Y. \tag{1.12}$$

Also we can find $0 < \delta < \delta_1$ such that, if $\|x - x_0\|_X \leq \delta$, then

$$\|F(x, y_0) - F(x_0, y_0)\|_Z \leq \frac{r}{2}. \tag{1.13}$$

Therefore, if $\|x - x_0\|_X \leq \delta$ and $\|y - y_0\|_Y \leq r$, then

$$\begin{aligned} \|F(x, y) - y_0\|_Z &= \|F(x, y) - F(x_0, y_0)\|_Z \\ &\leq \|F(x, y) - F(x, y_0)\|_Z + \|F(x, y_0) - F(x_0, y_0)\|_Z \\ &\leq \frac{1}{2}\|y - y_0\|_Y + \frac{r}{2} \quad \big(\text{see (1.12) and (1.13)}\big) \\ &\leq r. \end{aligned}$$

Hence $F(x, y)$ maps the closed ball $\bar{B}_r(y_0) = \{y \in Y : \|y - y_0\|_Y \leq r\}$ onto itself and similarly the open ball $B_r(y_0) = \{y \in Y : \|y - y_0\|_Y < r\}$ onto itself. Applying Proposition 1.1.22 to the parametric family of functions $\big\{y \longrightarrow F(x, y)\big\}_{x \in \bar{B}_\delta(x_0)}$ (see (1.12)), for every $x \in \bar{B}_\delta(x_0)$ we can find a unique $y = y(x) \in \bar{B}_r(y_0)$ such that $F(x, y) = y$; hence $f(x, y) = 0$ (which proves (1.9) and $x \longrightarrow g(x) = y(x)$ is continuous. Let

$$V = B_\delta(x_0) \quad \text{and} \quad W = B_r(y_0).$$

Choosing $\delta, r > 0$ small enough, we can guarantee that $V \times W \subseteq U$. We claim that $g \in C^1\big(B_\delta(x_0), Y\big)$. To show this let $(x_1, y_1) \in V \times W$, $y_1 = g(x_1)$ (recall that $G(x, \cdot)$ maps W onto itself). By virtue of the differentiability of f at (x_1, y_1), we have

$$f(x, y) = K(x - x_1) + L(y - y_1) + u(x, y) \quad \text{for all } (x, y) \in U, \tag{1.14}$$

where $K = f'_x(x_1, y_1)$ and $L = f'_y(x_1, y_1)$ and

$$\lim_{(x,y)\to(x_1,y_1)} \frac{\|u(x,y)\|_Z}{\|(x-x_1, y-y_1)\|_{X\times Y}} = 0. \tag{1.15}$$

Inasmuch as $f\big(x, g(x)\big) = 0$ for all $x \in V$, from (1.14) we obtain

$$g(x) = y_1 - L^{-1}K(x-x_1) - L^{-1}u\big(x, g(x)\big) \quad \text{for all } x \in V. \tag{1.16}$$

Because of (1.15), we can find $\eta > 0$ such that if $\|x-x_1\|_X \le \eta$ and $\|y-y_1\|_X \le \eta$, then

$$\begin{aligned} &\|u(x,y)\|_Z \le \frac{1}{2\|L^{-1}\|_{\mathcal{L}}}(\|x-x_1\|_X + \|y-y_1\|_Y) \\ \Rightarrow\ &\|u\big(x, g(x)\big)\|_Z \le \frac{1}{2\|L^{-1}\|_{\mathcal{L}}}(\|x-x_1\|_X + \|g(x)-g(x_1)\|_Y), \end{aligned} \tag{1.17}$$

for all $x \in V$.

From (1.16) and (1.17) it follows that

$$\begin{aligned} &\|g(x)-g(x_1)\|_Y \le \|L^{-1}K\|_{\mathcal{L}}\|x-x_1\|_X + \frac{1}{2}\|g(x)-g(x_1)\|_Y \\ \Rightarrow\ &\|g(x)-g(x_1)\|_Y \le \vartheta\|x-x_1\|_X \text{ with } \vartheta = 2\|L^{-1}K\|_{\mathcal{L}} + 1. \end{aligned} \tag{1.18}$$

Let $\xi(x) = L^{-1}u\big(x, g(x)\big)$, $x \in V$. From (1.16), we have

$$g(x) - g(x_1) = L^{-1}K(x-x_1) + \xi(x). \tag{1.19}$$

Note that $\|\xi(x)\|_Y \le \|L^{-1}K\|_{\mathcal{L}}\|u\big(x, g(x)\big)\|_Z$, $x \in V$ and

$$\lim_{x\to x_1} \frac{\|u\big(x, g(x)\big)\|_Z}{\|x-x_1\|_X} = 0 \quad \text{(see (1.15) and (1.16))}.$$

Hence we infer that

$$\lim_{x\to x_1} \frac{\|\xi(x)\|_Y}{\|x-x_1\|_X} = 0. \tag{1.20}$$

Combining (1.19) and (1.20), we conclude that g is Fréchet differentiable at $x_1 \in V$ and

$$g'(x_1) = -L^{-1}K \tag{1.21}$$

which proves (1.10). Moreover, it is clear from (1.21) that $g \in C^1(V, W)$. □

REMARK 1.1.24 A special case of interest is when $X = \mathbb{R}^n$ and $Y = Z = \mathbb{R}^m$. Then the function f has m-components $f_k\colon (x_1, \ldots, x_n, y_1, \ldots, y_m) \longrightarrow f_k(x_1, \ldots, x_n, y_1, \ldots, y_m)$, $1 \le k \le m$, which are C^1 $\mathbb{R}$-valued functions of $(n+m)$-variables $x_1, \ldots, x_n, y_1, \ldots, y_m$. Then

$$f'_y(x_0, y_0) = \left(\frac{\partial f_k}{\partial y_l}(x_0, y_0)\right)_{k,l=1}^m \in \mathbb{R}^{m\times m}.$$

Let $a_k = f_k(x_0, y_0)$, $k \in \{1, \dots, m\}$. Then according to Theorem 1.1.23, if $\det f'_y(x_0, y_0) \neq 0$, then the system of equations

$$f_k(x_1, \dots, x_n, y_1, \dots, y_m) = a_k, \; k \in \{1, \dots, m\}$$

for every value of the parameter $x = (x_i)_{i=1}^n \in V$ has a unique solution $y = g(x)$ with $(x, y) \in U$ and $g = (g_k)_{k=1}^m \in C^1(V, W)$. So we have

$$f_k\big(x_1, \dots, x_n, g_1(x_1, \dots, x_n), \dots, g_m(x_1, \dots, x_n)\big) = a_k, \; k \in \{1, \dots, m\}.$$

As a consequence of Theorem 1.1.23 we obtain the *inverse function theorem*. This theorem (as its name suggests), is about invertibility of functions. If $f\colon \mathbb{R}^n \longrightarrow \mathbb{R}^n$ has a nonzero Jacobian determinant at x (i.e., $\det\big(\partial f_i/\partial x_j\big)_{i,j=1}^n \neq 0$), then $f'(x) \in \mathcal{L}(\mathbb{R}^n, \mathbb{R}^n)$ is an isomorphism . From the fact that the best linear approximation is invertible, we would like to conclude the invertibility of f. In general this is not possible. Consider the simple case of $n = 1$; that is, $f\colon \mathbb{R} \longrightarrow \mathbb{R}$. If $f \in C^1(\mathbb{R})$ and $f'(x) \neq 0$, then we can only guarantee the invertibility of f near x, because f still has nonzero slope in a neighborhood of x. Therefore our main concern is local invertibility.

THEOREM 1.1.25 (*Inverse Function Theorem*) *If X, Y are Banach spaces, $U \subseteq Y$ is a nonempty open set, $f \in C^1(U, X)$, $y_0 \in U$, and $f'(y_0) \in \mathcal{L}(Y, X)$ is an isomorphism, then there exists a neighborhood U' of y_0, $U' \subseteq U$ and V' a neighborhood of $x_0 = f(y_0)$ such that $f\colon U' \longrightarrow V'$ is a C^1-diffeomorphism (i.e., both f and f^{-1} are C^1 maps) and $(f^{-1})'(x_0) = f'(y_0)^{-1}$.*

PROOF: Let $F(x, y) = f(y) - x$. Then $F'_y(x_0, y_0) = f'(y_0) \in \mathcal{L}(X, Y)$ which by hypothesis is an isomorphism. So we can apply Theorem 1.1.23 and obtain a neighborhood V' of x_0 and $g \in C^1(V', Y)$ such that $g(V') \subseteq U_0$ with U_0 a neighborhood of y_0, $F\big(x, g(x)\big) = 0$ for all $x \in V'$ (i.e., $f\big(x, g(x)\big) = x$ for all $x \in V'$) and $g(x_0) = y_0$. In what follows we consider f restricted on $g(V')$. Because $f\big(x, g(x)\big) = x$, we see that g is injective on V', hence a bijection from V' onto $g(V')$. Moreover, the set $g(V') = f^{-1}(V')$ is open in Y because $f \in C^1(U, X)$. Set $U' = g(V')$. We have that $f\colon U' \longrightarrow V'$ is a bijection. Also from Theorem 1.1.23 we have

$$g'(x_0) = -\Big(F'_y\big(x_0, g(x_0)\big)\Big)^{-1} F'_x\big(x_0, g(x_0)\big)$$

$$\text{hence} \quad f'(y_0) \circ g'(x_0) = \mathrm{Id}_X.$$

Therefore we conclude that $g'(x_0) = (f^{-1})'(x_0) = f'(y_0)^{-1}$. □

COROLLARY 1.1.26 *If X, Y are Banach spaces, $U \subseteq Y$ a nonempty open set, $f \in C^1(U, X)$, f is injective, and for every $y \in U$ $f'(y) \in \mathcal{L}(Y, X)$ is an isomorphism, then $f(U)$ is open in X and f is a C^1-diffeomorphism (i.e., both f and f^{-1} are C^1 maps) from U onto $f(U)$.*

REMARK 1.1.27 Suppose $X = \mathbb{R}^n$, $Y = \mathbb{R}^m$, $U \subseteq \mathbb{R}^m$ nonempty open and $f \in C^1(U, \mathbb{R}^n)$ with $f(y_0) = x_0$. Suppose that $m \leq n$ and $\operatorname{rank} f'(y_0) = m$ (note that $f'(y_0) \in \mathbb{R}^{n \times m}$). Then we can find $U' \subseteq U$ a neighborhood of y_0, V' a neighborhood of x_0, and a differentiable function $g\colon V' \longrightarrow U'$ such that $g \circ f = i$ with $i\colon \mathbb{R}^m \longrightarrow \mathbb{R}^n$

being the canonical injection map (i.e., $i(u_1,\ldots,u_m) = (u_1,\ldots,u_m,0,\ldots,0)$). If $m \geq n$ and $\operatorname{rank} f'(y_0) = n$, then we can find a $U' \subseteq U$ neighborhood of y_0 and a differentiable function $g\colon U' \longrightarrow U$ such that $g(y_0) = y_0$ and $f \circ g = p_{\mathbb{R}^n}$, where $p_{\mathbb{R}^n} : \mathbb{R}^m \longrightarrow \mathbb{R}^n$ is the canonical projection (i.e., $p_{\mathbb{R}^n}(u_1,\ldots,u_m) = (u_1,\ldots,u_n)$).

We conclude this section with a well-known result about the Nemytski operator (see, e.g., for example Gasiński–Papageorgiou [259]).

PROPOSITION 1.1.28 (a) *If (Ω,Σ,μ) is a complete σ-finite measure space, X,Y are separable Banach spaces, and $f\colon \Omega{\times}X \longrightarrow Y$ is a Carathéodory function (i.e., Σ-measurable in $\omega \in \Omega$ and continuous in $x \in X$) such that*

$$\|f(\omega,x)\|_Y \leq \alpha(\omega) + c\|x\|_X^{p/r} \quad p,\ r \in [1,\infty),\ \alpha \in L^r(\Omega),\ c{>}0$$

then $x \longrightarrow N_f(x)(\cdot) = f(\cdot,x(\cdot))$ is continuous and bounded from $L^p(\Omega,X)$ into $L^r(\Omega,Y)$.

(b) *If $f\colon \Omega{\times}\mathbb{R} \longrightarrow \mathbb{R}$ is a Carathéodory function*

$$F(\omega,x) = \int_0^x f(\omega,s)\mathrm{d}s \quad \textit{and} \quad \psi(u) = \int_\Omega F(\omega,u(\omega))\mathrm{d}\mu$$

for all $u \in L^p(\Omega)$, then $\psi \in C^1(L^p(\Omega))$ and $\psi'(u) = N_f(u)$.

1.2 Convex Functions

Throughout this section X is a locally convex vector space. Additional hypotheses are introduced as needed. Also we consider functions that also take the value $+\infty$. For this reason we introduce the set $\overline{\mathbb{R}} = \mathbb{R} \cup \{+\infty\}$.

DEFINITION 1.2.1 Let $\varphi\colon X \longrightarrow \overline{\mathbb{R}}$. The *effective domain* of φ is the set

$$\operatorname{dom} \varphi = \{x \in X \colon \varphi(x) < +\infty\}.$$

We say that φ is *proper* if $\operatorname{dom} \varphi \neq \emptyset$. The *epigraph* of φ is the set

$$\operatorname{epi} \varphi = \{(x,\lambda) \in X \times \mathbb{R} \colon \varphi(x) \leq \lambda\}.$$

The function φ is said to be *lower semicontinuous* (or *closed*) if for every $\lambda \in \mathbb{R}$,

$$L_\lambda = \{x \in X \colon \varphi(x) \leq \lambda\}$$

is closed. We say that φ is *convex* if for all $x_1, x_2 \in \operatorname{dom} \varphi$ and all $\lambda \in [0,1]$, we have

$$\phi(\lambda x_1 + (1-\lambda)x_2) \leq \lambda\varphi(x_1) + (1-\lambda)\varphi(x_2).$$

If this inequality is strict when $x_1 \neq x_2$ and $\lambda \in (0,1)$, then we say that φ is *strictly convex*.

The cone of proper, convex, and lower semicontinuous functions is denoted by $\Gamma_0(X)$.

If $C \subseteq X$, then the *indicator function* of C is defined by

$$i_C(x) = \begin{cases} 0 & \text{if } x \in C \\ +\infty & \text{otherwise} \end{cases}.$$

REMARK 1.2.2 It is easy to check that φ is lower semicontinuous if and only if epi φ is closed if and only if $f(x) = \liminf_{z \to x} f(z) = \sup_{U \in \mathcal{N}(x)} \inf_{z \in U} f(z)$, where $\mathcal{N}(x)$ is the filter of neighborhoods of x. Also φ is convex if and only if epi φ is convex in $X \times \mathbb{R}$. Moreover, if $\varphi \colon X \longrightarrow \overline{\mathbb{R}}$ is convex, then for every $\lambda \in \mathbb{R}$, the set $L_\lambda = \{x \in X \colon \varphi(x) \leq \lambda\}$ is convex. The converse is not true. Think, for example, of $\varphi(x) = \sqrt{|x|}$, $x \in \mathbb{R}$. If $C \subseteq X$ is nonempty, closed (and convex), then i_C is proper, lower semicontinuous (and convex, i.e., $i_C \in \Gamma_0(X)$).

What about continuity of convex functions? The next theorem summarizes the situation in this respect.

THEOREM 1.2.3 *If $\varphi \colon X \longrightarrow \overline{\mathbb{R}}$ is a proper, convex function, then the following statements are equivalent.*

(a) *φ is bounded above in a neighborhood of $x_0 \in X$.*

(b) *φ is continuous at $x_0 \in X$.*

(c) int epi $\varphi \neq \emptyset$.

(d) int dom $\varphi \neq \emptyset$ *and* $\varphi\big|_{\text{int dom } \varphi}$ *is continuous.*

Moreover, if one of the above statements holds, then

$$\text{int epi } \varphi = \{(x, \lambda) \in X \times \mathbb{R} \colon x \in \text{int dom } \varphi,\ \varphi(x) < \lambda\}.$$

PROOF: (a)$\Rightarrow$(b): Suppose that U is the neighborhood of x_0 such that $\varphi\big|_U \leq c$ for some $c > 0$. By translating things if necessary, without any loss of generality we may assume that $x_0 = 0$ and $\varphi(0) = 0$. Let $0 < \varepsilon \leq c$ and set $V_\varepsilon = \big((\varepsilon/c)U\big) \cap \big(-(\varepsilon/c)U\big)$. Then V_ε is a symmetric neighborhood of the origin. We show that

$$|\varphi(x)| \leq \varepsilon \quad \text{for all } x \in V_\varepsilon, \tag{1.22}$$

which of course implies the continuity of φ at $x_0 = 0$. So let $x \in U$. Then $(c/\varepsilon)x \in U$ and by virtue of the convexity of φ and because $\varphi(0) = 0$, we have

$$\varphi(x) \leq \frac{\varepsilon}{c}\varphi\Big(\frac{c}{\varepsilon}x\Big) + \Big(1 - \frac{\varepsilon}{c}\Big)\varphi(0) \leq \frac{\varepsilon}{c}c = \varepsilon. \tag{1.23}$$

Similarly $-(c/\varepsilon)x \in U$ and we have

$$\begin{aligned} 0 = \varphi(0) &\leq \frac{1}{1 + \frac{\varepsilon}{c}}\varphi(x) + \frac{\frac{\varepsilon}{c}}{1 + \frac{\varepsilon}{c}}\varphi\Big(- \frac{c}{\varepsilon}x\Big) \\ &\leq \frac{1}{1 + \frac{\varepsilon}{c}}\varphi(x) + \frac{\varepsilon}{1 + \frac{\varepsilon}{c}} \\ \Rightarrow \quad & -\varepsilon \leq \varphi(x). \end{aligned} \tag{1.24}$$

From (1.23) and (1.24), it follows that (1.22) holds and so φ is continuous at $x_0 = 0$.

(b)$\Rightarrow$(a): The continuity of φ at x_0 implies that φ is bounded above in a neighborhood of $x_0 \in X$.

(a)⇒(c): As before, let U be a neighborhood of x_0 such that $\varphi\big|_U \leq c$. Then $U \subseteq \text{int dom } \varphi$ and

$$\{(x, \lambda) \in X \times \mathbb{R} \colon x \in U,\ c < \lambda\} \subseteq \text{epi } \varphi$$

which implies that int epi $\varphi \neq \emptyset$.

(c)⇒(a): Let $(x, \lambda) \in \text{int epi } \varphi$. Then we can find U a neighborhood of x and $\varepsilon > 0$ such that

$$U \times [\lambda - \varepsilon, \lambda + \varepsilon] \subseteq \text{epi } \varphi.$$

Then $U \times \{\lambda\} \subseteq \text{epi } \varphi$ and so we have $\varphi(x) \leq \lambda$ for all $x \in U$, which is statement (a).

(a)⇒(d): Again without any loss of generality, we may assume that $x_0 = 0$. Let U be a neighborhood of x_0 such that $\varphi\big|_U$ is bounded above. Then $U \subseteq \text{dom } \varphi$ and so int dom $\varphi \neq \emptyset$. Clearly the set dom φ is convex. So if $x \in \text{int dom } \varphi$, we can find $\lambda > 1$ such that $v = \lambda x \in \text{dom } \varphi$. Set

$$V = x + \Big(1 - \frac{1}{\lambda}\Big)U,$$

which is a neighborhood of x. If $z \in V$, we have $z = x + \big(1 - \frac{1}{\lambda}\big)u$ with $u \in U$ and so exploiting the convexity of φ, we have

$$\begin{aligned} \varphi(z) = \varphi\Big(\frac{1}{\lambda}v + \Big(1 - \frac{1}{\lambda}\Big)u\Big) &\leq \frac{1}{\lambda}\varphi(v) + \Big(1 - \frac{1}{\lambda}\Big)\varphi(u) \\ &\leq \frac{1}{\lambda}\varphi(v) + \Big(1 - \frac{1}{\lambda}\Big)c = c_0. \end{aligned}$$

Therefore $\varphi\big|_V$ is bounded above and so it is continuous at $x \in \text{int dom } \varphi$ (recall that we have already proved (a)⇔(b)).

(d)⇒(a): Obvious.

Now let $E = \{(x, \lambda) \in X \times \mathbb{R} \colon x \in \text{int dom } \varphi,\ \varphi(x) < \lambda\}$. Evidently int epi $\varphi \subseteq E$. Let $v \in \text{int dom } \varphi$ such that $\varphi(v) < \lambda$. Choose $\varphi(v) < \mu < \lambda$. Because $\varphi\big|_{\text{int dom } \varphi}$ is continuous we can find U a neighborhood of v such that $U \subseteq \text{int dom } \varphi$ and $\varphi(v) < \mu$ for all $x \in U$. Therefore $(v, \lambda) \in U \times (\mu, +\infty) \subseteq \text{int epi } \varphi$ and so $E \subseteq \text{int epi } \varphi$. $\square$

REMARK 1.2.4 Note that int dom $\varphi = \{x \in X$: there exists $\lambda \in \mathbb{R}$ such that $(x, \lambda) \in \text{int epi } \varphi\}$.

In finite-dimensional spaces the situation simplifies.

PROPOSITION 1.2.5 *If X is finite-dimensional and $\varphi \colon X \longrightarrow \overline{\mathbb{R}}$ is convex, then φ is continuous on* int dom φ.

PROOF: If $x \in \text{int dom } \varphi$, then we can find $\delta > 0$ and $\{e_k\}_{k=0}^N \subseteq X$ ($N = \dim X$) such that

$$B_\delta(x) \subseteq \text{conv}\{e_k\}_{k=0}^N \subseteq \text{dom } \varphi.$$

So, if $v \in B_\delta(x)$, then we can find $\{t_k\}_{k=0}^N \subseteq [0, 1]$ such that

$$x = \sum_{k=0}^{N} t_k e_k \quad \text{with} \quad \sum_{k=0}^{N} t_k = 1.$$

Because of the convexity of φ, we have

$$\varphi(x) \leq \sum_{k=0}^{N} t_k \varphi(e_k) \leq \max_{0 \leq k \leq N} \varphi(e_k) = c.$$

Invoking Theorem 1.2.3, we conclude that $\varphi\big|_{\text{int dom } \varphi}$ is continuous. □

In infinite dimensions, we need an additional condition on φ in order to have continuity on int dom φ.

PROPOSITION 1.2.6 *If X is a Banach space and $\varphi\colon X \longrightarrow \overline{\mathbb{R}}$ is convex and lower semicontinuous, then $\varphi\big|_{\text{int dom } \varphi}$ is continuous.*

PROOF: Note that dom $\varphi = \bigcup_{n \geq 1} \{\varphi \leq n\}$. Due to the lower semicontinuity of φ, the sets $\{\varphi \leq n\}$ are closed. So, if $x \in$ int dom φ, then by the Baire category theorem we can find $n \geq 1$ such that $\text{int}\{\varphi < n\} \neq \emptyset$ and $\varphi(x) < n$. Let $u \in \text{int}\{\varphi < n\}$ and introduce the function

$$\xi(\lambda) = \varphi\big(x + \lambda(u - x)\big), \quad \lambda > 0.$$

Because $x \in$ int dom φ, we can find $\delta > 0$ such that $\bar{B}_{\delta \|u-x\|_X}(x) \subseteq$ dom φ. It follows that $[-\delta, \delta] \subseteq$ dom ξ and so $0 \in$ int dom ξ. Hence by Proposition 1.2.5 ξ is continuous at zero. Because $\xi(0) < n$, we can find $\varepsilon > 0$ such that $\xi(\lambda) < n$ for all $\lambda \in [-\varepsilon, 0]$. We set $y = x - \varepsilon(u - x)$. Then we have

$$y \in \{\varphi < n\} \quad \text{and} \quad u \in \text{int}\{\varphi < n\}.$$

Inasmuch as $x \in [u, y) = \big\{(1-t)u + ty\colon\ 0 \leq t < 1\big\}$, it follows that $x \in$ int dom $\{\varphi < n\}$ and so by Theorem 1.2.3 φ is continuous at x. □

In fact we can identify the kind of continuity of φ on int dom φ.

THEOREM 1.2.7 *If X is a Banach space and $\varphi\colon X \longrightarrow \overline{\mathbb{R}}$ is a convex function that is continuous at $x_0 \in$ dom φ, then there exists $r > 0$ such that φ is Lipchitz continuous on $\bar{B}_r(x_0)$.*

PROOF: The continuity of φ at x_0 implies that there exist $r, \delta, c > 0$ such that

$$|\varphi(y)| \leq c \qquad \text{for all } \ y \in \bar{B}_{r+\delta}(x_0).$$

Let $x, y \in \bar{B}_r(x_0)$ with $x \neq y$. We set

$$u = y + \delta \frac{y - x}{\|y - x\|} \qquad \text{and} \qquad \lambda = \frac{\|y - x\|}{\|y - x\| + \delta} \in (0, 1).$$

Then $u \in \bar{B}_{r+\delta}(x_0)$ and $y = \lambda u + (1 - \lambda)x$. So, because φ is convex

$$
\begin{aligned}
\varphi(y) &\leq \lambda\varphi(u) + (1-\lambda)\varphi(x) \\
\Rightarrow \varphi(y) - \varphi(x) &\leq \lambda\big(\varphi((u) - \varphi(x)\big) \\
&\leq \frac{\|y-x\|}{\|y-x\|+\delta}\, 2c \leq \frac{2c}{\delta}\, \|y-x\|.
\end{aligned}
$$

Interchanging the roles of x, y in the above argument, we conclude that

$$|\varphi(y) - \varphi(x)| \leq \frac{2c}{\delta}\, \|y-x\| \qquad \text{for all } x, y \in \bar{B}_r(x_0).$$

□

COROLLARY 1.2.8 *If X is a Banach space and $\varphi\colon X \longrightarrow \mathbb{R}$ is a continuous convex function, then φ is locally Lipschitz.*

Thus far we have concentrated our attention on the continuity properties of convex functions. Now we turn our attention to the differentiability properties that lead to the subdifferential.

DEFINITION 1.2.9 Let X be a normed space, $\varphi\colon X \longrightarrow \overline{\mathbb{R}}$ a convex function, and $x_0 \in \operatorname{dom}\varphi$. The *directional derivative* of φ at x_0 in the direction $h \in X$, denoted by $\varphi'(x_0; h)$, is defined by

$$\varphi'(x_0; h) = \lim_{\lambda \downarrow 0} \frac{\varphi(x_0 + \lambda h) - \varphi(x_0)}{\lambda}.$$

REMARK 1.2.10 Due to the convexity of φ, the quotient map $\lambda \longrightarrow \frac{\varphi(x_0+\lambda h)-\varphi(x_0)}{\lambda}$ is increasing on $(0, +\infty)$ and so $\varphi'(x_0; h)$ is well defined. In fact $\varphi'(x_0; h) = \inf\limits_{\lambda>0} \frac{\varphi(x_0+\lambda h)-\varphi(x_0)}{\lambda}$. It is easy to see that $\varphi'(x_0; \cdot)$ is convex and positively homogeneous. Of course, if $\varphi'(x_0; \cdot) \in X^*$, then $\varphi'(x_0; \cdot)$ is the Gâteaux derivative of φ at x_0.

The basic result on the Gâteaux differentiability of convex functions is the following theorem.

THEOREM 1.2.11 *If X is a separable Banach space, $U \subseteq X$ is a nonempty open convex set, and $\varphi\colon U \longrightarrow \mathbb{R}$ is a continuous convex function, then φ is Gâteaux differentiable on a dense G_δ-subset of U.*

PROOF: For every $h \in X$ and $n \geq 1$, we introduce the set

$$V(h,n) = \Big\{x \in U : \text{there exists } \delta = \delta(x,n) > 0 \text{ such that } \sup_{0<\lambda<\delta} \frac{\varphi(x+\lambda h) - \varphi(x-\lambda h) - 2\varphi(x)}{\lambda} < \frac{1}{n}\Big\}.$$

Because φ is continuous on U, given $n \geq 1$ for every $m \geq 1$ the set

$$W_m(h) = \Big\{x \in U : \frac{\varphi(x+\frac{1}{m}h) + \varphi(x-\frac{1}{m}h) - 2\varphi(x)}{\frac{1}{m}} < \frac{1}{n}\Big\}$$

is open in U. Note that $V(h,n) = \bigcup_{m\geq 1} W_m(h)$. So $V(h,n)$ is open.

Next we show that $V(h,n)$ is dense in U. If this is not the case, then we can find $x_0 \in U$ and $\delta > 0$ such that $V(h,n) \cap B_\delta(x_0) = \emptyset$. This means that the convex function $\lambda \longrightarrow \xi(\lambda) = \varphi(x_0 + \lambda h)$ is not differentiable on $(-\delta, \delta)$, which contradicts the classical result on the differentiability of real convex functions. So $V(h,n)$ is dense in U.

Because φ is continuous convex, from Corollary 1.2.8 we infer that $\varphi'(x;\cdot)$ is continuous for every $x \in U$. So if $\{h_k\}_{k\geq 1}$ is dense in $\partial B_1 = \{u \in X : \|u\|_X = 1\}$, then φ is Gâteaux differentiable on the set $\bigcap_{n,k\geq 1} V(h_k, n)$ which is dense in U (by Baire's theorem) and G_δ. □

There is a corresponding result for Fréchet differentiability due to Asplund [31].

THEOREM 1.2.12 *If X is a Banach space with a separable dual, U is a nonempty open convex set, and $\varphi \colon U \longrightarrow \mathbb{R}$ a continuous convex function, then φ is Fréchet differentiable on a dense G_δ-set in U.*

In Example 1.1.9 we saw that existence of partial derivatives does not in general imply Fréchet differentiability of a function. In the case of convex functions, the situation changes.

PROPOSITION 1.2.13 *If $U \subseteq \mathbb{R}^N$ is a nonempty open set, $\varphi \colon U \longrightarrow \mathbb{R}$ is convex, and all the partial derivatives of φ at $x \in U$ exist, then φ is Fréchet differentiable at $x \in U$.*

PROOF: Let $A(x) \in \mathbb{R}^N$ be defined by $\big(A(x), h\big)_{\mathbb{R}^N} = \sum_{k=1}^{N} (\partial\varphi/\partial x_k)(x) h_k$ for all $h = \big(h_k\big)_{k=1}^N \in \mathbb{R}^N$. Let $r > 0$ be such that $B_r(x) \subseteq U$. For each $h \in B_r(0)$, we set

$$\vartheta(h) = \varphi(x+h) - \varphi(x) - \big(A(x), h\big)_{\mathbb{R}^N}.$$

Clearly ϑ is convex on $B_r(0)$. For each $k \in \{1, \ldots, N\}$ we define $\eta_k : B_r(0) \longrightarrow \mathbb{R}$ by

$$\eta_k(h) = \begin{cases} \frac{\vartheta(h_k e_k)}{h_k} & \text{if } h_k \neq 0 \\ 0 & \text{if } h_k = 0 \end{cases} \qquad \text{for all } h = \big(h_k\big)_{k=1}^N \in B_r(0),$$

where $\{e_k\}_{k=1}^N$ is the usual orthonormal basis of $\mathbb{R}^N$. Note that $\eta_k(h) \longrightarrow 0$ as $\|h\|_{\mathbb{R}^N} \longrightarrow 0$. For each $h = \big(h_k\big)_{k=1}^N$ with $\|h\|_{\mathbb{R}^N} < r/N$, because ϑ is convex, we have

$$\begin{aligned} \vartheta(h) &= \vartheta\Big(\sum_{k=1}^{N} \frac{1}{N} N h_k e_k\Big) \leq \frac{1}{N} \sum_{k=1}^{N} \vartheta\Big(N h_k e_k\Big) \\ &= \sum_{k=1}^{N} h_k \eta_k(Nh) \leq \|h\|_{\mathbb{R}^N} \sum_{k=1}^{N} \big|\eta_k(Nh)\big|. \end{aligned}$$

Because $-\vartheta(-h) \leq \vartheta(h)$, we have

$$-\|h\|_{\mathbb{R}^N} \sum_{k=1}^{N} \big|\eta_k(-Nh)\big| \leq \vartheta h \leq \|h\|_{\mathbb{R}^N} \sum_{k=1}^{N} \big|\eta_k(Nh)\big|$$

$$\Rightarrow \quad \frac{\vartheta(h)}{\|h\|_{\mathbb{R}^N}} \longrightarrow 0 \quad \text{as} \quad \|h\|_{\mathbb{R}^N} \longrightarrow 0.$$

This proves that φ is Fréchet differentiable at $x \in U$ and $\varphi'(x) = A(x)$. □

COROLLARY 1.2.14 *If X is a finite-dimensional Banach space, $U \subseteq X$ is nonempty open, and $\varphi\colon U \longrightarrow \mathbb{R}$ is convex, then φ is Gâteaux differentiable at $x \in U$ if and only if it is Fréchet differentiable at $x \in U$.*

Before passing to the discussion of the subdifferential of convex functions, let us briefly mention a few things about the duality theory of such functions. Duality is a basic theme in the theory of convex analysis.

We work within a dual pair (X, X^*), where X is a Hausdorff locally convex space and X^* its dual. We supply X with the $w(X, X^*)$-topology (weak topology) and X^* with the $w(X^*, X)$-topology. The pairing of X and X^* is denoted by $\langle \cdot, \cdot \rangle_X$. In what follows $\mathbb{R}^* = \mathbb{R} \cup \{\pm\infty\}$.

DEFINITION 1.2.15 Given a function $\varphi\colon X \longrightarrow \mathbb{R}^*$, the *conjugate* (or *Legendre–Fenchel transform*) of φ is the function $\varphi^*\colon X^* \longrightarrow \mathbb{R}^*$ defined by

$$\varphi^*(x^*) = \sup\big[\langle x^*, x\rangle_X - \varphi(x) \colon x \in X\big].$$

Similarly the second conjugate of φ, is the function $(\varphi^*)^* = \varphi^{**}\colon X \longrightarrow \mathbb{R}^*$ defined by

$$\varphi^{**}(x) = \sup\big[\langle x^*, x\rangle_X - \varphi^*(x^*) \colon x^* \in X^*\big].$$

REMARK 1.2.16 If φ takes the value $-\infty$, then $\varphi^* \equiv +\infty$. Also if $\varphi \not\equiv +\infty$, then φ^* has values in $\overline{\mathbb{R}} = \mathbb{R} \cup \{+\infty\}$.

Directly from Definition 1.2.15, we have the following.

PROPOSITION 1.2.17 *If $\varphi\colon X \longrightarrow \overline{\mathbb{R}}$ is proper, then $\varphi^* \in \Gamma_0(X^*)$.*

PROPOSITION 1.2.18 *If $\varphi\colon X \longrightarrow \mathbb{R}^*$, then $\varphi(x) + \varphi^*(x^*) \geq \langle x^*, x\rangle_X$ for all $x^* \in X^*$, $x \in X$ (Young–Fenchel inequality).*

REMARK 1.2.19 From the Young–Fenchel inequality we deduce that we always have $\varphi^{**} \leq \varphi$. In the sequel we provide necessary and sufficient conditions for equality to hold.

PROPOSITION 1.2.20 *If $\varphi\colon X \longrightarrow \overline{\mathbb{R}}$ is proper, then φ admits a continuous affine minorant; that is there exists $(x_0^*, \vartheta_0) \in X^* \times \mathbb{R}$ such that*

$$\langle x_0^*, x\rangle - \vartheta_0 \leq \varphi(x) \qquad \textit{for all } x \in X.$$

PROOF: Let $x_0 \in X$ and $\beta \in \mathbb{R}$ such that $\beta < \varphi(x_0)$. Then $(x_0, \beta) \notin \operatorname{epi} \varphi$. Because $\operatorname{epi} \varphi \subseteq X \times \mathbb{R}$ is closed convex, by the strong separation theorem, we can find $(x_0^*, \xi_0) \in X^* \times \mathbb{R}$ and $\vartheta_0 \in \mathbb{R}$ such that

$$\begin{aligned} & \langle x_0^*, x\rangle_X + \xi_0\lambda \le \vartheta_0 < \langle x_0^*, x_0\rangle_X + \xi_0\beta \qquad \text{for all } (x,\lambda) \in \operatorname{epi} \varphi, \\ \Rightarrow\ & \langle x_0^*, x\rangle_X + \xi_0\varphi(x) \le \vartheta_0 < \langle x_0^*, x_0\rangle_X + \xi_0\beta. \end{aligned} \tag{1.25}$$

From (1.25) we see that $\xi_0 < 0$. Without any loss of generality we may assume that $\xi_0 = -1$. Then from (1.25) we have

$$\begin{aligned} \langle x_0^*, x\rangle_X - \varphi(x) \le \vartheta_0 \qquad & \text{for all } x \in X, \\ \langle x_0^*, x\rangle_X - \vartheta_0 \le \varphi(x) \qquad & \text{for all } x \in X. \end{aligned}$$

So φ is minorized by a continuous affine function. □

We use this proposition to produce necessary and sufficient conditions in order for the equality $\varphi^{**} = \varphi$ to hold. Recall that we always have $\varphi^{**} \le \varphi$ (see Remark 1.2.19).

THEOREM 1.2.21 *If $\varphi\colon X \longrightarrow \overline{\mathbb{R}}$ is proper, then $\varphi^{**} = \varphi$ if and only if $\varphi \in \Gamma_0(X)$.*

PROOF: $\Rightarrow$: Follows from Proposition 1.2.17.

$\Leftarrow$: From Remark 1.2.19 we know that $\varphi^{**} \le \varphi$. We show that the opposite inequality is also true. So let $x \in X$ and $\mu \in \mathbb{R}$ such that $\mu < \varphi(x)$. Then $(x, \mu) \notin \operatorname{epi} \varphi$ and because $\operatorname{epi} \varphi$ is closed convex, we can apply the strong separation theorem and obtain $(x^*, \xi) \in X^* \times \mathbb{R}$, $(x^*, \xi) \ne (0, 0)$ and $\varepsilon > 0$ such that

$$\langle x^*, y\rangle_X + \xi\lambda \le \langle x^*, x\rangle_X + \xi\mu - \varepsilon \qquad \text{for all } (y,\lambda) \in \operatorname{epi} \varphi. \tag{1.26}$$

Note that λ can increase to $+\infty$. So from (1.26) we infer that $\xi \le 0$.

Suppose $\xi < 0$. Then

$$\begin{aligned} & \langle x^*, y\rangle_X + \xi\varphi(y) < \langle x^*, x\rangle_X + \xi\mu \qquad \text{for all } y \in X, \\ \Rightarrow\ & (-\xi\varphi)^*(x^*) \le \langle x^*, x\rangle_X + \xi\mu. \end{aligned}$$

We can easily check that $(-\xi\varphi)^*(x^*) = -\xi\varphi^*(x^*/-\xi)$. So we have

$$\begin{aligned} & -\xi\varphi^*\Big(-\frac{x^*}{\xi}\Big) \le \langle x^*, x\rangle_X + \xi\mu \\ \Rightarrow\ & \mu \le \Big\langle -\frac{x^*}{\xi}, x\Big\rangle_X - \varphi^*\Big(-\frac{x^*}{\xi}\Big) \le \varphi^{**}(x). \end{aligned}$$

Because $\mu < \varphi(x)$ was arbitrary, it follows that $\varphi(x) \le \varphi^{**}(x)$.

Now suppose that $\xi = 0$. From (1.26) we have

$$\begin{aligned} & \langle x^*, y\rangle_X \le \langle x^*, x\rangle_X - \varepsilon \qquad \text{for all } y \in \operatorname{dom} \varphi, \\ \Rightarrow\ & x \notin \operatorname{dom} \varphi \quad \text{and so} \quad \varphi(x) = +\infty. \end{aligned}$$

Therefore we need to show that $\varphi^{**}(x) = +\infty$. Let $\eta \in \mathbb{R}$ be such that

$$\langle x^*, y\rangle_X < \eta < \langle x^*, x\rangle_X \qquad \text{for all } y \in \operatorname{dom} \varphi. \tag{1.27}$$

From Proposition 1.2.20 we know that φ is bounded below by a continuous affine function. So we can find $y^* \in X^*$ and $\vartheta_0 \in \mathbb{R}$ such that

$$\langle y^*, y\rangle_X - \vartheta_0 \leq \varphi(y) \qquad \text{for all } y \in X.$$

Then for all $\gamma > 0$ we have

$$\begin{aligned} &\langle y^*, y\rangle_X - \vartheta_0 + \gamma\big(\langle x^*, y\rangle_X - \eta\big) \leq \varphi(y) \qquad \text{for all } y \in X \quad (\text{see } (1.27)),\\ \Rightarrow\ &\langle y^* + \gamma x^*, y\rangle_X - \varphi(y) \leq \vartheta_0 + \gamma\eta \qquad \text{for all } y \in X,\\ \Rightarrow\ &\varphi^*(y^* + \gamma x^*) \leq \vartheta_0 + \gamma\eta. \end{aligned} \tag{1.28}$$

Therefore

$$\begin{aligned} \langle y^*, x\rangle_X - \vartheta_0 + \gamma\big(\langle x^*, x\rangle_X - \eta\big) &\leq \langle y^* + \gamma x^*, x\rangle_X - \varphi^*(y^* + \gamma x^*)\\ &\qquad (\text{see } (1.28))\\ &\leq \varphi^{**}(x). \end{aligned}$$

Recall that $\gamma > 0$ was arbitrary. So we let $\gamma \uparrow +\infty$ and because $\langle x^*, x\rangle - \eta > 0$, we conclude that $\varphi^{**}(x) = +\infty$. Therefore $\varphi^{**} = \varphi$. ☐

REMARK 1.2.22 It follows from Theorem 1.2.21 that $\varphi \in \Gamma_0(X)$ if and only if φ is the upper envelope of all continuous affine functions majorized by φ. Also it follows that φ^{**} is the lower semicontinuous, convex regularization φ (i.e., the biggest lower semicontinuous, convex function majorized by φ).

Next we introduce an operation that is basic in convex analysis.

DEFINITION 1.2.23 Let $\varphi, \psi\colon X \longrightarrow \overline{\mathbb{R}}$. The *infimal convolution* of φ and ψ denoted by $\varphi \oplus \psi$, is defined by

$$(\varphi \oplus \psi)(x) = \inf\big[\varphi(y) + \psi(x - y) : y \in X\big] = \inf\big[\varphi(u_1) + \psi(u_2) : u_1 + u_2 = x\big].$$

As it turns out this operation is dual to addition (see Theorem 1.2.25 below).

PROPOSITION 1.2.24 *If $\varphi, \psi\colon X \longrightarrow \overline{\mathbb{R}}$ are convex functions and there exists $x_0 \in \operatorname{dom}\psi$ such that φ is continuous at x_0, then* $\inf\big[\varphi(x) + \psi(x) : x \in X\big] = \max\big[-\varphi^*(x^*) - \psi^*(-x^*) : x^* \in X^*\big]$.

PROOF: Using Definition 1.2.15 we can check that for all $x \in X$ and all $x^* \in X^*$, we have

$$\begin{aligned} &-\varphi^*(x^*) - \psi^*(-x^*) \leq \varphi(x) + \psi(x)\\ \Rightarrow\quad & m^* = \sup\big[-\varphi^*(x^*) - \psi^*(-x^*) : x^* \in X^*\big]\\ &\quad\ \leq \inf\big[\varphi(x) + \psi(x) : x \in X\big] = m. \end{aligned}$$

Because $x_0 \in \operatorname{dom}\varphi \cap \operatorname{dom}\psi$, we have $-\infty \leq m < +\infty$. If $m = -\infty$, then we are done. So suppose that $m \in \mathbb{R}$. Let $E = \operatorname{int}\operatorname{epi}\varphi \neq \emptyset$ (see Theorem 1.2.3) and $G = \big\{(x, \lambda) \in X \times \mathbb{R}\colon \lambda \leq m - \psi(x)\big\} \neq \emptyset$. Both sets are convex and if $(x, \lambda) \in E$, then $\lambda > \varphi(x) \geq m - \psi(x)$, hence $E \cap G = \emptyset$. Applying the weak separation theorem, we can find $(x_0^*, \lambda_0) \in X^* \times \mathbb{R}$, $(x_0^*, \lambda_0) \neq (0, 0)$ and $\xi \in \mathbb{R}$ such that

$$\langle x_0^*, x\rangle_X + \lambda_0\lambda \le \xi \le \langle x_0^*, u\rangle_X + \lambda_0 t \qquad \text{for all } (x,\lambda)\in E \text{ and all } (u,t)\in G. \quad (1.29)$$

Because λ can grow to $+\infty$, from (1.29) we see that $\lambda_0 \le 0$. If $\lambda_0 = 0$, then $\langle x_0^*, x\rangle_X \le \xi \le \langle x_0^*, x_0\rangle_X$ for all $x \in \text{dom}\ \varphi$. But we can find $\delta > 0$ such that $B_\delta(x_0) \subseteq \text{dom}\ \varphi$. So it follows that $x_0^* = 0$, a contradiction to the fact that $(x_0^*, \lambda_0) \ne (0,0)$. Therefore $\lambda_0 < 0$ and we can assume that $\lambda_0 = -1$. We have

$$\begin{aligned} & -\langle x_0^*, u\rangle_X + t \le -\xi \qquad \text{for all } (u,t)\in G \quad (\text{see } (1.29)),\\ \Rightarrow\ & -\langle x_0^*, u\rangle_X + m - \psi(u) \le -\xi \qquad \text{for all } u \in X. \end{aligned} \quad (1.30)$$

Also we have

$$\langle x_0^*, x\rangle_X - \lambda \le \xi \qquad \text{for all } (x,\lambda) \in \bar{E} = \text{epi}\ \varphi. \quad (1.31)$$

From (1.30) we obtain $\psi^*(-x_0^*) \le -\xi - m$ and from (1.31) we have $\varphi^*(x^*) \le \xi$. Therefore

$$m \le -\varphi^*(x_0^*) - \psi^*(-x_0^*) \le m^* \le m.$$

□

Using this proposition we can establish the duality between addition and infimal convolution.

THEOREM 1.2.25 *If $\varphi, \psi\colon X \longrightarrow \overline{\mathbb{R}}$ are convex functions and there exists $x_0 \in$ dom ψ such that φ is continuous at x_0, then $(\varphi+\psi)^* = \varphi^* \oplus \psi^*$.*

PROOF: For every $x^* \in X^*$ we have $-(\varphi + \psi)^*(x^*) = \inf\big[g(x) + h(x)\colon x \in X\big]$ with $g(x) = \varphi(x) - \langle x^*, x\rangle_X$ and $h(x) = \psi(x)$. Evidently g, h are convex functions and g is continuous at $x_0 \in \text{dom}\ g = \text{dom}\ \psi$. So we can apply Proposition 1.2.24 and obtain

$$\begin{aligned} -(\varphi+\psi)^*(x^*) &= -\min\big[h^*(u^*) + g^*(-u^*)\colon u^* \in X^*\big] \\ &= -\min\big[\psi^*(u^*) + \varphi^*(x^* - u^*)\colon u^* \in X^*\big]. \end{aligned}$$

□

REMARK 1.2.26 It is straightforward to check that for any proper convex functions $\varphi, \psi\colon X \longrightarrow \overline{\mathbb{R}}$, we have $(\varphi \oplus \psi)^* = \varphi^* + \psi^*$.

EXAMPLE 1.2.27 (a) If $C \subseteq X$ is nonempty, and i_C is the indicator function, then $i_C^*(x^*) = \sigma(x^*; C) = \sup\big[\langle x^*, c\rangle_X : c \in C\big]$ (the *support function* of the set C). Moreover, if $C \subseteq X$ is nonempty, closed, and convex, then $x \in C$ if and only if $\langle x^*, x\rangle_X \le \sigma(x^*; C)$ for all $x^* \in X^*$.

(b) If X is a normed space, $C \subseteq X$ is nonempty, and $\varphi(x) = d(x, C) = \inf\big[\|x - c\|_X : c \in C\big]$, then $\varphi = i_C \oplus \|\cdot\|_X$ and so $\varphi^* = i_C^* + (\|\cdot\|_X)^* = \sigma(\cdot; C) + i_{\bar{B}_1^*}$ (see Remark 1.2.26).

(c) If X is a normed space and $h\colon \mathbb{R} \longrightarrow \mathbb{R}$ is an even convex function, we define $\varphi(x) = h(\|x\|)$ for all $x \in X$. Then $\varphi^*(x^*) = h^*(\|x^*\|_{X^*})$.

Now we pass to the study of the subdifferential of a convex function. The subdifferential characterizes the local behavior of the convex function in a way analogous to that in which derivatives determine the local behavior of smooth functions.

The mathematical setting remains unchanged. So X is a locally convex space, X^* its topological dual, and $\langle \cdot, \cdot \rangle_X$ denotes the duality brackets for the pair (X, X^*). We furnish X with the $w(X, X^*)$-topology and X^* with the $w(X^*, X)$-topology.

DEFINITION 1.2.28 Let $\varphi\colon X \longrightarrow \overline{\mathbb{R}}$ be a proper function and $x_0 \in \operatorname{dom} \varphi$. The *subdifferential* of φ at x_0 is the set $\partial\varphi(x_0) \subseteq X^*$ (possibly empty) defined by

$$\partial\varphi(x_0) = \big\{ x^* \in X^* : \langle x^*, h\rangle_X \leq \varphi(x_0 + h) - \varphi(x_0) \quad \text{for all } h \in X \big\}.$$

REMARK 1.2.29 It is clear from the above definition that $\partial\varphi(x_0)$ is $w(X^*, X)$-closed and convex. Moreover, $x_0 \in \arg\min \varphi$ (i.e., $\varphi(x_0) = \inf\limits_X \varphi$) if and only if $0 \in \partial\varphi(x_0)$ (*generalized Fermat's rule*).

The following propositions are straightforward consequences of Definition 1.2.28.

PROPOSITION 1.2.30 *If $\varphi, \psi\colon X \longrightarrow \overline{\mathbb{R}}$ are proper functions, $x_0 \in \operatorname{dom} \varphi \cap \operatorname{dom} \psi$ and $\lambda > 0$, then*

(a) $\partial\varphi(x_0) + \partial\psi(x_0) \subseteq \partial(\varphi + \psi)(x_0)$.
(b) $\partial(\lambda\varphi)(x_0) = \lambda\partial\varphi(x_0)$.

PROPOSITION 1.2.31 *If $\varphi\colon X \longrightarrow \overline{\mathbb{R}}$ is a proper function, then $x^* \in \partial\varphi(x)$ if and only if $\varphi(x) + \varphi^*(x^*) = \langle x^*, x\rangle_X$.*

REMARK 1.2.32 So for the elements of the subdifferential $\partial\varphi(x)$ (also known as subgradients of φ at $x \in X$), the Young–Fenchel inequality (see Proposition 1.2.18) becomes an equality.

Using this proposition, we obtain the following.

PROPOSITION 1.2.33 (a) *If $\varphi : X \longrightarrow \overline{\mathbb{R}}$ is a proper function and $x^* \in \partial\varphi(x)$, then $x \in \partial\varphi^*(x^*)$.*
(b) *If $\varphi \in \Gamma_0(X)$, then $x^* \in \partial\varphi(x)$ if and only if $x \in \partial\varphi^*(x^*)$.*

In general the set $\partial\varphi(x)$ may be empty (consider, e.g., the case $x \notin \operatorname{dom} \varphi$). The next theorem gives a situation where we have subdifferentiability of φ at $x \in X$ (i.e., $\partial\varphi(x) \neq \emptyset$).

THEOREM 1.2.34 *If X is a Banach space and $\varphi : X \longrightarrow \overline{\mathbb{R}}$ is a convex function that is continuous at $x \in X$, then $\partial\varphi(x) \neq \emptyset$ and it is a w^*-compact and convex set in X^*.*

PROOF: From Theorem 1.2.3 we have that int epi $\varphi \neq \emptyset$. Because $\big(x, \varphi(x)\big)$ is a boundary point of epi φ, we can apply the weak separation theorem and obtain $(x^*, \xi) \in X^* \times \mathbb{R}$, $(x^*, \xi) \neq (0, 0)$ such that

$$\xi\big(\varphi(x) - \lambda\big) \leq \langle x^*, y - x\rangle_X \qquad \text{for all } (y, \lambda) \in \text{epi } \varphi. \tag{1.32}$$

Because λ can increase to $+\infty$, from(1.32) we infer that $\xi \geq 0$. If $\xi = 0$, then

$$\langle x^*, x\rangle_X \leq \langle x^*, y\rangle_X \qquad \text{for all } y \in \operatorname{dom}\varphi.$$

But $x \in \operatorname{int}\operatorname{dom}\varphi$ and so for some $\delta > 0$ we have $\langle x^*, u\rangle_X \geq 0$ for all $u \in \bar{B}_\delta(0)$, which implies that $x^* = 0$, a contradiction to the fact that $(x^*, \xi) \neq (0,0)$. So $\xi > 0$ and we can take $\xi = 1$. Then from (1.32) with $\lambda = \varphi(y)$, we have

$$\begin{aligned} &\varphi(x) - \varphi(y) \leq \langle x^*, y - x\rangle_X \qquad \text{for all } y \in \operatorname{dom}\varphi, \\ \Rightarrow\ & -x^* \in \partial\varphi(x) \neq \emptyset. \end{aligned}$$

From Theorem 1.2.7 we know that there exists $r > 0$ such that $\varphi|_{\bar{B}_r(x)}$ is Lipschitz continuous, with Lipschitz constant, say, $k_r > 0$. Then for $x^* \in \partial\varphi(x)$, we have

$$\begin{aligned} &\langle x^*, h\rangle_X \leq \varphi(x+h) - \varphi(x) \leq k_r \|h\|_X \qquad \text{for all } h \in \bar{B}_r(0), \\ \Rightarrow\ & \|x^*\|_{X^*} \leq k_r. \end{aligned}$$

This proves that $\partial\varphi(x^*)$ is bounded hence w^*-compact (by Alaoglou's theorem) and of course convex. □

From Example 1.2.27(a) we know that closed convex sets are completely described by their support function. In the next theorem for a certain φ we identify the support function of $\partial\varphi(x)$.

THEOREM 1.2.35 *If X is a Banach space and $\varphi\colon X \longrightarrow \overline{\mathbb{R}}$ is a convex function that is continuous at $x \in X$, then $\sigma\big(h; \partial\varphi(x)\big) = \varphi'(x; h)$ for all $h \in X$ (see Definition* 1.2.9).

PROOF: Let $\psi(h) = \varphi'(x; h)$, $h \in X$. Because φ is continuous at $x \in X$, then ψ is continuous convex and $\partial\varphi(x) \neq \emptyset$ (see Theorem 1.2.34). So for all $x^* \in \partial\varphi(x)$ and all $h \in X$, we have

$$\langle x^*, h\rangle_X \leq \psi(h) \leq \varphi(x+h) - \varphi(x).$$

If $\psi(h) = (1/\lambda)\big[\varphi(x+h) - \varphi(x)\big]$, $\lambda > 0$, then we can easily calculate that

$$\psi_\lambda^*(x^*) = \frac{1}{\lambda}\big[\varphi^*(x^*) + \varphi(x) - \langle x^*, x\rangle_X\big], \quad \lambda > 0.$$

Because $\psi = \inf\limits_{\lambda > 0} \psi_\lambda$, then $\psi^* = \sup\limits_{\lambda > 0} \psi_\lambda^*$ and so

$$\psi^*(x^*) = \sup_{\lambda > 0} \frac{1}{\lambda}\big[\varphi^*(x^*) + \varphi(x) - \langle x^*, x\rangle_X\big] = \begin{cases} 0 & \text{if } x^* \in \partial\varphi(x) \\ +\infty & \text{otherwise.} \end{cases}$$

(see Propositions 1.2.31 and 1.2.18). Therefore $\psi^* = i_{\partial\varphi(x)}$, hence $\psi^{**} = \psi = \sigma_{\partial\varphi(x)}$ (see Theorem 1.2.21 and Example 1.2.27(a)). □

REMARK 1.2.36 Both Theorems 1.2.34 and 1.2.35 are still valid in the more general context of dual pairs of locally convex spaces. However, to avoid technical difficulties we have presented them in the more familiar framework of Banach spaces and their duals.

It is natural to ask what is the exact relation between the subdifferential and Gâteaux derivative when the latter exists.

THEOREM 1.2.37 *Let X be a Banach space and $\varphi\colon X \longrightarrow \overline{\mathbb{R}}$ a proper convex function.*

(a) *If φ is Gâteaux differentiable at $x \in X$, then $\partial\varphi(x) = \{\varphi'(x)\}$.*
(b) *If φ is continuous at $x \in X$ and $\partial\varphi(x)$ is a singleton, then φ is Gâteaux differentiable at $x \in X$ and $\partial\varphi(x) = \{\varphi'(x)\}$.*

PROOF: (a) Because φ is convex, we have

$$\begin{aligned}&\langle \varphi'(x), h\rangle_X \leq \frac{1}{\lambda}\big[\varphi(x+\lambda h) - \varphi(x)\big] \leq \varphi(x+h) - \varphi(x)\\ &\text{for all } 0<\lambda<1 \quad \text{and all } h \in X,\\ \Rightarrow\ &\varphi'(x) \in \partial\varphi(x) \quad \text{(see Definition 1.2.28).}\end{aligned}$$

Now let $x^* \in \partial\varphi(x)$. Then

$$\begin{aligned}&\langle x^*, h\rangle_X \leq \frac{1}{\lambda}\big[\varphi(x+h) - \varphi(x)\big] \quad \text{for all } \lambda>0 \text{ and all } h \in X,\\ \Rightarrow\ &\langle x^*, h\rangle_X \leq \langle \varphi'(x), h\rangle_X \qquad \text{for all } h \in X,\\ \Rightarrow\ &x^* = \varphi'(x);\ \text{that is, } \partial\varphi(x) = \{\varphi'(x)\}.\end{aligned}$$

(b) Due to convexity of φ,

$$\varphi(x) + \lambda\varphi'(x;h) \leq \varphi(x+h) \qquad \text{for all } \lambda \in \mathbb{R} \text{ and all } h \in X.$$

Then the line $L = \Big\{\big(x+\lambda h, \varphi(x) + \lambda\varphi'(x;h)\big) : \lambda \in \mathbb{R}\Big\}$ does not intersect the set int epi $\varphi \neq \emptyset$. So we can apply the weak separation theorem and produce a closed hyperplane H such that $L \subseteq H$ and $H \cap$int epi $\varphi = \emptyset$. We know that $H =$ Gr l where l is a continuous affine function on X and $l(x) = \varphi(x)$. By hypothesis $\partial\varphi(x) = \{x^*\}$. So $l(h) = \langle x^*, h\rangle_X + \xi$, $\xi \in \mathbb{R}$. Because $L \subseteq H$, we have

$$\begin{aligned}&\varphi'(x;h) = \langle x^*, h\rangle_X \qquad \text{for all } h \in X\\ \Rightarrow\ &\varphi'(x) = x^*;\ \text{that is, } \partial\varphi(x) = \{\varphi'(x)\}.\end{aligned}$$

□

The subdifferential of convex functions has a rich calculus which in many respects parallels that of smooth functions. Next we present two basic rules of this calculus.

THEOREM 1.2.38 *If $\varphi, \psi\colon X \longrightarrow \overline{\mathbb{R}}$ are proper convex functions and there exists $\hat{x} \in$ dom $\varphi \cap$ dom ψ where one of the two functions, say φ, is continuous, then $\partial(\varphi+\psi)(x) = \partial\varphi(x) + \partial\psi(x)$ for all $x \in X$.*

PROOF: From Proposition 1.2.30(a) we know that we always have $\partial\varphi + \partial\psi \subseteq \partial(\varphi+\psi)$, So it suffices to show that the opposite inclusion holds. To this end let $u^* \in \partial(\varphi+\psi)(x)$. Then $x \in$ dom $\varphi \cap$ dom ψ and

$$\psi(x) - \psi(y) \leq \varphi(y) - \varphi(x) - \langle u^*, y-x\rangle_X = h(y) \qquad \text{for all } y \in X. \tag{1.33}$$

We consider the following two sets

$$C_1 = \text{epi } h \quad \text{and} \quad C_2 = \{(y,\mu) \in X \times \mathbb{R} : \mu \leq \psi(x) - \psi(y)\}.$$

Note that both sets are convex and from Theorem 1.2.3 we have that int $C_1 \neq \emptyset$. Moreover, int $C_1 \cap C_2 = \emptyset$. To see this note that if $y \in \text{int } C_1 \cap C_2$, then

$$h(y) < \mu \leq \psi(x) - \psi(y) \quad \text{(see Theorem 1.2.3)},$$

which contradicts (1.33). So we can apply the weak separation theorem and obtain $(x^*,\xi) \in X^* \times \mathbb{R}$, $(x^*,\xi) \neq (0,0)$ such that

$$\langle x^*, u\rangle_X + \xi\lambda \leq \langle x^*, y\rangle_X + \xi\mu \quad \text{for all } (u,\lambda) \in C_1 \text{ and all } (y,\mu) \in C_2. \tag{1.34}$$

The inequality in (1.34) is strict if $(u,\lambda) \in \text{int } C_1$. Note that $(x,0) \in C_2$ and λ can increase to $+\infty$. So we must have $\xi \leq 0$. If $\xi = 0$, then

$$\langle x^*, u\rangle \leq \langle x^*, x\rangle \qquad \text{for all } u \in \text{dom } h. \tag{1.35}$$

But h is continuous at x and so dom h is a neighborhood of x. Therefore from (1.35) it follows that $x^* = 0$; hence $(x^*,\xi) = (0,0)$, a contradiction.

So $\xi < 0$ and we may take $\xi = -1$. Then from (1.34), we have

$$\langle x^*, u\rangle_X - h(u) \leq \langle x^*, x\rangle_X \leq \langle x^*, y\rangle_X - \big(\psi(x) - \psi(y)\big),$$

for all $u \in \text{dom} h$ and all $y \in \text{dom } \psi$. From the second inequality, we obtain $-x^* \in \partial\psi(x)$. From the first inequality and because $h(x) = 0$, we have $x^* \in \partial h(x)$, hence $x^* + u^* \in \partial\varphi(x)$. So finally

$$\begin{aligned} & u^* = u^* + x^* + (-x^*) \in \partial\varphi(x) + \partial\psi(x) \\ \Rightarrow\ & \partial(\varphi + \psi)(x) \subseteq \partial\varphi(x) + \partial\psi(x). \end{aligned}$$

□

REMARK 1.2.39 By induction it follows that Theorem 1.2.38 is true for any family $\{\varphi_k\}_{k=1}^n$ of proper convex functions $\varphi_k \colon X \longrightarrow \overline{\mathbb{R}} = \mathbb{R} \cup \{+\infty\}$, such that all but one of the φ_ks are continuous at a point $x \in \bigcap_{k=1}^{n} \text{dom } \varphi_k$.

THEOREM 1.2.40 *If Y is another locally convex space, $A \in \mathcal{L}(X,Y)$, and $\varphi\colon Y \longrightarrow \overline{\mathbb{R}}$ is a proper convex function, then $A^*\partial\varphi(Ax) \subseteq \partial(\varphi \circ A)(x)$ for all $x \in X$. Moreover, equality holds if there is a point in the range of A, where φ is continuous.*

PROOF: The inclusion follows at once from Definition 1.2.28. We show that equality holds when φ is continuous on the range of A. To this end let $x^* \in \partial(\varphi \circ A)(x)$. Then

$$\langle x^*, u - x\rangle_X + (\varphi \circ A)(x) \leq (\varphi \circ A)(u) \qquad \text{for all } u \in X. \tag{1.36}$$

Consider the affine space $L \subseteq Y \times \mathbb{R}$ defined by

$$L = \big\{\big(Au, \langle x^*, u - x\rangle_X + (\varphi \circ A)(x)\big) \in Y \times \mathbb{R} :\ u \in X\big\}.$$

Note that $L \cap \text{int epi } \varphi = \emptyset$ and have only boundary points in common. The weak separation theorem implies that there exists a closed hyperplane H such that

$$L \subseteq H \quad \text{and} \quad H \cap \text{int epi } \varphi = \emptyset.$$

We know that $H = \text{Gr } l$ where $l : Y \longrightarrow \mathbb{R}$ is the continuous affine function defined by

$$l(y) = \langle y^*, y \rangle_Y + \vartheta, \quad y^* \in Y^*, \quad \vartheta \in \mathbb{R}.$$

From the inclusion $L \subseteq H$, we have

$$\langle y^*, Au \rangle_Y + \vartheta = \langle x^*, u - x \rangle_X + (\varphi \circ A)(x) \qquad \text{for all } u \in X.$$

Let $u = 0$. Then

$$\begin{aligned} &\vartheta = (\varphi \circ A)(x) - \langle x^*, x \rangle_X, \\ \Rightarrow\ &\langle y^*, Au \rangle_Y = \langle x^*, u \rangle_X \qquad \text{for all } u \in X, \\ \Rightarrow\ &x^* = A^* y^*. \end{aligned}$$

Moreover, inasmuch as $H \cap \text{int epi } \varphi = \emptyset$, we have

$$\begin{aligned} &\langle y^*, y \rangle_Y + (\varphi \circ A)(x) - \langle A^* y^*, x \rangle_X \le \varphi(y) \qquad \text{for all } y \in Y, \\ \Rightarrow\ &y^* \in \partial\varphi(Ax); \text{ that is, } x^* \in A^* \partial\varphi(Ax). \end{aligned}$$

Therefore we have proved that $\partial(\varphi \circ A)(x) \subseteq A^* \partial\varphi(Ax)$ and so equality must hold. $\square$

EXAMPLE 1.2.41 (a) If $\varphi : \mathbb{R} \longrightarrow \overline{\mathbb{R}}$ is a proper convex function and $x \in \text{int dom } \varphi$, then

$$\partial\varphi(x) = \left[\varphi'_-(x), \varphi'_+(x)\right].$$

Here

$$\varphi'_-(x) = \lim_{u \uparrow x} \frac{\varphi(u) - \varphi(x)}{u - x}$$

and

$$\varphi'_+(x) = \lim_{u \downarrow x} \frac{\varphi(u) - \varphi(x)}{u - x}.$$

(b) If X is a Banach space and $\varphi(x) = \frac{1}{2}\|x\|^2$, then $\partial\varphi(x) = \mathcal{F}(x) = \left\{x^* \in X^* : \langle x^*, x \rangle_X = \|x\|_X^2 = \|x^*\|_{X^*}^2\right\}$. This is the duality map, which is important in the study of evolution equations and is examined in more detail in Section 3.2. At this point simply note that if $X = H$ a Hilbert space is identified with its dual, $\mathcal{F} = \text{Id}_H$.

(c) If X is a Banach space and $\varphi(x) = \|x\|$, then

$$\partial\varphi(x) = \begin{cases} \bar{B}_1^* & \text{if } x = 0 \\ \frac{\mathcal{F}(x)}{\|x\|} & \text{if } x \neq 0 \end{cases}.$$

Here $\bar{B}_1^* = \{x^* \in X^* : \|x^*\|_{X^*} \le 1\}$.

(d) If X is a locally convex space, $C \subseteq X$ is nonempty, closed, convex, and $\varphi(x) = i_C(x)$, then $\partial\varphi(x) = N_C(x) = \left\{x^* \in X^* : \langle x^*, c - x \rangle_X \le 0 \text{ for all } c \in C\right\} = \left\{x^* \in X^* : \langle x^*, x \rangle_X = \sigma(x^*, C)\right\}$. This set is a closed convex cone, known as the normal cone to C at x. In Section 1.4 we study tangent and normal cones in more detail.

1.3 Locally Lipschitz Functions

In this section we extend the subdifferential theory beyond the domain of convex functions. The starting point is Corollary 1.2.8. According to that result every continuous convex function is locally Lipschitz. So it is reasonable to focus our attention on locally Lipchitz functions.

Throughout this section X is a Banach space with topological dual X^*. By $\langle \cdot, \cdot \rangle_X$ we denote the duality brackets for the pair (X, X^*). Additional hypotheses are introduced as needed.

DEFINITION 1.3.1 A function $\varphi\colon X \longrightarrow \mathbb{R}$ is said to be *locally Lipschitz* if for every $x \in X$ we can find a neighborhood $U \subseteq X$ of x and a constant $k_U > 0$ (depending on U) such that

$$|\varphi(y) - \varphi(u)| \leq k_U \|y - u\|_X \qquad \text{for all}\, y, u \in U.$$

REMARK 1.3.2 If $\varphi\colon X \longrightarrow \mathbb{R}$ is Lipschitz continuous on bounded sets, then φ is locally Lipschitz and if $\dim X < +\infty$, then these two properties are equivalent.

Although the directional derivative need not exist, nevertheless the local Lipschitz property permits the introduction of the following substitute of it.

DEFINITION 1.3.3 Let $\varphi\colon X \longrightarrow \mathbb{R}$ be a locally Lipschitz function. The *generalized directional derivative* of φ at $x \in X$ in the direction $h \in X$, is defined by

$$\varphi^0(x; h) = \limsup_{\substack{x' \to x \\ \lambda \downarrow 0}} \frac{\varphi(x' + \lambda h) - \varphi(x')}{\lambda} = \inf_{\varepsilon, \delta > 0} \sup_{\substack{\|x' - x\| \leq \varepsilon \\ 0 < \lambda \leq \delta}} \frac{\varphi(x' + \lambda h) - \varphi(x')}{\lambda}.$$

This quantity has some useful properties that are summarized in the proposition that follows.

PROPOSITION 1.3.4 *If $\varphi\colon X \longrightarrow \mathbb{R}$ is locally Lipschitz, then*

(a) *For every $x \in X$, the function $h \longrightarrow \varphi^0(x; h)$ is sublinear and Lipschitz continuous.*

(b) *The function $(x, h) \longrightarrow \varphi^0(x; h)$ is upper semicontinuous.*

(c) $\varphi^0(x; -h) = (-\varphi)^0(x; h)$.

PROOF: (a) It is clear from Definition 1.3.3 that $\varphi^0(x; \cdot)$ is positively homogeneous. Also if $h_1, h_2 \in X$, we have

$$
\begin{aligned}
&\varphi^0(x;h_1+h_2)\\
=\ &\limsup_{\substack{x'\to x\\ \lambda\downarrow 0}} \frac{\varphi(x'+\lambda(h_1+h_2))-\varphi(x')}{\lambda}\\
=\ &\limsup_{\substack{x'\to x\\ \lambda\downarrow 0}} \frac{\varphi(x'+\lambda(h_1+h_2))-\varphi(x'+\lambda h_2)+\varphi(x'+\lambda h_2)-\varphi(x')}{\lambda}\\
\le\ &\limsup_{\substack{x'\to x\\ \lambda\downarrow 0}} \frac{\varphi(x'+\lambda(h_1+h_2))-\varphi(x'+\lambda h_2)}{\lambda}+\limsup_{\substack{x'\to x\\ \lambda\downarrow 0}} \frac{\varphi(x'+\lambda h_2)-\varphi(x')}{\lambda}\\
=\ &\varphi^0(x;h_1)+\varphi^0(x;h_2).
\end{aligned}
$$

This proves the subadditivity of $\varphi^0(x;\cdot)$, hence its sublinearity.

If x' is near x and $\lambda>0$ is near 0, then due to the local Lipschitzness of φ, we have

$$
\begin{aligned}
&\frac{\varphi(x'+\lambda h)-\varphi(x')}{\lambda}\le k\|h\|_X \qquad \text{for all } h\in X,\\
\Rightarrow\ &\varphi^0(x;h)\le k\|h\|_X \qquad \text{for all } h\in X,\\
\Rightarrow\ &|\varphi^0(x;h)|\le k\|h\|_X
\end{aligned}
\tag{1.37}
$$

for all $h\in X$ (by virtue of the sublinearity of $\varphi^0(x;\cdot)$). From this and the sublinearity of $\varphi^0(x;\cdot)$ follows easily the Lipschitz continuity of $h\longrightarrow\varphi^0(x;h)$.

(b) Suppose $(x_n,h_n)\longrightarrow(x,h)$ in $X\times X$ as $n\to\infty$. For every $n\ge 1$, we can find $u_n\in X$ and $\lambda_n>0$ such that $\|u_n\|_X+\lambda_n\le 1/n$ and

$$
\begin{aligned}
\varphi^0(x_n;h_n) &\le \frac{\varphi(x_n+u_n+\lambda_n h_n)-\varphi(x_n+u_n)}{\lambda_n}+\frac{1}{n}\\
&= \frac{\varphi(x_n+u_n+\lambda_n h_n)-\varphi(x_n+u_n+\lambda_n h)}{\lambda_n}\\
&\quad+\frac{\varphi(x_n+u_n+\lambda_n h)-\varphi(x_n+u_n)}{\lambda_n}+\frac{1}{n},\\
\Rightarrow\quad &\limsup_{n\to\infty}\varphi^0(x_n;h_n)\le\varphi^0(x;h).
\end{aligned}
$$

This proves the upper semicontinuity of $(x,h)\longrightarrow\varphi^0(x;h)$.

(c) From Definition 1.3.3, we have

$$
\begin{aligned}
\varphi^0(x;-h) &= \limsup_{\substack{x'\to x\\ \lambda\downarrow 0}} \frac{\varphi(x'-\lambda h)-\varphi(x')}{\lambda}\\
&= \limsup_{\substack{y\to x\\ \lambda\downarrow 0}} \frac{(-\varphi)(y+\lambda h)-(-\varphi)(y)}{\lambda} \qquad (\text{setting } y=x'-\lambda h)\\
&= (-\varphi)^0(x;h).
\end{aligned}
$$

□

The sublinearity and continuity of $\varphi^0(x;\cdot)$ lead naturally to the following definition.

DEFINITION 1.3.5 If $\varphi\colon X \longrightarrow \mathbb{R}$ is a locally Lipschitz function, the *generalized subdifferential* of φ at $x \in X$ is defined by

$$\partial\varphi(x) = \{x^* \in X^* : \langle x^*, h\rangle_X \leq \varphi^0(x;h) \text{ for all } h \in X\}.$$

REMARK 1.3.6 By virtue of Proposition 1.2.20, for every $x \in X$ the set $\partial\varphi(x)$ is nonempty and clearly it is convex and w^*-closed. Moreover, from (1.37) we see that $\partial\varphi(x)$ is also bounded; hence it is w^*-compact (Alaoglou's theorem). Also $\varphi^0(x;\cdot) = \sigma\big(\cdot;\partial\varphi(x)\big)$.

PROPOSITION 1.3.7 *If $\varphi\colon X \longrightarrow \mathbb{R}$ is locally Lipschitz, $x_n \longrightarrow x$ in X, $x_n^* \xrightarrow{w} x^*$ in X^*, and $x_n^* \in \partial\varphi(x_n)$ for all $n \geq 1$, then $x^* \in \partial\varphi(x)$.*

PROOF: For every $n \geq 1$ and every $h \in X$, we have $\langle x_n^*, h\rangle_X \leq \varphi^0(x_n;h)$. Passing to the limit as $n \to \infty$ and using Proposition 1.3.4(b), we obtain $\langle x^*, h\rangle_X \leq \varphi^0(x;h)$ for all $h \in X$; hence $x^* \in \partial\varphi(x)$. □

PROPOSITION 1.3.8 *Let $\varphi\colon X \longrightarrow \mathbb{R}$ be locally Lipschitz.*

(a) *If φ is Gâteaux differentiable at $x \in X$, then $\varphi'(x) \in \partial\varphi(x)$.*
(b) *If $\varphi \in C^1(X)$, then $\partial\varphi(x) = \{\varphi'(x)\}$.*

PROOF: (a) From Definition 1.3.3 we see that

$$\begin{aligned} &\big\langle \varphi'(x), h\big\rangle_X \leq \varphi^0(x;h) \qquad \text{for all } h \in X, \\ \Rightarrow\ &\varphi'(x) \in \partial\varphi(x) \quad \text{(see Definition 1.3.5)}. \end{aligned}$$

(b) Because $\varphi \in C^1(X)$, we have $\varphi^0(x;h) = \langle \varphi'(x), h\rangle_X$ for all $h \in X$; hence $\partial\varphi(x) = \{\varphi'(x)\}$. □

REMARK 1.3.9 It can happen that φ is Gâteaux differentiable at $x \in X$ but $\partial\varphi(x)$ contains elements other than $\varphi'(x)$. Consider the function $\varphi : \mathbb{R} \longrightarrow \mathbb{R}$ defined by

$$\varphi(x) = \begin{cases} x^2 \sin\frac{1}{x} & \text{if } x \neq 0 \\ 0 & \text{if } x = 0 \end{cases}.$$

Then φ is differentiable on $[0,1]$, Lipschitz continuous on $[0,1]$, and $\varphi'(0) = 0$. However, $\varphi^0(0;h) = |h|$ and so $\partial\varphi(0) = [-1,1]$. Also Proposition 1.3.8(b) is actually true under the weaker hypothesis that φ is strictly differentiable. Recall that φ is strictly differentiable, if for every $x \in X$, there exists $\varphi_s'(x) \in X^*$ such that for all $h \in X$ we have

$$\big\langle \varphi_s'(x), h\big\rangle_X = \lim_{\substack{x' \to x \\ \lambda \downarrow 0}} \frac{\varphi(x' + \lambda h) - \varphi(x')}{\lambda}.$$

A C^1-function is strictly differentiable and a strictly differentiable function is locally Lipschitz.

PROPOSITION 1.3.10 *If φ is continuous, and convex (hence locally Lipschitz, see Corollary 1.2.8), then $\partial\varphi(x)$ coincides with the subdifferential in the sense of convex analysis (see Definition 1.2.28).*

PROOF: For fixed $h \in X$, the function $(t,u) \longrightarrow \frac{\varphi(u+th)-\varphi(u)}{t}$ is continuous from $(0,+\infty) \times X$ into $\mathbb{R}$. So given $\varepsilon>0$, we can find $\delta>0$ such that

$$\frac{\varphi(u+th)-\varphi(u)}{t} \leq \frac{\varphi(x+\lambda h)-\varphi(x)}{\lambda} + \varepsilon$$

for all $(t,u) \in (0,+\infty) \times X$ such that $|t-\lambda| \leq \delta$ and $\|u-x\|_X \leq \delta$. Then

$$\sup_{\|u-x\|_X \leq \delta} \frac{\varphi\big(u+(\lambda+\delta)h\big)-\varphi(u)}{\lambda} \leq \frac{\varphi(x+\lambda h)-\varphi(x)}{\lambda} + \varepsilon.$$

Passing to the limit as $\lambda, \delta \downarrow 0$, we obtain

$$\varphi^0(x;h) \leq \varphi'(x;h) + \varepsilon.$$

Because $\varepsilon > 0$ was arbitrary let $\varepsilon \downarrow 0$, to conclude that $\varphi^0(x;h) \leq \varphi'(x;h)$. The opposite inequality is always true when $\varphi'(x;\cdot)$ exists, thus we conclude that $\varphi^0(x;\cdot) = \varphi'(x;\cdot)$. Then the proposition follows from Theorem 1.2.35. □

At this point let us recall the following fundamental result concerning vector-valued locally Lipschitz functions known as *Rademacher's theorem.* For a proof of it we refer to Evans–Gariepy [228].

THEOREM 1.3.11 *If* $\varphi\colon \mathbb{R}^n \longrightarrow \mathbb{R}^m$ *is locally Lipschitz, then* φ *is Fréchet differentiable* λ^n*-a.e.* (λ^n *is the Lebesgue measure on* $\mathbb{R}^n$).

Using this theorem, when X is finite-dimensional, we can have a definition of the generalized subdifferential, which is more intuitive and geometric than the one in terms of the generalized directional derivative (see Definition 1.3.5).

THEOREM 1.3.12 *If* $\varphi\colon \mathbb{R}^n \longrightarrow \mathbb{R}$ *is locally Lipschitz and* $E \subseteq \mathbb{R}^n$ *any Lebesgue-null set, then for every* $x \in \mathbb{R}^n$, $\partial\varphi(x) = \operatorname{conv}\{\lim_{n\to\infty} \nabla\varphi(x_n) : x_n \longrightarrow x,\ x_n \notin E \cup D_\varphi^c\}$ *(here by* D_φ *we denote the set of points of differentiability of* φ*; by Theorem 1.3.11,* $\lambda^n(D_\varphi^c) = 0$*).*

PROOF: By (1.37), $\partial\varphi(\cdot)$ is bounded (i.e., maps bounded sets to bounded sets). Because $\nabla\varphi(x_n) \in \partial\varphi(x_n)$, $n \geq 1$ (see Proposition 1.3.8(a)), we have $\{\nabla\varphi(x_n)\}_{n\geq 1} \in \mathbb{R}^n$ is bounded and so we may assume that $\nabla\varphi(x_n) \longrightarrow u^*$ in $\mathbb{R}^n$. Moreover, Proposition 1.3.7 implies that $u^* \in \partial\varphi(x)$. Hence

$$S(x) = \operatorname{conv}\{\lim \nabla\varphi(x_n) : x_n \longrightarrow x,\ x_n \notin E \cup D_\varphi^c\} \subseteq \partial\varphi(x). \tag{1.38}$$

Next for $h \neq 0$, let $\vartheta_h = \lim\limits_{\substack{x' \to x \\ x' \notin E \cup D_\varphi^c}} \big(\nabla\varphi(x'), h\big)_{\mathbb{R}^n}$. Given $\varepsilon>0$, we can find $\delta = \delta(\varepsilon)>0$ such that

$$\big(\nabla\varphi(x'), h\big)_{\mathbb{R}^n} \leq \vartheta_h + \varepsilon \qquad \text{for all } \|x'-x\|_{\mathbb{R}^n} \leq \delta,\ x' \notin E \cup D_\varphi^c.$$

If $0<\lambda<\delta/(2\|h\|_{\mathbb{R}^n})$, then for almost all $x' \in \mathbb{R}^n$ with $\|x'-x\|_{\mathbb{R}^n} \leq \delta/2$ we have

$$\varphi(x'+\lambda h)-\varphi(x')=\int_0^\lambda \big(\nabla\varphi(x'+th),h\big)_{\mathbb{R}^n}\mathrm{d}t\leq\lambda(\vartheta_h+\varepsilon),$$
$$\Rightarrow\ \varphi^0(x;h)\leq\vartheta_h+\varepsilon,$$
$$\Rightarrow\ \varphi^0(x;h)\leq\sigma\big(h,S(x)\big). \tag{1.39}$$

From (1.38) and (1.39), we conclude that $S(x)=\partial\varphi(x)$ (see Example 1.2.27(a)). □

COROLLARY 1.3.13 *If* $\varphi\colon\mathbb{R}^n\longrightarrow\mathbb{R}$ *is locally Lipschitz, then* $\varphi^0(x;h)=\limsup\limits_{\substack{x'\to x\\ x'\notin E\cup D_\varphi^c}}\big(\nabla\varphi(x'),h\big)_{\mathbb{R}^n}$.

The importance of the generalized subdifferential comes from the rich calculus that it has. We present a few basic results in this direction. We start with a definition.

DEFINITION 1.3.14 A locally Lipschitz function $\varphi\colon\mathbb{R}\longrightarrow\mathbb{R}$ is said to be *regular* at x if

(i) The directional derivative $\varphi'(x;h)$ in the sense of Definition 1.2.9 exists for all $h\in X$.

(ii) $\varphi^0(x;h)=\varphi'(x;h)$ for all $h\in X$.

REMARK 1.3.15 Continuous convex functions and strictly differentiable functions (in particular C^1-functions) are regular at $x\in X$.

The first calculus rule is a direct consequence of Definition 1.3.5 (see also Definition 1.3.3).

PROPOSITION 1.3.16 *If* $\varphi_k\colon X\longrightarrow\mathbb{R}$ $k\in\{1,\ldots,m\}$ *are locally Lipchitz functions and* $\{\lambda_k\}_{k=1}^m\subseteq\mathbb{R}$*, then* $\partial\Big(\sum\limits_{k=1}^m\lambda_k\varphi_k\Big)(x)\subseteq\sum\limits_{k=1}^m\lambda_k\varphi_k(x)$ *for all* $x\in X$*; equality holds if each* φ_k *is regular at* $x\in X$ *or if all but one of the* φ_k*s is strictly differentiable at* $x\in X$ *and if* $\lambda_k\geq 0$.

With the scalar multiples we can more precise.

PROPOSITION 1.3.17 *If* $\varphi\colon X\longrightarrow\mathbb{R}$ *is locally Lipchitz and* $\lambda\in\mathbb{R}$*, then* $\partial(\lambda\varphi)(x)=\lambda\partial\varphi(x)$ *for all* $x\in X$.

PROOF: If $\lambda\geq 0$, then $(\lambda\varphi)^0=\lambda\varphi^0$ and so $\partial(\lambda\varphi)=\lambda\partial\varphi$. Therefore we may assume that $\lambda<0$ and in fact we can take $\lambda=-1$. Then $x^*\in\partial(-\varphi)(x)$ if and only if $\langle x^*,h\rangle_X\leq(-\varphi)^0(x;h)$ (see Definition 1.3.5). But from Proposition 1.3.4(c), we have $(-\varphi)^0(x;h)=\varphi^0(x;-h)$. Hence $\langle x^*,h\rangle_X\leq\varphi^0(x;-h)$ for all $h\in X$ and this is equivalent to saying that $-x^*\in\partial\varphi(x)$; that is, $\partial(-\varphi)(x)=-\partial\varphi(x)$. □

We have the following chain rule.

PROPOSITION 1.3.18 *If Y is another Banach space, $f \in C^1(X,Y)$, and $\varphi\colon X \longrightarrow \mathbb{R}$ is locally Lipchitz, then $\varphi \circ f\colon X \longrightarrow \mathbb{R}$ is locally Lipchitz and*

$$\partial(\varphi \circ f)(x) \subseteq \partial\varphi\big(f(x)\big) \circ f'(x) \tag{1.40}$$

in the sense that for every $x^* \in \partial(\varphi \circ f)(x)$, we have

$$x^* = f'(x)^* u^* \qquad \text{for some } u^* \in \partial\varphi\big(f(x)\big). \tag{1.41}$$

Moreover, if φ or $(-\varphi)$ is regular at $f(x)$, then $\varphi \circ f$ (or $-\varphi \circ f$) is regular at x and equality holds in (1.40). Also if f maps every neighborhood of x onto a set that is dense in a neighborhood $f(x)$ (e.g., if $f'(x) \in \mathcal{L}(X,Y)$ is surjective), then equality holds in (1.40).

PROOF: Because φ is locally Lipschitz, we can find V a neighborhood of $f(x)$ such that $\varphi\big|_V$ is Lipschitz continuous (see Definition 1.3.1). Then $U = f^{-1}(V)$ is a neighborhood of $x \in X$ and because $f(U) \subseteq V$, we see that $\varphi \circ f\big|_U$ is Lipschitz continuous. Therefore $\varphi \circ f\colon X \longrightarrow \mathbb{R}$ is locally Lipschitz.

By virtue of Proposition 1.3.4(b), given $\varepsilon > 0$ we can find $0 < \delta \leq \varepsilon$ such that

$$\varphi^0\big(f(x); h'\big) \leq \varphi^0\big(f(x); h\big) + \varepsilon \qquad \text{for all } \|h' - h\|_Y < \delta.$$

Also from the definition of the generalized directional derivative (see Definition 1.3.3), we can find $0 < \xi, \beta$ such that

$$\frac{\varphi(y + \lambda h') - \varphi(y)}{\lambda} \leq \varphi^0\big(f(x); h'\big) + \varepsilon \leq \varphi^0\big(f(x); h\big) + 2\varepsilon \tag{1.42}$$

for all $\|y - f(x)\|_Y \leq \xi$, $\lambda \leq \beta$, $\|h' - h\|_Y < \delta$.

We set $h = f'(x)v$, $v \in X$. Because $f \in C^1(X,Y)$ we can find $0 < \eta \leq \beta$ such that

$$\Big\|\frac{f(u + \lambda v) - f(u)}{\lambda} - f'(x)v\Big\|_Y < \delta \quad \text{and} \quad \|f(u) - f(x)\|_Y \leq \xi$$

for all $\|u - x\|_X \leq \eta$, $\lambda \leq \eta$.

In (1.42), we set $y = f(u)$ and $h' = \frac{f(u+\lambda v) - \varphi(u)}{\lambda}$. We have

$$\frac{(\varphi \circ f)(u + \lambda v) - (\varphi \circ f)(u)}{\lambda} \leq \varphi^0\big(f(x); f'(x)v\big) + 2\varepsilon,$$

for all $\|u - x\|_X \leq \eta$, $\lambda \leq \eta$,

$$\Rightarrow\ (\varphi \circ f)^0(x; v) \leq \varphi^0\big(f(x); f'(x)v\big) = \max\big[\langle y^*, f'(x)v\rangle_Y : y^* \in \partial\varphi\big(f(x)\big)\big]. \tag{1.43}$$

Therefore (1.40) holds.

Next suppose that φ is regular at $f(x)$. The case where $-\varphi$ is regular at $f(x)$ can be derived from the former because $\partial(-\varphi)\big(f(x)\big) = -\partial\varphi\big(f(x)\big)$ (see Proposition 1.3.17).

Because of the regularity of φ at $f(x)$, we have

$$
\begin{aligned}
&\varphi^0\big(f(x);f'(x)v\big)=\varphi'\big(f(x);f'(x)v\big)\\
&=\lim_{\lambda\downarrow 0}\frac{\varphi\big(f(x)+\lambda f'(x)v\big)-\varphi\big(f(x)\big)}{\lambda}\\
&=\lim_{\lambda\downarrow 0}\Big[\frac{\varphi\big(f(x)+\lambda f'(x)v\big)-\varphi\big(f(x+\lambda v)\big)}{\lambda}+\frac{\varphi\big(f(x+\lambda v)\big)-\varphi\big(f(x)\big)}{\lambda}\Big]\\
&=(\varphi\circ f)'(x;v)\quad\text{(because } \varphi \text{ is locally Lipschitz and } f\in C^1(X,Y))\\
&\leq (\varphi\circ f)^0(x;v).
\end{aligned}
\tag{1.44}
$$

Comparing (1.43) and (1.44) we conclude that

$$
\begin{aligned}
&\varphi^0\big(f(x);f'(x)v\big)=(\varphi\circ f)^0(x;v)\qquad\text{for all } v\in X,\\
\Rightarrow\ &\partial(\varphi\circ f)(x)=\partial\varphi\big(f(x)\big)\circ f'(x).
\end{aligned}
\tag{1.45}
$$

If f maps any neighborhood of x onto a dense subset of a neighborhood of $f(x)$, then we can write

$$
\begin{aligned}
\varphi^0\big(f(x);f'(x)v\big)&=\limsup_{\substack{y\to f(x)\\ \lambda\downarrow 0}}\frac{\varphi\big(y+\lambda f'(x)v\big)-\varphi(y)}{\lambda}\\
&=\limsup_{\substack{u\to x\\ \lambda\downarrow 0}}\frac{\varphi\big(f(u)+\lambda f'(x)v\big)-\varphi\big(f(u)\big)}{\lambda}\\
&=\limsup_{\substack{u\to x\\ \lambda\downarrow 0}}\frac{\varphi\big(f(x+\lambda v)\big)-\varphi\big(f(u)\big)}{\lambda}\\
&\qquad\text{(because } f\in C^1(X,Y))\\
&=(\varphi\circ f)^0(x;v)\qquad\text{for all } v\in X.
\end{aligned}
$$

Therefore (1.45) must hold. □

REMARK 1.3.19 Using the adjoint of the linear operator $f'(x)\in\mathcal{L}(X,Y)$ we can write (1.40) as

$$\partial(\varphi\circ f)(x)\subseteq f'(x)^*\partial\varphi\big(f(x)\big)\ \text{(see (1.41))}.$$

COROLLARY 1.3.20 *If X is continuously and densely embedded in another Banach space Y, $\varphi\colon Y\longrightarrow\mathbb{R}$ is locally Lipschitz, and $\widehat{\varphi}=\varphi\big|_X$, then $\partial\widehat{\varphi}(x)=\partial\varphi(x)$ for all $x\in X$, which means that every element of $\partial\widehat{\varphi}(x)$ admits a unique extension to an element of $\partial\varphi(x)$.*

PROOF: Let $i\in\mathcal{L}(X,Y)$ be the canonical embedding. Then we apply Proposition 1.3.18 with $f=i$ to obtain the desired equality. □

Another result in the same direction as Proposition 1.3.18, is the following.

PROPOSITION 1.3.21 *If $T=[0,b]$, $f\in C^1(T,X)$, and $\varphi\colon X\longrightarrow\mathbb{R}$ is locally Lipschitz, then $g=\varphi\circ f\colon T\longrightarrow\mathbb{R}$ is differentiable a.e. on T and*

$$g'(t)\leq\max\big[\langle x^*,f'(t)\rangle_X : x^*\in\partial\varphi\big(f(t)\big)\big].$$

PROOF: The function $g\colon T \longrightarrow \mathbb{R}$ is locally Lipchitz, hence it is differentiable almost everywhere (see Theorem 1.3.11). Let $t_0 \in T$ be a point of differentiability of g. We have

$$\begin{aligned} g'(t_0) &= \lim_{\lambda\to 0} \frac{\varphi\big(f(t_0+\lambda)\big) - \varphi\big(f(t_0)\big)}{\lambda} \\ &= \lim_{\lambda\to 0} \frac{\varphi\big(f(t_0)+\lambda f'(t_0)+o(\lambda)\big) - \varphi\big(f(t_0)\big)}{\lambda} \\ &\qquad \Big(\text{here } \frac{o(\lambda)}{\lambda} \longrightarrow 0 \text{ as } \lambda \to 0\Big) \\ &= \lim_{\lambda\to 0} \Big[\frac{\varphi\big(f(t_0)+\lambda f'(t_0)\big) - \varphi\big(f(t_0)\big)}{\lambda} \\ &\qquad + \frac{\varphi\big(f(t_0)+\lambda f'(t_0)+o(\lambda)\big) - \varphi\big(f(t_0)+\lambda f'(t_0)\big)}{\lambda}\Big] \\ &= \lim_{\lambda\to 0} \frac{\varphi\big(f(t_0)+\lambda f'(t_0)\big) - \varphi\big(f(t_0)\big)}{\lambda} \\ &\le \varphi^0\big(f(t_0); f'(t_0)\big) = \max\big[\langle x^*, f'(t_0)\rangle_X : x^* \in \partial\varphi\big(f(t_0)\big)\big]. \end{aligned}$$

□

The next proposition produces an extension to the present nonsmooth setting of the classical Fermat's rule for local extrema.

PROPOSITION 1.3.22 *If $\varphi\colon X \longrightarrow \mathbb{R}$ is locally Lipschitz which attains a local extremum (local minimum or local maximum) at $x \in X$, then $0 \in \partial\varphi(x)$.*

PROOF: Because $\partial(-\varphi) = -\partial\varphi$ it suffices to prove the proposition when x is a local minimizer. Then directly from Definition 1.3.3 we see that $0 \le \varphi^0(x; h)$ for all $h \in X$, hence $0 \in \partial\varphi(x)$ (see Definition 1.3.5). □

Using this proposition, we can prove a mean value theorem for the generalized subdifferential.

THEOREM 1.3.23 *If $\varphi\colon X \longrightarrow \mathbb{R}$ is locally Lipschitz and $x, u \in X$, then we can find $0 < \lambda_0 < 1$ and $v^* \in \partial\varphi\big((1-\lambda_0)x + \lambda_0 u\big)$ such that*

$$\varphi(u) - \varphi(x) = \langle v^*, u - x\rangle_X .$$

PROOF: Let $\xi\colon \mathbb{R} \longrightarrow \mathbb{R}$ be defined

$$\xi(\lambda) = \varphi((1-\lambda)x + \lambda u) + \lambda\big(\varphi(x) - \varphi(u)\big), \quad \lambda \in \mathbb{R}.$$

Evidently ξ is locally Lipschitz on $\mathbb{R}$, hence Lipschitz continuous on $[0,1]$. Note that $\xi(0) = \xi(1) = \varphi(x)$. So we can find $\lambda_0 \in (0,1)$ such that ξ attains a local maximum or minimum at λ_0. By virtue of Proposition 1.3.22 we have $0 \in \partial\xi(\lambda_0)$. Moreover, an easy calculation gives

$$\partial\xi(\lambda_0) \subseteq \big\langle \partial\varphi\big(x + \lambda_0(u-x)\big), u - x\big\rangle_X + [\varphi(x) - \varphi(u)].$$

Therefore finally we have

$$\varphi(u) - \varphi(x) \in \langle \partial\varphi(x + \lambda_0(u - x)), u - x \rangle_X .$$

□

We conclude this section with a *Lagrange multipliers rule*. For a proof of it we refer to Clarke [149].

PROPOSITION 1.3.24 *If $\varphi, \psi \colon X \longrightarrow \mathbb{R}$ are locally Lipschitz functions and $x \in X$ solves the problem*

$$\inf[\varphi(x) \colon \psi(x) \le 0],$$

then there exist $\lambda, \mu \ge 0$ such that

$$0 \in \lambda\partial\varphi(x) + \mu\partial\psi(x) \quad \textit{and} \quad \mu\psi(x) = 0.$$

Moreover, if ψ is also convex and there exists $x_0 \in X$ such that $\psi(x_0) < 0$, then we have $\lambda > 0$ (and so we can take $\lambda = 1$, normal multipliers).

1.4 Variational Geometry

Differential geometry is built upon smooth manifolds where the sets of tangents are convenient vector spaces. But in many applications we encounter nonsmooth sets. The first step away from the smooth setting of differential geometry was to consider convex sets. Then the tangent and normal spaces of differential geometry are replaced by the tangent and normal cones which actually share many of the properties of their smooth counterparts. The second step is to get rid of the convexity requirement on the set using the generalized subdifferential of locally Lipschitz functions. We show that some interesting results can be obtained this way.

Throughout this section, X is a Banach space and X^* its topological dual. By $\langle \cdot, \cdot \rangle_X$ we denote the duality brackets for the pair (X, X^*).

DEFINITION 1.4.1 Let $C \subseteq X$ be a nonempty convex set and $x \in \bar{C}$. The *tangent cone* to C at x is the set

$$T_C(x) = \overline{\bigcup_{\lambda > 0} \frac{1}{\lambda}(C - x)}.$$

REMARK 1.4.2 Evidently $T_C(x) = T_{\bar{C}}(x)$ and so without any loss of generality we can always assume that C is closed. Also from Definition 1.4.1 it follows that $h \in T_C(x)$ if and only if there exist sequences $\lambda_n \longrightarrow 0^+$ in $\mathbb{R}$ and $h_n \longrightarrow h$ in X such that $x + \lambda_n h_n \in \bar{C}$ for all $n \ge 1$. Note that $C \subseteq x + T_C(x)$.

PROPOSITION 1.4.3 *If $C \subseteq X$ is a nonempty closed convex set and $x \in C$, then $T_C(x)$ is a closed convex cone.*

PROOF: Let $S_C(x) = \bigcup_{\lambda > 0} \frac{1}{\lambda}(C - x)$. First we show that for all $h \in S_C(x)$, we can find $\lambda_0 > 0$ such that for all $0 \le \lambda \le \lambda_0$ we have $x + \lambda h \in C$. Indeed note that from the definition of $S_C(x)$ we can find $\lambda_0 > 0$ such that $x + \lambda_0 h \in C$. Then for

$0 \le \lambda \le \lambda_0$ we have $x+\lambda h = \big(1-(\lambda/\lambda_0)\big)x+(\lambda/\lambda_0)(x+\lambda h)$ and the right-hand side is a convex combination of elements in C. Due to the convexity of C, we have that $x+\lambda h \in C$. Using this fact we can show that $S_C(x)$ is a convex cone. To this end let $h_1, h_2 \in S_C(x)$. Then we can find $\lambda_1, \lambda_2 > 0$ such that $x + \lambda_k h_k \in C$, $k = 1, 2$. Set $\lambda_0 = \min\{\lambda_1, \lambda_2\}$. From the first part of the proof we have $x + \lambda_0 h_k \in C$, $k = 1, 2$. Exploiting the convexity of C, we obtain $x + \lambda_0\big(th_1 + (1-t)h_2\big) \in C$, $t \in [0,1]$. This proves that $S_C(x)$ is a convex cone. It follows that $\overline{S_C(x)} = T_C(x)$ is a closed convex cone. $\square$

REMARK 1.4.4 We have $T_{\{x\}}(x) = \{0\}$ and if $x \in \operatorname{int} C$, then $T_C(x) = X$.

In differential geometry, the normal space to a smooth manifold embedded in $\mathbb{R}^N$ at a point x, is the orthogonal complement in $\mathbb{R}^N$ of the tangent space. In a similar way, replacing the orthogonal subspace by the negative polar cone, we can define the normal cone to C at x.

DEFINITION 1.4.5 Let $C \subseteq X$ be a nonempty, closed convex set. The *normal cone* to C at $x \in C$ is defined by

$$\begin{aligned} N_C(x) &= \big\{x^* \in X^* \colon \langle x^*, x\rangle_X = \sigma(x^*; C)\big\} \\ &= \big\{x^* \in X^* \colon \langle x^*, c - x\rangle \le 0 \qquad \text{for all } c \in C\big\}. \end{aligned}$$

REMARK 1.4.6 Evidently $N_C(x) = T_C(x)^-$ (= the negative polar cone of $T_C(x)$) and so it is closed and convex. Then $N_{\{x\}}(x) = X^*$ and if $x \in \operatorname{int} C$, we have $N_C(x) = \{0\}$ (see Remark 1.4.4). Moreover, if i_C is the indicator function of the set C; that is,

$$i_C(x) = \begin{cases} 0 & \text{if } x \in C \\ +\infty & \text{if } x \notin C \end{cases},$$

then $i_C \in \Gamma_0(X)$ and $\partial i_C(x) = N_C(x)$ (the subdifferential is taken in the sense of convex analysis; see Definition 1.2.28).

The following calculus rules for the tangent and normal cones follow easily from Definitions 1.4.1 and 1.4.5.

PROPOSITION 1.4.7 (a) *If $C_1 \subseteq C_2 \subseteq X$ are nonempty closed convex sets and $x \in C_1$, then $T_{C_1}(x) \subseteq T_{C_2}(x)$ and $N_{C_2}(x) \subseteq N_{C_1}(x)$.*

(b) *If $\{X_k\}_{k=1}^n$ are Banach spaces $C_k \subseteq X_k$, $k \in \{1, \ldots, n\}$ are nonempty, closed, and convex sets, $X = \prod_{k=1}^n X_k$, $C = \prod_{k=1}^n C_k \subseteq X$ and $x = (x_k)_{k=1}^n \in C$, then $T_C(x) = \prod_{k=1}^n T_{C_k}(x_k)$ and $N_C(x) = \prod_{k=1}^n N_{C_k}(x_k)$.*

(c) *If X, Y are Banach spaces, $C \subseteq X$ is a nonempty, closed, and convex set, and $A \in \mathcal{L}(X, Y)$, then $T_{\overline{A(C)}}(Ax) = \overline{AT_C(x)}$ and $N_{\overline{A(C)}}(Ax) = (A^*)^{-1}N_C(x)$.*

(d) *If C_1, $C_2 \subseteq X$ are two nonempty, closed convex sets and $x_1 \in C_1$, $x_2 \in C_2$, then $T_{\overline{C_1+C_2}}(x_1 + x_2) = \overline{T_{C_1}(x_1) + T_{C_2}(x_2)}$ and $N_{\overline{C_1+C_2}}(x_1 + x_2) = N_{C_1}(x_1) \cap N_{C_2}(x_2)$.*

For the next calculus rule, we need some auxiliary material which is actually of independent interest. We start with a minimax theorem, known as *Nikaido's theorem*, which we state here without a proof. We return to this subject in Section 2.2.

THEOREM 1.4.8 *If V, Z are locally convex spaces, $C \subseteq Z$ is nonempty, convex, and $w(Z, Z^*)$-compact, and $\eta\colon V \times Z \longrightarrow \mathbb{R}$ is a function such that*

(i) *For all $z \in Z$, $v \longrightarrow \eta(v, z)$ is convex and lower semicontinuous;*
(ii) *For all $v \in V$, $z \longrightarrow \eta(v, z)$ is concave and upper semicontinuous;*

then $\inf\limits_{v\in V} \max\limits_{z\in C} \eta(v, z) = \max\limits_{z\in C} \inf\limits_{v\in V} \eta(v, z)$.

Using this theorem, we can prove the following optimization result.

THEOREM 1.4.9 *If X, Y are reflexive Banach spaces, $A \in \mathcal{L}(X, Y)$ $\varphi \in \Gamma_0(X)$, $\psi \in \Gamma_0(Y)$, $0 \in \operatorname{int}\big(A(\operatorname{dom}\varphi) - \operatorname{dom}\psi\big)$, and $x_0 \in X$ satisfies*

$$\varphi(x_0) + \psi\big(A(x_0)\big) = \min\big[\varphi(x) + \psi\big(A(x)\big) : x \in X\big],$$

then $0 \in A^* \partial\psi\big(A(x_0)\big) + \partial\varphi(x_0)$.

PROOF: Let $\bar{B}_1^X = \{x \in X : \|x\|_X \leq 1\}$ and $\psi^* : Y^* \longrightarrow \overline{\mathbb{R}}$ be the conjugate of ψ (see Definition 1.2.15). Let

$$m = \sup_{n\geq 1} \inf_{y^*\in Y^*} \sup_{x\in n\bar{B}_1^X} \big[\psi^*(y^*) - \langle y^*, A(x)\rangle_Y - \varphi(x)\big].$$

Let $\eta(y^*, x) = \psi^*(y^*) - \langle y^*, A(x)\rangle_Y - \varphi(x)$. Evidently η satisfies hypotheses (i) and (ii) of Theorem 1.4.8 with $V = Y^*$ and $Z = X_w =$ the Banach space X furnished with the weak topology. Because $\bar{B}_1^X$ is w-compact in X (due to the reflexivity of X), by virtue of Theorem 1.4.8, we have

$$\begin{aligned} m &= \sup_{n\geq 1} \sup_{x\in n\bar{B}_1^X} \inf_{y^*\in Y^*} \big[\psi^*(y^*) - \langle y^*, A(x)\rangle_Y - \varphi(x)\big] \\ &= \sup_{x\in X} \inf_{y^*\in Y^*} \big[\psi^*(y^*) - \langle y^*, A(x)\rangle_Y - \varphi(x)\big]. \end{aligned} \tag{1.46}$$

From Propositions 1.2.18 and 1.2.31, we have that

$$\inf_{y^*\in Y^*} \big[\psi^*(y^*) - \langle y^*, A(x)\rangle_Y\big] = -\psi\big(A(x)\big). \tag{1.47}$$

Using this (1.47) in (1.46), we obtain

$$m = \sup_{x\in X} \big[-\psi\big(A(x)\big) - \varphi(x)\big] = -\inf_{x\in X} \big[\varphi(x) + \psi\big(A(x)\big)\big] = -\varphi(x_0) - \psi\big(A(x_0)\big). \tag{1.48}$$

Let $\{y_n^*\}_{n\geq 1} \subseteq Y^*$ be such that

$$\limsup_{n\to\infty} \big[\psi^*(y_n^*) - \langle y_n^*, A(x)\rangle_Y - \varphi(x)\big] \leq m \qquad \text{for all } x \in X. \tag{1.49}$$

By virtue of the hypothesis that $0 \in \mathrm{int}\big(A(\mathrm{dom}\ \varphi) - \mathrm{dom}\ \psi\big)$, we can find $r>0$ such that

$$r\bar{B}_1^Y \subseteq \mathrm{dom}\ \psi - A(\mathrm{dom}\ \varphi).$$

Every $u \in r\bar{B}_1^Y$ can be written as $u = y_u - A(x_u)$ with $y_u \in \mathrm{dom}\ \psi$ and $x_u \in \mathrm{dom}\ \varphi$. Then

$$\begin{aligned}\langle y_n^*, u\rangle_Y &= \langle y_n^*, y_u\rangle_Y - \langle A^*(y_n^*, x_u\rangle_X \\ &\leq \psi^*(y_n^*) + \psi(y_u) - \langle A^*(y_n^*, x_u\rangle_X\,, \quad \text{(see Proposition 1.2.18)},\end{aligned}$$
$$\Rightarrow \limsup_{n\to\infty} \langle y_n^*, y_u\rangle_Y \leq m + \varphi(x_u) + \psi(y_u) < +\infty \ \text{ for all } u \in r\bar{B}_1^Y \text{ (see(1.49))}.$$

Therefore $\{y_n^*\}_{n\geq 1} \subseteq Y^*$ is weakly bounded, thus bounded. Because Y is reflexive, it is relatively w-compact and so by the Eberlein–Smulian theorem and by passing to a subsequence if necessary, we may assume that $y_n^* \xrightarrow{w} y^*$ in Y^* as $n\to\infty$. Because $y^* \longrightarrow \psi^*(y^*) - \langle y^*, A(x)\rangle_Y$ is weakly lower semicontinuous, from (1.49) we have

$$\psi^*(y^*) - \langle y^*, A(x)\rangle_Y - \varphi(x) \leq m = -\varphi(x_0) - \psi\big(A(x_0)\big), \tag{1.50}$$

for all $x\in X$ (see (1.48)).

Taking $x = x_0$, we obtain

$$\begin{aligned}&\psi^*(y^*) + \psi\big(A(x_0)\big) \leq \langle y^*, A(x_0)\rangle_Y\,, \\ \Rightarrow\ & y^* \in \partial\psi\big(A(x_0)\big) \ \text{ (see Propositions 1.2.18 and 1.2.31)}.\end{aligned} \tag{1.51}$$

Returning to (1.50) and using (1.51) and Proposition 1.2.31, we obtain

$$\begin{aligned}&\varphi(x_0) - \varphi(x) \leq \langle -A^*(y^*), x_0 - x\rangle_X \qquad \text{for all } x\in X, \\ \Rightarrow\ & -A^*(y^*) \in \partial\varphi(x_0).\end{aligned}$$

So finally $0 = A^*(y^*) - A^*(y^*) \in A^*\big(\partial\psi\big(A(x_0)\big)$ (see (1.51)). ☐

THEOREM 1.4.10 *If X, Y are reflexive Banach spaces, $A \in \mathcal{L}(X,Y)$ is invertible, $C \subseteq X$, $D \subseteq Y$ are nonempty, closed, and convex sets such that $0 \in int\big(A(C) - D\big)$ and $E = \{x \in C : A(x) \in D\}$, then for all $x \in E$, we have*

$$T_E(x) = T_C(x) \cap A^{-1}T_D\big(A(x)\big) \quad \textit{and} \quad N_E(x) = N_C(x) + A^*N_D\big(A(x)\big).$$

PROOF: Let $x^* \in N_C(x)$ and $y^* \in N_D\big(A(x)\big)$. Set $u^* = x^* + A^*y^*$. If $u \in E$, then

$$\begin{aligned}\langle u^*, u\rangle_X &= \langle x^*, u\rangle_X + \langle y^*, A(u)\rangle_Y \leq \langle x^*, x\rangle_X + \langle y^*, A(x)\rangle_Y\,, \\ &\qquad \text{(see Definition 1.4.5)}, \\ \Rightarrow\ & \langle u^*, u\rangle_X \leq \langle u^*, x\rangle_X\,; \ \text{ that is, } \ u^* \in N_E(x).\end{aligned}$$

Conversely, if $u^* \in N_E(x)$, then $\langle x^*, u\rangle_X = \sup[\langle u^*, u\rangle_X : u \in C \text{ and } A(u) \in D]$. We apply Theorem 1.4.9 with $\varphi = i_C - u^* \in \Gamma_0(X)$ and $\psi = i_D \in \Gamma_0(Y)$. Note that dom $\varphi = C$ and dom $\psi = D$. Then because $\partial\varphi(x) = N_C(x)$ and $\partial\psi\big(A(x)\big) = N_D\big(A(x)\big) - u^*$, we have

$$0 \in A^*N_D\big(A(x)\big) + N_C(x).$$

Finally from Definition 1.4.5 we know that the tangent cone is a negative polar cone of the normal cone. So

$$\begin{aligned} T_E(x) &= \big(N_C(x) + A^* N_D\big(A(x)\big)\big)^- \\ &= T_C(x) \cap \big(A^* N_D\big(A(x)\big)\big)^- \\ &= T_C(x) \cap A^{-1} T_D\big(A(x)\big). \end{aligned}$$

□

This theorem permits the description of the tangent and normal cones for intersections.

PROPOSITION 1.4.11 *If X is a reflexive Banach space and $C, D \subseteq X$ are nonempty, closed, convex sets such that $0 \in int(C - D)$, then for all $x \in C \cap D$ we have*

$$T_{C\cap D}(x) = T_C(x) \cap T_D(x) \quad \textit{and} \quad N_{C\cap D}(x) = N_C(x) + N_D(x).$$

PROOF: Apply Theorem 1.4.10 with $Y = X$ and $A = \mathrm{Id}_X$. □

REMARK 1.4.12 The above result fails if we do not have the constraint qualification condition that $0 \in \mathrm{int}(C - D)$. Take, for example, two balls $C, D \subseteq \mathbb{R}^2$ that are tangent at $x \in \mathbb{R}^2$. Then $T_{C\cap D}(x) = T_{\{x\}}(x) = \{0\}$ but $T_C(x) \cap T_D(x) = \mathbb{R}$.

Recall that, if $X = H =$ a Hilbert space and $C \subseteq H$ is nonempty closed and convex, then every $x \in H$ has a unique best approximation in C; that is, there is a unique $c_0 \in C$ such that

$$\|x - c_0\|_H = \min\{\|x - c\|_H : x \in H\}.$$

This map, which to each $x \in H$ assigns its unique best approximation $p_C(x) \in C$, is called the *metric projection* map for C. Then directly from the definitions we have the following.

PROPOSITION 1.4.13 *If $X = H =$a Hilbert space, $C \subseteq H$ is nonempty, closed, and convex and $x \in C$, then $p_C^{-1}(x) = x + N_C(x)$ and $h \in T_C(x)$ if and only if $(h, u - x)_H \leq 0$ for all $u \in p_C^{-1}(x)$.*

PROPOSITION 1.4.14 *If $C \subseteq X$ is a nonempty, closed, and convex set such that* $\mathrm{int}\, C \neq \emptyset$, *then* $\mathrm{int}\, T_C(x) = \bigcup\limits_{\lambda>0} \frac{1}{\lambda}(\mathrm{int}\, C - x)$.

PROOF: Note that $\bigcup\limits_{\lambda>0} \frac{1}{\lambda}(\mathrm{int}\, C - x)$ is open as the union of open sets and so $\bigcup\limits_{\lambda>0} \frac{1}{\lambda}(\mathrm{int}\, C - x) \subseteq \mathrm{int}\, T_C(x)$. If $S_C(x) = \bigcup\limits_{\lambda>0} \frac{1}{\lambda}(C - x)$, then $\mathrm{int}\, T_C(x) = \mathrm{int}\, S_C(x)$. So to prove the proposition it suffices to show that if $h \in \mathrm{int}\, S_C(x)$, then for some $\lambda > 0$ $h \in \frac{1}{\lambda}(\mathrm{int}\, C - x)$.

Let $\delta > 0$ be such that $h + \delta B_1 \subseteq S_C(x)$ ($B_1 = \{x \in X : \|x\|_X < 1\}$. If $x + h \in C$, then we are done. Otherwise, let $u_0 \in \mathrm{int}\, C$ and set $h_0 = u_0 - x$.

Then $h - \delta(h_0/\|h_0\|_X) \in S_C(x)$ and so we can find $\lambda > 0$ such that $x + \lambda\Big(h - \delta(h_0/\|h_0\|_X)\Big) \in C$. Let $\mu = \lambda\delta/(\lambda\delta + \|h_0\|_X)$. We have

$$x + (1-\mu)\lambda h = \mu u_0 + (1-\mu)\Big(x + \lambda\Big(h - \delta\frac{h_0}{\|h_0\|_X}\Big)\Big).$$

Note that $u_0 \in \operatorname{int} C$ and $x + \lambda\Big(h - \delta(h_0/\|h_0\|_X)\Big) \in C$. Because $\mu \in (0,1)$, it follows that

$$x + (1-\mu)\lambda h \in \operatorname{int} C; \quad \text{that is,} \quad h \in \frac{1}{(1-\mu)\lambda}(\operatorname{int} C - x).$$

□

Let us give some simple examples of tangent cones.

EXAMPLE 1.4.15 (a) Let $C = \bar{B}_1 = \{x \in X; \|x\|_X \leq 1\}$ and $x \in \partial\bar{B}_1$. Then if $\mathcal{F}\colon X \longrightarrow 2^{X^*}$ is the duality map for the space; that is, $\mathcal{F}(u) = \{x^* \in X^* : \langle x^*, u\rangle = \|x^*\|^2_{X^*} = \|u\|^2_X\}$ for all $u \in X$ (see Example 1.2.41(b)), then

$$T_C(x) = \{\mathcal{F}(x)\}^- = \{u \in X : \langle \mathcal{F}(x), u\rangle_X \leq 0\} \quad \text{and} \quad N_C(x) = \bigcup_{\lambda \geq 0} \lambda\mathcal{F}(x).$$

(b) If $C \subseteq X$ is a nonempty, closed, and convex cone and $x \in C$, then

$$N_C(x) = C^- \cap \{x\}^\perp,$$

where $C^- = \{x^* \in X^* : \langle x^*, c\rangle_X \leq 0 \text{ for all } c \in C\}$ and $\{x\}^\perp = \{x^* \in X^* : \langle x^*, x\rangle_X = 0\}$. Also we have $T_C(x) = \overline{C + \mathbb{R}x}$.

If $C \subseteq X$ is a closed subspace and $x \in C$, then

$$T_C(x) = C \quad \text{and} \quad N_C(x) = C^\perp = \{x^* \in X^* : \langle x^*, c\rangle_X = 0 \text{ for all } c \in C\}.$$

(c) If Y is another Banach space, $A \in \mathcal{L}(X, Y)$, and $C = A^{-1}(\{y\})$, then if $A(x) = y$,

$$T_{A^{-1}(\{y\})}(x) = \ker A.$$

(d) If $C \subseteq \mathbb{R}^n$ is defined by $C = \{x = (x_k)_{k=1}^n \in \mathbb{R}^n_+ : \sum_{k=1}^n x_k = 1\}$ (the standard simplex in $\mathbb{R}^n$) and $I(x) = \{k = 1, \ldots, n : x_k = 0\}$, then

$$h = (h_k)_{k=1}^n \in T_{\mathbb{R}^n_+}(x) \text{ if and only if } h_k \geq 0 \quad \text{for all } k \in I(x)$$

and $$h = (h_k)_{k=1}^n \in T_C(x) \text{ if and only if } h_k \geq 0 \quad \text{for all } k \in I(x)$$

and $$\sum_{k=1}^n h_k = 0.$$

REMARK 1.4.16 A set $C \subseteq X$ is said to be star-shaped around $x \in C$, if for all $u \in C$ and all $\lambda \in [0,1]$, we have $x + \lambda(u - x) \in C$. For star-shaped around x sets C we still have $T_C(x) = \overline{S_C(x)} = \overline{\bigcup_{\lambda > 0} \frac{1}{\lambda}(C - x)}$ and $C \subseteq x + T_C(x)$.

Now let us see what can be said about nonconvex sets $C \subseteq X$. Let $d_C(\cdot) = \inf[\|\cdot - c\|; c \in C]$ (the distance function from $C \subseteq X$). For every $C \subseteq X$ nonempty $d_C(\cdot)$ is Lipschitz continuous with Lipschitz constant equal to 1. Note also that $d_C(\cdot) = (\|\cdot\| \oplus i_C)(\cdot)$. Therefore, if $C \subseteq X$ is convex, then $d_C(\cdot)$ is convex too.

DEFINITION 1.4.17 Let $C \subseteq X$ be nonempty and $x \in \bar{C}$. Then the *Clarke tangent cone* to C at x is defined by

$$T'_C(x) = \{h \in X : d^0_C(x;h) \leq 0\}.$$

The *Clarke normal cone* to C at x is defined by

$$N'_C(x) = [T'_C(x)]^- = \{x^* \in X^* : \langle x^*, h\rangle_X \leq 0 \text{ for all } h \in T'_C(x)\}.$$

REMARK 1.4.18 Note that both cones are closed and convex in X and X^*, respectively. Moreover, because $T'_C(x) = T'_{\bar{C}}(x)$, then we can always take without any loss of generality $C \subseteq X$ to be closed. Note that $T'_C(x) = [N'_C(x)]^-$ (bipolar theorem). Also if int $C \neq \emptyset$ and $x \in$int C, then $T'_C(x) = X$ and $N'_C(x) = \{0\}$. Finally in the definition of $T'_C(x)$ we actually have $d^0_C(x;h) = 0$.

PROPOSITION 1.4.19 *If $C \subseteq X$ is nonempty and $x \in \bar{C}$, then* $N'_C(x) = \overline{\bigcup_{\lambda \geq 0} \lambda \partial d_C(x)}^{w^*}$.

PROOF: Note that $h \in T'_C(x)$ if and only $\langle x^*, h\rangle_X \leq 0$ for all $x^* \in \partial d_C(x)$. So the negative polar cone of $T_C(x)$ is the weak*-closed convex cone generated by the set $\partial d_C(x)$. Hence $N'_C(x) = \overline{\bigcup_{\lambda \geq 0} \lambda \partial d_C(x)}^{w^*}$. □

We can have an intrinsic characterization of $T_C(x)$ independent of the norm used on X (hence of the distance function too).

PROPOSITION 1.4.20 *If $C \subseteq X$ is nonempty and $x \in \bar{C}$, then $h \in T_C(x)$ if and only if for every sequence $\{x_n\}_{n\geq 1} \subseteq C$ such that $x_n \longrightarrow x$ in X and every sequence $\lambda_n \downarrow 0$ in $\mathbb{R}$, we can find a sequence $\{h_n\}_{n\geq 1} \subseteq X$ such that $h_n \longrightarrow h$ in X and $x_n + \lambda_n h_n \in C$ for all $n \geq 1$.*

PROOF: $\Rightarrow$: Let $h \in T_C(x)$ and consider sequences $\{x_n\}_{n\geq 1} \subseteq C$ and $\{\lambda_n\}_{n\geq 1} \subseteq \mathbb{R}_+$ such that $x_n \longrightarrow x$ in X and $\lambda_n \downarrow 0$ as $n \to \infty$. Then by virtue of Definition 1.4.17 (see also Remark 1.4.18), we have

$$\lim_{n\to\infty} \frac{d_C(x_n + \lambda_n h) - d_C(x_n)}{\lambda_n} = \lim_{n\to\infty} \frac{d_C(x_n + \lambda_n h)}{\lambda_n} = 0. \tag{1.52}$$

Let $c_n \in C$ such that

$$\|x_n + \lambda_n h - c_n\|_X \leq d_C(x_n + \lambda_n h) + \frac{\lambda_n}{n} \tag{1.53}$$

and set $h_n = (1/\lambda_n)(c_n - x_n)$. Then from (1.52) and (1.53) it follows that $h_n \longrightarrow h$ in X as $n \to \infty$. Moreover, $x_n + \lambda_n h_n \in C$ for all $n \geq 1$.

$\Leftarrow$: Choose $\{u_n\}_{n\geq 1} \subseteq X$ with $u_n \longrightarrow x$ in X and $\{t_n\}_{n\geq 1} \subseteq \mathbb{R}_+$ with $t_n \downarrow 0$ such that

$$\lim_{n\to\infty} \frac{d_C(u_n + t_n h) - d_C(u_n)}{t_n} = d_C^0(x;h).$$

Let $c_n \in C$ such that $\|c_n - u_n\|_X \leq d_C(u_n) + (t_n/n)$, $n \geq 1$. Then $c_n \longrightarrow x$ in X. By hypothesis we can find $\{h_n\}_{n\geq 1} \subseteq X$ such that $h_n \longrightarrow h$ in X and $c_n + t_n h_n \in C$, $n \geq 1$. Because $d_C(\cdot)$ is Lipschitz continuous,

$$\begin{aligned} d_C(u_n + t_n h) &\leq d_C(c_n + t_n h) + \|u_n - c_n\|_X + t_n \|h - h_n\|_X \\ &\leq d_C(u_n) + t_n\Big(\frac{1}{n} + \|h - h_n\|_X\Big) \\ \Rightarrow \frac{d_C(u_n + t_n h) - d_C(u_n)}{t_n} &\leq \frac{1}{n} + \|h - h_n\|_X \\ \Rightarrow d_C^0(x;h) \leq 0. & \end{aligned}$$

□

Using this proposition, we can compare the two cones $T_C(x)$, $T'_C(x)$ when the set is convex.

PROPOSITION 1.4.21 *If $C \subseteq X$ is nonempty, closed, and convex and $x \in C$, then $T_C(x) = T'_C(x)$ and $N_C(x) = N'_C(x)$.*

PROOF: It is clear from Remark 1.4.2 and Proposition 1.4.20 that $T'_C(x) \subseteq T_C(x)$. So suppose that $h \in T_C(x)$. Then $d'_C(x;h) = d_C^0(x;h) = 0$ (see Remark 1.4.18). Hence $h \in T'_C(x)$ and we conclude that $T_C(x) = T'_C(x)$. Taking negative polar cones, we have $N_C(x) = N'_C(x)$. □

COROLLARY 1.4.22 *If $C \subseteq X$ is nonempty, closed, and convex and $x \in C$, then $h \in T'_C(x)$ if and only if $d'_C(x;h) = d_C^0(x;h) = 0$.*

The next variational principle underlines the significance of the distance function in variational problems. Roughly speaking, the result says that in many cases we can replace a constrained minimization problem by an unconstrained one with a suitably perturbed objective functional.

PROPOSITION 1.4.23 *If $\varphi\colon X \longrightarrow \mathbb{R}$ is a function that is Lipschitz continuous on $B \subseteq X$ with Lipschitz constant $k > 0$, $C \subseteq B$ is nonempty, and $\varphi(x) = \inf[\varphi(u) : u \in C]$, then for all $k' \geq k$ the function $u \longrightarrow \psi_{k'}(u) = \varphi(u) + k' d_C(u)$ attains its minimum over B at x. Moreover, if $k' > k$ and C is closed, then any other point minimizing $\psi_{k'}$ over B must also lie in C.*

PROOF: Let us prove the first assertion of the proposition. We argue indirectly. Suppose that the result is not true. We can find $v \in B$ and $\varepsilon > 0$ such that $\varphi(v) + k' d_C(v) < \varphi(x) - k'\varepsilon$. Choose $c \in C$ such that $\|v - c\|_X \leq d_C(v) + \varepsilon$. Then we have

$$\varphi(c) \leq \varphi(v) + k'\|c - v\|_X \leq \varphi(v) + k' d_C(v) + k'\varepsilon < \varphi(x),$$

a contradiction to the fact that $\inf\limits_C \varphi = \varphi(x)$.

Next suppose $k' > k$ and assume that $C \subseteq B$ is closed. Let $v \in B$ be an other minimizer of $\psi_{k'}$ on B. Using the first part of the proof (for $(k + k')/2 > k$), we have

$$\varphi(v) + k' d_C(v) = \varphi(x) \leq \varphi(v) + \frac{k+k'}{2} d_C(v)$$
$$\Rightarrow \quad d_C(v) = 0 \quad \text{(because } k' > k\text{) and so } v \in C \quad \text{(because } C \text{ is closed).}$$

□

COROLLARY 1.4.24 *If $\varphi\colon X \longrightarrow \mathbb{R}$ is Lipschitz continuous near $x \in X$, $C \subseteq X$ is nonempty, and $\varphi(x) = \inf_C \varphi$, then $0 \in \partial\varphi(x) + N'_C(x)$.*

PROOF: Let B be a neighborhood of $x \in X$ such that $\varphi\big|_B$ is Lipschitz continuous with constant $k > 0$. We may assume that $C \subseteq B$ (because the sets C and $C \cap B$ have the same normal cone at x). Then by Proposition 1.4.23 we know that $x \in C$ minimizes the function $u \longrightarrow \varphi(u) + k d_C(u)$ locally. Hence

$$0 \in \partial(\varphi + k d_C)(x) \quad \text{(see Proposition 1.3.22),}$$
$$\Rightarrow \quad 0 \in \partial\varphi(x) + k\partial d_C(x) \subseteq \partial\varphi(x) + N_C(x)$$

(see Propositions 1.3.16 and 1.4.19). □

We conclude this section with the introduction of two more geometric notions. They are useful in the analysis of noncoercive variational problems.

Let $C \subseteq X$ be a nonempty, closed, convex set. For $x \in C$, let

$$C_\infty(x) = \big\{h \in X : x + \lambda h \in C \ \text{ for all } \lambda > 0\big\} = \bigcap_{\lambda > 0} \frac{C - x}{\lambda}.$$

Clearly this set is a closed convex cone, which can be trivial (i.e., $C_\infty(x) = \{0\}$). Think of a bounded set in $\mathbb{R}^N$.

PROPOSITION 1.4.25 *The closed convex cone $C_\infty(x)$ does not depend on the base point $x \in C$.*

PROOF: Let $x, u \in C$. We need to show that $C_\infty(x) \subseteq C_\infty(u)$. To this end let $h \in C_\infty(x)$ and $\lambda > 0$. For $0 < \varepsilon < 1$ let $v_\varepsilon = x + \lambda h + (1 - \varepsilon)(u - x) = \varepsilon\big(x + (\lambda/\varepsilon)h\big) + (1 - \varepsilon)u$. Because $h \in C_\infty(x)$ and C is convex, we have $v_\varepsilon \in C$. Note that $u + \lambda h = \lim_{\varepsilon \downarrow 0} v_\varepsilon \in C$. □

This proposition suggests that the notation C_∞ is more appropriate.

DEFINITION 1.4.26 The *recession cone* (or *asymptotic cone*) of the closed convex set $C \subseteq X$, is the closed convex cone $C_\infty = \big\{h \in X : x + \lambda h \in C \ \text{ for all } \lambda > 0\big\} = \bigcap_{\lambda > 0}\big((C - x)/\lambda\big)$, for any $x \in C$.

REMARK 1.4.27 This cone describes the set of directions along which one can go straight to infinity without leaving C. So, if C is bounded, then $C_\infty = \{0\}$. Also we can view C_∞ as the maximal (with respect to inclusion) set $K \subseteq X$ such that $K + C \subseteq C$. If C is a closed convex cone, then $K = C$.

PROPOSITION 1.4.28 *A closed convex set $C \subseteq \mathbb{R}^N$ is compact if and only if $C_\infty = \{0\}$.*

PROOF: $\Rightarrow$: Because C is compact, it is bounded and so it cannot have direction receding to infinity.

$\Leftarrow$: Suppose that we can find $\{x_n\}_{n\geq 1} \subseteq C\setminus\{0\}$ such that $\|x_n\|_{\mathbb{R}^N} \longrightarrow +\infty$. Set $h_n = x_n/\|x_n\|_{\mathbb{R}^N}$, $n\geq 1$. We may assume that $h_n \longrightarrow h$ in $\mathbb{R}^N$ and $\|h\|_{\mathbb{R}^N} = 1$, hence $h \neq 0$. Given $x \in C$ and $\lambda > 0$, we can find $n \geq 1$ large such that $\|x_n\| \geq \lambda$. Then

$$x + \lambda h = \lim_{n\to\infty} \Big[(1 - \frac{\lambda}{\|x_n\|_{\mathbb{R}^N}}) x + \frac{\lambda}{\|x_n\|_{\mathbb{R}^N}} x_n \Big] \in C$$
$$\Rightarrow\ h \in C_\infty, \quad \text{a contradiction because } h \neq 0.$$

□

Given $\varphi \in \Gamma_0(X)$, we know that epi φ is a closed convex set in $X \times \mathbb{R}$. We can consider the set (epi $\varphi)_\infty$. This is the epigragh of a function φ_∞ and so we are led to the following definition.

DEFINITION 1.4.29 If $\varphi \in \Gamma_0(X)$, then the *recession function* (or *asymptotic function*) corresponding to φ is the function $\varphi_\infty \in \Gamma_0(X)$ such that epi $\varphi_\infty =$ (epi $\varphi)_\infty$. Then we have

$$\varphi_\infty(h) = \sup_{\lambda > 0} \frac{\varphi(x+\lambda h) - \varphi(x)}{\lambda} = \lim_{\lambda \to +\infty} \frac{\varphi(x+\lambda h) - \varphi(x)}{\lambda},$$

where x is arbitrary in dom φ.

Using a topological approach, we can extend the notions of recession cone and recession function to nonconvex sets and functionals. In what follows by τ we denote either the strong or the weak topology on X.

DEFINITION 1.4.30 (a) If $C \subseteq X$ is nonempty, then the *τ-recession cone* of C is the τ-closed cone (nonconvex in general) defined by

$$\tau\text{-}C_\infty = \Big\{ h \in X \text{:there exist } \{\lambda_n\}_{n\geq 1} \subseteq \mathbb{R}_+,\ \{h_n\}_{n\geq 1} \subseteq X \text{ with } \lambda_n \longrightarrow +\infty,\ h_n \xrightarrow{\tau} h \text{ and } \lambda_n h_n \in C \text{ for all } n \geq 1 \Big\}.$$

(b) If $\varphi\colon X \longrightarrow \overline{\mathbb{R}}$ is a proper function, then the *τ-recession function* of φ, is the function

$$\tau\text{-}\varphi_\infty(h) = \inf\Big[\liminf_{n\to\infty} \frac{\varphi(\lambda_n x_n)}{\lambda_n} : \lambda_n \longrightarrow +\infty,\ h_n \xrightarrow{\tau} h \Big]$$

PROPOSITION 1.4.31 *If $C \subseteq X$ is nonempty, then*

(a) *If C is bounded, we have $\tau\text{-}C_\infty = \{0\}$.*

(b) $\tau\text{-}C_\infty = \bigcap_{\mu>0} \overline{\bigcup_{\lambda>\mu} \frac{1}{\lambda}(C - x)}^{\tau}$ *where x is any point of X.*

(c) $\tau\text{-}C_\infty = \tau\text{-}(C\setminus B)_\infty = \tau\text{-}(C \cup B)_\infty$ *for any nonempty bounded set $B \subseteq X$.*

PROPOSITION 1.4.32 *If $\varphi, \psi\colon X \longrightarrow \overline{\mathbb{R}}$ are proper functions, then*

(a) *τ-φ_∞ is τ-lower semicontinuous and positively homogeneous of degree one.*

(b) $(\tau\text{-}\varphi_\infty) + (\tau\text{-}\psi_\infty) \leq \tau\text{-}(\varphi+\psi)_\infty$.

(c) *If φ is bounded below, then τ-$\varphi_\infty \geq 0$.*

(d) *If φ is weakly coercive (i.e., $\varphi(x) \longrightarrow +\infty$ as $\|x\| \longrightarrow \infty$), then φ_∞ is weakly coercive too and τ-$\varphi_\infty(h)>0$ for all $h \in X \setminus \{0\}$.*

1.5 Γ-Convergence

In optimization and in the calculus of variations, we are given an objective functional $\varphi\colon X \longrightarrow \mathbb{R}$ and a constraint set $C \subseteq X$ and we want to find the minimum value

$$m(\varphi, C) = \inf\big[\varphi(x) : x \in C\big] \tag{1.54}$$

and to determine the set of minimizers (solutions of (1.54))

$$M(\varphi, C) = \big\{x \in X : \varphi(x) = m(\varphi, C)\big\}.$$

The purpose of this section is to develop the tools that allow us to determine the dependence of the value $m(\varphi, C)$ and of the solution set $M(\varphi, C)$ on the data of the problem (φ, C).

The mathematical setting in this section is the following. We are given (X, τ) a Hausdorff topological space with topology τ and $\varphi_n\colon X \longrightarrow \overline{\mathbb{R}}$, $n \geq 1$, a sequence of proper functions. Additional hypotheses are introduced as needed. In what follows, given $x \in X$, by $\mathcal{N}(x)$ we denote the filter of neighborhoods of x.

DEFINITION 1.5.1 With the sequence $\{\varphi_n\}_{n\geq 1}$, we associate two limit functions

The Γ_τ-*lower limit* of the sequence $\{\varphi_n\}_{n\geq 1}$ defined by

$$\big(\Gamma_\tau\text{-}\liminf_{n\to\infty} \varphi_n\big)(x) = \sup_{U\in\mathcal{N}(x)} \liminf_{n\to\infty} \inf_{u\in U} \varphi_n(u)$$

and the Γ_τ-*upper limit* of the sequence $\{\varphi_n\}_{n\geq 1}$ defined by

$$\big(\Gamma_\tau\text{-}\limsup_{n\to\infty} \varphi_n\big)(x) = \sup_{U\in\mathcal{N}(x)} \limsup_{n\to\infty} \inf_{u\in U} \varphi_n(u).$$

If $\Gamma_\tau\text{-}\liminf\limits_{n\to\infty} \varphi_n = \Gamma_\tau\text{-}\limsup\limits_{n\to\infty} \varphi_n = \varphi$, then we say that the sequence $\{\varphi_n\}_{n\geq 1}$ Γ_τ-*converges* to φ and write $\Gamma_\tau\text{-}\lim\limits_{n\to\infty} \varphi_n = \varphi$.

REMARK 1.5.2 Evidently we always have that $\Gamma_\tau\text{-}\liminf\limits_{n\to\infty} \varphi_n \leq \Gamma_\tau\text{-}\limsup\limits_{n\to\infty} \varphi_n$ and both limit functions are τ-lower semicontinuous. In the above definition we can replace $\mathcal{N}(x)$ by a local base $\mathcal{B}(x)$. Also when the topology τ used is clearly understood and no confusion is possible, then we drop the subscript τ and simply write $\Gamma\text{-}\liminf_{n\to\infty} \varphi_n$, $\Gamma\text{-}\limsup_{n\to\infty} \varphi_n$ and $\Gamma\text{-}\lim_{n\to\infty} \varphi_n$. If $\varphi_n(x) = c_n \in \mathbb{R}$ for all $x \in X$, then $\Gamma_\tau\text{-}\liminf_{n\to\infty} \varphi_n = \liminf_{n\to\infty} c_n$ and $\Gamma_\tau\text{-}\limsup_{n\to\infty} \varphi_n = \limsup_{n\to\infty} c_n$. Also if $\varphi_n = \varphi$ for $n \geq 1$, then $\Gamma_\tau\text{-}\liminf_{n\to\infty} \varphi_n = \Gamma_\tau\text{-}\limsup_{n\to\infty} \varphi_n = \overline{\varphi}^\tau$, where $\overline{\varphi}^\tau$ is τ-lower semicontinuous regularization of φ; that is, $\overline{\varphi}^\tau = \sup_{U\in\mathcal{N}(x)} \inf_{u\in U} \varphi(u)$ or equivalently epi $\overline{\varphi}^\tau = \overline{\text{epi}\ \varphi}^\tau$.

In general Γ_τ-convergence and pointwise convergence are distinct modes of convergence.

EXAMPLE 1.5.3 (a) Let $X = \mathbb{R}$ and consider the sequence $\varphi_n(x) = nxe^{-n^2x^2}$, $n \geq 1$. Then

$$\Gamma\text{-}\lim_{n\to\infty} \varphi_n(x) = \begin{cases} -\frac{1}{\sqrt{2e}} & \text{if } x = 0 \\ 0 & \text{if } x \neq 0 \end{cases}$$

and $\lim \varphi_n(x) = 0$ for all $x \in \mathbb{R}$. So both the Γ and the pointwise limits exist but differ.

(b) Let $X=\mathbb{R}$ and consider the sequence $\varphi_n(x) = \sin(nx)$, $n \geq 1$. Then $\Gamma\text{-}\lim_{n\to\infty} \varphi_n = \varphi$ with $\varphi(x) = -1$ for all $x \in \mathbb{R}$, but the pointwise limit of the sequence $\{\varphi_n\}_{n\geq 1}$ does not exist.

(c) Let $X=\mathbb{R}$ and consider the sequence

$$\varphi_n(x) = \begin{cases} nxe^{-n^2x^2} & \text{if } n = \text{even} \\ 2nxe^{-n^2x^2} & \text{if } n = \text{odd} \end{cases}, \quad n \geq 1.$$

Then $\varphi_n(x) \longrightarrow 0$ for all $x \in \mathbb{R}$ but the Γ-limit does not exist. In fact we have

$$\big(\Gamma\text{-}\liminf_{n\to\infty} \varphi_n\big)(x) = \begin{cases} -\frac{\sqrt{2}}{\sqrt{e}} & \text{if } x = 0 \\ 0 & \text{if } x \neq 0 \end{cases} \quad \text{and}$$

$$\big(\Gamma\text{-}\limsup_{n\to\infty} \varphi_n\big)(x) = \begin{cases} -\frac{1}{\sqrt{2e}} & \text{if } x = 0 \\ 0 & \text{if } x \neq 0 \end{cases}.$$

Next we produce a characterization of the Γ-convergence in terms of the convergence of the epigraphs of the functions.For this purpose we need to introduce a mode of set convergence. The subject of set convergence is examined more systematically in Section 6.6.

DEFINITION 1.5.4 Let $\{C_n\}_{n\geq 1}$ be a sequence of subsets of X. The *K_τ-lower limit* of the sequence $\{C_n\}_{n\geq 1}$ denoted by $K_\tau\text{-}\liminf_{n\to\infty} C_n$ (or simply $K\text{-}\liminf_{n\to\infty} C_n$ when there is no ambiguity about the topology on X), is defined by

$$K_\tau\text{-}\liminf_{n\to\infty} C_n = \overline{\bigcup_{n\geq 1} \bigcap_{k\geq n} \overline{C}_k^\tau}^\tau .$$

So $x \in K_\tau\text{-}\liminf_{n\to\infty} C_n$ if and only if for every $U \in \mathcal{N}(x)$, we can find $n_0 = n_0(U) \geq 1$ such that $U \cap C_n \neq \emptyset$ for all $n \geq n_0$.

The K_τ*-upper limit* of the sequence $\{C_n\}_{n\geq 1}$ denoted by $K_\tau\text{-}\limsup_{n\to\infty} C_n$ (or $K\text{-}\limsup_{n\to\infty} C_n$) is defined by

$$K_\tau\text{-}\limsup_{n\to\infty} C_n = \bigcap_{n\geq 1} \overline{\bigcup_{k\geq n} C_k}^{\tau}.$$

So $x \in K_\tau\text{-}\limsup_{n\to\infty} C_n$ if and only if for all $U \in \mathcal{N}(x)$ and for every $n\geq 1$, we can find $k \geq n$ such that $U \cap C_k \neq \emptyset$.

If $K_\tau\text{-}\liminf_{n\to\infty} C_n = K_\tau\text{-}\limsup_{n\to\infty} C_n = C$, then we say that the sequence $\{C_n\}_{n\geq 1}$ converges to C in the sense of Kuratowski and we write $K_\tau\text{-}\lim_{n\to\infty} C_n = C$ (or $K\text{-}\lim_{n\to\infty} C_n = C$).

REMARK 1.5.5 Clearly we always have that $K_\tau\text{-}\liminf_{n\to\infty} C_n \subseteq K_\tau\text{-}\limsup_{n\to\infty} C_n = C$ and both sets are closed, possibly empty.

EXAMPLE 1.5.6 (a) Let $X = \mathbb{R}$ and

$$C_n = \begin{cases} [0, \frac{1}{n}] & \text{if } n = \text{even} \\ [1, 1+\frac{1}{n}] & \text{if } n = \text{odd} \end{cases},$$

then $K\text{-}\liminf_{n\to\infty} C_n = \emptyset$ and $K\text{-}\limsup_{n\to\infty} C_n = \{0, 1\}$.

(b) Let $X = \mathbb{R}$ and $C_n = [0, 1/n] \cup [n, +\infty)$. Then $C_n \xrightarrow{K} \{0\}$.

Recall that for $C \subseteq X$ we have the indicator function

$$i_C(x) = \begin{cases} 0 & \text{if } x \in C \\ +\infty & \text{if } x \notin C \end{cases}.$$

PROPOSITION 1.5.7 *If $\{C_n\}_{n\geq 1}$ is a sequence of nonempty subsets of X and we set*

$$C' = K_\tau\text{-}\liminf_{n\to\infty} C_n, \quad C'' = K_\tau\text{-}\limsup_{n\to\infty} C_n,$$

then $i_{C'} = \Gamma_\tau\text{-}\limsup_{n\to\infty} i_{C_n}$ and $i_{C''} = \Gamma_\tau\text{-}\liminf_{n\to\infty} i_{C_n}$.

PROOF: We show the first equality. The proof is similar for the second one.

Let $\varphi = \Gamma_\tau\text{-}\limsup_{n\to\infty} i_{C_n}$. Clearly φ takes only the values 0 and $+\infty$ (see Definition 1.5.1). So it suffices to show that $\varphi(x) = 0$ if and only if $x \in C'$. According to Definition 1.5.4, $x \in C'$ if and only if for every $U \in \mathcal{N}(x)$ we can find $n_0 \geq 1$ such that $C_n \cap U \neq \emptyset$ for all $n \geq n_0$. This is equivalent to $\inf_{u\in U} i_{C_n}(u) = 0$ for all $n \geq n_0$. Therefore $x \in C'$ if and only if $\limsup_{n\to\infty} \inf_{u\in U} i_{C_n}(u) = 0$ for all $U \in \mathcal{N}(x)$. So we conclude that $x \in C'$ if and only if $\varphi(x) = 0$; that is, $\varphi = i_{C'}$. $\square$

The next theorem relates Γ-convergence and the Kuratowski convergence of the epigraphs.

THEOREM 1.5.8 *If* $\varphi_n\colon X\longrightarrow\overline{\mathbb{R}}$, $n\geq 1$, *is a sequence of proper functions and*

$$\varphi' = \Gamma_\tau\text{-}\liminf_{n\to\infty}\varphi_n, \quad \varphi'' = \Gamma_\tau\text{-}\limsup_{n\to\infty}\varphi_n,$$

then $\operatorname{epi}\varphi' = K_\tau\text{-}\limsup_{n\to\infty}\operatorname{epi}\varphi_n$ *and* $\operatorname{epi}\varphi'' = K_\tau\text{-}\liminf_{n\to\infty}\operatorname{epi}\varphi_n$.

PROOF: Again we prove the first equality. The proof of the second is similar.

We know that $(x,\lambda)\in\operatorname{epi}\varphi'$ if and only if $\varphi'(x)\leq\lambda$. According to Definition 1.5.1, $\varphi'(x)\leq\lambda$ if and only if for every $\varepsilon>0$ and every $U\in\mathcal{N}(x)$ we have

$$\liminf_{n\to\infty}\inf_{u\in U}\varphi_n(u)<\lambda+\varepsilon.$$

This last inequality is equivalent to saying that for every $\varepsilon>0$, every $U\in\mathcal{N}(x)$ and every $k\geq 1$, we can find $n\geq k$ such that $\inf_{u\in U}\varphi_n(u)<\lambda+\varepsilon$. This in turn is equivalent to

$$U\times(\lambda-\varepsilon,\lambda+\varepsilon)\cap\operatorname{epi}\varphi_n\neq\emptyset.$$

Because the sets $U\times(\lambda-\varepsilon,\lambda+\varepsilon)$ with $\varepsilon>0$ and $U\in\mathcal{N}(x)$, form a local basis for (x,λ) in $X\times\mathbb{R}$ with the product topology; we conclude that $(x,\lambda)\in\operatorname{epi}\varphi'$ if and only if $(x,\lambda)\in K_\tau\text{-}\limsup_{n\to\infty}\operatorname{epi}\varphi_n$ (see Definition 1.5.4) □

REMARK 1.5.9 This is the reason why many authors call the Γ-convergence, *epigraphical convergence*, and denote it by $e_\tau\text{-}\lim_{n\to\infty}\varphi_n$, $e_\tau\text{-}\liminf_{n\to\infty}\varphi_n$ and $e_\tau\text{-}\limsup_{n\to\infty}\varphi_n$.

We have already seen that Γ-convergence and pointwise convergence are in general distinct notions. We determine what the precise relation is between them. First of all, directly from Definition 1.5.1 we have the following.

PROPOSITION 1.5.10 *If* $\varphi_n\colon X\longrightarrow\overline{\mathbb{R}}$, $n\geq 1$, *is a sequence of proper functions, then* $\Gamma_\tau\text{-}\liminf_{n\to\infty}\varphi_n\leq\liminf_{n\to\infty}\varphi_n$ *and* $\Gamma_\tau\text{-}\limsup_{n\to\infty}\varphi_n\leq\limsup_{n\to\infty}\varphi_n$. *In particular if* $\Gamma_\tau\text{-}\lim_{n\to\infty}\varphi_n=\widehat{\varphi}$ *and* $\lim\varphi_n=\widetilde{\varphi}$ *exist, then* $\widehat{\varphi}\leq\widetilde{\varphi}$.

PROPOSITION 1.5.11 *If* $\varphi_n\colon X\longrightarrow\overline{\mathbb{R}}$, $n\geq 1$, *is a sequence of proper functions that converge uniformly to* φ, *then* $\Gamma_\tau\text{-}\lim_{n\to\infty}\varphi_n=\overline{\varphi}^\tau$, *where* $\overline{\varphi}^\tau$ *is the* τ*-lower semicontinuous regularization of* φ.

PROOF: For every $U\subseteq X$ nonempty, and open, we have

$$\lim_{n\to\infty}\inf_{u\in U}\varphi_n(u)=\inf_{u\in U}\varphi(u).$$

Therefore for every $x\in X$,

$$\sup_{U\in\mathcal{N}(x)}\lim_{n\to\infty}\inf_{u\in U}\varphi_n(u)=\sup_{U\in\mathcal{N}(x)}\inf_{u\in U}\varphi(u)=\overline{\varphi}^\tau(x)$$
$$\Rightarrow\ \Gamma_\tau\text{-}\lim_{n\to\infty}\varphi_n=\overline{\varphi}^\tau.$$

□

REMARK 1.5.12 The uniform limit of τ-lower semicontinuous functions, is τ-lower semicontinuous too, therefore if each φ_n is τ-lower semicontinuous, then φ is τ-lower semicontinuous and so $\Gamma_\tau\text{-}\lim\limits_{n\to\infty} \varphi_n = \varphi$.

The next two propositions explain the importance of monotone methods in optimization.

PROPOSITION 1.5.13 *If $\varphi_n : X \longrightarrow \overline{\mathbb{R}}$, $n \geq 1$, is an increasing sequence of proper functions, then* $\Gamma_\tau\text{-}\lim\limits_{n\to\infty} \varphi_n = \lim\limits_{n\to\infty} \overline{\varphi}_n^\tau = \sup\limits_{n\geq 1} \overline{\varphi}_n^\tau$.

PROOF: If $U \subseteq X$ is nonempty open, then

$$\lim_{n\to\infty} \inf_{u\in U} \varphi_n(u) = \sup_{n\geq 1} \inf_{u\in U} \varphi_n(u).$$

So for every $x \in X$

$$\begin{aligned}\sup_{U\in\mathcal{N}(x)} \lim_{n\to\infty} \inf_{u\in U} \varphi_n(u) &= \sup_{U\in\mathcal{N}(x)} \sup_{n\geq 1} \inf_{u\in U} \varphi_n(u)\\ &= \sup_{n\geq 1} \sup_{U\in\mathcal{N}(x)} \inf_{u\in U} \varphi_n(u) = \sup_{n\geq 1} \overline{\varphi}_n^\tau(x).\end{aligned}$$

REMARK 1.5.14 If $\{\varphi_n\}_{n\geq 1}$ is an increasing sequence of τ-lower semicontinuous functions, then $\varphi_n \uparrow \varphi = \sup\limits_{n\geq 1} \varphi_n$ is τ-lower semicontinuous too and thus $\Gamma_\tau\text{-}\lim\limits_{n\to\infty} \varphi_n = \lim\limits_{n\to\infty} \varphi_n = \varphi$. So we see that in this context Γ_τ and pointwise convergence coincide. Without the τ-lower semicontinuity of the φ_ns this is no longer true.

PROPOSITION 1.5.15 *If $\varphi_n : X \longrightarrow \overline{\mathbb{R}}$, $n \geq 1$, is a decreasing sequence of proper functions and $\varphi_n(x) \longrightarrow \varphi(x)$ for all $x \in X$, then* $\Gamma_\tau\text{-}\lim\limits_{n\to\infty} \varphi_n = \overline{\varphi}^\tau$.

PROOF: If $U \subseteq X$ is nonempty open, we have

$$\lim_{n\to\infty} \inf_{u\in U} \varphi_n(u) = \inf_{u\in U} \inf_{n\geq 1} \varphi_n(u) = \inf_{u\in U} \varphi(u)$$

$$\Rightarrow \Gamma_\tau\text{-}\lim_{n\to\infty} \varphi_n(x) = \overline{\varphi}^\tau(x) \quad \text{(see Definition 1.5.1).}$$ □

To have a general result on the equivalence of the Γ_τ-convergence and of the pointwise convergence, we need the following notion.

DEFINITION 1.5.16 Let $\varphi_n : X \longrightarrow \overline{\mathbb{R}}$, $n \geq 1$, be a sequence of proper functions. We say that $\{\varphi_n\}_{n\geq 1}$ is *equi-τ-lower semicontinuous* at $x \in X$, if given any $\varepsilon > 0$, we can find $U \in \mathcal{N}(x)$ such that

$$\varphi_n(u) \geq \varphi_n(x) - \varepsilon \qquad \text{for all } u \in U \text{ and all } n \geq 1.$$

We say that $\{\varphi_n\}_{n\geq 1}$ is equi-τ-lower semicontinuous if it is equi-τ-lower semicontinuous at every $x \in X$.

PROPOSITION 1.5.17 *If $\varphi_n\colon X\longrightarrow\overline{\mathbb{R}}$, $n\geq 1$, is a sequence of proper functions that is equi-τ-lower semicontinuous at $x\in X$, then $(\Gamma_\tau\text{-}\liminf\limits_{n\to\infty}\varphi_n)(x)=\liminf\limits_{n\to\infty}\varphi_n(x)$ and $(\Gamma_\tau\text{-}\limsup\limits_{n\to\infty}\varphi_n)(x)=\liminf\limits_{n\to\infty}\varphi_n(x)$. So if $\{\varphi_n\}_{n\geq 1}$ is equi-τ-lower semicontinuous, then $\Gamma_\tau\text{-}\lim\limits_{n\to\infty}\varphi_n=\varphi$ if and only if $\varphi_n\longrightarrow\varphi$ pointwise.*

PROOF: As before we prove the first equality. The proof of the other is similar.

By virtue of Proposition 1.5.10 it suffices to show that

$$\liminf_{n\to\infty}\varphi_n(x)\leq(\Gamma_\tau\text{-}\liminf_{n\to\infty}\varphi_n)(x). \tag{1.55}$$

Exploiting the equi-τ-lower semicontinuity of $\{\varphi_n\}_{n\geq 1}$ at $x\in X$, given $\varepsilon>0$, we can find $U\in\mathcal{N}(x)$ such that $\varphi_n(x)-\varepsilon\leq\inf\limits_{u\in U}\varphi_n(u)$ for all $n\geq 1$. Hence

$$\liminf_{n\to\infty}\varphi_n(x)-\varepsilon\leq\sup_{U\in\mathcal{N}(x)}\liminf_{n\to\infty}\inf_{u\in U}\varphi_n(u).$$

Let $\varepsilon\downarrow 0$ to obtain (1.55). $\square$

If X is a Banach space and the φ_ns are convex functions, then we can use Theorem 1.2.7 to produce the following useful consequence of Proposition 1.5.17.

PROPOSITION 1.5.18 *If X is a Banach space and $\varphi_n\colon X\longrightarrow\overline{\mathbb{R}}$, $n\geq 1$, is a sequence of proper convex functions that is equibounded above in $U\in\mathcal{N}(x)$ (i.e., there exists $M>0$ such that $\sup\limits_{n\geq 1}\sup\limits_{u\in U}\varphi_n(u)\leq M<+\infty$), then $(\Gamma_\tau\text{-}\liminf\limits_{n\to\infty}\varphi_n)(x)=\liminf\limits_{n\to\infty}\varphi_n(x)$ and $(\Gamma_\tau\text{-}\limsup\limits_{n\to\infty}\varphi_n)(x)=\liminf\limits_{n\to\infty}\varphi_n(x)$. So, if $\{\varphi_n\}_{n\geq 1}$ is equibounded above in some neighborhood of every point $x\in X$, then $\Gamma_\tau\text{-}\lim\limits_{n\to\infty}\varphi_n=\varphi$ if and only if $\varphi_n\longrightarrow\varphi$ pointwise.*

REMARK 1.5.19 If X is finite-dimensional and the sequence $\{\varphi_n,\varphi\}_{n\geq 1}$ consists of convex functions with values in $\mathbb{R}$ (hence continuous; see Proposition 1.2.5), then $\Gamma_\tau\text{-}\lim\limits_{n\to\infty}\varphi_n=\varphi$ if and only if $\varphi_n\longrightarrow\varphi$ pointwise. This result fails if φ does not take values in $\mathbb{R}$. Let $X=\mathbb{R}$ and consider the sequence $\varphi_n(x)=\ln x-1$. Then $\varphi_n\xrightarrow{\Gamma_\tau}\widehat{\varphi}=I_{\{0\}}$ and $\varphi_n\longrightarrow\widetilde{\varphi}$ pointwise where

$$\widetilde{\varphi}=\begin{cases}1 & \text{if } x=0\\ +\infty & \text{if } x\neq 0\end{cases}.$$

Next by endowing with more structure the space X, we produce some convenient sequential characterizations of $\Gamma_\tau\text{-}\liminf\limits_{n\to\infty}\varphi_n$ and of $\Gamma_\tau\text{-}\limsup\limits_{n\to\infty}\varphi_n$.

PROPOSITION 1.5.20 *If X is first countable and $\varphi'=\Gamma_\tau\text{-}\liminf\limits_{n\to\infty}\varphi_n$, $\varphi''=\Gamma_\tau\text{-}\limsup\limits_{n\to\infty}\varphi_n$, then*

(a) *For every $x \in X$ and every sequence $x_n \longrightarrow x$, we have*

$$\varphi'(x) \le \liminf_{n\to\infty} \varphi_n(x_n); \tag{1.56}$$

and for every $x \in X$ we can find a sequence $x_n \longrightarrow x$ in X such that

$$\varphi'(x) = \liminf_{n\to\infty} \varphi_n(x_n). \tag{1.57}$$

(b) *For every $x \in X$ and every sequence $x_n \longrightarrow x$, we have*

$$\varphi''(x) \le \limsup_{n\to\infty} \varphi_n(x_n) \tag{1.58}$$

and for every $x \in X$ we can find a sequence $x_n \longrightarrow x$ in X such that

$$\varphi''(x) = \limsup_{n\to\infty} \varphi_n(x_n). \tag{1.59}$$

PROOF: (a) Let $U \in \mathcal{N}(x)$. Then we can find $n_0 = n_0(U) \ge 1$ such that $x_n \in U$ for all $n \ge n_0$. We have

$$\begin{aligned} &\inf_{u\in U} \varphi_n(u) \le \varphi_n(x_n) \qquad \text{for all } n \ge n_0,\\ \Rightarrow\ &\liminf_{n\to\infty} \inf_{u\in U} \varphi_n(u) \le \liminf_{n\to\infty} \varphi_n(x_n),\\ \Rightarrow\ &\varphi'(x) \le \liminf_{n\to\infty} \varphi_n(x_n). \end{aligned}$$

This proves (1.56). Next we prove (1.57). Let $x \in X$ such that $\varphi'(x) < +\infty$ and let $\{U_k\}_{k\ge1}$ be a local basis at $x \in X$ such that $U_{k+1} \subseteq U_k$ for all $k \ge 1$. Consider a sequence $\lambda_k \downarrow \varphi'(x)$ in $\mathbb{R}$ such that $\lambda_k > \varphi'(x)$ for all $k \ge 1$. We have

$$\liminf_{n\to\infty} \inf_{u\in U_k} \varphi_n(u) < \lambda_k \qquad \text{for all } k \ge 1.$$

So for all $k \ge 1$, we can find a strictly increasing sequence of integers $n(k) \ge 1$ such that

$$\inf_{u\in U_k} \varphi_{n(k)}(u) < \lambda_k \qquad \text{for all } k \ge 1.$$

For every $k \ge 1$ we can find $u_k \in U_k$ such that

$$\varphi_{n(k)}(u_k) < \lambda_k.$$

Then we introduce the sequence

$$x_n = \begin{cases} u_k & \text{if } n = n(k) \\ x & \text{if } n \ne n(k) \end{cases} \qquad \text{for all } k \ge 1.$$

Evidently $x_n \longrightarrow x$ and $\varphi'(x) = \lim_{k\to\infty} \lambda_k \ge \liminf_{k\to\infty} \varphi_{n(k)}(u_k) \ge \liminf_{n\to\infty} \varphi_n(x_n)$; that is, (1.57) holds.

From the first part of the proof we know that the opposite inequality holds. So

$$\varphi'(x) = \liminf_{n\to\infty} \varphi_n(x_n); \quad \text{that is, (1.57) holds.}$$

(b) Inequality (1.58) follows as (1.56). So we prove (1.59). Again let $x \in X$ such that $\varphi''(x) < +\infty$ and consider $\{\mu_k\}_{k\ge1} \subseteq \mathbb{R}$ such that $\mu_k \downarrow \varphi''(x)$ and $\mu_k > \varphi''(x)$

for $k \geq 1$. As before for every $k \geq 1$ we can find a strictly increasing sequence of integers $n(k) \geq 1$ such that

$$\inf_{u \in U_k} \varphi_n(u) < \mu_k \qquad \text{for all } n \geq n(k).$$

So for every $n \geq n(k)$ we can find $u_{n,k} \in U_k$ such that

$$\varphi_n(u_{n,k}) < \mu_k.$$

We introduce the sequence

$$x_n = \begin{cases} x & \text{if } n < n(1) \\ x_n & \text{if } n(k) \leq n < n(k+1) \end{cases}.$$

Then $x_n \longrightarrow x$ and so

$$\varphi''(x) = \lim_{k \to \infty} \mu_k \geq \limsup_{n \to \infty} \varphi_n(x_n).$$

Combining this with (1.58), we conclude that

$$\varphi''(x) = \limsup_{n \to \infty} \mu_k \varphi_n(x_n); \text{ that is, (1.59) holds.}$$

□

COROLLARY 1.5.21 *If X is first countable and $\varphi = \Gamma_\tau\text{-}\lim_{n \to \infty} \varphi_n$, then for every $x \in X$ and every sequence $x_n \longrightarrow x$ in X we have*

$$\varphi(x) \leq \liminf_{n \to \infty} \varphi_n(x_n). \tag{1.60}$$

and for every $x \in X$ we can find a sequence $x_n \longrightarrow x$ in X such that

$$\varphi(x) = \lim_{n \to \infty} \varphi_n(x_n). \tag{1.61}$$

PROOF: Use (1.56) to obtain (1.60) and combine (1.57) and (1.59) to get (1.61). □

REMARK 1.5.22 From Proposition 1.5.20 and Corollary 1.5.21 it follows that when X is first countable, then

$$(\Gamma_\tau\text{-}\liminf_{n \to \infty} \varphi_n)(x) = \min\left[\liminf_{n \to \infty} \varphi_n(x_n) : x_n \longrightarrow x \text{ in } X\right]$$

$$(\Gamma_\tau\text{-}\limsup_{n \to \infty} \varphi_n)(x) = \min\left[\limsup_{n \to \infty} \varphi_n(x_n) : x_n \longrightarrow x \text{ in } X\right]$$

and $$(\Gamma_\tau\text{-}\lim_{n \to \infty} \varphi_n)(x) = \min\left[\lim_{n \to \infty} \varphi_n(x_n) : x_n \longrightarrow x \text{ in } X\right].$$

Also note that in Corollary 1.5.21, (1.60) is a stronger assumption than the pointwise convergence because it is required to hold for every $x_n \longrightarrow x$ and not only for $x_n = x$. On the other hand, (1.61) is a weaker requirement than the pointwise convergence, where we must have $x_n = x$ for all $n \geq 1$. This explains why the two convergence notions are in general distinct.

In the case of a sequence $\{C_n\}_{n\geq 1} \subseteq 2^X$, then we can have the following convenient sequential characterizations of the *Kuratowki limits* (see Definitions 1.5.4 and Proposition 1.5.7).

PROPOSITION 1.5.23 *If X is first countable and $\{C_n\}_{n\geq 1} \subseteq 2^X$, then $\liminf\limits_{n\to\infty} C_n = \{x \in X : x = \lim x_n,\ x_n \in C_n \text{ for all } n \geq 1\}$ and*

$$\limsup_{n\to\infty} C_n = \{x \in X : x = \lim x_{n_k},\ x_{n_k} \in C_{n_k},\ n_1 < n_2 < \cdots < n_k < \cdots\}.$$

REMARK 1.5.24 So $\liminf\limits_{n\to\infty} C_n$ consists of all limit points of sequences with elements in C_n, $n \geq 1$ and $\limsup\limits_{n\to\infty} C_n$ consists of all subsequential limit points of sequences with elements in C_n, $n \geq 1$. When X is a metric space, we can write that $\liminf\limits_{n\to\infty} C_n = \{x \in X : \lim\limits_{n\to\infty} d(x, C_n) = 0\}$ and $\limsup\limits_{n\to\infty} C_n = \{x \in X : \liminf\limits_{n\to\infty} d(x, C_n) = 0\}$.

Using Proposition 1.5.23, we can have the following double sequence lemma which is a useful analytical tool on many occasions.

PROPOSITION 1.5.25 *If X is first countable, $\{x_{mn}\}_{m,n\geq 1} \subseteq X$ and $\lim\limits_{m\to\infty}\lim\limits_{n\to\infty} x_{mn} = x$, then we can find sequences of integers $\{m(n)\}_{n\geq 1}$ and $\{n(m)\}_{m\geq 1}$ increasing (not necessarily strictly) to $+\infty$ such that*

$$x_{mn(m)} \longrightarrow x \text{ as } m \to \infty \quad \text{and} \quad x_{m(n)n} \longrightarrow x \quad \text{as } n \to \infty.$$

PROOF: Let $A_n = \{x_{mn}\}_{m\geq 1}$ and $x_m = \lim\limits_{n\to\infty} x_{mn}$. By virtue of Proposition 1.5.23 we have $x_m \in \liminf\limits_{n\to\infty} A_n$ for all $m \geq 1$. Because $x_m \to x$ as $m \to \infty$ and $\liminf\limits_{n\to\infty} A_n$ is closed (see Remark 1.5.5), we have $x \in \liminf\limits_{n\to\infty} A_n$. Therefore $x = \lim\limits_{n\to\infty} u_n$ with $u_n \in A_n$, hence $u_n = x_{m(n)n}$ and so $x = \lim\limits_{n\to\infty} x_{m(n)n}$.

Next let $B_m = \{x_{mn}\}_{n\geq 1}$. Then $x_m \in \overline{B}_m$ for all $m \geq 1$ and so $x = \lim x_m \in \liminf\limits_{m\to\infty} \overline{B}_m = \liminf\limits_{m\to\infty} B_m$. Therefore we can find $v_m \in B_m$, $m \geq 1$, such that $x = \lim\limits_{m\to\infty} v_m$. Then $v_m = x_{mn(m)}$ and so $x = \lim\limits_{m\to\infty} x_{mn(m)}$. $\square$

Now let us check the variational properties of Γ-convergence. We start with a topological version of the notion of coercivity.

DEFINITION 1.5.26 Let (X, τ) be a Hausdorff topological space with τ being its topology

(a) A function $\varphi\colon X \longrightarrow \overline{\mathbb{R}}$ is said to be *coercive* (resp., *sequentially coercive*) if for every $\lambda \in \mathbb{R}$, the set $\overline{\{x \in X : \varphi(x) \leq \lambda\}}^{\tau}$ is τ-countably compact (resp., τ-sequentially compact).

(b) A sequence of functions $\varphi_n\colon X \longrightarrow \overline{\mathbb{R}}$, $n \geq 1$, is said to be *equicoercive* if for every $\lambda \in \mathbb{R}$, there exists a τ-closed, τ-countably compact set $K_\lambda \subseteq X$, such that

$$\{\varphi_n \leq \lambda\} = \{x \in X : \varphi_n(x) \leq \lambda\} \subseteq K_\lambda \qquad \text{for all } n \geq 1.$$

REMARK 1.5.27 Evidently every sequentially coercive function φ is coercive. If X is a reflexive Banach space and $\varphi\colon X \longrightarrow \overline{\mathbb{R}}$ is coercive for X with the weak topology, then $\varphi(x) \longrightarrow +\infty$ as $\|x\| \longrightarrow +\infty$. Some authors call this property weak coercivity and call φ coercive when $\big(\varphi(x)\big)/\|x\| \longrightarrow +\infty$ as $\|x\| \longrightarrow +\infty$ (i.e., φ exhibits superlinear growth at infinity).

PROPOSITION 1.5.28 *If $\varphi_n\colon X \longrightarrow \overline{\mathbb{R}}$ is a sequence such that $-\infty < \inf\limits_{n\geq 1} \varphi_n(x)$ for all $x \in X$, then the sequence $\{\varphi_n\}_{n\geq 1}$ is equicoercive if and only if there exists a τ-lower semicontinuous coercive function $g\colon X \longrightarrow \overline{\mathbb{R}}$ such that $g \leq \varphi_n$ for all $n \geq 1$.*

PROOF: $\Rightarrow$: Let $g_0(x) = \inf\limits_{n\geq 1} \varphi_n(x)$ and set $g = \overline{g}_0^{\tau}$ (the τ-lower semicontinuous regularization of g_0). Because $\{\varphi_n\}_{n\geq 1}$ is equicoercive, we can find a family $\{K_\lambda\}_{\lambda\in\mathbb{R}}$ of τ-closed, τ-countably compact sets such that $\{\varphi_n \leq \lambda\} \subseteq K_\lambda$ for all $\lambda \in \mathbb{R}$, $n \geq 1$. Given $x \in \{g \leq \lambda\}$ and $\varepsilon > 0$, we can find $n = n(\varepsilon) \geq 1$ such that $\varphi_n(x) \leq \lambda + \varepsilon$ and so $x \in K_{\lambda+\varepsilon}$. Therefore $\{g \leq \lambda\} \subseteq \bigcap\limits_{e>0} K_{\lambda+\varepsilon}$ and the latter set is τ-closed and τ-countably compact, which proves that g is τ-lower semicontinuous, coercive.

$\Leftarrow$: For every $\lambda \in \mathbb{R}$ and every $n \geq 1$, $\{\varphi_n \leq \lambda\} \subseteq \{g \leq \lambda\} = K_\lambda$ and K_λ of τ-closed (because g is τ-lower semicontinuous) and K_λ is τ-countably compact (because g is coercive). Therefore it follows that the sequence $\{\varphi_n\}_{n\geq 1}$ is equicoercive. $\square$

Equicoercive families exhibit interesting variational stability properties.

THEOREM 1.5.29 *If $\varphi_n\colon X \longrightarrow \overline{\mathbb{R}}$ $n \geq 1$ is an equicoercive sequence, $\inf\limits_{n\geq 1} \varphi_n(x) > -\infty$ for all $x \in X$ and as before $\varphi' = \Gamma_\tau\text{-}\liminf\limits_{n\to\infty} \varphi_n$, $\varphi'' = \Gamma_\tau\text{-}\limsup\limits_{n\to\infty} \varphi_n$, then φ' and φ'' are both coercive and*

$$\min_{x\in X} \varphi'(x) = \liminf_{n\to\infty} \inf_{x\in X} \varphi_n(x_n). \tag{1.62}$$

Moreover, if $\varphi = \Gamma_\tau\text{-}\lim\limits_{n\to\infty} \varphi_n$, then φ is coercive and

$$\min_{x\in X} \varphi(x) = \lim_{n\to\infty} \inf_{x\in X} \varphi_n(x_n). \tag{1.63}$$

PROOF: By virtue of Proposition 1.5.28, we can find $g\colon X \longrightarrow \overline{\mathbb{R}}$ a τ-lower semicontinuous and coercive function such that $g \leq \varphi_n$ for all $n \geq 1$. Then clearly $g \leq \varphi' \leq \varphi''$ and so both functions φ' and φ'' are coercive and τ-lower semicontinuous (see Remark 1.5.2).

Next we prove (1.62). First we show that because of coercivity, φ' attains its infimum on X. Indeed, if $\{x_n\}_{n\geq 1} \subseteq X$ is a minimizing sequence and $\lim \varphi'(x_n) = \inf\limits_X \varphi' = m < +\infty$, then we can find $n_0 \geq 1$ such that $x_n \in K_{m+1}$ for all $n \geq n_0$ with K_{m+1} τ-closed and τ-countably compact. So $\{x_n\}_{n\geq 1}$ has a cluster point $x \in X$ and because φ' is $\tau-$lower semicontinuous, we have

$$\begin{aligned} m \leq \varphi'(x) &\leq \liminf_{n\to\infty} \varphi'(x_n) = m, \\ \Rightarrow\ m = \varphi'(x);&\ \text{ that is, } \inf_X \varphi' = \min_X \varphi'. \end{aligned}$$

Directly from Definition 1.5.1, we have

$$\liminf_{n\to\infty} \inf_{x\in X} \varphi_n(x) \le \min_{x\in X} \varphi'(x). \tag{1.64}$$

Suppose that $\liminf_{n\to\infty} \inf_{x\in X} \varphi_n(x) < +\infty$ (or otherwise we are done, see (1.64)). We can find a subsequence $\{n_k\}_{k\ge 1}$ and $\lambda \in \mathbb{R}$ such that

$$\lim_{k\to\infty} \inf_{x\in X} \varphi_{n_k}(x) = \liminf_{n\to\infty} \inf_{x\in X} \varphi_n(x) < \lambda. \tag{1.65}$$

Evidently we may assume that

$$\inf_{x\in X} \varphi_{n_k}(x) < \lambda \qquad \text{for all } k \ge 1. \tag{1.66}$$

Because of the equicoercivity of the sequence $\{\varphi_n\}_{n\ge 1}$, we can find a τ-closed and τ-countably compact set K_λ such that $\{\varphi_{n_k} \le \lambda\} \subseteq K_\lambda$ for all $k \ge 1$. Note that because of (1.66) all the sets $\{\varphi_{n_k} \le \lambda\} \subseteq K$, $k \ge 1$, are nonempty. Hence

$$\inf_{x\in X} \varphi_{n_k}(x) = \inf_{x\in K} \varphi_{n_k}(x). \tag{1.67}$$

Let $x_{n_k} \in K$ such that $\varphi_{n_k}(x_{n_k}) = \inf_{x\in K} \varphi_{n_k}(x)$ (it exists because K is τ-closed, τ-countably compact, and φ_{n_k} is τ-lower semicontinuous by virtue of the equicoercivity of the sequence $\{\varphi_n\}_{n\ge 1}$). The sequence $\{x_{n_k}\}_{k\ge 1} \subseteq K$ has a cluster point $\widehat{x} \in K$. For every $U \in \mathcal{N}(\widehat{x})$, there exists $k_0 \ge 1$ such that for all $k \ge k_0$ we have $x_{n_k} \in U$. Hence

$$\begin{aligned} & \inf_{u\in U} \varphi_{n_k}(u) \le \varphi_{n_k}(x_{n_k}) \qquad \text{for all } k \ge k_0, \\ \Rightarrow\ & \liminf_{k\to\infty} \inf_{u\in U} \varphi_{n_k}(u) \le \lim_{k\to\infty} \varphi_{n_k}(x_{n_k}) \\ & \qquad = \liminf_{n\to\infty} \inf_{x\in X} \varphi_n(x) \end{aligned} \tag{1.68}$$

(see (1.67) and (1.65)).

Because $U \in \mathcal{N}(\widehat{x})$ was arbitrary from (1.68) it follows that

$$\begin{aligned} & \widehat{\varphi}'(\widehat{x}) \le \liminf_{n\to\infty} \inf_{x\in X} \varphi_n(x) \qquad (\text{with } \widehat{\varphi}' = \Gamma_\tau\text{-}\liminf_{k\to\infty} \varphi_{n_k}), \\ \Rightarrow\ & \min_{x\in X} \widehat{\varphi}'(x) \le \liminf_{n\to\infty} \inf_{x\in X} \varphi_n(x), \\ \Rightarrow\ & \min_{x\in X} \varphi'(x) \le \liminf_{n\to\infty} \inf_{x\in X} \varphi_n(x) \qquad (\text{because } \varphi' \le \widehat{\varphi}'). \end{aligned} \tag{1.69}$$

From (1.64) and (1.69) we obtain (1.62).

Finally if φ_n Γ_τ-converges to φ, then directly from Definition 1.5.1 we have

$$\limsup_{n\to\infty} \inf_{x\in X} \varphi_n(x) \le \inf_{x\in X} \varphi(x). \tag{1.70}$$

From (1.62) and (1.70) we obtain (1.63). □

To continue with the variational features of Γ-convergence, we introduce some relevant notation.

DEFINITION 1.5.30 Let $\varphi : X \longrightarrow \overline{\mathbb{R}}$ be a proper function. The set of all mininizers of φ in X, denoted by $\mathcal{M}(\varphi)$ is defined by

$$\mathcal{M}(\varphi) = \{x \in X : \varphi(x) = \inf_X \varphi\}.$$

If φ does not attain its infimum, we look for $\varepsilon > 0$ minimizers. So let $\varepsilon > 0$ be given. An *ε-minimizer* of φ in X is a point $x \in X$ such that

$$\varphi(x) \leq \max\left\{\inf_X \varphi + \varepsilon, -\frac{1}{\varepsilon}\right\}.$$

We denote the set of all ε-minimizers of φ in X, by $\mathcal{M}_\varepsilon(\varphi)$.

REMARK 1.5.31 If $\inf_X \varphi > -\infty$ and $\varepsilon > 0$ is small, then $x \in \mathcal{M}_\varepsilon(\varphi)$ if and only if $\varphi(x) \leq \inf_X \varphi + \varepsilon$. So in the above definition of ε-minimizer the term $-(1/\varepsilon)$ appears in order to take care of the case where $\inf_X \varphi = -\infty$. Note that $\mathcal{M}(\varphi) = \bigcap_{\varepsilon>0} \mathcal{M}_\varepsilon(\varphi)$. However, the set $\mathcal{M}_\varepsilon(\varphi)$ is nonempty for all $\varepsilon > 0$, but $\mathcal{M}(\varphi)$ may be empty.

PROPOSITION 1.5.32 *Suppose* $\varphi = \Gamma_\tau\text{-}\lim_{n\to\infty} \varphi_n$. *Then we have*

(a) $K_\tau\text{-}\limsup_{n\to\infty} \mathcal{M}(\varphi_n) \subseteq \bigcap_{\varepsilon>0} K_\tau\text{-}\limsup_{n\to\infty} \mathcal{M}_\varepsilon(\varphi_n) \subseteq \mathcal{M}(\varphi)$.

(b) *If* $\bigcap_{\varepsilon>0} K_\tau\text{-}\limsup_{n\to\infty} \mathcal{M}_\varepsilon(\varphi_n) \neq \emptyset$, *then* $\mathcal{M}(\varphi) \neq \emptyset$ *and* $\min_X \varphi = \limsup_{n\to\infty} \inf_X \varphi_n$.

(c) *If* $\bigcap_{\varepsilon>0} K_\tau\text{-}\liminf_{n\to\infty} \mathcal{M}_\varepsilon(\varphi_n) \neq \emptyset$, *then* $\mathcal{M}(\varphi) \neq \emptyset$ *and* $\min_X \varphi = \lim_{n\to\infty} \inf_X \varphi_n$.

PROOF: (a) For every $\varepsilon > 0$ and every $n \geq 1$, $\mathcal{M}(\varphi_n) \subseteq \mathcal{M}_\varepsilon(\varphi_n)$, thus first inclusion is obvious. Now let $x \in \bigcap_{\varepsilon>0} K_\tau\text{-}\limsup_{n\to\infty} \mathcal{M}_\varepsilon(\varphi_n)$. Then according to Definition 1.5.4, for every $\varepsilon > 0$, $U \in \mathcal{N}(x)$, and $k \geq 1$, we can find $n \geq k$ such that $U \cap \mathcal{M}_\varepsilon(\varphi_n) \neq \emptyset$. So we obtain

$$\begin{aligned} &\liminf_{n\to\infty} \inf_{u\in U} \varphi_n(u) \leq \max\left\{\limsup_{n\to\infty} \inf_X \varphi_n + \varepsilon, -\frac{1}{\varepsilon}\right\} \\ \Rightarrow\ &\varphi(x) \leq \limsup_{n\to\infty} \inf_X \varphi_n \quad \text{(because } \varepsilon > 0 \text{ was arbitrary).} \end{aligned}$$

But directly from Definition 1.5.1, we see that $\limsup_{n\to\infty} \inf_X \varphi_n \leq \inf_X \varphi$. So

$$\min_X \varphi = \limsup_{n\to\infty} \inf_X \varphi_n. \tag{1.71}$$

(b) Follows at once from (a) (see (1.71)).

(c) From the proof of (a) and Definition 1.5.1, we have

$$\begin{aligned} &\varphi(x) \leq \liminf_{n\to\infty} \inf_X \varphi_n \leq \inf_X \varphi \\ \Rightarrow\ &\inf_X \varphi = \liminf_{n\to\infty} \inf_X \varphi_n, \quad \text{which together with (b), proves (c).} \end{aligned}$$

□

If $\varphi = \Gamma_\tau\text{-}\lim_{n\to\infty} \varphi_n$ is not identically $+\infty$, then we can improve the conclusions of Proposition 1.5.32.

THEOREM 1.5.33 *Suppose that* $\varphi = \Gamma_\tau\text{-}\lim_{n\to\infty} \varphi_n$, φ *is not identically* $+\infty$ *and consider the following statements.*

(i) $\bigcap_{e>0} K_\tau\text{-}\limsup_{n\to\infty} \mathcal{M}_\varepsilon(\varphi_n) \neq \emptyset$.
(ii) $\mathcal{M}(\varphi) \neq \emptyset$ *and* $\min_X \varphi = \limsup_{n\to\infty} \inf_X \varphi_n$.
(iii) $\mathcal{M}(\varphi) = \bigcap_{e>0} K_\tau\text{-}\limsup_{n\to\infty} \mathcal{M}_\varepsilon(\varphi_n)$.
(iv) $\bigcap_{e>0} K_\tau\text{-}\liminf_{n\to\infty} \mathcal{M}_\varepsilon(\varphi_n) \neq \emptyset$.
(v) $\mathcal{M}(\varphi) \neq \emptyset$ *and* $\min_X \varphi = \lim_{n\to\infty} \inf_X \varphi_n$.
(vi) $\mathcal{M}(\varphi) = \bigcap_{e>0} K_\tau\text{-}\liminf_{n\to\infty} \mathcal{M}_\varepsilon(\varphi_n) = \bigcap_{e>0} K_\tau\text{-}\limsup_{n\to\infty} \mathcal{M}_\varepsilon(\varphi_n)$.

then

(a) (i)$\Leftrightarrow$ (ii)$\Rightarrow$(iii).
(b) (iv)$\Leftrightarrow$ (v)$\Rightarrow$(vi).

PROOF: (a) Note that (i) $\Rightarrow$ (ii) follows from Proposition 1.5.32(b). Let us prove that (ii) $\Rightarrow$ (i). For this purpose let $x \in \mathcal{M}(\varphi)$. Then by hypothesis $\varphi(x) = \min_X \varphi = \limsup_{n\to\infty} \inf_X \varphi_n < +\infty$. So given any $\varepsilon > 0$, we have

$$\varphi(x) - \frac{\varepsilon}{2} \leq \inf_X \varphi_n \qquad \text{for infinitely many } n\text{s}. \tag{1.72}$$

From Definition 1.5.1, we see that for all $U \in \mathcal{N}(x)$,

$$\begin{aligned} &\limsup_{n\to\infty} \inf_U \varphi_n < \max\left\{\varphi(x) + \frac{\varepsilon}{2}, -\frac{1}{\varepsilon}\right\}, \\ \Rightarrow\ &\inf_U \varphi_n < \max\left\{\varphi(x) + \frac{\varepsilon}{2}, -\frac{1}{\varepsilon}\right\} \qquad \text{for all } n \geq 1 \text{ large}, \\ \Rightarrow\ &\inf_U \varphi_n < \max\left\{\varphi(x) + \varepsilon, -\frac{1}{\varepsilon}\right\} \qquad \text{for infinitely many } n\text{s (see (1.72))}. \end{aligned} \tag{1.73}$$

Therefore $U \cap \mathcal{M}_\varepsilon(\varphi_n) \neq \emptyset$ for infinitely many ns and every $\varepsilon > 0$ and so $x \in \bigcap_{e>0} K_\tau\text{-}\limsup_{n\to\infty} \mathcal{M}_\varepsilon(\varphi_n)$. So we conclude that $\mathcal{M}(\varphi) = \bigcap_{e>0} K_\tau\text{-}\limsup_{n\to\infty} \mathcal{M}_\varepsilon(\varphi_n) \neq \emptyset$ and from this argument we also infer that (i) and (ii) both imply (iii).

(b) In this case (1.72) holds for $n \geq 1$ large and then arguing as above we obtain that (1.73) is valid for all $n \geq 1$ large. Hence $U \cap \mathcal{M}_\varepsilon(\varphi_n) \neq \emptyset$ for all $n \geq 1$ large and every $\varepsilon > 0$ and so $x \in \bigcap_{e>0} K_\tau\text{-}\liminf_{n\to\infty} \mathcal{M}_\varepsilon(\varphi_n)$. This combined with (a) above and Proposition 1.5.32, gives (iv) $\Leftrightarrow$ (v) $\Rightarrow$(vi). □

COROLLARY 1.5.34 *If* $\varphi = \Gamma_\tau\text{-}\lim_{n\to\infty} \varphi_n$, $x_n \in \mathcal{M}_\varepsilon(\varphi_n)$ *with* $\varepsilon_n \downarrow 0$ *and* x *is a cluster point of* $\{x_n\}_{n\geq 1}$, *then* $x \in \mathcal{M}(\varphi)$ *and* $\varphi(x) = \limsup_{n\to\infty} \varphi_n(x_n)$. *Moreover, if* $x_n \longrightarrow x$ *in* X, *then* $x \in \mathcal{M}(\varphi)$ *and*

$$\varphi(x) = \lim_{n\to\infty} \varphi_n(x_n).$$

PROOF: Note that $x \in K_\tau\text{-}\limsup\limits_{n\to\infty} \mathcal{M}_{\varepsilon_n}(\varphi_n)$ and $\bigcap\limits_{\varepsilon>0} K_\tau\text{-}\limsup\limits_{n\to\infty} \mathcal{M}_\varepsilon(\varphi_n) \neq \emptyset$. Then from Theorem 1.5.33(a) we obtain the first part of the corollary. If $x_n \longrightarrow x$ in X, then $x \in K_\tau\text{-}\liminf\limits_{n\to\infty} \mathcal{M}_{\varepsilon_n}(\varphi_n)$ and so $\bigcap\limits_{\varepsilon>0} K_\tau\text{-}\liminf\limits_{n\to\infty} \mathcal{M}_\varepsilon(\varphi_n) \neq \emptyset$. Then the second part of the corollary follows from Theorem 1.5.33(b). □

We have another result on the variational stability of equicoercive sequences.

THEOREM 1.5.35 *If* $\varphi = \Gamma_\tau\text{-}\lim\limits_{n\to\infty} \varphi_n$, *$\varphi$ is not identically $+\infty$, $\{\varphi_n\}_{n\geq 1}$ is equicoercive, and* $\inf\limits_{n\geq 1} \varphi_n(x) > -\infty$ *for all $x \in X$, then for every V neighborhood of $\mathcal{M}(\varphi)$, we can find $\varepsilon_0 > 0$ and $n_0 \geq 1$ such that*

$$\mathcal{M}(\varphi_n) \subseteq \mathcal{M}_{\varepsilon_0}(\varphi_n) \subseteq V \qquad \textit{for all } n \geq n_0.$$

Moreover, for every $x \in \mathcal{M}(\varphi)$, every $U \in \mathcal{N}(x)$, and every $\varepsilon > 0$, we can find $n_1 \geq 1$ such that $\mathcal{M}_\varepsilon(\varphi_n) \cap U \neq \emptyset$ for all $n \geq n_1$.

PROOF: Let $\lambda \in \mathbb{R}$ such that $\inf\limits_X \varphi < \lambda$. From Theorem 1.5.29, we know that φ is coercive and so we can find $K \subseteq X$ τ-closed, and τ-countably compact such that

$$\{\varphi_n \leq \lambda\} \subseteq K \qquad \text{for all } n \geq 1 \text{ and } \{\varphi \leq \lambda\} \subseteq K. \tag{1.74}$$

Let $\mu = \min\limits_{K\cap V^c} \varphi$. Note that $K \cap V^c$ is τ-closed and $\tau-$countably compact. This together with the τ-lower semicontinuity and coercivity of φ, imply that we can find $x \in K \cap V^c$ such that $\mu = \varphi(x)$. Because $x \notin \mathcal{M}(\varphi)$, we have $\inf\limits_X \varphi < \varphi(x) = \mu$. Also as in the proof of Theorem 1.5.29 we obtain

$$\mu = \min_{K\cap V^c} \varphi \leq \liminf_{n\to\infty} \inf_{K\cap V^c} \varphi_n.$$

From (1.74) we also have that $\lambda \leq \inf\limits_{X\cap V^c} \varphi_n$. Therefore

$$\min\{\lambda, \mu\} \leq \liminf_{n\to\infty} \inf_{X\cap V^c} \varphi_n.$$

On the other hand using Theorem 1.5.29, we have

$$\lim_{n\to\infty} \inf_X \varphi_n = \min_X \varphi < \min\{\lambda, \mu\} \leq \liminf_{n\to\infty} \inf_{X\cap V^c} \varphi_n.$$

So we can find $\varepsilon_0 > 0$ and $n_0 \geq 1$ such that

$$\begin{aligned} &\max\left\{\inf_X \varphi + \varepsilon_0, -\frac{1}{\varepsilon_0}\right\} < \inf_{X\cap V^c} \varphi_n \qquad \text{for all } n \geq n_0, \\ \Rightarrow\ &\mathcal{M}_{\varepsilon_0}(\varphi_n) \subseteq V \qquad \text{for all } n \geq n_0. \end{aligned}$$

Finally because $\{\varphi_n\}_{n\geq 1}$ is equicoercive, from Theorem 1.5.29 we see that statement (v) in Theorem 1.5.33 is satisfied. So by virtue of that theorem statement (vi) holds, which by virtue of Definition 1.5.4 implies the conclusion of the theorem. □

Sometimes it is suitable to consider different topologies when defining the two limit functions for a sequence $\{\varphi_n\}_{n\geq 1}$.

DEFINITION 1.5.36 (a) Let X be a Banach space and $\varphi_n\colon X \longrightarrow \overline{\mathbb{R}}$, $n \geq 1$, a sequence of proper functions. We say that φ_ns *Mosco converge* to a function $\varphi\colon X \longrightarrow \mathbb{R}$, denoted by $\varphi_n \xrightarrow{M} \varphi$, if the following two conditions hold.

(i) For every $x \in X$ and every $x_n \xrightarrow{w} x$, we have $\varphi(x) \leq \liminf\limits_{n\to\infty} \varphi_n(x_n)$.
(ii) For every $x \in X$, we can find a sequence $x_n \longrightarrow x$ such that $\varphi(x) = \lim\limits_{n\to\infty} \varphi_n(x_n)$.

There is a corresponding convergence of sets.

(b) Let $\{C_n\}_{n\geq 1} \subseteq 2^X \setminus \{\emptyset\}$. We set $s\text{-}\liminf\limits_{n\to\infty} C_n = \{x \in X : x = \lim x_n, x_n \in C_n \text{ for all } n \geq 1\}$ and $w\text{-}\limsup\limits_{n\to\infty} C_n = \{x \in X : x = w\text{-}\lim\limits_{k\to\infty} x_{n_k},\ x_{n_k} \in C_{n_k},\ n_1 < n_2 < \ldots < n_k < \ldots\}$. We say that the C_ns Mosco converge to a set C, denoted by $C_n \xrightarrow{M} C$, if

$$C = s\text{-}\liminf_{n\to\infty} C_n = w\text{-}\limsup_{n\to\infty} C_n.$$

REMARK 1.5.37 Clearly we always have $s\text{-}\liminf\limits_{n\to\infty} C_n \subseteq w\text{-}\limsup\limits_{n\to\infty} C_n$.

The following theorem justifies Definition 1.5.36. Its proof can be found in Attouch [34, p. 295], or Hu–Papageorgiou [313, p. 754].

THEOREM 1.5.38 *If X is a reflexive Banach space and $\{\varphi_n, \varphi\}_{n\geq 1} \subseteq \Gamma_0(X)$, then $\varphi_n \xrightarrow{M} \varphi$ in X if and only if $\varphi_n^* \xrightarrow{M} \varphi^*$ in X^*.*

REMARK 1.5.39 So convex conjugation (see Definition 1.2.15) is bicontinuous for the Mosco convergence from $\Gamma_0(X)$ onto $\Gamma_0(X^*)$, when X is reflexive.

An immediate consequence of this theorem is the following proposition.

PROPOSITION 1.5.40 *If X is a reflexive Banach space, $\{\varphi_n, \varphi\}_{n\geq 1} \subseteq \Gamma_0(X)$ and $\varphi_n \xrightarrow{M} \varphi$, then for every $x^* \in X^*$ we can find a sequence $x_n^* \longrightarrow x^*$ in X^* such that*

$$\inf_X (\varphi_n - x_n^*) \longrightarrow \inf_X (\varphi - x^*).$$

We conclude the section with some additional variational properties of the Mosco convergence. The proof of this proposition can be found in Denkowski–Migórski–Papageorgiou [195, p. 468].

PROPOSITION 1.5.41 *If X is a reflexive Banach space and $\{\varphi_n, \varphi\}_{n\geq 1} \subseteq \Gamma_0(X)$, then*

(a) *If $\varphi_n \xrightarrow{M} \varphi$ and $\inf\limits_X \varphi < \lambda$, then $\{\varphi_n \leq \lambda\} \xrightarrow{M} \{\varphi \leq \lambda\}$.*
(b) *If for every $\lambda \in \mathbb{R}$ $\{\varphi_n \leq \lambda\} \xrightarrow{M} \{\varphi \leq \lambda\}$, then $\varphi_n \xrightarrow{M} \varphi$.*

1.6 Remarks

1.1: Various aspects of the differential calculus in Banach spaces can be found in the books of Cartan [131], Denkowski–Migórski–Papageorgiou [194], Dieudonné [200], Vainberg [590], Zeidler [620], and the survey papers of Averbukh–Smolyanov [48] and Nashed [453].

1.2: There are several books treating the theory of convex functions. We mention the books of Rockafellar [522], Hiriart-Urruty–Lemarechal [308, 309], Webster [602], Rockafellar–Wets [530] (convex functions defined on $\mathbb{R}^N$), and of Laurent [370], Roberts–Varberg [516], Holmes [310], Barbu–Precupanu [59], Ekeland–Temam [222], Ioffe–Tichomirov [327], Giles [264], Aubin–Ekeland [39], Phelps [495], and Denkowski–Migórski–Papageorgiou [194] (convex functions defined on Banach spaces and on locally convex spaces). In the books of Holmes [310], Giles [264], and Phelps [495], the emphasis is on the relations between the theory of convex functions and the theory of Banach spaces. Theorem 1.2.11 dates back to Mazur [417]. The duality theory of convex functions was initiated by Fenchel [241] for functions defined on $\mathbb{R}^N$ and was extended to dual pairs of locally convex spaces by Brondsted [112], Moreau [442], and Rockafellar [528]. This duality theory can give the following useful result originally due to Hörmander [311].

PROPOSITION 1.6.1 *If X is a locally convex space, then there is a bijective correspondence between nonempty, closed, convex sets and sublinear $w(X^*,X)$-lower semicontinuous functions on X^* with values in $\overline{\mathbb{R}}$. This correspondence maps the set C to $\sigma(\cdot;C)$ its support function.*

The systematic study of the subdifferential starts with the works of Moreau [441, 442] and Rockafellar [517, 522, 523]. The subdifferential theory of convex integral functionals can be found in Rockafellar [519, 525]. The work of Rockafellar was extended by Levin [376, 377, 378], and Castaing–Valadier [134]. Interesting connections between the subdifferentiability of convex functions and Banach space theory can be found in Giles [264] and Phelps [495].

1.3: The subdifferential theory of locally Lipschitz functions started with the thesis of Clarke [148], who using Rademacher's theorem gave the description of the subdifferential included in Theorem 1.3.12 for functions in $\mathbb{R}^N$. Soon thereafter Clarke [149, 152] extended the results to locally Lipschitz functions defined on a Banach space. Proposition 1.3.24 is due to Lebourg [371]. A comprehensive presentation of the theory together with applications in mathematical programming, optimal control, and calculus of variations can be found in Clarke [153, 154].

We mention that Rademacher's theorem (see Theorem 1.3.11) can be extended to functions between Banach spaces as follows.

DEFINITION 1.6.2 Let $(G,+)$ be an Abelian Polish group and d an invariant metric on G compatible with the topology (therefore automatically complete). A universally measurable set $C \subseteq G$ is *Haar-null*, if there exists a probability measure μ on G (not unique), such that $\chi_C * \mu = 0$. Here χ_C is the characteristic function of the set C and $(\chi_C * \mu)(x) = \int_G \chi_C(x+y)\mathrm{d}\mu(y)$.

REMARK 1.6.3 So every translation of C is μ-null. The measure μ is usually called the *test measure*.

Then the infinite-dimensional extension of Theorem 1.3.11 reads as follows (see Christensen [145]).

THEOREM 1.6.4 *If X is a separable Banach space, Y is a Banach space with the RNP (Radon–Nikodym property; see Denkowski–Migórski–Papageorgiou* [194, p. 372]) *and $\varphi : X \longrightarrow Y$ is locally Lipschitz, then there exists a universally measurable set $D \subseteq X$ such that $X \setminus D$ is Haar-null and $\varphi\big|_D$ is Gâteaux differentiable.*

1.4: Tangent cones and normal cones have been introduced over the years in a variety of contexts and it is almost impossible today to give a full account of the various cones existing in the literature. We have concentrated on the two most profilic cones, the tangent cones $T_C(x)$ and $T'_C(x)$ and their respective normal cones. The tangent cone $T_C(x)$ (see Definition 1.4.1) was introduced by Bouligand [93] for general sets. It is usually called the *contingent cone* and it exhibits good properties in the context of convex sets. The other tangent cone $T'_C(x)$ (see Definition 1.4.17) was introduced by Clarke [148, 153] and is always convex without the set being convex. So duality techniques are readily available for this cone. The price we pay for this convenience is that often this cone reduces to the trivial one $\{0\}$. A detailed study of these cones and of related concepts can be found in Rockafellar–Wets [530] (finite-dimensional theory) and Aubin–Frankowska [40], and Hu–Papageorgiou [313] (infinite-dimensional theory).

1.5: Γ-convergence was introduced and studied by De Giorgi–Franzoni [190, 191]. The related K_τ-convergence of sets (see Definition 1.5.4), known as the Kuratowski convergence, was actually first used by Painlevé in 1902 in his lectures on analysis in the Université de Paris. However, because it was more systematically examined and popularized by Kuratowski [366, 368], it carries his name. The Γ-convergence is designed in such a way so as to be suitable in the study of the stability analysis of variational problems. For this reason it is said to be a *variational convergence* and can be found in Attouch [34], Dal Maso [171], Denkowski–Migórski–Papageorgiou [195], Dontchev–Zolezzi [202], Hu–Papageorgiou [313], and Rockafellar–Wets (only in $\mathbb{R}^N$) [530]. The Mosco convergence of sets and functions (see Definition 1.5.36) was introduced and studied by Mosco [445, 446], who proved the main result, which is Theorem 1.5.38 and which justifies the introduction of this alternative variational convergence. For more on the Mosco convergence we refer to Attouch [34], Dontchev–Zolezzi [202], and Hu–Papageorgiou [313]. We should also mention the multiple Γ-convergence which is defined on the product of two Hausdorff topological spaces and is convenient in the variational analysis of optimal control problems. We refer the reader to Buttazzo [126] for details.

2

Extremal Problems and Optimal Control

Summary. *This chapter is devoted to the study of various kinds of variational problems and of the mathematical tools needed to deal with them. We start with the so-called "direct method", that leads to a systematic study of lower semicontinuous functionals. Then we consider optimization problems with constraints using the Lagrange multipliers methods. This method is closely connected to nonlinear eigenvalue problems. We also develop a general duality theory for convex optimization problems and study saddle points, KKM-multimaps, coincidence theorems, variational inequalities and the Fenchel duality theory. Subsequently, we present the Ekeland variational principle and some of its most remarkable consequences. We also show that it is equivalent to a whole family of other important results of nonlinear analysis. With these mathematical tools, we then pass to the study of calculus of variations problems (Euler equation and canonical Hamiltonian equations) and of optimal control problems (existence theory, relaxation and Pontryagin's maximum principle).

Introduction

This chapter is devoted to the study of variational problems and of the mathematical tools that are necessary to do this.

So, in Section 2.1 we conduct a detailed investigation of the topological notion of lower semicontinuity. Since the work of Tonelli, the concept of lower semicontinuity has played a central role in the study of variational problems. Our study outlines the so-called direct method of the calculus of variations and also introduces the notion of relaxed functional which we encounter also in Section 2.6.

In Section 2.2 we deal with infinite-dimensional optimization problems with constraints and we analyze them by developing the so-called method of Lagrange multipliers. This method is closely connected to nonlinear eigenvalue problems.

Section 2.3 studies saddle points and develops a general duality theory for convex optimization problems. Saddle points appear in game theory and in general in control problems where the controllers exhibit conflicting interests. In connection with saddle points we also introduce and study KKM-maps which are useful in producing coincidence theorems for families of multifunctions and also lead to existence

N.S. Papageorgiou, S.Th. Kyritsi-Yiallourou, *Handbook of Applied Analysis*,
Advances in Mechanics and Mathematics 19, DOI 10.1007/b120946_2,

theorems for variational inequalities. Duality is at the core of convex analysis. In introductory functional analysis we encounter the first instances of duality with the separation theorems for convex sets, which was extended in Definition 1.2.15 with the introduction of the conjugate of a convex function. Here using a "perturbation" approach to a convex minimization problem, we associate a concave maximization one and investigate the precise relation between the two. As a special case, we present the so-called Fenchel duality theory.

In Section 2.4 first we present the Ekeland variational principle and discuss some remarkable consequences of it. Then we introduce some other fundamental results of nonlinear analysis, namely the Caristi fixed point theorem, the Takahashi variational principle, and the drop theorem, and we show that they are all equivalent to the Ekeland variational principle. Finally we prove a principle concerning ordered spaces, which we show generates all the aforementioned results.

In Section 2.5 we present some basic aspects of the calculus of variations. We consider problems with integral cost functional and fixed endpoints. We develop the Euler equation for an optimal state as well as the canonical Hamiltonian equations.

In Section 2.6 we deal with optimal control problems. We limit ourselves to systems monitored by ordinary differential equations (lumped parameter systems). First we develop the existence theory for such optimal control problems. This theory reveals the importance of convex structure in the problem. When this convexity is missing, we need to augment the system in order to assure existence of optimal pairs. This is the process of relaxation. We present four relaxation methods that we show are equivalent under reasonable conditions on the data. Finally we develop necessary conditions for optimality in the form of a maximum principle. Our approach is based on the Ekeland variational principle.

2.1 Lower Semicontinuity

In Section 1.2 we examined in some detail proper, lower semicontinuous, convex functions defined on a locally convex space. In this section we drop the convexity condition and concentrate on lower semicontinuous functions. The reason for this is that such functions are involved in the so-called direct method of the calculus of variations. Already in Section 1.5 we had some first results involving the notion of lower semicontinuity. Here we conduct a systematic study of this concept.

Our setting is purely topological. So let (X,τ) be a Hausdorff topological space, with τ denoting its topology. Additional conditions on X are introduced as needed. For $x \in X$, by $\mathcal{N}(x)$ we denote the filter of neighborhoods of x. Also recall that $\mathbb{R}^* = \mathbb{R} \cup \{\pm\infty\}$. We start by recalling the definition of τ-lower semicontinuity (both local and global) of a function $\varphi\colon X \longrightarrow \mathbb{R}^*$.

DEFINITION 2.1.1 A function $\varphi\colon X \longrightarrow \mathbb{R}^*$ is said to be *τ-lower semicontinuous at x*, if

$$\varphi(x) = \liminf_{u \to x} \varphi(u) = \sup_{U \in \mathcal{N}(x)} \inf_{u \in U} \varphi(u).$$

If φ is τ-lower semicontinuous at every $x \in X$, then we say that φ is τ-lower semicontinuous. If $-\varphi$ is τ-lower semicontinuous (at x), then we say that φ is *τ-upper semicontinuous* (at x).

REMARK 2.1.2 If the topology τ is clearly understood, then we drop τ from the definition and say that φ is lower semicontinuous (at $x \in X$). Also if $x \in X$ is a point where $\varphi(x) > -\infty$, then φ is τ-lower semicontinuous at x if and only if for every $\lambda \in \mathbb{R}$ with $\lambda < \varphi(x)$, we can find $U \in \mathcal{N}(x)$ such that $\lambda < \varphi(u)$ for all $u \in U$.

PROPOSITION 2.1.3 *For a function $\varphi \colon X \longrightarrow \mathbb{R}^*$ the following statements are equivalent.*

(a) *φ is τ-lower semicontinuous.*
(b) *For every $\lambda \in \mathbb{R}$, the set $L_\lambda = \{x \in X : \varphi(x) \leq \lambda\}$ is τ-closed.*
(c) epi $\varphi = \{(x, \lambda) \in X \times \mathbb{R} : \varphi \leq \lambda\}$ *is closed in $X \times \mathbb{R}$.*

PROOF: (a)$\Rightarrow$(b) We show that L_λ^c is open. Let $x \in L_\lambda$. Then $\lambda < \varphi(x)$ and by virtue of Remark 2.1.2 we can find $U \in \mathcal{N}(x)$ such that $\lambda < \varphi(u)$ for all $u \in U$, which shows that L_λ^c is open; hence L_λ is closed.

(b)$\Rightarrow$(a) If $\varphi(x) = -\infty$, then it is clear from Definition 2.1.1 that φ is τ-lower semicontinuous at $x \in X$. So suppose that $-\infty < \varphi(x)$ and let $\lambda < \varphi(x)$. Note that $x \in L_\lambda^c$ and because by hypothesis L_λ is closed, we can find $U \in \mathcal{N}(x)$ such that $U \subseteq L_\lambda^c$, hence $\lambda < \varphi(u)$ for all $u \in U$, which proves the τ-lower semicontinuity of φ at $x \in X$.

So we have established the equivalence of statements (a) and (b).

Next let $h \colon X \times \mathbb{R} \longrightarrow \mathbb{R}^*$ be defined by $h(x, \lambda) = \varphi(x) - \lambda$. Clearly φ is τ-lower semicontinuous on X if and only if h is lower semicontinuous on the product space $X \times \mathbb{R}$. Because epi $\varphi = \{(x, \lambda) \in X \times \mathbb{R} : h(x, \lambda) \leq 0\}$, the equivalence of (a) and (b) established above implies the equivalence of (b) and (c). $\square$

COROLLARY 2.1.4 *If $\varphi_i \colon X \longrightarrow \mathbb{R}^*$, $i \in I$, is a family of τ-lower semicontinuous functions, then*

(a) $\sup_{i \in I} \varphi_i$ *is τ-lower semicontinuous.*
(b) *If I is finite, then $\inf_{i \in I} \varphi_i$ is τ-lower semicontinuous.*

COROLLARY 2.1.5 *A set $C \subseteq X$ is τ-closed if and only if i_C is τ-lower semicontinuous.*

From Definition 2.1.1 we see that $\varphi \colon X \longrightarrow \mathbb{R}^*$ is τ-lower semicontinuous at $x \in X$; then for every $x_n \longrightarrow x$ in X, we have $\varphi(x) \leq \liminf_{n \to \infty} \varphi(x_n)$. The converse is not in general true.

EXAMPLE 2.1.6 Let X be an infinite-dimensional Banach space furnished with the weak topology denoted by w. Also let $C \subseteq X$ be a nonempty set which is sequentially w-closed but not w-closed. Then the indicator function i_C satisfies $i_C(x) \leq \liminf_{n \to \infty} i_C(x_n)$ for every sequence $x_n \xrightarrow{w} x$ in X, but it is not w-lower semicontinuous. For example in $X = l^1$, due to the Schur property the set $C = \partial B_1 = \{x \in l^1 : \|x\|_{l^1} = 1\}$ is sequentially w-closed but it is not w-closed (in fact, $\overline{C}^{w} = \overline{B}_1 = \{x \in l^1 : \|x\|_{l^1} \leq 1\}$).

DEFINITION 2.1.7 We say that $\varphi\colon X \longrightarrow \mathbb{R}^*$ is *sequentially τ-lower semicontinuous at $x \in X$*, if $\varphi(x) \leq \liminf_{n\to\infty} \varphi(x_n)$ for every sequence $x_n \longrightarrow x$ in X. We say that φ is *sequentially τ-lower semicontinuous*, if it is sequentially τ-lower semicontinuous at every $x \in X$.

REMARK 2.1.8 Let τ_{seq} denote the topology of X whose sets are sequentially τ-closed; then $\varphi\colon X \longrightarrow \mathbb{R}^*$ is sequentially τ-lower semicontinuous if and only if it is τ_{seq}-lower semicontinuous. In general the τ_{seq} topology is stronger than the τ-topology and so the notion of sequential τ-lower semicontinuous is more general than the τ-lower semicontinuity. Clearly $\tau = \tau_{\text{seq}}$ if and only if X is first countable. So we can state the following proposition.

PROPOSITION 2.1.9 *If (X, τ) is first countable, $\varphi\colon X \longrightarrow \mathbb{R}^*$ and $x \in X$, then φ is τ-lower semicontinuous at $x \in X$ if and only if φ is sequentially τ-lower semicontinuous at $x \in X$.*

The next theorem summarizes the direct method of the calculus of variations and underlines the importance of the notion of lower semicontinuity in variational analysis. The result is known as the *Weierstrass theorem.*

THEOREM 2.1.10 *If $\varphi\colon X \longrightarrow \mathbb{R}^*$ is τ-coercive and τ-lower semicontinuous (resp., sequentially τ-coercive and sequentially τ-lower semicontinuous), then*

(a) *There exists $x \in X$ such that $\varphi(x) = \inf_X \varphi$.*
(b) *If $\{x_n\}_{n\geq 1}$ is a minimizing sequence and x is a cluster point of $\{x_n\}_{n\geq 1}$ (resp., x is a subsequential limit of $\{x_n\}_{n\geq 1}$), then $\varphi(x) = \inf_X \varphi$.*
(c) *If φ is not identically $+\infty$, then every minimizing sequence has a cluster point (resp., a convergent subsequence).*

PROOF: If $\varphi = +\infty$, then all $x \in X$ minimize φ and so (a) and (b) hold.

So assume that φ is not identically $+\infty$. Let $\{x_n\}_{n\geq 1}$ be a minimizing sequence for φ. We have $\varphi(x_n) \downarrow \inf_X \varphi < +\infty$. Let $\lambda \in \mathbb{R}$ be such that $\inf_X \varphi < \lambda$. Then we can find $n_0 \geq 1$ such that $\varphi(x_n) \leq \lambda$ for all $n \geq n_0$. Due to the τ-coercivity (resp., sequential τ-coercivity) of φ, we have that $\{x_n\}_{n\geq 1}$ has a cluster point $x \in X$ (resp., a subsequential limit point $x \in X$); see Definition 1.5.26. Then due to the τ-lower semicontinuity (resp. sequential τ-lower semicontinuity) of φ, we have

$$\begin{aligned} &\inf_X \varphi \leq \varphi(x) \leq \lim_{n\to\infty} \varphi(x_n) = \inf_X \varphi, \\ \Rightarrow\ &\varphi(x) = \inf_X \varphi. \end{aligned}$$

∎

COROLLARY 2.1.11 *If X is a reflexive Banach space, $\varphi \in \Gamma_0(X)$, and $\varphi(x) \longrightarrow +\infty$ as $\|x\| \longrightarrow \infty$, then we can find $x \in X$ such that $-\infty < \varphi(x) = \inf_X \varphi$.*

PROOF: Note that due to the convexity of φ, the function is lower semicontinuous if and only if it is w-lower semicontinuous. Moreover, due to the reflexivity of X bounded sets are relatively w-compact (in fact relatively sequentially w-compact by the Eberlein–Smulian theorem). ∎

REMARK 2.1.12 If $\varphi\colon X \longrightarrow \mathbb{R}^*$ is lower semicontinuous convex and there exists $x_0 \in X$ such that $\varphi(x_0) = -\infty$, then φ is nowhere finite on X. For this reason, when convexity is present, we consider functions with values in $\overline{\mathbb{R}} = \mathbb{R} \cup \{+\infty\}$.

DEFINITION 2.1.13 Let $\varphi\colon X \longrightarrow \mathbb{R}^*$. The *$\tau$-lower semicontinuous regularization* (or the *τ-lower semicontinuous envelope*) of φ, is the function $\overline{\varphi}^{\,\tau}\colon X \longrightarrow \mathbb{R}^*$ defined by

$$\overline{\varphi}^{\,\tau}(x) = \liminf_{u \to x} \varphi(u) = \sup_{U \in \mathcal{N}(x)} \inf_{u \in U} \varphi(u).$$

REMARK 2.1.14 It is clear from this definition that $\overline{\varphi}^{\,\tau}$ is τ-lower semicontinuous.

In the next proposition we show that $\overline{\varphi}^{\,\tau}$ is the biggest τ-lower semicontinuous function majorized by φ.

PROPOSITION 2.1.15 *If $\varphi\colon X \longrightarrow \mathbb{R}^*$ and $\mathcal{S}(\varphi) = \big\{h : X \longrightarrow \mathbb{R}^* :$ h is τ-lower semicontinuous, $h \le \varphi\big\}$, then* $\operatorname{epi} \overline{\varphi}^{\,\tau} = \overline{\operatorname{epi} \varphi}$ *and* $\overline{\varphi}^{\,\tau}(x) = \sup\big[h(x) : h \in \mathcal{S}(\varphi)\big]$ *for all $x \in X$.*

PROOF: Evidently $\overline{\operatorname{epi} \varphi} \subseteq \operatorname{epi} \overline{\varphi}^{\,\tau}$. Let $(x, \lambda) \in \operatorname{epi} \overline{\varphi}^{\,\tau}$. Then $\overline{\varphi}^{\,\tau}(x) \le \lambda$. So for every $U \in \mathcal{N}(x)$ and for every $\varepsilon > 0$ we have $\inf_{u \in U} \varphi(u) \le \overline{\varphi}^{\,\tau}(x) \le \lambda < \lambda + \varepsilon$. So we can find $u \in U$ and $\mu \in (\lambda, \lambda+\varepsilon)$ such that $\varphi(u) < \mu$. Hence $\big(U \times (\lambda-\varepsilon, \lambda+\varepsilon)\big) \cap \operatorname{epi} \varphi \ne \emptyset$, from which it follows that $(x, \lambda) \in \overline{\operatorname{epi} \varphi}$. Therefore $\operatorname{epi} \overline{\varphi}^{\,\tau} = \overline{\operatorname{epi} \varphi}$.

Now note that $\overline{\varphi}^{\,\tau} \in \mathcal{S}(\varphi)$. Also if $h \in \mathcal{S}(\varphi)$, then $\operatorname{epi} \varphi \subseteq \operatorname{epi} h$ and so $\operatorname{epi} \overline{\varphi}^{\,\tau} = \overline{\operatorname{epi} \varphi} \subseteq \overline{\operatorname{epi} h} = \operatorname{epi} h$ (see Proposition 2.1.3). Therefore $h \le \overline{\varphi}^{\,\tau}$ and so we conclude that

$$\overline{\varphi}^{\,\tau}(x) = \sup\big[h(x) : h \in \mathcal{S}(\varphi)\big] \qquad \text{for all } x \in X.$$

□

COROLLARY 2.1.16 *If $\varphi\colon X \longrightarrow \mathbb{R}^*$ and $\lambda \in \mathbb{R}$, then* $\{\overline{\varphi}^{\,\tau} \le \lambda\} = \bigcap_{\mu > \lambda} \overline{\{\varphi \le \mu\}}^{\,\tau}$.

COROLLARY 2.1.17 *If $C \subseteq X$, then* $\overline{i_C}^{\,\tau} = i_{\overline{C}^{\,\tau}}$.

COROLLARY 2.1.18 *If $\varphi, \psi\colon X \longrightarrow \mathbb{R}^*$, then*

(a) $\overline{(\varphi + \psi)}^{\,\tau} \ge \overline{\varphi}^{\,\tau} + \overline{\psi}^{\,\tau}$.

(b) *If ψ is continuous and finite everywhere, then* $\overline{(\varphi + \psi)}^{\,\tau} = \overline{\varphi}^{\,\tau} + \psi$.

The following proposition characterizes $\overline{\varphi}^{\,\tau}$ in terms of sequences and follows from Proposition 1.5.20 (see also Remark 1.5.2).

PROPOSITION 2.1.19 *If (X, τ) is first countable, $\varphi\colon X \longrightarrow \mathbb{R}^*$ and $x \in X$, then* $\overline{\varphi}^{\,\tau}$.

(a) *For every sequence $x_n \longrightarrow x$ in X we have* $\overline{\varphi}^{\,\tau}(x) \le \liminf_{n \to \infty} \varphi(x_n)$.

(b) *There exists a sequence* $x_n \longrightarrow x$ *in* X *such that* $\overline{\varphi}^{\tau}(x) = \lim_{n\to\infty} \varphi(x_n)$.

The next theorem summarizes the relaxation method in the study of variational problems.

THEOREM 2.1.20 *If* $\varphi\colon X \longrightarrow \mathbb{R}^*$ *is* τ*-coercive, then*

(a) $\overline{\varphi}^{\tau}$ *is* τ*-coercive and* τ*-lower semicontinuous.*
(b) *There exists* $x \in X$ *such that* $\overline{\varphi}^{\tau}(x) = \inf_X \overline{\varphi}^{\tau}$.
(c) $\min_X \overline{\varphi}^{\tau} = \inf_X \varphi$.
(d) *Every cluster point of a minimizing sequence for* φ *is a minimizer of* $\overline{\varphi}^{\tau}$.
(e) *If* (X, τ) *is first countable,*
then the minimizers of $\overline{\varphi}^{\tau}$ *are the limits of minimizing sequences for* φ.

PROOF: (a) We already know that $\overline{\varphi}^{\tau}$ is τ-lower semicontinuous. The τ-coercivity of $\overline{\varphi}^{\tau}$ follows from Theorem 1.5.29 and Remark 1.5.2 (see also Proposition 2.1.19).

(b) Follows from part (a) and Theorem 2.1.10.

(c) Let $h(x) = \inf_X \varphi$ for all $x \in X$. Then $h \in \mathcal{S}(\varphi)$ and so $h \leq \overline{\varphi}^{\tau}$ (see Proposition 2.1.15). It follows that

$$\inf_X \varphi \leq \min_X \overline{\varphi}^{\tau}.$$

Because the opposite inequality is always true (recall that $\overline{\varphi}^{\tau} \leq \varphi$), we conclude that

$$\min_X \overline{\varphi}^{\tau} = \inf_X \varphi.$$

(d) Let $\{x_n\}_{n\geq 1}$ be a minimizing sequence for φ and x a cluster point of $\{x_n\}_{n\geq 1}$. Then

$$\overline{\varphi}^{\tau}(x) \leq \limsup_{n\to\infty} \overline{\varphi}^{\tau}(x_n) \leq \lim_{n\to\infty} \varphi(x_n) = \inf_X \varphi = \min_X \overline{\varphi}^{\tau}$$

(see part (c)),

$$\Rightarrow \overline{\varphi}^{\tau}(x) = \min_X \overline{\varphi}^{\tau}.$$

(e) Let $x \in X$ be such that $\overline{\varphi}^{\tau}(x) = \min_X \overline{\varphi}^{\tau}$. Then from Proposition 2.1.19 and (c), we can find $x_n \longrightarrow x$ such that

$$\inf_X \varphi = \overline{\varphi}^{\tau}(x) = \lim_{n\to\infty} \varphi(x_n).$$

□

REMARK 2.1.21 This theorem shows that the relaxed functional captures the asymptotic behavior of the minimizing sequences for φ.

For convex functions in a Banach space, the relaxation process is simpler because by Mazur's lemma for convex sets, the closures in the norm and weak topologies coincide. So we have the following.

PROPOSITION 2.1.22 *If X is a Banach space and $\varphi\colon X \longrightarrow \overline{\mathbb{R}}$ is convex, then $\overline{\varphi}^{\,s} = \overline{\varphi}^{\,w}$ (here by s we denote the strong topology on X and by w the weak topology on X).*

Next we prove a lower semicontinuity result for integral functionals defined on products of Lebesgue spaces. The mathematical setting is the following. Let (Ω, Σ, μ) be a complete, finite nonatomic measure space and $\varphi\colon \Omega \times \mathbb{R}^N \times \mathbb{R}^m \longrightarrow \overline{\mathbb{R}}$ a function that satisfies the following hypotheses.

H$_0$:

(i) φ is $\Sigma \times \mathbf{B}(\mathbb{R}^N) \times \mathbf{B}(\mathbb{R}^m)$-measurable with $\mathbf{B}(\mathbb{R}^N)$ (resp., $\mathbf{B}(\mathbb{R}^m)$) being the Borel σ-field of $\mathbb{R}^N$ (resp., of $\mathbb{R}^m$).

(ii) For μ-almost all $\omega \in \Omega$, $(x, u) \longrightarrow \varphi(\omega, x, u)$ is lower semicontinuous.

(iii) For μ-almost all $\omega \in \Omega$ and all $x \in \mathbb{R}^N$, $u \longrightarrow \varphi(\omega, x, u)$ is convex.

Let $1 \leq p, r \leq +\infty$ and consider the integral function $I_\varphi(x, u) : L^p(\Omega, \mathbb{R}^N) \times L^r(\Omega, \mathbb{R}^m) \longrightarrow \mathbb{R}^*$ defined by

$$I_\varphi(x, u) = \int_\Omega \varphi\big(\omega, x(\omega), u(\omega)\big) d\mu \qquad \text{for all } (x, u) \in L^p(\Omega, \mathbb{R}^N) \times L^r(\Omega, \mathbb{R}^m).$$

We want to establish a sequential lower semicontinuity result when we consider the strong topology on $L^p(\Omega, \mathbb{R}^N)$ and the weak topology on $L^r(\Omega, \mathbb{R}^m)$ (the weak* topology on $L^\infty(\Omega, \mathbb{R}^m)$ when $r = +\infty$). To do this we need the following approximation result, whose proof can be found in Buttazzo [126, p. 42].

PROPOSITION 2.1.23 *If $\varphi\colon \Omega \times \mathbb{R}^N \times \mathbb{R}^m \longrightarrow \overline{\mathbb{R}}$ satisfies H_0 above and one of the following conditions.*

(iv) *For μ-almost all $\omega \in \Omega$, we can find a continuous function $u_0\colon \mathbb{R}^N \longrightarrow \mathbb{R}^m$ such that the function $x \longrightarrow \varphi\big(\omega, x, u_0(x)\big)$ is finite and continuous; or*

(v) *There exists a function $\vartheta\colon \mathbb{R} \longrightarrow \mathbb{R}$ such that*

$$\lim_{s \to +\infty} \frac{\vartheta(s)}{s} = +\infty$$

and $\vartheta(\|u\|_{\mathbb{R}^m}) \leq \varphi(\omega, x, u)$ for μ-a.a. $\omega \in \Omega$, all $x \in \mathbb{R}^N$ and all $u \in \mathbb{R}^m$,

then we can find two sequences of Carathéodory functions $a_n\colon \Omega \times \mathbb{R}^N \longrightarrow \mathbb{R}^m$ and $c_n\colon \Omega \times \mathbb{R}^N \longrightarrow \mathbb{R}$ such that

$$\varphi(\omega, x, u) = \sup_{n \geq 1} \big[\big(a_n(\omega, x), u\big)_{\mathbb{R}^m}) + c_n(\omega, x)\big]$$

for μ-almost all $\omega \in \Omega$ and all $x \in \mathbb{R}^N$, all $u \in \mathbb{R}^m$.

REMARK 2.1.24 Recall that $h\colon \Omega \times \mathbb{R}^k \longrightarrow \mathbb{R}^s$ $(1 \leq s, k)$ is a Carathéodory function, if for all $x \in \mathbb{R}^k$, $\omega \longrightarrow h(\omega, x)$ is Σ-measurable and for μ-almost all $\omega \in \Omega$, $x \longrightarrow h(\omega, x)$ is continuous. Such a function is automatically $\Sigma \times \mathbf{B}(\mathbb{R}^k)$-measurable. If in the above proposition $\varphi \geq 0$, then it is easy to see the following approximation for φ holds

$$\varphi(\omega, x, u) = \sup_{n \geq 1} \big[\big(a_n(\omega, x), u\big)_{\mathbb{R}^m} + c_n(\omega, x)\big]^+$$

with $a_n\colon \Omega \times \mathbb{R}^N \longrightarrow \mathbb{R}^m$ and $c_n\colon \Omega \times \mathbb{R}^N \longrightarrow \mathbb{R}$ bounded Carathéodory functions.

We can prove the following sequential lower semicontinuity result for the integral functional $I_\varphi(x,u),(x,u)\in L^p(\Omega,\mathbb{R}^N)\times L^r(\Omega,\mathbb{R}^m)$.

THEOREM 2.1.25 *If $\varphi\colon \Omega\times\mathbb{R}^N\times\mathbb{R}^m\longrightarrow\overline{\mathbb{R}}$ satisfies H_0 and the following holds,*

"for every sequence $\{x_n\}_{n\geq 1}$ converging in $L^p(\Omega,\mathbb{R}^N)$ and every sequence $\{u_n\}_{n\geq 1}$ weakly converging in $L^r(\Omega,\mathbb{R}^m)$ (weakly if $r=+\infty$) such that*

$$\int_\Omega \varphi^+\big(\omega,x_n(\omega),u_n(\omega)\big)d\mu\leq c+\int_\Omega \varphi^-\big(\omega,x_n(\omega),u_n(\omega)\big)d\mu,$$

for some $c>0$ and all $n\geq 1$, the sequence $\Big\{\varphi^-\big(\cdot,x_n(\cdot),u_n(\cdot)\big)\Big\}_{n\geq 1}\subseteq L^1(\Omega)$ is relatively weakly compact,"

then the integral function I_φ is well-defined on $L^p(\Omega,\mathbb{R}^N)\times L^r(\Omega,\mathbb{R}^m)$ with values in $\overline{\mathbb{R}}$ and it is sequentially lower semicontinuous when we consider $L^p(\Omega,\mathbb{R}^N)$ with the norm topology and $L^r(\Omega,\mathbb{R}^m)$ with the weak topology (weak if $r=+\infty$).*

PROOF: Let $x\in L^p(\Omega,\mathbb{R}^N)$, $u\in L^r(\Omega,\mathbb{R}^m)$ and suppose that for some $c>0$ we have

$$\int_\Omega \varphi^+\big(\omega,x(\omega),u(\omega)\big)d\mu\leq c+\int_\Omega \varphi^-\big(\omega,x(\omega),u(\omega)\big)d\mu. \tag{2.1}$$

Then by hypothesis $\varphi^-\big(\cdot,x(\cdot),u(\cdot)\big)\subseteq L^1(\Omega)$ and so using (2.1) we conclude that $\varphi\big(\cdot,x(\cdot),u(\cdot)\big)\subseteq L^1(\Omega)$. If (2.1) is not satisfied, then

$$\int_\Omega \varphi^-\big(\omega,x(\omega),u(\omega)\big)d\mu<+\infty \quad\text{ and }\quad \int_\Omega \varphi^+\big(\omega,x(\omega),u(\omega)\big)d\mu=+\infty,$$
$$\Rightarrow \int_\Omega \varphi\big(\omega,x(\omega),u(\omega)\big)d\mu=+\infty.$$

So indeed I_φ is well-defined with values in $\overline{\mathbb{R}}=\mathbb{R}\cup\{+\infty\}$.

We prove the desired sequential lower semicontinuity of I_φ, in steps.

Step 1: We assume that there exists a function $\vartheta\colon\mathbb{R}\longrightarrow\mathbb{R}$ such that

$$\lim_{s\to+\infty}\frac{\vartheta(s)}{s}=+\infty \quad\text{ and }\quad \vartheta(\|u\|_{\mathbb{R}^m})\leq\varphi(\omega,x,u) \tag{2.2}$$

for μ-a.a. $\omega\in\Omega$, all $x\in\mathbb{R}^N$, and all $u\in\mathbb{R}^m$.

Because of (2.2) we can apply Proposition 2.1.23 (see also Remark 2.1.24) and obtain that

$$\varphi(\omega,x,u)=\sup_{n\geq 1}\Big[\big(a_n(\omega,x),u\big)_{\mathbb{R}^N}+c_n(\omega,x)\Big]^+ \tag{2.3}$$

with $a_n:\Omega\times\mathbb{R}^N\longrightarrow\mathbb{R}^m$ and $c_n:\Omega\times\mathbb{R}^N\longrightarrow\mathbb{R}$, $n\geq 1$, bounded Carathéodory functions. Recall that

$$\begin{aligned}&\int_\Omega\big[\big(a_n(\omega,x),u\big)_{\mathbb{R}^m}+c_n(\omega,x)\big]^+d\mu\\ =\;&\sup\Big[\int_C\big[\big(a_n(\omega,x),u\big)_{\mathbb{R}^m}+c_n(\omega,x)\big]d\mu : C\in\Sigma\Big].\end{aligned} \tag{2.4}$$

So in view of (2.3) and (2.4) it suffices to show that for all $n\geq 1$ and all $C\in\Sigma$

$$(x,u) \longrightarrow \int_C \left[\left(a_n(\omega, x(\omega)), u(\omega)\right)_{\mathbb{R}^N} + c_n\left(\omega, x(\omega)\right)\right] d\mu$$

is sequentially lower semicontinuous on $L^p(\Omega, \mathbb{R}^N) \times L^r(\Omega, \mathbb{R}^m)$ the first space in the product furnished with the norm topology and the second with the weak topology (weak* topology if $r = +\infty$). But this is immediate because a_n and c_n are bounded Carathéodory functions.

Step 2: Now we assume that φ is bounded below.
Without any loss of generality, we may assume that $\varphi \geq 0$. Suppose that $x_n \longrightarrow x$ in $L^p(\Omega, \mathbb{R}^N)$ and $u_n \xrightarrow{w} u$ in $L^r(\Omega, \mathbb{R}^m)$ (weakly* if $r = +\infty$). Because $u_n \xrightarrow{w} u$ in $L^r(\Omega, \mathbb{R}^m)$ (weakly* if $r = +\infty$) and μ is finite, we have that $u_n \xrightarrow{w} u$ in $L^1(\Omega, \mathbb{R}^m)$ and so by the Dunford–Pettis theorem the sequence $\{u_n\}_{n\geq 1} \subseteq L^1(\Omega, \mathbb{R}^m)$ is uniformly integrable. So according to the De La Vallée–Poussin theorem we can find $\vartheta \colon \mathbb{R} \longrightarrow \mathbb{R}_+$ such that

$$\lim_{s \to +\infty} \frac{\vartheta(s)}{s} = +\infty \qquad \text{and} \qquad \int_\Omega \vartheta(\|u_n(\omega)\|_{\mathbb{R}^m}) d\mu \leq 1 \qquad \text{for all } n \geq 1.$$

Let $\varepsilon > 0$ and set

$$\varphi_\varepsilon(\omega, x, u) = \varphi(\omega, x, u) + \varepsilon \vartheta(\|u\|_{\mathbb{R}^m}).$$

Because of Step 1, we have

$$\begin{aligned}
\int_\Omega \varphi\left(\omega, x(\omega), u(\omega)\right) d\mu &\leq \int_\Omega \varphi_\varepsilon\left(\omega, x(\omega), u(\omega)\right) d\mu \\
&\leq \liminf_{n \to \infty} \int_\Omega \varphi_\varepsilon\left(\omega, x_n(\omega), u_n(\omega)\right) d\mu \\
&\leq \liminf_{n \to \infty} \int_\Omega \varphi\left(\omega, x_n(\omega), u_n(\omega)\right) d\mu + \varepsilon.
\end{aligned}$$

Because $\varepsilon > 0$ was arbitrary, we let $\varepsilon \downarrow 0$ to conclude that

$$\int_\Omega \varphi\left(\omega, x(\omega), u(\omega)\right) d\mu \leq \liminf_{n \to \infty} \int_\Omega \varphi\left(\omega, x_n(\omega), u_n(\omega)\right) d\mu.$$

Step 3: General case.
Again suppose that $x_n \longrightarrow x$ in $L^p(\Omega, \mathbb{R}^N)$ and $u_n \xrightarrow{w} u$ in $L^r(\Omega, \mathbb{R}^m)$ (weakly* if $r = +\infty$). For every $k \geq 1$ we define

$$\varphi_k(\omega, x, u) = \max\{\varphi(\omega, x, u), -k\}, \quad g_n(\omega) = \varphi^-\left(\omega, x_n(\omega), u_n(\omega)\right)$$

and $C_{n,k} = \{\omega \in \Omega : g_n(\omega) > k\}$.

Without any loss of generality we may assume that $I_\varphi(x_n, u_n)$ tends to a finite limit as $n \to \infty$. So by virtue of the hypothesis of the theorem $\{g_n\}_{n\geq 1} \subseteq L^1(\Omega)$ is relatively weakly compact. Then for every $k \geq 1$, because of Step 2, we have

$$\begin{aligned}
I_\varphi(x, u) \leq I_{\varphi_k}(x, u) &\leq \liminf_{n \to \infty} I_{\varphi_k}(x_n, u_n) \\
&= \liminf_{n \to \infty} \left[I_\varphi(x_n, u_n) + \int_{C_{n,k}} \left(g_n(\omega) - k\right) \mathrm{d}\mu\right] \\
&\leq \liminf_{n \to \infty} I_\varphi(x_n, u_n) + \limsup_{n \to \infty} \int_{C_{n,k}} \left(g_n(\omega) - k\right) \mathrm{d}\mu.
\end{aligned}$$

The Dunford–Pettis theorem implies that $\{g_n\}_{\{x_n\}_{n\geq 1}\geq 1} \subseteq L^1(\Omega)$ is uniformly integrable. So if we pass to the limit as $k \longrightarrow \infty$, we conclude that

$$I_\varphi(x,u) \leq \liminf_{n\to\infty} I_\varphi(x_n,u_n).$$

□

COROLLARY 2.1.26 *If $\Omega \subseteq \mathbb{R}^k$ is a bounded domain with Lipschitz boundary and $\varphi\colon \Omega\times\mathbb{R}^N\times\mathbb{R}^{Nk} \longrightarrow \mathbb{R}_+$ satisfies*

(i) *φ is Borel measurable.*
(ii) *For almost all $z \in \Omega$, $(x,u) \longrightarrow \varphi(z,x,u)$ is lower semicontinuous.*
(iii) *For almost all $z \in \Omega$ and all $x \in \mathbb{R}^N$, $u \longrightarrow \varphi(z,x,u)$ is convex,*

then the integral functional $x \longrightarrow J_\varphi(x) = \int_\Omega \varphi\big(z,x(z),Dx(z)\big)\mathrm{d}z$ is sequentially w-lower semicontinuous on $W^{1,1}_{loc}(\Omega,\mathbb{R}^N)$.

REMARK 2.1.27 This result is optimal when $N=1$ (the so-called scalar case), in the sense that the convexity of $\varphi(z,x,\cdot)$ is a necessary condition for the sequential weak lower semicontinuity of $J_\varphi(\cdot)$ on $W^{1,p}(\Omega)$, $1 \leq p \leq \infty$. In the case $N>1$ (the so-called vector case), convex integrands produce only a small subclass of the sequentially weak lower semicontinuous functionals on $W^{1,p}(\Omega,\mathbb{R}^N)$ (see Section 2.5).

We can have a version of Theorem 2.1.25 for integrands defined on infinite-dimensional Banach spaces. The result is due to Balder [53] and a different proof based on Young measures can be found in Hu–Papageorgiou [316, p. 31].

THEOREM 2.1.28 *If X is a separable Banach space, Y is a separable reflexive Banach space, and $\varphi\colon \Omega\times X\times Y \longrightarrow \overline{\mathbb{R}}$ satisfies*

(i) *φ is $\Sigma \times \mathbf{B}(X) \times \mathbf{B}(Y)$ measurable (with $\mathbf{B}(X)$ (resp., $\mathbf{B}(Y)$) the Borel σ-field of X (resp., Y));*
(ii) *For μ-almost all $\omega \in \Omega$, $(x,y) \longrightarrow \varphi(\omega,x,y)$ is lower semicontinuous;*
(iii) *For μ-almost all $\omega \in \Omega$ and all $x \in X$, $y \longrightarrow \varphi(\omega,x,y)$ is convex;*
(iv) *There exist $M>0$ and $\vartheta \in L^1(\Omega)$ such that for μ–almost all $\omega \in \Omega$ and all $(x,y) \in X\times Y$ we have*

$$\varphi(\omega,x,y) \geq \vartheta(\omega) - M\big(\|x\|_X + \|y\|_Y\big),$$

then $(x,u) \longrightarrow I_\varphi(x,u) = \int_\Omega \varphi\big(\omega,x(\omega),u(\omega)\big)\mathrm{d}\mu$ is sequentially lower semicontinuous on $L^1(\Omega,X)\times L^1(\Omega,Y)_w$, where by $L^1(\Omega,Y)_w$ we denote the Lebesgue–Bochner space $L^1(\Omega,Y)$ furnished with the weak topology.

2.2 Constrained Minimization Problems

In many applied situations we are not just looking for the minimum of an objective functional on an open set U, but we want to determine the minimum of φ subject to certain restrictions on the points $x \in U$. In this section we examine such minimization problems with side conditions and develop the method of Lagrange multipliers.

DEFINITION 2.2.1 Let X, Y be Banach spaces, $U \subseteq X$ and $V \subseteq Y$ nonempty open sets and $\varphi\colon U \longrightarrow V$ a Fréchet differentiable map.

(a) We say that $x_0 \in U$ is a *critical point* of φ if and only if $\varphi'(x_0) \in \mathcal{L}(X, Y)$ is not surjective.

(b) We say that $x_0 \in U$ is a *regular point* of φ if and only if $\varphi'(x_0) \in \mathcal{L}(X, Y)$ is surjective.

REMARK 2.2.2 If $Y = \mathbb{R}$, then $x_0 \in U$ is a critical (resp., regular) point of the function $\varphi\colon U \longrightarrow \mathbb{R}$ only if $\varphi'(x_0) = 0$ (resp., $\varphi'(x_0) \neq 0$). If $X = \mathbb{R}$, then $x_0 \in (a, b)$ is a critical point of φ if and only if $\varphi'(x_0) = 0$.

EXAMPLE 2.2.3 Let $X = H =$ a Hilbert space, $A \in \mathcal{L}(H, H)$ is a self-adjoint isomorphism (i.e., $A^{-1} \in \mathcal{L}(H, H)$) and $\psi\colon H \longrightarrow \mathbb{R}$ is Fréchet differentiable. Let $\varphi\colon H \longrightarrow \mathbb{R}$ be defined by

$$\varphi(x) = \frac{1}{2} \langle Ax, x \rangle_H - \psi(x),$$

where by $\langle \cdot, \cdot \rangle_H$ we denote the inner product of H. Evidently φ is Fréchet differentiable and

$$\varphi'(x) = Ax - \psi'(x).$$

Therefore $x_0 \in H$ is a critical point of φ if and only if $x_0 = A^{-1}\psi'(x_0)$. If $\psi(x) = \lambda/2\|x\|_H^2$, $\lambda \in \mathbb{R}$, then $\psi'(x_0) = \lambda x_0$ and so $x_0 \in H$ is a critical point of the function $\varphi(x) = 1/2 \langle Ax, x \rangle_H - \lambda/2\|x\|_H^2$ if and only if $Ax_0 = \lambda x_0$; that is, x_0 is an eigenvector of the operator A with eigenvalue λ.

Next we extend Definition 1.4.1 to nonconvex sets C.

DEFINITION 2.2.4 Let X be a Banach space and $C \subseteq X$ nonempty.

(a) A vector $h \in X$ is said to be *tangent* to the set C at $x_0 \in \overline{C}$, if there exist $\varepsilon > 0$ and a map $r : [0, \varepsilon] \longrightarrow X$ such that

$$x_0 + \lambda h + r(\lambda) \in C \quad \text{for all } \lambda \in [0, \varepsilon] \quad \text{and} \quad \frac{\|r(\lambda)\|_X}{\lambda} \longrightarrow 0 \text{ as } \lambda \longrightarrow 0^+.$$

(b) The set of all vectors $h \in X$ that are tangent to the set C at $x_0 \in \overline{C}$ form a closed cone (which is nonempty because it contains the origin) denoted by $T_C(x_0)$. If this cone is a subspace, then it is called the *tangent space* to the set C at the point $x_0 \in \overline{C}$.

We also need the following notion from Banach space theory.

DEFINITION 2.2.5 Let X be a Banach space and Y a closed subspace of it. We say that a subspace V is a *topological complement* of Y if
(a) V is closed and
(b) $Y \cap V = \{0\}$ and $Y + V = X$. In this case we write $X = Y \oplus V$.

REMARK 2.2.6 A closed subspace Y of X admits a topological complement if and only if there exists a projection operator onto Y; that is, there exists $p_Y \in \mathcal{L}(X)$ such that $p_Y\big|_Y = \mathrm{id}_Y$. It is easy to see that if Y is finite-dimensional, then it admits a topological complement. Similarly, if Y has finite codimension (i.e., $\dim\big(X/Y\big) < +\infty$), then Y has a topological complement. In a Hilbert space every closed subspace admits a topological complement (consider the orthogonal complement of Y). In a Banach space that is not isomorphic to a Hilbert space, we can always find a closed subspace which has no topological complement.

Now let X, Y be two Banach spaces and $\psi\colon X \longrightarrow Y$ a map. We consider the following constraint set

$$C = \{x \in X : \psi(x) = 0\}.$$

The next theorem produces a very convenient characterization of the tangent space $T_C(x_0)$, $x_0 \in C$. The result is known as *Ljusternik's theorem.*

THEOREM 2.2.7 *If X, Y are Banach spaces, $U \subseteq X$ is a nonempty open set, $\psi\colon U \longrightarrow Y$ is continuously Fréchet differentiable, $C = \{x \in U : \psi(x) = 0\}$, and $x_0 \in C$ is a regular point of φ at which $N\big(\psi'(x_0)\big) = \ker \psi'(x_0)$ has a topological complement, then $T_C(x_0) = N\big(\psi'(x_0)\big)$.*

PROOF: Let $V = N\big(\psi'(x_0)\big)$. This is a closed subspace of X that by hypothesis has a topological complement $W \subseteq X$. So we can find $p_V,\ p_W \in \mathcal{L}(X)$ such that

$$V = p_V(X) = N(p_W),\ W = p_W(X) = N(p_V),$$
$$p_V^2 = p_V,\ p_W^2 = p_W$$

(i.e., they are linear projection operators)

$$\text{and}\quad p_V + p_W = \mathrm{id}_X$$

(i.e., $X = V \oplus W$).

Because $U \subseteq X$ is an open set, we can find $r > 0$ such that

$$x_0 + B_r(0) + B_r(0) \subseteq U$$

(recall that $B_r(0) = \big\{x \in X : \|x\|_X < r\big\}$). Let $f\colon \big(V \cap B_r(0)\big) \times \big(W \cap B_r(0)\big) \longrightarrow Y$ be defined by

$$f(v, w) = \psi(x_0 + v + w) \qquad \text{for all } v \in V \cap B_r(0) \text{ and all } w \in W \cap B_r(0).$$

The function f has the following properties.

$$f(0,0) = \psi(x_0) = 0 \tag{2.5}$$
$$\text{and}\quad f \in C^1(V \times W, Y) \quad \text{with} \quad f_1'(0,0) = \psi'(x_0)\big|_V = 0$$
$$\text{and}\quad f_2'(0,0) = \psi'(x_0)\big|_W \in \mathcal{L}(W, Y). \tag{2.6}$$

Because $f_2'(0,0) \in \mathcal{L}(W, Y)$ is bijective, from Banach's theorem we know that $f_2'(0,0)^{-1} \in \mathcal{L}(Y, W)$. These properties of f permit the use of the implicit function theorem (see Theorem 1.1.23), which gives a $0 < \delta < r$ and a uniquely determined continuously Fréchet differentiable $g\colon V \cap B_\delta(0) \longrightarrow W$ such that

$$f(v, g(v)) = 0 \qquad \text{for all } v \in V \cap B_\delta(0) \tag{2.7}$$

$$\text{and} \quad g(0) = 0,\ g'(0) = f_2'(0,0)^{-1} f_1'(0,0). \tag{2.8}$$

From (2.6) and (2.8), we have that $g'(0) = 0$ and so

$$\lim_{v \to 0} \frac{\|g(v)\|_Y}{\|v\|_X} = 0. \tag{2.9}$$

If we define $\xi\colon x_0 + V \cap B_\delta(0) \longrightarrow X$ by

$$\xi(x_0 + v) = x_0 + v + g(v),$$

then clearly ξ is continuous. Moreover, from (2.7) we have

$$0 = f(v, g(v)) = \psi(x_0 + v + g(v)),$$

hence ξ has values in C. Note that $v, g(v)$ are in complementary subspaces of X. Hence ξ is injective and so we can consider its inverse on $D = \{x_0 + v + g(v) : v \in V \cap B_\delta(0)\} \subseteq C$. So

$$\xi^{-1}(x_0 + v + g(v)) = x_0 + v = x_0 + p_V(v + g(v)),$$
$$\Rightarrow \ \xi^{-1} \text{ is continuous too.}$$

Therefore ξ is a homeomorphism of $U' = x_0 + V \cap B_\delta(0)$ onto $D \subseteq C$. This together with (2.9) imply that $T_C(x_0) = N(\psi'(x_0))$. □

REMARK 2.2.8 If every $x_0 \in C$ is a regular point and $N(\psi'(x_0))$ admits a topological complement, then $C = \{x \in X : \psi(x) = 0\}$ is a C^1-manifold. If in addition $\psi \in C^r(U, Y)$, then C is a C^r-manifold. Note that the rather technical hypothesis that $N(\psi'(x_0))$ admits a topological complement is automatically satisfied in the following three cases.

(a) Y is finite-dimensional.
(b) $X = H =$ a Hilbert space.
(c) For all $x \in U$ $\varphi'(x) \in \mathcal{L}(X, Y)$ is a Fredholm operator.

Cases (a) and (b) follow from Remark 2.2.6, and Case (c) is considered in Section 3.1.

COROLLARY 2.2.9 *If X, Y, Z are Banach spaces, $U \subseteq X$ is nonempty open, $\psi \in C^1(U, Y)$, $C = \{x \in U : \psi(x) = 0\}$, $x_0 \in C$ is a regular point of φ at which $N(\psi'(x_0))$ admits a topological complement, and $\varphi\colon U \longrightarrow Z$ is Fréchet differentiable at x_0, then there exists a neighborhood U_0 of the origin in $T_C(x_0)$, a neighborhood D_0 of x_0 in C, and a continuously Fréchet differentiable homeomorphism g of U_0 onto D_0 such that $\varphi(g(u)) = \varphi(x_0) + \varphi'(x_0)u + o(\|u\|_X)$ with $(o(\|u\|_X))/\|u\|_X \longrightarrow 0$ as $\|x\|_X \longrightarrow 0$.*

Using Theorem 2.2.7, we can produce necessary conditions for the existence of solutions for a constrained minimization problem. This theorem is the first result establishing the well-known *method of Lagrange multipliers*.

THEOREM 2.2.10 *If X, Y are Banach spaces, $U \subseteq X$ is a nonempty open set, $\psi \in C^1(U, Y)$, $\varphi \colon U \longrightarrow \mathbb{R}$ is Fréchet differentiable, $C = \{x \in U : \psi(x) = 0\}$, $x_0 \in C$ is a regular point of ψ at which $N(\psi'(x_0))$ admits a topological complement, and x_0 is a local minimizer of φ on the constraint set C, then there exists $y_0^* \in Y^*$ such that*

$$\varphi'(x_0) = y_0^* \circ \psi'(x_0).$$

PROOF: Let $V = N(\psi'(x_0))$ and let W be its topological complement (it exists by hypothesis). Because $x_0 \in C$ is a regular point of ψ,

$$K = \psi'(x_0)\Big|_W \in \mathcal{L}(W, Y)$$

is an isomorphism (i.e., $K^{-1} \in \mathcal{L}(Y, W)$ by Banach's theorem).

According to Theorem 2.2.7 we can find D_0 a neighborhood of x_0 in C such that for every $x_0 \in D_0$, we have

$$x = x_0 + u + g(u) \qquad \text{for all } u \in V \cap B_\delta(0),\ V = N(\psi'(x_0))$$

with $(\|g(u)\|_X)/\|u\|_X \longrightarrow 0$ as $u \longrightarrow 0$ in X. Because x_0 is a local minimizer of φ on C, we can find $0 < \delta_1 \leq \delta$ such that

$$\begin{aligned}
&\varphi(x_0) \leq \varphi(x_0 + u + g(u)) \qquad \text{for all } u \in V \cap B_\delta(0),\\
\Rightarrow\ & 0 \leq \langle \varphi'(x_0), u + g(u) \rangle_X + o_{x_0}(\|u + g(u)\|_X) \qquad \text{for all } u \in V \cap B_{\delta_1}(0),\\
\Rightarrow\ & \langle \varphi'(x_0), u \rangle_X = 0 \quad \text{for all } u \in V \cap B_{\delta_1}(0) \ \text{(recall g is a homeomorphism)},\\
\Rightarrow\ & V \subseteq N(\varphi'(x_0)) = \ker \varphi'(x_0).
\end{aligned} \tag{2.10}$$

Let $p_W \in \mathcal{L}(X)$ be the canonical projection on W, the topological complement of V. Because of (2.10) we can define $\widehat{\varphi}\,'(x_0) \in W^*$ by

$$\langle \widehat{\varphi}\,'(x_0), p_W x \rangle_W = \langle \varphi'(x_0), x \rangle_X \qquad \text{for all } x \in X.$$

Then $y_0^* = \widehat{\varphi}\,'(x_0) \circ K^{-1} \in Y^*$. If $p_V \in \mathcal{L}(X)$ is the canonical projection of X onto V, then for every $x \in X$ we have $x = p_V x + p_W x$ and so

$$\begin{aligned}
\langle y_0^* \circ \psi'(x_0), x \rangle_X &= \langle y_0^* \circ \psi'(x_0), p_W x \rangle_X\\
&= \langle y_0^* \circ K, p_W x \rangle_X\\
&= \langle \widehat{\varphi}\,'(x_0), p_W x \rangle_X = \langle \varphi'(x_0), x \rangle_X\\
\Rightarrow y_0^* \circ \psi'(x_0) &= \varphi'(x_0) \quad \text{in } X^*.
\end{aligned}$$

□

REMARK 2.2.11 According to Theorem 2.2.10, $x_0 \in C$ is a critical point of the function $L(x) = \varphi(x) - (y_0^* \circ \psi)(x)$, known as the *Lagrangian* function of the constrained minimization problem $\inf_C \varphi$. Also the element $y^* \in Y^*$ is a *Lagrange multiplier*.

COROLLARY 2.2.12 *If $U \subseteq \mathbb{R}^N$ is a nonempty open set, $\varphi \colon U \longrightarrow \mathbb{R}$ is Fréchet differentiable, $\psi \in C^1(U, \mathbb{R}^m)$, $C = \{x \in U : \psi(x) = 0\}$, $x_0 \in C$ is a regular point* *of* ψ*, and it is also a local minimizer of* φ *on* C*, then we can find* $\widehat{\lambda} = (\lambda_k)_{k=1}^m \in \mathbb{R}^m$ *such that*

$$\frac{\partial \varphi}{\partial x_i}(x_0) = \sum_{k=1}^{m} \lambda_k \frac{\partial \psi}{\partial x_i}(x_0) \qquad \textit{for all} \quad i = 1, \dots, N.$$

EXAMPLE 2.2.13 Let $X = H$ a Hilbert space and $A \in \mathcal{L}(H)$ a self-adjoint operator. Let $\varphi\colon H \longrightarrow \mathbb{R}$ be defined by $\varphi(x) = \langle Ax, x \rangle_H$. Also let $\psi\colon H \longrightarrow \mathbb{R}_+$ be defined by $\psi(x) = \|x\|_H^2 - 1$. Suppose that $x_0 \in H$, $\|x_0\|_H = 1$ is a minimizer of φ on $\partial B(0) = \{x \in H : \psi(x) = 0\}$. According to Theorem 2.2.10 (with $Y = \mathbb{R}$), we can find $\lambda \in \mathbb{R}$ such that $\varphi'(x_0) = \lambda \psi'(x_0)$, hence $2A(x_0) = 2\lambda x_0$ and so $A(x_0) = \lambda x_0$. In other words $x_0 \in H$ is an eigenvector of A for the eigenvalue λ. This example shows how we can establish the existence of eigenvalues for bounded self-adjoint linear operators on a Hilbert space.

Now we want to improve Theorem 2.2.10 by weakening the hypotheses on the constraint function ψ. To do this we need the following generalization of Ljusternik's theorem (see Theorem 2.2.7) due to Ioffe–Tichomirov [327, p. 34], where the interested reader can find its proof.

THEOREM 2.2.14 *If* X, Y *are Banach spaces,* $U \subseteq X$ *is a nonempty open set,* $\psi\colon U \longrightarrow \mathbb{R}$ *is Fréchet differentiable,* $C = \{x \in U : \psi(x) = 0\}$ *and* $x_0 \in C$ *is a regular point of* ψ *(i.e.,* $\psi'(x_0)(X) = Y$*), then* $T_C(x_0) = N(\psi'(x_0))$.

REMARK 2.2.15 Note that compared to Theorem 2.2.7, we have dropped the splitting property for $N(\psi'(x_0))$.

We also need the following general result from operator theory, which underlines the idea of the Lagrange multiplier method.

PROPOSITION 2.2.16 *If* X, Y *are Banach spaces,* $A \in \mathcal{L}(X, Y)$*,* $x_0^* \in X^*$*,* $\mathcal{R}(A)$ *is closed and* $N(A) \subseteq N(x_0^*)$*, then there exists* $y_0^* \in Y^*$ *such that* $x_0^* + y_0^* \circ A = 0$*. Moreover, if* $A(X) = Y$*, then* y_0^* *is unique.*

PROOF: We know that $\mathcal{R}(A^*) = N(A)^\perp = \{x^* \in X^* : \langle x^*, x \rangle_X = 0 \text{ for all } x \in N(A)\}$. By hypothesis $x_0^* \in N(A)^\perp$. So it follows that $x_0^* = -A^* y_0^*$ for some $y_0^* \in Y^*$. Therefore

$$x_0^* + y_0^* \circ A = 0. \tag{2.11}$$

If $\mathcal{R}(A) = Y$, then $N(A^*) = \mathcal{R}(A)^\perp = \{y^* \in Y^* : \langle y^*, y \rangle_Y = 0 \text{ for all } y \in \mathcal{R}(A)\} = \{0\}$. So A^* is injective and by (2.11) this means that y_0^* is uniquely determined by x_0^*. □

COROLLARY 2.2.17 *If* X, Y *are Banach spaces,* $A \in \mathcal{L}(X, Y)$*,* $x_0^* \in X^*$*,* $\mathcal{R}(A)$ *is closed, and* $\mathcal{R}(A) \neq Y$*, then there exists* $y_0^* \in Y^*$*,* $y_0^* \neq 0$ *such that* $y_0^* \circ A = 0$.

PROOF: Let $y \in Y$ such that $y \notin \mathcal{R}(A)$. Then we can find $y_0^* \in Y^*$ with $\|y_0^*\|_{Y^*} = 1$ such that

$$\langle y_0^*, y \rangle_Y = 0 \qquad \text{and} \qquad \langle y_0^*, u \rangle_Y = 0 \qquad \text{for all } \ u \in \mathcal{R}(A).$$

Therefore it follows that $y_0^* \neq 0$ and $y_0^* \circ A = 0$. $\square$

We are led to the following generalization of Theorem 2.2.10.

THEOREM 2.2.18 *If X, Y are Banach spaces, $U \subseteq X$ is a nonempty open set, $\varphi: U \longrightarrow \mathbb{R}$ is Fréchet differentiable, $\psi \in C^1(U,Y)$, $C = \{x \in U : \psi(x) = 0\}$, $x_0 \in C$, $\mathcal{R}(\psi'(x_0))$ is closed, and $x_0 \in C$ is a local minimizer of φ on C, then there exist $\lambda_0 \in \mathbb{R}$ and $y_0^* \in Y^*$ not both equal to zero such that*

$$\lambda_0 \varphi'(x_0) = y_0^* \circ \psi'(x_0). \tag{2.12}$$

PROOF: In the degenerate case $\mathcal{R}(\psi'(x_0)) \neq Y$, (2.12) is true with $\lambda_0 = 0$ by virtue of Corollary 2.2.17.

If $\mathcal{R}(\psi'(x_0)) = Y$, then provided that $N(\psi'(x_0))$ admits a topological complement, the validity of (2.12) with $\lambda_0 = 1$ is a consequence of Theorem 2.2.10.

Finally, if we do not have the splitting of $N(\psi'(x_0))$, then the result follows if we argue as in the proof of Theorem 2.2.10 using this time Theorem 2.2.14. $\square$

Theorems 2.2.10 and 2.2.18 indicate that there is a close relation between Lagrange multipliers and constrained critical points of a function φ.

DEFINITION 2.2.19 Let X be a Banach space, $U \subseteq X$ nonempty open, $\varphi: U \longrightarrow \mathbb{R}$ a Fréchet differentiable map, $C \subseteq X$ a nonempty closed set and $x_0 \in C$. We say that x_0 is *a critical point of φ subject to the constraint* C, if for all curves $u : (-\varepsilon, \varepsilon) \longrightarrow X$ such that $u(t) \in C$ for all $t \in (-\varepsilon, \varepsilon)$, $u(0) = x_0$, $u'(0)$ exists, we have

$$\frac{\mathrm{d}}{\mathrm{d}t} \varphi(u(t))\Big|_{t=0} = 0.$$

REMARK 2.2.20 If $x_0 \in \operatorname{int} C$ (the interior taken in X), then x_0 is a usual critical point of φ (also known as the *free critical point* of φ). A critical point that is not a local extremum of φ (i.e., it is not a local minimizer or a local maximizer of φ), is called a *saddle point.* So $x_0 \in C$ is a saddle point of φ subject to the constraint C, if for every $U \in \mathcal{N}(x_0) =$ filter of neighborhoods of x_0, we can find $y, v \in C \cap U$ such that $\varphi(y) < \varphi(x_0) < \varphi(v)$.

The next theorem establishes the connection between constrained critical points and Lagrange multipliers.

THEOREM 2.2.21 *If X, Y are Banach spaces, $U \subseteq X$ is nonempty open, $\varphi: U \longrightarrow \mathbb{R}$ is Fréchet differentiable, $\psi \in C^1(U,Y)$, $C = \{x \in X : \psi(x) = 0\}$, and $x_0 \in C$ is a regular point of φ at which $N(\varphi'(x_0))$ admits a topological complement, then x_0 is a critical point of φ subject to the constraint C if and only if we can find $y_0^* \in Y^*$ such that $\varphi'(x_0) = y_0^* \circ \psi'(x_0)$.*

PROOF: $\Rightarrow$: From Definition 2.2.19 it follows that $\langle \varphi'(x_0), h \rangle_X = 0$ for all $h \in N(\psi'(x_0))$. But from Theorem 2.2.10 we have that $N(\psi'(x_0)) = T_C(x_0)$. Invoking Proposition 2.2.16, we can find $y_0^* \in Y^*$ such that $\varphi'(x_0) = y_0^* \circ \psi'(x_0)$.

$\Leftarrow$: Let $u : (-\varepsilon, \varepsilon) \longrightarrow X$ be a curve as in Definition 2.2.19. Then $\psi(u(t)) = 0$ for all $t \in (-\varepsilon, \varepsilon)$ and $\psi'(x_0)u'(0) = 0$. It follows that $\langle \varphi'(x_0), u'(0) \rangle_X = 0$, hence $(\mathrm{d}/\mathrm{d}t)\, \varphi(u(t))\Big|_{t=0} = 0$ (by the chain rule). So $x_0 \in C$ is a critical point of φ subject to the constraint C. □

We conclude this section with the so-called *Dubovickii–Milyutin theorem*, which is crucial for the development of the Dubovickii–Milyutin formalism for the analysis of constrained optimization problems. For the proof we refer to Girsanov [266, p. 37].

THEOREM 2.2.22 *If X is a Banach space, $\{K_m\}_{m=1}^{n+1}$ are nonempty convex cones in X, and $\{K_m\}_{m=1}^{n}$ are open, then $\bigcap_{m=1}^{n+1} K_m = \emptyset$ if and only if we can find $x_m^* \in K_m^* = \{x^* \in X^* : \langle x^*, x \rangle_X \geq 0$ for all $x \in K_m\}$, $m = 1, \ldots, n+1$, not all zero, such that $\sum_{m=1}^{n+1} x_m^* = 0$.*

2.3 Saddle Points and Duality

Recall (see Remark 2.2.20), that a saddle point is a critical point of a function that is neither a local minimum nor local maximum. In this section we define a saddle point of a function on a product space $C \times D$. This new concept that we introduce is independent of the notion introduced in the previous section and it is global in nature.

DEFINITION 2.3.1 Let C, D be two nonempty sets and $\varphi\colon C \times D \longrightarrow \mathbb{R}$ a function. We say that (x_0, y_0) is a *saddle point* of φ, if we have

$$\varphi(x_0, y) \leq \varphi(x_0, y_0) \leq \varphi(x, y_0) \qquad \text{for all } (x, y) \in C \times D.$$

REMARK 2.3.2 The terminology of Definition 2.3.1 is justified by picturing the function $\varphi(x, y) = x^2 - y^2$ in $\mathbb{R}^3$ (a hyperbolic paraboloid). For this function $(0, 0)$ is a saddle point.

Note that we always have

$$\sup_{y \in D} \inf_{x \in C} \varphi(x, y) \leq \inf_{x \in C} \sup_{y \in D} \varphi(x, y). \tag{2.13}$$

If equality holds in (2.13), then the common value is called the *saddle value* of φ on $C \times D$.

DEFINITION 2.3.3 Let C, D be two nonempty sets and $\varphi\colon C \times D \longrightarrow \mathbb{R}$ a function. We say that φ satisfies a *minimax equality* on $C \times D$, if the following three conditions hold.

(a) φ has a saddle value (i.e., equality holds in (2.13)).

(b) There is $x_0 \in C$ such that $\sup_{y\in D} \varphi(x_0,y) = \inf_{x\in C} \sup_{y\in D} \varphi(x,y)$.

(c) There is $y_0 \in D$ such that $\inf_{x\in C} \varphi(x,y_0) = \sup_{y\in D} \inf_{x\in C} \varphi(x,y)$.

REMARK 2.3.4 In this case the inf and sup operations can be replaced by min and max, respectively, and we can write that $\max_{y\in D} \min_{x\in C} \varphi(x,y) = \min_{x\in C} \max_{y\in D} \varphi(x,y)$. This is the reason for the terminology minimax equality.

The following proposition is an immediate consequence of Definitions 2.3.1 and 2.3.3.

PROPOSITION 2.3.5 *If C, D are nonempty sets and $\varphi\colon C\times D \longrightarrow \mathbb{R}$, then φ has a saddle point in $C \times D$ if and only if φ satisfies a minimax equality on $C\times D$.*

DEFINITION 2.3.6 Let X be a vector space and $B \subseteq X$ a nonempty set. A multifunction (set-valued function) $F\colon B \longrightarrow 2^X \setminus \{\emptyset\}$ is said to be a *Knaster–Kuratowski–Mazurkiewicz map* (a KKM-map for short), if for every finite set $\{x_k\}_{k=1}^m$

$$\operatorname{conv}\{x_k\}_{k=1}^m \subseteq \bigcup_{k=1}^m F(x_k).$$

The basic property of KKM-maps is the following.

THEOREM 2.3.7 *If X is a Hausdorff topological vector space, $B \subseteq X$ is a nonempty set, and $F\colon B \longrightarrow 2^X \setminus \{\emptyset\}$ is a KKM-map with closed values, then the family $\{F(x)\}_{x\in B}$ of sets has the finite intersection property; that is, the intersection of each finite subfamily is nonempty.*

PROOF: We argue indirectly. So suppose that $\bigcap_{k=1}^m F(x_k)=\emptyset$. Let $C=\operatorname{conv}\{x_k\}_{k=1}^m$ and consider the function $\vartheta\colon C \longrightarrow \mathbb{R}_+$ defined by $\vartheta(u)=\sum_{k=1}^m d\big(u, F(x_k)\big)$. Evidently $\vartheta(u)>0$ for all $u \in C$ and so we can define the continuous map $h\colon C \longrightarrow C$ by

$$h(u) = \frac{1}{\vartheta(u)} \sum_{k=1}^m d\big(u, F(x_k)\big) x_k. \tag{2.14}$$

By Brouwer's theorem (see Theorem 3.5.3), we can find $u_0 \in C$ such that

$$h(u_0) = u_0.$$

Let $K = \big\{k \in \{1,\dots,m\} : d\big(u_0, F(x_k)\big) \neq 0\big\}$. Then $u_0 \notin \bigcup_{k\in K} F(x_k)$. On the other hand

$$u_0 = h(u_0) \in \operatorname{conv}\{x_k\}_{k\in K} \subseteq \bigcup_{k\in K} F(x_k),$$

(see (2.14) and recall that F is a KKM-map). So we have a contradiction, which finishes the proof. □

REMARK 2.3.8 The hypotheses of this theorem can be weakened in the following way. We can drop the requirement that F has closed values and only assume that each $F(x)$ is finitely closed. Namely, if V is a finite-dimensional flat in X, then $V \cap F(x)$ is closed in the Euclidean topology of V. Evidently in this case we need only assume that X is a vector space. The finitely closed sets define a topology on X, known as the finite topology. It is stronger than any Hausdorff linear topology on X.

COROLLARY 2.3.9 *If X is a Hausdorff topological vector space, $B \subseteq X$ is a nonempty set, and $F\colon B \longrightarrow 2^X \backslash \{\emptyset\}$ is a KKM-map with closed values one of which is compact, then $\bigcap\limits_{x \in B} F(x) \neq \emptyset$.*

Using this corollary we can prove the following useful coincidence theorem for set-valued maps.

PROPOSITION 2.3.10 *If X, Y are Hausdorff topological vector spaces, $C \subseteq X$ and $D \subseteq Y$ are nonempty compact and convex sets, and $F, G\colon C \longrightarrow 2^D$ are two multifunctions such that*

(i) *$F(x)$ is open in D and $G(x)$ is nonempty convex for every $x \in C$.*

(ii) *$G^{-1}(y) = \{x \in C : y \in G(x)\}$ is open in C and $F^{-1}(y) = \{x \in C : y \in F(x)\}$ is convex for every $y \in D$,*

then we can find $x_0 \in C$ such that $F(x_0) \cap G(x_0) \neq \emptyset$.

PROOF: Let $E = C \times D$ and let $H : C \times D \longrightarrow 2^{X \times Y}$ be defined by

$$H(x, y) = E \cap \big(G^{-1}(y) \times F(x)\big)^c.$$

Because of hypotheses (i) and (ii), for every $(x, y) \in E$ the set $H(x, y)$ is closed in E, hence compact. Also given any $(x_0, y_0) \in E$, choose $(x, y) \in F^{-1}(y_0) \times G(x_0)$ (by hypotheses (i) and (ii), $F^{-1}(y_0) \times G(x_0) \neq \emptyset$). Hence $(x_0, y_0) \in G^{-1}(y) \times F(x)$ and so we infer that

$$E = \bigcup_{(x,y) \in E} \big(G^{-1}(y) \times F(x)\big). \tag{2.15}$$

From (2.15) and the definition of the multifunction H, it follows that

$$\bigcap_{(x,y) \in E} H(x, y) = \emptyset. \tag{2.16}$$

Because of (2.16) H cannot be a KKM-map. So according to Definition 2.3.6, we can find elements $\big\{(x_k, y_k)\big\}_{k=1}^m \subseteq E$ such that

$$\begin{aligned} &\operatorname{conv}\big\{(x_k, y_k)\big\}_{k=1}^m \not\subseteq \bigcup_{k=1}^m H(x_k, y_k) \\ \Rightarrow\ & v = \sum_{k=1}^m \lambda_k (x_k, y_k) \notin \bigcup_{k=1}^m H(x_k, y_k) \qquad \text{with } \lambda_k \geq 0,\ \sum_{k=1}^m \lambda_k = 1. \end{aligned}$$

Note that due to the convexity of $E = C \times D$ we have $v \in E$ and so $v \in E \setminus \bigcup_{k=1}^{m} H(x_k, y_k) = \bigcap_{k=1}^{m} G^{-1}(y_k) \cap F(x_k)$. Hence

$$\sum_{k=1}^{m} \lambda_k x_k \in G^{-1}(y_k) \quad \text{and} \quad \sum_{k=1}^{m} \lambda_k y_k \in F(x_k) \quad \text{for all } k = 1, \ldots, m,$$

$$\Rightarrow y_k \in G\Big(\sum_{k=1}^{m} \lambda_k x_k\Big) \quad \text{and} \quad x_k \in F^{-1}\Big(\sum_{k=1}^{m} \lambda_k y_k\Big) \quad \text{for all } k = 1, \ldots, m,$$

$$\Rightarrow \sum_{k=1}^{m} \lambda_k y_k \in G\Big(\sum_{k=1}^{m} \lambda_k x_k\Big) \quad \text{and} \quad \sum_{k=1}^{m} \lambda_k x_k \in F^{-1}\Big(\sum_{k=1}^{m} \lambda_k y_k\Big),$$

$$\Rightarrow \sum_{k=1}^{m} \lambda_k y_k \in G\Big(\sum_{k=1}^{m} \lambda_k x_k\Big) \quad \text{and} \quad \sum_{k=1}^{m} \lambda_k y_k \in F\Big(\sum_{k=1}^{m} \lambda_k x_k\Big).$$

So, if we set $x_0 = \sum_{k=1}^{m} \lambda_k x_k$ and $y_0 = \sum_{k=1}^{m} \lambda_k y_k$, we see that $y_0 \in F(x_0) \cap G(x_0) \neq \emptyset$.

□

Using this proposition we can prove a basic minimax theorem. First we have a definition.

DEFINITION 2.3.11 Let V be a vector space, $B \subseteq V$ a nonempty convex set, and $\psi\colon B \longrightarrow \mathbb{R}$. We say that ψ is *quasiconvex* (resp., *quasiconcave*), if for all $\lambda \in \mathbb{R}$ the set $\{x \in B : \psi(x) \leq \lambda\}$ (resp., the set $\{x \in B : \psi(x) \geq \lambda\}$) is convex in B.

REMARK 2.3.12 Evidently a convex (resp., concave) function is quasiconvex (resp., quasiconcave). The converse is not in general true. Consider, for example, the function $x \longrightarrow \ln(|x| + 1)$ which is quasiconvex, but not convex.

THEOREM 2.3.13 *If X, Y are Hausdorff topological vector spaces, $C \subseteq X$ and $D \subseteq Y$ are nonempty, compact, and convex sets, and $\varphi\colon C \times D \longrightarrow \mathbb{R}$ satisfies*

(i) *For every $y \in D$, $x \longrightarrow \varphi(x, y)$ is lower semicontinuous and quasiconvex.*

(ii) *For every $x \in C$, $y \longrightarrow \varphi(x, y)$ is upper semicontinuous and quasiconcave,*

then φ has a saddle point on $C \times D$.

PROOF: By virtue of Proposition 2.3.5 it suffices to show that φ satisfies a minimax equality on $C \times D$, namely that

$$\min_{x \in C} \max_{y \in D} \varphi(x, y) = \max_{y \in D} \min_{x \in C} \varphi(x, y). \tag{2.17}$$

Because of the upper semicontinuity of $\varphi(x, \cdot)$ and the compactness of D, $\max_{y \in D} \varphi(x, y)$ exists for every $x \in C$ (see Theorem 2.1.10). Also if $h_1(x) = \max_{y \in D} \varphi(x, y)$, then we claim that $x \longrightarrow h_1(x)$ is lower semicontinuous. Indeed let $x_\alpha \longrightarrow x$ in C and suppose $h_1(x_\alpha) \leq \lambda$ for all $\alpha \in J$. We have $\varphi(x_\alpha, y) \leq \lambda$ for all $\alpha \in J$ and all $y \in D$. Because $\varphi(\cdot, y)$ is lower semicontinuous, we have

$\varphi(x,y) \leq \liminf_{\alpha \in J} \varphi(x_\alpha, y) \leq \lambda$ for all $y \in D$, hence $h_1(x) \leq \lambda$ and this by virtue of Proposition 2.1.3(b) implies that $h_1(\cdot)$ is lower semicontinuous. Therefore $\min_{x \in C} h_1(x) = \min_{x \in C} \max_{y \in D} \varphi(x,y)$ exists. Similarly we show that $\max_{y \in D} \min_{x \in C} \varphi(x,y)$ exists. We know that

$$\max_{y \in D} \min_{x \in C} \varphi(x,y) \leq \min_{x \in C} \max_{y \in D} \varphi(x,y) \quad \text{(see (2.13))}. \tag{2.18}$$

We show that in (2.18) we cannot have strict inequality. We proceed by contradiction. Suppose we can find $\lambda \in \mathbb{R}$ such that

$$\max_{y \in D} \min_{x \in C} \varphi(x,y) < \lambda < \min_{x \in C} \max_{y \in D} \varphi(x,y). \tag{2.19}$$

We introduce the multifunctions $F, G \colon C \longrightarrow 2^D$ defined by

$$F(x) = \big\{y \in D : \varphi(x,y) < \lambda\big\} \qquad \text{and} \qquad G(x) = \big\{y \in D : \varphi(x,y) > \lambda\big\}.$$

Note that because of hypothesis (ii) for each $x \in C$, $F(x)$ is open in D and $G(x)$ is convex and nonempty (see (2.19)). Also $G^{-1}(y) = \big\{x \in C : \varphi(x,y) > \lambda\big\}$ is open in C due to hypothesis (i), whereas $F^{-1}(y) = \big\{x \in C : \varphi(x,y) < \lambda\big\}$ is convex again due to hypothesis (i). So we can apply Proposition 2.3.10 and produce a point $(\widehat{x}_0, \widehat{y}_0) \in C \times D$ such that $\widehat{y}_0 \in F(\widehat{x}_0) \cap G(\widehat{x}_0)$. But then

$$\lambda < \varphi(\widehat{x}_0, \widehat{y}_0) < \lambda,$$

a contradiction. This proves that (2.17) holds and so φ admits a saddle point. □

Another saddle point theorem useful in applications is the following.

THEOREM 2.3.14 *If X, Y are reflexive Banach spaces, $C \subseteq X$ and $D \subseteq Y$ are nonempty, closed, and convex sets, and $\varphi \colon C \times D \longrightarrow \mathbb{R}$ is a function such that*

(i) *For every $y \in D$, $x \longrightarrow \varphi(x,y)$ is convex and lower semicontinuous.*

(ii) *For every $x \in C$, $y \longrightarrow \varphi(x,y)$ is concave and upper semicontinuous.*

(iii) *C is bounded or there exists $\widehat{y}_0 \in D$ such that $\varphi(x, \widehat{y}_0) \longrightarrow +\infty$ as $\|x\|_X \longrightarrow +\infty$.*

(iv) *D is bounded or there exists $\widehat{x}_0 \in C$ such that $\varphi(\widehat{x}_0, y) \longrightarrow -\infty$ as $\|y\|_Y \longrightarrow +\infty$,*

then φ has a saddle point on $C \times D$.

PROOF: Due to the reflexivity of X and Y, if $C \subseteq X$ and $D \subseteq Y$ are bounded, they are w-compact. Also due to convexity the functions $x \longrightarrow \varphi(x,y)$ and $y \longrightarrow -\varphi(x,y)$ are weakly lower semicontinuous. So in this case the result follows from Theorem 2.3.13.

So suppose that at least one of the sets $C \subseteq X$ and $D \subseteq Y$ is unbounded. For every $n \geq 1$ set

$$C_n = \big\{x \in C : \|x\|_X \leq n\big\} \qquad \text{and} \qquad D_n = \big\{y \in D : \|y\|_Y \leq n\big\}.$$

For $n \geq 1$ large $C_n \neq \emptyset$ and $D_n \neq \emptyset$. Then from the bounded case we know that we can find $(x_n, y_n) \in C_n \times D_n$ such that

$$\varphi(x_n, y) \leq \varphi(x_n, y_n) \leq \varphi(x, y_n) \qquad \text{for all } (x,y) \in C_n \times D_n,\ n \geq 1. \tag{2.20}$$

Let $x = \widehat{x}_0$ and $y = \widehat{y}_0$ (see hypotheses (iii) and (iv)). Using those hypotheses, from (2.20) we infer that $\{x_n\}_{n\geq 1} \subseteq X$, $\{y_n\}_{n\geq 1} \subseteq Y$, and $\{\varphi(x_n, y_n)\}_{n\geq 1} \subseteq \mathbb{R}$ are all bounded. So passing to a subsequence if necessary, we may assume that

$$x_n \xrightarrow{w} x_0 \text{ in } X, \qquad y_n \xrightarrow{w} y_0 \text{ in } Y \qquad \text{and} \qquad \varphi(x_n, y_n) \longrightarrow \xi \in \mathbb{R}.$$

Evidently $(x_0, y_0) \in C \times D$. Also because of hypotheses (i) and (ii) we have

$$\varphi(x_0, y) \leq \liminf_{n\to\infty} \varphi(x_n, y) \leq \xi \leq \limsup_{n\to\infty} \varphi(x, y_n) \leq \varphi(x, y_0)$$

for all $(x,y) \in C \times D$,

$$\Rightarrow \varphi(x_0, y) \leq \varphi(x_0, y_0) \leq \varphi(x, y_0),$$

that is, $(x_0, y_0) \in C \times D$ is a saddle point of φ.

□

REMARK 2.3.15 Functions $\varphi(x,y)$ such as those in Theorems 2.3.13 and 2.3.14 that are convex in $x \in C$ and concave in $y \in D$, are usually called *saddle functions* (or *convex–concave functions*).

Next we derive a duality theory for convex optimization. Saddle functions are helpful in this respect. The mathematical setting is the following. We are given two Banach spaces X, Y and a convex function $\varphi \colon X \times Y \longrightarrow \overline{\mathbb{R}} = \mathbb{R} \cup \{+\infty\}$. We know that $(X \times Y)^* = X^* \times Y^*$ and for the duality brackets we have

$$\langle (x^*, y^*), (x, y) \rangle_{X\times Y} = \langle x^*, x \rangle_X + \langle y^*, y \rangle_Y ,$$

for all $x \in X$, $x^* \in X^*$, $y \in Y$, $y^* \in Y^*$.

We can think of $y \in Y$ as a perturbation and $\varphi(x,y)$ as a perturbed functional with $\varphi(x,0)$ as the original unperturbed functional. Then the primal problem is:

$$(\text{P}): \inf_{x\in X} \varphi(x, 0). \tag{2.21}$$

We also consider the value function $m \colon Y \longrightarrow \mathbb{R}^*$ of the perturbed problem

$$m(y) = \inf_{x\in X} \varphi(x, y). \tag{2.22}$$

Clearly $m(\cdot)$ is convex. We have

$$\begin{aligned} m^*(y^*) &= \sup_{y\in Y} \big[\langle y^*, y \rangle_Y - \inf_{x\in X} \varphi(x,y) \big] \quad (\text{see } (2.22)) \\ &= \sup_{(x,y)\in X\times Y} \big[\langle 0, x \rangle_X + \langle y^*, y \rangle_Y - \varphi(x,y) \big] = \varphi^*(0, y^*). \end{aligned} \tag{2.23}$$

Then the dual problem associated to (P) (see (2.21)), is defined by

$$(\text{D}): \sup_{y^*\in Y^*} \big(-\varphi^*(0, y^*) \big). \tag{2.24}$$

PROPOSITION 2.3.16 *We always have* $\sup(\text{D}) \leq \inf(\text{P})$ (*weak duality*).

PROOF: For all $x \in X$ and all $y^* \in Y^*$, by the Young–Fenchel inequality (see Proposition 1.2.18), we have

$$\varphi(x,0) + \varphi^*(0,y^*) \geq \langle 0, x\rangle_X + \langle y^*, 0\rangle_Y = 0$$
$$\Rightarrow \inf(\mathrm{P}) \geq \sup(\mathrm{D}).$$

□

In the next proposition we provide conditions for strong duality to hold (i.e., to have equality of primal and dual problems).

PROPOSITION 2.3.17 *If $\partial m(0) \neq \emptyset$ and $y^* \in \partial m(0)$, then y^* is a solution of* (D) *and* $\inf(\mathrm{P}) = \max(\mathrm{D})$.

PROOF: Because $\partial m(0) \neq \emptyset$, we have $m(0) = \inf(\mathrm{P}) \in \mathbb{R}$ and for all $y^* \in Y^*$, $m^*(y^*) \geq -m(0) > -\infty$. From Proposition 1.2.31, we have

$$\inf(\mathrm{P}) + m^*(y^*) = m(0) + m^*(y^*) = \langle y^*, 0\rangle_Y = 0,$$
$$\Rightarrow \sup(\mathrm{D}) = \inf(\mathrm{P}) \quad \text{(see (2.23), (2.24), and Proposition 2.3.16).}$$

□

In order to analyze further the strong duality situation, we need the following lemma.

LEMMA 2.3.18 *If V is a Banach space, $f\colon V \longrightarrow \mathbb{R}^*$ is a convex function, and there exist $(v_0, \lambda_0) \in V \times \mathbb{R}$ such that $f(v_0) \in \mathbb{R}$ and $(v_0, \lambda_0) \in \operatorname{int} \operatorname{epi} f$, then $f(v) > -\infty$ for all $v \in V$.*

PROOF: Let $\mu_0 < f(v_0)$. Then $(v_0, \mu_0) \notin \operatorname{epi} f$ and so by the separation theorem for convex sets we can find $(v^*, t) \in V^* \times \mathbb{R}$, $(v^*, t) \neq (0,0)$ such that

$$\langle v^*, v_0\rangle_V + t\mu_0 \leq \langle v^*, v\rangle_V + t\lambda \quad \text{for all} \quad (v,\lambda) \in \operatorname{epi} f. \tag{2.25}$$

Because λ can increase up to $+\infty$, from (2.25) it follows that $t \geq 0$. If $t = 0$, then

$$\langle v^*, v_0\rangle_V \leq \langle v^*, v\rangle_V \quad \text{for all} \quad v \in \operatorname{dom} f. \tag{2.26}$$

But $\operatorname{int} \operatorname{dom} f \neq \emptyset$ (see Theorem 1.2.3(d)). So from (2.26) we infer that $v^* = 0$, a contradiction to the fact that $(v^*, t) \neq (0,0)$. Hence $t > 0$ and we may assume that $t = 1$. We have

$$\langle v^*, v_0 - v\rangle_V + \mu_0 \leq \lambda \qquad \text{for all} \ (v,\lambda) \in \operatorname{epi} f,$$
$$\Rightarrow \langle v^*, v_0 - v\rangle_V + \mu_0 \leq f(v) \qquad \text{for all} \ v \in V \ (\text{i.e., } f(v) > -\infty \quad \text{for all} \ v \in V).$$

□

Using this auxiliary result we can have a theorem on the duality of problems (P) and (D).

THEOREM 2.3.19 *If there exists $x_0 \in X$ such that $\varphi(x_0, \cdot)$ is finite and continuous at $y = 0$ and* $\inf(\mathrm{P}) \in \mathbb{R}$, *then*

(a) *m is continuous at $y = 0$.*

(b) $y^* \in \partial m(0)$ *if and only if* $y^* \in Y^*$ *is a solution of the dual problem* (D).
(c) $\inf(\mathrm{P}) = \max(\mathrm{D}) \in \mathbb{R}$ (*strong duality*).

PROOF: We can find $r>0$ and $M>0$ such that

$$\begin{aligned} \varphi(x_0, y) \le M \qquad &\text{for all } y \in \overline{B}_r(0), \\ \Rightarrow\ m(y) \le M \qquad &\text{for all } y \in \overline{B}_r(0). \end{aligned}$$

Then from Theorem 1.2.3 we have that m is continuous at $y=0$ and $(0,\lambda) \in \operatorname{int}\operatorname{epi} m$ for all $\lambda > M$. Because $m(0) = \inf(\mathrm{P}) \in \mathbb{R}$, from Lemma 2.3.18 we infer that $-\infty < m(y)$ for all $y \in Y$, hence m is a proper convex function continuous at $y=0$. By virtue of Theorem 1.2.34 we have that $\partial m(0) \ne \emptyset$. Invoking Proposition 2.3.17, we know that every $y^* \in \partial m(0)$ is a solution of (D) and $\inf(\mathrm{P}) = \max(\mathrm{P})$. Finally, if $y^* \in Y^*$ is a solution of (D), we have

$$m^*(y^*) \le m^*(u^*) \quad \text{for all } u^* \in Y^*. \tag{2.27}$$

Because m is continuous at $y = 0$, using Proposition 1.2.24 (with $\varphi = m$ and $\psi = i_{\{0\}}$) we obtain

$$\begin{aligned} m(0) = \sup_{u^* \in Y^*} \left[\langle u^*, 0\rangle_Y - m^*(u^*) \right] \le -m^*(y^*) \quad (\text{see } (2.27)), \\ \Rightarrow\ m(0) + m^*(y^*) \le 0 \quad (\text{i.e., } y^* \in \partial m(0);\ \text{see Proposition 1.2.31}). \end{aligned}$$

$\square$

REMARK 2.3.20 In the above theorem, the hypothesis that $\inf(\mathrm{P}) \in \mathbb{R}$ is important and cannot be dropped. To see this consider the case where $X = Y = \mathbb{R}$; let $C(y) = \{x \in \mathbb{R} : x \le y\}$ and

$$\varphi(x, y) = x + i_{C(y)}(x).$$

Then for every $x<0$, $\varphi(x,\cdot)$ is continuous at 0, but $m \equiv -\infty$ and so $m^* \equiv +\infty$.

DEFINITION 2.3.21 The *Lagrangian function* corresponding to the dual pair of problems $\{(\mathrm{P}),(\mathrm{D})\}$, is the function $L\colon X \times Y^* \longrightarrow \mathbb{R}^*$ defined by

$$L(x, y^*) = \inf \left[\varphi(x, y) - \langle y^*, y\rangle_Y : y \in Y \right] \qquad \text{for all } (x, y^*) \in X \times Y^*.$$

REMARK 2.3.22 Evidently $\inf\limits_{x \in X} L(x, y) = -\varphi^*(0, y^*)$ and L is a saddle function.

EXAMPLE 2.3.23 Consider the mathematical programming situation. Namely let

$$(\mathrm{P}):\ \inf_{x \in X} [f(x) : f_k(x) \le 0 \qquad \text{for all } k = 1, \ldots, m],$$

with $f, f_k\colon X \longrightarrow \overline{\mathbb{R}}$, $k = 1, \ldots, m$, convex functions. Then $\varphi\colon X \times \mathbb{R}^m \longrightarrow \overline{\mathbb{R}}$ is defined by

$$\varphi(x, y) = f(x) + i_{C(y)}(x) \qquad \text{for all } x \in X,\ y = (y_k)_{k=1}^m \in \mathbb{R}^m,$$

with $C(y) = \{x \in X : f_k(x) \le y_k \ \text{ for all } k = 1, \ldots, m\}$. Therefore

$$L(x, y^*) = \begin{cases} f(x) - \sum_{k=1}^m f_k(x) y_k^* & \text{if } y^* \in -\mathbb{R}_+^m \\ -\infty & \text{if } y^* \notin -\mathbb{R}_+^m \end{cases},$$

for all $x \in X$, $y^* = (y_k^*)_{k=1}^m \in \mathbb{R}^m$.

PROPOSITION 2.3.24 *If $\varphi \in \Gamma_0(X \times Y)$, then $\varphi(x,0) = \sup\limits_{y^* \in Y^*} L(x,y^*)$.*

PROOF: Because of Theorem 1.2.21, we have

$$\varphi(x,0) = \sup\left[\langle y^*, 0\rangle_Y - \big(\varphi(x,\cdot)\big)^*(y^*) : y^* \in Y^*\right] = \sup_{y^* \in Y^*} L(x,y^*).$$

□

We can characterize the situation of strong duality via the saddle points of the Lagrangian.

THEOREM 2.3.25 *If $\varphi \in \Gamma_0(X \times Y)$ and L is the corresponding Lagrangian function, then the following statements are equivalent.*

(a) *$(x_0, y_0^*) \in X \times Y^*$ is a saddle point of L.*
(b) *x_0 is a solution of* (P)*, y_0^* is a solution of* (D) *and* min(P)= max(D).
(c) *$\varphi(x_0, 0) = -\varphi^*(0, y_0^*)$.*

PROOF: (a)⇒**(b),(c):** We have

$$L(x_0,y_0^*) = \inf_{x \in X} L(x,y_0^*) = -\varphi^*(0,y_0^*) \quad \text{(see Remark 2.3.22)} \tag{2.28}$$

$$\text{and} \quad L(x_0,y_0^*) = \sup_{y^* \in Y^*} L(x_0,y^*) = \varphi(x_0,0) \quad \text{(see Proposition 2.3.24).} \tag{2.29}$$

So from (2.28) and (2.29), we have

$$\inf(\mathrm{P}) \le \varphi(x_0,0) = -\varphi^*(0,y_0^*) \le \inf(\mathrm{D}). \tag{2.30}$$

Combining (2.30) with Proposition 2.3.16, we conclude that

$$\min(\mathrm{P}) = \max(\mathrm{D}) = \varphi(x_0,0) = -\varphi^*(0,y_0^*).$$

(b)⇒(a), (c): We have

$$\varphi(x_0,0) = \min(\mathrm{P}), \quad -\varphi^*(0,y_0^*) = \max(\mathrm{D}) \quad \text{and} \quad \varphi(x_0,0) = -\varphi^*(0,y_0^*). \tag{2.31}$$

Note that

$$\varphi(x_0,0) = \sup_{y^* \in Y^*} L(x_0,y^*) \ge L(x_0,y^*) \quad \text{(see Proposition 2.3.24)} \tag{2.32}$$

$$\text{and} \quad -\varphi^*(0,y_0^*) = \inf_{x \in X} L(x,y_0^*) \le L(x,y_0^*) \quad \text{(see Remark 2.3.22).} \tag{2.33}$$

From (2.31) through (2.33) it follows that $(x_0, y_0^*) \in X \times Y^*$ is a saddle point of the Lagrangian.

(c)⇒(a), (b): We have

$$\inf{(\mathrm{P})} \leq \varphi(x_0, 0) = -\varphi^*(0, y_0^*) \leq \sup{(\mathrm{D})},$$
$$\Rightarrow \inf{(\mathrm{P})} = \sup{(\mathrm{D})} = \varphi(x_0, 0) = -\varphi^*(0, y_0^*) \quad \text{(see Proposition 2.3.16).}$$

□

Next we consider an important special case of the above general duality theory. So let $f \in \Gamma_0(X)$, $g \in \Gamma_0(Y)$, and $A \in \mathcal{L}(X,Y)$. The primal problem is:

$$(\mathrm{P})' : \inf[f(x) + g(A(x)) : x \in X]. \tag{2.34}$$

The convex perturbation functional $\varphi \colon X \times Y \longrightarrow \overline{\mathbb{R}} = \mathbb{R} \cup \{+\infty\}$ is given by

$$\varphi(x,y) = f(x) + g(A(x) + y) \qquad \text{for all } (x,y) \in X \times Y.$$

Then the dual problem is:

$$(\mathrm{D})' : \sup\big[-f^*(A^*(y^*)) - g^*(-y^*) : y^* \in Y^*\big]. \tag{2.35}$$

Also the Lagrangian function for the pair of problems $\{(\mathrm{P})', (\mathrm{D})'\}$ (see (2.34) and (2.35)) is given by

$$L(x,y^*) = \begin{cases} f(x) + \langle y^*, A(x)\rangle_Y - g^*(y^*) & \text{if } f(x) < +\infty \\ +\infty & \text{if } f(x) = +\infty \end{cases} \tag{2.36}$$

Using Theorem 2.3.25, we obtain the following.

PROPOSITION 2.3.26 *A pair $(x_0, y_0^*) \in X \times Y^*$ is a saddle point of the Lagrangian L defined by (2.36) if and only if*

$$y_0^* \in \partial g(A(x_0)) \qquad \textit{and} \qquad -A^*(y_0^*) \in \partial f(x_0).$$

Let us close this section with a simple application of duality theory on a calculus of variations problem.

EXAMPLE 2.3.27 Let $X = W_0^{1,2}(0,1)$, $Y = L^2(0,1)$, $A \in \mathcal{L}(X,Y)$ defined by $A(x) = x'$, and $f \in \Gamma_0(X)$, $g \in \Gamma_0(Y)$ defined by

$$f(x) = \langle x_0^*, x\rangle_X = -\int_0^1 h(t)x(t)dt = -(h,x)_{L^2(0,1)} \qquad \text{for some } h \in L^2(0,1)$$
$$\text{and} \quad g(y) = \frac{1}{2}\|y\|^2_{L^2(0,1)}.$$

Then identifying $L^2(0,1)$ with its dual (hence $X \subseteq Y \subseteq X^* = W^{-1,2}(0,1)$), we have

$$f^*(x^*) = i_{\{x_0^*\}}(x^*) \qquad \text{for all } x^* \in X^* \qquad \text{and} \qquad g^*(y^*) = \frac{1}{2}\|y^*\|^2_{L^2(0,1)}.$$

Note that $-A^*(y^*) = x_0^*$ if and only if

$$-\int_0^1 y^*(t)x'(t)dt = -\int_0^1 h(t)x(t)dt \qquad \text{for all } x \in W_0^{1,2}(0,1),$$
$$\Rightarrow y^* \in W^{1,2}(0,1) \qquad \text{and} \qquad (y^*)' = -h.$$

Then the primal problem is

$$(\mathrm{P})'' : \inf \Big[\frac{1}{2}\|x'\|^2_{L^2(0,1)} - (h,x)_{L^2(0,1)} : x \in W_0^{1,2}(0,1) \Big]$$

and the dual problem is

$$(\mathrm{D})'' : \sup \Big[-\frac{1}{2}\|y^*\|^2_{L^2(0,1)} : y^* \in W^{1,2}(0,1),\ (y^*)' = -h \Big].$$

Then problems (P)″ and (D)″ have solutions $x_0 \in W_0^{1,2}(0,1) \cap W^{2,2}(0,1)$ and $y_0^* \in W^{1,2}(0,1)$ (see Corollary 2.1.11) and they are unique due to the strict convexity of the functionals. Moreover, we have

$$-x_0''(t) = h(t) \text{ a.e. on } (0,1), \quad x_0(0) = x_0(1) = 0 \qquad \text{and} \qquad y_0^* = x_0'.$$

2.4 Variational Principles

In Theorem 2.1.10 and Corollary 2.1.11 we saw that if the objective functional $\varphi\colon X \longrightarrow \overline{\mathbb{R}}$ is lower semicontinuous in a topology and the domain exhibits some kind of (local) compactness in the same topology, then a minimizing point exists. This is the cornerstone of the so-called direct method of the calculus of variations. However, in many situations the above convenient setting is not present. Think, for example, of infinite-dimensional Banach space functionals that are strongly lower semicontinuous, but not weakly. An effective tool to approach such problems is provided by the so-called Ekeland variational principle. Since its appearance in 1974, this result found many significant applications in different parts of analysis. Also it turned out to be equivalent to some other important results of nonlinear analysis and also served to provide new and elegant proofs to known results. In addition it initiated the production of some other related variational principles.

In this section we survey this area exhibiting all those features that are significant for problems in nonlinear analysis. Of course our presentation is centered on the Ekeland variational principle.

We start by giving the general form of the Ekeland variational principle.

THEOREM 2.4.1 *If (X,d) is a complete metric space, $\varphi\colon X \longrightarrow \overline{\mathbb{R}}$ is a proper lower semicontinuous function that is bounded below, and $\varepsilon>0$, $\bar{x} \in X$ satisfy*

$$\varphi(\bar{x}) \leq \inf_X \varphi + \varepsilon,$$

then for any $\lambda>0$, we can find $x_\lambda \in X$ such that

$$\varphi(x_\lambda) \leq \varphi(\bar{x}), \quad d(x_\lambda, \bar{x}) \leq \lambda \quad \textit{and} \quad \varphi(x_\lambda) \leq \varphi(x) + \frac{\varepsilon}{\lambda} d(x, x_\lambda) \qquad \textit{for all } x \in X.$$

PROOF: In what follows $d_\lambda = (1/\lambda)d$. We introduce the following relation on $X \times X$.

$$x \leq u \qquad \text{if and only if } \varphi(x) \leq \varphi(u) - \varepsilon d_\lambda(x,u). \tag{2.37}$$

It is easy to see that $\leq$ is reflexive, antisymmetric, and transitive, hence a partial order on X. We define a sequence of nonempty sets $C_n \subseteq X$, $n \geq 1$ as follows. Let $x_1 = \bar{x}$ and set

$$C_1 = \{x \in X : x \leq x_1\}.$$

Choose $x_2 \in C_1$ such that $\varphi(x_2) \leq \inf_{C_1} \varphi + (\varepsilon/2^2)$ and define $C_2 = \{x \in X : x \leq x_2\}$. We continue this way and inductively define

$$C_n = \{x \in X : x \leq x_n\} \quad \text{and} \quad x_{n+1} \in C_n \quad \text{such that}$$
$$\varphi(x_{n+1}) \leq \inf_{C_n} \varphi + \frac{\varepsilon}{2^{n+1}}, \ n \geq 1. \tag{2.38}$$

Evidently $\{C_n\}_{n\geq 1}$ is a decreasing sequence and each set C_n is closed. Indeed, suppose $\{u_m\}_{m\geq 1} \subseteq C_n$ such that $u_m \longrightarrow u$. We have

$$\begin{aligned} &\varphi(u_m) \leq \varphi(x_n) - \varepsilon d_\lambda(u_m, x_n), \\ \Rightarrow \ &\varphi(u) \leq \liminf_{m\to\infty} \varphi(u_m) \leq \varphi(x_n) - \varepsilon d_\lambda(u, x_n) \quad \text{(i.e., } u \in C_n). \end{aligned}$$

We claim that diam $C_n \longrightarrow 0$ as $n \to \infty$. To this end let $u \in C_n$. We have $u \leq x_n$ and so

$$\varphi(u) \leq \varphi(x_n) - \varepsilon d_\lambda(u, x_n) \quad \text{(see (2.37))}. \tag{2.39}$$

Also note that $u \in C_{n-1}$, because $\{C_n\}_{n\geq 1}$ is a decreasing sequence. Therefore

$$\varphi(x_n) \leq \varphi(u) + \frac{\varepsilon}{2^n} \quad \text{(see (2.38))}. \tag{2.40}$$

Combining (2.39) and (2.40), we obtain

$$d_\lambda(u, x_n) \leq \frac{1}{2^n}. \tag{2.41}$$

Because $u \in C_n$ was arbitrary, from (2.41) and the triangle inequality it follows that

$$\begin{aligned} &\text{diam } C_n \leq \frac{1}{2^{n-1}}, \\ \Rightarrow \ &\text{diam } C_n \longrightarrow 0 \quad \text{as } n \to \infty. \end{aligned}$$

Then by Cantor's theorem, we have that $\bigcap_{n\geq 1} C_n = \{x_\lambda\}$. Because $x_\lambda \in C_1$, we have $x_\lambda \leq \bar{x}$ and so $\varphi(x_\lambda) \leq \varphi(\bar{x})$ (see (2.37)). Also from the triangle inequality and (2.41), we have

$$\begin{aligned} &d_\lambda(\bar{x}, x_n) \leq \sum_{k=1}^{n-1} d_\lambda(x_k, x_{k+1}) \leq \sum_{k=1}^{n-1} \frac{1}{2^k} \\ \Rightarrow \ &d_\lambda(\bar{x}, x_\lambda) = \lim_{n\to\infty} d_\lambda(\bar{x}, x_n) \leq 1 \quad \text{(i.e., } d_\lambda(\bar{x}, x_\lambda) \leq \lambda). \end{aligned}$$

Finally let $x \neq x_\lambda$. Evidently we cannot have $x \leq x_\lambda$ or otherwise $x \in \bigcap_{n\geq 1} C_n$, a contradiction. Then by virtue of (2.37)

$$\varphi(x_\lambda) < \varphi(x) + \frac{\varepsilon}{\lambda} d(x, x_\lambda).$$

∎

REMARK 2.4.2 In the conclusions of Theorem 2.4.1, the relations $d(\bar{x}, x_\lambda) \leq \lambda$ and $\varphi(x_\lambda) \leq \varphi(x) + (\varepsilon/\lambda)d(x, x_\lambda)$ for all $x \in X$, are in a sense complementary. Indeed, the choice of $\lambda > 0$ allows us to strike a balance between them. So if $\lambda > 0$ is large the inequality $d(x, x_\lambda) \leq \lambda$ gives little information on the whereabouts of x_λ, and the inequality $\varphi(x_\lambda) \leq \varphi(x) + (\varepsilon/\lambda)d(x, x_\lambda)$, $x \in X$ becomes sharper and says that x_λ is close to being a global minimizer of φ. The situation is reversed if $\lambda > 0$ is small. Two important cases are when $\lambda = 1$ (which means we do not care about the whereabouts of x_λ) and when $\lambda = \sqrt{\varepsilon}$ (which means that we need to have information from both inequalities). We state both cases as corollaries of Theorem 2.4.1.

COROLLARY 2.4.3 *If (X, d) is a complete metric space and $\varphi \colon X \longrightarrow \overline{\mathbb{R}}$ is a proper lower semicontinuous function that is bounded below, then for any $\varepsilon > 0$, we can find $x_\varepsilon \in X$ such that*

$$\varphi(x_\varepsilon) \leq \inf_X \varphi + \varepsilon \qquad \textit{and} \qquad \varphi(x_\varepsilon) \leq \varphi(x) + \varepsilon d(x, x_\varepsilon) \qquad \textit{for all } x \in X.$$

COROLLARY 2.4.4 *If (X, d) is a complete metric space, $\varphi \colon X \longrightarrow \overline{\mathbb{R}}$ is a proper lower semicontinuous function that is bounded below, and $\varepsilon > 0$, $x_\varepsilon \in X$ satisfy*

$$\varphi(x_\varepsilon) \leq \inf_X \varphi + \varepsilon,$$

then we can find $u_\varepsilon \in X$ such that

$$\varphi(u_\varepsilon) \leq \varphi(x_\varepsilon), \quad d(x_\varepsilon, u_\varepsilon) \leq \sqrt{\varepsilon} \quad \textit{and} \quad \varphi(u_\varepsilon) \leq \varphi(x) + \sqrt{\varepsilon} d(x, u_\varepsilon) \quad \textit{for all } x \in X.$$

Theorem 2.4.1 has some important consequences if we assume more structure on the space X and the function φ.

THEOREM 2.4.5 *If X is a Banach space, $\varphi \colon X \longrightarrow \overline{\mathbb{R}}$ is a function that is bounded below and Gâteaux differentiable, and $\varepsilon > 0$, $x_\varepsilon \in X$ satisfy*

$$\varphi(x_\varepsilon) \leq \inf_X \varphi + \varepsilon,$$

then there exists $u_\varepsilon \in X$ such that

$$\varphi(u_\varepsilon) \leq \varphi(x_\varepsilon), \quad \|u_\varepsilon - x_\varepsilon\|_X \leq \sqrt{\varepsilon} \qquad \textit{and} \qquad \|\varphi'(u_\varepsilon)\|_{X^*} \leq \sqrt{\varepsilon}.$$

PROOF: Apply Corollary 2.4.4 to obtain $u_\varepsilon \in X$ such that

$$\begin{aligned} &\varphi(u_\varepsilon) \leq \varphi(x_\varepsilon), \quad \|u_\varepsilon - x_\varepsilon\|_X \leq \sqrt{\varepsilon} && \text{and} \\ &-\sqrt{\varepsilon}\|x - u_\varepsilon\|_X \leq \varphi(x) - \varphi(u_\varepsilon) && \text{for all } x \in X. \end{aligned} \tag{2.42}$$

From the third inequality in (2.42), taking $x = u_\varepsilon + \lambda h$ with $\lambda > 0$ and $h \in X$ with $\|h\|_X \leq 1$, we have

$$\begin{aligned} & -\sqrt{\varepsilon} \leq \frac{\varphi(u_\varepsilon + \lambda h) - \varphi(u_\varepsilon)}{\lambda}, \\ \Rightarrow \quad & -\sqrt{\varepsilon} \leq \langle \varphi'(u_\varepsilon), h \rangle_X \qquad \text{for all } h \in \overline{B}_1(0), \\ \Rightarrow \quad & \|\varphi'(u_\varepsilon)\|_{X^*} \leq \sqrt{\varepsilon}. \end{aligned}$$

□

COROLLARY 2.4.6 *If X is a Banach space and $\varphi\colon X \longrightarrow \mathbb{R}$ is a function that is bounded below and Gâteaux differentiable, then for every minimizing sequence $\{x_n\}_{n\geq 1}$ of φ, we can find a minimizing sequence $\{u_n\}_{n\geq 1}$ of φ such that*

$$\varphi(u_n) \leq \varphi(x_n), \quad \|u_n - x_n\|_X \longrightarrow 0 \quad \text{and} \quad \|\varphi'(u_n)\|_{X^*} \longrightarrow 0 \quad \text{as } n\to\infty.$$

The following proposition is useful in critical point theory (see Section 4.1).

PROPOSITION 2.4.7 *If X is a Banach space and $\varphi\colon X \longrightarrow \mathbb{R}$ is a lower semicontinuous function that is bounded below, Gâteaux differentiable, and satisfies the following compactness-type condition*

> *every sequence $\{x_n\}_{n\geq 1} \subseteq X$ that satisfies $|\varphi(x_n)| \leq M$ for some $M>0$, all $n\geq 1$ and $\|\varphi'(x_n)\|_{X^*} \longrightarrow 0$, it admits a strongly convergent subsequence,*

then we can find $\widehat{x}_0 \in X$ such that $\varphi(\widehat{x}_0) = \inf\limits_X \varphi$ and $\varphi'(\widehat{x}_0) = 0$.

PROOF: Because of Theorem 2.4.5 for every $n\geq 1$, we can find $x_n \in X$ such that

$$\varphi(x_n) \leq \inf_X \varphi + \frac{1}{n} \qquad \text{and} \qquad \|\varphi'(x_n)\|_{X^*} \leq \frac{1}{n}.$$

Then by virtue of the compactness-type condition that φ satisfies, by passing to a subsequence if necessary, we may assume that $x_n \longrightarrow \widehat{x}_0$ in X. Clearly $\varphi(\widehat{x}_0) = \inf\limits_X \varphi$ and from this it follows that $\varphi'(\widehat{x}_0) = 0$. □

REMARK 2.4.8 In critical point theory (see Section 4.1), we encounter the compactness-type condition of Proposition 2.4.7 in the context of functions that are C^1 on X with values in $\mathbb{R}$. In such a setting it is called the *Palais–Smale condition* (*PS-condition* for short) and it is a basic tool in the derivation of minimax characterizations of the critical values of functions $\varphi \in C^1(X)$.

PROPOSITION 2.4.9 *If X is a Banach space and $\varphi\colon X \longrightarrow \mathbb{R}$ is a lower semicontinuous function that is bounded below, Gâteaux differentiable, and*

$$\varphi(x) \geq c_1\|x\| - c_2 \qquad \text{for some } c_1, c_2>0 \text{ and all } x\in X, \tag{2.43}$$

then $\varphi'(X)$ is dense in $c_1\overline{B}_1^*$ where $\overline{B}_1^* = \{x^* \in X^* : \|x^*\|_{X^*} \leq 1\}$.

PROOF: Let $x^* \in c_1\overline{B}_1^*$ and consider the function $\psi(x) = \varphi(x) - \langle x^*, x\rangle_X$ for all $x\in X$. Clearly ψ is lower semicontinuous, bounded below (because of (2.43)) and Gâteaux differentiable. So by virtue of Theorem 2.4.5 we can find $x_\varepsilon \in X$ such that $\|\psi'(x_\varepsilon)\|_{X^*} \leq \varepsilon$, hence $\|\varphi'(x_\varepsilon) - x^*\|_{X^*} < \varepsilon$, which proves the density of $\varphi'(X)$ in $c_1\overline{B}_1^*$. □

COROLLARY 2.4.10 *If X is a Banach space and $\varphi\colon X \longrightarrow \mathbb{R}$ is a lower semicontinuous function that is Gâteaux differentiable and satisfies $\varphi(x) \geq r(\|x\|)$ for all $x \in X$, with $r\colon \mathbb{R}_+ \longrightarrow \mathbb{R}$ a continuous function such that $\big(r(\lambda)/\lambda\big) \longrightarrow +\infty$ as $\lambda\longrightarrow +\infty$, then $\varphi'(X)$ is dense in X^*.*

PROOF: Let $c_1 > 0$ be given. Then we can find $\lambda_0 > 0$ such that for all $\lambda \geq \lambda_0$ we have $r(\lambda) \geq c_1 \lambda$. Hence $\varphi(x) \geq c_1 \|x\|$ for all $x \in X$ with $\|x\|_X > \lambda_0$. On the other hand if $\|x\|_X < \lambda_0$, then

$$\varphi(x) \geq m = \inf[r(\lambda) : 0 \leq \lambda \leq \lambda_0].$$

Let $c_2 = |m|$. Then we have

$$\varphi(x) \geq c_1 \|x\| - c_2$$

and we can apply Proposition 2.4.9 and obtain that $\varphi'(X)$ is dense in $c_1 \overline{B}_1^*$. Because $c_1 > 0$ was arbitrary we conclude that $\varphi'(X)$ is dense in X^*. □

Next we show that the Ekeland variational principle is actually equivalent to some other well-known results of nonlinear analysis. We start with *Caristi's fixed point theorem*.

THEOREM 2.4.11 *If (X, d) is a complete metric space, $\varphi \colon X \longrightarrow \overline{\mathbb{R}}$ is a proper, lower semicontinuous function that is bounded below, and $F : X \longrightarrow 2^X \backslash \{\emptyset\}$ is a multifunction such that*

$$\varphi(y) \leq \varphi(x) - d(x, y) \qquad \textit{for all } x \in X \textit{ and all } y \in F(x), \tag{2.44}$$

then there exists $x_0 \in X$ such that $x_0 \in F(x_0)$.

PROOF: Apply Corollary 2.4.3 with $\varepsilon = 1$, to obtain $x_0 \in X$ such that

$$\varphi(x_0) < \varphi(x) + d(x, x_0) \qquad \text{for all } x \neq x_0. \tag{2.45}$$

We claim that $x_0 \in F(x_0)$. If this is not the case, then for every $y \in F(x_0)$ we have $y \neq x_0$. Then from (2.44) and (2.45) we have $\varphi(y) \leq \varphi(x_0) - d(x_0, y) < \varphi(y)$, a contradiction. □

So we have seen that the Ekeland variational principle in the form of Corollary 2.4.3 implies Theorem 2.4.11 (Caristi's fixed point theorem). Next we show that the opposite is also true.

THEOREM 2.4.12 *Caristi's fixed point theorem (Theorem 2.4.11), implies the Ekeland variational principle in the form of Corollary 2.4.3.*

PROOF: We argue indirectly. Suppose that we cannot find $x_\varepsilon \in X$ satisfying $\varphi(x_\varepsilon) < \varphi(x) + \varepsilon d(x, x_\varepsilon)$ for all $x \neq x_\varepsilon$. Let $F(x) = \{y \in X : \varphi(x) \geq \varphi(y) + \varepsilon d(y, x),\ y \neq x\}$. Then $F(x) \neq \varnothing$ for all $x \in X$. Invoking Theorem 2.4.11 we can find $x_0 \in X$ such that $x_0 \in F(x_0)$, which is impossible. So the Ekeland variational principle holds. □

REMARK 2.4.13 Banach's fixed point theorem (see Section 3.4), in its existence part, can be deduced from Theorem 2.4.11. Indeed, if $F \colon X \longrightarrow X$ is a contraction, that is, $d(F(x), F(y)) \leq k\ d(x, y)$ for some $k \in [0, 1)$, then it satisfies (2.44) with $\varphi(x) = (1/1 - k) d(x, F(x))$. However, Banach's fixed point theorem includes more information than the mere existence of a fixed point (a computational scheme to find the fixed point, rate of convergence, error estimates, etc.).

Another result equivalent to the Ekeland variational principle and to the Caristi fixed point theorem, is the so-called *Takahashi variational principle.*

THEOREM 2.4.14 *If (X,d) is a complete metric space, $\varphi\colon X \longrightarrow \overline{\mathbb{R}} = \mathbb{R}\cup\{+\infty\}$ is a proper lower semicontinuous function that is bounded below, and for each $u \in X$ with $\varphi(u) > \inf\limits_X \varphi$ we can find $v \in X$ such that $v \neq u$ and $\varphi(v) + d(u,v) \leq \varphi(u)$, then there exists $x_0 \in X$ such that $\varphi(x_0) = \inf\limits_X \varphi$.*

PROOF: We proceed by contradiction. So suppose that $\inf\limits_X \varphi$ is not attained. We introduce the multifunction $F : X \longrightarrow 2^X$ defined by $F(x) = \{y \in X : \varphi(y) + d(x,y) \leq \varphi(x),\ y \neq x\}$. By hypothesis $F(x) \neq \varnothing$ for all $x \in X$. Invoking Theorem 2.4.11, we can find $x_0 \in F(x_0)$, a contradiction. So $\inf\limits_X \varphi$ is attained. □

Thus we have seen that Caristi's fixed point theorem (Theorem 2.4.11) implies Takahashi's variational principle (Theorem 2.4.14). Next we show that the converse is also true.

THEOREM 2.4.15 *Takahashi's variational principle (Theorem 2.4.14) implies Caristi's fixed point theorem (Theorem 2.4.11).*

PROOF: We proceed by contradiction. So suppose that in Caristi's fixed point theorem (Theorem 2.4.11), the multifunction F has no fixed point, (i.e., $x \notin F(x)$ for all $x \in X$). Thus from the property of the multifunction F, we see that for every $x \in X$ we can find $y \neq x$ such that

$$\varphi(y) + d(x,y) \leq \varphi(x).$$

This is the hypothesis in Theorem 2.4.14. By virtue of that result we can find $x_0 \in X$ such that

$$\varphi(x_0) = \inf_X \varphi.$$

Let $y_0 \in F(x_0)$ be such that $y_0 \neq x_0$ and

$$\varphi(y_0) + d(x_0,y_0) \leq \varphi(x_0).$$

Then we have

$$0 < d(x_0,y_0) \leq \varphi(x_0) - \varphi(y_0) \leq \varphi(y_0) - \varphi(y_0) = 0,$$

a contradiction. So F has a fixed point and Caristi's fixed point theorem (Theorem 2.4.11) holds. □

Another result of nonlinear analysis related to the Ekeland variational principle (in fact equivalent to it), is the so-called *drop theorem.*

THEOREM 2.4.16 *If X is a Banach space, $C \subseteq X$ is a nonempty closed set, $y \in X \setminus C$, $R = d(y,C)$, and $0 < r < R < \varrho$, then there exists $x_0 \in C$ such that $\|y - x_0\| \leq \varrho$ and $D(x_0;y,r) \cap C = \{x_0\}$, where $D(x_0;y,r) = \operatorname{conv}\big(\overline{B}_r(y) \cup \{x_0\}\big)$.*

PROOF: By translating things if necessary, without any loss of generality, we may assume that $y = 0$. Let $S = \overline{B}_r(0) \cap C$. Then S is a closed set in X, hence a complete metric space. Consider the function $\varphi\colon S \longrightarrow \mathbb{R}$ defined by

$$\varphi(x) = \frac{\varrho + r}{R - r}\|x\|_X.$$

By virtue of Corollary 2.4.3 with $\varepsilon = 1$, we can find $x_0 \in S$ such that

$$\varphi(x_0) < \varphi(x) + \|x - x_0\|_X \qquad \text{for all } x \in S,\ x \neq x_0. \tag{2.46}$$

Because we have assumed that $y = 0$, we have $\|y - x_0\|_X = \|x_0\|_X \leq \varrho$.

Next we show that $D(x_0; 0, r) \cap C = \{x_0\}$. Suppose that this is not the case and we can find $x \neq x_0$ such that $x \in D(x_0; 0, r) \cap C$. Then

$$x \in C \qquad \text{and} \qquad x = (1-\lambda)x_0 + \lambda u \quad \text{with } u \in \overline{B}_r(0) \text{ and } 0 \leq \lambda \leq 1. \tag{2.47}$$

Evidently $0 < \lambda < 1$ (recall that $\varrho < R$). From (2.47) we have

$$\begin{aligned} &\|x\|_X \leq (1-\lambda)\|x_0\|_X + \lambda\|u\|_X, \\ \Rightarrow\ &\lambda(R - r) \leq \lambda\big(\|x_0\|_X - \|x\|_X\big) \leq \|x_0\|_X - \|x\|_X. \end{aligned} \tag{2.48}$$

From (2.46) and (2.47) it follows that

$$\frac{\varrho + r}{R - r}\|x_0\|_X < \frac{\varrho + r}{R - r}\|x\|_X + \|x - x_0\|_X = \frac{\varrho + r}{R - r}\|x\|_X + \lambda\|x_0 - u\|_X. \tag{2.49}$$

Also from (2.48) we have

$$\lambda \leq \frac{\|x_0\|_X - \|x\|_X}{R - r}. \tag{2.50}$$

Using (2.50) in (2.49), we obtain

$$\begin{aligned} (\varrho + r)\|x_0\|_X &< (\varrho + r)\|x\|_X + \|x_0 - u\|_X\big(\|x_0\|_X - \|x\|_X\big) \\ &\leq (\varrho + r)\|x\|_X + (\varrho + r)\big(\|x_0\|_X - \|x\|_X\big) \end{aligned}$$

(because $\|x_0\|_X \leq \varrho$, $u \in \overline{B}_r(0)$),

$$\Rightarrow\ \|x_0\|_X < \|x\|_X + \|x_0\|_X - \|x\|_X = \|x_0\|_X, \qquad \text{a contradiction.}$$

Therefore we conclude that $D(x_0; 0, r) \cap C = \{x_0\}$ and the proof of the theorem is finished. □

REMARK 2.4.17 The set $D(x_0; y, r)$ is called a *drop*, in view of its evocative geometry.

In the above proof, we saw that the Ekeland variational principle (in the form of Corollary 2.4.3) implies the drop theorem (Theorem 2.4.16). In fact the converse is also true. For a proof of this we refer to Penot [494]. Thus we have the following.

THEOREM 2.4.18 *The drop theorem (Theorem 2.4.16) implies the Ekeland variational principle (Corollary 2.4.3).*

Summarizing we have the following.

THEOREM 2.4.19 *Corollary 2.4.3, Theorem 2.4.11, Theorem 2.4.14 and Theorem 2.4.16 are all equivalent.*

Now we present a general principle on partially ordered sets, which implies all the above equivalent theorems. Recall that on a set X, $\leq$ is a partial order if $x \leq x$ (reflexive) $x \leq y$ and $y \leq x$ imply $x = y$ (antisymmetric) and $x \leq y$, $y \leq z$ imply $x \leq z$ (transitive).

THEOREM 2.4.20 *If $(X, \leq)$ is a partially ordered set, every increasing sequence $\{x_n\}_{n\geq 1} \subseteq X$ has an upper bound in X (i.e. if $x_n \leq x_{n+1}$ for all $n \geq 1$, there exists $\overline{y} \in X$ such that $x_n \leq \overline{y}$ for all $n \geq 1$) and $\varphi\colon X \longrightarrow \mathbb{R}$ is an increasing function that is bounded above, then there exists $\widehat{x}_0 \in X$ such that $\widehat{x}_0 \leq y$ implies $\varphi(\widehat{x}_0) = \varphi(y)$.*

PROOF: Let $x_1 \in X$ be any element. Inductively we produce an increasing sequence $\{x_n\}_{n\geq 1} \subseteq X$. Suppose we have generated the element x_n. We define

$$C_n = \{x \in X : x_n \leq x\} \qquad \text{and} \qquad M_n = \sup_{C_n} \varphi.$$

If for x_n, we have that $x_n \leq y$ implies that $\varphi(x_n) = \varphi(y)$, then clearly we are done. Otherwise $\varphi(x_n) < M_n$ and so we can find $x_{n+1} \in C_n$ such that

$$M_n \leq \varphi(x_{n+1}) + \frac{1}{2}\big(M_n - \varphi(x_n)\big). \tag{2.51}$$

So by induction, we have produced an increasing sequence $\{x_n\}_{n\geq 1} \subseteq X$. By hypothesis we can find $\widehat{x}_0 \in X$ such that

$$x_n \leq \widehat{x}_0 \qquad \text{for all } n \geq 1. \tag{2.52}$$

We claim that $\widehat{x}_0 \in X$ is the desired solution. Suppose that this is not the case. Then we can find $y \in X$ such that $\widehat{x}_0 \leq y$ and $\varphi(\widehat{x}_0) < \varphi(y)$. The sequence $\big\{\varphi(x_n)\big\}_{n\geq 1} \subseteq \mathbb{R}$ is increasing and bounded above by $\varphi(y)$. Therefore it converges. We have

$$\lim_{n\rightarrow\infty} \varphi(x_n) \leq \varphi(\widehat{x}_0) \quad \text{(see (2.52))}. \tag{2.53}$$

Because of (2.52), $y \in C_n$ for all $n \geq 1$. So from (2.51) it follows that

$$\begin{aligned} & \varphi(y) \leq M_n \leq 2\varphi(x_{n+1}) - \varphi(x_n) \qquad \text{for all } n \geq 1 \\ \Rightarrow\ & \varphi(y) \leq \varphi(\widehat{x}_0) \qquad \text{(see (2.53))}, \end{aligned}$$

a contradiction to the choice of y. This proves the theorem. $\square$

REMARK 2.4.21 Theorem 2.4.20 can have a physical interpretation. We can think of φ as a function measuring the entropy of a system. The theorem guarantees the existence of a state of maximal entropy. To these states correspond stable equilibrium states of the system.

COROLLARY 2.4.22 *If X is a Hausdorff topological space equipped with a partial order $\leq$ and $\psi\colon X \longrightarrow \mathbb{R}$ a function that is bounded below such that*

(i) *For every $x \in X$ the set $\{y \in X : x \leq y\}$ is closed;*
(ii) *$x \leq y$ and $x \neq y$ imply $\psi(y) < \psi(x)$ (strictly decreasing ψ);*

(iii) *Any increasing sequence in X is relatively compact,*

then for each $x \in X$ we can find $\widehat{x}_0 \in X$ such that $x \leq \widehat{x}_0$ and $\widehat{x}_0$ is maximal.

PROOF: Let $\{x_n\}_{n\geq 1} \subseteq X$ be an increasing sequence in X. By virtue of hypothesis (iii) $\{x_n\}_{n\geq 1} \subseteq X$ is relatively compact and so we can find a subsequence $\{x_{n_k}\}_{k\geq 1}$ such that $x_{n_k} \longrightarrow y$ in X. We claim that $x_n \leq y$ for all $n \geq 1$. Indeed given $n \geq 1$, we have $n \leq n_k$ for $k \geq k_n$ and so $x_n \leq x_{n_k}$ for $k \geq k_n$. From this and hypothesis (i) it follows that $y \in C_n$ for all $n \geq 1$; that is, $x_n \leq y$ for all $n \geq 1$. Taking $\varphi = -\psi$, because of hypothesis (ii) we can apply Theorem 2.4.20 starting from $x_1 = x$ and obtain $\widehat{x}_0 \in X$ such that $x \leq \widehat{x}_0$ and $\widehat{x}_0$ is maximal. □

COROLLARY 2.4.23 *Theorem 2.4.20 implies the Ekeland variational principle in the form of Corollary 2.4.3.*

PROOF: Without any loss of generality we take $\varepsilon = 1$ and define a partial order $\leq$ by

$$x \leq u \qquad \text{if and only if } \varphi(u) - \varphi(x) \leq -d(x,u). \tag{2.54}$$

For any increasing sequence $\{x_n\}_{n\geq 1} \subseteq X$, $\{\varphi(x_n)\}_{n\geq 1}$ is decreasing and bounded below, so it converges. Therefore once more from (2.54), we infer that $\{x_n\}_{n\geq 1} \subseteq X$ is Cauchy, hence due to the completeness of X it converges to an element in X. Therefore we can apply Corollary 2.4.22 and finish the proof. □

REMARK 2.4.24 Thus Theorem 2.4.20 also implies Caristi's fixed point theorem, Takahashi's variational principle, and the drop theorem.

We conclude this section with a generalization of Theorem 2.4.1, which is used in Section 4.1. For a proof of it we refer to Zhong [626].

THEOREM 2.4.25 *If $h\colon \mathbb{R}_+ \longrightarrow \mathbb{R}_+$ is a continuous nondecreasing function such that $\int_0^{+\infty} \frac{1}{1+h(r)}\mathrm{d}r = +\infty$, (X,d) is a complete metric space, $x_0 \in X$ is fixed, $\varphi\colon X \longrightarrow \overline{\mathbb{R}} = \mathbb{R} \cup \{+\infty\}$ is a proper, lower semicontinuous, bounded below function, $\varepsilon > 0$, $\varphi(y) \leq \inf_X \varphi + \varepsilon$ and $\lambda > 0$, then there exists $x_\lambda \in X$ such that*

$$\varphi(x_\lambda) \leq \varphi(y),\ d(x_\lambda, x_0) \leq r_0 + \bar{r}$$

$$\text{and} \quad \varphi(x_\lambda) \leq \varphi(x) + \frac{\varepsilon}{\lambda\big(1 + h\big(d(x_0, x_\lambda)\big)\big)} d(x_\lambda, x) \qquad \text{for all } x \in X,$$

where $r_0 = d(x_0, y)$ and $\bar{r} > 0$ such that $\int_0^{r_0+\bar{r}} \frac{1}{1+h(r)}\,\mathrm{d}r \geq \lambda$.

REMARK 2.4.26 If $h \equiv 0$ and $x_0 = y$, then Theorem 2.4.25 reduces to Theorem 2.4.1.

2.5 Calculus of Variations

The calculus of variations deals with the minimization or maximization of functions defined on function spaces. It is as old as calculus itself and has important applications in mechanics and physics.

In this section we present some basic aspects of the theory concerning scalar problems. This means that the unknown functions are curves defined on closed intervals in $\mathbb{R}$ with values in a Banach space. For this reason we make the following definition.

DEFINITION 2.5.1 Let $T = [0, b]$ and X a Banach space. By $C^1(T, X)$ we denote the vector space of all functions $u \in C(T, X)$ which are differentiable on $(0, b)$ and the derivative $u' : (0, b) \longrightarrow X$ is bounded and uniformly continuous.

REMARK 2.5.2 This means that u' admits a unique continuous extension on $T = [0, b]$ and

$$u'(0) = \lim_{t \longrightarrow 0^+} u'(t), \quad u'(b) = \lim_{t \longrightarrow b^-} u'(t).$$

In fact $u'(0)$ is the right derivative of u at $t = 0$ and $u'(b)$ is the left derivative of u at $t = b$.

We furnish $C^1(T, X)$ with the norm

$$\|u\|_{C^1(T,X)} = \max_{t \in T} \|u(t)\|_X + \max_{t \in T} \|u'(t)\|_X. \tag{2.55}$$

It is easy to see that (2.55) indeed defines a norm on $C^1(T, X)$. Also we have the following.

PROPOSITION 2.5.3 *The space $C^1(T, X)$ equipped with the norm $\|\cdot\|_{C^1(T,X)}$ becomes a Banach space.*

Now we introduce the setting of the calculus of variations problem that we study. So let $T = [0, b]$, X be a Banach space, U be a nonempty open set in $\mathbb{R} \times X \times X$, and $L \colon U \longrightarrow \mathbb{R}$ a continuous function, usually called the Lagrangian. A map $u \in C^1(T, X)$ is said to be *admissible*, if for $t \in T$, we have $\big(t, u(t), u'(t)\big) \in U$. We introduce the integral functional $I_L \colon D \subseteq C^1(T, X) \longrightarrow \mathbb{R}$ defined by

$$I_L(u) = \int_0^b L\big(t, u(t), u'(t)\big) dt$$

for every admissible map $u \in C^1(T, X)$.

PROPOSITION 2.5.4 *The set $D \subseteq C^1(T, X)$ of all admissible maps is an open subset of $C^1(T, X)$.*

PROOF: We show that $D^c \subseteq C^1(T, X)$ is closed. To this end let $\{x_n\}_{n \geq 1} \subseteq D^c$ and suppose that $x_n \longrightarrow x$ in $C^1(T, X)$. Because for all $n \geq 1$ $x_n' \notin D$, we can find $t_n \in T$ such that $\big(t_n, x_n(t_n), x_n'(t_n)\big) \in U^c$. By passing to a subsequence if necessary, we may assume that $t_n \longrightarrow t \in T$. Because $x_n \longrightarrow x$ and $x_n' \longrightarrow x'$ in $C^1(T, X)$, we have that $x_n(t_n) \longrightarrow x(t)$ and $x_n'(t_n) \longrightarrow x'(t)$ in X as $n \to \infty$. So in the limit as $n \to \infty$, we have $\big(t, x(t), x'(t)\big) \in U^c$, hence $x \notin D^c$, which finishes the proof. $\square$

PROPOSITION 2.5.5 *If Y, Z are Banach spaces, $W \subseteq Y$ is a nonempty open set, $\vartheta \colon T \times W \longrightarrow Z$ is a continuous map, and we set $\xi(y) = \int_0^b \vartheta(t,y)dt$ for all $y \in W$, then*

(a) *$\xi \colon W \longrightarrow Z$ is a continuous map.*
(b) *If in addition for all $t \in T$, $y \longrightarrow \vartheta(t,y)$ is Fréchet differentiable and $(t,y) \longrightarrow \vartheta'_y(t,y)$ is continuous from $T \times W$ into $\mathcal{L}(Y,Z)$, then $\xi \in C^1(W,Z)$ and $\xi'(y) = \int_0^b \vartheta'_y(t,y)dt$.*

PROOF: (a) Let $\{y_n\}_{n\geq 1} \subseteq W$ and suppose that $y_n \longrightarrow y \in W$. Then $\{y_n, y\}_{n\geq 1} = C$ is compact in Y and $C \subseteq W$. So $h(T \times C)$ is compact in Z. Therefore we can find $M>0$ such that

$$\|h(t,y_n)\|_Z \leq M \qquad \text{for all } t \in T \text{ and all } n \geq 1.$$

Because $h(t,y_n) \longrightarrow h(t,y)$ in Z for all $t \in T$, from the dominated convergence theorem we have that

$$\xi(y_n) = \int_0^b h(t,y_n)dt \longrightarrow \int_0^b h(t,y)dt = \xi(y) \text{ in } Z \text{ as } n \to \infty,$$
$$\Rightarrow \xi \colon W \longrightarrow Z \text{ is continuous.}$$

(b) We show that ξ is continuously Gâteaux differentiable. Then it follows that $\xi \in C^1(W,Z)$ (see Proposition 1.1.10). So let $h \in Y$ and $\lambda \neq 0$. We have

$$\frac{\xi(y+\lambda h) - \xi(y)}{\lambda} = \int_0^b \frac{\vartheta(t,y+\lambda h) - \vartheta(t,y)}{\lambda} dt. \tag{2.56}$$

From Proposition 1.1.6 (the mean value theorem), we know that

$$\|\vartheta(t,y+\lambda h) - \vartheta(t,y)\| \leq \sup_{\mu \in [0,1]} \|\vartheta'_y(t,y+\mu\lambda h)\|_{\mathcal{L}(Y,Z)} \lambda \|h\|_Y. \tag{2.57}$$

Because by hypothesis $(t,y) \longrightarrow \vartheta'_y(t,y)$ is continuous from $T \times W$ into $\mathcal{L}(Y,Z)$, we can find $\overline{M}>0$ such that

$$\|\vartheta'_y(t,y+\mu\lambda h)\|_{\mathcal{L}(Y,Z)} \leq \overline{M} > 0 \qquad \text{for all } t \in T, \text{ all } \mu \in [0,1] \text{ and all } \lambda \in [-1,1]. \tag{2.58}$$

Therefore

$$\left\| \frac{\vartheta(t,y+\lambda h) - \vartheta(t,y)}{\lambda} \right\|_Z \leq \overline{M}\|h\|_Y \qquad \text{for all } t \in T \text{ and all } \lambda \in [-1,1] \quad \text{(see (2.57) and (2.58)).} \tag{2.59}$$

Recall that

$$\frac{\vartheta(t,y+\lambda h) - \vartheta(t,y)}{\lambda} \longrightarrow \vartheta'_y(t,y)h \qquad \text{for all } t \in T \text{ as } \lambda \longrightarrow 0. \tag{2.60}$$

Because of (2.59), (2.60), and the dominated convergence theorem, if we pass to the limit as $\lambda \longrightarrow 0$, we obtain

$$\xi'(y)h = \int_0^b \vartheta'_y(t,y)h \, dt \qquad \text{for all } h \in Y.$$

Therefore by virtue of part (a), $\xi \in C^1(W,Z)$ and

$$\xi'(y) = \int_0^b \vartheta'_y(t,y) \, dt \qquad \text{for all } y \in Y.$$

□

We use this proposition to establish the continuous Fréchet differentiability of the integral functional I_L and determine its derivative in terms of the partial derivatives of the Lagrangian L.

THEOREM 2.5.6 *If the Lagrangian $L\colon U \longrightarrow \mathbb{R}$ is continuously Fréchet differentiable, then $I_L : D \subseteq C^1(T,X) \longrightarrow \mathbb{R}$ is continuously Fréchet differentiable too and*

$$\begin{aligned}\langle I'_L(u), v\rangle_{C^1(T,X)} &= \int_0^b \big\langle L'_2\big(t,u(t),u'(t)\big), v(t)\big\rangle_X \, dt \\ &\quad + \int_0^b \big\langle L'_3\big(t,u(t),u'(t)\big), v'(t)\big\rangle_X \, dt\end{aligned}$$

for all $v \in C^1(T,X)$. Here by $\langle\cdot,\cdot\rangle_{C^1(T,X)}$ we denote the duality brackets for the pair $\big(C^1(T,X), C^1(T,X)^*\big)$, by $\langle\cdot,\cdot\rangle_X$ the duality brackets for the pair (X,X^*), and L'_k $k=2,3$ the partial derivative of $L(t,x,y)$ with respect to the second and third variables, respectively.

PROOF: From Proposition 2.5.4 we know that the set $D \subseteq C^1(T,X)$ of admissible maps is open. We consider the map $\vartheta\colon T \times D \longrightarrow \mathbb{R}$ defined by

$$\vartheta(t,u) = L\big(t,u(t),u'(t)\big) \qquad \text{for all } (t,u) \in T \times D.$$

Let $\vartheta_1\colon T \times D \longrightarrow U$ and $\vartheta_2\colon U \longrightarrow \mathbb{R}$ be defined by

$$\vartheta_1(t,u) = \big(t,u(t),u'(t)\big) \qquad \text{and} \qquad \vartheta_2(t,x,y) = L(t,x,y)$$

for all $(t,u) \in T \times D$ and all $(t,x,y) \in U$. Clearly ϑ_1, ϑ_2 are continuous and $\vartheta = \vartheta_2 \circ \vartheta_1$. So ϑ is continuous too.

Moreover, because for $t \in T$ fixed the maps $u \longrightarrow u(t)$ and $u \longrightarrow u'(t)$ are continuous linear, we see that for every $t \in T$, $\vartheta(t,\cdot)$ is continuously Fréchet differentiable. Then by virtue of Propositions 1.1.14 and 1.1.17, we have

$$\langle \vartheta'_2(t,u), v\rangle_{C^1(T,X)} = \big\langle L'_2\big(t,u(t),u'(t)\big), v(t)\big\rangle_X + \big\langle L'_3\big(t,u(t),u'(t)\big), v'(t)\big\rangle_X$$

for all $v \in C^1(T,X)$. It follows that $(t,u) \longrightarrow \vartheta'_2(t,u)$ is continuous from $T \times D$ into $C^1(T,X)^*$ and so we can apply Proposition 2.5.5 and conclude that

$$\begin{aligned}\langle I'_L(u), v\rangle_{C^1(T,X)} &= \int_0^b \big\langle L'_2\big(t,u(t),u'(t)\big), v(t)\big\rangle_X \, dt \\ &\quad + \int_0^b \big\langle L'_3\big(t,u(t),u'(t)\big), v'(t)\big\rangle_X \, dt\end{aligned}$$

for all $v \in C^1(T, X)$. $\square$

We consider the problem with fixed endpoints. Namely we require that all admissible maps satisfy $u(0)=x_0$ and $u(b) = x_b$, where $x_0, x_b \in X$ are given. So let

$$C^1_{x_0,x_b}(T, X) = \{u \in C^1(T, X) : u(0) = x_0, \; u(b) = x_b\}.$$

This is a closed affine subspace of $C^1(T, X)$ produced by translation of the closed subspace $C^1_0(T, X) = C^1_{0,0}(T, X)$.

Now let us formally introduce the concept of the local (relative) extremal point of a functional.

DEFINITION 2.5.7 Let Y be a Banach space, $C \subseteq Y$ a nonempty set, and $\varphi\colon C \longrightarrow \mathbb{R}$. We say that $y_0 \in C$ is a local (relative) extremum of φ, if we can find V a neighborhood of y_0 in Y such that one of the following two holds.

(a) $\varphi(y_0) \leq \varphi(y)$ for all $y \in V \cap C$ (*local minimum*).
(b) $\varphi(y_0) \geq \varphi(y)$ for all $y \in V \cap C$ (*local maximum*).

If the inequalities in (a) and (b) are strict for $y \neq y_0$, then we speak of a strict local extremum (strict local minimum and strict local maximum).

If the inequalities in (a) and (b) are true for all $y \in C$, then we speak of a (global) extremum on C ((global) minimum on C and (global) maximum on C).

In the next proposition we produce a necessary condition for I_L to admit a local extremum on $C^1_{x_0,x_b}(T, X)$. In what follows we assume that $D \cap C^1_{x_0,x_b}(T, X)$ is nonempty.

PROPOSITION 2.5.8 *If $L\colon U \longrightarrow \mathbb{R}$ is a continuously Fréchet differentiable Lagrangian and I_L admits a local relative extremum on $C^1_{x_0,x_b}(T, X)$ at the map $u \in C^1_{x_0,x_b}(T, X)$, then for every $v \in D \cap C^1_0(T, X)$ we have*

$$\begin{aligned} \langle I_L'(u), v\rangle_{C^1(T,X)} &= \int_0^b \big\langle L_2'\big(t, u(t), u'(t)\big), v(t)\big\rangle_X \, dt \\ &\quad + \int_0^b \big\langle L_3'\big(t, u(t), u'(t)\big), v'(t)\big\rangle_X \, dt. \end{aligned} \tag{2.61}$$

PROOF: From Proposition 2.5.4 we know that D is an open subset of $C^1(T, X)$. Hence $D \cap C^1_{x_0,x_b}(T, X)$ is an open subset of the closed affine subspace $C^1_{x_0,x_b}(T, X)$. By Theorem 2.5.6 $I_L\colon D \longrightarrow \mathbb{R}$ is continuously Fréchet differentiable, thus we have that its restriction on $C^1_{x_0,x_b}(T, X)$ is also continuously Fréchet differentiable. Because I_L attains its local relative minimum or local relative maximum at a map $u \in C^1_{x_0,x_b}(T, X)$, we have

$$\Big(I_L\big|_{C^1_{x_0,x_b}(T,X)}\Big)'(u) = 0.$$

But

$$\Big(I_L\big|_{C^1_{x_0,x_b}(T,X)}\Big)'(u) = I_L'(u)\big|_{C^1_0(T,X)}$$

and so (2.61) follows. $\square$

Equation (2.61) is not yet in a convenient form. To achieve this we need some auxiliary results that are basic in the theory of calculus of variations. The first auxiliary result is usually known in the literature as the *Lagrange lemma.*

LEMMA 2.5.9 *If $\xi \in C(T, X^*)$, then $\xi \equiv 0$ if and only if $\int_0^b \langle \xi(t), u(t) \rangle_X \, dt = 0$ for all $u \in C_0^1(T, X)$.*

PROOF: $\Rightarrow$: Obvious.

$\Leftarrow$: Suppose that ξ is not identically zero. Then we can find $t_0 \in (0, b)$ such that $\xi(t_0) \neq 0$. Hence we can find $x \in X$ such that $\langle \xi(t_0), x \rangle_X \neq 0$ and without any loss of generality we may assume that $\langle \xi(t_0), x \rangle_X > 0$. The function $t \longrightarrow \langle \xi(t), x \rangle_X$ is continuous on T and so we can find $\delta > 0$ that $[t_0 - \delta, t_0 + \delta] \subseteq (0, b)$ and $\langle \xi(t), x \rangle_X > 0$ for all $t \in [t_0 - \delta, t_0 + \delta]$. Consider a function $\vartheta \in C^1(T)$ such that

$$\vartheta \geq 0 \qquad \text{and} \qquad \vartheta(t) = 0 \qquad \text{for all } t \notin [t_0 - \delta, t_0 + \delta].$$

A possible such function (known as a cut-off function) is

$$\vartheta(t) = \begin{cases} 0 & \text{if } 0 \leq t \leq t_0 - \delta \\ \exp\left(-\frac{1}{\delta^2 - (t - t_0)^2}\right) & \text{if } t_0 - \delta \leq t \leq t_0 + \delta. \\ 0 & \text{if } t_0 + \delta \leq t \leq b \end{cases}$$

Then let $u_0(t) = \vartheta(t)x$. Evidently $u_0 \in C_0^1(T, X)$ and the function $t \longrightarrow \langle \xi(t_0), u_0(t) \rangle_X$ is continuous on T, nonnegative, and $\langle \xi(t_0), u_0(t) \rangle_X > 0$. Therefore it follows that

$$\int_0^b \langle \xi(t), u_0(t) \rangle_X \, dt > 0,$$

a contradiction to the hypothesis of the lemma. $\square$

The second auxiliary result is known as *Du Bois–Reymond's lemma.*

LEMMA 2.5.10 *If $\xi \in C(T, X^*)$, then ξ is constant if and only if for all $u \in C(T, X)$ with mean value zero (i.e., $\int_0^b u(t)dt = 0$) we have $\int_0^b \langle \xi(t), u(t) \rangle_X \, dt = 0$.*

PROOF: $\Rightarrow$: Suppose that $\xi(t) = x^* \in X^*$ for all $t \in T$. Also assume that $u \in C(T, X)$ has mean value zero. Then from the properties of the Bochner integral, we have

$$\int_0^b \langle \xi(t), u(t) \rangle_X \, dt = \int_0^b \langle x^*, u(t) \rangle_X \, dt = \left\langle x^*, \int_0^b u(t)dt \right\rangle_X = 0.$$

$\Leftarrow$: Suppose that ξ is not constant. Because $\xi \in C(T, X)$ we can find $0 < t_1 < t_2 < b$ such that $\xi(t_1) \neq \xi(t_2)$. So there exists $x \in X$ such that $\langle \xi(t_1), x \rangle_X \neq \langle \xi(t_2), x \rangle_X$. Without any loss of generality we may assume that

$$\langle \xi(t_1), x \rangle_X < \langle \xi(t_2), x \rangle_X \, .$$

By virtue of the continuity of the function $t \longrightarrow \langle \xi(t), x \rangle_X$ on T, we can find $\delta > 0$ such that $0 < t_1 - \delta < t_1 + \delta < t_2 - \delta < t_2 + \delta < b$ and

$$\max\left\{ \langle \xi(t), x\rangle_X : t \in [t_1-\delta, t_1+\delta]\right\} < \min\left\{ \langle \xi(t), x\rangle_X : t \in [t_2-\delta, t_2+\delta]\right\}.$$

From this inequality it follows that

$$\langle \xi(t_2+s) - \xi(t_1+s), x\rangle_X > 0 \qquad \text{for all } s \in [-\delta, \delta].$$

We consider a continuous cut-off function ϑ for the interval $[-\delta,\delta]$; that is, $\vartheta \geq 0$, $\vartheta(s) > 0$ for $s \in (-\delta,\delta)$ and $\vartheta(s) = 0$ for $s \notin (-\delta,\delta)$. A possible such function is given by

$$\vartheta(s) = \begin{cases} \frac{\delta-|s|}{\delta} & \text{if } |s| \leq \delta. \\ 0 & \text{if } |s| > \delta \end{cases}$$

Let $\eta(t) = \int_0^b \big(\vartheta(s-t_2) - \vartheta(s-t_1)\big)\mathrm{d}s$ for all $t \in T$ and set $u(t) = \eta(t)x$. Evidently $u \in C^1(T,X)$ and $u(0) = u(b) = 0$. Then

$$\begin{aligned} \int_0^b \langle \xi(t), u'(t)\rangle_X \,\mathrm{d}t &= \int_0^b \big(\vartheta(s-t_2) - \vartheta(s-t_1)\big)\, \langle \xi(t), x\rangle_X \,\mathrm{d}t \\ &= \int_\delta^\delta \vartheta(s)\, \langle \xi(t_2+s) - \xi(t_1+s), x\rangle_X \,\mathrm{d}s, \end{aligned}$$

a contradiction to the hypothesis. □

The previous two lemmata lead to the third auxiliary result which provides the tools to simplify the extremality equation (2.61).

LEMMA 2.5.11 *If* $\xi_1, \xi_2 \in C(T, X^*)$, *then* $\int_0^b \big[\langle \xi_1(t), u(t)\rangle_X + \langle \xi_2(t), u'(t)\rangle_X \big]\mathrm{d}t = 0$ *for all* $u \in C_0^1(T,X)$ *if and only if* ξ_2 *is differentiable on* T *and* $\xi_2'(t) = \xi_1(t)$ *for all* $t \in T$.

PROOF: Let $\beta(t) = \int_0^t \xi_1(s)\mathrm{d}s$ for all $t \in T$. Then $\beta \in C^1(T, X^*)$ and $\beta'(t) = \xi_1(t)$ for all $t \in T$. Let $u \in C_0^1(T,X)$. We have

$$\begin{aligned} \langle \xi_1(t), u(t)\rangle_X &= \frac{\mathrm{d}}{\mathrm{d}t}\langle \beta(t), u(t)\rangle_X - \langle \beta(t), u'(t)\rangle_X \qquad \text{for all } t \in T, \\ \Rightarrow \int_0^b \langle \xi_1(t), u(t)\rangle_X \,\mathrm{d}t &= \langle \beta(b), u(b)\rangle_X - \langle \beta(0), u(0)\rangle_X - \int_0^b \langle \beta(t), u'(t)\rangle_X \,\mathrm{d}t \\ &= -\int_0^b \langle \beta(t), u'(t)\rangle_X \,\mathrm{d}t \quad (\text{because } u(0) = u(b) = 0) \\ \Rightarrow \int_0^b \big[\langle \xi_1(t), u(t)\rangle_X + \langle \xi_2(t), u'(t)\rangle_X \big]\mathrm{d}t &= \int_0^b \langle \xi_2(t) - \beta(t), u'(t)\rangle_X \,\mathrm{d}t. \end{aligned}$$

Then by virtue of Lemma 2.5.10, we have

$$\int_0^b \langle \xi_2(t) - \beta(t), u'(t)\rangle_X \,\mathrm{d}t = 0 \quad \text{for all } u \in C_0^1(T,X)$$

if and only if

$$\xi_2(t) - \beta(t) = x_0^* \in X^* \qquad \text{for all } t \in T.$$

Therefore $\xi_2 \in C^1(T,X)$ and $\xi_2'(t) = \xi_1(t)$ for all $t \in T$. □

Using this lemma we can simplify equation (2.61).

THEOREM 2.5.12 *If* $L: U \longrightarrow \mathbb{R}$ *is a continuously differentiable Lagrangian and* I_L *admits a local relative extremum on* $C^1_{x_0,x_b}(T,X)$ *at* $u \in C^1_{x_0,x_b}(T,X)$*, then*

$$\frac{\mathrm{d}}{\mathrm{d}t} L_3'\big(t, u(t), u'(t)\big) = L_2'\big(t, u(t), u'(t)\big) \qquad \textit{for all } t \in T. \tag{2.62}$$

REMARK 2.5.13 Equation (2.62) is known as *Euler's equation* (in Lagrange form). The functions $u \in C^1_{x_0,x_b}(T,X)$ along which (2.62) is valid are called *extremals.*

Next we present some special cases where the Euler equation has easily determined integrals.

COROLLARY 2.5.14 *If the Lagrangian function* $L(t,x,y)$ *is actually independent of the third variable* $y \in X$*, then a necessary condition for the extremality of the map* $u \in C^1_{x_0,x_b}(T,X)$ *is given by*

$$L_2'\big(t, u(t)\big) = 0 \qquad \textit{for all } t \in T.$$

COROLLARY 2.5.15 *If the Lagrangian function* $L(t,x,y)$ *is actually independent of the second variable* $x \in X$*, then the Euler equation (2.62) admits the following solution*

$$p(t) = L_3'\big(t, u'(t)\big) = p_0^* \in X^* \qquad \textit{for all } t \in T.$$

REMARK 2.5.16 In mechanics $p(t)$ is the momentum function and Corollary 2.5.15 says that along an extremal the momentum is constant.

COROLLARY 2.5.17 *If the Lagrangian function* $L(t,x,y)$ *actually does not depend on the time variable* $t \in T$ *and the extremal* $u \in C^2(T,X) \cap C^1_{x_0,x_b}(T,X)$*, then the Euler equation (2.62) admits an energy integral*

$$\begin{aligned} H(t) &= \big\langle p(t), u'(t) \big\rangle_X - L\big(u(t), u'(t)\big) \\ &= \big\langle L_2'\big(u(t), u'(t)\big), u'(t) \big\rangle_X - L\big(u(t), u'(t)\big) = H_0 \in \mathbb{R} \qquad \textit{for all } t \in T. \end{aligned}$$

PROOF: From the chain rule we have

$$\begin{aligned} \frac{\mathrm{d}}{\mathrm{d}t} H(t) = &\left\langle \frac{\mathrm{d}}{\mathrm{d}t} L_2'\big(u(t), u'(t)\big), u'(t) \right\rangle + \big\langle L_2'\big(u(t), u'(t)\big), u''(t) \big\rangle_X \\ &- \big\langle L_1'\big(u(t), u'(t)\big), u'(t) \big\rangle_X - \big\langle L_2'\big(u(t), u'(t)\big), u''(t) \big\rangle_X = 0 \end{aligned}$$

(see (2.62)). Hence $H(t) = H_0 \in \mathbb{R}$ for all $t \in T$. □

REMARK 2.5.18 The above corollary says that along an extremal the energy is constant (conservation of energy).

To be able to analyze further the Euler equation (2.62), we need to introduce second-order derivatives.

DEFINITION 2.5.19 Let X, Y be Banach spaces, $U \subseteq X$ nonempty open, $x_0 \in U$, and $\varphi \colon U \longrightarrow Y$ a Fréchet differentiable map. If the Fréchet derivative $x \longrightarrow \varphi'(x)$ from U into $\mathcal{L}(X,Y)$ with the operator norm topology is differentiable at x_0, then φ is said to be *twice differentiable* at x_0 and this second-order derivative is denoted by $\varphi''(x_0)$.

REMARK 2.5.20 According to this definition $\varphi''(x_0) \in \mathcal{L}\big(X, \mathcal{L}(X,Y)\big)$.

DEFINITION 2.5.21 Let X_1, X_2 and Y be Banach spaces. A map $L \colon X = X_1 \times X_2 \longrightarrow Y$ is said to be *bilinear* if $x_1 \longrightarrow L(x_1, x_2)$ from X_1 into Y and $x_2 \longrightarrow L(x_1, x_2)$ from X_2 into Y are both linear. We say that L is *continuous*, if there exists $M \geq 0$ such that

$$\|L(x)\|_Y \leq M \|x_1\|_X \|x_2\|_X \qquad \text{for all } x = (x_1, x_2) \in X = X_1 \times X_2. \tag{2.63}$$

The infimum of all $M \geq 0$ for which (2.63) is true is the norm of L. Equipped with this norm the space of continuous bilinear maps $L \colon X_1 \times X_2 \longrightarrow Y$ is a Banach space denoted by $\mathcal{L}_2(X_1 \times X_2; Y)$.

REMARK 2.5.22 Note that on $X = X_1 \times X_2$, we consider the norm $\|x\|_X = \max\{\|x_1\|_{X_1}, \|x_2\|_{X_2}\}$ for all $x = (x_1, x_2) \in X = X_1 \times X_2$; then $\|L\|_{\mathcal{L}_2} = \sup\big\{\|L(x)\|_Y : x \in X,\ \|x\|_X \leq 1\big\}$.

PROPOSITION 2.5.23 *If X_1, X_2, and Y are Banach spaces, then $\mathcal{L}_2(X_1 \times X_2; Y)$, $\mathcal{L}\big(X_1, \mathcal{L}(X_2, Y)\big)$, and $\mathcal{L}\big(X_2, \mathcal{L}(X_1, Y)\big)$ are isometrically isomorphic Banach spaces.*

PROOF: Let $L \in \mathcal{L}_2(X_1 \times X_2; Y)$ and $x_1 \in X_1$. We consider the map $x_2 \longrightarrow K(x_1)(x_2) = L(x_1, x_2)$ from X_2 into Y. We have

$$\begin{aligned} &\|K(x_1)(x_2)\|_Y \leq \|L\|_{\mathcal{L}_2} \|x_1\|_{X_1} \|x_2\|_{X_2}, \\ \Rightarrow\ & K(x_1) \in \mathcal{L}(X_2, Y) \quad \text{and} \quad \|K(x_1)\|_{\mathcal{L}} \leq \|L\|_{\mathcal{L}_2} \|x_1\|_X. \end{aligned} \tag{2.64}$$

Also the map $x_1 \longrightarrow K(x_1)$ is linear and (2.64) implies that it is also continuous and satisfies

$$\|K\|_{\mathcal{L}(X_1, \mathcal{L}(X_2, Y))} \leq \|L\|_{\mathcal{L}_2}. \tag{2.65}$$

On the other hand, if $K \in \mathcal{L}\big(X_1, \mathcal{L}(X_2, Y)\big)$, for all $x_1 \in X_1$ we set

$$L(x_1, x_2) = K(x_1)x_2.$$

Evidently $L \colon X_1 \times X_2 \longrightarrow Y$ is a bilinear map and

$$\begin{aligned} &\|L(x_1, x_2)\|_Y \leq \|K(x_1)\|_{\mathcal{L}} \|x_2\|_{X_2} \leq \|K\|_{\mathcal{L}(X_1, \mathcal{L}(X_2, Y))} \|x_1\|_{X_1} \|x_2\|_{X_2}, \\ \Rightarrow\ & L \in \mathcal{L}_2(X_1 \times X_2; Y) \quad \text{and} \quad \|L\|_{\mathcal{L}_2} \leq \|K\|_{\mathcal{L}(X_1, \mathcal{L}(X_2, Y))}. \end{aligned} \tag{2.66}$$

From (2.65) and (2.66) we conclude that $\mathcal{L}_2(X_1 \times X_2; Y)$ and $\mathcal{L}\big(X_1, \mathcal{L}(X_2, Y)\big)$ are isometrically isomorphic and similarly for $\mathcal{L}\big(X_2, \mathcal{L}(X_1, Y)\big)$. □

By virtue of Proposition 2.5.23, if $\varphi \colon U \longrightarrow Y$ is a twice differentiable map and $x_0 \in U$, then

$$\varphi''(x_0) \in \mathcal{L}_2(X \times X; Y).$$

In fact we can say more.

PROPOSITION 2.5.24 *If X, Y are Banach spaces, $U \subseteq X$ is nonempty open, $x_0 \in X$, and $\varphi\colon U \longrightarrow X$ is a map which is twice differentiable at x_0 (see Definition 2.5.19), then $\varphi''(x_0) \in \mathcal{L}_2(X \times X; Y)$ is symmetric; that is, for all $x, u \in X$ we have*

$$\varphi''(x_0)(x,u) = \varphi''(x_0)(u,x).$$

PROOF: Let $r>0$ be such that $B_r(x_0) \subseteq U$ and consider the map $\psi\colon [0,1] \longrightarrow Y$ defined by

$$\psi(t) = \varphi(x_0 + tx + u) - \varphi(x_0 + tx) \qquad \text{for } \|x\|_X, \|u\|_Y \leq r.$$

From the mean value theorem (see Proposition 1.1.6), we have

$$\|\psi(1) - \psi(0) - \psi'(0)\|_X \leq \sup_{t \in [0,1]} \|\psi'(t) - \psi'(0)\|_X. \tag{2.67}$$

Also from the chain rule (see Proposition 1.1.14), we have

$$\begin{aligned}\psi'(t) &= \big(\varphi'(x_0 + tx + u) - \varphi'(x_0 + tx)\big)x \\ &= \big(\varphi'(x_0 + tx + u) - \varphi(x_0) - \varphi''(x_0)tx\big)x \\ &\quad -\big(\varphi'(x_0 + tx) - \varphi(x_0) - \varphi''(x_0)tx\big)x.\end{aligned}$$

Given $\varepsilon>0$, we can find $\delta>0$ such that if $\|x\|_X, \|u\|_X \leq \delta$ and $0 \leq t \leq 1$, then

$$\begin{aligned}&\|\varphi'(x_0 + tx + u) - \varphi'(x_0) - \varphi''(x_0)(tx + u)\|_{\mathcal{L}} \leq \varepsilon(\|x\|_X + \|u\|_X) \\ \text{and} \quad &\|\varphi'(x_0 + tx) - \varphi'(x_0) - \varphi''(x_0)(tx)\|_{\mathcal{L}} \leq \varepsilon\|x\|_X.\end{aligned}$$

These imply

$$\|\psi'(t) - \big(\varphi''(x_0)u\big)x\|_Y \leq 2\varepsilon(\|x\|_X + \|u\|_X) \tag{2.68}$$

$$\begin{aligned}\text{and} \quad \|\psi(1) - \psi(0) - \big(\varphi''(x_0)u\big)x\|_Y &\leq \|\psi(1) - \psi(0) - \varphi'(0)\|_Y \\ &\quad + \|\varphi'(0) - \big(\varphi''(x_0)u\big)x\|_Y \\ &\leq 6\varepsilon(\|x\|_X + \|u\|_X)\|x\|_X\end{aligned} \tag{2.69}$$

(see (2.67) and (2.68)).

But $\psi(1) - \psi(0) = \varphi(x_0 + tx + u) - \varphi(x_0 + x) - \varphi(x_0 + u) + \varphi(x_0)$ is symmetric in $x, u \in X$ and so in (2.68) we may interchange x and u and obtain

$$\|\big(\varphi''(x_0)u\big)x - \big(\varphi''(x_0)x\big)u\|_Y \leq 6\varepsilon(\|x\|_X + \|u\|_X)^2 \qquad \text{for all } \|x\|_X, \|u\|_X \leq \delta. \tag{2.70}$$

If we replace x, u by $\lambda x, \lambda u$ with $\lambda>0$, then both sides of (2.70) are multiplied with λ^2. So (2.70) holds also for $\|x\|_X = \|u\|_X = 1$. Hence

$$\|\varphi''(x_0)(u,x) - \varphi''(x_0)(x,u)\|_Y \leq 24\varepsilon \qquad \text{for all } \|x\|_X = \|u\|_X = 1. \tag{2.71}$$

Because $\varepsilon>0$ was arbitrary, from (2.71) we conclude that

$$\varphi''(x_0)(x,u) = \varphi''(x_0)(u,x) \qquad \text{for all } x, u \in X.$$

□

Now that we have the second-order derivative at our disposal, we can further analyze Euler's equation (2.62). Assume that the extremal u belongs in $C^2(T, X)$.

In order to keep track with respect to which variable we differentiate at each step, we consider the variables (t, x, y) of the Lagrangian and by L'_t (resp., L'_x, L'_y) we denote the partial derivative of L with respect to t (resp., with respect to x, y). Then we have

$$\begin{aligned}\frac{\mathrm{d}}{\mathrm{d}t} L'_3\big(t, u(t), u'(t)\big) &= \frac{\mathrm{d}}{\mathrm{d}t} L'_y\big(t, u(t), u'(t)\big) \\ &= L'_{yt}\big(t, u(t), u'(t)\big) + L'_{yx}\big(t, u(t), u'(t)\big)u'(t) \\ &\quad + L'_{yy}\big(t, u(t), u'(t)\big)u''(t) \text{ in } X^*.\end{aligned}$$

Therefore the Euler equation can be written as follows.

$$\begin{aligned}&L'_{yy}\big(t, u(t), u'(t)\big)u''(t) + L'_{yx}\big(t, u(t), u'(t)\big)u'(t) \\ &\quad + L'_{yt}\big(t, u(t), u'(t)\big) - L'_x\big(t, u(t), u'(t)\big) = 0 \qquad \text{in } X^*. \end{aligned} \tag{2.72}$$

This is a second-order differential equation with the function u unknown. Of course (2.72) is not in canonical form, because it is not solved for u''. In the sequel by imposing a suitable condition on $L'_{yy}\big(t, u(t), u'(t)\big) \in \mathcal{L}(X, X^*)$, we are able to put (2.72) in canonical form.

As this point we focus on a difficulty that we encounter. Equation (2.72) was derived under the assumption that $u \in C^2(T, X)$. On the other hand the Euler equation (2.62) in Theorem 2.5.12 was obtained under the assumption that the extremal u belongs in $C^1(T, \mathbb{R}^N)$. In what follows we show how we can overcome this difficulty.

PROPOSITION 2.5.25 *If $L\colon U \longrightarrow \mathbb{R}$ is a twice differentiable Lagrangian function (i.e., $L \in C^2(U)$), $(t_0, x_0, y_0) \in U$ with $0 < t_0 < b$, and $L''_{yy}(t_0, x_0, y_0)$ is an isomorphism from X into X^*, then*

(a) *There exists a solution u of the Euler equation (2.62) defined on an open interval I_0 containing t_0 such that $u(t_0) = x_0, u'(t_0) = y_0$ and u is C^2.*
(b) *Every other solution $\widehat{u}$ of the Euler equation (2.62) defined on an open interval $\widehat{I}$ containing t_0 and satisfying $\widehat{u}(t_0) = x_0$, $\widehat{u}'(t_0) = y_0$ coincides with u on an open interval containing t_0 and $\widehat{u}$ is C^2 on that interval.*

PROOF: Consider the map $\psi\colon U \longrightarrow \mathbb{R} \times X \times X$ defined by

$$\psi(t, x, y) = \big(t, x, L'_y(t, x, y)\big). \tag{2.73}$$

Because by hypothesis $L'_{yy}(t_0, x_0, y_0)$ is an isomorphism from X onto X^*, we can apply the inverse theorem on the function L'_y(see Theorem 1.1.25). So according to that theorem, we can find a neighborhood V of $(t_0, x_0, y_0) \in \mathbb{R} \times X \times X$ and a neighborhood W of $\big(t_0, x_0, p_0 = L'_y(t_0, x_0, y_0)\big) \in \mathbb{R} \times X \times X^*$, such that the function ψ defined by (2.73) is a C^1-diffeomorphism from V onto W. The inverse function ψ^{-1} (also a C^1-function) is of the form $(t, x, p) \longrightarrow \big(t, x, y = g(t, x, p)\big)$. We consider the following system on $X \times X^*$

$$\left\{\begin{array}{l} x'(t) = g\big(t, x(t), p(t)\big) \\ p'(t) = L'_x\Big(t, x(t), g\big(t, x(t), p(t)\big)\Big) \end{array}\right\}. \tag{2.74}$$

Note that (2.74) is in canonical form (it is solved with respect to the higher-order derivatives x', p' of the unknown functions x, p). The vector field of (2.74),

$$(t,x,p) \longrightarrow \big(g(t,x,p), L'_x(t,x,g(t,x,p))\big),$$

is a C^1-function from W into $X \times X^*$ and the triple (t_0, x_0, p_0) corresponds to the Cauchy data of (2.74). So by the basic existence and uniqueness theorem for initial value problems in Banach spaces, we can find a unique maximal solution $t \longrightarrow \big(x(t), p(t)\big)$ of (2.74) such that $x(t_0) = x_0$ and $p(t_0) = p_0$. Let I be the open interval containing t_0 on which the solution is defined. We know that on this interval the solution is C^2. Let $u(t) = x(t)$. Then $u \in C^2(I, X)$ satisfies the Euler equation (2.62) and the initial conditions $u(t_0) = x_0$, $u'(t_0) = y_0$.

Next let $\widehat{u}\colon \widehat{I} \longrightarrow X$ be another solution defined on an interval $\widehat{I}$ containing t_0 satisfying $\widehat{u}(t_0) = x_0$, $\widehat{u}'(t_0) = y_0$. For t close to t_0, $\big(t, \widehat{u}(t), \widehat{u}'(t)\big) \in V$. So we can consider $\big(\psi(t), \widehat{u}(t), \widehat{u}'(t)\big)$ (see (2.73)). Then the function $t \longrightarrow \big(\widehat{u}(t), \widehat{u}'(t)\big)$ solves (2.74) and satisfies the Cauchy data (t_0, x_0, y_0). By virtue of the uniqueness of the solution, we conclude that for all $t \in \widehat{I} \cap I$ we have $\widehat{u}(t) = u(t)$. $\square$

REMARK 2.5.26 This proposition shows that in order to hope to be able to transform (2.72) to a canonical first-order system we need to assume that the Lagrangian function L is C^2.

Before proceeding to the Hamiltonian formulation of the necessary conditions for extremality, let us illustrate the use of Euler's equation through some characteristic examples.

EXAMPLE 2.5.27 (a) Let $L(y) = \sqrt{1+y^2}$ and consider the problem $\inf\big[I_L(u)\colon u \in C^1([0,1]),\ u(0) = x_0,\ u(1) = x_1\big]$. Then the Euler equation is $(\mathrm{d}/\mathrm{d}t)(\partial/\partial y)\sqrt{1+u'(t)^2} = 0$, $u(0) = x_0$, $u(1) = x_1$, which gives $u(t) = (x_1 - x_0)t + x_0$ for all $t \in T$. This is a global minimum of $I_L(\cdot)$ over $C^1_{x_0,x_1}([0,1])$ and simply says that the curve with the minimum length joining x_0 and x_1 is the straight line connecting the two points.

(b) Let $L(y)=y^3$ and consider the problem $\inf[I_L(u) : u \in C^1([0,1]), u(0)=0,\ u(1)=1]$. Then the Euler equation is $\big(u'(t)^2\big)' = 0$, $u(0) = 0$, $u(1) = 1$. The unique solution of this boundary value problem is $u(t) = t$ for $t \in [0,1]$. This produces a local minimum for I_L on $C^1_{0,1}([0,1])$. Let $v \in C^1_0([0,1])$. Then $u+v$ is admissible and we have

$$\begin{aligned} I_L(u+v) &= \int_0^1 \big[\big(t+v(t)\big)'\big]^3 \mathrm{d}t \\ &= I_L(u) + 3\int_0^1 v'(t)\mathrm{d}t + \int_0^1 \big(3v'(t)^2 + v'(t)^3\big)\mathrm{d}t \\ &= I_L(u) + \int_0^1 \big(3v'(t)^2 + v'(t)^3\big)\mathrm{d}t \qquad \text{(because } v(0) = v(1) = 0\text{)}. \end{aligned}$$

Evidently if $3v'(t)^2 + v'(t)^3 \geq 0$ for all $t \in [0,1]$, then $I_L(u+v) \geq I_L(u)$. In particular if $\|v\|_{C^1([0,1])} \leq 3$, then $3v'(t)^2 + v'(t)^3 \geq 0$ for all $t \in [0,1]$ and so $I_L(u+v) \geq I_L(u)$ which proves that the extremal u produces a local minimum of I_L on $C^1_{0,1}([0,1])$.

(c) If L is not C^1, then the extremal function need not be C^1. Indeed let $L(t,y) = t^{2/3}y^2$ and consider the problem $\inf[I_L(u) : u \in C^1([0,1]),\ u(0) = 0,\ u(1) = 1]$. In this case the Euler equation has the form

$$\big(2t^{2/3}u'(t)\big)' = 0, \quad u(0) = 0, \quad u(1) = 1.$$

Solving this boundary value problem, we obtain $u(t) = t^{2/3}$ which does not belong in $C^1([0,1])$, but it is a global minimum of I_L over $C^1_{0,1}([0,1])$. This example is due to Hilbert and shows that a variational problem does not always have a solution in the given class of curves under consideration.

(d) In this example the Euler equation has no solutions and moreover, there is no solution even in the larger space of absolutely continuous functions. So let $L(t,y) = (ty)^2$ and consider the problem $\inf[I_L(u) : u \in C^1([0,1]),\ u(0) = 0,\ u(1) = 1]$. In this case the Euler equation has the form

$$\big(2t^2u'(t)\big)' = 0, \quad u(0) = 0, \quad u(1) = 1.$$

The general solution of the differential equation is $u(t) = (c/t) + d$. First we observe that u does not belong in $C^1([0,1])$ and none of these functions satisfies the boundary conditions $u(0) = 0,\ u(1) = 1$. In fact the variational problem has no solution in the space of absolutely continuous functions with the given boundary values. Indeed, if $v(t)$ is such a function, then $I_L(v) > 0$ and the value of the problem is zero. To see this consider

$$v_n(t) = \begin{cases} nt & \text{if } t \in [0, \frac{1}{n}] \\ 1 & \text{if } t \in [\frac{1}{n}, 1] \end{cases}.$$

Then we have $I_L(x_n) \longrightarrow 0$. This example is due to Weierstrass and was produced as an argument against Riemann's justification of the Dirichlet principle.

Now we introduce the Legendre transform, which is the "forefather" of the Legendre–Fenchel transform, which was introduced in Definition 1.2.15 and which is the basic tool in the duality theory of convex functions.

DEFINITION 2.5.28 Let X be a Banach space, $U \subseteq \mathbb{R} \times X \times X$ a nonempty open set, and $L\colon U \longrightarrow \mathbb{R}$ a C^2 Lagrangian function.

(a) The *Legendre transform* of L is the function $\mathcal{L}\colon U \longrightarrow \mathbb{R} \times X \times X^*$ defined by

$$\mathcal{L}(t,x,y) = \big(t, x, p = L'_y(t,x,y)\big).$$

(b) We say that the Lagrangian is *regular* if the corresponding Legendre transform $\mathcal{L}$ is a local diffeomorphism. We say that the Lagrangian is *hyperregular* if the corresponding Legendre transform $\mathcal{L}$ is a diffeomorphism of U on an open set $\mathcal{L}(U) \subseteq \mathbb{R} \times X \times X^*$.

REMARK 2.5.29 From the inverse theorem (see Theorem 1.1.25), the Lagrangian L is regular if and only if the Fréchet derivative of $\mathcal{L}$ at all $(t,x,y) \in U$ is an isomorphism from $T \times X \times X$ onto $T \times X \times X^*$. In what follows let D denote the Fréchet derivative with respect to all three variables $(t,x,y) \in U$ of the function involved. We have

$$D\mathcal{L}(t,x,y)(\lambda,v,w) = \big(\lambda, v, DL'_y(t,x,y)(\lambda,v,w)\big).$$

Note that

$$DL'_y(t,x,y)(\lambda,v,w) = \lambda L''_{yt}(t,x,y) + L''_{yx}(t,x,y)v + L''_{yy}(t,x,y)w \quad \text{in } X^*.$$

Then $D\mathcal{L}(t,x,y) \in \mathcal{L}(\mathbb{R}\times X\times X, \mathbb{R}\times X\times X^*)$ is an isomorphism if and only if $L''_{yy}(t,x,y) \in \mathcal{L}(X,X^*)$ is an isomorphism. In Proposition 2.5.25 we have assumed that $L''_{yy}(t_0,x_0,y_0) \in \mathcal{L}(X,X^*)$ is an isomorphism. Recall that in $\mathcal{L}(X,X^*)$ with the operator norm topology, the set of isomorphisms is open. So for all (t,x,y) near (t_0,x_0,y_0) in $\mathbb{R}\times X\times X$, we have that $L''_{yy}(t,x,y) \in \mathcal{L}(X,X^*)$ is an isomorphism. Therefore we see that the hypothesis in Proposition 2.5.25 implies that the Lagrangian L is regular in a neighborhood of the point (t_0,x_0,y_0).

THEOREM 2.5.30 *If $L\colon U \longrightarrow \mathbb{R}$ is a C^2 Lagrangian function that is hyperregular, $\mathcal{L}$ is the Legendre transform of L with $\mathcal{L}^{-1}$ its inverse $\big(\mathcal{L}^{-1}(t,x,y) = \big(t,x,y = g(t,x,p)\big)\big)$, and $H\colon \mathcal{L}(U) \longrightarrow \mathbb{R}\times X\times X^*$ is defined by*

$$H(t,x,p) = \langle p, g(t,x,p)\rangle_X - (L\circ\mathcal{L}^{-1})(t,x,p),$$

then the Euler equation

$$\frac{\mathrm{d}}{\mathrm{d}t} L'_y\big(t,u(t),u'(t)\big) = L'_x\big(t,u(t),u'(t)\big) \qquad \text{for all } t\in T=[0,b] \tag{2.75}$$

is equivalent to the following first-order canonical system

$$\left.\begin{cases} x'(t) = H'_p\big(t,x(t),p(t)\big) \\ p'(t) = -H'_x\big(t,x(t),p(t)\big) \qquad \text{for all } t\in T=[0,b] \end{cases}\right\}. \tag{2.76}$$

The function H is the *Hamiltonian* associated with the Lagrangian L and system (2.76) is called a *Hamiltonian system*. The equivalence of (2.75) and (2.76) is in the following sense: if $t \longrightarrow u(t)$ is a solution of the Euler equation (2.75), then $t \longrightarrow \big(u(t),\ L'_y\big(t,u(t),u'(t)\big)\big)$ is a solution of the Hamiltonian system (2.76). Conversely, if $t \longrightarrow \big(x(t),p(t)\big)$ is a solution of the Hamiltonian system (2.76), then $t \longrightarrow u(t) = x(t)$ is a solution of the Euler equation (2.75).

PROOF: In the proof of Proposition 2.5.25 we proved the local equivalence of (2.75) with the canonical system (2.74). In fact due to the hyperregularity of the Legendre transform $\mathcal{L}$, this equivalence is actually global. So it suffices to show the equivalence of (2.74) and (2.76).

By definition we have

$$H(t,x,p) = \langle p, g(t,x,p)\rangle_X - L\big(t,x,g(t,x,p)\big).$$

We have $H'_p(t,x,p) \in X^{**}$. So for all $x^*\in X^*$

$$\begin{aligned}\big\langle H'_p(t,x,p),x^*\big\rangle_{X^*} &= \langle x^*, g(t,x,p)\rangle_X + \big\langle p, g'_p(t,x,p)x^*\big\rangle_X \\ &\quad - \big\langle L'_y\big(t,x,g(t,x,p)\big), g'_p(t,x,p)x^*\big\rangle_X .\end{aligned} \tag{2.77}$$

But $p = L'_y\big(t,x,g(t,x,p)\big)$. Hence from (2.77) we infer that

$$\begin{aligned}&\big\langle H'_p(t,x,p),x^*\big\rangle_{X^*} = \langle x^*, g(t,x,p)\rangle_X \qquad \text{for all } x^*\in X^* \\ \Rightarrow\ & H'_p(t,x,p) = g(t,x,p) \in X.\end{aligned} \tag{2.78}$$

Then we see that the first equations in (2.74) and (2.76) are the same. Also if $v \in X$, then

$$\begin{aligned}\left\langle H'_x(t,x,p), v\right\rangle_X &= \left\langle p, h'_x(t,x,p)v\right\rangle_X - \left\langle L'_x\big(t,x,g(t,x,p)\big), v\right\rangle_X \\ &\quad - \left\langle L'_y\big(t,x,g(t,x,p)\big), h'_x(t,x,p)v\right\rangle_X. \end{aligned} \tag{2.79}$$

Again because $p = L'_x\big(t,x,g(t,x,p)\big)$, from (2.79) we obtain

$$\begin{aligned}&\left\langle H'_x(t,x,p), v\right\rangle_X = -\left\langle L'_x\big(t,x,g(t,x,p)\big), v\right\rangle_X \qquad \text{for all } v \in X, \\ \Rightarrow\ & H'_x(t,x,p) = L'_x\big(t,x,g(t,x,p)\big).\end{aligned}$$

□

EXAMPLE 2.5.31 *The Poincaré half-plane*: Let $P = \{(x,y) \in \mathbb{R}^2 : y > 0\}$ and $L\colon P \times \mathbb{R}^2 \longrightarrow \mathbb{R}_+$ is defined by

$$L(x,y,u,v) = \frac{1}{2}\frac{u^2+v^2}{y^2}.$$

The integral functional I_L is defined by

$$I_L(u) = \frac{1}{2}\int_0^b \frac{x'(t)^2 + y'(t)^2}{y(t)^2}\,dt,$$

where $u : T = [0,b] \longrightarrow P$ and $u(t) = \big(x(t), y(t)\big)$ for all $t \in T$. Then

$$\mathcal{L}(x,y,u,v) = (x,y,p,q)$$

with $p = u/y^2$, $q = v/y^2$. This is a diffeomorphism (i.e., $\mathcal{L}$ is hyperregular). Also

$$H(x,y,p,q) = \frac{1}{2}y^2(p^2+q^2).$$

Then the Hamiltonian system (2.76) takes the following form:

$$\begin{cases} x'(t) = y(t)^2\,p(t) \\ y'(t) = y(t)^2\,q(t) \end{cases} \qquad \begin{cases} p'(t) = 0 \\ q'(t) = -y(t)\big(p(t)^2 + q(t)^2\big) \end{cases}.$$

This system is a model for the hyperbolic geometry. For this reason its study is important in the understanding of non-Euclidean geometries.

2.6 Optimal Control

In this section we study some basic aspects of optimal control theory. Specifically we focus our attention on the existence theory, the relaxation methods, and the derivation of the maximum principle (a necessary condition for optimality of an admissible state-control pair). To avoid technical complications that require the development of a substantial mathematical background, we limit ourselves to finite dimensional systems (i.e., systems driven by ordinary differential equations, also known as lumped parameter systems).

So the mathematical setting is the following. The state space is $\mathbb{R}^N$ and the control space is a Polish space Y. Recall that a Polish space is a Hausdorff topological space which is separable and can be metrized by means of a complete metric. The time horizon is $T=[0,b]$. Let $\Gamma(T,Y)=\{u\colon T\longrightarrow Y : u \text{ is measurable}\}$. Any element of $\Gamma(T,Y)$ is called a control. Also $f\colon T\times\mathbb{R}^N\times Y\longrightarrow\mathbb{R}^N$ is a vector field and the controlled dynamical system we are considering is the following

$$x'(t)=f\big(t,x(t),u(t)\big) \qquad \text{a.e. on } T,\ x(0)=x_0\in\mathbb{R}^N. \tag{2.80}$$

Any solution $x(\cdot)$ of (2.80) is referred to as a state trajectory of the control system corresponding to the initial state $x_0\in\mathbb{R}^N$ and the control $u(\cdot)$. Note that we do not assume the uniqueness and/or the existence of solutions for (2.80). So for any control $u(\cdot)$ and any initial state x_0, we may have more than one or no response for the system. Moreover, in general we have some constraints on the control described by a multifunction (set-valued function) $U\colon T\longrightarrow 2^Y\setminus\{\emptyset\}$. For these reasons, the following definition is necessary.

DEFINITION 2.6.1 A pair $\big(x(\cdot),u(\cdot)\big)\in W^{1,1}\big((0,b),\mathbb{R}^N\big)\times\Gamma(T,Y)$ is said to be *admissible* (or *feasible*) state-control pair, if (2.80) is satisfied and $u(t)\in U(t)$ a.e. on T. In what follows by $\alpha_{\mathbf{ad}}$ we denote the set of all admissible state-control pairs.

We are also given an integral cost functional

$$J(x,u)=\int_0^b L\big(t,x(t),u(t)\big)dt \qquad \text{for all } (x,u)\in\alpha_{\mathbf{ad}}.$$

Then our optimal control problem can be stated as follows:

(P) Find $(\widehat{x},\widehat{u})\in\alpha_{\mathbf{ad}}$ such that

$$J(\widehat{x},\widehat{u})=\inf\big[J(x,u):(x,u)\in\alpha_{\mathbf{ad}}\big]=m. \tag{2.81}$$

DEFINITION 2.6.2 If we can find $(\widehat{x},\widehat{u})\in\alpha_{\mathbf{ad}}$ such that (2.81) is satisfied, then we say that $(\widehat{x},\widehat{u})$ is an *optimal admissible pair*. The function $\widehat{x}\in W^{1,1}\big((0,b),\mathbb{R}^N\big)$ is said to be an *optimal trajectory* and the function $u\in\Gamma(T,Y)$ is said to be an *optimal control*.

The hypotheses on the data of the optimal control problem (P), are the following.

H(f): $f\colon T\times\mathbb{R}^N\times Y\longrightarrow\mathbb{R}^N$ is a function, such that

(i) For all $(x,u)\in\mathbb{R}^N\times Y$, $t\longrightarrow f(t,x,u)$ is measurable.
(ii) For almost all $t\in T$, $(x,u)\longrightarrow f(t,x,u)$ is continuous.
(iii) For almost all $t\in T$, all $x\in\mathbb{R}^N$ and all $u\in U(t)$, we have

$$\|f(t,x,u)\|\leq\alpha(t)+c(t)\|x\| \qquad \text{with } \alpha,c\in L^1(T)_+.$$

H(U): $U\colon T\longrightarrow 2^Y\setminus\{\emptyset\}$ is a multifunction with compact values such that

$$\operatorname{Gr}U=\{(t,u)\in T\times Y : u\in U(t)\}\in\mathbf{B}(T)\times\mathbf{B}(Y),$$

with $\mathbf{B}(T)$ (resp., $\mathbf{B}(Y)$) being the Borel σ-field of T (resp., of Y).

REMARK 2.6.3 From measure theory we know that $\mathbf{B}(T)\times\mathbf{B}(Y)=\mathbf{B}(T\times Y)$(= the Borel σ-field of $T\times Y$).

H(L): $L\colon T\times\mathbb{R}^N\times Y\longrightarrow\overline{\mathbb{R}}=\mathbb{R}\cup\{+\infty\}$ is an integrand such that

(i) $(t,x,u)\longrightarrow L(t,x,u)$ is measurable.
(ii) For almost all $t\in T$, $(x,u)\longrightarrow L(t,x,u)$ is lower semicontinuous proper.
(iii) For almost all $t\in T$, all $x\in\mathbb{R}^N$, and all $u\in U(t)$

$$\alpha_0(t)-c_0\|x\|\le L(t,x,u)\qquad\text{with } \alpha_0\in L^1(T),\ c_0>0.$$

In the existence theory the following convexity-type hypothesis plays a central role.
$\mathbf{H_c}$: For all $(t,x)\in T\times\mathbb{R}^N$ the set $Q(t,x)=\{(v,\lambda)\in\mathbb{R}^N\times\mathbb{R} : v=f(t,x,u),\,u\in U(t),\ L(t,x,u)\le\lambda\}$ is convex.

REMARK 2.6.4 Hypotheses H(f), H(u), and H(L) imply that for all $(t,x)\in T\times\mathbb{R}^N$, $Q(t,x)$ is also closed. Hypothesis $\mathrm{H_c}$ means that the optimal control problem has enough "convex structure". Note that if the control variable enters linearly in the dynamics of the system (i.e., $f(t,x,u) = f_1(t,x) + f_2(t,x)u$), $Y=\mathbb{R}^m$, the control constraint multifunction has also convex values and the cost integrand $L(t,x,u)$ is in addition convex in $u\in\mathbb{R}^m$, then the hypothesis $\mathrm{H_c}$ is satisfied. Note that this hypothesis implies that for almost all $t\in T$ and all $x\in\mathbb{R}^N$, $F(t,x)=f\big(t,x,U(t)\big)$ is convex. In Chapter 6, we produce equivalent conditions for $\mathrm{H_c}$ to hold.

In what follows let $\alpha^1_{\mathbf{ad}}\subseteq W^{1,1}\big((0,b),\mathbb{R}^N\big)$ be the set of admissible states (trajectories). We start by establishing the nonemptiness and compactness in $C(T,\mathbb{R}^N)$ of $\alpha^1_{\mathbf{ad}}$. Recall that the Sobolev space $W^{1,1}\big((0,b),\mathbb{R}^N\big)$ is embedded continuously (but not compactly) in $C(T,\mathbb{R}^N)$. To determine the properties of $\alpha^1_{\mathbf{ad}}$, we study the following differential inclusion.

$$\left\{\begin{array}{l} x'(t)\in F\big(t,x(t)\big)\ \text{ a.e. on } T=[0,b],\\ x(0)=x_0\end{array}\right\}. \tag{2.82}$$

We impose the following conditions on the multifunction $F(t,x)$:

H(F): $F\colon T\times\mathbb{R}^N\longrightarrow 2^{\mathbb{R}^N}\setminus\{\emptyset\}$ is a multifunction with compact and convex values such that

(i) For all $x\in\mathbb{R}^N$, $\mathrm{Gr}F(\cdot,x)=\{(t,v)\in T\times\mathbb{R}^N : v\in F(t,x)\}\in\mathbf{B}(T)\times\mathbf{B}(\mathbb{R}^N)$.
(ii) For almost all $t\in T$, $\mathrm{Gr}F(t,\cdot)=\{(x,v)\in\mathbb{R}^N\times\mathbb{R}^N : v\in F(t,x)\}$ is closed.
(iii) For almost all $t\in T$, all $x\in\mathbb{R}^N$, and all $v\in F(t,x)$, we have

$$\|v\|\le\alpha(t)+c(t)\|x\|\qquad\text{with } \alpha,c\in L^1(T)_+.$$

REMARK 2.6.5 If $F(t,x)=f\big(t,x,U(t)\big)$, then by virtue of hypotheses H(f), H(U), H(L), and $\mathrm{H_c}$, the multifunction F satisfies hypotheses H(F).

DEFINITION 2.6.6 By a solution of (2.82), we mean a function $x\in W^{1,1}\big((0,b),\mathbb{R}^N\big)$ such that $x(0)=x_0$ and $x'(t)\in F\big(t,x(t)\big)$ for almost all $t\in T$. We denote the set of solutions of (2.82) by $S_F(x_0)$. Then $S_F(x_0)\subseteq W^{1,1}\big((0,b),\mathbb{R}^N\big)\subseteq C(T,\mathbb{R}^N)$.

To be able to solve (2.82) and eventually study the optimal control problem, we need some results from multivalued analysis, which we state here for easy reference and postpone their proof until Chapter 6, where we conduct a more systematic and detailed study of multifunctions.

The first result is known as the *Kakutani–Ky Fan fixed point theorem.*

THEOREM 2.6.7 *If X is a locally convex space, $C \subseteq X$ is nonempty, compact and convex, and $G\colon C \longrightarrow 2^C$ is a multifunction with nonempty, closed and convex values that has a closed graph (i.e., $Gr\ G = \{(x,y) \in C \times C : y \in G(x)\}$ is closed in $X \times X$), then there exists $x \in X$ such that $x \in G(x)$.*

The second result is known as the *Yankov–von Neumann–Aumann selection theorem*. It is stated in a less general form, which however suffices for our needs here. The general form of the result can be found in Theorem 6.3.20.

THEOREM 2.6.8 *If (Ω, Σ, μ) is a σ-finite measure space, X is a Polish space, and $G\colon \Omega \longrightarrow 2^X$ is a multifunction such that $Gr\ G = \{(\omega, x) \in \Omega \times X : y \in G(\omega)\} \in \Sigma \times \mathbf{B}(X)$, then there exists a Σ-measurable function $g\colon \Omega \longrightarrow X$ such that $g(\omega) \in G(\omega)$ μ-a.e. on Ω.*

It is well known that given a Borel set in $\mathbb{R}^2$, its projection on a coordinate axis need not be Borel. The next theorem provides conditions which guarantee that the projection is indeed Borel.

THEOREM 2.6.9 *If S, X are Polish spaces, $G \in \mathbf{B}(S \times X) = \mathbf{B}(S) \times \mathbf{B}(X)$, and for every $s \in S$ the section $G(s) = \{x \in X : (s,x) \in G\}$ is σ-compact, then* $\mathrm{proj}_S G \in \mathbf{B}(S)$.

Now we are ready to determine the properties of the solution set $S(x_0)$ of problem (2.82).

THEOREM 2.6.10 *If hypotheses $H(F)$ hold, then $S_F(x_0) \neq \emptyset$ and $S_F(x_0)$ is compact in $C(T, \mathbb{R}^N)$.*

PROOF: We start by establishing an a priori bound for the elements of $S_F(x_0)$. So suppose that $x \in S_F(x_0)$. Then for some $v \in L^1(T, \mathbb{R}^N)$ that satisfies $v(t) \in F\big(t, x(t)\big)$ a.e. on T, we have

$$\|x(t)\| \leq \|x_0\| + \int_0^t \|v(s)\| ds \qquad \text{for all } t \in T,$$

$$\Rightarrow \|x(t)\| \leq \|x_0\| + \int_0^t \big(\alpha(s) + c(s)\|x(s)\|\big) \qquad \text{for all } t \in T.$$

Invoking Gronwall's inequality, we can find $M_1 > 0$ such that

$$\|x(t)\| \leq M_1 \qquad \text{for all } t \in T \text{ and all } x \in S(x_0).$$

Let $p_{M_1} : \mathbb{R}^N \longrightarrow \mathbb{R}^N$ be the M_1-radial contraction; that is,

$$p_{M_1}(x) = \begin{cases} \frac{M_1 x}{\|x\|} & \text{if } \|x\| > M_1 \\ x & \text{if } \|x\| \le M_1 \end{cases}.$$

Clearly $p_{M_1}(\cdot)$ is Lipchitz continuous. Let $F_1(t,x) = F(t, p_{M_1}(x))$. Then F_1 still satisfies H(F)(i) and (ii) (with F replaced by F_1) and in addition for almost all $t \in T$, all $x \in \mathbb{R}^N$, and all $v \in F_1(t,x)$, we have

$$\|v\| \le \eta(t) \qquad \text{a.e. on } T, \text{ with } \eta(\cdot) = \alpha(\cdot) + c(\cdot)M \in L^1(T)_+.$$

Set $C = \{g \in L^1(T, \mathbb{R}^N) : \|g(t)\| \le \eta(t) \text{ a.e. on } T\}$. By virtue of the Dunford–Pettis theorem, C furnished with the relative weak topology is a compact convex set. Let $\xi\colon C \longrightarrow C(T, \mathbb{R}^N)$ be defined by

$$\xi(g)(t) = x_0 + \int_0^t g(s)\mathrm{d}s.$$

Via the Arzela–Ascoli theorem, we check that ξ is sequentially continuous from C with the relative weak topology into $C(T, \mathbb{R}^N)$ with the norm topology. Then let $G_1\colon C \longrightarrow 2^C$ be defined by

$$G_1(g) = \{h \in L^1(T, \mathbb{R}^N) : h(t) \in F_1(t, \xi(g)(t)) \qquad \text{a.e. on } T\}.$$

Let $\xi(g) \in C(T, \mathbb{R}^N)$ and let $\{s_n\}_{n \ge 1}$ be step functions from T into $\mathbb{R}^N$ such that $s_n(t) \longrightarrow x(t)$ a.e. on T as $n \to \infty$. Then because for all $y \in \mathbb{R}^N$, $t \longrightarrow F_1(t,y)$ has a measurable graph, applying Theorem 2.6.8 for every $n \ge 1$ we can find $h_n\colon T \longrightarrow \mathbb{R}^N$ a measurable map such that $h_n(t) \in F_1(t, s_n(t))$ a.e on T. Evidently $\|h_n(t)\| \le \eta(t)$ a.e on T and so by the Dunford–Pettis theorem and by passing to a suitable subsequence if necessary, we may assume that $h_n \xrightarrow{w} h$ in $L^1(T, \mathbb{R}^N)$. Then using Mazur's lemma and the fact that for almost all $t \in T$, $y \longrightarrow F_1(t,y)$ has closed graph, we have $h(t) \in F_1(t, x(t))$ a.e. on T (i.e., $h \in G_1(g)$). So G_1 has nonempty values, which clearly are closed and convex in C. Moreover, the above argument reveals that Gr $G \subseteq C \times C$ is sequentially closed in $C \times C$, when C is equipped with the relative weak topology. Recall that the relative weak topology on C is metrizable. So we can apply Theorem 2.6.7 and obtain $g \in C$ such that $g \in G_1(g)$. Then if $x = \xi(g)$, we have

$$\begin{aligned} & x'(t) \in F_1(t, x(t)) \qquad \text{a.e. on } T,\ x(0) = x_0, \\ \Rightarrow\ & \|x(t)\| \le \|x_0\| + \int_0^t \|v(s)\|\mathrm{d}s \qquad \text{for all } t \in T, \text{ with } v \in L^1(T, \mathbb{R}^N), \\ & v(s) \in F_1(s, x(s)) \qquad \text{a.e. on } T. \end{aligned} \tag{2.83}$$

But from the definition of F_1, we see that $\|v(s)\| \le \alpha(s) + c(s)\|x\|$ a.e. on T and so from (2.83) and Gronwall's inequality as before it follows that

$$\begin{aligned} & \|x(t)\| \le M_1 \qquad \text{for all } t \in T, \\ \Rightarrow\ & F_1(t, x(t)) = F(t, x(t)) \qquad \text{for all } t \in T \quad (\text{i.e., } x \in S_F(x_0)). \end{aligned}$$

Therefore we have proved the nonemptiness of $S_F(x_0)$.

Clearly $S(x_0) \subseteq C(T, \mathbb{R}^N)$ is closed, whereas from the Arzela–Ascoli theorem we have that $S(x_0)$ is relatively compact in $C(T, \mathbb{R}^N)$. Then $S_F(x_0)$ is compact in $C(T, \mathbb{R}^N)$. ◻

Now if $F(t,x) = f\big(t,x,U(t)\big)$ and $x \in S(x_0)$, then we set

$$V(t) = \{u \in U(t) : x'(t) = f\big(t,x(t),u\big)\},$$
$$\Rightarrow \text{Gr } V = \{(t,u) : x'(t) = f\big(t,x(t),u\big)\} \cap \text{Gr } U.$$

Because of hypotheses H(f)(i),(ii), $(t,u) \longmapsto x'(t) - f\big(t,x(t),u\big)$ is a Carathéodory function (i.e., for all $u \in Y$, $t \longmapsto x'(t) - f\big(t,x(t),u\big)$ is measurable, whereas for almost all $t \in T$, $u \longmapsto x'(t) - f\big(t,x(t),u\big)$ is continuous). Hence $(t,u) \longrightarrow x'(t) - f\big(t,x(t),u\big)$ is measurable and combining this with hypothesis H(U) we see that

$$\text{Gr } V \in \mathcal{L}(T) \times \mathbf{B}(Y)$$

with $\mathcal{L}(T)$ being the Lebesgue σ-field of T and $\mathbf{B}(Y)$ the Borel σ-field of Y. So we can apply Theorem 2.6.8 and obtain $u : T \longrightarrow Y$ a measurable map such that $u(t) \in V(t)$ a.e. on T. Then $(x,u) \in \alpha_{\mathbf{ad}}$ and so Theorem 2.6.10 implies that $\alpha^{\mathbf{1}}_{\mathbf{ad}} = S_F(x_0) \subseteq C(T,\mathbb{R}^N)$ is compact.

Next we solve the optimal control problem (P). The approach is the following. We use hypothesis H_c to transform (P) into a control-free variational problem with convex structure (calculus of variations problem) with the same values as (P). We solve the calculus of variations problem using the *direct method*. This method is known as the *reduction method*.

THEOREM 2.6.11 *If hypotheses H(f), H(U), H(L), and H_c hold and $m < +\infty$, then problem (P) admits an optimal state-control pair $(\widehat{x},\widehat{u}) \in \alpha^{\mathbf{1}}_{\mathbf{ad}}$.*

PROOF: We start implementing the reduction method described above. To this end let

$$A(t,x,v) = \{u \in U(t) : v = f(t,x,u)\}.$$

This is the set of all admissible controls which at time $t \in T$ and when the system is at state $x \in \mathbb{R}^N$, produce the "velocity" $v \in \mathbb{R}^N$. Evidently, by modifying f, U, and L on a Lebesgue-null set, we have

$$\begin{aligned}\text{Gr } A = \big\{(t,x,v,u) \in T \times \mathbb{R}^N \times \mathbb{R}^N \times Y : (t,u) \in \text{Gr } U,\ v - f(t,x,u) = 0\big\} \\ \in \mathbf{B}(T) \times \mathbf{B}(\mathbb{R}^N) \times \mathbf{B}(\mathbb{R}^N) \times \mathbf{B}(Y).\end{aligned}$$

Let $p \colon T \times \mathbb{R}^N \times \mathbb{R}^N \longrightarrow \overline{\mathbb{R}} = \mathbb{R} \cup \{+\infty\}$ be defined by

$$p(t,x,v) = \inf[L(t,x,u) : u \in A(t,x,v)] \tag{2.84}$$

which represents the minimum cost to generate "velocity" v, at time $t \in T$, when the state of the system is $x \in \mathbb{R}^N$ and we use admissible controls. As always $\inf \emptyset = +\infty$.

Claim 1: $(t,x,v) \longrightarrow p(t,x,v)$ is Borel-measurable.

For every $\lambda \in \mathbb{R}$ we have

$$\begin{aligned}&\{(t,x,v) \in T \times \mathbb{R}^N \times \mathbb{R}^N : p(t,x,v) \le \lambda\} \\ &= \text{proj}_{T \times \mathbb{R}^N \times \mathbb{R}^N} \big\{(t,x,v,u) \in T \times \mathbb{R}^N \times \mathbb{R}^N \times Y : L(t,x,u) \le \lambda, \\ &\quad u \in A(t,x,v)\big\} \in \mathbf{B}(T) \times \mathbf{B}(\mathbb{R}^N) \times \mathbf{B}(\mathbb{R}^N) \ \text{(see Theorem 2.6.9).}\end{aligned}$$

Claim 2: For every $t \in T$, the function $(x,v) \longrightarrow p(t,x,v)$ is lower semicontinuous.

We need to show that for every $\lambda \in \mathbb{R}$, the set $S(\lambda) = \{(x,v) \in \mathbb{R}^N \times \mathbb{R}^N : p(t,x,v) \leq \lambda\}$ is closed. So let $\{(x_n,v_n)\}_{n\geq 1} \subseteq S(\lambda)$ and assume that $x_n \longrightarrow x$, $v_n \longrightarrow v$ in $\mathbb{R}^N$ as $n \to \infty$. From (2.84) we know that we can find $u_n \in A(t,x_n,v_n)$ such that

$$p(t,x_n,v_n) = L(t,x_n,u_n).$$

We have $u_n \in U(t)$ and $v_n = f(t,x_n,u_n)$. By passing to a suitable subsequence if necessary, we may assume that $u_n \longrightarrow u \in U(t)$. Then $v = f(t,x,u)$ (see hypothesis H(f)(ii)) and so $u \in A(t,x,v)$. Therefore

$$p(t,x,v) \leq L(t,x,u) \leq \liminf_{n\to\infty} L(t,x_n,u_n) = \liminf_{n\to\infty} p(t,x_n,v_n) \leq \lambda$$

(see hypothesis H(L)(ii)),

$$\Rightarrow \ (x,v) \in S(\lambda), \quad \text{(i.e., } p(t,\cdot,\cdot) \text{ is lower semicontinuous).}$$

Claim 3: For all $(t,x) \in T \times \mathbb{R}^N$, $v \longrightarrow p(t,x,v)$ is convex.

Note that

$$\begin{aligned} \operatorname{epi}(t,x,\cdot) &= \{(v,\lambda) \in \mathbb{R}^N \times \mathbb{R} : p(t,x,v) \leq \lambda\} \\ &= \bigcap_{\varepsilon>0} \{(v,\lambda) \in \mathbb{R}^N \times \mathbb{R} : L(t,x,u) \leq \lambda + \varepsilon, \\ &\qquad\qquad u \in U(t),\ v = f(t,x,u)\}. \end{aligned}$$

So by virtue of hypothesis H$_c$, $\operatorname{epi}(t,x,\cdot)$ is convex; that is, $v \longrightarrow p(t,x,v)$ is convex as claimed.

Now let $\{(x_n,u_n)\}_{n\geq 1} \subseteq \alpha_{\mathbf{ad}}$ be a minimizing sequence for problem (P); that is, $J(x_n,u_n) \downarrow m$ as $n \to \infty$. Because of Theorem 2.6.10, we may assume that $x_n \longrightarrow \widehat{x}$ in $C(T,\mathbb{R}^N)$. Also from hypothesis H(f)(iii) we see that $\{x_n'(\cdot)\}_{n\geq 1} \subseteq L^1(T,\mathbb{R}^N)$ is uniformly integrable and so by the Dunford–Pettis theorem, we may assume that $x_n' \xrightarrow{w} v$ in $L^1(T,\mathbb{R}^N)$. Evidently $v = \widehat{x}'$. Using Claims 1, 2, and 3 and Theorem 2.1.28, we obtain

$$\begin{aligned} \int_0^b p\big(t,\widehat{x}(t),\widehat{x}'(t)\big)dt &\leq \liminf_{n\to\infty} \int_0^b p\big(t,x_n(t),x_n'(t)\big)dt \\ &\leq \lim_{n\to\infty} J(x_n,u_n) = m < +\infty. \end{aligned}$$

So due to hypothesis H(L)(iii) and by redefining $t \longrightarrow p\big(t,\widehat{x}(t),\widehat{x}'(t)\big)$ on a Lebesgue-null set, we can say that for all $t \in T$, $p\big(t,\widehat{x}(t),\widehat{x}'(t)\big)$ is finite. A straightforward application of Theorem 2.6.8 produces a measurable function $\widehat{u}\colon T \longrightarrow Y$ such that $\widehat{u}(t) \in A\big(t,\widehat{x}(t),\widehat{x}'(t)\big)$ for almost all $t \in T$ and

$$\begin{aligned} & p\big(t,\widehat{x}(t),\widehat{x}'(t)\big) = L\big(t,\widehat{x}(t),\widehat{u}'(t)\big) \text{ a.e. on } T, \\ \Rightarrow\ & J(\widehat{x},\widehat{u}) \leq m \quad \text{and } (\widehat{x},\widehat{u}) \in \alpha_{\mathbf{ad}}, \\ \Rightarrow\ & J(\widehat{x},\widehat{u}) = m \quad \text{(i.e., } (\widehat{x},\widehat{u}) \text{ is an optimal state–control pair).} \end{aligned}$$

□

Reviewing the proof of the above theorem, we see that hypothesis H$_c$ was crucial in the argument because it provided the needed convex structure for the direct method of the calculus of variations to work. If H$_c$ fails, we need not have a solution. Then, in order to capture the asymptotic behavior of the minimizing sequences, we embed the original problem in a "larger" system exhibiting the necessary convex structure that guarantees existence of optimal pairs. This method of augmenting the system is known as *relaxation*. There is no unique approach to relaxation. Nevertheless, we can agree that a reasonable relaxation method should meet the following three criteria.

(i) Every original state is also a relaxed state.

(ii) The set of original states is dense in the set of relaxed states.

(iii) The relaxed problem admits a solution and the values of the two problems (relaxed and original) are equal.

The first two requirements concern the dynamics of the system, and the third also involves the cost functional. The requirement that the values of the two problems are equal is known as *relaxability*. If relaxability fails, it can be said that the original system has been enlarged too much. A relaxation method that meets the three criteria is said to be *admissible*.

Next we present some different ways to relax (convexify) the original problem (P). The first relaxation method has its roots in the following observation. We recall that a sequence $\{u_n\}_{n\geq 1} \subseteq L^1(T, \mathbb{R}^m)$ that converges weakly but not strongly to a limit u oscillates violently around its limit. However, in the limit u a great deal of information about the faster and faster oscillations is forgotten and only an average value is registered in the limit. Clearly this is not satisfactory if the control enters in a nonlinear fashion in the dynamics. The idea then is to assign as a limit not a usual $\mathbb{R}^m$-valued function, but a probability-valued function, known as a transition measure (or parametrized measure or Young measure).

DEFINITION 2.6.12 Let (Ω, Σ, μ) be a finite measure space and Y a Polish space. Let $M_+^1(Y)$ be the space of all probability measures on Y. A *relaxed control* (or *transition probability* or *Young measure*) is a function $\lambda \colon \Omega \longrightarrow M_+^1(Y)$, such that for all $C \in \mathbf{B}(Y) =$ the Borel σ-field of Y, $\omega \longrightarrow \lambda(\omega)(C)$ is Σ-measurable. We denote the set of all relaxed controls from Ω into Y by $R(\Omega, Y)$.

REMARK 2.6.13 The "narrow topology" on $M_+^1(Y)$ is the weakest topology on $M_+^1(Y)$ that makes continuous all the maps $\psi_u \colon M_+^1(Y) \longrightarrow \mathbb{R}$ with $u \in C_b(Y) = \{u : Y \longrightarrow \mathbb{R} : u \text{ is continuous and bounded}\}$ defined by $\psi_u(\lambda) = \int_Y u(y)\lambda(dy)$. If we topologize $M_+^1(Y)$ this way, then $\lambda \in R(\Omega, Y)$ if and only if $\lambda \colon \Omega \longrightarrow M_+^1(Y)$ is Σ-measurable. So the narrow topology is the topology $w\big(M_+^1(Y), C_b(Y)\big)$. We can define an analogue of it on the space $R(\Omega, Y)$ of relaxed controls.

DEFINITION 2.6.14 Let $\mathrm{Car}(\Omega \times Y)$ be the space of L^1-Carathéodory integrands. Namely all functions $\varphi \colon \Omega \times Y \longrightarrow \mathbb{R}$ such that

(i) For all $y \in Y$, $\omega \longrightarrow \varphi(\omega, y)$ is Σ-measurable.
(ii) For μ-almost all $\omega \in \Omega$, $y \longrightarrow \varphi(\omega, y)$ is continuous.
(iii) There exists $h \in L^1(\Omega)_+$ such that for μ-almost all $\omega \in \Omega$ and all $y \in \mathbb{R}$, $|\varphi(\omega, y)| \leq h(\omega)$. The *narrow topology* on $R(\Omega, Y)$ is the weakest topology on $R(\Omega, Y)$ with respect to which the functionals

$$\lambda \longrightarrow I_\varphi(\lambda) = \int_\Omega \int_Y \varphi(\omega, y)\lambda(\omega)(\mathrm{d}y)\mathrm{d}\mu, \quad \varphi \in \mathrm{Car}(\Omega \times Y)$$

are continuous.
So the narrow topology on $R(\Omega, Y)$ is the topology $w\big(R(\Omega, Y), \mathrm{Car}(\Omega \times Y)\big)$.

REMARK 2.6.15 We say that an integrand $\varphi\colon \Omega \times Y \longrightarrow \overline{\mathbb{R}} = \mathbb{R} \cup \{+\infty\}$ is *normal* if

(i) φ is $\Sigma \times \mathbf{B}(Y)$-measurable.
(ii) For μ-almost all $\omega \in \Omega$, $y \longrightarrow \varphi(\omega, y)$ is proper and lower semicontinuous.

A positive normal integrand can be approximated pointwise from below by integrands in $\mathrm{Car}(\Omega \times Y)$. So equivalently the narrow topology on $R(\Omega, Y)$ can be defined as the weakest topology that makes all functionals $\lambda \longrightarrow I_\varphi(\lambda) = \int_\Omega \int_Y \varphi(\omega, y)\lambda(\omega)(\mathrm{d}y)\mathrm{d}\mu$ lower semicontinuous as φ ranges over all positive normal integrands. If Y is compact, then $R(\Omega, Y) \subseteq L^\infty\big(\Omega, M(Y)_{w^*}\big) = L^1\big(\Omega, C(Y)\big)^*$ (here $M(Y)$ is the space of all Radon measures on Y; recall that $C(Y)^* = M(Y)$) and the narrow topology on $R(\Omega, Y)$ coincides with the relative w^*-topology in $L^\infty\big(\Omega, M(Y)_{w^*}\big)$. If Σ is countably generated, then $L^1\big(\Omega, C(Y)\big)$ is separable and so on bounded subsets of $L^\infty\big(\Omega, M(Y)_{w^*}\big)$ (such as $R(\Omega, Y)$), the relative w^*−topology is metrizable. In the sequel this is the context in which we use the narrow topology on $R(\Omega, Y)$.

In what follows for simplicity we set $Y = \mathbb{R}^m$.

$\mathbf{H}(\mathbf{U})_1$: $U\colon T \longrightarrow 2^{\mathbb{R}^m} \setminus \{\emptyset\}$ is a multifunction with compact values such that Gr $U \in \mathbf{B}(T) \times \mathbf{B}(\mathbb{R}^m)$ and there exists $r > 0$ such that for almost all $t \in T$ and all $u \in U(t)$, $\|u\| \leq r$.

We set $\overline{B}_r = \{u \in \mathbb{R}^m : \|u\| \leq r\}$ and we introduce the constraint set for the relaxed controls, namely

$$\Sigma(t) = \big\{\mu \in M_+^1(\overline{B}_r) : \mu\big(U(t)\big) = 1\big\}.$$

Then given a state function $x \in W^{1,1}\big((0, b), \mathbb{R}^N\big) \subseteq C(T, \mathbb{R}^N)$, the set of admissible relaxed controls is given by

$$S_\Sigma = \{\lambda \in R(T, \overline{B}_r) : \lambda(t) \in \Sigma(t) \text{ a.e. on } T\}.$$

Now we have all the necessary tools to introduce and study the first relaxation of problem (P) which is based on Young measures (see Definition 2.6.12):

$$\left\{\begin{array}{ll} & J_r^1(x, \lambda) = \int_0^b \int_{\overline{B}_r} L\big(t, x(t), u\big)\lambda(t)(\mathrm{d}u)\mathrm{d}t \longrightarrow \inf = m_r^1 \\ \text{s.t.} & x'(t) = \int_{\overline{B}_r} f\big(t, x(t), u\big)\lambda(t)(\mathrm{d}u)\mathrm{d}t \text{ a.e. on } T \\ & x(0) = x_0, \ \lambda \in S_\Sigma \end{array}\right\}. \tag{2.85}$$

Note that in (2.85) in both the dynamics and the cost functional, the relaxed control enters linearly. This provides problem (2.85) with the necessary convex structure to obtain an optimal pair. For problem (2.85) every original control $u(\cdot)$ can be viewed as a relaxed control by considering the Dirac transition probability $\delta_{u(\cdot)}$. So the first requirement for admissibility is satisfied.

PROPOSITION 2.6.16 *If hypotheses $H(f)$, $H(U)_1$, and $H(L)$ hold and $m<+\infty$, then problem (2.85) has an optimal pair $(\widehat{x},\widehat{\lambda}) \in W^{1,1}\big((0,b),\mathbb{R}^N\big) \times S_\Sigma$; that is, $J_r^1(\widehat{x},\widehat{\lambda}) = m_r^1$.*

PROOF: Evidently $-\infty < m_r^1 \leq m < \infty$. Also let $\{(x_n,\lambda_n)\}_{n\geq 1}$ be a minimizing sequence for problem (2.85). Note that if $F(t,x)=\big\{\int_{\overline{B}_r} f(t,x,u)\lambda(t)(du) : \lambda\in S_\Sigma\big\}$, then F satisfies hypotheses H(F). On the other hand by virtue of Remark 2.6.15, $S_\Sigma \subseteq L^\infty\big(T, M(\overline{B}_r)_{w^*}\big)$ is w^*-compact. So we may assume that

$$x_n \longrightarrow x \text{ in } C(T,\mathbb{R}^N) \quad \text{(see Theorem 2.6.10) and}$$
$$\lambda_n \xrightarrow{w^*} \lambda \in S_\Sigma \quad \text{in } L^\infty\big(T, M(\overline{B}_r)_{w^*}\big).$$

Also from the dynamics of (2.85) and hypothesis H(f)(iii), we see that

$$x_n' \xrightarrow{w} x' \quad \text{in } L^1(T,\mathbb{R}^N).$$

We have

$$\begin{aligned}\|f(t,x_n(t),\cdot) - f(t,x(t),\cdot)\|_{C(T,\mathbb{R}^m)} &= \max_{u\in\overline{B}_r} \|f(t,x_n(t),u) - f(t,x(t),u)\| \\ &= \|f(t,x_n(t),u_n) - f(t,x(t),u_n)\|,\end{aligned}$$

with $u_n \in \overline{B}_r$.

We may assume that $u_n \longrightarrow u$ in $\mathbb{R}^m$. Then by virtue of hypothesis H(f)(ii)

$$\begin{aligned}&\|f(t,x_n(t),u_n) - f(t,x(t),u_n)\| \longrightarrow 0 \quad \text{as } n\rightarrow\infty,\\ \Rightarrow\ &\widehat{f}_n(t) = f(t,x_n(t),\cdot) \longrightarrow f(t,x(t),\cdot) = \widehat{f}(t) \text{ in } C(\overline{B}_r,\mathbb{R}^m) \quad \text{as } n\rightarrow\infty,\\ \Rightarrow\ &\widehat{f}_n \longrightarrow \widehat{f} \text{ in } L^1(T,C(\overline{B}_r,\mathbb{R}^m)) \text{ as } n\rightarrow\infty.\end{aligned}$$

Because $\lambda_n \xrightarrow{w^*} \lambda$ in $L^\infty\big(T, M(\overline{B}_r)_{w^*}\big)$, for every $C \in \mathbf{B}(T)$ we have

$$\int_C\int_{\overline{B}_r} f(t,x_n(t),u)\lambda_n(t)(du)dt \longrightarrow \int_C\int_{\overline{B}_r} f(t,x(t),u)\lambda(t)(du)$$

and $\int_C x_n'(t)dt \longrightarrow \int_C x'(t)dt$. So in the limit we have

$$\begin{aligned}&\int_C x'(t)dt = \int_C\int_{\overline{B}_r} f(t,x(t),u)\lambda(t)(du)dt \qquad \text{for all } C\in\mathbf{B},\\ \Rightarrow\ &x'(t) = \int_{\overline{B}_r} f(t,x(t),u)\lambda(t)(du) \qquad \text{for almost all } t\in T,\ x(0)=x_0.\end{aligned}$$

Therefore the pair (x,λ) is admissible for the relaxed problem (2.85).

Let $\widehat{L}_n(t,u) = L(t,x_n(t),u)$. Because of hypotheses H(L), this is a normal integrand (see Remark 2.6.15) and so we can find $\widehat{L}_n^k(t,u)$, $k\geq 1$, Carathéodory integrands such that

$$\begin{aligned}&\widehat{L}_n^k(t,u) \uparrow \widehat{L}_n(t,u) \qquad \text{for almost all } t\in T, \text{ all } u\in\overline{B}_r \text{ as } k\longrightarrow\infty\\ \text{and}\quad &\alpha_0(t) - c_0\|x_n(t)\| \leq \widehat{L}_n^k(t,u) \qquad \text{for almost all } t\in T, \text{ all } u\in\overline{B}_r.\end{aligned}$$

Then because $x_n \longrightarrow x$ in $C(T,\mathbb{R}^N)$ and $\lambda_n \xrightarrow{w^*} \lambda$ in $L^\infty\big(T, M(\overline{B}_r)_{w^*}\big)$, we have

$$\int_0^b \int_{\overline{B}_r} L_n^k(t, x_n(t), u)\lambda_n(t)(du)dt \longrightarrow \int_0^b \int_{\overline{B}_r} L^k(t, x(t), u)\lambda(t)(du)dt,$$

for all $k \geq 1$, as $n \to \infty$. Also from the monotone convergence theorem, we have

$$\int_0^b \int_{\overline{B}_r} L^k(t, x(t), u)\lambda(t)(du)dt \uparrow \int_0^b \int_{\overline{B}_r} L(t, x(t), u)\lambda(t)(du)dt$$
$$\text{as } k \longrightarrow \infty.$$

Therefore according to Proposition 1.5.25 we can find a sequence $\{k(n)\}_{n\geq 1}$ such that $k(n) \longrightarrow +\infty$ as $n \to \infty$ and

$$\int_0^b \int_{\overline{B}_r} L^{k(n)}(t, x_n(t), u)\lambda_n(t)(du)dt \longrightarrow \int_0^b \int_{\overline{B}_r} L(t, x(t), u)\lambda(t)(du)dt.$$

Because $\int_0^b \int_{\overline{B}_r} L^{k(n)}(t, x_n(t), u)\lambda_n(t)(du)dt \leq \int_0^b L(t, x_n(t), u)\lambda_n(t)(du)dt$ for all $n \geq 1$, we have

$$\int_0^b \int_{\overline{B}_r} L(t, x(t), u)\lambda(t)(du)dt \leq \liminf_{n\to\infty} \int_0^b \int_{\overline{B}_r} L(t, x_n(t), u)\lambda_n(t)(du)dt$$
$$\Rightarrow \ J_r^1(x, \lambda) \leq \liminf_{n\to\infty} J_r^1(x_n, \lambda_n) = m_r^1.$$

Because (x, λ) is admissible for the relaxed problem (2.85), we conclude that

$$J_r^1(x, \lambda) = m_r^1.$$

□

To establish the admissibility of the relaxed problem (2.85), we need to have that the original states are dense in the relaxed ones for the $C(T, \mathbb{R}^N)$-norm. This gives equality of the values of the original and relaxed problems. To do this we need one more result from multivalued analysis which we state here for convenience and postpone its proof until Section 6.4.

PROPOSITION 2.6.17 *If Ω, Σ, μ is a σ-finite nonatomic measure space, X is a separable Banach space, $F\colon \Omega \longrightarrow 2^X \setminus \{\emptyset\}$ a multifunction such that* Gr $F = \{(\omega, x) \in \Omega \times X : x \in F(\omega)\} \in \Sigma \times \mathbf{B}(X)$, *and for some $1 \leq p \leq \infty$, $S_F^p = \{u \in L^p(\Omega, X) : u(\omega) \in F(\omega)\ \mu\text{-a.e.}\} \neq \emptyset$, then denoting by w (resp., w^*) the weak (resp., weak*) topology on $L^p(\Omega, X), 1 \leq p < \infty$ (resp., on $L^\infty(\Omega, X)$) we have*

$$\overline{S_F^p}^{\,w} = S^p_{\overline{\mathrm{conv}}F} \quad 1 \leq p < \infty \qquad (\textit{resp., } \overline{S_F^p}^{\,w^*} = S^p_{\overline{\mathrm{conv}}F} \textit{ for } p = +\infty).$$

Here by $\overline{\mathrm{conv}}F$, we denote the multifunction $\omega \longrightarrow \overline{\mathrm{conv}}\, F(\omega)$ and

$$S^p_{\overline{\mathrm{conv}}F} = \big\{u \in L^p(\Omega, X) : u(\omega) \in \overline{\mathrm{conv}}\, F(\omega)\ \mu\text{-a.e. on } \Omega\big\}.$$

With the help of this abstract result for multifunctions, we are able to show the density for the $C(T, \mathbb{R}^N)$-norm topology of the original states in the relaxed ones (second requirement for the admissibility of a relaxation method). For this we need to strengthen our hypotheses on f as follows.

$\mathbf{H(f)_1}$: $f\colon T\times\mathbb{R}^N\times\mathbb{R}^m\longrightarrow\mathbb{R}^N$ is a function, such that

(i) For all $(x,u)\in\mathbb{R}^N\times\mathbb{R}^m$, $t\longrightarrow f(t,x,u)$ is measurable.
(ii) For almost all $t\in T$, all $x,y\in\mathbb{R}^N$, and all $u\in U(t)$, we have

$$\|f(t,x,u)-f(t,y,u)\|\leq k(t)\|x-y\| \qquad \text{with } k\in L^1(T)_+$$

and for almost all $t\in T$ and all $x\in\mathbb{R}^N$, $u\longrightarrow f(t,x,u)$ is continuous.
(iii) For almost all $t\in T$, all $x\in\mathbb{R}^N$ and all $u\in U(t)$

$$\|f(t,x,u)\|\leq \alpha(t)+c(t)\|x\| \qquad \text{with } \alpha,c\in L^1(T)_+.$$

In what follows by $\alpha^{\mathbf{1}}_{\mathbf{ad}}$ we denote the set of admissible states for the original system (see problem (P)) and by $\alpha^{\mathbf{1}}_{\mathbf{ad,r_1}}$ we denote the set of admissible states for the relaxed system (2.85).

PROPOSITION 2.6.18 *If hypotheses $H(f)_1$, $H(U)_1$ and $H(L)$ hold, then $\overline{\alpha^{\mathbf{1}}_{\mathbf{ad}}}=\alpha^{\mathbf{1}}_{\mathbf{ad,r_1}}$ the closure taken in the $C(T,\mathbb{R}^N)$-norm.*

PROOF: For every $(t,x)\in T\times\mathbb{R}^N$, let

$$F(t,x)=f\big(t,x,U(t)\big)$$
$$\text{and}\quad G(t,x)=\Big\{\int_{\overline{B}_r}f(t,x,u)\mu(du):\mu\in\Sigma(t)\Big\}.$$

We claim that conv $F(t,x)=G(t,x)$. Note that $f\big(t,x,U(t)\big)$ is compact and then so is conv $F(t,x)$. Also it is clear that $F(t,x)\subseteq G(t,x)$ and the latter is a convex set. Moreover, $M^1_+(\overline{B}_r)$ furnished with the narrow topology (see Remark 2.6.13), which due to the compactness of $\overline{B}_r$ coincides with the relative w^*-topology on $M(\overline{B}_r)$ = space of Radon measures on $\overline{B}_r$ (recall that $C(\overline{B}_r)^*=M(\overline{B}_r)$), is compact metrizable. Let $\{v_n\}_{n\geq1}\subseteq G(t,x)$ and suppose $v_n\longrightarrow v$ in $\mathbb{R}^N$. We have

$$v_n=\int_{\overline{B}_r}f(t,x,u)\mu_n(du) \qquad \text{with } \mu_n\in\Sigma(t).$$

We may assume that $\mu_n\longrightarrow\mu$ narrowly (equivalently in the relative w^*-topology) in $M^1_+(\overline{B}_r)$. So from the portmanteau theorem we have

$$\limsup_{n\to\infty}\mu_n\big(U(t)\big)\leq\mu\big(U(t)\big),$$
$$\Rightarrow\ 1\leq\mu\big(U(t)\big).$$

Because $\mu\in M^1_+(\overline{B}_r)$, we have $\mu\big(U(t)\big)=1$ and so $\mu\in\Sigma(t)$. Also

$$v_n=\int_{\overline{B}_r}f(t,x,u)\mu_n(du)\longrightarrow\int_{\overline{B}_r}f(t,x,u)\mu(du)=v\in G(t,x).$$

Therefore $G(t,x)$ is closed and so

$$\text{conv } F(t,x)\subseteq G(t,x). \tag{2.86}$$

On the other hand if $v \in G(t,x)$, we have $v = \int_{\overline{B}_r} f(t,x,u)\mu(du), \mu \in \Sigma(t)$. We can find $s_n = \sum_{k=1}^{n} r_k \delta_{u_k}$ with $u_k \in U(t)$ and $\{r_k\}_{n\geq 1} \subseteq [0,1]$, $\sum_{k=1}^{n} r_k = 1$ such that $s_n \xrightarrow{w^*} \mu$ in $M_+^1(\overline{B}_r)$. Then

$$v_n = \int_{\overline{B}_r} f(t,x,u)s_n(du) = \sum_{k=1}^{n} r_k f(t,x,u_k) \in \text{conv } F(t,x)$$

and $v_n \longrightarrow v = \int_{\overline{B}_r} f(t,x,u)\mu(du)$. So it follows

$$\text{conv } F(t,x) = G(t,x) \quad \text{(see (2.86))}. \tag{2.87}$$

Let $S_{\text{conv}F}(x_0)$ be the solution set of (2.82) when the multivalued right hand side is conv F and by $S_G(x_0)$ the solution set of (2.82) when the multivalued right hand is G. Then from (2.87) it follows that

$$S_{\text{conv}F}(x_0) = S_G(x_0). \tag{2.88}$$

Because of hypothesis H(f)(ii) for almost all $t \in T$, $x \longrightarrow \text{conv}F(t,x)$ is Lipchitz continuous for the Hausdorff metric. So invoking the relaxation theorem for differential inclusions (see Denkowski–Migórski–Papageorgiou [195, p. 262], we know that

$$S_{\text{conv}F}(x_0) = \overline{S_F(x_0)} \qquad \text{the closure taken in } C(T,\mathbb{R}^N). \tag{2.89}$$

Because $\alpha^{\mathbf{1}}_{\mathbf{ad}} = S_F(x_0)$ and $\alpha^{\mathbf{1}}_{\mathbf{ad,r_1}} = S_G(x_0)$, from (2.88) and (2.89) we conclude that

$$\alpha^{\mathbf{1}}_{\mathbf{ad,r_1}} = \overline{\alpha^{\mathbf{1}}_{\mathbf{ad}}} \qquad \text{the closure taken in } C(T,\mathbb{R}^N).$$

□

REMARK 2.6.19 It is well-known that the relaxation theorem for differential inclusions fails if the multivalued nonlinearity $F(t,x)$ is only continuous in the Hausdorff metric in the $x \in \mathbb{R}^N$-variable. For this reason we had to strengthen the hypothesis on the vector field $x \longrightarrow f(t,x,u)$ (see hypothesis H$(f)_1$(iii)).

The final step in the analysis of the relaxed problem (2.85) is to show that $m_r^1 = m$ (relaxability). This implies that the relaxation method of problem (2.85) is in fact admissible. To achieve this, we strengthen our hypotheses on the cost integrand L as follows.

H(L)$_1$: $L\colon T \times \mathbb{R}^N \times \mathbb{R}^m \longrightarrow \mathbb{R}$ is an integrand such that

(i) For all $(x,u) \in \mathbb{R}^N \times \mathbb{R}^m$, $t \longrightarrow L(t,x,u)$ is measurable.
(ii) For almost all $t \in T$, $(x,u) \longrightarrow L(t,x,u)$ is continuous.
(iii) For almost all $t \in T$, all $x \in \mathbb{R}^N$ with $\|x\| \leq n, n \geq 1$, and all $u \in U(t)$,

$$|L(t,x,u)| \leq \eta_n(t) \qquad \text{with } \eta_n \in L^1(T)_+.$$

THEOREM 2.6.20 *If hypotheses H$(f)_1$, H$(U)_1$, and H$(L)_1$ hold, then* $m_r^1 = m$.

PROOF: From Proposition 2.6.16 we know that we can find $(\widehat{x}, \widehat{\lambda})$ an admissible state-control pair for the relaxed problem (2.85) such that

$$J_r^1(\widehat{x}, \widehat{\lambda}) = m_r^1.$$

Also by virtue of Propositions 2.6.17 and 2.6.18 we can find $(x_n, u_n) \in \alpha_{\mathbf{ad}}$ such that

$$x_n \longrightarrow \widehat{x} \text{ in } C(T, \mathbb{R}^N) \qquad \text{and} \qquad \delta_{u_n} \xrightarrow{w^*} \widehat{\lambda} \text{ in } L^\infty\big(T, M(\overline{B}_r)_{w^*}\big) \text{ as } n \to \infty.$$

Set $\widehat{L}_n(t)(u) = L\big(t, x_n(t), u\big)$ and $\widehat{L}(t)(u) = L\big(t, \widehat{x}(t), u\big)$. Evidently $\widehat{L}_n \longrightarrow \widehat{L}$ in $L^1\big(T, C(\overline{B}_r)\big)$. So if by $\langle \cdot, \cdot \rangle$ we denote the duality brackets for the pair $\Big(L^1\big(T, C(\overline{B}_r)\big), L^\infty\big(T, M(\overline{B}_r)_{w^*}\big)\Big)$, we have

$$\begin{aligned} \int_0^b L\big(t, x_n(t), u_n(t)\big)dt &= \Big\langle \widehat{L}_n, \delta_{u_n} \Big\rangle \longrightarrow \Big\langle \widehat{L}, \widehat{\lambda} \Big\rangle \\ &= \int_0^b \int_{\overline{B}_r} L\big(t, \widehat{x}(t), u\big)\widehat{\lambda}(t)(du)dt = m_r^1, \\ &\Rightarrow m \le m_r^1. \end{aligned}$$

Because the opposite inequality is clearly true, we conclude that $m = m_r^1$. □

So we have established the admissibility of the relaxation method based on Young measures.

The next relaxation method is suggested by the reduction method used in the proof of Theorem 2.6.11. We introduce the following relaxed problem:

$$\left\{ \begin{array}{l} J_r^2(x, x') = \int_0^b p^{**}\big(t, x(t), x'(t)\big)dt \longrightarrow \inf = m_r^2 \\ \text{s.t.} \quad x'(t) \in \operatorname{conv} F\big(t, x(t)\big) \text{ a.e. on } T, \ \ x(0) = x_0 \end{array} \right\}. \tag{2.90}$$

Here p^{**} stands for the second conjugate (in the sense of Definition 1.2.15) of the function $v \longrightarrow p(t, x, v)$. Also as before $F(t, x) = f\big(t, x, U(t)\big)$. Note that problem (2.90) is control-free (a calculus of variations problem with multivalued dynamic constraints). In what follows by $\alpha^1_{\mathbf{ad,r_2}}$ we denote the set of admissible states for this new problem (2.90).

PROPOSITION 2.6.21 *If hypotheses $H(f)_1$, $H(U)_1$, and $H(L)$ hold, then $\alpha^1_{\mathbf{ad,r_1}} = \alpha^1_{\mathbf{ad,r_2}}$, $m_r^1 = m_r^2$, and there exists $\widehat{x} \in \alpha^1_{\mathbf{ad,r_2}}$ such that*

$$J_r^2(\widehat{x}, \widehat{x}') = m_r^2.$$

PROOF: From the proof of Proposition 2.6.18, we know that

$$\operatorname{conv} F(t, x) = \Big\{ \int_{\overline{B}_r} f(t, x, u)\mu(du) : \mu \in \Sigma(t) \Big\}. \tag{2.91}$$

From (2.91) it follows that $\alpha^1_{\mathbf{ad,r_1}} = \alpha^1_{\mathbf{ad,r_2}}$.

By definition we have

$$p^{**}(t,x,v) = \inf\left\{\eta \in \mathbb{R} : (v,\eta) \in \overline{\text{conv}}\ \text{epi}\ p(t,x,\cdot)\right\}. \tag{2.92}$$

As always $\inf \emptyset = +\infty$. Set

$$H(t,x) = \text{epi}\ p(t,x,\cdot) = \left\{\left(f(t,x,u),\eta\right) : u \in U(t),\ L(t,x,u) \leq \eta\right\}.$$

As in the proof of Proposition 2.6.18, we can show that

$$\begin{aligned}\overline{\text{conv}}\ H(t,x) &= \overline{\text{conv}}\ \text{epi}\ p(t,x,\cdot)\\ &= \left\{\Big(\int_{\overline{B}_r} f(t,x,u)\mu(du),\eta\Big) : \mu \in \Sigma(t),\ \int_{\overline{B}_r} L(t,x,u)\mu(du) \leq \eta\right\}.\end{aligned}$$

Hence we have

$$p^{**}(t,x,v) = \inf\Big\{\int_{\overline{B}_r} L(t,x,u)\mu(du) : \mu \in \Sigma(t),\ v = \int_{\overline{B}_r} f(t,x,u)\mu(du)\Big\}. \tag{2.93}$$

Because $\mu \longrightarrow \int_{\overline{B}_r} L(t,x,u)\mu(du)$ is narrowly lower semicontinuous and $M_+^1(\overline{B}_r)$ is narrowly compact, from Theorem 2.1.10, we see that the infimum in (2.93) is attained and an application of Theorem 2.6.8 implies that the minimizing measure depends measurable on $t \in T$. Combining this fact with Proposition 2.6.16, we infer that $m_r^1 = m_r^2$ and for $\widehat{x}$ the optimal state for problem (2.85), we also have

$$J_r^2(\widehat{x},\widehat{x}') = m_r^2.$$

□

Combining Propositions 2.6.18 and 2.6.21, we see the following.

PROPOSITION 2.6.22 *If hypotheses $H(f)$, $H(U)_1$, and $H(L)$ hold, then* $\overline{\alpha^1_{\mathbf{ad}}} = \alpha^1_{\mathbf{ad,r_2}}$*, the closure taken in the* $C(T,\mathbb{R}^N)$ *norm.*

Now we can establish the admissibility of the second relaxed problem (2.90).

THEOREM 2.6.23 *If hypotheses $H(f)$, $H(U)_1$, and $H(L)_1$ hold, then* $m_r^2 = m$.

PROOF: From Theorem 2.6.20 we know that $m_r^1 = m$. On the other hand Proposition 2.6.21 says that $m_r^1 = m_r^2$. So finally $m_r^2 = m$. □

So the second relaxation method, which was motivated from the reduction method, is also admissible.

The third relaxed problem has its roots in the well-known theorem of Carathéodory for the representation of convex sets in $\mathbb{R}^N$. Let us recall that theorem.

THEOREM 2.6.24 *If $C \subseteq \mathbb{R}^N$ is nonempty, then every point of the set* conv C *is a convex combination of no more than $n+1$ distinct points of the set C.*

Motivated by this result, we introduce the following relaxed problem. In what follows $\widehat{u} = (u_k)_{k=1}^{N+1}$ and $\widehat{r} = (r_k)_{k=1}^{N+1}$.

$$\left\{\begin{array}{l} J_r^3(x,\widehat{u},\widehat{r}) = \int_0^b \sum_{k=1}^{N+1} r_k(t)L\big(t,x(t),u_k\big)dt \longrightarrow \inf = m_r^3 \\ \text{s.t.} \quad x'(t) = \sum_{k=1}^{N+1} r_k(t)f\big(t,x(t),u_k(t)\big) \text{ a.e. on } T,\ x(0)=x_0 \\ u_k \in L^1(T,\mathbb{R}^m),\ u(t)\in U(t) \text{ a.e. on } T \\ r_k : T \longrightarrow [0,1] \text{ measurable } \sum_{k=1}^{N+1} r_k(t) = 1 \qquad \text{for all } t\in T \end{array}\right\}. \tag{2.94}$$

By $\alpha^{\mathbf{1}}_{\mathbf{ad,r_3}}$ we denote the admissible states for problem (2.94). From the proof of Proposition 2.6.21 it is clear that we have the following.

PROPOSITION 2.6.25 *If hypotheses $H(f)$, $H(U)_1$, and $H(L)$ hold, then $\alpha^{\mathbf{1}}_{\mathbf{ad,r_1}} = \alpha^{\mathbf{1}}_{\mathbf{ad,r_2}} = \alpha^{\mathbf{1}}_{\mathbf{ad,r_3}}$, $m_r^1 = m_r^2 = m_r^3$, $\overline{\alpha^{\mathbf{1}}_{\mathbf{ad}}} = \alpha^{\mathbf{1}}_{\mathbf{ad,r_3}}$ the closure taken in the $C(T,\mathbb{R}^N)$-norm, and there exists an optimal triple $(\widehat{x},\widehat{u}^*,\widehat{r}^*) = m_r^3$ from problem (2.94); that is,*

$$J_r^3(\widehat{x},\widehat{u}^*,\widehat{r}^*) = m_r^3.$$

Also we have the admissibility of the new relaxed problem (2.94).

THEOREM 2.6.26 *If hypotheses $H(f)$, $H(U)_1$, and $H(L)_1$ hold, then $m_r^3 = m$.*

Next we derive a fourth relaxation method based on semicontinuity techniques, which are different from the ones employed so far. The new approach uses the so-called multiple Γ-operators which are related to the Γ-convergence that we studied in Section 1.5.

DEFINITION 2.6.27 Let X_1, X_2 be two Hausdorff topological spaces, and let $\varphi\colon X_1\times X_2 \longrightarrow \overline{\mathbb{R}} = \mathbb{R}\cup\{+\infty\}$ be a proper function and $(x_1,x_2)\in X_1\times X_2$. In what follows by $\mathcal{Z}(+)$ we denote the *sup* operator and by $\mathcal{Z}(-)$ the *inf* operator. Also for $k=1,2$, let S_k be the set of sequences in X_k that converge to x_k and let β_k be one of the signs $+$ and $-$. We define

$$\Gamma_{\text{seq}}(X_1^{\beta_1}, X_2^{\beta_2})\varphi(x_1,x_2) = \underset{(x_n^1)\in S_1}{\mathcal{Z}(\beta_1)}\ \underset{(x_n^2)\in S_2}{\mathcal{Z}(\beta_2)}\ \underset{m\geq 1}{\mathcal{Z}(-\beta_1)}\ \underset{n\geq m}{\mathcal{Z}(\beta_1)}\ \varphi(x_n^1,x_n^2).$$

REMARK 2.6.28 Here for simplicity we have restricted ourselves to the sequential definition of multiple Γ-operators. Evidently we can have a more general topological version of the above notion, as we did for the Γ-convergence in Section 1.5.

EXAMPLE 2.6.29 By virtue of Definition 2.6.27, we have

$$\Gamma_{\text{seq}}(X_1^-, X_2^-)\varphi(x_1,x_2) = \inf_{x_n^1\to x_1}\ \inf_{x_n^2\to x_2}\ \liminf_{n\to\infty} \varphi(x_n^1,x_n^2)$$

$$\text{and} \quad \Gamma_{\text{seq}}(X_1^-, X_2^+)\varphi(x_1,x_2) = \inf_{x_n'\to x_1}\ \sup_{x_n^2\to x_2}\ \liminf_{n\to\infty} \varphi(x_n^1,x_n^2)$$

If the Γ_{seq}-limit is independent of the sign + or − associated with one of the spaces, then this sign is omitted. So if

$$\Gamma_{\text{seq}}(X_1^-, X_2^+)\varphi(x_1,x_2) = \Gamma_{\text{seq}}(X_1^+, X_2^+)\varphi(x_1,x_2)$$

then we simply write $\Gamma_{\text{seq}}(X_1, X_2^+)\varphi(x_1,x_2)$. Note that

$$\Gamma_{\text{seq}}(X_1^{\beta_1}, X_2^{\beta_2})\varphi(x_1,x_2) = -\Gamma_{\text{seq}}(X_1^{-\beta_1}, X_2^{-\beta_2})(-\varphi)(x_1,x_2)$$

and if both X_1, X_2 are metrizable, then

$$\Gamma_{\text{seq}}(X_1^-, X_2^-)\varphi(x_1,x_2) = \Gamma_{\text{seq}}\big((X_1 \times X_2)^-\big)\varphi(x_1,x_2).$$

Although in general multiple $\Gamma-$operators are not distributive with respect to addition, nevertheless we have some useful inequalities.

PROPOSITION 2.6.30 *If $\varphi, \psi \colon X_1 \times X_2 \longrightarrow \overline{\mathbb{R}} = \mathbb{R} \cup \{+\infty\}$ are proper functions, then*

$$\begin{aligned}\Gamma_{\text{seq}}(X_1^-, X_2^-)\varphi + \Gamma_{\text{seq}}(X_1^-, X_2^-)\psi &\le \Gamma_{\text{seq}}(X_1^-, X_2^-)(\varphi+\psi)\\ &\le \Gamma_{\text{seq}}(X_1^-, X_2^+)\varphi + \Gamma_{\text{seq}}(X_1^+, X_2^-)\psi.\end{aligned}$$

PROOF: It is a consequence of

$$\inf_{i\in I} \lambda_i + \inf_{i\in I} \mu_i \le \inf_{i\in I}(\lambda_i + \mu_i) \le \inf_{i\in I} \lambda_i + \sup_{i\in I} \mu_i$$

for all $\{\lambda_i\}_{i\in I}$, $\{\mu_i\}_{i\in I} \subseteq \overline{\mathbb{R}}$. □

COROLLARY 2.6.31 *If for $(x_1,x_2) \in X_1 \times X_2$, $\Gamma_{\text{seq}}(X_1^-, X_2)\varphi(x_1,x_2)$, and $\Gamma_{\text{seq}}(X_1, X_2^-)\psi(x_1,x_2)$ exist, then*

$$\Gamma_{\text{seq}}(X_1, X_2^-)(\varphi+\psi)(x_1,x_2) = \Gamma_{\text{seq}}(X_1^-, X_2)\varphi(x_1,x_2) + \Gamma_{\text{seq}}(X_1, X_2^-)\psi(x_1,x_2).$$

In abstract terms the idea of this new relaxation method is the following. Suppose X is the state space, Y the control space (both Hausdorff topological spaces), $\varphi \colon X \times Y \longrightarrow \overline{\mathbb{R}}$ is the cost functional, and C is the set of all admissible state-control pairs (in C we have incorporated all constraints of the problem dynamic and nondynamic). Then the optimal control problem is

$$\inf\big[(\varphi + i_C)(x,u) : (x,u) \in X \times Y\big], \tag{2.95}$$

where i_C is the indicator function of the set C; that is,

$$i_C(x,u) = \begin{cases} 0 & \text{if } (x,u) \in C \\ +\infty & \text{otherwise} \end{cases}.$$

Then the relaxed problem corresponding to (2.95) is

$$\inf\big[\Gamma_{\text{seq}}(X^-, U^-)(\varphi + i_C)(x,u) : (x,u) \in X \times Y\big];$$

in other words in the relaxed problem we minimize the Γ_{seq}-relaxation of the extended cost functional $\varphi + i_C$. So our goal is to determine $\Gamma_{\text{seq}}(X^-, U^-)(\varphi + i_C)$. We do this by employing the *auxiliary variable method*, which is outlined in the next proposition.

PROPOSITION 2.6.32 *If Z is a third Hausdorff topological space, $\xi\colon X\times Y\longrightarrow Z$ is a map that satisfies*

($\mathbf{C_0}$) *Every sequence $\{(x_n,u_n)\}_{n\geq 1}$ that is convergent in $X\times Y$ and $\{(\varphi+i_C)(x_n,u_n)\}_{n\geq 1}$ is bounded, the sequence $\{\xi(x_n,y_n)\}_{n\geq 1}$ has a convergent subsequence in Z,*

$$\textit{and}\quad \psi(x,u,v)=\begin{cases}(\varphi+i_C)(x,u) & \textit{if } v=\xi(x,u)\\ +\infty & \textit{otherwise}\end{cases}, \tag{2.96}$$

then $\Gamma_{\text{seq}}(X^-,Y^-)(\varphi+i_C)(x,u)=\inf\big[\Gamma_{\text{seq}}\big(X^-,(Y\times Z)^-\big)\psi(x,u,v):v\in Z\big]$.

PROOF: Let $(x,u,v)\in X\times Y\times Z$ and suppose that $x_n\to x$ in X, $u_n\longrightarrow u$ in Y, and $v_n\longrightarrow v$ in Z. We have

$$\begin{aligned}&(\varphi+i_C)(x_n,u_n)\leq\psi(x_n,u_n,v_n),\ n\geq 1 \qquad \text{(see (2.96))},\\ \Rightarrow\ &\Gamma_{\text{seq}}(X^-,Y^-)(\varphi+i_C)(x,u)\leq\Gamma_{\text{seq}}\big(X^-,(Y\times Z)^-\big)\psi(x,u,v)\\ &\quad\text{for all } v\in Z,\\ \Rightarrow\ &\Gamma_{\text{seq}}(X,^-,Y^-)(\varphi+i_C)(x,u)\\ &\leq\inf\big[\Gamma_{\text{seq}}\big(X^-,(X\times Y)^-\big)\psi(x,u,v):v\in Z\big].\end{aligned}\tag{2.97}$$

On the other hand let $(x,u)\in X\times Y$ and consider $x_n\longrightarrow x$ in X and $u_n\longrightarrow u$ in Y. Without any loss of generality we assume that $\lim\limits_{n\to\infty}(\varphi+i_C)(x_n,u_n)$ exists and it is finite. Then by virtue of condition ($\mathrm{C_0}$), $v_n=\xi(x_n,u_n)\longrightarrow v\in Z$. Hence using Definition 2.6.27 we have

$$\begin{aligned}&\Gamma_{\text{seq}}\big(X^-,(Y\times Z)^-\big)\psi(x,u,v)\leq\liminf_{n\to\infty}\psi(x_n,u_n,v_n)=\lim_{n\to\infty}\varphi(x_n,u_n)\\ \Rightarrow\ &\inf\big[\Gamma_{\text{seq}}\big(X^-,(Y\times Z)^-\big)\psi(x,u,v):v\in Z\big]\leq\Gamma_{\text{seq}}(X^-,Y^-)(\varphi+i_C)(x,u).\end{aligned}\tag{2.98}$$

From (2.97) and (2.98), we conclude that equality holds. □

The next proposition is helpful in the evaluation of $\Gamma_{\text{seq}}\big(X^-,(Y\times Z)^-\big)\psi$.

PROPOSITION 2.6.33 *If the cost functional φ satisfies*

$$\varphi(x,u)\leq\varphi(x',u)+w(x,x')h(x',u)\qquad \textit{for all } x,x'\in X,\ u\in Y \tag{2.99}$$

with $w\colon X\times X\longrightarrow\overline{\mathbb{R}}_+$ and $h\colon X\times Y\longrightarrow\overline{\mathbb{R}}_+$ satisfying

$$x_n\longrightarrow x\quad \textit{in } X \textit{ implies that } w(x,x_n),\ w(x_n,x)\longrightarrow 0 \tag{2.100}$$

and $\{(x_n,u_n)\}_{n\geq 1}$ is convergent in $X\times Y$ and $\{\varphi(x_n,u_n)\}_{n\geq 1}$ is bounded imply

$$\{h(x_n,u_n)\}_{n\geq 1}\ \textit{is bounded}, \tag{2.101}$$

then for all $(x,u)\in X\times Y$, $\Gamma_{\text{seq}}(X,Y^-)\,\varphi(x,u)$ exists and

$$\Gamma_{\text{seq}}(X,Y^-)\,\varphi(x,u)=\Gamma_{\text{seq}}(\delta_X,Y^-)\,\varphi(x,u)$$

with δ_X being the discrete topology on X.

PROOF: Let $(x_n, u_n) \longrightarrow (x, u)$ in $X \times Y$. Without any loss of generality we assume that $\{\varphi(x_n, u_n)\}_{n\geq 1}$ converges to a finite limit. From (2.99) we have

$$\varphi(x, u_n) \leq \varphi(x_n, u_n) + w(x, x_n)h(x_n, u_n), \quad n \geq 1.$$

Then by virtue of (2.100) and (2.101) we have

$$\begin{aligned} &\liminf_{n\to\infty} \varphi(x, u_n) \leq \liminf_{n\to\infty} \varphi(x_n, u_n), \\ \Rightarrow\ &\Gamma_{\text{seq}}(X, Y^-)\ \varphi(x, u) \leq \Gamma_{\text{seq}}(X, Y^-)\ \varphi(x, u). \end{aligned} \tag{2.102}$$

On the other hand let $(x_n, u_n) \longrightarrow (x, u)$ in $X \times Y$, and without any loss of generality assume that $\{\varphi(x, u_n)\}_{n\geq 1}$ converges to a finite limit. Then because of (2.99), we have

$$\varphi(x_n, u_n) \leq \varphi(x, u_n) + w(x_n, x)h(x, u_n), \quad n \geq 1.$$

Then by virtue of (2.100) and (2.101), we have

$$\begin{aligned} &\liminf_{n\to\infty} \varphi(x_n, u_n) \leq \lim \varphi(x, u_n), \\ \Rightarrow\ &\Gamma_{\text{seq}}(\delta_X, Y^-)\ \varphi(x, u) \leq \Gamma_{\text{seq}}(\delta_X, Y^-)\ \varphi(x, u). \end{aligned} \tag{2.103}$$

From (2.102) and (2.103) we conclude that equality must hold. $\square$

Now we introduce the specific optimal problem that we study. The state space X is $W^{1,1}\big((0,b),\mathbb{R}^N\big)$ endowed with the $C(T,\mathbb{R}^N)$-norm topology (recall that $W^{1,1}\big((0,b),\mathbb{R}^N\big)$ is embedded continuously in $C(T,\mathbb{R}^N)$) and the control space Y is $L^1(T,\mathbb{R}^m)$ endowed with the weak topology. The optimal control problem is the following.

$$\left\{ \begin{array}{ll} & J(x,u) = \int_0^b L\big(t, x(t), u(t)\big)dt \longrightarrow \inf = \widehat{m}, \\ \text{s.t.} & x'(t) = A\big(t, x(t)\big) + C\big(t, x(t)\big)g\big(t, u(t)\big) \quad \text{a.e. on } T \\ & x(0) = x_0, \ \ u \in L^1(T, \mathbb{R}^m), \ \ u(t) \in U(t) \quad \text{a.e. on } T \end{array} \right\}. \tag{2.104}$$

Here $A: T\times\mathbb{R}^N \longrightarrow \mathbb{R}^N$, $C: T\times\mathbb{R}^N \longrightarrow \mathbb{R}^{N\times k}$, and $g: T\times\mathbb{R}^m \longrightarrow \mathbb{R}^k$ are measurable functions, $U: T \longrightarrow 2^{\mathbb{R}^m}\setminus\{\emptyset\}$ is the control constraint multifunction, and $L: T \times \mathbb{R}^N \times \mathbb{R}^m \longrightarrow \overline{\mathbb{R}} = \mathbb{R}\cup\{+\infty\}$ is the cost integrand. We introduce the following hypotheses on the above data of problem (2.104).

H(A): $A: T\times\mathbb{R}^N \longrightarrow \mathbb{R}^N$ is a function such that

(i) For all $x \in \mathbb{R}^N$, $t \longrightarrow A(t,x)$ is measurable.
(ii) For every $r>0$, there exists $\beta_r \in L^1(T)_+$ such that for almost all $t\in T$ and all $\|x\|, \|y\| \leq r$, we have

$$\|A(t,x) - A(t,y)\| \leq \beta_r(t)\|x-y\|.$$

(iii) For almost all $t\in T$ and all $x \in \mathbb{R}^N$, we have

$$\|A(t,x)\| \leq \alpha(t) + c(t)\|x\| \quad \text{with } \alpha,\ c \in L^1(T)_+.$$

H(C): $C\colon T\times\mathbb{R}^N\longrightarrow\mathbb{R}^{N\times k}$ is a function such that

(i) For all $x\in\mathbb{R}^N$, $t\longrightarrow C(t,x)$ is measurable.
(ii) For every $r>0$, there exists $\gamma_r\in L^\infty(T)_+$ such that for almost all $t\in T$ and all $\|x\|,\|y\|\le r$, we have

$$\|C(t,x)-C(t,y)\|\le\gamma_r(t)\|x-y\|.$$

(iii) For almost all $t\in T$ and all $x\in\mathbb{R}^N$, we have

$$\|C(t,x)\|\le\alpha_1(t)+c_1(t)\|x\|\quad\text{with }\alpha_1,\ c_1\in L^\infty(T)_+.$$

H(g): $g\colon T\times\mathbb{R}^m\longrightarrow\mathbb{R}^k$ is a Borel function.

H(U)$_2$: $U\colon T\longrightarrow 2^{\mathbb{R}^m}\setminus\{\emptyset\}$ is a multifunction with compact values, such that Gr $U=\{(t,u)\in T\times\mathbb{R}^m: u\in U(t)\}\in\mathbf{B}(T)\times\mathbf{B}(\mathbb{R}^m)$ and there exists $u\in L^1(T,\mathbb{R}^m)$ such that $u(t)\in U(t)$ a.e. on T and $g(\cdot,u(\cdot))$ is integrable on T.

H(L)$_2$: $L\colon T\times\mathbb{R}^N\times\mathbb{R}^m\longrightarrow\overline{\mathbb{R}}_+=\mathbb{R}_+\cup\{+\infty\}$ is an integrand such that

(i) $(t,x,y)\longrightarrow L(t,x,y)$ is measurable.
(ii) For every $r>0$, almost all $t\in T$, all $\|x\|,\|y\|\le r$ and all $u\in U(t)$, we have

$$L(t,x,u)\le L(t,y,u)+\vartheta_r(t,\|x-y\|)+\sigma_r(\|x-y\|)L(t,y,u),$$

where $\vartheta_r: T\times\mathbb{R}_+\longrightarrow\mathbb{R}_+$ and $\sigma_r:\mathbb{R}_+\longrightarrow\mathbb{R}_+$ are such that $\vartheta_r(\cdot,\lambda)$ is integrable and $\vartheta_r(t,\cdot)$, $\sigma_r(\cdot)$ are continuous increasing with $\vartheta_r(t,0)=\sigma_r(0)=0$.

(iii) For almost all $t\in T$ and all $u\in U(t)$

$$\nu(\|g(t,x)\|+\|u\|)-\zeta(t)\le L(t,0,u)$$

with $\nu\colon\mathbb{R}_+\longrightarrow\mathbb{R}_+$ increasing convex function satisfying $\lim\limits_{r\to+\infty}\big(\nu(\tau)/\tau\big)=+\infty$ and $\zeta\in L^1(T)_+$.
(iv) There exist $u_0\in L^1(T,\mathbb{R}^m)$ such that $t\longrightarrow L\big(t,0,u_0(t)\big)$ is integrable on T.

We implement the method of the auxiliary variable, outlined earlier. Let $S=\{(x,u)\in W^{1,1}\big((0,b),\mathbb{R}^N\big)\times L^1(T,\mathbb{R}^m):(x,u)\text{ is an admissible pair for problem (2.104)}\}$ and set $\varphi=J+i_S$. The next lemma shows that the method of auxiliary variable indeed can be used within the context of problem (2.104).

LEMMA 2.6.34 *If $Z=L^1(T,\mathbb{R}^k)$ furnished with the weak topology and $\xi\colon X\times Y\longrightarrow Z$ is defined by*

$$\xi(x,u)(t)=\begin{cases}g\big(t,u(t)\big) & \text{if } g\big(\cdot,u(\cdot)\big)\text{ is integrable,}\\ 0 & \text{otherwise}\end{cases},$$

then φ satisfies the compactness condition (C_0) of Proposition 2.6.32.

PROOF: Suppose $(x_n, u_n) \longrightarrow (x, u)$ in $X \times Y$ and assume that $|\varphi(x_n, u_n)| \leq M$ for some $M > 0$, all $n \geq 1$. Let $v_n(t) = g\big(t, u_n(t)\big)$. If $\varepsilon_n = \|x_n - x\|_{C(T,\mathbb{R}^N)}$ and $r = \sup\limits_{n \geq 1} \|x_n\|_{C(T,\mathbb{R}^N)}$ (recall that on the state space $X = W^{1,1}\big((0,b), \mathbb{R}^N\big)$ we consider the $C(T,\mathbb{R}^N)$-norm topology), using hypothesis H(L), we have

$$\begin{aligned}\int_0^b \Big(\psi(\|v_n(t)\|) - \zeta(t)\Big)\mathrm{d}t &\leq \int_0^b L\big(t, 0, u_n(t)\big)\mathrm{d}t \\ &\leq \int_0^b \Big(L\big(t, x_n(t), u_n(t)\big)\mathrm{d}t + \vartheta_r(t, \varepsilon_n)\Big)\mathrm{d}t \\ &\quad + \int_0^b \sigma_r(\varepsilon_n)\, L\big(t, x_n(t), u_n(t)\big)\mathrm{d}t \\ &\leq M_1\big(1 + \sigma_r(\varepsilon_n)\big) + \int_0^b \vartheta_r(t, \varepsilon_n)\mathrm{d}t \end{aligned} \tag{2.105}$$

for some $M_1 > 0$, all $n \geq 1$.

Then from (2.105), hypothesis H$(L)_2$(iii), and the De La Vallée–Poussin theorem, we conclude that $\{v_n\}_{n\geq 1} \subseteq L^1(T, \mathbb{R}^k)$ has a weakly convergent subsequence. $\square$

This lemma permits the use of Proposition 2.6.32, which says the relaxation of the extended cost functional $\varphi = J + i_S$ is reduced to the relaxation in $X \times (Y \times Z)$ of

$$\psi(x, u, v) = \widehat{J}(x, u, v) + i_S(x, u, v),$$

where $\widehat{J}(x, u, v) = \int_0^b \widehat{L}\big(t, x(t), u(t), v(t)\big)\mathrm{d}t$ with $\widehat{L}(t, x, u, v) = L(t, x, u) + i_{\{v = g(t,u)\}}$ and $\widehat{S} = \big\{(x, u, v) \in X \times Y \times Z :\ x'(t) = A\big(t, x(t)\big) + C\big(t, x(t)\big)v(t)$ a.e. on $T,\ x(0) = x_0\big\}$.

We compute separately $\Gamma_{\text{seq}}\big(X, (Y \times Z)^-\big)\widehat{J}(x, u, v)$ and $\Gamma_{\text{seq}}\big(X^-, Y \times Z\big) i_{\widehat{C}}(x, u, v)$ and then use Corollary 2.6.31 to have $\Gamma_{\text{seq}}\big(X^-, (Y \times Z)^-\big)\psi(x, u, v)$, which is what we want.

PROPOSITION 2.6.35 *For all* $(x, u, v) \in X \times Y \times Z$ *we have*

$$\Gamma_{\text{seq}}\big(X, (Y \times Z)^-\big)\widehat{J}(x, u, v) = \int_0^b \widehat{L}^{**}\big(t, x(t), u(t), v(t)\big)\mathrm{d}t,$$

where $\widehat{L}^{**}(t, x, u, v)$ *denotes the second conjugate of* $\widehat{L}(t, x, \cdot, \cdot)$.

PROOF: From hypothesis H$(L)_2$(ii) we have

$$\begin{aligned}\widehat{J}(x, u, v) \leq \widehat{J}(y, u, v) &+ \int_0^b \vartheta_r(t, \|x - y\|_{C(T,\mathbb{R}^N)})\mathrm{d}t \\ &+\ \sigma_r(\|x - y\|_{C(T,\mathbb{R}^N)})\widehat{J}(y, u, v)\end{aligned}$$

for all $x, y \in X$, with $\|x\|_{C(T,\mathbb{R}^N)}$, $\|y\|_{C(T,\mathbb{R}^N)} \leq r, r > 0$ and all $u \in Y,\ v \in Z$. Set

$$w(x, y) = \int_0^b \vartheta_r(t, \|x - y\|_{C(T,\mathbb{R}^N)})\mathrm{d}t + \sigma_r(\|x - y\|_{C(T,\mathbb{R}^N)})$$

$$\text{and} \quad h(x, u, v) = 1 + \widehat{J}(x, u, v).$$

We can apply Proposition 2.6.33 and have

$$\Gamma_{\text{seq}}\big(X,(Y\times Z)^{-}\big)\widehat{J}(x,u,v)=\Gamma_{\text{seq}}\big(\delta_X,(Y\times Z)^{-}\big)\widehat{J}(x,u,v).$$

But $\Gamma_{\text{seq}}\big(\delta_X,(Y\times Z)^{-}\big)\widehat{J}(x,u,v)=\int_0^b \widehat{L}^{**}\big(t,x(t),u(t),v(t)\big)dt$ (see Buttazzo [126, p. 74], and Denkowski–Migórski–Papageorgiou [195, p. 588]). □

PROPOSITION 2.6.36 *For all* $(x,u,v)\in X\times Y\times Z$ *we have*

$$\Gamma_{\text{seq}}(X^{-},Y\times Z)\chi_{\widehat{S}}(x,u,v)=\chi_{\widehat{S}}(x,u,v).$$

PROOF: We need to show that

$$\begin{aligned}&\text{if } (x_n,u_n,v_n)\longrightarrow(x,u,v) \text{ in } X\times Y\times Z \text{ and } (x_n,u_n,v_n)\in\widehat{S}\\ &\quad\text{for all } n\geq 1, \text{ then } (x,u,v)\in\widehat{S}\end{aligned}\tag{2.106}$$

and

$$\begin{aligned}&\quad\text{if } (x,u,v)\in\widehat{S} \text{ and } (u_n,v_n)\longrightarrow(u,v) \text{ in } Y\times Z,\\ &\text{then there exists } x_n\longrightarrow x \text{ in } X \text{ with } (x_n,u_n,v_n)\in\widehat{S},\ n\geq 1.\end{aligned}\tag{2.107}$$

First we prove (2.106). We have

$$\left\{\begin{array}{l} x_n'(t)=A\big(t,x_n(t)\big)+C\big(t,x_n(t)\big)v_n(t) \text{ a.e. on } T\\ x_n(0)=x_0\end{array}\right\}.$$

By virtue of hypotheses H(A) and H(C) and because $v_n\xrightarrow{w}v$ in $L^1(T,\mathbb{R}^k)$ (recall that $Z=L^1(T,\mathbb{R}^k)$ is furnished with the weak topology), we have

$$A\big(\cdot,x_n(\cdot)\big)+C\big(\cdot,x_n(\cdot)\big)v_n(\cdot)\xrightarrow{w}A\big(\cdot,x(\cdot)\big)+C\big(\cdot,x(\cdot)\big)v(\cdot) \text{ in } L^1(T,\mathbb{R}^N).$$

Because $x_n\longrightarrow x$ in $C(T,\mathbb{R}^N)$ (recall that $X=W^{1,1}\big((0,b),\mathbb{R}^N\big)$ is furnished with the $C(T,\mathbb{R}^N)$-norm topology), in the limit as $n\to\infty$, we have

$$\left\{\begin{array}{l} x'(t)=A\big(t,x(t)\big)+C\big(t,x(t)\big)v(t) \text{ a.e. on } T\\ x(0)=x_0\end{array}\right\}.\tag{2.108}$$

So we have proved (2.106).

Next let $(x,u,v)\in X\times Y\times Z$ be such that it satisfies (2.108) and suppose $(u_n,v_n)\longrightarrow(u,v)$ in $Y\times Z$. For every $n\geq 1$, let $x_n\in W^{1,1}\big((0,b),\mathbb{R}^N\big)$ be the unique solution of the Cauchy problem

$$\left\{\begin{array}{l} x_n'(t)=A\big(t,x_n(t)\big)+C\big(t,x_n(t)\big)v_n(t) \text{ a.e. on } T\\ x_n(0)=x_0\end{array}\right\}.\tag{2.109}$$

Because $\sup\limits_{n\geq 1}\|v_n\|_{L^1(T,\mathbb{R}^k)}\leq M_2<+\infty$, then from (2.109), after integration over $[0,t]$ and using hypothesis H(A)(iii), H(C)(iii), and Gronwall's inequality, we obtain $r>0$ such that $\sup\limits_{n\geq 1}\|v_n\|_{L^1(T,\mathbb{R}^k)}\leq r$. From the Arzela–Ascoli theorem it follows that $\{x_n\}_{n\geq 1}\subseteq C(T,\mathbb{R}^N)$ is relatively compact and so we may assume that $x_n\longrightarrow\widehat{x}$ in $C(T,\mathbb{R}^N)$. In the limit we have

$$\left\{\begin{array}{l}\widehat{x}'(t)=A\big(t,\widehat{x}(t)\big)+C\big(t,\widehat{x}(t)\big)v(t) \text{ a.e. on } T\\ x(0)=x_0\end{array}\right\}.$$

Given the uniqueness of the solution of (2.108), it follows that $\widehat{x}=x$ and so we have proved (2.107) and also the proposition. □

Combining Propositions 2.6.35 and 2.6.36 with Corollary 2.6.31, we have

$$\Gamma_{\text{seq}}(X^-,Y^-)\varphi(x,u)=\inf\Big\{\int_0^b \widehat{L}^{**}\big(t,x(t),u(t),v(t)\big)\mathrm{d}t : v\in Z,$$
$$x'(t)=A\big(t,x(t)\big)+C\big(t,x(t)\big)v(t) \text{ a.e. on T}, \ x(0)=x_0\Big\}.$$

The final step in the derivation of the relaxed problem, is to eliminate the variable $v\in Z$ from the right-hand side expression. To this end let

$$G(t,u)=\big\{v\in\mathbb{R}^k : (u,v)\in\overline{\text{conv}}\{(u',v')\in\mathbb{R}^m\times\mathbb{R}^k : v'=g(t,u')\}\big\}$$
$$\overline{L}(t,x,u,w)=\inf\big\{\widehat{L}^{**}(t,x,u,v) : w=A(t,x)+C(t,x)v\big\}$$
$$\overline{S}=\big\{(x,u)\in X\times Y : \ x'(t)\in A\big(t,x(t)\big)+C\big(t,x(t)\big)G\big(t,u(t)\big)$$
$$\text{a.e. on } T, \ x(0)=x_0\big\}.$$

THEOREM 2.6.37 *If hypotheses $H(A)$, $H(C)$, $H(g)$, $H(U)_2$, and $H(L)_2$ hold, then* $\Gamma_{\text{seq}}(X^-,Y^-)\,\varphi(x,u)=\int_0^b \overline{L}\big(t,x(t),u(t),x'(t)\big)\mathrm{d}t+i_{\overline{S}}\,(x,u)$.

PROOF: Note that $v\in G(t,u)$, when $\widehat{L}^{**}(t,x,u,v)<+\infty$. Therefore

$$\Gamma_{\text{seq}}(X^-,Y^-)\,\varphi(x,u)=\inf\Big\{\int_0^b \widehat{L}^{**}\big(t,x(t),u(t),x'(t)\big)\mathrm{d}t : v\in Z, \ x'(t)$$
$$=A\big(t,x(t)\big)+C\big(t,x(t)\big)v(t) \text{ a.e. on } T\Big\}+i_{\overline{S}}\,(x,u).$$

Note that if $x'(t)=A\big(t,x(t)\big)+C\big(t,x(t)\big)v(t)$ a.e. on T, we have

$$\overline{L}\big(t,x(t),u(t),x'(t)\big)\le\widehat{L}^{**}\big(t,x(t),u(t),v(t)\big) \quad \text{a.e. on } T,$$
$$\Rightarrow \int_0^b \overline{L}\big(t,x(t),u(t),x'(t)\big)\mathrm{d}t+i_{\overline{S}}\,(x,u)\le\Gamma_{\text{seq}}(X^-,Y^-)\,\varphi(x,u). \tag{2.110}$$

On the other hand if $(x,u)\in\overline{S}$ and $\int_0^b \overline{L}\big(t,x(t),u(t),x'(t)\big)\mathrm{d}t<+\infty$, then $\overline{L}\big(t,x(t),u(t),x'(t)\big)<+\infty$ a.e. on T and so we must have

$$\overline{L}\big(t,x(t),u(t),x'(t)\big)=\widehat{L}^{**}\big(t,x(t),u(t),v(t)\big) \quad \text{a.e. on } T,$$

with $v\in L^1(T,\mathbb{R}^k)$ such that $x'(t)=A\big(t,x(t)\big)+C\big(t,x(t)\big)v(t)$ a.e. on T (see Theorem 2.6.8 and hypothesis $\text{H}(L)_2$). It follows that

$$\Gamma_{\text{seq}}(X^-,Y^-)\,\varphi(x,u)\le\int_0^b \widehat{L}^{**}\big(t,x(t),u(t),v(t)\big)\mathrm{d}t+i_{\overline{S}}\,(x,u)$$
$$=\int_0^b \overline{L}\big(t,x(t),u(t),x'(t)\big)\mathrm{d}t+i_{\overline{S}}\,(x,u). \tag{2.111}$$

Comparing (2.110) and (2.111), we obtain the desired equality. □

So the new relaxed problem corresponding to (2.104) is:

$$\inf\Big[\int_0^b \overline{L}(t,x(t),u(t),x'(t))dt + i_{\overline{S}}(x,u) : (x,u) \in X \times Y\Big] = \widehat{m}_r^4. \tag{2.112}$$

Then by virtue of Theorem 2.1.20, we have the following.

THEOREM 2.6.38 *If hypotheses $H(A)$, $H(L)$, $H(g)$, $H(U)_2$, and $H(L)_2$ hold, then problem (2.112) has a solution $(\widehat{x},\widehat{u}) \in X \times Y$ and $\widehat{m} = \widehat{m}_r^4$.*

To relate problem (2.112) with the relaxed problems introduced earlier we need to strengthen the hypotheses on g and U.

H(g)$_1$: $g\colon T \times \mathbb{R}^m \longrightarrow \mathbb{R}^k$ is a Carathéodory function (i.e., measurable in $t \in T$ and continuous in $x \in \mathbb{R}^m$) and for almost all $t \in T$ and all $\|u\| \leq r$, we have $\|g(t,u)\| \leq p(t)$ with $p \in L^1(T)_+$ (here $r>0$ as in $H(U)_1$).

For the control constraint set U we assume hypothesis $H(U)_1$. Then from (2.85) with $f(t,x,u) = A(t,x) + C(t,x)g(t,u)$, we have the following alternative relaxation of (2.104).

$$\left\{\begin{array}{ll} & J_r^1(x,\lambda) = \int_0^b \int_{\overline{B}_r} L\big(t,x(t),u(t)\big)\lambda(t)(du)dt \longrightarrow \inf = \widehat{m}_r^1 \\ \text{s.t.} & x'(t) = A\big(t,x(t)\big) + C\big(t,x(t)\big)\int_{\overline{B}_r} g(t,u)\lambda(t)(du) \quad \text{a.e. on } T \\ & x(0) = x_0,\ \lambda \in S_\Sigma \end{array}\right\}. \tag{2.113}$$

DEFINITION 2.6.39 Given an admissible original control $u(\cdot)$, the *barycenter* of $u(\cdot)$ is defined by

$$\operatorname{Bar}(u) = \Big\{\lambda \in S_\Sigma : u(t) = \int_{\overline{B}_r} u\lambda(t)(du) \text{ a.e. on } T\Big\}.$$

In what follows, for convenience we set

$$\overline{\varphi}(x,u) = \Gamma_{\text{seq}}(X^-,Y^-)\,\varphi(x,u). \tag{2.114}$$

PROPOSITION 2.6.40 *If hypotheses $H(A)$, $H(C)$, $H(g)$, $H(U)_1$, and $H(L)_2$ hold, then $\overline{\varphi}(x,u) = \min\big\{J_r^1(x,\lambda) : (x,\lambda)$ is admissible for (2.113) and $\lambda \in \operatorname{Bar}(u)\big\}$.*

PROOF: Let (x,λ) be an admissible state-control pair for problem (2.113) with $\lambda \in \operatorname{Bar}(u)$. Then by virtue of Proposition 2.6.17 we can find original controls $\{u_n\}_{n\geq 1}$ such that $\delta_{u_n} \longrightarrow \lambda$ narrowly in $R(T,\overline{B}_r)$. Let $x_n \in W^{1,1}\big((0,b),\mathbb{R}^N\big)$ be the unique state produced by control u_n. From Theorem 2.6.10 we know that $\{x_n\}_{n\geq 1} \subseteq C(T,\mathbb{R}^N)$ is relatively compact and so we may assume that $x_n \longrightarrow x$ in $C(T,\mathbb{R}^N)$. Also because of hypothesis $H(U)_1$, we may assume that $u_n \xrightarrow{w} \widehat{u}$ in $L^1(T,\mathbb{R}^m)$. For all $D \in \mathbf{B}(T)$ we have

$$\int_D u_n(t)dt \longrightarrow \int_D \widehat{u}(t)dt. \tag{2.115}$$

Moreover, because $\delta_{u_n} \longrightarrow \lambda$ narrowly and $\lambda \in \text{Bar } u$, we have

$$\int_D u_n(t)dt = \int_D \int_{\overline{B}_r} u\delta_{u_n(t)}(du)dt \longrightarrow \int_D \int_{\overline{B}_r} u\lambda(t)(du)dt = \int_D u(t)dt. \quad (2.116)$$

From (2.115), (2.116), and because $D \in \mathbf{B}(T)$ was arbitrary, we infer that $\widehat{u} = u$ and so $u_n \xrightarrow{w} u$ in $L^1(T, \mathbb{R}^m)$. Because (x_n, u_n) is admissible for the original optimal control problem, we have

$$\varphi(x_n, u_n) = J(x_n, u_n),$$

$$\Rightarrow \liminf_{n\to\infty} \varphi(x_n, u_n) = \liminf_{n\to\infty} J(x_n, u_n) = \liminf_{n\to\infty} J_r^1(x_n, \delta_{u_n}) = J_r^1(x, \lambda)$$

(see $\mathrm{H}(L)_2$),

$$\Rightarrow \overline{\varphi}(x, u) \leq \inf \left[J_r^1(x, \lambda) : (x, \lambda) \text{ is admissible for (2.113)}, \lambda \in \text{Bar } u\right]. \quad (2.117)$$

On the other hand, if $\varphi(x, u) < +\infty$, from the definition of $\overline{\varphi}$ (see (2.114)), given $\varepsilon > 0$, we can find $x_n \longrightarrow x$ in $C(T, \mathbb{R}^N)$ and $u_n \xrightarrow{w} u$ in $L^1(T, \mathbb{R}^m)$ such that

$$\liminf_{n\to\infty} J(x_n, u_n) \leq \overline{\varphi}(x, u) + \varepsilon. \quad (2.118)$$

By Alaoglu's theorem we may also assume that $\delta_{u_n} \xrightarrow{w^*} \lambda$ in $L^\infty(T, M(\overline{B}_r)_{w^*})$ (hence narrowly in $R(T, \overline{B}_r)$). Then for all $D \in \mathbf{B}(T)$ we have

$$\int_D u_n(t)dt = \int_D \int_{\overline{B}_r} u\delta_{u_n(t)}(du)dt \longrightarrow \int_D \int_{\overline{B}_r} u\lambda(t)(du)dt \quad (2.119)$$

and
$$\int_D u_n(t)dt \longrightarrow \int_D u(t)dt. \quad (2.120)$$

From (2.119), (2.120), and because $D \in \mathbf{B}(T)$ was arbitrary it follows that $u(t) = \int_{\overline{B}_r} u\lambda(t)(du)$ a.e. on T (i.e., $\lambda \in \text{Bar } u$). Also because of hypotheses $\mathrm{H}(L)_2$ we have $J(x_n, u_n) \longrightarrow J_r^1(x, \lambda)$. Therefore

$$J_r^1(x, \lambda) \leq \overline{\varphi}(x, u) + \varepsilon \quad \text{(see (2.118))}.$$

Let $\varepsilon \downarrow 0$, to conclude that

$$J_r^1(x, \lambda) \leq \overline{\varphi}(x, u). \quad (2.121)$$

Comparing (2.117) and (2.121), we obtain the desired equation. □

This proposition leads at once to the equivalence of relaxed problems (2.112) and (2.113).

THEOREM 2.6.41 *If hypotheses $H(A)$, $H(C)$, $H(g)$, $H(U)_1$, and $H(L)_2$ hold, then $\widehat{m} = \widehat{m}_r^1 = \widehat{m}_r^4$ and problems (2.112) and (2.113) are equivalent; that is, if $(\widehat{x}, \widehat{u})$ solves (2.112), then we can find $\widehat{\lambda} \in$ Bar(u) such that $(\widehat{x}, \widehat{\lambda})$ solves (2.113) and conversely if $(\widehat{x}, \widehat{\lambda})$ solves (2.113) we can find an admissible original control $\widehat{u}$ such that $(\widehat{x}, \widehat{u})$ solves (2.112). Both problems have optimal pairs.*

REMARK 2.6.42 Evidently (2.112) is also equivalent to the other two relaxed problems defined in (2.90) and (2.94) with $f(t,x,u) = A(t,x) + C(t,x)g(t,u)$.

In the last part of this section we derive the *Pontryagin maximum principle* for the following optimal control problem.

$$\left\{\begin{array}{ll} & J(x,u) = \int_0^b L\big(t,x(t),u(t)\big)dt \longrightarrow \inf = m \\ \text{s.t.} & x'(t) = f\big(t,x(t),u(t)\big) \quad \text{a.e. on } T=[0,b] \\ & \big(x(0),x(b)\big) \in C,\ u \in U_{\text{ad}} \end{array}\right\}. \tag{2.122}$$

Here $C \subseteq \mathbb{R}^N \times \mathbb{R}^N$ is nonempty and $U_{\text{ad}} = \{u \colon T \longrightarrow Y \text{ measurable}\}$, where Y is a separable metric space (the control space). Note the general endpoint constraints that include the periodic problem as a special case. Our derivation of the Pontryagin maximum principle is based on the Ekeland variational principle (see Theorem 2.4.1).

The precise mathematical setting for problem (2.122) is the following.

The state space is $X = \mathbb{R}^N$ and the control space Y is a separable metric space. We impose the following conditions on the data of (2.122).

$\mathbf{H(f)_2}$: $f \colon T \times \mathbb{R}^N \times Y$ is a function such that

(i) For all $(x,u) \in \mathbb{R}^N \times Y$, $t \longrightarrow f(t,x,u)$ is measurable.
(ii) For all $t \in T$ and all $u \in Y$, $x \longrightarrow f(t,x,u)$ is differentiable and $(x,u) \longrightarrow f'_x(t,x,u)$ is continuous;.
(iii) There exists $M > 0$ such that $\|f(t,0,u)\|$, $\|f'_x(t,x,u)\| \leq M$ for all $(t,x,u) \in T \times \mathbb{R}^N \times Y$.

$\mathbf{H(C)}$: $C \subseteq \mathbb{R}^N \times \mathbb{R}^N$ is nonempty, closed, and convex.

$\mathbf{H(L)_3}$: $L \colon T \times \mathbb{R}^N \times Y \longrightarrow \mathbb{R}$ is a function such that

(i) For all $(x,u) \in \mathbb{R}^N \times Y$, $t \longrightarrow L(t,x,u)$ is measurable.
(ii) For all $t \in T$ and all $u \in Y, x \longrightarrow L(t,x,u)$ is differentiable and $(x,u) \longrightarrow L'_x(t,x,u)$ is continuous.
(iii) There exists $M_1 > 0$ such that $\|L(t,0,u)\|$, $\|L'_x(t,x,u)\| \leq M_1$ for all $(t,x,u) \in T \times \mathbb{R}^N \times Y$.

The control space Y is only a separable metric space and does not have a linear structure, therefore we cannot talk about convexity. For this reason the variations of the control function $u(\cdot)$ are of special type and are called *spike perturbations* of u. For this reason we need to use measure-theoretic arguments when dealing with such perturbations. In particular we need to know how we can approximate the constant function 1 by an oscillatory function of the form $(1/\lambda)\,\chi_C$, $\lambda \in (0,1)$ (here χ_C denotes the characteristic function of C).

LEMMA 2.6.43 *If (Ω, Σ, μ) is a finite nonatomic measure space, X is a separable Banach space, and for any $\lambda \in (0,1)$, $\mathcal{S}_\lambda = \big\{C \in \Sigma : \mu(C) = \lambda\mu(\Omega)\big\}$, then for any $g \in L^1(\Omega, X)$ we have*

$$\inf_{C \in \mathcal{S}_\lambda} \Big\| \int_\Omega \big(\frac{1}{\lambda}\,\chi_C(\omega) - 1\big) g(\omega) d\mu \Big\| = 0. \tag{2.123}$$

PROOF: Given $\varepsilon > 0$, we can find a simple function $s_\varepsilon(\omega) = \sum\limits_{k=1}^{n} \chi_{A_k}(\omega)x_k$ with $A_k \in \Sigma$, $x_k \in X$ such that

$$\|g - s_\varepsilon\|_{L^1(\Omega,X)} < \varepsilon. \tag{2.124}$$

Due to the nonatomicity of μ, for each $k \in \{1, \ldots, n\}$ we can find $A_k^\lambda \subseteq A_k$ such that

$$\mu(A_k^\lambda) = \lambda\mu(A_k).$$

Let $A_\lambda = \bigcup\limits_{k=1}^{n} A_k^\lambda \in \mathcal{S}_\lambda$. Then

$$\int_\Omega \Big(\frac{1}{\lambda}\,\chi_{A_\lambda}(\omega) - 1\Big)s_\varepsilon(\omega)d\mu = \sum_{k=1}^{n}\Big(\frac{1}{\lambda}\mu(A_k^\lambda) - \mu(A_k)\Big)x_k = 0,$$

$$\begin{aligned}\Rightarrow \Big\|\int_\Omega \Big(\frac{1}{\lambda}\,\chi_{A_\lambda}(\omega) - 1\Big)g(\omega)d\mu\Big\|_X &\le \Big\|\int_\Omega \Big(\frac{1}{\lambda}\,\chi_{A_\lambda}(\omega) - 1\Big)s_\varepsilon(\omega)d\mu\Big\|_X + \\ &\quad + \Big(1 + \frac{1}{\lambda}\Big)\|s_\varepsilon - g\|_{L^1(\Omega,X)} \\ &\le \Big(1 + \frac{1}{\lambda}\Big)\,\varepsilon \quad \text{(see (2.124))}.\end{aligned}$$

Because $\varepsilon > 0$, was arbitrary, we conclude that (2.123) holds. □

In dealing with the endpoint constraints, we use the distance function from the set C. Recall that if Z is a Banach space and D is a nonempty, closed, and convex subset of Z, the distance function from D defined by $d_D(z) = \inf\{\|z - u\|_Z : u \in D\}$ is both convex and Lipschitz continuous with Lipschitz constant 1. So the convex subdifferential (see Definition 1.2.28) and the generalized subdifferential (see Definition 1.3.5) coincide and we have

$$\begin{aligned}\partial d_D(z) &= \{z^* \in Z^* : \langle z^*, z' - z\rangle_Z \le d_D(z') - d_D(z) \quad \text{for all } z' \in Z\} \\ &= \{z^* \in Z^* : \langle z^*, h\rangle_Z \le d_D^0(z; h) \quad \text{for all } h \in Z\}.\end{aligned} \tag{2.125}$$

LEMMA 2.6.44 *If Z is a Banach space and $D \subseteq Z$ is nonempty, closed, and convex, then for any $z \notin D$ and any $z^* \in \partial d_D(z)$ we have $\|z^*\|_{Z^*} = 1$.*

PROOF: Because $d_D(\cdot)$ is globally Lipschitz with constant 1, we have $\|z^*\|_{Z^*} \le 1$. Because $z \notin D$, given any $0 < \delta < 1$, we can find $u_\delta \in D$ such that

$$0 < (1 - \delta)\|z - u_\delta\|_Z \le d_D(z).$$

So by virtue of (2.125) we have

$$\begin{aligned}&\langle z^*, u_\delta - z\rangle_Z \le -d_D(z) \\ \Rightarrow\ &(1 - \delta)\|z - u_\delta\|_Z \le d_D(z) \le -\langle z^*, u_\delta - z\rangle_Z \le \|z^*\|_{Z^*}\|u_\delta - z\|_Z \\ \Rightarrow\ &(1 - \delta) \le \|z^*\|_{Z^*}.\end{aligned}$$

Because $\delta \in (0, 1)$ was arbitrary, we let $\delta \downarrow 0$, to conclude that $\|z^*\|_{Z^*} = 1$. □

COROLLARY 2.6.45 *If Z is a Banach space with a strictly convex dual Z^* and $D \subseteq Z$ is a nonempty, closed, and convex set, then for any $z \notin D$ and the set $\partial d_D(z)$ is a singleton.*

As we already mentioned, the derivation of the Pontryagin maximum principle is based on the Ekeland variational principle. We need to endow the set of admissible controls U_{ad} with a metric structure. So for $u,\ w \in U_{\text{ad}}$ we define

$$\widehat{d}(u,w) = \big|\{t \in T : u(t) \neq w(t)\}\big|_1.$$

Recall that by $|\cdot|_1$ we denote the one-dimensional Lebesgue measure. It is easy to check that $\widehat{d}$ is a metric and so $(U_{\text{ad}}, \widehat{d})$ is a metric space.

LEMMA 2.6.46 *$(U_{\text{ad}}, \widehat{d})$ is a complete metric space.*

PROOF: Let $\{u_n\}_{n\geq 1} \subseteq U_{\text{ad}}$ be a $\widehat{d}$-Cauchy sequence. By passing to a suitable subsequence, we may assume that

$$\widehat{d}(u_{n+1}, u_n) \leq \frac{1}{2^n}, \quad n \geq 1.$$

Set $D_{nm} = \{t \in T : u_n(t) \neq u_m(t)\}$, $n,m \geq 1$, and $C_k = \bigcup\limits_{n\geq k} D_{n(n+1)}$, $k \geq 1$. Clearly $\{C_k\}_{k\geq 1}$ is decreasing and we have

$$|C_k|_1 \leq \sum_{n\geq k} \frac{1}{2^n} = \frac{1}{2^{k-1}}, \quad k \geq 1.$$

It follows that $\Big|\bigcup\limits_{k\geq 1} C_k^c\Big|_1 = b$. We set $\widehat{u}(t) = u_k(t)$ if $t \in C_k^c$, $k \geq 1$. From the definition of C_k it is clear that $\widehat{u}$ is well-defined and $\widehat{u} \in U_{\text{ad}}$. Moreover,

$$\widehat{d}(u_k, \widehat{u}) \leq |C_k|_1 \longrightarrow 0 \quad \text{as } k \longrightarrow \infty.$$

$\square$

REMARK 2.6.47 A careful reading of the above proof reveals that the lemma remains valid if Y is only a measurable space without any metric structure. In addition T can be replaced by any nonatomic measure space.

Now we are ready to state and prove the Pontryagin maximum principle for problem (2.122).

THEOREM 2.6.48 *If hypotheses $H(f)_2$, $H(C)$ and $H(L)_3$ hold and $(\widehat{x}, \widehat{u})$ is an optimal pair for problem (2.122), then there exists $(\psi_0, \psi) \in \mathbb{R} \times W^{1,1}((0,b), \mathbb{R}^N)$, $(\psi_0, \psi) \neq (0,0)$ such that*

(a) $\psi_0 \leq 0$.

(b) $\psi'(t) = f'_x\big(t, \widehat{x}(t), \widehat{u}(t)\big)^* \psi(t) + \psi_0 L'_x\big(t, \widehat{x}(t), \widehat{u}(t)\big)$ *a.e. on T.*

(c) $\big(\psi(0), x_0 - \widehat{x}(0)\big)_{\mathbb{R}^N} \leq \big(\psi(b), x_1 - \widehat{x}(b)\big)_{\mathbb{R}^N}$ *for all* $(x_0, x_1) \in C$.

(d) *If* $H(t,x,u,v_0,v) = \big(v, f(t,x,u)\big)_{\mathbb{R}^N} + v_0 L(t,x,u)$ *for all* $(t,x,u,v_0,v) \in T \times \mathbb{R}^N \times Y \times \mathbb{R} \times \mathbb{R}^N$, *then* $H\big(t, \widehat{x}(t), \widehat{u}(t), \psi_0, \psi(t)\big) = \max\limits_{u \in Y} H\big(t, \widehat{x}(t), u, \psi_0, \psi(t)\big)$ *for almost all* $t \in T$.

PROOF: First let us briefly outline the strategy of the proof. It is divided into a series of steps.

(i) We introduce a penalty function that is used to define an approximation of problem (2.122) (called the approximate problem) which has no end point constraint.
(ii) We use the Ekeland variational principle to produce an optimal pair for the approximate problem. This pair is actually close to the original optimal pair.
(iii) We derive necessary conditions for the approximate optimal pair.
(iv) We pass to the limit to obtain necessary conditions for the original optimal pair.

On $\mathbb{R}^N \times U_{\text{ad}}$ we introduce the metric

$$\widehat{d}_0\big((x,u),(z,w)\big) = \|x - z\|_{\mathbb{R}^N} + \widehat{d}(u,w).$$

Clearly $(\mathbb{R}^N \times U_{\text{ad}}, \widehat{d}_0)$ is a complete metric space (see Lemma 2.6.46). In what follows by $x(x_0,u)(\cdot) \in W^{1,1}\big((0,b),\mathbb{R}^N\big)$ we denote the unique state generated by $u \in U_{\text{ad}}$ and the initial condition $x_0 \in \mathbb{R}^N$. Then we can write the cost functional $J(x_0,u)$. If $\widehat{x}(0) = \widehat{x}_0$, then by translating the cost functional if necessary (i.e., considering the cost functional $J(x_0,u) - J(\widehat{x}_0,\widehat{u})$), we may assume without any loss of generality that $J(\widehat{x}_0,\widehat{u}) = 0$.

We start executing the four steps of the method of proof outlined above.

Step i: We introduce the penalty function.

For $\varepsilon > 0$ and $(x_0,u) \in \mathbb{R}^N \times U_{\text{ad}}$, we set

$$J_\varepsilon(x_0,u) = \Big[d_C^2\big(x_0, x(x_0,u)(b)\big) + \big[\big(J(x_0,u)+\varepsilon\big)^+\big]^2\Big]^{1/2}. \tag{2.126}$$

Because of hypotheses $\mathrm{H}(f)_2$ and $\mathrm{H}(L)_3$ and Gronwall's inequality for all (x_0,u), $(\overline{x}_0,\overline{u}_0) \in \mathbb{R}^N \times U_{\text{ad}}$ we have

$$\|x(x_0,u)(\cdot) - x(\overline{x}_0,\overline{u}_0)(\cdot)\|_{\mathbb{R}^N} \leq c(1+\max\{\|x_0\|,\|\overline{x}_0\|\}) \times \widehat{d}_0\big((x_0,u),(\overline{x}_0,\overline{u}_0)\big) \tag{2.127}$$

and $$|J(x_0,u) - J(\overline{x}_0,\overline{u}_0)| \leq c\big(1+\max\{\|x_0\|,\|\overline{x}_0\|\}\widehat{d}_0\big((x_0,u),(\overline{x}_0,\overline{u}_0)\big), \quad \text{for some } c>0. \tag{2.128}$$

From (2.126) through (2.128) it follows that $J_\varepsilon(\cdot,\cdot)$ is continuous on $(\mathbb{R}^N \times U_{\text{ad}}, \widehat{d}_0)$.

Step ii: Solution (via the Ekeland variational principle) of the problem with the same dynamics and cost functional a new expression involving J_ε.

From (2.126) we see that $J_\varepsilon \geq 0$ and $J_\varepsilon(\widehat{x}_0,\widehat{u}) = \varepsilon \leq \inf\limits_{\mathbb{R}^N \times U_{\text{ad}}} J_\varepsilon(x_0,u) + \varepsilon$ (recall that $\widehat{x}(0) = \widehat{x}_0$ and that we have assumed that $J(\widehat{x}_0,\widehat{u}) = 0$). So we can apply Corollary 2.4.3 and obtain $(x_0^\varepsilon, u^\varepsilon) \in \mathbb{R}^N \times U_{\text{ad}}$ such that

$$\widehat{d}_0\big((x_0^\varepsilon, u^\varepsilon), (\widehat{x}_0, \widehat{u})\big) \le \sqrt{\varepsilon} \tag{2.129}$$

$$\text{and} \quad J_\varepsilon(x_0^\varepsilon, u^\varepsilon) \le J_\varepsilon(x_0, u) + \sqrt{\varepsilon}\, \widehat{d}_0\big((x_0, u), (x_0^\varepsilon, u^\varepsilon)\big) \quad \text{for all } (x_0, u) \in \mathbb{R}^N \times U_{\mathrm{ad}}. \tag{2.130}$$

So if $x^\varepsilon = x(x_0^\varepsilon, u^\varepsilon)$, then $(x^\varepsilon, u^\varepsilon)$ is a solution of the approximate optimal control problem with the same dynamics and cost functional as the right-hand side of (2.130). Note that because of (2.129) the optimal pair of the approximate problem is located close to the optimal pair of the original problem.

Step iii: We derive necessary conditions for the optimal pair $(x^\varepsilon, u^\varepsilon)$ of the approximate problem.

Fix $(e_0, w) \in \overline{B}_1 \times U_{\mathrm{ad}}$ ($\overline{B}_1 = \{e_0 \in \mathbb{R}^N : \|e_0\|_{\mathbb{R}^N} \le 1\}$) and $\lambda \in (0, 1]$ and set

$$g(s) = \begin{pmatrix} L\big(s, x^\varepsilon(s), w(s)\big) - L\big(s, x^\varepsilon(s), u^\varepsilon(s)\big) \\ f\big(s, x^\varepsilon(s), w(s)\big) - f\big(s, x^\varepsilon(s), u^\varepsilon(s)\big) \end{pmatrix}.$$

From Lemma 2.6.43 we know that given any $\delta > 0$. we can find $C_\lambda \subseteq T$ with $|C_\lambda|_1 = \lambda b$ such that

$$\Big\| \int_0^b g(s)\mathrm{d}s - \frac{1}{\lambda} \int_0^b \chi_{C_\lambda}(s) g(s)\mathrm{d}s \Big\|_{\mathbb{R}^N} \le \delta. \tag{2.131}$$

We introduce the following spike perturbation of the optimal control $u^\varepsilon(\cdot)$,

$$u_\lambda^\varepsilon(t) = \begin{cases} u^\varepsilon(t) & \text{if } t \in T \setminus C_\lambda \\ w(t) & \text{if } t \in C_\lambda \end{cases}. \tag{2.132}$$

Clearly $u_\lambda^\varepsilon \in U_{\mathrm{ad}}$ and $\widehat{d}(u_\lambda^\varepsilon, u^\varepsilon) \le |C_\lambda|_1 = \lambda b$. Set $x_\lambda^\varepsilon = x(x_0^\varepsilon + \lambda e_0, u_\lambda^\varepsilon)$. We define $(z^\varepsilon, z_0^\varepsilon) \in W^{1,1}\big((0, b), \mathbb{R}^N\big) \times \mathbb{R}$ as follows

$$(z^\varepsilon)'(t) = f'_x\big(t, x^\varepsilon(t), u^\varepsilon(t)\big) z^\varepsilon(t) + f\big(t, x^\varepsilon(t), w(t)\big) - f\big(t, x^\varepsilon(t), u^\varepsilon(t)\big)$$
$$\text{a.e. on } T, \; z^\varepsilon(0) = e_0 \quad \text{and}$$
$$z_0^\varepsilon = \int_0^b L'_x\big(t, x^\varepsilon(t), u^\varepsilon(t)\big) z^\varepsilon(t)\mathrm{d}t + \int_0^b \Big(L\big(t, x^\varepsilon(t), w(t)\big) - L\big(t, x^\varepsilon(t), u^\varepsilon(t)\big)\Big)\mathrm{d}t.$$

Exploiting hypotheses $\mathrm{H}(f)_2$ and $\mathrm{H}(L)_3$, via Gronwall's inequality, we can easily check that

$$\|x_\lambda^\varepsilon - x^\varepsilon - \lambda z^\varepsilon\|_{C(T, \mathbb{R}^N)} = o(\lambda) \tag{2.133}$$

$$\text{and} \quad \big| J(x_0^\varepsilon + \lambda e_0, u_\lambda^\varepsilon) - J(x_0^\varepsilon, u^\varepsilon) - \lambda z_0^\varepsilon \big| = o(\lambda), \tag{2.134}$$

where $\big(o(\lambda)/\lambda\big) \longrightarrow 0$ as $\lambda \downarrow 0$. Recalling that J_ε is continuous on $(\mathbb{R}^N \times U_{\mathrm{ad}}, \widehat{d}_0)$, from (2.133) and (2.133) and the definition of J_ε, we see that

$$J_\varepsilon(x_0^\varepsilon + \lambda e_0, u_\lambda^\varepsilon) = J_\varepsilon(x_0^\varepsilon, u^\varepsilon) + \varepsilon(\lambda) \quad \text{with } \varepsilon(\lambda) \downarrow 0. \tag{2.135}$$

From (2.130) we have

$$
\begin{aligned}
&-\sqrt{\varepsilon}(\|e_0\|_{\mathbb{R}^N}+b) \le \frac{1}{\lambda}\big(J_\varepsilon(x_0^\varepsilon+\lambda e_0,u_\lambda^\varepsilon)-J_\varepsilon(x_0^\varepsilon,u^\varepsilon)\big)\\
&=\frac{1}{J_\varepsilon(x_0^\varepsilon+\lambda e_0,u_\lambda^\varepsilon)+J_\varepsilon(x_0^\varepsilon,u^\varepsilon)}\Big[\frac{1}{\lambda}\big(J_\varepsilon(x_0^\varepsilon+\lambda e_0,u_\lambda^\varepsilon)^2-J_\varepsilon(x_0^\varepsilon,u^\varepsilon)^2\big)\Big]\\
&=\frac{1}{2J_\varepsilon(x_0^\varepsilon,u^\varepsilon)+\varepsilon(\lambda)}\Big[\frac{1}{\lambda}\big(d_C^2(x_0^\varepsilon+\lambda e_0,x_\lambda^\varepsilon(b))-d_C^2(x_0^\varepsilon,x^\varepsilon(b))\big)\Big]\\
&+\frac{1}{\lambda}\Big(\big[\big(J(x_0^\varepsilon+\lambda e_0,u^\varepsilon)+\varepsilon\big)^+\big]^2-\big[\big(J(x_0^\varepsilon,u^\varepsilon)+\varepsilon\big)^+\big]^2\Big) \quad \text{(see (2.133))}.
\end{aligned}
\tag{2.136}
$$

We know that $d_C^2 \in C^1(\mathbb{R}^N\times\mathbb{R}^N)$ and

$$
\nabla d_C^2(x_0,x_1)=\begin{cases}2\,d_C(x_0,x_1)\,\partial d_C(x_0,x_1) & \text{if } (x_0,x_1)\notin C\\ 0 & \text{if } (x_0,x_1)\in C\end{cases}.
$$

Recall that by virtue of Corollary 2.6.45 $\partial d_C(x_0,x_1)=(\alpha_0,\alpha_1)\in\mathbb{R}^N\times\mathbb{R}^N$ and $\|\alpha_0\|_{\mathbb{R}^N}^2+\|\alpha_1\|_{\mathbb{R}^N}^2=1$ (see Lemma 2.6.44), when $(x_0,x_1)\notin C$. On the other hand because d_C is Lipschitz continuous with Lipschitz constant 1, we always have $\partial d_C(x_0,x_1)\subseteq\overline{B}_1=\{(x,z)\in\mathbb{R}^N\times\mathbb{R}^N : \|x\|_{\mathbb{R}^N}^2+\|z\|_{\mathbb{R}^N}^2\le 1\}$. So we can write with no ambiguity that

$$
\begin{aligned}
\nabla d_C^2(x_0,x_1)=2\,d_C(x_0,x_1)(\alpha_0,\alpha_1) \quad &\text{with } (\alpha_0,\alpha_1)\in\partial d_C(x_0,x_1),\\
\|\alpha_0\|_{\mathbb{R}^N}^2+\|\alpha_1\|_{\mathbb{R}^N}^2&=1.
\end{aligned}
$$

Because of (2.133), we have

$$
\begin{aligned}
&\lim_{\lambda\downarrow 0}\frac{d_C^2\big(x_0^\varepsilon+\lambda e_0,x_\lambda^\varepsilon(b)\big)-d_C^2\big(x_0^\varepsilon,x^\varepsilon(b)\big)}{\lambda}\\
&=2\,d_C\big(x_0^\varepsilon,x^\varepsilon(b)\big)\Big((\alpha_0^\varepsilon,e_0)_{\mathbb{R}^N}+\big(\alpha_1^\varepsilon,z^\varepsilon(b)\big)_{\mathbb{R}^N}\Big)
\end{aligned}
$$

where

$$
(\alpha_0^\varepsilon,\alpha_1^\varepsilon)\in\partial d_C\big(x_0^\varepsilon,x^\varepsilon(b)\big) \text{ and } \|\alpha_0^\varepsilon\|_{\mathbb{R}^N}^2+\|\alpha_1^\varepsilon\|_{\mathbb{R}^N}^2=1. \tag{2.137}
$$

Also because of (2.133) we have

$$
\begin{aligned}
&\lim_{\lambda\downarrow 0}\frac{1}{\lambda}\Big[\big[\big(J(x_0^\varepsilon+\lambda e_0,u^\varepsilon)+\varepsilon\big)^+\big]^2-\big[\big(J(x_0^\varepsilon,u^\varepsilon)+\varepsilon\big)^+\big]^2\Big]\\
&=2\,\big(J(x_0^\varepsilon,u^\varepsilon)+\varepsilon\big)^+z_0^\varepsilon.
\end{aligned}
\tag{2.138}
$$

Returning to (2.136), passing to the limit as $\lambda\downarrow 0$ and using (2.137) and (2.138) we obtain

$$
-\sqrt{\varepsilon}(\|e_0\|_{\mathbb{R}^N}+b)\le(\widehat{\varphi}_\varepsilon,e_0)_{\mathbb{R}^N}+\big(\widehat{\psi}_\varepsilon,z_\varepsilon(b)\big)_{\mathbb{R}^N}+\widehat{\psi}_0^\varepsilon z_0^\varepsilon \tag{2.139}
$$

where $\widehat{\varphi}_\varepsilon=\frac{d_C\left(x_0^\varepsilon,x^\varepsilon(b)\right)}{J_\varepsilon(x_0^\varepsilon,u^\varepsilon)}\,\alpha_0^\varepsilon$, $\widehat{\psi}_\varepsilon=\frac{d_C\left(x_0^\varepsilon,x^\varepsilon(b)\right)}{J_\varepsilon(x_0^\varepsilon,u^\varepsilon)}\,\alpha_1^\varepsilon$, and

$$
\widehat{\psi}_0^\varepsilon=\frac{\big(J(x_0^\varepsilon,u^\varepsilon)+\varepsilon\big)^+}{J_\varepsilon(x_0^\varepsilon,u^\varepsilon)}. \tag{2.140}
$$

Because of (2.137) we have

$$\|\widehat{\varphi}_\varepsilon\|^2_{\mathbb{R}^N} + \|\widehat{\psi}_\varepsilon\|^2_{\mathbb{R}^N} + (\widehat{\psi}^\varepsilon_0)^2 = 1, \tag{2.141}$$

and from the definition of the convex subdifferential and because $\partial d_C\big(x^\varepsilon_0, x^\varepsilon(b)\big)$ is a cone we have

$$(\widehat{\varphi}_\varepsilon, x_0 - x^\varepsilon_0)_{\mathbb{R}^N} + \big(\widehat{\psi}_\varepsilon, x_1 - x^\varepsilon(b)\big)_{\mathbb{R}^N} \le 0 \qquad \text{for all } (x_0, x_1) \in C. \tag{2.142}$$

Conditions (2.138)⟶(2.142) are necessary conditions for optimality of $(x^\varepsilon, u^\varepsilon)$.

Step iv: Passage to the limit as $\varepsilon \downarrow 0$.

Let $(z, z^0) \in W^{1,1}\big((0,b), \mathbb{R}^N\big) \times \mathbb{R}$ be defined by

$$z'(t) = f'_x\big(t, \widehat{x}(t), \widehat{u}(t)\big)z(t) + f\big(t, \widehat{x}(t), w(t)\big) - f\big(t, \widehat{x}(t), \widehat{u}(t)\big)$$
$$\text{a.e. on } T, \; z(0) = e_0$$
$$z_0 = \int_0^b L'_x\big(t, \widehat{x}(t), \widehat{u}(t)\big)z(t)\mathrm{d}t + \int_0^b \Big(L\big(t, \widehat{x}(t), w(t)\big) - L\big(t, \widehat{x}(t), \widehat{u}(t)\big)\Big)\mathrm{d}t.$$

Then because of (2.129) and as before via Gronwall's inequality, we have

$$\lim_{\varepsilon \downarrow 0} \big[\|x^\varepsilon_0 - \widehat{x}_0\|_{\mathbb{R}^N} + \|z^\varepsilon - z\|_{C(T,\mathbb{R}^N)} + |z^\varepsilon_0 - z_0|\big] = 0.$$

So from (2.141) and (2.142), we have

$$\begin{aligned}(\widehat{\varphi}_\varepsilon, x_0 - \widehat{x}_0)_{\mathbb{R}^N} + \big(\widehat{\psi}_\varepsilon, x_1 - \widehat{x}(b)\big)_{\mathbb{R}^N} &\le \big(\|x^\varepsilon_0 - \widehat{x}_0\|^2_{\mathbb{R}^N} + \|x^\varepsilon(b) - \widehat{x}(b)\|^2_{\mathbb{R}^N} \\ &= \beta_\varepsilon \longrightarrow 0 \text{ as } \varepsilon \downarrow 0.\end{aligned}$$

Combining this inequality with (2.139), we obtain

$$\begin{aligned}&\big(\widehat{\varphi}_\varepsilon, e_0 - (x_0 - \widehat{x}_0)\big)_{\mathbb{R}^N} + \Big(\widehat{\psi}_\varepsilon, z(b) - \big(x_1 - \widehat{x}(b)\big)\Big)_{\mathbb{R}^N} + \widehat{\psi}^\varepsilon_0 z_0 \\ \ge\; & -\sqrt{\varepsilon}(\|e_0\|_{\mathbb{R}^N} + b) - \beta_\varepsilon - \|z^\varepsilon(b) - z(b)\|_{\mathbb{R}^N} - |z^\varepsilon_0 - z_0| \ge -\gamma_\varepsilon \\ & \text{for all } (x_0, x_1) \in C,\end{aligned} \tag{2.143}$$

where $\gamma_\varepsilon > 0$ independent of $u \in U_{\mathrm{ad}}$, $e_0 \in \overline{B}_1$, and $\gamma_\varepsilon \downarrow 0$ as $\varepsilon \downarrow 0$. Because of (2.141) we may assume that

$$(\widehat{\varphi}_\varepsilon, \widehat{\psi}_\varepsilon, \widehat{\psi}^\varepsilon_0) \longrightarrow (\widehat{\varphi}, \widehat{\psi}, \widehat{\psi}_0) \quad \text{in } \mathbb{R}^N \times \mathbb{R}^N \times \mathbb{R} \text{ as } \varepsilon \downarrow 0.$$

Clearly $(\widehat{\varphi}, \widehat{\psi}, \widehat{\psi}_0) \neq (0,0,0)$ (see (2.141)). Let $\varepsilon \downarrow 0$ in (2.143). We obtain

$$\big(\widehat{\varphi}, e_0 - (x_0 - \widehat{x}_0)\big)_{\mathbb{R}^N} + \Big(\widehat{\psi}, z(b) - \big(x_1 - \widehat{x}(b)\big)\Big)_{\mathbb{R}^N} + \widehat{\psi}_0 z_0 \ge 0 \tag{2.144}$$

for all $(x_0, x_1) \in C$, all $w \in U_{\mathrm{ad}}$, and all $e_0 \in \overline{B}_1$. Let $\psi \in W^{1,1}\big((0,b), \mathbb{R}^N\big)$ be the unique solution of the equation in statement (b) of the theorem $\psi(b) = -\widehat{\psi}$ and $\psi_0 = -\widehat{\psi}_0 \le 0$ (see (2.140)). We rewrite (2.144) as follows,

$$(\widehat{\varphi}, x_0 - \widehat{x}_0 - e_0)_{\mathbb{R}^N} - \big(\psi(b), x_1 - \widehat{x}(b) - z(b)\big)_{\mathbb{R}^N} + \psi_0 z_0 \le 0 \tag{2.145}$$

for all $(x_0, x_1) \in C$, all $w \in U_{\mathrm{ad}}$, and all $e_0 \in \overline{B}_1$.

A straightforward computation gives

$$\begin{aligned}&\big(\psi(b), z(b)\big)_{\mathbb{R}^N} - \big(\psi(0), e_0\big)_{\mathbb{R}^N} + \psi_0 z_0\\&= \int_0^b \Big(H\big(t, \widehat{x}(t), w(t), \psi_0, \psi(t)\big) - H\big(t, \widehat{x}(t), \widehat{u}(t), \psi_0, \psi(t)\big)\Big)\mathrm{d}t,\end{aligned} \tag{2.146}$$

for all $e_0 \in \overline{B}_1$ and all $w \in U_{\mathrm{ad}}$. In (2.145) we set $e_0 = 0$, $x_0 = \widehat{x}_b = \widehat{x}(0)$ and use the resulting inequality in (2.146) to obtain

$$\int_0^b \Big(H\big(t, \widehat{x}(t), w(t), \psi_0, \psi(t)\big) - H\big(t, \widehat{x}(t), \widehat{u}(t), \psi_0, \psi(t)\big)\Big)\mathrm{d}t \leq 0 \tag{2.147}$$

for all $w \in U_{\mathrm{ad}}$.

Recall that Y is separable. So we can find a countable dense set $\{u_m\}_{m\geq 1} \subseteq Y$. Then if

$$h_m(t) = H\big(t, \widehat{x}(t), u_m, \psi_0, \psi(t)\big) - H\big(t, \widehat{x}(t), \widehat{u}(t), \psi_0, \psi(t)\big), \quad m \geq 1,$$

we have $h_m \in L^1(T)$ and we can find $E_m \subseteq T$ with $|E_m|_1 = b$ such that

$$\lim_{\delta\downarrow 0} \frac{1}{2\delta}\int_{t-\delta}^{t+\delta} h_m(s)\mathrm{d}s = h_m(t) \qquad \text{for all } t \in E_m.$$

If $t \in E_m$, we define

$$w_\delta(s) = \begin{cases} \widehat{u}(s) & \text{if } |s-t|>\delta \\ u_m & \text{if } |s-t| \leq \delta \end{cases}.$$

From (2.147) we have

$$\int_{t-\delta}^{t+\delta} h_m(s)\mathrm{d}s \leq 0 \qquad \text{for all } \delta>0.$$

Divide with 2δ and let $\delta \downarrow 0$, to obtain $h_m(t) \leq 0$. Thus,

$$H\big(t, \widehat{x}(t), u_m, \psi_0, \psi(t)\big) - H\big(t, \widehat{x}(t), \widehat{u}(t), \psi_0, \psi(t)\big) \leq 0$$

for all $t \in E = \bigcap_{m\geq 1} E_m$, $|E|_1 = b$.

Exploiting the continuity of the Hamiltonian H, we have

$$H\big(t, \widehat{x}(t), \widehat{u}(t), \psi_0, \psi(t)\big) = \max_{u\in Y} H\big(t, \widehat{x}(t), u, \psi_0, \psi(t)\big) \text{ a.e. on } T.$$

Also, if in (2.145) we let $w = \widehat{u}$, $x_0 = \widehat{x}_0$, $x_1 = \widehat{x}(b)$ and we use (2.146) and (2.147), we obtain

$$\begin{aligned}&(\widehat{\varphi}, e_0)_{\mathbb{R}^N} \geq \big(\psi(b), z(b)\big)_{\mathbb{R}^N} + \psi_0 z_0 \geq \big(\psi(0), e_0\big)_{\mathbb{R}^N} \qquad \text{for all } e_0 \in \overline{B}_1,\\ \Rightarrow\ &\widehat{\varphi} = \psi(0).\end{aligned}$$

Therefore, if in (2.145) $e_0 = 0$ and $w = \widehat{u}$, we obtain

$$\big(\psi(0), x_0 - \widehat{x}(0)\big)_{\mathbb{R}^N} \leq \big(\psi(b), x_1 - \widehat{x}(b)\big)_{\mathbb{R}^N} \qquad \text{for all } (x_0, x_1) \in C.$$

Finally note that $(\psi_0, \psi) \neq (0,0)$, or otherwise we contradict the fact that $(\widehat{\varphi}, \widehat{\psi}, \widehat{\psi}_0) \neq (0,0,0)$. ∎

REMARK 2.6.49 The function $\psi(\cdot)$ is known as the *costate* or *adjoint state*. The system in (b) is known as the *adjoint system*. The inequality in (c) is known as the *transversality condition* and finally statement (d) is the *maximum condition.* In analogy with mechanics H is the *Hamiltonian*.

2.7 Remarks

2.1: The notion of lower semicontinuity for real functions was introduced by Borel [89], but systematic use of it in variational problems is due to Tonelli [585]. Lower semicontinuity is studied in the books of Barbu–Precupanu [59], Cesari [140], Dal Maso [171], Denkowski–Migórski–Papageorgiou [194], Ekeland–Temam [222], and Ioffe–Tichomirov [327]. Theorem 2.1.25 was first proved by De Giorgi [189] for the case φ is a nonnegative Carathéodory integrand. The version presented here is due to Ioffe [326]. Earlier results on the lower semicontinuity properties of integral functionals $J(x,u)=\int_\Omega \varphi\big(\omega,x(\omega),u(\omega)\big)d\mu$ on the Lebesgue spaces $L^p(\Omega,\mathbb{R}^N)\times L^r(\Omega,\mathbb{R}^m)$, were provided by Berkovitz [71], Cesari [138, 139], and Olech [467]. Corresponding results for the weak lower semicontinuity of integral functionals $J(x)=\int_\Omega \varphi\big(z,x(z),Dx(z)\big)dz$ ($\Omega\subseteq\mathbb{R}^N$ a bounded domain) on the Sobolev spaces $W^{1,p}(\Omega)$ were proved by Buttazzo [126], Dacorogna [170], Ekeland–Temam [222], Giaquinta [263], and Morrey [443]. Lower semicontinuity theorems in the weak topologies of Sobolev spaces when the function $x(\cdot)$ is vector-valued can be found in Acerbi–Fusco [1], Ball–Zhang [56], Dacorogna [170], and Morrey [443].

2.2: The method of Lagrange multipliers is discussed in Alexeev–Tichomirov–Fomin [7], Tichomirov [582] and Zeidler [620]. Theorem 2.2.7 is due to Ljusternik [392]. It can also be found in the book of Ljusternik–Sobolev [395]. Nonsmooth extensions of the Lagrange multipliers method can be found in Aubin [37], Clarke [149, 153], Rockafellar [522], and Rockafellar–Wets [530]. The Dubovitskii–Milyutin formalism for infinite dimensional optimization problems, was first formulated by Dubovitskii–Milyutin [208]. A comprehensive and readable presentation of the theory is contained in the monograph of Girsanov [266].

2.3: The first minimax theorem was formulated for bilinear functionals on finite-dimensional simplices by von Neumann [454]. It was a basic tool in the study of his model for the growth of an economy. Since then there have been several generalizations of the result of von Neumann. We mention the works of Fan [232], Sion [557], Browder [119, 121], Brezis–Nirenberg–Stampacchia [101], Terkelsen [580], and Simons [556]. Theorem 2.3.7 was first proved by Knaster–Kuratowski–Mazurkiewicz [358] in the special case in which B is the set of vertices of a simplex in $\mathbb{R}^N$. The form presented here is due to Dugundji–Granas [211, 212]. An earlier version was proved by Fan [233] who also gave numerous applications of it (see, e.g., Fan [234, 235]; in fact Fan [235] obtained the coincidence result proved in Proposition 2.3.10). Theorem 2.3.13 is due to Sion [557]. The duality theory for convex optimization, which uses the idea of perturbed problems and conjugate convex functionals, was developed for the finite-dimensional case by Rockafellar [522]. The infinite-dimensional case presented here is due to Ekeland–Temam [222]. The special case considered in (2.34) (problem (P)′) is due to Fenchel [241] when $A=I$ and to Rockafellar [518], the general situation with $A\in\mathcal{L}(X,Y)$.

2.4: Theorem 2.4.1 was proved by Ekeland [221]. A detailed discussion of the various applications that this theorem has can be found in Ekeland [223, 226]. Theorem 2.4.11 is due to Caristi [129] who proved it using transfinite induction. Takahashi [571] proved Theorem 2.4.14. Theorem 2.4.16 is due to Danes [175], who proved it using a result of Krasnoselskii–Zabreiko. Theorem 2.4.20 is due to Brezis–Browder [103]. A generalization of the Ekeland variational principle, in which the Lipschitz continuous perturbations are replaced by quadratic ones, was proved by Borwein–Preiss [92]. Additional results in this direction can be found in Deville–Godefroy–Zizler [197].

2.5: The calculus of variations is of course a much broader subject. Detailed presentation of the subject can be found in the books Bliss [80], Bolza [83], Cesari [140], Gelfand–Fomin [260], Hestenes [292], Ioffe–Tichomirov [327], Morrey [443], Morse [444], Troutman [586], and Young [617]. The more recent developments on vectorial problems (with significant applications in elasticity theory) can be found in Dacorogna [170].

2.6: Hypothesis H_c in the existence theory is related to the property (Q) of Cesari [138]. For a detailed discussion of the existence theory for finite dimensional optimal control problems and of the relevant property (Q), we refer to Cesari [140]. Similar results can also be found in Berkovitz [70], Fleming–Richel [249] (finite-dimensional problems) and Ahmed [4], Ahmed–Teo [3], Hu–Papageorgiou [316], Lions [388] and Tiba [581] (infinite-dimensional problems). Scrutinizing the proof of Theorem 2.6.11, we see that what makes it work is hypothesis H_c. So in order to be able to guarantee existence of an optimal pair in an optimal control problem, we need a convexity-type hypothesis. If such a condition is not present the problem need not have a solution. This difficulty was surmounted independently in the late fifties and early sixties by Filippov [247], Warga [601] and Gamkrelidze [255]. They realized that we need to suitably augment the system with the introduction of new admissible controls in order to be able to capture the asymptotic behavior of the minimizing sequences of the original problem. Gamkrelidze and Warga worked with parametrized measures (see problem (2.42)), whereas Filippov used a control-free problem (see problem (2.47)). Here we also present a third relaxation method based on Carathéodory's theorem for convex sets. The fourth relaxation method based on the Γ-regularization of the extended cost functional, is due to Buttazzo–Dal Maso [125] (see also Buttazzo [126]), who for that purpose developed the formalism of multiple Γ-limits. Further results on the relaxation of the calculus of variations and optimal control problems can be found in Buttazzo [126]), Dal Maso [171], Fattorini [239], and Roubicek [532]. Necessary conditions for optimality were first obtained by Pontryagin and his co-workers (see Pontryagin [499] and Pontryagin–Boltyanski–Gamkrelidze–Mischenko [500]. It has become a common terminology to call these necessary conditions the Pontryagin maximum principle. Proofs of the maximum principle for different types of optimal control problems can be found in Berkovitz [70], Cesari [140], Fleming–Richel [249], Hermes–La Salle [289], Ioffe–Tichomirov [327], Seierstadt–Sydsaeter [547] (finite-dimensional systems), and Ahmed [4], Ahmed–Teo [3], Fattorini [239], Li–Yong [381], Lions [388], and Tiba [581] (infinite-dimensional systems).

3

Nonlinear Operators and Fixed Points

Summary. *In this chapter, first we study certain broad classes of nonlinear operators that arise naturally in applications and then we develop some related degree theories and fixed point theorems. We start with compact operators, we conduct a detailed study of both linear and nonlinear operators. Then we pass to a broader class of nonlinear operators of monotone-type. We deal with operators from a reflexive Banach space X to its dual X^* (monotone operators and their generalizations) and with operators from X into itself (accretive operators). The latter, are related to the generation theory of linear and nonlinear semigroups. We deal with both. Then, we present the degree theories of Brouwer, (finite-dimensional) and Leray–Schauder and Browder–Skrypnik (infinite-dimensional) and several interesting topological applications of them. Finally we present the main aspects of fixed point theory. We consider metric fixed point theory, topologucal fixed point theory and the interplay between fixed point theory and partial order. In this direction we introduce and use the so-called "fixed point index".

Introduction

In this chapter, first we study certain classes on nonlinear operators that arise naturally in applications. Then we develop certain finite- and infinite-dimensional degree theories that are valuable tools in the study of boundary value problems. Degree theory leads in a natural way to fixed point theory, which is the topic of investigation in the last three sections of the chapter.

In Section 3.1 we study compact operators. Compact operators were the first attempt to deal with infinite-dimensional operator equations. By its nature, compactness is a suitable concept to achieve finite approximations of infinite objects. We present the basics of both the linear and nonlinear theories. In the linear case we focus on compact self-adjoint operators on a Hilbert space, which have remarkable spectral properties.

Compact operators have obvious limitations. A broader class producing a more general framework for the analysis of infinite-dimensional problems, is provided by the so-called monotone operators from a Banach space X into its topological dual X^*. They are the natural generalization of increasing real functions on $\mathbb{R}$ and their

N.S. Papageorgiou, S.Th. Kyritsi-Yiallourou, *Handbook of Applied Analysis*, Advances in Mechanics and Mathematics 19, DOI 10.1007/b120946_3,

definition does not require any order structure on the Banach space. Prominent among monotone operators are the so-called maximal monotone operators, which exhibit remarkable surjectivity properties. A detailed study of such operators appears in the first half of Section 3.2. In the second half we deal with accretive operators that are the corresponding of monotone operators, when we deal with maps from X into itself (and not into X^*). The importance of accretive operators comes from the fact that they are the generators of linear and nonlinear semigroups which play a central role in the study of evolution equations. In Section 3.2 we present the basic aspects of both the linear and nonlinear semigroup theories.

In Section 3.3 we present the basic finite- and infinite-dimensional degree theories. We start with Brouwer's degree theory which concerns all continuous functions $\varphi : \overline{U} \longrightarrow \mathbb{R}^N$ with $U \subseteq \mathbb{R}^N$ nonempty, bounded, and open. Then we present the Leray–Schauder degree theory which extends Brouwer's theory to an infinite-dimensional setting and specifically to maps of the form $I - \varphi$ with $\varphi : \overline{U} \longrightarrow X$ compact (compact perturbations of the identity). Using measures of noncompactness we extend the theory further to maps of the form $I - \varphi$ with φ being γ-condensing (Nussbaum–Sadovskii degree theory). Finally using Galerkin and Yosida approximations we produce degree maps for nonlinear operators of monotone type. We also have several interesting topological applications of these degree theories.

Degree theory leads smoothly to fixed points. So in Section 3.4 we present the metric fixed point theory. Starting with the celebrated Banach contraction principle, we prove several fixed point theorems in which the metric structure of the spaces and/or of the maps involved, play a crucial role.

In Section 3.5 we pass to topological fixed point theory. Now the topological properties of the spaces and/or of the maps involved are the important ones. The best-known result in this direction is Brouwer's fixed point theorem and its infinite-dimensional counterpart, the Schauder fixed point theorem. Other fixed point results of topological flavor are also presented.

Finally in Section 3.6 we deal with OBS (ordered Banach spaces) which have a partial order induced by an order cone and maps that exhibit certain positivity properties with respect to this partial order. Using structural properties of the order cones and degree-theoretic methods based on the so-called fixed point index, we prove theorems on the existence and multiplicity of fixed points.

3.1 Compact and Fredholm Operators

The first efforts to deal with nonlinear functional equations in infinite-dimensional spaces involved some kind of compactness. Moreover, compact operators provide the natural framework that allows a smooth transition from the finite-dimensional theory to the infinite-dimensional one.

DEFINITION 3.1.1 Let X, Y be Banach spaces, $C \subseteq X$ a nonempty set, and $f : C \longrightarrow Y$.

(a) We say that f is *compact* if it is continuous and maps bounded sets of C to relatively compact sets of Y. By $\mathcal{K}(C, Y)$ we denote the set of all compact maps from C into Y.

(b) We say that f is *completely continuous* if it is sequentially continuous from C with the relative weak topology into Y with the norm topology; that is, if $\{x_n\}_{n\geq 1} \subseteq C$ and $x_n \xrightarrow{w} x \in C$, then $f(x_n) \longrightarrow f(x)$ in Y.

REMARK 3.1.2 In general the two classes of mappings introduced above are not comparable. However, we have the following result.

PROPOSITION 3.1.3 *If X is a reflexive Banach space, Y is a Banach space, $C \subseteq X$ nonempty, closed, and convex and $f\colon C \longrightarrow Y$ is completely continuous, then f is also compact.*

PROOF: Evidently f is continuous. Let B be a bounded subset of X. We need to show that $f(B \cap C)$ is relatively compact in Y. So let $\{y_n\}_{n\geq 1} \subseteq f(B \cap C)$. Then $y_n = f(x_n)$ with $x_n \in B \cap C$. Because X is reflexive and $\{x_n\}_{n\geq 1}$ is bounded, it is relatively weakly sequentially compact (Eberlein–Smulian theorem) and so by passing to a suitable subsequence if necessary, we may assume that $x_n \xrightarrow{w} x$. Because f is completely continuous, we have that $y_n = f(x_n) \longrightarrow f(x) = y$ in Y and this proves the compactness of f. $\square$

PROPOSITION 3.1.4 *If X is a reflexive Banach space, Y is a Banach space, and $A \in \mathcal{L}(X,Y)$, then A is completely continuous if and only if A is compact.*

PROOF: $\Rightarrow$: Follows from Proposition 3.1.3.
$\Leftarrow$: Let $x_n \xrightarrow{w} x$ in X. Then $A(x_n) \xrightarrow{w} A(x)$ in Y. Also $\{x_n\}_{n\geq 1} \subseteq X$ is bounded and so $\{A(x_n)\}_{n\geq 1}$ is relatively compact in Y. Therefore $A(x_n) \longrightarrow A(x)$ in Y. $\square$

REMARK 3.1.5 It is clear from the second part of the above proof that for linear operators compactness always implies complete continuity, without any additional structure on the Banach spaces X, Y.

The set of all compact maps $f\colon X \longrightarrow Y$ is a linear space and we have the following.

PROPOSITION 3.1.6 *If X, Y are Banach spaces, $f_n\colon X \longrightarrow Y$, $n \geq 1$, is a sequence of compact mappings, and $f\colon X \longrightarrow Y$ is a mapping such that $f_n(x) \longrightarrow f(x)$ in Y uniformly on bounded subsets of X, then f is compact too.*

PROOF: Clearly f is continuous. We need to show that for every bounded set $B \subseteq X$, $f(B)$ is relatively compact. Given $\varepsilon > 0$, we can find $n \geq 1$ such that $\|f_n(x) - f(x)\|_Y \leq \varepsilon/2$ for all $x \in B$. Also, because $f_n(B)$ is relatively compact, we can find $\{x_m\}_{m=1}^M \subseteq B$ such that $\bigcup_{m=1}^{M} B_{\varepsilon/3}\big(f_n(x_m)\big) \supseteq f_n(B)$. Then if $x \in B$, there is $m \in \{1, \ldots, M\}$ such that

$$\begin{aligned}\|f(x) - f(x_m)\|_Y &\leq \|f(x) - f_n(x)\|_Y + \|f_n(x) - f_n(x_m)\|_Y + \\ &\quad + \|f_n(x_m) - f(x_m)\|_Y \\ &< \frac{\varepsilon}{3} + \frac{\varepsilon}{3} + \frac{\varepsilon}{3} = \varepsilon,\end{aligned}$$

$\Rightarrow$ $f(B)$ is totally bounded, hence relatively compact in Y. $\square$

REMARK 3.1.7 Note that the linear space $\mathcal{K}(X,Y)$ (see Definition 3.1.1(a)) is closed under composition with continuous bounded mappings.

DEFINITION 3.1.8 Let X, Y be Banach spaces, $C \subseteq X$ a nonempty set, and $f\colon C \longrightarrow Y$. We say that f is a *finite rank* mapping, if it is continuous, bounded (i.e., maps bounded sets to bounded ones), and $f(C)$ lies in a finite dimensional subspace of Y. We denote the linear space of finite rank mappings $f\colon C \longrightarrow Y$, by $\mathcal{K}_f(C,Y)$.

REMARK 3.1.9 By virtue of the Heine–Borel theorem we have $\mathcal{K}_f(C,Y) \subseteq \mathcal{K}(C,Y)$. In fact, we show that in a sense $\mathcal{K}(C,Y)$ is the closure of $\mathcal{K}_f(C,Y)$.

THEOREM 3.1.10 *If X, Y are Banach spaces and $C \subseteq X$ is nonempty bounded, then $f \in \mathcal{K}(C,Y)$ if and only if f is the uniform limit of elements in $\mathcal{K}_f(C,Y)$.*

PROOF: $\Rightarrow$: The set $f(C)$ is relatively compact in Y. So for every $n \geq 1$, we can find a finite set $\{y_k\}_{k=1}^{N_n} \subseteq f(C)$ such that

$$\min_{1 \leq k \leq N_n} \|f(x) - y_k\|_Y < \frac{1}{n} \quad \text{for all } x \in C. \tag{3.1}$$

Let $\xi_k(x) = \max\left\{(1/n) - \|f(x) - y_k\|_Y, 0\right\}$. Then $\xi_k \in C(C)$ and the ξ_ks do not all vanish simultaneously at $x \in C$ (see (3.1)). Moreover, if we set

$$f_n(x) = \frac{\sum_{k=1}^{N_n} \xi_k(x) y_k}{\sum_{k=1}^{N_n} \xi_k(x)}, \quad x \in C,$$

then we have

$$\|f(x) - f_n(x)\|_Y = \left\| \frac{\sum_{k=1}^{N_n} \xi_k(x)\big(y_k - f(x)\big)}{\sum_{k=1}^{N_n} \xi_k(x)} \right\| < \frac{1}{n} \quad \text{for all } x \in C \ \text{(see (3.1))}.$$

Then the boundedness of $f(C)$ implies the boundedness of $f_n(C)$ and so $f_n \in \mathcal{K}_f(C,Y)$.

$\Leftarrow$: It follows from Proposition 3.1.6 and Remark 3.1.9. $\square$

We recall the following generalization of Tietze's extension theorem due to Dugundji [209, p. 188].

THEOREM 3.1.11 *If X is a metric space, $A \subseteq X$ is a closed subset, Y is a locally convex space, and $f\colon A \longrightarrow Y$ is a continuous map, then there exists a continuous map $\widehat{f}\colon X \longrightarrow Y$ such that $\widehat{f}\big|_A = f$ and*

$$\widehat{f}(X) \subseteq \operatorname{conv} f(A).$$

REMARK 3.1.12 Evidently in the above theorem we can replace Y by any convex subset D of Y.

Using Theorem 3.1.11 and the fact that the closed convex hull of a norm compact set in a Banach space is again norm compact (Mazur's theorem), we can have the following theorem.

THEOREM 3.1.13 *If X is a Banach space, $C \subseteq X$ is nonempty, closed, and bounded, Y is a Banach space, and $f \in \mathcal{K}(C,Y)$, then there exists $\widehat{f} \in \mathcal{K}(X,Y)$ such that $\widehat{f}\big|_C = f$ and $\widehat{f}(X) \subseteq \overline{\text{conv}}\, f(C)$.*

DEFINITION 3.1.14 Let X, Y be metric spaces, $C \subseteq Y$ nonempty closed and $f\colon C \longrightarrow Y$ a continuous map. We say that f is *proper*, if for any compact set D in Y, the set $f^{-1}(D)$ is compact in C.

REMARK 3.1.15 The importance of this class of continuous maps stems from the fact that the property of properness restricts the size of the solution set $S(y) = \{x \in C : f(x) = y\}$ for any given $y \in Y$. Evidently, if X and Y are Banach spaces, then all proper operators $A \in \mathcal{L}(X,Y)$ are injective and have closed range.

PROPOSITION 3.1.16 *If X and Y are metric spaces and $f \in C(X,Y)$, then the following are equivalent.*

(a) *f is proper.*
(b) *f is a closed map (i.e., maps closed sets to closed sets) and for every $y \in Y$, the set $f^{-1}(y) = S(y)$ is compact in X.*

PROOF: (a)$\Rightarrow$(b): The singleton $\{y\}$ is compact in Y, thus from the properness of f it follows that $f^{-1}(y)$ is compact in X. Now let $C \subseteq X$ be nonempty and closed. We show that $f(C)$ is closed in Y. To this end let $\{y_n\}_{n\geq 1} \subseteq f(C)$ and suppose that $y_n \longrightarrow y$ in Y as $n \to \infty$. We have $y_n = f(x_n)$, $x_n \in C$ for all $n \geq 1$. Note that $D = \{y_n, y\}_{n\geq 1} \subseteq Y$ is compact and because f is proper, $f^{-1}(D)$ is compact in X. Because $\{x_n\}_{n\geq 1} \subseteq f^{-1}(D)$, by passing to a suitable subsequence if necessary, we may assume that $x_n \longrightarrow x$ in X. Evidently $x \in C$ and $f(x_n) \longrightarrow f(x)$. Hence $y = f(x) \in f(C)$ and so f is closed.

(b)$\Rightarrow$(a): Let $D \subseteq Y$ be nonempty compact. We have to show that $f^{-1}(D)$ is compact. For this purpose, let $\{x_n\}_{n\geq 1} \subseteq f^{-1}(D)$. Then $f(x_n) = y_n \in D$ and so we may assume that $y_n \longrightarrow y \in D$ in Y as $n \to \infty$. Let $C_n = \{x_k\}_{k\geq n}$. Because f is closed, $f(\overline{C}_n) = \overline{f(C_n)}$ and $y \in \overline{f(C_n)}$ for all $n \geq 1$. Therefore, if $x \in f^{-1}(y)$, then $f(x) \in \bigcap\limits_{n\geq 1} f(\overline{C}_n)$ and so $f^{-1}(y) \cap \overline{C}_n \neq \emptyset$. This means that $\{x_n\}_{n\geq 1}$ has a cluster point $x \in f^{-1}(y)$. So we conclude that $f^{-1}(D)$ is compact. $\square$

PROPOSITION 3.1.17 *If X and Y are finite-dimensional Banach spaces and $f \in C(X,Y)$, then the following are equivalent.*

(a) *f is proper.*
(b) *$\|f(x)\|_Y \longrightarrow +\infty$ as $\|x\|_X \longrightarrow +\infty$ (i.e., f is weakly coercive).*

PROOF: (a)$\Rightarrow$(b): Because both X and Y are finite-dimensional, the properness of f implies that the inverse image of a bounded set in Y is a bounded set in X. But this is simply a restatement of the property of weak coercivity.

(b)$\Rightarrow$(a): If $D \subseteq Y$ is nonempty compact, then by virtue of the weak coercivity of f, $f^{-1}(D)$ is bounded in X. It is also closed in X (because $f \in C(X,Y)$). So by the Heine–Borel theorem, we conclude that $f^{-1}(D)$ is compact, (i.e., f is proper).□

There is a partial generalization to an infinite-dimensional setting of the above proposition.

PROPOSITION 3.1.18 *If X, Y are Banach spaces, $f \in C(X,Y)$, f is weakly coercive (i.e., $\|f(x)\|_Y \longrightarrow +\infty$ as $\|x\|_X \longrightarrow +\infty$), and one of the following holds,*

(i) *$f = g + h$ with $g \in C(X,Y)$ proper and $h \in \mathcal{K}(X,Y)$, or*
(ii) *X is reflexive and $x_n \xrightarrow{w} x$ in X, $f(x_n) \longrightarrow y$ in Y imply $x_n \longrightarrow x$ in X,*

then f is proper.

PROOF: First assume that (i) holds.

Let $D \subseteq Y$ be nonempty compact and let $\{x_n\}_{n\geq 1} \subseteq f^{-1}(D)$. Then $y_n = f(x_n) \in D$ for all $n \geq 1$ and so we may assume that $y_n \longrightarrow y$ in Y. The weak coercivity of f implies that $\{x_n\}_{n\geq 1} \subseteq X$ is bounded. Because $h \in \mathcal{K}(X,Y)$, we may assume that $h(x_n) \longrightarrow \overline{y}$ in Y. Then $g(x_n) = y_n - h(x_n) \longrightarrow y - \overline{y}$ in Y as $n \to \infty$. Because g is proper, we must have, at least for a subsequence, that $x_n \longrightarrow x$ in X. Then $f(x_n) \longrightarrow f(x) = y \in D$ and so we conclude that f is proper.

Now suppose that (ii) holds.

Because X is reflexive and $f(x_n) \longrightarrow y$ in Y as $n \to \infty$, from the weak coercivity of f we have that $\{x_n\}_{n\geq 1} \subseteq X$ is bounded and so we may assume that $x_n \xrightarrow{w} x$ in X (Eberlein–Smulian theorem). Because of (ii), $x_n \longrightarrow x$ in X and so $f(x_n) \longrightarrow f(x) = y \in D$, which proves that f is proper. □

PROPOSITION 3.1.19 *If X is a Banach space, $C \subseteq X$ is nonempty, closed, and bounded, and $f \in \mathcal{K}(X,Y)$, then $g = I - f \in C(X,Y)$ is proper.*

PROOF: Let $D \subseteq X$ be nonempty and compact and let $\{x_n\}_{n\geq 1} \subseteq g^{-1}(D)$. Then $y_n = g(x_n) \in D$ for all $n \geq 1$ and so we may assume that $y_n \longrightarrow y \in D$ in X. Also because $f \in \mathcal{K}(X,Y)$, we have that $\{f(x_n)\}_{n\geq 1}$ is relatively compact in X and so we may assume that $f(x_n) \longrightarrow u$ in X as $n \to \infty$. Then $x_n \longrightarrow y + u$ in X as $n \to \infty$ and $y = g(y + u)$, which proves the properness of g. □

PROPOSITION 3.1.20 *If X, Y are Banach spaces, $U \subseteq X$ is nonempty open, and $f \in \mathcal{K}(U,Y)$ is Fréchet differentiable at $x_0 \in U$, then $f'(x_0) \in \mathcal{L}(X,Y)$ is a compact linear operator.*

PROOF: Suppose that $f'(x_0) \in \mathcal{L}(X,Y)$ is not compact. Then $f'(x_0)(\overline{B}_1^X)$ is not relatively compact in Y (recall that $\overline{B}_1^X = \{u \in X : \|u\|_X \leq 1\}$). So we can find $\{u_n\}_{n\geq 1} \subseteq \overline{B}_1^X$ and $\varepsilon > 0$ such that

$$\|f'(x_0)(u_n) - f'(x_0)(u_m)\|_Y \geq 3\varepsilon \qquad \text{for all } n, m \geq 1,\ n \neq m.$$

But from the Fréchet differentiability of f at x_0, we have

$$f(x_0 + h) - f(x_0) = f'(x_0)h + r(x_0, h) \qquad \text{for all } h \in X$$

and $\|r(x_0, h)\|_Y \leq \varepsilon \|h\|_X$ if $\|h\|_X \leq \delta$ for some $\delta > 0$, (see Definition 1.1.7). Therefore

$$\begin{aligned} &\|f(x_0 + \delta u_n) - f(x_0 + \delta u_m)\|_Y \\ &\geq \delta \|f'(x_0)u_n - f'(x_0)u_m\|_Y - \|r(x_0, \delta u_n)\|_Y - \|r(x_0, \delta u_n)\|_Y \\ &\geq 3\delta\varepsilon - \delta\varepsilon - \delta\varepsilon = \delta\varepsilon, \end{aligned}$$

which contradicts the fact that $f \in \mathcal{K}(U, Y)$. □

One of the fundamental notions in linear operator theory is that of the spectrum. It generalizes the concept of the eigenvalue of a linear mapping in a finite-dimensional space. Compact linear operators have remarkable spectral properties, culminating in the spectral resolution theorem for compact self-adjoint operators.

In what follows for X, Y Banach spaces, we set

$$\mathcal{L}_c(X, Y) = \mathcal{L}(X, Y) \cap \mathcal{K}(X, Y) \qquad \text{and} \qquad \mathcal{L}_f(X, Y) = \mathcal{L}(X, Y) \cap \mathcal{K}_f(X, Y).$$

DEFINITION 3.1.21 Let X be a complex Banach space and $A \in \mathcal{L}(X)$. The *spectrum* $\sigma(A)$ of A is defined by

$$\sigma(A) = \{\lambda \in \mathbb{C} : \lambda I - A \text{ is not invertible}\}$$

and the *resolvent set* $\varrho(A)$ is defined by $\varrho(A) = \mathbb{C} \setminus \sigma(A)$. If $\lambda \in \varrho(A)$, then the operator $R(\lambda) = (\lambda I - A)^{-1} \in \mathcal{L}(X)$ is called the *resolvent* of A.

REMARK 3.1.22 In the case of a real Banach space, the definition is analogous considering only real λ. However, in the real case some important results are not true. For example, if X is a real Banach space it is not true that for every $A \in \mathcal{L}(X)$, $\sigma(A) \neq \emptyset$, although this is true for X a complex Banach space. Nevertheless because we focus on self-adjoint operators in Hilbert spaces, we show that $\sigma(A) \subseteq \mathbb{R}$ and so we can assume that we have a real Hilbert space, as are all the vector spaces in this volume.

DEFINITION 3.1.23 Let X be a complex Banach space and $A \in \mathcal{L}(X)$. A number $\lambda \in \mathbb{C}$ is called an *eigenvalue* of A if there exists $x \in X \setminus \{0\}$ such that $A(x) = \lambda x$. Then such an $x \neq 0$ is called an *eigenvector* corresponding to the eigenvalue λ. The space $\ker(\lambda I - A)$ is called the *eigenspace* corresponding to the eigenvalue λ. The set of all eigenvalues of A is called the *point spectrum* of A and it is denoted by $\sigma_p(A)$.

REMARK 3.1.24 Again, there is an analogous definition for real Banach spaces. Evidently $\sigma_p(A) \subseteq \sigma(A)$. If $\dim X < +\infty$, then from linear algebra we know that the operator $\lambda I - A$ is invertible if and only if it is injective. So in this case $\sigma(A) = \sigma_p(A)$. On the other hand in an infinite-dimensional space it may happen that the point spectrum is empty. Indeed let $X = H = L^2[0,1]$ and let $A \in \mathcal{L}(H)$ be defined by $A(x)(t) = tx(t)$ for all $t \in [0,1]$. If $\lambda \in \mathbb{C}$ and $\lambda x(t) - tx(t) = 0$ for some $x \in L^2[0,1]$ and almost all $t \in [0,1]$, then $(\lambda - t)x(t) = 0$ a.e. on T and so $x(t) = 0$ a.e. on T. However, it can be shown that $[0,1] \subseteq \sigma(A)$. We leave the details to the reader.

PROPOSITION 3.1.25 *If H is a complex Hilbert space and $A \in \mathcal{L}(H)$ is self-adjoint, then $\sigma_p(A) \subseteq \mathbb{R}$ and eigenvectors corresponding to different eigenvalues are orthogonal.*

PROOF: A is self-adjoint therefore for every $x \in H$ we have

$$\begin{aligned} (Ax, x)_H &= (x, Ax)_H = \overline{(Ax, x)}_H, \\ \Rightarrow (Ax, x)_H &\in \mathbb{R} \qquad \text{for all } x \in H. \end{aligned}$$

If $\lambda \in \sigma_p(A)$ and $x \in H$ is an eigenvector corresponding to λ, then

$$\begin{aligned} (Ax, x)_H &= (\lambda x, x)_H = \lambda \|x\|^2 \\ \Rightarrow \lambda &= \frac{(Ax, x)_H}{\|x\|^2} \in \mathbb{R} \quad (\text{i.e., } \sigma_p(A) \subseteq \mathbb{R}). \end{aligned}$$

If $\lambda, \mu \in \sigma_p(A)$, $\lambda \neq \mu$ and x_1, x_2 are eigenvectors corresponding to λ, μ respectively, then

$$\begin{aligned} & Ax_1 = \lambda x_1 \text{ and } Ax_2 = \mu x_2, \\ \Rightarrow\ & (Ax_1, x_2)_H = \lambda(x_1, x_2)_H \text{ and } (Ax_1, x_2)_H = (x_1, Ax_2)_H = \mu(x_1, x_2)_H, \\ \Rightarrow\ & (\lambda - \mu)(x_1, x_2)_H = 0 \quad (\text{i.e., } x_1 \perp x_2, \text{ because } \lambda \neq \mu). \end{aligned}$$

□

REMARK 3.1.26 Similarly we can show that eigenvectors corresponding to different eigenvalues of an operator A on a Banach space X are linearly independent.

PROPOSITION 3.1.27 *If H is a complex Hilbert space, $A \in \mathcal{L}(H)$ is self-adjoint and $\lambda \in \mathbb{C}$, then $\lambda \in \sigma(A)$ if and only if* $\inf[\,(\|\lambda I - A)x\|_H : \|x\|_H = 1] = 0$.

PROOF: If $\lambda \in \varrho(A)$, then $(\lambda I - A)^{-1} \in \mathcal{L}(H)$ and for every $x \in \partial B_1 = \{x \in H : \|x\|_H = 1\}$, we have

$$\begin{aligned} & 1 = \|x\|_H = \|(\lambda I - A)^{-1}(\lambda I - A)x\|_H \leq \|(\lambda I - A)\|^{-1}_{\mathcal{L}(H)} \|(\lambda I - A)x\|_H, \\ \Rightarrow\ & \inf[\|(\lambda I - A)x\|_H : \|x\|_H = 1] \geq \|(\lambda I - A)\|_{\mathcal{L}(H)} > 0 \ \text{ for } \lambda \in \varrho(A). \end{aligned} \tag{3.2}$$

We show that if $\inf\left[\|(\lambda I - A)x\|_H : \|x\|_H = 1\right] > 0$, then $\lambda I - A$ is invertible. Clearly due to homogeneity, if $\inf[\|(\lambda I - A)x\|_H : \|x\|_H = 1] = c_0 > 0$, then

$$\|(\lambda I - A)x\|_H \geq c_0 \|x\|_H \qquad \text{for all } x \in H. \tag{3.3}$$

Hence $(\lambda I - A)x = 0$ if and only if $x = 0$, (i.e., $\lambda I - A$ is injective). Next we show that $\lambda I - A$ is also surjective, which by Banach's theorem implies that $(\lambda I - A)^{-1} \in \mathcal{L}(H)$.

First we show that $(\lambda I - A)(H)$ is dense in H. If this is not the case we can find $h \in H$, $h \neq 0$ such that

$$\begin{aligned} & \langle(\lambda I - A)x, h\rangle_H = 0 && \text{for all } x \in H, \\ \Rightarrow\ & \langle x, (\lambda I - A)h\rangle_H = 0 && \text{for all } x \in H \text{ (because } A \text{ is self-adjoint)}, \\ \Rightarrow\ & (\lambda I - A)h = 0, && \text{a contradiction.} \end{aligned}$$

Therefore $(\lambda I - A)(H)$ is dense in H.

Also we show that $(\lambda I - A)(H)$ is closed in H. To this end let $y_n = (\lambda I - A)x_n \longrightarrow y$ in H. From (3.3) it follows that $\{x_n\}_{n \geq 1} \subseteq H$ is bounded and so we may assume that $x_n \xrightarrow{w} x$ in H. Then

$$(\lambda I - A)x_n \xrightarrow{w} (\lambda I - A)x \qquad \text{in } H,$$
$$\Rightarrow \; y = (\lambda I - A)x \;\; \text{(i.e., } (\lambda I - A)(H) \text{ is closed).}$$

It follows that $\lambda I - A$ is surjective, hence $(\lambda I - A)^{-1} \in \mathcal{L}(H)$. So combining this with (3.2) we see that the proposition has been proved. □

PROPOSITION 3.1.28 *If H is a complex Hilbert space and $A \in \mathcal{L}(H)$, then $\sigma(A) \subseteq \mathbb{R}$.*

PROOF: Let $\lambda = a + i\beta$ with $\beta \neq 0$. We show that $\|(\lambda I - A)x\|_H \leq |\lambda| \|x\|_H$ for all $x \in H$. To this end, because A is self-adjoint, we have

$$\begin{aligned}
\langle \lambda x - Ax, x \rangle_H - \langle x, \lambda x - Ax \rangle_H &= \lambda \|x\|_H^2 - \langle Ax, x \rangle_H - \overline{\lambda} \|x\|_H^2 + \langle Ax, x \rangle_H \\
&= (\lambda - \overline{\lambda}) \|x\|_H^2 = 2i\beta \|x\|_H^2, \\
\Rightarrow \; 2|\beta| \|x\|_H^2 = \big| \langle \lambda x - Ax, x \rangle_H &- \langle x, \lambda x - Ax \rangle_H \big| \\
&\leq \big| \langle \lambda x - Ax, x \rangle_H \big| + \big| \langle x, \lambda x - Ax \rangle_H \big| \\
&\leq 2 \|(\lambda I - A)x\|_H \|x\|,
\end{aligned}$$
$$\Rightarrow \; |\beta| \|x\|_H \leq \|(\lambda I - A)x\|_H \qquad \text{for all } x \in H. \tag{3.4}$$

Invoking Proposition 3.1.27, from (3.4) we conclude that $\lambda \in \varrho(A)$. Therefore $\sigma(A) \subseteq \mathbb{R}$. □

REMARK 3.1.29 This proposition suggests that when dealing with the spectrum of self-adjoint operators, we can assume without any loss of generality that the Hilbert space H is real. In what follows we do this.

PROPOSITION 3.1.30 *If H is a Hilbert space, $A \in \mathcal{L}(H)$ is self-adjoint and $m = \inf_{\|x\|_H = 1} \langle Ax, x \rangle_H$, $M = \sup_{\|x\|_H = 1} \langle Ax, x \rangle_H$, then $\sigma(A) \subseteq [m, M]$ and $m, M \subseteq \sigma(A)$.*

PROOF: From Proposition 3.1.28 we know that $\sigma(A) \subseteq \mathbb{R}$. Let $\eta > 0$. We show that $M + \eta \in \varrho(A)$. From Proposition 3.1.27, we know that it suffices to show that

$$\inf \big[\|((M + \eta)I - A)x\|_H : \|x\|_H = 1 \big] > 0. \tag{3.5}$$

For every $x \in \partial B_1$, we have

$$\begin{aligned}
\langle (M + \eta)x - Ax, x \rangle_H &= (M + \eta) \|x\|_H^2 - \langle Ax, x \rangle_H \\
&\geq M + \eta - M = \eta > 0.
\end{aligned}$$

Hence (3.5) is true and then Proposition 3.1.27 implies that $M + \eta \in \varrho(A)$. Similarly $m - \eta \in \varrho(A)$. Therefore $\sigma(A) \subseteq [m, M]$.

Next note that $\sigma(A - \xi I) = \sigma(A) - \xi$. So by considering, if necessary, $A + \xi I$ instead of A, we may assume without any loss of generality that $0 \leq m \leq M$. Also using the parallelogram law, we can easily check that

$$\|A\|_{\mathcal{L}(H)} = \sup\left[\langle Ax, x\rangle_H : \|x\|_H = 1\right].$$

It follows that $M = \|A\|_{\mathcal{L}(H)}$. We show that

$$\inf\left[\ \|(MI - A)x\|_H : \|x\|_H = 1\right] = 0. \tag{3.6}$$

To this end let $\{x_n\}_{n\geq 1} \subseteq \partial B_1$ such that

$$\langle Ax_n, x_n\rangle \uparrow M = \|A\|_{\mathcal{L}(H)} \quad \text{as } n \to \infty.$$

Note that because $m \geq 0$, we have $\langle Ax, x\rangle_H \geq 0$ for all $x \in H$. Hence

$$\begin{aligned}
0 \leq \|(MI - A)x_n\|_H^2 &= \|Mx_n - Ax_n\|_H^2 \\
&= \langle Mx_n - Ax_n, Mx_n - Ax_n\rangle_H \\
&= M^2\|x_n\|_H^2 - 2M\langle Ax_n, x_n\rangle_H + \|Ax_n\|_H^2 \\
&\leq 2M^2 - 2M\langle Ax_n, x_n\rangle_H \longrightarrow 0 \quad \text{as } n \to \infty,
\end{aligned}$$

$$\Rightarrow \|(MI - A)x_n\|_H \longrightarrow 0 \quad \text{as } n \to \infty.$$

Therefore (3.6) holds and from Proposition 3.1.27 it follows that $M \in \sigma(A)$. Similarly we show that $m \in \sigma(A)$. □

PROPOSITION 3.1.31 *If H is a Hilbert space and $A \in \mathcal{L}(H)$ is self-adjoint compact, then every $\lambda \in \sigma(A)\setminus\{0\}$ is an eigenvalue; that is, $\lambda \in \sigma_p(A)$.*

PROOF: By virtue of Proposition 3.1.27, we can find $\{x_n\}_{n\geq 1} \subseteq \partial B_1$ such that

$$\|\lambda x_n - Ax_n\|_H \longrightarrow 0 \qquad \text{as } n \to \infty. \tag{3.7}$$

Due to the compactness of A and by passing to a subsequence if necessary, we may assume that $Ax_n \longrightarrow y$ in H. Because of (3.7) and because $\lambda \neq 0$, we have that $x_n \longrightarrow x$ in H. Then $y = Ax$ and in the limit as $n \to \infty$ we have

$$\lambda x = Ax \ \text{ with } \ \|x\| = 1 \ \ (\text{i.e., } \ \lambda \in \sigma_p(A)).$$

□

COROLLARY 3.1.32 *If H is a Hilbert space and $A \in \mathcal{L}(H)$ is self-adjoint compact, then $\sigma_p(A) \neq \emptyset$.*

PROOF: If $A = 0$, then $\lambda = 0$ is an eigenvalue of A. So suppose that $A \neq 0$. Then at least one of m and M defined in Proposition 3.1.30 is nonzero and belongs in $\sigma(A)$. By virtue of Proposition 3.1.31 it also belongs in $\sigma_p(A)$. □

PROPOSITION 3.1.33 *If X is a Banach space, $A \in \mathcal{L}_c(X)$ and $\lambda \in \sigma_p(A)\setminus\{0\}$, then* dim $\ker(\lambda I - A) < +\infty$*; that is, the eigenspace corresponding to λ is finite-dimensional.*

PROOF: Let $E_\lambda = \ker(\lambda I - A)$ and let $C \subseteq E_\lambda$ be a nonempty bounded set. We have

$$Ax = \lambda x \qquad \text{for all } \ x \in C. \tag{3.8}$$

Because A is compact, $A(C)$ is relatively compact. Therefore by (3.8) we have that $\lambda C = A(C)$ is relatively compact. Because C was an arbitrary bounded set in E_λ, we conclude that $\dim E_\lambda < +\infty$. □

PROPOSITION 3.1.34 *If H is a Hilbert space and $A \in \mathcal{L}_c(H)$ is self-adjoint, then $\sigma_p(A)$ is a countable set with 0 as the only possible limit point.*

PROOF: We show that for every $\varepsilon > 0, \sigma_p(A) \cap \{|\lambda| > \varepsilon\}$ is finite. We argue by contradiction. So suppose that there exists $\varepsilon > 0$ and a sequence of distinct eigenvalues $\{\lambda_n\}_{n\geq 1}$ of A such that $|\lambda_n| > \varepsilon$ for all $n \geq 1$. Let $\{x_n\}_{n\geq 1} \subseteq \partial B_1$ such that $Ax_n = \lambda_n x_n$ for all $n \geq 1$. If $n \neq m$, then from Proposition 3.1.25 we have that $(x_n, x_m)_H = 0$. Therefore

$$\begin{aligned}
&\|Ax_n - Ax_m\|_H^2 = \|\lambda_n x_n - \lambda_m x_m\|_H^2 = \lambda_n^2 + \lambda_m^2 \geq 2\varepsilon^2, \\
\Rightarrow\ & A(\overline{B}_1) \text{ is not relatively compact in } H.
\end{aligned}$$

But this contradicts the hypothesis that $A \in \mathcal{L}_c(H)$. □

PROPOSITION 3.1.35 *If H is a Hilbert space and $A \in \mathcal{L}_c(H)$ is self-adjoint, then there is an orthonormal basis of H formed by eigenvectors of A.*

PROOF: For every $\lambda \in \sigma_p(A)$, let $E_\lambda = \ker(\lambda I - A)$ and consider B_λ an orthonormal basis of E_λ. Then by virtue of Proposition 3.1.25 $\bigcup\limits_{\lambda \in \sigma_p(A)} B_\lambda$ is orthonormal set of H. Suppose $V = \overline{\text{span}} \bigcup\limits_{\lambda \in \sigma_p(A)} E_\lambda \neq H$. We consider the space $V^\perp$ (the orthogonal complement of V in H). Clearly $A(V) \subseteq V$ and so $A(V^\perp) \subseteq V^\perp$ (i.e., the subspace $V^\perp$ is A-invariant). By Corollary 3.1.32 $A\Big|_{V^\perp}$ has an eigenvalue and so an eigenvector $u \in V^\perp \setminus \{0\}$. Then $u \in V^\perp \cap V$, $u \neq 0$, a contradiction. So $V = H$ and $\bigcup\limits_{\lambda \in \sigma_p(A)} B_\lambda$ is an orthonormal basis of H. □

REMARK 3.1.36 This proposition essentially says that every self-adjoint, compact operator on a Hilbert space is diagonalizable.

Now we are ready for the *spectral resolution theorem* for self-adjoint, compact operators on a separable Hilbert space.

THEOREM 3.1.37 *If H is an infinite-dimensional separable Hilbert space and $A \in \mathcal{L}_c(H)$ is self-adjoint, then there is an orthonormal basis $\{e_n\}_{n\geq 1}$ of H consisting of eigenvectors of A such that for λ_n being the eigenvalue corresponding to e_n, we have*

$$A(x) = \sum_{n\geq 1} \lambda_n \langle x, e_n \rangle_H \qquad \textit{for all } x \in H.$$

PROOF: Due to Proposition 3.1.35 and the separability of H, we can find a countable orthonormal basis $\{e_n\}_{n\geq 1}$ of H consisting of eigenvectors of A. We have

$$\begin{aligned}
&\|\sum_{n=k}^{m} \lambda_n \langle x, e_n \rangle_H e_n\|_H^2 = \sum_{n=k}^{m} |\lambda_n \langle x, e_n \rangle_H|^2 \leq \|A\|_{\mathcal{L}(H)}^2 \sum_{n=k}^{m} |\langle x, e_n \rangle_H|^2 \\
&\quad \longrightarrow 0 \quad \text{as } k, m \longrightarrow \infty, \\
\Rightarrow\ & \sum_{n\geq 1} \lambda_n \langle x, e_n \rangle_H e_n \text{ is convergent in } H.
\end{aligned}$$

Moreover, if $\|x\| \le 1$, then for every $m \ge 1$ we have

$$\begin{aligned}\|\sum_{n=1}^{m} \lambda_n \langle x, e_n\rangle_H e_n\|_H^2 &= \sum_{n=1}^{m} \lambda_n^2 |\langle x, e_n\rangle_H|^2 \\ &\le \|A\|^2_{\mathcal{L}(H)} \sum_{n=1}^{m} |\langle x, e_n\rangle_H|^2 \\ &\le \|A\|^2_{\mathcal{L}(H)} \sum_{n\ge1} |\langle x, e_n\rangle_H|^2 \\ &= \|A\|^2_{\mathcal{L}(H)} \|x\|^2_H,\end{aligned}$$

$$\Rightarrow \sum_{n\ge1} \lambda_n \langle x, e_n\rangle_H e_n = F(x); \quad x \in H \text{ belongs in } \mathcal{L}(H).$$

Because $A(e_n) = \lambda_n e_n$, we infer that $A(e_n) = F(e_n)$ for all $n \ge 1$, hence $A = F$. □

PROPOSITION 3.1.38 *If H is a Hilbert space and $A \in \mathcal{L}_c(H)$ is self-adjoint, then $\sigma(A) = \overline{\sigma_p(A)}$.*

PROOF: If $\dim H < +\infty$, then $\sigma(A) = \sigma_p(A) =$ finite set (see Remark 3.1.24). If $\dim H = +\infty$ and $\lambda \in \sigma(A) \setminus \{0\}$, then $\lambda \in \sigma_p(A)$ (see Proposition 3.1.31). Due to the compactness of A, $0 \in \sigma(A)$. If 0 is not an eigenvalue, then by virtue of the proof of Proposition 3.1.34, we can find $\{\lambda_n\}_{n\ge1} \subseteq \sigma_p(A) \setminus \{0\}$ such that $\lambda_n \longrightarrow 0$. Therefore $\sigma(A) = \overline{\sigma_p(A)}$. □

Now we return to the general Banach space setting and prove some results that lead to the introduction of Fredholm operators.

PROPOSITION 3.1.39 *If X is a Banach space, $A \in \mathcal{L}_c(X)$, and $\lambda \ne 0$, then $\mathcal{R}(\lambda I - A)$ is closed.*

PROOF: We may assume that $\lambda = 1$. Let $u_n = x_n - A(x_n)$ and suppose that $u_n \longrightarrow u$ in X. We need to show that $u \in R(\lambda I - A)$. Let $d_n = d(x_n, \ker(I - A))$. Because $\ker(I - A)$ is finite-dimensional (see Proposition 3.1.33), we can find $y_n \in \ker(I - A)$ such that $d_n = \|x_n - y_n\|_X$. We have

$$u_n = x_n - y_n - A(x_n - y_n), \quad n \ge 1. \tag{3.9}$$

We show that $\{x_n - y_n\}_{n\ge1} \subseteq X$ is bounded. Suppose that this is not true. We may assume that $\|x_n - y_n\|_X \longrightarrow +\infty$ as $n \to \infty$. Set

$$v_n = \frac{x_n - y_n}{\|x_n - y_n\|_X}, \quad n \ge 1.$$

From (3.9) we have

$$v_n - A(v_n) = \frac{u_n}{\|x_n - y_n\|_X} \longrightarrow 0 \quad \text{as } n \to \infty. \tag{3.10}$$

Due to the compactness of A, we may assume that $A(v_n) \longrightarrow w$ in X. Then from (3.10) it follows that $v_n \longrightarrow w$ in X and $w - A(w) = 0$; that is, $w \in \ker(I - A)$. Note that

$$d\big(v_n, \ker(I-A)\big) = \frac{d\big(x_n, \ker(I-A)\big)}{\|x_n - y_n\|_X} = 1 \qquad \text{for all } n \geq 1$$
$$\Rightarrow\ d\big(w, \ker(I-A)\big) = 1, \quad \text{a contradiction.}$$

So indeed $\{x_n - y_n\}_{n\geq 1} \subseteq X$ is bounded and due to the compactness of A, we may assume that $A(x_n - y_n) \longrightarrow h$ in X. Therefore from (3.9), we have

$$x_n - y_n \longrightarrow u + h \quad \text{in } X \quad \text{as } n \to \infty.$$

Set $f = u + h$. Then $u = f - A(f)$ and so $u \in R(I-A)$, which proves the closedness of $\mathcal{R}(I-A)$. □

REMARK 3.1.40 From Proposition 3.1.39 it follows that $\mathcal{R}(I-A) = \ker(I - A^*)^{\perp}$ and $\mathcal{R}(I-A^*) = \ker(I-A)^{\perp}$.

The next result is known as *Schauder's theorem* and its proof is an easy application of the Arzela–Ascoli theorem. We leave the details to the reader.

THEOREM 3.1.41 *If X, Y are Banach spaces, then $A \in \mathcal{L}_c(X,Y)$ if and only if $A^* \in \mathcal{L}_c(Y^*, X^*)$.*

The theorem that follows is known as the *Fredholm alternative theorem* and is useful in the study of boundary value problems.

THEOREM 3.1.42 *If X is a Banach space, $A \in \mathcal{L}_c(X)$, and $\lambda \neq 0$, then $\ker(\lambda I - A) = \{0\}$ if and only if $\mathcal{R}(\lambda I - A) = X$.*

PROOF: Again without any loss of generality, we can assume that $\lambda = 1$.
$\Rightarrow$: We argue by contradiction. Suppose $X_1 = \mathcal{R}(I-A) \neq X$. From Proposition 3.1.39 we know that X_1 is a Banach space and $A(X_1) \subseteq X_1$ (i.e., X_1 is A-invariant). Then $A\big|_{X_1} \in \mathcal{L}_c(X_1)$ and $X_2 = (I - A)(X_1)$ is a Banach subspace of X_1. Note that because $I - A$ is injective, $X_2 \neq X_1$. We continue this way and produce $\{X_n = (I-A)^n(X)\}_{n\geq 1}$ a strictly decreasing sequence of Banach subspaces of X. Using the lemma of Riesz, we can have $x_n \in X_n$, $\|x_n\|_X = 1$ and $d(x_n, x_{n+1}) \geq \frac{1}{2}$, $n \geq 1$. We have

$$Ax_n - Ax_m = -(x_n - Ax_n) + (x_m - Ax_m) + (x_n - x_m).$$

If $n > m$, then $X_{n+1} \subseteq X_n \subseteq X_{m+1} \subseteq X_m$ (with strict inclusions) and so

$$-(x_n - Ax_n) + (x_m - Ax_m) + x_n \in X_{m+1},$$
$$\Rightarrow\ \|Ax_n - Ax_m\|_X \geq \frac{1}{2},$$

which contradicts the fact that $A \in \mathcal{L}_c(X)$. Therefore $\mathcal{R}(I-A) = X$.

$\Leftarrow$: Because $\mathcal{R}(I-A) = X$, we have $\{0\} = \mathcal{R}(I-A)^{\perp} = \ker(I - A^*)$. From Theorem 3.1.41, we know that $A^* \in \mathcal{L}_c(X^*)$. So from the first part of the proof, we know that $\mathcal{R}(I-A^*) = X^*$. Then $\{0\} = \mathcal{R}(I-A^*)^{\perp} = \ker(I-A)$. □

REMARK 3.1.43 This theorem says that if $A \in \mathcal{L}_c(X)$, then either for every $x \in H$ the equation $(\lambda I - A)x = h$ has a unique solution or otherwise the homogeneous equation $(\lambda I - A)x = 0$ has a nontrivial solution. If X is finite-dimensional, then every linear operator (which is automatically continuous) is injective if and only if it is surjective. In an infinite-dimensional space X, an operator $A \in \mathcal{L}(X)$ can be injective without being surjective and vice versa.

DEFINITION 3.1.44 Let X, Y be Banach spaces and $A \in \mathcal{L}(X, Y)$

(a) We say that A is a *Fredholm operator*, if ker A is finite-dimensional and $\mathcal{R}(A)$ is closed and finite-codimensional. Then

$$\text{ind}A = \dim(\ker A) - \text{codim}\big(\mathcal{R}(A)\big)$$

is called the *index* of A. We denote the set of Fredholm operators by $\Phi(X, Y)$.

(b) We say that A is a *semi-Fredholm operator*, if ker A is finite-dimensional and $\mathcal{R}(A)$ is closed. We denote the set of semi-Fredholm operators by $\Phi_+(X, Y)$.

EXAMPLE 3.1.45 (a) If $X = \mathbb{R}^n$ and $Y = \mathbb{R}^m$, then every linear operator $A \colon \mathbb{R}^n \longrightarrow \mathbb{R}^m$ is Fredholm with index ind $A = n - m$.

(b) Let $X = H = l^2$ and $A \in \mathcal{L}(H)$ is defined by

$$A\widehat{x} = (x_2, x_3, \ldots), \qquad \widehat{x} = (x_n)_{n \geq 1} \in l^2.$$

Then ker A is one-dimensional (ker $A = \{\widehat{x} = (x_1, 0, 0, \ldots)\}$) and $\mathcal{R}(A) = l^2$. If we consider $A' \in \mathcal{L}(H)$ defined by

$$A'\widehat{x} = (0, x_1, x_2, \ldots), \qquad \widehat{x} = (x_n)_{n \geq 1} \in l^2,$$

then ker $A' = \{0\}$ and codim $\mathcal{R}(A') = 1$. So both A, A' are Fredholm operators.

THEOREM 3.1.46 *If X is a Banach space and $A \in \mathcal{L}_c(X)$, then $\lambda I - A \in \Phi(X)$ for all $\lambda \neq 0$.*

PROOF: As before without any loss of generality, we can assume that $\lambda = 1$.

Note that $I\big|_{\ker(I-A)} = A\big|_{\ker(I-A)}$. So $I\big|_{\ker(I-A)}$ is compact which of course implies that dim ker$(I - A) < +\infty$.

Next we show that $\mathcal{R}(A)$ is closed. Let V be the topological complement of ker$(I - A)$, (i.e., $V \subseteq X$ is a closed subspace such that $X = \ker(I - A) \oplus V$). Let $G = I - A$ and consider $G\big|_V$ and $A\big|_V$. We see that $G\big|_V = \{0\}$. Then $G(V) = G(X)$ and so it suffices to show that $G\big|_V$ is closed. But this follows if we show that $(G\big|_V)^{-1} \subseteq \mathcal{L}\big(G(V), V\big)$. By linearity we have to show that $(G\big|_V)^{-1}$ is continuous at the origin. If $(G\big|_V)^{-1}$ is not continuous at $x = 0$, then we can find $\{x_n\}_{n \geq 1} \subseteq V$ such that $G(x_n) \longrightarrow 0$ but $\{x_n\}_{n \geq 1}$ does not converge to 0. We may assume that $\|x_n\|_X \geq \delta > 0$ for all $n \geq 1$. Then $(1/\|x_n\|_X) \leq 1/\delta$ for all $n \geq 1$ and so $\|G\big(x_n/\|x_n\|_X\big)\|_X \leq 1/\delta \|G(x_n)\|_X \longrightarrow 0$ as $n \to \infty$. Also because $A \in \mathcal{L}_c(X)$, we see that at least for a subsequence $A\big((x_n/\|x_n\|_X)\big) \longrightarrow u$ in X. Then because

$$G\Big(\frac{x_n}{\|x_n\|_X}\Big) = \frac{x_n}{\|x_n\|_X} - A\Big(\frac{x_n}{\|x_n\|_X}\Big),$$

we infer that

$$\frac{x_n}{\|x_n\|_X} \longrightarrow u \in V \quad \text{in } X,\ u \neq 0.$$

But $G(u) = 0$, which contradicts the fact that $\ker G \cap V = \{0\}$. Therefore $G(V) = G(X)$ is closed.

Finally we show that $G(X)$ has finite codimension. Suppose that this is not true. Then we can find a sequence $\{X_n\}_{n\geq 0}$ of closed subspaces of X such that

$$G(X) = X_0 \subseteq X_1 \subseteq \ldots \subseteq X_n \subseteq \ldots,$$

where X_n is closed and of codimension one in X_{n+1}, just by adding one-dimensional subspaces to $G(X) = X_0$. Using the Riesz lemma, for each X_n, we can find $x_n \in X_n$ with $\|x_n\|_X = 1$ and $d(x_n, x_{n-1}) \geq 1 - \varepsilon$, $\varepsilon > 0$. Then for all $m < n$ we have

$$\|Ax_n - Ax_m\|_X = \|x_n - G(x_n) - x_m + G(x_m)\|_X \geq 1 - \varepsilon$$

because $-Ax_n - x_m + Gx_m \in H_{n-1}$. But this contradicts the fact that $A \in \mathcal{L}_c(X)$. $\square$

THEOREM 3.1.47 *If X, Y are Banach spaces, then $\Phi(X,Y)$ is open in $\mathcal{L}(X,Y)$ and $S \longrightarrow \operatorname{ind} S$ is continuous.*

PROOF: Let $S \in \Phi(X,Y)$. Then $X = \ker S \oplus V$. Evidently S is an isomorphism of V on $S(V)$. Hence

$$Y = S(V) \oplus W$$

for some finite-dimensional subspace W. Consider the map $\xi \colon V \oplus W \longrightarrow S(V) \oplus W$ defined by

$$\xi(x,y) = Sx + y.$$

This is an isomorphism. But recall that $\operatorname{Isom}\big(V \oplus W, S(V) \oplus W\big)$ is open in $\mathcal{L}\big(V \oplus W, S(V) \oplus W\big)$. So if K is close to S, then the map $\eta \colon V \oplus W \longrightarrow K(V) \oplus W = Y$ defined by

$$\eta(x,y) = Kx + y$$

is still in $\operatorname{Isom}\big(V \oplus W, S(V) \oplus W\big)$. Hence $\ker K$ is finite-dimensional and $\mathcal{R}(K)$ has finite codimension and so it is closed; that is, $K \in \Phi(X,Y)$.

Note that we can find E a finite-dimensional space such that

$$\begin{aligned}
& X = V \oplus \ker K \oplus E, \\
\Rightarrow\ & K \in \operatorname{Isom}\big(V \oplus E, K(V) \oplus K(E)\big) \quad \text{and} \quad \dim E = \dim K(E), \\
\Rightarrow\ & \operatorname{ind} K = \dim \ker K - \dim W + \dim K(E) \\
& \qquad\quad = \dim \ker S - \dim W = \operatorname{ind} S, \\
\Rightarrow\ & S \longrightarrow \operatorname{ind} S \text{ is constant on connected components, hence continuous.}
\end{aligned}$$

$\square$

REMARK 3.1.48 The above theorem remains true if instead of $\Phi(X,Y)$ we consider $\Phi_+(X,Y)$. Also from this theorem, we see that if $X = Y$ and $A \in \mathcal{L}_c(X)$, then for all $\lambda \neq 0$ and $\operatorname{ind}(\lambda I - A) = 0$ (from Theorem 3.1.46 we know $\lambda I - A \in \Phi(X)$).

The following proposition summarizes some of the useful properties of Fredholm operators and can be proved using the basic theorems of introductory linear functional analysis.

PROPOSITION 3.1.49 *If X, Y are Banach spaces and $S \in \Phi_+(X,Y)$, then*

(a) *if* $\operatorname{ind} S = 0$ *and* $\ker S = \{0\}$, *we have* $S^{-1} \in \mathcal{L}(Y,X)$.

(b) $\mathcal{R}(S) = (\ker S^*)^\perp$.

(c) $S^* \in \Phi_+(Y^*, X^*)$, $\operatorname{ind} S^* = -\operatorname{ind} S$ *and* $\mathcal{R}(S^*) = (\ker S)^\perp$.

We conclude with a basic stability result for Fredholm operators. Its proof can be found in Kato [343, pp. 232–238].

THEOREM 3.1.50 *If X, Y are Banach spaces, $S \in \Phi(X,Y)$, and either $A \in \mathcal{L}_c(X,Y)$ or $A \in \mathcal{L}(X,Y)$ with $\|A\|_{\mathcal{L}} \leq \varepsilon$ for some $\varepsilon = \varepsilon(S) > 0$, then $S + A \in \Phi(X,Y)$ and* $\operatorname{ind}(S+A) = \operatorname{ind} S$.

REMARK 3.1.51 The result is still valid if $\Phi(X,Y)$ is replaced by $\Phi_+(X,Y)$.

3.2 Monotone and Accretive Operators

In this section we study the most important class of nonlinear maps from a Banach space X into its dual. These are the monotone maps. They constitute a manageable class of operators because of the very simple definition of monotonicity. They appear naturally in the calculus of variations and in the theory of nonlinear boundary value problems and provide an analytic framework broader than that of compact operators. When we deal with nonlinear maps from a Banach space into itself, then the corresponding notion is that of an accretive operator. When $X = H =$ a pivot Hilbert space, the two notions coincide. Accretive operators are closely connected with the generation theory of semigroups (linear and nonlinear). In the second half of this section we deal with accretive operators and the semigroups they generate.

Let X be a Banach space. By X^* we denote its topological dual and by $\langle \cdot, \cdot \rangle_X$ the duality brackets for the pair (X, X^*). Given a multivalued mapping $A \colon X \longrightarrow 2^{X^*}$, let us fix some notation associated with it.

(a) The *domain* of A is the set $\mathcal{D}(A) = \{x \in X : A(x) \neq \emptyset\}$.
(b) The *graph* of A is the set $\operatorname{Gr} A = \{(x, x^*) \in X \times X^* : x^* \in A(x)\}$.
(c) The *inverse* of A $A^{-1} : X^* \longrightarrow 2^X$ is defined by $(A^{-1})(x^*) = \{x \in X : x^* \in A(x)\}$ (hence $\operatorname{Gr} A^{-1} = \{(x^*, x) \in X^* \times X : (x^*, x) \in \operatorname{Gr} A\}$).

REMARK 3.2.1 The generality of our setting allows us to identify every subset $G \subseteq X \times X^*$ with a mapping $A \colon X \longrightarrow 2^{X^*}$ defined by $A(x) = \{x^* \in X^* : (x, x^*) \in G\}$. For this reason some authors identify a multivalued mapping with its graph. Note that in the present multivalued context the inverse mapping A^{-1} is always defined.

DEFINITION 3.2.2 Given a mapping $A \colon X \longrightarrow 2^{X^*}$, we say that:

(a) A is *monotone* if for any $x, y \in \mathcal{D}(A)$ and $x^* \in A(x)$, $y^* \in A(y)$ we have

$$\langle x^* - y^*, x - y \rangle \geq 0. \tag{3.11}$$

(b) A is *strictly monotone* if it is monotone and equality in (3.11) implies that $x = y$.
(c) A is *maximal monotone* if it is monotone and for $(y, y^*) \in X \times X^*$ the inequalities $\langle y^* - x^*, y - x \rangle_X \geq 0$ for all $(x, x^*) \in \mathrm{Gr}\, A$, imply $(y, y^*) \in \mathrm{Gr}\, A$.
(d) A is *locally bounded at* $x \in \mathcal{D}(A)$ if we can find a neighborhood U of x such that $A(U)$ is bounded. We say that A is *locally bounded* if it is locally bounded at every $x \in \mathcal{D}(A)$.
(e) A is *bounded* if it maps bounded sets in X to bounded sets in X^*.

REMARK 3.2.3 According to Definition 3.2.2(c), a mapping $A: X \longrightarrow 2^{X^*}$ is maximal monotone if and only if its graph Gr A is maximal with respect to inclusion among the graphs of all monotone mappings. This definition makes it necessary to consider set-valued mappings. Indeed a discontinuous nondecreasing function $f: \mathbb{R} \longrightarrow \mathbb{R}$ is not maximal monotone. To become maximal monotone, we need to fill in the jump discontinuities, namely introduce $\widehat{f}: \mathbb{R} \longrightarrow 2^{\mathbb{R}}$ defined by $\widehat{f}(x) = [f(x^-), f(x^+)]$ for all $x \in \mathbb{R}$. By Zorn's lemma, every monotone map has a maximal monotone extension.

Recall that $C \subseteq X$ is said to be absorbing if $0 \in C$ and $X = \bigcup_{\lambda > 0} \lambda C$ (equivalently for every $x \in X$, we can find $\lambda > 0$ such that $\lambda x \in C$). A point $x \in D \subseteq X$ is said to be the absorbing point of D if $D - x$ is absorbing. Evidently if int $D \neq \emptyset$, then every $x \in \mathrm{int}\, D$ is an absorbing point of D. But there can be absorbing points of D not belonging in int D. In fact it can happen that int $D = \emptyset$. Consider $D = \partial B_1 \cup \{0\}$. Then the origin is an absorbing point of D, but int $D = \emptyset$.

THEOREM 3.2.4 *If $A: X \longrightarrow 2^{X^*}$ is a monotone map and x_0 is an absorbing point of $\mathcal{D}(A)$, then A is locally bounded at x_0.*

PROOF: By picking any $x_0^* \in A(x_0)$ and replacing A by the monotone map $A_1(y) = A(y + x_0) - x_0^*$ we see that without any loss of generality, we may assume that $x_0 = 0$ and $(0, 0) \in \mathrm{Gr}\, A$. Let

$$\varphi(x) = \sup \left\{ \langle y^*, x - y \rangle_X : y \in \mathcal{D}(A),\ \|y\|_X \leq 1,\ y^* \in A(y) \right\}$$

for all $x \in X$ and $L_1^\varphi = \{x \in X : \varphi(x) \leq 1\}$.

Note that φ is the supremum of affine continuous functions. So φ is lower semicontinuous and convex, which means that L_1^φ is closed and convex. Note that because $(0, 0) \in \mathrm{Gr}\, A$, we have $\varphi \geq 0$. Also, if $y \in \mathcal{D}(A)$, $y^* \in A(y)$, then due to the monotonicity of A and because $(0, 0) \in \mathrm{Gr}\, A$, we have

$$\begin{aligned} & 0 \leq \langle y^*, y \rangle, \\ \Rightarrow\ & \varphi(0) = 0 \quad (\text{i.e., } 0 \in L_1^\varphi). \end{aligned} \tag{3.12}$$

We show that L_1^φ is also absorbing. Recall that by hypothesis $\mathcal{D}(A)$ is absorbing. So if $x \in X$, we can find $\lambda > 0$ such that $\lambda x \in \mathcal{D}(A)$ (i.e., $A(\lambda x) \neq \emptyset$). Let $u^* \in A(\lambda x)$. If $(y, y^*) \in \mathrm{Gr}\, A$, then by virtue of the monotonicity of A, we have

$$\begin{aligned}&\langle y^*, \lambda x - y\rangle_X \le \langle u^*, \lambda x - y\rangle_X\\ \Rightarrow\ & \varphi(\lambda x) \le \sup\left\{ \langle u^*, \lambda x - y\rangle_X : y \in \mathcal{D}(A),\ \|y\|_X \le 1\right\}\\ &\qquad \le \langle u^*, \lambda x\rangle_X + \|u^*\|_{X^*} < +\infty.\end{aligned}$$

Choose $t \in (0,1)$ such that $t\varphi(\lambda x) < 1$. Because of convexity of φ, we have

$$\begin{aligned}&\varphi(t\lambda x) \le t\varphi(\lambda x) + (1-t)\varphi(0) = t\varphi(\lambda x) < 1 \quad \text{(see (3.12))}\\ \Rightarrow\ & t\lambda x \in L_1^\varphi \quad \text{(i.e., } L_1^\varphi \text{ is absorbing).}\end{aligned}$$

Then $C{=}L_1^\varphi \cap (-L_1^\varphi)$ is nonempty (because $0 \in C$), closed, convex, symmetric, and absorbing, therefore $0 \in \operatorname{int} L_1^\varphi$. Hence we can find $\delta{>}0$ such that

$$\begin{aligned}&\varphi(x) \le 1 \qquad \text{for all } \|x\|_X \le 2\delta,\\ \Rightarrow\ & \langle y^*, x\rangle_X \le \langle y^*, y\rangle_X + 1 \qquad \text{for all } y \in \mathcal{D}(A),\ \|y\|_X \le 1,\end{aligned}$$

all $y^* \in A(y)$ and all $\|x\|_X \le 2\delta$,

$$\Rightarrow\ 2\delta\|y^*\|_{X^*} \le \|y^*\|_{X^*}\|y\|_X + 1 \le \delta\|y^*\|_{X^*} + 1$$

for all $y \in D(A) \cap B_\delta$ and all $y^* \in A(y)$,

$$\Rightarrow\ \|y^*\|_{X^*} \le \frac{1}{\delta} \qquad \text{for all } y^* \in A\big(\mathcal{D}(A) \cap B_\delta\big).$$

□

PROPOSITION 3.2.5 *If $A : X \longrightarrow 2^{X^*}$ is a maximal monotone map and $x \in \mathcal{D}(A)$, then $A(x)$ is nonempty, w^*-closed, and convex.*

PROOF: If $x_1^*, x_2^* \in A(x)$ and $\lambda \in [0,1]$, then for any $(y, y^*) \in \operatorname{Gr} A$, we have

$$\begin{aligned}\langle (1-\lambda)x_1^* + \lambda x_2^* - y^*, x - y\rangle_X &= (1-\lambda)\langle x_1^* - y^*, x - y\rangle_X\\ &\quad + \lambda\langle x_2^* - y^*, x - y\rangle_X \ge 0.\end{aligned}$$

From the maximality of A (see Definition 3.2.2(c)), it follows that

$$(1-\lambda)x_1^* + \lambda x_2^* \in A(x) \quad \text{(i.e., } A(x) \text{ is convex).}$$

Also if $\{x_\alpha^*\}_{\alpha\in J} \subseteq A(x)$ is a net such that $x_\alpha^* \xrightarrow{w^*} x^*$ in X^*, then for all $(y, y^*) \in \operatorname{Gr} A$ we have

$$\begin{aligned}&\langle x_\alpha^* - y^*, x - y\rangle_X \ge 0,\\ \Rightarrow\ & \langle x^* - y^*, x - y\rangle_X \ge 0,\end{aligned}$$

which by the maximality of A implies that $(x, x^*) \in \operatorname{Gr} A$; that is, $A(x)$ is w^*-closed.

□

Combining Theorem 3.2.4 and Proposition 3.2.5, we obtain the following.

PROPOSITION 3.2.6 *If $A : X \longrightarrow 2^{X^*}$ is a maximal monotone map and $x \in \operatorname{int} \mathcal{D}(A)$, then $A(x)$ is nonempty, w^*-compact, and convex.*

PROPOSITION 3.2.7 *If $A : X \longrightarrow 2^{X^*}$ is a maximal monotone map, then $\operatorname{Gr} A$ is closed in $X \times X_{w^*}^*$ and in $X_w \times X^*$. Here by $X_{w^*}^*$ (resp., X_w) we denote the space X^* (resp., X) furnished with the w^*-topology (resp., the w-topology).*

PROOF: Let $\{(x_\alpha, x_\alpha^*)\}_{\alpha\in J}$ be a net in Gr A such that $x_\alpha \longrightarrow x$ in X and $x_\alpha^* \xrightarrow{w^*} x^*$ in X^*. Then for any $(y,y^*) \in$ Gr A, we have

$$\begin{aligned} &\langle x_\alpha^* - y^*, x_\alpha - y\rangle_X \geq 0, \\ \Rightarrow\ &\langle x^* - y^*, x - y\rangle_X \geq 0. \end{aligned}$$

As before from the maximality of A we infer that $(x,x^*) \in$ Gr A, which proves that Gr A is closed in $X \times X^*_{w^*}$ and similarly for the closedness in $X_w \times X^*$. □

PROPOSITION 3.2.8 *If $A\colon X \longrightarrow 2^{X^*}$ is a monotone map with convex and w^*-closed values, $\mathcal{D}(A) = X$ and for all $x,u \in X$ the set-valued map $\lambda \longrightarrow A(x+\lambda u)$ from $[0,1]$ into X^* has a closed graph in $[0,1]\times X^*_{w^*}$, then A is maximal monotone.*

PROOF: Suppose that for $(x,x^*) \in X \times X^*$, we have

$$\langle x^* - y^*, x - y\rangle_X \geq 0 \qquad \text{for all } (y,y^*) \in \text{Gr } A.$$

We need to show that $(x,x^*) \in$ Gr A (see Definition 3.2.2(c)). We argue indirectly. So suppose that $x^* \notin A(x)$. By hypothesis $A(x)$ is w^*-closed and convex. So by the strong separation theorem, we can find $u \in X \setminus \{0\}$ and $\varepsilon > 0$ such that

$$\sup\big\{ \langle v^*, u\rangle_X : v^* \in A(x)\big\} \leq \langle x^*, u\rangle_X - \varepsilon. \tag{3.13}$$

In (3.13) we set $y = y_\lambda = x + \lambda u$, $\lambda > 0$ and $y^* = y_\lambda^* \in A(y_\lambda)$. Then

$$\langle y_\lambda^* - x^*, u\rangle_X \geq 0.$$

Because of Theorem 3.2.4 $\{y_\lambda^*\}_{\lambda\in[0,1]} \subseteq X^*$ is bounded and so we may assume that $y_\lambda^* \xrightarrow{w^*} v^*$ in X^*. Then by hypothesis $(x,v^*) \in$ Gr A. If we pass to the limit as $\lambda \downarrow 0$ we obtain

$$0 \leq \langle v^* - x^*, u\rangle_X$$

which contradicts (3.13). □

DEFINITION 3.2.9 Let $A\colon X \longrightarrow X^*$ be a single-valued map with $\mathcal{D}(A) = X$. Then we say:

(a) A is *demicontinuous* if $x_n \longrightarrow x$ in X, implies that $A(x_n) \xrightarrow{w^*} A(x)$ in X^*.

(b) A is *hemicontinuous* if for all $x,y \in X$, the map $\lambda \longrightarrow A(x+\lambda y)$ is continuous from $[0,1]$ into $X^*_{w^*}$.

REMARK 3.2.10 Evidently any demicontinuous map is hemicontinuous. Conversely, it is easy to check that if $A\colon X \longrightarrow X^*$ is monotone and hemicontinuous, it is demicontinuous.

COROLLARY 3.2.11 *If $A\colon X \longrightarrow X^*$ is monotone, and hemicontinuous, then A is maximal monotone.*

An important monotone map that is a basic tool in the study of maximal monotone maps and of nonlinear evolution equations, is the so-called duality map.

DEFINITION 3.2.12 Let X be a Banach space and X^* its dual.

(a) The *duality map* (or *normalized duality map*) is the map $\mathcal{F}\colon X \longrightarrow 2^{X^*}$ defined by

$$\mathcal{F}(x) = \{x^* \in X^* : \langle x^*, x\rangle_X = \|x\|_X^2 = \|x^*\|_{X^*}^2\}.$$

(b) Let $\xi\colon \mathbb{R}_+ \longrightarrow \mathbb{R}_+$ be a continuous, strictly increasing function such that $\xi(0) = 0$ and $\lim\limits_{n\to\infty} \xi(r) = +\infty$ (such a function is called the *gauge function*). The *duality map with gauge* ξ is the map $\mathcal{F}_\xi : X \longrightarrow 2^{X^*}$ defined by

$$\mathcal{F}_\xi(x) = \big\{x^* \in X^* : \langle x^*, x\rangle_X = \|x^*\|_{X^*}\|x\|_X,\ \|x^*\|_X = \xi(\|x\|_X)\big\}.$$

REMARK 3.2.13 By the Hahn–Banach theorem, it follows that $\mathcal{D}(\mathcal{F}) = \mathcal{D}(\mathcal{F}_\xi) = X$. Moreover, both $\mathcal{F}$ and $\mathcal{F}_\xi$ have closed and convex values. Note that if $\xi(r) = r$, that $\mathcal{F} = \mathcal{F}_\xi$. If ξ is a gauge function and $\eta(r) = \int_0^r \xi(s)ds$, then η is a convex function and $\mathcal{F}_\xi(x) = \partial\eta(\|x\|_X)$. In particular, if $\eta(x) = \frac{1}{2}\|x\|_X^2$, then $\mathcal{F}(x) = \partial\eta(x)$ for all $x \in X$. Here by $\partial\eta$ we denote the subdifferential of the convex function $\eta(\cdot)$ (see Definition 1.2.28).

In the sequel, for simplicity in the presentation, we limit ourselves to the normalized duality map $\mathcal{F}$.

DEFINITION 3.2.14 Let $A : X \longrightarrow 2^{X^*}$ be a map.

(a) We say that A is *coercive*, if either $\mathcal{D}(A)$ is bounded or $\mathcal{D}(A)$ is unbounded and

$$\frac{\inf[\langle x^*, x\rangle_X : x^* \in A(x)]}{\|x\|_X} \longrightarrow +\infty \quad \text{as } \|x\|_X \longrightarrow +\infty,\ x \in \mathcal{D}(A).$$

(b) We say that A is *weakly coercive* if either $\mathcal{D}(A)$ is bounded or $\mathcal{D}(A)$ is unbounded and $\inf\big\{\|x^*\|_{X^*} : x^* \in A(x)\big\} \longrightarrow +\infty$ as $\|x\|_X \longrightarrow +\infty$, $x \in \mathcal{D}(A)$.

REMARK 3.2.15 Evidently coercivity implies weak coercivity.

PROPOSITION 3.2.16 *If X is a reflexive Banach space and X^* is strictly convex, then $\mathcal{F}\colon X \longrightarrow 2^{X^*}$ is single-valued, demicontinuous, maximal monotone, bounded, coercive, and odd.*

PROOF: Suppose $x_k^* \in \mathcal{F}(x), k = 1, 2$. Then

$$\begin{aligned} 2\|x_k^*\|_{X^*}\|x\|_X \le \|x_1^*\|_{X^*}^2 + \|x_2^*\|_{X^*}^2 &= \langle x_1^* + x_2^*, x\rangle_X \\ &\le \|x_1^* + x_2^*\|_{X^*}\|x\|_X, \end{aligned}$$

$$\Rightarrow \quad \|x_1^*\|_{X^*} \le \frac{1}{2}\|x_1^* + x_2^*\|_{X^*}. \tag{3.14}$$

Because $\|x_1^*\|_{X^*} = \|x_2^*\|_{X^*}$ and X^* is strictly convex, then from (3.14) it follows that $x_1^* = x_2^*$.

Next we show that $\mathcal{F}$ is demicontinuous. To this end let $x_n \longrightarrow x$ in X. Then $\|\mathcal{F}(x_n)\|_{X^*} = \|x_n\|_X \longrightarrow \|x\|_X$ as $n \to \infty$. Because X^* is reflexive, we may assume (at least for a subsequence), that

$$\mathcal{F}(x_n) \xrightarrow{w} x^* \quad \text{in } X^* \quad \text{as } n \to \infty.$$

So for all $u \in X$, we have

$$\langle x^*, u \rangle_X = \lim_{n\to\infty} \langle \mathcal{F}(x_n), u \rangle_X \le \lim_{n\to\infty} \|x_n\|_X \|u\|_X = \|x\|_X \|u\|_X .$$

Also we have

$$\begin{aligned} & \langle x^*, x \rangle_X = \lim_{n\to\infty} \langle \mathcal{F}(x_n), x_n \rangle_X = \lim_{n\to\infty} \|x_n\|_X^2 = \|x\|_X^2 , \\ \Rightarrow \; & x^* = \mathcal{F}(x). \end{aligned}$$

Therefore by Urysohn's criterion for convergent sequences, we conclude that for the original sequence we have $\mathcal{F}(x_n) \xrightarrow{w} \mathcal{F}(x)$ in X^* as $n \to \infty$.

Note that for every $x, u \in X$, we have

$$\begin{aligned} & \langle \mathcal{F}(x) - \mathcal{F}(u), x - u \rangle_X = \|x\|_X^2 - 2\|x\|_X \|u\|_X + \|u\|_X^2 \ge 0, \\ \Rightarrow \; & \mathcal{F} \text{ is monotone, hence maximal monotone (see Corollary 3.2.11).} \end{aligned}$$

Finally it is clear from its definition that $\mathcal{F}$ is bounded, coercive, and odd. $\square$

PROPOSITION 3.2.17 *If X is a reflexive Banach space, X^* is strictly convex, and $\psi_0(x) = \|x\|_X$, then ψ_0 is Gâteaux differentiable at all $x \in X \setminus \{0\}$ and*

$$\psi_0'(x) = \frac{\mathcal{F}(x)}{\|x\|} \qquad \textit{for all } x \in X \setminus \{0\}.$$

PROOF: For all $x, y \in X$ we have

$$\begin{aligned} \langle \mathcal{F}(y), y - x \rangle_X & \ge \|y\|_X^2 - \|y\|_X \|x\|_X \\ & \ge \|y\|_X^2 - \frac{1}{2}(\|y\|_X^2 + \|x\|_X^2) \\ & = \frac{1}{2}\|y\|_X^2 - \frac{1}{2}\|x\|_X^2 \\ & \ge \|y\|_X \|x\|_X - \|x\|_X^2 \ge \langle \mathcal{F}(x), y - x \rangle_X . \end{aligned}$$

Let $y = x + \lambda h$, $h \in X$, and $\lambda \in \mathbb{R}$. We obtain

$$\lambda \langle \mathcal{F}(x), h \rangle_X \le \frac{1}{2}\|x + \lambda h\|_X^2 - \frac{1}{2}\|x\|_X^2 \le \lambda \langle \mathcal{F}(x + \lambda h), h \rangle_X . \tag{3.15}$$

From Proposition 3.2.16 we know that $\langle \mathcal{F}(x + \lambda h), h \rangle_X \longrightarrow \langle \mathcal{F}(x), h \rangle_X$ as $\lambda \to 0$. So from (3.15) it follows that

$$\mathcal{F}(x) = \eta'(x)$$

where $\eta'(x)$ is the Gâteaux derivative of $\eta(x) = \frac{1}{2}\|x\|_X^2$. Because $\psi_0(x) = \eta(x)^{1/2}$, from Proposition 3.2.16 we infer that ψ_0 is Gâteaux differentiable at all $x \in X \setminus \{0\}$ and $\psi_0'(x) = \big(\mathcal{F}(x)\big)/\|x\|$. $\square$

REMARK 3.2.18 Proposition 3.2.17 established the following duality implication. If X is a reflexive Banach space and X^* is strictly convex, then X is smooth. In fact the following stronger result is true. Namely, if X is reflexive, then X is strictly convex (resp., smooth) if and only if X^* is smooth (resp., strictly convex).

PROPOSITION 3.2.19 *If X is a reflexive Banach space and X^* is locally uniformly convex, then $\mathcal{F}\colon X \longrightarrow X^*$ is continuous and $\psi_0(x) = \|x\|$ is Fréchet differentiable at every $x \in X \setminus \{0\}$.*

PROOF: Let $x_n \longrightarrow x$ in X. Then $\|\mathcal{F}(x_n)\|_X \longrightarrow \|\mathcal{F}(x)\|_X$ and from Proposition 3.2.16 we also have that $\mathcal{F}(x_n) \xrightarrow{w} \mathcal{F}(x)$ in X^*. But X^* being locally uniformly convex it has the Kadec–Klee property and so $\mathcal{F}(x_n) \longrightarrow \mathcal{F}(x)$ in X^* (i.e., $\mathcal{F}$ is continuous). Combining this with Proposition 3.2.17, we see that $x \longrightarrow \eta(x) = \frac{1}{2}\|x\|_X^2$ is Fréchet differentiable (see Proposition 1.1.10). Therefore $x \longrightarrow \psi_0(x) = \|x\| = \big(2\eta(x)\big)^{1/2}$ is Fréchet differentiable at every $x \in X \setminus \{0\}$. $\square$

REMARK 3.2.20 In fact if X is reflexive and both X and X^* are locally uniformly convex, then $\mathcal{F}\colon X \longrightarrow X^*$ is a homeomorphism. Also if X^* is uniformly convex, then $\mathcal{F}$ is uniformly continuous on bounded sets of X.

EXAMPLE 3.2.21 (a) If (Ω, Σ, μ) is a σ-finite measure space and $1 \leq p < \infty$, then the duality map on $L^p(\Omega)$ corresponding to the gauge function $\xi(r) = r^{p-1}$, is the subdifferential of $x \longrightarrow (1/p)\|x\|_p^p$, $x \in L^p(\Omega)$. So

$$x^* \in \mathcal{F}_\xi(x) \quad \text{if and only if } x^*(z) \in |x(z)|^{p-1}\mathrm{sgn}\big(x(z)\big) \quad \text{a.e. on } \Omega.$$

(b) If $\Omega \in \mathbb{R}^N$ is a bounded domain and $1 < p < \infty$, then the duality map on the Sobolev space $W_0^{1,p}(\Omega)$ with norm $\|x\| = \|Dx\|_p$ corresponding to the gauge function $\xi(r) = r^{p-1}$ is

$$\mathcal{F}(x) = -\mathrm{div}(\|Dx\|^{p-2}Dx) \in W^{-1,p'}(\Omega) = W_0^{1,p}(\Omega)^*, \quad \frac{1}{p} + \frac{1}{p'} = 1.$$

The importance of maximal monotone operators stems from their remarkable surjectivity properties under coercivity conditions. Surjectivity results correspond to existence theorems for nonlinear boundary value problems. Next we prove the main surjectivity result for maximal monotone maps. To do this we need some preparation. We start with a finite-dimensional result.

PROPOSITION 3.2.22 *If X is a finite-dimensional Banach space, $C \subseteq X$ is a nonempty closed convex set, $A\colon X \longrightarrow 2^{X^*}$ is a monotone map with $\mathcal{D}(A) \subseteq C$, and $B\colon C \longrightarrow X^*$ is monotone, continuous, and coercive, then there exists $x_0 \in C$ such that*

$$\langle x^* + B(x_0), x - x_0\rangle_X \geq 0 \qquad \textit{for all } (x, x^*) \in \mathrm{Gr}\, A.$$

PROOF: Without any loss of generality, we may assume that $(0,0) \in \mathrm{Gr}\, A$. Indeed if this is not the case, we fix $\widehat{x}$, $\widehat{x}^* \in \mathrm{Gr}\, A$ and replace A and B, by $\widehat{A}$ and $\widehat{B}$ defined by

$$\widehat{A}(x) = A(x + \widehat{x}) - \widehat{x}^* \qquad \text{and} \qquad \widehat{B}(x) = B(x + \widehat{x}) - \widehat{x}^*.$$

Note that $\widehat{A}$ and $\widehat{B}$ still have all the properties of A and B, namely $\widehat{A}$ is monotone with $D(\widehat{A}) \subseteq C$ and $\widehat{B}$ is monotone, continuous, and coercive.

First suppose that $\mathcal{D}(A)$ is bounded. Let $K = \overline{\mathrm{conv}}\, \mathcal{D}(A)$. Then K is compact and convex. Suppose that the proposition is not true. Then for every $y \in K$ we can find $(x, x^*) \in \mathrm{Gr}\, A$ such that

$$\langle x^* + B(y), x - y \rangle_X < 0.$$

Therefore $K = \bigcup_{(x,x^*) \in \text{Gr } A} \{y \in K : \langle x^* + B(y), x - y \rangle_X < 0\}$. Clearly each set in the union is open. Because K is compact, we can find $\{(x_k, x_k^*)\}_{k=1}^n \subseteq \text{Gr } A$ such that

$$K = \bigcup_{k=1}^n \{y \in K : \langle x_k^* + B(y), x_k - y \rangle_X < 0\}.$$

Let $\{\vartheta_k\}_{k=1}^n$ be a continuous partition of unity subordinate to this cover and consider the map $\varphi: K \longrightarrow K$ defined by

$$\varphi(y) = \sum_{k=1}^n \vartheta_k(y) x_k.$$

By Brouwer's fixed point theorem (see Theorem 3.5.3), we can find $x_0 \in K$ such that $\varphi(x_0) = x_0$. For every $y \in K$, we have

$$\begin{aligned} h(y) &= \langle \sum_{k=1}^n \vartheta_k(y) x_k^* + B(y), \varphi(y) - y \rangle_X \\ &= \langle \sum_{k=1}^n \vartheta_k(y) x_k^* + B(y), \sum_{i=1}^n \vartheta_i(y)(x_i - y) \rangle_X \\ &= \sum_{k,i=1}^n \vartheta_k(y) \vartheta_i(y) \langle x_k^* + B(y), x_i - y \rangle_X . \end{aligned}$$

If $k = i$ and $\vartheta_k^2(y) \neq 0$, then $y \in K$ and so $\langle x_k^* + B(y), x_k - y \rangle_X < 0$. If $k \neq i$ and $\vartheta_k(y)\vartheta_i(y) \neq 0$, then $y \in \{y \in K : \langle x_k^* + B(y), x_k - y \rangle_X < 0\} \cap \{y \in K : \langle x_i^* + B(y), x_i - y \rangle_X < 0\}$ and so exploiting the monotonicity of A, we have

$$\begin{aligned} & \langle x_k^* + B(y), x_i - y \rangle_X + \langle x_i^* + B(y), x_k - y \rangle_X \\ = \ & \langle x_k^* + B(y), x_k - y \rangle_X + \langle x_i^* + B(y), x_i - y \rangle_X + \langle x_k^* - x_i^*, x_k - x_i \rangle_X < 0. \end{aligned}$$

Therefore $h(y) < 0$ for all $y \in K$. But

$$h(x_0) = \langle \sum_{k=1}^n \vartheta_k(x_0) x_k^* + B(x_0), \varphi(x_0) - x_0 \rangle_X = 0 \quad (\text{recall that } \varphi(x_0) = x_0),$$

a contradiction.

Next we remove the hypothesis that $\mathcal{D}(A)$ is bounded. From the first part of the proof, we can find $x_n \in C$ such that

$$\langle x^* + B(x_n), x - x_n \rangle_X \geq 0 \qquad \text{for all } (x, x^*) \in \text{Gr } A|_{\overline{B}_n(0)}.$$

Since $(0,0) \in \text{Gr } A|_{\overline{B}_n(0)}$, we obtain

$$\langle B(x_n), x_n \rangle_X \leq 0 \qquad \text{for all } n \geq 1.$$

Due to the coercivity of B, it follows that $\{x_n\}_{n \geq 1} \subseteq C$ is bounded. So we may assume that $x_n \longrightarrow x_0$ in X as $n \to \infty$. Then $x_0 \in C$ and

$$\langle x^* + B(x_0), x - x_0 \rangle_X \qquad \text{for all } (x, x^*) \in \text{Gr } A.$$

□

We extend this proposition to infinite dimensional Banach spaces and then we use the extension to prove surjectivity results for maximal monotone operators.

PROPOSITION 3.2.23 *If X is a reflexive Banach space, $C \subseteq X$ is a nonempty closed convex set, $A: X \longrightarrow 2^{X^*}$ is a monotone map with $\mathcal{D}(A) \subseteq C$ and $B: C \longrightarrow X^*$ is monotone, hemicontinuous, bounded, and coercive, then there exists $x_0 \in C$ such that*

$$\langle x^* + B(x_0), x - x_0 \rangle_X \geq 0 \qquad \textit{for all } (x, x^*) \in \text{Gr } A.$$

PROOF: We employ the method of finite-dimensional approximations (Galerkin method). So let $\{X_\alpha\}_{\alpha \in J}$ be a directed family of finite dimensional subspaces of X such that $\overline{\bigcup_{\alpha \in J} X_\alpha} = X$. Let $p_\alpha \in \mathcal{L}(X, X_\alpha)$ be the projection onto the X_α operator. Then $p_\alpha^* \in \mathcal{L}(X_\alpha^*, X^*)$ is the corresponding embedding map. We define

$$A_\alpha = p_\alpha^* \circ A \circ p_\alpha, \quad B_\alpha = p_\alpha^* \circ B \circ p_\alpha \quad \text{and} \quad C_\alpha = C \cap X_\alpha \qquad \text{for all } \alpha \in J.$$

Then A_α and B_α satisfy the hypotheses of Proposition 3.2.22 on X_α and so we can find $x_\alpha \in C_\alpha$ such that

$$\begin{aligned} &\langle \widehat{x}^* + B_\alpha(x_\alpha), \widehat{x} - x_\alpha \rangle_{X_\alpha} \geq 0 \qquad \text{for all } (\widehat{x}_\alpha, \widehat{x}_\alpha^*) \in \text{Gr } A_\alpha, \\ \Rightarrow\ &\langle \widehat{x}^* + B(x_\alpha), x - x_\alpha \rangle_X \geq 0 \qquad \text{for all } (x, x^*) \in \text{Gr } (A \circ p_\alpha). \end{aligned} \tag{3.16}$$

In particular, $\langle B(x_\alpha), x_\alpha \rangle_X \leq 0$ for all $\alpha \in J$ and due to the coercivity and boundedness of B, we can find $M > 0$ such that

$$\|x_\alpha\|_X \leq M \qquad \text{and} \qquad \|B(x_\alpha)\|_{X^*} \leq M \qquad \text{for all } \alpha \in J.$$

Due to the reflexivity of X, we can find a subsequence $\{x_{\alpha_n}\}_{n \geq 1} \subseteq \{x_\alpha\}_{\alpha \in J}$ such that

$$x_{\alpha_n} \xrightarrow{w} x_0 \qquad \text{and} \qquad B(x_{\alpha_n}) \xrightarrow{w} x_0^* \qquad \text{with } (x_0, x_0^*) \in C \times X^*.$$

Then from (3.16), we have

$$\limsup_{n \to \infty} \langle B(x_{\alpha_n}), x_{\alpha_n} \rangle_X \leq \langle x^*, x - x_0 \rangle_X + \langle x_0^*, x \rangle_X \qquad \text{for all } (x, x^*) \in \text{Gr } A. \tag{3.17}$$

By virtue of Zorn's lemma, without any loss of generality, we may assume that A is maximal on $\mathcal{D}(A)$. We claim that there exists $(x_1, x_1^*) \in \text{Gr } A$ such that

$$\langle x_1^*, x_1 - x_0 \rangle_X + \langle x_0^*, x_1 \rangle_X \leq \langle x_0^*, x_0 \rangle_X . \tag{3.18}$$

Indeed if (3.18) is not true, then we have

$$\langle x^* + x_0^*, x - x_0 \rangle_X > 0 \qquad \text{for all } (x, x^*) \in \text{Gr } A, \tag{3.19}$$

$$\Rightarrow\ (x_0, -x_0^*) \in \text{Gr } A \quad \text{(due to the maximality of } A).$$

So if in (3.19) we set $x = x_0$ and $x^* = x_0^*$, we reach a contradiction. Hence (3.18) holds and using this in (3.17), we obtain

$$\begin{aligned} & \limsup_{n\to\infty} \langle B(x_{\alpha_n}), x_{\alpha_n} \rangle_X \leq \langle x_0^*, x_0 \rangle_X , \\ \Rightarrow \; & \limsup_{n\to\infty} \langle B(x_{\alpha_n}), x_{\alpha_n} - x_0 \rangle_X \leq 0. \end{aligned} \tag{3.20}$$

Now let $x \in \mathcal{D}(A)$ and $x_\lambda = \lambda x_0 + (1-\lambda)x$, $\lambda \in [0,1]$. Then $x_\lambda \in C$ and because of the monotonicity of B, we have

$$\begin{aligned} & \langle B(x_{\alpha_n}) - B(x_\lambda), x_{\alpha_n} - x_\lambda \rangle_X \geq 0 \qquad \text{for all } n \geq 1 \text{ and all } \lambda \in [0,1], \\ \Rightarrow \; & \lambda \langle B(x_{\alpha_n}), x_{\alpha_n} - x_0 \rangle_X + (1-\lambda) \langle B(x_{\alpha_n}), x_{\alpha_n} - x \rangle_X \\ & \geq \lambda \langle B(x_\lambda), x_{\alpha_n} - x_0 \rangle_X + (1-\lambda) \langle B(x_\lambda), x_{\alpha_n} - x \rangle_X \\ \Rightarrow \; & \liminf_{n\to\infty} \langle B(x_{\alpha_n}), x_{\alpha_n} - x \rangle_X \geq \langle B(x_\lambda), x_0 - x \rangle_X , \end{aligned}$$

for all $\lambda \in [0,1]$ (see (3.20)).

Because B is hemicontinuous, we have

$$\begin{aligned} & \liminf_{n\to\infty} \langle B(x_{\alpha_n}), x_{\alpha_n} - x \rangle_X \geq \langle B(x_0), x_0 - x \rangle_X , \\ \Rightarrow \; & \liminf_{n\to\infty} \langle B(x_{\alpha_n}), x_{\alpha_n} \rangle_X \geq \langle B(x_0), x_0 - x \rangle_X + \langle x_0^*, x \rangle_X \qquad \text{for all } x \in \mathcal{D}(A) \end{aligned} \tag{3.21}$$

Then from (3.17) and (3.21), we obtain

$$\langle B(x_0), x_0 - x \rangle_X \leq \langle x^*, x - x_0 \rangle_X \qquad \text{for all } (x, x^*) \in \mathrm{Gr}\, A.$$ □

Using this proposition we can prove some remarkable surjectivity results for maximal monotone operators.

THEOREM 3.2.24 *If X is a reflexive Banach space, $C \subseteq X$ a nonempty, closed, convex set, $A: X \longrightarrow 2^{X^*}$ is a maximal monotone map with $\mathcal{D}(A) \subseteq C$ and $B: C \longrightarrow X^*$ is monotone, hemicontinuous, bounded, and coercive, then $A + B$ is surjective.*

PROOF: Let $x_0^* \in X^*$ and introduce the map

$$\widehat{A}(x) = A(x) - x_0^* \qquad \text{for all } x \in \mathcal{D}(A).$$

Note that $\widehat{A}$ is still maximal monotone and $D(\widehat{A}) = \mathcal{D}(A) \subseteq C$. We can apply Proposition 3.2.23 and obtain $x_0 \in C$ such that

$$\begin{aligned} & \langle \widehat{x}^* + B(x_0), x - x_0 \rangle_X \geq 0 \qquad \text{for all } (x, \widehat{x}^*) \in \mathrm{Gr}\, \widehat{A}, \\ \Rightarrow \; & \big\langle x^* - \big(x_0^* - B(x_0)\big), x - x_0 \big\rangle_X \geq 0 \qquad \text{for all } (x, x^*) \in \mathrm{Gr}\, A. \end{aligned} \tag{3.22}$$

Due to the maximal monotonicity of A, from (3.22), we infer that

$$\begin{aligned} & \big(x_0, x_0^* - B(x_0)\big) \subseteq \mathrm{Gr}\, A, \\ \Rightarrow \; & x_0^* \in A(x_0) + B(x_0). \end{aligned}$$

Because $x_0^* \in X^*$ was arbitrary, we conclude that $A + B$ is surjective. □

In what follows we use the following theorem, due to Troyanski [587], to equivalently renorm X and X^* in a way that it is helpful in our analysis.

THEOREM 3.2.25 *If X is a reflexive Banach space, then there exists an equivalent norm on X so that both X and X^* are locally uniformly convex.*

REMARK 3.2.26 From Proposition 3.2.16 and Remark 3.2.20, we know that the duality map $\mathcal{F}\colon X \longrightarrow X^*$ corresponding to this new equivalent norm, is monotone, a homeomorphism, bounded, and coercive.

THEOREM 3.2.27 *If X is a reflexive Banach space and $A\colon X \longrightarrow 2^{X^*}$ is maximal monotone, then A is surjective if and only if A^{-1} is locally bounded.*

PROOF: $\Rightarrow$: Note that A^{-1} is maximal monotone too. Moreover, $D(A^{-1}) = X^*$. Then from Theorem 3.2.4 it follows that A^{-1} is locally bounded.

$\Leftarrow$: We show that the range $\mathcal{R}(A)$ of A is both closed and open in X^*.

To this end let $x_n^* \in A(x_n)$, $n \geq 1$, and suppose that $x_n^* \longrightarrow x^*$. Then $x_n \in A^{-1}(x_n^*)$, $n \geq 1$. Because A^{-1} is locally bounded, $\{x_n\}_{n\geq 1} \subseteq X$ is bounded and so we may assume that $x_n \xrightarrow{w} x$ in X as $n \to \infty$. Because $(x_n, x_n^*) \in \mathrm{Gr}\, A$ for all $n \geq 1$, from Proposition 3.2.7, we infer that $(x, x^*) \in \mathrm{Gr}\, A$ and so we have that $\mathcal{R}(A)$ is closed in X.

Next we show that $\mathcal{R}(A)$ is open X. By virtue of Theorem 3.2.25, we may assume without any loss of generality that both X and X^* are locally uniformly convex. Let $x_0^* \in \mathcal{R}(A)$ and $r > 0$ such that $A^{-1}\big(\overline{B}_r(x_0^*)\big)$ is bounded in X. We can find $x_0 \in \mathcal{D}(A)$ such that $x_0^* \in A(x_0)$. By virtue of Theorem 3.2.24, for every $\lambda > 0$ $x \longrightarrow A(x) + \lambda\mathcal{F}(x - x_0)$ is surjective (here $\mathcal{F}\colon X \longrightarrow X^*$ is the duality map). Given $u^* \in B_{r/2}(x_0^*)$ we can find $x_\lambda \in \mathcal{D}(A)$ and $x_\lambda^* \in A(x_\lambda)$ such that

$$
\begin{aligned}
& x_\lambda^* + \lambda\mathcal{F}(x_\lambda - x_0) = u^* && (3.23)\\
\Rightarrow\ & \langle u^* - \lambda\mathcal{F}(x_\lambda - x_0) - x_0^*, x_\lambda - x_0\rangle_X \geq 0 \qquad \text{(because } A \text{ is monotone)} \\
\Rightarrow\ & \langle u^* - x_0^*, x_\lambda - x_0\rangle_X \geq \lambda\langle \mathcal{F}(x_\lambda - x_0), x_\lambda - x_0\rangle_X = \lambda\|x_\lambda - x_0\|_X^2, \\
\Rightarrow\ & \lambda\|x_\lambda - x_0\|_X \leq \|u^* - x_0^*\|_{X^*} < \frac{r}{2}. && (3.24)
\end{aligned}
$$

Also from (3.23), we have

$$
\begin{aligned}
& \|u^* - x_\lambda^*\|_{X^*} = \lambda\|\mathcal{F}(x_\lambda - x_0)\|_{X^*} = \lambda\|x_\lambda - x_0\|_X < \frac{r}{2} \\
\Rightarrow\ & \|x_\lambda^* - x_0^*\|_{X^*} \leq \|x_\lambda^* - u^*\|_{X^*} + \|u^* - x_0^*\|_{X^*} < r \qquad \text{(see (3.23))}.
\end{aligned}
$$

Because $A^{-1}|_{\overline{B}_r(x^*)}$ is bounded, then the net $\{x_\lambda\}_{\lambda>0}$ is bounded in X and so

$$
\begin{aligned}
& \|u^* - x_\lambda^*\|_{X^*} = \lambda\|x_\lambda - x\|_X \longrightarrow 0 \quad \text{as } \lambda \downarrow 0, \\
\Rightarrow\ & u^* \in \mathcal{R}(A) \quad \text{(because as we already proved } \mathcal{R}(A) \text{ is closed in } X^*\text{)}, \\
\Rightarrow\ & B_{r/2}(x_0^*) \subseteq \mathcal{R}(A) \quad \text{and so } \mathcal{R}(A) \text{ is open in } X^*.
\end{aligned}
$$

Because $\mathcal{R}(A)$ is both closed and open in X^*, we conclude that $\mathcal{R}(A) = X^*$. □

COROLLARY 3.2.28 *If X is a reflexive Banach space and $A : X \longrightarrow 2^{X^*}$ is maximal monotone, and weakly coercive, then A is surjective.*

PROOF: The weak coercivity of A implies that A^{-1} is locally bounded. Then from Theorem 3.2.27 it follows that A is surjective. □

Combining the above result with Corollary 3.2.11, we obtain the following.

COROLLARY 3.2.29 *If X is a reflexive Banach space and $A : X \longrightarrow 2^{X^*}$ is monotone, hemicontinuous, and weakly coercive, then A is surjective.*

The next theorem provides a useful criterion for maximal monotonicity in terms of the duality map.

THEOREM 3.2.30 *If X is a reflexive Banach space such that both X and X^* are strictly convex and $A : X \longrightarrow 2^{X^*}$ is a monotone map, then A is maximal monotone if and only if for every $\lambda > 0$ (equivalently for some $\lambda > 0$) $A + \lambda\mathcal{F}$ is surjective.*

PROOF: $\Rightarrow$: Follows from Theorem 3.2.24.

$\Leftarrow$: Suppose that for some $(y_0, y_0^*) \in X \times X^*$, we have

$$\langle x^* - y_0^*, x - y_0 \rangle_X \geq 0 \qquad \text{for all } (x, x^*) \in \text{Gr } A. \tag{3.25}$$

Then by hypothesis we can find $(x_1, x_1^*) \in \text{Gr } A$ such that

$$x_1^* + \lambda\mathcal{F}(x_1) = y_0^* + \lambda\mathcal{F}(y_0). \tag{3.26}$$

In (3.25) we set $x = x_1$ and $x^* = x_1^*$. We obtain

$$\lambda \langle \mathcal{F}(y_0) - \mathcal{F}(x_1), x_1 - y_0 \rangle_X \geq 0 \qquad (\text{see } (3.25)). \tag{3.27}$$

But because X, X^* are strictly convex, $\mathcal{F}$ is strictly monotone and so from (3.27) we infer that $x_1 = y_0$. Hence $(y_0, y_0^*) \in \text{Gr } A$ and this proves the maximality of A. □

COROLLARY 3.2.31 *If X is a reflexive Banach space, $A : X \longrightarrow 2^{X^*}$ is monotone and $B : X \longrightarrow X^*$ is strictly monotone, hemicontinuous, bounded, and coercive, then A is maximal monotone if and only if $A + B$ is surjective.*

COROLLARY 3.2.32 *If X is a reflexive Banach space, $A : X \longrightarrow 2^{X^*}$ is maximal monotone and $B : X \longrightarrow X^*$ is monotone, hemicontinuous, and bounded, then $A + B$ is maximal monotone.*

PROOF: By virtue of Theorem 3.2.25 we may assume without any loss of generality that both X and X^* are locally uniformly convex. Also if we replace A and B by

$$\widehat{A}(x) = A(x + x_0) + B(x_0) \qquad \text{and} \qquad \widehat{B}(x) = B(x + x_0) - B(x_0) \quad \text{for some } x_0 \in X,$$

then we see that we may assume that $0 = B(0)$. Then $B + \mathcal{F}$ is monotone, hemicontinuous, bounded, and coercive. Given $u^* \in X^*$, by Proposition 3.2.23 we can find $x_0 \in X$ such that

$$\langle x^* + B(x_0) + \mathcal{F}(x_0) - u^*, x - x_0 \rangle_X \geq 0 \qquad \text{for all } (x, x^*) \in \text{Gr } A.$$

Because A is maximal monotone, we have that

$$\begin{aligned} & u^* \in A(x_0) + B(x_0) + \mathcal{F}(x_0), \\ \Rightarrow\ & \mathcal{R}(A + B + \mathcal{F}) = X^* \end{aligned}$$

and this by Theorem 3.2.30 implies that $A + B$ is maximal monotone. □

Another result on the maximality of the sum is the following one whose proof can be found in Barbu [58, p. 46].

THEOREM 3.2.33 *If X is a reflexive Banach space and $A_1, A_2 \colon X \longrightarrow 2^{X^*}$ are two maximal monotone maps such that* $\operatorname{int} \mathcal{D}(A_1) \cap \mathcal{D}(A_2) \neq \emptyset$, *then $A_1 + A_2$ is maximal monotone.*

REMARK 3.2.34 If $X = H =$ a Hilbert space, then the domain condition in the above theorem can be replaced by a weaker one which says that $0 \in \operatorname{int}\big(\mathcal{D}(A_1) - \mathcal{D}(A_2)\big)$ (recall that $\operatorname{int}\big(\mathcal{D}(A_1) - \mathcal{D}(A_2)\big) \supseteq \operatorname{int} \mathcal{D}(A_1) - \mathcal{D}(A_2)$. This new condition is symmetric in A_1 and A_2 and can be satisfied even if $\mathcal{D}(A_1)$ and $\mathcal{D}(A_2)$ have an empty interior. For details we refer to Attouch [33].

It is easy to see that the subdifferential of a function $\varphi \in \Gamma_0(X)$ (see Definition 1.2.1) is monotone. Next we show that in fact it is maximal monotone.

THEOREM 3.2.35 *If X is a Banach space and $\varphi \in \Gamma_0(X)$, then $\partial\varphi \colon X \longrightarrow 2^{X^*}$ is maximal monotone.*

PROOF: We do the proof for the case when X is reflexive and for the general case we refer to Rockafellar [522]. Directly from the definition of the subdifferential (see Definition 1.2.28), we can see that $\partial\varphi$ is monotone. So we need to show that $\partial\varphi$ is maximal. By virtue of Theorem 3.2.30, it suffices to show that $\partial\varphi + \mathcal{F}$ is surjective. To this end, let $u^* \in X^*$ and consider the function $\psi \colon X \longrightarrow \overline{\mathbb{R}} = \mathbb{R} \cup \{+\infty\}$ defined by

$$\psi(x) = \varphi(x) + \frac{1}{2}\|x\|_X^2 - \langle u^*, x\rangle_X \qquad \text{for all } x \in X.$$

Clearly $\psi \in \Gamma_0(X)$. Also from Proposition 1.2.20 we know that we can find $(x_0^*, \vartheta_0) \in X^* \times \mathbb{R}$ such that $\varphi(x) \geq \langle x_0^*, x\rangle_X - \vartheta_0$ for all $x \in X$. So

$$\begin{aligned} & \psi(x) \geq -\|x_0^*\|_{X^*}\|x\|_X + \frac{1}{2}\|x\|_X^2 - \|u^*\|\|x\|_X, \\ \Rightarrow\ & \psi(x) \longrightarrow +\infty \quad \text{as } \|x\|_X \longrightarrow \infty. \end{aligned}$$

Invoking Corollary 2.1.11 we obtain $\widehat{x} \in X$ such that

$$\begin{aligned} & \psi(\widehat{x}) = \inf_X \psi, \\ \Rightarrow\ & 0 \in \partial\psi(\widehat{x}) = \partial\varphi(\widehat{x}) + \mathcal{F}(\widehat{x}) - u^* \\ \Rightarrow\ & u^* \in \partial\varphi(\widehat{x}) + \mathcal{F}(\widehat{x}). \end{aligned}$$

Because $u^* \in X^*$ was arbitrary, we conclude that $\partial\varphi + \mathcal{F}$ is surjective, hence $x \longrightarrow \partial\varphi(x)$ is maximal monotone. □

In order to identify those maximal monotone maps on X that are of the subdifferential type, we need to introduce the following specification of monotonicity.

DEFINITION 3.2.36 Let X be a Banach space and $A: X \longrightarrow 2^{X^*}$.

(a) We say that A is *cyclically monotone* if for any subset $\{x_0, \ldots, x_n\} \subseteq \mathcal{D}(A), n \geq 1$ and $x_k^* \in A(x_k)$, $0 \leq k \leq n$ we have

$$\sum_{k=0}^{n} \langle x_k^*, x_k - x_{k+1} \rangle_X \geq 0, \qquad \text{with } x_{n+1} = x_0.$$

(b) We say that A is *maximal cyclically monotone* if it is cyclically monotone and there is no cyclically monotone map $\widehat{A}: X \longrightarrow 2^{X^*}$ whose graph strictly contains the graph of A.

REMARK 3.2.37 Every monotone map $A: \mathbb{R} \longrightarrow 2^{\mathbb{R}}$ is in fact cyclically monotone. To see this let $\{x_k\}_{k=0}^n \subseteq \mathcal{D}(A)$ and $x_k^* \in A(x_k)$ for all $0 \leq k \leq n$. We may assume that $x_k < x_{k+1}$ for $0 \leq k \leq n-1$ and so $x_k^* \leq x_{k+1}^*$ for all $0 \leq k \leq n-1$. So if $x_{n+1} = x_0$, we have

$$\begin{aligned}
\sum_{k=0}^{n} x_k^*(x_k - x_{k+1}) &= \sum_{k=0}^{n-1} x_k^*(x_k - x_{k+1}) + x_n^*(x_n - x_0) \\
&= \sum_{k=0}^{n-1} (x_k^* - x_n^*)(x_k - x_{k+1}) \geq 0.
\end{aligned}$$

EXAMPLE 3.2.38 Not every monotone map is cyclically monotone. Consider the linear map in $\mathbb{R}^2$ defined by $A(x_1, x_2) = (x_2, -x_1)$. Consider the points $(1,1), (0,1)$, and $(1,0)$ to check that A is not cyclically monotone.

THEOREM 3.2.39 *If X is a Banach space and $A: X \longrightarrow 2^{X^*}$, then A is maximal cyclically monotone if and only if there exists $\varphi \in \Gamma_0(X)$ such that $A = \partial\varphi$.*

PROOF: $\Leftarrow$: Directly from Definition 1.2.28, we see that $\partial\varphi$ is cyclically monotone and this combined with Theorem 3.2.25 implies that $\partial\varphi$ is maximal cyclically monotone.

$\Rightarrow$: Let $(x_0, x_0^*) \in \operatorname{Gr} A$ and define

$$\varphi(x) = \sup\Big[\sum_{k=0}^{n-1} \langle x_k^*, x_{k+1} - x_k \rangle_X + \langle x_n^*, x - x_n \rangle_X : (x_k, x_k^*) \in \operatorname{Gr} A, \\ 0 \leq k \leq n,\ n \geq 1\Big]. \tag{3.28}$$

Evidently φ is convex and lower semicontinuous with values in $\overline{\mathbb{R}} = \mathbb{R} \cup \{+\infty\}$. Moreover, $\varphi(x_0) \leq 0$ because A is cyclically monotone. Hence φ is proper; that is, $\varphi \in \Gamma_0(X)$. Let $(x, x^*) \in \operatorname{Gr} A$ and $u \in X$. In the right-hand side of (3.28), we consider the $(n+1)$ pairs $\{(x_k, x_k^*)\}_{k=0}^n \cup \{(x, x^*)\} \subseteq \operatorname{Gr} A$. We obtain

$$\begin{aligned}
&\varphi(u) \geq \sum_{k=0}^{n-1} \langle x_k^*, x_{k+1} - x_k \rangle_X + \langle x_n^*, x - x_n \rangle_X + \langle x^*, u - x \rangle_X, \\
\Rightarrow\ &\varphi(u) \geq \varphi(x) + \langle x^*, u - x \rangle_X, \\
\Rightarrow\ &x^* \in \partial\varphi(x).
\end{aligned}$$

So Gr $A \subseteq$ Gr $\partial\varphi$ and from the first part of the proof we infer that Gr $A =$ Gr $\partial\varphi$ (i.e., $A = \partial\varphi$). $\square$

COROLLARY 3.2.40 *Every maximal monotone map $A: \mathbb{R} \longrightarrow 2^{\mathbb{R}}$ is of the subdifferential type.*

THEOREM 3.2.41 *If X is a reflexive Banach space, $\varphi \in \Gamma_0(X)$, and $0 \in \mathcal{D}(\partial\varphi)$, then*

$$\lim_{\substack{\|x\|_X \to \infty \\ x \in \mathrm{dom}\,\varphi}} \frac{\varphi(x)}{\|x\|_X} = +\infty$$

if and only if $A = \partial\varphi$ is coercive.

PROOF: $\Rightarrow$: Let $x_0 \in \mathcal{D}(\partial\varphi)$. Then

$$\varphi(x) - \varphi(x_0) \le \langle x^*, x - x_0\rangle \qquad \text{for all } (x, x^*) \in \text{Gr } \partial\varphi$$
$$\Rightarrow \lim_{\substack{\|x\|_X \to \infty \\ x \in \mathcal{D}(\partial\varphi)}} \frac{\inf\left[\langle x^*, x - x_0\rangle : x^* \in \partial\varphi(x)\right]}{\|x\|_X} = +\infty \text{ (i.e., } A = \partial\varphi \text{ is coercive).}$$

$\Leftarrow$: Because $A = \partial\varphi$ is coercive, maximal monotone (see Theorem 3.2.30) it is surjective (see Corollary 3.2.28). Also by virtue of Proposition 1.2.20, we may assume without any loss of generality that $\varphi \ge 0$. Then for every $x^* \in X^*$ with $\|x^*\|_{X^*} \le r$, we can find $x \in \mathcal{D}(\partial\varphi)$ and $M > 0$ such that $x^* \in \partial\varphi(x)$ and $\|x\|_X \le M$. By definition

$$\langle x^*, y - x\rangle_X \le \varphi(y) - \varphi(x) \qquad \text{for all } y \in X,$$
$$\Rightarrow \langle x^*, y\rangle_X \le \varphi(y) + Mr \qquad \text{for all } y \in X,$$
$$\Rightarrow \frac{\langle x^*, y\rangle_X}{\|y\|_X} \le \frac{\varphi(y)}{\|y\|_X} + \frac{Mr}{\|y\|_X},$$
$$\Rightarrow \lim_{\substack{\|y\|_X \to \infty \\ x \in \mathrm{dom}\,\varphi}} \frac{\varphi(y)}{\|y\|_X} = +\infty. \qquad \square$$

When we are in a Hilbert space we can introduce some useful single-valued approximations of a maximal monotone operator.

DEFINITION 3.2.42 Let H be a Hilbert space identified with its dual (i.e., $H = H^*$, pivot space) and $A: H \longrightarrow 2^H$ a maximal monotone map. For every $\lambda > 0$ we define the following well-known operators.

(a) $J_\lambda = (I + \lambda A)^{-1}$, the *resolvent* of A.

(b) $A_\lambda = (1/\lambda)(I - J_\lambda)$, the *Yosida approximation* of A.

REMARK 3.2.43 Note that $\mathcal{D}(J_\lambda) = \mathcal{D}(A_\lambda) = H$ for all $\lambda > 0$ and both maps J_λ and A_λ are single-valued.

Several important properties of J_λ and A_λ are gathered in the next proposition. They follow easily from Definition 3.2.42.

PROPOSITION 3.2.44 *If H is a Hilbert space, $H = H^*$, and $A: H \longrightarrow 2^H$ is maximal monotone, then*

(a) *J_λ is nonexpansive (i.e., Lipschitz continuous with constant 1) for all $\lambda > 0$.*

(b) *$A_\lambda(x) \in A\big(J_\lambda(x)\big)$ for all $x \in H$ and all $\lambda > 0$.*

(c) *A_λ is monotone and Lipschitz continuous with constant $1/\lambda$ for all $\lambda > 0$.*

(d) *For all $x \in \mathcal{D}(A)$, $\|A_\lambda(x)\|_H \leq \|A^0(x)\|_H = \min\big\{\|x^*\|_H : x^* \in A(x)\big\}$ for all $\lambda > 0$.*

(e) *$\lim\limits_{\lambda \to 0} A_\lambda(x) = A^0(x)$ for all $x \in \mathcal{D}(A)$.*

(f) *$\overline{\mathcal{D}(A)}$ is convex and $\lim\limits_{\lambda \to 0} J_\lambda(x) = \mathrm{proj}\big(x; \overline{\mathcal{D}(A)}\big)$ for all $x \in H$.*

REMARK 3.2.45 From (c) and Corollary 3.2.11 it follows that A_λ is maximal monotone. Also because A is maximal monotone, from Proposition 3.2.5 we know that $A(x)$ is nonempty, closed, and convex for every $x \in \mathcal{D}(A)$. Hence the element of minimum norm in $A(x)$ (the metric projection of the origin in $A(x)$) exists, is unique, and is denoted by $A^0(x)$. In fact in every reflexive Banach space X, every maximal monotone map $A : X \longrightarrow 2^{X^*}$ has domain $\mathcal{D}(A)$ and range $\mathcal{R}(A)$ which satisfy: $\overline{\mathcal{D}(A)}$ and $\overline{\mathcal{R}(A)}$ are both convex (virtual convexity). In (f) $\mathrm{proj}\big(x; \overline{\mathcal{D}(A)}\big)$ denotes the metric projection on the closed convex set $\overline{\mathcal{D}(A)}$.

The two operators J_λ and A_λ can also be defined when X is a reflexive Banach space that is not necessarily a Hilbert space. This is done as follows.

DEFINITION 3.2.46 Let X be a reflexive Banach space such that both X and X^* are locally uniformly convex and $A : X \longrightarrow 2^{X^*}$ a maximal monotone map. Then by Theorem 3.2.30 for every $\lambda > 0$ and $x \in X$, the operator inclusion

$$0 \in \mathcal{F}(y - x) + \lambda A(y)$$

has a solution. The hypothesis on X and X^* imply that this solution is unique and is denoted by $J_\lambda(x)$. Also we set $A_\lambda(x) = (1/\lambda)\mathcal{F}\big(x - J_\lambda(x)\big)$.

REMARK 3.2.47 If H is a pivot Hilbert space (i.e., $H = H^*$), then $\mathcal{F} = I$ and we recover Definition 3.2.42. Note that J_λ, A_λ are single-valued, and bounded and A_λ is also monotone, and demicontinuous (hence maximal monotone). Moreover, $\|A_\lambda(x)\|_{X^*} \leq \|A^0(x)\|_{X^*}$ for every $x \in \mathcal{D}(A)$ and $\lim\limits_{\lambda \downarrow 0} J_\lambda(x) = x$ for every $x \in \overline{\mathcal{D}(A)}$. For every $x \in \mathcal{D}(A)$, $A(x)$ is nonempty, closed convex. Therefore due to the local uniform convexity of X^*, $A^0(x)$ is well defined.

If A is of the subdifferential type, then as we show below, A_λ is also of this type. To show this we introduce the following notion.

DEFINITION 3.2.48 Let X be a Banach space and $\varphi \in \Gamma_0(X)$. Then for every $\lambda>0$, the *Moreau–Yosida approximation* φ_λ of φ, is defined by

$$\varphi_\lambda(x) = \inf\big\{\varphi(y) + (1/2\lambda)\|x-y\|_X^2 : y \in X\big\} \qquad \text{for all } x \in X.$$

REMARK 3.2.49 Note that $\varphi_\lambda = \varphi \oplus 1/(2\lambda)\|\cdot\|_X^2$ (see Definition 1.2.23). So φ_λ is convex and $\operatorname{dom}\varphi_\lambda = X$.

PROPOSITION 3.2.50 *If X is a reflexive Banach space, X and X^* are both locally uniformly convex, $\varphi \in \Gamma_0(X)$, and $A = \partial\varphi$, then*

(a) $\varphi_\lambda(x) = \varphi\big(J_\lambda(x)\big) + (1/2\lambda)\|x - J_\lambda(x)\|_X^2$ *for all $\lambda>0$ and all $x \in X$.*

(b) φ_λ *is Gâteaux differentiable and for all $\lambda>0, A_\lambda = (\partial\varphi)_\lambda = \partial\varphi_\lambda = \varphi'_\lambda$.*

(c) $\varphi(J_\lambda x) \le \varphi_\lambda(x) \le \varphi(x)$ *for all $\lambda>0$, all $x \in X$, and $\lim\limits_{\lambda\downarrow 0}\varphi_\lambda(x) = \varphi(x)$ for all $x \in X$.*

PROOF: (a) For fixed $x \in X$, we consider the function $y \longrightarrow \vartheta_\lambda(y) = \varphi(y) + (1/2\lambda)\|x-y\|_X^2$. Evidently $\vartheta_\lambda(\cdot)$ is proper, lower semicontinuous, strictly convex and coercive. So there exists unique element $x_\lambda \in X$ such that

$$\varphi_\lambda(x) = \vartheta_\lambda(x_\lambda).$$

The vector $x_\lambda \in X$ is the unique solution of the operator inclusion

$$0 \in -\frac{1}{\lambda}\mathcal{F}(x - x_\lambda) + \partial\varphi(x_\lambda), \tag{3.29}$$

$$\Rightarrow\ x_\lambda = J_\lambda(x) \ \text{(see Definition 3.2.46) and this proves (a).}$$

(b) From (3.29), we obtain

$$0 \le \varphi_\lambda(y) - \varphi_\lambda(x) - \langle A_\lambda(x), y-x\rangle_X \le \langle A_\lambda(y) - A_\lambda(x), y-x\rangle_X$$

for all $x, y \in X$. Let $y = x + th$ with $t>0$. Dividing with $t>0$, we obtain

$$\lim_{t\downarrow 0}\frac{\varphi_\lambda(x+th) - \varphi_\lambda(x)}{t} = \langle A_\lambda(x), h\rangle_X \qquad \text{for all } h \in X,$$

$$\Rightarrow\ \varphi_\lambda \text{ is Gâteaux differentiable and } A_\lambda = (\partial\varphi)_\lambda = \partial\varphi_\lambda = \varphi'_\lambda.$$

(c) In view of (a) we have

$$\varphi\big(J_\lambda(x)\big) \le \varphi_\lambda(x) \le \varphi(x) \qquad \text{for all } x \in X. \tag{3.30}$$

Let $x \in \operatorname{dom}\varphi$. Then from Remark 3.2.47 we know that $J_\lambda(x) \longrightarrow x$ as $\lambda \downarrow 0$ (because $\overline{\mathcal{D}(\partial\varphi)} = \overline{\operatorname{dom}\psi}$). So we have

$$\varphi(x) \le \liminf_{\lambda\downarrow 0}\varphi\big(J_\lambda(x)\big) \le \liminf_{\lambda\downarrow 0}\varphi_\lambda(x) \le \varphi(x) \quad \text{(see (3.30))},$$

$$\Rightarrow\ \varphi_\lambda(x) \uparrow \varphi(x) \quad \text{as } \lambda \downarrow 0.$$

Next suppose $x \notin \operatorname{dom}\varphi$. We show that $\varphi_\lambda(x) \longrightarrow +\infty$ as $\lambda \downarrow 0$. If this not the case, then we can find $\lambda_n \downarrow 0$ and $M>0$ such that

$$\varphi_{\lambda_n}(x) \le M \qquad \text{for all } n \ge 1.$$

From part (a) we infer that $J_{\lambda_n}(x) \longrightarrow x$ in X and $\varphi\big(J_{\lambda_n}(x)\big) \le M$. Hence $\varphi(x) \le M$, a contradiction to the fact that $x \notin \operatorname{dom} \varphi$. $\square$

In the context of a pivot Hilbert space, we can improve this proposition.

COROLLARY 3.2.51 *If $X = H =$ a pivot Hilbert space, $\varphi \in \Gamma_0(H)$, and $A = \partial\varphi$, then*

(a) *φ_λ is Fréchet differentiable and $\partial\varphi_\lambda$ is Lipschitz continuous with constant $1/\lambda$.*

(b) *If $\lambda_n \longrightarrow \lambda$, $x_n \longrightarrow x$ in H, $\partial\varphi_{\lambda_n}(x_n) \xrightarrow{w} x^*$ in H, then $(x, x^*) \in Gr\, \varphi$.*

(c) *If $\lambda_n \longrightarrow \lambda$ and $x_n \longrightarrow x$ in H, then $\varphi(x) \le \liminf\limits_{n\to\infty} \varphi_{\lambda_n}(x_n)$ and if in addition $\big\{\varphi_{\lambda_n}(x_n)\big\}_{n\ge1} \subseteq H$, is bounded, then $\varphi_{\lambda_n}(x_n) \longrightarrow \varphi(x)$.*

Before passing to generalizations of maximal monotonicity, let us mention the situation with linear operators. For a proof we refer to Brezis [102, p. 47].

PROPOSITION 3.2.52 *If H is a pivot Hilbert space and $A\colon H \longrightarrow H$ is linear, maximal monotone, then $A = \partial\varphi$ with*

$$\varphi(x) = \begin{cases} \frac{1}{2}\|A^{1/2}x\|_H^2 & \text{if } x \in \mathcal{D}(A^{1/2}) \\ +\infty & \text{if otherwise} \end{cases}$$

if and only if A is self-adjoint.

Next we introduce some generalizations of the notion of maximal monotonicity that are useful in applications.

DEFINITION 3.2.53 Let X be a reflexive Banach space and $A\colon X \longrightarrow 2^{X^*}$.

(a) We say that A is *pseudomonotone* if the following conditions are satisfied.
   (a_1) For every $x \in X$, $A(x)$ is nonempty, w-compact, and convex.

   (a_2) A is upper semicontinuous (usc for short) from each finite-dimensional subspace Y into X^* with the weak topology denoted by X^*_w (i.e., for all $U \subseteq X^*$ nonempty, w-open $A^+(U) = \{x \in Y : A(x) \subseteq U\}$ is open in Y).

   (a_3) If $x_n \xrightarrow{w} x$ in X and if $x_n^* \in A(x_n)$ satisfies $\limsup\limits_{n\to\infty} \langle x_n^*, x_n - x\rangle_X \le 0$, then for all $u \in X$ there exists $x^*(u) \in A(x)$ such that

$$\langle x^*(u), x - u\rangle_X \le \liminf_{n\to\infty} \langle x_n^*, x_n - u\rangle_X .$$

(b) We say that A is *generalized pseudomonotone* if for every sequence $x_n \xrightarrow{w} x$ in X and every sequence $x_n^* \xrightarrow{w} x^*$ in X^* with $x_n^* \in A(x_n)$ for all $n \ge 1$ such that

$$\limsup_{n\to\infty} \langle x_n^*, x_n - x\rangle_X \le 0,$$

then $x^* \in A(x)$ and $\langle x_n^*, x_n\rangle_X \longrightarrow \langle x^*, x\rangle_X$.

PROPOSITION 3.2.54 *If X is a reflexive Banach space and $A : X \longrightarrow 2^{X^*}$ is pseudomonotone, then A is generalized pseudomonotone.*

PROOF: Let $\{(x_n, x_n^*)\}_{n\geq 1} \subseteq \mathrm{Gr}\, A$ such that $x_n \xrightarrow{w} x$ in X, $x_n^* \xrightarrow{w} x^*$ in X^* and $\limsup_{n\to\infty} \langle x_n^*, x_n - x\rangle_X \leq 0$. Because A is pseudomonotone, for every $u \in X$ we can find $x^*(u) \in A(x)$ such that

$$\langle x^*(u), x - u\rangle_X \leq \liminf_{n\to\infty} \langle x_n^*, x_n - u\rangle_X .$$

The sequence $\{\langle x_n^*, x_n\rangle_X\}_{n\geq 1} \subseteq \mathbb{R}$ is bounded and so passing to a subsequence if necessary, we may assume that $\langle x_n^*, x_n\rangle_X \longrightarrow \beta \in \mathbb{R}$. Then

$$\limsup_{n\to\infty} \langle x_n^*, x_n - x\rangle_X = \beta - \langle x^*, x\rangle_X \leq 0. \tag{3.31}$$

On the other hand,

$$\begin{aligned} &\beta - \langle x^*, u\rangle_X \geq \liminf_{n\to\infty} \langle x_n^*, x_n - u\rangle_X \geq \langle x^*(u), x - u\rangle_X \\ \Rightarrow\ &\langle x^*, x - u\rangle_X \geq \langle x^*(u), x - u\rangle_X \qquad \text{for all } u \in X \quad \text{(see (3.31))}. \end{aligned} \tag{3.32}$$

We claim that $x^* \in A(x)$. If this is not true and since $A(x)$ is w-compact, convex in X^*, by the strong separation theorem we can find $v \in X \setminus \{0\}$ and $c \in \mathbb{R}$ such that

$$\langle x^*, v\rangle_X < c \leq \inf\big[\langle y^*, v\rangle_X : \text{ for all } y^* \in A(x)\big]. \tag{3.33}$$

If in (3.32) we choose $u = x - v$, we reach a contradiction to (3.33). So indeed $x^* \in A(x)$.

Next note that for some $x^*(x) \in X^*$ we have

$$\begin{aligned} &\liminf_{n\to\infty} \langle x_n^*, x_n - x\rangle_X \geq \langle x^*(x), x - x\rangle_X = 0 \\ \Rightarrow\ &\liminf_{n\to\infty} \langle x_n^*, x_n\rangle_X \geq \langle x^*, x\rangle_X . \end{aligned} \tag{3.34}$$

On the other hand from the choice of the sequences $\{x_n\}_{n\geq 1} \subseteq X$ and $\{x_n^*\}_{n\geq 1} \subseteq X^*$, we have

$$\limsup_{n\to\infty} \langle x_n^*, x_n\rangle_X \leq \langle x^*, x\rangle_X . \tag{3.35}$$

From (3.34) and (3.35) we conclude that $\langle x_n^*, x_n\rangle_X \longrightarrow \langle x^*, x\rangle_X$ and so A is generalized pseudomonotone. □

There is a converse to the previous proposition.

PROPOSITION 3.2.55 *If X is a reflexive Banach space and $A : X \longrightarrow 2^{X^*}$ is bounded, generalized pseudomonotone and for all $x \in X$ $A(x)$ is nonempty, closed, and convex in X^*, then A is pseudomonotone.*

PROOF: First we show Condition [a$_3$] in Definition 3.2.53(a), namely that $x_n \xrightarrow{w} x$ in X and $\limsup_{n\to\infty} \langle x_n^*, x_n\rangle_X \leq 0$, with $x_n^* \in A(x_n)$; for every $u \in X$ we can find $x^*(u) \in A(x)$ such that

$$\langle x^*(u), x-u\rangle_X \le \liminf_{n\to\infty} \langle x_n^*, x_n - u\rangle_X . \tag{3.36}$$

Because A is bounded, $\{x_n^*\}_{n\ge 1} \subseteq X^*$ is bounded and so by passing to subsequence if necessary, we may assume that $x_n^* \xrightarrow{w} x^*$ in X^*. Suppose that (3.36) is not true. So we can find $u \in X$ such that

$$\liminf_{n\to\infty} \langle x_n^*, x_n - u\rangle_X < \inf \left[\langle y^*, x-u\rangle_X : y^* \in A(x)\right].$$

By passing to a subsequence if necessary, we may assume that

$$\lim_{n\to\infty} \langle x_n^*, x_n - u\rangle_X < \inf \left[\langle y^*, x-u\rangle_X : y^* \in A(x)\right]. \tag{3.37}$$

Because A is generalized pseudomonotone, we have $x^* \in A(x)$ and

$$\langle x_n^*, x_n\rangle_X \longrightarrow \langle x^*, x\rangle_X .$$

Using this in (3.37), we obtain

$$\langle x^*, x-u\rangle_X < \inf \left[\langle y^*, x-u\rangle_X : y^* \in A(x)\right],$$

which contradicts the fact that $x^* \in A(x)$.

Finally we show that A is usc from every finite-dimensional subspace Y of X into X_w^* = the space X^* with the weak topology. But from Proposition 6.1.10, we know that it suffices to show that Gr A is sequentially closed in $Y \times X_w^*$. It follows at once from Definition 3.2.53(a). □

PROPOSITION 3.2.56 *If X is a reflexive Banach space and $A : X \longrightarrow 2^{X^*}$ is maximal monotone, then A is generalized pseudomonotone.*

PROOF: Consider a sequence $\{(x_n, x_n^*)\}_{n\ge 1} \subseteq$ Gr A such that $x_n \xrightarrow{w} x$ in X, $x_n^* \xrightarrow{w} x^*$ in X^*, and $\limsup_{n\to\infty} \langle x_n^*, x_n - x\rangle_X \le 0$. We need to show that $x^* \in A(x)$ and $\langle x_n^*, x_n\rangle_X \longrightarrow \langle x^*, x\rangle_X$ (see Definition 3.2.53(b)).

For every $(y, y^*) \in$ Gr A, we have

$$\langle x_n^*, x_n\rangle_X = \langle x_n^* - y^*, x_n - y\rangle_X + \langle y^*, x_n\rangle_X + \langle x_n^*, y\rangle_X - \langle y^*, y\rangle_X ,$$

for all $n \ge 1$

$$\begin{aligned} &\Rightarrow \ \langle x^*, x\rangle_X \ge \langle y^*, x\rangle_X + \langle x^*, y\rangle_X - \langle y^*, y\rangle_X , \\ &\Rightarrow \ \langle x^* - y^*, x - y\rangle_X \ge 0 \quad \text{for all } (y, y^*) \in \text{Gr } A. \end{aligned}$$

From the maximality of A, we infer that $(x^*, x) \in$ Gr A.

Also for every $n \ge 1$, we have

$$\begin{aligned} &\langle x_n^* - x^*, x_n - y\rangle_X \ge 0 \\ \Rightarrow \ &\liminf_{n\to\infty} \langle x_n^*, x_n\rangle_X \ge \langle x^*, x\rangle_X \ge \limsup_{n\to\infty} \langle x_n^*, x_n\rangle_X . \end{aligned}$$

Therefore $\langle x_n^*, x_n\rangle_X \longrightarrow \langle x^*, x\rangle_X$ and so we conclude that A is generalized pseudomonotone. □

Recalling that a maximal monotone operator is locally bounded on the interior of $\mathcal{D}(A)$ (see Theorem 3.2.4), we can easily prove the following result. We leave the details to the reader.

PROPOSITION 3.2.57 *If X is reflexive and $A\colon X \longrightarrow 2^{X^*}$ is maximal monotone with $\mathcal{D}(A) = X$, then A is pseudomonotone.*

The class of pseudomonotone maps is closed under addition. For a proof we refer to Browder–Hess [120].

PROPOSITION 3.2.58 *If X is reflexive and $A_1, A_2\colon X \longrightarrow 2^{X^*}$ are pseudomonotone maps, then $A_1 + A_2$ is pseudomonotone too.*

As was the case with maximal monotone maps, pseudomonotone maps exhibit remarkable surjectivity properties. The method of proof is based on Galerkin approximations, therefore we need the following finite-dimensional result whose proof uses fixed point results for multifunctions (see Section 6.5).

PROPOSITION 3.2.59 *If X is a finite-dimensional Banach space and $F : X \longrightarrow 2^{X^*}$ is usc, coercive, and has nonempty, compact, and convex values, then F is surjective.*

Using this result, we can prove the following surjectivity theorem for pseudomonotone maps.

THEOREM 3.2.60 *If X is a reflexive Banach space and $A\colon X \longrightarrow 2^{X^*}$ is pseudomonotone and coercive, then A is surjective.*

PROOF: Let $\mathcal{S}$ be the family of all finite-dimensional subspaces of X partially ordered by inclusion. If $Y \in \mathcal{S}$ and $i_Y \in \mathcal{L}(Y, X)$ is the embedding operator, then $i_Y^* = p_{Y^*} \in \mathcal{L}(X^*, Y^*)$ is the projection on Y^*. Let $A_Y = i_Y^* A i_Y : Y \longrightarrow 2^{Y^*}$. Then A_Y is usc, coercive, and is nonempty, compact, and convex.

Note that if $u^* \in X^*$, by considering $A_{u^*}(x) = A(x) - u^*$, we see that A_{u^*} has the same properties as A and so we see that it suffices to show that $0 \in A(X)$. From Proposition 3.2.59 we see that for every $Y \in \mathcal{S}$ we can find $x_Y \in Y$ such that $0 \in A_Y(x_Y)$, hence

$$0 = i_Y^* x_Y^* \qquad \text{for some } x_Y^* \in A(x_Y).$$

Because A is coercive $\{x_Y\}_{Y \in \mathcal{S}} \subseteq X$ is bounded.

For $Y \in \mathcal{S}$ let $C_Y = \bigcup_{\substack{Y' \supseteq Y \\ Y' \in \mathcal{S}}} \{x_{Y'}\}$. Then $C_Y \subseteq \overline{B}_M$ for some $M > 0$ large. From the reflexivity of X and the finite intersection property, we have

$$\bigcap_{Y \in \mathcal{S}} \overline{C_Y}^{\,w} \neq \emptyset.$$

Let $x_0 \in \bigcap_{Y \in \mathcal{S}} \overline{C_Y}^{\,w}$ and $u \in X$. Let $Y \in \mathcal{S}$ such that $\{x_0, u\} \subseteq Y$. We can find $\{x_{Y_k} = x_k\}_{k \geq 1} \subseteq C_V$ such that $x_k \xrightarrow{w} x_0$ in X. Then

$$\langle x_k^*, x_k - x_0 \rangle_X = 0 \quad \text{with } x_k^* \in A(x_k) \quad \text{for all } k \geq 1.$$

Because A is pseudomonotone, for every $u \in X$, we can find $u^*(u) \in A(x_0)$ such that

$$\langle u^*(u), x_0 - u\rangle_X \leq \lim_{k\to\infty} \langle x_k^*, x_k - u\rangle_X = 0. \tag{3.38}$$

If $0 \notin A(x_0)$, then by the strong separation theorem, we can find $u \in X$ such that

$$0 < \inf\left[\langle x^*, x_0 - u\rangle_X : x^* \in A(x_0)\right]. \tag{3.39}$$

Comparing (3.38) and (3.39), we reach a contradiction. So A is indeed surjective. □

We conclude our discussion of maps of monotone type, with one more surjectivity result. First we have a definition.

DEFINITION 3.2.61 Let X be a reflexive Banach space.

(a) A map $A: X \longrightarrow 2^{X^*}$ is said to be *quasibounded* if for each $M>0$, we can find $K(M)>0$ such that for all $(x,x^*) \in \mathrm{Gr}\, A$ with $\langle x^*, x\rangle_X \leq M\|x\|_X$ and $\|x\|_X \leq M$, we have $\|x^*\|_{X^*} \leq K(M)$.

(b) A map $C\colon X \longrightarrow X^*$ is said to be *smooth* if $\mathcal{D}(C) = X$ and it is bounded, coercive, and maximal monotone.

(c) A map $A: X \longrightarrow 2^{X^*}$ is said to be *regular* if it is generalized pseudomonotone, and for any smooth map $C\colon X \longrightarrow X^*$ we have $A + C$ is surjective.

(d) A map $A{:}X{\longrightarrow}X^*$ is said to be of *type $(S)_+$* if for any sequence $\{x_n\}_{n\geq 1}{\subseteq}\mathcal{D}(A)$ the conditions $x_n \xrightarrow{w} x$ in X and $\limsup\limits_{n\to\infty} \langle A(x_n), x_n - x\rangle \leq 0$, imply that $x_n \longrightarrow x$ in X.

REMARK 3.2.62 Clearly every bounded map $A : X \longrightarrow 2^{X^*}$ is quasibounded. Also a monotone map $A: X \longrightarrow 2^{X^*}$ with $0 \in \mathcal{D}(A)$ is regular if and only if it is maximal. A demicontinuous $(S)_+$ map, is pseudomonotone.

We have the following surjectivity result, whose proof can be found in Pascali–Sburlan [490, p. 139].

THEOREM 3.2.63 *If X is a reflexive Banach space, $A{:}X \longrightarrow 2^{X^*}$ is a maximal monotone map with $0 \in \mathcal{D}(A)$, and $C\colon X \longrightarrow X^*$ is a regular coercive map and either A or C is quasibounded, then $A + C$ is surjective.*

Thus far we have considered mappings from a Banach space into its dual. Next we deal with mappings from a Banach space into itself. Such operators are closely related to the generation theory of semigroups (linear and nonlinear), that play a central role in the theory of evolution equations.

In what follows X is a Banach space and $\mathcal{F}\colon X \longrightarrow 2^{X^*}$ is the normalized duality map. Additional hypotheses on X are introduced as needed.

DEFINITION 3.2.64 Let $A\colon X \longrightarrow 2^X$. Then

(a) We say that A is *accretive* if for any (x,u), $(y,u) \in \mathrm{Gr}\, A$, we can find $x^* \in \mathcal{F}(x-y)$ such that $\langle x^*, u - v\rangle_X \geq 0$.

(b) We say that A is *maximal accretive* if its graph is maximal with respect to inclusion among the graphs of all accretive mappings.

(c) We say that A is *m-accretive* if $\mathcal{R}(I+A)=X$.

(d) We say that A is *dissipative* (*m-dissipative*) if $-A$ is accretive (m-accretive).

REMARK 3.2.65 If $X=H=$ pivot Hilbert space (i.e., $H=H^*$), then $\mathcal{F}=I$ and so A is accretive if and only if A is monotone. Moreover, in this case the notions of maximal accretivity and m-accretivity coincide and they correspond to maximal monotonicity (see Theorem 3.2.30). Note that maximal accretivity is equivalent to saying that if $(x,u)\in X\times X$ is such that for all $(y,v)\in \operatorname{Gr} A$, we can find $x^*\in\mathcal{F}(x-y)$ with $\langle x^*,u-v\rangle_X\geq 0$, then $(x,u)\in \operatorname{Gr} A$.

PROPOSITION 3.2.66 *A mapping $A\colon X\longrightarrow 2^X$ is accretive if and only if $\|x-y+\lambda(u-v)\|_X\geq\|x-y\|_X$ for all $\lambda>0$ and all (x,u), $(y,v)\in \operatorname{Gr} A$.*

PROOF: $\Rightarrow$: By Definition 3.2.64(a), we can find $x^*\in\mathcal{F}(x-y)$ such that $\langle x^*,u-v\rangle_X\geq 0$. For $\lambda>0$ we have

$$\begin{aligned} \|x-y\|_X^2 &= \langle x^*,x-y\rangle_X \leq \langle x^*,x-y\rangle+\lambda\,\langle x^*,u-v\rangle_X \\ &\leq \|x^*\|_{X^*}\|x-y+\lambda(u-v)\|_X, \\ \Rightarrow\ \|x-y\|_X &\leq \|x-y+\lambda(u-v)\|_X \quad (\text{because } \|x^*\|_{X^*}=\|x-y\|_X). \end{aligned}$$

$\Leftarrow$: Let $x_\lambda^*\in\mathcal{F}\big(x-y+\lambda(u-v)\big)$ and set $y_\lambda^*=x_\lambda^*/\|x_\lambda^*\|_{X^*}$. We have

$$\begin{aligned} \|x-y\|_X\leq\|x-y+\lambda(u-v)\|_X &= \frac{1}{\|x_\lambda^*\|_{X^*}}\,\langle x_\lambda^*,x-y+\lambda(u-v)\rangle_X \\ &= \langle y_\lambda^*,x-y+\lambda(u-v)\rangle_X \\ &\leq \|x-y\|_X+\lambda\,\langle y_\lambda^*,u-v\rangle_X\,. \end{aligned} \tag{3.40}$$

By Alaoglou's theorem the net $\{y_\lambda^*\}_{\lambda>0}$ has a limit point y^* and because of (3.40), we see that

$$\begin{aligned} & \|y^*\|_{X^*}\leq 1,\ \ \langle y^*,u-v\rangle_X\geq 0 \ \text{ and } \ \langle y^*,x-y\rangle_X\geq\|x-y\|_X, \\ \Rightarrow\ & \|y^*\|_{X^*}=1 \text{ and } x^*=\|x\|_X\,y^*\in\mathcal{F}\big(x-y\big),\ \langle x^*,u-v\rangle_X\geq 0, \\ \Rightarrow\ & A \text{ is accretive.} \end{aligned}$$

□

As we did for monotone operators in Definition 3.2.42, we introduce the following single-valued approximations of the identity and of the operator itself.

DEFINITION 3.2.67 If $A\colon X\longrightarrow 2^X$ is an accretive map and $\lambda>0$, then we set

$$\begin{aligned} & J_\lambda(x)=(I+\lambda A)^{-1}(x),\ A_\lambda(x)=\frac{1}{\lambda}(I-J_\lambda)(x) \\ & \mathcal{D}(A_\lambda)=\mathcal{D}_\lambda=\mathcal{R}(I+\lambda A) \\ \text{and}\quad & |A(x)|=\inf\big\{\|u\|_X : u\in A(x)\big\}. \end{aligned}$$

Using Proposition 3.2.66 we check easily that the following is true.

PROPOSITION 3.2.68 *$A\colon X \longrightarrow 2^X$ is an accretive map if and only if for every $\lambda > 0$ J_λ is single-valued and $\|J_\lambda(x) - J_\lambda(y)\| \leq \|x - y\|_X$ for all $x, y \in \mathcal{D}_\lambda$.*

PROPOSITION 3.2.69 *If $A\colon X \longrightarrow 2^X$ is m-accretive, then A is maximal accretive and $\mathcal{R}(I + \lambda A) = X$ for every $\lambda > 0$.*

PROOF: Let $K\colon X \longrightarrow 2^X$ be an accretive operator such that Gr $A \subseteq$ Gr K and let $(x, u) \in$ Gr K. Because of the hypothesis A is m-accretive, thus we can find $(y, v) \in$ Gr A such that $x + u = y + v$. Because K is accretive, from Proposition 3.2.66 we obtain $x = y$ and so $u = v$. Therefore Gr $A =$ Gr K and this proves that A is maximal accretive. Next let $\lambda > 0$ and $\widehat{u} \in X$. Then $\widehat{u} \in \mathcal{R}(I + \lambda A)$ if and only if we can find $\widehat{x} \in \mathcal{D}(A)$ such that

$$\widehat{x} = (I + A)^{-1}\Big(\frac{1}{\lambda}\widehat{u} - \frac{1-\lambda}{\lambda}\widehat{x}\Big).$$

Let $K(x) = (I+A)^{-1}\Big((1/\lambda)\widehat{u} - \big((1-\lambda)/\lambda\big)x\Big)$. Then $\mathcal{D}(K)$ and using Proposition 3.2.66 we see that

$$\|K(x) - K(y)\|_X \leq \Big|\frac{1-\lambda}{\lambda}\Big| \, \|x - y\|_X. \tag{3.41}$$

If $\lambda > \frac{1}{2}$, then $\Big|(1-\lambda)/\lambda\Big| < 1$ and so because of (3.41) we can apply Banach's fixed point theorem (see Theorem 3.4.3) and obtain $\widehat{x} \in X$ such that $K(\widehat{x}) = \widehat{x}$. Now let $\beta > \frac{1}{2}$ and set $A_\beta = \beta A$. Then from the previous argument we have that $\mathcal{R}(I + \theta A_\beta) = X$ for all $\theta > \frac{1}{2}$, hence $\mathcal{R}(I + \theta\beta A) = X$ for all $\theta\beta > \frac{1}{4}$. Continuing this way, we have that $\mathcal{R}(I + \lambda A) = X$ for all $\lambda > 1/2^n$, $n \geq 1$. □

REMARK 3.2.70 When $X = H =$ a pivot Hilbert space, we recover a special case of Theorem 3.2.30 (with $\mathcal{F} = I$).

PROPOSITION 3.2.71 *If $A\colon X \longrightarrow 2^X$ is an accretive map, then*

(a) *For every $\lambda > 0$ A_λ is accretive, Lipschitz continuous on $\mathcal{D}_\lambda$ with constant $2/\lambda$, $\|A_\lambda(x)\|_X \leq |A(x)|$ for all $x \in \mathcal{D}_\lambda \cap \mathcal{D}(A)$ and $A_\lambda \in A\big(J_\lambda(x)\big)$ for all $x \in \mathcal{D}_\lambda$.*

(b) $\lim\limits_{\lambda \to 0^+} J_\lambda(x) = x$ *for all $x \in \mathcal{D}(A) \cap \big(\bigcap\limits_{\lambda > 0} \mathcal{D}_\lambda\big)$.*

PROOF: (a) Let $x, y \in \mathcal{D}_\lambda$ and $x^* \in \mathcal{F}(x - y)$. We have

$$\begin{aligned}
\langle x^*, A_\lambda(x) - A_\lambda(y)\rangle_X &= \frac{1}{\lambda}\langle x^*, x - y\rangle_X - \frac{1}{\lambda}\langle x^*, J_\lambda(x) - J_\lambda(y)\rangle_X \\
&\geq \frac{1}{\lambda}\|x - y\|_X^2 - \frac{1}{\lambda}\|x - y\|_X \|J_\lambda(x) - J_\lambda(y)\|_X \\
&\geq \frac{1}{\lambda}\|x - y\|_X \big(\|x - y\|_X - \|J_\lambda(x) - J_\lambda(y)\|_X\big) \geq 0
\end{aligned}$$

(see Proposition 3.2.69),

$\Rightarrow$ A_λ is an accretive map defined on $\mathcal{D}_\lambda$.

Also from Proposition 3.2.68, we have

$$\|A_\lambda(x) - A_\lambda(y)\|_X = \frac{1}{\lambda}\|x - y - \big(J_\lambda(x) - J_\lambda(y)\big)\|_X$$
$$\leq \frac{2}{\lambda}\|x - y\|_X \qquad \text{for all } x, y \in \mathcal{D}_\lambda.$$

If $x \in \mathcal{D}_\lambda \cap \mathcal{D}(A)$, then from Definition 3.2.67 we see that

$$J_\lambda(x + \lambda u) = x \quad \text{for every} \quad u \in A(x),$$
$$\Rightarrow \|A_\lambda(x)\|_X = \frac{1}{\lambda}\|x - J_\lambda(x)\|_X = \frac{1}{\lambda}\|J_\lambda(x + \lambda u) - J_\lambda(x)\|_X \leq \|u\|$$

(see Proposition 3.2.68),

$$\Rightarrow \|A_\lambda(x)\|_X \leq |A(x)| \quad \text{for all } \lambda > 0 \text{ and all } x \in \mathcal{D}_\lambda \cap \mathcal{D}(A).$$

Finally let $x \in \mathcal{D}_\lambda$ and put $y = J_\lambda(x)$. Then $x = y + \lambda v$ with $v \in A(y)$ (see Definition 3.2.67). So $A_\lambda(x) = (1/\lambda)\big(x - J_\lambda(x)\big) = (1/\lambda)(x - y) = v \in A(y)$. Therefore $A_\lambda(x) \in A\big(J_\lambda(x)\big)$ for all $\lambda > 0$ and all $x \in \mathcal{D}_\lambda$.

(b) If $x \in \mathcal{D}(A) \cap \big(\bigcap_{\lambda > 0} \mathcal{D}_\lambda\big)$, then from Part (a) we have

$$\|x - J_\lambda(x)\|_X = \lambda\|A_\lambda(x)\| \leq |A_\lambda(x)| \longrightarrow 0 \quad \text{as } \lambda \downarrow 0.$$

□

PROPOSITION 3.2.72 *If* $A \colon X \longrightarrow 2^X$ *is a maximal accretive map, then*

(a) Gr $A \subseteq X \times X$ *is closed and if* X^* *is uniformly convex, then* Gr A *is sequentially closed in* $X \times X_w$.

(b) *If* X^* *is strictly convex, then* $A(x)$ *is closed and convex for all* $x \in \mathcal{D}(A)$.

PROOF: (a) Let $\{(x_n, u_n\}_{n \geq 1} \subseteq$ Gr A and suppose that $x_n \longrightarrow \widehat{x}$ and $u_n \longrightarrow \widehat{u}$ in X as $n \to \infty$. Then from Proposition 3.2.66, we have

$$\|x_n - x\|_X \leq \|x_n - x + \lambda(u_n - u)\|_X \qquad \text{for all } n \geq 1,$$

all $\lambda > 0$, and all $(x, u) \in$ Gr A.

Passing to the limit as $n \to \infty$, we obtain

$$\|\widehat{x} - x\|_X \leq \|\widehat{x} - x + \lambda(\widehat{u} - u)\|_X \qquad \text{for all } \lambda > 0 \text{ and all } (x, u) \in \mathrm{Gr} A.$$

So Gr $A \cup \{(\widehat{x}, \widehat{u})\}$ is the graph of an accretive map. But by hypothesis A is maximal accretive. Hence $(\widehat{x}, \widehat{u}) \in$ Gr A.

Because X^* is uniformly convex, $\mathcal{F}$ is uniformly continuous on bounded sets of X (see Remark 3.2.20) and so if $x_n \longrightarrow \widehat{x}$, $u_n \xrightarrow{w} \widehat{u}$ in X, the inequalities $\langle \mathcal{F}(x_n - x), u_n - u \rangle_X \geq 0$ for all $(x, u) \in$ Gr A, imply $\langle \mathcal{F}(\widehat{x} - x), \widehat{u} - u \rangle_X \geq 0$ for all $(x, u) \in$ Gr A. Because of the maximality of A, we conclude that $(\widehat{x}, \widehat{u}) \in$ Gr A.

(b) From Proposition 3.2.16 we know that $\mathcal{F}$ is single-valued. The maximality of A implies that

$$A(x) = \bigcap_{(y,v) \in \mathrm{Gr} A} \{u \in X : \langle \mathcal{F}(x - y), u - v \rangle_X \geq 0\},$$
$$\Rightarrow A(x) \text{ is closed and convex for every } x \in \mathcal{D}(A).$$

□

DEFINITION 3.2.73 Let $A\colon X \longrightarrow 2^X$. The *minimal section* of A is the map $A^0\colon X \longrightarrow 2^X$ defined by

$$A^0(x) = \{u \in A(x) : \|u\|_X = |A(x)|\}.$$

PROPOSITION 3.2.74 *If X is reflexive, strictly convex, X^* is strictly convex too and $A\colon X \longrightarrow 2^X$ is maximal accretive, then $\mathcal{D}(A^0) = \mathcal{D}(A)$ and A^0 is single-valued.*

PROOF: Let $\{u_n\}_{n\geq 1} \subseteq A(x)$ be such that $\|u_n\|_X \downarrow |A(x)|$. Due to the reflexivity of X we may assume (at least for a subsequence), that $u_n \xrightarrow{w} u$ in X. But from Proposition 3.2.72(b) we know that $A(x)$ is closed, and convex. So $u \in A(x)$. We have $\|u\|_X \leq \liminf\limits_{n\to\infty} \|u_n\|_X = |A(x)|$, hence $\|u\|_X = |A(x)|$, which proves that $\mathcal{D}(A^0) = \mathcal{D}(A)$. Moreover, the strict convexity of X implies that A^0 is single-valued. □

The significance of the minimal map is explained by the next theorem, whose proof can be found in Barbu [58, p. 79].

THEOREM 3.2.75 *If X^* is uniformly convex and $A, K\colon X \longrightarrow 2^X$ are two m-accretive maps such that*

$$A^0(x) \cap K^0(x) \neq \emptyset \qquad \textit{for all } x \in \mathcal{D}(A) = \mathcal{D}(K),$$

then $A = K$. So if $A^0 = K^0$, then $A = K$.

Evidently the sum of two dissipative sets is a dissipative set on the intersection of the domains. But if the two operators are m-dissipative, it need not be m-dissipative. The next theorem provides such a criterion which is easy to check in applications. For a proof of it we refer to Barbu [58, p. 82].

THEOREM 3.2.76 *If X^* is uniformly convex and $A, K\colon X \longrightarrow 2^X$ are two m-accretive maps such that* (i) $\mathcal{D}(A) \cap \mathcal{D}(K) \neq \emptyset$ *and* (ii) $\langle \mathcal{F}(K_\lambda(x)), y\rangle \geq 0$ *for all $(x,y) \in Gr\,A$ and all $\lambda > 0$, then $A + K$ is m-accretive.*

The importance of accretive operators comes from their role in the generation theory of nonlinear semigroups. First let us briefly recall the very basic facts about linear semigroups. So suppose $T = \mathbb{R}_+$, A is an $N \times N$-matrix, $f \in L^1(T, \mathbb{R}^N)$, and $x_0 \in \mathbb{R}^N$ and we consider the following system,

$$x'(t) = Ax(t) + f(t),\ x(0) = x_0, \quad t \in T.$$

From the variation of constants formula, we know that

$$x(t) = S(t)x_0 + \int_0^b S(t-s)f(s)\mathrm{d}s \qquad \text{for all } t \in T, \tag{3.42}$$

where $S(t) = \mathrm{e}^{At}$, $t \in T$, is the fundamental solution of the equation $x'(t) = Ax(t)$, $x(0) = x_0$.

The theory of linear semigroups extends the concept of fundamental solutions to arbitrary Banach spaces X and gives precise meaning to (3.42).

Let X be a Banach space and $A \in \mathcal{L}(X)$. Given $x_0 \in X$, the function $x(t) = e^{At}x_0$ is the unique solution of $x'(t) = Ax(t)$, $t \in T$, $x(0) = x_0$, where $e^{At} = \sum_{k \geq 0} \frac{(At)^k}{k!}$. Let us briefly check the properties of $x_0 \longrightarrow x(t)$. This is a linear map on X and $\|x(t)\| \leq e^{t\|A\|_{\mathcal{L}}}\|x_0\|$. Hence this linear map is bounded. Moreover, as $t \downarrow 0$ we have $x(t) \longrightarrow x_0$ and also $x(0) = x_0$. Finally by the uniqueness of the solution to the problem, if the initial state is $x(t_0)$ and the system evolves for time $t > 0$, then we reach the state $x(t + t_0)$ (recall also that $e^{(t+t_0)A} = e^{tA}e^{t_0A}$). We formalize these properties in the following definition.

DEFINITION 3.2.77 Let X be a Banach space and $\{S(t)\}_{t \geq 0} \subseteq \mathcal{L}(X)$. We say that $\{S(t)\}_{t \geq 0}$ is a C_0-*semigroup*, if the following are true.

(a) $S(0) = I$, the identity operator on X.

(b) $S(t + s) = S(t)S(s)$ for all $t, s \geq 0$.

(c) For every $x \in X$, $S(t)x \longrightarrow x$ as $t \downarrow 0$.

REMARK 3.2.78 Property (b) is called the semigroup property. Property (c) implies continuity with respect to t in the strong operator topology. If $A \in \mathcal{L}(X)$, then we already saw that $S(t) = e^{At}$ defines a C_0-semigroup.

PROPOSITION 3.2.79 *If $\{S(t)\}_{t \geq 0}$ is a C_0-semigroup on X, then there exist $M \geq 1$ and $\omega \geq 0$ such that $\|S(t)\|_{\mathcal{L}} \leq Me^{\omega t}$ for all $t \geq 0$.*

PROOF: We claim that we can find $M \geq 1$ and $\delta > 0$ such that for $t \in [0, \delta]$, $\|S(t)\|_{\mathcal{L}} \leq M$. If not, then we can find $t_n \to 0$ such that $\{\|S(t_n)\|_{\mathcal{L}}\}_{n \geq 1}$ is unbounded. But from Definition 3.2.77(c), we have that $S(t_n)x \longrightarrow x$ in X for every $x \in X$. Hence $\{S(t_n)x\}_{n \geq 1}$ is bounded in X and so by the Banach–Steinhaus theorem, $\{\|S(t_n)\|_{\mathcal{L}}\}_{n \geq 1}$ is bounded, a contradiction.

Let $\omega = (\ln M)/\delta \geq 0$. Given $t \geq 0$ we can find an integer $n \geq 0$ and $0 \leq \theta \leq \delta$ such that $t = n\delta + \theta$. Then $S(t) = S(\delta)^n S(\theta)$ and so $\|S(t)\|_{\mathcal{L}} \leq \|S(\delta)\|_{\mathcal{L}}^n \|S(\theta)\|_{\mathcal{L}} \leq M^n M \leq Me^{\omega t}$ because $\ln M^n = n \ln M \leq n\omega\delta \leq \omega t$. ☐

COROLLARY 3.2.80 *For every $x \in X$, the map $t \longrightarrow S(t)x$ is continuous from $T = \mathbb{R}_+$ into X.*

DEFINITION 3.2.81 If in Proposition 3.2.79, $M = 1$ and $\omega = 0$, then $\|S(t)\|_{\mathcal{L}} \leq 1$ for all $t \geq 0$ and $\{S(t)\}_{t \geq 0}$ is called a *contraction semigroup*.

In the next definition we introduce an important concept in the theory of semigroups.

DEFINITION 3.2.82 Let $\{S(t)\}_{t \geq 0}$ be a C_0-semigroup on X. The *infinitesimal generator* of the semigroup is a linear operator A defined by

$$Ax = \lim_{t \downarrow 0} \frac{S(t)x - x}{t} \quad \text{for all } x \in \mathcal{D}(A) = \Big\{x \in X : \lim_{t \downarrow 0} \frac{S(t)x - x}{t} \text{ exists}\Big\}.$$

Next we derive some basic properties of the infinitesimal generator and for this reason we need the following simple lemma.

LEMMA 3.2.83 *If* $\{S(t)\}_{t\geq 0}$ *is a* C_0*-semigroup, then* $\lim_{h\downarrow 0} 1/h \int_t^{t+h} S(\tau)x\mathrm{d}\tau = S(t)x$ *for all* $t\geq 0$ *and all* $x\in X$.

PROOF: We have

$$\Big\|\frac{1}{h}\int_t^{t+h} S(\tau)x\mathrm{d}\tau - S(t)x\Big\| = \Big\|\frac{1}{h}\int_t^{t+h} \big(S(\tau)-S(t)\big)x\mathrm{d}\tau\Big\|$$
$$\leq \frac{1}{h}\int_t^{t+h} \|S(\tau)x - S(t)x\|\mathrm{d}\tau. \tag{3.43}$$

Given $\varepsilon>0$, we can find $h>0$ small such that $\|S(\tau)x-S(t)x\|<\varepsilon$ for all $|\tau-t|<h$ (see Corollary 3.2.80). So from (3.43) we obtain

$$\Big\|\frac{1}{h}\int_t^{t+h} S(\tau)x\mathrm{d}\tau - S(t)x\Big\|<\varepsilon$$

which proves the lemma. □

PROPOSITION 3.2.84 *If* $\{S(t)\}_{t\geq 0}$ *is a* C_0*-semigroup with infinitesimal generator* A *and* $x_0\in\mathcal{D}(A)$, *then*

(a) $S(t)x_0\in\mathcal{D}(A)$ *for each* $t\geq 0$.

(b) $(\mathrm{d}/\mathrm{d}t)S(t)x_0 = AS(t)x_0 = S(t)Ax_0$ *for each* $t>0$.

(c) $t\longrightarrow S(t)x_0$ *belongs in* $C^1\big((0,\infty),X\big)\cap C\big(\mathbb{R}_+,\mathcal{D}(A)\big)$.

PROOF: (a) We have

$$\lim_{h\to 0}\frac{1}{h}\big(S(h)S(t)x_0 - S(t)x_0\big) = \lim_{h\to 0}\frac{1}{h}\big(S(t)S(h)x_0 - S(h)x_0\big)$$
$$= S(t)\frac{1}{h}\big(S(h)x_0 - x_0\big) = S(t)Ax_0.$$

Therefore, $S(t)x_0\in\mathcal{D}(A)$ (see Definition 3.2.82) and $AS(t)x_0 = S(t)Ax_0$.

(b) For $h>0$, $t\geq 0$, we have

$$\frac{1}{h}\big(S(t+h)x_0 - S(t)x_0\big) = S(t)\frac{1}{h}\big(S(h)x_0 - x_0\big) = \frac{1}{h}\big(S(h)-I\big)S(t)x_0.$$

By virtue of Definition 3.2.82 we know that $\lim_{h\to 0}(1/h)\big(S(h)x_0 - x_0\big)$ exists and so the other two limits exist too. So it follows that $S(t)x_0\in\mathcal{D}(A)$. Also for $t>0$ and $h>0$ small, we have

$$\frac{1}{h}\big(S(t-h)x_0 - S(t)x_0\big) = -S(t-h)\frac{1}{h}\big(S(h)x_0 - x_0\big).$$

Hence we deduce that the strong derivative of $t\longrightarrow S(t)x_0$ exists at each $t>0$ and

$$\frac{\mathrm{d}}{\mathrm{d}t}S(t)x_0 = AS(t)x_0 = S(t)Ax_0.$$

(c) From (b) and Corollary 3.2.80 we infer that $t \longrightarrow S(t)x_0$ belongs in $C^1\big((0,\infty),X\big) \cap C\big(\mathbb{R}_+,\mathcal{D}(A)\big)$. $\square$

REMARK 3.2.85 From the above proposition it follows that $t \longrightarrow x(t) = S(t)x_0$ is the unique solution of the initial value problem $x'(t) = Ax(t)$ $t \geq 0$, $x(0) = x_0$, provided that the initial datum is in $\mathcal{D}(A)$. If $x_0 \notin \mathcal{D}(A)$, then $t \longrightarrow S(t)x_0$ is not differentiable with respect to t. We can, however, consider $t \longrightarrow x(t) = S(t)x_0$ as a generalized solution of the initial value problem.

PROPOSITION 3.2.86 *If $\{S(t)\}_{t\geq 0}$ is a C_0-semigroup with infinitesimal generator A, then A is closed and densely defined.*

PROOF: Let $x \in X$. Then for $h>0$ we have

$$\begin{aligned}\frac{1}{h}\big(S(h)-I\big)\int_0^t S(\tau)x\mathrm{d}\tau &= \frac{1}{h}\int_0^t \big(S(\tau+h)x - S(\tau)x\big)\mathrm{d}\tau \\ &= \frac{1}{h}\Big[\int_t^{t+h} S(\tau)x\mathrm{d}\tau - \int_0^h S(\tau)x\mathrm{d}\tau\Big] \longrightarrow S(t)x - x\end{aligned}$$

(see Lemma 3.2.83). Therefore $\int_0^t S(\tau)x\mathrm{d}\tau \in \mathcal{D}(A)$. Once again Lemma 3.2.83 implies that

$$\frac{1}{h}\int_0^h S(\tau)x\mathrm{d}\tau \longrightarrow S(0)x = x,$$
$$\Rightarrow \mathcal{D}(A) \text{ is dense in } X.$$

If $x_n \in \mathcal{D}(A)$ and $x_n \longrightarrow x$ in X, $A(x_n) \longrightarrow u$ in X, we need to show that $x \in \mathcal{D}(A)$ and $Ax = u$,

$$\frac{1}{h}\big(S(h)x - x\big) = \lim_{n\to\infty}\frac{1}{h}\big(S(h)x_n - x_n\big) = \lim_{n\to\infty}\frac{1}{h}\int_0^h S(\tau)Ax_n\mathrm{d}\tau.$$

Because $Ax_n \longrightarrow u$, we have

$$\frac{1}{h}\big(S(h)x - x\big) = \frac{1}{h}\int_0^h S(\tau)u\mathrm{d}\tau \longrightarrow u \quad \text{as } h \downarrow 0$$

(see Lemma 3.2.83). Therefore $x \in \mathcal{D}(A)$ and $Ax = u$. $\square$

Now let us reverse the problem and ask ourselves what conditions must be placed on A in order that A is the generator of a C_0-semigroup. If $A \in \mathcal{L}(X)$, then

$$S(t) = \mathrm{e}^{At} = \sum_{k\geq 0}\frac{(At)^k}{k!}.$$

For the unbounded case, we know that if A is the infinitesimal generator of a strongly continuous semigroup, then we require that

$$\overline{\mathcal{D}(A)} = X \quad \text{and } A \text{ is closed.}$$

The precise conditions are provided by the following classical result, known as the *Hille–Yosida theorem*. For a proof we refer to the book of Hille–Phillips [300, p. 360].

THEOREM 3.2.87 *Necessary and sufficient conditions for a linear unbounded operator A on a Banach space X to be the infinitesimal generator of a C_0-semigroup satisfying that $\|S(t)\|_{\mathcal{L}} \leq M e^{\omega t}$ for some $M \geq 1$, $\omega \geq 0$ are*

(a) *A is closed and densely defined.*

(b) *$(\lambda I - A)^{-1}$ exists for $\lambda > \omega$,*

(c) $\|(\lambda I - A)^{-m}\|_{\mathcal{L}} \leq \frac{M}{(\lambda-\omega)^m}$, $\lambda > \omega$, $m \geq 1$.

REMARK 3.2.88 The family of the operators $\mathcal{R}(\lambda) = (\lambda I - A)^{-1}$ for $\lambda > \omega$, is called the *resolvent* of A.

A byproduct of the proof of the Hille–Yosida theorem, is the following exponential formula for the C_0-semigroup.

THEOREM 3.2.89 *If $\{S(t)\}_{t\geq 0}$ is a contraction semigroup with infinitesimal generator A, then for any $x \in X$, $S(t)x = \lim_{n\to\infty} (I - (t/n)A)^{-n} = \lim_{n\to\infty} \big((n/t)\mathcal{R}(n/t)\big)^n x$.*

Let us give some characteristic examples of C_0-semigroups with their infinitesimal generators.

EXAMPLE 3.2.90 (a) Let $X = H = L^2(0, b)$ (a Hilbert space) and let us consider an operator $A = \mathrm{d}/\mathrm{d}x$ with domain

$$\mathcal{D}(A) = \{x \in W^{1,2}(0, b) : x(b) = 0\}.$$

The operator A is the generator of the so-called left-shift semigroup $\{S(t)\}_{t\geq 0}$

$$S(t)x(\tau) = \begin{cases} x(t+\tau) & \text{if } t+\tau \in (0, b) \\ 0 & \text{if } t+\tau \notin (0, b) \end{cases}.$$

(b) Let H be a Hilbert space with a complete and orthonormal basis $\{e_n\}_{n\geq 1}$ and let $\{\lambda_n\}_{n\geq 1}$ be a sequence in $\mathbb{R}$ such that $\lambda_n \longrightarrow -\infty$. We define

$$S(t)x = \sum_{n\geq 1} e^{\lambda_n t} \langle x, e_n \rangle_H e_n, \quad x \in H,\ t \geq 0. \tag{3.44}$$

It is easy to check that $\{S(t)\}_{t\geq 0}$ is a C_0-semigroup. The infinitesimal generator of this semigroup is given by

$$Ax = \sum_{n\geq 1} \lambda_n \langle x, e_n \rangle e_n$$

with domain $\mathcal{D}(A) = \Big\{x \in H : \sum_{n\geq 1} \big(\lambda_n \langle x, e_n \rangle\big)^2 < +\infty\Big\}$.

(c) Let $H = L^2[0,\pi]$ and $Ax = \frac{\mathrm{d}^2 x}{\mathrm{d}t^2}$ with domain $\mathcal{D}(A) = W_0^{1,2}(0,\pi) \cap W^{2,2}(0,\pi)$. Then A is the infinitesimal generator of a semigroup given by (3.44) with $e_n(t) = \sqrt{\frac{2}{\pi}}\sin(nt)$, $t \in [0,\pi]$, $\lambda_n = -n^2$, $n \geq 1$.

(d) The infinitesimal generator A of the semigroup $\{S(t)\}_{t\geq 0}$ on $C(\mathbb{R})$ given by

$$\big(S(t)x\big)(\tau) = \mathrm{e}^{-\lambda t}\sum_{k\geq 0}\frac{(\lambda t)^k}{k!}x(\tau - k\mu), \quad \lambda,\ \mu>0$$

is the difference operator $(Ax)(\tau) = \lambda[x(\tau-\mu) - x(\tau)]$.

Now we pass to nonlinear semigroups. So let X be a Banach space and $C \subseteq X$ be a nonempty subset of X.

DEFINITION 3.2.91 A family of maps $S(t)\colon C \longrightarrow C$, $t \geq 0$, is called a *semigroup of nonexpansive maps* on C if

(a) $S(t+\tau) = S(t)S(\tau)$ for all $t, \tau \geq 0$.

(b) $S(0) = I_C$.

(c) $\lim\limits_{t\downarrow 0} S(t)x = x$ for each $x \in C$.

(d) $\|S(t)x - S(t)u\| \leq \|x - u\|$ for all $t \geq 0$ and all $x, u \in C$.

REMARK 3.2.92 Given a semigroup of nonexpansive maps on C, the function $(t,x) \longrightarrow S(t)x$ is jointly continuous from $\mathbb{R}_+ \times C$ into C.

The main generation theorem for nonlinear semigroups, is the following theorem due to Crandall–Liggett [166], where the interested reader can find its proof (see also Barbu [58, p. 104]).

THEOREM 3.2.93 *If $A\colon \mathcal{D}(A) \subseteq X \longrightarrow 2^X$ is an m-accretive operator, then for every $x \in \overline{\mathcal{D}(A)}$ $S(t)x = \lim\limits_{n\to\infty}\big(I + (t/n)A\big)^{-n}x$ exists for each $x \in \overline{\mathcal{D}(A)}$ and uniformly in t in compact subsets of $\mathbb{R}_+$. The family of maps $S(t) : \overline{\mathcal{D}(A)} \longrightarrow \overline{\mathcal{D}(A)}$, $t \geq 0$, is a semigroup of nonexpansive maps and for each $x \in \mathcal{D}(A)$ and $t>0$, we have $\|S(t)x - x\| \leq t|A(x)|$, where $|A(x)| = \inf\big\{\|y\| : y \in A(x)\big\}$.*

Let $A\colon \mathcal{D}(A) \subseteq X \longrightarrow 2^X$ be an m-accretive operator. According to Theorem 3.2.93 it generates a semigroup of nonlinear contractions $S(t) : \overline{\mathcal{D}(A)} \longrightarrow \overline{\mathcal{D}(A)}$, $t \geq 0$. Recall that for each $\lambda > 0$, J_λ and A_λ are the resolvent and the Yosida approximation of A (i.e., $J_\lambda = (I + \lambda A)^{-1}$ and $A_\lambda = (1/\lambda)(I - J_\lambda)$; see Definition 3.2.67).

The next proposition relates the semigroup $S(t)$ with the resolvent J_λ. Its proof can be found in Vrabie [597, p. 46].

PROPOSITION 3.2.94 *If $x \in \overline{\mathcal{D}(A)}$, $t > 0$ and $\lambda > 0$, then $\|S(t)x - x\| \leq \big(2 + (t/\lambda)\big)\|J_\lambda(x) - x\|$ and $\|J_\lambda(x) - x\| \leq (2/t)\big(1 + (\lambda/t)\big)\int_0^t \|S(\tau)x - x\|\mathrm{d}\tau$.*

REMARK 3.2.95 If $X = H$ a Hilbert space and $A = \partial\varphi$ with $\varphi \in \Gamma_0(H)$, then the second inequality can be proved in the following way: $\|J_\lambda(x) - x\| \leq (1 + 1/\sqrt{2})\|S(t)x - x\|$ for each $x \in \overline{\mathcal{D}(\partial\varphi)}$ and $t, \lambda > 0$. Moreover, $\|J_t(x) - x\| \leq \sqrt{3/2}\,\|S(t)x - x\|$, with $\sqrt{3/2}$ being the best possible constant.

DEFINITION 3.2.96 A nonlinear semigroup $S(t)\colon C \longrightarrow C$, $t \geq 0$, of nonexpansive maps on $C \subseteq X$ is called *compact*, if for every $t > 0$, $S(t)$ is a compact operator.

REMARK 3.2.97 Recall that $S(0) = I_C$. So if $S(t)$ is compact for all $t \geq 0$, then C is compactly embedded in X. Moreover, if $C = X$, then X is finite dimensional.

DEFINITION 3.2.98 A nonlinear semigroup of nonexpansive maps $S(t)\colon C \longrightarrow C$, $t \geq 0$, is said to be *equicontinuous*, if for each bounded set $K \subseteq X$, the family $\{S(\cdot)x : x \in K\}$ is equicontinuous at every $t > 0$.

PROPOSITION 3.2.99 *If $S(t)\colon C \longrightarrow C$, $t \geq 0$, is a semigroup of nonlinear nonexpansive maps which is compact, then it is equicontinuous.*

PROOF: Let $K \subseteq X$ be bounded, $t > 0$, and choose $\lambda \in (0, t)$. Due to the compactness of the semigroup, $S(t - \lambda)K$ is totally bounded in X. So given $\varepsilon > 0$, we can find $\{x_k\}_{k=1}^N$ such that

$$S(t - \lambda)K \subseteq \bigcup_{k=1}^{N} B_{\varepsilon/3}\big(S(t - \lambda)x_k\big) \tag{3.45}$$

where $B_{\varepsilon/3}\big(S(t-\lambda)x_k\big) = \big\{y \in X : \|y - S(t-\lambda)x_k\| < \varepsilon/3\big\}$. Also from the continuity of $S(\cdot)x_k$ we can find $0 < \delta = \delta(\varepsilon, t) < \lambda$ such that

$$\|S(t + h)x_k - S(t)x_k\| < \frac{\varepsilon}{3} \tag{3.46}$$

for all $k \in \{1, \ldots, N\}$ and all $h \in \mathbb{R}$ with $|h| \leq \delta$. Because of (3.45) for each $x \in K$ there exists $k \in \{1, \ldots, N\}$ such that

$$\|S(t - \lambda)x - S(t - \lambda)x_k\| \leq \frac{\varepsilon}{3}. \tag{3.47}$$

Using (3.46), (3.47), and Definition 3.2.91, we obtain

$$\begin{aligned}
&\|S(t + h)x - S(t)x\| \\
&\leq \|S(t + h)x - S(t + h)x_k\| + \|S(t + h)x_k - S(t)x_k\| + \|S(t)x_k - S(t)x\| \\
&= \|S(\lambda + h)S(t - \lambda)x - S(\lambda + h)S(t - \lambda)x_k\| + \|S(t + h)x_k - S(t)x_k\| \\
&\quad + \|S(\lambda)S(t - \lambda)x_k - S(\lambda)S(t - \lambda)x\| \\
&\leq 2\,\|S(t - \lambda)x - S(t - \lambda)x_k\| + \|S(t + h)x_k - S(t)x_k\| \leq \varepsilon
\end{aligned}$$

for all $x \in K$ and all $h \in \mathbb{R}$ with $|h| \leq \delta$. This proves that $\{S(\cdot)x : x \in K\}$ is equicontinuous. ☐

Now we can characterize the compact nonlinear semigroups.

THEOREM 3.2.100 *If $A\colon \mathcal{D}(A) \subseteq X \longrightarrow 2^X$ is an m-accretive operator and $S(t) : \overline{\mathcal{D}(A)} \longrightarrow \overline{\mathcal{D}(A)}$, $t \geq 0$, is the semigroup generated by A, then the following statements are equivalent.*

(a) *The semigroup $\{S(t)\}_{t\geq 0}$ generated by A is compact.*

(b) (b_1) *For every $\lambda > 0$ J_λ is compact.*

(b_2) *$\{S(t)\}_{t\geq 0}$ is equicontinuous.*

PROOF: (a)$\Rightarrow$(b): From Proposition 3.2.99 we already know that $\{S(t)\}_{t\geq 0}$ is equicontinuous. We have:

$$\|S(t)J_\lambda(x) - J_\lambda(x)\| \leq t\|A_\lambda(x)\| = \frac{t}{\lambda}\|J_\lambda(x) - x\|. \tag{3.48}$$

Because J_λ is nonexpansive, from (3.48) it follows that

$$\lim_{t\downarrow 0} S(t)J_\lambda = J_\lambda$$

uniformly on bounded sets in X. For every $t > 0$ $S(t)J_\lambda$ is compact and so from Proposition 3.1.6 we infer that J_λ is compact.

(b)$\Rightarrow$(a): From Proposition 3.2.94, we see that for each $\lambda > 0$, $t > 0$, and $x \in \overline{\mathcal{D}(A)}$, we have

$$\|J_\lambda S(t)x - S(t)x\| \leq \frac{4}{\lambda}\int_0^\lambda \|S(t+\tau)x - S(t)x\| \mathrm{d}\tau. \tag{3.49}$$

Due to the equicontinuity of $\{S(t)\}_{t\geq 0}$ from (3.49) it follows that

$$\lim_{\lambda\downarrow 0} J_\lambda S(t) = S(t)$$

uniformly on bounded sets in $\overline{\mathcal{D}(A)}$. But $J_\lambda S(t)$ is compact. So Proposition 3.1.6 implies that $\{S(t)\}_{t\geq 0}$ is compact. □

COROLLARY 3.2.101 *If $A\colon \mathcal{D}(A) \subseteq X \longrightarrow 2^X$ is a densely defined, linear, m-accretive operator and $\{S(t)\}_{t\geq 0}$ is the nonexpansive semigroup generated by A, then the following statements are equivalent.*

(a) *The semigroup $\{S(t)\}_{t\geq 0}$ is compact.*

(b) (b_1) *For each $\lambda > 0$ J_λ is compact.*

(b_2) *$t \longrightarrow S(t)$ is continuous from $\mathbb{R}_+$ into $\mathcal{L}(X)$ with the uniform operator topology.*

REMARK 3.2.102 Recall the resolvent identity, which says that

$$J_\lambda(x) = J_\mu\Big(\frac{\mu}{\lambda}x + \frac{\lambda - \mu}{\lambda}J_\lambda(x)\Big)$$

for each $x \in X$ and $\lambda, \mu > 0$. So J_λ is compact for every $\lambda > 0$ if and only if it is compact for some $\lambda > 0$. So statement [b_1] is equivalent to saying that $(I + A)^{-1}$ is compact (i.e., $(I + A)^{-1} \in \mathcal{K}(X)$).

PROPOSITION 3.2.103 *If $A\colon \mathcal{D}(A) \subseteq X \longrightarrow 2^X$ is an m-accretive operator, then the following statements are equivalent.*

(a) *For each $\lambda > 0$ J_λ is compact.*

(b) *For each $k > 0$, the set $L_k = \{x \in \mathcal{D}(A) : \|x\| + |A(x)| \leq k\}$ is relatively compact in X.*

PROOF: (a)⇒(b): Recall that $\|A_\lambda(x)\| \leq |A(x)| \leq k$ for all $x \in L_k$ and $\lambda > 0$. So

$$\begin{aligned} & \|x - J_\lambda(x)\| \leq \lambda k \qquad \text{for each } x \in L_k \text{ and } \lambda > 0, \\ \Rightarrow\ & J_\lambda \longrightarrow I \qquad \text{as } \lambda \downarrow 0 \text{ uniformly on } L_k, \\ \Rightarrow\ & I_{L_k} \text{ is compact by Proposition 3.1.6 (i.e., } L_k \text{ is compact).} \end{aligned}$$

(b)⇒(a): We have

$$\begin{aligned} \|J_\lambda(x)\| + |A(J_\lambda(x))| &\leq \|J_\lambda(x)\| + \|A_\lambda(x)\| \quad (\text{because } A_\lambda(x) \in A(J_\lambda(x))) \\ &\leq \|J_\lambda(x)\| + \frac{1}{\lambda}\|x - J_\lambda(x)\| \qquad \text{for all } x \in X. \end{aligned} \tag{3.50}$$

If $K \subseteq X$ is bounded, then $J_\lambda(K)$ is bounded and so from (3.50) it follows that $J_\lambda(K) \subseteq L_k$ for $k > 0$ large. Therefore $J_\lambda(K)$ is relatively compact. □

Let $X = H =$ a Hilbert space and $\varphi \in \Gamma_0(H)$. In this case we can improve Proposition 3.2.103. First we have a definition.

DEFINITION 3.2.104 A convex function $\varphi\colon H \longrightarrow \overline{\mathbb{R}} = \mathbb{R} \cup \{+\infty\}$ is said to be of *compact type* if for each $k > 0$, the level set

$$L_k = \{x \in H : \|x\|^2 + \varphi(x) \leq k\}$$

is relatively compact in H.

REMARK 3.2.105 If $\varphi \in \Gamma_0(H)$, then L_k is compact for every $k > 0$.

Then we have the following improvement of Proposition 3.2.103. The proof of this result can be found in Vrabie [597, p. 58].

PROPOSITION 3.2.106 *If H is a Hilbert space, $\varphi \in \Gamma_0(H)$, and $A = \partial\varphi$, then the following statements are equivalent.*

(a) *For each $\lambda > 0$ J_λ is compact.*

(b) *φ is of compact type.*

(c) *The semigroup generated by A on $\overline{\mathcal{D}(A)}$ is compact.*

EXAMPLE 3.2.107 Let $Z\subseteq\mathbb{R}^N$ be a bounded domain with C^2-boundary ∂Z, $j\in\Gamma_0(\mathbb{R})$, $j\geq 0$, $j(0)=0$, $p\geq 2$, and $\lambda>0$. We consider the function $\varphi_\lambda\colon L^2(Z)\longrightarrow\overline{\mathbb{R}}_+=\mathbb{R}_+\cup\{+\infty\}$ defined by

$$\varphi_\lambda(x)=\begin{cases}\frac{1}{p}\|Dx\|_p^p+\frac{\lambda}{p}\|x\|_p^p+\int_{\partial Z} j\big(x(z)\big)\mathrm{d}\sigma & \text{if } x\in W^{1,p}(Z),\\ & j\big(x(\cdot)\big)\in L^1(\partial Z)\\ +\infty & \text{otherwise.}\end{cases}$$

Then $\varphi_\lambda\in\Gamma_0\big(L^2(Z)\big)$ and it is of compact type. Note that

$$\partial\varphi_\lambda(x)=-\Delta_p x+\lambda|x|^{p-2}x$$

for all $x\in\mathcal{D}(\partial\varphi_\lambda)=\big\{x\in W^{1,p}(Z):\big(\partial x/\partial n\big)(z)\in\partial j\big(x(z)\big)\text{ a.e. on }\partial Z\big\}$.

3.3 Degree Theory

Degree theory is a basic tool of nonlinear analysis and produces powerful existence and multiplicity results for nonlinear boundary value problems. It concerns operator equations of the form $\varphi(x)=y_0$, where φ is a map (often continuous) of $\overline{U}$, the closure of an open set U of the domain space X into the range space Y and $y_0\in Y$ satisfies $y_0\notin\varphi(\partial U)$. Then the degree of S at y_0 relative to U, written $d(\varphi,U,y_0)$, is an algebraic count of the number of solutions of the equation $\varphi(x)=y_0$. In particular the equation $\varphi(x)=y_0$ will have solutions in U, whenever $d(\varphi,U,y_0)\neq 0$. Here we deal with "classical" degree theories; that is, the value of the degree map is an integer. This integer can be both positive or negative and in the context of finite-dimensional problems positive counts (positive degree) correspond to solutions at which φ is orientation preserving, whereas negative counts correspond to solutions at which φ is not orientation preserving.

We start with a brief analytical presentation of Brouwer's degree theory, which is the first such theory and was introduced by Brouwer in 1912.

DEFINITION 3.3.1 Let $U\subseteq\mathbb{R}^N$ be a nonempty, bounded open set and $\varphi\in C^1(\overline{U},\mathbb{R}^N)$. We say that $x\in\overline{U}$ is a *critical point* of φ, if $J_\varphi(x)=\det\nabla\varphi(x)=0$ where $\nabla\varphi(x)=\Big(\big(\partial\varphi_i/\partial x_j\big)(x)\Big)_{i,j=1}^N$. We set $C_\varphi=\{x\in\overline{U}:J_\varphi(x)=0\}$ and $\varphi(C_\varphi)$ is called the *crease* of φ. If $y\notin\varphi(C_\varphi)$, then y is a *regular value* of φ.

Now we can give the first definition of degree.

DEFINITION 3.3.2 If $\varphi\in C^1(\overline{U},\mathbb{R}^N)$ and $y\notin[\varphi(C_\varphi)\cup\varphi(\partial U)]$, then the *degree* of φ at y with respect to U is defined by

$$d(\varphi,U,y)=\sum_{x\in\varphi^{-1}(y)}\operatorname{sgn}J_\varphi(x).$$

REMARK 3.3.3 Here sgn $u=1$ if $u>0$ and sgn $u=-1$ if $u<0$. Note that because y is not a critical value, the set $\varphi^{-1}(y)$ is discrete and so the summation in Definition 3.3.2 is finite. The determinant $J_\varphi(x)$ is positive or negative accordingly as φ is orientation preserving or orientation reversing at x.

Next the goal is to remove from Definition 3.3.2 the restrictions that y is a regular value of φ and that $\varphi \in C^1(\overline{U}, \mathbb{R}^N)$. The first can be removed with the help of the so-called *Sard's theorem.*

THEOREM 3.3.4 (Sard) *If $\varphi \in C^1(\overline{U}, \mathbb{R}^N)$, then $\varphi(C_\varphi)$ is a set of measure zero.*

We use Theorem 3.3.4 and suitable mollification techniques to replace the sum in Definition 3.3.2 by an appropriate integral. First let us recall the notion of mollifier.

DEFINITION 3.3.5 Consider the function $\psi_1 : \mathbb{R}^N \longrightarrow \mathbb{R}$ defined by

$$\psi_1(x) = \begin{cases} c \exp\left(\frac{1}{\|x\|^2 - 1}\right) & \text{for } \|x\| \leq 1 \\ 0 & \text{otherwise} \end{cases}$$

with $c > 0$ chosen so that $\int_{\mathbb{R}^N} \psi_1(x) \mathrm{d}x = 1$. For $\varepsilon > 0$, we define $\psi_\varepsilon(x) = 1/\varepsilon^N \psi_1(x/\varepsilon)$. We have $\psi_\varepsilon \in C^\infty(\mathbb{R}^N)$, $\int_{\mathbb{R}^N} \psi_\varepsilon(x) \mathrm{d}x = 1$, and $\operatorname{supp} \psi_\varepsilon = \overline{B}_\varepsilon(0) = \{x \in \mathbb{R}^N : \|x\| \leq \varepsilon\}$. The family of functions $\{\psi_\varepsilon\}_{e>0}$ is called *mollifiers.*

REMARK 3.3.6 If $U \subseteq \mathbb{R}^N$ is open and $f \in L^1_{\mathrm{loc}}(U)$, then we define $f_\varepsilon = \psi_\varepsilon * f = \int_U \psi_\varepsilon(x-y) f(y) \mathrm{d}y$ for all $x \in U_\varepsilon = \{x \in U : d(x, \partial U) > \varepsilon\}$. Evidently $f_\varepsilon \in C^\infty(U_\varepsilon)$ and if $f \in C(U)$, we have $f_\varepsilon \longrightarrow f$ uniformly on compact subsets of U. Also if $f \in L^p_{\mathrm{loc}}(U)$ for some $1 \leq p < \infty$, then $f_\varepsilon \longrightarrow f$ in $L^p_{\mathrm{loc}}(U)$.

PROPOSITION 3.3.7 *If (U, φ, y) are as in Definition 3.3.2 and $\{\psi_\varepsilon\}_{\varepsilon > 0}$ is a family of mollifiers, then there exists $\varepsilon_0 = \varepsilon_0(y, \varphi) > 0$ such that*

$$d(\varphi, U, y) = \int_U \psi_\varepsilon\big(\varphi(x) - y\big) J_\varphi(x) \mathrm{d}x \qquad \textit{for all } \varepsilon \in (0, \varepsilon_0).$$

PROOF: If $\varphi^{-1}(y) = \emptyset$, then the result is immediate because $\psi_\varepsilon\big(\varphi(x) - y\big) = 0$ for all $\varepsilon < d^*\big(y, \varphi(\overline{U})\big)$.

So suppose $\varphi^{-1}(y) = \{x_k\}_{k=1}^m$. Then by the inverse function theorem we can find disjoint balls $B_r(x_k)$ such that $\varphi|_{B_r(x_k)}$ is a homeomorphism onto a neighborhood V_k of y and $\operatorname{sgn} J_\varphi(x) = \operatorname{sgn} J_\varphi(x_k)$ for all $x \in B_r(x_k)$. Let $B_\varrho(y) \subseteq \bigcap_{k=1}^m V_k$ and $\Omega_k = B_r(x_k) \cap \varphi^{-1}\big(B_\varrho(y)\big)$. Then

$$\|\varphi(x) - y\| \geq \xi \text{ on } \overline{U} \setminus \bigcup_{k=1}^m U_k \qquad \text{for some } \xi > 0.$$

Therefore $\varepsilon < \xi$ implies that

$$\int_U \psi_\varepsilon\big(\varphi(x) - y\big) J_\varphi(x) \mathrm{d}x = \sum_{k=1}^m \operatorname{sgn} J_\varphi(x_k) \int_{U_k} \psi_\varepsilon\big(\varphi(x) - y\big) |J_\varphi(x)| \mathrm{d}x.$$

Because $J_\varphi(x) = J_{\varphi - y}(x)$ and $\varphi(U_k) - y = B_r(0)$, by change of variables, we obtain

$$\int_{U_k} \psi_\varepsilon\big(\varphi(x) - y\big) |J_{\varphi - y}(x)| \mathrm{d}x = \int_{B_r(0)} \psi_\varepsilon(x) \mathrm{d}x = 1 \qquad \text{for } \varepsilon \in (0, \min\{\xi, \varrho\}). \qquad \square$$

We use this proposition to pass from regular to critical values in the definition of the degree. For this purpose we need the following two lemmata.

LEMMA 3.3.8 *If $\psi \in C_c^1(\mathbb{R}^N)$, $K = \operatorname{supp} \psi$, $U \subseteq \mathbb{R}^N$ a bounded open set and $\gamma\colon [0,1] \longrightarrow \mathbb{R}^N$ a continuous path such that $A = \{k + \gamma(t) : k \in K,\ t \in [0,1]\} \subseteq U$, then we can find $u \in C_c^1(\overline{U})$ such that*

$$(\text{div } u)(x) = f\big(x - \gamma(0)\big) - f\big(x - \gamma(1)\big).$$

PROOF: First suppose that $\gamma(t) = t\bar{x}$ and set $\Psi(x) = \int_0^1 \psi(x - \vartheta\bar{x})\mathrm{d}\vartheta$, $u(x) = \bar{x}\Psi(x)$. Clearly $\Psi \in C^1(\overline{U})$. Also if $x \in \overline{U}$ is such that $\Psi(x) \neq 0$, then $\psi(x - \vartheta\bar{x}) \neq 0$ for some $\vartheta \in [0,1]$ and so $x \in K + \vartheta\bar{x} \subseteq A \subseteq U$. Therefore supp $\Psi \subseteq A$ and we have $u \in C_c^1(\overline{U}, \mathbb{R}^N)$. Note that

$$\begin{aligned}
\text{div } u(x) &= \sum_{k=1}^N \Big(\int_0^1 \sum_{j=1}^N \frac{\partial \psi}{\partial x_j}(x - \vartheta\bar{x})\frac{\partial x_j}{\partial x_k}\mathrm{d}\vartheta\Big)\bar{x}_k \\
&= \sum_{k=1}^N \Big(\int_0^1 \frac{\partial \psi}{\partial x_k}(x - \vartheta\bar{x})\mathrm{d}\vartheta\Big)\bar{x}_k \\
&= -\int_0^1 \frac{\mathrm{d}}{\mathrm{d}\vartheta}\psi(x - \vartheta\bar{x})\mathrm{d}\vartheta = \psi(x) - \psi(x - \bar{x}).
\end{aligned}$$

Now we consider a general path γ. For some $s, t \in [0,1]$ introduce the equivalence relation $s \sim t$ if and only if $\psi\big(x - \gamma(s)\big) - \psi\big(x - \gamma(t)\big) = \text{div } u(x)$ for some $u \in C_c^1(\overline{U}, \mathbb{R}^N)$. We show that each equivalence class is an open set and this by the compactness of $[0,1]$ implies that there is only one equivalence class, which is what we want. Set $y_t = \gamma(t) - \gamma(s)$, $t \in [0,1]$ and let E be an equivalence class for s. We introduce $\psi_s(x) = \psi\big(x - \gamma(s)\big)$, $K_s = \operatorname{supp}(\psi_s)$ and $\beta = d(K_s, \partial U) > 0$. We can find $\varepsilon > 0$ such that $|t - s| < \varepsilon$ implies that $\|y_t\| < \beta/2$. Therefore

$$K_s' = \{k + \vartheta y_t : k \in K_s,\ \vartheta \in [0,1]\} \subseteq U.$$

Then from the first part of the proof, we can find $u \in C_c^1(\overline{U}, \mathbb{R}^N)$ such that if $|t - s| < \varepsilon$

$$\text{div } u(x) = \psi_s(x) - \psi_s(x - y_t) = \psi\big(x - \gamma(s)\big) - \psi\big(x - \gamma(t)\big).$$

This means that $t \sim s$; that is, $t \in E$ and so E is open, hence $E = [0,1]$. □

LEMMA 3.3.9 *If $U \subseteq \mathbb{R}^N$ is a bounded open set, $\varphi \in C^2(\overline{U}, \mathbb{R}^N)$, and $u \in C_c^1(\mathbb{R}^N, \mathbb{R}^N)$ is such that*

$$\operatorname{supp} u \cap \varphi(\partial U) = \emptyset,$$

then we can find $w \in C_c^1(\mathbb{R}^N, \mathbb{R}^N)$ such that $\text{div } w(x) = J_\varphi(x)\text{div } u\big(\varphi(x)\big)$.

PROOF: Let $\big(A_{ij}(x)\big)_{i,j=1}^N$ be the matrix of cofactors of $\nabla\varphi(x)$. We set

$$w_i(x) = \sum_{j=1}^N u_j\big(\varphi(x)\big)A_{ji}(x).$$

It is easy to see that $w \in C_c^1(\overline{U}, \mathbb{R}^N)$. Recalling from the theory of differential forms that $\sum_{i=1}^{N} \frac{\partial A_{ji}}{\partial x_i}(x) = 0$, we obtain that

$$\begin{aligned}\frac{\partial w_i}{\partial x_i}(x) &= \sum_{j=1}^{N} \Big[\frac{\partial A_{ji}}{\partial x_i}(x) u_j\big(\varphi(x)\big) + A_{ji}(x) \sum_{k=1}^{N} \frac{\partial u_j}{\partial x_k}\big(\varphi(x)\big) \frac{\partial \varphi_k}{\partial x_i}(x)\Big] \\ &= \sum_{j,k=1}^{N} \frac{\partial u_j}{\partial x_k}\big(\varphi(x)\big) \delta_{jk} J_\varphi(x) = J_\varphi(x) \mathrm{div}\, u\big(\varphi(x)\big).\end{aligned}$$ □

The last auxiliary result before extending Definition 3.3.2 to critical values too, says that the definition of degree as it stands at this point (see Definition 3.3.2) is stable with respect to small changes in the function φ. For a proof of it we refer to Lloyd [396, p. 4].

LEMMA 3.3.10 *If $U \subseteq \mathbb{R}^N$ is a bounded open set, $\varphi \in C^1(\overline{U}, \mathbb{R}^N)$, and $y \notin \big[\varphi(C_\varphi) \cup \varphi(\partial U)\big]$, then there exists $\delta > 0$ such that if $\varphi_1 \in C^1(\overline{U}, \mathbb{R}^N)$ and $\|\varphi - \varphi_1\|_{C^1} \le \delta$, we have*

$$y \notin \big[\varphi_1(C_{\varphi_1}) \cup \varphi_1(\partial U)\big] \quad \text{and} \quad d(\varphi, U, y) = d(\varphi_1, U, y).$$

The next proposition allows us to admit critical values in Definition 3.3.2.

PROPOSITION 3.3.11 *If $U \subseteq \mathbb{R}^N$ is a bounded open set, $\varphi \in C^1(\overline{U}, \mathbb{R}^N)$, Z is a connected component of $\mathbb{R}^N \setminus \varphi(\partial U)$, and $y_1, y_2 \in Z \setminus \varphi(C_\varphi)$, then $d(\varphi, U, y_1) = d(\varphi, U, y_2)$.*

PROOF: First we suppose that $\varphi \in C^2(\overline{U}, \mathbb{R}^N)$. Recall that an open set in $\mathbb{R}^N$ is connected if and only if it is path-connected. So we can find a continuous path $\gamma\colon [0,1] \longrightarrow Z$ such that $\gamma(0) = y_1$, $\gamma(1) = y_2$. Let $\{\psi_\varepsilon\}_{\varepsilon > 0}$ be a family of mollifiers. By virtue of Proposition 3.3.7, we can find $\varepsilon_0 > 0$ such that $0 < \varepsilon \le \varepsilon_0$ implies that

$$d(\varphi, U, y_k) = \int_U \psi_\varepsilon\big(\varphi(x) - y_k\big) J_\varphi(x) \mathrm{d}x \qquad \text{for } k = 1, 2.$$

Let $0 < \varepsilon_1 < \varepsilon_0$ be such that the ε_1-neighborhood of the path γ is in Z. Set

$$A = \{k + \gamma(t) : k \in \mathrm{supp}\, \psi_{\varepsilon_1}, \quad t \in [0,1]\} \subseteq Z.$$

By Lemma 3.3.8, we can find $u \in C_c^1(\overline{Z}, \mathbb{R}^N)$ such that

$$\mathrm{div}\, u(x) = \varphi_{\varepsilon_1}(x - y_1) - \varphi_{\varepsilon_1}(x - y_2)$$

and $\mathrm{supp}\, u \cap \varphi(\partial U) \subseteq Z \cap \varphi(\partial U) = \emptyset$. So by Lemma 3.3.9 there exists $w \in C^1(\overline{U}, \mathbb{R}^N)$ such that

$$\mathrm{div}\, w(x) = \mathrm{div}\, u\big(\varphi(x)\big) J_\varphi(x) = \big[\psi_{\varepsilon_1}\big(\varphi(x) - y_1\big) - \psi_{\varepsilon_1}\big(\varphi(x) - y_2\big)\big] J_\varphi(x).$$

Using Green's formula, we conclude that

$$\begin{aligned}d(\varphi, U, y_1) - d(\varphi, U, y_2) &= \int_U \big[\psi_{\varepsilon_1}\big(\varphi(x) - y_1\big) - \psi_{\varepsilon_1}\big(\varphi(x) - y_2\big)\big] J_\varphi(x) \mathrm{d}x \\ &= \int_U \mathrm{div}\, u\big(\varphi(x)\big) J_\varphi(x) \mathrm{d}x = \int_U \mathrm{div}\, w(x) \mathrm{d}x = 0.\end{aligned}$$

Now we remove the restriction that $\varphi \in C^2(\overline{U}, \mathbb{R}^N)$. So if $\varphi \in C^1(\overline{U}, \mathbb{R}^N)$ we can find a sequence $\{\varphi_n\}_{n\geq 1} \subseteq C^2(\overline{U}, \mathbb{R}^N)$ such that $\|\varphi_n - \varphi\|_{C^1} \longrightarrow 0$ as $n \to \infty$. With the path γ being as above, let $\delta = d_1(\gamma, \varphi(\partial U)) > 0$. If $\|\varphi - \varphi_n\|_{C^1} < (\delta/2)$, then for $x \in \partial U$ and $t \in [0,1]$, we have

$$\|\gamma(t) - \varphi_n(x)\| \geq \|\gamma(t) - \varphi(x)\| - \|\varphi(x) - \varphi_n(x)\| > \frac{\delta}{2}.$$

Therefore y_1, y_2 are also in the same component of $\mathbb{R}^N \setminus \varphi_n(\partial U)$. Then using Lemma 3.3.10 and the first part of the proof, we obtain

$$d(\varphi, U, y_1) = d(\varphi_n, U, y_1) = d(\varphi_n, U, y_2) = d(\varphi, U, y_2).$$

□

This proposition leads to the following generalization of Definition 3.3.2.

DEFINITION 3.3.12 If $U \subseteq \mathbb{R}^N$ is a bounded open set, $\varphi \in C^1(\overline{U}, \mathbb{R}^N)$, and $y \notin \varphi(\partial U)$, then we define $d(\varphi, U, y)$ to be $d(\varphi, U, \widehat{y})$ for any $\widehat{y} \notin [\varphi(C_\varphi) \cup \varphi(\partial U)]$ such that $\|y - \widehat{y}\| < d(y, \varphi(\partial U))$.

REMARK 3.3.13 By Sard's theorem (see Theorem 3.3.4) every ball $B_r(y)$ contains vectors $\widehat{y}$ that are not critical values of φ. The component of y in $\mathbb{R}^N \setminus \varphi(\partial U)$ contains the ball $B_{d_1(y,\varphi(\partial U))}(y)$. So by Proposition 3.3.11, $d(\varphi, U, y)$ has the same value for all $\widehat{y}$ that is not a critical value of φ and $\|y - \widehat{y}\| < d(y, \varphi(\partial U))$.

DEFINITION 3.3.14 Let $U \subseteq \mathbb{R}^N$ be a bounded open set and $\varphi_0, \varphi_1 \in C^1(\overline{U}, \mathbb{R}^N)$. A map $h\colon [0,1] \times \overline{U} \longrightarrow \mathbb{R}^N$ is a C^1-homotopy between φ_0 and φ_1, if

(a) $h(0, \cdot) = \varphi_0(\cdot)$ and $h(1, \cdot) = \varphi_1(\cdot)$.

(b) $h(t, \cdot) \in C^1(\overline{U}, \mathbb{R}^N)$ for every $t \in [0,1]$.

(c) $\lim\limits_{t \to s} \|h(t, \cdot) - h(s, \cdot)\|_{C^1} = 0$ for every $s \in [0,1]$.

Using this notion and Sard's theorem, we can easily establish some basic properties of the degree $d(\varphi, U, y)$ for C^1-functions. Namely we have the following.

THEOREM 3.3.15 *If $U \subseteq \mathbb{R}^N$ is bounded open and $\varphi \in C^1(\overline{U}, \mathbb{R}^N)$, then*

(a) *$d(\varphi, U, \cdot)$ is constant on each connected component of $\mathbb{R}^N \setminus \varphi(\partial U)$.*

(b) *If $y \notin \varphi(\partial U)$, then we can find $\varepsilon > 0$ such that if $\|\widehat{\varphi} - \varphi\|_{C^1} \leq \varepsilon$; then $y \notin \widehat{\varphi}(\partial U)$ and $d(\varphi, U, y) = d(\widehat{\varphi}, U, y)$.*

(c) *If $h(t,x)$ is a C^1-homotopy between $\varphi_0, \varphi_1 \in C^1(\overline{U}, \mathbb{R}^N)$ and $y \notin h(t, \partial U)$ for all $t \in [0,1]$, then $d(\varphi_0, U, y) = d(\varphi_1, U, y)$.*

(d) *If $y \notin \varphi(\partial U)$, then $d(\varphi + z_0, U, y + z_0) = d(\varphi, U, y)$ for all $z_0 \in \mathbb{R}^N$.*

Property (c) (the homotopy invariance property) with a further approximation enables $d(\varphi, U, y)$ to be defined when φ is only continuous and not continuously differentiable. So let $\varphi \in C(\overline{U}, \mathbb{R}^N)$ and $y \notin \varphi(\partial U)$. Set $\beta = d\big(y, \varphi(\partial U)\big) > 0$. We can find $\varphi_0, \varphi_1 \in C^1(\overline{U}, \mathbb{R}^N)$ such that $\|\varphi - \varphi_k\|_{C^1} < \beta$, $k = 0, 1$. We consider the C^1-homotopy

$$h(t,x) = t\varphi_0(x) + (1-t)\varphi_1(x) \qquad \text{for all } t \in [0,1] \text{ and all } x \in \overline{U}.$$

We have

$$\|h(t,x) - \varphi(x)\| = \|t\big(\varphi_0(x) - \varphi(x)\big) + (1-t)\big(\varphi_1(x) - \varphi(x)\big)\| < \beta.$$

So if $x \in \partial U$, then

$$\|y - h(t,x)\| \geq \|y - \varphi(x)\| - \|\varphi(x) - h(t,x)\| > 0 \qquad \text{for all } t \in [0,1],$$
$$\Rightarrow \; y \notin h(t, \partial U) \qquad \text{for all } t \in [0,1].$$

So by virtue of Theorem 3.3.15(c), we have

$$d(\varphi_0, U, y) = d(\varphi_1, U, y).$$

It follows that $d(\widehat{\varphi}, U, y)$ is the same for all $\widehat{\varphi} \in C^1(\overline{U}, \mathbb{R}^N)$ within β of φ. This leads to a further generalization of Definition 3.3.2.

DEFINITION 3.3.16 Let $U \subseteq \mathbb{R}^N$ be a bounded open set, $\varphi \in C(\overline{U}, \mathbb{R}^N)$, and $y \notin \varphi(\partial U)$. We define $d(\varphi, U, y)$ to be equal to $d(\widehat{\varphi}, U, y)$, where $\widehat{\varphi} \in C^1(\overline{U}, \mathbb{R}^N)$ satisfies

$$\|\varphi(x) - \widehat{\varphi}(x)\| < d\big(y, \varphi(\partial U)\big) \qquad \text{for all } x \in \overline{U}.$$

REMARK 3.3.17 In fact it is easy to check that in the above definition we can choose $\widehat{\varphi} \in C^1(\overline{U}, \mathbb{R}^N)$ such that $y \notin \widehat{\varphi}(C_{\widehat{\varphi}})$.

This way we have concluded the definition of Brouwer's degree which has been defined for the class $C(\overline{U}, \mathbb{R}^N)$. In the next theorem we summarize the basic properties of Brouwer's degree.

THEOREM 3.3.18 *If $U \subseteq \mathbb{R}^N$ is a bounded open set, $\varphi \in C(\overline{U}, \mathbb{R}^N)$, and $y \notin \varphi(\partial U)$, then*

(a) $d(I, U, y) = 1$ *for all* $y \in U$.

(b) Additivity with respect to the domain: *If U_1, U_2 are disjoint open subsets of U and $y \notin \varphi\big(\overline{U} \setminus (U_1 \cup U_2)\big)$, then $d(\varphi, U, y) = d(\varphi, U_1, y) + d(\varphi, U_2, y)$.*

(c) Homotopy invariance: *If $h \colon [0,1] \times \overline{U} \longrightarrow \mathbb{R}^N$ is a continuous map and $y \notin h(t, \partial U)$ for all $t \in [0,1]$, then $d\big(h(t,\cdot), U, y\big)$ is independent of $t \in [0,1]$.*

(d) Dependence on the boundary values: *If $\widehat{\varphi} \in C(\overline{U}, \mathbb{R}^N)$ and $\varphi\big|_{\partial U} = \widehat{\varphi}\big|_{\partial U}$, then $d(\varphi, U, y) = d(\widehat{\varphi}, U, y)$.*

(e) Excision property: *If* $K \subseteq \overline{U}$ *is closed and* $y \notin \varphi(K)$, *then* $d(\varphi, U, y) = d(\varphi, U \setminus K, y)$.

(f) Continuity with respect to φ: *If* $\widehat{\varphi} \in C(\overline{U})$ *and* $\|\varphi - \widehat{\varphi}\|_\infty < d\big(y, \varphi(\partial U)\big)$, *then* $d(\widehat{\varphi}, U, y)$ *is defined and equals* $d(\varphi, U, y)$.

(g) Existence property: $d(\varphi, U, y) \neq 0$ *implies that* $\varphi^{-1}(y) \neq \emptyset$.

PROOF: (a) Follows at once from Definition 3.3.2.

(b) First we show that $\partial U_1, \partial U_2 \subseteq \partial U$. Clearly $\partial U_k \subseteq \overline{U}_k \subseteq \overline{U}$ for $k = 1, 2$. So if there is $x \in \partial U_1$ with $x \notin \partial U$, then we must have $x \in U$. So $x \in U_2$ and we find $r > 0$ such that $B_r(x) \subseteq U_2$, hence $B_r(x) \cap U_1 = \emptyset$. This contradicts the fact that $x \in \partial U_1$. So $\partial U_1 \subseteq \partial U$ and similarly $\partial U_2 \subseteq \partial U$.

We choose $\widehat{\varphi} \in C^1(\overline{U}, \mathbb{R}^N)$ such that

$$\|\varphi - \widehat{\varphi}\|_\infty < d\big(y, \varphi(\partial U)\big) \qquad \text{and} \qquad y \notin \widehat{\varphi}(C_{\widehat{\varphi}}).$$

Because $\partial U_k \subseteq \partial U$, it follows that $y \notin \widehat{\varphi}(\partial U_k)$ and

$$\begin{aligned} &\|\varphi(x) - \widehat{\varphi}(x)\| < d\big(y, \varphi(\partial U)\big) \qquad \text{for all } x \in \overline{U}_k,\ k = 1, 2, \\ \Rightarrow\ & d(\varphi, U_k, y) = d(\widehat{\varphi}, U_k, y), \ \ k = 1, 2. \end{aligned}$$

So we have

$$\begin{aligned} d(\varphi, U, y) = d(\widehat{\varphi}, U, y) &= \sum_{x \in \widehat{\varphi}^{-1}(y)} \operatorname{sgn} J_{\widehat{\varphi}}(x) \\ &= \sum_{x \in \widehat{\varphi}^{-1}(y) \cap U_1} \operatorname{sgn} J_{\widehat{\varphi}}(x) + \sum_{x \in \widehat{\varphi}^{-1}(y) \cap U_2} \operatorname{sgn} J_{\widehat{\varphi}}(x) \\ &= d(\widehat{\varphi}, U_1, y) + d(\widehat{\varphi}, U_2, y) \\ &= d(\varphi, U_1, y) + d(\varphi, U_2, y). \end{aligned}$$

(c) It follows from Theorem 3.3.15(c) and Definition 3.3.16.

(d) Consider the continuous map (homotopy) $h(t, x) = t\varphi(x) + (1 - t)\widehat{\varphi}(x)$ for all $t \in [0, 1]$ and all $x \in \overline{U}$. For $x \in \partial U$, we see that $h(t, x) = \varphi(x)$ and so $y \notin h(t, \partial U)$ for all $t \in [0, 1]$. Therefore (c) implies that $d(\varphi, U, y) = d(\widehat{\varphi}, U, y)$.

(e) Again we choose $\widehat{\varphi} \in C^1(\overline{U}, \mathbb{R}^N)$ such that

$$\|\varphi - \widehat{\varphi}\|_\infty < d\big(y, \varphi(\partial U)\big), \quad \|\varphi - \widehat{\varphi}\|_\infty < d\big(y, \varphi(K)\big) \quad \text{and} \quad y \notin \widehat{\varphi}(C_{\widehat{\varphi}}).$$

Then $y \notin \widehat{\varphi}(K)$ and from the choice of $\widehat{\varphi}$, we have

$$\begin{aligned} d(\varphi, U, y) = d(\widehat{\varphi}, U, y) &= \sum_{x \in \widehat{\varphi}^{-1}(y) \cap U} \operatorname{sgn} J_{\widehat{\varphi}}(x) = \sum_{x \in \widehat{\varphi}^{-1}(y) \cap (U \setminus K)} \operatorname{sgn} J_{\widehat{\varphi}}(x) \\ &= d(\widehat{\varphi}, U \setminus K, y) = d(\varphi, U \setminus K, y). \end{aligned}$$

(f) Because $\|\varphi - \widehat{\varphi}\|_\infty < d\big(y, \varphi(\partial U)\big)$ and $y \notin \varphi(\partial U)$, it follows that $y \notin \widehat{\varphi}(\partial U)$ and so $d(\widehat{\varphi}, U, y)$ is defined. Take $g \in C^1(\overline{U}, \mathbb{R}^N)$ such that

$$\begin{aligned} & \|g-\widehat{\varphi}\|_\infty + \|\widehat{\varphi}-\varphi\|_\infty < d\big(y,\varphi(\partial U)\big), && (3.51)\\ \Rightarrow\ & \|g-\varphi\|_\infty < d\big(y,\varphi(\partial U)\big),\\ \Rightarrow\ & d(g,U,y) = d(\varphi,U,y) \qquad \text{(see Definition 3.3.16)}. && (3.52)\end{aligned}$$

Because $d\big(y,\varphi(\partial U)\big) \le d\big(y,\widehat{\varphi}(\partial U)\big) + \|\varphi-\widehat{\varphi}\|_\infty$, we have

$$\|g-\widehat{\varphi}\|_\infty < d\big(y,\widehat{\varphi}(\partial U)\big) \qquad \text{(see (3.51))}.$$

So once more Definition 3.3.16 implies that

$$d(g,U,y) = d(\widehat{\varphi},U,y). \tag{3.53}$$

Therefore finally we have that if $\|\varphi-\widehat{\varphi}\|_\infty < d\big(y,\varphi(\partial U)\big)$, then

$$d(\widehat{\varphi},U,y) = d(\varphi,U,y) \qquad \text{(see (3.52) and (3.53))}.$$

(g) If $y \notin \varphi(\overline{U})$, we take $\widehat{\varphi} \in C^1(\overline{U},\mathbb{R}^N)$ such that

$$\|\varphi-\widehat{\varphi}\|_\infty < d\big(y,\varphi(\overline{U})\big).$$

Hence $y \notin \widehat{\varphi}(\overline{U})$ and so $d(\widehat{\varphi},U,y) = 0$. Therefore by Definition 3.3.16 we have $d(\varphi,U,y) = 0$. From this we infer that $y \in \varphi(\overline{U})$ if $d(\varphi,U,y) \neq 0$. □

Next we present some consequences of these properties and of Definition 3.3.16.

COROLLARY 3.3.19 *If $U \subseteq \mathbb{R}^N$ is bounded open and $\varphi \in C(\overline{U},\mathbb{R}^N)$, then $d(\varphi,U,\cdot)$ is constant on the connected components of $\mathbb{R}^N \setminus \varphi(\partial U)$.*

PROOF: Let Z be a connected component of $\mathbb{R}^N \setminus \varphi(\partial U)$ and let $y_1, y_2 \in Z$. As before we can find a continuous path $\gamma\colon [0,1] \longrightarrow Z$ such that $\gamma(0) = y_1$ and $\gamma(1) = y_2$. Choose $\widehat{\varphi} \in C(\overline{U},\mathbb{R}^N)$ such that

$$\|\varphi-\widehat{\varphi}\|_\infty < d\big(\gamma([0,1]),\varphi(\partial U)\big).$$

By virtue of Definition 3.3.16 we have that

$$d(\varphi,U,y_k) = d(\widehat{\varphi},U,y_k), \quad k=1,2$$

and y_1, y_2 are in the same component of $\mathbb{R}^N \setminus \widehat{\varphi}(\partial U)$. Then the result follows from Theorem 3.3.15(a). □

COROLLARY 3.3.20 *If $U \subseteq \mathbb{R}^N$ is bounded open, $\varphi \in C(\overline{U},\mathbb{R}^N)$, and $y \notin \varphi(\partial U)$, then $d(\varphi,U,y) = d(\varphi-v,U,y-v)$ for all $v \in \mathbb{R}^N$.*

PROOF: Let $\xi\colon [0,1] \longrightarrow \mathbb{Z}$ be defined by $\xi(t) = d(\varphi-tv,U,y-tv)$. Note that $y - tv \notin (\varphi-tv)(\partial U)$ and so ξ is well defined. Moreover, by virtue of Theorem 3.3.18(f) and Corollary 3.3.19, $\xi(\cdot)$ is continuous. Because it is $\mathbb{Z}$-valued, it is constant. □

This permits the following improvement of the homotopy invariance property (see Theorem 3.3.18(c)).

COROLLARY 3.3.21 *If $U \subseteq \mathbb{R}^N$ is bounded open, $h\colon [0,1]\times\overline{U}\longrightarrow\mathbb{R}^N$, $y\colon [0,1]\longrightarrow\mathbb{R}^N$ are continuous maps, and $y(t)\notin h(t,\partial U)$ for all $t\in[0,1]$, then $d\big(h(t,\cdot),U,y(t)\big)$ is independent of $t\in[0,1]$.*

PROOF: Consider the continuous map

$$\widehat{h}(t,x)=h(t,x)-y(t)\qquad\text{for all } (t,x)\in[0,1]\times\overline{U}.$$

Then from Corollary 3.3.20 we have

$$d\big(\widehat{h}(t,\cdot),U,0\big)=d\big(h(t,\cdot),U,y(t)\big)\qquad\text{for all } t\in[0,1].$$

But from Theorem 3.3.18(c) $d\big(\widehat{h}(t,\cdot),U,0\big)$ is independent of $t\in[0,1]$. Hence the same can be said for $d\big(h(t,\cdot),U,y(t)\big)$. □

COROLLARY 3.3.22 *If $U\subseteq\mathbb{R}^N$ is bounded open, $\varphi,\widehat{\varphi}\in C(\overline{U},\mathbb{R}^N)$ and for all $x\in\partial U$, $y\notin[\,\varphi(x),\widehat{\varphi}(x)\,]$, then $d(\varphi,U,y)=d(\widehat{\varphi},U,y)$.*

PROOF: Consider the homotopy

$$h(t,x)=t\varphi(x)+(1-t)\widehat{\varphi}(x)\qquad\text{for all } (t,x)\in[0,1]\times\overline{U}.$$

Note that by hypothesis $t\big(\varphi(x)-y\big)+(1-t)\big(\widehat{\varphi}(x)-y\big)\neq 0$ for all $t\in[0,1]$ and all $x\in\partial U$. So

$$h(t,x)\neq y\qquad\text{for all } (t,x)\in[0,1]\times\partial U.$$

Invoking Theorem 3.3.18(c), we infer that $d(\varphi,U,y)=d(\widehat{\varphi},U,y)$. □

Next we present some useful applications of Brouwer's degree. Additional applications are presented in Section 3.5.

PROPOSITION 3.3.23 *If $\overline{B}_R=\{x\in\mathbb{R}^N:\|x\|\leq R\}$, $\varphi\colon\overline{B}_R\longrightarrow\mathbb{R}^N$ is continuous and $\varphi(x)\neq-\lambda x$ for all $x\in\partial B_R$, $\lambda\geq 0$, then the equation $\varphi(x)=0$ has a solution $x\in B_R=\{x\in\mathbb{R}^N:\|x\|<R\}$.*

PROOF: By hypothesis $0\notin\varphi(\partial B_R)$ and so the degree $d(\varphi,B_R,0)$ is well defined. We consider the homotopy $h(t,x)=t\varphi(x)+(1-t)x$ for all $(t,x)\in[0,1]\times\overline{B}_R$. By hypothesis $0\notin h(t,\partial B_R)$ for all $t\in[0,1]$. So from the homotopy invariance and the normalization properties of the degree (see Theorem 3.3.18(a) and (c)),

$$d(\varphi,B_R,0)=d(I,B_R,0)=1.$$

So there exists $x_0\in B_R$ such that $\varphi(x_0)=0$ (see Theorem 3.3.18(g)). □

An interesting consequence of this proposition is the following result.

COROLLARY 3.3.24 *If $\varphi\in C(\mathbb{R}^N,\mathbb{R}^N)$ is such that $\big(\varphi(x),x\big)_{\mathbb{R}^N}\geq 0$ for all $x\in\partial B_R=\{x\in\mathbb{R}^N:\|x\|=R\}$, then we can find $x_0\in\overline{B}_R=\{x\in\mathbb{R}^N:\|x\|\leq R\}$ such that $\varphi(x_0)=0$.*

PROOF: If there is $x_0 \in \partial B_R$ such that $\varphi(x_0) = 0$, then we are done. So suppose that $0 \notin \varphi(\partial B_R)$. Then we cannot have that $\varphi(x) = -\lambda x$ for some $\lambda \geq 0$ and $x \in \partial B_R$ or otherwise we contradict the hypothesis that $\big(\varphi(x), x\big)_{\mathbb{R}^N} \geq 0$ for all $x \in \partial B_R$. Therefore we can apply Proposition 3.3.23 and conclude that $\varphi(x) = 0$ has a solution in $\overline{B}_R$. □

COROLLARY 3.3.25 *If* $\varphi \in C(\mathbb{R}^N, \mathbb{R}^N)$ *and* $\big((\varphi(x), x)_{\mathbb{R}^N}\big)/\|x\| \longrightarrow +\infty$ *as* $\|x\| \longrightarrow +\infty$, *then* φ *is surjective; that is,* $\varphi(\mathbb{R}^N) = \mathbb{R}^N$.

PROOF: Let $y \in \mathbb{R}^N$. Then

$$\frac{(\varphi(x) - y, x)_{\mathbb{R}^N}}{\|x\|} \longrightarrow +\infty \qquad \text{as} \quad \|x\| \longrightarrow +\infty.$$

So we can find $R > 0$ large such that $\big(\varphi(x) - y, x\big)_{\mathbb{R}^N} \geq 0$ for all $\|x\| = R$. Apply Corollary 3.3.24 to obtain $x_0 \in \overline{B}_R$ such that $\varphi(x_0) = y$. □

PROPOSITION 3.3.26 *If* N *is an odd integer,* $U \subseteq \mathbb{R}^N$ *is a bounded open set with* $0 \in U$, *and* $\varphi\colon \partial U \longrightarrow \mathbb{R}^N \setminus \{0\}$ *is continuous, then we can find* $x \in \partial U$ *and* $\lambda \neq 0$ *such that* $\varphi(x) = \lambda x$.

PROOF: By Tietze's extension theorem, without any loss of generality we may assume that $\varphi \in C(\mathbb{R}^N, \mathbb{R}^N)$. Because N is odd, then $d(-I, U, 0) = -1$. Suppose that $d(\varphi, U, 0) \neq -1$. Then the map $h(t, x) = (1-t)\varphi(x) + t(-x)$ must have a zero $(t_0, x_0) \in (0,1) \times \partial U$. Hence $\varphi(x_0) = \big(t_0/(1-t_0)\big)x_0$. On the other hand if $d(\varphi, U, 0) = -1$, then because $d(I, U, 0) = 1$, the map $\widehat{h}(t, x) = (1-t)\varphi(x) + tx$ must have a zero $(t_0, x_0) \in (0,1) \times \partial U$. Hence $\varphi(x_0) = -\big(t_0/(1-t_0)\big)x_0$. □

REMARK 3.3.27 In the above proposition the dimension of the space must be odd. If it is even it is no longer true. Consider, for example the map $\varphi\colon \mathbb{R}^2 \longrightarrow \mathbb{R}^2$ defined by $\varphi(x_1, x_2) = (-x_2, x_1)$ (rotation by $\pi/2$). If $U = B_1 = \{x \in \mathbb{R}^N : \|x\| < 1\}$, then the proposition says that there is no continuous vector field on $S = \partial B_1$ which is nowhere zero; that is, $\varphi\colon S \longrightarrow \mathbb{R}^N$ such that $\varphi(x) \neq 0$ and $\big(\varphi(x), x\big)_{\mathbb{R}^N} = 0$ on S. So you cannot comb a hedgehog.

Another useful topological application of Brouwer's degree is the following theorem known as *Borsuk's theorem.*

THEOREM 3.3.28 *If* $U \subseteq \mathbb{R}^N$ *is bounded, open, and symmetric (i.e.,* $U = -U$*) with* $0 \in U$ *and* $\varphi\colon \overline{U} \longrightarrow \mathbb{R}^N$ *is continuous, odd, and* $0 \notin \varphi(\partial U)$, *then* $d(\varphi, U, 0)$ *is odd.*

PROOF: First we assume that $\varphi \in C^1(U) \cap C(\overline{U})$ and that $\det \varphi'(0) \neq 0$. We construct an odd function $\psi \in C^1(U) \cap C(\overline{U})$ that is sufficiently close to φ and for which 0 is a regular value. The construction is done by induction. Let

$$U_k = \{x \in U : x_i \neq 0 \ \text{ for some } \ i \leq k\}$$

and let $\xi \in C(\mathbb{R})$ be odd and satisfy $\xi'(0) = 0$ and $\xi(t) = 0$ if and only if $t = 0$ (e.g., let $\xi(t) = t^3$). Let $\overline{\varphi}(x) = \big(1/(\xi(x_1))\big)\varphi(x)$ on the bounded open set U_1. Then by Sard's theorem (see Theorem 3.3.4) we can find $y^1 \notin \overline{\varphi}(C_{\overline{\varphi}}(U_1))$ with $\|y^1\|$ as small as we like. So if $\psi_1(x) = \varphi(x) - \xi(x_1)y^1$, $x \in U_1$, then $\psi_1'(x) = \xi(x_1)\overline{\varphi}'(x)$ for $x \in U_1$ such that $\psi_1'(x) = 0$ and so 0 is a regular value of ψ_1. Suppose that we already have an odd function $\psi_k \in C^1(U) \cap C(\overline{U})$ close to φ on U such that $0 \in \psi_k\big(C_{\psi_k}(U_k)\big)$ for some $k \leq N$. Then define $\psi_{k+1}(x) = \psi_k(x) - \xi(x_{k+1})y^{k+1}$ with $\|y^{k+1}\|$ small and such that 0 is a regular value of ψ_{k+1} on $\{x \in U : x_{k+1} \neq 0\}$. Evidently $\psi_{k+1} \in C^1(U) \cap C(\overline{U})$ is odd and close to φ on $\overline{U}$. Moreover, if $x \in U_{k+1}$ and $x_{k+1} = 0$, then $x \in U_k$, $\psi_{k+1}(x) = \psi_k(x)$, and $\psi_{k+1}'(x) = \psi_k'(x)$. Hence $\det \psi_{k+1}'(x) \neq 0$ and so $0 \notin \psi_{k+1}\big(C_{\psi_{k+1}}(U_k)\big)$. Continuing this way at the end we obtain $\psi = \psi_N \in C^1(U) \cap C(\overline{U})$ which is odd, close to φ on $\overline{U}$, and $0 \notin \psi(\{x \in U \setminus \{0\} : \det \psi'(x) = 0\})$ (because $U_N = U \setminus \{0\}$). Moreover, $\psi'(0) = \psi_1'(0) = \varphi'(0)$ and so 0 is a regular value of ψ.

Using Theorem 3.3.18(f), we have

$$d(\varphi, U, 0) = d(\psi, U, 0) = \operatorname{sgn} J_\psi(0) + \sum_{\substack{x \in \psi^{-1}(0) \\ x \neq 0}} \operatorname{sgn} J_\psi(x)$$

where the sum is even because $\psi(x) = 0$ if and only if $\psi(-x) = 0$ and $J_\psi(\cdot)$ is even. Therefore $d(\varphi, U, 0)$ is odd, .

Now we remove the restriction that $\varphi \in C^1(U) \cap C(\overline{U})$. Indeed if $\varphi \in C(\overline{U})$, we approximate φ by $\psi \in C^1(U) \cap C(\overline{U})$ and let $\psi_0(x) = \frac{1}{2}\big(\psi(x) - \psi(-x)\big)$ (the odd part of ψ). Choose $\delta > 0$ small such that it is not an eigenvalue of $\psi_0'(0)$ and set $\overline{\varphi} = \psi_0 - \delta I$. Then $\overline{\varphi} \in C^1(U) \cap C(\overline{U})$, it is odd, $\det \overline{\varphi}'(0) \neq 0$, and $\|\varphi - \overline{\varphi}\|_\infty$ can be as small as we like (just choose $\delta > 0$ small). Then by Theorem 3.3.18(f), we have

$$d(\varphi, U, 0) = d(\overline{\varphi}, U, 0) = \text{odd number.}$$

□

COROLLARY 3.3.29 *If $U \subseteq \mathbb{R}^N$ is bounded, open, and symmetric (i.e., $U = -U$) with $0 \in U$ and $\varphi\colon \partial U \longrightarrow \mathbb{R}^k$ is continuous with $k < N$, then there exists $x \in \partial U$ such that $\varphi(x) = \varphi(-x)$.*

PROOF: We proceed by contradiction. So suppose that $\varphi(x) \neq \varphi(-x)$ for all $x \in \partial U$. Let $\psi(x)$ be any odd extension of $\varphi(x) - \varphi(-x)$ on $\overline{U}$ and let

$$\widehat{\psi}(x) = \big(\psi(x), \overbrace{0, \ldots, 0}^{(N-k)-\text{elements}}\big) \qquad \text{for all } x \in U.$$

Then by virtue of Corollary 3.3.19 and Theorem 3.3.28, we have

$$d(\widehat{\psi}, U, y) = d(\widehat{\psi}, U, 0) \neq 0 \qquad \text{for all } y \in B_r(0) \text{ with } r > 0 \text{ small.}$$

But then Theorem 3.3.18 implies that $B_r(0) \subseteq \widehat{\psi}(U)$, a contradiction. □

REMARK 3.3.30 In particular Corollary 3.3.29 implies that if U is as above and $\varphi\colon \partial U \longrightarrow \mathbb{R}^k$, $k < N$, is a continuous odd map, then it must have a zero.

COROLLARY 3.3.31 *If $U \subseteq \mathbb{R}^N$ is bounded, open, and symmetric (i.e., $U = -U$) with $0 \in U$ and $\varphi \in C(\overline{U}, \mathbb{R}^N)$ is such that $0 \notin \varphi(\partial U)$ and $\varphi(-x) \neq \lambda\varphi(x)$ for all $\lambda \geq 1$ and all $x \in \partial U$, then $d(\varphi, U, 0)$ is odd.*

PROOF: Let $\psi(x) = \varphi(x) - \varphi(-x)$. Clearly $\psi \in C(\overline{U}, \mathbb{R}^N)$ and it is odd. Let $h(t,x) = (1-t)\varphi(x) + t\psi(x) = \varphi(x) - t\varphi(-x)$. Evidently, by hypothesis $0 \notin h(t, \partial U)$ for all $t \in [0,1]$ and so Theorem 3.3.18(c) and Theorem 3.3.28 imply

$$d(\varphi, U, 0) = d(\psi, U, 0) = \text{odd number}.$$

□

Using this corollary, we can prove something about coverings of the boundary ∂U. The result is known as the *Ljusternik–Schnirelmann–Borsuk theorem*. It is useful in the study of the so-called *Ljusternik–Schnirelmann category* (see Section 4.2).

THEOREM 3.3.32 *If $U \subseteq \mathbb{R}^N$ is bounded, open, and symmetric (i.e., $U = -U$) with $0 \in U$ and let $\{C_i\}_{i=1}^m$ be closed subsets of ∂U that cover ∂U and $C_i \cap (-C_i) = \emptyset$ for $i = 1, \ldots, m$, then $m \geq N+1$.*

PROOF: We argue indirectly. Suppose $m \leq N$. For $i \in \{1, \ldots, m-1\}$ let $\varphi_i(x) = 1$ if $x \in C_i$ and $\varphi_i(x) = -1$ if $x \in -C_i$. Also for $i \in \{m, \ldots, N\}$ let $\varphi_i(x) = 1$ for all $x \in \overline{U}$. For $i \in \{1, \ldots, m-1\}$ we extend φ_i to a continuous function on all of $\overline{U}$. Set $\varphi = (\varphi_i)_{i=1}^N$. We claim that φ satisfies the hypotheses of Corollary 3.3.31. So we need to show that

$$\varphi(-x) \neq \lambda\varphi(x) \qquad \text{for all } \lambda \geq 1 \text{ and all } x \in \partial U. \tag{3.54}$$

If $x \in C_m$, then $x \notin -C_m$ and so $x \in -C_i$ for some $i \in \{1, \ldots, m-1\}$. Therefore we have

$$\partial U \subseteq \bigcup_{i=1}^{m-1} (C_i \cap (-C_i)).$$

Let $x \in \partial U$. Then if $x \in C_i$, we have $\varphi_i(x) = 1$ and $\varphi_i(-x) = -1$, whereas if $x \in -C_k$, we have $\varphi_k(x) = -1$ and $\varphi_k(-x) = 1$. Therefore $\varphi_i(-x)$ does not point in the same direction as $\varphi_i(x)$ and so (3.54) holds. Apply Corollary 3.3.31 to obtain $d(\varphi, U, 0) \neq 0$ and so $\varphi_i(x) = 0$, for some $x \in U$, a contradiction to the fact that $\varphi_N(x) = 1$. □

REMARK 3.3.33 So if $\{C_i\}_{i=1}^N$ is a closed covering of ∂B_1 in $\mathbb{R}^N$, then at least one of the C_is must contain antipodal points.

A final topological application of Brouwer's degree is the so-called *invariance of domain theorem*, which provides sufficient conditions in order for a continuous map to be open.

THEOREM 3.3.34 *If $U \subseteq \mathbb{R}^N$ is open and $\varphi \in C(U, \mathbb{R}^N)$ is locally one-to-one, then φ is open.*

PROOF: We need to show that for every $V \subseteq U$ open, $\varphi(V)$ is open. Let $x_0 \in V$ and $y_0 = \varphi(x_0)$. By translating things if necessary, we may assume that $x_0 = y_0 = 0$. Choose $r > 0$ such that $B_r(0) \subseteq V$ and $\varphi|_{B_r(0)}$ is one-to-one. Define

$$h(t,x) = \varphi\Big(\frac{1}{1+t}x\Big) - \varphi\Big(-\frac{t}{1+t}x\Big) \qquad \text{for all } (t,x) \in [0,1] \times \overline{B}_r(0).$$

Clearly h is continuous, $h(0,\cdot) = \varphi$, and $h(1,x) = \varphi(x/2) - (-x/2)$ is odd. Moreover, if for some $(t,x) \in [0,1] \times \partial B_r(0)$, we have $h(t,x) = 0$, then $\varphi\big((1/(1+t))x\big) = \varphi\big(-(1/(1+t))x\big)$ and since $\varphi|_{B_r(0)}$ is one-to-one, we have $\big(1/(1+t)\big)x = -\big(1/(1+t)\big)x$, hence $x = 0$, a contradiction. So

$$d\big(\varphi, B_r(0), 0\big) = d\big(h(1,\cdot), B_r(0), 0\big) \neq 0 \quad \text{(see Theorem 3.3.28)},$$
$$\Rightarrow \; d\big(\varphi, B_r(0), y\big) \neq 0 \quad \text{for all } y \in B_\varrho(0) \text{ for some } \varrho > 0$$

(see Corollary 3.3.19),

$$\Rightarrow \; B_\varrho(0) \subseteq \varphi\big(B_r(0)\big); \text{ that is, } \varphi(V) \text{ is open and so } \varphi \text{ is an open map.}$$

□

Before passing to degree theory in infinite dimensional spaces, let us mention the so-called *multiplication formula*. It concerns the Brouwer degree of composition of maps and is useful in applications because it facilitates the computation of the degree in many concrete situations. The proof of the result is rather lengthy and can be found in Lloyd [396, p. 29].

THEOREM 3.3.35 *If $U \subseteq \mathbb{R}^N$ is bounded open, $\varphi \in C(\overline{U}, \mathbb{R}^N)$, V is a bounded open set containing $\varphi(U)$, $\{D_k\}_{k\geq 1}$ are the components of $V \setminus \varphi(\partial U)$, $\psi \in C(\overline{V}, \mathbb{R}^N)$, and $y \notin \psi\big(\varphi(\partial U)\big) \cup \psi(\partial V)$, then $d(\psi \circ \varphi, U, y) = \sum_{k\geq 1} d(\psi, D_k, y) d(\varphi, U, z_k)$ where z_k is any point in D_k.*

Now we extend the notion of degree to infinite-dimensional Banach spaces. However, because in an infinite-dimensional Banach space a closed and bounded set need not be norm compact, mere continuity of the function φ is not enough. Using the theory of compact maps (see Section 3.1), we can have a degree $\widehat{d}(\varphi, U, y)$ of a continuous map $\varphi\colon X \longrightarrow X$ (X an infinite-dimensional Banach space), which has the form

$$\varphi = I - f \tag{3.55}$$

with I being the identity operator on X and $f\colon X \longrightarrow X$ being a compact map (see Definition 3.1.1(a)) (i.e., $f \in \mathcal{K}(X,X)$).

So let X be a Banach space and $U \subseteq X$ a bounded open set. Also let $\varphi \in C(\overline{U}, X)$ be a map that admits the decomposition (3.55) with $f \in \mathcal{K}(\overline{U}, X)$ (i.e., φ is a compact perturbation of the identity map). From Propositions 3.1.18 and 3.1.19 we know that $\varphi(\partial U)$ is closed in X. So if $y \notin \varphi(\partial U)$, then

$$\vartheta = d\big(y, \varphi(\partial U)\big) > 0. \tag{3.56}$$

Invoking Theorem 3.1.10 we can find $f_\varepsilon \in \mathcal{K}_f(\overline{U}, X)$ with $\varepsilon \leq \vartheta/2$ such that

$$\|f(x) - f_\varepsilon(x)\| \leq \varepsilon \qquad \text{for all } x \in \overline{U}. \tag{3.57}$$

Let $Y_\varepsilon = \operatorname{span}\{\mathcal{R}(f_\varepsilon) \cup \{y\}\}$. This is a finite-dimensional subspace of X that contains the range of f_ε, $\mathcal{R}(f_\varepsilon)$, and the vector $y \in X$. Let $N = \dim Y_\varepsilon$ and choosing a basis of Y_ε, we can identify it with $\mathbb{R}^N$. Note that for all $x \in \partial(U \cap Y_\varepsilon) = \partial(U \cap \mathbb{R}^N)$, if $\varphi_\varepsilon = I - f_\varepsilon$, we have

$$\|\varphi_\varepsilon(x) - y\| = \|\varphi(x) + f(x) - f_\varepsilon(x) - y\|$$
$$\geq \|\varphi(x) - y\| - \|f(x) - f_\varepsilon(x)\| \geq \vartheta - \frac{\vartheta}{2} = \frac{\vartheta}{2} > 0$$

(see (3.56) and (3.57)),

$$\Rightarrow\ y \notin \varphi_\varepsilon\big(\partial(U \cap \mathbb{R}^N)\big) = (I - f_\varepsilon)\big(\partial(U \cap \mathbb{R}^N)\big).$$

This suggests that we can define the degree in an infinite-dimensional Banach space by means of the degree of the appropriate finite-dimensional approximation of φ; that is,

$$\widehat{d}(\varphi, U, y) = d(\varphi_\varepsilon, U \cap Y_\varepsilon, y). \tag{3.58}$$

We need some additional work to show that this degree is in fact well defined.

LEMMA 3.3.36 *If $U \subseteq \mathbb{R}^{N+k}$ $(k \geq 1)$ is a bounded open set, $f \in C(\overline{U}, \mathbb{R}^N)$, $y \in \mathbb{R}^N$ and $y \notin (I + f)(\partial U)$, then $d(I + f, U, y) = d(I + f, U \cap \mathbb{R}^N, y)$.*

PROOF: First assume that $f \in C^1(\overline{U}, \mathbb{R}^N)$. Let J_{N+k} be the Jacobian matrix in $\mathbb{R}^{N+k}$ of the map $I + \widehat{f}$ where $\widehat{f} = (f, \overbrace{0, \ldots, 0}^{k\text{–entries}})$. Then from Definition 3.3.2, we have

$$d(I + \widehat{f}, U, y) = \sum_{x \in (I+\widehat{f})^{-1}(y)} \operatorname{sgn} J_{N+k}(x) = \sum_{x \in (I+\widehat{f})^{-1}(y)} \operatorname{sgn} \det \begin{bmatrix} J_N(x) & 0 \\ 0 & I_k \end{bmatrix}$$
$$= d(I + f, U \cap \mathbb{R}^N, y).$$

Now consider the general case. For $i \in \{N+1, \ldots, N+k\}$, let $\widehat{f}_i \equiv 0$ and for $i \in \{1, \ldots, N\}$ choose $g_i \in C(\overline{U}, \mathbb{R})$ such that if $\widehat{g} = (g_i)_{i=1}^{N+k}$, then

$$\|\widehat{f}(x) - \widehat{g}(x)\| < d\big(y, \varphi(\partial U)\big) \qquad \text{for all } x \in \overline{U},$$
$$\Rightarrow\ \|\varphi(x) - \widehat{\psi}(x)\| < d\big(y, \varphi(\partial U)\big) \qquad \text{for all } x \in \overline{U}, \tag{3.59}$$

where $\widehat{\psi} = I + \widehat{g}$. Let $\psi = \widehat{\psi}|_{\overline{U} \cap \mathbb{R}^N}$.

Because $\widehat{\psi}^{-1}(y) \subseteq U \cap \mathbb{R}^N$ and $\psi(C_\psi)$ has $\mathbb{R}^N$-measure zero, by making a small translation on $\widehat{\psi}$ we can guarantee that in addition to (3.59), we have $y \notin \psi(C_\psi)$. Then

$$d(\varphi, U, y) = d(\psi, U, y)$$

and so the result of the lemma follows from the first part of the proof. $\square$

Using Lemma 3.3.36, we can show that the definition given in (3.58) is correct.

PROPOSITION 3.3.37 *For $\varepsilon > 0$ small $(\varepsilon < \vartheta = d\big(y, \varphi(\partial U)\big))$, $d(\varphi_\varepsilon, U \cap Y_\varepsilon, y)$ is independent of the ε-approximation $f_\varepsilon \in \mathcal{K}(\overline{U}, X)$ of f.*

PROOF: First we show that $d(\varphi_\varepsilon, U \cap Y_\varepsilon, y)$ is independent of the basis chosen in Y_ε. To see this, first assume that $\varphi_\varepsilon \in C^1(\overline{U \cap Y_\varepsilon}, Y_\varepsilon)$. Note that at a point the sign of the Jacobian determinant is invariant under a change of a basis in Y_ε. So the claim follows from Definition 3.3.2. The general case follows as before by approximating with C^1-functions.

Next using Lemma 3.3.36 we show that $d(\varphi_\varepsilon, U \cap Y_\varepsilon, y)$ stabilizes for $\varepsilon > 0$ small. To this end let $\varphi_\delta = I + f_\delta$ be the δ-approximation of φ and as before.

$$Y_\delta = \operatorname{span}\{\mathcal{R}(f_\delta) \cup \{y\}\}.$$

Let $Y = \operatorname{span}\{Y_\varepsilon \cup Y_\delta\}$. Then by virtue of Lemma 3.3.36 for $\varepsilon, \delta > 0$ small, we have

$$d(\varphi_\varepsilon, U \cap Y_\varepsilon, y) = d(\varphi_\varepsilon, U \cap Y, y) \tag{3.60}$$

$$\text{and} \quad d(\varphi_\delta, U \cap Y_\delta, y) = d(\varphi_\delta, U \cap Y, y). \tag{3.61}$$

Consider the homotopy $h(t, x) = t\varphi_\varepsilon(x) + (1 - t)\varphi_\delta(x)$ for all $(t, x) \in [0, 1] \times (U \cap Y)$. Evidently

$$y \notin h\big(t, \partial(U \cap Y)\big) \qquad \text{for all } t \in [0, 1]$$

and so from the homotopy invariance of Brouwer's degree, we have

$$d(\varphi_\varepsilon, U \cap Y, y) = d(\varphi_\delta, U \cap Y, y). \tag{3.62}$$

Comparing (3.60) through (3.62), we see that

$$d(\varphi_\varepsilon, U \cap Y_\varepsilon, y) = d(\varphi_\delta, U \cap Y_\delta, y) \qquad \text{for } \varepsilon, \delta > 0 \text{ small.}$$

□

This proposition leads to the following infinite-dimensional degree map.

DEFINITION 3.3.38 Let X be a Banach space, $U \subseteq X$ bounded open, $\varphi = I - f$ with $f \in \mathcal{K}(\overline{U}, X)$, and $y \notin \varphi(\partial U)$. We introduce $f_\varepsilon \in \mathcal{K}_f(\overline{U}, X)$ such that

$$\|f(x) - f_\varepsilon(x)\| < d\big(y, \varphi(\partial U)\big) \qquad \text{for all } x \in \overline{U}.$$

Choose a finite-dimensional subspace Y_ε of X such that $Y_\varepsilon \supseteq f_\varepsilon(\overline{U})$ and $y \in Y_\varepsilon$. We set $U_\varepsilon = U \cap Y$, $\varphi_\varepsilon = I\big|_{Y_\varepsilon} - f_\varepsilon$ and we define

$$\widehat{d}(\varphi, U, y) = d(\varphi_\varepsilon, U_\varepsilon, y).$$

REMARK 3.3.39 It is clear from the above definition that U need not be bounded. We only need that $U \cap V$ is bounded for every finite-dimensional subspace V of Y. So we only need to know that U is finitely bounded. The degree $\widehat{d}$ is known as the *Leray–Schauder degree*.

So from Definition 3.3.38 and the corresponding properties of Brouwer's degree we obtain the following theorem.

THEOREM 3.3.40 *If X is a Banach space, $U \subseteq X$ is bounded open, $\varphi = I - f$ with $f \in \mathcal{K}(\overline{U}, X)$, and $y \notin \varphi(\partial U)$, then*

(a) $\widehat{d}(I, U, y) = 1$ *for all* $y \in U$ *and* $\widehat{d}(I, U, y) = 0$ *if* $y \notin \overline{U}$.

(b) Additivity with respect to the domain: *If U_1, U_2 are disjoint open subsets of U such that $y \notin \varphi\big(\overline{U} \setminus (U_1 \cup U_2)\big)$, then*

$$\widehat{d}(I - f, U, y) = \widehat{d}(I - f, U_1, y) + \widehat{d}(I - f, U_2, y).$$

(c) Homotopy invariance: *If $h\colon [0,1] \times \overline{U} \longrightarrow X$ is compact, $y\colon [0,1] \longrightarrow X$ is continuous and $y(t) \notin \big(I - h(t,\cdot)\big)(\partial U)$ for all $t \in T$, then $\widehat{d}\big(I - h(t,\cdot), U, y(t)\big)$ is independent of $t \in [0,1]$.*

(d) Dependence on the boundary values: *If $\psi = I - g$ with $g \in \mathcal{K}(\overline{U}, X)$ and $\varphi = \psi$ on ∂U, then $\widehat{d}(I - f, U, y) = \widehat{d}(I - g, U, y)$.*

(e) Excision property: *If $K \subseteq \overline{U}$ is closed and $y \notin \varphi(K)$, then $\widehat{d}(\varphi, U, y) = \widehat{d}(\varphi, U \setminus K, y)$.*

(f) Continuity with respect to φ: *If $\psi = I - g$ with $g \in \mathcal{K}(\overline{U}, X)$ and $\|\varphi(x) - \psi(x)\| < d\big(y, \varphi(\partial U)\big)$ for all $x \in \overline{U}$, then $y \notin \psi(\partial U)$ and $\widehat{d}(\varphi, U, y) = \widehat{d}(\psi, U, y)$.*

(g) Existence property: *$\widehat{d}(\varphi, U, y) \neq 0$ implies that there exists $x \in U$ such that $x - f(x) = y$.*

(h) Dependence on y: *$\widehat{d}(\varphi, U, \cdot)$ is constant on the connected components of $X \setminus \varphi(\partial U)$.*

(i) *If $u \in X$, then $\widehat{d}(\varphi, U, y) = \widehat{d}(\varphi - u, U, y - u)$.*

Next as we did for Brouwer's degree, we present some topological applications of the Leray–Schauder degree. We start with the infinite-dimensional version of Borsuk's theorem (see Theorem 3.3.28).

THEOREM 3.3.41 *If X is a Banach space, $U \subseteq X$ is bounded, open, and symmetric (i.e., $U = -U$) with $0 \in U$ $\varphi = I - f$ with $f \in \mathcal{K}(\overline{U}, X)$, $0 \notin \varphi(\partial U)$, and $\varphi(-x) \neq \lambda \varphi(x)$ for all $\lambda \geq 1$ and all $x \in \partial U$, then $\widehat{d}(I - f, U, 0)$ is odd and in particular if $f\big|_{\partial U}$ is odd, then $\widehat{d}(I - f, U, 0)$ is odd.*

PROOF: Consider the compact homotopy

$$h(t,x) = \frac{1}{1+t} f(x) - \frac{t}{1+t} f(-x) \qquad \text{for all } (t,x) \in [0,1] \times \overline{U}.$$

Because by hypothesis $\varphi(-x) \neq \lambda \varphi(x)$ for all $\lambda \geq 1$ and all $x \in \partial U$, it follows that $x \neq h(t,x)$ for all $(t,x) \in [0,1] \times \partial U$. Thus by virtue of Theorem 3.3.40(c) we have

$$\widehat{d}(I - f, U, 0) = \widehat{d}(I - g, U, 0)$$

where $g(x) = h(1,x) = \frac{1}{2}\big(f(x) - f(-x)\big)$. Note that g is the odd part of the map f. We choose $g_1 \in \mathcal{K}_f(\overline{U}, X)$ such that $\|g_1(x) - g(x)\| < d\big(0, \psi(\partial U)\big)$ where $\psi = I - g$ and let $g_2(x) = \frac{1}{2}\big(g_1(x) - g_1(-x)\big)$ (the odd part of g_1). Then $g_2 \in \mathcal{K}_f(\overline{U}, X)$ is odd and

$$\sup_{x\in\overline{U}}\|g_2(x)-g(x)\| \leq \sup_{x\in\overline{U}}\frac{1}{2}\|g_1(x)-g(x)\| + \sup_{x\in\partial U}\frac{1}{2}\|g_1(-x)+g(x)\|$$
$$= \sup_{x\in\overline{U}}\frac{1}{2}\|g_1(x)-g(x)\| + \sup_{x\in\overline{U}}\frac{1}{2}\|g_1(-x)-g(-x)\|,$$
$$\text{(because } g \text{ is odd)}$$
$$= \sup_{x\in\overline{U}}\|g_1(x)-g(x)\| < d\big(0,\psi(\partial U)\big).$$

Therefore by virtue of Definition 3.3.38, we have

$$\widehat{d}(I-g,U,0)=\widehat{d}(I-g_2,U\cap Y_2,0)=\text{odd} \qquad \text{(by Theorem 3.3.28).}$$

□

Also with a proof similar to that of Theorem 3.3.34, we can have the following *invariance of domain theorem.*

THEOREM 3.3.42 *If X is a Banach space, $U\subseteq X$ is open, and $\varphi=I-f$ with $f\in\mathcal{K}(\overline{U},X)$ and it is locally one-to-one, then φ is an open map.*

The next result is an infinite-dimensional generalization of Lemma 3.3.36 and provides one more property of the Leray–Schauder degree, namely the *reduction property*.

THEOREM 3.3.43 *If X is a Banach space, $U\subseteq X$ is bounded open, Y is a closed linear subspace of X, $f\in\mathcal{K}(\overline{U},X)$, $\varphi=I-f$, $y\in Y$, and $y\notin\varphi(\partial U)$, then $\widehat{d}(\varphi,U,y)=\widehat{d}(\varphi|_{\overline{U\cap Y}},U\cap Y,y)$.*

PROOF: By hypothesis $\vartheta=d\big(y,\varphi(\partial U)\big)>0$. We can find $g\in\mathcal{K}_f(\overline{U},X)$ such that

$$\sup_{x\in\overline{U}}\|g(x)-f(x)\|<\vartheta.$$

Let Y_1 be a finite-dimensional linear subspace of X such that $g(\overline{U})\subseteq Y_1$, $y\in Y$, and let $U_0=U\cap Y$ and $U_1=U\cap Y_1$. Because $\partial U_1\subseteq\partial U$, we have $\sup_{x\in\overline{U}_0}\|g(x)-f(x)\|<d^*\big(y,\varphi(\partial U_0)\big)$. Hence by Lemma 3.3.36 and Definition 3.3.38, we have

$$\widehat{d}(\varphi,U,y)=d\big((I-g)|_{\overline{U}_1},U_1,y\big)=\widehat{d}(\varphi|_{\overline{U}_0},U_0,y).$$

□

The next theorem, known as *Birkoff–Kellogg theorem*, is characteristic of infinite-dimensional spaces and it is no longer true in a finite-dimensional setting.

THEOREM 3.3.44 *If X is an infinite-dimensional Banach space, $U\subseteq X$ is bounded open with $0\in U$, and $f\in\mathcal{K}(\overline{U},X)$ satisfies $\inf\big[\,\|f(x)\|:x\in\partial U\big]>0$, then there exists $x\in\partial U$ such that $\varphi(x)=\mu x$ for some $\mu>0$ (i.e., φ has an invariant direction on ∂U).*

PROOF: We argue indirectly. So suppose that $\mu x \neq \varphi(x)$ for all $\mu>0$ and all $x \in \partial U$. Let $r = \inf_{x\in\partial U} \|\varphi(x)\| > 0$. The set $K = \overline{\varphi(\overline{U})}$ is compact, and the ball $B_{r/2}(0)$ is noncompact (since X is infinite dimensional). So we can find $y_0 \in B_{r/2}(0) \setminus K$. For each $v \in B_{r/2}(0) \setminus \{y_0\}$ we can write uniquely $v = (1-t)y_0 + t\widehat{v}$ with $t \in (0,1)$ and $\widehat{v} \in \partial B_{r/2}(0)$. Then we define

$$h(v) = \begin{cases} 0 & \text{if } v = y_0 \\ t\widehat{v} & \text{if } v \in B_{r/2}(0)\setminus\{y_0\} \\ v & \text{if } v \in X\setminus B_{r/2}(0) \end{cases} .$$

Evidently this is a homeomorphism on X and on $B_{r/2}(0)$, which is the identity on $X\setminus B_{r/2}(0)$ and sends y_0 to zero. Let $g = h \circ f$. Then $g \in \mathcal{K}(\overline{U}, X)$ and $g\big|_{\partial U} = f\big|_{\partial U}$. Moreover, by hypothesis $\varphi(x) = x - f(x) \neq 0$ for all $x \in \partial U$. So by Theorem 3.3.40(d) we have $\widehat{d}(\varphi, U, 0) = \widehat{d}(\psi, U, 0)$, where $\varphi = I - f$, $\psi = I - g$.

Choose $R>0$ large such that $\overline{U} \subseteq B_R(0)$ and let $g_1(x) = R\big(g(x)\big)/\|g(x)\|$, $x \in \overline{U}$. Clearly $g_1 \in \mathcal{K}(\overline{U}, X)$ and $g_1(\overline{U}) \cap \overline{U} = \emptyset$. So $\widehat{d}(\psi_1, U, 0) = 0$ for $\psi_1 = I - g_1$. Moreover, if $h(t,x) = tR\big(g(x)\big)/\|g(x)\|$, for all $t \in [0,1]$, $x \in \overline{U}$, then h is a compact homotopy and so

$$1 = \widehat{d}\big(I - h(0,\cdot), U, 0\big) = \widehat{d}\big(I - h(1,\cdot), U, 0\big) = \widehat{d}\big(\psi_1, U, 0\big) = 0,$$

a contradiction. This proves that f has an invariant direction on ∂U. □

EXAMPLE 3.3.45 As we already said Theorem 3.3.44 is no longer true if $\dim X < +\infty$. Indeed, let $X = \mathbb{R}^N$, $U = B_1(0)$, and $f\colon \overline{B}_1(0) \longrightarrow \mathbb{R}^N$ defined by $\varphi(x) = -x$. Then all the hypotheses of Theorem 3.3.44 are satisfied, but there is no invariant direction for φ on $\partial B_1(0)$. Because if for some $\mu>0$ and some $x \in \partial B_1(0)$ we have $\mu x = \varphi(x)$, then $(\mu+1)x = 0$ and so $x = 0$, a contradiction.

Finally there is a multiplication formula for the Leray–Schauder degree analogous to the one for Brouwer's degree which was stated in Theorem 3.3.35.

THEOREM 3.3.46 *If X is a Banach space, $U \subseteq X$ is bounded open, $f \in \mathcal{K}(\overline{U}, X)$, $\varphi = I - f$, V is a bounded open set containing $\varphi(U)$, $\{D_k\}_{k\geq 1}$ are the components of $V \setminus \varphi(\partial U)$, $g \in \mathcal{K}(\overline{V}, X)$, $\psi = I - g$, and $y \notin \big(\psi(\varphi(\partial U)\big) \cup \psi(\partial U)$, then $\widehat{d}(\psi\circ\varphi, U, y) = \sum_{k\geq 1} d(\psi, D_k, y)\, d(\varphi, U, z_k)$ where z_k is any point in D_k.*

For Brouwer's degree we consider the triplets

$$\mathcal{S}_B = \big\{(\varphi, U, y) : U \subseteq \mathbb{R}^N,\ \varphi \in C(\overline{U}, \mathbb{R}^N),\ y \notin \varphi(\partial U)\big\},$$

whereas for the Leray–Schauder degree the admissible triplets are

$$\mathcal{S}_{LS} = \big\{(\varphi, U, y) : U \subseteq X \text{ bounded open, } \varphi = I - f \text{ with } f \in \mathcal{K}(\overline{U}, X),\ y \notin \varphi(\partial U)\big\}.$$

We look for $\mathbb{Z}$-valued functions on $\mathcal{S}_B$ and $\mathcal{S}_{LS}$, respectively, that satisfy the normalization, additivity, and homotopy invariance properties.

THEOREM 3.3.47 (a) *The Brouwer's degree is the unique topological degree on* $\mathcal{S}_B$.
(b) *The Leray–Schauder degree is the unique topological degree on* $\mathcal{S}_{LS}$.

Next we extend the Leray–Schauder degree to cover wider classes of maps of the form $I - f$. We require that f exhibit some kind of set-contracting. For this reason we introduce the following notion.

DEFINITION 3.3.48 Let X be a Banach space and $\mathcal{B}$ the family of bounded subsets of X.

(a) The *Kuratowski measure of noncompactness* $\alpha : \mathcal{B} \longrightarrow \mathbb{R}_+$ is defined by

$$\alpha(B) = \inf\{d{>}0 : B \text{ admits a finite cover by sets of diameter} \leq d\}.$$

(b) The *Hausdorff or ball measure of noncompactness* $\beta : \mathcal{B} \longrightarrow \mathbb{R}_+$ is defined by

$$\beta(B) = \inf\{r{>}0 : B \text{ admits a finite cover by balls of radius } r\}.$$

REMARK 3.3.49 It is clear from the above definition that $\beta(B) \leq \alpha(B) \leq 2\beta(B)$ for all $B \in \mathcal{B}$.

PROPOSITION 3.3.50 *If X is a Banach space and $\gamma : \mathcal{B} \longrightarrow \mathbb{R}_+$ is either α or β (see Definition 3.3.48), then*

(a) $A \subseteq B \Rightarrow \gamma(A) \leq \gamma(B)$.

(b) $\gamma(A) = \gamma(\overline{A})$.

(c) $\gamma(A) = 0$ *if and only if* $\overline{A}$ *is compact.*

(d) $\gamma(A \cup B) = \max\{\gamma(A), \gamma(B)\}$, $\gamma(A + B) \leq \gamma(A) + \gamma(B)$ *and* $\gamma(tA) = |t|\gamma(A)$, $t \in \mathbb{R}$*:*

(e) $\gamma(A) = \gamma(\text{conv}A)$.

(f) *If* $\{A_n\}_{n\geq 1}$ *is a decreasing sequence of nonempty closed sets in* $\mathcal{B}$ *such that* $\gamma(A_n) \longrightarrow 0$ *as* $n \to \infty$, *then* $A = \bigcap_{n\geq 1} A_n$ *is nonempty and compact.*

PROOF: (a) If $A \subseteq B$, then any cover of B is also a cover of A and so we have the monotonicity property of γ.

(b) From (a) we have $\gamma(A) \leq \gamma(\overline{A})$. If $\{U_n\}_{n=1}^N$ is a cover of A, then $\{\overline{U}_n\}_{n=1}^N$ is a cover of $\overline{A}$. So because diam U_n = diam $\overline{U}_n$, directly from Definition 3.3.48 it follows that $\gamma(\overline{A}) \leq \gamma(A)$. Therefore $\gamma(A) = \gamma(\overline{A})$.

(c) Clear from Definition 3.3.48 and the definition of compactness.

(d) Evidently $\gamma(A \cup B) \leq \max\{\gamma(A), \gamma(B)\}$. The opposite inequality is true by (a), therefore equality must hold. The other two relations are direct consequences of Definition 3.3.48.

(e) It is enough to show that $\gamma(\text{conv } A) \leq \gamma(A)$, because the opposite inequality is always true by part (a). Let $\xi > \gamma(A)$ and $A \subseteq \bigcup_{n=1}^{N} C_n$ where diam $C_n \leq \xi$ if $\gamma = \alpha$ and $C_n = B_\xi(x_n)$ if $\gamma = \beta$. Since diam(conv $C_n) \leq \xi$ and $B_\xi(x_n)$ is convex, without any loss of generality we may assume that C_n is convex. Note that

$$\begin{aligned} \text{conv } A &\subseteq \text{conv}\Big[C_1 \cup \text{conv}\big(\bigcup_{n=2}^{N} C_n\big)\Big] \\ &= \text{conv}\Big[C_1 \cup \text{conv}\big[C_2 \cup \text{conv}\big(\bigcup_{n=3}^{N} C_n\big)\big]\Big] \subseteq \dots . \end{aligned}$$

So it suffices to show that for convex sets C_1, C_2, we have

$$\gamma\big(\text{conv}(C_1 \cup C_2)\big) \leq \max\big\{\gamma(C_1), \gamma(C_1)\big\}.$$

Note that $\text{conv}(C_1 \cup C_2) \subseteq \bigcup_{0 \leq t \leq 1} \big[tC_1 + (1-t)C_2\big]$ and because $C_1 - C_2$ is bounded we can find $r > 0$ such that $C_1 - C_2 \subseteq \overline{B}_r(0)$. Given $\varepsilon > 0$, let $\{t_k\}_{k=1}^m \subseteq [0,1]$ such that $[0,1] \subseteq \overset{m}{\underset{k=1}{\cup}} B_{\frac{\varepsilon}{r}}(t_k)$ and so

$$\begin{aligned} &\text{conv}(C_1 \cup C_2) \subseteq \overset{m}{\underset{k=1}{\cup}} \big[t_k C_1 + (1-t_k)C_2 + \overline{B}_\varepsilon(0)\big] \\ \Rightarrow\ &\gamma\big(\text{conv}(C_1 \cup C_2)\big) \leq \max\big\{\gamma(C_1), \gamma(C_1)\big\} + 2\varepsilon \quad \text{(see part (d))}. \end{aligned}$$

Let $\varepsilon \downarrow 0$ to finish the proof.

(f) Let $x_n \in A_n$, $n \geq 1$. Then $\gamma\big(\{x_n\}_{n \geq k}\big) \leq \gamma(A_k) \longrightarrow 0$ as $k \to \infty$. So $x_n \longrightarrow x$ in X and $x \in A \neq \emptyset$. Moreover, $\gamma(A) \leq \gamma(A_n) \longrightarrow 0$ as $n \to \infty$, hence $\gamma(A) = 0$; that is, A being closed is compact; see part (c). ☐

REMARK 3.3.51 Part (f) of the above theorem generalizes a well-known theorem of Cantor which characterizes complete metric spaces.

Using the measures of noncompactness, we can introduce some important classes of nonlinear operator that generalize the compact ones.

DEFINITION 3.3.52 Let X be a Banach space, $C \subseteq X$, and $f: C \longrightarrow X$ a continuous map.

(a) We say that f is γ-*Lipschitz* if there is a $k \geq 0$ such that

$$\gamma\big(f(A)\big) \leq k\gamma(A) \qquad \text{for all } A \subseteq X \text{ bounded.}$$

(b) If f is γ-Lipschitz with constant $k < 1$, then we say that f is a γ-*contraction*. The family of such maps is denoted by $\mathcal{SC}_\gamma(C)$.

(c) We say that f is γ-*condensing* if $\gamma\big(f(A)\big) < \gamma(A)$ for all $A \subseteq C$ bounded with $\gamma(A) > 0$. The family of such maps is denoted by $\mathcal{C}_\gamma(C)$.

REMARK 3.3.53 Evidently γ-Lipschitz and γ-condensing maps are bounded (i.e., map bounded sets to bounded sets). In addition γ-contractions and γ-condensing maps are proper (hence closed). Note that $\mathcal{SC}_\gamma(C) \subseteq \mathcal{C}_\gamma(C)$ and compact maps are of course γ-Lipschitz with constant $k = 0$.

Let X be a Banach space and $U \subseteq X$ bounded, open. We fix $k<1$ and define a degree for the class

$$\Sigma_k(\overline{U}) = \{\varphi = I - f : f \in \mathcal{SC}_\gamma(\overline{U})\}.$$

We set $D_0 = \overline{U}$ and define inductively:

$$D_n = \overline{\text{conv}}\, f(D_{n-1} \cap \overline{U}) \qquad \text{for all } n \geq 1.$$

Clearly $\{D_n\}_{n\geq 1}$ is a decreasing sequence of closed convex sets. We set

$$D = \bigcap_{n \geq 1} D_n.$$

Evidently D is closed and convex. Also using Proposition 3.3.50, we have

$$\begin{aligned} \gamma(D_n) &= \gamma\big(\overline{\text{conv}}\, f(D_{n-1} \cap \overline{U})\big) = \gamma\big(f(D_{n-1} \cap \overline{U})\big) \\ &\leq \gamma\big(f(D_{n-1})\big) \leq k\gamma(D_{n-1}), \\ \Rightarrow\ \gamma(D_n) &\leq k^n \gamma(\overline{D}). \end{aligned}$$

Because $k<1$, we see that $\gamma(D_n) \longrightarrow 0$ and so by virtue of Proposition 3.3.50(f), D is nonempty and compact. Using Theorem 3.1.13 we can find $\widehat{f} : \overline{U} \longrightarrow D$ which is continuous and $\widehat{f}\big|_{D\cap\overline{U}} = f$. Because the range of $\widehat{f}$ is compact and if we assume that $0 \notin \varphi(\partial U)$, then we can define $\widehat{d}(I - \widehat{f}, U, 0)$. In fact we can show that this degree is actually independent of the extension $\widehat{f}$ of f that we use. Indeed, let $\widetilde{f} : \overline{U} \longrightarrow D$ be another continuous map such that $\widetilde{f}\big|_{D\cap U} = f$. Consider the compact homotopy

$$h(t,x) = t\widehat{f}(x) + (1-t)\widetilde{f}(x) \qquad \text{for all } t \in [0,1] \text{ and all } x \in \overline{U}.$$

Because $\widehat{f}$ and $\widetilde{f}$ both map $\overline{U}$ in D, then $h(t,x) = x$ only if $x \in D$. But $\widehat{f} = \widetilde{f} = f$ on D. So $h(t,x) = x$ only if $f(x) = x$. Therefore $x \neq h(t,x)$ for all $(t,x) \in [0,1] \times \partial U$ and we can use the homotopy invariance of the Leray–Schauder degree to conclude that $\widehat{d}(I - \widetilde{f}, U, 0) = \widehat{d}(I - \widehat{f}, U, 0)$. This then makes the next definition meaningful.

DEFINITION 3.3.54 Let X be a Banach space, $U \subseteq X$ bounded open, $\varphi \in \Sigma_k(\overline{U})$ $(k < 1)$, and $0 \notin \varphi(\partial U)$. We define the *degree* $\widehat{d}_S(\varphi, U, 0)$ to be the *Leray–Schauder degree* $\widehat{d}(\widehat{\varphi}, U, 0)$ where $\widehat{\varphi} = I - \widehat{f}$ and $\widehat{f} : \overline{U} \longrightarrow D$ is any extension of $f\big|_{D\cap\overline{U}}$ (see the previous arguments). If $y \notin \varphi(\partial U)$, then $\widehat{d}_S(\varphi, U, y)$ is defined to be equal to $\widehat{d}_S(\varphi - y, U, 0)$.

This definition defines a topological degree, in the sense that it satisfies the normalization property (for the identity), the additivity with respect to the domain and the homotopy invariance property. The admissible homotopies in this case are $I - h(t, \cdot)$, where $h \in C([0,1] \times \overline{U}, X)$ and $\gamma\big(h(t,B)\big) \leq k\gamma(B)$ for all $t \in [0,1]$ and all $B \subseteq \overline{U}$. In fact exploiting the uniqueness of the Leray–Schauder degree (see Theorem 3.3.47(c)), we can have the following result.

THEOREM 3.3.55 *The degree $\widehat{d}_S(\varphi, U, 0)$ is the unique topological degree defined on $\Sigma_k(\overline{U})$.*

Next, we extend the definition of degree to cover the class

$$\widehat{\Sigma}(\overline{U}) = \{\varphi = I - f : f \in \mathcal{C}_\gamma(\overline{U})\}.$$

But this is done easily by approximating the condensing map f with γ-contractions. To this end let $\varphi \in \widehat{\Sigma}(U)$. Then $\varphi = I - f$ with $f \in \mathcal{C}_\gamma(\overline{U})$. Suppose that $0 \notin \varphi(\partial U) > 0$. Set $\vartheta = d\big(0, \varphi(\partial U)\big) > 0$ (note that $\varphi(\partial U)$ is a closed set because φ is a closed map; see Remark 3.3.53). We choose $f_0 \in \mathcal{SC}_\gamma(\overline{U})$ such that $\|f(x) - f_0(x)\| < \vartheta/3$ for all $x \in \overline{U}$. Such a choice is possible, because we can take $f_0 = \lambda f$ where $\lambda \in (0,1)$ with $1-\lambda$ sufficiently small. Let $\varphi_0 = I - f_0$. Then $\varphi_0(\partial U)$ is closed and $d\big(0, \varphi_0(\partial U)\big) \geq \frac{2}{3}\vartheta$. Suppose that f_1 is another element in $\mathcal{SC}_\gamma(\overline{U})$ such that $\|f(x) - f_1(x)\| < \vartheta/3$ for all $x \in \overline{U}$. Suppose f_0 is a γ-contraction with constant $k_0 < 1$ and f_1 is a γ-contraction with constant $k_1 < 1$. If $k = \max\{k_0, k_1\}$, then both f_0 and f_1 are γ-contractions with constant k. Consider the homotopy

$$h(t,x) = tf_0(x) + (1-t)f_1(x) \qquad \text{for all } (t,x) \in [0,1] \times \overline{U}.$$

If $h(t,x) = x$ with $(t,x) \in [0,1] \times \partial U$, then because $\|f_0(x) - f_1(x)\| < \frac{2}{3}\vartheta$ for all $x \in \overline{U}$, we have

$$\|x - f_1(x)\| < \frac{2}{3}\vartheta,$$

a contradiction to the fact that $\|x - f_1(x)\| \geq \frac{2}{3}\vartheta$. So $x \neq h(t,x)$ for all $(t,x) \in [0,1] \times \partial U$. Invoking the homotopy invariance of the degree $\widehat{d}_S$, we have

$$\widehat{d}_S(I - f_0, U, 0) = \widehat{d}_S(I - f_1, U, 0).$$

This then justifies the following definition.

DEFINITION 3.3.56 Let X be a Banach space, $U \subseteq X$ bounded open, $\varphi \in \widehat{\Sigma}(\overline{U})$, and $0 \notin \varphi(\partial U)$. We define the *degree* $\widehat{d}_0(\varphi, U, 0)$ to be *equal to* $\widehat{d}_S(I - \overline{f}, U, 0)$ where $\overline{f}$ is any γ-contraction such that $\|f(x) - \overline{f}(x)\| < \delta$ for all $x \in \overline{U}$, where $\delta > 0$ is small enough. If $y \notin \varphi(\partial U)$, then we define $\widehat{d}_0(\varphi, U, y) = \widehat{d}_0(\varphi - y, U, 0)$.

REMARK 3.3.57 In this definition we may take $\delta = \frac{1}{3}\, d\big(0, \varphi(\partial U)\big)$.

Again this definition defines a topological degree, in the sense that it satisfies the normalization property (for the identity), the additivity with respect to the domain, and the homotopy invariance property. The admissible homotopies in this case are $I - h(t,\cdot)$, where $h \in C([0,1] \times \overline{U}, X)$ and $\gamma\big(h(t,B)\big) < \gamma(B)$ for all $t \in [0,1]$ and all $B \subseteq \overline{U}$. Moreover, Theorem 3.3.55 leads to the following result.

THEOREM 3.3.58 *The degree $\widehat{d}_0(\varphi, U, 0)$ is the unique topological degree on $\widehat{\Sigma}(\overline{U})$.*

REMARK 3.3.59 The degree $\widehat{d}_0$ is usually called in the literature the *Nussbaum–Sadovskii degree*.

It is clear from the presentation of the Nussbaum–Sadovskii degree, that the basic task of finding degree theorems more general than the Leray–Schauder degree, is therefore in part a problem of characterizing suitable classes of nonlinear maps that have enough structure and are sufficiently narrowed down from the broad class of all continuous maps so that a degree theory can exist. In the sequel we focus on maps $f \colon \overline{U} \subseteq X \longrightarrow X^*$, where X is a reflexive Banach space and X^* is its topological dual. Such an extension of degree theory permits the introduction of degree-theoretic methods to nonlinear boundary value problems.

First we present an extension of the degree to operators of type $(S)_+$ (see Definition 3.2.61(d)). Note that if $X = X^*$ is a Hilbert space, the mappings of type $(S)_+$ contain as a special case the Leray–Schauder maps $I - f$, with f compact. This follows from the fact that each strongly monotone map is of type $(S)_+$ and the class $(S)_+$ is closed under compact perturbations.

So let X be a reflexive Banach space with X^* being its dual and $\langle \cdot, \cdot \rangle$ denoting the duality brackets for the pair (X, X^*). Let $U \subseteq X$ be bounded open and $f \colon \overline{U} \longrightarrow X^*$ a demicontinuous map of type $(S)_+$ (see Definition 3.2.61(d)). Let $\{X_\alpha\}_{\alpha \in J}$ be the family of all finite-dimensional subspaces of X such that $U_\alpha = U \cap X_\alpha \neq \emptyset$. We order J by inclusion. Also by i_α we denote the injection of X_α into X and by i_α^* the corresponding projection of X^* onto X_α^*. In addition in order to simplify things, we assume that the reference point $y^* \in X^*$ is the origin. We can always do this without any loss of generality, because we can replace f by $f - y^*$. We consider f_α the Galerkin approximation of f with respect to X_α; that is,

$$\langle f_\alpha(x), y \rangle_{X_\alpha} = \langle f(x), y \rangle \qquad \text{for all } x \in \overline{U} \cap X_\alpha \text{ and all } y \in X_\alpha.$$

Here by $\langle \cdot, \cdot \rangle_{X_\alpha}$ we denote the duality brackets for the pair (X_α, X_α^*). Evidently $f_\alpha = i_\alpha^* \circ f \circ i_\alpha$.

An essential tool in producing a degree for $(S)_+$–maps is the following result due to Browder [122].

PROPOSITION 3.3.60 *If Y is a finite-dimensional subspace of X, a degree map $\deg_{(S)_+}(\cdot)$ is defined on the demicontinuous $(S)_+$-maps defined on the closures of bounded open sets U in X with values in X^*, the degree $\deg_{(S)_+}$ is invariant under affine homotopies (i.e., homotopies $h(t,x) = tf_0(x) + (1-t)f_1(x)$ with $f_0, f_1 \colon \overline{U} \longrightarrow X^*$ demicontinuous, $(S)_+$-maps) and it is normalized by a duality map $\mathcal{F}$ of type $(S)_+$; if $f \colon \overline{U} \longrightarrow X^*$ is such a demicontinuous, $(S)_+$-map, and f_Y is the Galerkin approximation of f with respect to Y, suppose that $\deg_{(S)_+}(f, U, 0)$ is not defined or $d(f_Y, U \cap Y, 0)$ is not defined or both are well defined but $\deg_{(S)_+}(f, U, 0) \neq d(f_Y, U \cap Y, 0)$, then there exists $x \in \partial U$ such that $\langle f(x), x \rangle \leq 0$ and $\langle f(x), y \rangle = 0$ for all $y \in Y$.*

REMARK 3.3.61 By virtue of Theorem 3.2.25, we can always equivalently renorm X so that both X and X^* are locally uniformly convex, in which case the corresponding duality map $\mathcal{F}$ is of type $(S)_+$.

Using Proposition 3.3.60 we can make the decisive step towards the stated goal of extending degree theory to demicontinuous $(S)_+$-maps from the closure of a bounded open set $U \subseteq X$ in a reflexive Banach space X into its dual X^*.

PROPOSITION 3.3.62 *If $\{f, U, X\}$ are as above and $0 \notin f(\partial U)$, then there exists $\alpha_0 \in J$ such that for all $\alpha \geq \alpha_0$ (i.e., $X_\alpha \supseteq X_{\alpha_0}$), we have $0 \notin f_\alpha(\partial U_\alpha)$ and $d(f_\alpha, U_\alpha, 0)$ is independent of $\alpha \in J$, $\alpha \geq \alpha_0$.*

PROOF: We argue indirectly. So suppose that for each $\alpha \in J$, we can find $\beta \geq \alpha$ (i.e., $X_\beta \supseteq X_\alpha$) such that

$$d(f_\beta, U_\beta, 0) \neq d(f_\alpha, U_\alpha, 0)$$

or one of the degrees is not defined. In all these cases we can apply Proposition 3.3.60 with X replaced by X_β and $Y = X_\alpha$. Evidently f_α is the Galerkin approximation of f_β with respect to X_α. Both X_β and X_α have degree functions defined (Brouwer's degree because they are finite-dimensional), thus we can apply Proposition 3.3.60 and obtain $x \in \partial U_\beta$ such that

$$\langle f(x), x\rangle = \langle f_\beta(x), x\rangle_{X_\beta} \leq 0 \qquad \text{and} \qquad \langle f(x), y\rangle = \langle f_\beta(x), y\rangle_{X_\beta} = 0$$

for all $y \in X_\alpha$. We set

$$V_\alpha = \big\{x \in \partial U : \langle f(x), x\rangle \leq 0 \text{ and } \langle f(x), y\rangle = 0 \quad \text{for all } y \in X_\alpha\big\}.$$

Evidently if $\alpha_1 \leq \alpha_2$, then $V_{\alpha_2} \subseteq V_{\alpha_1}$. So the family $\{V_\alpha\}_{\alpha \in J}$ is contained in a fixed bounded set ∂U and has the finite intersection property. The set $\overline{V}_\alpha^{\,w}$ is weakly compact in X and also $\bigcap_{\alpha \in J} \overline{V}_\alpha^{\,w} \neq \emptyset$. Let $x_0 \in \bigcap_{\alpha \in J} \overline{V}_\alpha^{\,w}$ and $y \in X$. We choose a finite-dimensional subspace X_α of X that contains both x_0 and y. Because $x_0 \in \overline{V}_\alpha^{\,w}$ and the space X is reflexive, we can find a sequence $\{v_n\}_{n \geq 1} \subseteq V_\alpha$ such that $v_n \xrightarrow{w} x_0$ in X (see Denkowski–Migórski–Papageorgiou [194, p. 306]). By definition

$$\langle f(v_n), v_n\rangle \leq 0 \quad \text{and} \quad \langle f(v_n), y\rangle = 0 \qquad \text{for all } y \in X_\alpha \text{ and all } n \geq 1.$$

So it follows that

$$\limsup_{n \to \infty} \langle f(v_n), v_n - x_0\rangle \leq 0.$$

Because f is of type $(S)_+$, it follows that $v_n \longrightarrow x_0$ in X and $x_0 \in \partial U$. Due to the demicontinuity of f we have $f(v_n) \xrightarrow{w} f(x_0)$ in X^*. Hence $\langle f(x_0), y\rangle = 0$ for all $y \in X$, so $f(x_0) = 0$, a contradiction to the hypothesis that $0 \notin f(\partial U)$. □

Let X be a reflexive Banach space and consider the family

$$\begin{aligned}\mathcal{S}_{(S)_+} = \{&(f, U, y^*) : U \subseteq X \text{ bounded, open, } f \colon \overline{U} \longrightarrow X^* \text{ demicontinuous,}\\ &(S)_+,\ y^* \notin f(\partial U)\}.\end{aligned}$$

Proposition 3.3.62 leads to a degree map defined on $\mathcal{S}_{(S)_+}$.

DEFINITION 3.3.63 We can define a *degree map* on $\mathcal{S}_{(S)_+}$ by setting

$$\deg_{(S)_+}(f, U, y^*) = d(f_\alpha, U \cap X_\alpha, y^*) \qquad \text{for } X_\alpha \text{ sufficiently large.}$$

REMARK 3.3.64 For this definition only we used a general reference point y^*. In the sequel we return to our initial simplifying convention that $y^* = 0$.

As admissible homotopies, we could use the affine homotopies $h(t,x) = tf_0(x) + (1-t)f_1(x)$ with $(t,x) \in [0,1] \times \overline{U}$ and $f_0, f_1 : \overline{U} \longrightarrow X^*$ demicontinuous $(S)_+$-maps. The degree map $\deg_{(S)_+}$ is invariant under these homotopies. However, affine homotopies are too weak for some important applications. For this reason we introduce the following broader class of homotopies.

DEFINITION 3.3.65 Let X be a reflexive Banach space and $U \subseteq X$ bounded open. The map $h\colon [0,1] \times \overline{U} \longrightarrow X^*$ is said to be *homotopy of class* $(S)_+$, if it satisfies the following conditions. For any sequence $\{t_n\}_{n\geq 1} \subseteq [0,1]$ converging to t and any sequence $\{x_n\}_{n\geq 1} \subseteq \overline{U}$ such that $x_n \xrightarrow{w} x$ in X for which

$$\limsup_{n\to\infty} \langle h(t_n, x_n), x_n - x\rangle \leq 0$$

we have that $x_n \longrightarrow x$ in X and $h(t_n, x_n) \xrightarrow{w} h(t,x)$ in X^*.

Using this broader class of admissible homotopies, we have the following.

THEOREM 3.3.66 *The degree map* $\deg_{(S)_+} : \mathcal{S}_{(S)_+} \longrightarrow \mathbb{Z}$ *defined in Definition 3.3.63 is the unique degree map such that*

(a) $\deg_{(S)_+}(\mathcal{F}, U, y^*) = 1$, *where* $\mathcal{F}\colon X \longrightarrow X^*$ *is the duality map corresponding to an equivalent norm on* X *for which both* X *and* X^* *are locally uniformly convex (see Theorem 3.2.25).*

(b) Additivity with respect to the domain: *If* U_1, U_2 *are disjoint open subsets of* U *and* $y^* \notin f\big(\overline{U} \setminus (U_1 \cup U_2)\big)$, *then* $\deg_{(S)_+}(\mathcal{F}, U, y^*) = \deg_{(S)_+}(\mathcal{F}, U_1, y^*) + \deg_{(S)_+}(\mathcal{F}, U_2, y^*)$.

(c) Homotopy invariance: *If* $h(t,x)$ *is a homotopy of type* $(S)_+$ *and* $y^* : I \longrightarrow X^*$ *is continuous such that* $y^*(t) \notin h(t, \partial U)$ *for all* $t \in [0,1]$, *then* $\deg_{(S)_+}\big(h(t,\cdot), U, y^*(t)\big)$ *is independent of* $t \in [0,1]$.

REMARK 3.3.67 Of course the degree map $\deg_{(S)_+}$ has other properties too, such as the excision property, the solution property, and the dependence on boundary values.

Combining the Galerkin approximations technique (see the proof of Proposition 3.3.62) with a corresponding finite-dimensional result (see Amann [18]), we can have the following theorem.

THEOREM 3.3.68 *If* X *is a reflexive Banach space,* $U \subseteq X$ *is open,* $F\colon U \longrightarrow \mathbb{R}$ *is Gâteaux differentiable with* $f = F' : U \longrightarrow X^*$ *a demicontinuous,* $(S)_+$*-map, and there exist* $a, b \in \mathbb{R}$, $a < b$, *and* $x_0 \in X$ *such that*

(i) $V = \{F < b\}$ *is bounded and* $\overline{V} \subseteq U$.
(ii) *If* $x \in \{F \leq a\}$, *then* $tx + (1-t)x_0 \in V$ *for all* $t \in [0,1]$.
(iii) $f(x) = F'(x) \neq 0$ *for all* $x \in \{a \leq F \leq b\}$,

then $\deg_{(S)_+}(f, V, 0) = 1$.

This theorem has some noteworthy consequences.

COROLLARY 3.3.69 *If X is a reflexive Banach space, $F\colon X \longrightarrow \mathbb{R}$ is a bounded, Gâteaux differentiable functional such that $f = F'\colon X \longrightarrow X^*$ is of type $(S)_+$, $F(x) \longrightarrow +\infty$ as $\|x\| \longrightarrow +\infty$, and there exists $r_0 > 0$ such that for all $\|x\| \geq r_0$ we have $f(x) \neq 0$, then there exists $r_1 \geq r_0$ such that $\deg_{(S)_+}\big(f, B_r(0), 0\big) = 1$ for all $r \geq r_1$.*

PROOF: Let $a = \sup\{F(x) : x \in B_{r_0}(0)\}$ and $r_1 = \sup\big\{\|x\| : x \in \{F \leq a\}\big\}$. Moreover, given $r \geq r_1$, let $b > \sup\{F(x) : x \in B_r(0)\}$. Then the corollary follows from Theorem 3.3.68 with $x_0 = 0$ and the excision property of the degree (see Remark 3.3.67). □

COROLLARY 3.3.70 *If X is a reflexive Banach space, $U \subseteq X$ is open convex, $F\colon U \longrightarrow \mathbb{R}$ is a C^1-functional on U such that $f = F'\colon U \longrightarrow X^*$ is of type $(S)_+$, and F has a local minimum at x_0 and also x_0 is an isolated critical point of F, then $\deg_{(S)_+}\big(f, B_r(x_0), 0\big) = 1$ for some $r > 0$.*

PROOF: First we show that F is weakly sequentially lower semicontinuous on U. We argue by contradiction. So suppose that we can find $\{x_n, x\}_{n \geq 1} \subseteq U$ such that $x_n \xrightarrow{w} x$ in X and

$$\liminf_{n \to \infty} F(x_n) < F(x). \tag{3.63}$$

We consider a subsequence $\{x_{n_k}\}_{k \geq 1}$ of $\{x_n\}_{n \geq 1}$ such that

$$\lim_{k \to \infty} F(x_{n_k}) = \liminf_{n \to \infty} F(x_n). \tag{3.64}$$

From the mean value theorem (see Proposition 1.1.6), we can find $t_{n_k} \in (0,1)$ such that

$$\begin{aligned} & F(x_{n_k}) - F(x) = \big\langle f\big(x + t_{n_k}(x_{n_k} - x)\big), x_{n_k} - x\big\rangle \\ \Rightarrow\ & \limsup_{k \to \infty} \big[t_{n_k}\big(F(x_{n_k}) - F(x)\big)\big] \\ =\ & \limsup_{k \to \infty} \big\langle f\big(x + t_{n_k}(x_{n_k} - x)\big), x + t_{n_k}(x_{n_k} - x) - x\big\rangle. \end{aligned} \tag{3.65}$$

From (3.63) and (3.64), we infer that

$$\limsup_{k \to \infty} \big[t_{n_k}\big(F(x_{n_k}) - F(x)\big)\big] \leq 0. \tag{3.66}$$

Note that $x + t_{n_k}(x_{n_k} - x) \xrightarrow{w} x$ in X. So from (3.65), (3.66), and the fact that f is of type $(S)_+$, we have that $x + t_{n_k}(x_{n_k} - x) \longrightarrow x$ in X. Because by hypothesis f is continuous, we obtain

$$\lim_{k \to \infty} \big[F(x_{n_k}) - F(x)\big] = \lim_{k \to \infty} \big\langle f\big(x + t_{n_k}(x_{n_k} - x)\big), x_{n_k} - x\big\rangle = 0,$$

a contradiction to the choice of $\{x_{n_k}\}_{k \geq 1}$. So indeed f is weakly sequentially lower semicontinuous on U.

By hypothesis x_0 is an isolated critical point of F and also a local minimizer, therefore we can find $r_0 > 0$ such that

$$F(x_0) < F(y) \quad \text{and} \quad f(y) = F'(y) \neq 0 \qquad \text{for all } y \in \overline{B_{r_0}(x_0)} \setminus \{x_0\}. \tag{3.67}$$

We show that for all $r \in (0, r_0)$, the following is true.

$$\inf\left[F(y) : y \in \overline{B_{r_0}(x_0)} \setminus B_r(x_0)\right] > F(x_0). \tag{3.68}$$

Suppose that (3.68) is not true. We can find $r > 0$ and $\{x_n\}_{n\geq 1} \subseteq \overline{B_{r_0}(x_0)} \setminus B_r(x_0)$ such that

$$F(x_n) \downarrow F(x_0) \ \text{ as } \ n \to \infty. \tag{3.69}$$

Due to reflexivity of the space, we may assume that $x_n \xrightarrow{w} x$ in X. Because F is weakly sequentially lower semicontinuous we have

$$F(x) \leq \liminf_{n\to\infty} F(x_n) = F(x_0) \ \ \text{(i.e., } x = x_0\text{)}.$$

By the mean value theorem we have

$$F(x_n) - F\Big(\frac{x_n + x_0}{2}\Big) = \Big\langle f\Big(t_n x_n + (1 - t_n)\frac{x_n + x_0}{2}\Big), \frac{x_n - x_0}{2}\Big\rangle$$

with $t_n \in (0, 1)$. Passing to the limit as $n \to \infty$ and using the weak sequential lower semicontinuity of F and (3.69), we obtain

$$\begin{aligned}
&\limsup_{n\to\infty} \Big\langle f\Big(t_n x_n + (1 - t_n)\frac{x_n + x_0}{2}\Big), x_n - x_0\Big\rangle \leq 0 \\
\Rightarrow\ &\limsup_{n\to\infty} \Big\langle f\Big(t_n x_n + (1 - t_n)\frac{x_n + x_0}{2}\Big), t_n x_n + (1 - t_n)\frac{x_n - x_0}{2} - x_0\Big\rangle \\
=\ &\limsup_{n\to\infty} \Big[\frac{1 + t_n}{2}\Big\langle f\Big(t_n x_n + (1 - t_n)\frac{x_n + x_0}{2}\Big), x_n - x_0\Big\rangle\Big] \leq 0.
\end{aligned} \tag{3.70}$$

Because f is of type $(S)_+$, from (3.70) it follows that

$$t_n x_n + (1 - t_n)\frac{x_n + x_0}{2} \longrightarrow x_0 \ \text{ in } X.$$

But

$$\|t_n x_n + (1 - t_n)\frac{x_n + x_0}{2} - x_0\| = (1 - t_n)\|\frac{x_n - x_0}{2}\| \geq \frac{r}{2}$$

for all $n \geq 1$, a contradiction. So (3.68) is valid.

Let

$$b = \inf\left[F(x) : x \in \overline{B_{r_0}(x_0)} \setminus B_{r_0/2}(x_0)\right] - F(x_0).$$

From (3.68) we see that $b > 0$. Also let

$$V = \big\{x \in B_{r_0/2}(x_0) : F(x) - F(x_0) < \mu\big\}.$$

Evidently V is nonempty and open. Fix $r \in (0, r_0/2)$ such that $\overline{B_r(x_0)} \subseteq V$. Choose $a \in \mathbb{R}$ satisfying

$$0 < a < \inf\left[F(x) : x \in B_{r_0}(x_0) \setminus B_r(x_0)\right] - F(x_0).$$

Note that

$$\big\{x \in B_{r_0}(x_0) : F(x) - F(x_0) \leq a\big\} \subseteq B_r(x_0) \subseteq \overline{B_r(x_0)} \subseteq V \subseteq \overline{V} \subseteq B_{r_0}(x_0).$$

So we can apply Theorem 3.3.68 with $U = B_{r_0}(x_0)$, F replaced by $F|_{B_{r_0}(x_0)} - F(x_0)$, and a, b as above. We obtain

$$\deg_{(S)_+}(f, V, 0) = 1.$$

By the excision property of the degree, we conclude that

$$\deg_{(S)_+}(f, B_r(x_0), 0) = 1.$$

$\square$

Finally let us extend the degree to maps of the form $f + A$, with f bounded, demicontinuous, $(S)_+$, and A is maximal monotone with $(0,0) \in \mathrm{Gr}\, A$. We assume that the reflexive Banach space is equipped with a norm such that both X and X^* are locally uniformly convex (see Theorem 3.2.25). Let $U \subseteq X$ be a bounded open set, $f : \overline{U} \longrightarrow X^*$ a bounded, demicontinuous, $(S)_+$-map and $A : \mathcal{D}(A) \subseteq X \longrightarrow 2^{X^*}$ a maximal monotone operator such that $(0,0) \in \mathrm{Gr}\, A$. We assume that $y^* \notin (f + A)(\partial U)$. We consider the Yosida approximation A_λ $(\lambda > 0)$ of the operator A (see Definition 3.2.46). We know that A_λ is demicontinuous, monotone everywhere defined, hence it is maximal monotone (see Remark 3.2.47 and Corollary 3.2.11). Then $f + A_\lambda$ is a bounded, demicontinuous, and $(S)_+$-map and it can be shown (see Browder [122]) that for $\lambda > 0$ small

$$y^* \notin (f + A_\lambda)(\partial U) \qquad \text{and} \qquad \deg_{(S)_+}(f + A_\lambda, U, y^*) \text{ stabilizes.} \tag{3.71}$$

So for X a reflexive Banach renormed so that both X and X^* are locally uniformly convex, we consider the family

$$\begin{aligned} \mathcal{S}_{\mathrm{SM}} = \Big\{ (f + A, U, y^*) : \; & U \subseteq X \text{ bounded open, } f : \overline{U} \longrightarrow X^* \text{ bounded, demi-} \\ & \text{continuous, } (S)_+, A : \mathcal{D}(A) \longrightarrow 2^{X^*} \text{maximal monotone} \\ & \text{with } (0,0) \in \mathrm{Gr}\, A \text{ and } y^* \notin (f + A)(\partial U) \Big\}. \end{aligned}$$

Then (3.71) makes the following definition possible.

DEFINITION 3.3.71 We can define a *degree map* on $\mathcal{S}_{\mathrm{SM}}$ by setting

$$\deg(f + A, U, y^*) = \deg_{(S)_+}(f + A_\lambda, U, y^*) \qquad \text{for } \lambda > 0 \text{ small.}$$

In this case the admissible homotopies are of the form

$$h(t, x) = f_t + A^t(x) \qquad \text{for all } (t, x) \in [0,1] \times \overline{U},$$

where $\{f_t\}_{t \in [0,1]}$ is an $(S)_+$-homotopy (see Definition 3.3.65) and $\{A^t\}_{t \in [0,1]}$ is a pseudomonotone homotopy defined below.

DEFINITION 3.3.72 Let $\{A^t\}_{t \in [0,1]}$ be a family of maximal monotone maps such that $(0,0) \in \mathrm{Gr}\, A^t$ for all $t \in [0,1]$. We say that $\{A^t\}_{t \in [0,1]}$ is a *pseudomonotone homotopy* of maximal monotone operators if it satisfies the following mutually equivalent conditions.

(a) If $t_n \to t$ in $[0,1]$, $x_n \xrightarrow{w} x$ in X, $x_n^* \xrightarrow{w} x^*$ in X^*, $(x_n, x_n^*) \in \operatorname{Gr} A^{t_n}$ for all $n \geq 1$ and

$$\limsup_{n\to\infty} \langle x_n^*, x_n \rangle \leq \langle x^*, x \rangle ,$$

then $(x,x^*) \in \operatorname{Gr} A^t$ and $\langle x_n^*, x_n \rangle \longrightarrow \langle x^*, x \rangle$.

(b) For every $\lambda > 0$ (equivalently for some $\lambda > 0$), the map $(t,x) \longrightarrow J_\lambda^{A^t}(x)$ is continuous from $[0,1] \times X$ into X (X furnished with the strong topology).

(c) For every $\lambda > 0$ (equivalently for some $\lambda > 0$) and every $x \in X$, the map $t \longrightarrow J_\lambda^{A^t}(x)$ is continuous from $[0,1]$ to X with the strong topology.

(d) If $t_n \longrightarrow t$ in $[0,1]$ and $(x,x^*) \in \operatorname{Gr} A^t$, then there exist sequences $\{x_n\}_{n\geq 1} \subseteq X$, $\{x_n^*\}_{n\geq 1} \subseteq X^*$ such that $(x_n, x_n^*) \in \operatorname{Gr} A^{t_n}$, $x_n \longrightarrow x$ in X and $x_n^* \xrightarrow{w} x^*$ in X^*.

REMARK 3.3.73 Anticipating some terminology from Section 6.6, we can rewrite Part (d) of the above definition as $\operatorname{Gr} A^t \subseteq s - \liminf_{n\to\infty} \operatorname{Gr} A^{t_n}$. In general affine homotopies are not pseudomonotone homotopies; that is, if A^0, A^1 are two maximal monotone operators from X into X^* such that $(0,0) \in \operatorname{Gr} A^k$ $(k=0,1)$ and $\overline{\mathcal{D}(A^0)} \neq \overline{\mathcal{D}(A^1)}$, then $A^t = tA^0 + (1-t)A^1$, $t \in [0,1]$ is not a pseudomonotone homotopy. However, if A^0 is a continuous maximal monotone operator with $\mathcal{D}(A^0) = X$, then $\{A^t\}_{t\in[0,1]}$ is a pseudomonotone homotopy.

THEOREM 3.3.74 *The degree map* $\deg : \mathcal{S}_{\mathrm{SM}} \longrightarrow \mathbb{Z}$ *defined in Definition 3.3.71 is the unique degree map such that*

(a) $\deg(\mathcal{F}, U, y^*) = 1$, *where* $\mathcal{F}\colon X \longrightarrow X^*$ *is the duality map.*

(b) Additivity with respect to the domain: *If* U_1, U_2 *are disjoint open subsets of* U *and* $y^* \notin (f+A)\big(\overline{U} \setminus (U_1 \cup U_2)\big)$, *then* $\deg(f+A,U,y^*) = \deg(f+A,U_1,y^*) + \deg(f+A,U_2,y^*)$.

(c) Homotopy invariance: *If* $\{f_t\}_{t\in[0,1]}$ *is an* $(S)_+$*-homotopy such that each* f_t *is bounded,* $\{A^t\}_{t\in[0,1]}$ *is a pseudomonotone homotopy of maximal monotone operators, and* $y^* : [0,1] \longrightarrow X^*$ *is a continuous map such that* $y^*(t) \notin (f_t + A^t)(\partial U)$ *for all* $t \in [0,1]$, *then* $\deg\big(f_t + A^t, U, y^*(t)\big)$ *is independent of* $t \in [0,1]$.

REMARK 3.3.75 Of course deg has other properties too such as excision property, solution property, and dependence on boundary values.

3.4 Metric Fixed Points

Fixed point theory is concerned with the conditions which guarantee that a map $\varphi\colon X \longrightarrow X$ from a topological space X into itself, has one or more fixed points; that is, we can find $x \in X$ such that $\varphi(x) = x$. Fixed point theory is one of the basic tools in nonlinear analysis and in particular in the study of nonlinear boundary value problems.

There is an informal classification of fixed point theorems to metric fixed points, topological fixed points, and order fixed points. Metric fixed point theorems are always formulated in a metric space setting (usually in a Banach space setting) and the methods involved in their study exploit the metric structure and geometry of the spaces involved together with the metric properties of the maps. The typical representative of this family of fixed point theorems is the celebrated Banach contraction principle. In topological fixed point theorems, the emphasis is on the topological properties of the spaces and maps involved. For this family, the typical representatives are the well-known Brouwer's fixed point theorem and its infinite-dimensional extension, the Schauder–Tychonov fixed point theorem. Finally there is a third class of fixed point theorems, the so-called order fixed point theorems, in which the order structure of the ambient space (usually an ordered Banach space) and the monotonicity properties of the maps with respect to this order, play a crucial role. Here degree theory, via the so-called fixed point index is a valuable tool. The forerunner of results of this type is the Tarski fixed point theorem.

In this section we deal with the metric fixed point theory, starting with the Banach contraction principle. Banach's fixed point theorem is essentially an abstraction of the Liouville–Picard method of successive approximations to produce the solution of some integral equations. From that moment, because of its simplicity and remarkable power, it has found applications in many different branches of mathematical analysis. As we show, the importance of the theorem, derives from the fact that in addition to the existence of solutions, it provides much other valuable information, such as uniqueness of the solution, stability of the solution under small perturbations of the equation, a convergent approximation method to produce the solution, a priori error estimates and estimates of the rate of convergence of the approximation method, and finally stability of this method.

DEFINITION 3.4.1 Let (X, d) be a metric space and $\varphi\colon C \subseteq X \longrightarrow X$ a map

(a) We say that φ is a *k-contraction* if and only if

$$d\big(\varphi(x), \varphi(y)\big) \leq kd(x, y) \qquad \text{for all } x, y \in C \text{ with } k \in [0, 1).$$

(b) We say that φ is *nonexpansive* if and only if

$$d\big(\varphi(x), \varphi(y)\big) \leq d(x, y) \qquad \text{for all } x, y \in C.$$

(c) We say that φ is *contractive* if and only if

$$d\big(\varphi(x), \varphi(y)\big) < d(x, y) \qquad \text{for all } x, y \in C,\ x \neq y.$$

REMARK 3.4.2 Of course if $d\big(\varphi(x), \varphi(y)\big) \leq kd(x, y)$ for all $x, y \in C$ and some $k \geq 0$, we simply say that φ is k-Lipschitz continuous (or simply Lipschitz continuous) and $k \geq 0$ is the Lipschitz constant of φ. If $\varphi, \psi\colon X \longrightarrow X$ are two Lipschitz continuous maps with Lipschitz constants $k(\varphi)$ and $k(\psi)$, respectively, then $\psi \circ \varphi$ is Lipschitz continuous too with constant $k(\psi \circ \varphi) \leq k(\psi)k(\varphi)$. In particular then $k(\varphi^{(n)}) \leq k(\varphi)^n$ for all $n \geq 1$ (recall that $\varphi^{(n)} = \varphi \circ \ldots \circ \varphi$ n-times). If we are in the setting of a linear space, then $k(\varphi + \psi) \leq k(\varphi) + k(\psi)$ and $k(\lambda\varphi) = \lambda k(\varphi)$ for all $\lambda \geq 0$.

The next theorem is the celebrated *Banach fixed point theorem* (or *Banach contraction principle*).

THEOREM 3.4.3 *If (X,d) is a complete metric space (d denoting the metric of X) and $\varphi\colon X \longrightarrow X$ is a k-contraction, then φ has a unique fixed point (i.e., there exists unique $x \in X$ such that $\varphi(x) = x$).*

PROOF: Choose any point $x_0 \in X$ and define the sequence $\{x_{n+1} = \varphi(x_n)\}_{n\geq 0}$. This sequence satisfies

$$d(x_n, x_{n+1}) = d\big(\varphi^{(n)}(x_0), \varphi^{(n+1)}(x_0)\big) \leq k\, d\big(\varphi^{(n-1)}(x_0), \varphi^{(n)}(x_0)\big).$$

Hence by induction we obtain

$$d(x_n, x_{n+1}) \leq k^n d(x_0, x_1). \tag{3.72}$$

If $m \geq n$, from the triangle inequality we have

$$\begin{aligned} d\big(\varphi^{(n)}(x_0), \varphi^{(m)}(x_0)\big) &\leq d\big(\varphi^{(n)}(x_0), \varphi^{(n+1)}(x_0)\big) + \dots \\ &\quad + d\big(\varphi^{(m-1)}(x_0), \varphi^{(m)}(x_0)\big) \\ &\leq (k^n + \dots + k^{m-1})\, d(x_0, x_1) \qquad \text{(see (3.72))} \\ &\leq \frac{k^n}{1-k} d(x_0, x_1), \end{aligned} \tag{3.73}$$

$$\Rightarrow d(x_n, x_m) = d\big(\varphi^{(n)}(x_0), \varphi^{(m)}(x_0)\big) \longrightarrow 0 \quad \text{as } n, m \to \infty;$$

that is, $\{x_n\}_{n\geq 1} \subseteq C$ is a Cauchy sequence. Because (X,d) is by hypothesis complete, we have $x_n = \varphi^{(n)}(x_0) \longrightarrow x \in X$ in X. Because φ is continuous, we must have $\varphi(x_n) = \varphi\big(\varphi^{(n)}(x_0)\big) \longrightarrow \varphi(x)$, hence $x_{n+1} \longrightarrow \varphi(x)$ and so $\varphi(x) = x$.

Suppose $y \in C$ is another fixed point of φ. We have

$$d(x,y) = d\big(\varphi(x), \varphi(y)\big) \leq k d(x,y),$$

a contradiction unless $d(x,y) = 0$; that is, $x = y$. □

Some useful consequence of Theorem 3.4.3 and of its proof is collected in the next proposition.

PROPOSITION 3.4.4 *If (X,d) is a complete metric space and $\varphi\colon X \longrightarrow X$ is a k-contraction, then*

(a) *The unique fixed point x of φ is obtained as $\lim_{n\to\infty} \varphi^{(n)}(x_0)$ for any $x_0 \in X$.*

(b) $d\big(x, \varphi^{(n)}(x_0)\big) \leq \big(k^n/(1-k)\big)\, d\big(x_0, \varphi(x_0)\big)$ *for all $n \geq 0$.*

(c) *For any $x_0 \in X$, $d(x_0, x) \leq \big(1/(1-k)\big)\, d\big(x_0, \varphi(x_0)\big)$.*

(d) *For all $n \geq 0$ we have $d(x_{n+1}, x) \leq k\, d(x_n, x)$ where $x_k = \varphi^{(k)}(x_0)$ for all $k \geq 0$.*

PROOF:

(a) Follows from the proof of Theorem 3.4.3.
(b) If in (3.73) we let $m \to \infty$, we obtain $d\big(\varphi^{(n)}(x_0), x\big) \leq \big(k^n/(1-k)\big)\, d\big(x_0, \varphi(x_0)\big)$ for all $n \geq 0$.
(c) Let $n = 0$ in (b).

(d) Note that $d(x_{n+1}, x) = d\big(\varphi(x_n), \varphi(x)\big) \leq k\, d(x_n, x)$. □

REMARK 3.4.5 Part (b) gives an a priori error estimate of the successive approximations process starting from x_0. Part (d) determines the rate of convergence of the successive approximations process.

There is a parametric version of Theorem 3.4.3 that illustrates the stability of the successive approximations method.

PROPOSITION 3.4.6 *If (X, d) is a complete metric space, (T, ϱ) is another metric space (the parameter space with ϱ its metric), for each $t \in T$ $\varphi_t \colon X \longrightarrow X$ is a k-contraction with $k \in [0, 1)$ independent of $t \in T$ and for each $x \in X$ $t \longrightarrow \varphi_t(x)$ is continuous, then for each $t \in T$ there exists a unique $x_t \in X$ such that $\varphi_t(x_t) = x_t$ and the map $t \longrightarrow x_t$ is continuous from T into X.*

PROOF: Theorem 3.4.3 guarantees the existence and uniqueness of x_t. If $t_n \to t$, we have

$$\begin{aligned} d(x_{t_n}, x_t) = d\big(\varphi_{t_n}(x_{n_t}), \varphi_t(x_t)\big) &\leq d\big(\varphi_{t_n}(x_{t_n}), \varphi_{t_n}(x_t)\big) + d\big(\varphi_{t_n}(x_t), \varphi_t(x_t)\big) \\ &\leq k\, d(x_{t_n}, x_t) + d\big(\varphi_{t_n}(x_t), \varphi_t(x_t)\big), \end{aligned}$$

$$\Rightarrow d(x_{t_n}, x_t) \leq \frac{1}{1-k}\, d\big(\varphi_{t_n}(x_t), \varphi_t(x_t)\big) \longrightarrow 0 \ \text{ as } n \to \infty,$$

$$\Rightarrow t \longrightarrow x_t \ \text{ is continuous from } T \text{ into } X.$$

□

There is also a useful local version of Theorem 3.4.3.

PROPOSITION 3.4.7 *If (X, d) is a complete metric space, $B = B_r(x_0) = \{x \in X : d(x, x_0) < r\}$ and $\varphi \colon B \longrightarrow X$ is a k-contraction such that $d\big(\varphi(x_0), x_0\big) < (1-k)\, r$, then φ has a fixed point.*

PROOF: We choose $\varepsilon < r$ such that $d\big(\varphi(x_0), x_0\big) \leq (1-k)\, \varepsilon < (1-k)\, r$. Consider the set $C = \{x \in X : d(x, x_0) \leq \varepsilon\}$. We claim that $\varphi(C) \subseteq C$. We have

$$d\big(\varphi(x), x_0\big) \leq d\big(\varphi(x), \varphi(x_0)\big) + d\big(\varphi(x_0), x_0\big) \leq k\, d(x, x_0) + (1-k)\, \varepsilon \leq \varepsilon.$$

Because $C \subseteq X$ is closed and $\varphi \colon C \longrightarrow C$ is a k-contraction, we can apply Theorem 3.4.3 and obtain a fixed point. □

We can improve Theorem 3.4.3 by assuming that some "power" $\varphi^{(n)}$ of φ is a k-contraction.

THEOREM 3.4.8 *If (X, d) is a complete metric space and $\varphi \colon X \longrightarrow X$ is a map that for some $n \geq 1$, $\varphi^{(n)}$ is a k-contraction, then φ has a unique fixed point.*

PROOF: By Theorem 3.4.3 we can find a unique $x \in X$ such that $\varphi^{(n)}(x) = x$. Then

$$\varphi(x) = \varphi\big(\varphi^{(n)}(x)\big) = \varphi^{(n+1)}(x) = \varphi^{(n)}\big(\varphi(x)\big),$$

$$\Rightarrow \varphi(x) \ \text{ is the unique fixed point of } \varphi^{(n)}(\cdot) \ \text{ (i.e., } \varphi(x) = x\text{)}.$$

Suppose $y \in X$ is another fixed point of φ. Then

$$y = \varphi(y) = \varphi(\varphi(y)) = \cdots = \varphi^{(n)}(\varphi(y))$$
$$\Rightarrow \; y \text{ is also a fixed point of } \varphi^{(n)}.$$

So by the uniqueness of the fixed point of $\varphi^{(n)}$, we have $y = x$. □

REMARK 3.4.9 A function $\varphi: X \longrightarrow X$ satisfying the hypotheses of Theorem 3.4.8 need not be continuous. To see this, consider the function $\varphi: \mathbb{R} \longrightarrow \mathbb{R}$ defined by

$$\varphi(x) = \begin{cases} 1 & \text{if } x = \text{rational} \\ 0 & \text{if } x = \text{irrational} \end{cases}.$$

Evidently φ is not continuous but $\varphi^{(2)}(x) \equiv 1$ for all $x \in \mathbb{R}$.

In the next example we present a continuous map that is not a k-contraction but $\varphi^{(n)}$ is a contraction for some $n \geq 1$.

EXAMPLE 3.4.10 Let $\varphi: C[0,2] \longrightarrow C[0,2]$ be defined by

$$\varphi(x)(t) = \int_0^t x(s)ds \qquad \text{for all } t \in [0,2].$$

It can be shown that

$$\varphi^{(n)}(x)(t) = \frac{1}{(n-1)!} \int_0^t (t-s)^{n-1} x(s)ds, \quad n \geq 1.$$

So for $n \geq 1$, the map $\varphi^{(n)}$ is a k-contraction, but for $n = 1$ is not.

We can also relax the hypothesis that φ is a k-contraction, provided we strengthen the topological structure of (X, d).

THEOREM 3.4.11 *If (X, d) is a compact metric space and $\varphi: X \longrightarrow X$ is a contractive map, then φ has a unique fixed point and for any $x_0 \in X$ the sequence $\{\varphi^{(n)}(x_0)\}_{n \geq 1}$ converges to this unique fixed point.*

PROOF: Consider the function $\vartheta: X \longrightarrow \mathbb{R}_+$ defined by $\vartheta(x) = d(x, \varphi(x))$. Clearly ϑ is continuous and so we can find $x \in X$ such that $\vartheta(x) = \inf_X \vartheta$. We claim that $x = \varphi(x)$. Indeed, if this is not the case, then

$$\vartheta(\varphi(x)) = d(\varphi(x), \varphi^{(2)}(x)) < d(x, \varphi(x)) = \vartheta(x),$$

a contradiction to the fact that $x \in X$ is a minimizer of ϑ. So $x = \varphi(x)$. Clearly this fixed point is unique.

Next let $x_0 \in X$ and set $\xi_n = d(x, \varphi^{(n)}(x_0))$. We have

$$\xi_{n+1} = d(x, \varphi^{(n+1)}(x_0)) < d(x, \varphi^{(n)}(x_0)) = \xi_n, \tag{3.74}$$
$$\Rightarrow \; \{\xi_n\}_{n \geq 1} \subseteq \mathbb{R}_+ \text{ is decreasing.}$$

So we must have $\xi_n \downarrow \xi \geq 0$. Also because $\{\varphi^{(n)}(x_0)\}_{n\geq 1} \subseteq X$ and X is compact, we may assume (by passing to a subsequence if necessary) that $\varphi^{(n)}(x_0) \longrightarrow u \in X$ as $n \to \infty$. Using (3.74) we have

$$\begin{aligned}\xi = \lim_{n\to\infty} d\big(x, \varphi^{(n+1)}(x_0)\big) &= d\big(x, \varphi(u)\big)\\ &= d\big(\varphi(x), \varphi(u)\big) < d(x,u) = \xi \quad \text{if } \xi > 0,\end{aligned}$$

a contradiction. So $x = u$. By Urysohn's criterion for convergent sequences, we conclude that the original sequence $\{\varphi^{(n)}(x_0)\}_{n\geq 1}$ converges to x. $\square$

EXAMPLE 3.4.12 The compactness of (X, d) cannot be replaced by completeness and boundedness. To see this let $X = \{x \in C[0,1] : 0 = x(0) \leq x(t) \leq x(1) = 1$ for all $t \in [0,1]\}$ and $\varphi\colon X \longrightarrow X$ be defined by $\varphi(x)(t) = tx(t)$ for all $x \in X$ and all $t \in [0,1]$. The map φ is contractive and fixed point free.

Next we present a useful consequence of Banach's fixed point theorem. It is analogous to the invariance of domain results in Section 3.3 (see Theorems 3.3.34 and Theorem 3.3.42).

THEOREM 3.4.13 *If X is a Banach space, $U \subseteq X$ is nonempty open, $\varphi\colon U \longrightarrow X$ is a k-contraction, and $h(x) = x - \varphi(x)$, then*

(a) *h is an open map (so $h(U)$ is open in X).*

(b) *$h\colon U \longrightarrow h(U)$ is a homeomorphism.*

PROOF: We show that for any $x \in U$ if $B_r(x) \subseteq U$, then $B_{(1-k)r}\big(h(x)\big) \subseteq h\big(B_r(x)\big)$. To this end let $u_o \in B_{(1-k)r}\big(h(x)\big)$ and define $g\colon B_r(x) \longrightarrow X$ by $g(y) = u_0 + \varphi(y)$. Then clearly g is a k-contraction and

$$\|g(x) - x\| = \|u_0 + \varphi(x) - x\| = \|u_0 - h(x)\| < (1-k)r.$$

Invoking Proposition 3.4.7 we obtain $x_0 \in B_r(x)$ such that $x_0 = g(x_0) = u_0 + \varphi(x_0)$, hence $u_0 = h(x_0)$. Therefore $B_{(1-k)r}\big(h(x)\big) \subseteq h\big(B_r(x)\big)$ and this proves that h is an open map. In particular $h(U) \subseteq X$ is open. This proves (a). To prove (b), note that if $y, x \in U$, then

$$\begin{aligned}&\|h(x) - h(y)\| \geq \|x - y\| - \|\varphi(x) - \varphi(y)\| \geq (1-k)\|x - y\|,\\ \Rightarrow\ & h \text{ is one-to-one.}\end{aligned}$$

Therefore $h\colon U \longrightarrow h(U)$ is a continuous bijection, thus a homeomorphism. $\square$

COROLLARY 3.4.14 *If (X, d) is a Banach space and $\varphi\colon X \longrightarrow X$ is a k-contraction, then $h = I - \varphi$ has a homeomorphism.*

PROOF: We need only to show that h is surjective. So given $x_0 \in X$ let $g(x) = x_0 + \varphi(x)$, $x \in X$. Then g is a k-contraction and so by Theorem 3.4.3 we can find $\widehat{x} \in X$ such that $\widehat{x} = x_0 + \varphi(\widehat{x})$, hence $h(\widehat{x}) = x_0$. $\square$

Before moving to other metric fixed point theorems let us state one more descendant of the Banach contraction principle.

PROPOSITION 3.4.15 *If (X,d) is a complete metric space and $\varphi\colon X \longrightarrow X$ is a map such that*

$$d(\varphi(x),\varphi(y)) \leq kd(x,\varphi(x)) + kd(y,\varphi(y)) \qquad \text{with } k<\frac{1}{2} \tag{3.75}$$

then φ has a unique fixed point x_0 and $\varphi^{(n)}(x) \longrightarrow x_0$ for every $x\in X$.

PROOF: The function φ need not be continuous (see Example 3.4.20 below). However, if we set $\xi(x) = d(x,\varphi(x))$ because of (3.75) we have

$$\begin{aligned}
&\xi(\varphi^{(n+1)}(x)) \leq k\xi(\varphi^{(n)}(x)) + k\xi(\varphi^{(n+1)}(x)),\\
\Rightarrow\ &\xi(\varphi^{(n+1)}(x)) \leq \Big(\frac{k}{1-k}\Big)^{n+1}\xi(x),\\
\Rightarrow\ &\xi(\varphi^{(n+1)}(x)) \longrightarrow 0 \qquad \text{as } n\rightarrow\infty\\
\Rightarrow\ &\{\varphi^{(n)}(x)\}_{n\geq 1} \quad \text{is a Cauchy sequence.}
\end{aligned}$$

Due to the completeness of X, we have $\varphi^{(n)}(x) \longrightarrow x_0$ in X. Then

$$\xi(x_0) = \lim_{n\rightarrow\infty} d(\varphi^{(n)}(x),\varphi(x_0)) \leq k\lim_{n\rightarrow\infty}\xi(\varphi^{(n-1)}(x)) + k\xi(x_0) = k\xi(x_0),$$
$$\Rightarrow\ \xi(x_0) = 0 \quad (\text{because } k<1;\ \text{i.e., } x_0 = \varphi(x_0)).$$

Clearly from (3.75) we infer that this fixed point is unique. □

In Section 2.4 we proved another metric fixed point theorem under very general hypotheses. This was Caristi's fixed point theorem (see Theorem 2.4.11). For completeness we repeat here the statement of that theorem which is a far-reaching generalization of the Banach contraction principle.

THEOREM 3.4.16 *If (X,d) is a complete metric space, $\varphi\colon X \longrightarrow \overline{\mathbb{R}}=\mathbb{R}\cup\{+\infty\}$ is a proper, lower semicontinuous function that is bounded below and $F : X \longrightarrow 2^X\setminus\{\emptyset\}$ is a multifunction such that*

$$\varphi(y) \leq \varphi(x) - d(x,y) \quad \text{for all } x\in X \text{ and all } y \in F(x),$$

then there exists $x_0 \in X$ such that $x_0 \in F(x_0)$.

REMARK 3.4.17 In Section 2.4 we proved that Theorem 3.4.16 is in fact equivalent to some basic variational principles of nonlinear analysis, most notable of which is the Ekeland variational principle (see Theorem 2.4.12).

Next we present a generalization of Theorem 3.4.16.

THEOREM 3.4.18 *If $(X,d),(Y,\varrho)$ are complete metric spaces, $g\colon X \longrightarrow X$, $f\colon X\longrightarrow Y$ is a closed map, and $\varphi\colon f(X)\longrightarrow\mathbb{R}_+$ is a lower semicontinuous function such that for each $x\in X$,*

$$\begin{aligned}
&d(x,g(x)) \leq \varphi(f(x)) - \varphi(f(g(x)))\\
\text{and}\quad &c\varrho(f(x),f(g(x))) \leq \varphi(f(x)) - \varphi(f(g(x))) \quad \text{for some } c>0,
\end{aligned} \tag{3.76}$$

then g has a fixed point in X.

PROOF: On X we define a relation $\leq$ by

$$\begin{aligned} & x \leq u \qquad \text{if and only if } \ d(x,u) \leq \varphi\big(f(x)\big) - \varphi\big(f(u)\big) \\ \text{and} \quad & c\varrho\big(f(x), f(u)\big) \leq \varphi\big(f(x)\big) - \varphi\big(f(u)\big). \end{aligned}$$

It is easy to verify that $\leq$ is a partial order on X. We consider a chain $\{x_\alpha\}_{\alpha\in J}$ in X (i.e., a totally ordered set in $(X,\leq)$ with $x_{\alpha_1} \leq x_{\alpha_2}$ if and only if $\alpha_1 \preceq \alpha_2$ in J). Then $\big\{\varphi\big(f(x_\alpha)\big)\big\}_{\alpha\in J}$ is a decreasing net in $\mathbb{R}_+$. So $\varphi\big(f(x_\alpha)\big) \downarrow \eta$, $\eta \geq 0$. Given $\varepsilon > 0$, we can find $\alpha_0 \in J$ such that

$$\eta \leq \varphi\big(f(x_\alpha)\big) \leq \eta + \varepsilon \qquad \text{for all } \ \alpha \succeq \alpha_0.$$

So if $\alpha_0 \preceq \alpha \preceq \beta$, we have

$$\begin{aligned} & d(x_\alpha, x_\beta) \leq \varphi\big(f(x_\alpha)\big) - \varphi\big(f(x_\beta)\big) \leq \varepsilon \\ \text{and} \quad & c\varrho\big(f(x_\alpha), f(x_\beta)\big) \leq \varphi\big(f(x_\alpha)\big) - \varphi\big(f(x_\beta)\big) \leq \varepsilon. \end{aligned}$$

Therefore $\{x_\alpha\}_{\alpha\in J} \subseteq X$ and $\big\{f(x_\alpha)\big\}_{\alpha\in J} \subseteq Y$ are both Cauchy nets. So we can find $\widehat{x} \in X$ and $\widehat{y} \in Y$ such that $x_\alpha \longrightarrow \widehat{x}$ in X and $f(x_\alpha) \longrightarrow \widehat{y}$ in Y. Since by hypothesis f is closed we have that $f(\widehat{x}) = \widehat{y}$. Moreover, due to the lower semicontinuity of φ, we have $\varphi\big(f(\widehat{x})\big) \leq \eta$. Also if $\alpha \preceq \beta$, then

$$\begin{aligned} & d(x_\alpha, x_\beta) \leq \varphi\big(f(x_\alpha)\big) - \varphi\big(f(x_\beta)\big) \leq \varphi\big(f(x_\alpha)\big) - \eta \\ \text{and} \quad & c\varrho\big(f(x_\alpha), f(x_\beta)\big) \leq \varphi\big(f(x_\alpha)\big) - \eta. \end{aligned}$$

Taking the limit with respect to β, we obtain

$$\begin{aligned} & d(x_\alpha, \widehat{x}) \leq \varphi\big(f(x_\alpha)\big) - \eta \leq \varphi\big(f(x_\alpha)\big) - \varphi\big(f(\widehat{x})\big) \\ \text{and} \quad & c\varrho\big(f(x_\alpha), f(\widehat{x})\big) \leq \varphi\big(f(x_\alpha)\big) - \varphi\big(f(\widehat{x})\big), \\ \Rightarrow \quad & x_\alpha \leq \widehat{x} \qquad \text{for all } \ \alpha \in J. \end{aligned}$$

This proves that every chain in $(X,\leq)$ has an upper bound. So by Zorn's lemma there exists a maximal element $x \in X$. Then from (3.76) we have $x \leq g(x)$ and so finally $x = g(x)$. $\square$

REMARK 3.4.19 If $X = Y$, $f = I$, and $c = 1$, then Theorem 3.4.18 reduces to Theorem 3.4.16.

Next we pass to the investigation of fixed points for nonexpansive maps (see Definition 3.4.1(b)). In general a nonexpansive map in a Banach space need not have fixed points.

EXAMPLE 3.4.20 Let X be a Banach space and $\varphi\colon X \longrightarrow X$ be defined by $\varphi(x) = x + u$ with $u \neq 0$. This map is nonexpansive and clearly fixed point free. Moreover, if $\varphi = I$ (the identity map on X), then we see that the fixed point of a nonexpansive map need not be unique.

PROPOSITION 3.4.21 *If X is a Banach space, $C \subseteq X$ a nonempty, bounded, closed set, and $\varphi\colon C \longrightarrow C$ a nonexpansive map such that $(I - \varphi)(C) \subseteq X$ is a closed set, then φ has a fixed point.*

PROOF: By translating things if necessary, we may assume that $0 \in C$. Also because by hypothesis C is bounded, we can find $R>0$ large such that $C \subseteq B_R(0)$. Let $\lambda_n \uparrow 1$, $\lambda_n \in (0,1)$, $n \geq 1$ and introduce $\varphi_n = \lambda_n \varphi$. For each $n \geq 1$ φ_n is a λ_n-contraction and by Theorem 3.4.3 we can find $x_n \in C$ such that $\varphi_n(x_n) = x_n$. We have

$$\begin{aligned}
& \|\varphi_n(x_n) - x_n\| = \|\varphi_n(x_n) - \lambda_n \varphi(x_n)\| \leq (1-\lambda_n)\, R, \\
\Rightarrow\ & \lim_{n\to\infty} \|\varphi_n(x_n) - x_n\| = 0, \\
\Rightarrow\ & 0 \in (I-\varphi)(C) \quad \text{(because by hypothesis } (I-\varphi)(C) \text{ is closed).}
\end{aligned}$$

So we can find $x \in C$ such that $\varphi(x) = x$. □

Using this proposition, we are led to the first general fixed point theorem for nonexpansive maps.

THEOREM 3.4.22 *If H is a Hilbert space, $C \subseteq H$ is a nonempty, bounded, closed, and convex set, and $\varphi: C \longrightarrow C$ is nonexpansive, then φ has a fixed point.*

PROOF: Let $r: H \longrightarrow C$ be a retraction map for C (i.e., r is continuous and $r|_C = I_C =$ the identity map on C, in fact $r(x) = p_C(x)$ with $p(\cdot)$ being the metric projection to the set C). Then r is nonexpansive and so is $\varphi \circ r$. Therefore, we can easily check that $I - \varphi \circ r$ is continuous, and monotone, hence maximal monotone (see Corollary 3.2.11). Using Proposition 3.2.7 we see that $(I-\varphi\circ r)(C) = (I-\varphi)(C)$ is closed. So we can apply Proposition 3.4.21 and establish the existence of a fixed point for φ. □

We can improve this theorem by replacing the Hilbert space H with a general uniformly convex Banach space X. Recall that parallelogram law we verify that a Hilbert space is uniformly convex.

THEOREM 3.4.23 *If X is a uniformly convex Banach space, $C \subseteq X$ is nonempty, bounded, closed, and convex, and $\varphi: C \longrightarrow C$ is a nonexpansive map, then φ has a fixed point.*

PROOF: Let $\mathcal{S} = \{K \subseteq C:$ nonempty, closed, and φ-invariant (i.e., $\varphi(K) \subseteq K)\}$. Evidently $\mathcal{S} \neq \emptyset$ since $C \in \mathcal{S}$. We partially order $\mathcal{S}$ by reverse inclusion (i.e. $K_1 \preceq K_2$ if and only if $K_2 \subseteq K_1$). If $\mathcal{D}$ is a chain in $(\mathcal{S}, \leq)$ (i.e., a totally ordered subset), then the set $\bigcap_{K\in\mathcal{D}} K$ is closed, convex φ-invariant, and nonempty because all sets $K \in \mathcal{D}$ are weakly compact in X (finite intersection property). So $\bigcap_{K\in\mathcal{D}} K \in \mathcal{S}$ and is an upper bound of $\mathcal{D}$. By Zorn's lemma $\mathcal{S}$ has a maximal element $\widehat{K}$ and due to its maximality we have that $\widehat{K} = \overline{\text{conv}}\ \varphi(\widehat{K})$. Clearly we finish the proof if we can show that $\widehat{K}$ is a singleton. If $\widehat{K}$ is not a singleton, because $\widehat{K} \neq \emptyset$, we have that $r = \text{diam}\ \widehat{K} > 0$. Choose $x_1, x_2 \in \widehat{K}$ such that $\|x_1 - x_2\| \geq r/2$. Let $x \in \widehat{K}$ be the midpoint of the segment joining x_1 and x_2 (i.e., $\{x' = \lambda x_1 + (1-\lambda)x_2 : \lambda \in [0,1]\}$). Then $x - y$ is the midpoint of the segment joining $x_1 - y$ and $x_2 - y$, $y \in \widehat{K}$ and $\|x_1 - y\| < r$, $\|x_2 - y\| < r$. By virtue of the uniform convexity of the space X, we can find $\vartheta > 0$ such that $\|x - y\| \leq (1-\vartheta)\, r = r_0 < r$. Let $K_0 = \bigcap_{u\in\widehat{K}} \{v \in \widehat{K} : \|v - u\| \leq r_0\}$.

Then K_0 is a nonempty, closed, convex subset of $\widehat{K}$ and $x \in K_0$. The set K_0 is a strict subset of $\widehat{K}$ since $r_0 < r = \operatorname{diam} \widehat{K}$. We show that K_0 is φ-invariant. To this end let $v \in K_0$ and $y \in \widehat{K}$. Given any $\varepsilon > 0$, we can find $\{w_k\}_{k=1}^N \subseteq \widehat{K}$ and $\{\lambda_k\}_{k=1}^N \subseteq [0,1]$ such that

$$\|y - \sum_{k=1}^N \lambda_k \varphi(w_k)\| < \varepsilon \qquad \text{with} \qquad \sum_{k=1}^N \lambda_k = 1.$$

Then we have

$$\begin{aligned} \|\varphi(v) - y\| &\leq \|\varphi(v) - \sum_{k=1}^N \lambda_k \varphi(w_k)\| + \varepsilon \\ &\leq \sum_{k=1}^N \lambda_k \|\varphi(v) - \varphi(w_k)\| + \varepsilon \\ \text{and} \quad \|\varphi(v) - \varphi(w_k)\| &\leq \|v - w_k\| \leq r_0 \quad (\text{recall } v \in K_0). \end{aligned}$$

Therefore it follows that

$$\|\varphi(v) - y\| \leq \sum_{k=1}^N \lambda_k r_0 + \varepsilon = r_0 + \varepsilon.$$

Because $\varepsilon > 0$ was arbitrary, we let $\varepsilon \downarrow 0$ and obtain $\|\varphi(v) - y\| \leq r_0$ for all $y \in \widehat{K}$ (i.e., $\varphi(v) \in K_0$). Therefore K_0 is a maximal element of $\mathcal{S}$, a contradiction (because K_0 is a strict subset of $\widehat{K}$). This implies that $\widehat{K}$ is a singleton and so we are done. □

To further generalize this fixed point theorem, we need the following geometrical notion.

DEFINITION 3.4.24 Let X be a Banach space and $C \subseteq X$ nonempty. A point $x \in C$ is said to be the *diametral point of* **C**, if $\sup\{\|x - y\| : y \in C\} = \operatorname{diam} C$. A convex set $C \subseteq X$ is said to have *normal structure*, if for each bounded, convex subset K of C with $\operatorname{diam} K > 0$, there exists some $x \in K$ that is not a diametral point of K (i.e., there exists $x \in K$ such that $\sup\{\|x - u\| : u \in K\} < \operatorname{diam} K$).

REMARK 3.4.25 It is clear from the above definition that sets with normal structure have no convex subsets K which consist entirely of diametral points, except singletons.

PROPOSITION 3.4.26 *Every compact convex subset C of a Banach space X has normal structure.*

PROOF: Suppose that K does not have the normal structure. Without any loss of generality we assume that $\operatorname{diam} C > 0$. For every $x_1 \in C$ there exists $x_2 \in C$ such that $\operatorname{diam} C = \|x_2 - x_1\|$. Since C is convex we have $\frac{1}{2}(x_1 + x_2) \in C$. We can find $x_3 \in C$ such that

$$\operatorname{diam} C = \left\|x_3 - \frac{1}{2}(x_1 + x_2)\right\|.$$

Proceeding this way, we obtain a sequence $\{x_n\}_{n\geq 1}$ such that

$$\text{diam } C = \Big\|x_{n+1} - \frac{1}{n}\sum_{k=1}^{n} x_k\Big\|, \ n \geq 2,$$

$$\Rightarrow \text{diam } C \leq \frac{1}{n}\sum_{k=1}^{n} \|x_{n+1} - x_k\| \leq \text{diam } C,$$

$$\Rightarrow \text{diam } C = \|x_{n+1} - x_k\| \ \text{ for } \ 1 \leq k \leq n.$$

So the sequence $\{x_n\}_{n\geq 1}$ has no convergent subsequence, a contradiction to the fact that K is compact. □

PROPOSITION 3.4.27 *If X is a uniformly convex Banach space and $C \subseteq X$ is closed, convex, and bounded, then C has normal structure.*

PROOF: Without any loss of generality, we assume that $C \subseteq \overline{B}_1(0)$. Let K be a closed and convex subset of C. Let $x_1 \in K$ and $\varepsilon = \frac{1}{2}$. We can find $x_2 \in K$ such $\|x_2 - x_1\| \geq \frac{1}{2}K_1$. Then for any $x \in K$ we have

$$\begin{aligned}\Big\|x - \frac{1}{2}(x_1 + x_2)\Big\| &= \Big\|\frac{1}{2}(x - x_1) + \frac{1}{2}(x - x_2)\Big\| \\ &\leq \text{diam } K\Big(1 - \delta\big(\frac{\frac{1}{2}\text{diam } K}{\text{diam } K}\big)\Big) \\ &\leq \text{diam } K\Big(1 - \delta\big(\frac{1}{2}\big)\Big),\end{aligned}$$

where $\delta(\varepsilon) > 0$ is the modulus of convexity of X. This proves that C has normal structure (see Definition 3.4.24). □

DEFINITION 3.4.28 Let X be a Banach space and $C \subseteq X$ a nonempty, bounded, closed, and convex set. We introduce the following geometric quantities.

(a) $r_x(C) = \sup\{\|x - y\| : y \in C\}$ = the *radius* of C relative to $x \in X$.

(b) $r(C) = \inf\{r_x(C) : x \in X\}$ = the *Chebyshev radius* of C.

(c) $C_0(C) = \{x \in X : r_x(C) = r(C)\}$ = the *Chebyshev center* of C.

LEMMA 3.4.29 *If X is reflexive, then $C_0(C)$ is nonempty and convex.*

PROOF: Let $C_n(x) = \{u \in C : \|u - x\| \leq r(C) + 1/n\}$. Evidently $\big\{D_n = \bigcap_{x\in C} C_n(x)\big\}_{n\geq 1}$ is a decreasing sequence of nonempty, closed, and convex sets. Because X is reflexive by Smulian's theorem, we have that $C_0(C) = \bigcap_{n\geq 1} D_n$ is nonempty and of course convex. □

LEMMA 3.4.30 *If X is a Banach space, $C \subseteq X$ a closed convex set with* diam $C > 0$, *and C has normal structure, then* diam $\big(C_0(C)\big) <$ diam C.

PROOF: Because by hypothesis C has normal structure, it has at least a nondiametral point x. Then $r_x(C) < \text{diam}\, C$. Let $y, v \in C_0(C)$, then $\|y - v\| \leq r_y(C) = r(C)$. Therefore

$$\text{diam}\big(C_0(C)\big) = \sup\{\|y - v\| : y, v \in C_0(C)\} \leq r(C) \leq r_x(C) < \text{diam}\, C.$$

□

Now we are ready for the main fixed point theorem for nonexpansive maps.

THEOREM 3.4.31 *If X is a reflexive Banach space, $C \subseteq X$ is a nonempty, bounded, closed, and convex set with normal structure and $\varphi\colon C \longrightarrow C$ is nonexpansive, then φ has a fixed point.*

PROOF: Let $\mathcal{S}$ be the collection of all nonempty, closed, and convex subsets of C, which are mapped to themselves by the map φ. From Smulian's theorem and Zorn's lemma, as in the proof of Theorem 3.4.23, we infer that $\mathcal{S}$ has a maximal element K. We finish the proof by showing that K is a singleton. So suppose that is not true. Then by Lemma 3.4.29 we can take $x \in C_0(K)$. We have

$$\begin{aligned} &\|\varphi(x) - \varphi(y)\| \leq \|x - y\| \leq r(K) \qquad \text{for all } y \in K,\\ \Rightarrow\ &\varphi(K) \subseteq B_{r(K)}\big(\varphi(x)\big). \end{aligned}$$

Because $\varphi\Big(B_{r(K)}\big(\varphi(x)\big) \cap K\Big) \subseteq B_{r(K)}\big(\varphi(x)\big) \cap K$, the minimality of K implies that

$$K \subseteq \overline{B}_{r(K)}\big(\varphi(x)\big).$$

Therefore $\varphi(x) \in C_0(K)$; that is, φ maps $C_0(K)$ into itself and so $C_0(K) \in \mathcal{S}$. If diam $K > 0$, then by Lemma 3.4.30 we have that $C_0(K)$ is a proper subset of K, a contradiction to the minimality of K. So K is a singleton and so φ has a fixed point. □

Another formulation of the above theorem is the following.

THEOREM 3.4.32 *If X is a Banach space, $C \subseteq X$ is nonempty, weakly compact, and convex with normal structure, and $\varphi\colon C \longrightarrow C$ is nonexpansive, then φ has a fixed point.*

EXAMPLE 3.4.33 In Theorem 3.4.31 the reflexivity of X (or Theorem 3.4.32 the weak compactness of the set C) can not be dropped. Let $X = C[0,1]$ and let $C = \{x \in X : x(0) = 0, x(1) = 1, 0 \leq x(t) \leq 1 \text{ for all } t \in [0,1]\}$. Then C is bounded, closed, and convex, but C is not weakly compact in X. Let $\varphi\colon C \longrightarrow C$ be defined by $\varphi(x)(t) = tx(t)$ for all $t \in [0,1]$. Clearly φ is nonexpansive, but has no fixed points.

Next we consider a certain class of maps that need not map the set into itself. For this reason, we introduce the following definition.

DEFINITION 3.4.34 (a) Let X be a Banach space, $C \subseteq X$ is nonempty and $\{x_n\}_{n\geq 1}$ a bounded sequence in X. The *asymptotic center* of $\{x_n\}_{n\geq 1}$ relative to C is defined by

$$A(C, \{x_n\}) = \{y \in C : \limsup_{n\to\infty} \|y - x_n\| = \inf_{z\in C}\{\limsup_{n\to\infty} \|z - x_n\|\}\}.$$

(b) Let X be a Banach space; $C \subseteq X$ is nonempty, closed, and convex. Then

$$I_C(x) = \{(1-\lambda)x + \lambda y : \lambda \geq 0,\ y \in C\}$$

is the *inward set of* $x \in C$ *with respect to* C. Also a map $\varphi\colon C \longrightarrow X$ is said to be *inward* if $\varphi(x) \in I_C(x)$ for all $x \in C$ and *weakly inward* if $\varphi(x) \in \overline{I_C(x)}$ for every $x \in C$.

REMARK 3.4.35 If we set $r(y) = \limsup_{n\to\infty} \|y - x_n\|$, then $r(\cdot)$ is convex and nonexpansive. Also, if the space X is uniformly convex and $C \subseteq X$ is nonempty, bounded, closed, and convex, then $A(C, \{x_n\})$ is a singleton. Finally it is easy to check that $\varphi\colon C \longrightarrow X$ is weakly inward if and only if $\lim_{\lambda\downarrow 0}(1/\lambda)\big(x + \lambda(\varphi(x) - x), C\big) = 0$ for all $x \in C$.

PROPOSITION 3.4.36 *If X is a uniformly convex Banach space, $C \subseteq X$ is a nonempty, closed, and convex set, $\{x_n\}_{n\geq 1}$ is a bounded sequence in C, and x is the asymptotic center of $\{x_n\}$ with respect to C (see Remark 3.4.35), then x is also the asymptotic center of $\{x_n\}_{n\geq 1}$ with respect to $\overline{I_C(x)}$.*

PROOF: Let v be the asymptotic center of $\{x_n\}_{n\geq 1}$ with respect to $\overline{I_C(x)}$ and assume that $v \neq x$. Because $C \subseteq \overline{I_C(x)}$, we have $v \in \overline{I_C(x)} - C$ and $r(v) < r(x)$ (by the uniqueness of the asymptotic center). Due to the continuity of r (see Remark 3.4.35), we can find $u \in I_C(x) - C$ such that $r(u) < r(x)$. Therefore $u = (1-\lambda)x + \lambda y$ for some $y \in C$ and $\lambda > 1$. Exploiting the convexity of the function $r(\cdot)$ we have

$$\begin{aligned} r(y) = r\Big(\frac{1}{\lambda} + \big(1 - \frac{1}{\lambda}\big)x\Big) &\leq \frac{1}{\lambda} r(u) + \big(1 - \frac{1}{\lambda}\big) r(x) \\ &< \frac{1}{\lambda} r(x) + \big(1 - \frac{1}{\lambda}\big) r(x) = r(x), \end{aligned}$$

a contradiction to the definition of x (see Definition 3.4.34). Hence $v = x$. □

The following result is due to Caristi [129] and is used in the proof of the fixed point theorem for weakly inward nonexpansive maps.

PROPOSITION 3.4.37 *If X is a Banach space, $C \subseteq X$ is a nonempty, closed, and convex set, and $\varphi\colon C \longrightarrow C$ is weakly inward and a k-contraction, then φ has a unique fixed point.*

This proposition is used to prove the following fixed point theorem.

THEOREM 3.4.38 *If X is a uniformly Banach space, $C \subseteq X$ is a nonempty, bounded, closed, and convex set, and $\varphi\colon C \longrightarrow C$ is weakly inward and nonexpansive, then φ has a unique fixed point.*

PROOF: Let $u \in C$ and define φ_n by

$$\varphi_n(x) = (1-\lambda_n)u + \lambda_n \varphi(x), \qquad 0<\lambda_n<1 \text{ and } \lambda_n \longrightarrow 1.$$

Evidently φ_n is a λ_n-contraction and so by Proposition 3.4.37 φ_n has a unique fixed point x_n. Moreover,

$$\begin{aligned} \|x_n - \varphi_n(x_n)\| &= \Big\|x_n - \frac{1}{\lambda_n}x_n - \Big(\frac{1}{\lambda_n} - 1\Big)u\Big\| \\ &= \Big(\frac{1}{\lambda_n} - 1\Big)\|x_n - u\| \longrightarrow 0 \qquad \text{as } n \to \infty \end{aligned}$$

(recall that C is bounded). Let x be the asymptotic center with respect to C. We have

$$\begin{aligned} r\big(\varphi(x)\big) &= \limsup_{n\to\infty} \|\varphi(x) - x_n\| \\ &\leq \limsup_{n\to\infty} \|\varphi(x) - \varphi(x_n)\| \\ &\leq \lim_{n\to\infty} \|x - x_n\| = r(x) \end{aligned} \tag{3.77}$$

(recall that φ is nonexpansive).

Because φ is also weakly inward, we have $\varphi(x) \in \overline{I_C(x)}$. Also from Proposition 3.4.36 we have that x is the asymptotic center with respect to $\overline{I_C(x)}$. So from (3.77) it follows that $\varphi(x) = x$ (recall that the asymptotic center is unique). Moreover, it is clear from the above argument that the fixed point of φ is unique. $\square$

We have one more fixed point theorem for weakly inward mappings. The result is due to Deimling [187], where the interested reader can find its proof (see also Deimling [188, p. 211]).

THEOREM 3.4.39 *If X is a Banach space, $C \subseteq X$ a nonempty, bounded, closed, and convex set, and $\varphi\colon C \longrightarrow X$ a weakly inward and γ-condensing map (with γ equal to α or β, see Definition 3.3.48), then φ has a fixed point.*

We conclude this section with a structural result concerning the set of fixed points $\mathrm{Fix}(\varphi)$ a map φ. The result is due to Vidossich [594].

THEOREM 3.4.40 *If X is a Banach space, $U \subseteq X$ is bounded open, $\varphi\colon \overline{U} \longrightarrow X$ is compact and nonexpansive, $\varphi(x) \neq x$ on ∂U, and there exists $x_0 \in U$ such that $\varphi(x) - x_0 \neq \lambda(x - x_0)$ on ∂U for all $\lambda>1$ (Leray–Schauder condition), then* $\mathrm{Fix}(\varphi)$ *is connected.*

3.5 Topological Fixed Points

In this section we present some basic topological fixed point theorems. Now crucial are the topological properties of the spaces and/or of the maps involved. In particular the notion of compactness is important in our considerations here.

The grandfather of all topological fixed point theorems is, of course, the famous fixed point theorem of Brouwer. To state and prove it in its general form, we need the following topological result.

PROPOSITION 3.5.1 *If $C \subseteq \mathbb{R}^N$ is a nonempty, bounded, closed, and convex set with nonempty interior, then C is homeomorphic to the closed unit ball $\overline{B}_1$ in $\mathbb{R}^N$.*

PROOF: Let $x_0 \in \operatorname{int} C$. Suppose that $x \neq x_0$. Then the segment $[x, x_0] = \{\lambda x_0 + (1-\lambda)x : \lambda \in [0,1]\}$ or its continuation beyond x meets the boundary of C in a uniquely determined point $\xi(x)$. We define

$$\vartheta(x) = \begin{cases} 0 & \text{if } x = x_0 \\ \frac{x - x_0}{\|\xi(x) - x_0\|} & \text{if } x \neq x_0 \end{cases}, \qquad x \in \mathbb{R}^N.$$

Then ϑ maps C onto $\overline{B}_1$ and is a homeomorphism of $\mathbb{R}^N$ onto itself. □

REMARK 3.5.2 A convex set C in $\mathbb{R}^N$ may have empty interior. However, we always have that $\operatorname{r\,int} C \neq \emptyset$, where $\operatorname{r\,int} C$ is the interior of C with respect to span C.

Now we can state and prove *Brouwer's fixed point theorem.*

THEOREM 3.5.3 *If $U \subseteq \mathbb{R}^N$ is a bounded, open, and convex set, and $\varphi \colon \overline{U} \longrightarrow \overline{U}$ is a continuous map, then φ has a fixed point.*

PROOF: First we assume that $U = B_1(0) = \{x \in \mathbb{R}^N : \|x\| < 1\}$. Clearly we may assume that $\varphi(x) \neq x$ for all $x \in \partial B_1(0) = \{x \in \mathbb{R}^N : \|x\| = 1\}$ or otherwise we are done. Consider the continuous homotopy $h(t,x) = x - t\varphi(x)$ for all $t \in [0,1] \times \overline{B}_1(0)$. Then for all $(t,x) \in [0,1] \times \partial B_1(0)$, we have

$$\|h(t,x)\| \geq \|x\| - t\|\varphi(x)\| > (1-t)r > 0$$

since $\|x\| = 1$ and $\varphi(x) \neq x$ on $\partial B_1(0)$. Therefore by virtue of the homotopy invariance and normalization properties of Brouwer's degree, we have

$$d(I - \varphi, U, 0) = d(I, U, 0) = 1,$$

$$\Rightarrow \; x = \varphi(x) \text{ for some } x \in B_1(0) \text{ (existence property of Brouwer's degree).}$$

Now for the general case, we use Proposition 3.5.1 to find a homeomorphism $\theta \colon \overline{U} \longrightarrow \overline{B}_1$. We set $\psi = \theta \circ \varphi \circ \theta^{-1}$. Clearly $\psi \colon \overline{B}_1(0) \longrightarrow \overline{B}_1(0)$ is continuous. So from the first part of the proof we obtain $u \in \overline{B}_1(0)$ such that $\psi(u) = u$. Let $x \in \overline{U}$ such that $u = \theta(x)$. Then

$$\theta \circ \varphi(x) = (\theta \circ \varphi \circ \theta^{-1})(u) = \psi(u) = u = \theta(x).$$

Because θ is a homeomorphism, we have $\varphi(x) = x$; that is, $x \in \overline{B}_1(0)$ is a fixed point of φ. □

DEFINITION 3.5.4 A Hausdorff topological space X has the *fixed point property* (*fpp*), if every continuous map $\varphi \colon X \longrightarrow X$ has a fixed point.

REMARK 3.5.5 Theorem 3.5.3 says that every compact convex set in $\mathbb{R}^N$ has the fpp. More generally, it is easy to prove that if the space X has the fpp and C is a retract of X, then C has the fpp too. Recall that C is a retract of X, if $C \subseteq X$ and there exists a continuous map $r : X \longrightarrow C$ such that $r\big|_C = I\big|_C$ (the identity map restricted on C). The map r is said to be a *retraction.*

Next we present two remarkable consequences of Brouwer's fixed point theorem (see Theorem 3.5.3). The first is the so-called *Perron–Frobenius theorem* from linear algebra.

THEOREM 3.5.6 *If $A = (a_{ij})_{i,j=1}^N$ is an $N \times N$-matrix which is positive; that is, $a_{ij} \geq 0$ for all $i, j \in \{1, \ldots, N\}$, then A has a nonnegative eigenvalue and a corresponding nonnegative eigenvector; that is, there is a $\lambda \geq 0$ and $x \neq 0$ with $x_k \geq 0$ for all $k \in \{1, \ldots, N\}$ such that $Ax = \lambda x$.*

PROOF: Let $C = \{x \in \mathbb{R}^N : x = (x_k)_{k=1}^N, x_k \geq 0 \text{ and } \sum_{k=1}^N x_k = 1\}$. If for some $x \in C$ we have $Ax = 0$, then the theorem holds for $\lambda = 0$. If $Ax \neq 0$ on C, then we can find $\xi > 0$ such that $\sum_{k=1}^N (Ax)_k \geq \xi$ for all $x \in C$. Then the map

$$\varphi(x) = \frac{1}{\sum_{k=1}^N (Ax)_k} Ax, \quad x \in C,$$

is continuous and $\varphi(C) \subseteq C$ (because $a_{ij} \geq 0$). So we can apply Theorem 3.5.3 and obtain $x_0 \in C$ such that $\varphi(x_0) = x_0$, hence $Ax_0 = \lambda x_0$ with $\lambda = \sum_{k=1}^N (Ax_0)_k \geq 0$. □

The next theorem is the so-called *no-retraction theorem* and it says that it is impossible to retract the whole unit ball of $\mathbb{R}^N$ onto its boundary, keeping the boundary fixed.

THEOREM 3.5.7 *If $B_r(0) = \{x \in \mathbb{R}^N : \|x\| < 1\}$, then there is no continuous map $\varphi : \overline{B}_1(0) \longrightarrow \partial B_1(0)$ such that $\varphi|_{\partial B_1(0)} = I|_{\partial B_1(0)}$.*

PROOF: We proceed by contradiction and assume that such a retraction exists. Then from the dependence on the boundary values of Brouwer's degree (see Theorem 3.3.18) we have

$$\begin{aligned} & d(\varphi, B_1(0), 0) = d(I, B_1(0), 0) = 1, \\ \Rightarrow\ & 0 = \varphi(x) \quad \text{for some } x \in B_1(0), \end{aligned}$$

a contradiction to the fact that $\varphi(\overline{B}_1(0)) \subseteq \partial B_1(0)$. □

Isolating the essential parts of the proof of Theorem 3.5.3, we can prove the following fixed point theorem.

THEOREM 3.5.8 *If $U \subseteq \mathbb{R}^N$ is bounded open, $\varphi \in C(\overline{U}, \mathbb{R}^N)$, and there is a $u \in U$ such that*

$$\varphi(x) - u \neq \lambda(x - u) \qquad \textit{for all } x \in \partial U \textit{ and all } \lambda > 1, \tag{3.78}$$

then φ has a fixed point in $\overline{U}$.

PROOF: Again we assume that $\varphi(x) \neq x$ for all $x \in \partial U$ or otherwise there is nothing to prove.

Consider the continuous homotopy $h(t,x)$ defined by

$$h(t,x) = x - u - t\big(\varphi(x) - u\big) \qquad \text{for all } (t,x) \in [0,1] \times \overline{U}.$$

Suppose that $h(t,x) = 0$ for some $t \in (0,1)$ and $x \in \partial U$. Then

$$\varphi(x) - u = \frac{1}{t}(x - u)$$

and because $1/t > 1$ this contradicts (3.78). So by the homotopy invariance of Brouwer's degree we have

$$d(I - \varphi, U, 0) = d(I - u, U, 0) = d(I, U, u) = 1 \qquad (\text{because } u \in U).$$

So we can find $x \in U$ such that $\varphi(x) = x$. □

REMARK 3.5.9 The geometric interpretation of (3.78), is that for some point $u \in U$, for no $x \in \partial U$ does $\varphi(x)$ lie on the continuation of the segment $[u, x]$ beyond x.

The next result is known as *Borsuk's fixed point theorem.*

THEOREM 3.5.10 *If $U \subseteq \mathbb{R}^N$ is bounded, open, and symmetric with $0 \in U$ and $\varphi \in C(\overline{U}, \mathbb{R}^N)$ is odd, then φ has a fixed point.*

PROOF: Assume that $x \neq \varphi(x)$ for all $x \in \partial U$, or otherwise we are done. From Theorem 3.3.28 we know that $d(I - \varphi, U, 0) \neq 0$ (being odd) and so φ has a fixed point. □

Can we have an infinite-dimensional version of Theorem 3.5.3? The next theorem shows that this is not possible. Its proof can be found in Dugundji–Granas [212, p. 47].

THEOREM 3.5.11 *If X is a normed space, then each continuous map $\varphi\colon \overline{B}_1(0) \longrightarrow \overline{B}_1(0)$ has a fixed point if and only if $\dim X < +\infty$.*

This shows that in order to extend the Brouwer fixed point theorem to infinite-dimensional spaces, we need to introduce additional hypotheses.

EXAMPLE 3.5.12 Let $H = l^2 = \big\{\widehat{x} = (x_n)_{n \geq 1} : \sum_{n \geq 1} x_n^2 < +\infty\big\}$. We know that this is a Hilbert space with inner product $(x, y) = \sum_{n \geq 1} x_n y_n$ and norm $\|x\| = \big(\sum_{n \geq 1} x_n^2\big)^{1/2}$. By $\overline{B}_1(0)$ we denote the closed unit ball in H. Let $\varphi\colon \overline{B}_1(0) \longrightarrow \overline{B}_1(0)$ be defined by

$$\varphi(\widehat{x}) = \big((1 - \|x\|^2)^{1/2}, x_1, x_2, \ldots\big).$$

Then $\|\varphi(\widehat{x})\| = \big[(1 - \|x\|^2) + \|x\|^2\big]^{1/2} = 1$; that is, $\varphi\big(\overline{B}_1(0)\big) \subseteq \partial B_1(0)$ and φ is continuous. However, φ is fixed point free. Indeed, if $\varphi(\widehat{x}) = \widehat{x}$, then $\|\widehat{x}\| = \|\varphi(\widehat{x})\| = 1$ and so

$$\begin{aligned} &\varphi(\widehat{x}) = (0, x_1, x_2, \dots, x_n, \dots) = \widehat{x} = (x_1, x_2, \dots, x_n, \dots), \\ \Rightarrow\ & 0 = x_1, \quad x_1 = x_2, \dots, \ x_{n-1} = x_n, \dots, \\ \Rightarrow\ & x_n = 0 \qquad \text{for all } n \geq 1, \text{ a contradiction (because } \|\widehat{x}\| = 1). \end{aligned}$$

REMARK 3.5.13 The difficulty is that in an infinite-dimensional Banach space a bounded closed set need not be compact. For this reason as we did with the infinite-dimensional extension of Brouwer's degree, we look to maps exhibiting some kind of compactness.

The next theorem is known as *Schauder's fixed point theorem.*

THEOREM 3.5.14 *If X is a Banach space, $C \subseteq X$ is nonempty, bounded, closed, and convex, and $\varphi\colon C \longrightarrow C$ is compact, then φ has a fixed point.*

PROOF: Let $\varepsilon_n \downarrow 0$ and let $\varphi_{\varepsilon_n} \in \mathcal{K}_f(C, X)$ be the finite-dimensional ε_n-approximation of φ guaranteed by Theorem 3.1.10. Let X_{ε_n} be the finite-dimensional range space of φ_{ε_n}. Then $\varphi_{\varepsilon_n}\colon C \cap X_{\varepsilon_n} \longrightarrow C \cap X_{\varepsilon_n}$ is continuous and so by Theorem 3.5.3 we can find $x_n \in C \cap X_{\varepsilon_n}$ such that $\varphi_{\varepsilon_n}(x_n) = x_n$. Because $\|\varphi(y) - \varphi_{\varepsilon_n}(y)\| \leq \varepsilon_n$ for all $y \in C$ and $\varphi(C)$ is compact, it follows that $\{\varphi_{\varepsilon_n}(x_n)\}_{n \geq 1} \subseteq X$ is relatively compact. So we may assume that

$$\varphi_{\varepsilon_n}(x_n) = x_n \longrightarrow \widehat{x} \quad \text{in } x \in X.$$

But $\|\varphi(x_n) - x_n\| = \|\varphi(x_n) - \varphi_{\varepsilon_n}(x_n)\| \leq \varepsilon_n \longrightarrow 0$. Hence $\varphi(\widehat{x}) = \widehat{x}$; that is, φ has a fixed point. □

REMARK 3.5.15 We have given a nondegree-theoretic proof of Theorem 3.5.14. It is easy to see that we can have a degree-theoretic proof of the result by mimicking the proof of Theorem 3.5.3.

The next result is known as the *nonlinear alternative theorem.*

THEOREM 3.5.16 *If $U \subseteq X$ is a bounded open set with $0 \in U$ and $\varphi\colon \overline{U} \longrightarrow X$ is a compact map, then we have the following alternative; either*

(a) *φ has a fixed point in $\overline{U}$, or*
(b) *There exist $\lambda \in (0,1)$ and $x \in \partial U$, such that $x = \lambda\varphi(x)$.*

PROOF: Let $h(t, x)$ be the compact homotopy defined by $h(t, x) = t\varphi(x)$ for all $(t, x) \in [0,1] \times \overline{U}$. If $h(t, x) \neq x$ for all $(t, x) \in [0,1] \times \partial U$, then from the homotopy invariance of the Leray–Schauder degree we have

$$\begin{aligned} &\widehat{d}(I - \varphi, U, 0) = \widehat{d}(I, U, 0) = 1, \\ \Rightarrow\ &\varphi(x) = x \qquad \text{for some } x \in U. \end{aligned}$$

If $h(1, x) = x$ for some $x \in \partial U$, then $\varphi(x) = x$ and so φ has a fixed point. If $h(t, x) = x$ for some $(t, x) \in (0,1) \times \partial U$, then $x = t\varphi(x)$ and so (b) is satisfied. □

This theorem suggests that we have to impose conditions on φ which do not allow (b) to happen in Theorem 3.5.16. Then automatically φ has a fixed point.

COROLLARY 3.5.17 *If X is a Banach space, $\xi\colon X \longrightarrow \mathbb{R}_+$ is a function such that $\xi^{-1}(0) = 0$, and $\xi(\lambda x) = \lambda\xi(x)$ for all $\lambda > 0$, $x \in X$, $U \subseteq X$ is a bounded open set with $0 \in U$, $\varphi\colon \overline{U} \longrightarrow X$ a compact map, and one of the following conditions is satisfied,*

(i) $\xi\big(\varphi(x)\big) \leq \xi(x)$ *for all* $x \in \partial U$ (Röthe's condition), *or*
(ii) $\xi\big(\varphi(x)\big)^2 \leq \xi\big(\varphi(x) - x\big)^2 + \xi(x)^2$ *for all* $x \in \partial U$ (Altman's condition,)

then φ has a fixed point.

PROOF: We show that both Röthe's condition and Altman's condition exclude alternative (b) in Theorem 3.5.16. Arguing by contradiction, suppose that there is $(\lambda, x) \in (0,1) \times \partial U$ such that $x = \lambda\varphi(x)$. Because $\xi(x) = \lambda\xi\big(\varphi(x)\big)$, Röthe's condition is not satisfied. Also

$$\xi\big(\varphi(x) - x\big)^2 + \xi(x)^2 = (1-\lambda)^2\xi\big(\varphi(x)\big)^2 + \lambda^2\xi\big(\varphi(x)\big)^2 < \xi\big(\varphi(x)\big)^2,$$

$\Rightarrow$ Altman's condition is not satisfied. □

The next consequence of Theorem 3.5.16 is known in the literature as *Leray–Schauder alternative principle* or *Schäfer's fixed point theorem*.

COROLLARY 3.5.18 *If X is a Banach space, $\varphi\colon X \longrightarrow X$ is a compact map, and we set*

$$S(\varphi) = \{x \in X : \textit{there exists } \lambda \in (0,1) \textit{ such that } x = \lambda\varphi(x)\},$$

then either $S(\varphi)$ is unbounded or φ has a fixed point.

PROOF: Suppose that $S(\varphi)$ is bounded. Then we can find $r > 0$ such that $S(\varphi) \subseteq B_r(0)$. Because $\lambda\varphi|_{\overline{B}_r(0)}\colon \overline{B}_r(0) \longrightarrow X$, $\lambda \in (0,1)$, is a compact map without a fixed point on $\partial\overline{B}_r(0)$, then in Theorem 3.5.16 alternative (b) is excluded and so φ must have a fixed point. □

We can have an infinite-dimensional version of Theorem 3.5.10 (Borsuk's fixed point theorem). The proof is identical.

THEOREM 3.5.19 *If X is a Banach space, $U \subseteq X$ a bounded, open, convex, and symmetric set with $0 \in U$ and $\varphi\colon \overline{U} \longrightarrow X$ a compact odd map such that $0 \notin \varphi(\partial U)$, then φ has a fixed point.*

All these results for compact maps can be extended to γ-condensing maps. The proofs remain the same. We simply replace the Leray–Schauder degree by the Nussbaum–Sadovskii degree. So we omit the proofs. We only prove the *Sadovskii fixed point theorem* that follows and which is an extension of Theorem 3.5.14 (Shauder's fixed point theorem).

THEOREM 3.5.20 *If X is a Banach space, $C \subseteq X$ is nonempty, bounded, closed, and convex, and $\varphi\colon C \longrightarrow C$ is a γ-condensing map, then φ has a fixed point.*

PROOF: First suppose that φ is a γ-contraction with constant $k<1$. We define

$$C_0 = C \quad \text{and inductively} \quad C_n = \overline{\text{conv}}\, \varphi(C_{n-1}) \qquad \text{for all } n \geq 1.$$

We have

$$\begin{aligned} & \gamma(C_n) \leq k\,\gamma(C_{n-1}) \leq \ldots \leq k^n \gamma(C_0) \quad \text{for all } n \geq 1 \\ \Rightarrow\ & \gamma(C_n) \longrightarrow 0 \qquad \text{as } n \to \infty \\ \Rightarrow\ & \widehat{C} = \bigcap_{n \geq 1} C_n \qquad \text{is nonempty, compact, and convex} \end{aligned}$$

(see Proposition 3.3.50).

Note that $\varphi\colon \widehat{C} \longrightarrow \widehat{C}$. So by Theorem 3.5.14 we can find $x \in \widehat{C}$ such that $\varphi(x) = x$.

Now we remove the restriction that φ is a γ-contraction. Let $k_n \uparrow 1$ and set $\varphi_n = k_n \varphi$. Then φ_n is a γ-contraction and so from the first part of the proof we have $x_n \in C$ such that $\varphi_n(x_n) = x_n$. Then $x_n - \varphi(x_n) = (k_n - 1)\varphi(x_n) \longrightarrow 0$ and because $(I - \varphi)(C)$ is closed, we conclude that for some $x \in C$ we have $\varphi(x) = x$. $\square$

The nonlinear alternative theorem is still valid for γ-condensing maps.

THEOREM 3.5.21 *If X is a Banach space, $U \subseteq X$ is bounded open with $0 \in U$, and $\varphi\colon \overline{U} \longrightarrow X$ is a γ-condensing map, then the following alternative holds; either*

(a) *φ has a fixed point in $\overline{U}$, or*
(b) *There exists $(\lambda, x) \in (0,1) \times \partial U$ such that $x = \lambda\varphi(x)$.*

This alternative theorem has some interesting consequences.

COROLLARY 3.5.22 *If X is a Banach space, $\xi\colon X \longrightarrow \mathbb{R}_+$ is a function as in Corollary 3.5.17, $U \subseteq X$ is a bounded open set with $0 \in U$ and $\varphi\colon \overline{U} \longrightarrow X$ is a γ-condensing map satisfying one of the following conditions,*

(i) *$\xi\big(\varphi(x)\big) \leq \xi(x)$ for all $x \in \partial U$ (Röthe's condition), or*
(ii) *$\xi\big(\varphi(x)\big)^2 \leq \xi\big(\varphi(x) - x\big)^2 + \xi(x)^2$ for all $x \in \partial U$ (Altman's condition),*

then φ has a fixed point.

COROLLARY 3.5.23 *If X is a Banach space, $\varphi\colon X \longrightarrow X$ is a γ-condensing map, and we set*

$$S(\varphi) = \{x \in X : \textit{there exists } \lambda \in (0,1) \textit{ such that } x = \lambda\varphi(x)\},$$

then either $S(\varphi)$ is unbounded or φ has a fixed point.

Borsuk's fixed point theorem (see Theorem 3.5.19), is still valid for γ-condensing maps.

THEOREM 3.5.24 *If X is a Banach space, $U \subseteq X$ a bounded, open, convex, and symmetric set with $0 \in U$, and $\varphi\colon \overline{U} \longrightarrow X$ is a γ-condensing odd map such that $0 \notin \varphi(\partial U)$, then φ has a fixed point.*

We prove two more fixed point theorems, this time directly for γ-condensing maps. We start with a definition.

DEFINITION 3.5.25 Let X be a Banach space and $\varphi\colon X \longrightarrow X$ a map.

(a) We say that φ is *quasibounded* if

$$|\varphi|_\infty = \limsup_{\|x\| \longrightarrow \infty} \frac{\|\varphi(x)\|}{\|x\|} = \inf_{M>0} \sup_{\|x\| \geq M} \frac{\|\varphi(x)\|}{\|x\|} < +\infty.$$

The number $|\varphi|_\infty$ is called the *quasinorm* of φ.

(b) We say that φ is *asymptotically linear* if there exists $T \in \mathcal{L}(X)$ such that

$$\lim_{\|x\| \longrightarrow \infty} \frac{\|Tx - \varphi(x)\|}{\|x\|} = 0.$$

The operator $T \in \mathcal{L}(X)$ is uniquely determined; it is called the *asymptotic derivative* of φ and it is denoted by $\varphi'(\infty)$.

THEOREM 3.5.26 *If X is a Banach space and $\varphi\colon X \longrightarrow X$ is a quasibounded γ-condensing map, then for every $y \in X$ and every $\lambda \in [-1,1]$ such that $|\lambda| < 1/\big(|\varphi|_\infty\big)$ the operator equation $y = x - \lambda\varphi(x)$ has a solution; in particular then the γ-condensing map $\lambda\varphi$ has a fixed point.*

PROOF: Let $\psi_\lambda = y + \lambda\varphi(x)$. Because $|\lambda| \leq 1$, we see that ψ_λ is γ-condensing. It is easy to check that $|\psi_\lambda|_\infty = |\lambda||\varphi|_\infty < 1$. So we can find $M > 0$ such that

$$\frac{\|\psi_\lambda(x)\|}{\|x\|} < 1 \qquad \text{for all } \|x\| \geq M.$$

Then we can apply Corollary 3.5.23 with Rothe's condition and $\xi(\cdot) = \|\cdot\|$ to obtain a fixed point for φ. □

THEOREM 3.5.27 *If X is a Banach space, $\varphi\colon X \longrightarrow X$ is asymptotically linear, γ-condensing, and*

$$\|\varphi'(\infty)\| < 1,$$

then for every $y \in X$ and every $\lambda \in [-1,1]$, the equation $y = x - \lambda\varphi(x)$ has a solution; in particular the γ-condensing map $\lambda\varphi$ has a fixed point.

PROOF: It is easy to check that φ is quasibounded and $|\varphi|_\infty \leq \|\varphi'(\infty)\|_{\mathcal{L}}$. Hence $|\varphi|_\infty < 1$ and so we can apply Theorem 3.5.27 and finish the proof. □

We conclude this section with an extension of Theorem 3.5.14 to locally convex spaces. The result is known as *Tychonov's fixed point theorem.*

THEOREM 3.5.28 *If X is a locally convex space, $C \subseteq X$ a nonempty, closed, and convex set, $\varphi\colon C \longrightarrow C$ is continuous, and $\overline{\varphi(C)}$ is compact, then φ has a fixed point.*

3.6 Order and Fixed Points

In many applied fields, nonnegativity of the solution of the boundary value problem is a minimal requirement in order for a function to qualify as the state of the model. What we normally understand by nonnegativity can be described using order cones. Using order cones we can extend the notion of monotonicity of operators beyond the framework of dual pairs (see Section 3.2). Also the theory of cones can be coupled with the notion of a fixed point index in order to investigate positive fixed points of various classes of nonlinear operators. Cone analysis is a valuable tool in the study of nonlinear problems and in this section we survey some important parts of this theory.

We start with a discussion of order cones.

DEFINITION 3.6.1 Let X be a Banach space. By an *order cone*, we understand a closed convex set $K \subseteq X$ such that $\lambda K \subseteq K$ for all $\lambda \geq 0$ and $K \cap (-K) = \{0\}$.

REMARK 3.6.2 Given an order cone $K \subseteq X$, we can introduce a partial ordering $\leq$ on X by $x \leq u$ if and only if $u - x \in K$. Usually we write $x < u$ if and only if $u - x \in K \setminus \{0\}$ and $x \ll u$ if and only $u - x \in \operatorname{int} K$, provided of course that $\operatorname{int} K \neq \emptyset$. Using this partial ordering we can speak of monotone sequences $\{x_n\}_{n \geq 1}$; that is, of sequences satisfying $x_n \leq x_{n+1}$, $n \geq 1$ (increasing) or $x_{n+1} \leq x_n$ (decreasing). Also a set $C \subseteq X$ is said to be bounded above (resp., bounded below), if there exists $u \in X$ such that $c \leq u$ (resp., $u \leq c$) for all $c \in C$. By $\sup C$ (resp., $\inf C$) we denote the least upper bound (resp., the greatest lower bound) of C with respect to $\leq$ if it exists. The space X with the order $\leq$ induced by K is an ordered Banach space (OBS for short).

Next we introduce some special kinds of order cones that we encounter in practice.

DEFINITION 3.6.3 Let X be a Banach space, $K \subseteq X$ an order cone and, $\leq$ the partial ordering induced by K.

(a) We say that K is *solid* if $\operatorname{int} K \neq \emptyset$.
(b) We say that K is *generating* (resp., *total*) if $X = K - K$ (resp., $X = \overline{K - K}$).
(c) We say that K is *normal* if there exists $\beta > 0$ such

$$\|x + u\| \geq \beta \qquad \text{for all } \|x\| = \|u\| = 1.$$

(d) We say that K is *regular* if every increasing sequence which is bounded above is convergent.
(e) We say that K is *fully regular* if every increasing sequence which is norm bounded is convergent.
(f) We say that K is *minihedral* if $\sup\{x, u\}$ exist for all $x, u \in X$.
(g) We say that K is *strongly minihedral* if every set which is bounded above has a supremum.

REMARK 3.6.4 Normality means that the angle between two positive unit vectors is bounded away from π; that is, a normal cone cannot be too large. Also it is clear that K is regular if and only if every decreasing and bounded below sequence is convergent.

In the next proposition we present some useful criteria for normality of a cone.

PROPOSITION 3.6.5 *If X is a Banach space and $K \subseteq X$ is an order cone, then the following statements are equivalent.*

(a) *K is normal.*

(b) *There exists $\gamma > 0$ such that $\|x+u\| \geq \gamma \max\{\|x\|, \|u\|\}$ for all $x, u \in X$.*

(c) *There exists $M > 0$ such that $0 \leq x \leq u$ implies $\|x\| \leq M\|y\|$ (i.e., $\|\cdot\|$ is semimonotone).*

(d) *There exists an equivalent norm $\|\cdot\|_1$ on X such that $0 \leq x \leq u$ implies $\|x\|_1 \leq \|u\|_1$ (i.e., the new norm $\|\cdot\|_1$ is monotone).*

(e) *If $x_n \leq v_n \leq u_n$, $n \geq 1$ and $\|x_n - x\| \longrightarrow 0$, $\|u_n - x\| \longrightarrow 0$, then $\|v_n - x\| \longrightarrow 0$.*

(f) *$\big(\overline{B}_1(0) + K\big) \cap \big(\overline{B}_1(0) - K\big)$ is bounded, where $\overline{B}_1(0) = \{x \in X : \|x\| \leq 1\}$.*

(g) *Every order interval $[x, u] = \{v \in X : x \leq v \leq u\}$ is bounded (in norm).*

PROOF: (a)$\Rightarrow$(b): We assume without any loss of generality that $\|x\| = 1$ and $\|u\| \leq 1$. We have

$$1 = \|x\| \leq \|x+u\| + \|-u\| = \|x+u\| + \|u\|. \tag{3.79}$$

Also because K is normal, from Definition 3.6.3(c), we have

$$\|x+u\| = \|x + \frac{u}{\|u\|} - \frac{1-\|u\|}{\|u\|}u\| \geq \|x + \frac{u}{\|u\|}\| - 1 + \|u\| \geq \beta - 1 + \|u\|. \tag{3.80}$$

Combining (3.79) and (3.80) we obtain

$$\gamma = \frac{\beta}{2} \leq \|x+u\|.$$

(b)$\Rightarrow$(c): If the implication is not true, then we can find sequences $\{x_n\}_{n\geq 1}$, $\{u_n\}_{n\geq 1} \subseteq K$ such that $x_n \leq u_n$ and $\|x_n\| > n\|u_n\|$ for all $n \geq 1$. We set

$$y_n = \frac{x_n}{\|x_n\|} + \frac{u_n}{n\|u_n\|} \quad \text{and} \quad v_n = -\frac{x_n}{\|x_n\|} + \frac{u_n}{n\|u_n\|} \quad \text{for all } n \geq 1.$$

Then $\{y_n\}_{n\geq 1}$, $\{v_n\}_{n\geq 1} \subseteq K$ and we have

$$\|y_n\| \geq 1 - \frac{1}{n} \quad \text{and} \quad \|v_n\| \geq 1 - \frac{1}{n} \quad \text{for all } n \geq 1,$$

$$\Rightarrow \frac{2}{n} = \|y_n + v_n\| \geq \gamma\left(1 - \frac{1}{n}\right) \quad \text{for all } n \geq 1 \text{ (because of (b))}.$$

For $n \geq 1$ large the above inequality leads to a contradiction.

(c)$\Rightarrow$(d): We define

$$\|x\|_1 = \inf_{u \leq x} \|u\| + \inf_{v \geq x} \|v\|. \tag{3.81}$$

We show that $\|\cdot\|_1$ is the desired monotone norm on X. Clearly $\|0\|_1 = 0$. Also suppose that $\|x\|_1 = 0$. From (3.81) we see that given $\varepsilon > 0$ we can find $u \leq x \leq v$ such that $\|u\|, \|v\| < \varepsilon$. Then because of (c), we have

$$\|x\| \leq \|x - u\| + \|u\| \leq M\|v - u\| + \|u\| \leq (2N + 1)\varepsilon.$$

Because $\varepsilon > 0$ was arbitrary, we let $\varepsilon \downarrow 0$ to get that $\|x\| = 0$; that is, $x = 0$.

Clearly $\|\mu x\|_1 = |\mu| \|x\|_1$ for all $\mu \in \mathbb{R}$ and all $x \in X$. We check that $\|\cdot\|_1$ satisfies the triangle inequality in order to establish that $\|\cdot\|_1$ is a norm on X. So suppose $x, y \in X$. Given $\varepsilon > 0$ we can find $u_1, v_1, u_2, v_2 \in X$ such that $u_1 \leq x \leq v_1$, $u_2 \leq y \leq v_2$, $\|u_1\| + \|v_1\| \leq \|x\|_1 + \varepsilon$ and $\|u_2\| + \|v_2\| \leq \|y\|_1 + \varepsilon$. We have $u_1 + u_2 \leq x + y \leq v_1 + v_2$ and so it follows that $\|x + y\|_1 \leq \|u_1 + u_2\| + \|v_1 + v_2\|$. Hence

$$\|x + y\|_1 \leq \|x\|_1 + \|y\|_1 + 2\varepsilon.$$

Because $\varepsilon > 0$ was arbitrary, we let $\varepsilon \downarrow 0$, to obtain $\|x + y\|_1 \leq \|x\|_1 + \|y\|_1$.

From (3.81) it is clear that $0 \leq x \leq y$, then

$$\|x\|_1 = \inf_{v \geq x} \|v\| \leq \inf_{v \geq y} \|v\| = \|y\|_1,$$
$$\Rightarrow \quad \|\cdot\|_1 \text{ is monotone.}$$

Finally we need to show that the two norms $\|\cdot\|$, $\|\cdot\|_1$ are equivalent. Clearly from (3.81), we have $\|\cdot\|_1 \leq 2\|\cdot\|$. On the other hand for any $u \leq x \leq v$, we have

$$\|x\| \leq \|x - u\| + \|u\| \leq M\|v - u\| + \|u\| \leq (M + 1)(\|u\| + \|v\|)$$
$$\Rightarrow \quad \|x\| \leq (N + 1)\|x\|_1 \quad \text{(see (3.81))}.$$

So the two norms are equivalent and we are done.

(d)$\Rightarrow$(e): We have $0 \leq v_n - x_n \leq u_n - x_n$ and so from (d) $\|v_n - x_n\|_1 \leq \|u_n - x_n\|_1$. Because of the equivalence of the norms $\|\cdot\|$ and $\|\cdot\|_1$ and since $\|x_n - x\| \longrightarrow 0$, $\|u_n - x\| \longrightarrow 0$, we conclude that $\|v_n - x\| \longrightarrow 0$.

(e)$\Rightarrow$(f): If $D = \big(\overline{B}_1(0) + K\big) \cap \big(\overline{B}_1(0) - K\big)$ is unbounded, then we can find $\{y_n\}_{n \geq 1} \subseteq D$ such that $\|y_n\| \longrightarrow \infty$. Then $x_n \leq y_n \leq u_n$ with x_n, $u_n \in \overline{B}_1(0)$ for all $n \geq 1$. We set

$$\widehat{x}_n = \frac{x_n}{\|y_n\|}, \quad \widehat{y}_n = \frac{y_n}{\|y_n\|} \quad \text{and} \quad \widehat{u}_n = \frac{u_n}{\|y_n\|} \quad \text{for all } n \geq 1$$

and we have $\widehat{x}_n \leq \widehat{y}_n \leq \widehat{u}_n$ for all $n \geq 1$. Note that $\widehat{x}_n$, $\widehat{u}_n \longrightarrow 0$. Hence because of (e) it follows that we must have $\widehat{y}_n \longrightarrow 0$, a contradiction to the fact that $\|\widehat{y}_n\| = 1$ for all $n \geq 1$.

(f)$\Rightarrow$(g): Suppose that $\big(\overline{B}_1(0) + K\big) \cap \big(\overline{B}_1(0) - K\big) \subseteq r\overline{B}_1(0)$, $r > 0$. Let $\varrho = \max\{\|x\|, \|u\|\}$. Then for any $x \in [x, u]$, we have $(1/\varrho)v \in \big(\overline{B}_1(0) + K\big) \cap \big(\overline{B}_1(0) - K\big) \subseteq r\overline{B}_1(0)$, hence $[x, u] \subseteq \varrho r\overline{B}_1(0)$; that is, $[x, u]$ is bounded.

(g)⇒(a): Suppose K is not normal. Then according to Definition 3.6.3(c), we can find $\{x_n\}_{n\geq 1}$, $\{u_n\}_{n\geq 1} \subseteq \partial B_1(0)\cap K$ such that $\|x_n+u_n\|<(1/2^{2n})$, $n\geq 1$. We set

$$y_n = \frac{x_n}{\|x_n+u_n\|^{1/2}} \qquad \text{and} \qquad v_n = \frac{x_n+u_n}{\|x_n+u_n\|^{1/2}} \qquad \text{for all } n\geq 1.$$

Then $0 \leq y_n \leq v_n$ and $\sum\limits_{n\geq 1}\|v_n\| \leq \sum\limits_{n\geq 1}(1/2^n) < +\infty$. So the series $\sum\limits_{n\geq 1} v_n$ is converging strongly to $v\in X$. We have $0\leq y_n\leq v_n\leq v$ and

$$\|y_n\| = \frac{1}{\|x_n+u_n\|^{1/2}} > 2^n \qquad \text{for all } n\geq 1,$$
$$\Rightarrow \ [0,v] \text{ is unbounded, a contradiction.}$$

□

REMARK 3.6.6 So by virtue of this proposition, in an ordered Banach space with a normal order cone, every order bounded set C (i.e., there is an order interval $[x,u]$ containing C) is bounded.

PROPOSITION 3.6.7 *If X is an OBS with order cone K and* int $K\neq\emptyset$, *then K is generating.*

PROOF: Let $x_0\in \text{int}\, K$ and choose $r>0$ such that $\overline{B}_r(x_0)\subseteq K$. If $x\in X\setminus\{0\}$, then $x_0 + r\,(x/\|x\|)\in K$. Hence

$$x = \frac{\|x\|}{r}\Big(x_0 + \frac{r}{\|x\|}x\Big) - \frac{\|x\|}{r}x_0 \in K - K;$$

that is, $X = K-K$. □

From the above proof it is also clear that the following result is true.

PROPOSITION 3.6.8 *If X is an OBS with order cone K,* int $K\neq\emptyset$ *and $x_0\in$* int K, *then for every $x\in X$ we can find $\lambda=\lambda(x)>0$ such that $\lambda x_0 - x\in$* int K.

PROPOSITION 3.6.9 *If X is an OBS with order cone K and we consider the following conditions,*

(i) *K is fully regular,*
(ii) *K is regular,*
(iii) *K is normal,*

then (i)⇒(ii)⇒(iii).

PROOF: First we show that if K is not normal, then it is not regular, nor fully regular. Because K is not normal, by virtue of Proposition 3.6.5(c), we can find $\{x_n\}_{n\geq 1}$, $\{u_n\}_{n\geq 1}\subseteq K$ such that $0\leq x_n\leq u_n$ and $\|x_n\|>2^n\|u_n\|$ for all $n\geq 1$. We set

$$y_n = \frac{x_n}{\|x_n\|} \qquad \text{and} \qquad v_n = \frac{u_n}{2^n\|u_n\|} \qquad \text{for all } n\geq 1.$$

Then we have

$$0 \leq y_n \leq \frac{x_n}{2^n \|u_n\|} \leq v_n \qquad \text{for all } n \geq 1. \tag{3.82}$$

Also

$$\sum_{n \geq 1} \|v_n\| = \sum_{n \geq 1} \frac{1}{2^n} = 1. \tag{3.83}$$

Hence $\sum\limits_{n \geq 1} v_n = v \in X$. We define

$$w_n = \begin{cases} \sum\limits_{k=1}^{n} v_k & \text{if } n = 2m,\ m \geq 1 \\ \\ \sum\limits_{k=1}^{n} v_k + y_{2m+1} & \text{if } n = 2m+1,\ m \geq 1 \end{cases}.$$

Then because of (3.82) and (3.83) it follows that

$$0 \leq w_1 \leq \cdots \leq w_n \leq w_{n+1} \leq \cdots \leq v. \tag{3.84}$$

$$\sup_{n \geq 1} \|w_n\| \leq 2. \tag{3.85}$$

On the other hand $\|w_{2m+1} - w_{2m}\| = \|y_{2m+1}\| = 1$. So $\{w_n\}_{n \geq 1}$ does not converge and this means that K is not regular (see (3.84)) and K is not fully regular (see (3.85)).

So it remains to show that (i)⇒(ii). Because K is fully regular, then from the above argument it follows that K is normal. So let $\{x_n\}_{n \geq 1} \subseteq X$ be an increasing, order bounded sequence. We have

$$x_1 \leq x_2 \leq \cdots \leq x_n \leq x_{n+1} \leq \cdots \leq u,$$

$$\Rightarrow\ 0 \leq u - x_n \leq u - x_1 \qquad \text{for all } n \geq 1,$$

$$\Rightarrow\ \|u - x_n\| \leq M\|u - x_1\| \ \text{ for some } M > 0 \text{ and all } n \geq 1,$$

(see Proposition 3.6.5(c))

$$\Rightarrow\ \{x_n\}_{n \geq 1} \subseteq X \ \text{ is norm bounded and by full regularity of } K \text{ converges.}$$

□

PROPOSITION 3.6.10 *If X is a reflexive OBS with order cone K, then K is normal ⇔ K is regular ⇔ K is fully regular.*

PROOF: By virtue of Proposition 3.6.9 it suffices to show that normality of K implies full regularity of K. So suppose that $\{x_n\}_{n \geq 1} \subseteq X$ is an increasing, norm bounded sequence. Since X is reflexive we can find $\{x_{n_k}\}_{k \geq 1}$ a subsequence of $\{x_n\}_{n \geq 1}$ such that $x_{n_k} \xrightarrow{w} \widehat{x} \in X$. We claim that $x_{n_k} \leq \widehat{x}$ for all $k \geq 1$. If this is not the case, we can find $k_0 \geq 1$ such that $\widehat{x} - x_{n_{k_0}} \notin K$. Then the strong separation theorem for convex sets implies that there exist $x^* \in X^* \setminus \{0\}$ and $\varepsilon > 0$ such that

$$\left\langle x^*, \widehat{x} - x_{n_{k_0}} \right\rangle + \varepsilon \leq \left\langle x^*, x_{n_k} - x_{n_{k_0}} \right\rangle \qquad \text{for all } k \geq k_0. \tag{3.86}$$

If we pass to the limit as $k \longrightarrow \infty$ in (3.86), we reach a contradiction. So $x_{n_k} \leq \widehat{x}$ for all $k \geq 1$.

We show that $\{x_{n_k}\}_{k\geq 1}$ has a subsequence which strongly converges to $\widehat{x}$ in X. Indeed if this is not the case, we can find $\varepsilon > 0$ and $k_0 \geq 1$ such that

$$\|\widehat{x} - x_{n_k}\| \geq \varepsilon \qquad \text{for all } k \geq k_0. \tag{3.87}$$

Let $C_k = \{x \in X : x \leq x_{n_k}\}$, $k \geq k_0$, and $C = \bigcup_{k\geq k_0} C_k$. Evidently each set is convex and $\{C_k\}_{k\geq 1}$ is increasing. So C is convex and thus so is $\overline{C}$. If $x \in C$, then $x \in C_k$ for some $k \geq k_0$ and so $x \leq x_{n_k}$, hence $0 \leq \widehat{x} - x_{n_k} \leq \widehat{x} - x$ and so $\|\widehat{x} - x_{n_k}\| \leq M\|\widehat{x} - x\|$. Because of (3.87), we obtain

$$\frac{\varepsilon}{M} \leq \|\widehat{x} - x\| \qquad \text{for all } x \in C, \quad (\text{i.e., } \widehat{x} \notin \overline{C}).$$

Then by the strong separation theorem for convex set we can find $u^* \in X^* \setminus \{0\}$ and $\delta > 0$ such that

$$\langle x^*, \widehat{x}\rangle + \delta \leq \langle x^*, c\rangle \qquad \text{for all } c \in \overline{C}. \tag{3.88}$$

But clearly $\widehat{x} \in C$ and this contradicts (3.88). Therefore (3.87) can not happen and so we can find a subsequence $\{x_{n_{k_m}}\}_{m\geq 1}$ of $\{x_{n_k}\}_{k\geq 1}$ such that $x_{n_{k_m}} \longrightarrow \widehat{x}$ in X as $m \to \infty$. Since for every $n \geq 1$, we can find $m \geq 1$ large enough such that $x_n \leq x_{n_{k_m}}$, we deduce that $x_n \leq \widehat{x}$. Then

$$\begin{aligned}
& 0 \leq \widehat{x} - x_n \leq \widehat{x} - x_{n_{k_m}} \qquad \text{for all } n \geq n_{k_m}, \\
\Rightarrow\ & \|\widehat{x} - x_n\| \leq M\, \|\widehat{x} - x_{n_{k_m}}\| \quad (\text{see Proposition 3.6.5(c)}), \\
\Rightarrow\ & x_n \longrightarrow \widehat{x} \text{ in } X.
\end{aligned}$$

□

PROPOSITION 3.6.11 *If X is a separable OBS with order cone K that is regular and minihedral, then K is strongly minihedral.*

PROOF: Let $C \subseteq X$ be bounded above. Let $\{x_n\}_{n\geq 1} \subseteq D$ be dense in D. Set $u_n = \sup\{x_k\}_{k=1}^n$ which exist because by hypothesis K is minihedral. We have

$$u_1 \leq u_2 \leq \cdots \leq u_n \leq \cdots \leq w \quad \text{for some } w \in X \ \ (\text{since } D \text{ is bounded above}).$$

Because K is regular, we have $u_n \longrightarrow u$ in X. We show that $u = \sup C$. Clearly

$$x_n \leq u_n \leq u \qquad \text{for all } n \geq 1. \tag{3.89}$$

Let $x \in C$.Then we can find a subsequence $\{x_{n_k}\}_{k\geq 1}$ of $\{x_n\}_{n\geq 1}$ such that $x_{n_k} \longrightarrow x$ in X. Because of (3.89) it follows that $x \leq u$ and so u is an upper bound of C. If $v \in X$ is any other upper bound of C, then $u_n \leq v$ for all $n \geq 1$ and so $u \leq v$. Therefore $u = \sup C$. □

COROLLARY 3.6.12 *If X is a separable reflexive OBS and the order cone $K \subseteq X$ is normal and minihedral, then K is strongly minihedral.*

Let us see some characteristic examples of ordered Banach spaces.

EXAMPLE 3.6.13 (a) $X=\mathbb{R}^N$ with order (positive) cone $K=\mathbb{R}_+^N=\{\widehat{x}=(x_k)_{k=1}^N : x_k \geq 0$ for all $k\in\{1,\dots,N\}\}$. This cone is solid, generating, and because the norm in $\mathbb{R}^N$ is monotone, K is also normal (see Proposition 3.6.5(d)). Then Proposition 3.6.10 implies that K also regular and fully regular. Then because K is clearly minihedral, Proposition 3.6.10 implies that K is strongly minihedral.

(b) $X=C(D)$ where $D\subseteq\mathbb{R}^N$ is a closed subset, with positive cone $K=\{x\in C(D) : x(t)\geq 0 \quad \text{for all } t\in D\}$. Then K is solid and normal (note that order intervals are bounded; see Proposition 3.6.5(g)). However, it is not regular. To see this, consider $x_n(t)=1-t^n$, $t\in[0,1]$, $n\geq 1$. Then $\{x_n\}_{n\geq 1}$ is an increasing sequence, order bounded by $u\equiv 1$, but it does not converge uniformly on $[0,1]$. Moreover, K is minihedral but not strongly minihedral. Indeed, if $E=\{x\in C[0,2] : x(t)<1$ for $t\in (0,1)$ and $x(t)<2$ for $t\in(0,2)\}$, then $\sup E$ does not exist.

Another positive cone for $X=C(D)$ is the following.

$$\widehat{K}=\{x\in C(D) : x(t)\geq 0 \quad \text{for all } t\in D \text{ and } \min_{t\in D_0} x(t)\geq \varepsilon_0\|x\|_\infty\}$$

with D_0 a closed subset of D and $0<\varepsilon_0<1$. This is a solid normal cone.

(c) $X=L^p(Z)$, $1\leq p<\infty$, where $Z\subseteq\mathbb{R}^N$ is a Lebesgue measurable set with finite Lebesgue measure, with positive cone $K=\{x\in L^p(Z) : x(z)\geq 0$ a.e. on $Z\}$. Clearly K is generating but not solid (i.e., $\operatorname{int} K=\emptyset$). Also it is normal because it has bounded order intervals. Moreover, if $1<p<\infty$, Proposition 3.6.10 implies that K is regular and fully regular. If $p=1$, by the monotone convergence theorem and the Lebesgue dominated convergence theorem, we infer that K is fully regular, hence regular too. Also K is minihedral and because it is separable and regular, it is also strongly minihedral (see Proposition 3.6.11).

(d) $X=W^{1,p}(Z)$, $1<p<\infty$, and $Z\subseteq\mathbb{R}^N$ a bounded open set with a C^1-boundary ∂Z. The positive cone of this OBS is $K=\{x\in W^{1,p}(Z) : x(z)\geq 0$ a.e. on $Z\}$. In general K is not solid. However, if $p>N$, by the Sobolev embedding theorem $W^{1,p}(Z)$ is embedded in $C(\overline{Z})$ and so in this case K is solid (i.e., $\operatorname{int} K\neq 0$, see Example 3.6.13(b)). Note that order intervals are not bounded, so K is not normal (see Proposition 3.6.5).

(e) $X=C_0(\mathbb{R}_+)=\{x\in C(\mathbb{R}_+) : x(t)\longrightarrow 0$ as $t\to\infty\}$ with positive cone $K=\{x\in C_0(\mathbb{R}_+) : x(t)\geq 0 \quad \text{for all } t\in\mathbb{R}_+\}$ (the norm on X is the usual supremum norm $\|x\|_\infty=\max_{t\geq 0}|x(t)|$. This cone is regular (hence normal too), but not fully regular.

Next let us say a few things about positive linear functionals. For this purpose we make the following definition.

DEFINITION 3.6.14 Let X be an OBS with positive cone K. An element $x^*\in X^*$ is said to be *positive* if $\langle x^*,x\rangle\geq 0$ for all $x\in K$. We consider the set of all positive elements of X^*

$$K^*=\{x^*\in X^* : \langle x^*,x\rangle\geq 0 \quad \text{for all } x\in K\}.$$

Then K^* is nonempty, closed, and convex but need not satisfy $K^*\cap(-K^*)=\{0\}$. Nevertheless we call K^* the *dual cone*.

REMARK 3.6.15 Clearly if K is generating, then $K^* \cap (-K^*) = \{0\}$ and so K^* is an order cone too. We say that $x^* \in K^*$ is strictly positive, if $\langle x^*, x\rangle > 0$ for all $x \in K \setminus \{0\}$ and uniformly positive if $\langle x^*, x\rangle \geq c\|x\|$ for some $c > 0$ and all $x \in K$.

In the next proposition we have collected some useful properties of dual cones and their elements.

PROPOSITION 3.6.16 *If X is an OBS with positive cone $K \neq \{0\}$ and K^* its dual cone, then*

(a) $K^* \neq \{0\}$ *and* $x \in K$ *if and only if* $\langle x^*, x\rangle \geq 0$ *for all* $x^* \in K^*$.

(b) *For every* $x \in K \setminus \{0\}$ *we can find* $x^* \in K^*$ *such that* $\langle x^*, x\rangle > 0$.

(c) *If* $x \notin K$, *then we can find* $x^* \in K^*$ *such that* $\langle x^*, x\rangle < 0$.

(d) $x \in \operatorname{int} K$ *if and only if* $\langle x^*, x\rangle > 0$ *for all* $x^* \in K^* \setminus \{0\}$.

(e) *If X is separable, then we can find* $x_0 \in X^*$ *such that* $\langle x_0^*, x\rangle > 0$ *for all* $x \in K \setminus \{0\}$.

PROOF: (a),(b),(c): If $x \notin K$, then by the strong separation theorem for convex sets we can find $x^* \in X^* \setminus \{0\}$ and $\varepsilon > 0$ such that

$$\begin{aligned} \langle x^*, x\rangle + \varepsilon \leq \langle x^*, c\rangle \qquad &\text{for all } c \in K \qquad (3.90)\\ \Rightarrow \ \langle x^*, x\rangle \leq -\varepsilon < 0 \ \ &(\text{i.e., } x^* \neq 0). \end{aligned}$$

Moreover, because $\lambda K \subseteq K$ for all $\lambda \geq 0$, from (3.90) it follows that $\langle x^*, c\rangle \geq 0$ for all $c \in K$ and so $x^* \in K^* \setminus \{0\}$. So (a), (b), and (c) follow.

(d): If $x_0 \in \partial K$, then by the weak separation theorem (applied to the sets $\{x_0\}$ and $\operatorname{int} K$, both convex), we can find $x^* \in K^* \setminus \{0\}$ such that $\langle x^*, x_0\rangle = 0$. On the other hand if $x \in \operatorname{int} K$, we can find $r > 0$ such that $B_r(x) \subseteq K$. Suppose that $\langle x^*, x\rangle = 0$ for some $x^* \in K^* \setminus \{0\}$. Then $\langle x^*, ru\rangle \geq 0$ for all $u \in B_r(0)$ and so $x^* = 0$.

(e): Because X is separable, the set $K^* \cap \overline{B}_1^*(0)$ furnished with the weak*-topology is compact metrizable. Let $\{x_n^*\}_{n\geq 1}$ be a dense sequence and set $u^* = \sum_{n\geq 1} (1/n^2)x_n^*$. Clearly $u^* \in K^*$ and if $\langle u^*, x\rangle = 0$ for some $x \in K$, then $\langle x^*, x\rangle = 0$ for all $x^* \in K^*$ and so $x = 0$ (see part (b)). □

There is a duality relation between being normal and being generating for an order cone. The result is known as *Krein's theorem* and its proof can be found in Deimling [188, p. 223].

PROPOSITION 3.6.17 *If X is an OBS, K is the positive cone, and K^* the dual cone, then*

(a) *K is generating if and only if K^* is normal.*
(b) *K is normal if and only if K^* is generating.*

Using the machinery of order cones, we can easily produce fixed point theorems for increasing maps.

DEFINITION 3.6.18 Let X be an OBS with positive cone K and $\leq$ the partial ordering defined by K. Let $D \subseteq X$ and $\varphi\colon D \longrightarrow X$. We say that

(a) φ is *increasing* (resp., *decreasing*), if $x \leq u$, then $\varphi(x) \leq \varphi(u)$ (resp. $\varphi(u) \leq \varphi(x)$).

(b) φ is *strictly increasing* (resp., *strictly decreasing*), if $x < u$, then $\varphi(x) < \varphi(u)$ (resp., $\varphi(u) < \varphi(x)$).

(c) φ is *strongly increasing* (resp., *strongly decreasing*), if $x < u$, then $\varphi(x) \ll \varphi(u)$ (resp., $\varphi(u) \ll \varphi(x)$), provided $\operatorname{int} K \neq \emptyset$;

THEOREM 3.6.19 *If X is an OBS with a normal positive cone K and $\varphi\colon [x_0, u_0] \longrightarrow [x_0, u_0]$ is increasing and γ-condensing, then φ has a maximal fixed point $\overline{x}$ and a minimal fixed point $\underline{x}$ in $[x_0, u_0]$; moreover, if $u_n = \varphi(u_{n-1})$ and $x_n = \varphi(x_{n-1})$ for all $n \geq 1$, then*

$$\overline{x} = \lim_{n\to\infty} u_n \qquad \textit{and} \qquad \underline{x} = \lim_{n\to\infty} x_n.$$

PROOF: Because φ is increasing, we have

$$x_0 \leq x_1 \leq \cdots \leq x_n \leq \cdots \leq u_n \leq \cdots \leq u_0.$$

The sequence $S = \{x_n\}_{n\geq 0}$ is bounded and $S = \varphi(S) \cup \{x_0\}$. Therefore

$$\gamma(S) = \gamma\big(\varphi(S)\big) \qquad \text{(see Proposition 3.3.50(d)).} \tag{3.91}$$

Because φ is γ-condensing, from (3.91) we infer that $\gamma(S)=0$. So $S=\{x_n\}_{n\geq 0}$ is relatively compact. Hence we can find a subsequence $\{x_{n_k}\}_{k\geq 1}$ of $\{x_n\}_{n\geq 1}$ such that $x_{n_k} \longrightarrow \underline{x}$. Then $x_n \leq \underline{x} \leq u_n$ for all $n \geq 1$. For $n \geq n_k$ we have $0 \leq \underline{x} - x_n \leq \underline{x} - x_{n_k}$ and because K is normal, by Proposition 3.6.5(c) we have $\|\underline{x} - x_n\| \leq M\|\underline{x} - x_{n_k}\|$, hence $x_n \longrightarrow \underline{x}$ in X. Recall that $x_n = \varphi(x_{n-1})$, $n \geq 1$. So passing to the limit as $n \to \infty$, we obtain $\underline{x} = \varphi(\underline{x})$. Similarly we show that $u_n \longrightarrow \overline{x}$ in X and $\overline{x} = \varphi(\overline{x})$.

It remains to show that $\overline{x}$ is the maximal fixed point of φ in $[x_0, u_0]$ and $\underline{x}$ is the minimal fixed point of φ in $[x_0, u_0]$. Let $v \in [x_0, u_0]$ be a fixed point of φ; that is, $v = \varphi(v)$. Because φ is increasing, we see that $x_n \leq v \leq u_n$ for all $n \geq 1$, hence $\underline{x} \leq v \leq \overline{x}$. □

REMARK 3.6.20 The above theorem produces an iteration process that generates the maximal and minimal fixed points of φ in $[x_0, u_0]$. So, if φ has a unique fixed point in $[x_0, u_0]$, then the successive iterates converge to it.

In the next fixed point theorem we do not require the map φ to be continuous.

THEOREM 3.6.21 *If X is an OBS with a strongly minihedral positive cone and $\varphi\colon [x_0, u_0] \longrightarrow [x_0, u_0]$ is an increasing map, then φ has a greatest fixed point $\overline{x}$ and a smallest fixed point $\underline{x}$ in $[x_0, u_0]$.*

PROOF: Let $C = \{x \in [x_0, u_0] : x \leq \varphi(x)\}$. Evidently $x_0 \in C$ and u_0 is an upper bound of C. Because K is strongly minihedral $\overline{x} = \sup C$ exists. We claim that $\overline{x}$ is the maximal fixed point of φ in $[x_0, u_0]$. For any $x \in C$ we have $x_0 \leq x \leq \overline{x} \leq u_0$ and because φ is increasing, we obtain $x_0 \leq \varphi(x_0) \leq \varphi(x) \leq \varphi(\overline{x}) \leq \varphi(u_0) \leq u_0$. Since $x \in C$ we have $x \leq \varphi(x)$ and so $x \leq \varphi(\overline{x})$ for all $x \in C$. Recalling that $\overline{x} = supC$, we obtain $\overline{x} \leq \varphi(\overline{x})$. Also since $\overline{x} \leq \varphi(\overline{x})$, we have $\varphi(\overline{x}) \leq \varphi(\varphi(\overline{x}))$ (recall that φ is increasing) and so $\varphi(\overline{x}) \in C$. Therefore $\varphi(\overline{x}) \leq \overline{x}$, hence $\overline{x} = \varphi(\overline{x})$. Now if x is any other fixed point of φ in $[x_0, u_0]$, then $x \in C$ and $x \leq \overline{x}$, which proves that $\overline{x}$ is the greatest fixed point of φ.

Similarly if $D = \{x \in [x_0, u_0] : \varphi(x) \leq x\}$, we can show that $\underline{x} = \inf D$ exists and is the smallest fixed point of φ in $[x_0, u_0]$. □

We continue with fixed point theorems for maps φ that are not necessarily continuous.

THEOREM 3.6.22 *If X is an OBS with positive cone K and $\varphi\colon [x_0, u_0] \longrightarrow [x_0, u_0]$ is an increasing map such that $\varphi([x_0, u_0])$ is relatively compact in X, then φ has a fixed point in $[x_0, u_0]$.*

PROOF: Let $C = \{x \in \varphi([x_0, u_0]) : x \leq \varphi(x)\}$. Because $x_0 \leq \varphi(x_0)$ and φ is increasing, we have $\varphi(x_0) \leq \varphi(\varphi(x_0))$ and so $\varphi(x_0) \in C$, which means that the set C is nonempty. Evidently C is a partially ordered set with the partial ordering that it inherits from X. Let S be a chain (i.e., a totally ordered subset) of C. Because $S \subseteq C \subseteq \varphi([x_0, u_0])$ and the latter is by hypothesis relatively compact, it follows that S is relatively compact in X. So S is separable. Let $E = \{v_n\}_{n \geq 1} \subseteq S$ be a dense sequence. Since S is a chain $y_n = \sup\{v_k\}_{k=1}^{n}$, $n \geq 1$, exist and $y_n \in S$ (actually $y_n = v_{k_0}$ for some $k_0 \in \{1, \ldots, n\}$. Because S is relatively compact, by passing to a subsequence if necessary, we may assume that $y_n \longrightarrow \widehat{y}$ in X. Clearly $v_n \leq y_n \leq \widehat{y}$ for all $n \geq 1$ and $\widehat{y} \in \overline{S} \subseteq \overline{C} \subseteq \overline{\varphi([x_0, u_0])} \subseteq [x_0, u_0]$. Hence $v \leq \widehat{y}$ for all $v \in S$ and so $v \leq \varphi(v) \leq \varphi(\widehat{y})$ for all $v \in S$, which means that $\varphi(\widehat{y})$ is an upper bound of S. In addition since $y_n \leq \varphi(\widehat{y})$, $n \geq 1$, we obtain $\widehat{y} \leq \varphi(\widehat{y})$ and so $\varphi(\widehat{y}) \leq \varphi(\varphi(\widehat{y}))$, hence $\varphi(\widehat{y}) \in C$. Therefore $\varphi(\widehat{y})$ is an upper bound of C. So we can apply Zorn's lemma and obtain a maximal element $\widehat{x}$ of C. Because $\widehat{x} \leq \varphi(\widehat{x})$, we have $\varphi(\widehat{x}) \leq \varphi(\varphi(\widehat{x}))$ and so $\varphi(\widehat{x}) \in C$. By virtue of the maximality of $\widehat{x}$, we conclude that $\widehat{x} = \varphi(\widehat{x})$. □

COROLLARY 3.6.23 *If X is an OBS with a normal positive cone K and $\varphi\colon [x_0, u_0] \longrightarrow [x_0, u_0]$ is an increasing compact map, then φ has a fixed point.*

PROOF: Because K is normal, the order interval $[x_0, u_0]$ is bounded (see Proposition 3.6.5). Because φ is compact, it follows that $\varphi([x_0, u_0])$ is relatively compact. So we can apply Theorem 3.6.22. □

THEOREM 3.6.24 *If X is an OBS with a minihedral positive cone K and $\varphi\colon [x_0, u_0] \longrightarrow [x_0, u_0]$ is increasing with $\varphi([x_0, u_0])$ relatively compact in X, then φ has a greatest fixed point $\overline{x}$ and a smallest point $\underline{x}$ in $[x_0, u_0]$.*

PROOF: Let C be as in the proof of Theorem 3.6.22. In that proof via Zorn's lemma, we proved that C has a maximal element $\overline{x}$ and $\overline{x} = \varphi(\overline{x})$. We claim that $\overline{x}$ is the greatest fixed point of φ in $[x_0, u_0]$. To this end let y be any fixed point of φ

in $[x_0, u_0]$. Because K is minihedral $v = \sup\{y, \overline{x}\}$ exists. We have $y = \varphi(y) \leq \varphi(v)$ and $\overline{x} = \varphi(\overline{x}) \leq \varphi(v)$. Hence $v \leq \varphi(v)$ and so $y \leq \overline{x}$, which proves that $\overline{x}$ is the greatest fixed point of φ in $[x_0, u_0]$.

In a similar fashion, working with $D = \{x \in \varphi([x_0, u_0]) : \varphi(x) \leq x\}$, we produce a minimal element $\underline{x}$, which is a fixed point of φ and in fact is the smallest fixed point of φ in $[x_0, u_0]$. □

COROLLARY 3.6.25 *If X is an OBS with a normal, minihedral positive cone K and $\varphi\colon [x_0, u_0] \longrightarrow [x_0, u_0]$ is an increasing compact map, then φ has a greatest fixed point $\overline{x}$ and a smallest point $\underline{x}$ in $[x_0, u_0]$.*

Thus far we have assumed that the map $\varphi\colon D \subseteq X \longrightarrow X$ is increasing. Now we drop this monotonicity requirement and instead use tools and methods from degree theory. For this purpose we consider a Banach space X and $C \subseteq X$ a retract of X (see Remark 3.5.5). Let $r\colon X \longrightarrow C$ be an arbitrary retraction. Let $U \subseteq X$ be nonempty, bounded, and open in C and $\varphi\colon \overline{U} \longrightarrow C$ a compact map such that $x \neq \varphi(x)$ for all $x \in \partial U$. We choose $R>0$ such that $U \subseteq B_R(0)$. We define

$$i_C(\varphi, U) = \widehat{d}\big(I - \varphi \circ r, B_R(0) \cap r^{-1}(U), 0\big). \tag{3.92}$$

We show that this definition is independent of the choices of r and $R>0$.

First let $r_1\colon X \longrightarrow C$ be another retraction of X onto C and set $V = B_R(0) \cap r^{-1}(U) \cap r_1^{-1}(U)$. Then V is a bounded open set of X and $U \subseteq V$. It is easy to see that $\varphi \circ r$ has no fixed points in $\overline{B_R(0) \cap r^{-1}(U)} \setminus V$ and $\varphi \circ r$ has no fixed points in $\overline{B_R(0) \cap r_1^{-1}(U)} \setminus V$. So from the excision property of the Leray–Schauder degree, we have

$$\begin{aligned} &\widehat{d}\big(I - \varphi \circ r, B_R(0) \cap r^{-1}(U), 0\big) = \widehat{d}\big(I - \varphi \circ r, V, 0\big) \\ \text{and} \quad &\widehat{d}\big(I - \varphi \circ r_1, B_R(0) \cap r_1^{-1}(U), 0\big) = \widehat{d}\big(I - \varphi \circ r_1, V, 0\big). \end{aligned} \tag{3.93}$$

We introduce the compact homotopy $h(t,x) = r\Big(t\varphi\big(r(x)\big) + (1-t)\varphi\big(r_1(x)\big)\Big)$ for all $(t,x) \in [0,1] \times \overline{V}$. We claim that $x \neq h(t,x)$ for all $t \in [0,1]$ and all $x \in \partial V$. Indeed suppose that we can find $(t,x) \in [0,1] \times \partial V$ such that $x = h(t,x)$. We have

$$\begin{aligned} &x = r\Big(t\varphi\big(r(x)\big) + (1-t)\varphi\big(r_1(x)\big)\Big) \in C, \\ \Rightarrow\ & r(x) = x,\ r_1(x) = x \qquad \text{and so } x = \varphi(x), \end{aligned}$$

hence $x \in U \subseteq V$, a contradiction.

Invoking the homotopy invariance of the Leray–Schauder degree, we obtain

$$\widehat{d}\big(I - \varphi \circ r, V, 0\big) = \widehat{d}\big(I - \varphi \circ r_1, V, 0\big). \tag{3.94}$$

From (3.93) and (3.94) it follows that

$$\widehat{d}\big(I - \varphi \circ r, B_R(0) \cap r^{-1}(U), 0\big) = \widehat{d}\big(I - \varphi \circ r_1, B_R(0) \cap r_1^{-1}(U), 0\big). \tag{3.95}$$

From (3.95) we see that the definition made in (3.92) is in fact independent of the particular retraction we use.

Next we show that the definition made in (3.92) is also independent of $R > 0$. So let $R_1 > R$. We have

$$U \subseteq B_R(0) \cap r^{-1}(U) \subseteq B_{R_1}(0) \cap r^{-1}(U)$$

and so $\varphi \circ r$ has no fixed points in $\overline{B_{R_1}(0) \cap r^{-1}(U)} \setminus \big(B_R(0) \cap r^{-1}(U)\big)$. So once again from the excision property of the Leray–Schauder degree, we have

$$\widehat{d}\big(I - \varphi \circ r, B_{R_1}(0) \cap r^{-1}(U), 0\big) = \widehat{d}\big(I - \varphi \circ r, B_R(0) \cap r^{-1}(U), 0\big)$$

which proves that the definition made in (3.92) is also independent of $R > 0$. So we can now formally introduce the notion of the fixed point index of φ with respect to a retract C.

DEFINITION 3.6.26 Let X be a Banach space, $C \subseteq X$ a retract of X and consider the family

$$\mathcal{S}_I = \big\{(\varphi, U) : U \subseteq C \text{ bounded open in } C,\ \varphi\colon \overline{U} \longrightarrow X \text{ a compact map}$$
$$\text{such that } x \neq \varphi(x) \text{ for all } x \in \partial U\big\}.$$

We can define a map $i_C\colon \mathcal{S}_I \longrightarrow \mathbb{Z}$ by

$$i_C(\varphi, U) = \widehat{d}\big(I - \varphi \circ r, B_R(0) \cap r^{-1}(U), 0\big),$$

where $r\colon X \longrightarrow C$ is an arbitrary retraction of X onto C and $R > 0$ is such that $U \subseteq B_R(0)$. The integer $i_C(\varphi, U)$ is called the *fixed point index* of φ on U with respect to C.

From this definition and the properties of the Leray–Schauder degree (see Theorem 3.3.40) we obtain the following.

THEOREM 3.6.27 *If X is a Banach space, $C \subseteq X$ a retract, and $(\varphi, U) \in \mathcal{S}_I$ (see Definition 3.6.26), then*

(a) Normalization: *$i_C(\varphi, U) = 1$ if $\varphi(x) = u_0 \in U$ for all $x \in \overline{U}$ and $i_C(\varphi, U) = 0$ if $u_0 \notin \overline{U}$.*

(b) Additivity with respect to the domain: *If U_1, U_2 are disjoint open subsets of U and φ has no fixed points on $\overline{U} \setminus (U_1, U_2)$, then $i_C(\varphi, U) = i_C(\varphi, U_1) + i_C(\varphi, U_2)$.*

(c) Homotopy invariance: *If $h\colon [0,1] \times \overline{U} \longrightarrow C$ is a compact map such that $x \neq h(t, x)$ for all $(t, x) \in [0,1] \times \partial U$, then $i_C\big(h(t, \cdot), U\big)$ is independent of $t \in [0,1]$.*

(d) Reduction property: *If D is a retract of C and $\varphi(\overline{U}) \subseteq D$, then $i_C(\varphi, U) = i_D(\varphi, U \cap D)$.*

(e) Excision property: *If U_0 is an open subset of U and φ has no fixed points in $\overline{U} \setminus U_0$, then $i_C(\varphi, U) = i_C(\varphi, U_0)$.*

(f) Solution property: *If $i_C(\varphi, U) \neq 0$, then there exists $x \in U$ such that $x = \varphi(x)$.*

REMARK 3.6.28 The notion of the fixed point index can be extended to γ-condensing maps, using the Nussbaum–Sadovskii degree $\widehat{d_0}$ (see Definition 3.3.56). We leave the straightforward details to the reader. We still denote it by $i_C(\varphi, U)$ and Theorem 3.6.27 remains valid.

Recall that every nonempty, closed, and convex subset of a Banach space X is retract of X (see Theorem 3.1.10 and Remark 3.1.12).

PROPOSITION 3.6.29 *If X is a Banach space, $C \subseteq X$ is nonempty, closed, and convex, $(\varphi, U) \in \mathcal{S}_I$, and there exists $u \in U$ such that $\varphi(x) - x \neq \lambda(x-u)$ for all $\lambda \geq 0$ and all $x \in \partial U$, then $i_C(\varphi, U) = 1$.*

PROOF: Let $h(t,x) = t\varphi(x) + (1-t)u$ for all $(t,x) \in [0,1] \times \overline{U}$. If for some $(t,x) \in [0,1] \times \partial U$ we have $x = h(t,x)$, then clearly $t \neq 0$ (because $u \in U$) and $\varphi(x) - x = \big((1-t)/t\big)(x-u)$, so $\varphi(x) - x = \lambda(x-u)$ with $\lambda = (1-t)/t$, $t \neq 0$, a contradiction to the hypothesis. So invoking the homotopy invariance and normalization properties of the fixed point index (see Theorem 3.6.27(c) and (a)), we have

$$i_C(\varphi, U) = i_C(u, U) = 1.$$

$\square$

PROPOSITION 3.6.30 *If X is a Banach space, $C \subseteq X$ is nonempty, closed, and convex, $(\varphi, U) \in \mathcal{S}_I$, and there exists $u \in C \setminus U$ such that $\varphi(x) - x = \lambda(x-u)$ for all $\lambda \geq 0$ and all $x \in \partial U$, then $i_C(\varphi, U) = 0$.*

PROOF: We consider the compact homotopy $h(t,x) = t\varphi(x) + (1-t)u$. If for some $(t,x) \in [0,1] \times \partial U$, $x = h(t,x)$, then clearly $t \neq 0$ and $\varphi(x) - x = \big((1-t)/t\big)(x-u)$, with $(1-t)/t \geq 0$ and this contradicts the assumption. So from the homotopy invariance and normalization properties of the fixed point index, we have

$$i_C(\varphi, U) = i_C(u, U) = 0.$$

$\square$

COROLLARY 3.6.31 *If X is an OBS with order cone K, $U \subseteq X$ is a bounded open set, $\varphi\colon K \cap \overline{U} \longrightarrow K$ is a compact map, and there exists $u \in K \setminus \{0\}$ such that $x - \varphi(x) \neq \lambda u$ for all $\lambda \geq 0$, and all $x \in K \cap \partial U$, then $i_K(\varphi, K \cap U) = 0$.*

REMARK 3.6.32 If X is a Banach space and $C \subseteq X$ is a nonempty, closed, and convex set (hence a retract), then for any $U \subseteq X$ bounded open, the set $K \cap U$ is bounded open in K and $\partial(K \cap U) = K \cap \partial U$ and $\overline{K \cap U} = K \cap \overline{U}$.

PROPOSITION 3.6.33 *If X is an OBS with order cone K, $U \subseteq X$ is a bounded open set, and $\varphi\colon K \cap \overline{U} \longrightarrow K$ is a compact map such that*

(i) $\inf\big[\, \|\varphi(x)\| : x \in K \cap \partial U \big] > 0$, *and*
(ii) $\varphi(x) \neq \lambda x$ *for all* $\lambda \in (0,1]$ *and all* $x \in K \cap \partial U$,

then $i_K(\varphi, K \cap U) = 0$.

PROOF: Consider the compact homotopy $h(t,x) = (1+t)\varphi(x)$ for all $t \geq 0$ and all $x \in K \cap \overline{U} = \overline{K \cap U}$. Because of hypothesis (ii), we see that $x \neq h(t,x)$ for all $t \geq 0$ and all $x \in K \cap \partial U = \partial(K \cap U)$. So by the homotopy invariance of the fixed point index we have $i_C(\varphi, K \cap U) = i_C\big((1+t)\varphi, K \cap U\big)$ for all $t \geq 0$. If $i_C(\varphi, K \cap U) \neq 0$, then we can find $x = x(t) \in U$ such that $(1+t)\varphi(x) = x$. But let $t^* > \big(\xi_1 + \xi_2\big)/\xi_3$ where $\xi_1 = \sup\{\|x\| : x \in K \cap \overline{U}\}$, $\xi_2 = \sup\{\|\varphi(x)\| : x \in K \cap \overline{U}\}$

and $\xi_3 = \inf\{\|\varphi(x)\| : x \in K \cap \overline{U}\}$. Note that because of hypothesis (i), the dependence of the Leray–Schauder degree on the boundary values and Theorem 3.1.13 we have $\xi_3 > 0$ and so $(\xi_1 + \xi_2)/\xi_3$ is finite. From the solution property of the fixed point index we can find $x \in K \cap U$ such that

$$\begin{aligned} &\varphi(x) + t^* \varphi(x) = x \\ \Rightarrow\ & t^* = \frac{\|x - \varphi(x)\|}{\|\varphi(x)\|} \leq \frac{\xi_1 + \xi_2}{\xi_3}, \quad \text{a contradiction.} \end{aligned}$$

Therefore we conclude that $i_K(\varphi, K \cap U) = 0$. □

PROPOSITION 3.6.34 *If X is an OBS with order cone K, $U \subseteq X$ is a bounded open set, and $\varphi\colon K \cap \overline{U} \longrightarrow X$ is a compact map such that*

(i) *$\varphi(x) \neq \lambda x$ for all $\lambda \in [0,1]$ and all $x \in K \cap \partial U = \partial(K \cap U)$,*
(ii) *The set $\left\{\varphi(x)/\|\varphi(x)\| : x \in K \cap \partial U\right\}$ is relatively compact,*

then $i_K(\varphi, U) = 0$.

PROOF: Let $\xi = \sup\{\|\varphi(x)\| : x \in K \cap \partial U\} > 0$ and consider the map $\psi\colon K \cap \partial U \longrightarrow K$ defined by

$$\psi(x) = \xi \, \frac{\varphi(x)}{\|\varphi(x)\|}.$$

By hypothesis (ii), ψ is compact and so by Theorem 3.1.13 we can find a compact extension $\widehat{\psi}$ of ψ with values in K. It is easy to see that $\widehat{\psi}$ satisfies the hypotheses of Proposition 3.6.33. So we have $i_K(\widehat{\psi}, K \cap U) = 0$. We consider the compact homotopy

$$h(t,x) = t\varphi(x)(1-t) + \widehat{\psi}(x) \qquad \text{for all } t \in [0,1] \text{ and all } x \in K \cap \overline{U} = \overline{K \cap U}.$$

Then using hypothesis (i) we see $x \neq h(t,x)$ for all $t \in [0,1]$ and $K \cap \partial U = \partial(K \cap U)$. Hence by the homotopy invariance of the degree, we have

$$i_K(\varphi, K \cap U) = i_K(\widehat{\psi}, K \cap U) = 0.$$

□

Now we use the above degree theoretic results involving the fixed point index in order to prove fixed point theorems of the cone expansion-compression form.

THEOREM 3.6.35 *If X is an OBS with order cone K, $U_1, U_2 \subseteq X$ are bounded open sets with $0 \in U_1$ $\overline{U}_1 \subseteq U_2$, and $\varphi\colon K \cap (\overline{U}_2 \setminus U_1) \longrightarrow K$ is a compact map that satisfies one of the following two hypotheses,*

(i) *$\varphi(x) - x \notin K$ for all $x \in K \cap \partial U_1$ and $x - \varphi(x) \notin K$ for all $x \in K \cap \partial U_2$, or*
(ii) *$x - \varphi(x) \notin K$ for all $x \in K \cap \partial U_1$ and $\varphi(x) - x \notin K$ for all $x \in K \cap \partial U_2$,*

then φ has at least one fixed point in $K \cap (\overline{U}_2 \setminus U_1)$.

PROOF: By virtue of Theorem 3.1.13 we can assume that φ is defined on $K \cap \overline{U}_2$.

If hypothesis (i) is in effect, then $\varphi(x) \neq \lambda x$ for all $\lambda \geq 1$ and all $x \in K \cap \partial U_1$ and so we can apply Theorem 3.6.29 and obtain

$$i_K(\varphi, K \cap U_1) = 1. \tag{3.96}$$

On the other hand choosing arbitrary $u \in K \setminus \{0\}$, because of the second part of hypothesis (i), we have that $x - \varphi(x) \neq \lambda u$ for all $\lambda \geq 0$ and all $x \in K \cap \partial U_2$. Therefore from Corollary 3.6.31, we have

$$i_K(\varphi, K \cap U_2) = 1. \tag{3.97}$$

From (3.96), (3.97), and the additivity property of the fixed point index, we have

$$\begin{aligned} & i_K\big(\varphi, K \cap (U_2 \backslash \overline{U}_1)\big) + i_K(\varphi, K \cap U_1) = i_K(\varphi, K \cap U_2) \\ \Rightarrow\ & i_K\big(\varphi, K \cap (U_2 \backslash \overline{U}_1)\big) = -1. \end{aligned}$$

So by the solution property of the index, φ has a fixed point in $U_2 \backslash \overline{U}_1$.

If hypothesis (ii) is in effect, then $i_K(\varphi, K \cap U_1) = 0$ and $i_K(\varphi, K \cap U_2) = 1$ and so

$$i_K\big(\varphi, K \cap (U_2 \setminus U_1)\big) = 1$$

from which we infer that φ has a fixed point in $U_2 \setminus \overline{U}_1$. □

In the previous theorem the expansion-compression of the cone was in terms of the order of the space. In the next theorem is with respect to the norm. The proof is similar and so is omitted.

THEOREM 3.6.36 *If X is an OBS with order cone K, $U_1, U_2 \subseteq X$ are bounded open sets with $0 \in U_1$ $\overline{U}_1 \subseteq U_2$, and $\varphi\colon K \cap (\overline{U}_2 \setminus U_1) \longrightarrow K$ is a compact map that satisfies one of the following two hypotheses,*

(i) *$\|\varphi(x)\| \leq \|x\|$ for all $x \in K \cap \partial U_1$ and $\|\varphi(x)\| \geq \|x\|$ for all $x \in K \cap \partial U_2$, or*
(ii) *$\|\varphi(x)\| \geq \|x\|$ for all $x \in K \cap \partial U_1$ and $\|\varphi(x)\| \leq \|x\|$ for all $x \in K \cap \partial U_2$,*

then φ has at least one fixed point in $K \cap (\overline{U}_2 \setminus U_1)$.

Finally we present some theorems on the existence of multiple fixed point theorems.

DEFINITION 3.6.37 Let X be a Banach space and $C \subseteq X$ a nonempty closed convex set. Given a continuous convex function $\xi\colon C \longrightarrow \mathbb{R}$, a continuous concave function $\vartheta\colon C \longrightarrow \mathbb{R}$, and real numbers η, μ, we set

$$\begin{aligned} & C(\xi,\eta) = \{x \in C : \xi(x) \leq \eta\}, \quad C(\vartheta,\mu) = \{x \in C : \vartheta(x) \geq \mu\} \quad \text{and} \\ & C(\xi,\vartheta,\eta,\mu) = C(\xi,\eta) \cap C(\vartheta,\mu). \end{aligned}$$

PROPOSITION 3.6.38 *If X is a Banach space, $C \subseteq X$ is a nonempty, closed, and convex set, $\xi\colon C \longrightarrow \mathbb{R}$ is continuous convex, $\vartheta\colon C \longrightarrow \mathbb{R}$ is continuous concave, $\eta, \mu \in \mathbb{R}$, $\varphi : \overline{\{x \in C : \vartheta(x) < \mu\}} \longrightarrow C$ is compact, and the following hypotheses hold,*

(i) $\{x \in C : \vartheta(x) < \mu\} \neq \emptyset$ *and it is bounded.*
(ii) $\{x \in C(\xi, \vartheta, \eta, \mu) : \vartheta(x) > \mu\} \neq \emptyset$ *and* $\vartheta(\varphi(x)) > \mu$ *for all* $x \in C(\xi, \vartheta, \eta, \mu)$.
(iii) $\vartheta(\varphi(x)) < \eta$ *for all* $x \in C(\vartheta, \mu)$ *with* $\xi(\varphi(x)) > \eta$,

then setting $U = \{x \in C : \vartheta(x) < \mu\}$, *we have* $i_C(\varphi, U) = 0$.

PROOF: By hypothesis $U \subseteq C$ is bounded open in C. Using hypothesis (ii) we can find $u \in C$ such that

$$\vartheta(u) > \mu \qquad \text{and} \qquad \xi(u) \leq \eta.$$

Hence $u \in C \setminus \overline{U}$. We show that $\varphi(x) - x \neq \lambda(x - u)$ for all $\lambda \geq 0$ and all $x \in \partial U$. Indeed, if there exist $\lambda \geq 0$ and $x \in \partial U$ such that $\varphi(x) - x = \lambda(x - u)$, then $\vartheta(x) = \mu$ and $x = (1/(1+\lambda))\varphi(x) + (\lambda/(1+\lambda))u$. If $\xi(\varphi(x)) \leq \eta$, then exploiting the convexity of ξ, we have $\xi(x) \leq (1/(1+\lambda))\xi(\varphi(x)) + (\lambda/(1+\lambda))\xi(u) \leq \eta$. Hence from hypothesis (ii) it follows that $\vartheta(\varphi(x)) > \mu$. Exploiting the concavity of ϑ, we have $\vartheta(x) \geq (1/(1+\lambda))\vartheta(\varphi(x)) + (\lambda/(1+\lambda))\vartheta(u) > \mu$, a contradiction. Now suppose $\xi(\varphi(x)) > \eta$. Then from hypothesis (iii) we have $\vartheta(\varphi(x)) > \mu$, again a contradiction. Therefore we can apply Proposition 3.6.30 and conclude that $i_C(\varphi, U) = 0$. □

PROPOSITION 3.6.39 *If X is a Banach space, $C \subseteq X$ is nonempty, closed, and convex, $\xi \colon C \longrightarrow \mathbb{R}$ is continuous convex, $\vartheta \colon C \longrightarrow \mathbb{R}$ is continuous concave, $\eta, \mu \in \mathbb{R}$, $\varphi \colon \overline{\{x \in C : \xi(x) < \eta\}} \longrightarrow C$ is compact, and the following hypotheses hold,*

(i) $\{x \in C : \xi(x) < \eta\}$ *is bounded;*
(ii) $\{x \in C(\xi, \vartheta, \eta, \mu) : \xi(x) < \eta\} \neq \emptyset$ *and* $\xi(\varphi(x)) < \eta$ *for all* $x \in C(\xi, \vartheta, \eta, \mu)$;
(iii) $\xi(\varphi(x)) < \eta$ *for all* $x \in C(\xi, \eta)$ *with* $\vartheta(\varphi(x)) < \mu$,

then setting $U = \{x \in C : \xi(x) < \eta\}$, *we have* $i_C(\varphi, U) = 1$.

PROOF: By hypothesis $U \subseteq C$ is bounded open in C. Using hypothesis (ii) we can find $u \in C$ such that

$$\vartheta(u) \geq \mu \qquad \text{and} \qquad \xi(u) < \eta.$$

So we have $u \in C$. We show that $\varphi(x) - x \neq \lambda(x - u)$ for all $\lambda \geq 0$ and all $x \in \partial U$. Indeed, if for some $\lambda \geq 0$ and $x \in \partial U$ we have $\varphi(x) - x = \lambda(x - u)$, then $\xi(x) = \eta$ and $x = (1/(1+\lambda))\varphi(x) + (\lambda/(1+\lambda))u$. First suppose that $\vartheta(\varphi(x)) \geq \mu$. Then due to the concavity of ϑ, we have $\vartheta(x) \geq \frac{1}{1+\lambda}\vartheta(\varphi(x)) + \frac{\lambda}{1+\lambda}\vartheta(u) \geq \mu$ and so from hypothesis (ii) we have $\xi(\varphi(x)) < \eta$. So from the convexity of ξ, we have $\eta = \xi(x) \leq (1/(1+\lambda))\xi(\varphi(x)) + (\lambda/(1+\lambda))\xi(u) < \eta$, a contradiction. Now suppose that $\vartheta(\varphi(x)) < \mu$. Then we have

$$\mu = \xi(x) \leq \frac{1}{1+\lambda}\xi(\varphi(x)) + \frac{\lambda}{1+\lambda}\varphi(u) < \mu,$$

again a contradiction. Therefore $\varphi(x) - x \neq \lambda(x - u)$ for all $\lambda \geq 0$ and all $x \in \partial U$. So we can apply Proposition 3.6.29 and conclude that $i_C(\varphi, U) = 1$. □

Next we present some multiplicity results for the fixed points of a compact map on an order cone.

THEOREM 3.6.40 *If X is a Banach space, $C \subseteq X$ a nonempty, bounded, closed, and convex set, $U_1, U_2 \subseteq X$ nonempty open, $\overline{U}_1 \cap \overline{U}_2 = \emptyset$, and $\varphi: C \longrightarrow C$ a compact map satisfying*

(i) *There exists $u_1 \in U_1$ such that $\varphi(x) - x \neq \lambda(x - u_1)$ for all $\lambda \geq 0$ and all $x \in \partial U_1$;*

(ii) *There exists $u_2 \in U_2$ such that $\varphi(x) - x \neq \lambda(x - u_2)$ for all $\lambda \geq 0$ and all $x \in \partial U_2$,*

then φ has at least three fixed points $x_1, x_2, x_3 \in C$ such that $x_1 \in U_1, x_2 \in U_2, x_3 \in C \setminus (\overline{U_1 \cup U_2})$.

PROOF: Clearly $i_C(\varphi, C) = 1$. Also from Proposition 3.6.29, we have

$$i_C(\varphi, U_1) = i_C(\varphi, U_2) = 1.$$

Then from the additivity property of the fixed point index (see Theorem 3.6.27(b)), we have

$$\begin{aligned} & i_C\big(\varphi, C \setminus (\overline{U_1 \cup U_2})\big) + i_C(\varphi, U_1) + i_C(\varphi, U_2) = i_C(\varphi, X) = 1, \\ \Rightarrow\ & i_C\big(\varphi, C \setminus (\overline{U_1 \cup U_2})\big) = -1. \end{aligned}$$

□

COROLLARY 3.6.41 *If X is a Banach space, $C \subseteq X$ is nonempty, bounded, closed, and convex, $\xi: C \longrightarrow \mathbb{R}$ is continuous convex, $\vartheta: C \longrightarrow \mathbb{R}$ is continuous concave, $\eta, \mu \in \mathbb{R}$, $U_1 = \{x \in C : \xi(x) < \eta\}$ and $U_2 = \{x \in C : \vartheta(x) > \mu\}$ are nonempty, $\overline{U}_1 \cap \overline{U}_2 = \emptyset$, $\varphi: C \longrightarrow C$ is compact, and*

(i) *$\xi(\varphi(x)) < \eta$ for all $x \in C$ with $\xi(x) = \eta$;*
(ii) *$\vartheta(\varphi(x)) > \mu$ for all $x \in C$ with $\vartheta(x) = \mu$,*

then φ has at least three fixed points $x_1, x_2, x_3 \in C$ such that $\xi(x_1) < \eta$, $\mu < \vartheta(x_2)$, and $\eta < \xi(x_3)$, $\vartheta(x_3) < \mu$.

PROOF: First let $u_1 \in U_1$. Then $\xi(u_1) < \eta$. If we can find $\lambda \geq 0$ and $x \in \partial U_1$ such that $\varphi(x) - x = \lambda(x - u_1)$, then $\xi(x) = \eta$ and $x = \big(1/(1+\lambda)\big)\varphi(x) + \big(\lambda/(1+\lambda)\big)u_1$. From the convexity of ξ and hypothesis (i) we have

$$\eta = \xi(x) \leq \frac{1}{1+\lambda}\xi\big(\varphi(x)\big) + \frac{\lambda}{1+\lambda}\xi(u_1) < \eta,$$

a contradiction. Therefore $\varphi(x) - x \neq \lambda(x - u_1)$ for all $\lambda \geq 0$ and $x \in \partial U_1$.

Similarly let $u_2 \in U_2$. Then $\vartheta(u_2) > \mu$ and arguing as above we establish that $\varphi(x) - x \neq \lambda(x - u_2)$ for all $\lambda \geq 0$ and $x \in \partial U_2$. So we can apply Theorem 3.6.40 and produce the three fixed points of φ with the stated properties. □

THEOREM 3.6.42 *If X is an OBS with order cone K, $\varphi: K \longrightarrow K$ is a compact map, $\xi: K \longrightarrow \mathbb{R}$ is continuous, convex, $\vartheta: K \longrightarrow \mathbb{R}$ is continuous concave, η, $\mu \in \mathbb{R}$, and*

(i) *$0 \in \{x \in K : \vartheta(x) < \mu\}$ and the set $\{x \in K : \vartheta(x) < \mu\}$ is bounded;*

(ii) $\{x\in K(\xi,\vartheta,\eta,\mu):\vartheta(x)>\mu\}\neq\emptyset$ *and* $\vartheta(\varphi(x))>\mu$ *for all* $x\in K(\xi,\vartheta,\eta,\mu)\}$;

(iii) $\vartheta(\varphi(x))>\mu$ *for all* $x\in K(\vartheta,\mu)$ *with* $\xi(\varphi(x))>\eta$;

(iv) *There exist* $r_0>0$ *small and* $R_0>r_0$ *large such that for* $r\leq r_0<R_0\leq R$ *we have*

$$i_K(\varphi,K_r)=i_K(\varphi,K_R)=1 \qquad \textit{where } K_\varrho=K\cap B_\varrho(0) \quad \textit{for all } \varrho>0,$$

then φ *has at least three fixed points* $x_1,x_2,x_3\in K$.

PROOF: Let $U=\{x\in K:\vartheta(x)<\mu\}$. From Proposition 3.6.38, we have $i_K(\varphi,U)=0$. Because $0\in\{x\in K:\vartheta(x)<\mu\}$ and the set $\{x\in K:\vartheta(x)<\mu\}$ is bounded (see hypothesis (i)), we can find $r>0$ small and $R>0$ large such that $\overline{K}_r\subseteq U\subseteq\overline{U}\subseteq K_R$. Then using the additivity property of the fixed point index, we obtain

$$i_K(\varphi,\overline{K}_r)=-1,\ i_K(\varphi,K_R\setminus\overline{U})=1 \qquad \text{and} \qquad i_K(\varphi,K_r)=1.$$

So we obtain three fixed points $x_1,x_2,x_3\in K$ of φ such that $x_1\in K_r$, $x_2\in U\setminus\overline{K}_r$, and $x_3\in K_R\setminus\overline{U}$. □

In a similar fashion using this time Proposition 3.6.39, we obtain the following.

THEOREM 3.6.43 *If* X *is an OBS with order cone* K, $\varphi\colon K\longrightarrow K$ *is a compact map,* $\xi\colon K\longrightarrow\mathbb{R}$ *is continuous, convex,* $\vartheta\colon K\longrightarrow\mathbb{R}$ *is continuous concave,* $\eta,\ \mu\in\mathbb{R}$, *and*

(i) $0\in\{x\in K:\xi(x)<\eta\}$ *and the set* $\{x\in K:\xi(x)<\eta\}$ *is bounded;*

(ii) $\{x\in K(\xi,\vartheta,\eta,\mu):\xi(x)<\eta\}\neq\emptyset$ *and* $\xi(\varphi(x))<\eta$ *for all* $x\in K(\xi,\vartheta,\eta,\mu)\}$;

(iii) $\xi(\varphi(x))<\eta$ *for all* $x\in K(\xi,\eta)$ *with* $\vartheta(\varphi(x))<\mu$;

(iv) *There exist* $r_0>0$ *small and* $R_0>r_0$ *large such that for* $r\leq r_0<R_0\leq R$ *we have*

$$i_K(\varphi,K_r)=i_K(\varphi,K_R)=0,$$

then φ *has at least two fixed points.*

We conclude this section with Amann's *three fixed points theorem*. For a proof of this result we refer to Amann [16].

THEOREM 3.6.44 *If* X *is an OBS with order cone* K *that is solid (i.e.,* $\operatorname{int}K\neq\emptyset$*) and normal,* $y_1,y_2,v_1,v_2\in X$, $y_1<v_1<y_2<v_2$, $\varphi\colon[y_1,v_2]\longrightarrow[y_1,v_2]$ *is compact and strongly increasing (see Definition 3.6.18(c)) and* $\varphi(v_1)<v_1$, $y_2\leq\varphi(y_2)$, *then* φ *has at least three fixed points* $x_1,x_2,x_3\in[y_1,v_2]$ *with*

$$y_1\leq x_1\ll v_1,\ y_2\ll x_2\leq v_2, \quad \textit{and} \quad y_2\nleq x_3\nleq v_1.$$

3.7 Remarks

3.1: The first attempts to deal with nonlinear operator equations in infinite-dimensional Banach spaces, involved compactness properties of the operator involved. Proposition 3.1.6 is due to Schauder [540]. Theorem 3.1.10 is due to Dugundji [209] (see also Dugundji [210, p.188]). Proper maps (see Definition 3.1.14) are discussed in Berger [68]. A thorough discussion of compact operators and their spectral properties can be found in the books of Kato [343], Megginson [424], and Reid–Simon [513]. For Fredholm operators the main reference is the book of Kato [343].

3.2: Compact operators limit the class of nonlinear boundary value problems that we can study. Monotone operators, and in particular maximal monotone operators, are a systematic effort to broaden the problems that we can study. Monotone operators is a rather natural class of nonlinear operators, which is rooted in the calculus of variations. The systematic study of monotone nonlinear operators started in the early 1960s and coincided with the advent of the so-called nonsmooth analysis, with which for a long period it developed in parallel. The first result on monotone operators, was obtained by Kachurovski [336], who proved that the gradient of a convex function on a Banach space is a monotone operator. In fact he is the one responsible for the term monotone operator. Then first came Minty [430] (for Hilbert spaces) and then Browder [121] (for general reflexive Banach spaces), who illustrated that the notion of monotonicity is a powerful tool in proving the existence theorem for nonlinear operator equations. Theorem 3.2.4 is essentially due to Rockafellar [520]. Here we state a slightly more general version of Rockafellar's result. The duality map (see Definition 3.2.12) was introduced by Beurling–Livingstone [73] and is a basic tool in the study of evolution equations as well as in the investigation of the geometric properties of Banach spaces. A detailed study of the duality map can be found in the books of Browder [121], Cioranescu [146], Gasiński–Papageorgiou [258], and Zeidler [622]. Theorems 3.2.24 and 3.2.25 are due to Browder [119] and are useful in the study of nonlinear operator equations. Theorem 3.2.30 was first proved for pivot Hilbert spaces (in which case $\mathcal{F}=I$) by Minty [430]. The maximal monotonicity of the sum of monotone operators was considered by Rockafellar [522, 523] who also proved Theorems 3.2.35 and 3.2.39. The notion of pseudomonotonicity (see Definition 3.2.53) was first introduced by Brezis [98] using nets and soon thereafter Browder [121] provided the sequential definition. The importance of this class of nonlinear operators comes from Theorem 3.2.60 which is due to Browder–Hess [120]. The basic works on pseudomonotone operators are those by Browder–Hess [120] and Kenmochi [345, 346]. Accretive operators (see Definition 3.2.64) were introduced by Kato [341, 342]. For a comprehensive introduction of their theory, we refer to the books of Barbu [58] and Miyadera [436]. Semigroups (linear and nonlinear alike), are important in the study of evolution equations. The linear semigroup theory can be found in the books of Hille–Phillips [300] and Pazy [491]. Theorem 3.2.87 is the main result of the linear theory and was proved independently (for $N = 1, \omega = 0$, i.e., contraction semigroups) by Hille [299] and Yosida [613]. The general case (stated in Theorem 3.2.89) is independently due to Feller [240], Miyadera [435], and Phillips [496]. For the nonlinear theory we refer to Barbu [58] and Vrabie [597]. The main result here is Theorem 3.2.93 proved by Crandall–Liggett [166].

3.3: The classical degree theory, was introduced by Brouwer [115] (foreshadowed by earlier work of Kronecker) and is an algebraic count of the solutions of $y_0 = f(x)$ where $f : \overline{U} \longrightarrow \mathbb{R}^N$ is a continuous map, $U \subseteq \mathbb{R}^N$ a bounded open set, and $y_0 \in \mathbb{R}^N$. Here we present the analytical approach due to Heinz [288] in the derivation of Brouwer's degree. Other analytical versions of the theory can be found in Nagumo [450] and Schwartz [545]. The proof of Borsuk's Theorem (see Theorem 3.3.28) is due to Gromes [274]. Theorem 3.3.32 is independently due to Ljusternik–Schnirelmann [393] and Borsuk [91] and is useful in the study of the so-called Ljusternik–Schnirelmann category, which in turn plays a basic role in the critical point theory developed by Ljusternik–Schnirelmann. The classical degree theory of Brouwer, was extended by Leray–Schauder [375] to a class of mappings in an infinite-dimensional Banach space. Their class of mapping had the form $I - \varphi$ with φ compact (compact perturbations of the identity). The uniqueness of Brouwer's and of course of the Leray–Schauder degree, was established by Amann–Weiss [15] and Führer [253]. Since 1934 there have been various efforts to move beyond the Leray–Schauder framework (i.e., beyond compact perturbations of the identity). The extension to maps $I - \varphi$ with φ γ-condensing is due to Nussbaum [464]. The extensions of degree theory to operators of monotone type, are due to Browder [122, 123].

Some of the results of the Leray–Schauder degree theory can be casted in terms of essential and inessential maps (see Granas [273] and Dugundji–Granas [212]).

DEFINITION 3.7.1 Let X be a Banach space, $C \subseteq X$ a nonempty convex set, $E \subseteq C$ a nonempty bounded set and $A \subseteq E$ closed in E. By $\mathcal{K}_A(E, C)$ we denote the set of all compact maps $\varphi : E \longrightarrow C$ such that $\varphi|_A$ is fixed point free. A map $\varphi \in \mathcal{K}_A(E, C)$ is said to be *traverse* or *essential* provided every $\psi \in \mathcal{K}_A(E, C)$ such that $\varphi|_A = \psi|_A$ has a fixed point. A map that is not essential is called inessential.

REMARK 3.7.2 In geometric terms, a compact map $\varphi : E \longrightarrow C$ is essential if the graph of $\varphi|_A$ does not intersect the diagonal $D \subseteq E \times C$, but the graph of every compact map $\psi : E \longrightarrow C$ such that it coincides with ϕ on A must cross the diagonal D.

The books of Deimling [188], Denkowski–Migòrski–Papageorgiou [195], Fonseca–Gangbo [252], Lloyd [396], and Zeidler [620], have detailed presentations of the Brouwer, Leray–Schauder and Nussbaum–Sadovskii degree theories.

3.4: Theorem 3.4.3 is due to Banach [57]. In applying the theorem, sometimes we need to equivalently renorm the Banach space, in order to make the appropriate map a k-contraction. This renorming trick was first used by Bielecki [75] (see also Gasiński–Papageorgiou [258], Example 7.1.4(b)). Theorem 3.4.11 is due to Edelstein [217]. Theorem 3.4.23 was proved independently by Browder [116] and Göhde [271] and was the first fixed point theorem for nonexpansive maps. Immediately after that Kirk [354] observed that their result could be extended using the notion of normal structure (see Definition 3.4.24) and proved Theorem 3.4.31. The normal structure of a convex set $C \subseteq X$ was introduced by Brodskii–Milman [113], who used it to study fixed points of isometries. The asymptotic center of $\{x_n\}_{n \geq 1}$ relative to C (see Definition 3.4.34(a)) was introduced by Edelstein [218], and the notions of inward set and weakly inward map (see Definition 3.4.34(b)), were introduced by Halpern in his thesis (see Halpern–Bergman [283]), who also established Theorem 3.4.38. Additional results can be found in Caristi–Kirk [128].

Detailed expositions of metric fixed point theory can be found in Dugundji–Granas [212], Goebel–Kirk [269], Smart [558], and Zeidler [619].

3.5: Theorem 3.5.3 is one of the oldest and best known results of topology. It was proved for $N = 3$ by Brouwer [114]. The case of general $N \geq 1$ was proved by Hadamard [281], who in his proof used the Kronecker index. A proof of the theorem based on a combinatorial technique was provided by Knaster–Kuratowski–Mazurkiewicz [358]. The infinite dimensional generalization of Theorem 3.5.6, (the Peron–Frobenius theorem), is the so-called Krein–Rutman theorem (see Zeidler [619, p. 290]).

THEOREM 3.7.3 *If X is an OBS with a solid order cone K (i.e., int $K \neq \emptyset$) which is total (i.e., $X = \overline{K - K}$), $A \in \mathcal{L}_c(X)$, A is positive (i.e., $A(K) \subseteq K$) and the spectral radius $r(A) = \lim_{n\to\infty} \|A^n\|_{\mathcal{L}}^{1/n} > 0$, then $r(A) > 0$ is an eigenvalue of A with positive eigenvector.*

Concerning Theorem 3.5.7, we should mention that in contrast in infinite-dimensional Banach spaces such a retraction exists. Example 3.5.12 is due to Kakutani [338], whereas Theorem 3.5.14 was proved by Schauder [540] and Theorem 3.5.10 is due to Borsuk [90]. Theorem 3.5.16 (the nonlinear alternative theorem), formalizes the well-known informal principle which roughly speaking says that existence of a priori bounds implies existence. A direct elementary proof of it was given by Schaeffer [538] (see Corollary 3.5.18). Theorem 3.5.20 is due to Sadovskii [534]. Historical forerunners are the fixed point theorems of Darbo [176] and Krasnoselskii [361, 362].

The topological fixed point theory can be found in the books of Brown [124], Dugundji–Granas [212], Istratescu [329], Smart [558], and Zeidler [619].

3.6: The various types of order cones were first considered by Krasnoselkii [361]. Deimling [188] and Guo–Lakshmikantham [277] presented them in detail. The fixed point index was first used systematically by Amann [16]. Fixed point theorems using the order structure can be found in Amann [16], Guo [276], Guo–Sun [278], Leggett–Wiliams [374], Sun [567], and Sun–Sun [566]. Also we refer to the lecture notes of Amann [16] and the books of Deimling [188], Guo–Lakshmikantham [277] and Heikkila–Lakshmikantham [287].

4

Critical Point Theory and Variational Methods

Summary. *The variational method in the study of nonlinear boundary value problems is based on the critical point theory, that provides minimax characterizations of the critical values over certain homotopically stable families of sets. Using the deformation method, we derive the main results of the smooth critical point theory and we also present results guaranteeing the existence of multiple critical points. We present an extension of the theory (perturbation of C^1-functionals by proper, convex, lsc functionals) which is suitable in the study of problems with unilateral constraints (variational inequalities). Then we present the Ljusternik–Schnirelmann theory which leads naturally to nonlinear eigenvalue problems and to the study of the Ljusternik–Schnirelmann category and of the Krasnoseleskii genus (two instances of topological indices). The Ljusternik–Schnirelmann theory allows us to develop the spectral properties of the Dirichlet and Neumann Laplacian and p-Laplacian and of the periodic scalar p-Laplacian. Also, we examine further certain abstract nonlinear eigenvalue problems. We conclude with a look at the main features of bifurcation theory.

Introduction

In this chapter we develop the abstract results that are necessary in order to implement the variational method in the study of nonlinear boundary value problems.

In Section 4.1 we develop the critical point theory for smooth functionals. This is the main tool in the variational method. This method consists of trying to find solutions of a given equation, by looking for stationary points of a real functional defined on the function space in which the solution of the equation is to lie. The given equation is the Euler–Lagrange equation satisfied by a stationary point. This functional (called energy or Euler functional of the problem) is often unbounded (indefinite functional) and so one cannot look for (global) maxima and minima. Instead one seeks saddle points characterized by a minimax expression. The approach that we follow to characterize the critical values of the functional is based on deformation arguments. We derive all the classical results such as the mountain pass theorem, the saddle point theorem, and the generalized mountain pass theorem. We also prove results on the existence of multiple critical points under the local linking

N.S. Papageorgiou, S.Th. Kyritsi-Yiallourou, *Handbook of Applied Analysis*,
Advances in Mechanics and Mathematics 19, DOI 10.1007/b120946_4,

condition or under symmetry conditions. Finally we present an extension of the theory to functionals that are perturbations of a smooth functional by a proper, lower semicontinuous and convex functional.

In Section 4.2 we present the Ljusternik–Schnirelmann theory for smooth functionals defined on a Banach space. When the functional is quadratic, then the corresponding eigenvalue problem is linear and for such problems, the eigenvalues have minimax characterizations, known as Courant's minimax principle. The Ljusternik–Schnirelmann theory is a method to advance critical point theory beyond quadratic functionals and independent of nondegeneracy considerations, which are present in Morse theory (not covered in this volume). To produce topological analogue of the minimax principles of the linear theory, we need to introduce substitutes of the notion of dimension. This is achieved using topological indices. We present two such indices, the Ljusternik–Schnirelmann category and the genus. Using them we state the basic results of the Ljusternik–Schnirelmann theory.

Section 4.3 investigates the spectral properties of the Laplacian and of the p-Laplacian under both Dirichlet and Neumann boundary conditions). We show that for the negative Laplacian we can have a full description of its eigenvalues and variational characterizations of them (via Courant's minimax principle). The situation is different with the negative p-Laplacian. Using the Ljusternik–Schnirelmann theory (Section 4.2), we produce an increasing sequence of eigenvalues converging to $+\infty$. However, we do not know if these eigenvalues exhaust all the eigenvalues. Nevertheless, we are able to fully describe the first eigenvalue and also show that the second eigenvalue and the second variational eigenvalue coincide.

In Section 4.4, we use the Ljusternik–Schnirelmann theory to study abstract nonlinear eigenvalue problems. The notion of a sphere-like constraint is important in these considerations and so we investigate it further.

Finally Section 4.5 deals with bifurcation theory. Bifurcation theory is concerned with the structure of the solutions of the parametrized equation $\varphi(\lambda, x) = 0$ as a function on the parameter λ near a solution (λ_0, x_0) that is also a singular point of the map $x \longrightarrow \varphi(\lambda_0, x)$ (hence $f'_x(\lambda_0, x_0)$ is not invertible and so we can not use the implicit function theorem). Using some degree-theoretic arguments we prove local and global bifurcation theorems.

4.1 Critical Point Theory

One of the typical features of modern physics is the characterization of phenomena by variational principles. Thus many phenomena can be understood in terms of the minimization of an energy functional $\varphi(x)$ over an appropriate class of functionals. This leads naturally to the study of critical points of a functional φ (i.e., the solutions of the gradient operator equation $\nabla\varphi(x) = \varphi'(x) = 0$). There are two types of critical points of φ. The local extrema (local minima and local maxima) and the saddle points. It is well known that in a finite-dimensional Banach space X, any continuous functional defined on a compact subset of X attains its infimum. This is not in general true in infinite-dimensional Banach spaces, because bounded closed subsets are not necessarily compact. In fact this lack of minimizers for such functionals was the starting point of the criticism of Weierstrass for Riemann's approach to potential theory (see Courant [163]). To guarantee minimizers (or maximizers) we need two types of properties of the functional φ. One is quantitative and requires that the sublevel sets $\{x \in X : \varphi(x) \leq \lambda\}$ are relatively compact in some useful topology of X (coercivity property) and the other qualitative, which requires that φ is lower semicontinuous (or upper semicontinuous) for the same topology on X. A coercive functional is bounded below. But many natural functionals φ that we may encounter may not be bounded at all, neither from above or from below. Thus we need to look for results where other types of critical points may be identified.

DEFINITION 4.1.1 Let X be a Banach space and $\varphi \in C^1(X)$. We say that $x \in X$ is a *critical point* of φ, if $\varphi'(x) = 0$. We say that $c \in \mathbb{R}$ is a *critical value* of φ, if there is a critical point $x \in X$ such that $c = \varphi(x)$. We say that $c \in \mathbb{R}$ is a *regular value* of φ, if it is not a critical value of φ.

We introduce the following notation which is used throughout this chapter. For every $\lambda \in \mathbb{R} \cup \{+\infty\}$ and every $c \in \mathbb{R}$, we set

$$\begin{aligned} \varphi^\lambda &= \{x \in X : \varphi(x) \leq \lambda\} \\ K &= \{x \in X : \varphi'(x) = 0\} \qquad \text{(the set of critical points of } \varphi) \\ \text{and} \quad K_c &= \{x \in K : \varphi(x) = c\} \qquad \text{(the set of critical points of } \varphi \text{ at level } c \in \mathbb{R}). \end{aligned}$$

We want to get information about the set K. To achieve this we employ the so-called *deformation method*.

DEFINITION 4.1.2 A continuous function $h\colon [0,1] \times X \longrightarrow X$ such that $h(0,\cdot) = I$, (i.e., $h(0,x) = x$ for all $x \in X$) is called a *deformation of* X. A family $\mathcal{S} \subseteq 2^X$ is said to be *deformation invariant*, if $h(1,A) \in \mathcal{S}$ for every $A \in \mathcal{S}$ and every $h(\cdot,\cdot)$ is a deformation of X.

We look at deformations h which effectively decrease the values of φ on $X \setminus K$ and try to determine critical values of φ by consideration of minimax expressions of the form

$$c = \inf_{A \in \mathcal{S}} \sup_{x \in A} \varphi(x)$$

for various deformation invariant classes $\mathcal{S}$. The construction of suitable deformations is the most technical part of this method. The guiding condition to produce the appropriate deformations is the following one.

(D) : for $h : [0,1] \times X \longrightarrow X$ a deformation of X, we have:

(1) For all $-\infty < a < b < +\infty$ with $\varphi^{-1}([a,b]) \cap K = \emptyset$, there exists $t_0 > 0$ such that

$$h(t_0, \varphi^b) \subseteq \varphi^a.$$

(2) If $c \in \mathbb{R}$ and U is a neighborhood of K_c, there exist t_0 and $a, b \in \mathbb{R}$ such that $a < c < b$ for which we have $h(t_0, \varphi^b) \subseteq U \cup \varphi^a$.

REMARK 4.1.3 Condition (D)(1) says that h effectively decreases the values of φ in $X \setminus K$. So nothing can happen topologically between the levels a, b, if the interval $[a,b]$ does not contain any critical values. On the other hand condition (D)(2) says that if we start a little above a critical level c, then we will either bypass the "critical" neighborhood U and as before reach a harmless level $a < c$ or we will end up in U, where topologically interesting things may happen.

The deformation theorem which we prove in the sequel, shows how to construct deformations satisfying condition (D) above. The idea is to use the gradient or gradient-like flows on X; that is, some kind of method of steepest descent.

To compensate for the lack of compactness in the ambient space X, we introduce the following compactness-type condition.

DEFINITION 4.1.4 Let X be a Banach space and $\varphi \in C^1(X)$.

(a) We say that φ satisfies the *Palais–Smale condition* at level $c \in \mathbb{R}$ (the PS_c-condition for short), if every sequence $\{x_n\}_{n\geq 1} \subseteq X$ such that

$$\varphi(x_n) \longrightarrow c \qquad \text{and} \qquad \varphi'(x_n) \longrightarrow 0,$$

has a strongly convergent subsequence. If this is true at every level $c \in \mathbb{R}$, then we simply say that φ satisfies the PS-condition.

(b) We say that φ satisfies the *Cerami condition* at level $c \in \mathbb{R}$ (the C_c-condition for short), if every sequence $\{x_n\}_{n\geq 1} \subseteq X$ such that

$$\varphi(x_n) \longrightarrow c \qquad \text{and} \qquad (1 + \|x_n\|)\varphi'(x_n) \longrightarrow 0,$$

has a strongly convergent subsequence. If this is true at every level $c \in \mathbb{R}$, then we simply say that φ satisfies the C-condition.

REMARK 4.1.5 Clearly the C_c-condition is weaker that the PS_c-condition. These conditions are fairly strong and are not satisfied even for some "nice" functions.

EXAMPLE 4.1.6 (a) Let $X = \mathbb{R}$ and $\varphi(x) = e^x$. Because $\varphi'(x) = e^x$, we see that φ does not satisfy the PS_0 and C_0 conditions. It trivially satisfies the PS_c and C_c conditions for $c \neq 0$.

(b) Let $X = \mathbb{R}$ and $\varphi(x) = c \in \mathbb{R}$. Then clearly φ does not satisfy the PS and C conditions.

(c) Let $X = \mathbb{R}$ and $\varphi(x) = \sin x$. If $\{x_n = (4n+1)(\pi/2)\}_{n\geq 1}$, then $\varphi(x_n) = 1$ and $\varphi'(x_n) = 0$ for all $n \geq 1$. However, $\{x_n\}_{n\geq 1} \subseteq \mathbb{R}$ has no convergent subsequence.

When the functional φ is bounded below, then as we show the situation simplifies and the compactness conditions imply weak coercivity (i.e., $\varphi(x) \longrightarrow +\infty$ as $\|x\| \longrightarrow \infty$; see Definition 3.2.14(b)).

PROPOSITION 4.1.7 *If X is a Banach space, $\varphi \in C^1(X)$, and is bounded below, $-\infty < m = \inf_X \varphi < +\infty$ and φ satisfies the C_m-condition, then the set $\varphi^{m+\vartheta}$ is bounded for some $\vartheta > 0$.*

PROOF: We argue indirectly. Suppose that $\varphi^{m+\vartheta}$ is unbounded for all $\vartheta > 0$. Then we can find a sequence $\{y_n\}_{n\geq 1} \subseteq X$ such that

$$m \leq \varphi(y_n) \leq m + \frac{1}{n} \qquad \text{and} \qquad \|y_n\| \geq n.$$

Apply Theorem 2.4.25 (with $\varepsilon = 1/n$, $h(r) = r$, $x_0 = y_n$, and $\lambda = 1/\sqrt{n}$) to obtain a sequence $\{x_n\}_{n\geq 1} \subseteq X$ such that

$$\varphi(x_n) \downarrow 0, \quad (1+\|x_n\|)\varphi'(x_n) \longrightarrow m \qquad \text{and} \qquad \|x_n - y_n\| \leq \overline{r}_n,$$

where $\overline{r}_n > 0$ is such that $\ln(1+\overline{r}_n) = 1/\sqrt{n}$. Clearly $\overline{r}_n \longrightarrow 0$ and $n \leq \|y_n\| \leq \overline{r}_n + \|x_n\|$, hence $\|x_n\| \longrightarrow +\infty$. But this contradicts the fact that φ satisfies the C_m-condition. □

PROPOSITION 4.1.8 *If X is a Banach space, $\varphi \in C^1(X)$, and $c \in \mathbb{R}$ is such that φ^{ϑ} is unbounded for $\vartheta > c$ and φ^{ϑ} is bounded for $\vartheta < c$, then there exists $\{x_n\}_{n\geq 1} \subseteq X$ such that*

$$\varphi(x_n) \longrightarrow c, \quad (1+\|x_n\|)\varphi'(x_n) \longrightarrow 0 \quad \text{in} \quad X^* \qquad \text{and} \qquad \|x_n\| \longrightarrow +\infty. \tag{4.1}$$

PROOF: By hypothesis for every $n \geq 1$, we can find $R_n \geq n$ such that

$$\varphi^{c-(1/n)} \subseteq B_{R_n}(0) = \{x \in X : \|x\| < R_n\}. \tag{4.2}$$

Set $C_n = X \setminus B_{R_n}(0), \widehat{\varphi}_n = \varphi\big|_{C_n} : C_n \longrightarrow \mathbb{R}$ and note that

$$m_n = \inf_{C_n} \widehat{\varphi}_n \geq c - \frac{1}{n} \qquad \text{(see (4.2))}. \tag{4.3}$$

Because by hypothesis $\varphi^{c+(1/n)}$ is unbounded, we can find $y_n \in X$ such that

$$\varphi(y_n) \leq c + \frac{1}{n} \qquad \text{and} \qquad \|y_n\| \geq R_n + 1 + \frac{1}{\sqrt{n}} \qquad \text{for all } n \geq 1. \tag{4.4}$$

Therefore we have $y_n \in C_n$ and from (4.3) and (4.4) it follows that

$$\varphi(y_n) \leq c + \frac{1}{n} \leq m_n + \frac{2}{n} \qquad \text{for all } n \geq 1. \tag{4.5}$$

Applying Theorem 2.4.25 (with $\varepsilon = 2/n, h(r) = r$, $x_0 = y_n$, and $\lambda = 1/\sqrt{n}$) we obtain a sequence $\{x_n\}_{n\geq 1}$ such that $x_n \in C_n$ and

$$c - \frac{1}{n} \leq m_n \leq \varphi(x_n) \leq \varphi(y_n) \leq c + \frac{1}{n} \leq m_n + \frac{2}{n} \qquad \text{for all } n \geq 1,$$

(see (4.5)), (4.6)

$$(1+\|x_n\|)\|\varphi'(x_n)\| \leq \frac{2}{\sqrt{n}} \qquad \text{for all } n \geq 1 \tag{4.7}$$

and $\|y_n - x_n\| \leq \overline{r}_n$ for all $n \geq 1$, where $\overline{r}_n > 0$ is such that

$$\ln(1+\overline{r}_n) = \frac{1}{\sqrt{n}}. \tag{4.8}$$

From (4.4) and (4.6) through (4.8) we conclude that the sequence $\{x_n\}_{n\geq 1} \subseteq X$ satisfies (4.1). □

COROLLARY 4.1.9 *If X is a Banach space and $\varphi \in C^1(X)$, satisfies the PS_c-condition for some $c \in \mathbb{R}$ and for every $\vartheta < c, \varphi^{\vartheta}$ is bounded, then $\varphi^{c+\eta}$ is also bounded for some $\eta > 0$.*

REMARK 4.1.10 Evidently φ^c is bounded too. We can restate the corollary by assuming that φ^c is bounded. Of course this corollary generalizes Proposition 4.1.7.

THEOREM 4.1.11 *If X is a Banach space and $\varphi \in C^1(X)$, is bounded below but not coercive, then φ does not satisfy the C_ξ-condition with $\xi = \sup\{\vartheta \in \mathbb{R} : \varphi^\vartheta$ is bounded$\}$.*

PROOF: Let $C = \{\vartheta \in \mathbb{R} : \varphi^\vartheta$ is bounded$\}$. Because by hypothesis φ is bounded from below, we have that $(-\infty, m) \subseteq C$ where $m = \inf_X \varphi$ (recall that the empty set is bounded; see Kuratowki [368, p. 107]). So $C \neq \emptyset$ and we can define $\xi = \sup C$. Because φ is not coercive, $\xi < +\infty$. Then for all $\eta > \xi$ we have that φ^η is unbounded. Invoking Proposition 4.1.8, we conclude that φ does not satisfy the C_ξ-condition. □

A remarkable consequence of Theorem 4.1.11 is the following result.

THEOREM 4.1.12 *If X is a Banach space, $\varphi \in C^1(X)$, it is bounded below and satisfies the C-condition, then φ is weakly coercive (i.e., $\varphi(x) \longrightarrow +\infty$ as $\|x\| \longrightarrow \infty$).*

REMARK 4.1.13 In Theorem 4.1.11, we can alternatively define $\xi = \inf\{\eta \in \mathbb{R} : \varphi^\eta$ is unbounded$\}$. In general for any functional $\varphi: X \longrightarrow \mathbb{R}$, the set $C = \{\vartheta \in \mathbb{R} : \varphi^\vartheta$ is bounded$\}$ is a left half-line either open or closed and we may have $C = \emptyset$ or $C = \mathbb{R}$, the latter case being equivalent to the weak coercivity of φ.

THEOREM 4.1.14 *If X is a Banach space and $\varphi \in C^1(X)$ is bounded below, then the PS and C conditions are equivalent.*

PROOF: Evidently we only need to show that the C-condition implies the PS-condition. To this end, let $\{x_n\}_{n\geq 1} \subseteq X$ be such that $|\varphi(x_n)| \leq M$ for some $M > 0$, all $n \geq 1$, and $\varphi'(x_n) \longrightarrow 0$. From Theorem 4.1.12 we know that φ is weakly coercive. Therefore it follows that $\{x_n\}_{n\geq 1} \subseteq X$ is bounded. Hence $(1 + \|x_n\|)\varphi'(x_n) \longrightarrow 0$. But φ satisfies the C-condition. So $\{x_n\}_{n\geq 1}$ admits a strongly convergent subsequence and this proves the theorem. □

As we already indicated, the deformation approach will be implemented by considering the negative gradient or gradientlike flows. If we are in a pivot Hilbert space (i.e. $H = H^*$), the negative gradient flow is a suitable choice, because it occurs within H. However, if we are in a general Banach space, the gradient has values in X^* and so the negative gradient flow will occur there hence it is not helpful. In addition, even in the Hilbert space situation, for the steepest descent method to work we need more than a $C^1(H)$-functional (additional smoothness on the functional φ). For this reason we introduce the following substitute of the gradient vector field of φ.

DEFINITION 4.1.15 We say that $u : X \setminus K \longrightarrow X$ is a *pseudogradient vector field* for φ, if u is locally Lipschitz and for every $x \in X \setminus K$, we have

$$\|\varphi'(x)\|^2_{X^*} \leq \langle \varphi'(x), u(x) \rangle \qquad \text{and} \qquad \|u(x)\|_X \leq 2\|\varphi'(x)\|_{X^*}.$$

REMARK 4.1.16 More generally for $0 < \alpha < \beta$, we can require that

$$\alpha\|\varphi'(x)\|^2_{X^*} \leq \langle \varphi'(x), u(x) \rangle \qquad \text{and} \qquad \|u(x)\|_X \leq \beta\|\varphi'(x)\|_{X^*} \quad \text{for all } x \in X \setminus K.$$

However, to simplify things we have decided to take $\alpha = 1$ and $\beta = 2$ as it is done in most cases in the literature. If $X = H$ is a Hilbert space with $H = H^*$ and $\varphi \in C^2(H)$, we can take

$$u(x) = \frac{\alpha + \beta}{2} \frac{1}{\|\varphi'(x)\|^2} \varphi'(x).$$

First we establish the existence of a pseudogradient vector field. To do this, we need the following lemma.

LEMMA 4.1.17 *If X is a Banach space, Y is a metric space, and $F : Y \longrightarrow 2^X \setminus \{\emptyset\}$ is a multifunction with convex values such that for every $y \in Y$ we can find a neighborhood U of y such that $\bigcap_{y' \in U} F(y') \neq \emptyset$, then we can find a locally Lipschitz map $u \colon Y \longrightarrow X$ such that $u(y) \in F(y)$ for all $y \in Y$.*

PROOF: By hypothesis, for every $y \in Y$ we can find $U(y)$ an open neighborhood of y such that

$$\bigcap_{y' \in U(y)} F(y') \neq \emptyset.$$

The family $\{U(y)\}_{y \in Y}$ is an open cover of Y and Y being a metric space it is paracompact. So we can find a locally finite open refinement $\{V_i\}_{i \in I}$. There is a partition of unity $\{g_i\}_{i \in I}$ subordinated to the open covering $\{V_i\}_{i \in I}$. For every $i \in I$, we choose $z_i \in \bigcap_{y' \in V_i} F(y')$ and we define a locally Lipschitz map $u \colon Y \longrightarrow X$ by

$$u(y) = \sum_{i \in I} g_i(y) z_i.$$

Now given $y \in Y$, we can find $\{V_{i_k}\}_{k=1}^n \subseteq \{V_i\}_{i \in I}$ such that $y \in V_{i_k}$, $k \in \{1, \ldots, n\}$ (because the cover is locally finite) and so we have

$$u(y) = \sum_{k=1}^n g_{i_k}(y) z_{i_k} \quad \text{with} \quad \sum_{k=1}^n g_{i_k}(y) = 1.$$

Because $y \in V_{i_k}$ for all $k \in \{1, \ldots, n\}$, $z_{i_k} \in F(y)$ and the latter is convex. So we conclude that $u(y) \in F(y)$ and u is a locally Lipschitz selector of F. □

THEOREM 4.1.18 *If X is a Banach space and $\varphi \in C^1(X)$, then there exists a pseudogradient vector field for φ.*

PROOF: We let $Y = X \setminus K = \{x \in X : \varphi'(x) \neq 0\}$ (the set of regular points of φ). For every $y \in Y$, let $F(y) = \{v \in X : \|v\|_X \leq 2\|\varphi'(y)\|_{X^*} \text{ and } \|\varphi'(y)\|_{X^*}^2 \leq \langle \varphi'(y), v \rangle\}$. Clearly $F(y)$ is a convex set. For every $y \in Y$ from the definition of the dual norm, we can find $x \in X$ with $\|x\|_X \leq 1$ such that

$$\frac{4}{5}\|\varphi'(y)\|_{X^*} \leq \langle \varphi'(y), x \rangle . \tag{4.9}$$

We set $v = 4/5\|\varphi'(y)\|_{X^*} x$. Then $\|v\|_X \leq 5/3\|\varphi'(y)\|_{X^*}$ and $\langle \varphi'(y), v \rangle \leq 4/3\|\varphi'(y)\|_{X^*}^2$ (see (4.9)). Because $\varphi \in C^1(X)$, we can find an open neighborhood U of y such that

$$\|v\|_X < 2\|\varphi'(y')\|_{X^*} \quad \text{and} \quad \langle \varphi'(y'), v\rangle > \|\varphi'(y')\|_{X^*} \qquad \text{for all } y' \in U,$$
$$\Rightarrow v \in \bigcap_{y' \in U} F(y').$$

So we can use Lemma 4.1.17 and produce a locally Lipschitz map $u\colon Y \longrightarrow X$ such that $u(y) \in F(y)$ for all $y \in Y$. Evidently this is the pseudogradient vector field for φ. □

Now we can prove the so-called *deformation theorem* that leads to minimax characterizations of the critical values of φ.

THEOREM 4.1.19 *If X is a Banach space and $\varphi \in C^1(X)$ and satisfies the C_c-condition for some $c \in \mathbb{R}$, then for every $\varepsilon_0 > 0$ and every neighborhood U of K_c (if $K_c = \emptyset$, then $U = \emptyset$) and every $\lambda > 0$, we can find $\varepsilon \in (0, \varepsilon_0)$ and a continuous map $h : [0,1] \times X \longrightarrow X$ (a continuous homotopy) such that for all $(t,x) \in [0,1] \times X$, we have*

(a) $\|h(t,x) - x\| \leq \lambda(1 + \|x\|)t$.
(b) $\varphi(h(t,x)) \leq \varphi(x)$.
(c) $h(t,x) \neq x \Rightarrow \varphi(h(t,x)) < \varphi(x)$.
(d) $|\varphi(x) - c| \geq \varepsilon_0 \Rightarrow h(t,x) = x$.
(e) $h(1, \varphi^{c+\varepsilon}) \subseteq \varphi^{c-\varepsilon} \cup U$.

PROOF: Because φ satisfies the C_c-condition, we readily verify that K_c is compact (it may be empty). So we can find $r > 0$ such that $B_{3r}(K_c) = \{x \in X : d(x, K_c) < 3r\} \subseteq U$.

We show that there exist $\varepsilon_1 > 0$ and $\vartheta > 0$ such that

$$c - 2\varepsilon_1 \leq \varphi(x) \leq c + 2\varepsilon_1 \qquad \text{for all } x \notin B_r(K_c) \Rightarrow (1 + \|x\|)\|\varphi'(x)\|_{X^*} \geq \vartheta. \tag{4.10}$$

Suppose that this is not possible. Then we can find $\{x_n\}_{n\geq 1} \subseteq X$ such that

$$\varphi(x_n) \longrightarrow c \quad \text{and} \quad (1 + \|x_n\|)\|\varphi'(x_n)\|_{X^*} \longrightarrow 0 \quad \text{as } n \to \infty. \tag{4.11}$$

Because φ satisfies the C_c-condition, from (4.11) and by passing to a subsequence if necessary, we may assume that $x_n \longrightarrow x$ in X. Then $\varphi(x) = c$, $x \notin B_r(K_c)$ and $\varphi(x) = 0$, a contradiction.

Let $A = \{x \in X : |\varphi(x) - c| \geq 2\varepsilon_1\} \cap \overline{B_r(K_c)}$ and $B = \{x \in X : |\varphi(x) - c| \leq \varepsilon_1\} \cap \big(B_{2r}(K_c)\big)^c$. These are two nonempty disjoint closed sets. So we can find $\beta\colon X \longrightarrow [0,1]$ a locally Lipschitz function such that $\beta\big|_A = 0$ and $\beta\big|_B = 1$ (consider, e.g., $\beta(x) = \big(d(x,B)\big)/\big(d(x,A) + d(x,B)\big)$). Choose $0 < \eta < 1$ such that $e^\eta \leq \lambda$. If $u\colon X \setminus K \longrightarrow X$ is the pseudogradient vector field for φ obtained in Theorem 4.1.18, we define

$$V(x) = \begin{cases} -\vartheta\eta\beta(x)\frac{u(x)}{\|u(x)\|^2} & \text{if } |\varphi(x) - c| \leq 2\varepsilon_1 \text{ and } x \notin B_r(K_c) \\ 0 & \text{otherwise.} \end{cases}$$

Evidently $V\colon X \longrightarrow X$ is a locally Lipschitz vector field. Also if $|\varphi(x) - c| \leq 2\varepsilon_1$ and $x \notin B_r(K_c)$, from (4.11) and Definition 4.1.15, we have

$$\|V(x)\| \leq \vartheta\eta \frac{1}{\|u(x)\|} \leq \vartheta\eta \frac{1}{\|\varphi'(x)\|_{|xd}} \leq \eta(1 + \|x\|) \qquad \text{for all } x \in X \tag{4.12}$$

and
$$\langle \varphi'(x), V(x) \rangle = -\vartheta\eta\beta(x) \frac{\langle \varphi'(x), u(x) \rangle}{\|u(x)\|^2} \leq -\vartheta\eta\beta(x) \frac{\|\varphi'(x)\|^2_{X^*}}{\|u(x)\|^2}$$
$$\leq -\frac{1}{4}\vartheta\eta\beta(x) \qquad \text{for all } x \in X. \tag{4.13}$$

We consider the following Cauchy problem in the Banach space X:

$$\left\{ \begin{array}{ll} \frac{d\xi(t)}{dt} = V\big(\xi(t)\big) & \text{a.e. on } [0,1] \\ \xi(0) = x & \end{array} \right\}. \tag{4.14}$$

Because V is a locally Lipschitz vector field on X with sublinear growth (see (4.12)), problem (4.14) admits a unique global solution $\xi(x)\colon [0,1] \longrightarrow X$. Then if we set

$$h(t,x) = \xi(x)(t),$$

because of the continuous dependence of the solution $\xi(x)$ on the initial condition x, we see that the map $h\colon [0,1] \times X \longrightarrow X$ is a continuous homotopy. Note that from the definition of V we see that if $|\varphi(x) - c| \geq 2\varepsilon_0$, then $h(t,x) = x$ for all $t \in [0,1]$. So we have proved statement (d) of the theorem. Also for all $(t,x) \in [0,1] \times X$, we have

$$\frac{\partial}{\partial t}\varphi\big(h(t,x)\big) = \left\langle \varphi'\big(h(t,x)\big), \frac{\partial h(t,x)}{\partial t} \right\rangle$$
$$= \big\langle \varphi'\big(h(t,x)\big), V\big(h(t,x)\big) \big\rangle \leq 0 \qquad \text{(see (4.13) and (4.14))}, \tag{4.15}$$
$$\Rightarrow \quad \varphi\big(h(t,x)\big) \leq \varphi(x) \quad \text{for all } (t,x) \in [0,1] \times X.$$

So we have proved statement (b) of the theorem. Moreover, note that

$$\frac{\partial}{\partial t}\varphi\big(h(t,x)\big) \leq \big\langle \varphi'\big(h(t,x)\big), V\big(h(t,x)\big) \big\rangle \leq -\frac{1}{4}\vartheta\eta\beta\big(h(t,x)\big)$$

see (4.13) and (4.15)). Therefore, if $h(t,x) \neq x$, we see that

$$\varphi\big(h(t,x)\big) < \varphi(x),$$

which proves statement (c) of the theorem. Integrating (4.14), we obtain

$$\|h(t,x) - x\| \leq \int_0^t \|V\big(h(s,x)\big)\| \mathrm{d}s$$
$$\leq \eta \int_0^t \big(1 + \|h(s,x)\|\big) \mathrm{d}s \qquad \text{(see (4.13))}$$
$$\leq \eta \int_0^t \|h(s,x) - x\| \mathrm{d}s + \eta(1 + \|x\|)\, t. \tag{4.16}$$

Invoking Gronwall's inequality, from (4.16) we infer that

$$
\begin{aligned}
\|h(t,x)-x\| &\le \eta(1+\|x\|)\,t+\eta^2(1+\|x\|)\,t\int_0^t e^{\eta s}ds\\
&= \eta(1+\|x\|)\,t\,e^{\eta t}\\
&\le \lambda(1+\|x\|)\,t
\end{aligned}
$$

(recall the choice of $0<\eta<1$ and that $t\in[0,1]$). This proves statement (a) of the theorem.

Moreover, because for $0\le t_1\le t_2\le 1$, we have $h(t_2,x_1)=h\big(t_2-t_1,h(t_1,x)\big)$, we also have

$$
\|h(t_2,x)-h(t_1,x)\|\le\lambda\big(1+\|h(t_1,x)\|\big)(t_2-t_1). \tag{4.17}
$$

Finally let $R>0$ and $0<\varepsilon\le\varepsilon_1$ be such that

$$
\overline{B_{2r}(K_c)}\subseteq B_R(0),\ 8\varepsilon\le\vartheta\eta \qquad\text{and}\qquad 8\lambda(1+R)\varepsilon\le\vartheta\eta r. \tag{4.18}
$$

To prove statement (e) of the theorem we proceed by contradiction. So let $x\in\varphi^{c+\varepsilon}$ and assume that $\varphi\big(h(1,x)\big)>c-\varepsilon$ and $h(1,x)\notin U$. Note that because the homotopy h is φ-decreasing (see statement (b) of the theorem), we have

$$
c-\varepsilon<\varphi\big(h(t,x)\big)\le c+\varepsilon \qquad\text{for all } t\in[0,1].
$$

Also $h([0,1],x)\cap B_{2r}(K_c)\neq\emptyset$. Indeed, if this is not the case, then from (4.13) we have

$$
\frac{1}{4}\vartheta\eta\le\varphi(x)-\varphi\big(h(1,x)\big)<2\varepsilon. \tag{4.19}
$$

Because $h(1,x)\notin U$ and $B_{3r}(K_c)\subseteq U$, we can find $0\le t_1 \xrightarrow{w}{}_2\le 1$ such that

$$
d\big(h(t_1,x),K_c\big)=2r,\ d\big(h(t_2,x),K_c\big)=3r\ \text{ and }\ 2r<d\big(h(t,x),K_c\big)<3r,
$$

for all $t\in(t_1,t_2)$

$$
\begin{aligned}
&\Rightarrow\ \frac{1}{4}\vartheta\eta(t_2-t_1)\le\varphi\big(h(t_1,x)\big)-\varphi\big(h(t_2,x)\big)<2\varepsilon\quad\text{(see (4.19))}\\
&\Rightarrow\ r\le\|h(t_2,x)-h(t_1,x)\|\le\lambda\big(1+\|h(t_1,x)\|\big)(t_2-t_1)<\lambda(1+R)\frac{8\varepsilon}{\vartheta\eta}\le r
\end{aligned}
$$

see (4.17) and (4.18)), a contradiction. This proves statement (e) of the theorem. □

Now we use this deformation theorem to produce minimax characterizations of the critical values of $\varphi\in C^1(X)$. For this we need the following basic topological notion.

DEFINITION 4.1.20 Let Y be a Hausdorff topological space, $E_0\subseteq E$ and D be nonempty subsets of Y with D closed, and $\gamma^*\in C(E_0,X)$. We say that the sets $\{E_0,E\}$ and D *link* in Y via γ^* if and only if the following conditions are satisfied.

(a) $E_0\cap D=\emptyset$.
(b) For any $\gamma\in C(E,X)$ such that $\gamma|_{E_0}=\gamma^*|_{E_0}$, we have $\gamma(E)\cap D\neq\emptyset$.

The sets $\{E_0,E,D\}$ are said to be *linking sets* via $\gamma^*\in C(E_0,X)$ and if $\gamma^*=I|_{E_0}$, then we simply say that $\{E_0,E,D\}$ are *linking sets*.

EXAMPLE 4.1.21 (a) Let $E_0=\{0,x\}, E=\{tx : t \in [0,1]\}$, and $D=\partial B_r(0)$ with $0 < r < \|x\|$. Then the sets $\{E_0, E, D\}$ are linking sets. Indeed, first note that $E_0 \cap D=\emptyset$. Also let $\gamma \in C(E,X)$ be such that $\gamma|_{E_0}=I_{E_0}$. Hence $\gamma(0)=0, \gamma(x)=x$. Consider the function $\widehat{\gamma} \in C([0,1])$ defined by $\widehat{\gamma}(t)=\|\gamma(tx)\|$, $t \in [0,1]$. Then $\widehat{\gamma}(0)=0$ and $\widehat{\gamma}(1)=\|x\|>r$. So by the intermediate value theorem, we can find $t_0 \in (0,1)$ such that $\widehat{\gamma}(t_0)=r$, hence $\|\gamma(t_0x)\|=r$ and so we conclude that $\gamma(E) \cap D \neq \emptyset$.

(b) Suppose $X=Y \oplus V$ with $\dim Y<+\infty$. Let $E_0=\{x \in Y : \|x\| = R\}=\partial B_R(0) \cap Y$, $E=\{x \in Y : \|x\| \leq R\} = \overline{B_R(0)} \cap Y$ and $D=V$ as in the decomposition. We claim that the sets $\{E_0, E, D\}$ are linking. To this end, let $p_Y \in \mathcal{L}(X)$ be the projection of X onto Y. It exists because Y is finite-dimensional. Let $\gamma \in C(E,X)$ be such that $\gamma|_{E_0} = I_{E_0}$. In order to prove that $\gamma(E) \cap D \neq \emptyset$, it suffices to show that $0 \in p_Y\big(\gamma(E)\big)$. We consider the continuous homotopy $h\colon [0,1] \times Y \longrightarrow Y$ defined by

$$h(t,x) = tp_Y\big(\gamma(x)\big) + (1-t)x.$$

Note that $h(0,\cdot) = I\big|_Y$, $h(1,\cdot) = p_Y \circ \gamma$, and $h(t,\cdot)\big|_{E_0} = I_{E_0}$ for all $t \in [0,1]$. So from the homotopy invariance and normalization properties of Brouwer's degree, we have

$$\begin{aligned} & d\big(p_Y \circ \gamma, B_R(0) \cap Y, 0\big) = d\big(I\big|_Y, B_R(0) \cap Y, 0\big) = 1 \\ \Rightarrow\ & 0 \in p_Y\big(\gamma(E)\big) \qquad \text{(existence property of Brouwer's degree).} \end{aligned}$$

(c) Let $X=Y \oplus V$ with $\dim Y<+\infty$ and let $v_0 \in V$ with $\|v_0\|=1$. Suppose $0<r<R_1$ and $0<R_2$ and set

$$\begin{aligned} E_0 &= \big\{\lambda v_0 + y : y \in Y \text{ and } (0 \leq \lambda \leq R_1, \|y\| = R_2) \text{ or} \\ &\qquad (\lambda \in \{0, R_1\},\ \|y\| \leq R_2)\big\} \\ E &= \big\{\lambda v_0 + y : y \in Y,\ 0 \leq \lambda \leq R_1 \text{ and } \|y\| \leq R_2\big\} \\ \text{and}\quad D &= \partial B_r(0) \cap V. \end{aligned}$$

Note that E is a cylinder in the space $Y \oplus \mathbb{R}v_0$ and E_0 is the boundary of E in $Y \oplus \mathbb{R}v_0$ (the upper and lower bases corresponding to $\lambda \in \{0, R_1\}$ and $\|y\| \leq R_2$ and the lateral surface corresponding to $0 \leq \lambda \leq R_1$ and $\|y\| = R_2$). We claim that the sets $\{E_0, E, D\}$ are linking. To show this, as before let $p_Y \in \mathcal{L}(X)$ be the projection operator of X onto Y. Also let $\gamma \in C(E,X)$ be such that $\gamma\big|_{E_0} = I\big|_{E_0}$. To establish the claimed linking, we need to show that there exists $x \in E$ such that $\|\gamma(x)\| = r$ and $p_Y\big(\gamma(x)\big) = 0$. Consider the continuous homotopy $h\colon [0,1] \times (Y \times \mathbb{R}) \longrightarrow Y \times \mathbb{R}$ defined by

$$h\big(t,(y,\lambda)\big) = \Big(tp_Y\big(\gamma(x)\big) + (1-t)y, t\|\gamma(x) - p_Y\big(\gamma(x)\big)\| + (1-t)\lambda - r\Big),$$

where $x = \lambda v_0 + y \in Y \oplus \mathbb{R}v_0 \simeq Y \times \mathbb{R}$. If $x = \lambda v_0 + y \in E_0$, then

$$h\big(t,(y,\lambda)\big) = \big(y, t\|x-y\| + (1-t)\lambda - r\big) = (y, \lambda - r).$$

So if we identify E with a subset of $Y \times \mathbb{R}$ (using the decomposition $x = \lambda v_0 + y$), we see that the Brouwer degree $d\big(h(t,\cdot), \text{int}E, 0\big)$ (the interior taken in $Y \times \mathbb{R}$) is well defined and from the homotopy invariance property of the degree we have

$$d\big(h(0,\cdot),\mathrm{int}\ E,0\big) = d\big(h(1,\cdot),\mathrm{int}\ E,0\big).$$

Note that $h(0,(y,\lambda)) = (\lambda - r, y)$. So $d\big(h(0,\cdot),\mathrm{int}\ E,0\big) = 1$, hence $d\big(h(1,\cdot),\mathrm{int}\ E,0\big) = 1$. So there exists $x = \lambda v_0 + y \in E$ such that $h\big(1,(y,\lambda)\big) = 0$, which implies

$$p_Y\big(\gamma(x)\big) = 0 \qquad \text{and} \qquad \|\gamma(x) - p_Y\big(\gamma(x)\big)\| = \|\gamma(x)\| = r$$

which is what we wanted. Therefore the sets $\{E_0, E, D\}$ are linking.

(d) Let $X=Y \oplus V$ with $\dim Y<+\infty$, $0<r<R$ and let $v_0 \in V$ be such that $\|v_0\|=r$ and set

$$\begin{aligned} E_0 &= \big\{x = \lambda v_0 + y : y \in Y \text{ and } (\lambda \geq 0,\ \|x\| = R) \text{ or } (\lambda = 0,\ \|x\| \leq R)\big\} \\ E &= \big\{x = \lambda v_0 + y : y \in Y,\ \lambda \geq 0,\ \|x\| \leq R\big\} \\ \text{and} \quad D &= \partial B_r(0) \cap V. \end{aligned}$$

Note that E_0 is the north hemisphere of a sphere of radius R in $Y \times \mathbb{R}v_0$ plus the circle of radius R in Y and it is the boundary of E in $Y \times \mathbb{R}v_0 \simeq Y \times \mathbb{R}$, Arguing as in the previous example, we check that the sets $\{E_0, E, D\}$ are linking.

THEOREM 4.1.22 *If X is a Banach space, $\varphi \in C^1(X)$, the sets $\{E_0, E, D\}$ are linking via γ^*, $\alpha = \sup\limits_{E_0} < \inf\limits_{D} \varphi = \beta$, $\Gamma = \big\{\gamma \in C(E,X) : \gamma\big|_{E_0} = \gamma^*\big\}$, $c = \inf\limits_{\gamma\in\Gamma} \sup\limits_{x\in E} \varphi\big(\gamma(x)\big)$, and φ satisfies the C_c-condition, then $c \geq \beta$ and c is a critical value of φ.*

PROOF: Because by hypothesis the sets $\{E_0, E, D\}$ are linking via $\gamma^* \in C(E_0, X)$, for every $\gamma \in \Gamma$ we have $\gamma(E) \cap D \neq \emptyset$ and so $\beta \leq c$.

To show that c is a critical value of φ, we argue indirectly. So suppose by contradiction that $K_c = \emptyset$. Set $\varepsilon_0 = \beta - \alpha$ and $U = \emptyset$. Then by virtue of Theorem 4.1.19 (the deformation theorem), we can find $0<\varepsilon<\varepsilon_0$ and a continuous homotopy $h\colon [0,1] \times X \longrightarrow X$ that satisfies statements (a)$\longrightarrow$(e) of Theorem 4.1.19. We choose $\gamma \in \Gamma$ such that

$$\varphi\big(\gamma(x)\big) < c + \varepsilon \qquad \text{for all } x \in E. \tag{4.20}$$

Set $\gamma_0(x) = h\big(1, \gamma_0(x)\big)$ for all $x \in E$. From the choice of $\varepsilon_0 > 0$ and statement (d) in Theorem 4.1.19 we see that $\gamma_0\big|_{E_0} = \gamma^*$ and so $\gamma_0 \in \Gamma$. For all $x \in E$, we have

$$\varphi\big(\gamma_0(x)\big) = \varphi\Big(h\big(1, \gamma(x)\big)\Big). \tag{4.21}$$

Combining (4.20), (4.21), and statement (e) of Theorem 4.1.19, we infer that

$$\varphi\big(\gamma_0(x)\big) \leq c - \varepsilon \qquad \text{for all } x \in E,$$

a contradiction to the definition of c. □

REMARK 4.1.23 It can be shown that if $c = \beta$, then there exists a critical point of φ in D. For details we refer to Gasiński–Papageorgiou [259, p. 633].

Now with suitable choices of the linking sets, we can have the *mountain pass*, *saddle point* and *generalized mountain pass theorems.* We start with the *mountain pass theorem.*

THEOREM 4.1.24 *If X is a Banach space, $\varphi \in C^1(X)$, there exist $x_0, x_1 \in X$ and $r>0$ such that $\|x_1 - x_0\| > r$, $\max\{\varphi(x_0), \varphi(x_0)\} < \inf\{\varphi(x) : \|x - x_0\| = r\}$, $\Gamma = \{\gamma \in C([0,1], X) : \gamma(0) = x_0,\ \gamma(1) = x_1\}$, $c = \inf\limits_{\gamma \in \Gamma} \max\limits_{t \in [0,1]} \varphi(\gamma(t))$, and φ satisfies the C_c-condition, then $c \geq \inf\{\varphi(x) : \|x - x_0\| = r\}$ and c is a critical value of φ.*

PROOF: We consider the linking sets $E_0 = \{x_0, x_1\}$, $E = \{(1-t)x_0 + tx_1 : t \in [0,1]\}$, $D = \partial B_r(x_0)$ (see Example 4.1.21(a)) and we apply Theorem 4.1.22. $\square$

Next we state the *saddle point theorem.*

THEOREM 4.1.25 *If X is a Banach space, $X = Y \oplus V$ with $\dim Y < +\infty$, $\varphi \in C^1(X)$, there exists $R>0$ such that $\max\{\varphi(x) : x \in \partial B_R(0) \cap Y\} < \inf\{\varphi(x) : x \in V\} = \beta$, $\Gamma = \{\gamma \in C(\overline{B}_R(0) \cap Y, X) : \gamma|_{\partial \overline{B}_R(0) \cap Y} = I_{\partial \overline{B}_R(0) \cap Y}$, $c = \inf\limits_{\gamma \in \Gamma} \sup\limits_{x \in E} \varphi(\gamma(x))$, and φ satisfies the C_c-condition, then $c \geq \beta$ and c is a critical value of φ.*

PROOF: We consider the linking sets $E_0 = \partial B_R(0) \cap Y$, $E = \overline{B}_R(0) \cap Y$, $D = V$ (see Example 4.1.21(b)), and we apply Theorem 4.1.22. $\square$

Finally we state the *generalized mountain pass theorem.*

THEOREM 4.1.26 *If X is a Banach space, $X = Y \oplus V$ with $\dim Y < +\infty$, $\varphi \in C^1(X)$ and there exist $0 < r < R$ and $v \in V$ with $\|v\| = 1$ such that if E_0, E, D are the sets from Example 4.1.21(c), have*

$$\max_{x \in E_0} \varphi(x) < \inf_{x \in D} \varphi(x) = \beta,$$

$\Gamma = \{\gamma \in C(E, X) : \gamma|_{E_0} = I_{E_0}\}$, $c = \inf\limits_{\gamma \in \Gamma} \sup\limits_{x \in E} \varphi(\gamma(x))$, and φ satisfies the C_c-condition, then $c \geq \beta$ and c is a critical value of φ.

PROOF: Let $\{E_0, E, D\}$ be the linking triple of Example 4.1.21(c) and apply Theorem 4.1.22. $\square$

REMARK 4.1.27 An alternative version of Theorem 4.1.25 can be obtained using the linking triple $\{E_0, E, D\}$ from Example 4.1.21(d).

If φ is bounded below (definite functional), the situation is much simpler.

THEOREM 4.1.28 *If X is a Banach space, $\varphi \in C^1(X)$ and it is bounded below, $m = \inf\limits_X \varphi$ and φ satisfies the C_m-condition, then m is a critical value of φ.*

PROOF: Invoking Theorem 2.4.25, we can find a sequence $\{x_n\}_{n \geq 1} \subseteq X$ such that

$$\varphi(x_n) \downarrow m \quad \text{as } n \to \infty \quad \text{and} \quad (1 + \|x_n\|)\varphi'(x_n) \longrightarrow 0.$$

By hypothesis φ satisfies the C_m-condition, thus by passing to a suitable subsequence if necessary, we may assume that $x_n \longrightarrow x$ in X. Then $\varphi(x_n) \downarrow \varphi(x)$ and so $\varphi(x) = m = \inf\limits_X \varphi$. Hence $\varphi'(x) = 0$ and so m is a critical value of φ. $\square$

Next we look for multiple critical points of a smooth functional φ. For this purpose we introduce the following notion.

DEFINITION 4.1.29 Let X be a Banach space, $X=Y\oplus V$ and $\varphi\in C^1(X)$. We say that φ has a *local linking* at 0, if there exists $r>0$ such that

$$\begin{cases}\varphi(x)\leq 0 & \text{if } x\in Y,\ \|x\|\leq r\\ \varphi(x)\geq 0 & \text{if } x\in V,\ \|x\|\leq r\end{cases}.$$

REMARK 4.1.30 Evidently $x=0$ is a critical point of φ. We produce two more, distinct nontrivial critical points of φ.

We start with an auxiliary result. Here $u(\cdot)$ is the pseudogradient vector field (see Theorem 4.1.18).

LEMMA 4.1.31 *If X is a Banach space, $\varphi\in C^1(X)$ and satisfies the PS-condition, $x_0\in X$ is the unique global minimum of φ on X, $y\in X$ is such that $\varphi'(y)\neq 0$, and φ has no critical value in the interval $\big(\varphi(x_0),\varphi(y)\big)$, then the negative pseudogradient flow defined by*

$$\frac{\mathrm{d}\overline{x}(t)}{\mathrm{d}t}=-\frac{u\big(\overline{x}(t)\big)}{\|u\big(\overline{x}(t)\big)\|^2},\quad \overline{x}(0)=y, \tag{4.22}$$

exists for a maximal finite time $T(y)>0$ and $\overline{x}\big(T(y)\big)=x_0$.

PROOF: Without any loss of generality we assume that $x_0=0$ and $\varphi(x_0)=0$. Evidently problem (4.22) has a unique local solution $\overline{x}(t)$ and we have

$$\begin{aligned}\frac{\mathrm{d}\varphi\big(\overline{x}(t)\big)}{\mathrm{d}t}&=\left\langle\varphi'\big(\overline{x}(t)\big),\frac{\mathrm{d}\overline{x}(t)}{\mathrm{d}t}\right\rangle\\&=\left\langle\varphi'\big(\overline{x}(t)\big),-\frac{u\big(\overline{x}(t)\big)}{\|u\big(\overline{x}(t)\big)\|_X^2}\right\rangle\qquad\text{(see (4.22))}\\&\leq-\frac{\|\varphi'\big(\overline{x}(t)\big)\|_{X^*}^2}{\|u\big(\overline{x}(t)\big)\|_X^2}\leq-\frac{1}{4}\qquad\text{(see (4.1.15)).}\end{aligned}\tag{4.23}$$

Let $(0,b)$ be the maximal open interval of existence of the solution $\overline{x}(\cdot)$. From (4.23), we have

$$\begin{aligned}&\varphi(y)\geq b+\varphi\big(\overline{x}(b)\big)\geq b\qquad\text{(recall }\varphi\geq 0)\\\text{and}\quad&0<\varphi\big(\overline{x}(t)\big)<\varphi(y)\qquad\text{for all } t\in(0,b).\end{aligned}$$

We show that $\varphi\big(\overline{x}(t)\big)\longrightarrow 0=\varphi(x_0)$ as $t\longrightarrow b$.

First suppose that we can find $\delta>0$ such that $\|\varphi'\big(\overline{x}(t)\big)\|_{X^*}\geq\delta$ for all $t\in(0,b)$. From Definition 4.1.15, we have

$$\|u\big(\overline{x}(t)\big)\|_X\geq\|\varphi'\big(\overline{x}(t)\big)\|_{X^*}\geq\delta>0\qquad\text{for all } t\in[0,b]$$

and so

$$\begin{aligned}&\int_0^b\|\frac{\mathrm{d}\overline{x}(t)}{\mathrm{d}t}\|\mathrm{d}t=\int_0^b\frac{1}{\|u\big(\overline{x}(t)\big)\|}\mathrm{d}t<+\infty,\\\Rightarrow\ &\lim_{t\to b}\overline{x}(t)=\overline{x}(b)\quad\text{exists in } X.\end{aligned}$$

Note that $\overline{x}(b)=0$ or otherwise we could continue $\overline{x}(t)$ beyond $t=b$, a contradiction to the maximality of b.

Now suppose that we can find a sequence $t_n \longrightarrow b$ and $\|\varphi'(\overline{x}(t_n))\|_{X^*} \longrightarrow 0$. But by hypothesis φ satisfies the PS-condition. So by passing to a subsequence if necessary, we may assume that $\overline{x}(t_n) \longrightarrow \widehat{x}_0$ in X and $\widehat{x}_0$ is a critical point of φ. Because by hypothesis φ has no critical values in $\big(\varphi(x_0)=0, \varphi(y)\big)$, we must have $\widehat{x}_0 = x_0 = 0$. Then $\lim\limits_{t\rightarrow b} \varphi\big(\overline{x}(t)\big) = 0$ and since φ satisfies the PS-condition, by the Ekeland variational principle (see Corollary 2.4.6), we conclude that $\overline{x}(t) \longrightarrow 0 = x_0$ as $t \longrightarrow b$. □

Using this lemma, we can prove the following multiplicity result.

THEOREM 4.1.32 *If X is a Banach space, $X = Y \oplus V$ with $\dim Y < +\infty$, $\varphi \in C^1(X)$ and satisfies the PS-condition, φ is bounded below, $m = \inf\limits_X < \varphi(0) = 0$, and there exists $r > 0$ such that*

$$\begin{cases} \varphi(x) \leq 0 & \text{if } x \in Y,\ \|x\| \leq r \\ \varphi(x) \geq 0 & \text{if } x \in V,\ \|x\| \leq r \end{cases},$$

(i.e., φ has a local linking at 0), then φ has at least two nontrivial critical points.

PROOF: From Theorem 4.1.28, we know that φ attains its infimum at some point $x_0 \in X$. Then

$$\begin{aligned} & m = \inf_X \varphi = \varphi(x_0) < \varphi(0) = 0, \\ \Rightarrow\ & x_0 \neq 0. \end{aligned}$$

Suppose that $\{0, x_0\}$ are the only critical points of φ.

Case I: $\dim Y > 0$ and $\dim V > 0$.

Without any loss of generality, we assume that $r = 1 < \|x_0\|$. We show that there exists $\delta > 0$ such that

$$\big\{x \in X : \varphi(x) < \varphi(x_0) + \delta\big\} \subseteq \big\{x \in X : \|x - x_0\| < \frac{1}{2}\|x_0\|\big\}. \tag{4.24}$$

Indeed, if no such $\delta > 0$ can be found, then there is a minimizing sequence $\{x_n\}_{n\geq 1} \subseteq X$ (i.e., $\varphi(x_n) \downarrow m = \varphi(x_0)$) such that $\|x_n - x_0\| \geq \frac{1}{2}\|x_0\|$ for all $n \geq 1$. Because of Corollary 2.4.6 we can find $\{y_n\}_{n\geq 1} \subseteq X$ another minimizing sequence (i.e., $\varphi(y_n) \downarrow m = \varphi(x_0)$) such that $\|y_n - x_n\| \leq 1/n$ and $\varphi'(y_n) \longrightarrow 0$. Because by hypothesis φ satisfies the PS-condition, we may assume that $y_n \longrightarrow \widehat{x}_0$ in X. We have $\|y_n - x_0\| \geq \|x_n - x_0\| - \|y_n - x_n\| \geq \frac{1}{2}\|x_0\| - \frac{1}{n}$, hence $\|\widehat{x}_0 - x_0\| \geq \frac{1}{2}\|x_0\| > 0$ and so $\widehat{x}_0 \neq x_0$. Also $\varphi'(y_n) \longrightarrow \varphi'(\widehat{x}_0) = 0$ and so $\widehat{x}_0 \in X$ is a third critical point of φ, a contradiction to our hypothesis. So we can find $\delta > 0$ for which (4.24) is true.

For every $y \in Y$ with $\|y\| = 1 = r < \|x_0\|$, we have $\varphi'(y) \neq 0$ and so we can apply Lemma 4.1.31 and have that the flow generated by (4.22) exists on a maximal open interval $\big(0, b(y)\big)$ with $b(y) \leq -4\varphi(x_0)$ and $\overline{x}(t) \longrightarrow x_0$ as $t \longrightarrow b(y)$. Moreover, by choosing $\delta > 0$ small in (4.24), we can find unique $t = t(y) \in \big(0, b(y)\big)$ such that

$\varphi(\overline{x}(t(y))) = \varphi(x_0) + \delta$. The uniqueness of $t(y)$ also implies easily the continuity of the map $y \longrightarrow t(y)$. Also $\varphi \le 0$ on $\{x \in X : \varphi(x) \le \varphi(x_0) + \delta\}$.

Let $v_0 \in V$ be such that $\|v_0\| = 1$ and let

$$\begin{aligned} E &= \{x = \lambda v_0 + y : y \in Y,\ \lambda \ge 0,\ \|x\| \le 1\} \\ E_0 &= \{x = \lambda v_0 + y : y \in Y \text{ and } (\lambda \ge 0,\ \|x\| = 1) \text{ or } (\lambda = 0,\ \|x\| \le 1)\}. \end{aligned}$$

We define $\gamma^* \in C(E_0, X)$ as follows.

$$\gamma^*(y) = y \quad \text{if } y \in Y,\ \|y\| \le 1 \qquad \text{and} \qquad \gamma^*(v_0) = x_0. \tag{4.25}$$

Also for any $x \in E_0$ with $x \ne v_0$ and $\|x\|=1$, we have the unique representation

$$x = \lambda v_0 + \eta y \tag{4.26}$$

with $0 \le \lambda \le 1$, $y \in Y$, $\|y\| = 1$, and $0<\eta \le 1$ (the uniqueness of this representation means that λ, η, y in (4.26) are uniquely defined). Then for such a $x \in E_0$, we set

$$\gamma^*(x) = \gamma^*(\lambda v_0 + \eta y) = \overline{x}(2\lambda t(y)) \quad \text{when } \lambda \in \left[0, \frac{1}{2}\right] \tag{4.27}$$

with $\overline{x}(t)$ being the solution of (4.22). So

$$\gamma^*\left(\frac{1}{2} v_0 + \eta y\right) = \overline{x}(t(y)) \in \{x \in X : \|x - x_0\| \le \frac{1}{2}\|x_0\|\} \qquad \text{(see (4.24))}.$$

Finally set

$$\gamma^*(\lambda v_0 + \eta y) = (2\lambda - 1) v_0 + (2 - 2\lambda)\overline{x}(t(y)) \quad \text{when } \frac{1}{2} \le \lambda \le 1. \tag{4.28}$$

Note that as λ moves from $\frac{1}{2}$ to 1, the right-hand side of the above equation traverses the line segment from $\overline{x}(t(y))$ to x_0 and so for $\lambda \in \left[\frac{1}{2}, 1\right]$ we have

$$\|\gamma^*(\lambda v_0 + \eta y) - x_0\| \le \frac{1}{2}\|x_0\|.$$

The map $\gamma^* \colon E_0 \longrightarrow X$ is clearly continuous and $\varphi \circ \gamma^* \le 0$ (recall the choice of $\delta>0$). In addition note that for some $0<\vartheta \le 1$, we have

$$\|\gamma^*(x)\| \ge \vartheta > 0 \qquad \text{for all } x \in E_0,\ \|x\| = 1$$

(see (4.25), (4.27), and (4.28)).

Next we show that if $D = \partial B_\varrho(0) \cap V$ with $\varrho<\vartheta$, then the sets $\{E_0, E, D\}$ are linking via γ^*. Let γ be any continuous extension of γ^* on E. Let $W = \mathbb{R}v_0 \oplus Y$ and consider the map $S\colon E \longrightarrow W$ defined by

$$S(x) = p_Y(\gamma(x)) + \|(I - p_Y)\gamma(x)\| v_0, \tag{4.29}$$

where $p_Y \in \mathcal{L}(X)$ is the projection operator onto Y (it exists because Y is finite-dimensional). In order to show that $\gamma(E) \cap D \ne \emptyset$, it suffices to show that there exists $\widehat{x} \in E$ such that $S(\widehat{x}) = \varrho\widehat{x}$. Because $\varrho<\vartheta$, we see that $S(x) \ne \varrho v_0$ for all $x \in E_0$ and so the degree $d(S, \text{int}\, E, \varrho v_0)$ (the $\text{int}E$ taken in W) is well-defined. Recall that the degree depends only on the boundary values. So we may consider only $S\big|_{E_0 = \partial E}$ (the boundary ∂E considered in W). Set

$$B_1 = \{y \in Y : \|y\| \leq 1\} \qquad \text{and} \qquad B_2 = \{x \in E : \|x\| = 1\}.$$

Evidently $E_0 = \partial E = B_1 \cup B_2$ (B_1 is the flat basis of the half-ball E and B_2 is the hemisphere of the half-ball E). It is clear from (4.29) that $S\big|_{B_1} = I\big|_{B_1}$ and $\|S(x)\| \geq \vartheta' > 0$ for $x \in B_2$. So on $E_0 = \partial E$ we define

$$\widehat{S}(x) = \begin{cases} x & \text{if } x \in B_1 \\ \frac{S(x)}{\|S(x)\|} & \text{if } x \in B_2 \end{cases}.$$

Let $h : [0,1] \times E_0 \longrightarrow W \setminus \{\varrho v_0\}$ be defined by

$$h(t,x) = tS(x) + (1-t)\widehat{S}(x).$$

This is a continuous homotopy. Note that $\widehat{S}(B_2) \subseteq B_2$ and if we consider the boundary ∂B_2 of the manifold B_2 in W, we see that $\widehat{S}\big|_{\partial B_2} = I\big|_{\partial B_2}$. Since B_2 is homeomorphic to a ball, there is a continuous homotopy $\widehat{h}(t,x)$ connecting $\widehat{S}$ to $I\big|_{\partial B_2}$ with $\widehat{h}(t,\cdot)\big|_{\partial B_2} = I\big|_{\partial B_2}$. So $S\big|_{E_0 = \partial E}$ is homotopic to the identity in $W \setminus \{\varrho v_0\}$ and thus $d(S, \operatorname{int} E, \varrho v_0) = d(I, \operatorname{int} E, \varrho v_0) = 1$. This means that we can find $\widehat{x} \in E$ such that $S(\widehat{x}) = \varrho v_0$ and so the sets $\{E_0, E, D\}$ are linking via γ^*. Invoking Theorem 4.1.22 we deduce that

$$c = \inf_{\gamma \in \Gamma} \max_{x \in E} \varphi(\gamma(x)),$$

where $\Gamma = \{\gamma \in C(E,X) : \gamma\big|_{E_0} = \gamma^*\}$ is a critical value of φ and $c \geq \inf_D \varphi \geq 0$ (by the local linking at 0). If $c > 0$, then we have a second nontrivial critical point because $\varphi(x_0) < \varphi(0) = 0$. If $c = 0$, then by Remark 4.1.23, we can find $x \in D$ such that $\varphi(x) = c$ and so $x \neq 0$ is the other nontrivial critical point of φ.

Case II: $\dim Y = 0$ and $\dim V > 0$.

In this case by virtue of the local linking at 0, we see that the origin is a local minimizer of φ. Recall that we have assumed that $\{0, x_0\}$ are the only critical points of φ. So we see that we can find $\varrho > 0$ small such that

$$\varphi(x) > 0 \qquad \text{for all } x \in \overline{B}_\varrho(0) \setminus \{0\}.$$

We claim that there exists $0 < \varrho_0 < \varrho$ such that

$$\inf\left[\varphi(x) : x \in \partial B_{\varrho_0}(0)\right] > 0 = \varphi(0). \tag{4.30}$$

If not, for any given fixed $0 < \varrho_0 < \varrho$, we have

$$\inf\left[\varphi(x) : x \in \partial B_{\varrho_0}(0)\right] = 0. \tag{4.31}$$

Let $\delta > 0$ be such that $0 < \varrho_0 - \delta < \varrho_0 + \delta < \varrho$ and consider the ring

$$R = \{x \in X : \varrho_0 - \delta \leq \|x\| \leq \varrho_0 + \delta\}.$$

We can find $\{x_n\}_{n \geq 1} \subseteq R$ such that

$$\|x_n\| = \varrho_0 \qquad \text{and} \qquad \varphi(x_n) \leq \frac{1}{n} \qquad \text{(see (4.31))}.$$

Invoking the Ekeland variational principle (see Theorem 2.4.1) we can find $\{y_n\}_{n\geq 1} \subseteq R$ such that

$$\varphi(y_n) \leq \varphi(x_n), \ \|y_n - x_n\| \leq \frac{1}{n} \ \text{ and } \ \varphi(y_n) \leq \varphi(x) + \frac{1}{n}\|y_n - x\| \quad \text{for all } x \in R. \tag{4.32}$$

From the second assertion in (4.32), we see that for $n \geq 1$ large we have that $y_n \in \text{int } R$. So in the third assertion in (4.32), we can take $x = y_n + \lambda u$ with $u \in X$, $\|u\| = 1$, and $\lambda > 0$ small. With this choice and by letting $\lambda \longrightarrow 0$, we obtain

$$-\frac{1}{n}\|u\| \leq \langle \varphi'(y_n), u \rangle .$$

Because $u \in \partial B_1(0)$ was arbitrary it follows that

$$\|\varphi'(y_n)\|_{X^*} \leq \frac{1}{n}.$$

This together with the first assertion in (4.32) and the fact that φ satisfies the PS-condition, implies that we may assume (at least for a subsequence) that $y_n \longrightarrow y$ in X.

Evidently $\varphi(y) = 0$, $\varphi'(y) = 0$ and $\|y\| = \varrho$, a contradiction. So (4.30) is true. If $\varrho > 0$ is small enough we have

$$0 < \varrho < \|x_0\| \qquad \text{and} \qquad \varphi(x_0) < 0 = \varphi(0) < \inf\big[\varphi(x) : x \in \partial B_\varrho(0)\big].$$

So we can apply Theorem 4.1.24 and obtain a critical point $x_1 \in X$ of φ such that

$$\begin{aligned} &\varphi(x_1) \geq \inf\big[\varphi(x) : x \in \partial B_\varrho(0)\big] > 0 = \varphi(0) > \varphi(x_0) \\ \Rightarrow\ & x_1 \neq 0, \quad x_1 \neq x_0 \quad \text{and is the second nontrivial critical point of } \varphi. \end{aligned}$$

Case III: $\dim Y > 0$ and $\dim V = 0$.

In this case we may assume that $\dim Y = +\infty$. Now the origin is a local minimizer of $-\varphi$. Because φ is bounded below and satisfies the PS-condition, from Theorem 4.1.12 we have that φ is weakly coercive. So $-\varphi(x) \longrightarrow -\infty$ as $\|x\| \longrightarrow \infty$ and we can find $u \in X$ such that

$$-\varphi(u) < -\varphi(0) = 0 < \inf\big[-\varphi(x) : x \in \partial B_\varrho(0)\big] < -\varphi(x_0)$$

for $\varrho > 0$ small so that $\|u\| > \varrho > 0$ (see Case II). Once again Theorem 4.1.24 gives the second nontrivial critical point of φ. $\square$

Hidden in the argument for Case II in the proof of Theorem 4.1.32 is the following result.

THEOREM 4.1.33 *If X is a Banach space, $\varphi \in C^1(X)$, φ satisfies the PS-condition, and it has two local minima, then φ has at least one more critical point.*

Borrowing the introductory concept of Morse theory, we make the following definition.

DEFINITION 4.1.34 Let H be a Hilbert space and $\varphi \in C^2(H)$. Let $u \in H$ and consider the operator $S_u \in \mathcal{L}(H)$ defined by

$$\langle S_u(x), y\rangle_H = \varphi''(u)(x,y) \qquad \text{for all } (x,y) \in H \times H.$$

Here by $\langle \cdot, \cdot \rangle_H$ we denote the inner product of H. The operator $S_u \in \mathcal{L}(H)$ is self-adjoint and is identified with $\varphi''(u)$. Suppose $u \in H$ is a critical point of φ. We say that φ is a *nondegenerate critical point* if $\varphi''(u) = S_u$ is invertible. The *Morse index* of u is defined as the supremum of the dimensions of the linear subspaces of H on which $\varphi''(u) = S_u \in \mathcal{L}(H)$ is negative definite.

We state the following result, known as the *Morse lemma* and for its proof we refer to Chang [143, p. 46].

LEMMA 4.1.35 *If H is a Hilbert space, $\varphi \in C^2(H)$, and x_0 is a nondegenerate critical point of φ, then there exists a Lipschitz homeomorphism h from a neighborhood U of the origin to a neighborhood W of x_0 such that $h(0) = x_0$ and*

$$\varphi\big(h(x)\big) = \varphi(x_0) + \|x_+\|^2 - \|x_-\|^2$$

where $x = x_+ + x_-$ corresponds to the decomposition of $H = H_+ \oplus H_-$ with respect to the operator $\varphi''(x_0) \in \mathcal{L}(H)$.

Using this lemma, we deduce the following useful consequence of Theorem 4.1.32.

COROLLARY 4.1.36 *If H is a Hilbert space, $\varphi \in C^2(H)$ is bounded below, satisfies the PS-condition, and x_0 is a nondegenerate critical point of φ with a finite Morse index, and $\inf_X \varphi < \varphi(x_0)$, then φ has at least two critical points that are distinct from x_0.*

PROOF: By considering $\widehat{\varphi}(x) = \varphi(x + x_0) - \varphi(x_0)$, we may assume without any loss of generality that $x_0 = 0$ and $\varphi(x_0) = 0$. Then from Lemma 4.1.35 we see that φ has a local linking at 0 with $Y = H_-$ ($\dim H_- < \infty$ because φ has a finite Morse index at $x_0 = 0$) and $V = H_+$. So we can apply Theorem 4.1.32 and produce two nontrivial critical points of φ. □

We can have a multiplicity result under symmetry conditions on the functional φ. The result is sometimes called in the literature the *symmetric mountain pass theorem*. We state the result and for a proof we refer to Rabinowitz [508, p. 55].

THEOREM 4.1.37 *If X is a Banach space, $\varphi \in C^1(X)$, satisfies the PS-condition, it is even, and also the following hold*

(i) *There exists a linear subspace V of X of finite codimension and numbers $\beta, r > 0$ such that $\varphi\big|_{\partial B_r(0) \cap V} \geq \beta$.*

(ii) *There exists a finite-dimensional linear subspace Y of X with $\dim Y > \operatorname{co\,dim} V$ such that $\varphi\big|_Y$ is weakly anticoercive; that is, $\varphi(y) \longrightarrow -\infty$ as $\|y\| \longrightarrow \infty$, $y \in Y$,*

then φ has at least $k = \dim Y - \operatorname{co\,dim} V$ distinct pairs of nontrivial critical points.

COROLLARY 4.1.38 *If the hypotheses of Theorem 4.1.37 are satisfied with* (ii) *replaced by* (ii)′ *for any integer $k \geq 1$, there is a k-dimensional linear subspace Y of X such that $\varphi|_Y$ is weakly anticoercive; that is, $\varphi(y) \longrightarrow -\infty$ as $\|y\| \longrightarrow \infty$, $y \in Y$, then φ has infinitely many distinct pairs of nontrivial critical points.*

Continuing with functionals exhibiting symmetry, we present the *principle of symmetric criticality*. First we have a definition.

DEFINITION 4.1.39 Let G be a topological group. A *representation* of G over a Banach space X is a family $\{S(g)\}_{g\in G} \subseteq \mathcal{L}(X)$ such that

(a) $S(e) = I$ (e is the unit of G).
(b) $S(g_1 * g_2) = S(g_1)\, S(g_2)$ for all $g_1, g_2 \in G$ (here $*$ denotes the operation of G).
(c) $(g, x) \longrightarrow S(g)x$ is continuous from $G \times X$ into X.

A set $C \subseteq X$ is *invariant* (under the representation), if $S(g)(C) \subseteq C$ for all $g \in G$. A function $\varphi\colon X \longrightarrow \mathbb{R}$ is *invariant* (under the representation), if $\varphi \circ S(g) = \varphi$ for all $g \in G$. A map $h\colon X \longrightarrow X$ is *equivariant* (under the representation), if $S(g) \circ h = h \circ S(g)$ for all $g \in G$. A representation $\{S(g)\}_{g\in G}$ of G over X is said to be *isometric*, if $\|S(g)x\| = \|x\|$ for all $g \in G$ and all $x \in X$. Finally the set of *invariant points* of X under the representation, is defined by

$$\mathrm{Fix}(S) = \mathrm{Fix}(G) = \{x \in X : S(g)x = x \ \text{ for all } g \in G\}.$$

REMARK 4.1.40 Very often to simplify the notation we identify $S(g)$ with g. We do this here too. Also we speak about the *action* of G on X, to mean the representation of G over X.

THEOREM 4.1.41 *If H is a Hilbert space, G is a topological group that has an isometric representation over H, $\varphi \in C^1(H)$ is invariant, and x_0 is a critical point of $\varphi|_{\mathrm{Fix}(G)}$, then x_0 is also a critical point of φ.*

PROOF: For every $x, y \in H$ we have

$$\begin{aligned}
\left(\varphi'(gx), y\right)_H &= \lim_{\lambda\to 0} \frac{\varphi(gx+\lambda y)-\varphi(gx)}{\lambda}\\
&= \lim_{\lambda\to 0} \frac{\varphi(gx+\lambda g g^{-1} y)-\varphi(gx)}{\lambda}\\
&= \lim_{\lambda\to 0} \frac{\varphi(x+\lambda g^{-1} y)-\varphi(x)}{\lambda} \qquad \text{(because } \varphi \text{ is invariant)}\\
&= \left(\varphi'(x), g^{-1}y\right)_H.
\end{aligned} \tag{4.33}$$

Moreover, because the representation is isometric (hence it preserves inner products), we have

$$\begin{aligned}
&\left(\varphi'(x), g^{-1}y\right)_H = \left(g\,\varphi'(x), y\right)_H,\\
\Rightarrow\ &\left(\varphi'(gx), y\right)_H = \left(g\,\varphi'(x), y\right)_H \qquad \text{for all } y \in H \quad \text{(see (4.33))},\\
\Rightarrow\ &\varphi'(gx) = g\,\varphi'(x); \ \text{ that is, } \varphi' \text{ is equivariant.}
\end{aligned}$$

Now suppose that x_0 is a critical point of $\varphi|_{\mathrm{Fix}(G)}$. Then clearly

$$g\,\varphi'(x_0) = \varphi'(gx_0) = \varphi'(x_0)$$
$$\Rightarrow \ \varphi'(x_0) \in \mathrm{Fix}(G).$$

Therefore we have $\varphi'(x_0) \in \mathrm{Fix}(G) \cap \mathrm{Fix}(G)^{\perp} = \{0\}$; that is, $\varphi'(x_0) = 0$. $\square$

We conclude by briefly mentioning a generalization of this theory to a class of functionals that arise when dealing with problems with unilateral constraints.

So let X be a Banach space. The generalization is for functionals

$$\widehat{\varphi} = \varphi + \psi \tag{4.34}$$

with $\widehat{\varphi} \in C^1(X)$ and $\psi \in \Gamma_0(X)$ (see Definition 1.2.1).

DEFINITION 4.1.42 Let X be a Banach space and $\widehat{\varphi}\colon X \longrightarrow \overline{\mathbb{R}} = \mathbb{R} \cup \{+\infty\}$ a functional satisfying the decomposition (4.34).

(a) A point $x \in X$ is said to be a *critical point* of $\widehat{\varphi}$, if $x \in \mathrm{dom}\,\psi$ and

$$\big\langle \varphi'(x), y - x \big\rangle + \psi(y) - \psi(x) \geq 0 \qquad \text{for all } y \in X. \tag{4.35}$$

(b) We say that $\widehat{\varphi}$ satisfies the *generalized PS-condition* if every sequence $\{x_n\}_{n\geq 1} \subseteq X$ such that $\{\widehat{\varphi}(x_n)\}_{n\geq 1}$ is bounded in $\mathbb{R}$ and

$$\big\langle \varphi'(x_n), y - x_n \big\rangle + \psi(y) - \psi(x_n) \geq -\varepsilon_n \|y - x_n\| \qquad \text{for all } y \in X$$

with $\varepsilon_n \downarrow 0$, has a strongly convergent subsequence.

REMARK 4.1.43 Clearly $x \in X$ is a critical point of $\widehat{\varphi}$ if and only if $-\varphi'(x) \in \partial\psi(x)$.

LEMMA 4.1.44 *If X is a Banach space, $\psi \in \Gamma_0(X)$ with $\psi(0) = 0$ and $\psi(x) \geq -\|x\|$ for all $x \in X$, then there exists $x^* \in X^*$ with $\|x^*\| \leq 1$ such that $\psi(x) \geq \langle x^*, x\rangle$ for all $x \in X$.*

PROOF: Let $\xi(x) = \psi(x) + \|x\|$, $x \in X$. Then $\xi \in \Gamma_0(X)$, $\xi \geq 0$, and $\xi(0) = 0$. So $0 \in \partial\xi(0) = \partial\psi(0) + \partial\|\cdot\|(0)$ (see Proposition 1.2.38). Because $\partial\|\cdot\|(0) = \overline{B}_1^*(0) = \{x^* \in X^* : \|x^*\| \leq 1\}$, we obtain $x^* \in \partial\psi(0)$ with $\|x^*\| \leq 1$ such that $\langle x^*, x\rangle \leq \psi(x)$ for all $x \in X$. $\square$

Using this lemma we can equivalently reformulate the generalized PS-condition in a more convenient form.

PROPOSITION 4.1.45 *If X is a Banach space and $\widehat{\varphi}\colon X \to \overline{\mathbb{R}} = \mathbb{R} \cup \{+\infty\}$ a functional satisfying the decomposition (4.34), then the generalized PS-condition (see Definition 4.1.42* (b)) *is equivalent to the following condition,*

every sequence $\{x_n\}_{n\geq 1} \subseteq X$ such that $\{\varphi(x_n)\}_{n\geq 1}$ is bounded and

$$\big\langle \varphi'(x_n), y - x_n \big\rangle + \psi(y) - \psi(x_n) \geq \langle x_n^*, y - x_n\rangle \qquad \text{for all } y \in X$$

with $x_n^* \longrightarrow 0$, has a strongly convergent subsequence.

Using the Ekeland variational principle (see Theorem 2.4.1), we can prove a *deformation theorem* for this new class of functionals and then use it to produce minimax characterizations of the critical values of $\widehat{\varphi}$. The details can be found in Szulkin [568]. Here we simply state without proofs the formulations of the *mountain pass*, *saddle point* and *generalized mountain pass* theorems in this context.

First we have the *mountain pass theorem*.

THEOREM 4.1.46 *If X is a Banach space, $\widehat{\varphi}: X \longrightarrow \overline{\mathbb{R}} = \mathbb{R} \cup \{+\infty\}$ is a functional satisfying the decomposition (4.34) and the generalized PS-condition, and the following hold,*

(i) *$\widehat{\varphi}(0) = 0$ and there exist $\beta, r > 0$ such that $\widehat{\varphi}|_{\partial B_r(0)} \geq \beta$,*
(ii) *$\widehat{\varphi}(u) \leq 0$ for some $u \in X$ with $\|u\| > r$,*

then $\widehat{\varphi}$ has a critical value $c \geq \beta$ that can be characterized by

$$c = \inf_{\gamma \in \Gamma} \sup_{t \in [0,1]} \widehat{\varphi}(\gamma(t)),$$

where $\Gamma = \{\gamma \in C([0,1], X) : \gamma(0) = 0,\ \gamma(1) = u\}$.

Next we state the *saddle point theorem*.

THEOREM 4.1.47 *If X is a Banach space, $X = Y \oplus V$ with $\dim Y < +\infty$, $\widehat{\varphi}: X \longrightarrow \overline{\mathbb{R}} = \mathbb{R} \cup \{+\infty\}$ is a functional satisfying the decomposition (4.34) and the generalized PS-condition and the following hold,*

(i) *There exist constants $r > 0$ and $\alpha \in \mathbb{R}$ such that $\widehat{\varphi}|_{\partial B_r(0) \cap Y} \leq \alpha$*
(ii) *There exists a constant $\beta > \alpha$ such that $\widehat{\varphi}|_V \geq \beta$*

then $\widehat{\varphi}$ has a critical value $c \geq \beta$ that can be characterized by

$$c = \inf_{\gamma \in \Gamma} \sup_{x \in \overline{B}_r(0) \cap Y} \widehat{\varphi}(\gamma(x)),$$

where $\Gamma = \{\gamma \in C(\overline{B}_r(0) \cap Y, X) : \gamma|_{\partial \overline{B}_r(0) \cap Y} = I|_{\partial \overline{B}_r(0) \cap Y}\}$.

Finally we state the *generalized mountain pass theorem*.

THEOREM 4.1.48 *If X is a Banach space, $X = Y \oplus V$ with $\dim Y < +\infty$, $\widehat{\varphi}: X \longrightarrow \overline{\mathbb{R}} = \mathbb{R} \cup \{+\infty\}$ is a functional satisfying the decomposition (4.34) and the generalized PS-condition, and the following hold,*

(i) *There exist constants $r, \beta > 0$ such that $\widehat{\varphi}|_{\partial B_r(0) \cap V} \geq \beta$,*
(ii) *There exists a constant $R > r$ and $v_0 \in V$ with $\|v_0\| = 1$ such that*

$$E = \{x = \lambda v_0 + y : y \in Y,\ \|y\| \leq R,\ 0 \leq \lambda \leq R\}$$

and on $E_0 = \partial E$ in $W = \mathbb{R}v_0 \oplus Y$, we have $\widehat{\varphi}|_{E_0} \leq 0$,

then $\widehat{\varphi}$ has a critical value $c \geq \beta$ that can be characterized by

$$c = \inf_{\gamma \in \Gamma} \sup_{x \in E} \widehat{\varphi}(\gamma(x)),$$

where $\Gamma = \{\gamma \in C(E, X) : \gamma|_{E_0} = I|_{E_0}\}$.

Also the following is a straightforward consequence of the Ekeland variational principle (see Theorem 2.4.1).

THEOREM 4.1.49 *If X is a Banach space and $\widehat{\varphi}\colon X \longrightarrow \overline{\mathbb{R}} = \mathbb{R} \cup \{+\infty\}$ is a functional satisfying the decomposition (4.34), the generalized PS-condition, and also is bounded below, then $m = \inf_X \widehat{\varphi} = \widehat{\varphi}(x_0)$ for some $x_0 \in X$ and it is a critical value of $\widehat{\varphi}$.*

REMARK 4.1.50 Multiplicity results under symmetry, similar to Theorem 4.1.37 and Corollary 4.1.38, can be found in Szulkin [568].

4.2 Ljusternik–Schnirelmann Theory

Let H be an infinite-dimensional Hilbert space. In what follows we identify H with H^* and by $\langle \cdot, \cdot \rangle_H$ we denote the inner product of H. In Section 3.1 we studied the spectral properties of a self-adjoint operator $A \in \mathcal{L}_c(H)$ (i.e., $A : H \longrightarrow H$ is linear, compact, and self-adjoint). In particular we proved that A has a sequence $\{\lambda_n\}_{n\geq 1}$ of eigenvalues such that $\lambda_n \longrightarrow 0$ as $n \to \infty$. Moreover, there is an orthonormal basis $\{e_n\}_{n\geq 1} \subseteq H$ corresponding to the eigenvalues $\{\lambda_n\}_{n\geq 1}$ and we have the spectral resolution $A(x) = \sum\limits_{n\geq 1} \lambda_n \langle x, e_n \rangle_H e_n$. There is a variational characterization of the eigenvalues. In particular, let $\{\lambda_n^+\}_{n\geq 1}$ be the positive eigenvalues ordered in decreasing order (with multiplicities repeated), $\mathcal{L}_n$ the collection of all linear subspaces of H of dimension n, and $\mathcal{C}_n$ the collection of all linear subspaces of H of codimension $n-1$. Then we have the following variational characterizations of the eigenvalues. The result is known as the Courant minimax principle. Sometimes the names of Weyl and Fischer are attached to it too.

PROPOSITION 4.2.1 *If everything is as described above, then*

$$\lambda_n^+ = \sup_{Y \in \mathcal{L}_n} \min_{x \in Y} \frac{\langle A(x), x \rangle_H}{\|x\|^2} = \inf_{V \in \mathcal{C}_n} \sup_{x \in V} \frac{\langle A(x), x \rangle_H}{\|x\|^2}. \tag{4.36}$$

REMARK 4.2.2 An analogous variational characterization is also true for the negative eigenvalues $\{\lambda_n^-\}_{n\geq 1}$. Namely,

$$-\lambda_n^- = \sup_{Y \in \mathcal{L}_n} \min_{x \in Y} -\frac{\langle A(x), x \rangle_H}{\|x\|^2} = \inf_{V \in \mathcal{C}_n} \sup_{x \in V} -\frac{\langle A(x), x \rangle_H}{\|x\|^2}. \tag{4.37}$$

Of course, if A is monotone, then only nonnegative eigenvalues exist (such as in the case of the Laplacian differential operator; see Section 4.3). Note that if $S = \partial B_1(0) = \{x \in H : \|x\| = 1\}$, then we can equivalently rewrite (4.36) and (4.37) as

$$\lambda_n^+ = \sup_{Y \in \mathcal{L}_n} \min_{x \in S \cap Y} \langle A(x), x \rangle_H = \inf_{V \in \mathcal{C}_n} \sup_{x \in S \cap V} \langle A(x), x \rangle_H \tag{4.38}$$

$$-\lambda_n^- = \sup_{Y \in \mathcal{L}_n} \min_{x \in S \cap Y} -\langle A(x), x \rangle_H = \inf_{V \in \mathcal{C}_n} \sup_{x \in S \cap V} -\langle A(x), x \rangle_H. \tag{4.39}$$

The eigenvalues of A are precisely the critical values of the quadratic functional $\varphi(x) = \langle A(x), x \rangle_H$ on the unit sphere $S = \partial B_1(0)$ in H, therefore it is natural to extend (4.38) and (4.39) to general smooth functionals φ by finding "topological" analogues of the sets $S \cap Y$ and of the families $\{S \cap Y : Y \in \mathcal{L}_n\}_{n \geq 1}$. This will produce a far-reaching generalization of Proposition 4.2.1 to nonlinear eigenvalue problems. This is the goal of the Ljusternik–Schnirelmann theory, which we briefly outline in this section.

A crucial step in extending the theory to nonquadratic functionals (i.e., to nonlinear eigenvalue problems in a Banach space X) is the determination of an interesting (i.e., nontrivial), computable $\mathbb{N}_0 \cup \{+\infty\}$-valued function ξ on a class of closed subsets of X that exhibits the following properties.

(C) (1) $\xi(\emptyset) = 0$ and $\xi(A) = 1$ if A is a finite nonempty set.
(2) $A \subseteq B \Rightarrow \xi(A) \leq \xi(B)$ (monotonicity),
(3) If $h(t, x)$ is a deformation of Y (i.e. $h: [0,1] \times Y \longrightarrow Y$ is a continuous map such that $h(0, x) = x$ for all $x \in Y$), then $\xi(A) \leq \xi\big(h(1, A)\big)$.
(4) $\xi(A \cup B) \leq \xi(A) + \xi(B)$ (subadditivity).
(5) There is a neighborhood U of A such that $\xi(U) = \xi(A)$ (continuity).

REMARK 4.2.3 Such functions ξ are usually called *topological indices*. If $\mathcal{S}$ is the family of all closed sets in a Banach space X, the trivial function $\xi: \mathcal{S} \longrightarrow \mathbb{N}_0 \cup \{+\infty\}$ defined by

$$\xi(A) = \begin{cases} 1 & \text{if } A \subseteq X \text{ is nonempty, closed} \\ 0 & \text{if } A = \emptyset, \end{cases}$$

satisfies properties (C). Fortunately we can find more interesting indices than this one.

DEFINITION 4.2.4 Let Y be a Hausdorff topological space and $C \subseteq Y$ a nonempty set. We say that C is *contractible* in Y, if there exist a continuous map $h: [0,1] \times C \longrightarrow Y$ and $y_0 \in Y$ such that $h(0, x) = x$ and $h(1, x) = y_0$ for all $x \in C$.

REMARK 4.2.5 Intuitively, regarding $t \in [0,1]$ as a time parameter, $h(t, x)$ describes a continuous deformation of the set C down to the singleton $\{y_0\}$, with $h(t_0, C)$ expressing the state of the deformation at time $t_0 \in [0,1]$. Another way to describe contractible sets C is to say that the identity map is null-homotopic.

Next we give the definition of a nontrivial function $\mathrm{cat}_Y(A)$ which admits properties (C) and which is also maximal in the sense that if $\xi(A)$ is another function defined on the closed subsets of X and possessing properties (C), then $\xi(A) \leq \mathrm{cat}_Y(A)$.

DEFINITION 4.2.6 Let Y be a Hausdorff topological space. The *Ljusternik–Schnirelmann category* (LS-category for short), denoted by cat_Y, is defined on the closed subsets of Y by $\mathrm{cat}_Y(\emptyset) = 0, \mathrm{cat}_Y(A) = \min\{n \in \mathbb{N} : A \subseteq \bigcup_{k=1}^{n} S_k\}$, S_k is closed, contractible for all $k \in \{1, \ldots, n\}$ and $\mathrm{cat}_Y(A) = +\infty$ if $A \neq \emptyset$ does not admit such a finite cover.

REMARK 4.2.7 The properties of closedness and contractibility are preserved by homeomorphisms, thus it is clear that we get the same values for homeomorphic Y or homeomorphic closed set C. However, the index Y in $\operatorname{cat}_Y(C)$ is essential if Y lies in a larger space $\widehat{Y}$, because a set may be contractible in $\widehat{Y}$ but not in the smaller space Y. If Y is embedded continuously in $\widehat{Y}$, then $\operatorname{cat}_{\widehat{Y}}(C) \leq \operatorname{cat}_Y(C)$ for all closed $C \subseteq Y$.

Next we verify that for a large class of metric spaces Y, $\operatorname{cat}_Y(C)$ satisfies properties (C). For this purpose, we need to recall the following definition from topology.

DEFINITION 4.2.8 A metric space Y is said to be an *absolute neighborhood retract* (*ANR* for short), if for any metric space Z, any closed set $A \subseteq Z$ and any continuous function $f\colon A \longrightarrow Y$, we can find a neighborhood U of A (U depends on f) and a continuous function $\widehat{f}\colon U \longrightarrow Y$ such that $\widehat{f}\big|_A = f$. If f can always be extended to all of Z, then we say that Y is an *absolute retract* (*AR* for short).

This property is not that restrictive as the following list of examples illustrates.

EXAMPLE 4.2.9 (a) It is straightforward to check that a finite product of ANR's is again an ANR.

(b) From Theorem 3.1.11 (Dugundji's extension theorem), we know that every closed and convex set of a locally convex space is an AR (hence a ANR too).

(c) If $S^1 = \{x \in \mathbb{R}^2 : \|x\| = 1\}$ (the circle), then S^1 is an ANR. To see this let Z be a metric space, $A \subseteq Z$ a closed subset and $f \in C(A, S^1)$. By virtue of the Tietze extension theorem we can find $\widehat{f} \in C(Z, \mathbb{R}^2)$ such that $\widehat{f}\big|_A = f$. On the other hand we can find a neighborhood U of S^1 in $\mathbb{R}^2$ and a retraction $r\colon U \longrightarrow S^1$ (recall that no global retraction on S^1 exists, see Theorem 3.5.7). Then the map $r \circ \widehat{f}$ defined on $\widehat{f}^{-1}(U)$ is the desired extension of f.

PROPOSITION 4.2.10 *If Y is a path-connected ANR, then cat_Y satisfies properties* (C).

PROOF: Property (C)(1) holds if and if Y is path-connected.
Property (C)(2) (monotonicity property) is obvious.
For property (C)(3) we argue as follows. Suppose that $h\colon [0,1] \times Y \longrightarrow Y$ is a continuous map such that $h(0,\cdot) = I$ (a deformation of Y) and that $h(1, A) \subseteq \bigcup_{k=1}^{n} A_k$ with $A_k \subseteq Y$ closed, contractible for all $k \in \{1, \ldots, n\}$. So we can find continuous maps $h_k\colon [0,1] \times A_k \longrightarrow Y$ such that $h_k(0, x) = x$ for all $x \in A_k$ and $h(1, A_k) = \{x_k\}$ with $x_k \in Y$ (see Definition 4.2.4) and let $B_k = h(1,\cdot)^{-1}(A_k)$ for all $k \in \{1, \ldots, n\}$. Evidently B_k is closed and if we set

$$\widehat{h}_k(t,x) = \begin{cases} h(2t, x) & \text{if } t \in \left[0, \frac{1}{2}\right] \\ h_k\big(2t-1, h(t,x)\big) & \text{if } t \in \left[\frac{1}{2}, 1\right] \end{cases},$$

then we see that $\widehat{h}_k$ is a continuous homotopy between $I\big|_{B_k}$ and the constant map equal to x_k. So B_k is contractible to $\{x_k\}$. We have $A\subseteq \bigcup\limits_{k=1}^{n} B_k$ and so $\mathrm{cat}_Y(A) \leq \mathrm{cat}_Y\big(h(1,A)\big)$.

Property (C)(4) is obvious.

To prove property (C)(5), we proceed as follows. First suppose that A is closed with $\mathrm{cat}_Y(A) = 1$. Let $h\colon [0,1] \times A \longrightarrow Y$ be a deformation of A onto $\{y_0\}$ with $y_0 \in Y$. Set

$$Z = [0,1] \times Y \qquad \text{and} \qquad D = (\{0\} \times Y) \cup ([0,1] \times A) \cup (\{1\} \times Y)$$

and define $f\colon C \longrightarrow Y$ by $f(0,x) = x$ for all $x \in Y$, $f = h$ on $[0,1] \times A$ and $f(1,x) = y_0$ for all $x \in Y$. Since by hypothesis Y is an ANR, we can find $\widehat{f}$ a continuous extension of f on some neighborhood V of D and $V \supseteq [0,1] \times U$ for some neighborhood U of A. Then $\widehat{f}\big|_{[0,1]\times U}$ deforms U onto $\{x_0\}$ and so $\mathrm{cat}_Y(U) = \mathrm{cat}_Y(A) = 1$. If $\mathrm{cat}_Y(A) = n$, then $A \subseteq \bigcup\limits_{k=1}^{n} A_k$ with A_k closed and $\mathrm{cat}_Y(A_k) = 1$. Therefore $A\subseteq \bigcup\limits_{k=1}^{n} U_k = U$ with $A_k \subseteq U_k$ and $\mathrm{cat}_Y(U_k) = 1$. Hence $\mathrm{cat}_Y(U) = n$. If $\mathrm{cat}_Y(A) = +\infty$, then clearly $\mathrm{cat}_Y(Y) = +\infty$. □

Now we show that cat_Y is maximal among all $\mathbb{N}_0 \cup \{+\infty\}$-valued functions $\xi(\cdot)$ on the closed subsets of Y having properties (C).

PROPOSITION 4.2.11 *If Y is a path connected ANR and ξ is an $\mathbb{N}_0 \cup \{+\infty\}$-valued function on the closed subsets of Y that satisfies properties* (C)*, then $\xi \leq \mathrm{cat}_Y$.*

PROOF: If $\mathrm{cat}_Y(A) = 1$, then according to Definition 4.2.6, A is deformable to a point $y \in Y$. So from property (C)(3), we have $\xi(A) \leq \xi(\{y\}) = 1$ (property (C)(3)), hence $\xi(A) \leq \mathrm{cat}_Y(A)$. If $\mathrm{cat}_Y(A) = n < +\infty$, then $A = \bigcup\limits_{k=1}^{n} A_k$, where each A_k is closed and deformable to a point $y_k \in Y$. So $\xi(A_k) = \xi(\{y_k\}) = 1$ for all $k \in \{1, \ldots, n\}$ (see property (C)(3)) and then using the subadditivity property (C)(4) we obtain

$$\xi(A) = \xi\Big(\bigcup_{k=1}^{n} A_k\Big) \leq \sum_{k=1}^{n} \xi(A_k) = n = \mathrm{cat}_Y(A).$$

Therefore we conclude that $\xi \leq \mathrm{cat}_Y$. □

EXAMPLE 4.2.12 (a) If $Y = T^n = \mathbb{R}^n/\mathbb{Z}_2$ is an n-torus, then $\mathrm{cat}_{T^n}(T^n) = n+1$. For a proof of this we refer to Schwartz [545, p. 161].

(b) If $Y = P^n = S^n/\mathbb{Z}_2$ where $S^n = \{x \in \mathbb{R}^{n+1} : \|x\| = 1\}$ (the real n-projective space, it arises from S^n by identifying antipodal points), then $\mathrm{cat}_{P^n}(P^n) = n+1$. Also $\mathrm{cat}_{P^n}(P^k) = k+1$ for all $n \geq k$. If X is an infinite-dimensional Banach space and $P^\infty(X)$ is the infinite -dimensional projective space obtained by identifying the antipodal points of the unit sphere $\partial B_1(0) = \{x \in X : \|x\| = 1\}$ (i.e., $P^\infty(X) = \partial B_1(0)/\mathbb{Z}_2$), then $\mathrm{cat}_{P^\infty}(P^\infty) = +\infty$.

(c) If $Y = \mathbb{R}^n$ and A is a closed ball in $\mathbb{R}^n$, then $\mathrm{cat}_{\mathbb{R}^n}(A) = 1$.

(d) If $Y = \mathbb{R}^n$ and $A = S^{n-1} = \{x \in \mathbb{R}^n : \|x\| = 1\}$, then $\mathrm{cat}_{\mathbb{R}^n}(S^{n-1}) = \mathrm{cat}_{S^{n-1}}(S^{n-1}) = 2$ (take as covering sets $\{A_1, A_2\}$, two slightly overlapping "northern" and "southern" hemispheres).

(e) If $Y = X =$ an infinite-dimensional Banach space and $A = S = \{x \in X : \|x\| = 1\}$. Then $\mathrm{cat}_X(S) = \mathrm{cat}_S(S) = 1$ (recall that due to the infinite-dimensionality of X, S is contractible in itself).

The determination of the LS-category of a set $A \subseteq X$, is in general a difficult and complicated task, which involves various results from cohomology and homotopy theories. So it is desirable to have another interesting function ξ satisfying properties (C), which is more elementary than the LS-category. This is provided by the so-called *genus* of a set.

As before X is a Banach space. In this case we define an $\mathbb{N}_0 \cup \{+\infty\}$-valued function γ, on the closed and symmetric with respect to the origin, subsets of $X \backslash \{0\}$.

DEFINITION 4.2.13 Let X be a Banach space and $\mathrm{Sym}(X) = \{A \subseteq X \setminus \{0\} : A \text{ is closed and } A = -A\}$. The *genus* $\gamma\colon \mathrm{Sym}(X) \longrightarrow \mathbb{N}_0 \cup \{+\infty\}$ is defined by

$$\gamma(\emptyset) = 0, \quad \gamma(A) = \min\big\{n \in \mathbb{N} : \text{ there exists odd } f \in C(A, \mathbb{R}^n \setminus \{0\})\big\}$$
$$\gamma(A) = +\infty \quad \text{if no such continuous odd map exists.}$$

REMARK 4.2.14 For any $A \in \mathrm{Sym}(X)$, by the Tietze extension theorem any odd map $f \in C(A, \mathbb{R}^n \setminus \{0\})$ admits an extension $\widehat{f} \in C(X, \mathbb{R}^n)$. In fact if we let $\widehat{f}_0(x) = \frac{1}{2}\big(\widehat{f}(x) - \widehat{f}(-x)\big)$, the extension can be assumed to be odd too. The function γ is often called *Krasnoselskii genus*.

The notion of genus generalizes the concept of dimension of a linear space.

PROPOSITION 4.2.15 *If X is a Banach space and U is a bounded symmetric neighborhood of the origin in X, then $\gamma(\partial U) = \dim X$.*

PROOF: First suppose that $0 < \dim X < +\infty$. If in Definition 4.2.13 we use $f = I$, we see that $\gamma(\partial U) \leq \dim X$. If $k < \dim X$, then by Corollary 3.3.29 we know that there is no odd function $f \in C(\partial U, \mathbb{R}^k \setminus \{0\})$. Hence we must have $\gamma(\partial U) = \dim X$.

Next suppose that $\dim X = +\infty$. Let X_n be an n-dimensional linear subspace of X. Clearly from Definition 4.2.13 we have $\gamma(\partial U \cap X_n) \leq \gamma(\partial U)$. Also from the first part of the proof we know that $\gamma(\partial U \cap X_n) = n$. So $n \leq \gamma(\partial U)$ for all $n \geq 1$, hence $\gamma(\partial U) = +\infty = \dim X$.

Finally if $X = \{0\}$, then $\partial U = \emptyset$ and so $\gamma(\partial U) = 0 = \dim X$. $\square$

There is a sort of converse to this proposition.

PROPOSITION 4.2.16 *If H is a Hilbert space, $A \in \mathrm{Sym}(H)$ and it is compact, and $\gamma(A) = k < +\infty$, then A contains at least k-mutually orthogonal vectors $\{x_i\}_{i=1}^k$.*

PROOF: Let $\{x_i\}_{i=1}^m$ be a maximal set of mutually orthogonal vectors in A. Let $Y = \operatorname{span}\{x_i\}_{i=1}^m$ which is isomorphic to $\mathbb{R}^m$. Let $p_Y : H \longrightarrow Y$ be the orthogonal projection onto Y. Then $0 \notin p_Y(A)$ and $f = p_Y\big|_A : A \longrightarrow \mathbb{R}^m \setminus \{0\}$ is continuous, odd, hence $\gamma(A) = k < m$. □

Next we state the relation between the two indices cat_Y and γ. The result is due to Rabinowitz [503], where the interested reader can find its proof.

PROPOSITION 4.2.17 *If X is a Banach space, $P = \{(x, -x) : x \in X \setminus \{0\}\}$ and $A \in \operatorname{Sym}(X)$, then $\operatorname{cat}_P(\widehat{A}) = \gamma(A)$ where $\widehat{A} = \{(x, -x) : x \in A\}$.*

Using this result we can now prove that γ satisfies properties (C).

PROPOSITION 4.2.18 *If X is a Banach space then the genus γ: $\operatorname{Sym}(X) \longrightarrow \mathbb{N}_0 \cup \{+\infty\}$ satisfies properties (C) provided the deformations $h(t, x)$ are understood to be odd in $x \in X$ and $h(1, A) \in \operatorname{Sym}(X)$ for all $A \in \operatorname{Sym}(X)$.*

PROOF: Properties (C)(1) and (2) follow at once from Definition 4.2.13. Also let $A \in \operatorname{Sym}(X)$ and suppose that $h(t, x)$ is a deformation that is odd in $x \in X$ and $h(1, A) \in \operatorname{Sym}(X)$. Then if $f \in C(h(1, A), \mathbb{R}^m \setminus \{0\})$, we have that $f \circ h(1, \cdot) : A \longrightarrow \mathbb{R}^m \setminus \{0\}$ is continuous and odd and so $\gamma(A) \leq m$, hence $\gamma(A) \leq \gamma(h(1, A))$. This proves property (C)(3). Next we prove property (C)(4) (the subadditivity property). Clearly we may assume that both $\gamma(A) = m$ and $\gamma(B) = n$ are finite, or otherwise we are done. So we can find $f \in C(A, \mathbb{R}^m \setminus \{0\})$ and $g \in C(B, \mathbb{R}^n \setminus \{0\})$ odd maps (see Definition 4.2.13). By the Tietze extension theorem, we can find $\widehat{f} \in C(X, \mathbb{R}^m)$ and $\widehat{g} \in C(X, \mathbb{R}^n)$ extensions of these maps. Let $u = (\widehat{f}, \widehat{g}) \in C(X, \mathbb{R}^{m+n})$. This is an odd map and $u(x) \neq 0$ for all $x \in A \cup B$. Hence $\gamma(A \cup B) \leq m + n$. This proves property (C)(4). Property (C)(5) follows from Propositions 4.2.17 and 4.2.10. □

PROPOSITION 4.2.19 *If X is a Banach space and $A \in \operatorname{Sym}(X)$ is compact, then $\gamma(A) < +\infty$.*

PROOF: Because $A \in \operatorname{Sym}(X)$, we have $0 \notin A$ and so we find $r > 0$ such that $A \cap B_r(0) = \emptyset$. The family $\{B_r(x) \cup B_r(-x)\}_{x \in A}$ is an open cover of A and because A is compact, we can find a finite subcover $\{B_r(x_k) \cup B_r(-x_k)\}_{k-1}^N$. Let $\{\psi_k\}_{k-1}^N$ be a continuous partition of unity subordinated to the open cover $\{B_r(x_k) \cup B_r(-x_k)\}_{k-1}^N$. Replacing ψ_k by $\widehat{\psi}_k(x) = \frac{1}{2}[\psi_k(x) + \psi_k(-x)]$ if necessary, we may assume that each ψ_k, $k = 1, \ldots, N$, is even. From the choice of $r > 0$ we see that $B_r(x_k) \cap B_r(-x_k) = \emptyset$ for all $k = 1, \ldots, N$. Consider the map $u : X \longrightarrow \mathbb{R}^N$ with kth-component defined by

$$u_k(x) = \begin{cases} \psi_k(x) & \text{if } x \in B_r(x_k) \\ -\psi_k(x) & \text{if } x \in B_r(-x_k) \end{cases}.$$

Evidently $u \in C(X, \mathbb{R}^m)$, u is odd, and $0 \notin u(A)$. Therefore $\gamma(A) \leq N < +\infty$. □

PROPOSITION 4.2.20 *If X is a Banach space, Y is a finite-dimensional linear subspace of X, $p_Y \in \mathcal{L}(X)$ is the projection operator onto Y, and $A \in \operatorname{Sym}(X)$ with $\gamma(A) > k = \dim Y$, then $A \cap (I - p_Y)(X) \neq \emptyset$.*

PROOF: We have $X = Y \oplus V$ with $V = (I - p_Y)(X)$. If $A \cap (I - p_Y)(X) = \emptyset$, then $A \cap V = \emptyset$ and so $A \subseteq Y$. Note that $p_Y \in C(A, Y \setminus \{0\})$ and it is odd. Hence $\gamma(A) \leq k$, a contradiction to the hypothesis. Therefore $A \cap V = A \cap (I - p_Y)(X) \neq \emptyset$. □

The next result is an immediate consequence of Definition 4.2.13.

PROPOSITION 4.2.21 *If X is a Banach space and $A, B \in \mathrm{Sym}(X)$ are homeomorphic with respect to an odd homeomorphism, then $\gamma(A) = \gamma(B)$.*

Now that we have two interesting and computable topological indices (namely the LS-category and the (Krasnoselskii) genus), we can state the main result of the Ljusternik–Schnirelmann theory, which justifies all the effort to define topological indices satisfying properties (C).

So let X be a Banach, $\mathcal{S}$ a deformation invariant family of closed sets of X, and ξ a topological index on $\mathcal{S}$ (i.e., $\xi \colon \mathcal{S} \longrightarrow \mathbb{N}_0 \cup \{+\infty\}$ and satisfies properties (C)). For example we can have $\xi = \mathrm{cat}_X$ defined on all closed subsets of X or we can have $\xi = \gamma$ defined on $\mathrm{Sym}(X)$. For every $k \geq 1$, we set

$$\mathcal{S}_k = \{A \in \mathcal{S} : \xi(A) \geq k\}. \tag{4.40}$$

Clearly property (C)(3) implies that $\mathcal{S}_k$ is deformation invariant; that is, if $h(t,x)$ is a deformation of X and $A \in \mathcal{S}_k$, then $h(1, A) \in \mathcal{S}_k$. This fact in conjunction with the next proposition suggests the right way to proceed.

PROPOSITION 4.2.22 *If X is a Banach space, $\varphi \in C(X)$, $K \subseteq X$, and $\varphi(K) \subseteq \mathbb{R}$ are both closed, condition* (D) *from the beginning of Section 4.1 is satisfied and $\widehat{\mathcal{S}} \subseteq 2^X$ is a nonempty deformation invariant family, then $\widehat{c} = \inf\limits_{A \in \widehat{\mathcal{S}}} \sup\limits_{x \in A} \varphi(x) < +\infty$ implies that $\widehat{c} \in \varphi(K)$.*

PROOF: Suppose that $\widehat{c} \notin \varphi(K)$. Because by hypothesis $\varphi(K)$ is closed, we can find $a < \widehat{c} < b$ such that $\varphi^{-1}([a,b]) \cap K = \emptyset$. Let $A \in \widehat{\mathcal{S}}$ such that $\sup\limits_{x \in A} \varphi(x) \leq b$. Then from condition (D)(1) we know that we can find $t_0 > 0$ such that $\varphi\big(h(t_0, x)\big) \leq a$ for all $x \in A$. Since $\widehat{h}(t,x) = h(tt_0, x)$, $t \in [0,1]$, $x \in X$, defines a deformation of X with $\widehat{h}(1, A) = h(t_0, A)$, we have that $h(t_0, A) \in \widehat{\mathcal{S}}$ (recall that by hypothesis $\widehat{\mathcal{S}}$ is deformation invariant). Therefore $\widehat{c} \leq a < \widehat{c}$, a contradiction. □

So we are led to consider

$$c_k = \inf_{A \in \mathcal{S}_k} \sup_{x \in A} \varphi(x) \quad \text{if } \mathcal{S}_k \neq \emptyset. \tag{4.41}$$

THEOREM 4.2.23 *If X is a Banach space, $\mathcal{S}_k \neq \emptyset$ for all $k \geq 1$ (see (4.40)), $\varphi \in C^1(X)$ is bounded below and satisfies the PS-condition, then*

(a) *c_k (see (4.41)) if finite it is attained and it is a critical value of φ.*
(b) *If $c_k = c_{k+1} = c < +\infty$, then* card *$K_c = +\infty$, where recall that $K_c = \{x \in X : \varphi'(x) = 0$ and $\varphi(x) = c\}$.*
(c) *If $c_k = c_{k+1} = \cdots = c_{k+m} = c < +\infty$, then $\xi(K_c) \geq m + 1$.*

(d) *If* $c_k = +\infty$ *for some* $k \geq 1$, *then* $\sup\limits_K \varphi = +\infty$, *where recall that* $K = \{x \in X : \varphi'(x) = 0\}$.

PROOF: (a) If c_k is finite and it is not a critical value of φ, then by virtue of Theorem 4.1.19(e) we may find $\varepsilon > 0$ small such that $\varphi^{c_k+\varepsilon}$ can be deformed to $\varphi^{c_k-\varepsilon}$. So every $A \in \mathcal{S}_k$ such that $A \subseteq \varphi^{c_k+\varepsilon}$, is deformed into a set $B \subseteq \varphi^{c_k-\varepsilon}$ and so by property (C)(3) we have $\xi(B) \geq k$ (i.e., $B \in \mathcal{S}_k$; recall that $\mathcal{S}_k$ is deformation invariant). Then $c_k \leq \sup\limits_{x \in B} \varphi(x) \leq c_k - \varepsilon$, a contradiction. Therefore c_k is a critical value of φ.

(b) Suppose that $\text{card}(K_c) < +\infty$. Then from properties (C)(3) and (5), we can find a neighborhood U of K_c such that $\xi(U) = 1$. Moreover, from Theorem 4.1.19(e) we know that we can find $\varepsilon > 0$ and a deformation $h(t, x)$, such that $h(1, \varphi^{c+\varepsilon}) \subseteq \varphi^{c-\varepsilon} \cup U$. Because $c = c_{k+1} < c + \varepsilon$, we can find $A \in \mathcal{S}_{k+1}$ such that $\sup\limits_{x \in A} \varphi(x) \leq c + \varepsilon$, hence $h(1, A) \subseteq \varphi^{c-\varepsilon} \cup U$ and so from properties (C) we have

$$\begin{aligned} & k + 1 \leq \xi(A) \leq \xi\big(h(1, A)\big) \leq \xi(\varphi^{c-\varepsilon}) + \xi(U) = \xi(\varphi^{c-\varepsilon}) + 1 \\ \Rightarrow\ & k \leq \xi(\varphi^{c-\varepsilon});\ \text{that is, } \varphi^{c-\varepsilon} \in \mathcal{S}_k. \end{aligned}$$

So we have

$$\sup_{x \in \varphi^{c-\varepsilon}} \varphi(x) \geq c_k = c > c - \varepsilon \geq \varphi(x) \qquad \text{for all } x \in \varphi^{c-\varepsilon},$$

a contradiction. This proves that $\text{card}(K_c) = +\infty$.

(c) Again property (C)(5) implies the existence of a neighborhood U of K_c such that $\xi(U) = \xi(K_c)$. Also Theorem 4.1.19(e) implies the existence of an $\varepsilon > 0$ and of a deformation $h(t, x)$ such that $h(1, \varphi^{c+\varepsilon}) \subseteq \varphi^{c-\varepsilon} \cup U$. So from properties (C), we have

$$\xi(\varphi^{c+\varepsilon}) \leq \xi(\varphi^{c-\varepsilon} \cup U) \leq \xi(\varphi^{c-\varepsilon}) + \xi(U) = \xi(\varphi^{c-\varepsilon}) + \xi(K_c). \tag{4.42}$$

Note that $\varphi^{c+\varepsilon}$ contains subsets A with $\xi(A) \geq k + m$. So

$$\xi(\varphi^{c+\varepsilon}) \geq k + m. \tag{4.43}$$

Also because $c = c_k$, we have

$$\xi(\varphi^{c-\varepsilon}) \leq k - 1. \tag{4.44}$$

Combining (4.42) through (4.44), we obtain $m + 1 \leq \xi(K_c)$.

(d) Suppose that $\beta = \sup\limits_K \varphi < +\infty$. Then by virtue of Theorem 4.1.19(e) for some $\varepsilon > 0$, the space X is deformable to $\varphi^{\beta+\varepsilon}$. Hence $\xi(X) = \xi(\varphi^{\beta+\varepsilon})$. Therefore $\varphi^{\beta+\varepsilon} \in \mathcal{S}_k$ and so

$$c_k = \inf_{A \in \mathcal{S}_k} \sup_{x \in A} \varphi(x) \leq \beta + \varepsilon,$$

a contradiction to the fact that $c_k = +\infty$. □

In fact this theorem can be restated in a global setting using the following notion.

DEFINITION 4.2.24 A C^1-manifold M (which is assumed to be a metric space), modelled on a Banach space X is said to be a *Finsler manifold* if and only if the following hold.

(a) For every $z \in M$, the tangent space $T_z M$ has a norm $\|\cdot\|_z$ that is equivalent to the norm $\|\cdot\|$ of X.
(b) If $z \in M$ and U is an open neighborhood of z locally trivializing the tangent bundle TM (i.e., $TM\big|_U = U \times X$), then for every $k > 1$, there exists $V_k \subseteq U$ another open neighborhood of z such that

$$\frac{1}{k}\|x\|_v \leq \|x\|_z \leq l\|x\|_v \qquad \text{for all } (v,x) \in V_k \times X.$$

REMARK 4.2.25 Trivially every Banach space X is itself a complete C^∞-Finsler manifold. Note that the cotangent bundle T^*M also carries a natural Finsler structure by letting

$$\|x^*\|_z = \sup\{|\langle x^*, x\rangle| : x \in T_z M,\ \|x\|_z \leq 1\} \qquad \text{for any } x^* \in T_z^* M.$$

Finally, if M is also connected, then the Finsler structure introduces a metric on M, known as the *Finsler metric*, defined by

$$d(z,u) = \inf\Big\{\int_0^1 \|\frac{\mathrm{d}\gamma(t)}{\mathrm{d}t}\|_{\gamma(t)}\mathrm{d}t : \gamma \in C^1([0,1], M),\ \gamma(0) = z,\ \gamma(1) = u\Big\}.$$

The existence of C^1-paths joining z and u follows from the work of Palais [471].

The deformation approach of critical point theory (see Section 4.1 and in particular Theorem 4.1.19) and the Ljusternik–Schnirelmann theory (in particular Theorem 4.2.23) can be recast in the more general framework of a connected, metric, C^1-Finsler manifold. In this context the next theorem provides useful information. For a proof of it we refer to Ghoussoub [262, p. 78].

THEOREM 4.2.26 *If M is a connected, metric, C^1-Finsler manifold modelled on a Banach space X, and $\varphi \in C^1(M)$ is bounded below and satisfies the PS-condition, then φ has at least $\mathrm{cat}_M(M)$-distinct critical points.*

4.3 Spectrum of the Laplacian and of the p-Laplacian

In this section we determine some important spectral properties of the Laplacian and p-Laplacian differential operators.

We start with the Laplacian (linear eigenvalue problem). So let $Z \subseteq \mathbb{R}^N$ be a bounded open set and first consider the Dirichlet eigenvalue problem

$$\left\{\begin{array}{l} -\triangle u(z) = \lambda u(z) \text{ a.e. on } Z, \\ u\big|_{\partial Z} = 0,\ \lambda \in \mathbb{R} \end{array}\right\}. \tag{4.45}$$

Recall that $\lambda \in \mathbb{R}$ is an *eigenvalue* of $\big(-\triangle, H_0^1(Z)\big)$, provided problem (4.45) has a nontrivial solution $u \in H_0^1(Z)$. Then such a nontrivial solution is called an *eigenfunction* of $\big(-\triangle, H_0^1(Z)\big)$. Also the pair (λ, u) is said to be an *eigenelement* of $\big(-\Delta, H_0^1(Z)\big)$. To fully determine the properties of the eigenelements of $\big(-\Delta, H_0^1(Z)\big)$, we need to recall the basic regularity results for linear second order elliptic equations. We state the main regularity results without proofs which can be found in Evans [229, Section 6.3].

So let L be the following second order differential operator in divergence form,

$$Lu = -\sum_{i,j=1}^{N} D_j\big(a_{ij}(z)D_i u\big) + \sum_{i=1}^{N} D_j b_i(z) D_i u + c(z)u, \tag{4.46}$$

where $D_k = \partial/\partial z_k$, $z = (z_k)_{k=1}^N$ being the generic element of Z, $a_{ij}, b_i, c \in L^\infty(Z)$, $1 \leq i,j \leq N$.

We require that L is uniformly elliptic; namely there exists $\vartheta > 0$ such that

$$\sum_{i,j=1}^{N} a_{ij}(z)\xi_i\xi_j \geq \vartheta \|\xi\|_{\mathbb{R}^N}^2 \tag{4.47}$$

for a.a. $z \in Z$ and all $\xi = (\xi_k)_{k=1}^N \in \mathbb{R}^N$. We will also assume without any loss of generality that $a_{ij} = a_{ji}$ for all $1 \leq i,j \leq N$ (symmetry condition). Uniform ellipticity means that for almost all $z \in Z$, the symmetric $N \times N$-matrix $A(z) = \big(a_{ij}(z)\big)_{i,j=1}^N$ is positive definite, with the smallest eigenvalue greater or equal to $\vartheta > 0$ (see (4.47)).

First we state the interior regularity results and then pass to the boundary regularity ones.

THEOREM 4.3.1 *If $a_{ij} \in C^1(Z)$, $b_i, c \in L^\infty(Z)$ for $i,j \in \{1,\ldots,N\}$, $f \in L^2(Z)$ and $x \in H^1(Z)$ is a weak solution of the problem $Lx = f$ in Z (i.e., $\int_Z \sum_{i,j=1}^{N} a_{ij}(z) D_j x(z) D_i \psi(z) dz + \int_Z \sum_{i=1}^N b_i(z)x(z)\psi(z)dz + \int_Z c(z)x(z)\psi(z)dz = \int_Z f(z)\psi(z)dz$ for all $\psi \in C_c^\infty(Z)$), then $x \in H^2_{\text{loc}}(Z)$ (i.e., $x \in H^2(Z')$ for every $Z' \subseteq Z$ open with $\overline{Z}' \subseteq Z$). Moreover, for every $Z' \subseteq Z$ open with $\overline{Z}' \subseteq Z$ (denoted by $Z' \subset\subset Z$), we have the estimate*

$$\|x\|_{H^2(Z')} \leq c\,\big(\|f\|_{L^2(Z)} + \|x\|_{L^2(Z)}\big),$$

with $c > 0$ depending only on Z', Z, and the coefficients of L.

REMARK 4.3.2 Note that in the above theorem we do not assume any Dirichlet conditions for the problem; that is, we do not require that $x \in H_0^2(Z)$. Moreover, because finally $x \in H^2_{\text{loc}}(Z)$, then a weak solution x is in fact a strong solution; namely $Lx(z) = f(z)$ a.e. on Z.

Increasing the regularity of the coefficient functions, we can iterate the argument used to prove Theorem 4.3.1 and have higher interior regularity.

THEOREM 4.3.3 *If $m \geq 1$ is an integer, $a_{ij}, b_j, c \in C^{m+1}(Z)$ for all $i,j \in \{1,\ldots,N\}$, $f \in H^m(Z)$, and $x \in H^1(Z)$ is a weak solution $Lx = f$ in Z, then $x \in H^{m+2}_{\text{loc}}(Z)$ and for each $Z' \subset\subset Z$ (see Theorem 4.3.1), we have the estimate*

$$\|x\|_{H^{m+2}(Z')} \leq c\,\big(\|f\|_{H^m(Z)} + \|x\|_{L^2(Z)}\big),$$

with the constant $c > 0$ depending only on m, Z', Z, and the coefficients of L.

We can now repeatedly apply Theorem 4.3.3 for $m \geq 1$, increasing every time the regularity of the solution by 2, to obtain the following infinite interior regularity result.

THEOREM 4.3.4 *If $a_{ij}, b_i, c \in C^\infty(Z)$ for all $i,j \in \{1,\dots,N\}$, $f \in C^\infty(Z)$, and $x \in H^1(Z)$ is a weak solution $Lx = f$ in Z, then $x \in C^\infty(Z)$.*

REMARK 4.3.5 Again note that we are not making any assumptions about the boundary values (in the sense of trace) of the solution x. So the theorem says that any possible singularities of x on the boundary, do not "propagate" in the interior.

The next regularity results establish regularity of the weak solutions up to the boundary.

THEOREM 4.3.6 *If $a_{ij} \in C^1(\overline{Z}), b_i, c \in L^\infty(Z)$ for all $i,j \in \{1,\dots,N\}, f \in L^2(Z), \partial Z$ is a C^2-manifold and $x \in H_0^1(Z)$ is a weak solution of the Dirichlet problem*

$$Lx = f \quad \text{in } Z, \quad x\big|_{\partial Z} = 0, \tag{4.48}$$

then $x \in H^2(Z)$ and we have the estimate

$$\|x\|_{H^2(Z)} \leq c\left(\|f\|_{L^2(Z)} + \|x\|_{L^2(Z)}\right) \tag{4.49}$$

with the constant $c > 0$ depending only on Z and the coefficients of L.

REMARK 4.3.7 If problem (4.48) has a unique solution $x \in H^2(Z)$, then the estimate simplifies as follows,

$$\|x\|_{H^2(Z)} \leq c\|f\|_{L^2(Z)}.$$

Also note that in this case in contrast to Theorem 4.3.1 we impose Dirichlet boundary conditions (in the sense of trace).

Proceeding as in the interior regularity, by improving the hypotheses on the coefficients we can conclude higher boundary regularity of the weak solution.

THEOREM 4.3.8 *If $m \geq 1$ is an integer, $a_{ij}, b_i, c \in C^{m+1}(\overline{Z})$ for all $i,j \in \{1,\dots,N\}$, $f \in H^m(Z)$, ∂Z is a C^{m+2}-manifold, and $x \in H_0^1(Z)$ is a weak solution of the Dirichlet problem*

$$Lx = f \quad \text{in } Z, \quad x\big|_{\partial Z} = 0, \tag{4.50}$$

then $x \in H^{m+2}(Z) \cap H_0^1(Z)$ and we have the estimate

$$\|x\|_{H^{m+2}(Z)} \leq c\left(\|f\|_{H^m(Z)} + \|x\|_{L^2(Z)}\right) \tag{4.51}$$

with the constant $c > 0$ depending only on m, Z, and the coefficients of L.

REMARK 4.3.9 Again problem (4.50) has a unique solution; the estimate (4.51) becomes

$$\|x\|_{H^{m+2}(Z)} \leq c\,\|f\|_{H^m(Z)}.$$

We conclude with an infinite boundary regularity result.

THEOREM 4.3.10 *If $a_{ij}, b_i, c \in C^\infty(\overline{Z})$ for all $i,j \in \{1,\ldots,N\}$, $f \in C^\infty(\overline{Z})$, ∂Z is a C^∞-manifold, and $x \in H_0^1(Z)$ is a weak solution of the Dirichlet problem*

$$Lx = f \quad \text{in } Z, \ x\big|_{\partial Z} = 0,$$

then $x \in C^\infty(\overline{Z})$.

Now we return to the linear eigenvalue problem (4.45) and prove the following fundamental result on the spectrum of $\big(-\Delta, H_0^1(Z)\big)$.

THEOREM 4.3.11 *Problem (4.45) has a sequence $\big\{(\lambda_n, u_n)\big\} \subseteq \mathbb{R}_+ \times H_0^1(Z)$ of eigenelements such that $0 < \lambda_1 < \lambda_2 \leq \cdots \leq \lambda_n \leq \cdots$, $\lambda_n \longrightarrow +\infty$ as $n \to \infty$, $\{u_n\}_{n \geq 1}$ is an orthonormal basis of $L^2(Z)$, $\Big\{(1/\sqrt{\lambda_n})u_n\Big\}_{n\geq 1}$ is an orthonormal basis of $H_0^1(Z)$ and $\{u_n\} \in H_0^1(Z) \cap C^\infty(Z)$ for all $n \geq 1$.*

PROOF: For every $h \in L^2(Z)$ we consider the Dirichlet problem

$$-\Delta x = h \quad \text{in } Z, \quad x\big|_{\partial Z} = 0. \tag{4.52}$$

If we consider the operator $A \in \mathcal{L}\big(H_0^1(Z), H^{-1}(Z)\big)$ defined by

$$\langle Ax, y\rangle = \int_Z \big(Dx(z), Dy(z)\big)_{\mathbb{R}^N} dz,$$

then clearly A is strongly monotone and weakly coercive (use Poincarè's inequality) and so by virtue of Corollary 3.2.29 we can find a unique $x \in H_0^1(Z)$ such that $Ax = h$. Evidently this is the unique weak (in fact strong) solution of problem (4.52). So we can define the solution map $K\colon L^2(Z) \longrightarrow H_0^1(Z)$ that to each $h \in L^2(Z)$ assigns the unique solution of problem (4.52). Clearly K is linear. We show that K is also completely continuous, namely if $h_n \xrightarrow{w} h$ in $L^2(Z)$, then $K(h_n) \longrightarrow K(h)$ in $H_0^1(Z)$ as $n \to \infty$. Note that for every $n \geq 1$, we have

$$A(x_n) = h_n \quad \text{in } H^{-1}(Z), \tag{4.53}$$

$$\Rightarrow \ \langle A(x_n), x_n\rangle = \|Dx_n\|_2^2 = \int_Z h_n x_n dz \leq \|h_n\|_2 \|x_n\|_2 \leq c_1 \|Dx_n\|_2,$$

for some $c_1 > 0$,

$$\Rightarrow \ \|Dx_n\|_2 \leq c_1, \ \text{ i.e., } \{x_n\}_{n\geq 1} \subseteq H_0^1(Z) \text{ is bounded,}$$

(by Poincaré's inequality).

Therefore by passing to a suitable subsequence if necessary, we may assume that $x_n \xrightarrow{w} x$ (i.e., $x = K(h)$). By Urysohn's criterion for convergent sequences, we conclude that for the original sequence we have $x_n \xrightarrow{w} x = K(h)$ in $H_0^1(Z)$, hence $x_n \longrightarrow x$ in $L^2(Z)$ (recall that $H_0^1(Z)$ is embedded compactly in $L^2(Z)$). We have

$$\langle A(x_n), x_n - x\rangle = \langle h_n, x_n - x\rangle = \int_Z h_n(x_n - x) \longrightarrow 0,$$

$$\Rightarrow \ \|Dx_n\|_2 \longrightarrow \|Dx\|_2.$$

But $Dx_n \xrightarrow{w} Dx$ in $L^2(Z, \mathbb{R}^N)$ and a Hilbert space has the Kadec–Klee property, thus $x_n \longrightarrow x$ in $H_0^1(Z)$. This proves the complete continuity of K. In particular then K viewed as a map from $L^2(Z)$ into itself is linear compact (i.e., $K \in \mathcal{L}_c\big(L^2(Z)\big)$; see Proposition 3.1.3) and it is also self-adjoint, because for every $f \in L^2(Z)$ we have

$$\int_Z K(h) f \mathrm{d}z = \int_Z \big(DK(h), DK(f)\big)_{\mathbb{R}^N} \mathrm{d}z = \int_Z hK(f)\mathrm{d}z.$$

Finally note that

$$\int_Z K(h) h \mathrm{d}z = \int_Z \big(DK(h), DK(h)\big)_{\mathbb{R}^N} \mathrm{d}z = \|DK(h)\|_2^2 > 0,$$

for all $h \in L^2(Z) \setminus \{0\}$,

$\Rightarrow$ $K \in \mathcal{L}_c\big(L^2(Z)\big)$ is positive definite.

So we can apply Theorem 3.1.37 and produce a sequence $\{\mu_n\}_{n \geq 1}$ of eigenvalues of K with $\mu_n \longrightarrow 0$ and $\{u_n\}_{n \geq 1} \subseteq L^2(Z)$ a sequence of corresponding eigenfunctions that form an orthonormal basis of $L^2(Z)$. We have

$$K(u_n) = \mu_n u_n.$$

Because clearly $K = A^{-1}$, we have

$$A(u_n) = \lambda_n u_n \qquad \text{with } \lambda_n = \frac{1}{\mu_n}.$$

So $0 < \lambda_1 \leq \lambda_2 \leq \cdots \leq \lambda_n \leq \cdots$, $\lambda_n \longrightarrow +\infty$, $\{u_n\}_{n \geq 1} \subseteq H_0^1(Z)$ and are eigenfunctions of $\big(-\Delta, H_0^1(Z)\big)$. Moreover, from Theorem 4.3.4 we have that $u_n \in H_0^1(Z) \cap C^\infty(Z)$ for all $n \geq 1$.

Finally, if in $H_0^1(Z)$ we consider the inner product

$$\langle x, y\rangle_{H_0^1(Z)} = \int_Z (Dx, Dy)_{\mathbb{R}^N} \mathrm{d}z \qquad \text{for all } x, y \in H_0^1(Z),$$

then we have

$$\begin{aligned}\Big\langle \frac{1}{\sqrt{\lambda_n}} u_n, \frac{1}{\sqrt{\lambda_k}} u_k \Big\rangle_{H_0^1(Z)} &= \frac{1}{\sqrt{\lambda_n \lambda_k}} \int_Z (Du_n, Du_k)_{\mathbb{R}^N} \mathrm{d}z \\ &= \frac{\sqrt{\lambda_n}}{\sqrt{\lambda_k}} \int_Z u_n u_k \mathrm{d}z = \delta_{nk}.\end{aligned}$$

Also if $y \in H_0^1(Z)$ is such that $\langle y, u_n\rangle_{H_0^1(Z)} = 0$ for all $n \geq 1$, then

$$0 = \int_Z (Dy, Du_n)_{\mathbb{R}^N} \mathrm{d}z = \lambda_n \int_Z y u_n \mathrm{d}z,$$

$$\Rightarrow \int_Z y u_n \mathrm{d}z = 0 \qquad \text{for all } n \geq 1,$$

$$\Rightarrow y = 0 \quad (\text{because } \{u_n\}_{n \geq 1} \subseteq L^2(Z) \text{ is a basis}).$$

Therefore we conclude that $\big\{(1/\sqrt{\lambda_n})u_n\big\}_{n \geq 1}$ is an orthonormal basis of $H_0^1(Z)$. □

REMARK 4.3.12 If ∂Z is a C^∞-manifold, then from Theorem 4.3.10 we have that $u_n \in C^\infty(\overline{Z})$ for all $n \geq 1$.

Next we produce variational characterizations for the eigenelements $\{(\lambda_n, u_n)\}_{n\geq 1}$. For this purpose we introduce the so-called *Rayleigh quotient*.

$$R(x) = \frac{\|Dx\|_2^2}{\|x\|_2^2} \quad \text{for all } x \in H_0^1(Z),\ x \neq 0. \tag{4.54}$$

THEOREM 4.3.13 *We have*

$$\lambda_1 = \min\{R(x) : x \in H_0^1(Z),\ x \neq 0\} = R(u_1) \tag{4.55}$$

$$n \geq 2,\ \lambda_n = \max\left\{R(x) : x \in \operatorname{span}\{u_k\}_{k=1}^n,\ x \neq 0\right\} \tag{4.56}$$

$$= \min\left\{R(x) : x \in \left[\operatorname{span}\{u_k\}_{k=1}^{n-1}\right]^\perp,\ x \neq 0\right\} \tag{4.57}$$

$$= \min_{\substack{Y \subseteq H_0^1(Z) \\ \dim Y = n}} \max_{\substack{y \in Y \\ y \neq 0}} R(y) \tag{4.58}$$

$$= \max_{\substack{V \subseteq H_0^1(Z) \\ \dim V^\perp = n-1}} \min_{\substack{v \in V \\ v \neq 0}} R(v) = R(u_n). \tag{4.59}$$

PROOF: Let $Y_n = \operatorname{span}\{u_k\}_{k=1}^n$, $n \geq 1$. Clearly for all $n \geq 1$, $\lambda_n = R(u_n)$ (see (4.45)). If $v \in Y_{n-1}^\perp$, then the expansion of v in $L^2(Z)$ with respect to the orthonormal basis $\{u_n\}_{n\geq 1}$, is given by

$$v = \sum_{k \geq n} \vartheta_k u_k \quad \text{with} \quad \vartheta_k = \int_Z v u_k dz.$$

Let $v_m = \sum_{k=n}^m \vartheta_k u_k$. Then $v_m \longrightarrow v$ in $L^2(Z)$ and in $H_0^1(Z)$ (see Theorem 4.3.11). Hence

$$R(v_m) \longrightarrow R(v).$$

We have

$$R(v_m) = \frac{\sum\limits_{k=n}^m \vartheta_k^2 \|Du_k\|_2^2}{\sum\limits_{k=n}^m \vartheta_k^2} = \frac{\sum\limits_{k=n}^m \lambda_k \vartheta_k^2}{\sum\limits_{k=n}^m \vartheta_k^2} \geq \lambda_n,$$

$$\Rightarrow R(v) \geq \lambda_n.$$

Because $R(u_n) = \lambda_n$, we conclude that (4.57) holds, hence (4.55) holds too (let $n = 1$).

On the other hand if $y \in Y_n$, then $y = \sum\limits_{k=1}^n \vartheta_k u_k$ and so

$$R(y) \leq \frac{\sum\limits_{k=1}^n \lambda_k \vartheta_k^2}{\sum\limits_{k=1}^n \vartheta_k^2} \leq \lambda_n.$$

Because $R(u_n) = \lambda_n$, we deduce that (4.56) holds.

Now let Y be an n-dimensional linear subspace of $H_0^1(Z)$. Let $y_0 \in Y$ such that

$$y_0 \perp \operatorname{span}\{u_k\}_{k=1}^{n-1}.$$

Then from (4.57) we have that

$$\begin{aligned} & R(y_0) \geq \lambda_n, \\ \Rightarrow\ & \max_{y \in Y} R(y) \geq \lambda_n. \end{aligned}$$

Because Y was an arbitrary n-dimensional linear subspace of $H_0^1(Z)$, we obtain

$$\min_{\substack{Y \subseteq H_0^1(Z) \\ \dim Y = n}} \max_{y \in Y} R(y) \geq \lambda_n. \tag{4.60}$$

Then because of (4.56), we see that equality must hold in (4.60). Dually we obtain (4.59). □

REMARK 4.3.14 Characterizations (4.58) and (4.59) are more satisfactory than (4.56) and (4.57), since they do not require a priori knowledge of the spectrum of $\big(-\triangle, H_0^1(Z)\big)$. Moreover, we can say that (4.58) is more satisfactory than (4.59), because (4.58) involves a finite dimensional maximization and (4.59) has an infinite dimensional minimization. Expressions (4.58) and (4.59) are also known as *Courant's minimax principles* (see also Section 4.2).

Next we focus on the eigenelement $\{(\lambda_1, u_1)\}$.

PROPOSITION 4.3.15 *If for some $x \in H_0^1(Z)$, $x \neq 0$ we have $R(x) = \lambda_1$, then $x \in H_0^1(Z)$ is an eigenfunction corresponding to the eigenvalue $\lambda_1 > 0$.*

PROOF: Let $y \in H_0^1(Z)$. Then we have

$$R(x + ry) \geq R(x) = \lambda_1 \qquad \text{for all } r > 0.$$

Without any loss of generality, we may assume that $\|x\|_2 = 1$. Then we have

$$\begin{aligned} & \frac{\|D(x+ry)\|_2^2}{\|x+ry\|_2^2} \geq \|Dx\|_2^2 = \lambda_1 \\ \Rightarrow\ & r^2\|Dy\|_2^2 + 2r\int_Z (Dx, Dy)_{\mathbb{R}^N} dz \geq \lambda_1\Big(2r\int_Z xy dz + r^2\|y\|_2^2\Big). \end{aligned}$$

Dividing with $r > 0$ and then letting $r \longrightarrow 0$, since $y \in H_0^1(Z)$ was arbitrary, we obtain

$$\begin{aligned} & \int_Z (Dx, Dy)_{\mathbb{R}^N} dz = \lambda_1 \int_Z xy dz, \\ \Rightarrow\ & -\Delta x(z) = \lambda_1 x(z) \quad \text{a.e. on } Z,\ x\big|_{\partial Z} = 0. \end{aligned}$$

□

To fully describe the first eigenelement (λ_1, u_1), we need the so-called *strong maximum principle*. The maximum principles for second-order linear partial differential equations are based upon the observation that if $u \in C^2(Z)$ attains its maximum over an open Z at a point $z_0 \in U$, then

$$Dx(z_0) = 0 \qquad \text{and} \qquad D^2u(z_0) \leq 0, \tag{4.61}$$

where the second -order condition means that the symmetric $N{\times}N$ matrix $D^2u(z_0) = \big(D_jD_iu(z_0)\big)_{i,j=1}^N$ is negative definite. Evidently arguments based on (4.61) are pointwise in nature, in contrast to the integral-based energy methods that we employed so far.

Although we use the strong maximum principle in the context of the Laplacian differential operator, we state and prove the strong maximum principle for general second-order elliptic differential operators in *nondivergence form*

$$Lu = -\sum_{i,j=1}^N a_{ij}(z)D_jD_iu + \sum_{i=1}^N b_i(z)D_iu + cu. \tag{4.62}$$

We assume that $a_{ij}, b_i, c \in C(Z)$, the uniform ellipticity condition (4.47) holds and without any loss of generality, we also have the symmetry condition $a_{ij} = a_{ji}$ for all $i, j \in \{1, \ldots, N\}$.

First let us state the so-called *weak maximum principle*, which is used in the proof of the strong maximum principle. For its proof we refer to Evans [229, p. 333].

THEOREM 4.3.16 *If L is as in (4.62) with $c \geq 0$ and $u \in C^2(Z) \cap C(\overline{Z})$, then*

(a) *If $Lu(z) \leq 0$ for all $z \in Z$, we have $\max\limits_{\overline{Z}} u \leq \max\limits_{\partial Z} u^+$ $(u^+ = \max\{u, 0\})$.*
(b) *If $Lu(z) \geq 0$ for all $z \in Z$, we have $\min\limits_{\overline{Z}} u \geq -\max\limits_{\partial Z} u^-$ $(u^- = \max\{-u, 0\})$.*

REMARK 4.3.17 If $Lu(z) \leq 0$ for all $z \in Z$, we say that u is a lower solution of the equation $Lu(z) = 0$. Similarly if $Lu(z) \geq 0$ for all $z \in Z$, we say that u is an upper solution of the equation $Lu(z) = 0$. In Section 5.2 we generalize these concepts. Note that if $Lu(z) = 0$ for all $z \in U$, then $\max\limits_{\overline{Z}} |u| = \max\limits_{\partial Z} |u|$. Finally if $c \equiv 0$, then in (a) we have $\max\limits_{\overline{Z}} u = \max\limits_{\partial Z} u$ and in (b) $\min\limits_{\overline{Z}} u = \min\limits_{\partial Z} u$.

The strong maximum principle is essentially the following lemma, known in the literature as *Hopf's lemma*.

LEMMA 4.3.18 *If $u \in C^2(Z) \cap C(\overline{Z})$, L is as in (4.62), $Lu(z) \leq 0$ for all $z \in Z$, there exists $z_0 \in \partial Z$ such that $u(z) < u(z_0)$ for all $z \in Z$, and Z satisfies the interior ball condition at $z_0 \in \partial Z$ (i.e., there exists an open ball $B \subseteq Z$ such that $z_0 \in \partial B$; this is the case if ∂Z in C^2), then*

(a) *If $c = 0$, we have $(\partial u/\partial n)(z_0) > 0$ with n being the outward unit normal at $z_0 \in \partial Z$.*
(b) *If $c \geq 0$, then the same conclusion holds provided $u(z_0) \geq 0$.*

PROOF: We assume $c \geq 0$ and cover both (a) and (b). By translating things if necessary, we may assume without any loss of generality that $B = B_r(0)$, $r > 0$. Let

$$v(z) = e^{-\lambda\|z\|^2} - e^{-\lambda r^2} \quad \text{for all } z \in B_r(0),\ \lambda > 0 \text{ to be chosen later.}$$

We have

$$\begin{aligned}
Lv(z) &= -\sum_{i,j=1}^{N} a_{ij}(z)D_jD_iv(z) + \sum_{i=1}^{N} b_i(z)D_iv(z) + c(z)v(z) \\
&= e^{-\lambda\|z\|^2}\sum_{i,j=1}^{N} a_{ij}(z)\big(-4\lambda^2 z_iz_j + 2\lambda\delta_{ij}\big) \\
&\quad -e^{-\lambda\|z\|^2}\sum_{i=1}^{N} b_i(z)2\lambda z_i + c(z)\big(e^{-\lambda\|z\|^2} - e^{-\lambda r^2}\big) \\
&\le e^{-\lambda\|z\|^2}\big(-4\vartheta\lambda^2\|z\|^2 + 2\lambda tr A(z) + 2\lambda\|b(z)\|\|z\| + c(z)\big) \quad \text{(see (4.47))}
\end{aligned}$$

for $A(z) = \big(a_{ij}(z)\big)_{ij=1}^{N}$ and $b(z) = \big(b_i(z)\big)_{i=1}^{N}$. Consider the open ring $R = B_r(0)\setminus \overline{B}_{r/2}(0)$. We have

$$Lv(z) \le e^{-\lambda\|z\|^2}\big(-\vartheta\lambda^2 r^2 + 2\lambda tr A(z) + 2\lambda\|b(z)\|r + c(z)\big) \le 0 \tag{4.63}$$

for all $z \in R$, provided we choose $\lambda>0$ large enough.

By virtue of the hypothesis about $z_0 \in \partial Z$, we see that we can find $\varepsilon>0$ small such that

$$u(z) + \varepsilon v(z) \le u(z_0) \qquad \text{for all } z \in \partial B_{r/2}(0). \tag{4.64}$$

Moreover, because $v\big|_{\partial B_r(0)} \le 0$, we have that

$$u(z) + \varepsilon v(z) \le u(z) \qquad \text{for all } z \in \partial B_r(0). \tag{4.65}$$

Using (4.63) and the fact that $Lu(z) \le 0$ for all $z\in Z$, we obtain

$$L\big(u(z) + \varepsilon v(z) - u(z_0)\big) \le -cu(z_0) \le 0 \qquad \text{for all } z \in R,$$

(recall either $c = 0$ or $u(z_0)>0$).

Also from (4.64) and (4.65), we have that

$$u + \varepsilon v - u(z_0)\big|_{\partial R} \le 0.$$

Invoking Theorem 4.3.16 (the weak maximum principle), we infer that

$$u(z) + \varepsilon v(z) - u(z_0) \le 0 \qquad \text{for all } z \in R.$$

Note that $u(z_0) + \varepsilon v(z_0) - u(z_0) = 0$. Hence $(\partial/\partial n)\big(u + \varepsilon v - u(z_0)\big)(z_0) = \big(\partial u/\partial n\big)(z_0) + \varepsilon\big(\partial v/\partial n\big)(z_0) \ge 0$. Consequently

$$\frac{\partial u}{\partial n}(z_0) \ge -\varepsilon\frac{\partial v}{\partial n}(z_0) = -\frac{\varepsilon}{r}\big(Dv(z_0), z_0\big)_{\mathbb{R}^N} = 2\lambda\varepsilon r e^{-\lambda r^2} > 0. \qquad \square$$

Using this lemma, we can prove the strong maximum principle.

THEOREM 4.3.19 *If L is as in (4.62) with c=0, $Z\subseteq\mathbb{R}^N$ is bounded, open, and connected, and $u \in C^2(Z)\cap C(\overline{Z})$, then*

(a) *If $Lu(z) \le 0$ for all $z\in Z$ and u attains its maximum on $\overline{Z}$ at an interior point, then u is constant on $\overline{Z}$.*

(b) *If $Lu(z) \geq 0$ for all $z \in Z$ and u attains its minimum on $\overline{Z}$ at an interior point, then u is constant on $\overline{Z}$.*

PROOF: (a) Let $M = \max\limits_{\overline{Z}} u$ and $E = \{z \in U : u(z) = M\}$. Suppose u is not constant (i.e., $u \neq M$), and let $V = \{z \in Z : u(z) < M\} \neq \emptyset$. This is an open set and we can choose $x \in V$ such that $d(x, E) < d^*(x, \partial Z)$. Let B be the largest open ball with center at x located in V. Then we can find $z_0 \in E$ such that $z_0 \in \partial B$. Evidently V satisfies the interior ball condition at z_0. So by virtue of Lemma 4.3.18 $(\partial u/\partial n)(z_0) > 0$. But because u attains its supremum at z_0, $Du(z_0) = 0$ and so $(\partial u/\partial n)(z_0) = 0$, a contradiction.

(b) Apply part (a) on $-u$. ◻

Using part (b) of Lemma 4.3.18, we can have the following variant of the strong maximum principle.

THEOREM 4.3.20 *If L is as in (4.62) with $c \geq 0$, $Z \subseteq \mathbb{R}^N$ is bounded, open, and connected, and $u \in C^2(Z) \cap C(\overline{Z})$, then*

(a) *If $Lu(z) \leq 0$ for all $z \in Z$ and u attains a nonnegative maximum on $\overline{Z}$ at an interior point, then u is constant on $\overline{Z}$.*
(b) *If $Lu(z) \geq 0$ for all $z \in Z$ and u attains a nonpositive minimum on $\overline{Z}$ at an interior point, then u is constant on $\overline{Z}$.*

We return to the linear eigenvalue problem (4.45) and using Theorem 4.3.19, we fully describe the first eigenelement (λ, u_1).

THEOREM 4.3.21 *If $Z \subseteq \mathbb{R}^N$ is a bounded domain (i.e., a bounded, open, and connected set) with a C^∞-boundary ∂Z, then $\lambda_1 > 0$ is simple and the corresponding eigenfunction $u_1 \in C^\infty(\overline{Z})$ does not change sign and does not vanish on Z (so we may assume that $u_1(z) > 0$ for all $z \in Z$).*

PROOF: We know that $u_1^+, u_1^- \in H_0^1(Z)$ and we have

$$\int_Z (Du_1, Du_1^+)_{\mathbb{R}^N} dz = \lambda_1 \int_Z u_1 u_1^+ dz \Rightarrow \|Du_1^+\|_2^2 = \lambda_1 \|u_1^+\|_2^2 \tag{4.66}$$

and

$$\int_Z (Du_1, Du_1^-)_{\mathbb{R}^N} dz = \lambda_1 \int_Z u_1 u_1^- dz \Rightarrow \|Du_1^-\|_2^2 = \lambda_1 \|u_1^-\|_2^2. \tag{4.67}$$

If u_1 does change sign, then $u_1^+, u_1^- \neq 0$ and from (4.66) and (4.67), we have

$$R(u_1^+) = \lambda_1 = R(u_1^-). \tag{4.68}$$

Invoking Proposition 4.3.15 from (4.68) we deduce that u_1^+, u_1^- are eigenfunctions for $\lambda_1 > 0$. Because $u_1^+, u_1^- \geq 0$, from Theorem 4.3.19(b), we deduce that

$$u_1^+(z) > 0 \text{ and } u_1^-(z) > 0 \quad \text{for all } z \in Z,$$

a contradiction. So u_1 must have constant sign on Z and we can always assume $u_1(z) > 0$ for all $z \in Z$.

If $\lambda_1 > 0$ is not simple, then we can find another eigenfunction $\widehat{u}_1 \in C^\infty(\overline{Z})$ such that

$$\int_Z u_1 \widehat{u}_1 dz = 0. \tag{4.69}$$

From the first part of the proof we know that $u_1, \widehat{u}_1$ do not change sign on Z and we can assume $u_1, \widehat{u}_1 > 0$ on Z. This contradicts (4.69) and so $\lambda_1 > 0$ is simple. □

REMARK 4.3.22 The hypothesis that ∂Z is C^∞ was made for convenience. As we show later in this section, in our discussion of the p-Laplacian, it suffices to assume that ∂Z is C^2 (see Theorem 4.3.35). What we need in the proof of Theorem 4.3.21 is to guarantee that $u_1 \in C(\overline{Z})$ which allows us to use Theorem 4.3.19. Note that if ∂Z is Lipschitz and $N \leq 3$, then since $H^2(Z) \subseteq C(\overline{Z})$ (Sobolev embedding theorem) and $u_1 \in H^2(Z)$ (see Theorem 4.3.6), we have $u_1 \in C(\overline{Z})$. For $N > 3$, note that we can use Theorem 4.3.8, in order through more smoothness of the manifold ∂Z to achieve higher regularity of u_1 and eventually guarantee that $u_1 \in C(\overline{Z})$. But as we already indicated we can do better.

Now we consider the Neumann linear eigenvalue problem (i.e., we determine the spectrum of $\big(-\triangle, H^1(Z)\big)\big)$. So we consider the following problem.

$$\left\{ \begin{array}{l} -\triangle x(z) = \lambda x(z) \ \text{ a.e. on } Z, \\ \frac{\partial x}{\partial n}(z) = 0 \quad \text{for all } \ z \in \partial Z \end{array} \right\}. \tag{4.70}$$

Here $\partial x / \partial n$ denotes the derivative in the direction of the exterior normal on ∂Z. Working as for the Dirichlet linear eigenvalue problem, we can have the following fundamental theorem for problem (4.70).

THEOREM 4.3.23 *If $Z \subseteq \mathbb{R}^N$ is a bounded domain with a C^∞-boundary ∂Z, then problem (4.70) has countably many eigenvalues $\{\lambda_n\}_{n\geq 0}$ such that*

$$0 = \lambda_0 < \lambda_1 \leq \lambda_2 \leq \cdots \leq \lambda_n \leq \cdots, \quad \text{with } \lambda_n \longrightarrow +\infty$$

and the corresponding eigenfunctions $\{u_n\}_{n\geq 0} \subseteq H^1(Z)$ form an orthonormal basis for $L^2(Z)$, $\Big\{|Z|_N^{-(1/2)}, \big(u_n/\sqrt{\lambda_n}\big)\Big\}_{n\geq 0} \subseteq H^1(Z)$ form an orthonormal basis for $H^1(Z)$ $\lambda_0 = 0$ simple with corresponding eigenspace $\mathbb{R}$ (i.e., the constant functions) and we have the following variational characterizations of the eigenvalues

$$\begin{aligned} \lambda_0 &= \min\{R(u) : u \in H^1(Z),\ u \neq 0\}, \\ n \geq 1,\ \ \lambda_n &= \max\big\{R(u) : u \in \operatorname{span}\{u_k\}_{k=0}^n,\ u \neq 0\big\} \\ &= \min\big\{R(u) : u \in \big[\operatorname{span}\{u_k\}_{k=0}^{n-1}\big]^\perp,\ u \neq 0\big\} \\ &= \min_{\substack{Y \subseteq H^1(Z) \\ \dim Y = n+1}} \max_{\substack{y \in Y \\ y \neq 0}} R(y) \\ &= \max_{\substack{V \subseteq H^1(Z) \\ \dim V^\perp = n}} \min_{\substack{v \in V \\ v \neq 0}} R(v) \end{aligned} \tag{4.71}$$

where recall $R(u) = \|Du\|_2^2 / \|u\|_2^2$ (the Rayleigh quotient).

REMARK 4.3.24 Because $H_0^1(Z) \subseteq H^1(Z)$, we deduce that if by $\left\{\lambda_n^D\right\}_{n\geq 1}$ we denote the Dirichlet eigenvalues and by $\left\{\lambda_n^N\right\}_{n\geq 0}$ the Neumann eigenvalues, for all $n\geq 1$ we have $\lambda_{n-1}^N \leq \lambda_n^D$.

PROPOSITION 4.3.25 *If $Z_1 \subseteq Z_2 \subseteq \mathbb{R}^N$ are bounded open sets and $\{\lambda_n(Z_1)\}_{n\geq 1}$ (resp., $\{\lambda_n(Z_2)\}_{n\geq 1}$) are the eigenvalues of $(-\Delta, H_0^1(Z_1))$ (resp., $(-\Delta, H_0^1(Z_2))$), then $\lambda_n(Z_2) \leq \lambda_n(Z_1)$ for all $n\geq 1$.*

PROOF: Note that any $u \in H_0^1(Z_1)$ can be extended to a function $\widehat{u} \in H_0^1(Z_2)$, by

$$\widehat{u}(z) = \begin{cases} u(z) & \text{if } z \in Z_1 \\ 0 & \text{if } z \in Z_2 \setminus Z_1 \end{cases}.$$

So $H_0^1(Z_1) \subseteq H_0^1(Z_2)$ and this by virtue of (4.58) implies that $\lambda_n(Z_2) \leq \lambda_n(Z_1)$ for all $n\geq 1$. □

REMARK 4.3.26 This proposition is also valid for the Neumann eigenvalues but the proof is more involved.

The next theorem is known as *Courant's nodal set theorem* and its proof can be found in Courant–Hilbert [164].

THEOREM 4.3.27 *If $Z\subseteq\mathbb{R}^N$ is a bounded open set, $\{\lambda_n\}_{n\geq 1} \subseteq \mathbb{R}_+$ are the eigenvalues of $(-\Delta, H_0^1(Z))$ (i.e., of problem (4.45)) and $\{u_n\}_{n\geq 1} \subseteq H_0^1(Z) \cap C^\infty(Z)$ are the corresponding eigenfunctions, and for every $n\geq 1$ $Z_n = \{z\in Z : u_n(z) = 0\}$ (the nodal set of u_n), then $Z \setminus Z_n$ has at most n components.*

Before passing to nonlinear problems and in particular to the study of the spectral properties of the p–Laplacian, let us briefly summarize the situation with the scalar eigenvalue problems (i.e., $N=1$, ordinary differential operators).

So first consider the following eigenvalue problem:

$$\left\{ \begin{array}{l} -x''(t) = \lambda x(t) \text{ a.e. on } (0,b), \\ x(0) = x(b) = 0 \end{array} \right\}. \tag{4.72}$$

PROPOSITION 4.3.28 *The eigenvalues of (4.72) are $\left\{\lambda_n=(n\pi/b)^2\right\}_{n\geq 1}$ and the eigenfunctions are $\left\{u_n(t)=\sqrt{2/b}\sin\left((n\pi/b)t\right)\right\}_{n\geq 1}$.*

Next consider the Neumann problem:

$$\left\{ \begin{array}{l} -x''(t) = \lambda x(t) \text{ a.e. on } (0,b), \\ x'(0) = x'(b) = 0 \end{array} \right\}. \tag{4.73}$$

PROPOSITION 4.3.29 *The eigenvalues of (4.73) are $\left\{\lambda_n = (n\pi/b)^2\right\}_{n\geq 0}$ and the eigenfunctions are $\left\{u_0(t)=1/\sqrt{b},\ u_n(t)=\sqrt{2/b}\cos\left((n\pi/b)t\right)\right\}_{n\geq 1}$.*

Finally we consider the periodic problem:

$$\left\{\begin{array}{l} -x''(t) = \lambda x(t) \ \text{a.e. on } (0,b), \\ x(0) = x(b), \ x'(0) = x'(b) \end{array}\right\}. \tag{4.74}$$

PROPOSITION 4.3.30 *The eigenvalues of (4.74) are* $\left\{\lambda_n = (2n\pi/b)^2\right\}_{n\geq 0}$ *and the eigenfunctions are* $\left\{u_0(t) = 1/\sqrt{b}, u_n(t) = (2/\sqrt{b})\Big(\sin\big((2n\pi/b)t\big) + \cos\big((2n\pi/b)t\big)\Big)\right\}_{n\geq 1}$.

Now we turn our attention to nonlinear eigenvalue problems. Namely we study the spectral properties of the p-Laplacian differential operator

$$-\triangle_p x = -\mathrm{div}(\|Dx\|^{p-2}Dx), \quad p>1. \tag{4.75}$$

We start with some regularity results for elliptic equations involving the p-Laplacian (see (4.75)). In what follows $Z \subseteq \mathbb{R}^N$ is a bounded domain with a C^2–boundary ∂Z. Also for $1 < p < \infty$, by p^* we denote the critical Sobolev exponent, namely

$$p^* = \left\{\begin{array}{ll} \frac{Np}{N-p} & \text{if } p<N \\ +\infty & \text{if } p \geq N \end{array}\right. . \tag{4.76}$$

We start with a result on Sobolev functions, whose proof can be found in Brezis–Browder [103].

LEMMA 4.3.31 *If $x \in W_0^{m,p}(Z)$ with $m \geq 1, 1 < p < +\infty$, $S \in W^{-m,p'}(Z) \cap L^1_{\mathrm{loc}}(Z)$ $\big((1/p) + (1/p') = 1\big)$, and for some $g \in L^1(Z)$ we have $g(z) \leq S(z)x(z)$ a.e. on Z, then $S(x) \in L^1(Z)$ and $\langle S, x\rangle = \int_Z S(z)x(z)dz$ (by $\langle\cdot,\cdot\rangle$ we denote the duality brackets for the pair $\big(W_0^{m,p}(Z), W^{-m,p'}(Z)\big)$).*

LEMMA 4.3.32 *If $x \in W_0^{1,p}(Z), \triangle_p x \in L^1_{\mathrm{loc}}(Z)$, $c>0$, $\vartheta \in [1, p^*)$ (see (4.76)), $q \in \big[1, (p^*/p)\big)$, and $\alpha \in L^{q'}(Z)_+$, $\big((1/q) + (1/q') = 1\big)$ are such that*

$$-x(z)\triangle_p x(z) \leq \alpha(z)|x(z)| + c|x(z)|^{\vartheta} \quad \text{a.e. on } Z \tag{4.77}$$

and

$$p_0 = \left\{\begin{array}{ll} p^* & \text{if } p^* < +\infty \\ 2\max\{pq, \vartheta\} & \text{if } p^* = +\infty \end{array}\right. \quad (\text{see (4.76)}),$$

$$p_{n+1} = p_0 + \frac{p_0}{p}\min\left\{p_n - \vartheta, \frac{p_n}{q} - 1\right\} \quad \text{for all } n \geq 0,$$

then $x \in L^{p_n}(Z)$ for all $n \geq 0$.

PROOF: From the Sobolev embedding theorem, we have that $x \in L^{p_0}(Z)$. We proceed by induction. So suppose that $x \in L^{p_n}(Z)$. We show that $x \in L^{p_{n+1}}(Z)$.

Consider $\{y_k\}_{k\geq 1}$ the truncations of u at the levels $-k$ and k; that is,

$$y_k(z) = \left\{\begin{array}{ll} k & \text{if } \quad k \leq x(z) \\ x(z) & \text{if } -k \leq x(z) \leq k \\ -k & \text{if } \quad x(z) \leq -k \end{array}\right. .$$

Set $a = (p(p_{n+1} - p_0))/p_0$. Clearly $a \geq 0$ because we have that $p_0 \leq p_n$ for all $n \geq 0$. Then we have

$$-|y_k(z)|^a y_k(z)\Delta_p x(z) \leq \alpha(z)|x(z)|^{a+1} + c|x(z)|^{\vartheta+a} \quad \text{a.e. on } Z \text{ (see (4.77)).} \tag{4.78}$$

By Hölder's inequality, we have

$$\int_Z \left(\alpha(z)|x(z)|^{a+1} + c|x(z)|^{\vartheta+a}\right)dz \leq \|\alpha\|_{q'}\|x\|_{q(a+1)}^{a+1} + c\|x\|_{\vartheta+a}^{\vartheta+a}. \tag{4.79}$$

Because $p_n = \max\{\vartheta + a,\ q(a+1)\}$, from (4.79) it follows that

$$\int_Z \left(\alpha(z)|x(z)|^{a+1} + c|x(z)|^{\vartheta+a}\right)dz \leq c_1\left(\|x\|_{p_n}^{p_n} + 1\right) \quad \text{for some } c_1 > 0. \tag{4.80}$$

Because $|y_k|^a y_k \in W_0^{1,p}(Z)$, $\Delta_p\, x \in W^{-1,p'}(Z) \cap L^1_{\text{loc}}(Z)$, and $|y_k|^a y_k \Delta_p x$ is bounded below by an $L^1(Z)$-function (see (4.78) and (4.80)), from Lemma 4.3.31 we have that $|y_k|^a y_k \Delta_p x \in L^1(Z)$ and

$$-\int_Z |y_k|^a y_k \Delta_p x dz = \langle -\Delta_p u, |y_k|^a y_k\rangle\,,$$

where by $\langle\cdot,\cdot\rangle$ we denote the duality brackets for the pair $\left(W_0^{1,p}(Z), W^{-1,p'}(Z)\right)$. Hence after integration by parts, we obtain

$$\begin{aligned}
-\int_Z |y_k|^a y_k \Delta_p x dz &= (a+1)\int_Z |y_k|^a \|Dy_k\|^{\vartheta} dz \\
&= (a+1)\left(\frac{p}{a+p}\right)^p \int_Z \left\|D(|y_k|^{a/p} y_k)\right\|^p dz \\
&\geq \frac{1}{c_0^p}\left(\frac{p}{a+p}\right)\|y_k\|^{a+p}_{p_0\left(\frac{p}{a+p}\right)},
\end{aligned} \tag{4.81}$$

where $c_0 > 0$ is the embedding constant of $W_0^{1,p}(Z)$ into $L^{p_0}(Z)$ (i.e., $\|u\|_{p_0} \leq c_0\|u\|_{1,p}$ for all $u \in W_0^{1,p}(Z)$). Combining (4.78), (4.80) and (4.81), we obtain

$$\|y_k\|_{p_{n+1}}^{p(p_{n+1})/p_0} \leq c_2 p_{n+1}^p\left(\|x\|_{p_n}^{p_n} + 1\right)$$

with $c_2 = c_1\left(c_0/p_0\right)^p$. Because $y_k(z) \longrightarrow x(z)$ a.e. on Z, we have

$$\begin{aligned}
&\|x\|_{p_{n+1}}^{p(p_{n+1})/p_0} \leq \liminf_{k\to\infty}\|y_k\|_{p_{n+1}}^{p(p_{n+1})/p_0} \leq c_2 p_{n+1}^p\left(\|x\|_{p_n}^{p_n} + 1\right), \\
\Rightarrow\ & x \in L^{p_{n+1}}(Z).
\end{aligned} \tag{4.82}$$

So we have completed the induction and we can say that $x \in L^{p_n}(Z)$ for all $n \geq 1$. □

Next we consider the sequence $\{q_n\}_{n\geq 0}$ defined by

$$q_0 = p_0 \qquad \text{and} \qquad q_{n+1} = \left(q_n + q(p-1)\right)\delta \quad \text{with} \quad \delta = \frac{p_0}{pq}, \qquad n \geq 0. \tag{4.83}$$

LEMMA 4.3.33 *If all the hypotheses of Lemma 4.3.32 are satisfied and $\{q_n\}_{n\geq 0} \subseteq \mathbb{R}_+$ is defined by (4.83), then $x \in L^{q_n}(Z)$ for all $n \geq 0$ and*

$$\|x\|_{q_{n+1}}^{\frac{q_{n+1}}{\delta q}} \leq c_3 q_{n+1}^p \|x\|_{q_n}^{\frac{q_n}{q}},$$

where $c_3 > 0$ depends only on $\big(\|\alpha\|_q,\ c,\ \vartheta,\ N,\ p, q$ and $\|x\|_{p_0}\big)$.

PROOF: If $r_0 = \min\{q(p_0 - \vartheta),\ p_0 - q\}$, for every $n \geq 1$ we have

$$p_0 + r_0 \sum_{k=1}^{n} \delta^k = p_0 + r_0 \delta\, \frac{\delta^n - 1}{\delta - 1} \leq p_n. \tag{4.84}$$

Because $r_0 > 0$ and $\delta > 1$, from (4.84) we see that $p_n \longrightarrow +\infty$. So from Lemma 4.3.32 we have that $x \in L^r(Z)$ for all $1 \leq r < \infty$. Then $|x|^{\vartheta-1} \in L^{q'}(Z)$ (see (4.77)) and from (4.82) we have that

$$\|\,|x|^{\vartheta-1}\|_{q'} \leq \widehat{c}_3$$

with $\widehat{c}_3 > 0$ depending only on $\big(\|\alpha\|_q,\ c,\ \vartheta,\ N,\ p$ and $\|x\|_{p_0}\big)$. We have

$$\int_Z \big(\alpha(z)|x|^{q_n/q} + c|x|^{\vartheta-1}\big)dz \leq \big(c\widehat{c}_3 + \|\alpha\|_{q'}\big)\|x\|_{q_n}^{q_n/q}.$$

If in (4.78) we replace a by $(q_n/q) - 1$ and argue as in the proof of Lemma 4.3.32, we reach the conclusion of the lemma. $\square$

Now we are ready for the first nonlinear regularity result.

THEOREM 4.3.34 *If $x \in W_0^{1,p}(Z), \Delta_p x \in L^1_{\text{loc}}(Z), c > 0,\ \vartheta \in [1, p^*),\ q \in [1, (p^*/p))$ and $\alpha \in L^{q'}(Z)_+$ are such that*

$$-x(z)\Delta_p x(z) \leq \alpha(z)|x(z)| + c|x(z)|^{\vartheta} \quad \text{a.e. on } Z,$$

then $x \in L^\infty(Z)$ and $\|x\|_\infty \leq c_4$ with $c_4 > 0$ depending only on $\big(\|\alpha\|_{q'},\ c, \vartheta, N, p, q$ and $\|x\|_{p_0}\big)$ where $p_0 = \begin{cases} p^* & \text{if } p^* < +\infty \\ 2\max\{pq, \vartheta\} & \text{if } p^* = +\infty \end{cases}$.

PROOF: Let $\xi_n = q_n \ln(\|x\|_{q_n})$ and $\beta_n = q\ln(c_3 q_{n+1}^p)$ with $c_3 > 0$ as in Lemma 4.3.33. Then from Lemma 4.3.33, we have

$$\xi_{n+1} \leq \delta(\xi_n + \beta_n) \qquad \text{for all } n \geq 1,$$
$$\Rightarrow\ \xi_n \leq \delta^n \xi_0 + \sum_{k=1}^{n} \delta^k \beta_{n-k} \quad \text{for all } n \geq 1.$$

Note that $q_n = \delta^n p_0 + \delta q(p-1)\big((\delta^n - 1)/(\delta - 1)\big)$. So we have $\delta^n p_0 \leq q_n \leq \delta^n \widehat{c}_4$ with $\widehat{c}_4 = p_0 + \delta q\big((p-1)/(\delta - 1)\big)$, hence

$$\beta_n \leq \gamma_1 + (n+1)\gamma_2$$

with $\gamma_1 = q\ \ln(c_3 \widehat{c}_4^{\,p})$ and $\gamma_2 = pq \ln \delta$. Then through some elementary calculations, we obtain

$$\sum_{k=1}^{n} \delta^k \beta_{n-k} \leq \gamma_3 \delta^n$$

with $\gamma_3 = \Big(\gamma_1 + \gamma_2\big(\delta/(\delta-1)\big)\Big)\big(\delta/(\delta-1)\big)$. So finally

$$\begin{aligned} &\|x\|_{q_n} \leq \exp\Big(\xi_n/\delta^n p_0\Big) \leq \exp\Big((\xi_0+\gamma_3)/p_o\Big) \\ \Rightarrow\ &\|x\|_\infty \leq \limsup_{n\to\infty} \|x\|_{q_n} \leq \exp\Big((\xi_0+\gamma_3)/p_o\Big) = c_4. \end{aligned}$$

□

Using this theorem, we can state a basic nonlinear regularity result, which is actually a particular case of a result of Lieberman [382].

THEOREM 4.3.35 *If $x \in W_0^{1,p}(Z) \cap L^\infty(Z)$ and $\triangle_p x \in L^\infty(Z)$, then $x \in C^{1,\alpha}(\overline{Z})$ and $\|x\|_{C^{1,\alpha}(\overline{Z})} \leq c_5$ with $0<\alpha<1$ and $c_5>0$ depending only on $\big(N,\ p,\ \|x\|_\infty, \|\triangle_p x\|_\infty\big)$.*

REMARK 4.3.36 In fact the above regularity theorem is still valid if instead of the Dirichlet boundary condition, we have the Neumann boundary condition (so $W_0^{1,p}(Z)$ is replaced by $W^{1,p}(Z)$); see Lieberman [382, Theorem 2].

In addition to the nonlinear regularity result, we also need the following nonlinear strong maximum principle which generalizes Lemma 4.3.18 (Hopf's lemma) and Theorem 4.3.19. The result is due to Vazquez [593] and the interested reader can look there for its proof.

THEOREM 4.3.37 *If $x \in C^1(Z)$, $x(z) \geq 0$ for all $z \in Z$, $x \neq 0$, $\triangle_p x \in L^2_{\mathrm{loc}}(Z)$, and there exists continuous nondecreasing function $\beta\colon \mathbb{R}_+ \longrightarrow \mathbb{R}$ such that $\beta(0)=0$ and either $\beta(r_0)=0$ for some $r_0>0$ or*

$$\int_0^1 \frac{1}{\big(r\beta(r)\big)^{1/p}}\, dr = +\infty,$$

for which we have

$$\triangle_p x(z) \leq \beta\big(x(z)\big) \quad \text{a.e. on } Z,$$

then $x(z)>0$ for all $z\in Z$; moreover, if $z_0 \in \partial Z$, $x \in C^1(Z\cup\{z_0\})$ and $x(z_0)=0$, then $(\partial x/\partial n)(z_0)<0$.

REMARK 4.3.38 In fact the result is true for any bounded domain $Z \subseteq \mathbb{R}^N$. In this case, as in Lemma 4.3.18, we need to assume that Z satisfies the interior ball condition at $z_0 \in \partial Z$. Recall that this condition is automatically satisfied at every boundary point if ∂Z is a C^2-manifold.

Now we have at our disposal all the necessary analytical tools to start investigating the spectral properties of $\big(-\triangle_p, W_0^{1,p}(Z)\big)$. So let $m \in L^\infty(Z)$ and assume that

$$\big|\{m>0\}\big|_N > 0.$$

We consider the following nonlinear weighted eigenvalue problem with weight $m \in L^\infty(Z) \setminus \{0\}$.

$$\left\{ \begin{array}{l} -\operatorname{div}\big(\|Du(z)\|^{p-2}Du(z)\big) = \lambda m(z)|u(z)|^{p-2}u(z) \text{ a.e. on } Z, \\ u\big|_{\partial Z} = 0, \ \lambda \in \mathbb{R}, \ 1<p<\infty \end{array} \right\}. \tag{4.85}$$

We say that $\lambda \in \mathbb{R}$ is an eigenvalue of $\big(-\triangle_p, W_0^{1,p}(Z), m\big)$, if problem (4.85) has a nontrivial solution $u \in W_0^{1,p}(Z)$. This solution is an eigenfunction corresponding to the eigenvalue $\lambda \in \mathbb{R}$. The pair (λ, u) is an eigenelement of $\big(-\triangle_p, W_0^{1,p}(Z), m\big)$.

The first observation about problem (4.85) is a straightforward consequence of the previous discussion on nonlinear regularity theory and the nonlinear strong maximum principle.

PROPOSITION 4.3.39 *If $u \in W_0^{1,p}(Z)$ is an eigenfunction of $\big(-\triangle_p, W_0^{1,p}(Z)\big)$, then $u \in C^{1,\alpha}(\overline{Z})$ for some $0<\alpha<1$; moreover, if $u \geq 0$, then $u(z)>0$ for all $z \in Z$ and $(\partial u/\partial n)(z)<0$ for all $z \in \partial Z$.*

PROOF: The fact that $u \in C^{1,\alpha}(\overline{Z})$ for some $\alpha \in (0,1)$ is a consequence of Theorem 4.3.35. Also if $u \geq 0$, then because $u \neq 0$, $\triangle_p u \in L^\infty(Z)$, and $\triangle_p u(z) \leq |\lambda|\|m\|_\infty u(z)^{p-1}$ a.e. on Z, we can apply Theorem 4.3.37 and conclude that $u(z)>0$ for all $z \in Z$ and $(\partial u/\partial n)(z)<0$ for all $z \in \partial Z$. □

REMARK 4.3.40 If we consider the Banach space

$$C_0^1(\overline{Z}) = \{x \in C^1(\overline{Z}) : x\big|_{\partial Z} = 0\},$$

then this becomes an ordered Banach space (OBS), with order cone

$$C_0^1(\overline{Z})_+ = \{x \in C_0^1(\overline{Z}) : x(z) \geq 0 \quad \text{for all } z \in \overline{Z}\}.$$

It is well known that int $C_0^1(\overline{Z})_+$ is nonempty and more precisely, we have

$$\begin{aligned} \operatorname{int} C_0^1(\overline{Z})_+ = \Big\{x \in C_0^1(\overline{Z})_+ : x(z)>0 \text{ for all } z \in Z \text{ and } \frac{\partial x}{\partial n}(z)<0 \\ \text{for all } z \in \partial Z\Big\}. \end{aligned}$$

So Proposition 4.3.39 says that, if $u \in W_0^{1,p}(Z)$ is a nonnegative eigenfunction of $\big(-\triangle_p, W_0^{1,p}(Z), m\big)$, then $u \in \operatorname{int} C_0^1(\overline{Z})_+$. When $m \equiv 1$, we simply write $\big(-\triangle_p, W_0^{1,p}(Z)\big)$.

Now consider the functions $\xi, \eta : W_0^{1,p}(Z) \longrightarrow \mathbb{R}_+$ defined by

$$\xi(x) = \frac{1}{p}\|Dx\|_p^p \qquad \text{and} \qquad \eta(x) = \frac{1}{p}\int_Z m|x|^p dz.$$

Also for the needs of the argument that follows, we introduce the function $\widehat{\eta}: L^p(Z,\mathbb{R}^N) \longrightarrow \mathbb{R}_+$ defined by

$$\widehat{\eta}(y) = \frac{1}{p}\|y\|^p_{L^p(Z,\mathbb{R}^N)}, \quad y \in L^p(Z,\mathbb{R}^N).$$

Evidently $\xi, \eta \in C^1\big(W_0^{1,p}(Z)\big)$ and $\widehat{\eta} \in C^1\big(L^p(Z,\mathbb{R}^N)\big)$. Note that

$$\frac{1}{p}m(z)|x|^p = \int_0^x m(z)|r|^{p-2}r\mathrm{d}r \qquad \text{for all } z\in Z, \text{ all } x\in\mathbb{R}.$$

So from Proposition 1.1.14 we have

$$\eta'(x)(\cdot) = m(\cdot)|x(\cdot)|^{p-2}x(\cdot). \tag{4.86}$$

Similarly from Proposition 3.2.17 and Example 3.2.21(a), we have

$$\eta'(y)(\cdot) = \|y(\cdot)\|_{\mathbb{R}^N}^{p-2}y(\cdot) \qquad \text{for all } y\in L^p(Z,\mathbb{R}^N). \tag{4.87}$$

Note that $\xi = \widehat{\eta}\circ D$, $D\in\mathcal{L}\big(W_0^{1,p}(Z), L^p(Z,\mathbb{R}^N)\big)$ being the gradient operator. Hence from the chain rule (see Proposition 1.1.14), we have

$$\begin{aligned} &\xi'(x) = \widehat{\eta}\,'(Dx)\circ D \\ \Rightarrow\ &\langle \xi'(x), u\rangle = \int_Z \|Dx\|^{p-2}(Dx, Du)_{\mathbb{R}^N}\mathrm{d}z \qquad \text{(see (4.87))}. \end{aligned}$$

So, if we define $A\colon W_0^{1,p}(Z)\longrightarrow W^{-1,p'}(Z)$ by

$$\langle A(x), u\rangle = \int_Z \|Dx\|^{p-2}(Dx, Du)_{\mathbb{R}^N}\mathrm{d}z \qquad \text{for all } x,u\in W_0^{1,p}(Z), \tag{4.88}$$

then $\xi'(x) = A(x)$. We recall the following elementary inequalities:

$$\begin{aligned} &\frac{p-1}{2}\|v_1-v_2\|^2(\|v_1\|+\|v_2\|)^{p-2} \le (\|v_1\|^{p-2}v_1 - \|v_2\|^{p-2}v_2, v_1-v_2)_{\mathbb{R}^N} \\ &\le 2^{2-p}\|v_1-v_2\|^p \qquad \text{for all } v_1,v_2\in\mathbb{R}^N,\ 1<p\le 2, \end{aligned} \tag{4.89}$$

and

$$\begin{aligned} &2^{2-p}\|v_1-v_2\|^p \le (\|v_1\|^{p-2}v_1 - \|v_2\|^{p-2}v_2, v_1-v_2)_{\mathbb{R}^N} \\ &\le c\|v_1-v_2\|^2(\|v_1\|+\|v_2\|)^{p-2} \qquad \text{for all } v_1,v_2\in\mathbb{R}^N,\ p\ge 2. \end{aligned} \tag{4.90}$$

From (4.88) and using (4.89) and (4.90) it follows easily that A is demicontinuous, strictly monotone (strongly monotone if $p\ge 2$), and coercive. Hence A is maximal monotone (see Corollary 3.2.11) and surjective (see Corollary 3.2.28). Moreover, recalling that $D^* = -\mathrm{div}\in\mathcal{L}\big(L^{p'}(Z,\mathbb{R}^N), W^{-1,p'}(Z)\big)$, then we see that $A(x) = -\Delta_p x$ for all $x\in W_0^{1,p}(Z)$.

PROPOSITION 4.3.41 *If $A\colon W_0^{1,p}(Z)\longrightarrow W^{-1,p'}(Z)$ is defined by (4.88), then A is of type $(S)_+$ (see Definition 3.2.61(d)).*

PROOF: Let $x_n \xrightarrow{w} x$ in $W_0^{1,p}(Z)$ and suppose that $\limsup\limits_{n\to\infty}\langle A(x_n), x_n - x\rangle \le 0$. A is maximal monotone, therefore it is generalized pseudomonotone (see Proposition 3.2.56) and so

$$\begin{aligned} &\langle A(x_n), x_n\rangle \longrightarrow \langle A(x), x\rangle \quad \text{(see Definition 3.2.53(b))}, \\ \Rightarrow\ &\|Dx_n\|_p \longrightarrow \|Dx\|_p. \end{aligned}$$

We know that $Dx_n \xrightarrow{w} Dx$ in $L^p(Z,\mathbb{R}^N)$ and $L^p(Z,\mathbb{R}^N)$, being uniformly convex, has the Kadec–Klee property. Therefore we conclude that $Dx_n \longrightarrow Dx$ in $L^p(Z,\mathbb{R}^N)$ and so $x_n\longrightarrow x$ in $W_0^{1,p}(Z)$, which proves that A is of type $(S)_+$. □

We define

$$\lambda_1 = \inf\big\{\xi(u) : x \in W_0^{1,p}(Z),\ \eta(u) = 1\big\}$$
$$= \inf\big\{\frac{1}{p}\|Du\|_p^p : x \in W_0^{1,p}(Z),\ \frac{1}{p}\int_Z m|u|^p dz = 1\big\}. \tag{4.91}$$

PROPOSITION 4.3.42 *If λ_1 is defined by (4.91), then $\lambda_1>0$ is an eigenvalue of $\big(-\Delta_p, W_0^{1,p}(Z), m\big)$; moreover $u \in W_0^{1,p}(Z)$ is an eigenfunction corresponding to $\lambda_1>0$ if and only if*

$$\xi(u) - \lambda_1\eta(u) = 0 = \inf\big\{\xi(x) - \lambda_1\eta(x) : x \in W_0^{1,p}(Z)\big\}$$

and $u \in \operatorname{int} C_0^1(\overline{Z})_+$ or $u \in -\operatorname{int} C_0^1(\overline{Z})_+$.

PROOF: Let $\{u_n\}_{n\geq 1} \subseteq W_0^{1,p}(Z)$, and $\|u_n\|_p = 1$ be a minimizing sequence for problem (4.91). Then by Poincaré's inequality we have that $\{u_n\}_{n\geq 1} \subseteq W_0^{1,p}(Z)$ is bounded and so by passing to a suitable subsequence if necessary, we may assume that

$$u_n \xrightarrow{w} u \quad \text{in} \quad W_0^{1,p}(Z) \quad \text{and} \quad u_n \longrightarrow u \quad \text{in} \quad L^p(Z).$$

From the weak lower semicontinuity of the norm functional in a Banach space, we have

$$\frac{1}{p}\|Du\|_p^p \leq \liminf_{n\to\infty} \frac{1}{p}\|Du_n\|_p^p = \lambda_1. \tag{4.92}$$

Also because $u_n \longrightarrow u$ in $L^p(Z)$, we have that $(1/p)\int_Z m|x|^p dz{=}1$. Therefore from (4.92) it follows that

$$\frac{1}{p}\|Du\|_p^p = \lambda_1.$$

Then by the Lagrange multiplier rule we can find $\lambda \in \mathbb{R}$ such that

$$A(u) = \lambda\eta'(u),$$
$$\Rightarrow \ \|Du\|_p^p = \lambda\int_Z m|u|^p dz \ \ \text{(i.e., } \lambda = \lambda_1).$$

From the equation $A(u) = \lambda_1\eta'(u)$, it follows that $\lambda_1 > 0$ is an eigenvalue of $\big(-\Delta_p, W_0^{1,p}(Z), m\big)$. Moreover, exploiting the p-homogeneity of the functions ξ and η, we have

$$\xi(u) - \lambda_1\eta(u) = 0 = \inf\big\{\xi(x) - \lambda_1\eta(x) : x \in W_0^{1,p}(Z)\big\}.$$

Since $u \neq 0$ we must have $u^+ \neq 0$ or $u^- \neq 0$. We assume without any loss of generality that $u^+ \neq 0$ (the other case is treated similarly). Then using as a test function $u^+ \in W_0^{1,p}(Z)$ we obtain

$$\big\langle A(u), u^+\big\rangle = \lambda_1\big\langle \eta'(u), u^+\big\rangle$$
$$\Rightarrow \ \|Du^+\|_p^p = \lambda_1\int_Z m|u^+|^p dz,$$
$$\Rightarrow \ u^+ \text{ is an eigenfunction of } \big(-\Delta_p, W_0^{1,p}(Z), m\big).$$

Then by virtue of Proposition 4.3.39, we have that $u^+(z){>}0$ for all $z{\in}Z$, hence $u = u^+ \in \operatorname{int} C_0^1(\overline{Z})_+$. □

REMARK 4.3.43 The previous proposition gives a variational characterization of the eigenelement (λ_1, u), namely $\|Du\|_p^p = \lambda_1 \|u\|_p^p$ and $\|Dx\|_p^p \geq \lambda_1 \|x\|_p^p$ for all $x \in W_0^{1,p}(Z)$.

Next we establish the simplicity of $\lambda_1 > 0$.

PROPOSITION 4.3.44 *The eigenvalue $\lambda_1 > 0$ is simple; that is, the corresponding eigenspace is one-dimensional.*

PROOF: Suppose that $u \geq 0$ and $v \geq 0, u \neq v$ are both eigenfunctions corresponding to $\lambda_1 > 0$. Then we have $u, v \in \operatorname{int} C_0^1(\overline{Z})_+$ (see Proposition 4.3.39) and

$$A(u) = \lambda_1 \eta'(u) \tag{4.93}$$

$$\text{and} \quad A(v) = \lambda_1 \eta'(v). \tag{4.94}$$

Note that $u/v,\ v/u \in L^\infty(Z)$ and set $\varphi = (u^p - v^p)/u^{p-1},\ \psi = (v^p - u^p)/v^{p-1}$. Evidently $\varphi,\ \psi \in W_0^{1,p}(Z)$ and we have

$$D\varphi = \Big(1 + (p-1)\big(\frac{v}{u}\big)^p\Big) Du - p\big(\frac{v}{u}\big)^{p-1} Dv \tag{4.95}$$

$$\text{and} \quad D\psi = \Big(1 + (p-1)\big(\frac{u}{v}\big)^p\Big) Dv - p\big(\frac{u}{v}\big)^{p-1} Du. \tag{4.96}$$

We use φ as a test function in (4.93) and ψ as a test function in (4.94). We obtain

$$\int_Z \|Du\|^{p-2}(Du, D\varphi)_{\mathbb{R}^N} dz = \lambda_1 \int_Z m|u|^{p-2} u\varphi dz \tag{4.97}$$

$$\text{and} \quad \int_Z \|Dv\|^{p-2}(Dv, D\psi)_{\mathbb{R}^N} dz = \lambda_1 \int_Z m|v|^{p-2} v\psi dz. \tag{4.98}$$

Using (4.95) and (4.96) in (4.97) and (4.98), respectively, and then adding the two equalities, we obtain

$$\begin{aligned} &\int_Z \Big[\Big(1 + (p-1)\big(\frac{v}{u}\big)^p\Big)\|Du\|^p + \Big(1 + (p-1)\big(\frac{u}{v}\big)^p\Big)\|Dv\|^p\Big] dz \\ = &\int_Z \Big(p\big(\frac{v}{u}\big)^{p-1}\|Du\|^{p-2}(Du, Dv)_{\mathbb{R}^N} + p\big(\frac{u}{v}\big)^{p-1}\|Dv\|^{p-2}(Dv, Du)_{\mathbb{R}^N}\Big) dz. \end{aligned} \tag{4.99}$$

Note that $\|D\ln u\| = \|Du\|/u$ and $\|D\ln v\| = \|Dv\|/v$. Therefore from (4.99) we have

$$\begin{aligned} &\int_Z (u^p - v^p)(\|D\ln u\|^p - \|D\ln v\|^p) dz \\ = &\int_Z pv^p \|D\ln u\|^{p-2}(D\ln u, D\ln v - D\ln u)_{\mathbb{R}^N} dz \\ &+ \int_Z pu^p \|D\ln v\|^{p-2}(D\ln v, D\ln u - D\ln v)_{\mathbb{R}^N} dz. \end{aligned} \tag{4.100}$$

If $p \geq 2$, we recall the elementary inequality which says that

$$\|\xi_2\|^2 - \|\xi_1\|^2 \geq p\, \|\xi_1\|^{p-2}(\xi_1, \xi_2 - \xi_1)_{\mathbb{R}^N} + \frac{1}{2^{p-1}-1}\|\xi_2 - \xi_1\|^p \quad \text{for all } \xi_1, \xi_2 \in \mathbb{R}^N. \tag{4.101}$$

Therefore when $p \geq 2$, we use (4.101) in (4.100) and have

$$\begin{aligned} &\frac{1}{2^{p-1}-1}\int_Z \left(\frac{1}{v^p} + \frac{1}{u^p}\right)\|vDu - uDv\|^p dz = 0, \\ \Rightarrow\ & \|v(z)Du(z) - u(z)Dv(z)\| = 0, \ \text{ a.e. on } Z, \\ \Rightarrow\ & u = kv \quad \text{for some } k>0. \end{aligned}$$

If $1<p<2$, then we recall the elementary inequality which says that

$$\|\xi_2\|^p - \|\xi_1\|^p \geq p\, \|\xi_1\|^{p-2}(\xi_1, \xi_2 - \xi_1)_{\mathbb{R}^N} + \frac{3p(p-1)}{16}\frac{\|\xi_2 - \xi_1\|^2}{(\|\xi_1\| + \|\xi_2\|)^{2-p}}, \tag{4.102}$$

for all $\xi_1, \xi_2 \in \mathbb{R}^N$. Therefore when $1<p<2$, we use (4.102) in (4.100) and have

$$\begin{aligned} &\frac{3p(p-1)}{16}\int_Z \left(\frac{1}{v^p} + \frac{1}{u^p}\right)\frac{\|vDu - uDv\|^2}{(v\|Du\| + u\|Dv\|)^{2-p}} dz = 0, \\ \Rightarrow\ & \|v(z)Du(z) - u(z)Dv(z)\| = 0 \quad \text{a.e. on } Z, \\ \Rightarrow\ & u = kv \quad \text{for some } k>0. \end{aligned}$$

We conclude that the eigenspace corresponding to $\lambda_1 > 0$ is one-dimensional; that is, $\lambda_1 > 0$ is simple. □

In Proposition 4.3.42 we saw that the eigenfunctions corresponding to $\lambda_1 > 0$ do not change sign. Next we show that these are the only eigenfunctions of $\left(-\triangle_p, W_0^{1,p}(Z), m\right)$ with a fixed sign.

PROPOSITION 4.3.45 *If $u \in C_0^1(\overline{Z})$ is an eigenfunction of $\left(-\triangle_p, W_0^{1,p}(Z), m\right)$ corresponding to an eigenvalue $\lambda > \lambda_1$, then u must change sign (i.e., $u^+ \neq 0$ and $u^- \neq 0$) and we have*

$$\min\{|Z^+|_N,\ |Z^-|_N\} \geq (\lambda\|m\|_\infty c)^\eta, \tag{4.103}$$

where $Z^+ = \{u > 0\}$, $Z^- = \{u < 0\}$, $c > 0$ is a constant independent of the eigenelement (λ, u) and

$$\eta = \begin{cases} -\frac{N}{p} & \text{if } p<N \\ -1 & \text{if } p \geq N \end{cases}.$$

PROOF: Let $u_1 \in \operatorname{int} C_0^1(\overline{Z})_+$ be the normalized (i.e., $\|Du\|_p = 1$) eigenfunction corresponding to $\lambda_1 > 0$. We argue indirectly. So suppose that u has a constant sign on Z, say $u \geq 0$ (the case $u \leq 0$ is treated similarly). We may assume that $\|Du\|_p = 1$. Note that by Proposition 4.3.44, we obtain

$$\begin{aligned} 0 &\leq \left\langle A(u_1), \frac{u_1^p - u^p}{u_1^{p-1}}\right\rangle + \left\langle A(u), \frac{u^p - u_1^p}{u^{p-1}}\right\rangle \\ &= \int_Z (\lambda_1 - \lambda)m(u_1^p - u^p)dz = (\lambda_1 - \lambda)\left(\frac{1}{\lambda_1} - \frac{1}{\lambda}\right) < 0, \end{aligned}$$

a contradiction.

Now we prove estimate (4.103). Note that

$$\begin{aligned} & A(u^+) = \lambda\eta'(u^+) \\ \Rightarrow\ & \|Du^+\|_p^p = \lambda\int_Z m|u^+|^p dz \leq \lambda\|m\|_\infty\|u^+\|_{p^*}^p|Z^+|_N^{1-(p/p^*)}, \end{aligned} \tag{4.104}$$

(by Hölder's inequality). Similarly we obtain

$$\|Du^-\|_p^p \leq \lambda\|m\|_\infty\|u^-\|_{p^*}^p|Z^-|_N^{1-(p/p^*)}. \tag{4.105}$$

From the continuous embedding of $W_0^{1,p}(Z)$ into $L^{p^*}(Z)$ (Sobolev embedding theorem), we can find a constant $c>0$ (independent of the eigenelement (u,λ)) such that

$$\|u^+\|_{p^*}^p \leq c\|Du^+\|_p^p \quad \text{and} \quad \|u^-\|_{p^*}^p \leq c\|Du^-\|_p^p. \tag{4.106}$$

Combining (4.104) through (4.106) we obtain

$$\begin{aligned} & 1 \leq \lambda\|m\|_\infty c|Z^\pm|_N^{1-(p/p^*)} \\ \Rightarrow\ & (\lambda\|m\|_\infty c)^\eta \leq \min\{|Z^+|_N, |Z^-|_N\} \end{aligned}$$

with

$$\eta = \begin{cases} -\frac{N}{p} & \text{if } p<N \\ -1 & \text{if } p\geq N \end{cases}.$$

□

Using this proposition we show that $\lambda_1>0$ is isolated.

PROPOSITION 4.3.46 *The eigenvalue $\lambda_1>0$ of $\big(-\Delta_p, W_0^{1,p}(Z), m\big)$ is isolated.*

PROOF: Clearly from Proposition 4.3.42 (see also Remark 4.3.43), we have that $\lambda_1>0$ is the smallest positive eigenvalue.

Suppose we could find $\{\widehat{\lambda}_n\}\subseteq\mathbb{R}_+$ eigenvalues, $\widehat{\lambda}_n\neq\lambda_1$, and $\widehat{\lambda}_n\downarrow\lambda_1$ as $n\to\infty$. Then we can find $\{\widehat{u}_n\}\subseteq W_0^{1,p}(Z)$ such that $\|D\widehat{u}_n\|_p=1$ and

$$\begin{aligned} & A(\widehat{u}_n) = \widehat{\lambda}_n\eta'(\widehat{u}_n) \\ \Rightarrow\ & \|D\widehat{u}_n\|_p^p = \widehat{\lambda}_n\int_Z m|\widehat{u}_n|^p dz. \end{aligned} \tag{4.107}$$

We may assume that

$$\widehat{u}_n \xrightarrow{w} \widehat{u} \ \text{ in } \ W_0^{1,p}(Z) \ \text{ and } \ \widehat{u}_n \longrightarrow \widehat{u} \ \text{ in } \ L^p(Z).$$

Passing to the limit as $n\to\infty$ in (4.107) and using the weak lower semicontinuity of the norm functional in a Banach space, we obtain

$$\begin{aligned} & \|D\widehat{u}\|_p^p \leq \lambda_1\int_Z m|\widehat{u}|^p dz, \\ \Rightarrow\ & \|D\widehat{u}\|_p^p = \lambda_1\int_Z m|\widehat{u}|^p dz \ \text{(see Remark 4.3.43)}. \end{aligned} \tag{4.108}$$

Note that

$$\begin{aligned} &\langle A(\widehat{u}_n), \widehat{u}_n - \widehat{u}\rangle = \lambda_n \int_Z m|\widehat{u}_n|^{p-2}\widehat{u}_n(\widehat{u}_n - \widehat{u})dz, \\ \Rightarrow\ &\lim_{n\to\infty} \langle A(\widehat{u}_n), \widehat{u}_n - \widehat{u}\rangle = 0, \\ \Rightarrow\ &\widehat{u}_n \longrightarrow \widehat{u} \text{ in } W_0^{1,p}(Z) \quad \text{(see Proposition 4.3.41).} \end{aligned}$$

Therefore we infer that $\widehat{u} = u_1$. We have $\widehat{u}_n \longrightarrow u_1$ in $W_0^{1,p}(Z)$ and by Egorov's theorem given $\varepsilon > 0$, we can find $Z_\varepsilon \subseteq Z$ a closed subset such that $|Z_\varepsilon|_N \geq |Z|_N - \varepsilon$ and $\widehat{u}_n(z) \longrightarrow u_1(z)$ uniformly on Z_ε. If $Z_n^+ = \{\widehat{u}_n > 0\}$ and $Z_n^- = \{\widehat{u}_n < 0\}$, $n \geq 1$, we have

$$\liminf_{n\to\infty} |Z_n^+|_N \geq \liminf_{n\to\infty} \left|Z_n^+ \cap Z_\varepsilon\right|_N \geq |Z|_N - \varepsilon \qquad (\text{recall } u_1 > 0). \tag{4.109}$$

On the other hand from Proposition 4.3.45, we have

$$|Z_n^-|_N \geq \xi > 0 \qquad \text{for all } n \geq 1. \tag{4.110}$$

Because $|Z_n^+|_N + |Z_n^-|_N \leq |Z|_N$, by choosing $\varepsilon > 0$ small and using (4.109) and (4.110) we reach a contradiction. □

So summarizing the situation with the beginning of the positive part of the spectrum of $\left(-\Delta_p, W_0^{1,p}(Z), m\right)$ (see problem (4.85)), we can state the following theorem.

THEOREM 4.3.47 *Problem (4.85) has a smallest positive eigenvalue $\lambda_1 > 0$. This eigenvalue is simple (i.e., the corresponding eigenspace is one-dimensional), isolated, it has a variational characterization*

$$\lambda_1 = \inf\left\{\frac{\|Dx\|_p^p}{\int_Z m|x|^p dz} : x \in W_0^{1,p}(Z),\ x \neq 0\right\}, \tag{4.111}$$

this infimum is realized at the eigenfunction $u_1 \in \operatorname{int} C_0^1(\overline{Z})_+$ with $\int_Z m u_1^p dz = 1$, and this is the only eigenfunction (up to scalar multiplication) that does not change sign.

REMARK 4.3.48 If in problem (4.85) we replace m by $-m$, we see that the above analysis gives a first negative eigenvalue $\lambda_{-1} < 0$, which is simple, isolated, and is the only negative eigenvalue with a positive eigenfunction. It is easy to check that the set of eigenvalues of $\left(-\Delta_p, W_0^{1,p}(Z), m\right)$ is closed in $\mathbb{R}$.

What can be said about other eigenvalues and eigenfunctions of $\left(-\Delta_p, W_0^{1,p}(Z), m\right)$? Here is where the Ljusternik–Schnirelmann theory from Section 4.2 enters into the picture. So let $\varphi\colon W_0^{1,p}(Z) \longrightarrow \mathbb{R}$ be defined by

$$\varphi(x) = \xi(x)^2 - \eta(x), \quad x \in W_0^{1,p}(Z).$$

Recall that

$$\xi(x) = \frac{1}{p}\|Dx\|_p^p \quad \text{and} \quad \eta(x) = \frac{1}{p}\int_Z m|x|^p dz, \quad x \in W_0^{1,p}(Z).$$

Evidently $\varphi \in C^1\big(W_0^{1,p}(Z)\big)$. Suppose $x \neq 0$ is a critical point of φ (i.e., $\varphi'(x) = 0$) with critical value c (i.e., $\varphi(x) = c$), then we have

$$\begin{aligned}
&0 = \varphi'(x) = 2\xi(x)\xi'(x) - \eta'(x),\\
\Rightarrow\ &2\xi(x)\left\langle \xi'(x), x\right\rangle = \left\langle \eta'(x), x\right\rangle,\\
\Rightarrow\ &2\xi(x)\|Dx\|_p^p = \int_Z m|x|^p dz,\\
\Rightarrow\ &2p\xi(x)^2 - p\eta(x) = 0\\
\Rightarrow\ &p\big(2\xi(x)^2 - \eta(x)\big) = 0\\
\Rightarrow\ &p\big(\xi(x)^2 + c\big) = 0 \ \ (\text{i.e., } \ c = -\xi(x)^2 < 0).
\end{aligned}$$

Then we see that

$$\xi(x) = \frac{1}{2\sqrt{-c}}\,\eta(x)$$

and so $\lambda = 1/(2\sqrt{-c}) > 0$ is an eigenvalue of $\big(-\triangle_p, W_0^{1,p}(Z), m\big)$ with associated eigenfunction x.

Conversely if $u \neq 0$ is an eigenfunction of $\big(-\triangle_p, W_0^{1,p}(Z), m\big)$ corresponding to an eigenvalue $\lambda > 0$, then $v = 1/\big(2\lambda\xi(x)\big)^{1/p}\, u$ is also an eigenfunction corresponding to $\lambda > 0$. Exploiting the p-homogeneity of ξ, we see that $\lambda = 1/\big(2\xi(v)\big)$. Because $\varphi'(v) = 2\xi(v)\xi'(v) - \eta'(v)$, we see that v is a critical point of φ with corresponding critical value $c = -(1/4\lambda^2)$.

So the eigenelements of $\big(-\triangle_p, W_0^{1,p}(Z), m\big)$ are completely determined by those of the functional φ. Hence we may work with φ and apply on it the Ljusternik–Schnirelmann theory. To this end we introduce

$$\mathcal{S}_n^c = \big\{K \subseteq W_0^{1,p}(Z) : K \ \text{is compact, symmetric (i.e., } K = -K) \ \text{and} \ \gamma(K) \geq n\big\}, \quad n \geq 1. \tag{4.112}$$

Here γ is the genus introduced in Definition 4.2.13. Then we define

$$c_n = \inf_{K \in \mathcal{S}_n^c} \max_{x \in K} \varphi(x), \quad n \geq 1. \tag{4.113}$$

We proceed with the genus as our topological index, because as we explained in Section 4.2 it is in general easier to calculate than the LS-category ξ. Also note that in the family (4.112), without any loss of generality we assume that the sets are compact (see Proposition 4.2.19 and Theorem 4.2.23).

LEMMA 4.3.49 *The sequence $\{c_n\}_{n \geq 1}$ defined by (4.113) consists of critical points of φ and*

$$-\infty < \inf_{W_0^{1,p}(Z)} \varphi = c_1 \leq c_2 \leq \cdots \leq c_n \leq \cdots < 0 = \varphi(0).$$

PROOF: First we show that φ is bounded below.

Indeed we have

$$\begin{aligned}
\varphi(x) = \xi(x)^2 - \eta(x) &= \frac{1}{p^2}\|Dx\|_p^{2p} - \frac{1}{p}\int_Z m|x|^p dz\\
&\geq \frac{1}{p}\Big(\|Dx\|_p^p - \frac{1}{\lambda_1}\Big)\|Dx\|_p^p \quad (\text{see } (4.111)),
\end{aligned}$$

$\Rightarrow$ $\varphi(x)$ is coercive, hence in particular bounded below.

Next we show that φ satisfies the PS-condition. To this end let $\{x_n\}_{n\geq 1} \subseteq W_0^{1,p}(Z)$ be a sequence such that $\{\varphi(x_n)\}_{n\geq 1} \subseteq \mathbb{R}$ is bounded and $\varphi'(x_n) \longrightarrow 0$. Because φ is coercive, it follows that $\{x_n\}_{n\geq 1} \subseteq W_0^{1,p}(Z)$ is bounded and so by passing to a suitable subsequence if necessary, we may assume that

$$x_n \xrightarrow{w} x \text{ in } W_0^{1,p}(Z) \qquad \text{and} \qquad x_n \longrightarrow x \text{ in } L^p(Z). \tag{4.114}$$

Then we have

$$\langle \eta'(x_n), x_n - x \rangle = \int_Z m|x_n|^{p-2} x_n (x_n - x) dz \longrightarrow 0$$

and because $\varphi'(x_n) = 2\xi(x_n)A(x_n) - \eta'(x_n) \longrightarrow 0$ in $W^{-1,p'}(Z)$, it follows that

$$\xi(x_n) \langle A(x_n), x_n - x \rangle \longrightarrow 0. \tag{4.115}$$

Because of (4.114), we may assume (at least for a subsequence), that $\xi(x_n) \longrightarrow \vartheta$. If $\vartheta = 0$, then $x_n \longrightarrow x = 0$ in $W_0^{1,p}(Z)$ and we are done. Otherwise, from (4.115), we have

$$\begin{aligned} &\langle A(x_n), x_n - x \rangle = 0, \\ \Rightarrow\ & x_n \longrightarrow x \text{ in } W_0^{1,p}(Z) \quad \text{(see Proposition 4.3.41).} \end{aligned}$$

This proves that φ satisfies the PS-condition.

Finally for every $n \geq 1$, we show there exists $K \subseteq W_0^{1,p}(Z)$ compact, and symmetric with $\gamma(K) = n$ such that $\sup\limits_{x\in K} \varphi(x) < 0$. Consider functions $\{y_k\}_{k=1}^n \subseteq W_0^{1,p}(Z)$ such that

$$\begin{aligned} &\|y_k\|_p = 1, \qquad \int_Z m|y_k|^p dz > 0 \\ \text{and} \quad &\operatorname{supp} y_k \cap \operatorname{supp} y_m = \emptyset \quad \text{if } k \neq m,\ k, m \in \{1, \ldots, n\}. \end{aligned}$$

Let $Y_n = \operatorname{span}\{y_k\}_{k=1}^n$. Because Y_n is a finite-dimensional subspace of $W_0^{1,p}(Z)$, all norms are equivalent and so we can find $c > 0$ such that

$$c\xi(y) \leq \eta(y) \leq \frac{1}{c}\xi(y) \qquad \text{for all } y \in Y_n \qquad \text{(by Poincaré's inequality).}$$

We consider the compact set

$$K = \Big\{ y \in Y_n : \frac{c^2}{4} \leq \eta(y) \leq \frac{c^2}{3} \Big\}.$$

We have $\sup\limits_{y\in K} \varphi(y) \leq -(c^2/8) < 0$. Because Y_n is isomorphic in $\mathbb{R}^n$, we can identify K with a ring K^1 in $\mathbb{R}^n$ such that $\partial B_1(0) = S^{n-1} = \{y \in \mathbb{R}^n : \|y\| = 1\} \subseteq K^1 \subseteq \mathbb{R}^n \setminus \{0\}$. Then by virtue of Propositions 4.2.15 and 4.2.18, we have $\gamma(K) = n$. ☐

REMARK 4.3.50 By virtue of Theorem 4.2.23(b), we see that φ has infinitely many critical points.

LEMMA 4.3.51 *If $\{c_n\}_{n\geq 1} \subseteq \mathbb{R}_-$ is defined by (4.113), then $c_n \longrightarrow 0$ as $n \to \infty$.*

PROOF: Let $G=\{x\in W_0^{1,p}(Z):\varphi(x)\leq 0\}$. Because φ is coercive (see the proof of Lemma 4.3.49), we see that $G\subseteq W_0^{1,p}(Z)$ is bounded. We show that given $\varepsilon>0$ we can find $n_\varepsilon\geq 1$ and $K\in\mathcal{S}^c_{n_\varepsilon}$ with $K\subseteq G$ such that $-\varepsilon\leq\sup\limits_{x\in K}\varphi(x)$. Exploiting the compact embedding of $W_0^{1,p}(Z)$ into $L^p(Z)$, given $\vartheta>0$ we can find $Y_\vartheta\subseteq L^p(Z)$ finite-dimensional and $i_\vartheta\colon G\longrightarrow Y_\vartheta$ continuous such that

$$\sup_{x\in G}\|x-i_\vartheta(x)\|_p\leq\vartheta \qquad \text{(see Theorem 3.1.10).}$$

We consider the odd component of i_ϑ defined by

$$\widehat{i}_\vartheta(x)=\frac{1}{2}\big(i_\vartheta(x)-i_\vartheta(-x)\big) \qquad \text{for all } x\in G.$$

Clearly we have

$$\sup_{x\in G}\|x-\widehat{i}_\vartheta(x)\|_p\leq\vartheta. \tag{4.116}$$

The set G is relatively compact in $L^p(Z)$ (because of the compact embedding of $W_0^{1,p}(Z)$ into $L^p(Z)$). Hence given $\varepsilon>0$, we can find $\vartheta_\varepsilon>0$ such that

$$|\eta(x)-\eta(\widehat{i}_{\vartheta_\varepsilon}(x))|\leq\frac{\varepsilon}{2} \qquad \text{for all } x\in G \qquad \text{(see (4.116)).} \tag{4.117}$$

Let $\delta_\varepsilon>0$ be such that $\eta(x)\leq\varepsilon/2$ if $\|x\|_p\leq\delta_\varepsilon$. Then for $x\in G$ with $\|\widehat{i}_{\vartheta_\varepsilon}(x)\|_p\leq\delta_\varepsilon$, we have

$$\eta(x)\leq|\eta(x)-\eta(\widehat{i}_{\vartheta_\varepsilon}(x))|+\eta(\widehat{i}_\vartheta(x))\leq\varepsilon \qquad \text{(see (4.117)).} \tag{4.118}$$

From (4.118) it follows that for every K compact, and symmetric, $K\subseteq G\cap\{x\in W_0^{1,p}(Z):\eta(x)\geq\varepsilon\}$, we have

$$\widehat{i}_{\vartheta_\varepsilon}(K)\subseteq\{y\in Y_{\vartheta_\varepsilon}:\|y\|_p\geq\delta_\varepsilon\}. \tag{4.119}$$

The set $\widehat{i}_{\vartheta_\varepsilon}(K)$ is compact and symmetric in $L^p(Z)$ and so if by $\widehat{\gamma}$ we denote the genus on $L^p(Z)$, from (4.119) we have

$$\begin{aligned}&\widehat{\gamma}\big(\widehat{i}_{\vartheta_\varepsilon}(K)\big)\leq\dim Y_{\vartheta_\varepsilon},\\ \Rightarrow\ &\gamma(K)\leq\widehat{\gamma}\big(\widehat{i}_{\vartheta_\varepsilon}(K)\big)\leq\dim Y_{\vartheta_\varepsilon}.\end{aligned}$$

Therefore for all compact and symmetric $K\subseteq G$ satisfying $\gamma(K)\geq\dim Y_{\vartheta_\varepsilon}+1$, we can find $x_0\in K$ such that $\inf\limits_{x\in K}\eta(x)\leq\eta(x_0)<\varepsilon$. Then because $\varphi(x)\geq-\eta(x)$, we have

$$\sup_{x\in K}\varphi(x)\geq-\inf_{x\in K}\eta(x)>-\varepsilon.$$

□

So combining Lemma 4.3.49 and Lemma 4.3.51, we can have the following result concerning the spectrum of $\big(-\Delta_p,W_0^{1,p}(Z),m\big)$.

THEOREM 4.3.52 *There exists a sequence* $\Big\{\lambda_n=1/\big(2\sqrt{-c_n}\big)\Big\}_{n\geq1}$ *of positive eigenvalues of* $\big(-\Delta_p,W_0^{1,p}(Z),m\big)$ *such that* $\lambda_n\longrightarrow+\infty$.

REMARK 4.3.53 So in addition to $\lambda_1>0$ (see (4.91)) we have produced a whole sequence $\{\lambda_n\}_{n\geq 1}$ of eigenvalues of $\big(-\Delta_p, W_0^{1,p}(Z), m\big)$ such that $\lambda_n \longrightarrow +\infty$. If $p = 2$ (linear eigenvalue problem), then these are all the positive eigenvalues of $\big(-\Delta, H_0^1(Z), m\big)$ (see Theorem 4.3.13 which treats the case $m \equiv 1$). If $p \neq 2$ (nonlinear eigenvalue problem), then we do not know if this in this case. What we know is that $\lambda_2>\lambda_1$ (see Proposition 4.3.46).

Because the set of eigenvalues of $\big(-\Delta_p, W_0^{1,p}(Z), m\big)$ is clearly closed, we have that

$$\lambda_2^* = \inf\big\{\lambda>0 : \lambda \text{ is an eigenvalue of } \big(-\Delta_p, W_0^{1,p}(Z), m\big),\ \lambda\neq\lambda_1\big\} \tag{4.120}$$

is the second positive eigenvalue of $\big(-\Delta_p, W_0^{1,p}(Z), m\big)$. Evidently $\lambda_2 \geq \lambda_2^*$ and it is natural to ask whether $\lambda_2^* = \lambda_2$. In what follows we show that indeed this is the case. So the second eigenvalue and the second positive variational eigenvalue of $\big(-\Delta_p, W_0^{1,p}(Z), m\big)$ coincide.

Let $(\lambda, u)\in\mathbb{R}_+\times C_0^1(\overline{Z})$ be an eigenelement of $\big(-\Delta_p, W_0^{1,p}(Z), m\big)$ (see Proposition 4.3.39). We introduce the set

$$Z(\lambda) = \{z\in Z : u(z) = 0\}.$$

Let $\mathcal{N}(\lambda)$ be the number of components (maximal connected open sets) of the open set $Z\backslash Z(\lambda)$. Generalizing Courant's nodal theorem, Anane–Tsouli [26] proved the following.

PROPOSITION 4.3.54 *If $(\lambda, u) \in \mathbb{R}_+\times C_0^1(\overline{Z})$ is an eigenelement of $\big(-\Delta_p, W_0^{1,p}(Z), m\big)$, then $\lambda_{\mathcal{N}(\lambda)} \leq \lambda$.*

COROLLARY 4.3.55 *If $\lambda_n<\lambda_{n+1}$, then $\mathcal{N}(\lambda_n)\leq n$.*

PROOF: If $\mathcal{N}(\lambda_n)>n$, then $\lambda_{\mathcal{N}(\lambda_n)} \geq \lambda_{n+1}$. On the other hand Proposition 4.3.54 implies that $\lambda_{\mathcal{N}(\lambda_n)} \leq \lambda_n$. Therefore it follows that $\lambda_n=\lambda_{n+1}$, a contradiction to the hypothesis. Hence $\mathcal{N}(\lambda_n)\leq n$. $\square$

THEOREM 4.3.56 $\lambda_2^* = \lambda_2$.

PROOF: Because λ_1 is isolated (see Proposition 4.3.46) from (4.120) we see that $\lambda_2^*>\lambda_1$. If $u \in C_0^1(\overline{Z})$ is an eigenfunction corresponding to $\lambda_2^*>0$, u must change sign on Z (see Proposition 4.3.45). Hence $\mathcal{N}(\lambda_2^*) \geq 2$ and so $\lambda_2 \leq \lambda_{\mathcal{N}(\lambda_2^*)}$. On the other hand from Proposition 4.3.54 we have that $\lambda_{\mathcal{N}(\lambda_2^*)} \leq \lambda_2^*$ and so from (4.120) we conclude that $\lambda_2^* = \lambda_2$. $\square$

Next we investigate the dependence of λ_1 and λ_2 on the weight function m. Recall that $m \in L^\infty$ and $|\{m>0\}|_N>0$. In what follows in order to stress the dependence of λ_1 and λ_2 on the weight function m, we write $\lambda_1(m)$ and $\lambda_2(m)$.

The first proposition follows at once from (4.111).

PROPOSITION 4.3.57 *If $m_1, m_2\in L^\infty(Z)$, $m_1(z)\leq m_2(z)$ a.e. on Z with strict inequality on a set of positive measure and $|\{m_1>0\}|_N>0$, then $\lambda_1(m_2)<\lambda_1(m_1)$.*

For the second eigenvalue we have the following result corresponding to the monotone dependence on the weight m. In what follows $Z_m^+ = \{m>0\}$.

PROPOSITION 4.3.58 *If $m_1, m_2 \in L^\infty(Z)$, $m_1(z) \leq m_2(z)$ a.e. on Z, $|Z_m^+|_N > 0$, and $m_1(z) < m_2(z)$ a.e. on $Z_{m_1}^+$, then $\lambda_2(m_2) < \lambda_2(m_1)$.*

PROOF: Let u_2 be an eigenfunction corresponding to $\lambda_2 > 0$ of $\left(-\Delta_p, W_0^{1,p}(Z), m_1\right)$. We know that u_2 changes sign (see Proposition 4.3.45). So we can define

$$v_1(z) = \begin{cases} \frac{u_2^+(z)}{(\frac{1}{p}\int_Z m_1|u_2^+|^p dz)^{1/p}} & \text{if } z \in \{u_2 > 0\} \\ 0 & \text{if otherwise} \end{cases} \quad \text{and}$$

$$v_2(z) = \begin{cases} \frac{u_2^-(z)}{(\frac{1}{p}\int_Z m_1|u_2^-|^p dz)^{1/p}} & \text{if } z \in \{u_2 < 0\} \\ 0 & \text{if otherwise} \end{cases}.$$

Evidently $v_1, v_2 \in W_0^{1,p}(Z)$. Let $Y_2 = \operatorname{span}\{v_1, v_2\}$ and $K = \left\{y \in Y_2 : (1/p)\int_Z m_1|y|^p dz = -2c_2\right\}$ where $c_2 = -1/\left(4\lambda_2^2(m_1)\right)$. Clearly K is compact and since Y_2 is isomorphic to $\mathbb{R}^2$, K can be identified with the boundary of the unit ball in $\mathbb{R}^2$. Therefore from Proposition 4.2.15 we have $\gamma(K) = 2$. Note that $|\{m_2 > 0\} \cap \{u_2 > 0\}|_N > 0$ and $|\{m_1 < 0\} \cap \{u_2 < 0\}|_N > 0$. So

$$\int_{\{m_1>0\}\cap\{u_2>0\}} m_1|v_k|^2 dz < \int_{\{m_1>0\}\cap\{u_2>0\}} m_2|v_k|^2 dz \quad \text{for } k = 1, 2$$

$$\text{and} \int_{\{m_1<0\}\cap\{u_2<0\}} m_1|v_k|^2 dz < \int_{\{m_1<0\}\cap\{u_2<0\}} m_2|v_k|^2 dz \quad \text{for } k = 1, 2.$$

It follows that

$$\begin{aligned}
&\varphi_{m_1}(v) = \xi(v)^2 - \eta_{m_1}(v) = c_2(m_1) \qquad \text{for all } v \in K, \\
\Rightarrow\ &\varphi_{m_2}(v) < \varphi_{m_1}(v) = c_2(m_1) \qquad \text{for all } v \in K, \\
\Rightarrow\ &\max_{v\in K} \varphi_{m_2}(v) < c_2(m_1), \\
\Rightarrow\ &c_2(m_2) < c_2(m_1), \\
\Rightarrow\ &\lambda_2(m_2) < \lambda_2(m_1).
\end{aligned}$$

□

We can have another sequence of variational eigenvalues of $\left(-\Delta_p, W_0^{1,p}(Z)\right)$ (i.e., now we take $m = 1$). We consider the set $B_1^{L^p(Z)} = \{x \in L^p(Z) : \|x\|_p = 1\}$ and we set

$$M = W_0^{1,p}(Z) \cap \partial B_1^{L^p(Z)}.$$

This is a C^1-manifold (if $p \geq 2$ it is a $C^{1,1}$-manifold). For such manifolds we still have the deformation theorem (see Theorem 4.1.19). The result is due to Ghoussoub [262, p. 55], where the reader can find its proof.

THEOREM 4.3.59 *If M is a complete, connected, C^1-Finsler manifold (see Definition 4.2.24), $\widehat{\varphi} \in C^1(M)$, it satisfies the PS-condition, $c \in \mathbb{R}$ is a regular value of $\widehat{\varphi}$, and $\varepsilon_0 > 0$, then there exist $0 < \varepsilon < \varepsilon_0$ and a continuous one-parameter family of homeomorphisms $h : [0,1] \times M \longrightarrow M$ such that*

(a) *$h(0,x) = x$ and $|\widehat{\varphi}(x) - c| \geq \varepsilon_0 \Rightarrow h(t,x) = x$ for all $t \in [0,1]$.*

(b) *For any* $x \in M$ $t \longrightarrow \widehat{\varphi}\big(h(t,x)\big)$ *is nonincreasing.*
(c) $h(1, \widehat{\varphi}^{\,c+\varepsilon}) \subseteq \widehat{\varphi}^{\,c-\varepsilon}$.
(d) *If* $\widehat{\varphi}$ *is even, then for all* $t \in [0,1]$ $h(t,\cdot)$ *is odd.*

We consider the functional $\widehat{\varphi}\colon W_0^{1,p}(Z) \longrightarrow \mathbb{R}$ defined by

$$\widehat{\varphi}(x) = \|Dx\|_p^p \qquad \text{for all } x \in W_0^{1,p}(Z).$$

We restrict $\widehat{\varphi}$ on the C^1-manifold M; that is, we consider $\widehat{\varphi}\big|_M$. Then $\widehat{\varphi}\big|_M \in C^1(M)$. Note that

$$\big(\widehat{\varphi}\big|_M\big)'(x) = p\big(A(x) - \widehat{\varphi}(x)|x|^{p-2}x\big) \qquad \text{for all } x \in M.$$

PROPOSITION 4.3.60 $\widehat{\varphi}\big|_M$ *satisfies the PS-condition.*

PROOF: Suppose that $\{x_n\}_{n\geq 1} \subseteq M$ is a sequence such that $|\widehat{\varphi}(x_n)| \leq c$ for some $c>0$, all $n\geq 1$, and $\big(\widehat{\varphi}\big|_M\big)'(x_n) = p\big(A(x_n) - \widehat{\varphi}(x_n)|x_n|^{p-2}x_n\big) \longrightarrow 0$ in $W^{-1,p'}(Z)$. Evidently $\{x_n\}_{n\geq 1} \subseteq W_0^{1,p}(Z)$ is bounded and so we may assume that $x_n \xrightarrow{w} x$ in $W_0^{1,p}(Z)$ and $x_n \longrightarrow x$ in $L^p(Z)$. Also we may assume that $\widehat{\varphi}(x_n) \longrightarrow \eta$ in $\mathbb{R}$. Note that $\widehat{\varphi}(x_n)\int_Z |x_n|^{p-2}x_n(x_n - x)dz \longrightarrow 0$ and so $\langle A(x_n), x_n - x\rangle \longrightarrow 0$, which by virtue of Proposition 4.3.41 implies that $x_n \longrightarrow x$ in $W_0^{1,p}(Z)$. ☐

Let $\mathcal{D}_n = \{C \subseteq M : C$ is compact, symmetric, and there exists $h \in C(S^{n-1}, C)$, odd, surjective$\}$, $n\geq 1$ and $S^{n-1} = \{y \in \mathbb{R}^n : \|y\| = 1\}$. We define

$$\mu_n = \inf_{C\in\mathcal{D}^n} \sup_{x\in C} \widehat{\varphi}(x), \quad n\geq 1. \tag{4.121}$$

Then $\mu_n \in \mathbb{R}$ and we have the following counterpart of Theorem 4.3.52.

THEOREM 4.3.61 *For every integer* $n\geq 1$, μ_n *is a critical value of* $\widehat{\varphi}\big|_M$.

PROOF: We argue indirectly. So suppose that μ_n is a regular value of $\widehat{\varphi}\big|_M$. We apply Theorem 4.3.59 with $\varepsilon_0 = 1$ and $c = \mu_n$. We can find $0 < \varepsilon < \varepsilon_0$ and a one-parameter family of homeomorphisms $h(t,\cdot) : M \longrightarrow M$, $t \in [0,1]$, which satisfy statements (a)→(d) of Theorem 4.3.59. From (4.121) we see that we can find $C \in \mathcal{D}_n$ such that $\sup_{x\in C} \widehat{\varphi}(x) \leq \mu_k + \varepsilon$. If $\psi \in C(S^{n-1}, C)$ is odd and surjective, then so is $h\big(1, \psi(\cdot)\big) : S^{n-1} \longrightarrow h(1,C)$ (see Theorem 4.3.59(d)). Hence $h(1,C) \in \mathcal{D}_n$ and

$$\sup_{x\in h(1,C)} \widehat{\varphi}(x) \leq \mu_k - \varepsilon \quad \text{(see Theorem 4.3.59(c))},$$

a contradiction. ☐

Because of this proposition we have $v_n \in M$ such that

$$\begin{aligned} &\big(\widehat{\varphi}\big|_M\big)'(v_n) = 0, \\ \Rightarrow\ &(\mu_n, v_n) \text{ is an eigenelement of } \big(-\Delta_p, W_0^{1,p}(Z)\big). \end{aligned}$$

Note that $\mathcal{D}_n$ is a subfamily of the family of sets with genus $n\geq 1$, we infer that

$$\lambda_n \le \mu_n \qquad \text{for all } n\ge 1. \tag{4.122}$$

Clearly $\lambda_1=\mu_1$ and also $\lambda_2=\mu_2$. To see this let u_2 be a normalized eigenfunction corresponding to the eigenvalue λ_2. We know that u_2 must change sign (see Proposition 4.3.45). So $u_2^+ \neq 0$, $u_2^- \neq 0$. Set $C = \{\xi u_2^+ + \eta u_2^- : \xi,\eta \in \mathbb{R}$ and $|\xi|^p\|u_2^+\|_p + |\eta|^p\|u_2^-\|_p = 1\}$. Clearly $C \in \mathcal{D}_2$ and for any $u \in C$, we have $\widehat{\varphi}(u) = \lambda_2$. Therefore from (4.121) it follows that $\mu_2 \le \lambda_2$. This combined with (4.122) implies that $\mu_2 = \lambda_2$. For $n>2$ we do not know that if $\lambda_n = \mu_n$.

We return to the nonlinear eigenvalue problem (4.85) and we introduce the C^1-manifold

$$M_0 = \{x \in W_0^{1,p}(Z) : \int_Z m|x|^p dz = 1\}$$

and then the open set U in M_0, defined by

$$U = \{x \in M_0 : \|Dx\|_p^p < \lambda_2\}.$$

Let $u_1 \in M_0 \cap \operatorname{int} C_0^1(\overline{Z})_+$ be the eigenfunction corresponding to $\lambda_1>0$. Evidently $u_1, -u_1 \in U$.

LEMMA 4.3.62 *u_1 and $-u_1$ belong to different connected components of U.*

PROOF: We proceed by contradiction. So suppose that u_1 and $-u_1$ belong to the same connected component of U. Because such a component is also path-connected, we can find a continuous curve $\gamma\colon [-1,1] \longrightarrow U$ joining u_1 and $-u_1$. Set $E = \gamma([-1,1])$ and $C_0 = E \cup (-E)$. Clearly $0 \notin C_0, C_0$ is compact and symmetric, and $\gamma(C_0)>1$. Therefore $C_0 \in \mathcal{S}_2$ and we have

$$\begin{aligned} &\max\big\{\|Dx\|_p^p : x \in C_0\big\}<\lambda_2 \qquad \text{(recall the definition of } U),\\ \Rightarrow\ &\inf_{C\in\mathcal{S}_2}\max_{x\in C}\|Dx\|_p^p<\lambda_2. \end{aligned}$$

Then by virtue of Theorem 4.3.52 we have $\lambda_2<\lambda_2$, a contradiction. □

Let U^+ be the connected component of U containing u_1 and U^- the connected component of U containing $-u_1$. Evidently $U^+ = -U^-$. Set

$$E_+ = \{tU^t : t>0\} \ \ E_- = -E_+ \qquad \text{and} \qquad E = E_+ \cup E_-.$$

We have

$$\|Dx\|_p^p < \lambda_2 \int_Z m|x|^p dz \qquad \text{for all } x \in E \tag{4.123}$$

$$\text{and} \quad \|Dx\|_p^p = \lambda_2 \int_Z m|x|^p dz \qquad \text{for all } x \in \partial E. \tag{4.124}$$

Here ∂E is the boundary of E in $W_0^{1,p}(Z)$. For $r>0$, let

$$(\partial E)_r = \{x\in\partial E : \|x\| = r\}.$$

Let u_2 be a normalized eigenfunction of $\big(-\Delta_p, W_0^{1,p}(Z), m\big)$ corresponding to $\lambda_2>0$. We set

$$Y_1 = \mathbb{R}u_1, \quad Y_2 = \operatorname{span}\{u_1,u_2\}$$

$$\text{and} \quad D = \Big\{x \in W_0^{1,p}(Z) : \|Dx\|_p^p = \lambda_2 \int_Z m|x|^p dz\Big\}.$$

Note that (4.124) implies that $\partial E \subseteq D$.

PROPOSITION 4.3.63 *If $u \in \mathbb{R}_+ u_1, u \neq 0, S=[-u,u]=\{v = t(-u) + (1-t)u : t \in [0,1]\}$, and $S_0 = \{-u, u\}$, then the pair $\{S_0, S\}$ and D link in $W_0^{1,p}(Z)$ via $\gamma^* = I$ (see Definition 4.1.20).*

PROOF: Clearly $S_0 \cap D = \emptyset$. Also let $\gamma \in C(S, W_0^{1,p}(Z))$ be such that $\gamma|_{S_0} = I|_{S_0}$. Note that E has two connected components E_+ and $E_- = -E_+$ (see Lemma 4.3.62 and recall the definition of E). Then if for $x = t(-u) + (1-t)u \in S$, $t \in [0,1]$, we set $\widehat{\gamma}(t) = \gamma(x)$, then $\widehat{\gamma}\colon [0,1] \longrightarrow W_0^{1,p}(Z)$ is a continuous curve joining $-u$ and u and so we must have $\gamma(S) \cap \partial E \neq \emptyset$. Because $\partial E \subseteq D$, we obtain that $\gamma(S) \cap D \neq \emptyset$. This proves that the pair $\{S_0, S\}$ and D link in $W_0^{1,p}(Z)$ via $\gamma^* = I$. □

PROPOSITION 4.3.64 *If $0 < r < R < \infty$, $\widehat{u}_1 = (1/\lambda_1^{1/p})\, u_1$, $\widehat{u}_2 = (1/\lambda_2^{1/p})\, u_2$ and*

$$
\begin{aligned}
C &= \{u = t\widehat{u}_1 + s\widehat{u}_2 : \|u\| \leq R,\ s \geq 0\},\\
C_0 &= \{u = t\widehat{u}_1,\ \|u\| = |t| \leq R\} \cup \{u \in C : \|u\| = R\},\\
\text{and}\quad D_r &= \{u \in D : \|u\| = r\},
\end{aligned}
$$

then $\{C_0, C\}$ and D_r link in $W_0^{1,p}(Z)$ through $\gamma^* = I$.

PROOF: Clearly $C_0 \cap D_r = \emptyset$. Let $\gamma\colon C \longrightarrow W_0^{1,p}(Z)$ be a continuous map that $\gamma|_{C_0} = I|_{C_0}$. Let $d_1 = d(\widehat{u}_1, \partial K)$ and define the map $k\colon W_0^{1,p}(Z) \longrightarrow Y_2$ by

$$
k(u) = \begin{cases} \min\{d(u,\partial E), Rd_1\}\widehat{u}_1 + (\|u\| - r)\widehat{u}_2 & \text{if } u \notin E_- = -E_+ \\ -\min\{d(u,\partial E), Rd_1\}\widehat{u}_1 + (\|u\| - r)\widehat{u}_2 & \text{if } u \in E_- \end{cases}.
$$

Evidently k is continuous. If $g = k \circ \gamma \in C(C, Y_2)$, then we can easily check that $0 \notin g(C_0)$ and $d(g, \text{int}\, C, 0) = 1$ (the int C is taken in Y_2). So we can find $u \in C$ such that $g(u) = k(\gamma(u)) = 0$. This implies that $\gamma(u) \in \partial E$ and also $\|\gamma(u)\| = r$. Hence $\gamma(u) \in (\partial E)_r$ and $\gamma(C) \cap (\partial E)_r \neq \emptyset$. Because $(\partial E)_r \subseteq D_r$, we have $\gamma(C) \cap D_r \neq \emptyset$ and so the sets $\{C_0, C\}$ and D_r link in $W_0^{1,p}(Z)$ through $\gamma^* = I$. □

Finally before turning our attention to the Neumann nonlinear eigenvalue problem for the negative p-Laplacian, let us mention what happens if in problem (4.85) the weight function m does not belong in $L^\infty(Z)$, but we have $m \in L^s(Z)$ for some $s > N/p$ if $1 < p \leq N$ and $m \in L^1(Z)$ if $p > N$. Details can be found in Cuesta [168]. In what follows we always assume $m^+ \neq 0$.

If this is the case with the weight function m, then for $u \in W_0^{1,p}(Z)$ an eigenfunction corresponding to an eigenvalue λ, we can still say that it belongs in $L^\infty(Z)$ but we no longer have that $u \in C_0^1(\overline{Z})$.

PROPOSITION 4.3.65 *If $m \in L^s(Z)$ for some $s > N/p$ if $1 < p \leq N$ and $m \in L^1(Z)$ if $p > N$ and (λ, u) an eigenelement of $(-\Delta_p, W_0^{1,p}(Z), m)$, then $u \in L^\infty(Z)$ and u is locally Hölder continuous; that is, there exists $a \in (0,1)$ which depends on $(p, N, \lambda, \|m\|_s)$ such that for every $Z' \subset\subset Z$, we can find $c > 0$, c depending on $(p, N, \lambda, \|m\|_s, d(Z', \partial Z))$, such that*

$$
|u(z) - u(z')| \leq c\|u\|_\infty \|z - z'\|_{\mathbb{R}^N}^a \qquad \text{for all } z, z' \in Z'.
$$

The lack of global regularity of the eigenfunction u is a source of difficulties. Nevertheless we can still conduct an analysis of the eigenvalue problem and produce for $\lambda_2 > \lambda_1 > 0$ (the first two eigenvalues) results similar to the ones we had for the case when $m \in L^\infty(Z)$ with $|\{m>0\}|_N>0$.

THEOREM 4.3.66 *If $m \in L^s(Z)$ with $s > N/p$ if $1 < p \leq N$ and $m \in L^1(Z)$ if $p > N$, then problem (4.85) has a smallest positive eigenvalue $\lambda_1 > 0$, which is simple and isolated and the corresponding eigenfunction does not change sign on Z; moreover*

$$\lambda_1 = \inf\Big\{\frac{\|Du\|_p^p}{\int_Z m|u|^p dz} : u \in W_0^{1,p}(Z),\ u \neq 0\Big\}$$

and the infimum is attained at the normalized principal eigenfunction $u_1 \in L^\infty(Z) \cap C^{1,\gamma}_{\mathrm{loc}}(Z)$ for some $\gamma \in (0,1)$.

In deriving Theorem 4.3.66 the analysis differs from that of the $m \in L^\infty(Z)$ case in two respects. First, in the proof of Proposition 4.3.44, we used the function

$$J(u,v) = \Big\langle A(u), \frac{u^p - v^p}{u^{p-1}}\Big\rangle + \Big\langle A(v), \frac{v^p - u^p}{v^{p-1}}\Big\rangle \tag{4.125}$$

for all $(u,v) \in D(J) = \big\{(u,v) \in W_0^{1,p}(Z) \times W_0^{1,p}(Z) : u,v \geq 0,\ (u/v),\ (v/u) \in L^\infty(Z)\big\}$. We exploited the fact that

$$J(u,v) \geq 0 \quad \text{for all } (u,v) \in D(J) \quad \text{and} \quad J(u,v) = 0 \text{ if and only if } u = \beta v,\ \beta \in \mathbb{R}. \tag{4.126}$$

This is sometimes called in the literature the *Diaz–Saa inequality*. Note that if u,v are eigenfunctions corresponding to $\lambda_1>0, u,v \geq 0$, then $(u,v) \in D(J)$ (see Proposition 4.3.39). This is no longer available in the present $m \in L^s(Z)$ setting due to the lack of global regularity of the eigenfunctions. Then (4.126) is replaced by *Picone's identity*, proved by Allegretto–Huang [11].

THEOREM 4.3.67 *If $u,v \in C(Z)$, are differentiable a.e. on Z, $u \geq 0$, $v>0$, and*

$$L(u,v) = \|Du\|^p + (p-1)\frac{u^p}{v^p}\|Dv\|^p - p\frac{u^p}{v^p}\|Dv\|^{p-2}(Dv, Du)_{\mathbb{R}^N}$$
$$R(u,v) = \|Du\|^p - \|Dv\|^{p-2}\big(D\big(\frac{u^p}{v^{p-1}}\big), Dv\big)_{\mathbb{R}^N},$$

then

(a) $L(u,v) = R(u,v)$.
(b) $L(u,v) \geq 0$ *a.e. on* Z.
(c) $L(u,v) = 0$ *a.e. on Z if and only if $u = kv$ for some $k \in \mathbb{R}$.*

Second, in obtaining Proposition 4.3.39, which justified the use of the function $J(u,v)$ (see (4.125)) and of (4.126), we used the nonlinear strong maximum principle (see Theorem 4.3.37). This result is not available in the $m \in L^s(Z)$ case due to the lack of global regularity of the eigenfunctions. So the nonlinear strong maximum principle (Theorem 4.3.37) is replaced by the following nonlinear *Harnack's inequality*, due to Serrin [548].

THEOREM 4.3.68 *If $u \in W_0^{1,p}(Z) \cap L^\infty(Z)$ is a nonnegative eigenfunction of $\big(-\Delta_p, W_0^{1,p}(Z), m\big)$ and for some $z_0 \in Z$ and $r>0$ we have $B_{3r}(z_0) \subseteq Z$, then for some $c>0$ depending on (p, N, r, λ, m, Z) we have*

$$\max_{\overline{B}_r(z_0)} u \leq c \min_{\overline{B}_r(z_0)} u. \tag{4.127}$$

REMARK 4.3.69 Harnack's inequality (see (4.127)), says that the values of nonnegative eigenfunctions of $\big(-\Delta_p, W_0^{1,p}(Z), m\big)$ are comparable, at least in any subregion away from the boundary ∂Z.

Also as before for the second eigenvalue $\lambda_2>0$, we have the following.

THEOREM 4.3.70 *If $m \in L^s(Z)$ with $s > N/p$ if $1<p \leq N$ and $m \in L^1(Z)$ if $p > N$, then $\lambda_2^* = \lambda_2$.*

There is an alternative variational characterization of $\lambda_2>0$ due to Cuesta–De Figueiredo–Gossez [167].

THEOREM 4.3.71 *If $S = W_0^{1,p}(Z) \cap \partial B_1^{L^p(Z)}$ where $\overline{B}_1^{L^p(Z)} = \{x \in L^p(Z) : \|x\|_p \leq 1\}$ and*

$$\Gamma_0 = \big\{\gamma_0 \in C([-1,1], S) : \gamma_0(-1) = -u_1, \gamma_0(1) = u_1\big\},$$

then $\lambda_2 = \inf\ \max\big[\|Dx\|_p^p : x \in \gamma([-1,1])\big]$.

Now we turn our attention to the Neumann nonlinear eigenvalue problem.

$$\left\{\begin{array}{l} -\operatorname{div}\big(\|Dx(z)\|^{p-2}Dx(z)\big) = \lambda|x(z)|^{p-2}x(z) \ \text{ a.e. on } Z, \\ \\ \frac{\partial x}{\partial n_p} = 0 \quad \text{on } \partial Z,\ \lambda \in \mathbb{R} \end{array}\right\}. \tag{4.128}$$

Here for $x \in C^1(\overline{Z})$, $(\partial x/\partial n_p)(z) = \|Dx(z)\|^{p-2}\big(Dx(z), n(z)\big)_{\mathbb{R}^N}$, $z \in \partial Z$, where $n(z)$ is the outward unit normal on ∂Z. First we recall a nonlinear Green's formula, which is a useful tool in the analysis of problem (4.128). So consider the following space,

$$V^q(Z, \operatorname{div}) = \big\{v \in L^q(Z, \mathbb{R}^N) : \operatorname{div} v \in L^q(Z)\big\}, \quad 1<q<+\infty.$$

On $V^q(Z, \operatorname{div})$ we consider the norm

$$\|v\|_{V^q(Z,\operatorname{div})} = \big(\|v\|^q_{L^q(Z,\mathbb{R}^N)} + \|\operatorname{div} v\|_q^q\big)^{1/q}.$$

Furnished with this norm $V^q(Z, \operatorname{div})$ becomes a separable, reflexive Banach space in which $C^\infty(\overline{Z}, \mathbb{R}^N)$ is dense. The next theorem extends to a nonlinear setting the classical Green's identity. For its proof we refer to Casas–Fernadez [132] and Kenmochi [346].

THEOREM 4.3.72 *If $1<p<\infty$ and $(1/p)+(1/p') = 1$, then there exists a unique bounded linear operator*

$$\gamma_n : V^{p'}(Z, \operatorname{div}) \longrightarrow W^{-(1/p'),p'}(\partial Z) = W^{(1/p'),p}(\partial Z)^*$$

such that $\gamma_n(v) = (v, n)_{\mathbb{R}^N}$ *for all* $v \in C^\infty(\overline{Z}, \mathbb{R}^N)$ *and*

$$\int_Z u \operatorname{div} v dz + \int_Z (Du, v)_{\mathbb{R}^N} dz = \langle \gamma_0(u), \gamma_n(v) \rangle_{\partial Z}$$

for all $(u, v) \in W^{1,p}(Z) \times V^{p'}(Z, \operatorname{div})$, *where* γ_0 *is the usual trace map on* $W^{-1,p'}(Z)$ *and* $\langle \cdot, \cdot \rangle_{\partial Z}$ *denotes the duality brackets for the pair*

$$\left(W^{(1/p'),p}(\partial Z), W^{-(1/p'),p'}(\partial Z)\right).$$

If for $u \in W^{1,p}(Z)$, we set $\Delta_p u = \operatorname{div}(\|Du\|^{p-2} Du)$, then from Theorem 4.3.71, we obtain the following nonlinear generalization of the so-called *second Green's identity*.

THEOREM 4.3.73 *If* $1 < p < \infty$, $(1/p) + (1/p') = 1$, $u \in W^{1,p}(Z)$, *and* $\Delta_p u \in L^{p'}(Z)$, *then there exists a unique element of* $W^{-(1/p'),p'}(\partial Z)$, *which by extension we denote by* $\partial u / \partial n_p$ *satisfying*

$$\int_Z (\Delta_p u) v dz + \int_Z \|Du\|^{p-2} (Du, Dv)_{\mathbb{R}^N} dz = \left\langle \frac{\partial u}{\partial n_p}, \gamma_0(v) \right\rangle_{\partial Z}.$$

Clearly $\lambda = 0$ is an eigenvalue of problem (4.128) $\big($i.e., of $\left(-\Delta_p, W^{1,p}(Z)\right)\big)$, with the constant functions as eigenfunctions (i.e., the corresponding eigenspace is $\mathbb{R}$). In fact we can say more.

PROPOSITION 4.3.74 $\lambda = 0$ *is the first eigenvalue of* $\left(-\Delta_p, W^{1,p}(Z)\right)$ *and it is isolated and simple.*

PROOF: First note that problem (4.128) cannot have negative eigenvalues. Indeed, if $\lambda < 0$ is an eigenvalue for (4.128) with corresponding eigenfunction $u \in W^{1,p}(Z)$, then

$$\|Du\|_p^p = \lambda \|u\|_p^p \qquad \text{(see Theorem 4.3.73)},$$

which is a contradiction because $\|u\|_p > 0$.

The simplicity of $\lambda = 0$ is a direct consequence of the fact that

$$0 = \inf \left\{ \frac{\|Dx\|_p^p}{\|x\|_p^p} : x \in W^{1,p}(Z),\ x \neq 0 \right\}.$$

Finally suppose that $\lambda = 0$ is not isolated. We can find a sequence $\{\lambda_n\}_{n \geq 1}$ of nonzero eigenvalues such that $\lambda_n \downarrow 0$. Let $\{u_n\}_{n \geq 1} \subseteq W^{1,p}(Z)$ be a sequence of corresponding eigenfunctions. We know that $u_n \in C^1(\overline{Z})$ for all $n \geq 1$ (see Remark 4.3.36). We may assume that $\|u_n\|_p = 1$. Hence

$$\lambda_n = \|Du_n\|_p^p \longrightarrow 0 \quad \text{as } n \to \infty,$$
$$\Rightarrow \{u_n\}_{n \geq 1} \subseteq W^{1,p}(Z) \text{ is bounded.}$$

So we may assume that

$$u_n \xrightarrow{w} u \quad \text{in} \quad W^{1,p}(Z) \quad \text{and} \quad u_n \longrightarrow u \quad \text{in} \quad L^p(Z).$$

We have $\|u\|_p = 1$ and $\|Du\|_p = 0$. So

$$u = \pm \frac{1}{|Z|_N^{1/p}}. \tag{4.129}$$

Using as test function $v = 1$ in (4.128), we obtain

$$\int_Z |u_n|^{p-2} u_n dz = 0.$$

Passing to the limit as $n \to \infty$, we have

$$\int_Z |u|^{p-2} u dz = 0,$$

which contradicts (4.129). □

Next we characterize the first nonzero eigenvalue of $\left(-\triangle_p, W_0^{1,p}(Z)\right)$. Suppose that $\lambda > 0$ is an eigenvalue of (4.128) with corresponding eigenfunction $u \in C^1(\overline{Z})$. Integrate (4.128) over Z and then apply Theorem 4.3.73 (the nonlinear second Green's identity). We obtain

$$\int_Z |u|^{p-2} u dz = 0.$$

So we are naturally led to the consideration of the following nonempty, pointed, closed, and symmetric cone

$$C(p) = \left\{x \in W^{1,p}(Z) : \int_Z |x(z)|^{p-2} x(z) dz = 0\right\}. \tag{4.130}$$

We consider the set $C_1(p) = C(p) \cap \partial B_1^{L_p(Z)}$; that is,

$$C_1(p) = \left\{x \in W^{1,p}(Z) : \int_Z |x(z)|^{p-2} x(z) dz = 0,\ \|x\|_p = 1\right\}. \tag{4.131}$$

Let $\vartheta_p \colon W^{1,p}(Z) \longrightarrow \mathbb{R}$ be the strictly convex C^1-map defined by

$$\vartheta_p(x) = \|Dx\|_p^p,\ x \in W^{1,p}(Z).$$

We consider the following minimization problem

$$\lambda_1(p) = \inf\left\{\vartheta_p(x) : x \in C_1(p)\right\}. \tag{4.132}$$

PROPOSITION 4.3.75 *Problem (4.132) has a value $\lambda_1 = \lambda_1(p) > 0$ that is attained in $C_1(p)$.*

PROOF: Let $\{x_n\}_{n\geq 1} \subseteq C_1(p)$ be a minimizing sequence for problem (4.130) (i.e., $\vartheta_p(x_n) \downarrow \lambda_1$). Evidently $\{x_n\}_{n\geq 1} \subseteq W^{1,p}(Z)$ is bounded and so we may assume that

$$x_n \xrightarrow{w} x \quad \text{in} \quad W^{1,p}(Z),\ x_n \longrightarrow x \quad \text{in} \quad L^p(Z),\ x_n(z) \longrightarrow x(z) \text{ a.e. on } Z$$
$$\text{and} \quad |x_n(z)| \leq k(z) \text{ a.e. on } Z, \quad \text{for all } n \geq 1 \text{ and with } k \in L^p(Z).$$

Using these relations and the Lebesgue dominated convergence theorem, in the limit as $n \to \infty$, we obtain

$$\int_Z |x(z)|^{p-2}x(z)\mathrm{d}z = 0 \text{ and } \|x\|_p = 1 \text{ (i.e., } x \in C_1(p); \quad \text{see (4.131)).}$$

Also exploiting the weak lower semicontinuity of the norm functional in a Banach space we have $\|Dx\|_p^p \leq \lambda_1$, hence $\|Dx\|_p^p = \lambda_1$. Because $x_1 \in C(p)$, it follows that x is not constant and so $\lambda_1 > 0$. □

An interesting consequence of Proposition 4.3.74, is the following Poincaré–Wirtinger type inequality.

COROLLARY 4.3.76 *If $x \in C(p)$ (see (4.130)), then $\lambda_1 \|x\|_p^p \leq \|Dx\|_p^p$.*

In fact for $p \geq 2$, we show that $\lambda_1 > 0$ is the first nonzero eigenvalue of $(-\Delta_p, W_0^{1,p}(Z))$.

PROPOSITION 4.3.77 *If $p \geq 1$, then $\lambda_1 > 0$ is the first nonzero eigenvalue of $(-\Delta_p, W_0^{1,p}(Z))$.*

PROOF: Let $x \in C_1(p)$ be a solution of problem (4.132) (see Proposition 4.3.74). By the Lagrange multiplier rule (see Theorem 2.2.10), we can find $a, b, c \in \mathbb{R}$, not all of them equal to zero, such that for all $y \in W^{1,p}(Z)$, we have

$$ap \int_Z \|Dx\|^{p-2}(Dx, Dy)_{\mathbb{R}^N} \mathrm{d}z + bp \int_Z |x|^{p-2}xy\mathrm{d}z + c(p-1) \int_Z |x|^{p-2}y\mathrm{d}z = 0. \tag{4.133}$$

Let $y = c$ and recall that because $x \in C_1(p)$, we have $\int_Z |x|^{p-2}x\mathrm{d}z = 0$ (see (4.131)). So

$$c^2(p-1) \int_Z |x|^{p-2}\mathrm{d}z = 0,$$
$$\Rightarrow \quad c = 0.$$

Using this in (4.133), we obtain

$$ap \int_Z \|Dx\|^{p-2}(Dx, Dy)_{\mathbb{R}^N} \mathrm{d}z + bp \int_Z |x|^{p-2}xy\mathrm{d}z = 0 \qquad \text{for all } y \in W^{1,p}(Z).$$

If $a = 0$, then

$$bp \int_Z |x|^{p-2}xy\mathrm{d}z = 0 \qquad \text{for all } y \in W^{1,p}(Z).$$

Select $y = x \in W^{1,p}(Z)$. We have

$$bp\|x\|_p^p = 0 \text{ (i.e., } b = 0),$$

a contradiction, because we cannot have $a = b = c = 0$. Hence $a \neq 0$ and so without any loss of generality, we may assume that $a = 1$. Then we have

$$\int_Z \|Dx\|^{p-2}(Dx, Dy)_{\mathbb{R}^N} \mathrm{d}z + b \int_Z |x|^{p-2}xy\mathrm{d}z = 0 \qquad \text{for all } y \in W^{1,p}(Z).$$

Once again we use $y = x \in W^{1,p}(Z)$. Then

$$\|Dx\|_p^p + b\|y\|_p^p = 0,$$

and $\quad b = -\lambda_1$ because $x \in C_1(p)$ is a solution of (4.132).

Via Theorem 4.3.73, we establish that x solves (4.128) with $\lambda = \lambda_1$. Clearly from the definition of $\lambda_1 > 0$, we see that we cannot find an eigenvalue of $\left(-\Delta_p, W_0^{1,p}(Z)\right)$ in $(0, \lambda_1)$. ☐

REMARK 4.3.78 The eigenfunctions of $-\Delta$ (Dirichlet or Neumann), have the so-called *unique continuation property*, namely if u is an eigenfunction that vanishes on a set of positive Lebesgue measure, then $u \equiv 0$. In fact because the eigenfunctions belong in $C(\overline{Z})$ (recall that we have assumed that $Z \subseteq \mathbb{R}^2$ has a C^2-boundary ∂Z), the unique continuation property can be equivalently restated as follows. If u is a nontrivial eigenfunction, then the set $\{z \in Z : u(z) = 0\}$ has an empty interior. The existence of such property is an open problem for $p \neq 2$. There is a counterexample due to Martio [409], which suggests that the unique continuation property is difficult to extend to the p-Laplacian.

Let us briefly mention the situation with the eigenvalue problems for the scalar ordinary p-Laplacian (i.e., $N = 1$). For details see Drabek–Manasevich [205] and Gasiński–Papageorgiou [259, Section 6.3].

We start with the Dirichlet eigenvalue problem:

$$\left\{ \begin{array}{l} -\left(|x'(t)|^{p-2}x'(t)\right)' = \lambda |x(t)|^{p-2}x(t) \text{ a.e. on } T = [0,b], \\ x(0) = x(b) = 0, \ \lambda \in \mathbb{R}, \ 1<p<\infty \end{array} \right\}. \tag{4.134}$$

Following the $p = 2$ case (linear eigenvalue problem), we set

$$\pi_p = 2\int_0^{(p-1)^{1/p}} \frac{ds}{\left(1 - s^p/(p-1)\right)^{1/p}} = \frac{2\pi(p-1)^{1/p}}{p \sin \frac{\pi}{p}}.$$

Note that $\pi_2 = \pi$. Also define the function $\sin_p : \mathbb{R} \longrightarrow \mathbb{R}$ by

$$\int_0^{\sin_p t} \frac{ds}{\left(1 - s^p/(p-1)\right)^{1/p}} = t \qquad \text{for all } t \in \left[0, \pi_p/2\right]$$

and then extend $\sin_p(\cdot)$ to all of $\mathbb{R}$ in a similar way as for $\sin(\cdot)$. We produce a $2\pi_p$-periodic function.

THEOREM 4.3.79 *Problem (4.134) has a sequence of eigenelements* $\left\{(\lambda_n^D, u_n^D)\right\}_{n\geq 1}$, *where*

$$\lambda_n^D = \left(\frac{n\pi_p}{b}\right)^p \quad \textit{and} \quad u_n^D(t) = \mu \sin_p \frac{n\pi_p}{b} t \quad \mu \in \mathbb{R}, \ n \geq 1.$$

These are all the eigenvalues of the scalar ordinary p-Laplacian with Dirichlet boundary conditions.

Next we consider the Neumann eigenvalue problem:

$$\left\{ \begin{array}{l} -\left(|x'(t)|^{p-2}x'(t)\right)' = \lambda |x(t)|^{p-2}x(t) \text{ a.e. on } T = [0,b], \\ x'(0) = x'(b) = 0, \ \lambda \in \mathbb{R}, \ 1<p<\infty \end{array} \right\}. \tag{4.135}$$

THEOREM 4.3.80 *Problem (4.135) has a sequence of eigenelements* $\{(\lambda_n^N, u_n^N)\}_{n\geq 0}$, *where*

$$\lambda_0^N = 0,\ \lambda_n^N = \lambda_n^D \quad \text{and} \quad u_0^N(t) = c \in \mathbb{R}\setminus\{0\},\ u_n^N(t) = u_n^D\big(t - \frac{b}{2n}\big),$$

for all $t\in T$, $n\geq 1$. *These are all the eigenvalues of the scalar ordinary p-Laplacian with Neumann boundary conditions.*

Finally we deal with the periodic eigenvalue problem:

$$\left\{\begin{array}{l} -\big(|x'(t)|^{p-2}x'(t)\big)' = \lambda|x(t)|^{p-2}x(t) \ \text{ a.e. on } T=[0,b] \\ x(0)=x(b),\ x'(0)=x'(b)=0,\ \lambda\in\mathbb{R},\ 1<p<\infty \end{array}\right\}. \tag{4.136}$$

THEOREM 4.3.81 *Problem (4.136) has a sequence of eigenelements* $\{(\lambda_n^P, u_n^P)\}_{n\geq 0}$, *where*

$$\lambda_0^P = 0,\ \lambda_n^P = \lambda_{2n}^D \quad \text{and} \quad u_n^P(t) = u_{2n}^D(t) \qquad \text{for all } t\in T,\ n\geq 1.$$

These are all the eigenvalues of the scalar ordinary p-Laplacian with periodic boundary conditions.

REMARK 4.3.82 The situation changes if instead of the scalar ordinary p-Laplacian, we consider the vector ordinary p-Laplacian; that is, $-\big(\|x'(t)\|^{p-2}x'(t)\big)'$ with $x\in W^{1,p}\big((0,b),\mathbb{R}^N\big)$. In this case for the Dirichlet problem, Theorem 4.3.79 remains unchanged, but for the periodic problem the spectrum contains more than the sequence $\{\lambda_n^P\}_{n\geq 0}$, for example, the infinite sequence $\{\big(n(2\pi/b)\big)^p\}_{n\geq 1}$. The spectrum of the vectorial ordinary p-Laplacian with periodic boundary conditions is far from being understood.

We also mention the following sharp Poincaré–Wirtinger inequality for the scalar ordinary p-Laplacian.

THEOREM 4.3.83 *If* $u \in W^{1,p}_{\text{per}}(0,b) = \{u\in W^{1,p}(0,b) : u(0)=u(b)\}$ *and* $\int_0^b u(t)dt = 0$, *then* $\lambda_1^P\|u\|_p^p \leq \|u'\|_p^p$.

REMARK 4.3.84 In the definition of $W^{1,p}_{\text{per}}(0,b)$ the evaluations at $t=0$ and $t=b$ make sense because $W^{1,p}(0,b)$ is embedded continuously (in fact compactly) in $C[0,b]$.

4.4 Abstract Eigenvalue Problems

We have seen in the previous section in the context of eigenvalue problems involving the partial p-Laplacian differential operator, that the Ljusternik–Schnirelmann theory (see Section 4.2), can be carried over to C^1-manifolds modelled on a reflexive Banach space. In this section we go beyond the p-Laplacian, consider general nonlinear eigenvalue problems and see how the Ljusternik–Schnirelmann theory can be used to study them.

So let X be a separable reflexive Banach space. By the Troyanski renorming theorem we assume without any loss of generality that both X and X^* are locally uniformly convex. By $\langle\cdot,\cdot\rangle$ we denote the duality brackets for (X, X^*). Let $\varphi, \psi \colon X \longrightarrow \mathbb{R}$ be two C^1-functions. For fixed $\beta > 0$, we consider the nontrivial solutions of the equation (eigenvalue problem)

$$\varphi'(x) = \lambda\psi'(x) \quad \text{with } \varphi(x) = \beta. \tag{4.137}$$

The solutions of (4.137) are contained in the set of critical points of the functional ψ on the surface $M_\beta = \{x \in X : \varphi(x) = \beta\}$. We assume that β is a regular value of φ and so M_β is a C^1-manifold of codimension 1 modelled on the Banach space X.

Now we introduce the precise hypotheses on the functionals φ and ψ.

H: (i) $\varphi, \psi \in C^1(X)$ are even (hence φ', ψ' are odd potential operators), $\varphi(0) = \psi(0) = 0$, and $\varphi'(0) = 0$.
(ii) φ is weakly coercive; that is, $\varphi(x) \longrightarrow +\infty$ as $\|x\| \longrightarrow \infty$.
(iii) For every $x \neq 0$, the function $s \longrightarrow \langle\varphi'(sx), x\rangle$ is strictly increasing on $\mathbb{R}_+$.
(iv) If $x_n \xrightarrow{w} x$ in X, $\varphi'(x_n) \longrightarrow y$ in X^*, then $x_n \longrightarrow x$ in X and φ' is bounded.
(v) For every $x \in X$, $\psi'(x) \in X^*$ is compact and $\psi'(x) = 0$ if and only if $x = 0$.

We have the following existence result concerning problem (4.137).

THEOREM 4.4.1 *If hypotheses (H) hold, then problem (4.137) has a sequence of solutions $(\lambda_n, x_n) \in \mathbb{R} \times M_\beta$ such that*

$$|\lambda_n| \longrightarrow +\infty \quad \textit{and} \quad x_n \xrightarrow{w} 0 \quad \textit{in } X.$$

PROOF: Note that hypotheses (H)(i) and (ii) imply that $M_\beta \neq \emptyset$ and it is bounded. Also for $x \in M_\beta$ we have

$$\begin{aligned} &\langle\varphi'(x), x\rangle > \int_0^1 \langle\varphi'(sx), x\rangle\, \mathrm{d}s = \varphi(x) = \beta > 0, \\ \Rightarrow\ &\|\varphi'(x)\|_{X^*} > \frac{\beta}{\sup\limits_{x \in M_\beta} \|x\|} > 0. \end{aligned} \tag{4.138}$$

Therefore β is a regular value of φ and so M_β is a C^1-manifold of codimension 1 modelled on the Banach space X.

The manifold M_β is star-shaped with respect to the origin, because every ray through the origin $\{rx \in X : r \in \mathbb{R},\ \|x\| = 1\}$, intersects M_β at precisely two points $\pm r(x)x$ (see (4.138)). Therefore there is a bijection from $P^\infty(X) = \partial B_1(0)/\mathbb{Z}_2$ (see Example 4.2.12(b)) onto $M_\beta/\mathbb{Z}_2$ defined by

$$\vartheta(x) = r(x)x \qquad \text{for all } x \in P^\infty(X). \tag{4.139}$$

Note that for $x \in M_\beta$, we have

$$\frac{\mathrm{d}}{\mathrm{d}s}\varphi(sx) = \langle\varphi'(x), x\rangle > 0 \qquad \text{for all } s > 0 \quad \text{(see hypothesis (H)(iii))}.$$

So the implicit function theorem (see Theorem 1.1.23) implies that $x \longrightarrow r(x)$ is a differentiable function, consequently $x \longrightarrow \vartheta(x)$ is continuous (see (4.139)). So we have proved that $M_\beta/\mathbb{Z}_2$ (i.e., the manifold M_β with the antipodal points identified), is homeomorphic to the projective space $P^\infty(X)$ and the homeomorphism is ϑ (see (4.139)).

Now we produce deformations of subsets of $M_\beta/\mathbb{Z}_2$ analogous to the pseudogradient deformations (see Theorem 4.1.19). So consider the duality map $\mathcal{F}^*: X^* \longrightarrow X$ of X^*. Because X and X^* are both locally uniformly convex, we have that $\mathcal{F}^*$ is locally Lipschitz. We consider the following Cauchy problem in X.

$$\left\{ \begin{array}{l} \frac{\mathrm{d}x(t)}{\mathrm{d}t} = v + \xi(x,v)\mathcal{F}^*\big(\varphi'(x)\big) \\ x(0) = x_0 \in M_\beta \end{array} \right\}, \tag{4.140}$$

where $v \in X$ and $\xi(x,v) \in \mathbb{R}$ are chosen so that

(a) $x(t;x_0) \in M_\beta$ for all $t \geq 0$ (by $x(\cdot;x_0)$ we denote the unique solution of (4.140)).
(b) $t \longrightarrow \psi\big(x(t;t_0)\big)$ is decreasing.

Note that, if $x(t) = x(t;x_0)$ is the unique solution of (4.140), then if we want (a) to be true, then $\varphi\big(x(t)\big) = \beta$, hence $(\mathrm{d}/\mathrm{d}t)\, \varphi\big(x(t)\big) = \big\langle \varphi'\big(x(t)\big), x'(t)\big\rangle = 0$ and this by (4.140) leads to the following determination of $\xi(x,v)$.

$$\begin{aligned} &\big\langle \varphi'(x), v + \xi(x,v)\mathcal{F}^*\big(\varphi'(x)\big)\big\rangle = 0, \\ \Rightarrow\ &\xi(x,v) = -\frac{\langle \varphi'(x), v\rangle}{\|\varphi'(x)\|_{X^*}^2}. \end{aligned} \tag{4.141}$$

We also determine $v \in X$. We have

$$\begin{aligned} \psi\big(x(t)\big) - \psi(x_0) &= \int_0^t \Big\langle \psi'\big(x(s)\big), v + \xi\big(x(s),v\big)\mathcal{F}^*\Big(\varphi'\big(x(s)\big)\Big)\Big\rangle\, \mathrm{d}s \\ &= \int_0^t \left\langle \psi'\big(x(s)\big) - \frac{\Big\langle \psi'\big(x(s)\big), \mathcal{F}^*\Big(\varphi'\big(x(s)\big)\Big)\Big\rangle}{\|\varphi'\big(x(s)\big)\|_{X^*}^2}\, \varphi'\big(x(s)\big), v \right\rangle \mathrm{d}s \end{aligned}$$

(see (4.141)).

We set

$$\psi'\big(x(s)\big) - \frac{\Big\langle \psi'\big(x(s)\big), \mathcal{F}^*\Big(\varphi'\big(x(s)\big)\Big)\Big\rangle}{\|\varphi'\big(x(s)\big)\|_{X^*}^2} = D\big(\psi|_{M_\beta}\big)\big(x(s)\big),$$

the gradient of $\varphi|_{M_\beta}$. So if $v = -D\big(\psi|_{M_\beta}\big)\big(x(s)\big)$, we have

$$\varphi\big(x(t)\big) - \varphi(x_0) = -\int_0^t \big\|D\big(\psi|_{M_\beta}\big)\big(x(s)\big)\big\|_{X^*}^2 \mathrm{d}s.$$

We define

$$\mathcal{S}_n = \big\{C : C \subseteq M_\beta/\mathbb{Z}_2 = \widehat{M},\ \ \mathrm{cat}_{\widehat{M}}(C) \geq n\big\}.$$

Because $\widehat{M} \simeq P^\infty(X)$, we have that $\{\mathcal{S}_n\}_{n\geq 1}$ is a strictly decreasing family. We define

$$c_n^+ = \inf_{C \in \mathcal{S}_n} \sup_{x \in C} \psi(x) \ \text{ and } \ c_n^- = \sup_{C \in \mathcal{S}_n} \inf_{x \in C} \psi(x),\ n \geq 1.$$

We show that $\{c_n^+\}_{n\geq 1}$, $\{c_n^-\}_{n\geq 1}$ are critical values of $\psi|_{M_\beta}$. To this end we need to verify the PS-condition. So we show that

$$\text{if } c \neq 0,\ \varepsilon > 0 \text{ is sufficiently small, } x_n \in \{c - \varepsilon \leq \psi \leq c + \varepsilon\} \cap M_\beta \text{ and } D(\psi|_{M_\beta})(x_n) \longrightarrow 0, \text{ then } \{x_n\}_{n\geq 1} \text{ has a strongly convergent subsequence.} \tag{4.142}$$

Because M_β is bounded in X, we may assume that $x_n \xrightarrow{w} x$ in X. Also because of hypothesis (H)(iv), we have that $\{\varphi'(x_n)\}_{n\geq 1} \subseteq X^*$ is bounded and so we may assume that $\varphi'(x_n) \xrightarrow{w} u^*$ in X^* as $n \to \infty$. Finally recall that $\|\varphi'(x_n)\|_{X^*} \geq \eta_1 > 0$ for all $n \geq 1$ (see (4.138)). From the choice of the sequence $\{x_n\}_{n\geq 1}$, we have

$$D(\psi|_{M_\beta})(x_n) = \psi'(x_n) - \frac{\langle \psi'(x_n), \mathcal{F}^*(\varphi'(x_n))\rangle}{\|\varphi'(x_n)\|^2_{X^*}}\, \varphi'(x_n) \longrightarrow 0 \quad \text{as } n \to \infty. \tag{4.143}$$

Because of hypothesis (H)(v), we have that $\psi'(x_n) \longrightarrow v^*$ in X^* and so

$$\langle \psi'(x_n), \mathcal{F}^*(\varphi'(x_n))\rangle\, \varphi'(x_n) \longrightarrow v^* \quad \text{in } X^* \text{ as } n \to \infty.$$

We may assume $\langle \psi'(x_n), \mathcal{F}^*(\varphi'(x_n))\rangle \longrightarrow a$. Note that $a \neq 0$ or otherwise from (4.143) we have

$$\begin{aligned} &\psi'(x_n) \longrightarrow 0 \text{ in } X^* \text{ as } n \to \infty \\ \Rightarrow\ &\psi'(x) = 0 \text{ and so } x = 0 \quad \text{(see hypothesis (H)(v)).} \end{aligned} \tag{4.144}$$

Using hypothesis (H)(v) and the mean value theorem, we can easily check that the set $\{c - \varepsilon \leq \psi \leq c + \varepsilon\}$ is weakly closed and for $\varepsilon > 0$ small does not contain the origin. Since $x \in \{c - \varepsilon \leq \psi \leq c + \varepsilon\}$ from (4.144) we have a contradiction. This proves that (4.142) holds; that is, $\psi|_{M_\beta}$ satisfies the PS–condition.

Now we prove that each $c_n^+ \neq 0$ is a critical value of $\psi|_{M_\beta}$. Because of the PS-condition, the set of critical points of $\psi|_{M_\beta}$ is closed. So if c_n^+ is not a critical value of $\psi|_{M_\beta}$, we can find $\varepsilon > 0$ small such that $\{c_n^+ - \varepsilon \leq \psi \leq c_n^+ + \varepsilon\} \cap M_\beta$ contains no critical points of $\psi|_{M_\beta}$. Consequently, as in the proof of Theorem 4.1.19(e), we can find a deformation $h(t,x)$ and a set $C \in \mathcal{S}_n$ such that $\sup\limits_{x\in A} \psi(x) \leq c_n^+ + \varepsilon$ and $\sup\limits_{x\in h(1,C)} \varphi(x) \leq c_n^+ - \varepsilon$. But $\operatorname{cat}_{\widehat{M}}(h(1,C)) \geq n$ and so

$$c_n^+ \leq \sup_{x\in h(1,C)} \psi(x) \leq c_n^+ - \varepsilon,$$

a contradiction. Similarly, using deformations so that $t \longrightarrow \psi(x(t))$ is increasing, we establish that c_n^-, $n \geq 1$, are critical values of $\psi|_{M_\beta}$.

Let x_n be a critical point with critical value c_n^+ or c_n^-, $n \geq 1$. We have

$$\psi'(x_n) = \frac{\langle \psi'(x_n), \mathcal{F}^*(\varphi'(x_n))\rangle}{\|\varphi'(x_n)\|^2_{X^*}}\, \varphi'(x_n)$$

$$\text{and} \quad \frac{\langle \psi'(x_n), \mathcal{F}^*(\varphi'(x_n))\rangle}{\|\varphi'(x_n)\|^2_{X^*}} \neq 0 \qquad \text{for all } n \geq 1.$$

So finally we can say

$$\varphi'(x_n) = \lambda_n \psi'(x_n)$$

$$\text{and} \quad |\lambda_n| \longrightarrow +\infty, \ x_n \xrightarrow{w} 0 \quad \text{in } X \quad \text{as } n \to \infty.$$

∎

REMARK 4.4.2 Because of hypotheses (H)(ii) and (iv), we see that $M_\beta \subseteq X$ is compact.

If we drop the evenness hypothesis on the functionals φ, ψ, we have the following.

COROLLARY 4.4.3 *If hypotheses (H) (without the evenness of φ and ψ) hold and we can find $x_0^\pm \in M_\beta$ such that $\pm\psi(x_0^\pm)>0$, then problem (4.137) has a solution $(\lambda^\pm, x^\pm)$ such that*

$$\pm\psi(x^\pm) = \max_{x\in M_\beta} \pm\psi(x).$$

Motivated from the proof of Theorem 4.4.1, we make the following definition.

DEFINITION 4.4.4 A compact C^1-manifold M in a Banach space X is said to be *spherelike*, if it is C^1-diffeomorphic to the unit sphere $\partial B_1(0) = \{x\in X : \|x\| = 1\}$ in X.

The next proposition, resulting from the proof of Theorem 4.4.1, answers the question of when the level surface $M_\beta = \{x\in X : \varphi(x) = \beta\}$ of a function $\varphi\colon X \longrightarrow \mathbb{R}$ is spherelike.

PROPOSITION 4.4.5 *If X is a Banach space, $\varphi \in C^1(X)$, and*

(i) $\langle\varphi'(x), x\rangle \neq 0$ *for all* $x \in M_\beta$.
(ii) *Every ray from the origin intersects* M_β *at exactly one point,*

then M_β is spherelike; the C^1-diffeomorphism $r_\beta\colon M_\beta \longrightarrow \partial B_1(0)$ is given by

$$r_\beta = r\big|_{M_\beta},$$

where $r : X\setminus\{0\} \longrightarrow \partial B_1(0)$ is the radial retraction defined by

$$r(x) = \frac{x}{\|x\|};$$

also we have $\|Dr_\beta(x)\|_{\mathcal{L}} \leq (2/\|x\|)$ for all $x \in M_\beta$.

PROOF: Note that r is Fréchet differentiable on $X\setminus\{0\}$ and $r'(x) \in \mathcal{L}(X)$ is given by

$$r'(x)(h) = \frac{1}{\|x\|}h - \frac{\langle\mathcal{F}(x), h\rangle}{\|x\|^3}\,x \qquad \text{for all } h \in X, \tag{4.145}$$

$$\Rightarrow \|r'(x)\|_{X^*} \leq \frac{2}{\|x\|} \qquad \text{for all } x\in X\setminus\{0\}.$$

Because $\langle\mathcal{F}(x), x\rangle = \|x\|^2$, from (4.145) we also have that

$$r'(x)(x) \subseteq \ker n'(x) \qquad \text{for all } x\in X\setminus\{0\},$$

where $n\colon X \longrightarrow \mathbb{R}$ is the norm function $n(x) = \|x\|$ which is Gâteaux differentiable at every $x\in X \setminus \{0\}$ because by hypothesis the space X is smooth.

Recall that M_β is a C^1-manifold in X, which does not contain the origin of X. Also

$$T_x(M_\beta) = \ker \varphi'(x) \qquad \text{for all } x \in M_\beta \quad \text{(see Theorem 2.2.7).} \tag{4.146}$$

Because of hypothesis (ii) $r_\beta = r\big|_{M_\beta}$ is a bijection from M_β into $\partial B_1(0)$. We show that the derivative of this bijection is an isomorphism between the corresponding tangent spaces; that is,

$$Dr_\beta : T_x(M_\beta) \longrightarrow T_{r_\beta(x)}\big(\partial B_1(0)\big) \tag{4.147}$$

is an isomorphism. To this end, it suffices to show that $r'(x)$ is an isomorphism from $\ker \varphi'(x)$ into $\ker n'\big(r_\beta(x)\big)$ (see (4.146), (4.147), and recall that $\ker n'(x) = T_x\big(\partial B_1(0)\big)$ if $\|x\| = 1$). Recall that $r'(x)$ maps X into $\ker n'(x)$ for all $x \in X \setminus \{0\}$. Because $n'(x) = n'(\lambda x)$ for all $\lambda > 0$ and all $x \in X \setminus \{0\}$ we see that $r'(x)$ maps linearly into $\ker n'\big(r_\beta(x)\big)$ and this operator is also continuous. For $x \in M_\beta$ and $u \in \ker n'\big(r_\beta(x)\big) = \ker n'(x)$, we set

$$v = \|x\| u - \|x\| \, \frac{\langle \varphi'(x), u\rangle}{\langle \varphi'(x), x\rangle}\, x.$$

The map $u \longrightarrow v$ is linear and continuous. Moreover, note that

$$\big\langle \varphi'(x), v\big\rangle = \|x\| \, \big\langle \varphi'(x), u\big\rangle - \|x\| \, \frac{\langle \varphi'(x), u\rangle}{\langle \varphi'(x), x\rangle} \, \big\langle \varphi'(x), x\big\rangle = 0,$$

$$\Rightarrow \quad v \in \ker \varphi'(x).$$

Also we have $r(x)(v) = u$. Therefore $r'(x) \in \mathcal{L}\big(\ker \varphi'(x), \ker n'\big(r_\beta(x)\big)\big)$ is surjective. It is also injective, because

$$r'(x)(h) = 0 \quad \text{if and only if} \quad h = \frac{\langle \mathcal{F}(x), h\rangle}{\|x\|^2}\, x \qquad \text{(see (4.146)).}$$

Because x does not belong to $\ker n'(x)$, this equality holds only for $h = 0$, which proves the desired injectivity. Thus

$$Dr_\beta(x) : T_x(M_\beta) \longrightarrow T_{r_\beta(x)}\big(\partial B_1(0)\big)$$

is an isomorphism and so $r_\beta : M_\beta \longrightarrow \partial B_1(0)$ is a C^1-diffeomorphism. □

DEFINITION 4.4.6 We say that $\varphi \in C^1(X)$ defines a *spherelike constraint* at $\beta \in \mathbb{R}$, if $M_\beta = \varphi^{-1}(\{\beta\})$ is a spherelike manifold (see Definition 4.4.4).

We would like to have concrete and easily verifiable conditions on $\varphi \in C^1(X)$, which imply that for certain levels β, the manifold M_β is spherelike. The next proposition provides such conditions.

PROPOSITION 4.4.7 *If X is a Banach space and $\varphi \in C^1(X)$ satisfies the following conditions.*

(i) *φ and $\varphi' : X \longrightarrow X^*$ are bounded (i.e., map bounded sets to bounded sets);*
(ii) *$\langle \varphi'(x), x\rangle \longrightarrow +\infty$ as $\|x\| \longrightarrow \infty$,*

then there exists a $\beta_0 > 0$ such that for all $\beta \geq \beta_0$ the manifolds $M_\beta = \varphi^{-1}(\{\beta\})$ are spherelike.

PROOF: Because of hypothesis (ii), given $M>0$, we can find $R=R(M)>0$ such that

$$\langle \varphi'(x), x\rangle \geq M \qquad \text{for all } x\in X \text{ with } \|x\|\geq R. \tag{4.148}$$

For $r\geq 1$ and $x\in X$ with $\|x\|=R$, we have

$$\begin{aligned} \varphi(rx)-\varphi(x) &= \int_1^r \frac{\mathrm{d}}{\mathrm{d}t}\varphi(tx)\mathrm{d}t \\ &= \int_1^r \langle \varphi'(tx), x\rangle \,\mathrm{d}t \\ &= \int_1^r \frac{1}{t}\langle \varphi'(tx), tx\rangle \,\mathrm{d}t \\ &\geq M\int_1^r \frac{\mathrm{d}t}{x} = M\ln r. \end{aligned} \tag{4.149}$$

Because of hypothesis (i), we have

$$m=\sup\big\{|\varphi(x)| : \|x\|=R\big\}<+\infty.$$

So from (4.149) we have

$$\begin{aligned} &\varphi(rx)\geq -m+M\ln r, \\ \Rightarrow\ &\varphi(x)\longrightarrow +\infty \text{ as } \|x\|\longrightarrow\infty \quad \text{(i.e., } \varphi \text{ is weakly coercive).} \end{aligned} \tag{4.150}$$

In particular (4.150) implies that for every $\beta\in\mathbb{R}$, M_β is bounded. Moreover, (4.150) shows that there exist $R<R_0<R_1$ and $M_0<M_1$ such that

$$M_0\leq\varphi(x)\leq M_1 \qquad \text{for all } x\in X \text{ with } R_0\leq\|x\|\leq R_1.$$

Fix a $y\in\partial B_1(0)$, otherwise arbitrary, and consider the ray $\{x_t=ty\}_{t\geq 0}$. For $M\in[M_0,M_1]$ there exists a minimal $\widehat{t}\geq R_0$ such that the ray intersects M_β. As above for $t>\widehat{t}\geq R$, we have

$$\varphi(ty)-\varphi(y)\geq M\ln\frac{t}{\widehat{t}}>0.$$

So this ray intersects M_β at exactly one point. □

4.5 Bifurcation Theory

Bifurcation theory is concerned with the structure of branch points of nonlinear parametric operator equations in a Banach space. So suppose that X,Y are Banach spaces and $G\colon\mathbb{R}\times X\longrightarrow Y$ is a continuously Fréchet differentiable map such that $G(\lambda_0,x_0)=0$. We are interested in constructing all solutions of the nonlinear operator equation

$$G(\lambda,x)=0 \tag{4.151}$$

in a neighborhood of (λ_0,x_0). If $G'_x(\lambda_0,x_0)\in\mathcal{L}(X,Y)$ is an isomorphism, then according to the implicit function theorem (see Theorem 1.1.23), there is a continuously Fréchet differentiable function $x(\lambda)$ defined for $\lambda\in\mathbb{R}$ such that $G\big(\lambda,x(\lambda)\big)=0$.

Moreover, all solutions of (4.151) in a sufficiently small neighborhood of (λ_0, x_0) in $\mathbb{R} \times X$ lie on the curve $\big(\lambda, x(\lambda)\big)$.

Bifurcation theory is concerned with what happens when $G'_x(\lambda_0, x_0)$ is no longer an isomorphism. This simplest case occurs when $L_0 = G'_x(\lambda_0, x_0)$ is a Fredholm operator (see Definition 3.1.44(a)). In that case the bifurcation problem may be reduced, via an alternative method, to a system of n algebraic equations with n unknowns

$$F_k(\lambda, y) = 0, \quad k \in \{1, \ldots, n\} \quad \text{and} \quad y = (y_k)_{k=1}^n \in \mathbb{R}^n. \tag{4.152}$$

For such problems, the main task is the analysis of system (4.152), which is often a quite complex endeavor especially if $n \geq 1$ is big.

Bifurcation theory is an important topic in applied mathematics, because the phenomenon of bifurcation is intimately associated with loss of stability of the system under consideration. A complete resolution of the branching problem therefore requires a study of the stability of the bifurcation solutions.

The bifurcation problem that we examine in this section usually arises from the steady-state solutions of dynamical systems. So let X be a Banach space, $U \subseteq X$ an open set containing the origin, and $T = (\lambda_0 - \delta, \lambda_0 + \delta)$ for some $\lambda_0 \in \mathbb{R}$ and $\delta > 0$. It is also given a compact map $G \colon T \times U \longrightarrow X$ that satisfies $G(\lambda, 0) = 0$ for all $\lambda \in T$. We consider the following operator system

$$x = G(\lambda, x). \tag{4.153}$$

Clearly $(\lambda, 0)$ for $\lambda \in T$ satisfies (4.153). We call such solutions of (4.153) trivial solutions. All other solutions of (4.153) are called nontrivial solutions.

DEFINITION 4.5.1 A point $(\lambda_0, 0) \in T \times U$ is a *bifurcation point* of (4.153) if every neighborhood of $(\lambda_0, 0)$ contains a nontrivial solution of (4.153); that is,

$$(\lambda_0, 0) \in \operatorname{cl} \{(\lambda, x) \in T \times U : x = G(\lambda, x), \;\; x \neq 0\}.$$

REMARK 4.5.2 According to this definition $(\lambda_0, 0)$ is a bifurcation point of (4.153) if and only if we can find a sequence $\{(\lambda_n, x_n)\}_{n \geq 1} \subseteq T \times U$ such that $x_n = G(\lambda_n, x_n)$, $x_n \neq 0$, and $\lambda_n \longrightarrow \lambda_0$, $x_n \longrightarrow 0$ as $n \to \infty$. Also, although strictly speaking $(\lambda_0, 0)$ is the bifurcation point, we usually say (for simplicity) that λ_0 is the bifurcation point (in fact if we identify $\mathbb{R} \times \{0\}$ with $\mathbb{R}$, we see that this abuse of the terminology is in fact justified).

We focus our attention on maps $G(\lambda, x)$ of the form

$$G(\lambda, x) = \lambda K(x) + \vartheta(\lambda, x) \tag{4.154}$$

with $K \in \mathcal{L}_c(X)$ and $\|\vartheta(\lambda, x)\| / \|x\| \longrightarrow 0$ as $x \longrightarrow 0$ uniformly in $\lambda \in T$. Clearly (4.154) is obtained from (4.153) via linearization. Note that, if $G \colon T \times U \longrightarrow X$ is a compact map which is Fréchet differentiable at $x = 0$ for all $\lambda \in T$ with Fréchet derivative λK, $K \in \mathcal{L}(X)$, then from Proposition 3.1.20, we know that $K \in \mathcal{L}_c(X)$ and so (4.154) is valid.

The first result that we prove is a rather negative result, because it says that most numbers are not bifurcation points. Nevertheless, it is a useful result because it clarifies the situation and allows us to focus on the possible locations for bifurcation points.

PROPOSITION 4.5.3 *If $G(\lambda,x)$ has the form (4.154) and $(\lambda_0,0)$ is a bifurcation point of (4.153), then $1/\lambda_0$ is an eigenvalue of $K\in\mathcal{L}_c(X)$.*

PROOF: Suppose that $1/\lambda_0$ is not an eigenvalue of K. Then $I-\lambda_0K$ is invertible. So if $c=1/\|(I-\lambda_0K)^{-1}\|_{\mathcal{L}}$, we have

$$\|(I-\lambda_0K)(x)\|\geq c\,\|x\| \qquad \text{for all } x\in X.$$

Hence we have

$$\begin{aligned}\|x-G(\lambda,x)\| &= \|x-\lambda K(x)-\vartheta(\lambda,x)\|\\ &= \|x-\lambda K(x)-\lambda_0K(x)+\lambda_0K(x)-\vartheta(\lambda,x)\|\\ &\geq \|(I-\lambda_0K)(x)\|-|\lambda-\lambda_0|\,\|K\|_{\mathcal{L}}\|x\|-\|\vartheta(\lambda,x)\|\\ &\geq c\|x\|-|\lambda-\lambda_0|\,\|K\|_{\mathcal{L}}\|x\|-\|\vartheta(\lambda,x)\|. \end{aligned} \tag{4.155}$$

Recall that by hypothesis $\|\vartheta(\lambda,x)\|/\|x\|\longrightarrow 0$ as $x\longrightarrow 0$ uniformly in $\lambda\in T$. So from (4.155) we infer that if $|\lambda-\lambda_0|$, $\|x\|$ are small and $x\neq 0$, we have

$$\|x-G(\lambda,x)\|>0.$$

So for all λ near λ_0 and all $x\neq 0$ with $\|x\|$ small $x\neq G(\lambda,x)$, which shows that $(\lambda_0,0)$ is not a bifurcation point (see Definition 4.5.1). □

COROLLARY 4.5.4 *If $G(\lambda,x)$ has the form (4.154) and is defined on all $\mathbb{R}\times X$, then the set of bifurcation points is a discrete set in $\mathbb{R}$.*

PROOF: If $\lambda\neq 0$ is a bifurcation point, then by virtue of Proposition 4.5.3 $1/\lambda$ belongs to the point spectrum of $K\in\mathcal{L}_c(X)$, which is either a finite set or a sequence converging to zero. Therefore Proposition 4.5.3 implies that the set of bifurcation points is discrete. □

We have seen in Proposition 4.5.3 that a necessary condition for $(\lambda_0,0)$ to be a bifurcation point is the noninvertibility of $\lambda_0K=G'_x(\lambda_0,0)$. However, this condition is not sufficient in general, as shown by the following simple example.

EXAMPLE 4.5.5 Let $T=\mathbb{R}$, $X=\mathbb{R}^2$, and consider

$$G(\lambda,x,y)=(-\lambda x+y^3,-\lambda y-x^3) \qquad \text{for all } v=\begin{pmatrix}x\\y\end{pmatrix}\in\mathbb{R}^2.$$

Then $G'_v(-1,0,0)=I=(-1)(-I)$ and $K=-I$ has an eigenvalue equal to -1, but $(-1,0)$ is not a bifurcation point, because

$$\begin{aligned} & G(\lambda,x,y)=(x,y),\\ \Rightarrow\ & y(-\lambda x+y^3)-x(-\lambda y-x^3)=0,\end{aligned}$$

(taking inner product with $\begin{pmatrix}x\\-y\end{pmatrix}\in\mathbb{R}^2$),

$$\Rightarrow\ x^4+y^4=0 \ \text{(i.e., } x=y=0).$$

In what follows extending a standard notation for invertible $N \times N$-matrices to isomorphisms of X onto itself, we set

$$GL(X) = \{L \in \mathcal{L}(X) : L \text{ is an isomorphism}\}.$$

Motivated from Proposition 4.5.3, we introduce the following definition.

DEFINITION 4.5.6 The set of *singular points* of (4.153) is the set

$$D_s = \{(\lambda, 0) \in T \times U : I - \lambda K \notin GL(X)\}.$$

We want to derive sufficient conditions for $(\lambda_0, 0)$ to be a bifurcation point of (4.153). According to Corollary 4.5.4 we focus on isolated critical points $(\lambda_0, 0)$ and by virtue of Definition 4.5.1, $(\lambda_0, 0)$ is a bifurcation point if and only if in every neighborhood of $(\lambda, 0)$ problem (4.153) has a nontrivial solution. In order to distinguish nontrivial solutions from trivial ones, we use a function $\alpha \in C(T \times U)$ that is negative on the set $T \times \{0\}$ of trivial solutions of (4.153). This function is called the *auxiliary function*. Hence the search for nontrivial solutions of problem (4.153) is reduced to the resolution of the following system,

$$\left\{ \begin{array}{c} x = G(\lambda, x) \\ \alpha(\lambda, x) = 0 \end{array} \right\}. \tag{4.156}$$

To solve system (4.156), we can use degree-theoretic methods. For this purpose we introduce the maps $\varphi \colon T \times U \longrightarrow X$ and $\varphi_\alpha \colon T \times U \longrightarrow \mathbb{R} \times X$ defined by

$$\begin{aligned} &\varphi(\lambda, x) = x - G(\lambda, x) \\ \text{and} \quad &\varphi_\alpha(\lambda, x) = \big(\alpha(\lambda, x), \varphi(\lambda, x)\big) \qquad \text{for all } (\lambda, x) \in T \times U. \end{aligned}$$

Let U be a small neighborhood of $(\lambda_0, 0)$. If $\varphi_\alpha(\lambda, x) \neq (0, 0)$ for all $(\lambda, x) \in \partial U$, then we can consider the Leray–Schauder degree $\widehat{d}(\varphi_\alpha, U, 0)$ and to solve (4.156), try to show that $\widehat{d}(\varphi_\alpha, U, 0) \neq 0$. To pursue this idea, we introduce the following notion.

DEFINITION 4.5.7 Let $(\lambda_0, 0)$ be an isolated singular point of (4.153) and let $\varepsilon > 0$, $r > 0$ be such that

(a) $c < |\lambda - \lambda_0| \leq \varepsilon \Rightarrow \lambda \notin D_s$.
(b) $x \neq G(\lambda, x)$ for all $|\lambda - \lambda_0| = \varepsilon$ and $0 < \|x\| \leq r$.

We set $V(\varepsilon, r) = (\lambda_0 - \varepsilon, \lambda_0 + \varepsilon) \times B_r(0)$ and we call $V(\varepsilon, r)$ a *special neighborhood* of $(\lambda_0, 0)$. Also a function $\xi \colon \overline{V(\varepsilon, r)} \times \mathbb{R} \longrightarrow \mathbb{R}$ satisfying

(c) $\xi(\lambda, 0) = -|\lambda - \lambda_0|$,
(d) $\xi(\lambda, x) = r$ for $\|x\| = r$

is called a *complementing function* (or *Ize's function*) for $(\lambda_0, 0)$.

REMARK 4.5.8 It is easy to see that a special neighborhood of an isolated critical point $(\lambda_0, 0)$ exists. Indeed, let $\varepsilon > 0$ be small so that $I - \lambda K$ is invertible for all $|\lambda - \lambda_0| = \varepsilon$ and set

$$c = \min\left\{\frac{1}{\|(I-\lambda_0 K)^{-1}\|_{\mathcal{L}}} : |\lambda - \lambda_0| = \varepsilon\right\}.$$

Then choose $r > 0$ small enough so that $\|\vartheta(\lambda, x)\|/\|x\| \leq c$ for all $0 < \|x\| \leq r$. We assume without any loss of generality that this inequality is always satisfied by some special neighborhood of $(\lambda_0, 0)$. A complementing function is then defined by $\xi(\lambda, x) = |\lambda - \lambda_0|\,\big((\|x\| - r)/r\big) + \|x\|$. Note that for every $\eta > 0$, $\alpha_\eta(\lambda, x) = \xi(\lambda, x) - \eta$ is an auxiliary function. Exploiting the homotopy invariance of the Leray–Schauder degree we have $\widehat{d}\big(\varphi_\xi, V(\varepsilon, r), 0\big) = \widehat{d}\big(\varphi_{\alpha_\eta}, V(\varepsilon, r), 0\big)$ for $\eta > 0$ small. So the existence of nontrivial solutions in $V(\varepsilon, r)$ can be deduced from the relation $\widehat{d}\big(\varphi_\xi, V(\varepsilon, r), 0\big) \neq 0$.

PROPOSITION 4.5.9 *If $G\colon T \times U \longrightarrow X$ is a compact map given by (4.154), $(\lambda_0, 0)$ is an isolated singular point of (4.153), $V(\varepsilon, r)$ is a special neighborhood of $(\lambda_0, 0)$, and $\xi(\lambda, x)$ a complementing function defined on $V(\varepsilon, r)$, then for every special neighborhood $V(\varepsilon', r')$ of $(\lambda_0, 0)$ with $0 < \varepsilon' < \varepsilon, 0 < r' < r$ and for every complementing function $\xi'(\lambda, x)$ defined by $V(\varepsilon', r')$ we have*

$$\widehat{d}\big(\varphi_\xi, V(\varepsilon, r), 0\big) = \widehat{d}\big(\varphi_{\xi'}, V(\varepsilon', r'), 0\big).$$

PROOF: Let $h\colon [0,1] \times \overline{V(\varepsilon, r)} \longrightarrow X$ be defined by

$$h(t, \lambda, x) = x - \lambda K(x) - t\vartheta(\lambda, x)$$

and $h_\xi\colon [0,1] \times \overline{V(\varepsilon, r)} \longrightarrow \mathbb{R} \times X$ be defined by

$$h_\xi(t, \lambda, x) = \big(\xi(\lambda, x), h(t, \lambda, x)\big).$$

If we set $k(\lambda, x) = x - \lambda K(x)$ and $\varphi(\lambda, x) = x - G(\lambda, x)$, then because

$$c = \min\left\{\frac{1}{\|(I-\lambda_0 K)^{-1}\|_{\mathcal{L}}} : |\lambda - \lambda_0| = \varepsilon\right\} \geq \frac{\|\vartheta(\lambda, x)\|}{\|x\|} \qquad \text{for all } 0 < \|x\| \leq r$$

(see Remark 4.5.8), we see that h_ξ is an admissible homotopy (for compact perturbations of the identity) between k_ξ and φ_ξ. So from the homotopy invariance of the Leray–Schauder degree, we have

$$\widehat{d}\big(k_\xi, V(\varepsilon, r), 0\big) = \widehat{d}\big(\varphi_\xi, V(\varepsilon, r), 0\big). \tag{4.157}$$

Here $k_\xi(\lambda, x) = \big(\xi(\lambda, x); x - k(\lambda, x)\big)$. In a similar fashion we show that

$$\widehat{d}\big(k_{\xi'}, V(\varepsilon, r'), 0\big) = \widehat{d}\big(\varphi_{\xi'}, V(\varepsilon, r'), 0\big). \tag{4.158}$$

Note that $k_\xi^{-1}(0) = (\lambda_0, 0)$ and so from (4.157), (4.158), and the excision property of the Leray–Schauder degree, we conclude that

$$\widehat{d}\big(\varphi_\xi, V(\varepsilon, r), 0\big) = \widehat{d}\big(\varphi_{\xi'}, V(\varepsilon, r'), 0\big).$$

□

Recall that if we consider the general linear group $GL(N, \mathbb{R})$, we can define

$$GL^+(N, \mathbb{R}) = \big\{A \in GL(N, \mathbb{R}) : \det A > 0\big\}$$
$$\text{and} \quad GL^-(N, \mathbb{R}) = \big\{A \in GL(N, \mathbb{R}) : \det A < 0\big\}.$$

Note that $GL(N,\mathbb{R})$ can be regarded as a subset of $\mathbb{R}^{N\times N}$ and so $GL(N,\mathbb{R})$ is a topological space with the topology induced from $\mathbb{R}^{N\times N}$. It is well known that $GL^+(N,\mathbb{R})$ and $GL^-(N,\mathbb{R})$ are two open, connected components of $GL(N,\mathbb{R})$. In fact this has an infinite-dimensional analogue. Namely let $GL_c(X) = \{I+K : K \in \mathcal{L}_c(X) \text{ and } (I+K)^{-1} \in \mathcal{L}(X)\}$. Then $GL_c(X)$ has two connected components $GL^+(X)$ (which contains I) and $GL^-(X)$.

DEFINITION 4.5.10 Let $L \in GL_c(X)$. Then

$$\operatorname{sgn} L = \begin{cases} 1 & \text{if } L \in GL^+(X) \\ -1 & \text{if } L \in GL^-(X) \end{cases}.$$

REMARK 4.5.11 As for Brouwer's degree, if $\varphi = I - f$, with $f \in K(U,X) \cap C^1(U,X)$ and $\varepsilon = \inf\{\|\varphi(x)\| : x \in \partial U\} > 0$, then there exists $y_0 \in B_\varepsilon(0)$ a regular value of $\varphi|_U$ such that $\widehat{d}(\varphi,U,0) = \widehat{d}(\varphi - y_0,U,0) = \sum\limits_{x\in(\varphi-y_0)^{-1}(0)} \operatorname{sgn} \varphi'(x)$.

Now we can produce a useful sufficient condition for $(\lambda_0,0)$ to be a bifurcation point.

THEOREM 4.5.12 *If $G\colon T\times U \longrightarrow X$ is a compact map given by (4.154), $(\lambda_0,0)$ is an isolated singular point of (4.153), $V(\varepsilon,r)$ is a special neighborhood of $(\lambda_0,0)$, $\xi\colon \overline{V(\varepsilon,r)} \longrightarrow \mathbb{R}$ is a complementing function, and* $\operatorname{sgn}\big(I-(\lambda_0-\varepsilon)K\big) \neq \operatorname{sgn}\big(I-(\lambda_0+\varepsilon)K\big)$, *then $(\lambda_0,0)$ is a bifurcation point.*

PROOF: Let $k(\lambda,x)=x-\lambda K(x)$ and $k_\xi(\lambda,x)=\big(\xi(\lambda,x),k(\lambda,x)\big)$ for all $(\lambda,x) \in \overline{V(\varepsilon,r)}$. Then as in the proof of Proposition 4.5.9 we can show that

$$\widehat{d}\big(\varphi_\xi,V(\varepsilon,r),0\big) = \widehat{d}\big(k_\xi,V(\varepsilon,r),0\big). \tag{4.159}$$

By virtue of the homotopy invariance of the Leray–Schauder degree the map $\lambda \longrightarrow \lambda K$ is constant on some neighborhoods J_+ and J_- of $\lambda_0+\varepsilon$ and $\lambda_0-\varepsilon$, respectively. We define the $\alpha(t,\lambda,x) = \xi(\lambda,x)+t\eta$ with $\eta \in (0,\varepsilon)$ such that $\lambda_0+\eta \in J_+$ and $\lambda_0-\varepsilon \in J_-$. Then we set

$$h(t,\lambda,x) = \big(\alpha(t,\lambda,x),k(\lambda,x)\big).$$

We have

$$\begin{aligned} &\widehat{d}\big(h(1,\cdot,\cdot),V(\varepsilon,r),0\big) \\ &= \operatorname{sgn}\big(I-(\lambda_0-\varepsilon)K\big) - \operatorname{sgn}\big(I-(\lambda_0+\varepsilon)K\big) \neq 0 \quad \text{(see Remark 4.5.11)} \\ \Rightarrow\ &\widehat{d}\big(\varphi_\xi,V(\varepsilon,r),0\big) \neq 0 \qquad \text{(see (4.159))}, \\ \Rightarrow\ &(\lambda_0,0) \text{ is a bifurcation point.} \end{aligned}$$

□

DEFINITION 4.5.13 Let $G\colon T\times U \longrightarrow X$ be a compact map given by (4.154), $(\lambda_0,0)$ an isolated singular point of (4.153), $V(\varepsilon,r)$ a special neighborhood of $(\lambda_0,0)$, and $\xi\colon \overline{V(\varepsilon,r)} \longrightarrow \mathbb{R}$ a complementing function. The number $\gamma(\lambda_0) = \operatorname{sgn}\big(I-(\lambda_0-\varepsilon)K\big) - \operatorname{sgn}\big(I-(\lambda_0+\varepsilon)K\big)$ is called a *crossing number* at $(\lambda_0,0)$.

Recall that the algebraic multiplicity of an eigenvalue μ, denoted by $m_\alpha(\mu)$, is the dimension of the space $\bigcup\limits_{n\geq 1} N(\mu I-K)^n=\bigcup\limits_{n\geq 1}\ker N(\mu I-K)^n$.

THEOREM 4.5.14 *If $G\colon T\times U\longrightarrow X$ is a compact map given by (4.154), and $1/\lambda_0$ is an eigenvalue of K which has odd algebraic multiplicity m_α, then $(\lambda_0,0)$ is a bifurcation point.*

PROOF: By virtue of Definition 4.5.7, $1/\lambda$ is not an eigenvalue of K if $\lambda\in[\lambda_0-\varepsilon,\lambda_0+\varepsilon]\setminus\{\lambda_0\}$. Then from Remark 4.5.11 we know that as λ moves from $\lambda_0-\varepsilon$ to $\lambda_0+\varepsilon$, the degree $\widehat{d}\big(I-\lambda K,B_r(0),0\big)=\operatorname{sgn}(I-\lambda K)$ changes by a multiplicative factor $(-1)^{m_\alpha}$. Therefore Theorem 4.5.12 implies that $(\lambda_0,0)$ is a bifurcation point.

□

So this theorem relates an eigenvalue of odd algebraic multiplicity to the occurrence of a bifurcation point. Next we show that bifurcation from eigenvalues of odd multiplicity is a global rather than a local phenomenon. We start with a topological result.

LEMMA 4.5.15 *If (Y,d) is a compact metric space, $A\subseteq Y$ is a component (i.e., a maximal connected closed set), and $B\subseteq Y$ is a closed set such that $A\cap B=\emptyset$, then there exist compact sets $K_1\supseteq A$, $K_2\supseteq B$ such that $K_1\cap K_2=\emptyset$ and $K_1\cup K_2=Y$.*

PROOF: Given $\varepsilon>0$, we say that two points $a,b\in Y$ are ε-chainable, if we can find a finite sequence $\{y_k\}_{k=1}^n$ such that $y_1=\alpha, y_n=b$ and $d(y_k,y_{k+1})<\varepsilon$ for all $k\in\{1,\dots,n-1\}$. The finite sequence $\{y_k\}_{k=1}^n$ is called an ε-chain joining a and b.

We set

$$A_\varepsilon=\big\{y\in Y:\text{there exists } a\in A \text{ such that } a \text{ and } y \text{ are } \varepsilon\text{-chainable}\big\}.$$

Evidently $A\subseteq A_\varepsilon$ and A_ε is both closed and open in Y (a clopen set in Y). So it suffices to show that $A_\varepsilon\cap B=\emptyset$ for some $\varepsilon>0$.

We argue indirectly. So suppose that $A_\varepsilon\cap B\neq\emptyset$ for every $\varepsilon>0$. Then we can find sequences $\{\varepsilon_n\}_{n\geq 1}\subseteq\mathbb{R}_+\setminus\{0\}$, $\{a_n\}_{n\geq 1}\subseteq A$ and $\{b_n\}_{n\geq 1}\subseteq B$ such that

$$\varepsilon_n\longrightarrow 0 \quad\text{and}\quad a_n \text{ and } b_n \text{ are } \varepsilon_n\text{-chainable for every } n\geq 1.$$

Both sets A and B are compact and so we may assume that $a_n\longrightarrow\widehat{a}\in A$ and $b_n\longrightarrow\widehat{b}\in B$. Therefore for every $n\geq 1$, we can find an ε_n-chain C_n joining $\widehat{a}$ and $\widehat{b}$. Consider the set

$$Y_0=\big\{y\in Y: y=\lim_{k\to\infty} y_{n_k} \text{ with } y_{n_k}\in C_{n_k}\big\}.$$

We have

(i) Y_0 is a closed, hence compact set in Y.
(ii) $\widehat{a},\widehat{b}\in Y_0$.
(iii) For any $u,v\in Y_0$ and any $n\geq 1$, the points u and v are ε_n-chainable.

Suppose that Y_0 is not connected. We can find compact sets C_1 and C_2 and $\delta > 0$ such that $Y_0 = C_1 \cup C_2$ and $d(u,v) > \delta$ for all $u,v \in Y$ with $d(u,C_1) < \delta$, $d(v,C_2) < \delta$. Choose $\varepsilon_n < \delta$, $\widehat{u} \in C_1$, and $\widehat{v} \in C_2$. Then $\widehat{u}$ and $\widehat{v}$ are ε_n-chainable, but this contradicts the previously mentioned property of C_1 and C_2. Therefore Y_0 is connected. Because $\widehat{a} \in Y_0$, we have $Y_0 \subseteq A$ (recall that by hypothesis A is a component of Y). But $\widehat{b} \in Y_0$, hence $\widehat{b} \in A \cap B$, a contradiction. □

THEOREM 4.5.16 *If $U \subseteq \mathbb{R} \times X$ is a bounded neighborhood of $(\lambda_0, 0)$, $G : \overline{U} \longrightarrow X$ is a compact map given by (4.154), and the points in $D_s = \{(\lambda, x) \in \overline{U} : I - \lambda K \notin GL(X)\}$ are isolated, then if we set $S = \{(\lambda, x) \in \overline{U} : x = G(\lambda, x),\ x \neq 0\}$, the component S_0 of $\overline{S}$ that contains $(\lambda_0, 0)$ has at least one of the following properties,*

(a) *$S_0 \cap \partial U \neq \emptyset$.*
(b) *S_0 contains a finite number of trivial solutions $(\lambda_k, 0)$ $k \in \{0, 1, \ldots, n\}$ and $\sum\limits_{k=0}^{n} \gamma(\lambda_k)$ where $\gamma(\lambda_k)$ denotes the crossing number of $(\lambda_k, 0)$ (see Definition 4.5.13).*

PROOF: First note that $\overline{S}$ is compact in $\mathbb{R} \times X$. Suppose that $S_0 \cap \partial U = \emptyset$.

Claim: There exists a bounded open set U_0 such that $S_0 \subseteq U_0 \subseteq \overline{U}$ and $\overline{S} \cap \partial U_0 = \emptyset$.

Set $W_\delta = \{(\lambda, x) \in U : d\big((\lambda, x), S_0\big) < \delta\}$. Clearly $\overline{S} \cap \overline{W}_\delta$ is compact and so $S_0 \cap \partial W_\delta = \emptyset$. First suppose that $\overline{S} \cap \overline{W}_\delta$ is connected. Then $S_0 = \overline{S} \cap \overline{W}_\delta$ because S_0 is a component of $\overline{S}$. Therefore we can choose $U_0 = W_\delta$ for $\delta > 0$ small enough. Next suppose that $\overline{S} \cap \overline{W}_\delta$ is not connected. By Lemma 4.5.15 we can find compact sets $K_1 \supseteq S_0$ and $K_2 \supseteq \overline{S} \cap \partial W_\delta$ such that $K_1 \cap K_2 = \emptyset$ and $K_1 \cup K_2 = \overline{S} \cap \overline{W}_\delta$. Let $a = d(K_1, K_2) > 0$ and set $U_0 = (K_1)_{\alpha/2} \cap \overline{W}_\delta$ where $(K_1)_{\alpha/2} = \{(\lambda, x) \in \overline{U} : d\big((\lambda, x), K_1\big) < a/2\}$. Then for $\delta > 0$ small, we have

$$S_0 \subseteq U_0 \subseteq \overline{U} \quad \text{and} \quad \overline{S} \cap \partial U_0 = \emptyset.$$

Because by hypothesis the singular set D_s consist of isolated points and S_0 is compact, S_0 must contain a finite number of trivial solutions $\{(\lambda_k, 0)\}_{k=0}^{n}$. We may assume that

$$\overline{U}_0 \cap \big[(\mathbb{R} \times \{0\}) \cap \overline{U}\big] = \bigcup_{k=0}^{n} \big([\lambda_k - \varepsilon, \lambda_k + \varepsilon] \times \{0\}\big),$$

where the intervals are disjoint. Also without any loss of generality we may assume that there is an open set $V_0 \subseteq U_0$ such that $U_0 \backslash \overline{V}_0$ is a union of special neighborhoods $\{V(\varepsilon_k, r_k)\}_{k=0}^{n}$ of isolated points $\{(\lambda_k, 0)\}_{k=0}^{n}$.

Taking $\varepsilon = \min\limits_{1 \le k \le n} \varepsilon_k$ and $r = \min\limits_{1 \le k \le n} r_k$, we may assume that

$$V(\varepsilon_k, r_k) = (\lambda_k - \varepsilon, \lambda_k + \varepsilon) \times B_r(0),\ k \in \{0, \ldots, n\}.$$

Also let $\xi_k : \overline{V(\varepsilon_k, r_k)} \longrightarrow \mathbb{R}$ be a corresponding complementing function for $(\lambda_k, 0)$; that is,

$$\xi_k\big((\lambda, 0)\big) = -|\lambda - \lambda_k| \quad \text{and} \quad \xi_k\big((\lambda, x)\big) = r \quad \text{for } \|x\| = r.$$

We can define a function $\xi : \overline{U}_0 \longrightarrow \mathbb{R}$ by

$$\xi(\lambda,x)=\begin{cases}\xi_k(\lambda,x) & \text{if } (\lambda,x)\in V(\varepsilon_k,r_k)\\ r & \text{if } (\lambda,x)\in V_0\end{cases}.$$

Clearly ξ is continuous and because $\varphi_\xi(\lambda,x)\neq 0$ for all $(\lambda,x)\in\partial U_0$, the Leray–Schauder degree $\widehat{d}(\varphi_\xi,U_0,0)$ is well-defined.

Let $R>\varepsilon>0$ and consider the following compact homotopy

$$h(t,\lambda,x)=\big((1-t)\xi(\lambda,x)-tR,\ \varphi(\lambda,x)\big)\qquad\text{for all } t\in[0,1] \text{ and all } (\lambda,x)\in\overline{U}_0.$$

If for some $(t,\lambda,x)\in[0,1]\times\partial U_0$, we have $h(t,\lambda,x)=0$, then because $\varphi(\lambda,x)\neq 0$ for all $(\lambda,x)\in\partial U_0\setminus\overline{V}_0$, we have that $(\lambda,x)\in\partial U_0\cap\overline{V(\varepsilon_k,r_k)}$. In this case $\varphi(\lambda,x)=0$ implies $x=0$ and so $\xi(\lambda,x)=\xi_k(\lambda,x)=-|\lambda-\lambda_k|=-\varepsilon$. Hence $(1-t)\xi(\lambda,x)-tR<0$, which contradicts the fact that $h(t,\lambda,x)=0$. So the homotopy h is admissible.

Note that $h(0,\cdot,\cdot)=\varphi_\xi$. Then by virtue of the homotopy invariance of the Leray–Schauder we have

$$\widehat{d}(\varphi_\xi,U_0,0)=\widehat{d}\big(h(1,\cdot,\cdot),U_0,0\big). \tag{4.160}$$

Because $h(1,\lambda,x)=\big(-R,\varphi(\lambda,x)\big)\neq 0$ for all $(\lambda,0)\in U_0$, we must have

$$\widehat{d}\big(h(1,\cdot,\cdot),U_0,0\big)=0. \tag{4.161}$$

From (4.160) and (4.161) it follows that

$$\widehat{d}(\varphi_\xi,U_0,0)=0.$$

Because $\varphi_\xi(\lambda,x)\neq 0$ for all $(\lambda,x)\in V_0$, from the excision and additivity with respect to the domain properties of the degree map, we have

$$\begin{aligned}0=\widehat{d}(\varphi_\xi,U_0,0)&=\widehat{d}\big(\varphi_\xi,\bigcup_{k-0}^{n}V(\varepsilon_k,r_k),0\big)\\&=\sum_{k-0}^{n}\widehat{d}\big(\varphi_\xi,V(\varepsilon_k,r_k),0\big)\\&=\sum_{k-0}^{n}\gamma(\lambda_k).\end{aligned}$$

□

As a consequence of this theorem, we obtain the following.

THEOREM 4.5.17 *If $U\subseteq\mathbb{R}\times X$ is a bounded neighborhood of $(\lambda_0,0)$, $G:\overline{U}\longrightarrow X$ is a compact map given by (4.154), and $1/\lambda_0$ is an eigenvalue of odd algebraic multiplicity of K, then if we set $S=\{(\lambda,x)\in U: x=G(\lambda,x),\ x\neq 0\}$, the component S_0 of $\overline{S}$ that contains $(\lambda_0,0)$ has at least one the following properties,*

(a) $S_0\cap\partial U\neq\emptyset$.

(b) *S_0 contains an odd number of trivial solutions $(\lambda_k,0)\neq(\lambda_0,0)$ with $1/\lambda_k$ an eigenvalue of K of odd algebraic multiplicity.*

4.6 Remarks

4.1: In the literature we can find two approaches to the critical point theory. One is based on deformation techniques along the negative gradient flow or a suitable substitute of it, namely the pseudogradient flow (see Definition 4.1.15). This is the approach that we follow here. An alternative approach and an outline of it can be found in de Figueiredo [244] (see also Cuesta [169] for critical point theory on C^1-Banach manifolds). The Palais–Smale condition and the slightly more general Cerami condition (see Definition 4.1.4), are crucial in both approaches and were introduced by Palais–Smale [470] (the PS-condition) and by Cerami [136] (the C-condition). The relation between these compactness-type conditions and weak coercivity (see Theorem 4.1.12) was investigated by many authors. We mention the works by Caklovic–Li–Willem [127], Costa–Silva [158], Brezis–Nirenberg [105], and Goeleven [270]. The concept of pseudogradient vector field (see Definition 4.1.15) was introduced by Palais [471] in order to extend the classical Ljusternik–Schnirelmann theory (see Section 4.2) to infinite-dimensional Banach manifolds. He also proved that the pseudogradient vector field exists (see Theorem 4.1.18). Various forms of the deformation theorem (see Theorem 4.1.19) were first obtained by Browder [116], Palais [471], Schwartz [545], and Clark [147]. Of course this result is the main step in the deformation approach. The notion of linking sets (see Definition 4.1.20), is due to Benci–Rabinowitz [64]. A more restrictive version of Theorem 4.1.22 can be found in Struwe [564, p. 118]. The mountain pass theorem (see Theorem 4.1.24), is due to Ambrosetti–Rabinowitz [20]. The saddle point theorem (see Theorem 4.1.25), is due to Rabinowitz [504]. The generalized mountain pass theorem (see Theorem 4.1.26), is due to Rabinowitz [505]. The notion of local linking was introduced first by Brezis–Nirenberg [105]. Extensions of it can be found in Li–Willem [380]. The principle of symmetric criticality (see Theorem 4.1.41) is due to Palais [472] (see also Willem [606, p. 18]). The extension of the critical point theory to functions of the form $\widehat{\varphi} = \varphi + \psi$ with $\varphi \in C^1(X)$, $\psi \in \Gamma_0(X)$ is due to Szulkin [568]. In fact there are extensions of the theory to nonsmooth functions. We refer to Degiovanni–Marzocchi [192], Motreanu–Radulescu [439], and Gasiński–Papageorgiou [258].

Finally in addition to Theorem 4.1.19, there is the so-called second deformation theorem due to Röthe [531] Marino–Prodi [408] and Chang [142]. For a proof of it we refer to Gasiński–Papageorgiou [258, p. 617]. The result is useful in proving multiplicity results for nonlinear boundary value problems.

THEOREM 4.6.1 *If $\varphi \in C^1(X)$, $a \in \mathbb{R}$, $a < b \leq +\infty$ (if $b = +\infty$, then $\varphi^b \setminus K_b = X$), φ satisfies the PS_c-condition for every $c \in [a, b)$, φ has no critical values in (a, b), and $\varphi^{-1}(a)$ contains at most a finite number of critical points of φ, then there exists a φ-decreasing homotopy $h : [0,1] \times (\varphi^b \setminus K_b) \longrightarrow \varphi^b$ such that*

$$\begin{aligned} & h(1, \varphi^b \setminus K_b) \subseteq \varphi^a \\ \text{and} \quad & h(t, x) = x \qquad \text{for all } (t, x) \in [0,1] \times \varphi^a. \end{aligned}$$

REMARK 4.6.2 So according to this theorem φ^a is a strong deformation retract of $\varphi^b \setminus K_b$.

The critical point theory for smooth functionals (i.e., $\varphi \in C^1(X)$), using the deformation approach can be found in the books by Chang [143], Struwe [564], and Willem [606].

4.2: The study of critical points of not necessarily quadratic functionals, started with the work of Ljusternik [391], who considered C^2-functionals on finite dimensional manifolds and introduced for that purpose the notion of category (see Definition 4.2.6). Soon thereafter these ideas were pursued further by Ljusternik–Schnirelmann [393, 394]. They exploited the fact that a compact set has a neighborhood of the same category, to calculate categories by means of elementary concepts of combinatorial topology. A notion closely related to the genus (see Definition 4.2.13), was first introduced by Yang [610], under the name B-index and denoted by $i(A)$. In fact $i(A) \leq \gamma(A)$ and $i(S^n) = n$, where $S^n = \partial B_1(0)$ in $\mathbb{R}^{n+1}$. The genus was first considered by Krasnoselskii [361], although the definition we use here is due to Coffmann [156]. Discussions of the Ljusternik–Schnirelmann theory and of the notions of category and genus can be found in Deimling [188], Gasiński–Papageorgiou [258], Schwartz [545], Struwe [564], and Zeidler [619]. Extensions to Banach manifolds with Finsler structure (see Definition 4.2.24) were established by Palais [471], Szulkin [569], and Ghoussoub [262].

4.3: The spectral properties of the negative Laplacian with Dirichlet or Neumann conditions are standard and various parts of them can be found in Evans [229] and Jost [335] (see also Gasiński–Papageorgiou [258] for more general linear elliptic operators). The p-Laplacian differential operator has been the object of intense research for the last fifteen years. Here in developing the spectral properties of $-\triangle_p$ we follow the works of Anane [25] and Lindqvist [384, 385]. Theorem 4.3.56 is due to Anane–Tsouli [26]. Theorem 4.3.61 was proved by Drabek–Robinson [206], and Theorem 4.3.66 is due to Cuesta [168]. For the eigenvalues of the vector ordinary p-Laplacian, we refer to Manasevich–Mawhin [406, 407], and Mawhin [416].

4.4: Theorem 4.4.1 is essentially due to Amann [16]. Discussion of abstract nonlinear eigenvalue problems can be found in Berger [68], Blanchard–Brüning [79], and Zeidler [620].

4.5: Bifurcation theory has its roots in the work of Poincaré [498]. His work was dealing with the determination of equilibrium forms for a rotating ideal fluid. The local bifurcation theory (Theorem 4.5.14) is due to Krasnoselskii [362], and the global bifurcation theory (Theorem 4.5.17) is due to Rabinowitz [502]. However, here we follow the approach of Ize [325] (see also Krawcewicz–Wu [364]). Many aspects of local and global bifurcation bifurcation theory are discussed in Berger [68], Krasnoselskii–Zabreiko [363], Deimling [188], Zeidler [619], Dancer [174], Nussbaum [465], and Krawcewicz–Wu [364].

5

Boundary Value Problems–Hamiltonian Systems

Summary. *Having the general abstract theories developed in Chapter 4, in this chapter we deal with concrete boundary value problems (both for ordinary and partial differential equations). In the first section, we illustrate the uses of the so-called "variational method", which is based on the critical point theory. We consider Dirichlet and Neumann problems for nonlinear problems driven by the p-Laplacian as well as semilinear periodic systems with indefinite linear part. We prove existence and multiplicity results. Then we illustrate the method of "upper-lower solutions". In fact combining this method with variational arguments, we prove three solutions theorems for p-Laplacian pde's. We also deal with nonlinear nonvariational boundary value and provide at unifying framework that enables us to deal at the same time with the Dirichlet, Neumann, Sturm–Liouville, and periodic problems. Subsequently we illustrate the "degree-theoretical approach", by proving multiplicity results for both ode's and pde's problems. In Section 5.4, we deal with perturbed elliptic eigenvalue problems and we prove existence and nonexistence results using Pohozaev identity. Then we prove maximum and comparison principles for the p-Laplacian. These results are useful tools when studying the existence of multiple nontrivial solutions. Finally, we examine Hamiltonian systems and deal with the minimal period and prescribed energy level problems.

Introduction

In this chapter we use the general theories of the previous two chapters in order to solve certain boundary value problems for nonlinear ordinary and elliptic partial differential equations.

In Section 5.1, we use the critical point theory of Chapter 4 in order to illustrate the variational method for the study of boundary value problems. We consider Dirichlet and Neumann problems for nonlinear elliptic equations driven by the p-Laplacian differential operator and semilinear periodic systems with indefinite linear part. We prove existence and multiplicity results.

In Section 5.2 we present the method of upper and lower solutions. First we consider a perturbed eigenvalue problem with the p-Laplacian and a nonlinear perturbation, which is p-superlinear. Combining the method of upper and lower solutions, with some variational arguments, we show that the problem has at least

N.S. Papageorgiou, S.Th. Kyritsi-Yiallourou, *Handbook of Applied Analysis*, Advances in Mechanics and Mathematics 19, DOI 10.1007/b120946_5, three nontrivial solutions when $\lambda > \lambda_2$(= the second eigenvalue of $(-\Delta_p, W_0^{1,p}(Z))$). In the second part of the section, we use the method to solve an ordinary differential equation with the scalar p-Laplacian, a nonlinearity that may be discontinuous and nonlinear and with multivalued boundary conditions which provide a unified treatment for the Dirichlet, Neumann, and Sturm–Liouville problems. Moreover, our method of proof also applies to the periodic problem. We also establish the existence of extremal solutions in the order interval formed by an ordered pair of upper and lower solutions.

In Section 5.3, we illustrate the degree-theoretical approach in the study of nonlinear boundary value problems. First using the degree map for $(S)_+$-operators (see Section 5.3), we prove the existence of multiple solutions of constant sign for nonlinear elliptic equations with the p-Laplacian. We also consider a periodic problem with the scalar ordinary p-Laplacian and we prove two existence theorems. In the first we impose a uniform nonresonance condition between two successive eigenvalues and in the second we employ certain Landesman–Lazer type conditions.

In Section 5.4 we study two perturbed elliptic eigenvalue problems. The first is nonlinear with the p-Laplacian and the second semilinear with the Laplacian. For the nonlinear one, using the Ambrosetti–Rabinowitz condition and the mountain pass theorem (see Theorem 4.1.24), we prove the existence of a nontrivial smooth solution. For the semilinear problem, we assume a critical nonlinear perturbation. We establish the so-called Pohozaev identity, which we subsequently use to show that the semilinear problem cannot have a positive solution when $\lambda \notin (0, \lambda_1)$ and the domain is star-shaped. For the case $\lambda \in (0, \lambda_1)$ we state a result that reveals an interesting dependence on the dimension N of the domain.

In Section 5.5 we present some maximum and comparison principles involving the p-Laplacian. These results extend well-known results concerning the Laplacian and are useful in proving positivity and multiplicity results for the solutions of nonlinear boundary value problems.

Finally in Section 5.6, we study periodic Hamiltonian systems. We prove two existence theorems. One for a prescribed minimal period and the other for a prescribed energy level. The approach is variational.

5.1 Variational Method

In this section we use the variational method to solve some characteristic second-order boundary value problems (both for ordinary and partial differential equations). So our tools come from critical point theory (see Section 4.1) and the spectral properties of the Laplacian and of the p-Laplacian (see Section 4.3).

Throughout this section $Z \subseteq \mathbb{R}^N$ is a bounded domain with a C^2-boundary ∂Z. We consider the following problem,

$$\left\{ \begin{array}{l} -\operatorname{div}\big(\|Dx(z)\|^{p-2}Dx(z)\big) = f\big(z, x(z)\big) \ \text{ a.e. on } Z, \\ x\big|_{\partial Z} = 0, \ \ 1<p<\infty \end{array} \right\}. \tag{5.1}$$

We look for nontrivial positive solutions of problem (5.1) under conditions that make the Euler (energy) functional of problem (5.1) noncoercive, in fact indefinite. More precisely the hypotheses on the nonlinearity $f(z, x)$ are the following,

$\mathbf{H(f)_1}$: $f\colon Z \times \mathbb{R} \longrightarrow \mathbb{R}$ is a function such that $f(z,0)=0$ a.e. on Z and

(i) For all $x \in \mathbb{R}$, $z \longrightarrow f(z,x)$ is measurable.

(ii) For almost all $z \in Z$, $x \longrightarrow f(z,x)$ is continuous.

(iii) $|f(z,x)| \leq \alpha(z) + c|x|^{p-1}$ for a.a. $z \in Z$, all $x \in \mathbb{R}$, and with $\alpha \in L^\infty(Z)_+$, $c>0$.

(iv) There exists $\vartheta \in L^\infty(Z)_+$ such that $\vartheta(z) \geq \lambda_1$ a.e. on Z with strict inequality on a set of positive measure and $\vartheta(z) \leq \liminf\limits_{x \to +\infty} \frac{f(z,x)}{x^{p-1}}$ uniformly for a.a. $z \in Z$.

(v) There exists $\eta \in L^\infty(Z)_+$ such that $\eta(z) \leq \lambda_1$ a.e. on Z with strict inequality on a set of positive measure and $\limsup\limits_{x \to 0^+} \frac{f(z,x)}{x^{p-1}} \leq \eta(z)$ uniformly for a.a. $z \in Z$.

REMARK 5.1.1 Hypotheses $(H)(f)_1$(iv) and (v) are nonuniform nonresonance conditions near $+\infty$ and 0^+, respectively, at the principal eigenvalue λ_1 of the $\left(-\Delta_p, W_0^{1,p}(Z)\right)$. When $p=2$ (semilinear problem), these hypotheses incorporate in our framework the so-called *asymptotically linear problems*. Note that because $f(z,0) = 0$ a.e., problem (5.1) has a trivial solution. Here we are interested in nontrivial positive solutions.

We consider the Lipschitz truncation function $\tau\colon \mathbb{R} \longrightarrow \mathbb{R}_+$ defined by

$$\tau(x) = \begin{cases} 0 & \text{if } x \leq 0 \\ x & \text{if } x > 0 \end{cases}$$

and set $f_1(z,x) = f\big(z,\tau(x)\big)$. Note that $f_1(z,x)$ is still a Carathéodory function (i.e., f_1 is measurable in $z \in Z$, for all $x \in \mathbb{R}$ and continuous in $x \in \mathbb{R}$ for a.a. $z \in Z$). We consider the Euler functional for problem (5.1) when the right-hand side nonlinearity is $f_1(z,x)$. So $\varphi_1\colon W_0^{1,p}(Z) \longrightarrow \mathbb{R}$ is defined by

$$\varphi_1(x) = \frac{1}{p}\|Dx\|_p^p - \int_Z F_1\big(z,x(z)\big)dz,$$

where $F_1(z,x) = \int_0^x f_1(z,r)dr$ for all $(z,x) \in Z \times \mathbb{R}$ (the potential (or primitive) function corresponding to $f_1(z,x)$). We know that $\varphi_1 \in C^1\big(W_0^{1,p}(Z)\big)$.

PROPOSITION 5.1.2 *If hypotheses $H(f)_1$ hold, then φ_1 satisfies the PS-condition.*

PROOF: Let $\{x_n\}_{n\geq 1} \subseteq W_0^{1,p}(Z)$ be a sequence such that

$$|\varphi_1(x_n)| \leq M_1 \quad \text{for some } M_1>0, \text{ all } n \geq 1 \text{ and } \varphi'(x_n) \longrightarrow 0 \text{ as } n \to \infty.$$

Let $A\colon W_0^{1,p}(Z) \longrightarrow W^{-1,p'}(Z) = W_0^{1,p}(Z)^*$, $(1/p)+(1/p')=1$ be the nonlinear operator defined by

$$\langle A(x), y\rangle = \int_Z \|Dx\|^{p-2}(Dx, Dy)_{\mathbb{R}^N}\,\mathrm{d}z \qquad \text{for all } x, y \in W_0^{1,p}(Z).$$

Hereafter by $\langle\cdot,\cdot\rangle$ we denote the duality brackets for the pair $\big(W_0^{1,p}(Z), W^{-1,p'}(Z)\big)$. We know that this operator is strictly monotone, maximal monotone and of type $(S)_+$ (see Proposition 4.3.41). Also let $N_{f_1}: W_0^{1,p}(Z) \longrightarrow L^{p'}(Z)$ be the Nemitsky operator corresponding to the function f_1; that is, $N_{f_1} = f_1\big(\cdot, x(\cdot)\big)$ for all $x \in W_0^{1,p}(Z)$. We know that this map is bounded continuous (see Proposition 1.1.28).

We have

$$\varphi_1'(x_n) = A(x_n) - N_{f_1}(x_n) \quad \text{for all } n \geq 1.$$

We claim that $\{x_n\}_{n\geq 1} \subseteq W_0^{1,p}(Z)$ is bounded. Suppose that this is not the case. We may assume that $\|x_n\| \longrightarrow \infty$. Set $y_n = x_n/\|x_n\|$, $n\geq 1$. We can say (at least for a subsequence) that

$$y_n \xrightarrow{w} y \ \text{ in } W_0^{1,p}(Z) \ \text{ and } \ y_n \longrightarrow y \ \text{ in } L^p(Z) \text{ as } n\to\infty.$$

Because of hypothesis $\mathrm{H}(f)_1$(iii), we have

$$\frac{|f_1(z, x_n(z))|}{\|x_n\|^{p-1}} \leq \frac{\alpha(z)}{\|x_n\|^{p-1}} + c\,|y_n(z)|^{p-1} \quad \text{a.e. on } Z, \tag{5.2}$$

$$\Rightarrow \Big\{\frac{f_1(\cdot, x_n(\cdot))}{\|x_n\|^{p-1}}\Big\}_{n\geq 1} \subseteq L^{p'}(Z) \text{ is bounded.}$$

So we may assume that

$$\frac{f_1(\cdot, x_n(\cdot))}{\|x_n\|^{p-1}} = \frac{N_{f_1}(x_n)}{\|x_n\|^{p-1}} \xrightarrow{w} h \quad \text{in } L^{p'}(Z) \text{ as } n\to\infty.$$

Given $\varepsilon>0$ and $n\geq 1$, we consider the set

$$C_{\varepsilon,n} = \Big\{z\in Z : x_n(z)>0,\ \frac{u_n(z)}{x_n(z)^{p-1}} \geq \vartheta(z) - \varepsilon\Big\}, \quad n\geq 1.$$

Note that for a.a. $z \in \{y>0\}$, we have $x_n(z) \longrightarrow +\infty$ as $n\to\infty$. Then because of hypothesis $\mathrm{H}(f)_1$(iv), we have $\chi_{C_{\varepsilon,n}}(z) \longrightarrow 1$ a.e. on $\{y>0\}$. By the dominated convergence theorem, we have

$$\Big\|(1-\chi_{C_{\varepsilon,n}})\frac{N_{f_1}(x_n)}{\|x_n\|^{p-1}}\Big\|_{L^1(\{y>0\})} \longrightarrow 0,$$

$$\Rightarrow \chi_{C_{\varepsilon,n}}\frac{N_{f_1}(x_n)}{\|x_n\|^{p-1}} \xrightarrow{w} h \quad \text{in } L^{p'}(Z) \quad \text{as } n\to\infty.$$

From the definition of the truncated nonlinearity $f_1(z,x)$ and (5.2), we see that $h(z) = 0$ a.e. on $\{y>0\}$. We have

$$\begin{aligned}\chi_{C_{\varepsilon,n}}(z)\,\frac{f_1(z, x_n(z))}{\|x_n\|^{p-1}} &= \frac{f_1(z, x_n(z))}{x_n(z)^{p-1}}\,y_n(z)^{p-1}\\ &\geq \chi_{C_{\varepsilon,n}}(z)\big(\vartheta(z)-\varepsilon\big)y_n(z)^{p-1} \quad \text{a.e. on } Z.\end{aligned}$$

Taking weak limits in $L^1(\{y>0\})$, using Mazur's lemma and letting $\varepsilon\downarrow 0$, we obtain

$$h(z) \geq \vartheta(z)\, y^+(z)^{p-1} \qquad \text{a.e. on } Z.$$

Also note that $h(z) \leq c\, y^+(z)^{p-1}$ a.e on Z. So it follows that

$$h(z) = g(z)\, y^+(z)^{p-1} \quad \text{a.e. on } Z \tag{5.3}$$

with $g \in L^\infty(Z)_+, g(z) \geq \vartheta(z)$ a.e. on Z. From the choice of the sequence $\{x_n\}_{n\geq 1} \subseteq W_0^{1,p}(Z)$ we have

$$|\langle A(x_n), y_n - y\rangle - \int_Z N_{f_1}(x_n)(y_n - y)\mathrm{d}z| \leq \varepsilon_n \|y_n - y\| \quad \text{with } \varepsilon_n \downarrow 0.$$

Dividing by $\|x_n\|^{p-1}$, we obtain

$$\big|\langle A(y_n), y_n - y\rangle - \int_Z \frac{N_{f_1}(x_n)}{\|x_n\|^{p-1}}(y_n - y)\mathrm{d}z\big| \leq \frac{\varepsilon_n}{\|x_n\|^{p-1}}\|y_n - y\|. \tag{5.4}$$

Evidently we have

$$\int_Z \frac{N_{f_1}(x_n)}{\|x_n\|^{p-1}}(y_n - y)\mathrm{d}z \longrightarrow 0,$$
$$\Rightarrow \ \langle A(x_n), y_n - y\rangle \longrightarrow 0 \quad (\text{see } (5.4)). \tag{5.5}$$

Because A is of type $(S)_+$, from (5.5) we infer that $y_n \longrightarrow y$ in $W_0^{1,p}(Z)$. Recall that

$$\big|\langle A(y_n), v\rangle - \int_Z \frac{N_{f_1}(x_n)}{\|x_n\|^{p-1}} v\mathrm{d}z\big| \leq \varepsilon_n\|v\| \qquad \text{for all } v \in W_0^{1,p}(Z).$$

In the limit as $n \to \infty$, we obtain

$$\langle A(y), v\rangle = \int_Z g(y^+)^{p-1} v\mathrm{d}z \qquad \text{for all } v \in W_0^{1,p}(Z). \tag{5.6}$$

From (5.6) it follows that

$$\left\{ \begin{array}{ll} -\mathrm{div}\big(\|Dy(z)\|^{p-2}Dy(z)\big) = g(z)\big(y^+(z)\big)^{p-1} & \text{a.e. on } Z, \\ y^+\big|_{\partial Z} = 0 & \end{array} \right\}. \tag{5.7}$$

Using the test function $-y^-$, we obtain $y = y^+ \geq 0$. Then by virtue of Proposition 4.3.57

$$\lambda_1(g) \leq \lambda_1(\vartheta) < \lambda_1(\overline{\lambda}_1) = 1, \tag{5.8}$$

with $\overline{\lambda}_1$ being the principal eigenvalue of $\big(-\Delta_p, W_0^{1,p}(Z), m = 1\big)$.

Combining (5.7), (5.8), and Proposition 4.3.45, we infer that $y = 0$. Because

$$\big|\langle A(y_n), y_n\rangle - \int_Z \frac{N_{f_1}(x_n)}{\|x_n\|^{p-1}} y_n\mathrm{d}z\big| \leq \varepsilon_n' \quad \text{with } \varepsilon_n' \downarrow 0$$

and clearly $\int_Z \frac{N_{f_1}(x_n)}{\|x_n\|^{p-1}} y_n\mathrm{d}z \longrightarrow 0$, we obtain $\|Dy_n\|_p \longrightarrow 0$, hence $y_n \longrightarrow 0$ in $W_0^{1,p}(Z)$, a contradiction to the fact that $\|y_n\| = 1$ for all $n \geq 1$. So $\{x_n\}_{n\geq 1} \subseteq W_0^{1,p}(Z)$ is bounded and we may assume that $x_n \xrightarrow{w} x$ in $W_0^{1,p}(Z)$. Because

$$\big|\langle A(x_n), x_n - x\rangle - \int_Z N_{f_1}(x_n)(x_n - x)\mathrm{d}z\big| \leq \varepsilon\|x_n - x\| \qquad \text{for all } n \geq 1 \tag{5.9}$$

and $\int_Z N_{f_1}(x_n)(x_n - x)dz \longrightarrow 0$, from (5.9) and the type $(S)_+$ property of A, we obtain that $x_n \longrightarrow x$ in $W_0^{1,p}(Z)$. This proves that φ_1 satisfies the PS-condition. □

Next we show that φ_1 satisfies the mountain pass geometry (see Theorem 4.1.24). We start with a lemma that highlights the significance of the nonuniform nonresonance condition at 0^+ (see hypothesis $\mathrm{H}(f)_1$(v)).

LEMMA 5.1.3 *If $\eta \in L^\infty(Z)_+$ satisfies $\eta(z) \leq \lambda_1$ a.e. on Z with strict inequality on a set of positive measure, then there exists $\xi > 0$ such that $\psi(x) = \|Dx\|_p^p - \int_Z \eta(z)|x(z)|^p dz \geq \xi \|Dx\|_p^p$ for all $x \in W_0^{1,p}(Z)$.*

PROOF: Note that $\psi \geq 0$. Suppose that the Lemma is not true. Exploiting the p-homogeneity of ψ, we can find $\{x_n\}_{n\geq 1} \subseteq W_0^{1,p}(Z)$ with $\|Dx_n\|_p = 1$ such that $\psi(x_n) \downarrow 0$. We may assume that

$$x_n \xrightarrow{w} x \text{ in } W_0^{1,p}(Z) \quad \text{and} \quad x_n \longrightarrow x \text{ in } L^p(Z).$$

Clearly ψ is w-lower semicontinuous on $W_0^{1,p}(Z)$ and so we have $\psi(x) \leq 0$. Hence

$$\|Dx\|_p^p \leq \int_Z \eta |x|^p dz \leq \overline{\lambda}_1 \|x\|_p^p, \tag{5.10}$$

$$\Rightarrow \ x = 0 \text{ or } x = \pm u_1 \text{ (the principal eigenfunction of } \big(-\Delta_p, W_0^{1,p}(Z)\big)).$$

If $x=0$, then $\|Dx_n\|_p \longrightarrow 0$, which is a contradiction to the fact that $\|Dx_n\|_p = 1$ for all $n \geq 1$.

So $x = \pm u_1$. Recall that $|x(z)| > 0$ for all $z \in Z$ (see Proposition 4.3.42). Then from the first inequality in (5.10) and the hypothesis in $\eta \in L^\infty(Z)_+$, we obtain

$$\|Dx\|_p^p < \overline{\lambda}_1 \|x\|_p^p,$$

which contradicts the variational characterization of $\lambda_1 > 0$ (see Theorem 4.3.47) with $m = 1$. □

PROPOSITION 5.1.4 *If hypotheses $H(f)_1$ hold, then there exists $\varrho > 0$ such that $\varphi_1(x) \geq \beta > 0$ for all $x \in W_0^{1,p}(Z)$ with $\|x\| = \varrho$.*

PROOF: By virtue of hypothesis $\mathrm{H}(f)_1$(v), given $\varepsilon > 0$ we can find $\delta = \delta(\varepsilon) > 0$ such that

$$f_1(z,x) = f(z,x) \leq \big(\eta(z) + \varepsilon\big) x^{p-1} \quad \text{for a.a. } z \in Z \text{ and all } x \in (0, \delta],$$

$$\Rightarrow F_1(z,x) \leq (1/p)\big(\eta(z) + \varepsilon\big) x^p \quad \text{for a.a. } z \in Z \text{ and all } x \in (0, \delta]. \tag{5.11}$$

On the other hand, from hypothesis $\mathrm{H}(f)_1$(iii) and the mean value theorem, we have

$$F_1(z,x) \leq c_1 x^r \quad \text{for a.a. } z \in Z, \text{ all } x \geq \delta \text{ and with } c_1 > 0,$$

$$p < r \leq p^* = \begin{cases} \frac{pN}{N-p} & \text{if } p < N \\ +\infty & \text{if } p \geq N \end{cases}. \tag{5.12}$$

Because $F_1(z,x) = 0$ for a.a. $z \in Z$ and all $x \leq 0$, from (5.11) and (5.12) it follows that

$$F_1(z,x) \leq \frac{1}{p}\big(\eta(z)+\varepsilon\big)|x|^p + c_1|x|^r \quad \text{for a.a. } z\in Z, \text{ and all } x\in\mathbb{R}. \tag{5.13}$$

Hence if $x \in W_0^{1,p}(Z)$, we have

$$\begin{aligned}\varphi_1(x) &= \frac{1}{p}\|Dx\|_p^p - \int_Z F_1\big(z,x(z)\big)\mathrm{d}z \\ &\geq \frac{1}{p}\|Dx\|_p^p - \frac{1}{p}\int_Z \eta|x|^p\mathrm{d}z - \frac{\varepsilon}{p}\|x\|_p^p - c_1\|x\|^r \quad \text{(see (5.13))} \\ &\geq \frac{1}{p}\Big(\xi - \frac{\varepsilon}{\lambda_1}\Big)\|Dx\|_p^p - c_2\|Dx\|_p^r \quad \text{for some } c_2>0 \end{aligned} \tag{5.14}$$

(see Lemma 5.1.3).

Choose $\varepsilon<\lambda_1\xi$. Because $p<r$, from (5.14) and Poincaré's inequality it follows that if we choose $\varrho>0$ small, then

$$\varphi_1(x) \geq \beta>0 \qquad \text{for all } x \in W_0^{1,p}(Z), \text{ with } \|x\| = \varrho.$$

□

PROPOSITION 5.1.5 *If hypotheses $H(f)_1$ hold, then $\varphi_1(tu_1) \longrightarrow -\infty$ as $t \longrightarrow +\infty$.*

PROOF: Using hypotheses $H(f)_1$(iii), (iv), and the mean value theorem, we see that given $\varepsilon>0$ we can find $c_\varepsilon>0$ such that

$$F_1(z,x) \geq \frac{1}{p}\big(\vartheta(z)-\varepsilon\big)x^p - c_\varepsilon \quad \text{for a.a. } z\in Z, \text{ and all } x\geq 0. \tag{5.15}$$

Then for $t>0$, we have

$$\begin{aligned}\varphi_1(tu_1) &= \frac{t^p}{p}\|Du_1\|_p^p - \int_Z F_1\big(z,tu_1(z)\big)\mathrm{d}z \\ &\leq \frac{t^p}{p}\|Du_1\|_p^p - \frac{t^p}{p}\int_Z \vartheta|u_1|^p\mathrm{d}z + \frac{\varepsilon t^p}{p}\|u_1\|_p^p + c_\varepsilon|Z|_N \quad \text{(see (5.15))} \\ &= \frac{t^p}{p}\Big[\int_Z(\lambda_1-\vartheta)|u_1|^p\mathrm{d}z + \varepsilon\Big] + c_\varepsilon|Z|_N \\ &\qquad \text{(because } \|Du_1\|_p^p = \lambda_1\|u_1\|_p^p \text{ and } \|u_1\|_p = 1). \end{aligned} \tag{5.16}$$

Because $u_1(z)>0$ for all $z\in Z$, because of hypothesis $H(f)_1$(iv) we have

$$\int_Z(\lambda_1-\vartheta)|u_1|^p\mathrm{d}z = \gamma<0.$$

So if we choose $\varepsilon<-\gamma$, then $\gamma+\varepsilon<0$ and from (5.16) we conclude that $\varphi_1(tu_1) \longrightarrow -\infty$ as $t \longrightarrow +\infty$. □

THEOREM 5.1.6 *If hypotheses $H(f)_1$ hold, then problem (5.1) has a nontrivial solution $x \in C_0^1(\overline{Z})$ such that $x(z) \geq 0$ for all $z \in \overline{Z}$.*

PROOF: Because of Proposition 5.1.5, we can find $t>0$ large such that

$$\varphi_1(tu_1) \le 0 = \varphi_1(0) < \beta \le \inf[\varphi_1(v) : \|v\| = \varrho] \quad \text{and} \quad \varrho < \|tu_1\|. \tag{5.17}$$

Then (5.17) together with Proposition 5.1.2 permit the use of Theorem 4.1.24 (the mountain pass theorem) which gives $x \in W_0^{1,p}(Z)$ such that

$$\varphi_1(0) = 0 < \beta \le \varphi_1(x) \quad \text{and} \quad \varphi_1'(x) = 0. \tag{5.18}$$

From the inequality in (5.18), we infer that $x \neq 0$. Also from the equation $\varphi_1'(x) = 0$, we obtain

$$A(x) = N_{f_1}(x). \tag{5.19}$$

We act with the test function $-x^- \in W_0^{1,p}(Z)$. Recalling that $Dx^+(z) = 0$ a.e. on $\{x \le 0\}$ and that $f_1(z,x) = 0$ for a.a. $z \in Z$ and all $x \le 0$, we obtain

$$\|Dx^-\|_p = 0 \ \ (\text{i.e.,} \ \ x \ge 0).$$

Also from (5.19) it follows that $x \in W_0^{1,p}(Z)_+$ solves problem (5.1). Finally Theorem 4.3.35 implies that $x \in C_0^1(\overline{Z})$. □

We can improve the conclusion of this theorem, by strengthening a little the hypotheses on the nonlinearity $f(z,x)$.

H(f)$_2$: $f\colon Z \times \mathbb{R} \longrightarrow \mathbb{R}$ is a function such that $f(z,0) = 0$ a.e. on Z, it satisfies hypotheses $H(f)_1$(i)$\longrightarrow$(v), and

(vi) There exists $c_0 > 0$ such that $-c_0 x^{p-1} \le f(z,x)$ for a.a. $z \in Z$ and $x \ge 0$.

THEOREM 5.1.7 *If hypotheses $H(f)_2$ hold, then problem (5.1) has a nontrivial solution $x \in \operatorname{int} C_0^1(\overline{Z})_+$.*

PROOF: Theorem 5.1.6 gives a solution $x \in C_0^1(\overline{Z})_+ \setminus \{0\}$. Because of hypothesis $H(f)_2$(vi) we have

$$\operatorname{div}\big(\|Dx(z)\|^{p-2} Dx(z)\big) \le c_0 x(z)^{p-1} \quad \text{a.e. on } Z.$$

Invoking Theorem 4.3.37, we conclude that $x \in \operatorname{int} C_0^1(\overline{Z})_+$. □

It is clear from the above analysis that if the asymptotic conditions on the slopes $\big(f(z,x)\big)/\big(|x|^{p-2}x\big)$ are symmetric for $\pm\infty$ and for $0^\pm$, we can have a multiplicity result for problem (5.1). So we assume that:

H(f)$_3$: $f\colon Z \times \mathbb{R} \longrightarrow \mathbb{R}$ is a function such that $f(z,0) = 0$ a.e. on Z, it satisfies hypotheses $H(f)_3$(i),(ii),(iii), and

(iv) $\liminf\limits_{|x| \to \infty} \frac{f(z,x)}{|x|^{p-2}x} \ge \vartheta(z)$ uniformly for a.a. $z \in Z$ with $\vartheta \in L^\infty(Z)_+$ as in hypothesis $H(f)_1$(iv).

(v) $\limsup\limits_{x \to 0} \frac{f(z,x)}{|x|^{p-2}x} \le \eta(z)$ uniformly for a.a. $z \in Z$ with $\eta \in L^\infty(Z)_+$ as in hypothesis $H(f)_1$(v).

THEOREM 5.1.8 *If hypotheses $H(f)_3$ hold, then problem (5.1) has two nontrivial solutions $x, u \in C_0^1(\overline{Z})$ such that $u(z) \leq 0 \leq x(z)$ for all $z \in Z$.*

Also if we strengthen hypotheses $\mathrm{H}(f)_3$ we can have a stronger conclusion.

H(f)$_4$: $f: Z \times \mathbb{R} \longrightarrow \mathbb{R}$ is a function such that $f(z,0) = 0$ a.e. on Z, it satisfies hypotheses $\mathrm{H}(f)_3$(i)$\longrightarrow$(v), and

(vi) for almost all $z \in Z$ and all $x \in \mathbb{R}$, $f(z,x)x \geq 0$ (sign condition).

THEOREM 5.1.9 *If hypotheses $H(f)_4$ hold, then problem (5.1) has two nontrivial solutions x, u such that $x, -u \in \mathrm{int} C_0^1(\overline{Z})_+$.*

Next turn our attention to the Neumann problem:

$$\left\{ \begin{array}{l} -\mathrm{div}\big(\|Dx(z)\|^{p-2}Dx(z)\big) = f\big(z, x(z)\big) + h(z) \;\; \text{a.e. on } Z, \\ \\ \frac{\partial x}{\partial n} = 0 \text{ on } \partial Z, \; 1<p<\infty, \; h \in L^\infty(Z) \end{array} \right\}. \tag{5.20}$$

In this case the hypotheses on the nonlinearity $f(z,x)$ are the following.

H(f)$_5$: $f: Z \times \mathbb{R} \longrightarrow \mathbb{R}$ is a function such that

(i) For all $x \in \mathbb{R}$, $z \longrightarrow f(z,x)$ is measurable.

(ii) For almost all $z \in Z$, $x \longrightarrow f(z,x)$ is continuous.

(iii) $|f(z,x)| \leq \alpha(z) + c|x|^{r-1}$ for a.a. $z \in Z$, all $x \in \mathbb{R}$ and with $\alpha \in L^\infty(Z)_+$, $c > 0$, $1 \leq r < p^*$.

(iv) If $F(z,x) = \int_Z f(z,r)\mathrm{d}r$, then the limits $\lim\limits_{x \to \pm\infty} F(z,x) = F_\pm(z)$ exist for almost all $z \in Z$, and there is $M > 0$ such that $F(z,x) \geq F_+(z)$ for a.a. $z \in Z$, all $x \geq M$ and $F(z,x) \geq F_-(z)$ for a.a. $z \in Z$, all $x \leq -M$.

(v) There exist $\xi \in L^1(Z)$ and $x_0 \neq 0$ such that $F(z,x) \leq \xi(z)$ for a.a. $z \in Z$ and also $\int_Z F(z,x_0)\mathrm{d}z > 0$.

Let $h \in L^\infty(Z)$ be such that $\int_Z h(z)\mathrm{d}z = 0$. We consider the following auxiliary problem.

$$\left\{ \begin{array}{l} -\mathrm{div}\big(\|Dx(z)\|^{p-2}Dx(z)\big) = h(z) \;\; \text{a.e. on } Z, \\ \\ \frac{\partial x}{\partial n} = 0 \text{ on } \partial Z \end{array} \right\}. \tag{5.21}$$

PROPOSITION 5.1.10 *If $h \in L^\infty(Z)$ and $\int_Z h(z)\mathrm{d}z = 0$, then problem (5.21) has a unique solution $v_0 \in C_0^1(\overline{Z})$ such that $\int_Z v_0(z)\mathrm{d}z = 0$.*

PROOF: Let $\psi\colon W^{1,p}(Z)\longrightarrow\mathbb{R}$ be the C^1–function defined by

$$\psi(x)=\frac{1}{p}\|Dx\|_p^p-\frac{1}{p}\int_Z hx\mathrm{d}z \qquad \text{for all } x\in W^{1,p}(Z).$$

We consider the direct sum decomposition

$$W^{1,p}(Z)=\mathbb{R}\oplus V,$$

where $V=\{v\in W^{1,p}(Z):\int_Z v(z)\mathrm{d}z=0\}$. Because of the hypothesis on h we have $\psi|_{\mathbb{R}}=0$. Let $\widehat{\psi}=\psi|_V$ (the restriction of ψ on V). From the Poincaré–Wirtinger inequality, we see that $\widehat{\psi}$ is coercive and also it is clear that $\widehat{\psi}\colon V\longrightarrow\mathbb{R}$ is weakly lower semicontinuous. So by the Weierstrass theorem (see Theorem 2.1.10(a)), we can find $v_0\in V$ such that

$$\begin{aligned}&-\infty<m_0=\inf_V\widehat{\psi}=\widehat{\psi}(v_0),\\ \Rightarrow\ &\widehat{\psi}'(v_0)=0 \ \text{ in } \ V^*.\end{aligned} \tag{5.22}$$

Note that if $x\in W^{1,p}(Z)$, then $x=\overline{x}+\widehat{x}$ uniquely with $\overline{x}\in\mathbb{R}$, $\widehat{x}\in V$, and

$$\psi(x)=\psi(\overline{x}+\widehat{x})=\frac{1}{p}\|Dx\|_p^p-\frac{1}{p}\int_Z h\widehat{x}\mathrm{d}z=\psi(\widehat{x}).$$

Hence if $p_V:W^{1,p}(Z)\longrightarrow V$ is a projection map (i.e., $p_V(x)=p_V(\overline{x}+\widehat{x})=\widehat{x}\in V$), then

$$\psi=\widehat{\psi}\circ p_V.$$

Thus by the chain rule, we obtain

$$\psi'(x)=p_V^*\widehat{\psi}\,'(p_V(x)) \qquad \text{for all } x\in W^{1,p}(Z).$$

Let $\langle\cdot,\cdot\rangle_V$ denote the duality brackets for the pair (V,V^*) and $\langle\cdot,\cdot\rangle$ the duality brackets for the pair $(W^{1,p}(Z),W^{1,p}(Z)^*)$. For any $x,y\in W^{1,p}(Z)$, we have

$$\begin{aligned}\langle\psi'(x),y\rangle&=\left\langle p_V^*\widehat{\psi}\,'(p_V(x)),y\right\rangle=\left\langle\widehat{\psi}\,'(p_V(x)),p_V(y)\right\rangle_V\\ \Rightarrow\langle\psi'(v_0),y\rangle&=\left\langle\widehat{\psi}\,'(v_0),p_V(y)\right\rangle_V=\left\langle\widehat{\psi}\,'(v_0),\widehat{y}\right\rangle_V=0 \quad (\text{see (5.22)}).\end{aligned}$$

Because $y\in W^{1,p}(Z)$ was arbitrary, it follows that $\psi'(v_0)=0$ in $W^{1,p}(Z)^*$. Hence

$$A(v_0)=h, \tag{5.23}$$

where $A\colon W^{1,p}(Z)\longrightarrow W^{1,p}(Z)^*$ is defined by

$$\langle A(x),y\rangle=\int_Z\|Dx\|^{p-2}(Dx,Dy)_{\mathbb{R}^N}\mathrm{d}z \qquad \text{for all } x,y\in W^{1,p}(Z).$$

For every $\vartheta\in C_c^\infty(Z)$ we have

$$\int_Z\|Dv_0\|^{p-2}(Dv_0,D\vartheta)_{\mathbb{R}^N}\mathrm{d}z=\int_Z h\vartheta\mathrm{d}z \qquad (\text{see (5.23)}). \tag{5.24}$$

From the representation theorem for the elements of $W^{-1,p'}(Z)=W_0^{1,p}(Z)^*$, we infer that $\operatorname{div}(\|Dx\|^{p-2}Dx)\in W^{-1,p'}(Z)$. Let $\langle\cdot,\cdot\rangle_0$ denote the duality brackets for the pair $\big(W_0^{1,p}(Z), W^{-1,p'}(Z)\big)$. From (5.24), we have

$$\big\langle -\operatorname{div}(\|Dv_0\|^{p-2}Dv_0), \vartheta\big\rangle_0 = \langle h, \vartheta\rangle_0 \qquad \text{for all } \vartheta \in C_c^\infty(Z). \tag{5.25}$$

But $C_c^\infty(Z)$ is dense in $W_0^{1,p}(Z)$. So from (5.25) it follows that

$$-\operatorname{div}\big(\|Dv_0(z)\|^{p-2}Dv_0(z)\big) = h(z) \qquad \text{a.e. on } Z.$$

Also using Theorem 4.3.72 (nonlinear Green's identity), for every $y\in W^{1,p}(Z)$, we have

$$\int_Z \|Dv_0\|^{p-2}(Dv_0, Dy)_{\mathbb{R}^N}\,dz + \int_Z y\operatorname{div}(\|Dv_0\|^{p-2}Dv_0)dz = \left\langle \frac{\partial v_0}{\partial n_p}, \gamma_0(y)\right\rangle_{\partial Z}, \tag{5.26}$$

where $\partial v_0/\partial n_p = \|Dv_0\|^{p-2}(Dx,n)_{\mathbb{R}^N}$ on ∂Z, γ_0 is the trace map and $\langle\cdot,\cdot\rangle_{\partial Z}$ denotes the duality brackets for the pair $\big(W^{(1/p),p}(\partial Z), W^{-(1/p),p'}(\partial Z)\big)$. Combining (5.24) through (5.26), we infer that

$$\left\langle \frac{\partial v_0}{\partial n_p}, \gamma_0(y)\right\rangle_{\partial Z} = 0. \tag{5.27}$$

Recall that $\gamma_0\big(W^{1,p}(Z)\big)=W^{-(1/p'),p}(\partial Z)$. So from (5.27) we obtain

$$\frac{\partial v_0}{\partial n_p} = 0 \quad \text{in the sense of traces.}$$

Therefore $x \in W^{1,p}(Z)$ solves problem (5.21). Moreover, from Theorem 4.3.35 and Remark 4.3.36, we have that $x \in C^1(\overline{Z})$ and so the Neumann boundary condition is understood in a pointwise sense. □

Using this auxiliary result, we can prove an existence theorem for the nonlinear Neumann problem (5.20).

THEOREM 5.1.11 *If hypotheses $H(f)_5$ hold and $h\in L^\infty(Z)$ satisfies $\int_Z h(z)dz = 0$, then problem (5.20) has a nontrivial solution $x \in C^1(\overline{Z})$.*

PROOF: We consider the C^1-Euler functional $\varphi\colon W^{1,p}(Z)\longrightarrow\mathbb{R}$ for problem (5.20), defined by

$$\begin{aligned}\varphi(x) &= \frac{1}{p}\|Dx\|_p^p - \int_Z F\big(z,x(z)\big)dz - \int_Z h(z)x(z)dz\\ &\geq \frac{1}{p}\|D\widehat{x}\|_p^p - \|\xi\|_1 - c_1\|D\widehat{x}\|_p \text{ for some } c_1>0,\end{aligned} \tag{5.28}$$

(see hypothesis $\mathrm{H}(f)_5$(v)).

Here we have used the fact that $x=\overline{x}+\widehat{x}$ with $\overline{x}\in\mathbb{R}$, $\widehat{x}\in V$, that $\int_Z h(z)dz = 0$, and the Poincaré–Wirtinger inequality. From (5.28) it is clear that φ is bounded below.

Let $-\infty < m = \inf\limits_{W^{1,p}(Z)} \varphi$ and consider a minimizing sequence $\{x_n\}_{n\geq 1} \subseteq W^{1,p}(Z)$; that is, $\varphi(x_n) \downarrow m$ as $n \to \infty$. If $\{x_n\}_{n\geq 1} \subseteq W^{1,p}(Z)$ is bounded in $W^{1,p}(Z)$, then we may assume that $x_n \xrightarrow{w} x$ in $W^{1,p}(Z)$ and $x_n \longrightarrow x$ in $L^p(Z)$. Exploiting the weak lower semicontinuity of the Euler functional φ, we have $m = \varphi(x)$ and $\varphi'(x) = 0$. Note that $\varphi(x) \leq \varphi(x_0) < 0 = \varphi(0)$ and so $x \neq 0$. From the equation $\varphi'(x) = 0$, as in the proof of Proposition 5.1.10, we infer that x solves problem (5.20) and $x \in C^1(\overline{Z})$. Therefore, if m is not attained, then $\{x_n\}_{n\geq 1} \subseteq W^{1,p}(Z)$ is unbounded. We have $x_n = \overline{x}_n + \widehat{x}_n$, with $\overline{x}_n \in \mathbb{R}$, $\widehat{x}_n \in V$, $n \geq 1$. From (5.28) and Young's inequality, we obtain

$$\begin{aligned} &\varphi(x_n) \geq c_2 \|D\widehat{x}_n\|_p^p - c_3 \quad \text{for some } c_2, c_3 > 0, \text{ all } n \geq 1, \\ \Rightarrow\ &\{\widehat{x}_n\}_{n\geq 1} \subseteq W^{1,p}(Z) \text{ is bounded (Poincaré–Wirtinger inequality).} \end{aligned}$$

Because $\{x_n\}_{n\geq 1} \subseteq W^{1,p}(Z)$ is unbounded, we must have $|\overline{x}_n| \longrightarrow \infty$ as $n \to \infty$. Suppose that $\overline{x}_n \longrightarrow +\infty$ as $n \to \infty$. Because $\{\widehat{x}_n\}_{n\geq 1} \subseteq W^{1,p}(Z)$ is bounded, we may assume that

$$\widehat{x}_n \xrightarrow{w} \widehat{x} \ \text{ in } W^{1,p}(Z), \quad \widehat{x}_n \longrightarrow \widehat{x} \ \text{ in } L^p(Z), \quad \widehat{x}_n(z) \longrightarrow \widehat{x}(z) \quad \text{a.e. on } Z$$

and $|\widehat{x}_n(z)| \leq k(z)$ for a.a. $z \in Z$, all $n \geq 1$, with $k \in L^p(Z)$. Hence

$$\begin{aligned} &x_n(z) = \overline{x}_n + \widehat{x}_n(z) \geq \overline{x}_n - k(z) \qquad \text{a.e. on } Z, \\ \Rightarrow\ &x_n(z) \longrightarrow +\infty \quad \text{a.e. on } Z \text{ as } n \to \infty. \end{aligned}$$

Therefore we have

$$\begin{aligned} m &= \lim_{n\to\infty} \Big[\frac{1}{p}\|Dx_n\|_p^p - \int_Z F\big(z, x_n(z)\big)dz - \int_Z h(z)x_n(z)dz\Big] \\ &\geq \liminf_{n\to\infty} \frac{1}{p}\|Dx_n\|_p^p - \limsup_{n\to\infty} \int_Z F\big(z, x_n(z)\big)dz - \lim_{n\to\infty} \int_Z h(z)\widehat{x}_n(z)dz \\ &\geq \frac{1}{p}\|Dx\|_p^p - \int_Z F_+(z)dz - \int_Z h(z)\widehat{x}(z)dz \quad \text{(by Fatou's lemma)} \\ &= \psi(\widehat{x}) - \int_Z F_+(z)dz. \end{aligned} \tag{5.29}$$

By virtue of Proposition 5.1.10 and because $\overline{x}_n \longrightarrow +\infty$, for $n \geq 1$ large we have

$$\overline{x}_n + v_0(z) \geq M \quad \text{for all } z \in \overline{Z}.$$

So by hypothesis $H(f)_5$(iv) we have

$$F\big(z, \overline{x}_n + v_0(z)\big) \geq F_+(z) \quad \text{for a.a. } z \in Z \text{ and for all } n \geq 1 \text{ large.}$$

Using this in (5.29), we obtain for all $n \geq 1$ large

$$\begin{aligned} m &\geq \psi(\widehat{x}) - \int_Z F\big(z, \overline{x}_n + v_0(z)\big)dz \\ &\geq \psi(v_0) - \int_Z F\big(z, \overline{x}_n + v_0(z)\big)dz \quad \text{(see the proof of Proposition 5.1.10)} \\ &= \varphi(\overline{x}_n + v_0) \qquad \Big(\text{because } \int_Z h(z)dz = 0\Big), \\ \Rightarrow m &= \varphi(\overline{x}_n + v_0). \end{aligned}$$

Therefore m is attained, a contradiction. Similarly, if $\overline{x}_n \longrightarrow -\infty$, then again we reach a contradiction.

It follows that m is attained; that is, we can find $x \in W^{1,p}(Z)$ such that $m = \varphi(x)$. Because $\varphi(x) \leq \varphi(x_0) < 0 = \varphi(0)$, we have $x \neq 0$. Also $\varphi'(x) = 0$ from which as we already mentioned we infer that x solves (5.20) and $x \in C^1(\overline{Z})$ (see the proof of Proposition 5.1.10). □

Next we consider a semilinear periodic system. Using the second deformation theorem (see Theorem 4.6.1), we prove a multiplicity result for such systems.

So the problem under consideration is the following:

$$\left\{ \begin{array}{l} -x''(t) - A(t)x(t) = \nabla F\big(t, x(t)\big) \ \text{a.e. on } T = [0, b], \\ x(0) = x(b),\ x'(0) = x'(b) \end{array} \right\}. \tag{5.30}$$

Let $GL_s(\mathbb{R}^N)$ be the group of $N \times N$ symmetric invertible matrices. In what follows for any $A_1, A_2 \in GL_s(\mathbb{R}^N)$, we set $A_1 \leq A_2$ if and only if $A_2 - A_1$ is positive definite and $A_1 < A_2$ if and only if $A_2 - A_1$ is strictly positive definite. In problem (5.30) $A \in L^\infty\big(T, GL_s(\mathbb{R}^N)\big)$, but note that we do not require that $A(t) \geq 0$ for almost all $t \in T$. Also $F(t, x)$ is a Carathéodory function on $T \times \mathbb{R}^N$ into $\mathbb{R}$ and for almost all $t \in T$, $F(t, \cdot) \in C^1(\mathbb{R}^N)$. By $\nabla F(t, x)$ we denote the gradient of $F(t, \cdot)$.

We consider the Sobolev space

$$W^{1,2}_{\text{per}}\big((0, b), \mathbb{R}^N\big) = \big\{x \in W^{1,2}\big((0, b), \mathbb{R}^N\big) : x(0) = x(b)\big\}.$$

Because $W^{1,2}\big((0, b), \mathbb{R}^N\big)$ is embedded continuously (in fact compactly) in $C(T, \mathbb{R}^N)$, we see that the evaluations at $t = 0$ and $t = b$ make sense and so $W^{1,2}_{\text{per}}\big((0, b), \mathbb{R}^N\big)$ is well-defined. This is a Hilbert space with inner product

$$(x, y)_{1,2} = \int_0^b \big(x(t), y(t)\big)_{\mathbb{R}^N} dt + \int_0^b \big(x'(t), y'(t)\big)_{\mathbb{R}^N} dt,$$

for all $x, y \in W^{1,2}_{\text{per}}\big((0, b), \mathbb{R}^N\big)$. The corresponding norm is of course the usual Sobolev norm

$$\|x\|^2 = \|x\|^2_{1,2} = \|x\|^2_2 + \|x'\|^2_2.$$

PROPOSITION 5.1.12 *If $A \in L^\infty\big(T, GL_s(\mathbb{R}^N)\big)$, then there exists a sequence $\big\{\lambda_n = \lambda_n(A)\big\}_{n \geq 1} \subseteq \mathbb{R}$ such that*

$$\lambda_1 \leq \lambda_2 \leq \cdots \leq \lambda_n \leq \cdots,\ \lambda_n = \lambda_n(A) \longrightarrow +\infty$$

and

$$\left\{ \begin{array}{l} -x''(t) - A(t)x(t) = \lambda_n x(t) \ \text{a.e. on } T = [0, b], \\ x(0) = x(b),\ x'(0) = x'(b) \end{array} \right\} \tag{5.31}$$

has a nontrivial solution. Moreover, we have

$$W^{1,2}_{\text{per}}\big((0, b), \mathbb{R}^N\big) = H_- \oplus H_0 \oplus H_+ \tag{5.32}$$

with
$$\begin{aligned} H_- &= \operatorname{span}\big\{x \in W^{1,2}_{\text{per}}\big((0, b), \mathbb{R}^N\big) : -x'' - A(\cdot)x = \lambda x,\ \lambda < 0\big\} \\ H_0 &= \ker\big(-x'' - A(\cdot)x\big) \\ H_+ &= \operatorname{span}\big\{x \in W^{1,2}_{\text{per}}\big((0, b), \mathbb{R}^N\big) : -x'' - A(\cdot)x = \lambda x,\ \lambda > 0\big\}. \end{aligned}$$

and the subspaces H_- and H_0 are finite-dimensional.

PROOF: Let $\lambda^* > \|A\|_\infty$ and for $h \in L^2(T, \mathbb{R}^N)$ we consider the following auxiliary periodic problem.

$$\left\{ \begin{array}{l} -x''(t) + (\lambda^* I - A(t))x(t) = h(t) \text{ a.e. on } T = [0, b], \\ x(0) = x(b), \ x'(0) = x'(b) \end{array} \right\}. \tag{5.33}$$

It is easy to see that (5.33) has a unique solution in $C(T, \mathbb{R}^N) \cap W^{2,2}((0,b), \mathbb{R}^N)$. We denote this solution by $K(h)$. So we have defined a map $K\colon L^2(T, \mathbb{R}^N) \longrightarrow L^2(T, \mathbb{R}^N)$, which clearly is linear, monotone, and self-adjoint, and because of the compact embedding of $W^{1,2}_{\text{per}}((0,b), \mathbb{R}^N)$ into $L^2(T, \mathbb{R}^N)$, it is also compact. Hence by virtue of Theorem 3.1.37, we can find $\{\mu_n\}_n \geq 1 \subseteq \mathbb{R}_+$ such that $\mu_n \longrightarrow 0$ and the μ_ns are eigenvalues of K. Set $\lambda_n = (1/\mu_n) - \lambda^*$, $n \geq 1$. Then the λ_ns with the corresponding eigenfunctions of K satisfy (5.31). If by $E_n(A)$ we denote the corresponding eigenspace and

$$H_- = \bigoplus_{k=1}^{n} E_k(A) \ \text{ if } \lambda_1 \leq \lambda_2 \leq \cdots \leq \lambda_n < \lambda_{n+1} = 0,$$

$$H_0 = \bigoplus_{k=n+1}^{m} E_k(A) \ \text{ if } \lambda_{n+1} = \cdots = \lambda_m = 0 < \lambda_{m+1}$$

$$\text{and } H_+ = \overline{\bigoplus_{k \geq n+1} E_k(A)},$$

then we have the orthogonal direct sum decomposition (5.32). Note that if

$$q_A(x, y) = \int_0^b \left[\left(x'(t), y'(t)\right)_{\mathbb{R}^N} - \left(A(t)x(t), y(t)\right)_{\mathbb{R}^N} \right] dt$$

for all $x, y \in W^{1,2}_{\text{per}}((0,b), \mathbb{R}^N)$, then H_- is the subspace where q_A is strictly negative definite, H_0 is the subspace where q_A is null and H_+ is the subspace where q_A is strictly positive definite.

Finally because $\mu \downarrow 0$, then only a finite number of $\lambda_n = (1/\mu_n) - \lambda^*$ is less than or equal to zero and so we conclude that $\dim H_-$, $\dim H_0 < +\infty$. ☐

We have the following unique continuation property for the nontrivial solutions of the eigenvalue problem (5.31).

LEMMA 5.1.13 *If $A \in L^\infty(T, GL_s(\mathbb{R}^N))$ and $x \in C^1(T, \mathbb{R}^N)$ is a nontrivial solution of (5.31), then x has finite number of zeros.*

PROOF: Suppose x has an infinite number of distinct zeros. Without any loss of generality, we may assume that the sequence of zeros $\{t_n\}_{n \geq 1} \subseteq T$ satisfies $t_n < t_{n+1}$ for all $n \geq 1$ and $t_n \longrightarrow t^*$. Let $x(t) = \left(x_k(t)\right)_{k=1}^N$. By Rolle's theorem, we can find $r_n \in (t_n, t_{n+1})$, $n \geq 1$, such that

$$x_k'(r_n) = 0 \quad \text{for all } n \geq 1.$$

Evidently $r_n \longrightarrow t^*$ as $n \to \infty$. Hence $x_k'(t^*) = 0$ for all $k \in \{1, \ldots, N\}$. Because $x_k(t^*) = x_k'(t^*) = 0$ for all $k \in \{1, \ldots, N\}$, from the theory of linear differential equations, we conclude that $x \equiv 0$, a contradiction. ☐

The next lemma gives some useful inequalities satisfied by the elements in the component spaces H_- and H_+.

LEMMA 5.1.14 (a) *There exists* $\beta_1 > 0$ *such that*

$$\|x'\|_2^2 - \int_0^b \big(A(t)x(t), x(t)\big)_{\mathbb{R}^N} dt \geq \beta_1 \|x\|^2 \qquad \textit{for all } x \in H_+.$$

(b) *There exists* $\beta_2 > 0$ *such that*

$$\|x'\|_2^2 - \int_0^b \big(A(t)x(t), x(t)\big)_{\mathbb{R}^N} dt \leq -\beta_2 \|x\|^2 \qquad \textit{for all } x \in H_-.$$

PROOF: (a) Let $\psi \colon H_+ \longrightarrow \mathbb{R}$ be the C^1–function defined by

$$\psi(x) = \|x'\|_2^2 - \int_0^b \big(A(t)x(t), x(t)\big)_{\mathbb{R}^N} dt.$$

Evidently, if $x \in H_+$, then $x'' \in L^2(T, \mathbb{R}^N)$ and so by Green's identity, we have

$$\|x'\|_2^2 = \langle -x'', x \rangle .$$

We denote for the pair $\big(W^{1,2}_{\text{per}}((0,b), \mathbb{R}^N), W^{1,2}_{\text{per}}((0,b), \mathbb{R}^N)^*\big)$ the duality brackets $\langle \cdot, \cdot \rangle$. So from the definition of H_+ (see Proposition 5.1.12), we have

$$\psi(x) = \langle -x'', x \rangle - \left\langle \widehat{A}(x), x \right\rangle \geq 0, \tag{5.34}$$

where $\widehat{A} \in \mathcal{L}\big(L^2(T, \mathbb{R}^N), L^2(T, \mathbb{R}^N)\big)$ is defined by $\widehat{A}(x)(\cdot) = A(\cdot)x(\cdot)$ (the Nemitsky operator corresponding to $A(\cdot)$). If the inequality in (a) was not true, then we could find $\{x_n\}_{n \geq 1} \subseteq H_+$ with $\|x_n\| = 1$ such that $\psi(x_n) \downarrow 0$ as $n \to \infty$. By passing to a suitable subsequence if necessary, we may assume that $x_n \xrightarrow{w} x$ in $W^{1,2}_{\text{per}}((0,b), \mathbb{R}^N)$ and $x_n \longrightarrow x$ in $C(T, \mathbb{R}^N)$. So in the limit as $n \to \infty$, we obtain

$$\psi(x) = \|x'\|_2^2 - \int_0^b \big(A(t)x(t), x(t)\big)_{\mathbb{R}^N} dt \leq 0,$$

which in conjunction with (5.34), implies

$$\psi(x) = \|x'\|_2^2 - \int_0^b \big(A(t)x(t), x(t)\big)_{\mathbb{R}^N} dt = 0. \tag{5.35}$$

Because $x \in H_+$, from (5.35) it follows that $x = 0$. Then $x_n', x_n \longrightarrow 0$ in $L^2(T, \mathbb{R}^N)$ and so $x_n \longrightarrow 0$ in $W^{1,2}_{\text{per}}((0,b), \mathbb{R}^N)$ as $n \to \infty$, a contradiction to the fact that $\|x_n\| = 1$ for all $n \geq 1$.

(b) The proof of the inequality for the elements in H_- is done similarly. $\square$

We assume that $\dim H_- = 0$ and $\dim H_0 \neq 0$.

The hypotheses on the potential function $F(t,x)$ are the following.

H(F): $F \colon T \times \mathbb{R}^N \longrightarrow \mathbb{R}$ is a function such that $F(t,0) = 0$ a.e. on T and

(i) For every $x \in \mathbb{R}^N$, $t \longrightarrow F(t,x)$ is measurable.

(ii) For almost all $t \in T$, $F(t,\cdot) \in C^1(\mathbb{R}^N)$.

(iii) $\|\nabla F(t,x)\| \leq \alpha(t)$ for a.a. $t \in T$, all $x \in \mathbb{R}^N$ and with $\alpha \in L^1(T)_+$.

(iv) There exists a function $F_\infty \in L^1(T)$ such that $\int_0^b F_\infty(t)dt \leq 0$ and

$$F(t,x) \longrightarrow F_\infty(t) \quad \text{a.e. on } T \quad \text{as } \|x\| \longrightarrow \infty$$

and if $x_n \in C^1(T,\mathbb{R}^N)$ and $\|x_n(t)\| \longrightarrow \infty$ for almost all $t \in T$, then

$$\int_0^b \big(\nabla F(t,x_n(t)), x_n(t)\big)_{\mathbb{R}^N} dt \longrightarrow 0 \quad \text{as } n \to \infty.$$

(v) There exists $\delta > 0$ such that for a.a. $t \in T$ and all $\|x\| \leq \delta$, we have

$$F(t,x) > 0 \qquad \text{for } x \neq 0.$$

(vi) If $\lambda_m > 0$ is the first strictly positive eigenvalue for problem (5.31), then

$$F(t,x) \leq \frac{\lambda_m}{2}\|x\|^2 \quad \text{for a.a. } t \in T \text{ and all } x \in \mathbb{R}^N.$$

EXAMPLE 5.1.15 The following functions satisfy hypotheses H(F). For simplicity we drop the t-dependence:

$$F_1(x) = \begin{cases} \frac{\lambda_m}{2}(\|x\|^2 - \|x\|^3) & \text{if } \|x\| \leq 1 \\ \frac{\lambda_m}{2}\frac{1}{\|x\|} - \frac{\lambda_m}{2} & \text{if } \|x\| > 1 \end{cases}, \quad F_2(x) = c\,\frac{\ln(\|x\|^2+1)}{\|x\|^2+1}, \text{ and}$$

$$F_3(x) = c\|x\|^2 e^{-\|x\|^2} \text{ with } c \leq (\lambda_m/2).$$

We consider the Euler functional $\varphi\colon W^{1,2}_{\text{per}}((0,b),\mathbb{R}^N) \longrightarrow \mathbb{R}$ for problem (5.31), defined by

$$\varphi(x) = \frac{1}{2}\|x'\|_2^2 - \int_0^b \big(A(t)x(t), x(t)\big)_{\mathbb{R}^N} dt - \int_0^b F\big(t,x(t)\big)dt$$

for all $x \in W^{1,2}_{\text{per}}((0,b),\mathbb{R}^N)$.

PROPOSITION 5.1.16 *If $A \in L^\infty(T, GL_s(\mathbb{R}^N))$, hypotheses $H(F)$ hold and $c \neq -\int_0^b F_\infty(t)dt$, then φ satisfies the C_c-condition.*

PROOF: Let $\{x_n\}_{n\geq 1} \subseteq W^{1,2}_{\text{per}}((0,b),\mathbb{R}^N)$ be a sequence such that

$$\varphi(x_n) \longrightarrow c \qquad \text{and} \qquad (1+\|x_n\|)\varphi'(x_n) \longrightarrow 0 \quad \text{as } n \to \infty.$$

Let $V \in \mathcal{L}\big(W^{1,2}_{\text{per}}((0,b),\mathbb{R}^N), W^{1,2}_{\text{per}}((0,b),\mathbb{R}^N)^*\big)$ be defined by

$$\langle V(x), y\rangle = \int_0^b \big(x'(t), y'(t)\big)_{\mathbb{R}^N} dt \qquad \text{for all } x,y \in W^{1,2}_{\text{per}}((0,b),\mathbb{R}^N).$$

Evidently V is monotone, hence maximal monotone. Also let $N\colon L^1(T,\mathbb{R}^N) \longrightarrow L^1(T,\mathbb{R}^N)$ be the Nemitsky operator corresponding to the nonlinearity ∇F; that is,

$$N(x)(\cdot) = \nabla F(\cdot, x(\cdot)) \qquad \text{for all } x \in L^1(T, \mathbb{R}^N).$$

Clearly N is continuous and bounded. Moreover, exploiting the compact embedding of $W^{1,2}_{\text{per}}\big((0,b), \mathbb{R}^N\big)$ into $C(T, \mathbb{R}^N)$ (hence into $L^1(T, \mathbb{R}^N)$ too), we deduce that $N\big|_{W^{1,2}_{\text{per}}((0,b),\mathbb{R}^N)}$ is completely continuous, thus compact (see Proposition 3.1.3). We have

$$\varphi'(x_n) = V(x_n) - \widehat{A}(x_n) - N(x_n) \qquad \text{for all } n \geq 1,$$

where as before $\widehat{A} \in \mathcal{L}\big(L^2(T, \mathbb{R}^N), L^2(T, \mathbb{R}^N)\big)$ is defined by $\widehat{A}(x)(\cdot) = A(\cdot)x(\cdot)$. We show that the sequence $\{x_n\}_{n\geq 1} \subseteq W^{1,2}_{\text{per}}\big((0,b), \mathbb{R}^N\big)$ is bounded. We proceed by contradiction. So suppose that (by passing to a subsequence if necessary), we have $\|x_n\| \longrightarrow +\infty$. We set $y_n = x_n/\|x_n\|$ for all $n \geq 1$. We may assume that

$$y_n \xrightarrow{w} y \text{ in } W^{1,2}_{\text{per}}\big((0,b), \mathbb{R}^N\big) \qquad \text{and} \qquad y_n \longrightarrow y \text{ in } C(T, \mathbb{R}^N).$$

From the choice of the sequence $\{x_n\}_{n\geq 1} \subseteq W^{1,2}_{\text{per}}\big((0,b), \mathbb{R}^N\big)$, we have

$$\begin{aligned} & \big|\big\langle \varphi'(x_n), y_n - y\big\rangle\big| \leq \varepsilon_n \text{ with } \varepsilon_n \downarrow 0, \\ \Rightarrow\ & \Big|\langle V(x_n), y_n - y\rangle - \int_0^b \big(A(t)x_n(t), y_n(t) - y(t)\big)_{\mathbb{R}^N} dt \\ & - \int_0^b \big(N(x_n)(t), y_n(t) - y(t)\big)_{\mathbb{R}^N} dt\Big| \leq \varepsilon_n. \end{aligned}$$

We divide by $\|x_n\|$ and we obtain

$$\begin{aligned} & \Big|\langle V(y_n), y_n - y\rangle - \int_0^b \big(A(t)y_n(t), y_n(t) - y(t)\big)_{\mathbb{R}^N} dt \\ & - \int_0^b \Big(\frac{N(x_n)(t)}{\|x_n\|}, y_n(t) - y(t)\Big)_{\mathbb{R}^N} dt\Big| \leq \varepsilon_n. \end{aligned} \tag{5.36}$$

Note that

$$\int_0^b \big(A(t)y_n(t), y_n(t) - y(t)\big)_{\mathbb{R}^N} dt \longrightarrow 0 \quad \text{as } n \to \infty$$

$$\text{(because } A \in L^\infty\big(T, GL_s(\mathbb{R}^N)\big))$$

$$\text{and} \quad \int_0^b \Big(\frac{N(x_n)(t)}{\|x_n\|}, y_n(t) - y(t)\Big)_{\mathbb{R}^N} dt \longrightarrow 0 \quad \text{as } n \to \infty$$

(see hypothesis H(F)(iii)).

So from (5.36), we obtain

$$\langle V(y_n), y_n - y\rangle \longrightarrow 0.$$

Because V is maximal monotone, it follows that

$$\begin{aligned} & \langle V(y_n), y_n\rangle \longrightarrow \langle V(y), y\rangle, \\ \Rightarrow\ & \|y_n'\|_2 \longrightarrow \|y'\|_2. \end{aligned}$$

Because $y_n \xrightarrow{w} y$ in $W^{1,2}_{\text{per}}\big((0,b), \mathbb{R}^N\big)$, we infer that $y_n' \longrightarrow y'$ in $L^2(T, \mathbb{R}^N)$ and so finally

$$y_n \longrightarrow y \quad \text{in } W^{1,2}_{\text{per}}\big((0,b),\mathbb{R}^N\big). \tag{5.37}$$

Recall that for all $v \in W^{1,2}_{\text{per}}\big((0,b),\mathbb{R}^N\big)$, we have

$$\Big|\langle V(y_n),v\rangle - \int_0^b \big(A(t)y_n(t),v(t)\big)_{\mathbb{R}^N}\,dt$$
$$-\int_0^b \big(\frac{Nx_n(t)}{\|x_n\|},v(t)\big)_{\mathbb{R}^N}\,dt\Big| \le \varepsilon_n \quad \text{with } \varepsilon_n \downarrow 0.$$

Passing to the limit as $n\to\infty$ and using (5.37) and hypothesis H(F)(iii), we obtain

$$\langle V(y),v\rangle = \int_0^b \big(A(t)y(t),v(t)\big)_{\mathbb{R}^N}\,dt \qquad \text{for all } v \in W^{1,2}_{\text{per}}\big((0,b),\mathbb{R}^N\big)$$
$$\Rightarrow\ V(y) = \widehat{A}(y) \ \text{ in } W^{1,2}_{\text{per}}\big((0,b),\mathbb{R}^N\big)^* \ \text{ (i.e., } y \in H_0).$$

Because $\|y_n\| = 1$ for all $n \ge 1$, from (5.37), we have $\|y\| = 1$ (i.e., $y \ne 0$). Moreover, from Lemma 5.1.13, we have $y(t) \ne 0$ a.e. on T and so $\|x_n(t)\| \longrightarrow \infty$ a.e. on T. It follows that

$$F\big(t,x_n(t)\big) \longrightarrow F_\infty(t) \qquad \text{a.e. on } T. \tag{5.38}$$

From the choice of the sequence $\{x_n\}_{n\ge1} \subseteq W^{1,2}_{\text{per}}\big((0,b),\mathbb{R}^N\big)$, we have

$$\Big|\langle V(x_n),x_n\rangle - \int_0^b \big(A(t)x_n(t),x_n(t)\big)_{\mathbb{R}^N}\,dt - \int_0^b \big(\nabla F\big(t,x_n(t)\big),x_n(t)\big)_{\mathbb{R}^N}\,dt\Big|$$
$$\le \varepsilon_n,$$
$$\Rightarrow\ \Big|\,\|x_n'\|_2^2 - \int_0^b \big(A(t)x_n(t),x_n(t)\big)_{\mathbb{R}^N}\,dt - \int_0^b \big(\nabla F\big(t,x_n(t)\big),x_n(t)\big)_{\mathbb{R}^N}\,dt\Big|$$
$$\le \varepsilon_n.$$

Because of hypothesis H(F)(iv), we have

$$\int_0^b \big(\nabla F\big(t,x_n(t)\big),x_n(t)\big)_{\mathbb{R}^N}\,dt \longrightarrow 0 \quad \text{as } n\to\infty,$$
$$\Rightarrow\ \|x_n'\|_2^2 - \int_0^b \big(A(t)x_n(t),x_n(t)\big)_{\mathbb{R}^N}\,dt \longrightarrow 0 \quad \text{as } n\to\infty. \tag{5.39}$$

Hypotheses H(F)(iii) and (iv) imply that we can find $\vartheta \in L^1(T)_+$ such that

$$|F(t,x)| \le \vartheta(t) \quad \text{for a.a. } t\in T, \text{ all } x \in \mathbb{R}^N. \tag{5.40}$$

From (5.38), (5.40), and the dominated convergence theorem, it follows that

$$\int_0^b F\big(t,x_n(t)\big)dt \longrightarrow \int_0^b F_\infty(t)dt \quad \text{as } n\to\infty. \tag{5.41}$$

By hypothesis $\varphi(x_n) \longrightarrow c$ and so given $\varepsilon>0$, we can find $n_0 = n_0(\varepsilon) \ge 1$ such that for all $n \ge n_0$ we have

$$c-\varepsilon \le \varphi(x_n) \le c+\varepsilon,$$
$$\Rightarrow\ c-\varepsilon \le \frac{1}{2}\|x_n'\|_2^2 - \frac{1}{2}\int_0^b \big(A(t)x_n(t),x_n(t)\big)_{\mathbb{R}^N}\,dt - \int_0^b F\big(t,x_n(t)\big)dt \le c+\varepsilon.$$

If we pass to the limit as $n \to \infty$ and we use (5.39) and (5.41), we obtain

$$c - \varepsilon \leq -\int_0^b F_\infty(t)dt \leq c + \varepsilon.$$

But $\varepsilon > 0$ was arbitrary. So we let $\varepsilon \downarrow 0$ to conclude that

$$c = -\int_0^b F_\infty(t)dt,$$

a contradiction. Therefore $\{x_n\}_{n\geq 1} \subseteq W^{1,2}_{\text{per}}\big((0,b),\mathbb{R}^N\big)$ is bounded. Hence we may assume that

$$x_n \xrightarrow{w} x \text{ in } W^{1,2}_{\text{per}}\big((0,b),\mathbb{R}^N\big) \qquad \text{and} \qquad x_n \longrightarrow x \text{ in } C(T,\mathbb{R}^N).$$

Then as above, we obtain

$$\lim_{n\to\infty} \langle V(x_n), x_n - x\rangle = 0,$$

from which because V is maximal monotone, we conclude that $x_n \longrightarrow x$ in $W^{1,2}_{\text{per}}\big((0,b),\mathbb{R}^N\big)$. Therefore φ satisfies the C_c-condition for $c \neq -\int_0^b F_\infty(t)dt$. □

REMARK 5.1.17 Using hypotheses H(F)(iii) and (iv) and because dim $H_- = 0$, we can easily check that φ is bounded below.

Using Lemma 5.1.14, we obtain the following.

PROPOSITION 5.1.18 *If $A \in L^\infty\big(T, GL_s(\mathbb{R}^N)\big)$, hypotheses H(F) hold and $c < 0$, then φ satisfies the PS_c-condition.*

Now we are ready for the multiplicity result concerning problem (5.30).

THEOREM 5.1.19 *If $A \in L^\infty\big(T, GL_s(\mathbb{R}^N)\big)$ and hypotheses H(F) hold, then problem (5.30) has at least two nontrivial solutions $\widehat{x}_0, \widehat{x}_1 \in C^1(T,\mathbb{R}^N)$.*

PROOF: We know that φ is bounded below (see Remark 5.1.17). So

$$-\infty < m = \inf\{\varphi(x) : x \in W^{1,2}_{\text{per}}\big((0,b),\mathbb{R}^N\big)\}.$$

By virtue of hypothesis H(F)(v), if $x \in H_0$ and $\|x(t)\| \leq \delta$ for all $t \in T$, then

$$\varphi(x) < 0 = \varphi(0) \leq -\int_0^b F_\infty(t)dt$$
$$\Rightarrow\ m < 0 = \varphi(0) \leq -\int_0^b F_\infty(t)dt.$$

Because of Proposition 5.1.18, φ satisfies the PS_m-condition and so we can find $\widehat{x}_0 \in W^{1,2}_{\text{per}}\big((0,b),\mathbb{R}^N\big)\}$ such that

$$\varphi(\widehat{x}_0) = m < 0 = \varphi(0) \qquad \text{and} \qquad \varphi'(\widehat{x}_0) = 0 \qquad \text{(see Theorem 4.1.28)}. \tag{5.42}$$

From the first relation in (5.42) we infer that $\widehat{x}_0 \neq 0$. Suppose that $\widehat{x}_0$ and 0 are the only critical points of φ. By virtue of hypothesis H(F)(v), we can find $r > 0$ small such that

$$\xi = \sup\{\varphi(x) : x \in \partial B_r(0) \cap H_0\} < 0, \tag{5.43}$$

where $\overline{B}_r(0) = \{x \in W^{1,2}_{\text{per}}((0,b),\mathbb{R}^N) : \|x\| \leq r\}$. On the other hand, if $x \in H_+$, then

$$\|x'\|_2^2 - \int_0^b \big(A(t), x_n(t), x(t)\big)_{\mathbb{R}^N} dt \geq \lambda_m \|x\|_2^2,$$

$$\Rightarrow \ \varphi(x) \geq 0 \quad \text{for all } x \in H_+ \qquad \text{(see hypothesis H}(F)\text{(vi))}.$$

Let $\Gamma = \big\{\gamma \in C\big(\overline{B}_r(0), W^{1,2}_{\text{per}}((0,b),\mathbb{R}^N)\big) : \gamma|_{\partial B_r(0)\cap H_0} = I|_{\partial B_r(0)\cap H_0}\big\}$ and consider the map $\gamma_0 : \overline{B}_r(0) \longrightarrow W^{1,2}_{\text{per}}((0,b),\mathbb{R}^N)$ defined by

$$\gamma_0(x) = \begin{cases} \widehat{x}_0 & \text{if } \|x\| < \frac{r}{2} \\ h\big(\frac{2(r-\|x\|)}{r}, \frac{rx}{\|x\|}\big) & \text{if } \|x\| \geq \frac{r}{2} \end{cases},$$

where $h(t,x)$ is the homotopy postulated by Theorem 4.6.1 (note that φ satisfies the PS_c-condition for all $c \in [m,0)$, see Proposition 5.1.18). Note that

$$\gamma_0(x) = h(1,2x) \qquad \text{for all } x \in \overline{B}_r(0) \cap H_0 \text{ with } \|x\| = \frac{r}{2}.$$

We have $\|2x\| = r$ and so $\varphi(2x) < 0$ (see (5.43)). Because by hypothesis $\widehat{x}_0$ is the only minimizer of φ on $W^{1,2}_{\text{per}}((0,b),\mathbb{R}^N)$ (recall that $m < 0 = \varphi(0)$), from Theorem 4.6.1 we infer that

$$\gamma_0(x) = h(1,2x) = \widehat{x}_0,$$

which proves the continuity of γ_0; that is, $\gamma_0 \in C\big(\overline{B}_r(0) \cap H_0, W^{1,2}_{\text{per}}((0,b),\mathbb{R}^N)\big)$. Moreover, for all $x \in \partial B_r \cap H_0$, we have

$$\gamma_0(x) = h(0,x) = x \quad \text{(see Theorem 4.6.1)}, \tag{5.44}$$

$$\Rightarrow \ \gamma_0|_{\partial B_r \cap H_0},$$

$$\Rightarrow \ \gamma_0 \in \Gamma.$$

Because h is φ-decreasing (see Theorem 4.6.1), we have

$$\varphi\big(h(s,x)\big) \leq \varphi\big(h(t,x)\big) \qquad \text{for all } t < s \text{ and all } x \in \varphi^0 \setminus \{0\}.$$

So from (5.43) and (5.44), it follows that

$$\varphi\big(\gamma_0(x)\big) < 0 \qquad \text{for all } x \in \overline{B}_r(0) \cap H_0. \tag{5.45}$$

The sets $\{\partial B_r \cap H_0, \overline{B}_r(0) \cap H_0, H_+\}$ are linking in $W^{1,2}_{\text{per}}((0,b),\mathbb{R}^N)$ via I (see Definition 4.1.20 and Example 4.1.21(b)). So

$$\gamma\big(\overline{B}_r(0) \cap H_0\big) \cap H_+ \neq \emptyset \qquad \text{for all } \gamma \in \Gamma.$$

Because $0 = \inf_{H_+} \varphi$, we have

$$\sup\big\{\varphi(\gamma(x)) : x \in \overline{B}_r(0) \cap H_0\big\} \geq 0 \qquad \text{for all } \gamma \in \Gamma,$$

$$\Rightarrow \ \sup\big\{\varphi(\gamma_0(x)) : x \in \overline{B}_r(0) \cap W\big\} \geq 0. \tag{5.46}$$

Comparing (5.45) and (5.46), we reach a contradiction. So we can find $\widehat{x}_1 \in W^{1,2}_{\text{per}}((0,b),\mathbb{R}^N)$ such that

$$\widehat{x}_1 \neq \widehat{x}_0,\ \widehat{x}_1 \neq 0 \qquad \text{and} \qquad \varphi'(\widehat{x}_1) = 0.$$

Now let $x \in W^{1,2}_{\mathrm{per}}\big((0,b),\mathbb{R}^N\big)$ be a critical point of φ; that is, $\varphi'(x) = 0$. We have

$$V(x) - \widehat{A}(x) = N(x). \tag{5.47}$$

Let $v \in C^1\big((0,b),\mathbb{R}^N\big)$. Because $-x'' \in W^{-1,2}\big((0,b),\mathbb{R}^N\big) = W^{1,2}_0\big((0,b),\mathbb{R}^N\big)^*$ and if by $\langle\cdot,\cdot\rangle_0$ we denote the duality brackets for the pair $\big(W^{1,2}_0\big((0,b),\mathbb{R}^N\big), W^{-1,2}_0\big((0,b),\mathbb{R}^N\big)\big)$, we have

$$\begin{aligned} &\langle V(x), v\rangle - \int_0^b \big(A(t)x(t), v(t)\big)_{\mathbb{R}^N} dt = \int_0^b \big(\nabla F\big(t,x(t)\big), v(t)\big)_{\mathbb{R}^N} dt, \\ \Rightarrow\ & \langle -x'', v\rangle_0 - \big\langle \widehat{A}(x), x\big\rangle_0 = \langle N(x), x\rangle_0 . \end{aligned} \tag{5.48}$$

Because $C_c\big((0,b),\mathbb{R}^N\big)$ is dense in $W^{1,2}_0\big((0,b),\mathbb{R}^N\big)$, from (5.48) we infer that

$$\left\{ \begin{array}{l} -x''(t) - A(t)x(t) = \nabla F\big(t,x(t)\big) \ \text{ a.e. on } T, \\ \\ x(0) = x(b) \end{array} \right\}, \tag{5.49}$$

$$\begin{aligned} &\Rightarrow\ x'' \in L^1(T,\mathbb{R}^N) \ \ (\text{i.e., } x' \in W^{1,1}\big((0,b),\mathbb{R}^N\big) \subseteq C(T,\mathbb{R}^N)), \\ &\Rightarrow\ x \in C^1(T,\mathbb{R}^N). \end{aligned}$$

Also if $y \in W^{1,2}_{\mathrm{per}}\big((0,b),\mathbb{R}^N\big)$, via integration by parts, we have

$$\begin{aligned} &\langle V(x), y\rangle = \int_0^b \big(x'(t), y'(t)\big)_{\mathbb{R}^N} dt \\ &= \big(x'(b), y(b)\big)_{\mathbb{R}^N} - \big(x'(0), y(0)\big)_{\mathbb{R}^N} - \int_0^b \big(x''(t), y(t)\big)_{\mathbb{R}^N} dt, \\ &\Rightarrow\ -\big\langle \widehat{A}(x), y\big\rangle - \langle N(x), y\rangle \\ &= \big(x'(b), y(b)\big)_{\mathbb{R}^N} - \big(x'(0), y(0)\big)_{\mathbb{R}^N} - \big\langle \widehat{A}(x), y\big\rangle - \langle N(x), y\rangle \end{aligned}$$

(see (5.47) and (5.49)),

$$\begin{aligned} &\Rightarrow\ \big(x'(0), y(0)\big)_{\mathbb{R}^N} = \big(x'(b), y(b)\big)_{\mathbb{R}^N} \qquad \text{for all } y \in W^{1,2}_{\mathrm{per}}\big((0,b),\mathbb{R}^N\big), \\ &\Rightarrow\ x'(0) = x'(b). \end{aligned}$$

Therefore $x \in C^1(T,\mathbb{R}^N)$ solves problem (5.30). So we conclude that both $\widehat{x}_0, \widehat{x}_1 \in C^1(T,\mathbb{R}^N)$ are nontrivial distinct solutions of (5.30). □

Continuing with multiplicity results, we use Theorem 4.1.32, to produce two nontrivial solutions for problem (5.1).

If we consider the nonlinear eigenvalue problem

$$\left\{ \begin{array}{l} -\mathrm{div}\big(\|Dx(z)\|^{p-2} Dx(z)\big) = \lambda |x(z)|^{p-2} x(z) \ \text{ a.e. on } Z, \\ \\ x\big|_{\partial Z} = 0, \ \ \lambda > 0, \ \ 1 < p < \infty \end{array} \right\}, \tag{5.50}$$

then (5.50) has a first eigenvalue $\lambda_1 > 0$, which is isolated (see Proposition 4.3.46) and simple (see Proposition 4.3.44). If by $u_1 \in C^1_0(\overline{Z})$ we denote a normalized eigenfunction corresponding to $\lambda_1 > 0$, then $\mathbb{R}u_1$ is the eigenspace corresponding to $\lambda_1 > 0$. If V is a topological complement of $\mathbb{R}u_1$ (i.e., $W_0^{1,p}(Z) = \mathbb{R}u_1 \oplus V$), then

$$\lambda_V = \inf\big\{\|Du\|_p^p : u \in V,\ \|u\|_p = 1\big\} > \lambda_1. \tag{5.51}$$

We consider the following topological complement of $\mathbb{R}u_1$,

$$V = \Big\{u \in W_0^{1,p}(Z) : \int_Z u(z)u_1(z)^{p-1}\mathrm{d}z = 0\Big\}.$$

Evidently if $p = 2$, then $V = (\mathbb{R}u_1)^{\perp}$.

The hypotheses on the nonlinearity $f(z,x)$ of (5.1) are the following.

H(f)$_6$: $f\colon Z \times \mathbb{R} \longrightarrow \mathbb{R}$ is a function such that $f(z,x) = 0$ a.e. on Z and

(i) For all $x \in \mathbb{R}$, $z \longrightarrow f(z,x)$ is measurable.

(ii) For almost all $z \in Z$, $x \longrightarrow f(z,x)$ is continuous.

(iii) For almost all $z \in Z$ and all $x \in \mathbb{R}$, we have

$$|f(z,x)| \leq \alpha(z) + c|x|^{r-1} \quad \text{with } \alpha \in L^\infty(Z)_+,\ c>0$$

$$\text{and} \quad 1 \leq r < p^* = \begin{cases} \frac{Np}{N-p} & \text{if } N>p \\ +\infty & \text{if } N \leq p \end{cases}.$$

(iv) There exist $\delta > 0$ and $\lambda \in (\lambda_1, \lambda_V)$ such that for a.a. $z \in Z$ and all $|x| \leq \delta$, we have

$$\lambda_1|x|^p \leq pF(z,x) \leq \lambda|x|^p,$$

where $F(z,x) = \int_0^x f(z,r)\mathrm{d}r$ (the potential function corresponding to $f(z,x)$);

(v) $\lim\limits_{|x|\to\infty} \frac{pF(z,x)}{|x|^p} = \lambda_1$ and $\lim\limits_{|x|\to\infty} \big(pF(z,x) - \lambda_1|x|^p\big) = -\infty$ uniformly for almost all $z \in Z$.

REMARK 5.1.20 Hypothesis H$(f)_6$(iv) implies that problem (5.1) is resonant near zero at the first eigenvalue $\lambda_1 > 0$ from the right. On the other hand by virtue of hypothesis H$(f)_6$(v), problem (5.1) is resonant near infinity from the left. Therefore we are dealing with a problem which is resonant both near zero and near infinity and we cross λ_1 as we move from 0 to infinity.

The Euler functional $\varphi\colon W_0^{1,p}(Z) \longrightarrow \mathbb{R}$ is given by

$$\varphi(x) = \frac{1}{p}\|Dx\|_p^p - \int_Z F\big(z, x(z)\big)\mathrm{d}z.$$

We know that $\varphi \in C^1\big(W_0^{1,p}(Z)\big)$.

PROPOSITION 5.1.21 *If hypotheses $H(f)_6$ hold, then φ is weakly coercive; that is, $\varphi(x) \longrightarrow +\infty$ as $\|x\| \longrightarrow +\infty$.*

PROOF: We proceed by contradiction. So suppose that φ is not weakly coercive. Then we can find a sequence $\{x_n\}_{n\geq 1} \subseteq W_0^{1,p}(Z)$ and $M>0$ such that

$$\|x_n\| \longrightarrow \infty \qquad \text{and} \qquad \varphi(x_n) \leq \qquad \text{for all } n\geq 1. \tag{5.52}$$

Set $y_n = x_n/\|x_n\|$, $n\geq 1$. We may assume that

$$y_n \xrightarrow{w} y \text{ in } W_0^{1,p}(Z) \qquad \text{and} \qquad y_n \longrightarrow y \text{ in } L^p(Z).$$

From (5.52) we have

$$\frac{1}{p}\|Dy_n\|_p^p - \int_Z \frac{F\big(z,x_n(z)\big)}{\|x_n\|^p}\mathrm{d}z \leq \frac{M}{\|x_n\|^p} \qquad \text{for all } n\geq 1. \tag{5.53}$$

From hypotheses $\mathrm{H}(f)_6$(iii), (v), and the mean value theorem, we have

$$|F(z,x)| \leq \alpha_1(z) + c_1|x|^p \quad \text{for a.a. } z\in Z, \quad \text{with } \alpha_1 \in L^\infty(Z)_+,\ c_1>0,$$

$$\Rightarrow \quad \frac{F\big(z,x_n(z)\big)}{\|x_n\|^p} \leq \frac{\alpha_1(z)}{\|x_n\|^p} + c_1|y_n(z)|^p \qquad \text{a.e. on } Z \quad \text{for all } n\geq 1. \tag{5.54}$$

Because of (5.54) and the Dunford–Pettis theorem, we may assume that

$$\frac{F\big(\cdot,x_n(\cdot)\big)}{\|x_n\|^p} \xrightarrow{w} h \quad \text{in } L^1(Z) \text{ as } n\to\infty. \tag{5.55}$$

For every $\varepsilon>0$ and every $n\geq 1$, we consider the set

$$C_{\varepsilon,n} = \Big\{z\in Z : x_n(z)\neq 0,\ \frac{1}{p}(\lambda_1-\varepsilon) \leq \frac{F\big(z,x_n(z)\big)}{|x_n(z)|^p} \leq \frac{1}{p}(\lambda_1+\varepsilon)\Big\}.$$

Due to hypothesis $\mathrm{H}(f)_6$(v), we have

$$\chi_{C_{\varepsilon,n}}(z) \longrightarrow 1 \qquad \text{a.e. on } \{y\neq 0\}.$$

Note that

$$\Big\|(1-\chi_{C_{\varepsilon,n}})\frac{F\big(\cdot,x_n(\cdot)\big)}{\|x_n\|^p}\Big\|_{L^1(\{y\neq 0\})} \longrightarrow 0 \quad \text{as } n\to\infty,$$

$$\Rightarrow \chi_{C_{\varepsilon,n}}(\cdot)\frac{F\big(\cdot,x_n(\cdot)\big)}{\|x_n\|^p} \xrightarrow{w} h \text{ in } L^1(\{y\neq 0\}) \quad (\text{see } (5.55)).$$

From the definition of the set $C_{\varepsilon,n}$, we have

$$\begin{aligned}\frac{1}{p}(\lambda_1-\varepsilon)|y_n(z)|^p &\leq \chi_{C_{\varepsilon,n}}(z)\frac{F\big(z,x_n(z)\big)}{\|x_n\|^p} = \chi_{C_{\varepsilon,n}}(z)\frac{F\big(z,x_n(z)\big)}{|x_n(z)|^p}|y_n(z)|^p\\ &\leq \frac{1}{p}(\lambda_1+\varepsilon)|y_n(z)|^p \qquad \text{a.e. on } Z.\end{aligned}$$

Taking weak limits in $L^1(\{y\neq 0\})$ and using Mazur's lemma, we obtain

$$\frac{1}{p}(\lambda_1-\varepsilon)|y_n(z)|^p \leq h(z) \leq \frac{1}{p}(\lambda_1+\varepsilon)|y_n(z)|^p \qquad \text{a.e. on } \{y\neq 0\}.$$

Because $\varepsilon>0$ was arbitrary, it follows that

$$h(z) = \frac{\lambda_1}{p}|y(z)|^p \qquad \text{a.e. on } \{y \neq 0\}. \tag{5.56}$$

On the other hand it is clear from (5.54) that

$$h(z) = 0 \qquad \text{a.e. on } \{y \neq 0\}. \tag{5.57}$$

From (5.56) and (5.57) it follows that

$$h(z) = \frac{\lambda_1}{p}|y(z)|^p \qquad \text{a.e. on } Z. \tag{5.58}$$

We return to (5.53) and pass to the limit as $n \to \infty$. Using (5.55) and (5.58), we obtain

$$\begin{aligned} &\|Dy\|_p^p \leq \lambda_1 \|y\|_p^p, \\ \Rightarrow \ & y = 0 \text{ or } y = \pm u_1 \qquad \text{(see Remark 4.3.43)}. \end{aligned}$$

If $y = 0$, then we have $y_n \longrightarrow 0$ in $W_0^{1,p}(Z)$, a contradiction to the fact that $\|y_n\| = 1,\ n \geq 1$.

If $y = \pm u_1$, then $|x_n(z)| \longrightarrow +\infty$ for a.a. $z \in Z$ (see Theorem 4.3.47). By virtue of hypothesis $H(f)_6$(v), we have

$$pF\big(z, x_n(z)\big) - \lambda_1 |x_n(z)|^p \longrightarrow -\infty \qquad \text{a.e. on } Z \text{ as } n \to \infty. \tag{5.59}$$

We have

$$\begin{aligned} \varphi(x_n) &= \frac{1}{p}\|Dx_n\|_p^p - \int_Z \Big(F\big(z, x_n(z)\big) - \frac{\lambda_1}{p}|x_n(z)|^p\Big)dz - \frac{\lambda_1}{p}\|x_n\|_p^p \\ &\geq - \int_Z \Big(F\big(z, x_n(z)\big) - \frac{\lambda_1}{p}|x_n(z)|^p\Big)dz \qquad \text{(see Remark 4.3.43)}. \end{aligned}$$

Using Fatou's lemma and (5.59), we obtain

$$\varphi(x_n) \longrightarrow +\infty \quad \text{as } n \to \infty,$$

which contradicts (5.52). This proves the weak coercivity of φ. □

COROLLARY 5.1.22 *If hypotheses $H(f)_6$ hold, then φ satisfies the PS-condition and it is bounded below.*

PROOF: Let $\{x_n\}_{n \geq 1} \subseteq W_0^{1,p}(Z)$ be a PS-sequence, namely $\{\varphi(x_n)\}_{n \geq 1} \subseteq \mathbb{R}$ is bounded and $\varphi'(x_n) \longrightarrow 0$ in $W^{-1,p'}(Z)$ as $n \to \infty$ $\big(\frac{1}{p} + \frac{1}{p'} = 1\big)$. From Proposition 5.1.21 we know that φ is weakly coercive. So it follows that $\{x_n\}_{n \geq 1} \subseteq W_0^{1,p}(Z)$ is bounded. Then arguing as in the last part of Proposition 5.1.2, we obtain $x_n \longrightarrow x$ in $W_0^{1,p}(Z)$. This proves that φ satisfies the PS-condition.

Next suppose that $\{x_n\}_{n \geq 1} \subseteq W_0^{1,p}(Z)$ satisfies

$$\varphi(x_n) \longrightarrow -\infty \quad \text{as } n \to \infty. \tag{5.60}$$

Proposition 5.1.21 implies that $\{x_n\}_{n \geq 1} \subseteq W_0^{1,p}(Z)$ is bounded and so we may assume that $x_n \xrightarrow{w} x$ in $W_0^{1,p}(Z)$. Exploiting the compact embedding of $W_0^{1,p}(Z)$

into $L^p(Z)$, we can easily see that φ is weakly lower semicontinuous on $W_0^{1,p}(Z)$. So we have

$$-\infty < \varphi(x) \leq \liminf_{n\to\infty} \varphi(x_n),$$

which contradicts (5.60). This proves that φ is bounded below. $\square$

Now we can have a multiplicity result for problem (5.1).

THEOREM 5.1.23 *If hypotheses $H(f)_6$ hold, then problem (5.1) has at least two nontrivial solutions $\widehat{x}_0, \widehat{x}_1 \in C_0^1(\overline{Z})$.*

PROOF: Recall the direct sum decomposition

$$W_0^{1,p}(Z) = \mathbb{R}u_1 \oplus V$$

with $V = \left\{u \in W_0^{1,p}(Z) : \int_Z u(z)u_1(z)^{p-1}\mathrm{d}z\right\}$.

We know that $u_1 \in C_0^1(\overline{Z})$ (see Proposition 4.3.39). So we can find $t_0 > 0$ such that

$$|tu_1(z)| \leq \delta \qquad \text{for all } z \in \overline{Z} \text{ and all } |t| \leq t_0.$$

Then hypothesis $H(f)_6$(iv) implies that

$$\lambda_1 |t|^p u_1(z)^p \leq p\, F\big(z, tu_1(z)\big) \leq \lambda |t|^p u_1(z)^p \qquad \text{a.e. on } Z, \text{ with } \lambda_1 < \lambda < \lambda_V. \tag{5.61}$$

So for $|t| \leq t_0$, we have

$$\begin{aligned} \varphi(tu_1) &= \frac{|t|^p}{p}\|Du_1\|_p^p - \int_Z F\big(z, tu_1(z)\big)\mathrm{d}z \\ &\leq \frac{|t|^p}{p}\|Du_1\|_p^p - \frac{|t|^p}{p}\lambda_1\|u_1\|_p^p \quad \text{(see (5.61))} \\ &= 0 \qquad \text{(recall that } \|Du_1\|_p^p = \lambda_1\|u_1\|_p^p, \text{ see Remark 4.3.43).} \end{aligned} \tag{5.62}$$

Note that for a.a. $z \in Z$ and all $x \in \mathbb{R}$, we have

$$F(z,x) \leq \frac{\lambda}{p}x^p + c_2|x|^r \quad \text{with } c_2 > 0 \text{ and } p < r < p^*. \tag{5.63}$$

So if $u \in V$, we have

$$\begin{aligned} \varphi(u) &= \frac{1}{p}\|Du\|_p^p - \int_Z F\big(z, u(z)\big)\mathrm{d}z \\ &\geq \frac{1}{p}\|Du\|_p^p - \frac{\lambda}{p}\|u\|_p^p - c_2\|u\|_r^r \qquad \text{(see (5.63))}, \\ &\geq \frac{1}{p}\Big(1 - \frac{\lambda}{\lambda_V}\Big)\|Du\|_p^p - c_3\|Du\|_r^r \qquad \text{for some } c_3 > 0 \end{aligned} \tag{5.64}$$

(see (5.51) and recall $r < p^*$). Recall that $\lambda < \lambda_V$ and $r > p$. So we can find $\varrho > 0$ small such that

$$\varphi(u) \geq 0 \qquad \text{for all } u \in V,\ \|u\| \leq \varrho \quad \text{(see (5.64))}. \tag{5.65}$$

On the other hand from (5.62), if $\varrho>0$ is small, we have

$$\varphi(u) \leq 0 \qquad \text{for all } u \in \mathbb{R}u_1, \text{ with } \|u\| \leq \varrho. \tag{5.66}$$

If $\inf \varphi = 0$, then because of (5.66) all the elements of $\mathbb{R}u_1 \cap \overline{B}_\varrho(0) \setminus \{0\}$ are critical points of φ and so we have a continuum of nontrivial solutions for problem (5.1), which by virtue of Theorems 4.3.34 and 4.3.35 belong in $C_0^1(\overline{Z})$.

If $\inf \varphi < 0$, then because of (5.65), (5.66), and Corollary 5.1.22, we can apply Theorem 4.1.32 and produce two nontrivial critical points $\widehat{x}_0, \widehat{x}_1 \in W_0^{1,p}(Z)$ of φ. These are solutions of (5.1) and as above the nonlinear regularity theory implies that $\widehat{x}_0, \widehat{x}_1 \in C_0^1(\overline{Z})$. □

We prove a multiplicity theorem for semilinear problems (i.e. $p = 2$). In this multiplicity theorem, we combine the variational method with the method of monotone iterations. This way we are well placed to consider the method of upper-lower solutions, examined in the next section.

The problem under consideration is the following.

$$\left\{ \begin{aligned} & -\triangle x(z) = x(z)^2 + h(z) \text{ a.e. on } Z, \\ & x\big|_{\partial Z} = 0, \ h \in L^\infty(Z) \end{aligned} \right\}. \tag{5.67}$$

First using the monotone iteration method, we produce a nontrivial solution for problem (5.67).

PROPOSITION 5.1.24 *If $h \in L^\infty(Z)$, $h \neq 0$ and $h(z) \leq 0$ a.e. on Z, then problem (5.67) has a solution $u \in -\mathrm{int}\, C_0^1(\overline{Z})_+$.*

PROOF: Let $x_0 \in H_0^1(Z)$ be the unique solution of $-\triangle x_0(z) = h(z)$ a.e. on Z. From Theorems 4.3.34 and 4.3.35 we have that $x_0 \in C_0^1(\overline{Z})$. Moreover, Theorem 4.3.37 implies that $x_0 \in -\mathrm{int}\, C_0^1(\overline{Z})_+$. We set

$$M = 2\big|\min_{z \in \overline{Z}} x_0(z)\big| > 0.$$

Given $y \in H_0^1(Z)$, we consider the problem

$$\left\{ \begin{aligned} & -\triangle u(z) + Mu(z) = My(z) + y(z)^2 + h(z) \text{ a.e. on } Z, \\ & y\big|_{\partial Z} = 0 \end{aligned} \right\}. \tag{5.68}$$

Problem (5.68) has a unique solution in $C_0^1(\overline{Z})$ denoted by $G(y)$. So we define a map $G\colon H_0^1(Z) \longrightarrow H_0^1(Z)$. Evidently a fixed point of G solves problem (5.67).

Claim: If $y \in H_0^1(Z)$, $x_0 \leq y \leq 0$, then $x_0 \leq G(y) \leq 0$.

We have $My + y^2 = y(M + y) \leq 0$. So from (5.68) and Theorem 4.3.37 it follows that

$$u = G(y) \leq 0.$$

Then

$$-\triangle(u - x_0) + M(u - x_0) = My + y^2 + h - h - Mx_0 = M(y - x_0) + y^2$$
$$\geq 0,$$
$$\Rightarrow \ u \geq x_0 \qquad \text{(see Theorem 4.3.37).}$$

In a similar fashion, we show that G is increasing; that is,

$$x_0 \leq y_1 \leq y_2 \leq 0 \Rightarrow G(y_1) \leq G(y_2). \tag{5.69}$$

Now let $u_0 = 0$ and define $u_{n+1} = G(u_n)$, $n \geq 0$. By virtue of the claim and (5.69), we have

$$x_0 \leq \cdots \leq u_n \leq u_{n-1} \leq \cdots \leq u_0 = 0, \quad n \geq 1. \tag{5.70}$$

We have

$$|u_n| \leq |x_0| \qquad \text{for all } n \geq 1,$$
$$\Rightarrow \ \{u_n\}_{n\geq 1} \subseteq L^2(Z) \quad \text{is bounded.} \tag{5.71}$$

Moreover, by definition we have

$$-\triangle u_{n+1} + Mu_{n+1} = Mu_n + u_n^2 + h,$$
$$\Rightarrow \ \|u_{n+1}\|^2 + M\|u_{n+1}\|_2^2 = M\int_Z u_n u_{n+1} \mathrm{d}z + \int_Z u_n^2 u_{n+1} \mathrm{d}z + \int_Z h u_{n+1} \mathrm{d}z$$
$$\leq (M\|u_n\|_2 + \|h\|_2)\|u_{n+1}\|$$
$$\leq M_1\|u_{n+1}\| \text{ for some } M_1 > 0, \text{ all } n \geq 1$$

(see (5.71)),

$$\Rightarrow \ \{u_n\}_{n\geq 1} \subseteq H_0^1(Z) \text{ is bounded.}$$

From this and (5.70), it follows that for the whole sequence $\{u_n\}_{n\geq 1}$, we have $u_n \xrightarrow{w} u$ in $H_0^1(Z)$ with $u(z) = \inf_{n\geq 1} u_n(z)$. If $A \in \mathcal{L}(H_0^1(Z), H^{-1}(Z))$, is defined by

$$\langle A(v), y\rangle = \int_Z (Dv, Dy)_{\mathbb{R}^N} \mathrm{d}z \qquad \text{for all } v, y \in H_0^1(Z),$$

then A is maximal monotone and

$$Au_{n+1} + Mu_{n+1} = Mu_n + u_n^2 + h, \tag{5.72}$$
$$\Rightarrow \ \lim \langle Au_{n+1}, u_{n+1} - u\rangle = 0,$$
$$\Rightarrow \ u_{n+1} \longrightarrow u \text{ in } H_0^1(Z).$$

So if we pass to the limit as $n \to \infty$ in (5.72), we obtain

$$Au = u^2 + h,$$
$$\Rightarrow \ u \text{ solves problem (5.67) and } u \in C_0^1(\overline{Z}).$$

Finally, it is clear that $u \neq 0$ and from Theorem 4.3.37 we have $u \in -\mathrm{int}\, C_0^1(\overline{Z})$. ∎

If $h \geq 0$, $h \neq 0$, it can happen that problem (5.67) has no solution.

EXAMPLE 5.1.25 Let $u_1 \in C_0^1(\overline{Z})$ be the principal eigenfunction of $\big(-\triangle, H_0^1(Z)\big)$. We know that $u_1(z) > 0$ for all $z \in Z$. Let $h(z) = tu_1(z)$ and $t > 0$ to be specified in the sequel. Suppose that we could find $u \in H_0^1(Z)$ a solution of (5.67) with h as above. We have

$$\lambda_1 \int_Z uu_1 dz = \int_Z (Du, Du_1)_{\mathbb{R}^N} dz = \int_Z u^2 u_1 dz + \int_Z hu_1 dz$$
$$= \int_Z u^2 u_1 dz + t \qquad (\text{because } \|u_1\|_2 = 1). \tag{5.73}$$

Note that

$$\int_Z uu_1 dz = \int_Z (u\sqrt{u_1}\,)\sqrt{u_1} dz$$
$$\leq \Big(\int_Z u_1 dz\Big)^{1/2} \Big(\int_Z u^2 u_1 dz\Big)^{1/2}.$$

Using the estimate in (5.73), we obtain

$$\int_Z u^2 u_1 dz + t \leq \lambda_1 c \Big(\int_Z u^2 u_1 dz\Big)^{1/2} \quad \text{with } c = \Big(\int_Z u_1 dz\Big)^{1/2} = \sqrt{\|u_1\|_1}.$$

But this is possible only if $4t \leq \lambda_1^2 c^2$. So if $t > \left(\frac{\lambda_1 c}{2}\right)^2$, then problem (5.67) with $h = tu_1$ has no solution.

Now using the variational method we produce a second solution for problem (5.67). So we have the following multiplicity theorem.

THEOREM 5.1.26 *If $h \in L^\infty(Z)$, $h \neq 0$, and $h(z) \leq 0$ a.e. on Z, then problem (5.67) has at least two nontrivial solutions $u, v \in C_0^1(\overline{Z})$.*

PROOF: The first solution $u \in C_0^1(\overline{Z})$ was obtained in Proposition 5.1.24.

Next using the mountain pass theorem (see Theorem 4.1.24), we produce a second nontrivial solution for problem (5.67).

Let $y = u + \varepsilon w$, $w \in H_0^1(Z)$, $\|w\| = 1$, $\varepsilon > 0$. We have

$$\varphi(u + \varepsilon w) - \varphi(u) = \varepsilon \int_Z (Du, Dw)_{\mathbb{R}^N} dz + \frac{\varepsilon^2}{2}\|Dw\|_2^2 - \varepsilon \int_Z u^2 w dz$$
$$-\varepsilon \int_Z uw^2 dz - \frac{\varepsilon^3}{3}\int_Z w^3 dz - \varepsilon \int_Z hw dz$$
$$= \frac{\varepsilon^2}{2}\|Dw\|_2^2 - \varepsilon \int_Z uw^2 dz - \frac{\varepsilon^3}{3}\int_Z w^3 dz, \tag{5.74}$$

the second equality following from the fact that $u \in C_0^1(\overline{Z})$ solves (5.67). Because $u \in -\mathrm{int}\, C_0^1(\overline{Z})_+$ (see Proposition 5.1.24), we have $\int_Z uw^2 dz \leq 0$. Also $\left|\int_Z w^2 dz\right| \leq c_1$ for some $c_1 > 0$ and all $\|w\| = 1$. Therefore

$$\varphi(u + \varepsilon w) - \varphi(u) \geq \varepsilon^2 \Big(\frac{1}{2} - \frac{c_1 \varepsilon}{3}\Big) \qquad (\text{see (5.74)}). \tag{5.75}$$

Choosing $0<\varepsilon<\frac{3}{4}c_1$, from (5.75) we have

$$\varphi(y)-\varphi(u)\geq\frac{\varepsilon^2}{4}\qquad\text{for all } y\in H_0^1(Z) \text{ with } \|y-u\|=\varepsilon. \tag{5.76}$$

Next let $y=tu_1$. Then

$$\begin{aligned}\varphi(y)&=\frac{t^2}{2}\lambda_1-\frac{t^3}{3}\|u_1\|_3^3-t\int_Z hu_1\mathrm{d}z,\\ \Rightarrow\ \varphi(y)&\longrightarrow-\infty\quad\text{as } t\longrightarrow+\infty.\end{aligned}$$

Therefore we can find $y\in H_0^1(Z)$ such that

$$\varphi(y)<\varphi(u)\qquad\text{and}\qquad\|y-u\|>\varepsilon. \tag{5.77}$$

Claim: φ satisfies the PS-condition.

Suppose $\{x_n\}_{n\geq1}$ is a PS-sequence; that is,

$$|\varphi(x_n)|\leq c_2\quad\text{for some } c_2>0 \text{ and all } n\geq1 \tag{5.78}$$
$$\text{and}\quad \varphi'(x_n)\longrightarrow0\quad\text{in } H^{-1}(Z). \tag{5.79}$$

Because of (5.79), we have

$$\begin{aligned}&\langle\varphi'(x_n),x_n\rangle=\|Dx_n\|_2^2-\int_Z x_n^3\mathrm{d}z-\int_Z hx_n\mathrm{d}z\leq\varepsilon_n\|x_n\|\quad\text{with } \varepsilon_n\downarrow0,\\ \Rightarrow\ &3\varphi(x_n)-\frac{1}{2}\|Dx_n\|_2^2+2\int_Z hx_n\mathrm{d}z\leq\varepsilon_n\|x_n\|\quad\text{(see (5.78))}.\end{aligned}$$

From Poincaré's inequality and (5.78), it follows that $\{x_n\}_{n\geq1}\subseteq H_0^1(Z)$ is bounded. So we may assume that

$$x_n\xrightarrow{w}x\ \text{in } H_0^1(Z)\qquad\text{and}\qquad x_n\longrightarrow x\ \text{in } L^2(Z).$$

Then as before, due to the maximal monotonicity of $A\in\mathcal{L}\big(H_0^1(Z),H^{-1}(Z)\big)$, we conclude that $x_n\longrightarrow x$ in $H_0^1(Z)$ and so φ satisfies the PS-condition.

Because of (5.76), (5.77), and the claim, we can apply the mountain pass theorem (see Theorem 4.1.24) and obtain $v\in H_0^1(Z)$ such that

$$\varphi(v)\geq\varphi(u)+\frac{\varepsilon^2}{4},\ \text{i.e. } v\neq w,\ v\neq0$$

and $\varphi'(v)=0$.

So v solves problem (5.67) and also $v\in C_0^1(\overline{Z})$ (regularity theory). $\square$

We conclude this section with a remarkable theorem, which provides a bridge between variational methods and methods based on the nonlinear strong maximum principle. The result is a valuable tool in the study of nonlinear boundary value problems involving the p-Laplacian. For a proof of it we refer to Gasiński–Papageorgiou [258, p. 655].

So let $f:Z\times\mathbb{R}\longrightarrow\mathbb{R}$ be a Carathéodory function such that

$$|f(z,x)|\leq\alpha(z)+c|x|^{r-1}\quad\text{for a.a. } z\in Z \text{ and all } x\in\mathbb{R},$$

with $\alpha \in L^\infty(Z)_+$, $c>0$ and $1 \leq r<p^*$.

We introduce the C^1-functional $\varphi : W_0^{1,p}(Z) \longrightarrow \mathbb{R}$ defined by

$$\varphi(x) = \frac{1}{p}\|Dx\|_p^p - \int_Z F\big(z,x(z)\big)\mathrm{d}z \qquad \text{for all } x \in W_0^{1,p}(Z),$$

where $F(z,x) = \int_0^x f(z,r)\mathrm{d}r$.

THEOREM 5.1.27 *If $\varphi \in C^1(W_0^{1,p}(Z))$ is as above and $x_0 \in W_0^{1,p}(Z)$ is a local $C_0^1(\overline{Z})$-minimizer of φ; that is, there exists $\varrho>0$ such that*

$$\varphi(x_0) \leq \varphi(x_0+y) \qquad \textit{for all } \|y\|_{C_0^1(\overline{Z})} \leq \varrho,$$

then $x_0 \in C_0^1(\overline{Z})$ and it is also a local $W_0^{1,p}(Z)$-minimizer of φ; that is, there exists $\varrho_0>0$ such that

$$\varphi(x_0) \leq \varphi(x_0+y) \qquad \textit{for all } \|y\| \leq \varrho_0.$$

5.2 Method of Upper–Lower Solutions

In this section we present the method of upper and lower solutions in the study of nonlinear boundary value problems. With this method, via truncation and penalization techniques, we produce solutions that are located in the order interval determined by an ordered pair of upper and lower solutions, which serve as upper and lower bounds, respectively. Under appropriate monotonicity conditions on the data, this method leads to monotone iterative processes that are amenable to numerical treatment. For a given problem one may try several different methods in order to produce the necessary ordered pair of upper and lower solutions. There is no specific methodology and in general one should try simple functions (such as constants, linear, quadratic, exponential, eigenfunctions of simpler operators, solutions of simpler auxiliary equations, etc.).

Here we illustrate this method by analyzing two nonlinear boundary value problems. The first deals with elliptic differential equations, whereas the second is a boundary value problem for ordinary differential equations with general boundary conditions which unify the Dirichlet, Neumann, and Sturm–Liouville problems.

We start with a nonlinear eigenvalue problem driven by the p-Laplacian. So let $Z \subseteq \mathbb{R}^N$ be a bounded domain with a C^2-boundary ∂Z. We consider the following problem,

$$\left\{\begin{array}{l} -\mathrm{div}\big(\|Dx(z)\|^{p-2}Dx(z)\big) = \lambda|x(z)|^{p-2}x(z) - f\big(z,x(z)\big) \ \text{ a.e. on } Z, \\ x\big|_{\partial Z} = 0, \ \ \lambda \in \mathbb{R}, \ \ 1<p<\infty \end{array}\right\}. \tag{5.80}$$

We produce a multiplicity result for problem (5.80) under the hypothesis that the right-hand side nonlinearity $f(z,x)$ exhibits a p-superlinear growth both near 0 and $\pm\infty$. For this purpose our hypotheses on $f(z,x)$ are the following.

$\mathbf{H(f)_1}$: $f\colon Z \times \mathbb{R} \longrightarrow \mathbb{R}$ is a function such that $f(z,0) = 0$ a.e. on Z and

(i) For all $x \in \mathbb{R}$, $z \longrightarrow f(z,x)$ is measurable.

(ii) For almost all $z \in Z$, $x \longrightarrow f(z,x)$ is continuous and $f(z,x)x \geq 0$.

(iii) For almost all $z \in Z$ and all $x \in \mathbb{R}$, we have

$$|f(z,x)| \leq \alpha(z) + c|x|^{r-1} \quad \text{with } \alpha \in L^\infty(Z)_+,\ c>0$$

$$\text{and} \qquad 1 \leq r < p^* = \begin{cases} \frac{Np}{N-p} & \text{if } N>p \\ +\infty & \text{if } N \leq p \end{cases}.$$

(iv) $\lim\limits_{x \to 0} \frac{f(z,x)}{|x|^{p-2}x} = 0$ uniformly for almost all $z \in Z$.

(v) $\lim\limits_{|x| \to \infty} \frac{f(z,x)}{|x|^{p-2}x} = +\infty$ uniformly for almost all $z \in Z$.

REMARK 5.2.1 When $p=2$ (semilinear problem), hypotheses $H(f)_1$(iv) and (v) imply that the nonlinearity $f(z,x)$ is superlinear both at zero and at infinity.

From hypotheses $H(f)_1$(iii), (iv), and (v), we see that given $\mu > 0$, we can find $\vartheta_\mu \in L^\infty(Z)_+$, $\vartheta_\mu \neq 0$, such that

$$f(z,x) > \mu|x|^{p-2}x - \vartheta_\mu \quad \text{for a.a. } z \in Z \text{ and all } x \geq 0 \tag{5.81}$$

$$\text{and} \quad f(z,x) < \mu|x|^{p-2}x + \vartheta_\mu \quad \text{for a.a. } z \in Z \text{ and all } x \leq 0. \tag{5.82}$$

We use (5.81) and (5.82) to produce an ordered pair of upper and lower solutions for problem (5.80). Let us start by recalling the definitions of upper and lower solutions for (5.80).

DEFINITION 5.2.2 (a) A function $\overline{x} \in W^{1,p}(Z)$ is an *upper solution* for problem (5.80), if $\overline{x}\big|_{\partial Z} \geq 0$ and

$$\int_Z \|D\overline{x}\|^{p-2}(D\overline{x}, Du)_{\mathbb{R}^N}\,dz \geq \lambda \int_Z |\overline{x}|^{p-2}\overline{x}u\,dz - \int_Z f(z,\overline{x})u\,dz$$

for all $u \in W_0^{1,p}(Z)$, $u \geq 0$.

(b) A function $\underline{x} \in W^{1,p}(Z)$ is a *lower solution* for problem (5.80), if $\underline{x}\big|_{\partial Z} \leq 0$ and

$$\int_Z \|D\underline{x}\|^{p-2}(D\underline{x}, Du)_{\mathbb{R}^N}\,dz \leq \lambda \int_Z |\underline{x}|^{p-2}\underline{x}u\,dz - \int_Z f(z,\underline{x})u\,dz$$

for all $u \in W_0^{1,p}(Z)$, $u \geq 0$.

Motivated by inequality (5.81), we consider the following auxiliary boundary value problem.

$$\left\{ \begin{array}{l} -\operatorname{div}\big(\|Dx(z)\|^{p-2}Dx(z)\big) = (\lambda-\mu)|x(z)|^{p-2}x(z) + \vartheta_\mu(z) \ \text{ a.e. on } Z, \\ x\big|_{\partial Z} = 0 \end{array} \right\}. \tag{5.83}$$

PROPOSITION 5.2.3 *If $\lambda > \lambda_1$ and we choose $\mu > \lambda - \lambda_1$, then problem (5.83) has a solution $\overline{x} \in \operatorname{int} C_0^1(\overline{Z})_+$.*

PROOF: In what follows by $\langle\cdot,\cdot\rangle$ we denote the duality brackets for the pair $(W_0^{1,p}(Z), W^{-1,p'}(Z))$, $(1/p)+(1/p')=1$. Let $A: W_0^{1,p}(Z)\longrightarrow W^{-1,p'}(Z)$ be the nonlinear operator defined by

$$\langle A(x),y\rangle=\int_Z \|Dx\|^{p-2}(Dx,Dy)_{\mathbb{R}^N}\,dz \qquad \text{for all } x,y\in W_0^{1,p}(Z).$$

We know that A is maximal monotone. Also let $K: W_0^{1,p}(Z)\longrightarrow L^{p'}(Z)$ be defined by

$$K(x)(\cdot)=(\lambda-\mu)|x(\cdot)|^{p-2}x(\cdot).$$

Clearly K is completely continuous (recall that $W_0^{1,p}(Z)$ is embedded compactly into $L^p(Z)$). It follows then that $A-K: W_0^{1,p}(Z)\longrightarrow W^{-1,p'}(Z)$ is pseudomonotone. Moreover, we have

$$\langle A(x)-K(x),x\rangle=\|Dx\|_p^p-(\lambda-\mu)\|x\|_p^p. \tag{5.84}$$

If $\lambda\leq\mu$, then it is clear from (5.84) that $A-K$ is coercive. If $\lambda>\mu$, then by hypothesis $\lambda-\mu=\lambda_1-\varepsilon$ with $0<\varepsilon<\lambda_1$ and so

$$\langle A(x)-K(x),x\rangle\geq\|Dx\|_p^p-\frac{\lambda_1-\varepsilon}{\lambda_1}\|Dx\|_p^p=\frac{\varepsilon}{\lambda_1}\|Dx\|_p^p,$$
$$\Rightarrow\ A-K \text{ is coercive.}$$

So in both cases $A-K$ is coercive. Then by virtue of Theorem 3.2.60 we can find $\overline{x}\in W_0^{1,p}(Z)$ such that

$$A(\overline{x})-K(\overline{x})=\vartheta_\mu,$$
$$\Rightarrow\ \overline{x}\in W_0^{1,p}(Z) \text{ is a solution of problem (5.83) and belongs in } C_0^1(\overline{Z})_+.$$

Because $\vartheta_\mu\neq 0$, it follows that $\overline{x}\neq 0$. Moreover, using as a test function $\overline{x}^-\in W_0^{1,p}(Z)$, we obtain

$$\|D\overline{x}^-\|_p^p\leq(\lambda-\mu)\|\overline{x}^-\|_p^p \qquad (\text{recall that } \vartheta_\mu\geq 0).$$

If $\lambda\leq\mu$, then $\|D\overline{x}^-\|_p=0$ and so we have $\overline{x}=0$. If $\lambda-\mu>0$, then $\lambda-\mu=\lambda_1-\varepsilon$ and so

$$\|D\overline{x}^-\|_p^p\leq(\lambda_1-\varepsilon)\|\overline{x}^-\|_p^p,$$

which contradicts Theorem 4.3.37, unless $\overline{x}\geq 0$.

Therefore we have established that $\overline{x}\geq 0$, $\overline{x}\neq 0$. Also from (5.83), we have

$$\operatorname{div}\big(\|D\overline{x}(z)\|^{p-2}D\overline{x}(z)\big)\leq|\lambda-\mu|\,|\overline{x}(z)|^{p-2}\overline{x}(z) \qquad \text{a.e. on } Z$$

and this by virtue of Theorem 4.3.37 implies that $\overline{x}\in\operatorname{int}C_0^1(\overline{Z})_+$. □

COROLLARY 5.2.4 *If $\lambda>\lambda_1$ and we have $\mu>\lambda-\lambda_1$, then the solution $\overline{x}\in\operatorname{int}C_0^1(\overline{Z})_+$ of problem (5.83) produced in Proposition 5.2.3 is an upper solution for problem (5.80).*

Hypothesis $H(f)_1$(iv) implies that given $\varepsilon>0$, we can find $\delta=\delta(\varepsilon)>0$ such that

$$f(z,x)<\varepsilon|x|^{p-2}x \quad \text{for a.a. } z\in Z \text{ and all } x\in[0,\delta] \tag{5.85}$$
$$\text{and}\quad f(z,x)>\varepsilon|x|^{p-2}x \quad \text{for a.a. } z\in Z \text{ and all } x\in[-\delta,0]. \tag{5.86}$$

The next lemma helps us generate a lower solution for problem (5.80).

LEMMA 5.2.5 *If X is an ordered Banach space with order cone K such that* int $K \neq \emptyset$ *and $x_0 \in$* int K*, then given $y \in X$, we can find $\delta > 0$ such that $\beta x_0 - y \in$* int K*.*

PROOF: Because $x_0 \in \text{int}\, K$, we can find $\delta > 0$ such that

$$\overline{B}_\delta(x_0) = \{x \in X : \|x - x_0\| \leq \delta\} \subseteq \text{int}\, K.$$

Let $y \in X$, $y \neq 0$ (if $y = 0$, then the lemma is trivially true for all $\beta > 0$). We have

$$\begin{aligned} & x_0 \pm \delta \frac{y}{\|y\|} \in \overline{B}_\delta(x_0) \subseteq \text{int}\, K, \\ \Rightarrow \quad & \frac{\|y\|}{\delta} x_0 - y \in \text{int}\, K. \end{aligned}$$

So if $\beta = \beta(y) = \|y\|/\delta$, then we have $\beta x_0 - y \in \text{int}\, K$. □

Now let $u_1 \in \text{int}\, C_0^1(\overline{Z})_+$ be the principal eigenfunction of $\big(-\triangle_p, W_0^{1,p}(Z)\big)$. Using Proposition 5.2.3 and Lemma 5.2.5, we can find $\xi > 0$ small such that

$$\xi u_1(z) \in [0, \delta] \qquad \text{for all } z \in \overline{Z} \text{ and } \xi u_1 \leq \overline{x}. \tag{5.87}$$

We set $\underline{x} = \xi u_1$. Evidently $\underline{x} \in \text{int}\, C_0^1(\overline{Z})_+$ (of course $\underline{x}$ depends on $\varepsilon > 0$).

PROPOSITION 5.2.6 *If $\lambda > \lambda_1$, then $\underline{x} \in$* int $C_0^1(\overline{Z})_+$ *defined above is a lower solution for problem (5.80).*

PROOF: Let $\varepsilon > 0$ be such that $\lambda = \lambda_1 + \varepsilon$. Then we have

$$\begin{aligned} -\text{div}\big(\|D\underline{x}(z)\|^{p-2} D\underline{x}(z)\big) &= \lambda_1 |\underline{x}(z)|^{p-2} \underline{x}(z) \\ &= (\lambda - \varepsilon)|\underline{x}(z)|^{p-2}\underline{x}(z) \\ &< \lambda |\underline{x}(z)|^{p-2}\underline{x}(z) - f\big(z, \underline{x}(z)\big) \qquad \text{a.e. on } Z \end{aligned}$$

(see (5.85) and (5.87)),

$\Rightarrow$ $\underline{x} \in \text{int}\, C_0^1(\overline{Z})_+$ is a lower solution of (5.80) (see Definition 5.2.2(b)).

□

Therefore we have produced an ordered pair $\{\underline{x}, \overline{x}\}$ of lower and upper solutions. We introduce the truncation map $\tau_+ : \mathbb{R} \longrightarrow \mathbb{R}$ defined by

$$\tau_+(x) = \begin{cases} 0 & \text{if } x \leq 0 \\ x & \text{if } x \geq 0 \end{cases}.$$

We set

$$f_+(z, x) = f\big(z, \tau_+(x)\big), \; F_+(z, x) = \int_0^x f_+(z, r)\text{d}r \quad \text{and} \quad F(z, x) = \int_0^x f(z, r)\text{d}r.$$

We also consider the Euler functionals $\varphi_+, \varphi : W_0^{1,p}(Z) \longrightarrow \mathbb{R}$ defined by

$$\begin{aligned} \varphi_+(x) &= \frac{1}{p}\|Dx\|_p^p - \frac{\lambda}{p}\|x\|_p^p + \int_Z F_+\big(z, x(z)\big)\text{d}z \\ \text{and} \quad \varphi(x) &= \frac{1}{p}\|Dx\|_p^p - \frac{\lambda}{p}\|x\|_p^p + \int_Z F\big(z, x(z)\big)\text{d}z \qquad \text{for all } x \in W_0^{1,p}(Z). \end{aligned}$$

We have $\varphi_+, \varphi \in C^1\big(W_0^{1,p}(Z)\big)$ and the critical points of φ are solutions of (5.80). Also we introduce the order interval

$$E_+ = [\,\underline{x}, \overline{x}\,] = \{x \in W_0^{1,p}(Z) : \underline{x}(z) \leq x(z) \leq \overline{x}(z) \quad \text{a.e. on } Z\}.$$

In the proof of the next proposition, we need the following nonlinear strong comparison principle, whose proof is postponed until Section 5.5 (see Theorem 5.5.11).

THEOREM 5.2.7 *If $x, y \in C_0^1(\overline{Z})$, $x \neq 0$, $f, g \in L^\infty(Z)$, $f \geq 0$,*

$$-\operatorname{div}\big(\|Dx(z)\|^{p-2}Dx(z)\big) = f(z) \qquad a.e.\ on\ Z$$

and

$$-\operatorname{div}\big(\|Dy(z)\|^{p-2}Dy(z)\big) = g(z) \qquad a.e.\ on\ Z$$

$$f(z) \geq g(z) \qquad a.e.\ on\ Z\ and\ the\ set$$

$$C = \{z \in Z : f(z) = g(z)\}$$

has an empty interior, then $x(z) > y(z)$ for all $z \in Z$ and $(\partial x/\partial n)(z) < (\partial y/\partial n)(z)$ for all $z \in \partial Z$, that is $x - y \in \operatorname{int} C_0^1(\overline{Z})_+$.

PROPOSITION 5.2.8 *If hypotheses $H(f)_1$ hold and $\lambda > \lambda_1$, then we can find $x_0 \in E_+$ which is a local minimizer of φ.*

PROOF: Because of (5.81), we have

$$F_+(z,x) \geq \frac{\mu}{p}|x|^p - \vartheta_\mu(z)|x| \qquad \text{for a.a. } z \in Z \text{ and all } x \in \mathbb{R}_+. \tag{5.88}$$

Choose $\mu > 0$ such that $\lambda - \mu < \lambda_1$. Then

$$\begin{aligned} \varphi_+(x) &= \frac{1}{p}\|Dx\|_p^p - \frac{\lambda}{p}\|x\|_p^p + \int_Z F_+\big(z, x(z)\big)\mathrm{d}z \\ &\geq \frac{1}{p}\|Dx\|_p^p - \frac{\lambda-\mu}{p}\|x\|_p^p - c_1\|Dx\|_p \quad \text{for some } c_1 > 0 \ \text{(see (5.88))} \\ &= \frac{1}{p}\Big(1 - \frac{\lambda-\mu}{\lambda_1}\Big)\|Dx\|_p^p - c_1\|Dx\|_p, \end{aligned}$$

$\Rightarrow \varphi_+$ is coercive on $W_0^{1,p}(Z)_+$ (because $\lambda - \mu < \lambda_1$).

Because of compact embedding of $W_0^{1,p}(Z)$ into $L^p(Z)$, we can easily see that φ_+ is weakly lower semicontinuous. So we can find $x_0 \in E_+$ such that

$$\varphi_+(x_0) = \inf_{E_+} \varphi_+. \tag{5.89}$$

Now let $y \in E_+$ and set $\eta(t) = \varphi_+\big(ty + (1-t)x_0\big)$ for all $t \in [0,1]$. Then

$$\begin{aligned} &\eta(0) \leq \eta(t) \qquad \text{for all } t \in [0,1] \ \text{(see (5.10))}, \\ \Rightarrow\ & 0 \leq \eta'(0^+), \\ \Rightarrow\ & 0 \leq \langle A(x_0), y - x_0\rangle - \lambda\int_Z |x_0|^{p-2}x_0(y - x_0)\mathrm{d}z + \int_Z f_+\big(z, x_0(z)\big)(y - x_0)\mathrm{d}z \end{aligned}$$

$$\text{for all } y \in E_+. \tag{5.90}$$

Given $v \in W_0^{1,p}(Z)$ and $\varepsilon > 0$, we define

$$y(z)=\begin{cases}\underline{x}(z) & \text{if } z\in\{x_0+\varepsilon v\le\underline{x}\}\\ x_0(z)+\varepsilon v(z) & \text{if } z\in\{\underline{x}<x_0+\varepsilon v<\overline{x}\}\\ \overline{x}(z) & \text{if } z\in\{\overline{x}\le x_0+\varepsilon v\}\end{cases}.$$

Evidently $y\in E_+$. We return to (5.90) and use this $y\in E_+$. We obtain

$$\begin{aligned}
0\le&\int_{\{\underline{x}<x_0+\varepsilon v<\overline{x}\}}\|Dx_0\|^{p-2}(Dx_0,Dv)_{\mathbb{R}^N}\mathrm{d}z-\lambda\varepsilon\int_Z|x_0|^{p-2}x_0v\mathrm{d}z\\
&+\varepsilon\int_{\{\underline{x}<x_0+\varepsilon v<\overline{x}\}}f_+(z,x_0)v\mathrm{d}z+\int_{\{x_0+\varepsilon v\le\underline{x}\}}\|Dx_0\|^{p-2}(Dx_0,D\underline{x}-Dv)_{\mathbb{R}^N}\mathrm{d}z\\
&-\lambda\int_{\{x_0+\varepsilon v\le\underline{x}\}}|x_0|^{p-2}x_0(\underline{x}-x_0)\mathrm{d}z+\int_{\{x_0+\varepsilon v\le\underline{x}\}}f_+(z,x_0)(\underline{x}-x_0)\mathrm{d}z\\
&+\int_{\{x_0+\varepsilon v\ge\overline{x}\}}\|Dx_0\|^{p-2}(Dx_0,D\overline{x}-Dx_0)_{\mathbb{R}^N}\mathrm{d}z-\lambda\int_{\{x_0+\varepsilon v\ge\overline{x}\}}|x_0|^{p-2}x_0(\overline{x}-x_0)\mathrm{d}z\\
&+\int_{\{x_0+\varepsilon v\ge\overline{x}\}}f_+(z,x_0)(\overline{x}-x_0)\mathrm{d}z=\\
=&\,\varepsilon\int_Z\|Dx_0\|^{p-2}(Dx_0,Dv)_{\mathbb{R}^N}\mathrm{d}z-\lambda\varepsilon\int_Z|x_0|^{p-2}x_0v\mathrm{d}z\\
&+\varepsilon\int_Z f_+(z,x_0)v\mathrm{d}z-\int_{\{x_0+\varepsilon v\ge\overline{x}\}}\|D\overline{x}\|^{p-2}\big(D\overline{x},D(x_0+\varepsilon v-\overline{x})\big)_{\mathbb{R}^N}\mathrm{d}z\\
&+\lambda\int_{\{x_0+\varepsilon v\ge\overline{x}\}}|\overline{x}|^{p-2}\overline{x}(x_0+\varepsilon v-\overline{x})\mathrm{d}z-\int_{\{x_0+\varepsilon v\ge\overline{x}\}}f_+(z,\overline{x})(x_0+\varepsilon v-\overline{x})\mathrm{d}z\\
&+\int_{\{x_0+\varepsilon v\le\underline{x}\}}\|D\underline{x}\|^{p-2}\big(D\underline{x},D(\underline{x}-x_0-\varepsilon v)\big)_{\mathbb{R}^N}\mathrm{d}z\\
&-\lambda\int_{\{x_0+\varepsilon v\le\underline{x}\}}|\underline{x}|^{p-2}\underline{x}(\underline{x}-x_0-\varepsilon v)\mathrm{d}z+\int_{\{x_0+\varepsilon v\le\underline{x}\}}f_+(z,\underline{x})(\underline{x}-x_0-\varepsilon v)\mathrm{d}z\\
&-\int_{\{x_0+\varepsilon v\le\underline{x}\}}\big(f_+(z,x_0)-f_+(z,\underline{x})\big)(x_0+\varepsilon v-\underline{x})\mathrm{d}z\\
&-\int_{\{x_0+\varepsilon v\ge\overline{x}\}}\big(f_+(z,\overline{x})-f_+(z,x_0)\big)(\overline{x}-x_0-\varepsilon v)\mathrm{d}z\\
&+\int_{\{x_0+\varepsilon v\ge\overline{x}\}}(\|D\overline{x}\|^{p-2}D\overline{x}-\|Dx_0\|^{p-2}Dx_0,Dx_0-D\overline{x})_{\mathbb{R}^N}\mathrm{d}z\\
&-\lambda\int_{\{x_0+\varepsilon v\ge\overline{x}\}}(|\overline{x}|^{p-2}\overline{x}-|x_0|^{p-2}x_0)(x_0-\overline{x})\mathrm{d}z\\
&-\int_{\{x_0+\varepsilon v\le\underline{x}\}}(\|D\underline{x}\|^{p-2}D\underline{x}-\|Dx_0\|^{p-2}Dx_0,D\underline{x}-Dx_0)_{\mathbb{R}^N}\mathrm{d}z\\
&+\lambda\int_{\{x_0+\varepsilon v\le\underline{x}\}}(|\underline{x}|^{p-2}\underline{x}-|x_0|^{p-2}x_0)(\underline{x}-x_0)\mathrm{d}z\\
&+\varepsilon\int_{\{x_0+\varepsilon v\ge\overline{x}\}}(\|D\overline{x}\|^{p-2}D\overline{x}-\|Dx_0\|^{p-2}Dx_0,Dv)_{\mathbb{R}^N}\mathrm{d}z
\end{aligned}$$

$$+ \varepsilon \int_{\{x_0+\varepsilon v\leq \underline{x}\}} (\|D\underline{x}\|^{p-2}D\underline{x} - \|Dx_0\|^{p-2}Dx_0, Dv)_{\mathbb{R}^N}\mathrm{d}z$$
$$- \lambda\varepsilon \int_{\{x_0+\varepsilon v\geq \overline{x}\}} (|\overline{x}|^{p-2}\overline{x} - |x_0|^{p-2}x_0)v\mathrm{d}z$$
$$- \lambda\varepsilon \int_{\{x_0+\varepsilon v\leq \underline{x}\}} (|\underline{x}|^{p-2}\underline{x} - |x_0|^{p-2}x_0)v\mathrm{d}z. \tag{5.91}$$

If $u = (x_0+\varepsilon v-\overline{x})^+ \in W_0^{1,p}(Z)_+$, then because $\overline{x}$ is an upper solution for problem (5.80), we have

$$- \int_{\{x_0+\varepsilon v\geq \overline{x}\}} \|D\overline{x}\|^{p-2}\big(D\overline{x}, D(x_0+\varepsilon v-\overline{x})\big)_{\mathbb{R}^N}\mathrm{d}z + \lambda \int_{\{x_0+\varepsilon v\geq \overline{x}\}} |\overline{x}|^{p-2}\overline{x}(x_0+\varepsilon v-\overline{x})\mathrm{d}z$$
$$- \int_{\{x_0+\varepsilon v\geq \overline{x}\}} f_+(z,\overline{x})(x_0+\varepsilon v-\overline{x})\mathrm{d}z \leq 0. \tag{5.92}$$

If $u = (\underline{x}-x_0-\varepsilon v)^+ \in W_0^{1,p}(Z)_+$, then because $\underline{x}$ is a lower solution for problem (5.80), we have

$$\int_{\{x_0+\varepsilon v\leq \underline{x}\}} \|D\underline{x}\|^{p-2}\big(D\underline{x}, D(\underline{x}-x_0-\varepsilon v)\big)_{\mathbb{R}^N}\mathrm{d}z - \lambda \int_{\{x_0+\varepsilon v\leq \underline{x}\}} |\underline{x}|^{p-2}\underline{x}(\underline{x}-x_0-\varepsilon v)\mathrm{d}z$$
$$+ \int_{\{x_0+\varepsilon v\leq \underline{x}\}} f_+(z,\underline{x})(\underline{x}-x_0-\varepsilon v)\mathrm{d}z \leq 0. \tag{5.93}$$

Recall that the map $\psi_p\colon \mathbb{R}^N \longrightarrow \mathbb{R}^N$ defined by

$$\psi_p(y) = \begin{cases} \|y\|^{p-2}y & \text{if } y \neq 0 \\ 0 & \text{if } y = 0 \end{cases},$$

is a strictly monotone homeomorphism. Therefore

$$\int_{\{x_0+\varepsilon v\geq \overline{x}\}} (\|D\overline{x}\|^{p-2}D\overline{x} - \|Dx_0\|^{p-2}Dx_0, Dx_0 - D\overline{x})_{\mathbb{R}^N}\mathrm{d}z \leq 0 \tag{5.94}$$

$$\text{and} \quad \int_{\{x_0+\varepsilon v\leq \underline{x}\}} (\|D\underline{x}\|^{p-2}D\underline{x} - \|Dx_0\|^{p-2}Dx_0, D\underline{x} - Dx_0)_{\mathbb{R}^N}\mathrm{d}z \leq 0. \tag{5.95}$$

Because $x_0 \leq \overline{x}$ and $\overline{x} - x_0 \leq \varepsilon v$ on $\{x_0+\varepsilon v \geq \overline{x}\}$, we have

$$-\lambda \int_{\{x_0+\varepsilon v\geq \overline{x}\}} (|\overline{x}|^{p-2}\overline{x} - |x_0|^{p-2}x_0)(x_0-\overline{x})\mathrm{d}z \leq \lambda\varepsilon \int_{\{x_0+\varepsilon v\leq \underline{x}\}} (|\underline{x}|^{p-2}\underline{x} - |x_0|^{p-2}x_0)v\mathrm{d}z. \tag{5.96}$$

Similarly because $\underline{x} \leq x_0$ and $\underline{x} - x_0 \geq \varepsilon v$ on $\{x_0+\varepsilon v \leq \underline{x}\}$, we have

$$\lambda \int_{\{x_0+\varepsilon v\leq \underline{x}\}} \big(|\underline{x}|^{p-2}\underline{x} - |x_0|^{p-2}x_0\big)(\underline{x}-x_0)\mathrm{d}z \leq \lambda\varepsilon \int_{\{x_0+\varepsilon v\leq \underline{x}\}} (|\underline{x}|^{p-2}\underline{x} - |x_0|^{p-2}x_0)v\mathrm{d}z. \tag{5.97}$$

Because $\underline{x}, \overline{x} \in C_0^1(\overline{Z})_+$, $\underline{x} \leq x_0$ and using hypothesis $H(f)_1$(iii), we have

$$-\int_{\{x_0+\varepsilon v\leq \underline{x}\}} \big(f_+(z,x_0) - f_+(z,\underline{x})\big)(x_0+\varepsilon v-\underline{x})\mathrm{d}z \leq c_2\varepsilon \int_{\{x_0+\varepsilon v\leq \underline{x}<x_0\}} (-v)\mathrm{d}z \quad \text{for some } c_2>0. \tag{5.98}$$

Similarly we have

$$-\int_{\{x_0+\varepsilon v\geq\overline{x}\}}\big(f_+(z,\overline{x})-f_+(z,x_0)\big)(\overline{x}-x_0-\varepsilon v)dz\leq c_3\varepsilon\int_{\{x_0+\varepsilon v\geq\overline{x}>x_0\}}vdz \quad \text{for some } c_3>0. \tag{5.99}$$

We return to (5.91) and use (5.92)⟶(5.99). This way we have

$$0\leq\varepsilon\int_Z\|Dx_0\|^{p-2}(Dx_0,Dv)_{\mathbb{R}^N}dz-\lambda\varepsilon\int_Z\|x_0\|^{p-2}x_0vdz+\varepsilon\int_Z f_+(z,x_0)vdz$$
$$+\,c_2\varepsilon\int_{\{x_0+\varepsilon v\leq\underline{x}<x_0\}}(-v)dz+c_3\varepsilon\int_{\{x_0+\varepsilon v\geq\overline{x}>x_0\}}vdz.$$

Divide with $\varepsilon>0$ and then let $\varepsilon\downarrow 0$. If by $|\cdot|_N$ we denote the Lebesgue measure in $\mathbb{R}^N$, then

$$|\{x_0+\varepsilon v\leq\underline{x}<x_0\}|_N\longrightarrow 0\quad\text{and}\quad|\{x_0+\varepsilon v\geq\overline{x}>x_0\}|_N\longrightarrow 0\quad\text{as }\varepsilon\downarrow 0.$$

Therefore in the limit as $\varepsilon\downarrow 0$, we obtain

$$0\leq\int_Z\|Dx_0\|^{p-2}(Dx_0,Dv)_{\mathbb{R}^N}dz-\lambda\int_Z\|x_0\|^{p-2}x_0vdz+\int_Z f_+(z,x_0)vdz. \tag{5.100}$$

Because $x_0\in W_0^{1,p}(Z)$ was arbitrary, from (5.99) we infer that

$$\left\{\begin{array}{l}-\mathrm{div}\big(\|Dx_0(z)\|^{p-2}Dx_0(z)\big)=\lambda|x_0(z)|^{p-2}x_0(z)-f\big(z,x_0(z)\big)\ \text{ a.e. on } Z,\\ x\big|_{\partial Z}=0\end{array}\right\}, \tag{5.101}$$

(note that $f_+\big(z,x_0(z)\big)=f\big(z,x_0(z)\big)$). From (5.101) we have that $x_0\in C_0^1(\overline{Z})$. Moreover, from (5.81) and Proposition 5.2.6, we have

$$\overline{x}-x_0\in\mathrm{int}\,C_0^1(\overline{Z})_+.$$

Similarly because of (5.85) and Proposition 5.2.6, we have

$$x_0-\underline{x}\in\mathrm{int}\,C_0^1(\overline{Z})_+.$$

From the definition of f_+, it follows that x_0 is a $C_0^1(\overline{Z})$-local minimizer of φ. Invoking Theorem 5.1.27, we infer that x_0 is a $W_0^{1,p}(Z)$-local minimizer of φ. □

From the above proof we have that $x_0\in C_0^1(\overline{Z})_+$ is a solution of problem (5.80). We repeat a similar argument on the negative semiaxis of $\mathbb{R}$. In this case motivated by (5.82), we consider the following auxiliary problem.

$$\left\{\begin{array}{l}-\mathrm{div}\big(\|Dx(z)\|^{p-2}Dx(z)\big)=(\lambda-\mu)|x(z)|^{p-2}x(z)-\vartheta_\mu(z)\ \text{ a.e. on } Z,\\ x\big|_{\partial Z}=0\end{array}\right\}. \tag{5.102}$$

Following the reasoning of Proposition 5.2.3, we solve problem (5.102) and obtain a solution $\underline{v}\in-\mathrm{int}\,C_0^1(\overline{Z})_+$, which we can show is a lower solution for problem (5.80). Moreover, taking $\xi>0$ as in the definition of $\underline{x}$, we introduce $\overline{v}=\xi(-u_1)$ which is an upper solution for problem (5.80). Using the ordered pair $\{\underline{v},\overline{v}\}$, we consider the corresponding Carathéodory truncation $\tau_-(z,x)$ and then set $f_-(z,x)=f\big(z,\tau_-(z,x)\big)$, $F_-(z,x)=\int_0^x f_-(z,r)dr$, and

$$\varphi_-(x) = \frac{1}{p}\|Dx\|_p^p - \frac{\lambda}{p}\|x\|_p^p + \int_Z F_-\big(z, x(z)\big) \qquad \text{for all } x \in W_0^{1,p}(Z).$$

We work as above, this time on the order interval $E_- = [\,\underline{v}, \overline{v}\,]$ and obtain the following.

PROPOSITION 5.2.9 *If hypotheses $H(f)_1$ hold and $\lambda > \lambda_1$, then there exists $v_0 \in E_-$ which is a local minimizer of φ.*

Again $v_0 \in -\text{int}\, C_0^1(\overline{Z})_+$ is a solution of problem (5.80). Therefore we have produced two nontrivial solutions $x_0, v_0 \in C_0^1(\overline{Z})$ of problem (5.80) which have constant sign. Next we show that if we restrict further the parameter $\lambda > 0$, namely we require that $\lambda > \lambda_2$, then we have a third nontrivial solution of problem (5.80), distinct from the other two.

THEOREM 5.2.10 *If hypotheses $H(f)_1$ hold and $\lambda > \lambda_2$, then problem (5.80) has at least three distinct nontrivial solutions $x_0, v_0, y_0 \in C_0^1(\overline{Z})$.*

PROOF: We have produced two solutions $x_0, v_0 \in C_0^1(\overline{Z})$ that have constant sign and are also local minimizers of φ. We assume that these are the only nontrivial critical points of φ^λ, or otherwise we are done. Also clearly without any loss of generality, we can say that both x_0, v_0 are strict local minimizers of φ^λ. We choose $\delta > 0$ small such

$$\varphi(v_0) < \inf\big[\varphi(x) : x \in \partial B_\delta(v_0)\big] \qquad \text{and} \qquad \varphi(x_0) < \inf\big[\varphi(x) : x \in \partial B_\delta(x_0)\big].$$

Let $\varphi(v_0) \le \varphi(x_0)$, $V = \partial B_\delta(x_0)$, $W_0 = \{v_0, x_0\}$, and $W = [v_0, x_0] = \{x \in W_0^{1,p}(Z) : v_0(z) \le x(z) \le x_0(z) \text{ a.e. on } Z\}$. The pair $\{W_0, W\}$ links with V in $W_0^{1,p}(Z)$ via the identity map. Using hypothesis $\text{H}(f)_1$(v) it is easy to verify that φ is coercive from which it follows that φ satisfies the PS-condition. So by virtue of Theorem 4.1.28, we can find $y_0 \in W_0^{1,p}(Z)$ such that

$$\varphi'(y_0) = 0 \qquad (\text{i.e., } y_0 \text{ is a critical point of } \varphi) \tag{5.103}$$

$$\text{and} \quad \varphi(v_0), \varphi(x_0) < \varphi(y_0) = \inf_{\gamma \in \Gamma} \max_{t \in [-1,1]} \varphi\big(\gamma(t)\big), \tag{5.104}$$

where $\Gamma = \big\{\gamma \in C\big([-1,1], W_0^{1,p}(Z)\big) : \gamma(-1) = v_0,\ \gamma(1) = x_0\big\}$.

Our goal is to produce a path $\gamma \in \Gamma$ such that $\varphi\big|_\gamma < 0$. Then $y_0 \ne 0$ and of course it is distinct from v_0, x_0. To this end let $S = W_0^{1,p}(Z) \cap \partial B_1^{L^p(Z)}$ where $\overline{B}_1^{L^p(Z)} = \big\{x \in L^p(Z) : \|x\|_p \le 1\big\}$ and $S_c = W_0^{1,p}(Z) \cap C_0^1(\overline{Z}) \cap \partial B_1^{L^p(Z)}$. Evidently S_c is dense in S. From Theorem 4.1.28, we know that we can find $\gamma_0 \in \Gamma_0 = \big\{\gamma_0 \in C([-1,1], S) : \gamma(-1) = -u_1,\ \gamma_0(1) = u_1\big\}$ such that $\gamma_0([-1,1]) \subseteq S_c$ and

$$\max\big\{\|Du\|_p^p : u \in \gamma_0([-1,1])\big\} \le \lambda_2 + \delta_0. \tag{5.105}$$

We choose $\delta_0 > 0$ such that $\lambda_2 + \delta_0 < \lambda$. Also because of hypothesis $\text{H}(f)_1$(iv), given $\varepsilon > 0$ we can find $\delta = \delta(\varepsilon) > 0$ such that

$$F(z, x) \le \frac{\varepsilon}{p}|x|^p \qquad \text{for a.a. } z \in Z \text{ and all } |x| \le \delta. \tag{5.106}$$

Recall that $\gamma_0([-1,1]) \subseteq S_c$ and $-v_0, x_0 \in \text{int}\, C_0^1(\overline{Z})_+$. So we can choose $\varepsilon > 0$ small so that

$$
\begin{aligned}
&|\varepsilon u(z)| \leq \delta \qquad \text{for all } z \in \overline{Z}, \text{ all } u \in \gamma_0([-1,1]), \\
&\lambda_2 + \delta_0 + \varepsilon < \lambda \qquad \text{and} \qquad \varepsilon u \in W = [v_0, x_0] \qquad \text{for all } u \in \gamma_0([-1,1]).
\end{aligned}
\tag{5.107}
$$

Therefore, if $u \in \gamma_0([-1,1])$, we have

$$
\begin{aligned}
\varphi(\varepsilon u) &= \frac{\varepsilon^p}{p}\|Du\|_p^p - \frac{\lambda\varepsilon^p}{p}\|u\|_p^p + \int_Z F\big(z, \varepsilon u(z)\big)\mathrm{d}z \\
&\leq \frac{\varepsilon^p}{p}(\lambda_2+\delta_0) - \frac{\lambda\varepsilon^p}{p} + \frac{\varepsilon^{p+1}}{p} \qquad \text{(see (5.105) and (5.106)}
\end{aligned}
$$

and recall that $\|u\|_p = 1$),

$$
= \frac{\varepsilon^p}{p}(\lambda_2+\delta_0+\varepsilon-\lambda) < 0 \qquad \text{(see (5.107))}.
$$

Hence the path $\varepsilon\gamma_0$ joins $-\varepsilon u_1$ and εu_1 and we have

$$
\varphi\big|_{\varepsilon\gamma_0} < 0. \tag{5.108}
$$

Next we produce another continuous path $\gamma_+ : [0,1] \longrightarrow W_0^{1,p}(Z)$ which joins εu_1 and x_0 and along which φ is negative. For this purpose, we consider the truncation map

$$
\widehat{\tau}_+(x) = \begin{cases} x & \text{if } x \geq 0 \\ 0 & \text{if } x < 0 \end{cases}.
$$

We set $\widehat{f}_+(z,x) = f\big(z, \widehat{\tau}_+(x)\big)$, $\widehat{F}_+(z,x) = \int_0^x \widehat{f}_+(z,r)\mathrm{d}r$, and

$$
\widehat{\varphi}_+(x) = \frac{1}{p}\|Dx\|_p^p - \frac{\lambda}{p}\|x^+\|_p^p + \int_Z \widehat{F}_+\big(z, x(z)\big)\mathrm{d}z \qquad \text{for all } x \in W_0^{1,p}(Z).
$$

Because of hypothesis $\mathrm{H}(f)_1$(v), $\widehat{\varphi}_+$ is coercive and so it satisfies the *PS*-condition. Also $\{0, x_0\}$ are the only critical points of $\widehat{\varphi}_+$ and $\widehat{\varphi}_+(x_0) = \inf\limits_{W_0^{1,p}(Z)} \widehat{\varphi}_+$. We set

$$
\widehat{b}_+ = \widehat{\varphi}_+(\varepsilon u_1) \qquad \text{and} \qquad \widehat{a}_+ = \widehat{\varphi}_+(x_0) = \inf_{W_0^{1,p}(Z)} \widehat{\varphi}_+.
$$

Invoking the second deformation theorem (see Theorem 4.6.1), we can find $h \in C\big([0,1] \times \widehat{\varphi}_+^{\widehat{b}_+}, \widehat{\varphi}_+^{\widehat{b}_+}\big)$ such that

$$
\begin{aligned}
&h(t,x) = x \qquad \text{for all } t \in [0,1] \text{ and all } x \in \widehat{\varphi}_+^{\widehat{a}_+}, \\
&h(0,x) = x \qquad \text{for all } \widehat{\varphi}_+^{\widehat{b}_+} \\
\text{and} \quad &h(1,x) = x_0 \qquad \text{for all } \widehat{\varphi}_+^{\widehat{b}_+}.
\end{aligned}
$$

Then consider the continuous path $\gamma_+(t) = h(t, \varepsilon u_1)$ for all $t \in [0,1]$, which joins εu_1 and x_0. We have

$$
\begin{aligned}
&\varphi\big(\gamma_+(t)\big) \leq \widehat{\varphi}_+\big(\gamma_+(t)\big) \leq \widehat{b}_+ = \widehat{\varphi}_+(\varepsilon u_1) < 0 \\
\Rightarrow\ &\varphi\big|_{\gamma_+} < 0.
\end{aligned}
\tag{5.109}
$$

In a similar fashion, we produce a continuous path $\gamma_-\colon [0,1] \longrightarrow W_0^{1,p}(Z)$ joining $-\varepsilon u_1$ and v_0 such that

$$\varphi\big|_{\gamma_-} < 0. \tag{5.110}$$

Concatenating the paths $\varepsilon\gamma_0$, γ_+, and γ_- we generate a path $\gamma \in \Gamma$ such that

$$\begin{aligned} &\varphi\big|_{\gamma} < 0 \qquad \text{(see (5.108) through (5.110))} \\ \Rightarrow\ &\varphi(y_0) < 0 = \varphi(0) \ \ \text{(i.e. } y_0 \neq 0). \end{aligned}$$

Finally, nonlinear regularity theory implies that $y_0 \in C_0^1(\overline{Z})$. □

As a second example of the method of upper–lower solutions, we consider the following nonlinear boundary value problem.

$$\left\{ \begin{array}{l} -\big(|x'(t)|^{p-2}x'(t)\big)' = f\big(t, x(t), x'(t)\big) + \vartheta\big(t, x(t)\big) \ \text{ a.e. on } T = [0,b], \\ x'(0) \in \xi_1\big(x(0)\big),\ -x'(b) \in \xi_2\big(x(b)\big) \end{array} \right\}. \tag{5.111}$$

In this problem ξ_1 and ξ_2 are maximal monotone graphs in $\mathbb{R}^2$.

First let us specify the notions of upper and lower solutions for problem (5.111).

DEFINITION 5.2.11 (a) A function $\varphi \in C^1(T)$ such that $|\varphi'|^{p-2}\varphi' \in W^{1,p'}\big((0,b)\big)$, $(1/p) + (1/p') = 1$ is said to be *an upper solution* of problem (5.111), if

$$\left\{ \begin{array}{ll} -\big(|\varphi'(t)|^{p-2}\varphi'(t)\big)' \geq f\big(t, \varphi(t), \varphi'(t)\big) + \vartheta\big(\varphi(t)\big) & \text{a.e. on } T, \\ \varphi'(0) \in \xi_1\big(\varphi(0)\big) - \mathbb{R}_+,\ -\varphi'(b) \in \xi_2\big(\varphi(b)\big) - \mathbb{R}_+ & \end{array} \right\}.$$

(b) A function $\psi \in C^1(T)$ such that $|\psi'|^{p-2}\psi' \in W^{1,p'}\big((0,b)\big)$, $(1/p) + (1/p') = 1$ is said to be *a lower solution* of problem (5.111), if

$$\left\{ \begin{array}{ll} -\big(|\psi'(t)|^{p-2}\psi'(t)\big)' \leq f\big(t, \psi(t), \psi'(t)\big) + \vartheta\big(\psi(t)\big) & \text{a.e. on } T, \\ \psi'(0) \in \xi_1\big(\psi(0)\big) + \mathbb{R}_+,\ -\psi'(b) \in \xi_2\big(\psi(b)\big) + \mathbb{R}_+ & \end{array} \right\}.$$

We make the following hypothesis.

H₀: There exist a lower solution $\psi \in C^1(T)$ and an upper solution $\varphi \in C^1(T)$ such that $\psi(t) \leq \varphi(t)$ for all $t \in T$.

The hypotheses on the data of problem (5.111) are the following.

$\mathbf{H(f)_2}$: $f\colon T \times \mathbb{R} \times \mathbb{R} \longrightarrow \mathbb{R}$ is a function such that

(i) For all $x, y \in \mathbb{R}$, $t \longrightarrow f(t,x,y)$ is measurable.

(ii) For almost all $t \in T$, $(x,y) \longrightarrow f(t,x,y)$ is continuous.

(iii) For almost all $t \in T$, all $x \in [\psi(t), \varphi(t)]$ and all $y \in \mathbb{R}$, we have

$$|f(t,x,y)| \leq \eta(|y|^{p-1})\big(\alpha(t) + c|y|\big)$$

with $\alpha \in L^1(T)_+$, $c > 0$ and $\eta \colon \mathbb{R}_+ \longrightarrow \mathbb{R}_+ \setminus \{0\}$ a Borel measurable nondecreasing function such that

$$\int_{\lambda^{p-1}}^{\infty} \frac{ds}{\eta(s)} > \|\alpha\|_1 + c(\max_T \varphi - \min_T \psi) + \sup \frac{b}{\eta(\lambda)} \big[|\vartheta(z)| : |z| \leq \max\{\|\varphi\|_\infty, \|\psi\|_\infty\}\big]$$

with $\lambda = (1/b) \max \big[|\psi(0) - \varphi(b)|, |\psi(b) - \varphi(0)|\big]$.

(iv) For every $r > 0$, we can find $\gamma_r \in L^{p'}(T)$ such that for almost all $x, y \in \mathbb{R}$ with $|x|, |y| \leq r$ we have $|f(t,x,y)| \leq \gamma_r(t)$.

REMARK 5.2.12 Hypothesis $H(f)_2$(iii) is known as a *Bernstein–Nagumo–Wintner growth condition* and produces a uniform a priori bound for the derivatives of the solutions of problem (5.111). It is clear that if for almost all $t \in T$, all $x \in [\ \psi(t), \varphi(t)\ ]$, and all $y \in \mathbb{R}$ we have $|f(t,x,y)| \leq \alpha(t) + c|y|^p$ with $\alpha \in L^1(T)_+$, $c > 0$, then hypothesis $H(f)_2$(iii) is satisfied. This is the growth condition initially used by Bernstein in his existence theory for second-order boundary value problems.

H(ξ): $\xi_1, \xi_1 \colon \mathbb{R} \longrightarrow 2^{\mathbb{R}}$ are maximal monotone maps such that $0 \in \xi_1(0)$ and $0 \in \xi_2(0)$.

REMARK 5.2.13 We know that there exist functions $k_1, k_2 \in \Gamma_0(\mathbb{R})$ such that $\xi_1 = \partial k_1$ and $\xi_2 = \partial k_2$.

H(ϑ): $\vartheta \colon \mathbb{R} \longrightarrow \mathbb{R}$ is a function that maps bounded sets to bounded sets and there exists $M > 0$ such that $x \longrightarrow \vartheta(x) + Mx$ is increasing.

REMARK 5.2.14 We emphasize that ϑ need not be continuous.

We start with a lemma that produces a uniform bound for the derivatives of the solutions of problem (5.111). As we already mentioned, hypothesis $H(f)_2$(iii) is the crucial tool for this.

LEMMA 5.2.15 *If hypothesis $H(f)_2$(iii) holds and $x \in C^1(T)$ satisfies*

$$-\big(|x'(t)|^{p-2}x'(t)\big)' = f\big(t, x(t), x'(t)\big) + \vartheta\big(x(t)\big) \quad \text{a.e. on } T$$
$$\text{and} \quad \psi(t) \leq x(t) \leq \varphi(t) \quad \text{for all } t \in T,$$

then there exists $M_1 > 0$ (depending only on $\psi, \varphi, \eta, \alpha, c$) such that

$$|x'(t)| \leq M_1 \quad \text{for all } t \in T.$$

PROOF: Set $\mu = \big((1/\eta(\lambda)\big) \sup \big[|\vartheta(z)| : |z| \leq \max\{\|\varphi\|_\infty, \|\psi\|_\infty\}\big] < +\infty$ (see Hypothesis H(ϑ)). By virtue of hypothesis H$(f)_2$(iii), we can find $M_1 > \lambda$ such that

$$\int_{\lambda^{p-1}}^{M_1^{p-1}} \frac{ds}{\eta(s)} > \|\alpha\|_1 + c(\max_T \varphi - \min_T \psi) + \mu b.$$

We claim that $|x'(t)| \leq M_1$ for all $t \in T$. Suppose that this is not the case. Then we can find $t' \in T$ such that $|x'(t')| > M_1$. By the mean value theorem, we can find $t_0 \in (0, b)$ such that $x(b) - x(0) = x'(t_0)b$. Without any loss of generality, we assume that $t_0 \leq t'$ (the analysis is similar if $t_0 > t'$). Note that $\psi(b) - \varphi(0) \leq x(b) - x(0) \leq \varphi(b) - \psi(0)$. So we have

$$\begin{aligned} &|x'(t_0)| = \frac{1}{b}|x(b) - x(0)| \leq \frac{1}{b} \max\big\{|\psi(0) - \varphi(b)|,\ |\psi(b) - \varphi(0)|\big\}, \\ \Rightarrow\ & |x'(t_0)| \leq \lambda > M_1 \ \text{ and } \ |x'(t')| < M_1 \quad (\text{i.e., } t_0 < t'). \end{aligned}$$

Because $x \in C^1(T)$, by the intermediate value theorem we can find $t_1, t_2 \in [t_0, t']$ with $t_1 < t_2$ such that $|x'(t_1)| = \lambda$ and $|x'(t_2)| = M_1$. By hypothesis we have

$$\begin{aligned} & -\big(|x'(t)|^{p-2}x'(t)\big)' = f\big(t, x(t), x'(t)\big) + \vartheta\big(x(t)\big) \quad \text{a.e. on } T, \\ \Rightarrow\ & \big(|x'(t)|^{p-1}\big)' \leq |\big(|x'(t)|^{p-2}x'(t)\big)'| \leq |f\big(t, x(t), x'(t)\big)| + |\vartheta\big(x(t)\big)| \\ & \qquad \leq \eta\big(|x'(t)|^{p-1}\big)\big(\alpha(t) + c|x'(t)|\big) + |\vartheta\big(x(t)\big)| \qquad \text{a.e. on } T, \\ \Rightarrow\ & \frac{\big(|x'(t)|^{p-1}\big)'}{\eta\big(|x'(t)|^{p-1}\big)} \leq \alpha(t) + c|x'(t)| + \frac{|\vartheta\big(x(t)\big)|}{\eta(\lambda)} \qquad \text{a.e. on } [t_1, t_2] \\ \Rightarrow\ & \int_{t_1}^{t_2} \frac{\big(|x'(t)|^{p-1}\big)'}{\eta\big(|x'(t)|^{p-1}\big)} dt \leq \|\alpha\|_1 + c(\max_T \varphi - \min_T \psi) + \mu b \\ \Rightarrow\ & \int_{\lambda^{p-1}}^{M_1^{p-1}} \frac{ds}{\eta(s)} \leq \|\alpha\|_1 + c(\max_T \varphi - \min_T \psi) + \mu b, \end{aligned}$$

which contradicts the choice of $M_1 > 0$. □

The upper–lower solutions method employs *truncation* and *penalization* techniques. So we introduce a truncation map $u \colon T \times \mathbb{R} \times \mathbb{R} \longrightarrow \mathbb{R}^2$ and a penalty function $\beta \colon T \times \mathbb{R} \longrightarrow \mathbb{R}$ defined by

$$u(t, x, y) = \begin{cases} \big(\psi(t), \psi'(t)\big) & \text{if } x < \psi(t) \\ \big(\varphi(t), \varphi'(t)\big) & \text{if } x > \varphi(t) \\ (x, M_0) & \text{if } \psi(t) \leq x \leq \varphi(t),\ y > M_0 \\ (x, -M_0) & \text{if } \psi(t) \leq x \leq \varphi(t),\ y < -M_0 \\ (x, y) & \text{if otherwise,} \end{cases} \tag{5.112}$$

where $M_0 > \max\{M_1, \|\varphi'\|_\infty, \|\psi'\|_\infty\}$ and

$$\beta(t, x) = \begin{cases} |\psi(t)|^{p-2}\psi(t) - |x|^{p-2}x & \text{if } x < \psi(t) \\ 0 & \text{if } \psi(t) \leq x \leq \varphi(t) \\ |\varphi(t)|^{p-2}\varphi(t) - |x|^{p-2}x & \text{if } \varphi(t) < x. \end{cases} \tag{5.113}$$

We set $f_1(t,x,y)=f\big(t_1,u(t,x,y)\big)$. Note that for almost all $x\in[\psi(t),\varphi(t)]$ and all $|y|\le M$, we have $f_1(t,x,y)=f(t,x,y)$. Moreover, for almost all $t\in T$ and all $x,y\in\mathbb{R}$, we have $|f_1(t,x,y)|\le\alpha_r(t)$ with $r=\max\{M_0,\|\varphi\|_\infty,\|\psi\|_\infty\}$. For every $x\in W^{1,p}\big((0,b)\big)$, let $N_1(x)(\cdot)=f_1\big(\cdot,x(\cdot),x'(\cdot)\big)$ and $\widehat{\beta}(x)(\cdot)=\beta\big(\cdot,x(\cdot)\big)$ (i.e., the Nemitsky operators corresponding to f_1 and β, respectively). Let $G(x)=N_1(x)+\widehat{\beta}(x)$ for every $x\in W^{1,p}\big((0,b)\big)$. The next proposition is an immediate consequence of the continuity of the truncation map $(x,y)\longrightarrow u(t,x,y)$ (see (5.112)).

PROPOSITION 5.2.16 *If hypotheses $H(f)_2$ hold, then $G\colon W^{1,p}\big((0,b)\big)\longrightarrow L^{p'}(T)$, $(1/p)+(1/p')=1$ is continuous.*

Next we introduce the set

$$D=\big\{x\in C^1(T):|x'|^{p-2}x'\in W^{1,p'}\big((0,b)\big),\ x'(0)\in\xi_1\big(x(0)\big)\\ \text{and }-x'(b)\in\xi_2\big(x(b)\big)\big\}$$

and then define the nonlinear operator $U:D\subseteq L^p(T)\longrightarrow L^{p'}(T)$ by

$$U(x)(\cdot)=-\big(|x'(\cdot)^{p-2}x'(\cdot)|\big)'\qquad\text{for all }x\in D.$$

PROPOSITION 5.2.17 *If hypotheses $H(\xi)$ hold, then $U:D\subseteq L^p(T)\longrightarrow L^{p'}(T)$ is maximal monotone.*

PROOF: Given $h\in L^{p'}(T)$, we consider the following nonlinear boundary value problem,

$$\left\{\begin{array}{l}-\big(|x'(t)|^{p-2}x'(t)\big)'+|x(t)|^{p-2}x(t)=h(t)\ \text{ a.e. on }T,\\ x'(0)\in\xi_1\big(x(0)\big),\ -x'(b)\in\xi_2\big(x(b)\big)\end{array}\right\}.\tag{5.114}$$

We show that problem (5.114) has a unique solution $x\in C^1(T)$. To this end, given $v,w\in\mathbb{R}$, we consider the following two-point boundary value problem.

$$\left\{\begin{array}{l}-\big(|x'(t)|^{p-2}x'(t)\big)'+|x(t)|^{p-2}x(t)=h(t)\ \text{ a.e. on }T,\\ x(0)=v,\ x(b)=w\end{array}\right\}.\tag{5.115}$$

Let us set $\gamma(t)=\big(1-(t/b)\big)v+(t/b)w$. Then $\gamma(0)=v$ and $\gamma(b)=w$. We consider the function $y(t)=x(t)-\gamma(t)$ and rewrite problem (5.115) in terms of this function

$$\left\{\begin{array}{l}-\big(|(y+\gamma)'(t)|^{p-2}(y+\gamma)'(t)\big)'+|(y+\gamma)(t)|^{p-2}(y+\gamma)(t)=h(t)\ \text{ a.e. on }T,\\ y(0)=0,\ y(b)=0\end{array}\right\}.\tag{5.116}$$

This is a homogeneous Dirichlet problem for y. To solve (5.116), we argue as follows.

Let $V_1:W_0^{1,p}\big((0,b)\big)\longrightarrow W^{-1,p'}\big((0,b)\big)$, be the nonlinear operator defined by

$$\begin{aligned}\langle V_1(u),z\rangle_0&=\int_0^b|(u+\gamma)'(t)|^{p-2}(u+\gamma)'(t)z'(t)dt\\ &\quad+\int_0^b|(u+\gamma)(t)|^{p-2}(u+\gamma)(t)z(t)dt\qquad\text{for all }u,z\in W_0^{1,p}\big((0,b)\big).\end{aligned}$$

It is easy to check that V_1 is strictly monotone, demicontinuous, hence maximal monotone too. Moreover, we have

$$\begin{aligned}&\langle V_1(u), u\rangle_0 \geq \|u+\gamma\|^p - c_3\|u+\gamma\|^{p-1} \quad \text{for some } c_3>0,\\ \Rightarrow\ & V_1 \text{ is coercive.}\end{aligned}$$

So from Corollary 3.2.28, we infer that there exists $y \in W_0^{1,p}\big((0,b)\big)$ such that $V_1(y)=h$ and due to the strict monotonicity of V_1, this solution is unique. From the equation $V_1(y) = h$ it follows easily that $y \in C^1(T)$ and it solves problem (5.116). Therefore $x=y+\gamma \in C^1(T)$ and it is the unique solution of problem (5.115).

We can define the solution map $s\colon \mathbb{R}\times\mathbb{R}\longrightarrow C^1(T)$ which to each pair (v,w) assigns the unique solution of problem (5.115). Then let $\varrho\colon \mathbb{R}\times\mathbb{R}\longrightarrow\mathbb{R}\times\mathbb{R}$ be defined by

$$\varrho(v,w) = \big(s(v,w)'(0), -s(v,w)'(b)\big). \tag{5.117}$$

Using integration by parts, we check easily that ϱ is monotone.

Also let $\alpha_n \longrightarrow \alpha$ and $\beta_n \longrightarrow \beta$ in $\mathbb{R}$. Set

$$\begin{aligned}& x_n = s(\alpha_n,\beta_n),\ x = s(\alpha,\beta), \quad \gamma_n(t) = \Big(1-\frac{t}{b}\Big)\alpha_n + \frac{t}{b}\beta_n\\ \text{and}\quad & \gamma(t) = \Big(1-\frac{t}{b}\Big)\alpha + \frac{t}{b}\beta \qquad \text{for all } n\geq 1.\end{aligned}$$

Then directly from (5.115) we obtain that $\{x_n\}_{n\geq1}\subseteq W^{1,p}\big((0,b)\big)$ is bounded. It follows that $\big\{|x_n|^{p-2}x_n\big\}_{n\geq1}\subseteq L^{p'}(T)$ and $\big\{|x_n'|^{p-2}x_n'\big\}_{n\geq1}\subseteq W^{1,p'}\big((0,b)\big)$ are both bounded. So we may assume that

$$\begin{aligned}& x_n \xrightarrow{w} u \text{ in } W^{1,p}\big((0,b)\big),\ |x_n|^{p-2}x_n \xrightarrow{w} w \text{ in } L^{p'}(T),\\ \text{and}\quad & |x_n'|^{p-2}x_n' \xrightarrow{w} v \text{ in } W^{1,p'}\big((0,b)\big).\end{aligned}$$

Due to the compact embeddings of $W^{1,p}\big((0,b)\big)$ and $W^{1,p'}\big((0,b)\big)$ into $C(T)$, we have

$$x_n \longrightarrow u \text{ in } C(T) \qquad \text{and} \qquad |x_n'|^{p-2}x_n' \longrightarrow v \text{ in } C(T).$$

The map $\sigma\colon C(T)\longrightarrow C(T)$ defined by

$$\sigma(y)(\cdot) = \vartheta_p^{-1}\big(y(\cdot)\big) = \vartheta_{p'}\big(y(\cdot)\big),$$

where for any $1<r<\infty$,

$$\vartheta_r(y) = \begin{cases} |y|^{r-2}y & \text{if } y\neq 0\\ 0 & \text{if } y=0\end{cases},$$

is continuous and maps bounded sets to bounded sets. Therefore

$$\begin{aligned}& x_n' \longrightarrow \sigma(v) \text{ in } C(T)\\ \Rightarrow\ & u' = \sigma(v) \ \text{(i.e., } v = |u'|^{p-2}u').\end{aligned}$$

Therefore in the limit as $n\to\infty$, we obtain

$$\left\{\begin{array}{l} -\big(|u'(t)|^{p-2}u'(t)\big)' + |u(t)|^{p-2}u(t) = h(t) \ \text{ a.e. on } T,\\ u(0)=\alpha,\ u(b)=\beta \end{array}\right\}.$$

$$\Rightarrow\ x = s(\alpha,\beta) \ \text{(i.e., } s:\mathbb{R}\times\mathbb{R}\longrightarrow C^1(T) \text{ is continuous).}$$

From the continuity of s it follows at once the continuity of ϱ (see (5.117)).

Finally it is straightforward to check that ϱ is coercive. So ϱ is maximal monotone (being monotone, continuous) and coercive. Evidently so is $k = \varrho + \xi : \mathbb{R}\times\mathbb{R} \longrightarrow 2^{\mathbb{R}\times\mathbb{R}}$. So we can find $(\alpha,\beta) \in \mathbb{R}\times\mathbb{R}$ such that $(0,0) \in k(\alpha,\beta)$. Then $x_0 = s(\alpha,\beta)$ is the unique solution of the auxiliary problem (5.114).

If $K : L^p(T)\longrightarrow L^{p'}(T)$ is the maximal monotone operator defined by

$$K(x)(\cdot) = |x(\cdot)|^{p-2}x(\cdot),$$

then from the previous argument we have that

$$\mathcal{R}(U+K) = L^{p'}(T), \quad \text{(i.e., } U+K \text{ is surjective).} \tag{5.118}$$

We claim that this surjectivity implies the maximal monotonicity of U. For this purpose suppose that for some $y \in L^p(T)$ and some $v \in L^{p'}(T)$, we have

$$\langle U(x) - v, x - y\rangle_p \geq 0 \qquad \text{for all } x \in D. \tag{5.119}$$

Here by $\langle\cdot,\cdot\rangle_p$ we denote the duality brackets for the pair $\big(L^p(T), L^{p'}(T)\big)$. Because of (5.118), we can find $x_1 \in D$ such that

$$U(x_1) + K(x_1) = v + K(y).$$

We use this in (5.119) with $x = x_1$. So we obtain

$$\begin{aligned} &\langle U(x_1) - U(x_1) - K(x_1) + K(y), x_1 - y\rangle_p \geq 0, \\ \Rightarrow\ &\langle K(y) - K(x_1), x_1 - y\rangle_p \geq 0. \end{aligned} \tag{5.120}$$

Because K is strictly monotone, from (5.120) we conclude that $y = x_1 \in D$ and $v = U(x_1)$. Therefore U is maximal monotone. ☐

The operator $U+K : D \subseteq L^p(T)\longrightarrow L^{p'}(T)$ is a maximal monotone and strictly monotone operator that is coercive (indeed note that $\langle U(x)+K(x), x\rangle_p \geq \|x\|_p^p$, because $\langle U(x), x\rangle_p \geq 0$ and $\langle K(x), x\rangle_p = \|x\|_p^p$). Therefore $U+K$ is surjective. This means that the operator $L : (U+K)^{-1} : L^{p'}(T)\longrightarrow D \subseteq W^{1,p}\big((0,b)\big)$ is well-defined, single-valued, and maximal monotone (from $L^{p'}(T)$ into $L^p(T)$).

PROPOSITION 5.2.18 *If hypothesis $H(\xi)$ holds, then $L : L^{p'}(T) \longrightarrow D \subseteq W^{1,p}\big((0,b)\big)$ is completely continuous.*

PROOF: Suppose that $v_n \xrightarrow{w} v$ in $L^{p'}(T)$. We need to show that $L(v_n) \longrightarrow L(v)$ in $L^p(T)$. We set $x_n = L(v_n)$ for all $n \geq 1$. Then $x_n \in D$ and

$$\begin{aligned} &U(x_n) + K(x_n) = v_n, \\ \Rightarrow\ &\langle U(x_n), x_n\rangle_p + \langle K(x_n), x_n\rangle_p = \langle v_n, x_n\rangle_p. \end{aligned} \tag{5.121}$$

Because of the integration by parts formula, we have

$$\langle U(x_n), x_n\rangle_p = -|x_n'(b)|^{p-2}x_n'(b)x_n(b) + |x_n'(0)|^{p-2}x_n'(0)x_n(0) + \|x_n'\|_p^p. \tag{5.122}$$

Because $x_n \in D$, we have $x_n(0) \in \xi_1\big(x_n(0)\big)$ and $-x_n'(b) \in \xi_2\big(x_n(b)\big)$ for all $n\geq 1$. Then recalling that $(0,0) \in \operatorname{Gr}\xi_i,\ i=1,2$, we have

$$\begin{aligned} & x_n'(0)x_n(0) \geq 0 \qquad \text{and} \qquad -x_n'(b)x_n(b) \geq 0, \\ \Rightarrow\ & -|x_n'(b)|^{p-2}x_n'(b)x_n(b) \geq 0 \qquad \text{and} \qquad |x_n'(0)|^{p-2}x_n'(0)x_n(0) \geq 0 \\ \Rightarrow\ & \langle U(x_n), x_n\rangle_p \geq \|x_n'\|_p^p \qquad \text{(see (5.122))}. \end{aligned} \tag{5.123}$$

Using (5.123) in (5.121), we obtain

$$\begin{aligned} & \|x_n'\|_p^p + \|x_n\|_p^p = \|x_n\|^p \leq \|v_n\|_{p'}\|x_n\|_p \\ \Rightarrow\ & \|x_n\|^{p-1} \leq c_3 \quad \text{for some } c_3>0 \text{ and all } n\geq 1, \\ \Rightarrow\ & \{x_n\}_{n\geq 1} \subseteq W^{1,p}\big((0,b)\big) \text{ is bounded.} \end{aligned}$$

Therefore we may assume that $x_n \xrightarrow{w} x$ in $W^{1,p}\big((0,b)\big)$ and $x_n \longrightarrow x$ in $C(T)$. Also directly from the equation $U(x_n)+K(x_n)=v_n$ for all $n\geq 1$, it follows that $\big\{|x_n'|^{p-2}x_n'\big\}_{n\geq 1} \subseteq W^{1,p'}\big((0,b)\big)$ is bounded. So by passing to a suitable subsequence if necessary, we may assume that

$$|x_n'|^{p-2}x_n' \xrightarrow{w} h \text{ in } W^{1,p'}\big((0,b)\big) \quad \text{and} \quad |x_n'|^{p-2}x_n' \longrightarrow h \text{ in } C(T). \tag{5.124}$$

If $\vartheta_p\colon \mathbb{R}\longrightarrow\mathbb{R}$ is the homeomorphism

$$\vartheta_p(y) = \begin{cases} |y|^{p-2}y & \text{if } y\neq 0 \\ 0 & \text{if } y=0 \end{cases}$$

(see also the proof of Proposition 5.2.17), then

$$\begin{aligned} & \vartheta_p^{-1}\big(|x_n'(t)|^{p-2}x_n'(t)\big) \longrightarrow \vartheta_p^{-1}\big(h(t)\big) \qquad \text{for all } t\in T, \\ \Rightarrow\ & x_n'(t) \longrightarrow \vartheta_p^{-1}\big(h(t)\big) \qquad \text{for all } t\in T, \\ \Rightarrow\ & x_n' \longrightarrow \vartheta_p^{-1}\big(h(\cdot)\big) \quad \text{in } L^p(T) \text{ (by the dominated convergence theorem).} \end{aligned}$$

But recall that $x_n' \xrightarrow{w} x'$ in $L^p(T)$. So we infer that

$$\begin{aligned} & x' = \vartheta_p^{-1}\big(h(\cdot)\big), \\ \Rightarrow\ & h = \vartheta_p\big(|x'(\cdot)|^{p-2}x'(\cdot)\big), \\ \Rightarrow\ & |x_n'|^{p-2}x_n' \longrightarrow |x'|^{p-2}x' \quad \text{in } C(T) \quad \text{(see (5.124))}, \\ \Rightarrow\ & x_n' \longrightarrow x' \quad \text{in } C(T), \\ \Rightarrow\ & x_n \longrightarrow x \quad \text{in } W^{1,p}\big((0,b)\big). \end{aligned}$$

This proves the complete continuity of the operator L. □

We introduce the order interval

$$E=[\psi,\varphi]=\{x\in W^{1,p}\big((0,b)\big) : \psi(t)\leq x(t)\leq\varphi(t) \quad \text{for all } t\in T\}.$$

Also let $\tau : W^{1,p}\big((0,b)\big)\longrightarrow W^{1,p}\big((0,b)\big)$, be the truncation operator defined by

$$\tau(x)(t)=\begin{cases} \psi(t) & \text{if } x(t)<\psi(t) \\ x(t) & \text{if } \psi(t)\leq x(t)\leq\varphi(t) \\ \varphi(t) & \text{if } \varphi(t)\leq x(t) \end{cases}.$$

Clearly τ is continuous and bounded (i.e., maps bounded sets to bounded ones). Given $w \in E$, we consider the following auxiliary boundary value problem:

$$\left\{\begin{array}{l} -\big(|x'(t)|^{p-2}x'(t)\big)' = f_1\big(t, x(t), x'(t)\big) + \beta\big(t, x(t)\big) - M\tau(x)(t) + \vartheta\big(w(t)\big) \\ \qquad\qquad + Mw(t) \text{ a.e. on } T, \\ x'(0) \in \xi_1\big(\psi(0)\big), \ -x'(b) \in \xi_2\big(\psi(b)\big) \end{array}\right\}. \tag{5.125}$$

PROPOSITION 5.2.19 *If hypotheses H_0, $H(f)_2$, $H(\xi)$, and $H(\vartheta)$ hold, then problem (5.125) has a solution $x \in C^1(T) \cap E$.*

PROOF: Let $G_1 : W^{1,p}\big((0,b)\big) \longrightarrow L^{p'}(T)$ be the nonlinear operator defined by

$$G_1(x) = G(x) + K(x) - M\tau(x) + \widehat{\vartheta}(w) + Mw \quad \text{for all } x \in W^{1,p}\big((0,b)\big).$$

From Proposition 5.2.16 and the continuity of the operators K and τ, we infer that G_1 is continuous. For every $x \in W^{1,p}\big((0,b)\big)$, we have

$$\begin{aligned} \|G_1(x)\|_{p'} &\leq \|\gamma_r\|_{p'} + b^{1/p'} \max\big\{\|\varphi\|_\infty^{p-1}, \|\psi\|_\infty^{p-1}\big\} + Mb^{1/p'} \max\big\{\|\varphi\|_\infty, \|\psi\|_\infty\big\} \\ &\quad + \|\widehat{\vartheta}(w)\|_{p'} + M\|w\|_{p'} \\ &= M_2, \ M_2 > 0. \end{aligned}$$

Recall that $r = \max\big\{M_0, \|\varphi\|_\infty, \|\psi\|_\infty\big\}$. Set $C = \{g \in L^{p'}(T) : \|g\|_{p'} \leq M_2\}$. Clearly G_1 maps bounded sets to bounded ones and $LG_1\big(W^{1,p}\big((0,b)\big)\big) \subseteq L(E)$ which is relatively compact in $W^{1,p}\big((0,b)\big)$ (see Proposition 5.2.18). Therefore we can find $x \in D \subseteq W^{1,p}\big((0,b)\big)$ such that

$$\begin{aligned} & x = LG_1(x), \\ \Rightarrow \ & U(x) + K(x) = G(x) + K(x) - M\tau(x) + \widehat{\vartheta}(w) + Mw, \\ \Rightarrow \ & U(x) = G(x) - M\tau(x) + \widehat{\vartheta}(w) + Mw, \\ \Rightarrow \ & x \in D \subseteq C^1(T) \text{ solves problem (5.125)).} \end{aligned}$$

It remains to show that $x \in E$. Because $\psi \in C^1(T)$ is a lower solution for problem (5.111), we have

$$\left\{\begin{array}{l} -\big(|\psi'(t)|^{p-2}\psi'(t)\big)' \leq f\big(t, \psi(t), \psi'(t)\big) + \vartheta\big(\psi(t)\big) \text{ a.e. on } T, \\ \psi'(0) \in \xi_1\big(\psi(0)\big) + \mathbb{R}_+, \ -x'(b) \in \xi_2\big(\psi(b)\big) + \mathbb{R}_+ \end{array}\right\}. \tag{5.126}$$

Subtracting (5.126) from (5.125), we obtain

$$\begin{aligned} \big(|\psi'(t)|^{p-2}\psi'(t)\big)' - \big(|x'(t)|^{p-2}x'(t)\big)' \geq\ & f\big(t, x(t), x'(t)\big) + \beta\big(t, x(t)\big) - M\tau(x)(t) \\ & + \vartheta\big(w(t)\big) + Mw(t) - f\big(t, \psi(t), \psi'(t)\big) \\ & - \vartheta\big(\psi(t)\big) \quad \text{a.e. on } T. \end{aligned} \tag{5.127}$$

We multiply (5.127) with $(\psi - x)^+ \in W^{1,p}\big((0,b)\big)$ and then integrate on $[0,b]$ the resulting inequality. We obtain

$$\begin{aligned}&\int_0^b \left[\left(|\psi'(t)|^{p-2}\psi'(t)\right)' - \left(|x'(t)|^{p-2}x'(t)\right)'\right](\psi - x)^+(t)dt \\ \geq &\int_0^b \left[f_1\big(t, x(t), x'(t)\big) - f\big(t, \psi(t), \psi'(t)\big)\right](\psi - x)^+(t)dt + \int_0^b \beta\big(t, x(t)\big)(\psi - x)^+(t)dt \\ &+ \int_0^b \left[\vartheta\big(w(t)\big) + Mw(t) - \vartheta\big(\psi(t)\big) - M\tau(x)(t)\right](\psi - x)^+(t)dt.\end{aligned} \tag{5.128}$$

First we estimate the left-hand side in inequality (5.128). Performing integration by parts, we obtain

$$\begin{aligned}&\int_0^b \left[\left(|\psi'(t)|^{p-2}\psi'(t)\right)' - \left(|x'(t)|^{p-2}x'(t)\right)'\right](\psi - x)^+(t)dt \\ = &\left(|\psi'(b)|^{p-2}\psi'(b) - |x'(b)|^{p-2}x'(b)\right)(\psi - x)^+(b) \\ &- \left(|\psi'(0)|^{p-2}\psi'(0) - |x'(0)|^{p-2}x'(0)\right)(\psi - x)^+(0) \\ &- \int_0^b \left(|\psi'(t)|^{p-2}\psi'(t) - |x'(t)|^{p-2}x'(t)\right)\left[(\psi - x)^+\right]'(t)dt.\end{aligned} \tag{5.129}$$

Recall that

$$\left[(\psi - x)^+\right]'(t) = \begin{cases} (\psi - x)(t) & \text{if } \psi(t) > x(t) \\ 0 & \text{if } \psi(t) \leq x(t) \end{cases}. \tag{5.130}$$

Also from the boundary conditions in (5.125) and (5.126), we have

$$-x'(b) \in \xi_2\big(x(b)\big) \quad \text{and} \quad -\psi'(b) \in \xi_2\big(\psi(b)\big) + e_b \quad \text{with } e_b \geq 0.$$

If $\psi(b) \geq x(b)$, then from the monotonicity of ξ_2 (see hypothesis H(ξ)), we have $\psi'(b) \leq x'(b) \Rightarrow \vartheta_p\big(\psi'(b)\big) \leq \vartheta_p\big(x'(b)\big) \Rightarrow |\psi'(b)|^{p-2}\psi'(b) \leq |x'(b)|^{p-2}x'(b)$. So it follows that

$$\left(|\psi'(b)|^{p-2}\psi'(b) - |x'(b)|^{p-2}x'(b)\right)(\psi - x)^+(b) \leq 0. \tag{5.131}$$

In a similar fashion using the boundary conditions $x'(0) \in \xi_1\big(x(0)\big)$ and $\psi'(0) \in \xi_1\big(\psi(0)\big) + e_0$ with $e_0 \geq 0$, we obtain that

$$\left(|\psi'(0)|^{p-2}\psi'(0) - |x'(0)|^{p-2}x'(0)\right)(\psi - x)^+(0) \geq 0. \tag{5.132}$$

Also we have

$$\begin{aligned}&\int_0^b \left(|\psi'(t)|^{p-2}\psi'(t) - |x'(t)|^{p-2}x'(t)\right)\left[(\psi - x)^+\right]'(t)dt \\ = &\int_{\{\psi > x\}} \left(|\psi'(t)|^{p-2}\psi'(t) - |x'(t)|^{p-2}x'(t)\right)(\psi' - x')(t)dt \geq 0.\end{aligned} \tag{5.133}$$

Using (5.131) through (5.133) in (5.129), we see that

$$\int_0^b \left[\left(|\psi'(t)|^{p-2}\psi'(t)\right)' - \left(|x'(t)|^{p-2}x'(t)\right)'\right](\psi - x)^+(t)dt \leq 0. \tag{5.134}$$

Next we estimate the right-hand side of (5.128). We have

$$f_1\big(t,x(t),x'(t)\big)-f\big(t,\psi(t),\psi'(t)\big) = f\big(t,\psi(t),\psi'(t)\big)-f\big(t,\psi(t),\psi'(t)\big) = 0 \quad \text{a.e. on } \{\psi \geq x\},$$

$$\Rightarrow \int_0^b \big[f_1\big(t,x(t),x'(t)\big) - f\big(t,\psi(t),\psi'(t)\big)\big](\psi - x)^+(t)\mathrm{d}t = 0. \tag{5.135}$$

Also from the definition of the penalty function β (see (5.113)), if $|\{\psi > x\}| > 0$ (by $|\cdot|$ we denote the Lebesgue measure on $\mathbb{R}$), then

$$\begin{aligned}&\int_0^b \beta\big(t,x(t)\big)(\psi - x)^+(t)\mathrm{d}t\\ &= \int_{\{\psi>x\}}\big(|\psi(t)|^{p-2}\psi(t) - |x(t)|^{p-2}x(t)\big)(\psi - x)(t)\mathrm{d}t > 0.\end{aligned} \tag{5.136}$$

Finally by virtue of hypothesis H(ϑ) and since $w \in E$, we see that

$$\int_0^b \Big(\vartheta\big(w(t)\big) + Mw(t) - \vartheta\big(\psi(t)\big) - M\tau(x)(t)\Big)(\psi - x)^+(t)\mathrm{d}t \geq 0. \tag{5.137}$$

Using (5.135) through (5.137), we have

$$\begin{aligned}&\int_0^b\big[f\big(t,x(t),x'(t)\big)-f\big(t,\psi(t),\psi'(t)\big)\big](\psi-x)^+(t)\mathrm{d}t+\int_0^b\beta\big(t,x(t)\big)(\psi-x)^+(t)\mathrm{d}t\\ &+\int_0^b\Big(\vartheta\big(w(t)\big)+Mw(t) - \vartheta\big(\psi(t)\big)-M\tau(x)(t)\Big)(\psi-x)^+(t)\mathrm{d}t>0,\end{aligned} \tag{5.138}$$

provided $|\{\psi > x\}| > 0$.

Returning to (5.128) and using (5.134) and (5.138), we reach a contradiction when $|\{\psi > x\}| > 0$. So it follows that $\psi(t) \leq x(t)$ for all $t \in T$. In a similar fashion we show that $x(t) \leq \varphi(t)$ for all $t \in T$; that is, $x \in E$. □

We use the solvability of the auxiliary problem (5.125) in order to produce a solution for the original problem (5.111). To do this, we need a fixed point theorem for multifunctions in an ordered Banach space. More on the fixed point theory for multifunctions can be found in Section 6.5.

THEOREM 5.2.20 *If X is a separable, reflexive, ordered Banach space, $E \subseteq X$ is a nonempty and weakly closed set, and $S: E \longrightarrow 2^E \setminus \{\emptyset\}$ is a multifunction with weakly closed values, $S(E)$ is bounded and*

(i) *The set $M = \{x \in E : x \leq y$ for some $y \in S(x)\}$ is nonempty.*
(ii) *If $x_1 \leq y_1$, $y_1 \in S(x_1)$ and $x_1 \leq x_2$, then we can find $y_2 \in S(x_2)$ such that $y_1 \leq y_2$,*

then S has a fixed point; that is, there exists $x \in E$ such that $x \in S(x)$.

We apply this theorem, using the following data. The separable, reflexive, ordered Banach space $X = W^{1,p}\big((0,b)\big)$, $E = [\psi, \phi]$ and $S: E \longrightarrow 2^E \setminus \{\emptyset\}$ is the solution multifunction for the auxiliary problem (5.125); that is, for every $w \in E$, $S(w)$ is the set of solutions for problem (5.125). From Proposition 5.2.19, we know that $S(w) \neq \emptyset$ and $S(w) \subseteq E$.

THEOREM 5.2.21 *If hypotheses $H(f)_2$, $H(\xi)$, and $H(\vartheta)$ hold, then problem (5.111) has a solution $x \in C^1(T)$.*

PROOF: Let $X = W^{1,p}\big((0,b)\big)$, $E = [\psi, \phi] \subseteq W^{1,p}\big((0,b)\big)$ and $S: E \longrightarrow 2^E \setminus \{\emptyset\}$ the solution multifunction for problem (5.125). From Proposition 5.2.19 we know that for every $w \in E$, $S(w) \neq \emptyset$ and $S(w) \subseteq E$. Moreover, it is routine to check that $S(w)$ is weakly closed in $W^{1,p}\big((0,b)\big)$. In addition, from the proof of Proposition 5.2.19 we know that $S(E) \subseteq W^{1,p}\big((0,b)\big)$ is bounded. Therefore it remains to verify conditions (i) and (ii) of Theorem 5.2.20.

Note that if $w = \psi$, then by Proposition 5.2.19, $S(\psi) \neq \emptyset$ and $S(\psi) \subseteq E$. So we have satisfied condition (i) of Theorem 5.2.20.

Next we verify condition (ii) of Theorem 5.2.20. So let $w_1, w_2 \in E$, $w_1 \leq w_2$, and $x_1 \in S(w_1)$ with $w_1 \leq x_1$. Because $x_1 \in S(w_1) \subseteq E$, we have $\beta\big(t, x_1(t)\big) = 0$ for all $t \in T$ (see (5.113)) and $\tau(x_1) = x_1$. Also note that if $p_{M_0}: \mathbb{R} \longrightarrow \mathbb{R}$ is the truncation function defined by

$$p_{M_0}(y) = \begin{cases} -M_0 & \text{if } y < -M_0 \\ y & \text{if } -M_0 \leq y \leq M_0 \\ M_0 & \text{if } M_0 < y \end{cases},$$

then from the definition of the function u (see (5.112)), we have

$$u\big(t, x_1(t), x_1'(t)\big) = \big(\tau(x_1)(t), p_{M_0}\tau(x_1)'(t)\big)\big) = \big(x_1(t), p_{M_0}\big(x_1'(t)\big)\big)$$

and so it follows that

$$f_1\big(t, x_1(t), x_1'(t)\big) = f\big(t, x_1(t), x_1'(t)\big) \qquad \text{for all } t \in T.$$

Therefore the auxiliary problem (5.125) becomes

$$\left\{ \begin{array}{l} -\big(|x_1'(t)|^{p-2} x_1'(t)\big)' = f\big(t, x_1(t), x_1'(t)\big) + \vartheta\big(w_1(t)\big) + M w_1(t) - M x_1(t) \\ \quad \text{a.e. on } T, \\ x_1'(0) \in \xi_1\big(x_1(0)\big), \ -x_1'(b) \in \xi_2\big(x_1(b)\big) \end{array} \right\}. \tag{5.139}$$

Because $w_1 \leq w_2$, by virtue of hypothesis $\mathrm{H}(\vartheta)$, we have

$$\vartheta\big(w_1(t)\big) + M w_1(t) \leq \vartheta\big(w_2(t)\big) + M w_2(t) \qquad \text{for all } t \in T. \tag{5.140}$$

Using (5.141) in (5.140), it follows that $x_1 \in C^1(T)$ is a lower solution for the problem

$$\left\{ \begin{array}{l} -\big(|x'(t)|^{p-2} x'(t)\big)' = f\big(t, x(t), x'(t)\big) + \vartheta\big(w_2(t)\big) + M w_2(t) \\ \qquad\qquad -Mx(t) \ \text{ a.e. on } T, \\ x'(0) \in \xi_1\big(x(0)\big), \ -x'(b) \in \xi_2\big(x(b)\big) \end{array} \right\} \tag{5.141}$$

(see Definition 5.2.11(b)). Also because $\varphi \in C^1(T)$ is an upper solution for problem (5.111), we have

$$\left\{ \begin{array}{l} -\big(|\varphi'(t)|^{p-2} \varphi'(t)\big)' \geq f\big(t, \varphi(t), \varphi'(t)\big) + \vartheta\big(\varphi(t)\big) \ \text{ a.e. on } T, \\ \varphi'(0) \in \xi_1\big(\varphi(0)\big) - \mathbb{R}_+, \ -\varphi'(b) \in \xi_2\big(\varphi(b)\big) - \mathbb{R}_+ \end{array} \right\}. \tag{5.142}$$

But $f\big(t,\varphi(t),\varphi'(t)\big)+\vartheta\big(\varphi(t)\big)+M\varphi(t)\geq f\big(t,\varphi(t),\varphi'(t)\big)+\vartheta\big(w_2(t)\big)+Mw_2(t)$ a.e. on T (see hypothesis H(ϑ) and recall that $w_2 \in E$). Using this inequality in (5.142), we obtain

$$\left\{\begin{array}{l} -\big(|\varphi'(t)|^{p-2}\varphi'(t)\big)' \geq f\big(t,\varphi(t),\varphi'(t)\big)+\vartheta\big(w_2(t)\big)+Mw_2(t) \\ -M\varphi(t) \quad \text{a.e. on } T, \\ \\ \varphi'(0) \in \xi_1\big(\varphi(0)\big)-\mathbb{R}_+, \ -\varphi'(b) \in \xi_2\big(\varphi(b)\big)-\mathbb{R}_+ \end{array}\right\}.$$

This means that $\varphi \in C^1(T)$ is an upper solution for problem (5.142) (see Definition 5.2.11(a)). Then for problem (5.142), using truncation and penalization techniques based on the ordered upper–lower solution pair $\{\varphi, x_1\}$, as in Proposition 5.2.19 we obtain a solution $x_2 \in E_1 = [x_1,\varphi]$ for problem (5.142). Evidently $x_2 \in S(w_2)$ and $x_1 \leq x_2$. So we have satisfied condition (ii) of Theorem 5.2.20. Therefore we can apply that theorem and obtain $x \in E$ such that $x \in S(x)$. Clearly $x \in C^1(T)$ solves problem (5.111). □

Next we establish the existence of a greatest and of a smallest solution in the order interval E. These solutions are called *extremal solutions*.

So let $C_1 = \big\{x \in C^1(T) : x \text{ be a solution of (5.111) and } x \in E\big\}$. On $L^\infty(T)$ we consider the partial order structure induced by the order cone $L^\infty(T)_+ = \{x \in L^\infty(T) : x(t) \geq 0 \text{ a.e. on } T\}$. So $x \leq y$ in $L^\infty(T)$ if and only if $x(t) \leq y(t)$ a.e. on T. From Theorem 5.2.21, we know that under hypotheses H_0, $H(f)_2$, $H(\xi)$, and $H(\vartheta)$, the set C_1 is nonempty. Recall that a set C in a partially ordered set is a chain (or totally ordered subset), if for every $x,y \in C$, either $x \leq y$ or $y \leq x$.

PROPOSITION 5.2.22 *If hypotheses H_0, $H(f)_2$, $H(\xi)$, and $H(\vartheta)$ hold, then every chain C in C_1 has an upper bound.*

PROOF: Because $C \subseteq L^\infty(T)$ is bounded and $L^\infty(T)$ is a complete lattice, we can find $\{x_n\}_{n\geq1} \subseteq C$ such that $\sup\limits_{n\geq1} x_n = \sup C \in L^\infty(T)$. Moreover, because of the lattice structure of $L^\infty(T)$, we can assume that the sequence $\{x_n\}_{n\geq1}$ is increasing. Invoking the monotone convergence theorem, we have that $x_n \longrightarrow x$ in $L^p(T)$. Because of Lemma 5.2.15 and $H(f)_2$(iv) and $H(\vartheta)$, we have

$$\begin{aligned} &\big|\big(|x_n'(t)|^{p-2}x_n'(t)\big)'\big| \leq \gamma(t) \qquad \text{a.e. on } T, \ \text{ for all } n\geq1 \text{ with } \gamma \in L^{p'}(T), \\ \Rightarrow\ &\big\{|x_n'|^{p-2}x_n'\big\}_{n\geq1} \subseteq W^{1,p'}\big((0,b)\big) \quad \text{is bounded.} \end{aligned}$$

Also $\|x_n\|_\infty, \|x_n'\|_\infty \leq r = \max\{M_1, \|\varphi\|_\infty, \|\psi\|_\infty\}$ for all $n \geq 1$, hence $\{x_n\}_{n\geq1} \subseteq W^{1,p}\big((0,b)\big)$ is bounded. Therefore we may assume that

$$|x_n'|^{p-2}x_n' \xrightarrow{w} v \ \text{ in } W^{1,p'}\big((0,b)\big) \ \text{ and } \ x_n \xrightarrow{w} u \ \text{ in } W^{1,p}\big((0,b)\big).$$

Evidently $u = x$ and as in proof of Proposition 5.2.18, we can check that $v = |x'|^{p-2}x'$. Note that $x_n \longrightarrow x$ in $C(T)$ and $|x_n'|^{p-2}x_n' \longrightarrow |x'|^{p-2}x'$ in $C(T)$. So we have $x_n(t) \longrightarrow x(t)$ and $x_n'(t) \longrightarrow x'(t)$ for all $t \in T$. From the dominated convergence theorem, we have

$$f\big(\cdot,x_n(\cdot),x_n'(\cdot)\big) \longrightarrow f\big(\cdot,x(\cdot),x'(\cdot)\big) \ \text{ in } L^{p'}(T).$$

Also from the monotone convergence theorem we have

$$\widehat{\vartheta}(x_n) + Mx_n \longrightarrow \widehat{\vartheta}(x) + Mx \ \text{ in } \ L^{p'}(T).$$

Because $Mx_n \longrightarrow Mx$ in $C(T)$, it follows that $\widehat{\vartheta}(x_n) \longrightarrow \widehat{\vartheta}(x)$ in $L^{p'}(T)$. Therefore in the limit as $n \to \infty$, we have

$$-\big(|x'(t)|^{p-2}x'(t)\big)' = f\big(t, x(t), x'(t)\big) + \vartheta\big(x(t)\big) \quad \text{a.e. on } T.$$

Moreover, exploiting the fact that Gr ξ_1, Gr ξ_2 are closed in $\mathbb{R} \times \mathbb{R}$, we also have

$$x'(0) \in \xi_1\big(x(0)\big) \quad \text{and} \quad -x'(b) \in \xi_2\big(x(b)\big).$$

Therefore $x \in C^1(T)$ is a solution of problem (5.111) and in addition $x \in E$. Hence $x \in C_1$ and clearly is an upper bound of C. □

Recall that if $(C_0, \leq)$ is a partially ordered set, we say that C_0 is directed, if for each pair $u_1, u_2 \in C_0$, we can find $u_3 \in C_0$ such that $u_1 \leq u_3$ and $u_2 \leq u_3$.

PROPOSITION 5.2.23 *If hypotheses H_0, $H(f)_2$, $H(\xi)$, and $H(\vartheta)$ hold, then $C_1 \subseteq W^{1,p}\big((0,b)\big)$ is directed, when $W^{1,p}\big((0,b)\big)$ is endowed with the pointwise order (order induced by $C(T)$).*

PROOF: Let $x_1, x_2 \in C_1$ and set $x_3 = \max\{x_1, x_2\}$. We have $x_3 = (x_1 - x_2)^+ + x_2$ and so $x_3 \in W^{1,p}\big((0,b)\big)$. We introduce the following truncation and penalty functions.

$$u_0(t,x,y) = \begin{cases} \big(x_3(t), x_3'(t)\big) & \text{if } x < x_3(t) \\ \big(\varphi(t), \varphi'(t)\big) & \text{if } x > \varphi(t) \\ (x, M_1) & \text{if } x_3(t) \leq x \leq \varphi(t),\ y > M_1 \\ (x, -M_1) & \text{if } x_3(t) \leq x \leq \varphi(t),\ y < -M_1 \\ (x, y) & \text{if otherwise} \end{cases}$$

and
$$\beta_0(t,x) = \begin{cases} |\varphi(t)|^{p-2}\varphi(t) - |x|^{p-2}x & \text{if } x > \varphi(t) \\ 0 & \text{if } x_3(t) \leq x \leq \varphi(t)\,. \\ |x_3(t)|^{p-2}x_3(t) - |x|^{p-2}x & \text{if } x < x_3(t). \end{cases}$$

Then we introduce the following modification of the nonlinearity f,

$$f_0(t,x,y) = f\big(t, u_0(t,x,y)\big).$$

Note that for almost all $t \in T$, all $x \in [x_3(t), \varphi(t)]$, and all $|y| \leq M_1$ we have

$$f_0(t,x,y) = f(t,x,y).$$

Moreover, for almost all $t \in T$ and all $x, y \in \mathbb{R}$,

$$|f_0(t,x,y)| \leq \gamma_r(t) \quad \text{with } \gamma_r \in L^{p'}(T), \quad r = \max\{M_1, \|\varphi_\infty, \|\psi\|_\infty\}.$$

Also we introduce the truncation operator $\tau_0 : W^{1,p}\big((0,b)\big) \longrightarrow W^{1,p}\big((0,b)\big)$ defined by

$$\tau(x)(t)=\begin{cases} x_3(t) & \text{if } x(t)<x_3(t) \\ x(t) & \text{if } x_3(t)\le x(t)\le \varphi(t) \\ \varphi(t) & \text{if } \varphi(t)<x(t) \end{cases}.$$

Let $E_0 = [x_3, \varphi] = \{x \in W^{1,p}((0,b)) : x_3(t) \le x(t) \le \varphi(t) \text{ for all } t \in T\}$. Given $w \in E_0$, we consider the following auxiliary problem

$$\left\{ \begin{array}{l} -\big(|x'(t)|^{p-2}x'(t)\big)' = f_0\big(t,x(t),x'(t)\big) + \beta_0\big(t,x(t)\big) - M\tau_0(x)(t) \\ \qquad\qquad +\vartheta\big(w(t)\big) + Mw(t) \text{ a.e. on } T, \\ \\ x'(0) \in \xi_1\big(x(0)\big),\ -x'(b) \in \xi_2\big(x(b)\big) \end{array} \right\}. \tag{5.143}$$

Following the reasoning of Proposition 5.2.19, we can show that problem (5.143) has a solution $x \in C^1(T) \cap E_0$. So we can define the solution multifunction $S: E_0 \longrightarrow 2^{W^{1,p}((0,b))} \setminus \{\emptyset\}$ and then as in Theorem 5.2.21, via the use of Theorem 5.2.20 we can produce $x \in C^1(T) \cap E_0$ such that it solves problem (5.111). Therefore $x \in C_1$ and clearly $x_1 \le x$, $x_2 \le x$ which proves that the set C_1 is directed. □

Now we use this proposition to establish the existence of extremal solutions for problem (5.111) in the order interval $E = [\psi, \varphi]$.

THEOREM 5.2.24 *If hypotheses H_0, $H(f)_2$, $H(\xi)$, and $H(\vartheta)$ hold, then problem (5.111) has extremal solutions in the order interval $E = [\psi, \varphi]$.*

PROOF: By virtue of Proposition 5.2.22 and Zorn's lemma, we can find $x^* \in C_1$ a maximal element for the pointwise ordering on $W^{1,p}((0,b))$. If $x \in C_1$, then because of Proposition 5.2.23, we can find $y \in C_1$ such that $x \le y, x^* \le y$. Because $x^* \in C_1$ is maximal, we must have $x^* = y$. Then $x \le x^*$ and because $x \in C_1$ was arbitrary, it follows that $x^* \in C_1$ is the greatest element of C_1.

If on $W^{1,p}((0,b))$ we use the partial order $\le_1$ defined by

$$x \le_1 y \qquad \text{if and only if } y(t) \le x(t) \quad \text{a.e. on } T,$$

then by the same argument we produce $x_* \in C_1$, which is the smallest element of C_1. Hence $\{x_*, x^*\}$ are the extremal solutions of (5.111) in $E = [\psi, \varphi]$. □

The framework of problem (5.111) is general and it incorporates as special cases standard boundary value problems.

EXAMPLE 5.2.25 (a) Let $I_1, I_2 \subseteq \mathbb{R}$ be nonempty closed intervals containing zero. For $i = 1, 2$, we set

$$i_{I_i}(x)=\begin{cases} 0 & \text{if } x \in I_i \\ +\infty & \text{if otherwise} \end{cases} \qquad \text{(the indicator function of } I_i\text{)}.$$

Set $\xi_i = \partial i_{I_i}$, $i = 1, 2$ (the subdifferential in the sense of convex analysis. Then problem (5.111) becomes

$$\left\{ \begin{array}{l} -\big(|x'(t)|^{p-2}x'(t)\big)' = f\big(t,x(t),x'(t)\big) + \vartheta\big(x(t)\big) \text{ a.e. on } T, \\ x(0) \in I_1,\ x(b) \in I_2 \\ x'(0)x(0) = \sup[vx'(0) : v \in I_1] \\ -x'(b)x(b) = \sup[-wx'(b) : w \in I_2] \end{array} \right\}. \tag{5.144}$$

(b) If $I_1 = I_2 = \{0\}$, then $\xi_i(x) = \mathbb{R}$ for all $x \in \mathbb{R}$ and for $i = 1, 2$. Hence problem (5.144) is the classical Dirichlet problem.

(c) If $I_1 = I_2 = \mathbb{R}$, then $\xi_i(x) = \{0\}$ for all $x \in \mathbb{R}$ and for $i = 1, 2$. Hence problem (5.144) is the classical Neumann problem.

(d) If $\xi_1(x) = (1/\beta)x$ and $\xi_2(x) = (1/\gamma)x$, $\beta, \gamma > 0$, then problem (5.111) becomes the following Sturm–Liouville problem.

$$\left\{ \begin{array}{ll} -\big(|x'(t)|^{p-2}x'(t)\big)' = f\big(t, x(t), x'(t)\big) + \vartheta\big(x(t)\big) & \text{a.e. on } T, \\ x(0) - \beta x'(0) = 0, \ x(b) + \gamma x'(b) = 0 & \end{array} \right\}.$$

(e) If $u_1, u_2 : \mathbb{R} \longrightarrow \mathbb{R}$ are two contractions and $\xi_1 = u_1 - \mathrm{id}$, $\xi_2 = u_2 - \mathrm{id}$ (both maximal monotone functions), then problem (5.111) becomes

$$\left\{ \begin{array}{ll} -\big(|x'(t)|^{p-2}x'(t)\big)' = f\big(t, x(t), x'(t)\big) + \vartheta\big(x(t)\big) & \text{a.e. on } T, \\ x(0) + x'(0) = u_1\big(x(0)\big), \ x(b) - x'(b) = u_2\big(x(b)\big) & \end{array} \right\}.$$

REMARK 5.2.26 The boundary conditions of problem (5.111) do not include the periodic ones. However, a careful analysis of the methods of the proof, reveals that they are also valid for the periodic problem. So Theorems 5.2.21 and 5.2.24 are also true for the periodic problem

$$\left\{ \begin{array}{ll} -\big(|x'(t)|^{p-2}x'(t)\big)' = f\big(t, x(t), x'(t)\big) + \vartheta\big(x(t)\big) & \text{a.e. on } T, \\ x(0) = x(b), \ x'(0) = x'(b) & \end{array} \right\}. \tag{5.145}$$

5.3 Degree-Theoretic Methods

In this section we use degree theoretic methods to study second-order nonlinear boundary value problems. We consider two examples. The first is a nonlinear elliptic equation driven by the p-Laplacian with Dirichlet boundary conditions. Using degree-theoretic methods based on the degree map for operators of monotone type (see Section 3.3), we establish the existence of multiple nontrivial solutions. In the second example, we deal with a second-order scalar equation driven by the ordinary p-Laplacian with periodic boundary conditions and using the Leray–Schauder degree, we produce solutions under conditions of nonuniform nonresonance and of Landesman–Lazer type for the nonlinearity.

We start with the elliptic Dirichlet boundary value problem. So let $Z \subseteq \mathbb{R}^N$ be a bounded domain with a C^2-boundary ∂Z. The problem under consideration is the following.

$$\left\{ \begin{array}{l} -\mathrm{div}\big(\|Dx(z)\|^{p-2}Dx(z)\big) = f\big(z, x(z)\big) \ \text{ a.e. on } Z, \\ \\ x\big|_{\partial Z} = 0, \ \ 1 < p < \infty \end{array} \right\}. \tag{5.146}$$

The hypotheses on the nonlinearity $f(z, x)$ are the following.

H(f)$_1$: $f \colon Z \times \mathbb{R} \longrightarrow \mathbb{R}$ is a function such that $f(z, 0) = 0$ a.e. on Z, $f(z, x) \geq 0$ a.e. on Z for all $x \geq 0$ and

(i) For all $x \in \mathbb{R}$, $z \longrightarrow f(z,x)$ is measurable.

(ii) For almost all $z \in Z$, $x \longrightarrow f(z,x)$ is continuous.

(iii) For almost all $z \in Z$ and all $x \in \mathbb{R}$, we have

$$|f(z,x)| \leq \alpha(z) + c|x|^{p-1} \quad \text{with } \alpha \in L^\infty(Z)_+,\ c>0.$$

(iv) There exists $\vartheta \in L^\infty(Z)_+$ such that $\vartheta(z) \leq \lambda_1$ a.e. on Z with strict inequality on a set of positive measure and

$$\limsup_{x \rightarrow +\infty} \frac{f(z,x)}{x^{p-1}} \leq \vartheta(z) \quad \text{uniformly for a.a. } z \in Z,$$

and there exist $\eta_1, \eta_2 \in L^\infty(Z)_+$ such that $\eta_1(z) \geq \lambda_1$ a.e. on Z with strict inequality on a set of positive measure and

$$\eta_1(z) \leq \liminf_{x \rightarrow -\infty} \frac{f(z,x)}{|x|^{p-2}x} \leq \limsup_{x \rightarrow -\infty} \frac{f(z,x)}{|x|^{p-2}x} \leq \eta_2(z)$$

uniformly for a.a. $z \in Z$.

(v) There exist $\eta, \widehat{\eta} \in L^\infty(Z)_+$ such that $\eta(z) \geq \lambda_1$ a.e. on Z with strict inequality on a set of positive measure and

$$\eta(z) \leq \liminf_{x \rightarrow 0^+} \frac{f(z,x)}{x^{p-1}} \leq \limsup_{x \rightarrow 0^+} \frac{f(z,x)}{x^{p-1}} \leq \widehat{\eta}(z) \quad \text{uniformly for a.a. } z \in Z$$

and $\lim_{x \rightarrow 0^-} \frac{f(z,x)}{|x|^{p-2}x} = 0$ uniformly for a.a. $z \in Z$.

REMARK 5.3.1 Hypotheses $H(f)_1$(iv) and (v) are nonresonance conditions at $+\infty$ and 0^+, respectively, with respect to $\lambda_1 > 0$ the principal eigenvalue of $\left(-\Delta_p, W_0^{1,p}(Z)\right)$. In both conditions we allow partial interaction with $\lambda_1 > 0$ (nonuniform nonresonance). Note that when $p = 2$, these conditions incorporate in the example the so-called *asymptotically linear* problems.

Let $\tau_+ : \mathbb{R} \longrightarrow \mathbb{R}_+$ be the positive truncation function defined by

$$\tau_+(x) = \begin{cases} 0 & \text{if } x \leq 0 \\ x & \text{if } x \geq 0 \end{cases}.$$

We set $f_+(z,x) = f\big(z, \tau_+(x)\big)$ and $F_+(z,x) = \int_0^x f_+(z,r)dr$. Then we produce the energy functional $\varphi_+ : W_0^{1,p}(Z) \longrightarrow \mathbb{R}$ defined by

$$\varphi_+(x) = \frac{1}{p}\|Dx\|_p^p - \int_Z F_+\big(z, x(z)\big)dz \qquad \text{for all } x \in W_0^{1,p}(Z).$$

We know that $\varphi_+ \in C^1\big(W_0^{1,p}(Z)\big)$. Also $\varphi(x) = (1/p)\|Dx\|_p^p - \int_Z F\big(z, x(z)\big)dz$ for all $x \in W_0^{1,p}(Z)$. Again $\varphi \in C^1(W_0^{1,p}(Z))$.

PROPOSITION 5.3.2 *If $H(f)_1$ hold, then there exists $\widehat{x}_0 \in \operatorname{int} C_0^1(\overline{Z})$ a local minimizer of φ.*

PROOF: By virtue of hypothesis $H(f)_1$(iv), given $\varepsilon>0$, we can find $M=M(\varepsilon)>0$ such that

$$f_+(z,x) \le \big(\vartheta(z)+\varepsilon\big)x^{p-1} \quad \text{for a.a. } z\in Z \text{ and all } x\ge M. \tag{5.147}$$

On the other hand from hypothesis $H(f)_1$(iii) and because $f_+(z,x)=0$ for a.a. $z\in Z$ and all $x\le 0$, we have

$$f_+(z,x) \le \widehat{\alpha}_\varepsilon(z) \quad \text{for a.a. } z\in Z \text{ all } x\le M, \text{ with } \widehat{\alpha}_\varepsilon\in L^\infty(Z)_+. \tag{5.148}$$

Combining (5.147) and (5.148), we obtain

$$\begin{aligned} & f_+(z,x) \le \big(\vartheta(z)+\varepsilon\big)|x|^{p-1}+\widehat{\alpha}_\varepsilon(z) && \text{for a.a. } z\in Z, \text{ all } x\in\mathbb{R}, \\ \Rightarrow\ & F_+(z,x) \le \frac{1}{p}\big(\vartheta(z)+\varepsilon\big)|x|^{p}+\widehat{\alpha}_\varepsilon(z)|x| && \text{for a.a. } z\in Z, \text{ all } x\in\mathbb{R}. \end{aligned} \tag{5.149}$$

Then for all $x\in W_0^{1,p}(Z)$ we have

$$\begin{aligned} \varphi_+(x) &= \frac{1}{p}\|Dx\|_p^p - \int_Z F_+\big(z,x(z)\big)dz \\ &\ge \frac{1}{p}\|Dx\|_p^p - \frac{1}{p}\int_Z \vartheta(z)|x(z)|^p - \frac{\varepsilon}{p}\|x\|_p^p - c_1\|Dx\|_p \end{aligned}$$

for some $c_1>0$, (see (5.149))

$$\ge \frac{1}{p}\Big(\xi-\frac{\varepsilon}{\lambda_1}\Big)\|Dx\|_p^p - c_1\|Dx\|_p \tag{5.150}$$

(see Lemma 5.1.3 and Theorem 4.3.47).

We choose $\varepsilon<\lambda_1\xi$. Then from (5.150) it follows that φ_+ is coercive. Exploiting the compact embedding of $W_0^{1,p}(Z)$ into $L^p(Z)$, we can see that φ_+ is weakly lower semicontinuous. So we can find $\widehat{x}_0\in W_0^{1,p}(Z)$ such that $\varphi_+(\widehat{x}_0)=\inf\big[\varphi_+(x): x\in W_0^{1,p}(Z)\big]$.

Using hypothesis $H(f)_1$(v), we see that $\varphi_+(tu_1)<0$ for $t>0$ small. Hence $m=\varphi_+(x_0)<0=\varphi_+(0)$ and so $\widehat{x}_0\neq 0$. As before from nonlinear regularity theory and the nonlinear maximum principle (see Theorems 4.3.35 and 4.3.37), we obtain that $\widehat{x}_0\in \operatorname{int} C_0^1(\overline{Z})$. Then by virtue of Theorem 5.1.27, it follows that $\widehat{x}_0$ is a local minimizer of φ. □

In what follows by N we denote the Nemitsky operator corresponding to the Carathéodory function $f(z,x)$; that is,

$$N(x)(\cdot) = f\big(\cdot,x(\cdot)\big) \qquad \text{for all } x\in L^p(Z).$$

We know that $N: L^p(Z)\longrightarrow L^{p'}(Z)$, $(1/p)+(1/p')=1$ is continuous and bounded (see hypotheses $H(f)_1$(i), (ii), and (iii)). Also let $A: W_0^{1,p}(Z)\longrightarrow W^{-1,p'}(Z)=W_0^{1,p}(Z)^*$ be the nonlinear operator defined by

$$\langle A(x),y\rangle = \int_Z \|Dx\|^{p-2}(Dx,Dy)_{\mathbb{R}^N}dz \qquad \text{for all } x,y\in W_0^{1,p}(Z).$$

Here by $\langle\cdot,\cdot\rangle$ we denote the duality brackets for the pair $\big(W_0^{1,p}(Z),W^{-1,p'}(Z)\big)$.

We know that A is of type $(S)_+$ (see Proposition 4.3.41). Due to the compact embedding of $W_0^{1,p}(Z)$ into $L^p(Z)$, we can see that N is completely continuous. So it follows at once that the operator $x \longrightarrow (A-N)(x)$ is of $(S)_+$–type and we can consider the degree in the sense of Definition 3.3.63.

PROPOSITION 5.3.3 *If hypotheses $H(f)_1$ hold, then there exists $R_0>0$ such that for all $R \geq R_0$ we have*

$$d_{(S)_+}(A-N, B_R, 0) = 0.$$

PROOF: Let $K_- : W_0^{1,p}(Z) \longrightarrow W^{-1,p'}(Z)$ be defined by

$$K_-(x) = \big(x^-(\cdot)\big)^{p-1} \qquad \text{for all } x \in W_0^{1,p}(Z).$$

Fix $h \in L^\infty(Z)_+$, $h(z) \geq \lambda_1$ a.e. on Z with strict inequality on a set of positive measure.

We consider the $(S)_+$-homotopy $h_1 : [0,1] \times W_0^{1,p}(Z) \longrightarrow W^{-1,p'}(Z)$ defined by

$$h_1(t,x) = A(x) - tN(x) + (1-t)hK_-(x).$$

Claim: There exists $R_0 > 0$ such that for all $t \in [0,1]$, all $x \in \partial B_R(0)$ and all $R \geq R_0$, $0 \neq h_1(t,x)$.

Suppose that the claim is not true. Then we can find $\{t_n\}_{n\geq 1} \subseteq [0,1]$ and $x_n \subseteq W_0^{1,p}(Z)$ such that

$$\begin{aligned} & t_n \longrightarrow t \quad \text{in} \quad [0,1], \ \|x_n\| \longrightarrow \infty \\ \text{and } & A(x_n) = t_n N(x_n) - (1-t_n)hK_-(x_n) \qquad \text{for all } n \geq 1. \end{aligned}$$

Acting with the test function x_n^+ and using hypothesis H(f)(iv), we obtain that $\{x_n^+\}_{n\geq 1} \subseteq W_0^{1,p}(Z)$ is bounded. Then we must have $\|x_n^-\| \longrightarrow \infty$. We set $y_n = (x_n^- / \|x_n^-\|)$ and we may assume that

$$\begin{aligned} & y_n \xrightarrow{w} y \ \text{ in } W_0^{1,p}(Z), \quad y_n \longrightarrow y \text{ in } L^p(Z),\ y_n(z) \longrightarrow y(z) \text{ a.e. on } Z \\ \text{and} \quad & |y_n(z)| \leq k(z) \text{ a.e. on } Z \quad \text{for all } n \geq 1, \text{ with } k \in L^p(Z)_+. \end{aligned}$$

We have

$$\frac{1}{\|x_n^-\|^{p-1}} A(x_n^+) - A(y_n) = t_n \frac{N(x_n)}{\|x_n^-\|^{p-1}} - (1-t_n)hy_n^{p-1}.$$

Acting with $y_n - y$ and passing to the limit as $n \to \infty$, we obtain

$$\lim \langle A(y_n), y_n - y\rangle = 0.$$

Because A is of type $(S)_+$, it follows that $y_n \longrightarrow y$ in $W_0^{1,p}(Z)$ and so $\|y\| = 1$. Also arguing in the proof of Proposition 5.1.2, we see that

$$\frac{N(x_n)}{\|x\|^{p-1}} \xrightarrow{w} \widehat{h} \quad \text{in } L^{p'}(Z),$$

with $\widehat{h} = -gy^{p-1}$ where $g \in L^\infty(Z)_+$, $\eta_1(z) \leq g(z) \leq \eta_2(z)$ a.e on Z. Hence

$$A(y) = \widehat{g}y^{p-1} \tag{5.151}$$

with $\widehat{g} = tg + (1-t)h$. From (5.151) we have

$$\left\{\begin{array}{l} -\operatorname{div}\big(\|Dy(z)\|^{p-2}Dy(z)\big) = \widehat{g}(z)|y(z)|^{p-2}y(z) \text{ a.e. on } Z, \\ y|_{\partial Z} = 0 \end{array}\right\}. \tag{5.152}$$

Note that $1=\widehat{\lambda}_1(\lambda_1)>\widehat{\lambda}_1(\widehat{g})$ and so from (5.152) it follows that y must change sign, a contradiction. This proves the claim. So we have

$$d_{(S)_+}(A-N, B_R, 0) = d_{(S)_+}(A+hK_-, B_R, 0) \qquad \text{for all } R \geq R_0. \tag{5.153}$$

But from Godoy–Gossez–Paczka [268], we have that

$$d_{(S)_+}(A+hK_-, B_R, 0) = 0 \qquad \text{for all } R>0. \tag{5.154}$$

Using (5.154) in (5.153), we conclude that

$$d_{(S)_+}(A-N, B_R, 0) = 0 \qquad \text{for all } R \geq R_0.$$

□

Next we prove an analogous result for small balls.

PROPOSITION 5.3.4 *If hypotheses $H(f)_1$ hold, then there exists $\varrho_0 > 0$ such that*

$$d_{(S)_+}(A-N_+, B_\varrho, 0) = 0 \qquad \textit{for all } 0<\varrho \leq \varrho_0.$$

PROOF: In what follows $K_+ : L^p(Z) \longrightarrow L^{p'}(Z)$ is a continuous monotone (hence maximal monotone too) map defined by

$$K_+(x)(\cdot) = \big(x^+(\cdot)\big)^{p-1}.$$

Let $h_3 : [0,1] \times W_0^{1,p}(Z) \longrightarrow W^{-1,p'}(Z)$ be the homotopy defined by

$$h_3(t,x) = A(x) - (1-t)\eta K_+(x) - tN(x),$$

where $\eta \in L^\infty(Z)_+$ is as in hypothesis $H(f)_1$(v). Because of the complete continuity of the maps $K_+, N : W_0^{1,p}(Z) \longrightarrow W^{-1,p'}(Z)$, we can easily verify that $h_3(t,x)$ is a homotopy of type $(S)_+$.

Claim: There exists $\varrho_0>0$ such that for all $t \in [0,1]$, all $x \in \partial B_\varrho(0)$ and all $0<\varrho \leq \varrho_0$, $x \neq h_3(t,x)$.

As in proof of Proposition 5.3.3, we argue indirectly. So suppose that the claim is not true. Then we can find $\{t_n\}_{n\geq 1} \subseteq [0,1]$ and $\{x_n\}_{n\geq 1} \subseteq W_0^{1,p}(Z)$ such that

$$t_n \longrightarrow t \text{ in } [0,1],\ \|x_n\| \longrightarrow 0 \text{ as } n\rightarrow\infty \text{ and } h_3(t_n,x_n) = 0 \qquad \text{for all } n\geq 1.$$

From the last equality, we have

$$A(x_n) = (1-t_n)\eta K_+(x_n) + t_n N(x_n) \qquad \text{for all } n\geq 1.$$

We set $y_n = x_n/\|x_n\|$, $n\geq 1$. By passing to a suitable subsequence if necessary, we may assume that

$$y_n \xrightarrow{w} y \text{ in } W_0^{1,p}(Z) \quad \text{and} \quad y_n \longrightarrow y \text{ in } L^p(Z) \quad \text{as } n \to \infty.$$

We have

$$A(y_n) = (1 - t_n)\eta K_+(y_n) + t_n \frac{N(x_n)}{\|x_n\|^{p-1}} \qquad \text{for all } n \geq 1. \tag{5.155}$$

By virtue of hypotheses $\mathrm{H}(f)_1$, we deduce that

$$|f(z,x)| \leq c_1|x|^{p-1} \text{ for a.a. } z \in Z, \text{ all } x \in \mathbb{R}, \tag{5.156}$$

with $c_1 > 0$. From (5.156) we infer that

$$\int_Z \Big(\frac{N(x_n)}{\|x_n\|^{p-1}}\Big)^{p/(p-1)} dz \leq c_1^{p/(p-1)} \|y\|_p^p \qquad \text{for all } n \geq 1,$$

$$\Rightarrow \Big\{\frac{N(x_n)}{\|x_n\|^{p-1}}\Big\}_{n\geq 1} \subseteq L^{p'}(Z) \qquad \text{is bounded.} \tag{5.157}$$

By passing to a subsequence if necessary, we may assume that

$$\frac{N(x_n)}{\|x_n\|^{p-1}} \xrightarrow{w} h_0 \quad \text{in } L^{p'}(Z) \text{ as } n \to \infty, \text{ with } h_0 \in L^{p'}(Z). \tag{5.158}$$

As before, using hypothesis $\mathrm{H}(f)_1(\mathrm{v})$, we can check that

$$\eta(z)y^+(z) \leq h_0(z) \leq \widehat{\eta}(z)y^+(z) \qquad \text{a.e. on } Z$$
$$\Rightarrow h_0(z) = g(z)y^+(z)^{p-1} \qquad \text{a.e. on } Z,$$

with $g \in L^\infty(Z)_+$ such that $\eta(z) \leq g(z) \leq \widehat{\eta}(z)$ a.e. on Z. Acting on (5.155) with the test function $y_n - y \in W_0^{1,p}(Z)$, we have

$$\langle A(y_n), y_n - y\rangle = \int_Z \Big[(1 - t_n)\eta (y_n^+)^{p-1} + t_n \frac{N(x_n)}{\|x_n\|^{p-1}}\Big](y_n - y)(z)dz \longrightarrow 0,$$

as $n \to \infty$, (see (5.156)). Because A is of type $(S)_+$, it follows that

$$y_n \longrightarrow y \quad \text{in } W_0^{1,p}(Z). \tag{5.159}$$

Passing to the limit as $n \to \infty$ in (5.155) and using (5.158) and (5.159), we obtain

$$A(y) = (1-t)\eta K_+(y) + tgK_+(y). \tag{5.160}$$

Recall that

$$Dy^+(z) = \begin{cases} Dy(z) & \text{a.e. on } \{y > 0\} \\ 0 & \text{a.e. on } \{y \leq 0\} \end{cases}.$$

So we can write that

$$A(y^+) = \sigma K_+(y), \tag{5.161}$$

where $\sigma \in L^\infty(Z)_+$ is given by $\sigma = (1-t)\eta + tg$. From (5.161) we have

$$\left\{ \begin{array}{l} -\mathrm{div}\big(\|Dy(z)\|^{p-2}Dy(z)\big) = \sigma(z)|y(z)|^{p-2}y(z) \text{ a.e. on } Z, \\ y|_{\partial Z} = 0 \end{array} \right\}. \tag{5.162}$$

Once again, the monotonicity of the principal eigenvalue on the weight function, implies that

$$\widehat{\lambda}_1(\sigma) \leq \widehat{\lambda}_1(\eta) < \widehat{\lambda}_1(\lambda_1) = 1.$$

Hence because of (5.162), we see that $y^+ \in W_0^{1,p}(Z)$ cannot be the principal eigenfunction of $\left(-\Delta_p, W_0^{1,p}(Z), \sigma\right)$ and so it must change sign, which is a contradiction, unless $y^+ = 0$. Therefore $y \leq 0$ and so $K_+(y) = 0$, from which we have

$$\begin{aligned} & A(y) = 0 \quad \text{(see (5.161))}, \\ \Rightarrow \; & y = 0, \; \text{ a contradiction to the fact that } \|y\| = 1, \; \text{ see (5.160)}. \end{aligned}$$

This proves the claim.

Then the homotopy invariance property implies that

$$d_{(S)_+}(A - N, B_\varrho, 0) = d_{(S)_+}(A - \eta K_+, B_\varrho, 0) \qquad \text{for all } 0 < \varrho \leq \varrho_0. \tag{5.163}$$

As in the proof of Proposition 5.3.3, we have

$$d_{(S)_+}(A - \eta K_+, B_\varrho, 0) = 0. \tag{5.164}$$

From (5.163) and (5.164), we conclude that

$$d_{(S)_+}(A - N_+, B_\varrho, 0) = 0 \qquad \text{for all } 0 < \varrho \leq \varrho_0.$$

□

Now we can prove the existence of multiple nontrivial solutions of constant sign for problem (5.146).

THEOREM 5.3.5 *If hypotheses $H(f)_1$ hold, then problem (5.146) has at least two distinct nontrivial solutions*

$$\widehat{x}_0 \in \operatorname{int} C_0^1(\overline{Z})_+ \quad \textit{and} \quad \widehat{x}_1 \in C_0^1(\overline{Z}).$$

PROOF: Let $\widehat{x}_0 \in W_0^{1,p}(Z)$ be the local minimizer of φ obtained in Proposition 5.3.2. We have

$$\begin{aligned} & \varphi'(\widehat{x}_0) = 0, \\ \Rightarrow \; & A(\widehat{x}_0) = N(\widehat{x}_0). \end{aligned}$$

Assume that $\widehat{x}_0$ is an isolated critical point of φ_+ (otherwise we have infinitely many critical points of φ_+, hence infinitely many nontrivial positive solutions of (5.146)). So we can find $r_0 > 0$ such that

$$\varphi(\widehat{x}_0) < \varphi(y) \quad \text{and} \quad \varphi'(y) \neq 0 \qquad \text{for all } y \in \overline{B}_{r_0}(\widehat{x}_0) \setminus \{\widehat{x}_0\}, \tag{5.165}$$

where $\overline{B}_{r_0}(\widehat{x}_0) = \left\{y \in W_0^{1,p}(Z) : \|y - \widehat{x}_0\| \leq r_0\right\}$. We show that for all $r \in (0, r_0)$, the following property holds

$$\inf\left\{\varphi(y) : y \in \overline{B}_{r_0}(\widehat{x}_0) \setminus B_r(\widehat{x}_0)\right\} > \varphi(\widehat{x}_0). \tag{5.166}$$

We argue by contradiction. So suppose that there exists $r > 0$ and a sequence $\{x_n\}_{n \geq 1} \subseteq \overline{B}_{r_0}(\widehat{x}_0) \setminus B_r(\widehat{x}_0)$ such that

$$\varphi(x_n) \downarrow \varphi(\widehat{x}_0) \quad \text{as } n \to \infty.$$

Evidently, we may assume that

$$x_n \xrightarrow{w} \overline{x} \quad \text{and} \quad x_n \longrightarrow \overline{x} \quad \text{in } L^p(Z) \text{ as } n \to \infty. \tag{5.167}$$

Note that φ is weakly lower semicontinuous. So

$$\varphi(\overline{x}) \leq \liminf_{n\to\infty} \varphi(x_n) = \varphi(\widehat{x}_0).$$

Because $\overline{x} \in \overline{B}_{r_0}(\widehat{x}_0)$, it follows that $\varphi(\overline{x}) = \varphi(\widehat{x}_0)$ and by virtue of (5.165), we have $\overline{x} = \widehat{x}_0$. From the mean value theorem, we have

$$\varphi(\widehat{x}_0) - \varphi\Big(\frac{x_n + \widehat{x}_0}{2}\Big) = \Big\langle w_n^*, \frac{x_n - \widehat{x}_0}{2}\Big\rangle, \tag{5.168}$$

where $w_n^* = \varphi'\Big(\lambda_n x_n + (1-\lambda_n)\frac{x_n+\widehat{x}_0}{2}\Big)$, $\lambda_n \in [0,1]$. We know that

$$w_n^* = A\Big(\lambda_n x_n + (1-\lambda_n)\frac{x_n + \widehat{x}_0}{2}\Big) + u_n^*, \tag{5.169}$$

where $u_n^* = N\Big(\frac{x_n+\widehat{x}_0}{2}\Big)$, $n \geq 1$. Using (5.169) in (5.168) and then passing to the limit as $n \to \infty$, we obtain

$$\limsup_{n\to\infty} \Big\langle A\Big(\lambda_n x_n + (1-\lambda_n)\frac{x_n + \widehat{x}_0}{2}\Big), x_n - \widehat{x}_0\Big\rangle \leq 0.$$

We have

$$\limsup_{n\to\infty} \Big\langle A\Big(\lambda_n x_n + (1-\lambda_n)\frac{x_n + \widehat{x}_0}{2}\Big), \lambda_n x_n + (1-\lambda_n)\frac{x_n + \widehat{x}_0}{2} - \widehat{x}_0\Big\rangle$$
$$\limsup_{n\to\infty} \Big[\frac{\lambda_n + 1}{2} \limsup_{n\to\infty} \Big\langle A\Big(\lambda_n x_n + (1-\lambda_n)\frac{x_n + \widehat{x}_0}{2}\Big), x_n - \widehat{x}_0\Big\rangle\Big] \leq 0.$$

Because A is of type $(S)_+$, it follows that

$$\lambda_n x_n + (1-\lambda_n)\frac{x_n + \widehat{x}_0}{2} \longrightarrow \widehat{x}_0 \quad \text{in } W_0^{1,p}(Z). \tag{5.170}$$

But note that

$$\Big\|\lambda_n x_n + (1-\lambda_n)\frac{x_n + \widehat{x}_0}{2} - \widehat{x}_0\Big\| = (1+\lambda_n)\Big\|\frac{x_n - \widehat{x}_0}{2}\Big\| \geq \frac{r}{2}. \tag{5.171}$$

Comparing (5.170) and (5.171), we reach a contradiction. So (5.166) is valid.

Let $\mu = \inf\big[\varphi(x) : x \in B_{r_0}(\widehat{x}_0) \setminus B_{r_0/2}(\widehat{x}_0)\big] - \varphi(\widehat{x}_0)$. By (5.171) we have $\mu > 0$. Also we introduce the set

$$V = \big\{x \in B_{\frac{r_0}{2}}(\widehat{x}_0) : \varphi(x) - \varphi(\widehat{x}_0) < \mu\big\}.$$

Because $\widehat{x}_0 \in V$, the set V is nonempty. It is clear that V is also bounded open. Fix $r \in (0, r_0/2)$ with $\overline{B}_r(\widehat{x}_0) \subseteq V$. Choose a number $\lambda \in \mathbb{R}$ such that

$$0 < \lambda < \inf\big[\varphi(x) : x \in B_{r_0}(\widehat{x}_0) \setminus B_r(\widehat{x}_0)\big] - \varphi(\widehat{x}_0) \qquad (\text{see } (5.171)).$$

Note that $\lambda<\mu$ because $r<r_0/2$. Obviously $\overline{V}\subseteq B_{r_0}(\widehat{x}_0)$ and

$$V=\{x\in B_{r_0}(\widehat{x}_0):\varphi(x)-\varphi(\widehat{x}_0)<\mu\}.$$

From the definition of $\lambda>0$, we see that

$$\{x\in B_{r_0}(\widehat{x}_0):\varphi(x)-\varphi(\widehat{x}_0)\leq\lambda\}\subseteq B_r(\widehat{x}_0)\subseteq V.$$

Moreover, from (5.170) we see that

$$\varphi'(x)\neq 0 \qquad \text{for all } x\in B_{r_0}(\widehat{x}_0) \text{ with } \lambda\leq\varphi(x)-\varphi(\widehat{x}_0)\leq\mu.$$

Therefore we can apply Corollary 3.3.70 and obtain

$$d_{(S)_+}(A-N,V,0)=1 \qquad (\text{recall } \varphi'=A-N). \tag{5.172}$$

From the excision property of the degree map, we have

$$\begin{aligned} & d_{(S)_+}(A-N,V,0)=d_{(S)_+}(A-N,B_{r_0}(\widehat{x}_0),0) \\ \Rightarrow\ & d_{(S)_+}(A-N,B_{r_0}(\widehat{x}_0),0)=1 \qquad (\text{see } (5.172)). \end{aligned} \tag{5.173}$$

Now we fix $R_0>0$ in Proposition 5.3.3 sufficiently large and $\varrho_0>0$ in Proposition 5.3.4 sufficiently small in order to have $\widehat{x}_0\in B_{R_0}\setminus\overline{B}_{\varrho_0}$. Then we can find $r>0$ such that

$$B_r(\widehat{x}_0)\subseteq B_{R_0} \quad \text{and} \quad B_r(\widehat{x}_0)\cap B_{\varrho_0}=\emptyset.$$

Let $R>R_0$ and $\varrho\in(0,\varrho_0)$. We claim that there exists $\widehat{x}_1\in\overline{B}_R\backslash\big(B_r(\widehat{x}_0)\cup B_\varrho\big)$ with $A(\widehat{x}_1)=N(\widehat{x}_1)$. Indeed, if this is not the case, then $0\notin(A-N)\Big(\overline{B}_R\backslash\big(B_r(\widehat{x}_0)\cup B_\varrho\big)\Big)$ and so invoking the additivity property of the degree map, we obtain

$$\begin{aligned} & d_{(S)_+}(A-N,B_R,0)=d_{(S)_+}(A-N,B_r(\widehat{x}_0),0)+d_{(S)_+}(A-N,B_\varrho,0), \\ \Rightarrow\ & 1=1+(-1) \quad (\text{see } (5.172) \text{ and Propositions 5.3.3 and 5.3.4}), \end{aligned}$$

which is a contradiction. So we can find $\widehat{x}_1\in W_0^{1,p}(Z)$ such that

$$\widehat{x}_1\neq\widehat{x}_0,\ \widehat{x}_1\neq 0 \quad \text{and} \quad A(\widehat{x}_1)=N(\widehat{x}_1).$$

Therefore $\widehat{x}_1>0$ is a nontrivial solution of problem (5.146) and from nonlinear regularity $\widehat{x}_1\in C_0^1(\overline{Z})_+$. □

The second boundary value problem that we study in this section is the following nonlinear second-order periodic differential equation.

$$\left\{\begin{array}{l} -\big(|x'(t)|^{p-2}x'(t)\big)'=f\big(t,x(t)\big)+h(t) \quad \text{a.e. on } T, \\ x(0)=x(b),\ x'(0)=x'(b),\ h\in L^1(T),\ 1<p<\infty \end{array}\right\}. \tag{5.174}$$

We use degree-theoretic methods to prove two existence theorems for problem (5.174). In the first we assume conditions of nonuniform nonresonance between two successive eigenvalues of the negative scalar ordinary p-Laplacian with periodic boundary conditions. In the second existence theorem, the asymptotic at infinity nonuniform nonresonance conditions for the "slopes" $\left\{\frac{f(z,x)}{|x|^{p-2}x}\right\}_{x\neq 0}$ are replaced by certain Landesman–Lazer conditions.

From Section 4.3, we know that the nonlinear eigenvalue problem

$$\left\{\begin{array}{l} -\left(|x'(t)|^{p-2}x'(t)\right)' = \lambda|x(t)|^{p-2}x(t) \quad \text{a.e. on } T, \\ x(0)=x(b),\ x'(0)=x'(b),\ \lambda\in\mathbb{R},\ 1<p<\infty \end{array}\right\}, \tag{5.175}$$

has a sequence of eigenvalues

$$\left\{\lambda_{2n}=\left(\frac{2n\pi_p}{b}\right)^p\right\}_{x\neq 0}$$

where

$$\pi_p=\frac{2\pi(p-1)^{\frac{1}{p}}}{p\sin\frac{\pi}{p}}.$$

These are all the eigenvalues of (5.175). Also if we consider the nonlinear weighted eigenvalue problem

$$\left\{\begin{array}{l} -\left(|x'(t)|^{p-2}x'(t)\right)' = \left(\lambda+m(t)\right)|x(t)|^{p-2}x(t) \quad \text{a.e. on } T \\ x(0)=x(b),\ x'(0)=x'(b),\quad \lambda\in\mathbb{R},\quad m\in L^1(T),\quad 1<p<\infty, \end{array}\right\}, \tag{5.176}$$

we know (see Zhang [624]) that (5.176) has a double sequence of eigenvalues $\{\underline{\lambda}_{2n}(m)\}_{n\geq 1}$ and $\{\overline{\lambda}_{2n}(m)\}_{n\geq 0}$ such that

$$-\infty<\overline{\lambda}_0(m)<\underline{\lambda}_2(m)\leq\overline{\lambda}_2(m)<\cdots<\underline{\lambda}_{2n}(m)\leq\overline{\lambda}_{2n}(m)<\cdots$$

and $\underline{\lambda}_{2n}(m) \longrightarrow +\infty$ as $n\rightarrow\infty$. If $p=2$ (linear case), the two sequences $\{\underline{\lambda}_{2n}(m)\}_{n\geq 1}$ and $\{\overline{\lambda}_{2n}(m)\}_{n\geq 0}$ are all the eigenvalues of (5.176). If $p\neq 2$ (nonlinear case), we do not know if this is true.

For the first existence result under the nonuniform nonresonance conditions in an arbitrary spectral interval, our hypotheses on the nonlinearity $f(t,x)$, are the following.

H(f)₂: $f\colon T\times\mathbb{R}\longrightarrow\mathbb{R}$ is a function such that

(i) For all $x\in\mathbb{R}$, $t\longrightarrow f(t,x)$ is measurable.

(ii) For almost all $t\in T$, $x\longrightarrow f(t,x)$ is continuous.

(iii) For every $r>0$, there exists $\alpha_r\in L^1(T)_+$ such that for almost all $t\in T$ and all $x\in\mathbb{R}$ with $|x|\leq r$, we have $|f(t,x)|\leq\alpha_r(t)$.

(iv) There exist functions $\vartheta_1,\vartheta_2\in L^\infty(T)_+$ such that for some $n\geq 0$, we have

$$\lambda_{2n}\leq\vartheta_1(t)\leq\vartheta_2(t)\leq\lambda_{2n+2}\qquad\text{a.e. on } T$$

with the first and third inequalities strict on sets (not necessarily the same) of positive measure and

$$\vartheta_1(t)\leq\liminf_{|x|\to\infty}\frac{f(t,x)}{|x|^{p-2}x}\leq\limsup_{|x|\to\infty}\frac{f(t,x)}{|x|^{p-2}x}\leq\vartheta_2(t)$$

uniformly for almost all $t\in T$.

REMARK 5.3.6 Hypothesis $H(f)_2$(iv) is the nonuniform nonresonance condition in the spectral interval $[\lambda_{2n}, \lambda_{2n+2}]$.

We start with a simple observation about the eigenvalues of problem (5.176).

PROPOSITION 5.3.7 *If $\vartheta_1, \vartheta_2 \in L^\infty(T)_+$ are as in hypothesis $H(f)_2$(iv) and $m \in L^1(T)_+$ satisfies*

$$\vartheta_1(t) \leq m(t) \leq \vartheta_2(t) \qquad \text{a.e. on } T,$$

then all the eigenvalues of problem (5.176) are nonzero and do not have zero as a limit point.

PROOF: By virtue of the monotonicity of the eigenvalues $\{\underline{\lambda}_{2n}(m)\}_{n\geq 1}$ and $\{\overline{\lambda}_{2n}(m)\}_{n\geq 0}$ on the weight function $m \in L^1(T)_+$, we have

$$\overline{\lambda}_{2n}(m) \leq \overline{\lambda}_{2n}(\vartheta_1) < \overline{\lambda}_{2n}(\lambda_{2n}) = 0 \quad \text{and} \tag{5.177}$$

$$0 = \underline{\lambda}_{2n}(\lambda_{2n}) < \underline{\lambda}_{2n}(\vartheta_2) \leq \underline{\lambda}_{2n}(m). \tag{5.178}$$

Also we know that if $\lambda \in \mathbb{R}$ is an eigenvalue of (5.176), then

$$\lambda \in \bigcup_{k\geq 1} [\underline{\lambda}_{2k}(m), \overline{\lambda}_{2k}(m)] \cup \big(-\infty, \overline{\lambda}_0(m)\big] \tag{5.179}$$

(see Zhang [624]). Combining (5.179) with (5.177) and (5.178), we see that $\lambda \neq 0$. So if by $\sigma(p)$ we denote the spectrum of the nonlinear weighted eigenvalue problem (5.176), we have that $0 \notin \sigma(p)$.

Suppose that we can find $\{\lambda_n\}_{n\geq 1} \subseteq \sigma(p)$ such that $\lambda_n \longrightarrow 0$. We can find $u_n \in C^1_{\text{per}}(T) = \{x \in C^1(T) : x(0) = x(b), x'(0) = x'(b)\}$ such that $u_n \neq 0$ and

$$-\big(|u_n'(t)|^{p-2}u_n'(t)\big)' = \big(\lambda + m(t)\big)|u_n(t)|^{p-2}u_n(t) \qquad \text{a.e. on } T. \tag{5.180}$$

Because of the $(p-1)$-homogeneity of problem (5.180), we may assume that $\|u_n\|=1$ for all $n\geq 1$ ($\|\cdot\|$ denotes the norm of the Sobolev space $W^{1,p}_{\text{per}}\big((0,b)\big) = \{x \in W^{1,p}\big((0,b)\big) : x(0)=x(b)\}$. So by passing to a suitable subsequence if necessary, we may assume that

$$u_n \xrightarrow{w} u \quad \text{in } W^{1,p}_{\text{per}}\big((0,b)\big) \quad \text{and} \quad u_n \longrightarrow u \quad \text{in } C(T) \text{ as } n \to \infty.$$

By $\langle \cdot, \cdot \rangle$ we denote the duality brackets for the pair $\Big(W^{1,p}_{\text{per}}\big((0,b)\big), W^{1,p}_{\text{per}}\big((0,b)\big)^*\Big)$ and we let $A: W^{1,p}_{\text{per}}\big((0,b)\big) \longrightarrow W^{1,p}_{\text{per}}\big((0,b)\big)^*$ be the nonlinear operator defined by

$$\langle A(x), y \rangle = \int_0^b |x'(t)|^{p-2}x'(t)y'(t)dt \qquad \text{for all } x, y \in W^{1,p}_{\text{per}}\big((0,b)\big).$$

We know that A is monotone demicontinuous, hence it is maximal monotone.

Also by $K: W^{1,p}_{\text{per}}\big((0,b)\big) \longrightarrow L^{p'}(T)$, $(1/p) + (1/p') = 1$ we denote the operator

$$K(x)(\cdot) = |x(\cdot)|^{p-2}x(\cdot).$$

Evidently K is bounded continuous. Then in terms of A and K, we can equivalently rewrite (5.180) as the following abstract operator equation

$$\begin{aligned} & A(u_n)=(\lambda_n+m)K(u_n), \quad n\geq 1, \\ \Rightarrow\ & \langle A(u_n),u_n-u\rangle=\int_0^b(\lambda_n+m(t))|u_n(t)|^{p-2}u_n(t)(u_n-u)dt, \quad n\geq 1. \end{aligned} \tag{5.181}$$

Note that

$$\int_0^b(\lambda_n+m(t))|u_n(t)|^{p-2}u_n(t)(u_n-u)(t)dt \longrightarrow 0.$$

So we obtain

$$\lim_{n\to\infty}\langle A(u_n),u_n-u\rangle=0. \tag{5.182}$$

But A being maximal monotone, it is generalized pseudomonotone and so from (5.182) it follows that

$$\begin{aligned} & \langle A(u_n),u_n\rangle \longrightarrow \langle A(u),u\rangle, \\ \Rightarrow\ & \|u_n'\|_p \longrightarrow \|u'\|_p. \end{aligned}$$

Because $u_n' \xrightarrow{w} u'$ in $L^p(T)$ and $L^p(T)$ being uniformly convex it has the Kadec–Klee property, it follows that $u_n' \longrightarrow u'$ in $L^p(T)$ and so $u_n' \longrightarrow u'$ in $W^{1,p}_{\text{per}}((0,b))$. Hence $\|u\|=1$ and so $u\neq 0$. Passing to the limit as $n\to\infty$ in (5.181), we obtain

$$A(u)=mK(u). \tag{5.183}$$

From (5.183) it follows that

$$\left\{\begin{array}{ll} -\big(|u'(t)|^{p-2}u'(t)\big)'=m(t)|u(t)|^{p-2}u(t) & \text{a.e. on } T, \\ u(0)=u(b),\ u'(0)=u'(b) & \end{array}\right\}. \tag{5.184}$$

Because $u\neq 0$, from (5.184) it follows that $0\in\sigma(p)$, which contradicts the first part of the proof. $\square$

From the above proposition we have that $0\notin\sigma(p)$ and we can find $\varepsilon_0\in(0,1)$ such that $(-\varepsilon_0,\varepsilon_0)\cap\sigma(p)=\emptyset$. We fix $\varepsilon\in(0,\varepsilon_0)$ and we consider the following periodic problem.

$$\left\{\begin{array}{ll} -\big(|x'(t)|^{p-2}x'(t)\big)'+\varepsilon|x(t)|^{p-2}x(t)=h(t) & \text{a.e. on } T, \\ x(0)=x(b),\ x'(0)=x'(b),\ \ h\in L^1(T),\ \ 1<p<\infty & \end{array}\right\}. \tag{5.185}$$

PROPOSITION 5.3.8 *For every $h\in L^1(T)$ problem (5.185) has a unique solution $S_\varepsilon(h)\in C^1_{\text{per}}(T)$ and the solution map $S_\varepsilon: L^1(T)\longrightarrow W^{1,p}_{\text{per}}((0,b))$ is completely continuous; that is, if $h_n\xrightarrow{w} h$ in $L^1(T)$, then $S_\varepsilon(h_n)\longrightarrow S_\varepsilon(h)$ in $W^{1,p}_{\text{per}}((0,b))$.*

PROOF: Let $A,K:W^{1,p}_{\text{per}}((0,b))\longrightarrow W^{1,p}_{\text{per}}((0,b))^*$ be as in the proof of Proposition 5.3.7 (recall that $L^{p'}(T)$ is embedded compactly in $W^{1,p}_{\text{per}}((0,b))^*$). We consider the nonlinear operator $L_\varepsilon:W^{1,p}_{\text{per}}((0,b))\longrightarrow W^{1,p}_{\text{per}}((0,b))^*$ defined by

$$L_\varepsilon(x) = A(x) + \varepsilon K(x) \qquad \text{for all } x \in W^{1,p}_{\text{per}}\big((0,b)\big).$$

Clearly L_ε is strictly monotone and demicontinuous, hence it is maximal monotone. Moreover, for every $x \in W^{1,p}_{\text{per}}\big((0,b)\big)$, we have

$$\langle L_\varepsilon(x), x\rangle = \|x'\|_p^p + \varepsilon\|x\|^p$$
$$\Rightarrow \; L_\varepsilon \text{ is coercive.}$$

So L_ε is surjective and we find $x \in W^{1,p}_{\text{per}}\big((0,b)\big)$ such that

$$L_\varepsilon(x) = h. \tag{5.186}$$

Due to the strict monotonicity of L_ε, the solution $x \in W^{1,p}_{\text{per}}\big((0,b)\big)$ of (5.191) is unique and we denote it by $S_\varepsilon(h)$. Thus we have defined the solution map $S_\varepsilon : L^1(T) \longrightarrow W^{1,p}_{\text{per}}\big((0,b)\big)$, which to each forcing term $h \in L^1(T)$ assigns the unique solution of problem (5.185). We show that S_ε is completely continuous. To this end let $h_n \xrightarrow{w} h$ in $L^1(T)$ and set $x_n = S_\varepsilon(h_n) \in C^1_{\text{per}}(T)$, $n \geq 1$. We have

$$\left\{\begin{array}{ll} -\big(|x_n'(t)|^{p-2}x_n'(t)\big)' + \varepsilon|x_n(t)|^{p-2}x_n(t) = h_n(t) & \text{a.e. on } T, \\ x_n(0) = x_n(b),\; x_n'(0) = x_n'(b) & \end{array}\right\},$$

$$\Rightarrow \; A(x_n) + \varepsilon K(x_n) = h_n, \quad n \geq 1. \tag{5.187}$$

Taking duality brackets in $\Big(W^{1,p}_{\text{per}}\big((0,b)\big), W^{1,p}_{\text{per}}\big((0,b)\big)^*\Big)$ with $x_n \in C^1_{\text{per}}(T)$, we obtain

$$\varepsilon\|x_n\|^p \leq \|x_n'\|_p^p + \varepsilon\|x_n\|_p^p = \int_0^b h_n(t)x_n(t)dt \leq c_1\|x_n\|,$$

for some $c_1 > 0$, all $n \geq 1$,

$$\Rightarrow \; \{x_n\}_{n\geq 1} \subseteq W^{1,p}_{\text{per}}\big((0,b)\big) \quad \text{is bounded.}$$

So we may assume that

$$x_n \xrightarrow{w} x \;\text{ in }\; W^{1,p}_{\text{per}}\big((0,b)\big) \qquad \text{and} \qquad x_n \longrightarrow x \;\text{ in }\; C(T).$$

As before, acting on (5.187) with the test function $x_n - x \in W^{1,p}_{\text{per}}\big((0,b)\big)$ and using the Kadec–Klee property of the space $L^p(T)$, we can show that $x_n \longrightarrow x$ in $W^{1,p}_{\text{per}}\big((0,b)\big)$. Passing to the limit as $n \to \infty$ in (5.187), we obtain

$$\begin{aligned} & A(x) + \varepsilon K(x) = h, \\ \Rightarrow \; & -\big(|x'(t)|^{p-2}x'(t)\big)' + \varepsilon|x(t)|^{p-2}x(t) = h(t) \qquad \text{a.e. on } T, \\ & x(0) = x(b),\; x'(0) = x'(b), \end{aligned}$$

$$\Rightarrow \quad x = S_\varepsilon(h).$$

From Urysohn's criterion for the convergence of sequences, we conclude for the original sequence $\big\{x_n = S_\varepsilon(h_n)\big\}_{n\geq 1} \subseteq W^{1,p}_{\text{per}}\big((0,b)\big)$, that

$$x_n \longrightarrow x = S_\varepsilon(h) \;\text{ in }\; W^{1,p}_{\text{per}}\big((0,b)\big)$$
$$\Rightarrow \; S_\varepsilon \text{ is indeed completely continuous.}$$

□

THEOREM 5.3.9 *If hypotheses $H(f)_2$ hold, then for every $h \in L^1(T)$ problem (5.174) has a solution $x \in C^1_{\text{per}}(T)$.*

PROOF: Let $N_f : W^{1,p}_{\text{per}}\big((0,b)\big) \longrightarrow L^1(T)$ be the Nemitsky operator corresponding to the function $f(t,x)$; that is,

$$N_f(x)(\cdot) = f\big(\cdot, x(\cdot)\big).$$

Because of the compact embedding of $W^{1,p}_{\text{per}}\big((0,b)\big)$ into $C(T)$, we see that N_f is compact. Let $g \in L^1(T)$ such that $\vartheta_1(t) \leq g(t) \leq \vartheta_2(t)$ a.e. on T and consider the compact homotopy $h : [0,1] \times W^{1,p}_{\text{per}}\big((0,b)\big) \longrightarrow W^{1,p}_{\text{per}}\big((0,b)\big)$ defined by

$$h(\beta, x) = S_\varepsilon\big(\varepsilon K(x) + \beta N_f(x) + \beta h + (1-\beta) g K(x)\big).$$

Claim: There exists $R_0 > 0$ such that $h(\beta, x) \neq x$ for all $\beta \in [0,1]$ and all $x \in W^{1,p}_{\text{per}}\big((0,b)\big)$ with $\|x\| = R$ and all $R \geq R_0$.

We argue indirectly. Suppose that the claim is not true. Then we can find $\{\beta_n\}_{n\geq 1} \subseteq [0,1]$ and $\{x_n\}_{n\geq 1} \subseteq W^{1,p}_{\text{per}}\big((0,b)\big)$ such that

$$\beta_n \longrightarrow \beta \quad \text{in } [0,1], \quad \|x_n\| \longrightarrow \infty \quad \text{and} \quad x_n = h(\beta_n, x_n) \qquad \text{for all } n \geq 1.$$

We have

$$A(x_n) = \beta_n N_f(x_n) + \beta_n h + (1-\beta_n) g K(x_n) \qquad \text{for all } n \geq 1. \tag{5.188}$$

Set $y_n = x_n / \|x_n\|$, $n \geq 1$. We may assume that

$$y_n \xrightarrow{w} y \quad \text{in } W^{1,p}_{\text{per}}\big((0,b)\big) \quad \text{and} \quad y_n \longrightarrow y \quad \text{in } C(T) \text{ as } n \to \infty.$$

Using hypothesis $H(f)_2$(iii) and (iv), we can check that

$$\left\{\frac{N_f(x_n)}{\|x_n\|^{p-1}}\right\}_{n\geq 1} \subseteq L^1(T) \text{ is uniformly integrable.}$$

So by the Dunford–Pettis theorem, we may assume (at least for a subsequence), that

$$\frac{N_f(x_n)}{\|x_n\|^{p-1}} \xrightarrow{w} w \quad \text{in } L^1(T). \tag{5.189}$$

As in previous proofs, we can check that

$$w(t) = g_0(t)|y(t)|^{p-2} y(t) \qquad \text{a.e. on } T, \tag{5.190}$$

with $g_0 \in L^1(T)$ such that $\vartheta_1(t) \leq g_0(t) \leq \vartheta_2(t)$ a.e. on T. Recall that

$$A(x_n) = \beta_n N_f(x_n) + \beta_n h + (1-\beta_n) g K(x_n), \quad n \geq 1.$$

Dividing with $\|x_n\|^{p-1}$, we obtain

$$A(y_n) = \beta_n \frac{N_f(x_n)}{\|x_n\|^{p-1}} + \beta_n \frac{h}{\|x_n\|^{p-1}} + (1-\beta_n) g K(y_n), \quad n \geq 1. \tag{5.191}$$

As before, acting with the test function $y_n - y \in W^{1,p}_{\text{per}}\big((0,b)\big)$ and using the Kadec–Klee property, we can conclude that $y_n \longrightarrow y$ in $W^{1,p}_{\text{per}}\big((0,b)\big)$. So if we pass to the limit as $n \to \infty$ in (5.191), we obtain

$$A(y) = \big(\beta g_0 + (1-\beta)g\big)K(y).$$

Let $\widehat{g} = \beta g_0 + (1-\beta)g$. Then $\widehat{g} \in L^1(T)_+$ and $\vartheta_1(t) \leq \widehat{g}(t) \leq \vartheta_2(t)$ a.e. on T. Also

$$\left\{ \begin{array}{l} -\big(|y'(t)|^{p-2}y'(t)\big)' = \widehat{g}(t)|y(t)|^{p-2}y(t) \quad \text{a.e. on } T, \\ y(0) = y(b),\ y'(0) = y'(b) \end{array} \right\}. \tag{5.192}$$

Note that $\|y\| = 1$ and so $y \neq 0$. Hence from (5.192), we infer that $0 \in \sigma(p)$ which contradicts Proposition 5.3.8. This proves the validity of the claim. Then from the homotopy invariance of the Leray–Schauder degree (see Section 3.3), we have

$$\widehat{d}\big(I - S_\varepsilon \circ (\varepsilon + g)K, B_R, 0\big) = \widehat{d}\big(I - S_\varepsilon \circ (\varepsilon K + N_f + h), B_R, 0\big) \qquad \text{for all } R \geq R_0. \tag{5.193}$$

From the choice of $\varepsilon > 0$ and Proposition 5.3.8, we see that

$$x \neq S_\varepsilon \circ (\varepsilon + g)K(x) \qquad \text{for all } \|x\| = R \geq R_0.$$

Moreover, it is clear that the map $x \longrightarrow S_\varepsilon \circ (\varepsilon + g)K(x)$ is odd. So invoking Borsuk's theorem (see Theorem 3.3.41), we have

$$\begin{aligned} & \widehat{d}\big(I - S_\varepsilon \circ (\varepsilon + g)K, B_R, 0\big) \neq 0 \quad \text{for } R \geq R_0, \\ \Rightarrow\ & \widehat{d}\big(I - S_\varepsilon \circ (\varepsilon K + N_f + h), B_R, 0\big) \neq 0 \quad \text{for } R \geq R_0 \qquad (\text{see } (5.193)). \end{aligned}$$

Therefore by the solution property of the Leray–Schauder degree, we can find $x \in B_R$ such that

$$\begin{aligned} & x = S_\varepsilon \circ (\varepsilon K + N_f + h)(x), \\ \Rightarrow\ & A(x) = N_f(x) + h, \\ \Rightarrow\ & -\big(|x'(t)|^{p-2}x'(t)\big)' = f\big(t, x(t)\big) + h(t) \qquad \text{a.e. on } T, \\ & x(0) = x(b),\ x'(0) = x'(b), \\ \Rightarrow\ & x \in C^1_{\text{per}}(T) \text{ is a solution of problem (5.174).} \end{aligned}$$

□

In the second existence theorem for problem (5.174), we employ a Landesman–Lazer type condition instead of the nonuniform nonresonance condition in the spectral interval $[\lambda_{2n}, \lambda_{2n+2}]$. More precisely, our hypotheses on the nonlinearity $f(t,x)$ are the following.

H(f)$_3$: $f\colon T \times \mathbb{R} \longrightarrow \mathbb{R}$ is a function such that

(i) For all $x \in \mathbb{R}$, $t \longrightarrow f(t,x)$ is measurable.

(ii) For almost all $t \in T$, $x \longrightarrow f(t,x)$ is continuous.

(iii) For every $r > 0$, there exists $\alpha_r \in L^1(T)_+$ such that for almost all $t \in T$ and all $x \in \mathbb{R}$ with $|x| < r$, we have $|f(t,x)| \leq \alpha_r(t)$;.

(iv) There exist functions $\eta_-, \eta_+ \in L^1(T)$ such that

$$\eta_-(t) = \liminf_{x \to -\infty} f(t,x) \quad \text{and} \quad \eta_+(t) = \limsup_{x \to +\infty} f(t,x),$$

uniformly for a.a. $t \in T$.

THEOREM 5.3.10 *If hypotheses $H(f)_3$ hold and $h \in L^1(T)$ satisfies*

$$\int_0^b \eta_+(t)\mathrm{d}t < -\int_0^b h(t)\mathrm{d}t < \int_0^b \eta_-(t)\mathrm{d}t,$$

then problem (5.174) has a solution $x \in C^1_{\mathrm{per}}(T)$.

PROOF: We keep the notation introduced in the proof of Theorem 5.3.9. We consider the compact homotopy $\widehat{h}: [0,1] \times W^{1,p}_{\mathrm{per}}((0,b)) \longrightarrow W^{1,p}_{\mathrm{per}}((0,b))$ defined by

$$\widehat{h}(\beta, x) = S_\varepsilon \circ \big(\beta(\varepsilon K + N_f + h)\big)(x).$$

Again we show that there exists $R > 0$ large enough so that

$$\widehat{h}(\beta, x) \neq x$$

for all $\beta \in [0,1]$ and all $x \in W^{1,p}_{\mathrm{per}}((0,b))$ with $\|x\| = R$.

We proceed by contradiction. So suppose that we can find $\{\beta_n\}_{n \geq 1} \subseteq [0,1]$ and $\{x_n\}_{n \geq 1} \subseteq W^{1,p}_{\mathrm{per}}((0,b))$ such that

$$\beta_n \longrightarrow \beta \quad \text{in } [0,1], \ \|x_n\| \longrightarrow \infty \quad \text{and} \quad x_n = S_\varepsilon \circ \big(\beta_n(\varepsilon K + N_f + h)\big)(x_n),$$

for all $n \geq 1$.

We set $y_n = x_n / \|x_n\|$, $n \geq 1$. We may assume that

$$y_n \xrightarrow{w} y \text{ in } W^{1,p}_{\mathrm{per}}((0,b)) \quad \text{and} \quad y_n \longrightarrow y \text{ in } C(T) \text{ as } n \to \infty.$$

For every $n \geq 1$, we have

$$A(x_n) + \varepsilon K(x_n) = \varepsilon \beta_n K(x_n) + \beta_n N_f(x_n) + \beta_n h, \quad n \geq 1.$$

Dividing by $\|x_n\|^{p-1}$, we obtain

$$A(y_n) + \varepsilon K(y_n) = \varepsilon \beta_n K(y_n) + \beta_n \frac{N_f(x_n)}{\|x_n\|^{p-1}} + \beta_n \frac{h}{\|x_n\|^{p-1}}, \quad n \geq 1. \tag{5.194}$$

From hypotheses $\mathrm{H}(f)_3$(iii) and (iv), we have that

$$\frac{N_f(x_n)}{\|x_n\|^{p-1}} \longrightarrow 0 \quad \text{in } L^1(T).$$

Also we have

$$\frac{h}{\|x_n\|^{p-1}} \longrightarrow 0 \quad \text{in } L^1(T)$$

$$\text{and} \quad \int_0^b y_n(t)(y_n - y)(t)\mathrm{d}t \longrightarrow 0.$$

So if in (5.194) we act with the test function $y_n - y \in W^{1,p}_{\text{per}}\big((0,b)\big)$ and then pass to the limit as $n \to \infty$, we obtain

$$\begin{aligned} & \lim \langle A(y_n), y_n - y \rangle = 0, \\ \Rightarrow\ & y_n \longrightarrow y \quad \text{in } W^{1,p}_{\text{per}}\big((0,b)\big). \end{aligned}$$

Therefore from (5.194) in the limit as $n \to \infty$, we obtain

$$A(y) = (\beta - 1)\varepsilon K(y), \tag{5.195}$$

with $\|y\| = 1$, $(y \neq 0)$. Because $\beta \in [0,1]$, $\varepsilon > 0$, from (5.195) it follows that $\beta = 1$ and so $A(y) = 0$, hence $y = \xi \in \mathbb{R}$, $\xi \neq 0$ (recall that $y \neq 0$).

First suppose that $\xi > 0$. If in (5.194) we act with the test function $y = \xi$, we obtain

$$(1 - \beta_n)\varepsilon \|x_n\|^{p-1}\xi \int_0^b |y_n|^{p-2} y_n \mathrm{d}t = \beta_n \xi \int_0^b \big(N_f(x_n) + h\big)\mathrm{d}t. \tag{5.196}$$

Because $\beta_n \in [0,1]$, $\beta_n \longrightarrow \beta$, $y_n \longrightarrow \xi > 0$ in $C(T)$, from (5.196) it follows that we can find $n_0 \geq 1$ such that

$$\int_0^b \big(N_f(x_n) + h\big)\mathrm{d}t > 0 \qquad \text{for all } n \geq n_0. \tag{5.197}$$

Note that because we have assumed that $\xi > 0$, for all $t \in T$ we have that $x_n(t) \longrightarrow +\infty$ as $n \to \infty$. We claim that this convergence is uniform in $t \in T$. To this end let $0 < \delta < \xi$. Because $y_n \longrightarrow \xi$ in $C(T)$, we can find $n_1 = n_1(\delta) \geq 1$ such that for all $t \in T$ and all $n \geq n_1$ we have $|y_n(t) - \xi| < \delta$, hence $0 < \xi - \delta < y_n(t)$. Also because $\|x_n\| \longrightarrow \infty$, given $\widehat{M} > 0$ we can find $n_2 = n_2(\widehat{M}) \geq 1$ such that $\|x_n\| \geq \widehat{M}$ for all $n \geq n_2$. So for all $n \geq n_2$ and all $t \in T$, we have

$$\begin{aligned} & \frac{x_n(t)}{\widehat{M}} \geq \frac{x_n(t)}{\|x_n\|} = y_n(t) > \xi - \delta = \gamma > 0, \\ \Rightarrow\ & x_n(t) \geq \gamma \widehat{M} \qquad \text{for all } n \geq n_2 \text{ and all } t \in T. \end{aligned}$$

Because $\widehat{M} > 0$ was arbitrary, we conclude that $x_n(t) \longrightarrow +\infty$ uniformly in $t \in T$ as $n \to \infty$. This then by virtue of hypothesis $\mathrm{H}(f)_3$(iv) implies that

$$\limsup_{n \to \infty} f\big(t, x_n(t)\big) = \eta_+(t) \qquad \text{uniformly for a.a. } t \in T.$$

Then by Fatou's lemma and using (5.197), we obtain

$$-\int_0^b h(t)\mathrm{d}t \leq \limsup_{n \to \infty} \int_0^b N_f(x_n)\mathrm{d}t \leq \int_0^b \eta_+(t)\mathrm{d}t,$$

a contradiction to the hypotheses of the theorem.

If we assume $\xi < 0$, then arguing in a similar fashion we reach the contradiction

$$\int_0^b \eta_-(t)\mathrm{d}t \leq -\int_0^b h(t)\mathrm{d}t.$$

Therefore we can find $R > 0$ large enough such that

$$\widehat{h}(\beta, x) \neq x \qquad \text{for all } \beta \in [0,1] \text{ and all } \|x\| = R.$$

Invoking the homotopy invariance property of the Leray–Schauder degree, we have

$$\widehat{d}\big(I - \widehat{h}(0,\cdot), B_R, 0\big) = \widehat{d}\big(I - \widehat{h}(1,\cdot), B_R, 0\big).$$

Note that $\widehat{h}(0,\cdot) = 0$ and so $\widehat{d}\big(I - \widehat{h}(0,\cdot), B_R, 0\big) = \widehat{d}(I, B_R, 0) = 1$, hence

$$\widehat{d}\big(I - \widehat{h}(1,\cdot), B_R, 0\big) = 1.$$

From the solution property of the Leray–Schauder degree, it follows that thee exists $x \in W^{1,p}_{\text{per}}\big((0,b)\big)$ with $\|x\| \leq R$ such that

$$\begin{aligned}
& x = \widehat{h}(1,x) = S_\varepsilon \circ (\varepsilon K + N_f + h)(x), \\
\Rightarrow \quad & A(x) = N_f(x) + h, \\
\Rightarrow \quad & -\big(|x'(t)|^{p-2}x'(t)\big)' = f\big(t, x(t)\big) + h(t) \qquad \text{a.e. on } T, \\
& x(0) = x(b), \ x'(0) = x'(b), \\
\Rightarrow \quad & x \in C^1_{\text{per}}(T) \ \text{ is a solution of problem (5.174).}
\end{aligned}$$

□

REMARK 5.3.11 Theorems 5.3.9 and 5.3.10 are still valid if we consider the more general situation in which instead of the ordinary p-Laplacian differential operator, we have an operator of the form $x \longrightarrow \big(\varphi(x')\big)'$, with $\varphi\colon \mathbb{R} \longrightarrow \mathbb{R}$ a suitable homeomorphism.

5.4 Nonlinear Eigenvalue Problems

In this section we study two perturbed eigenvalue problems, in which the perturbation is nonlinear. The first problem is driven by the p-Laplacian and the second by the Laplacian differential operator. For both problems we establish the existence of nontrivial smooth solutions, as the parameter $\lambda \in \mathbb{R}$ moves in a certain interval. In fact for the second problem (the semilinear one), we also have a negative result, showing that no positive solution exists, when λ is out of the aforementioned interval. This is done using the so-called Pohozaev's identity, which we also prove here.

In this section $Z \subseteq \mathbb{R}^N$ is a bounded domain with C^2-boundary ∂Z. We start with the following perturbed weighted eigenvalue problem.

$$\left\{\begin{aligned}
& -\operatorname{div}\big(\|Dx(z)\|^{p-2}Dx(z)\big) - \lambda m(z)|x(z)|^{p-2}x(z) = f\big(z, x(z)\big) \ \text{ a.e. on } Z, \\
& x\big|_{\partial Z} = 0, \ \ 1<p<\infty
\end{aligned}\right\}. \tag{5.198}$$

The hypotheses on the weight function $m(z)$ and the perturbation term $f(z,x)$ are the following.

$\mathbf{H(m)}$: $m \in L^\infty(Z)_+ = \{m \in L^\infty(Z)_+ : m(z) \geq 0 \ \text{ a.e. on } Z\}$, $m \neq 0$.

H(f)$_1$: $f\colon Z\times\mathbb{R}\longrightarrow\mathbb{R}$ is a function such that $f(z,0)=0$ a.e. on Z and

(i) For all $x\in\mathbb{R}$, $z\longrightarrow f(z,x)$ is measurable.

(ii) For almost all $z\in Z$, $x\longrightarrow f(z,x)$ is continuous.

(iii) For almost all $z\in Z$ and all $x\in\mathbb{R}$, we have

$$|f(z,x)|\leq\alpha(z)+c|x|^{r-1}\quad\text{with }\alpha\in L^{\infty}(Z)_+,\ c>0$$
$$\text{and}\qquad p<r<p^*=\begin{cases}\frac{Np}{N-p} & \text{if } p<N\\ +\infty & \text{if } p\geq N\end{cases}.$$

(iv) If $F(z,x)=\int_0^x f(z,r)dr$ (the potential function corresponding to the nonlinearity $f(z,x)$), then there exist constants $\mu>p$ and $M>0$ such that for almost all $z\in Z$ and all $|x|\geq M$, we have

$$\mu F(z,x)\leq xf(z,x)$$
$$\text{and}\qquad \operatorname{ess\,inf} F(\cdot,M),\ \operatorname{ess\,inf} F(\cdot,-M)>0.$$

(v) $\limsup_{x\to 0}\frac{f(z,x)}{x^{p-2}x}\leq 0$ uniformly for almost all $z\in Z$.

REMARK 5.4.1 Hypothesis H$(f)_1$(iv) is the so-called Ambrosetti–Rabinowitz condition (AR-condition for short). However, note that in contrast to the standard AR-condition, we do not require that $F(z,x)>0$ for almost all $z\in Z$ and all $|x|\geq M$. Instead we only assume that $\operatorname{ess\,inf} F(\cdot,M)$, $\operatorname{ess\,inf} F(\cdot,-M)>0$. The following function satisfies all the hypotheses in H$(f)_1$. For simplicity we drop the z-dependence and so we have

$$f(x)=|x|^{r-2}x-c|x|^{p-2}x$$

with $c\geq 0$ and $p<r<p^*$. In this example the AR-condition (see hypothesis H$(f)_1$(iv)), is satisfied for $\mu=r$ and all $M>0$.

We consider the functional $\varphi_\lambda : W_0^{1,p}(Z)\longrightarrow\mathbb{R}$ defined by

$$\varphi_\lambda(x)=\frac{1}{p}\|Dx\|_p^p-\frac{\lambda}{p}\int_Z m(z)|x(z)|^p dz-\int_Z F\big(z,x(z)\big)dz\qquad\text{for all } x\in W_0^{1,p}(Z).$$

We know that this functional is C^1 on $W_0^{1,p}(\mathbb{R})$. Next we show that it satisfies the Cerami condition (C-condition).

PROPOSITION 5.4.2 *If hypotheses H(m) and H$(f)_1$ hold, then φ_λ satisfies the C-condition.*

PROOF: Let $\{x_n\}_{n\geq 1}\subseteq W_0^{1,p}(Z)$ be a sequence such that

$$|\varphi_\lambda(x_n)|\leq\overline{M}\quad\text{for some }\overline{M}>0,\text{ all } n\geq 1$$
$$\text{and}\qquad (1+\|x_n\|)\varphi_\lambda'(x_n)\longrightarrow 0\ \text{ as } n\to\infty. \tag{5.199}$$

In the sequel by $\langle\cdot,\cdot\rangle$ we denote the duality brackets for the pair $(W_0^{1,p}(Z), W^{-1,p'}(Z))$, $(1/p)+(1/p')=1$. Let $A\colon W_0^{1,p}(Z)\longrightarrow W^{-1,p'}(Z)$ be the nonlinear operator defined by

$$\langle A(x),y\rangle = \int_Z \|Dx(z)\|^{p-2}\big(Dx(z),Dy(z)\big)_{\mathbb{R}^N}\mathrm{d}z \qquad \text{for all } x,y\in W_0^{1,p}(Z).$$

We know that A is of type $(S)_+$ (see Proposition 4.3.41). Moreover, we have

$$\varphi_\lambda'(x_n) = A(x_n) - \lambda m|x_n|^{p-2}x_n - N_f(x_n),$$

where $N_f(x_n)(\cdot)=f(\cdot,x_n(\cdot))$ (the Nemitsky operator corresponding to the nonlinearity f). Note that $N_f\colon W_0^{1,p}(Z)\longrightarrow L^{r'}(Z)\subseteq W^{-1,p'}(Z)$, where $\frac{1}{r}+\frac{1}{r'}=1$ (recall $p<r<p^*$).

From the choice of the sequence $\{x_n\}_{n\geq 1}\subseteq W_0^{1,p}(Z)$(see (5.199)), we have

$$\begin{aligned} &\big|\big\langle\varphi_\lambda'(x_n),x_n\big\rangle\big| \leq \varepsilon_n \quad \text{with } \varepsilon_n\downarrow 0,\\ \Rightarrow\ & -\|Dx_n\|_p^p + \lambda\int_Z m|x_n|^p\mathrm{d}z + \int_Z x_n N_f(x_n)\mathrm{d}z \leq \varepsilon_n. \end{aligned} \tag{5.200}$$

Also

$$\begin{aligned} &p\,\varphi_\lambda(x_n) \leq p\overline{M} \qquad \text{for all } n\geq 1,\\ \Rightarrow\ & \|Dx_n\|_p^p - \lambda\int_Z m|x_n|^p\mathrm{d}z - \int_Z pF\big(z,x_n(z)\big)\mathrm{d}z \leq p\overline{M}. \end{aligned} \tag{5.201}$$

Adding (5.200) and (5.201), we obtain

$$\begin{aligned} &-\int_Z\Big(pF\big(z,x_n(z)\big) - x_n(z)f\big(z,x_n(z)\big)\Big)\mathrm{d}z \leq \varepsilon_n + p\overline{M} \qquad \text{for all } n\geq 1,\\ \Rightarrow\ & -\int_Z\Big(\mu F\big(z,x_n(z)\big) - x_n(z)f\big(z,x_n(z)\big)\Big)\mathrm{d}z\\ &+(\mu-p)\int_Z F\big(z,x_n(z)\big)\mathrm{d}z \leq \varepsilon_n + p\overline{M}. \end{aligned} \tag{5.202}$$

Note that

$$\begin{aligned} &-\int_Z\Big(\mu F\big(z,x_n(z)\big) - x_n(z)f\big(z,x_n(z)\big)\Big)\mathrm{d}z\\ &= -\int_{\{|x_n|\geq M\}}\big(\mu F\big(z,x_n(z)\big) - x_n(z)f\big(z,x_n(z)\big)\big)\mathrm{d}z\\ &\quad -\int_{\{|x_n|<M\}}\big(\mu F\big(z,x_n(z)\big) - x_n(z)f\big(z,x_n(z)\big)\big)\mathrm{d}z. \end{aligned} \tag{5.203}$$

From hypothesis $H(f)_1$(iv), we have

$$-\int_{\{|x_n|\geq M\}}\big(\mu F\big(z,x_n(z)\big) - x_n(z)f\big(z,x_n(z)\big)\big)\mathrm{d}z \geq 0. \tag{5.204}$$

Also from hypothesis $H(f)_1$(iii) and the mean value theorem, we have

$$-\int_{\{|x_n|<M\}} \big(\mu F(z,x_n(z)) - x_n(z) f(z,x_n(z))\big)\mathrm{d}z \geq -c_1 \quad \text{for some } c_1>0, \text{ for all } n\geq 1. \tag{5.205}$$

Using (5.204) and (5.205) in (5.203) we obtain

$$-\int_Z \big(\mu F(z,x_n(z)) - x_n(z) f(z,x_n(z))\big)\mathrm{d}z \geq -c_1 \qquad \text{for all } n\geq 1. \tag{5.206}$$

For almost all $z\in Z$ and all $|x|\geq M$, the function $s \longrightarrow (1/s^{\mu})F(z,sx)$ is C^1 on $\mathbb{R}_+ \setminus \{0\}$. So we have

$$\frac{\mathrm{d}}{\mathrm{d}s}\Big(\frac{1}{s^{\mu}}F(z,sx)\Big) = -\frac{\mu}{s^{\mu+1}}F(z,sx) + \frac{1}{s^{\mu}}\Big(\frac{\mathrm{d}}{\mathrm{d}x}F(z,sx)\Big)x.$$

By virtue of the mean value theorem, for $s>1$ we can find $\eta\in(1,s)$ such that

$$\begin{aligned}\frac{1}{s^{\mu}}F(z,sx) - F(z,x) &= \Big(-\frac{\mu}{\eta^{\mu+1}}F(z,\eta x) + \frac{1}{\eta^{\mu}}\Big(\frac{\mathrm{d}}{\mathrm{d}x}F(z,\eta x)\Big)x\Big)(s-1)\\ &= \frac{s-1}{\eta^{\mu+1}}\Big(-\mu F(z,\eta x) + \Big(\frac{\mathrm{d}}{\mathrm{d}x}F(z,\eta x)\Big)\eta x\\ &\geq 0 \quad \text{for a.a. } z\in Z \text{ and all } |x|\geq M,\end{aligned}$$

(see hypothesis $H(f)_1$(iv)),

$$\Rightarrow s^{\mu}F(z,x) \leq F(z,sx) \ \text{ for a.a. } z\in Z, \ \text{ all } |x|\geq M \text{ and all } s\geq 1. \tag{5.207}$$

Using (5.207), we see that for almost all $z\in Z$, we have

$$F(z,x) = F\Big(z,\frac{x}{M}M\Big) \geq \frac{x^{\mu}}{M^{\mu}}F(z,M) \quad \text{if } x\geq M$$

$$\text{and} \quad F(z,x) = F\Big(z,\frac{|x|}{M}(-M)\Big) \geq \frac{|x|^{\mu}}{M^{\mu}}F(z,-M) \quad \text{if } x\leq M.$$

We set $\xi = (1/M^{\mu})\min\big\{\operatorname{ess\,inf} F(\cdot,M), \ \operatorname{ess\,inf} F(\cdot,-M)\big\} > 0$ (see hypothesis $H(f)_1$(iv)). So we have

$$F(z,x) \geq \xi|x|^{\mu} \quad \text{for a.a. } z\in Z \text{ and all } |x|\geq M. \tag{5.208}$$

From (5.208), hypothesis $H(f)_1$(iii), and the mean value theorem, we conclude that

$$F(z,x) \geq \xi|x|^{\mu} - c_2 \quad \text{for some } c_2>0, \ \text{almost all } z\in Z, \ \text{and all } x\in\mathbb{R}. \tag{5.209}$$

Using (5.209) and because $\mu>p$, we obtain

$$(\mu-p)\int_Z F(z,x_n(z))\mathrm{d}z \geq c_3\|x_n\|_{\mu}^{\mu} - c_4 \quad \text{for some } c_3,c_4>0, \ \text{all } n\geq 1. \tag{5.210}$$

Returning to (5.202) and using (5.206) and (5.210), we obtain

$$\begin{aligned}& c_3\|x_n\|_{\mu}^{\mu} \leq c_5 \qquad \text{for some } c_5>0, \ \text{all } n\geq 1,\\ \Rightarrow\ & \{x_n\}_{n\geq 1} \subseteq L^{\mu}(Z) \ \text{ is bounded},\\ \Rightarrow\ & \{x_n\}_{n\geq 1} \subseteq L^{p}(Z) \ \text{ is bounded} \quad (\text{because } p<\mu).\end{aligned} \tag{5.211}$$

From (5.199) we have

$$
\begin{aligned}
&\mu\, \varphi_\lambda(x_n) - \left\langle \varphi_\lambda'(x_n), x_n \right\rangle \leq \mu \overline{M} + \varepsilon_n \qquad \text{for all } n \geq 1, \\
\Rightarrow\ & \left(\frac{\mu}{p} - 1\right) \|Dx_n\|_p^p - \lambda \left(\frac{\mu}{p} - 1\right) \int_Z m|x_n|^p \mathrm{d}z \\
& - \int_Z \Big(\mu F\big(z, x_n(z)\big) - x_n(z) f\big(z, x_n(z)\big)\Big) \mathrm{d}z \leq \mu \overline{M} + \varepsilon_n, \\
\Rightarrow\ & \left(\frac{\mu}{p} - 1\right) \|Dx_n\|_p^p \leq c_6 \quad \text{for some } c_5 > 0, \text{ all } n \geq 1,
\end{aligned} \tag{5.212}
$$

(see (5.206) and (5.211)). Recall that $\mu > p$. So from (5.212) and Poincaré's inequality, it follows that $\{x_n\}_{n \geq 1} \subseteq W_0^{1,p}(Z)$ is bounded. Hence we may assume that

$$
\begin{aligned}
& x_n \xrightarrow{w} x \quad \text{in } W_0^{1,p}(Z), \ x_n \longrightarrow x \text{ in } L^p(Z) \\
\text{and} \quad & x_n \longrightarrow x \quad \text{in } L^r(Z) \quad (\text{because } r < p^*).
\end{aligned}
$$

From (5.199), we have

$$
\langle A(x_n), x_n - x \rangle - \lambda \int_Z m|x_n|^{p-2} x_n (x_n - x) \mathrm{d}z - \int_Z f(z, x_n)(x_n - x) \mathrm{d}z \leq \varepsilon_n \tag{5.213}
$$

for all $n \geq 1$. Note that

$$
\begin{aligned}
& \lambda \int_Z m|x_n|^{p-2} x_n (x_n - x) \mathrm{d}z \longrightarrow 0 \quad (\text{because } x_n \longrightarrow x \text{ in } L^p(Z) \text{ and} \\
& \qquad \text{because of (5.211)}), \\
\text{and} \quad & \int_Z f(z, x_n)(x_n - x) \mathrm{d}z \longrightarrow 0 \quad (\text{because } x_n \longrightarrow x \text{ in } L^r(Z) \text{ and} \\
& \qquad \big\{N_f(x_n)\big\}_{n \geq 1} \subseteq L^{r'}(Z) \text{ is bounded; see hypothesis H}(f)_1\text{(iii)}).
\end{aligned}
$$

So from (5.213) it follows that

$$
\limsup \langle A(x_n), x_n - x \rangle \leq 0. \tag{5.214}
$$

Because A is of type $(S)_+$ (see Proposition 4.3.41), from (5.214) it follows that

$$
x_n \longrightarrow x \text{ in } W_0^{1,p}(Z).
$$

Thus we conclude that φ_λ satisfies the C-condition. □

Let $u_2 \in C_0^1(\overline{Z})$ be an eigenfunction corresponding to the second eigenvalue $\lambda_2 > 0$ of $\big(-\triangle_p, W_0^{1,p}(Z)\big)$ (see Section 4.3). Let

$$
\begin{aligned}
& Y = \operatorname{span}\{u_1, u_2\} \\
\text{and} \quad & H_R = \big\{y \in Y : y = t_1 u_1 + t_2 u_2,\ t_1 \in \mathbb{R},\ t_2 \geq 0,\ \|y\| \leq R\big\} \text{ for } R > 0.
\end{aligned} \tag{5.215}
$$

Evidently H_R is a hemisphere in Y. We consider the boundary of this hemisphere in Y. So we have

$$
H_R^0 = \big\{y \in H_R : y = t_1 u_1,\ t_1 \in \mathbb{R},\ \|y\| \leq R\big\} \cup \big\{y \in H_R : \|y\| = R\big\}. \tag{5.216}
$$

The boundary H_R^0 is the union of two sets. The first set in this union is a circle and its interior and the second set is the dome of the hemisphere. Also we set

$$K = \{x \in W_0^{1,p}(Z) : \|Dx\|_p^p = \lambda_2 \int_Z m|x|^p dz\}. \tag{5.217}$$

This is a closed, pointed, symmetric cone in $W_0^{1,p}(Z)$. For $\varrho > 0$, we set

$$D^\varrho = K \cap \partial B_\varrho$$

where $\partial B_\varrho = \{x \in W_0^{1,p}(Z) : \|x\| = \varrho\}$.

PROPOSITION 5.4.3 *If hypotheses $H(m)$, $H(f)_1$ hold and $\lambda \in [\lambda_1, \lambda_2)$, then we can find $R > 0$ large enough such that $\varphi_\lambda\big|_{H_R^0} \leq 0$.*

PROOF: Let $y \in Y$. Then we have

$$\begin{aligned} \varphi_\lambda(y) &= \frac{1}{p}\|Dy\|_p^p - \frac{\lambda}{p}\int_Z m|y|^p dz - \int_Z F\big(z, y(z)\big)dz \\ &\leq \frac{1}{p}\|Dy\|_p^p - \xi\|y\|_\mu^\mu + c_7 \quad \text{for some } c_7 > 0 \text{ independent of } y \in Y \end{aligned} \tag{5.218}$$

(see (5.209)).

Because $Y \subseteq C_0^1(\overline{Z})$ is finite-dimensional, all the L^p and Sobolev norms on it are equivalent. So we can find $c_8 > 0$ such that

$$c_8\|Dy\|_p^\mu \leq \|y\|_\mu^\mu \qquad \text{for all } y \in Y.$$

Using this in (5.218), we obtain

$$\varphi_\lambda(y) \leq \frac{1}{p}\|Dy\|_p^p - c_9\|Dy\|_p^\mu + c_7 \quad \text{for some } c_9 > 0 \text{ and all } y \in Y. \tag{5.219}$$

Because $\mu > p$, from (5.219), using Poincaré's inequality, we deduce that

$$\varphi_\lambda(y) \longrightarrow -\infty \text{ as } \|y\| \longrightarrow \infty, \tag{5.220}$$

that is, $\varphi_\lambda\big|_Y$ is weakly anticoercive. Also for $t > 0$ and because $u_1 \in \operatorname{int} C_0^1(\overline{Z})_+$, we have

$$\begin{aligned} \varphi_\lambda(tu_1) &= \frac{t^p}{p}\|Du_1\|_p^p - \frac{t^p}{p}\lambda\int_Z m|u_1|^p dz - \int_Z F(z, tu_1)dz \\ &\leq \frac{t^p}{p}\Big(1 - \frac{\lambda}{\lambda_1}\Big)\|Du_1\|_p^p - t^\mu\|u_1\|_\mu^\mu + c_{10} \quad \text{for some } c_{10} > 0 \end{aligned}$$

(see (5.209))

$$\leq \frac{t^p}{p}\Big(1 - \frac{\lambda}{\lambda_1}\Big)\|Du_1\|_p^p - t^\mu c_{11}\|Du_1\|_p^\mu + c_{10} \quad \text{for some } c_{11} > 0. \tag{5.221}$$

From (5.220) and (5.221) and because $\lambda_1 \leq \lambda < \lambda_2$ and $p < \mu$, we deduce that there exists $R > 0$ large such that

$$\varphi_\lambda\big|_{H_R^0} \leq 0 \qquad \text{(see (5.216))}.$$

∎

PROPOSITION 5.4.4 *If $H(m)$, $H(f)_1$ hold and $0<\lambda<\lambda_2$, then we can find $\varrho>0$ small such that $\varphi_\lambda\big|_{D^\varrho} \geq \xi_0>0$.*

PROOF: By virtue of hypothesis $\mathrm{H}(f)_1$(v), given $\varepsilon>0$, we can find $\delta=\delta(\varepsilon)>0$ such that

$$\begin{aligned} & f(z,x) \leq \varepsilon x^{p-1} \quad \text{for a.a. } z\in Z \text{ and all } x\in[0,\delta] \\ \text{and} \quad & f(z,x) \geq \varepsilon |x|^{p-2}x \quad \text{for a.a. } z\in Z \text{ and all } x\in[-\delta,0]. \end{aligned}$$

From these inequalities, after integration we obtain

$$F(z,x) \leq \frac{\varepsilon}{p}|x|^p \quad \text{for a.a. } z\in Z \text{ and all } x\in[-\delta,\delta]. \tag{5.222}$$

On the other hand from hypothesis $\mathrm{H}(f)_1$(iii) and the mean value theorem, we have

$$F(z,x) \leq c_\varepsilon |x|^r \quad \text{for some } c_\varepsilon>0, \text{ a.a. } z\in Z \text{ and all } |x|>\delta. \tag{5.223}$$

Combining (5.222) and (5.223), we obtain

$$F(z,x) \leq \frac{\varepsilon}{p}|x|^p + c_\varepsilon |x|^r \quad \text{for a.a. } z\in Z \text{ and all } x\in\mathbb{R}. \tag{5.224}$$

Let $x\in K$. Then

$$\begin{aligned} \varphi_\lambda(x) &= \frac{1}{p}\|Dx\|_p^p - \frac{\lambda}{p}\int_Z m|x|^p dz - \int_Z F\big(z,x(z)\big)dz \\ &\geq \frac{1}{p}\|Dx\|_p^p - \frac{\lambda}{p}\int_Z m|x|^p dz - \frac{\varepsilon}{p}\|x\|_p^p - c_\varepsilon\|x\|_r^r \quad \text{(see (5.224))} \\ &\geq \frac{1}{p}\Big(1-\frac{\lambda}{\lambda_2}\Big)\|Dx\|_p^p - \frac{\varepsilon}{p}c_{14}\|Dx\|_p^p - \widehat{c}_\varepsilon\|Dx\|_p^r \end{aligned} \tag{5.225}$$

for some $c_{14}, \widehat{c}_\varepsilon>0$ (see (5.217)).

Because $\lambda<\lambda_2$, choosing $\varepsilon>0$ small from (5.225) we have

$$\varphi_\lambda(x) \geq c_{15}\|Dx\|_p^p - \widehat{c}_\varepsilon\|Dx\|_p^r \qquad \text{for some } c_{14}>0 \text{ and all } x\in K. \tag{5.226}$$

Recall that $r>p$. So from (5.226) and Poincaré's inequality, it follows that we can find $\varrho>0$ small such that

$$\varphi_\lambda\big|_{D^\varrho} \geq \xi_0>0,$$

where $D^\varrho=K\cap\partial B_\varrho(0)$. □

Now we can establish the existence of a nontrivial solution for problem (5.198).

THEOREM 5.4.5 *If $H(m)$, $H(f)_1$ hold and $0<\lambda<\lambda_2$, then problem (5.198) has a nontrivial solution $x\in C_0^1(\overline{Z})$.*

PROOF: First assume that $\lambda_1 \leq \lambda<\lambda_2$.

We choose $\varrho>0$ small and $R>0$ big so that Propositions 5.4.3 and 5.4.4 hold. From Proposition 4.3.63, we know that the sets $\{H_R^0, H_R, D^\varrho\}$ link in $W_0^{1,p}(Z)$ via the identity map. Then Proposition 5.4.2 allows the use of Theorem 4.1.22, which gives $x\in W_0^{1,p}(Z)$ such that

$$\varphi_\lambda(x) \geq \xi_0 > 0 = \varphi_\lambda(0) \ \ \text{(i.e., } x \neq 0)$$
$$\text{and} \quad \varphi'(x) = 0.$$

From the last equation, we have

$$A(x) - \lambda m|x|^{p-2}x = N_f(x),$$

$$\Rightarrow \left\{ \begin{array}{l} -\mathrm{div}\big(\|Dx(z)\|^{p-2}Dx(z)\big) - \lambda m(z)|x(z)|^{p-2}x(z) = f\big(z, x(z)\big) \\ \text{a.e. on } Z, \\ x\big|_{\partial Z} = 0, \ \ 1<p<\infty \end{array} \right\}. \tag{5.227}$$

From nonlinear regularity theory (see Theorem 4.3.35), we have that $x \in C_0^1(\overline{Z})$ and of course it solves problem (5.198) (see (5.227)).

Now assume that $0<\lambda<\lambda_1$. From the proof of Proposition 5.4.3, we know that

$$\varphi_\lambda(tu_1) \longrightarrow -\infty \quad \text{as } t \longrightarrow \infty \quad \text{(see (5.221))}. \tag{5.228}$$

Moreover, as in the proof of Proposition 5.4.4, using (5.224), we obtain

$$\varphi_\lambda(x) \geq \frac{1}{p}\Big(1 - \frac{\lambda}{\lambda_1}\Big)\|Dx\|_p^p - \frac{\varepsilon}{p}\, c_{14}\|Dx\|_p^p - \widehat{c}_\varepsilon\|Dx\|_p^r.$$

Because $\lambda<\lambda_1$, choosing $\varepsilon>0$ small, we have

$$\varphi_\lambda(x) \geq c_{16}\|Dx\|_p^p - \widehat{c}_\varepsilon\|Dx\|_p^r \quad \text{for some } c_{16}>0, \text{ all } x \in W_0^{1,p}(Z). \tag{5.229}$$

Because $p<r$, we can find $\varrho>0$ small, we have

$$\varphi_\lambda\big|_{\partial B_\varrho(0)} \geq \xi_0 > 0 \tag{5.230}$$

(see (5.229) and use Poincaré's inequality). From (5.228), (5.230), and Proposition 5.4.2 and because $\varphi_\lambda(0) = 0$, we see that we can apply the mountain pass theorem (see Theorem 4.1.24) and obtain $x \in W_0^{1,p}(Z)$ such that

$$\varphi_\lambda(x) \geq \xi_0 > 0 = \varphi_\lambda(0) \ \ \text{(i.e., } x \neq 0)$$
$$\text{and} \quad \varphi'(x) = 0.$$

As above it follows that $x \in C_0^1(\overline{Z})$ and it is a nontrivial solution of problem (5.198). □

Next we consider the following perturbed eigenvalue problem

$$\left\{ \begin{array}{l} -\Delta x(z) = \lambda x(z) + |x(z)|^{2^*-2}x(z) \ \text{ a.e. on } Z, \\ x\big|_{\partial Z} = 0 \end{array} \right\}. \tag{5.231}$$

Here

$$2^* = \left\{ \begin{array}{ll} \frac{2N}{N-2} & \text{if } 2<N \\ +\infty & \text{if } N = 1,2 \end{array} \right.$$

(the critical Sobolev exponent for 2). So we see that in this case the nonlinear perturbation in problem (5.231) has a critical growth and this causes serious difficulties dealing with the problem, because the embedding of the Sobolev space $H_0^1(Z)$ into $L^{2^*}(Z)$ is no longer compact.

We are looking for positive solutions of problem (5.231). First let us consider the special case $\lambda = 0$. One way to approach problem (5.231) variationally is to try to obtain positive solutions of (5.231), as relative minima of the functional.

$$\varphi_\lambda(x) = \frac{1}{2}\|Dx\|_2^2 - \frac{\lambda}{2}\|x\|_2^2 - \frac{1}{2^*}\|x\|_{2^*}^{2^*}$$

on the C^1-manifold

$$M = \{x \in H_0^1(Z) : \|x\|_{2^*} = 1\}.$$

So we want to find a solution of the following minimization problem

$$\inf\Big[\frac{\|Dx\|_2^2 - \lambda\|x\|_2^2}{\|x\|_{2^*}^2} : x \in H_0^1(Z),\ x \neq 0\Big] = S_\lambda(Z).$$

If $\lambda = 0$, then we have

$$S_0(Z) = \inf\Big[\frac{\|Dx\|_2^2}{\|x\|_{2^*}^2} : x \in H_0^1(Z),\ x \neq 0\Big] = S(Z),$$

where $S(Z)$ is the best Lipschitz constant for the embedding of $H_0^1(Z)$ into $L^{2^*}(Z)$. We know that $S(Z) = S$ is independent of the domain Z and is never attained on a domain $Z \subseteq \mathbb{R}^N$, $Z \neq \mathbb{R}^N$ (see, for example, Willem [606, Section 1.9]). So when $\lambda = 0$, problem (5.231) has no positive solution. More generally we show that this true for all $\lambda \leq 0$ and all $\lambda \geq \lambda_1$, provided that the domain Z has certain geometric structure. So we start with a definition.

DEFINITION 5.4.6 A domain $Z \subseteq \mathbb{R}^N$ containing the origin is said to be *star-shaped* with respect to the origin, if $\big(z, n(z)\big)_{\mathbb{R}^N} > 0$ for all $z \in \partial Z$ (by $n(z)$ we denote the unit outward normal at $z \in \partial Z$).

EXAMPLE 5.4.7 Any open ball in $\mathbb{R}^N$ centered at 0, is star-shaped with respect to the origin. To see this let $z \in \partial Z$. Then $n(z) = z/R$ where $R>0$ is the radius of the ball. So $\big(z, n(z)\big)_{\mathbb{R}^N} = \|z\|^2/R = R>0$.

The negative result on the existence of positive solutions for problem (5.231) when $\lambda \notin (0, \lambda_1)$, is based on the following *Pohozaev's identity*.

PROPOSITION 5.4.8 *If* $g\colon \mathbb{R} \longrightarrow \mathbb{R}$ *is continuous,* $G(x) = \int_0^x g(r)dr$, *and* $x \in C_0^1(\overline{Z})$ *is a solution of*

$$\left\{\begin{array}{l} -\Delta x(z) = g\big(x(z)\big) \ \ a.e.\ on\ Z, \\ x\big|_{\partial Z} = 0 \end{array}\right\}, \tag{5.232}$$

then

$$\Big(1-\frac{N}{2}\Big)\int_Z g\big(x(z)\big)x(z)dz + N\int_Z G\big(x(z)\big)x(z)dz = \frac{1}{2}\int_{\partial G}\big(z, n(z)\big)_{\mathbb{R}^N}\frac{\partial x}{\partial n}(z)d\sigma. \tag{5.233}$$

PROOF: We multiply the equation in (5.232) with $z_i(\partial x/\partial z_i)$ and then integrate over Z. So we have

$$-\int_Z \triangle x(z) z_i \frac{\partial x}{\partial z_i}(z)\mathrm{d}z = \int_Z g\big(x(z)\big) z_i \frac{\partial x}{\partial z_i}(z)\mathrm{d}z = \int_Z \frac{\partial G\big(x(z)\big)}{\partial z_i} z_i \mathrm{d}z$$

$$\Rightarrow -\int_Z \triangle x(z) z_i \frac{\partial x}{\partial z_i}(z)\mathrm{d}z = -\int_Z G\big(x(z)\big)\mathrm{d}z \qquad \text{(by integration by parts).}$$

So we have

$$\begin{aligned}
-\int_Z G\big(x(z)\big)\mathrm{d}z &= -\int_Z \Big(\sum_{k=1}^N \frac{\partial^2 x}{\partial z_k^2}\Big)\frac{\partial x}{\partial z_i} z_i \mathrm{d}z \\
&= \int_Z \sum_{k=1}^N \frac{\partial x}{\partial z_k}\frac{\partial^2 x}{\partial z_k \partial z_i} z_i \mathrm{d}z + \int_Z \sum_{k=1}^N \frac{\partial x}{\partial z_k}\frac{\partial x}{\partial z_i}\delta_{ik}\mathrm{d}z \\
&\quad - \int_{\partial Z} \sum_{k=1}^N \frac{\partial x}{\partial z_k}\frac{\partial x}{\partial z_i} z_i n_k \mathrm{d}\sigma \\
&= \frac{1}{2}\int_Z \sum_{k=1}^N \frac{\partial}{\partial z_i}\Big(\big(\frac{\partial x}{\partial z_k}\big)^2\Big) z_i \mathrm{d}z + \int_Z \big(\frac{\partial x}{\partial z_i}\big)^2 \mathrm{d}z \\
&\quad - \int_{\partial Z} \frac{\partial x}{\partial z_i} z_i \frac{\partial x}{\partial n}\mathrm{d}\sigma \\
&= -\frac{1}{2}\int_Z \sum_{k=1}^N \Big(\frac{\partial x}{\partial z_k}\Big)^2 \mathrm{d}z + \int_Z \big(\frac{\partial x}{\partial z_i}\big)^2 \mathrm{d}z \\
&\quad + \frac{1}{2}\int_{\partial Z} \sum_{k=1}^N \big(\frac{\partial x}{\partial z_k}\big)^2 z_i n_i \mathrm{d}\sigma - \frac{1}{2}\int_{\partial Z} \frac{\partial x}{\partial z_i} z_i \frac{\partial x}{\partial n}\mathrm{d}\sigma.
\end{aligned}$$

Summing over i, we obtain

$$\begin{aligned}
-N\int_Z G\big(x(z)\big)\mathrm{d}z = \Big(1-\frac{N}{2}\Big)\|Dx\|_2^2 + \frac{1}{2}\int_{\partial Z} \|Dx\|^2 (z,n)_{\mathbb{R}^N}\mathrm{d}\sigma \\
- \int_{\partial Z} \frac{\partial x}{\partial n}(z, Dx)_{\mathbb{R}^N}\mathrm{d}\sigma. \qquad (5.234)
\end{aligned}$$

From (5.232) we have

$$\|Dx\|_2^2 = \int_Z g\big(x(z)\big)x(z)\mathrm{d}z. \tag{5.235}$$

Moreover, because $x\big|_{\partial Z} = 0$, we have

$$\|Dx(z)\|_{\mathbb{R}^N}^2 = \Big(\frac{\partial x}{\partial n}(z)\Big)^2 \qquad \text{for all } z \in \partial Z \tag{5.236}$$

$$\text{and} \quad \big(z, Dx(z)\big)_{\mathbb{R}^N} = \big(z, n(z)\big)_{\mathbb{R}^N}\frac{\partial x}{\partial n}(z) \qquad \text{for all } z \in \partial Z. \tag{5.237}$$

Using (5.235) through (5.237) in (5.234), we obtain (5.233) (Pohozaev's identity). ☐

Using Proposition 5.4.8, we can now prove a negative result concerning problem (5.231). Let $Z \subseteq \mathbb{R}^N$ be a bounded domain with a C^2-boundary ∂Z, which is star-shaped with respect to the origin. If $x \in H_0^1(Z)$ is a solution of (5.231), then by the strong maximum principle we have that either $x = 0$ or $x(z) > 0$ for all $z \in Z$. We show that when $\lambda \notin (0, \lambda_1)$, the second possibility can not occur.

THEOREM 5.4.9 *If $Z \subseteq \mathbb{R}^N$, $N \geq 3$ is a bounded domain with a C^2-boundary, which is star-shaped with respect to the origin, and $\lambda \notin (0, \lambda_1)$, then problem (5.231) has no nontrivial solution $x \in H_0^1(Z)_+$.*

PROOF: First suppose that $\lambda \geq \lambda_1$.

If $x \in H_0^1(Z) \setminus \{0\}$ is a solution of (5.231), then using the eigenfunction $u_1 > 0$ corresponding to $\lambda_1 > 0$ as a test function, we obtain

$$\begin{aligned}\lambda_1 \int_Z x u_1 \mathrm{d}z = \int_Z (Dx, Du_1)_{\mathbb{R}^N} \mathrm{d}z &= \lambda \int_Z x u_1 \mathrm{d}z + \int_Z |x|^{2^*-2} x u_1 \mathrm{d}z \\ &> \lambda \int_Z x u_1 \mathrm{d}z \\ &\geq \lambda_1 \int_Z x u_1 \mathrm{d}z,\end{aligned}$$

a contradiction.

Next suppose that $\lambda \leq 0$.

We set $g(r) = \lambda r + |r|^{2^*-2} r$. Then $G(r) = (\lambda/2) r^2 + (1/2^*) |r|^{2^*}$. Using these functions in Pohozaev's identity (5.233) and because $x \in C_0^1(\overline{Z})$ (regularity theory), we have

$$\begin{aligned}&\left(1 - \frac{N}{2}\right) \int_Z (\lambda x^2 + |x|^{2^*}) \mathrm{d}z + \frac{N\lambda}{2} \int_Z x^2 \mathrm{d}z + \frac{N}{2^*} \int_Z |x|^{2^*} \mathrm{d}z = \\ &\frac{1}{2} \int_{\partial Z} (z, n)_{\mathbb{R}^N} \left(\frac{\partial x}{\partial n}\right)^2 \mathrm{d}\sigma \\ \Rightarrow\ &\left(\frac{N}{2^*} + 1 - \frac{N}{2}\right) \int_Z |x|^{2^*} \mathrm{d}z + \lambda \int_Z \lambda x^2 \mathrm{d}z = \frac{1}{2} \int_{\partial Z} (z, n)_{\mathbb{R}^N} \left(\frac{\partial x}{\partial n}\right)^2 \mathrm{d}\sigma.\end{aligned}$$

But

$$\frac{N}{2^*} + 1 - \frac{N}{2} = \frac{N}{2} - \frac{N}{2} = 0.$$

So we obtain

$$\lambda \int_Z x^2 \mathrm{d}z = \frac{1}{2} \int_{\partial Z} (z, n)_{\mathbb{R}^N} \left(\frac{\partial x}{\partial n}\right)^2 \mathrm{d}\sigma$$

which is a contradiction if $\lambda < 0$. If $\lambda = 0$ then because Z is star-shaped with respect to the origin, we must have $(\partial x / \partial n)(z) = 0$ for $z \in \partial Z$. Hence

$$\int_Z |x|^{2^*} \mathrm{d}z = -\int_Z \Delta x(z) = 0 \qquad \text{(by Green's formula)}$$

and so $x = 0$. $\square$

REMARK 5.4.10 If $Z \subseteq \mathbb{R}^N$ is not star-shaped with respect to the origin, then we can have solutions even when $\lambda \leq 0$. In fact if $Z \subseteq \mathbb{R}^N$ is an annular domain, then we know that (5.231) admits a radial solution for all $\lambda \in (-\infty, \lambda_1)$. Recall that $x \in H_0^1(Z)$ is a radial solution, if $x(z) = x(\|z\|_{\mathbb{R}^N})$ for all $z \in Z$.

When $0<\lambda<\lambda_1$, problem (5.231) may admit nontrivial solutions $x \in H_0^1(Z)_+$. In this case we have an interesting dependence on the dimension N. More precisely, we have the following theorem due to Brezis–Nirenberg [104]. As always by $\lambda_1 > 0$ we denote the principal eigenvalue of $\big(-\Delta, H_0^1(Z)\big)$.

THEOREM 5.4.11 *If $Z \subseteq \mathbb{R}^N$, $(N \geq 3)$ is a bounded domain with a C^2-boundary ∂Z, then*

(a) *If $N \geq 4$, then for any $\lambda \in (0, \lambda_1)$, there exists a positive solution for problem (5.231).*
(b) *If $N = 3$, then there exists $\lambda_* \in [0, \lambda_1)$ such that for any $\lambda \in (\lambda_*, \lambda_1)$ problem (5.231) has a positive solution.*
(c) *If $N = 3$ and $Z = B_1(0)$, then $\lambda_* = \lambda_1/4$ and for $\lambda \leq \lambda_1/4$ problem (5.231) has no nontrivial solution in $H_0^1(Z)_+$.*

5.5 Maximum and Comparison Principles

The main goal of this section is to present some maximum and comparison principles for certain nonlinear differential operators involving the p-Laplacian.

So let $Z \subseteq \mathbb{R}^N$ be a bounded domain with a C^2-boundary ∂Z. We start by considering the nonlinear differential operator

$$V_p(x) = -\mathrm{div}(\|Dx\|^{p-2}Dx) + m\widehat{\vartheta}_p(x), \quad x \in W_0^{1,p}(Z), \tag{5.238}$$

where $1<p<\infty, m\in L^\infty(Z)$ (the weight function), $\vartheta_p:\mathbb{R}\longrightarrow\mathbb{R}$ is the homeomorphism defined by

$$\vartheta_p(x) = \begin{cases} |x|^{p-2}x & \text{if } x \neq 0 \\ 0 & \text{if } x = 0 \end{cases}$$

and $\widehat{\vartheta}_p:L^p(Z)\longrightarrow L^{p'}(Z)$, $(1/p)+(1/p') = 1$ is the Nemitsky operator corresponding to ϑ_p; that is, $\widehat{\vartheta}_p(x)(\cdot)=\vartheta_p\big(x(\cdot)\big)$. Evidently $\widehat{\vartheta}_p$ is continuous and bounded (i.e., maps bounded sets to bounded sets).

We start by recalling the following basic notions.

DEFINITION 5.5.1 (a) A function $x\in W_{\mathrm{loc}}^{1,p}(Z)$ is said to be *a weak solution* of

$$-\mathrm{div}(\|Dx\|^{p-2}Dx) + m|x|^{p-2}x = g, \quad g \in W^{-1,p'}(Z) \tag{5.239}$$

if and only if

$$\int_Z \|Dx\|^{p-2}(Dx, Dv)_{\mathbb{R}^N}\mathrm{d}z + \int_Z m|x|^{p-2}xv\mathrm{d}z = \langle g, v\rangle \quad \text{for all } v\in C_c^\infty(Z)$$

(here by $\langle\cdot,\cdot\rangle$ we denote the duality brackets for the pair $\big(W_0^{1,p}(Z), W^{-1,p'}(Z)\big)$).

(b) A function $\overline{x}\in W_{\mathrm{loc}}^{1,p}(Z)$ is said to be an *upper solution* of (5.239), if

$$\int_Z \|Dx\|^{p-2}(Dx, Dv)_{\mathbb{R}^N}\mathrm{d}z + \int_Z m|x|^{p-2}xv\mathrm{d}z \geq \langle g, v\rangle \quad \text{for all } v\in C_c^\infty(Z)_+. \tag{5.240}$$

(c) A function $\overline{x}\in W^{1,p}_{\text{loc}}(Z)$ is said to be a *lower solution* of (5.239), if

$$\int_Z \|D\overline{x}\|^{p-2}(D\overline{x}, Dv)_{\mathbb{R}^N}\,dz + \int_Z m|\overline{x}|^{p-2}\overline{x}v\,dz \leq \langle g, v\rangle \quad \text{for all } v\in C_c^\infty(Z)_+. \tag{5.241}$$

REMARK 5.5.2 Recall that $C_c^\infty(Z)$ is dense in $W_0^{1,p}(Z)$ and $W^{-1,p'}(Z) = W_0^{1,p}(Z)^*$. So relations (5.239) through (5.241) are in fact valid for all $v \in W_0^{1,p}(Z)$.

DEFINITION 5.5.3 (a) We say that the operator V_p (see (5.238)) satisfies the *maximum principle* (MP, for short), if every weak solution $x \in W^{1,p}(Z)$ of

$$\left\{\begin{array}{l} -\operatorname{div}\big(\|Dx\|^{p-2}Dx\big) + m|x|^{p-2}x = g \\ x\big|_{\partial Z} \geq 0, \;\; g \in W^{-1,p'}(Z) \end{array}\right\} \tag{5.242}$$

satisfies $x(z) \geq 0$ a.e. on Z when $g \geq 0$.

(b) We say that the operator V_p (see (5.238)) satisfies the *strong maximum principle* (SMP, for short), if every weak solution $x \in W^{1,p}(Z)$ of (5.242) satisfies $x(z)>0$ a.e. on Z when $g \geq 0,\ g \neq 0$.

(c) We say that the operator V_p (see (5.238)) satisfies the *weak comparison principle* (WCP, for short), if $V_px_1 \leq V_px_2$ in Z and $x_1\big|_{\partial Z} \leq x_2\big|_{\partial Z}$ with $x_1, x_2 \in W^{1,p}(Z)$, imply that $x_1(z) \leq x_2(z)$ a.e. on Z.

To establish the maximum and comparison principles for the operator V_p, we need to know its spectrum. For this purpose the relevant nonlinear eigenvalue problem is the following one.

$$\left\{\begin{array}{l} -\operatorname{div}\big(\|Dx(z)\|^{p-2}Dx(z)\big) + m|x(z)|^{p-2}x(z) = \lambda|x(z)|^{p-2}x(z) \;\text{ a.e. on } Z, \\ x\big|_{\partial Z} = 0, \;\; \lambda \in \mathbb{R} \end{array}\right\}. \tag{5.243}$$

Following step by step the arguments used to establish the existence and properties of the first eigenvalue of $\big(-\Delta_p, W_0^{1,p}(Z)\big)$ with a weight $m \in L^\infty(Z)$ (see Section 4.3), we obtain the following result.

PROPOSITION 5.5.4 *The nonlinear eigenvalue problem (5.243) has a unique eigenvalue $\lambda = \lambda_1(m)$ defined by*

$$\lambda_1(m) = \inf\Big[\|Dx\|_p^p + \int_Z m|x|^p dz : x \in W_0^{1,p}(Z),\;\; \|x\|_p = 1\Big] \tag{5.244}$$

with the property that it has a positive eigenfunction $\widehat{u}_1 \in W_0^{1,p}(Z)$. Moreover, the eigenvalue $\lambda_1(m)$ is simple and isolated and the eigenfunction $\widehat{u}_1$ satisfies $\widehat{u}_1 \in \operatorname{int}C_0^1(\overline{Z})_+$.

The next theorem is in the direction of the nonlinear regularity results proved in Section 4.3) and is due to Di Benedetto [198] and Tolksdorf [583]. So suppose $f\colon Z \times \mathbb{R} \longrightarrow \mathbb{R}$ is a Carathéodory function (i.e., for all $x \in \mathbb{R},\ z \longrightarrow f(z,x)$ is measurable and for almost all $z\in Z,\ x \longrightarrow f(z,x)$ is continuous). We consider the following nonlinear elliptic equation,

$$-\operatorname{div}\big(\|Dx(z)\|^{p-2}Dx(z)\big) = f\big(z, x(z)\big) \;\text{ in } Z,\ 1<p<\infty. \tag{5.245}$$

In analogy to Definition 5.5.1(a), we make the following definition.

DEFINITION 5.5.5 A function $x \in W^{1,p}_{\mathrm{loc}}(Z)$ is said to be a *weak solution* of problem (5.245), if

$$\int_Z \|Dx\|^{p-2}(Dx, Dv)_{\mathbb{R}^N}\mathrm{d}z = \int_Z f(z,x)v\mathrm{d}z \qquad \text{for all } v \in C_c^\infty(Z).$$

The regularity result of Di Benedetto–Tolksdorf, reads as follows.

THEOREM 5.5.6 *If $Z \subseteq \mathbb{R}^N$ is a bounded domain, $x \in W^{1,p}(Z)$ is a weak solution of (5.245), and $z \longrightarrow f(z, x(z))$ belongs in $L^r(Z)$ with $p/(p-1)N < r \leq \infty$, then $x \in C^{1,\alpha}(Z)$ for some $0<\alpha<1$.*

Another such regularity result that we need is the next theorem, which is essentially Theorem 4.3.35 but with nonhomogeneous Dirichlet boundary conditions. So we consider the following problem:

$$\left\{\begin{array}{l} -\mathrm{div}\big(\|Dx\|^{p-2}Dx\big) + m|x|^{p-2}x = g \ \text{ in } Z \\ x\big|_{\partial Z} = h, \ 1<p<\infty, \ g \in L^\infty(Z) \end{array}\right\}. \tag{5.246}$$

The next theorem is a particular case of a more general result due to Lieberman [382].

THEOREM 5.5.7 *If $x \in W^{1,p}(Z) \cap L^\infty(Z)$ is a weak solution of (5.246) and $h \in C^2(\partial Z)$, then $x \in C^1(\overline{Z})$.*

Now we can pass to the analysis of operator V_p.

THEOREM 5.5.8 *The following statements are equivalent.*

(a) *The operator V_p satisfies the MP.*
(b) *The operator V_p satisfies the SMP.*
(c) *$\lambda_1(m)>0$ (see (5.244)).*
(d) *There exists a positive strict upper solution $\overline{x} \in W_0^{1,p}(Z)$ of the equation $V_p(x) = 0$ and $V_p(\overline{x}) \in L^\infty(Z)$ (i.e., $V_p(\overline{x}) = g$ in Z, $g \in L^\infty(Z)_+$, $g \neq 0$).*
(e) *For every $g \in L^\infty(Z)_+$, the problem $V_p(x) = g$, $g\big|_{\partial Z} = 0$, has a unique weak solution $x \in W_0^{1,p}(Z)_+$.*

PROOF: (a)$\Rightarrow$(b). Let $x \in W_0^{1,p}(Z)$ be a weak solution of

$$-\Delta_p x + m|x|^{p-2}x = g \quad \text{in } Z, \ x\big|_{\partial Z} \geq 0, \tag{5.247}$$

where $g \in L^\infty(Z)_+$, $g \neq 0$. Because by hypothesis V_p satisfies the MP, we have $x(z) \geq 0$ a.e. on Z. Because $g \neq 0$, we have that $x \neq 0$. Moreover, taking into account the fact that $x \in L^\infty(Z)$ (see Theorem 4.3.34), from Theorem 5.5.6 we infer that $x \in C^1(\overline{Z})$. We have

$$-\Delta_p x + \|m\|_\infty |x|^{p-2}x \geq 0 \tag{5.248}$$

(see (5.247) and recall that $x \geq 0$, $g \geq 0$). Then from (5.248) and the nonlinear strong maximum principle (see Theorem 4.3.37), we deduce that $x(z) > 0$ for all $z \in Z$. So V_p satisfies the SMP.

(b)$\Rightarrow$(c). Suppose that $\lambda_1(m) \leq 0$ and let $\widehat{u}_1 \in \operatorname{int} C_0^1(\overline{Z})_+$ be a positive eigenfunction for the operator $\big(V_p, W_0^{1,p}(Z)\big)$ (see Theorem 5.5.6). Then we have

$$V_p(-\widehat{u}_1) = -\Delta_p(-\widehat{u}_1) + m|-\widehat{u}_1|^{p-2}(-\widehat{u}_1) \geq 0.$$

Because $V_p(\widehat{u}_1) \in L^\infty(Z)$ and by hypothesis V_p satisfies the SMP, we must have $\widehat{u}_1 \leq 0$, a contradiction.

(c)$\Rightarrow$(d). Just take $\overline{x} = \widehat{u}_1$ (= a positive eigenfunction of $\big(V_p, W_0^{1,p}(Z)\big)$).

(d)$\Rightarrow$(c). As in Proposition 4.3.44 we introduce the set

$$\mathcal{D}(J) = \Big\{(x,y) \in W_0^{1,p}(Z) \times W_0^{1,p}(Z) : x \geq 0,\ y \geq 0,\ \frac{x}{y}, \frac{y}{x} \in L^\infty(Z)\Big\}.$$

Let $\widehat{u}_1 \in \operatorname{int} C_0^1(\overline{Z})_+$ be a positive eigenfunction of $\big(V_p, W_0^{1,p}(Z)\big)$. Note that because $\overline{x} \in W_0^{1,p}(Z)$ is by hypothesis a strict upper solution of $V_p(x) = 0$ and $V_p(\overline{x}) \in L^\infty(Z)$, we have $\overline{x} \in \operatorname{int} C_0^1(\overline{Z})_+$. So it follows that $(\widehat{u}_1, \overline{x}) \in \mathcal{D}(J)$. In particular we have $\overline{x}/\widehat{u}_1 \in L^\infty(Z)$. Set $\overline{u}_1 = c\widehat{u}_1$ with $c>0$ such that $c \geq \|\overline{x}/\widehat{u}_1\|_\infty$. Suppose that $\lambda_1(m) \leq 0$. Then if by $\langle \cdot,\cdot \rangle$ we denote the duality brackets for the pair $\big(W_0^{1,p}(Z), W^{-1,p'}(Z)\big)$, $(1/p) + (1/p') = 1$, we have

$$\begin{aligned}
&\Big\langle -\Delta_p \overline{x}, \frac{\overline{u}_1^p - \overline{x}^p}{\overline{x}^{p-1}} \Big\rangle \geq \int_Z -m(\overline{u}_1^p - \overline{x}^p)dz,\\
\Rightarrow\ & J(\overline{u}_1, \overline{x}) \leq \int_Z \lambda_1(m)(\overline{u}_1^p - \overline{x}^p)dz \leq 0, \qquad \text{(see (4.124) in Section 4.3)}\\
\Rightarrow\ & J(\overline{u}_1, \overline{x}) = 0 \ \text{ and so } \ \overline{u}_1 = \theta\overline{x} \quad \text{for some } \theta>0 \qquad \text{(see (4.125) in Section 4.3).}
\end{aligned}$$

So it follows that

$$-\Delta_p \overline{u} + m|\overline{u}|^{p-2}\overline{u} = \theta^{p-1} g, \quad g \in L^\infty(Z)_+,\ g \neq 0.$$

This contradicts the hypothesis that $\lambda_1(m) \leq 0$. Hence we must have $\lambda_1(m)>0$.

(c)$\Rightarrow$(a). Suppose that $x \in W^{1,p}(Z)$ satisfies

$$-\Delta_p x + m|x|^{p-2}x = g \quad \text{in } Z,\ x\big|_{\partial Z} \geq 0,$$

with $g \in W^{-1,p'}(Z)$, $g \geq 0$ (i.e., $\langle g, v\rangle \geq 0$ for all $v \in W_0^{1,p}(Z)_+$). Using as a test function $-x^- \in W_0^{1,p}(Z)$ and recalling that

$$Dx^-(z) = \begin{cases} -Dx(z) & \text{if } x(z)<0 \\ 0 & \text{if } x(z)\geq 0 \end{cases},$$

we obtain

$$\begin{aligned}
&\|Dx^-\|_p^p + \int_Z m|x^-|^p dz = \langle g, -x^-\rangle \leq 0\\
\Rightarrow\ & \lambda_1(m) \leq 0,
\end{aligned}$$

a contradiction, unless $x^- = 0$ in which case $x \geq 0$. Thus far we have proved the equivalence of statements (a), (b), (c), (d).

(e)⇒(d). Immediate.

So we are done if we show that one of the statements (a), (b), (c), or (d) implies (e). More precisely we show the following.

(c)⇒(e). We consider the Euler functional $\varphi\colon W_0^{1,p}(Z)\longrightarrow\mathbb{R}$ defined by

$$\varphi(x)=\frac{1}{p}\|Dx\|_p^p+\frac{1}{p}\int_Z m|x|^p dz-\int_Z gx dz. \tag{5.249}$$

We claim that φ is weakly coercive. Suppose that this is not the case. Then we can find $\{x_n\}_{n\geq1}\subseteq W_0^{1,p}(Z)$ such that

$$\|x_n\|\longrightarrow\infty \text{ as } n\rightarrow\infty \quad\text{and}\quad \varphi(x_n)\leq M \quad\text{for some } M>0,\text{ all } n\geq1.$$

We have

$$\begin{aligned}&\frac{\lambda_1(m)}{p}\|x_n\|_p^p-\int_Z gx_n dz\leq\varphi(x_n)\leq M\quad\text{for all } n\geq1,\\ \Rightarrow\ &\frac{\lambda_1(m)}{p}\|x_n\|_p^p\leq M+\widehat{c}\|x_n\|_p\quad\text{for some } \widehat{c}>0,\text{ all } n\geq1,\\ \Rightarrow\ &\{x_n\}_{n\geq1}\subseteq L^p(Z)\text{ is bounded.}\end{aligned}$$

Then from the definition of φ (see (5.249)), we infer that

$$\begin{aligned}&\{Dx_n\}_{n\geq1}\subseteq L^p(Z,\mathbb{R}^N)\text{ is bounded,}\\ \Rightarrow\ &\{x_n\}_{n\geq1}\subseteq W_0^{1,p}(Z)\text{ is bounded (by Poincaré's inequality),}\end{aligned}$$

a contradiction to the fact that $\|x_n\|\longrightarrow\infty$. Therefore φ is weakly coercive. Moreover, exploiting the compact embedding of $W_0^{1,p}(Z)$ into $L^p(Z)$, we can see that φ is weakly lower semicontinuous on $W_0^{1,p}(Z)$. By the Weierstrass theorem, we can find $\widehat{x}\in W_0^{1,p}(Z)$ such that

$$\begin{aligned}&\varphi(\widehat{x})=\inf_{W_0^{1,p}(Z)}\varphi,\\ \Rightarrow\ &\varphi'(\widehat{x})=0,\\ \Rightarrow\ &A(\widehat{x})+m|\widehat{x}|^{p-2}\widehat{x}=g,\end{aligned}\tag{5.250}$$

where $A:W_0^{1,p}(Z)\longrightarrow W^{-1,p'}(Z)$ as before is defined by

$$\langle A(x),y\rangle=\int_Z\|Dx\|^{p-2}(Dx,Dy)_{\mathbb{R}^N}dz\qquad\text{for all } x,y\in W_0^{1,p}(Z).$$

From (5.250) it follows that

$$\left\{\begin{array}{l}-\mathrm{div}\big(\|D\widehat{x}(z)\|^{p-2}D\widehat{x}(z)\big)+m(z)|\widehat{x}(z)|^{p-2}\widehat{x}(z)=g(z)\ \text{ a.e. on } Z,\\ \widehat{x}\big|_{\partial Z}=0\end{array}\right\}$$

$$\Rightarrow\ \widehat{x}\in C_0^1(\overline{Z})_+\ \text{ is a strong solution of the problem }\ V_p(x)=g,\ x\big|_{\partial Z}=0.$$

Moreover, acting on (5.250) with the test function $-\widehat{x}^-\in W_0^{1,p}(Z)$ and since by hypothesis $\lambda_1(m)>0$, it follows that $\widehat{x}\geq0$. In fact if $g\neq0$, by the nonlinear strong maximum principle we have $\widehat{x}\in\operatorname{int}C_0^1(\overline{Z})_+$.

Finally if $\widehat{x}, \widehat{y} \in \operatorname{int} C_0^1(\overline{Z})_+$ are two such solutions, when $g \neq 0$, then

$$\begin{aligned} 0 \leq J(\widehat{x}, \widehat{y}) &= \left\langle -\Delta_p \widehat{x}, \frac{\widehat{x}^p - \widehat{y}^p}{\widehat{x}^{p-1}} \right\rangle - \left\langle -\Delta_p \widehat{y}, \frac{\widehat{x}^p - \widehat{y}^p}{\widehat{y}^{p-1}} \right\rangle \\ &= \int_Z \frac{\widehat{y}^{p-1} - \widehat{x}^{p-1}}{\widehat{x}^{p-1}\widehat{y}^{p-1}} (\widehat{x}^p - \widehat{y}^p) dz \leq 0, \\ &\Rightarrow J(\widehat{x}, \widehat{y}) = 0 \ \text{ (i.e., } \widehat{x} = c\widehat{y} \text{ for some } c > 0), \\ &\Rightarrow c^{p-1} g = g \ \text{ (i.e., } c = 1). \end{aligned}$$

Therefore $\widehat{x} = \widehat{y}$ and we have proved the uniqueness of the solution.

If $g = 0$, then we have

$$A(\widehat{x}) + m|\widehat{x}|^{p-2}\widehat{x} = 0 \quad \text{(see (5.250))}.$$

Acting with the test function $\widehat{x} \in W_0^{1,p}(Z)$, we obtain

$$\begin{aligned} & \|D\widehat{x}\|_p^p + \int_Z m|\widehat{x}|^p dz = 0, \\ \Rightarrow\ & \lambda_1(m)\|\widehat{x}\|_p^p \leq 0 \quad \text{(see (5.244))}, \\ \Rightarrow\ & \widehat{x} = 0 \quad \text{(because by hypothesis } \lambda_1(m) > 0). \end{aligned}$$

So when $g = 0$, the only possible solution is the trivial one. □

THEOREM 5.5.9 *If $\lambda_1(m) > 0$, $x_k \in W^{1,p}(Z) \cap L^\infty(Z)$ satisfy*

$$V_p(x_k) \in L^\infty(Z),\ x_k\big|_{\partial Z} \in C^2(\partial Z) \quad k = 1, 2,$$

also we have

$$\left\{ \begin{array}{l} V_p(x_1) \leq V_p(x_2) \ \text{ in } Z \\ x_1\big|_{\partial Z} \leq x_2\big|_{\partial Z} \end{array} \right\} \tag{5.251}$$

and finally $V_p(x_2) \geq 0$ in Z with $x_2\big|_{\partial Z} \geq 0$, then $x_1(z) \leq x_2(z)$ for every $z \in Z$. Moreover, if in (5.251) $x_1\big|_{\partial Z} = x_2\big|_{\partial Z} = 0$, then the same conclusion holds under the following less restrictive assumptions,

$$x_k \in W_0^{1,p}(Z),\ V_p(x_k) \in L^\infty(Z) \quad \text{for } k = 1, 2 \text{ and } V_p(x_2) \geq 0 \text{ in } Z.$$

PROOF: If $x_2 = 0$, then clearly then result is true, with $x_1 = 0$.

So suppose that $x_2 \neq 0$. From Theorem 5.5.7 we have that $x_k \in C^1(\overline{Z})$, $k = 1, 2$. Then since $x_2 \geq 0$ (recall $\lambda_1(m) > 0$), from the strong maximum principle, we have that $x_2 \in \operatorname{int} C_0^1(\overline{Z})_+$. Therefore we can find $c > 1$ such that $x_1 < cx_2$. Let $g = V_p(x_2)$, $h = x_2\big|_{\partial Z}$ and consider the following problem

$$\left\{ \begin{array}{l} -\Delta_p y + m|y|^{p-2}y = g \ \text{ in } Z, \\ y\big|_{\partial Z} = h \end{array} \right\}. \tag{5.252}$$

We can easily check that

$$\underline{y} = x_1 \quad \text{and} \quad \overline{y} = cx_2$$

are lower and upper solutions, respectively, of (5.252). Thus employing the truncation and penalization techniques developed in Section 5.2, we can have a solution $y \in C^1(\overline{Z})$ (see Theorem 5.5.7) such that

$$x_1(z) \le y(z) \le cx_2(z) \qquad \text{for all } z \in \overline{Z}.$$

Because $\lambda_1(m) > 0$, $y \ge 0$. But from the proof of Theorem 5.5.6 we know that (5.252) has a unique solution. So $y = x_2$ and we have $x_1 \le x_2$.

Finally if we have homogeneous Dirichlet boundary conditions, then by Theorem 4.3.35, it suffices to assume that $V_p(x_k) \in L^\infty(Z)$, $x_k \in W_0^{1,p}(Z)$, $k = 1, 2$, and $V_p(x_2) \ge 0$. □

The classical maximum principle (see Theorem 4.3.19), asserts that any superharmonic C^2-function x on a smooth domain $Z \subseteq \mathbb{R}^N$ can not achieve its minimum in the interior of Z, unless x is a constant. We saw that this result is still true if $x \in W^{1,p}_{\text{loc}}(Z)$ is a p-superharmonic function such that $\operatorname{div}(\|Dx\|^{p-2}Dx) \in L^2_{\text{loc}}(Z)$ (see Theorem 4.3.37). From the classical maximum principle one immediately derives the strong comparison principle between two C^2-functions x_k $(k = 1, 2)$ satisfying $-\triangle x_1 \le -\triangle x_2$. Next we prove corresponding results for the p-Laplacian differential operator. Recall that $Z \subseteq \mathbb{R}^N$ is a bounded domain with a C^2-boundary ∂Z.

PROPOSITION 5.5.10 *If $x, y \in W^{1,p}_{\text{loc}}(Z)$ $(1 < p < \infty)$, $f, g \in L^\infty(Z)$,*

$$\begin{aligned} & -\operatorname{div}\big(\|Dx(z)\|^{p-2}Dx(z)\big) = f(z) \\ \textit{and} \quad & -\operatorname{div}\big(\|Dy(z)\|^{p-2}Dy(z)\big) = g(z) \quad \textit{a.e. on } Z \\ & x(z) \ge y(z) \qquad \textit{for all } z \in Z,\ f(z) \ge g(z) \qquad \textit{a.e. on } Z \\ \textit{and} \quad & K = \{x \in Z : x(z) = y(z)\} \end{aligned}$$

is compact, then $K = \emptyset$.

PROOF: From nonlinear regularity theory (see Theorems 4.3.34 and 4.3.35), we know that $x, y \in C^{1,\alpha}(\overline{Z})$. Assume that K is nonempty compact. We can find a relatively compact open set Z_1 such that $K \subseteq Z_1 \subseteq \overline{Z}_1 \subseteq Z$ and

$$y(z) < x(z) \qquad \text{for all } z \in Z \setminus K.$$

Given $\varepsilon > 0$, let $x_\varepsilon, y_\varepsilon \in W^{1,p}(Z_1)$ be solutions of the equations

$$\left\{ \begin{array}{l} -\operatorname{div}\big((\varepsilon + \|Dx_\varepsilon(z)\|^2)^{(p-2)/2}Dx_\varepsilon(z)\big) = f(z) \ \text{ a.e. on } Z_1, \\ x_\varepsilon\big|_{\partial Z_1} = x \end{array} \right\} \tag{5.253}$$

$$\text{and} \quad \left\{ \begin{array}{l} -\operatorname{div}\big((\varepsilon + \|Dy_\varepsilon(z)\|^2)^{(p-2/2}Dy_\varepsilon(z)\big) = g(z) \ \text{ a.e. on } Z_1, \\ y_\varepsilon\big|_{\partial Z_1} = y \end{array} \right\}. \tag{5.254}$$

Such solutions can be easily obtained by direct minimization of the Euler functionals corresponding to problems (5.253) and (5.254). Moreover, $\{Dx_\varepsilon\}_{\varepsilon \in (0,1]}$ and $\{Dy_\varepsilon\}_{\varepsilon \in (0,1]} \subseteq L^p(Z_1)$ are both bounded. Again we have $x_\varepsilon, y_\varepsilon \in C^{1,\alpha}(\overline{Z}_1)$ (we can always assume that $\partial \overline{Z}_1$ is C^2) and

$$\lim_{\varepsilon \downarrow 0} x_\varepsilon = x \quad \text{and} \quad \lim_{\varepsilon \downarrow 0} y_\varepsilon = y$$

weakly in $W^{1,p}(Z_1)$ and strongly in $C^{1,\beta}_{\text{loc}}(Z_1)$ for any $\beta \in (0, \alpha)$. We choose Z_2 an open subset of Z such that $K \subseteq Z_2 \subseteq \overline{Z}_2 \subseteq Z_1$ and set

$$\xi = \min_{\partial Z_2}(x-y) > 0 \quad \text{and} \quad u_\varepsilon = x_\varepsilon - y_\varepsilon.$$

We choose $\varepsilon > 0$ small so that

$$\|x - x_\varepsilon\|_{L^\infty(Z_2)} < \frac{\xi}{4} \quad \text{and} \quad \|y - y_\varepsilon\|_{L^\infty(Z_2)} < \frac{\xi}{4}.$$

So it follows that

$$x_\varepsilon(z) - y_\varepsilon(z) > \frac{\xi}{2} \quad \text{for all } z \in \partial Z_2 \quad \text{and} \quad x_\varepsilon(z) - y_\varepsilon(z) < \frac{\xi}{2} \qquad \text{for all } z \in K.$$

Invoking the mean value theorem, we have

$$-\sum_{i,j=1}^N \frac{\partial}{\partial z_i}\Big(\alpha_{ij}^\varepsilon \frac{\partial u_\varepsilon}{\partial z_j}\Big) = f - g \geq 0 \quad \text{in } Z_2,$$

where

$$\alpha_{ij}^\varepsilon = \big(\varepsilon + \|t_i Dx_\varepsilon + (1-t_i)Dy_\varepsilon\|^2\big)^{(p-4)/2}\Big[\delta_{ij}\big(\varepsilon + \|t_i Dx_\varepsilon + (1-t_i)Dy_\varepsilon\|^2\big) + (p-2)\Big(t_i\frac{\partial x_\varepsilon}{\partial z_i} + (1-t_i)\frac{\partial y_\varepsilon}{\partial z_i}\Big)\Big(t_i\frac{\partial x_\varepsilon}{\partial z_j} + (1-t_i)\frac{\partial y_\varepsilon}{\partial z_j}\Big)\Big]$$

with $t_i \in (0,1)$. Set $\eta_\varepsilon = \min_{Z_2}(x_\varepsilon - y_\varepsilon)$ and $K_{\eta_\varepsilon} = \big\{z \in Z_2 : (x_\varepsilon - y_\varepsilon)(z) = \eta_\varepsilon\big\}$. Evidently K_{η_ε} is a nonempty compact subset of Z_2 and $Dx_\varepsilon(z) = Dy_\varepsilon(z)$ for all $z \in K_{\eta_\varepsilon}$. So we have

$$\alpha_{ij}^\varepsilon = \big(\varepsilon + \|Dx_\varepsilon\|^2\big)^{(p-4)/2}\Big[\delta_{ij}\big(\varepsilon + \|Dx_\varepsilon\|^2\big) + (p-2)\frac{\partial x_\varepsilon}{\partial z_i}\frac{\partial x_\varepsilon}{\partial z_j}\Big] \quad \text{on } K_{\eta_\varepsilon}. \tag{5.255}$$

Moreover, because α_{ij}^ε are the entries of the Hessian matrix of the strongly convex function $v = (v_k)_{k=1}^N \longrightarrow (1/p)\big(\varepsilon + \|v\|^2\big)^{p/2}$, we have

$$\sum_{i,j=1}^N \alpha_{ij}^\varepsilon \vartheta_i \vartheta_j \geq \gamma \|\vartheta\|^2 \tag{5.256}$$

for some $\gamma > 0$ and all $\vartheta = (\vartheta_k)_{k=1}^N \in \mathbb{R}^N$. We choose an open neighborhood Z_{η_ε} of K_{η_ε} such that $\overline{Z}_{\eta_\varepsilon} \subseteq Z_2$, $x_\varepsilon - y_\varepsilon > \eta_\varepsilon$ and (5.256) still holds with γ replaced by $\frac{1}{2}\gamma$. Note that such a choice is possible because $Dx_\varepsilon, Dy_\varepsilon \in C(\overline{Z}_2, \mathbb{R}^N)$. Then by the strong maximum principle (see Theorem 4.3.19), it follows that $u_\varepsilon = x_\varepsilon - y_\varepsilon$ is constant on $\overline{Z}_{\eta_\varepsilon}$, a contradiction. This proves that $K \neq \emptyset$. □

This proposition leads to the following strong comparison principle.

THEOREM 5.5.11 *If $x, y \in C_0^1(\overline{Z})$, $x \neq 0$, $f, g \in L^\infty(Z)$, $f \geq 0$,*

$$-\text{div}\big(\|Dx(z)\|^{p-2}Dx(z)\big) = f(z) \qquad \text{a.e. on } Z$$

and

$$-\text{div}\big(\|Dy(z)\|^{p-2}Dy(z)\big) = g(z) \qquad \text{a.e. on } Z$$

$$f(z) \geq g(z) \qquad \text{a.e. on } Z \text{ and the set}$$

$$C = \big\{z \in Z : f(z) = g(z)\big\}$$

has an empty interior, then $x(z) > y(z)$ for all $z \in Z$ and $(\partial x/\partial n)(z) < (\partial y/\partial n)(z)$ for all $z \in \partial Z$; that is, we have that $x - y \in \text{int}\, C_0^1(\overline{Z})_+$.

PROOF: From the nonlinear strong maximum principle (see Theorem 4.3.37), we have

$$x(z)>0 \qquad \text{for all } z\in Z \quad \text{and} \quad \frac{\partial x}{\partial n}(z)<0 \qquad \text{for all } z\in \partial Z.$$

Acting with the test function $-(x-y)^-$, we can see that

$$x(z)\geq y(z) \qquad \text{for all } z\in \overline{Z}.$$

We set $K=\{z\in Z : x(z)=y(z)\}$ (the coincidence set in Z). From Proposition 5.5.10 we know that K cannot be compact unless K is empty. Suppose K is nonempty. Then we can find $\{z_n\}_{n\geq 1}\subseteq K$ and $\widehat{z}\in\partial Z$ such that $z_n \longrightarrow \widehat{z}$. Since $Dx, Dy\in C(\overline{Z},\mathbb{R}^N)$ (see Theorems 4.3.34 and 4.3.35) and $x,y\in C_0^1(\overline{Z})$, we have

$$\frac{\partial x}{\partial n}(\widehat{z}) = \frac{\partial y}{\partial n}(\widehat{z}) = -\beta^2<0. \tag{5.257}$$

Set $u=x-y$. Then $u\in C_0^1(\overline{Z})$ and it satisfies

$$-\sum_{i,j=1}^{N}\frac{\partial}{\partial z_j}\Big(\alpha_{ij}\frac{\partial u}{\partial x_i}\Big) = f-g\geq 0$$

with $\alpha_{ij}=\alpha_{ij}^0$ (see (5.255) with $\varepsilon=0$). In particular, we have

$$\alpha_{ij}(\widehat{z}) = \Big|\frac{\partial x}{\partial n}(\widehat{z})\Big|^{p-4}\Big[\delta_{ij}\Big|\frac{\partial x}{\partial n}(\widehat{z})\Big|^2+(p-2)\frac{\partial x}{\partial z_i}(\widehat{z})\frac{\partial x}{\partial z_j}(\widehat{z})\Big].$$

Exploiting the continuity of the gradient vectors Dx, Dy, we can find a ball $B\subseteq Z$ such that $\widehat{z}\in\partial B$ and the elliptic operator defined by the α_{ij} is strictly elliptic on B. It follows that either $u=0$ in B, which is impossible because by hypothesis int $C\neq\emptyset$ for $u>0$ on B and $(\partial u/\partial n)(\widehat{z})<0$, which contradicts (5.257). Therefore we infer that

$$x(z)>y(z) \qquad \text{for all } z\in Z. \tag{5.258}$$

Moreover, from Lemma 4.3.18 we have

$$\frac{\partial x}{\partial n}(z)<\frac{\partial y}{\partial n} \qquad \text{for all } z\in\partial Z. \tag{5.259}$$

Hence from (5.258) and (5.259), we conclude that

$$x-y\in \text{int}\, C_0^1(\overline{Z})_+.$$

□

5.6 Periodic Solutions for Hamiltonian Systems

The Hamiltonian equations in mechanics (classical and celestial), can be written in the compact form

$$-Jx'(t)=\nabla H\big(t,x(t)\big) \qquad \text{a.e. on } T=[0,b], \tag{5.260}$$

where

$$J=\begin{pmatrix}0_N & -I_N\\ I_N & 0_N\end{pmatrix}$$

is the standard $2N\times 2N$ simplectic matrix and $\nabla H(t,x)$ is the gradient of the Hamiltonian function $x\longrightarrow H(t,x)$ defined on $\mathbb{R}^{2N}$ which is assumed to be C^1. Recall that the simplectic matrix has the following properties,

$$J^2=-I_{2N} \quad \text{and} \quad (Jx,y)_{\mathbb{R}^{2N}}=-(x,Jy)_{\mathbb{R}^{2N}} \qquad \text{for all } x,y\in\mathbb{R}^{2N}. \tag{5.261}$$

A basic problem in mechanics is to prove the existence of periodic trajectories for system (5.260). So first we establish the existence of periodic solutions for problem (5.260), when the Hamiltonian function is superquadratic, but does not necessarily satisfy the Ambrosetti–Rabinowitz condition (AR-condition, for short). Recall that the AR-condition on $H(t,\cdot)$, says the following.

(AR): there exist $\mu>0$ and $R>0$ such for almost all $t\in T$ and all $\|x\|\geq R$

$$0<\mu H(t,x)\leq \big(\nabla H(t,x),x\big)_{\mathbb{R}^{2N}}. \tag{5.262}$$

Condition (AR) implies that the growth of $x\longrightarrow H(t,x)$ is superquadratic. Here instead, we make the following hypotheses concerning the Hamiltonian function $H(t,x)$.

$\mathbf{H}_1$: $H\colon T\times\mathbb{R}^{2N}\longrightarrow\mathbb{R}$ is a function such that

(i) For all $x\in\mathbb{R}^{2N}$, $t\longrightarrow H(t,x)$ is measurable.

(ii) For almost all $t\in T$, $x\longrightarrow H(t,x)$ is a C^1-function.

(iii) $H(t,x)\geq 0$ for a.a. $t\in T$, all $x\in\mathbb{R}^{2N}$ and

$$\lim_{\|x\|\to\infty}\frac{H(t,x)}{\|x\|^2}=+\infty \quad \text{uniformly for a.a. } t\in T.$$

(iv) $\lim\limits_{x\to 0}\frac{H(t,x)}{\|x\|^2}=0$ uniformly for a.a. $t\in T$.

(v) There exist constants $1<r$ and $1<\vartheta<1+\frac{r-1}{r}$, $c_1,c_2>0$, and $M>0$ such that

$$c_1\|x\|^r\leq \big(\nabla H(t,x),x\big)_{\mathbb{R}^{2N}}-2H(t,x)$$
$$\text{and} \quad \|\nabla H(t,x)\|\leq c_2\|x\|^{\vartheta} \quad \text{for a.a. } t\in T \text{ and all } \|x\|\geq M.$$

EXAMPLE 5.6.1 The function $H(x)=\|x\|^2\big(\ln(1+\|x\|^p)\big)^q$ with $p,q>1$ satisfies hypotheses H_1 but not the AR-condition (see (5.262)).

Let $S^1=\mathbb{R}\big/(2\pi/b)\mathbb{Z}$. We consider the Hilbert space $V=W^{1/2,2}(S^1,\mathbb{R}^{2N})$ of all b-periodic functions

$$x(t)=\sum_{k\in\mathbb{Z}}\exp\Big(\frac{2\pi k}{b}tJ\Big)x_k\in L^2\big([0,1],\mathbb{R}^{2N}\big)$$

with Fourier coefficients $x_k\in\mathbb{R}^{2N}$ that satisfy

$$\|x\|_V^2 = \|x_0\|_{\mathbb{R}^{2N}}^2 + \frac{2\pi}{b}\Big(\sum_{k\in\mathbb{Z}} |k| \|x_k\|_{\mathbb{R}^{2N}}^2\Big).$$

The inner product on the Hilbert space V is given by

$$(x, y)_V = (x_0, y_0)_{\mathbb{R}^{2N}} + \frac{2\pi}{b}\Big(\sum_{k\in\mathbb{Z}} |k|(x_k, y_k)_{\mathbb{R}^{2N}}\Big).$$

Evidently the inner product $(\cdot,\cdot)_V$ generates the norm $\|\cdot\|_V$.

We define the operator $A\in\mathcal{L}(V)$ by

$$(Ax, y)_V = \int_0^b \big(-Jx'(t), y(t)\big)_{\mathbb{R}^N} dt \qquad \text{for all } x, y \in V.$$

Using the properties of the simplectic matrix (see (5.261)), we see that A is a bounded self-adjoint operator and $\ker A = \mathbb{R}^{2N}$. Note that $W^{1/2,2}(S^1, \mathbb{R}^{2N}) = \mathcal{D}(|A|^{1/2})$, where $|A| = \sqrt{A^2}$ (since A is self-adjoint). Also we consider the energy functional $\varphi\colon W^{1/2,2}(S^1, \mathbb{R}^{2N}) \longrightarrow \mathbb{R}$ defined by

$$\varphi(x) = \frac{1}{2}(Ax, x)_V - \int_0^b H\big(t, x(t)\big)dt \quad \text{for all } x \in W^{1/2,2}(S^1, \mathbb{R}^{2N}).$$

Clearly $\varphi \in C^1(V)$ and the critical points of φ are the solutions of

$$\left\{\begin{array}{ll} -Jx'(t) = \nabla H\big(t, x(t)\big) \text{ a.e. on } [0, b], \\ x(0) = x(b),\ x'(0) = x'(b) \end{array}\right\}. \tag{5.263}$$

We consider the orthogonal direct sum decomposition $V = V^- \oplus V_0 \oplus V^+$, where

$$\begin{aligned} V^- &= \{x \in V : x_k = 0 \quad \text{for } k \geq 0\} \\ V_0 &= \{x \in V : x_k = 0 \quad \text{for } k \neq 0\} \\ \text{and} \quad V^+ &= \{x \in V : x_k = 0 \quad \text{for } k \leq 0\}. \end{aligned}$$

So every $x \in V$ may be uniquely written as

$$x = \overline{x} + x^0 + \widehat{x}$$

with $\overline{x} \in V^-$, $x^0 \in V_0$ and $\widehat{x} \in V^+$. Note that the self-adjoint operator $A \in \mathcal{L}(V)$ has eigenspaces

$$E_k = \Big\{\exp\Big(\frac{2\pi k}{b} tJ\Big)x_k : x_k \in \mathbb{R}^{2N}\Big\}$$

with corresponding eigenvalues $2\pi k$, $k\in\mathbb{Z}$. Moreover, $\dim E_k = 2N$. Hence

$$\begin{aligned} V^- &= \text{the negative definite subspace of } A \\ V_0 &= \ker A = \mathbb{R}^{2N} \\ \text{and} \quad V^+ &= \text{the positive definite subspace of } A. \end{aligned}$$

PROPOSITION 5.6.2 *If hypotheses H_1 hold, then φ satisfies the PS-condition.*

PROOF: Let $\{x_n\}_{n\geq 1} \subseteq V$ be a sequence such that

$$|\varphi(x_n)| \leq M \quad \text{for some } M>0, \text{ all } n\geq 1 \text{ and } \varphi'(x_n) \longrightarrow 0 \text{ as } n\to\infty. \tag{5.264}$$

Because of hypotheses H_1, we have that

$$|H(t,x)| \leq c_3 + c_4\|x\|^{\vartheta+1} \quad \text{for a.a. } t\in T, \text{ all } x\in\mathbb{R}^{2N} \text{ with } c_3,c_4>0.$$

This growth condition combined with hypothesis H_1(v), imply that

$$c_1\|x\|^r - c_5 \leq \big(\nabla H(t,x),x\big)_{\mathbb{R}^{2N}} - 2H(t,x) \tag{5.265}$$

for a.a. $t\in T$, all $x\in\mathbb{R}^{2N}$, with $c_5>0$.

Then we have

$$\begin{aligned} 2\varphi(x_n) - \big(\varphi'(x_n),x_n\big)_V &= \int_0^b \big[\big(\nabla H(t,x_n),x_n\big)_{\mathbb{R}^{2N}} - 2H(t,x_n)\big]\mathrm{d}t \\ &\geq c_1\int_0^b \|x_n\|^r \mathrm{d}t - c_5 b \quad \text{(see (5.265))}. \end{aligned} \tag{5.266}$$

We claim that $\{x_n\}_{n\geq 1}\subseteq V$ is bounded. Suppose that this is not the case. We may assume that $\|x_n\|_V \longrightarrow \infty$. Then from (5.264) and (5.266), it follows that

$$\frac{1}{\|x_n\|_V}\int_0^b \|x_n\|^r \mathrm{d}t \longrightarrow 0 \quad \text{as } n\to\infty. \tag{5.267}$$

By hypothesis H_1(v), we have $1<\vartheta<1+(r-1)/r$. Let $\xi = (r-1)/\big(r(\vartheta-1)\big)$. We have

$$\xi>1 \quad \text{and} \quad \xi\vartheta - 1 = \xi - \frac{1}{r}. \tag{5.268}$$

Hypothesis H_1(v) implies that

$$\|\nabla H(t,x)\|^\xi \leq c_6\|x\|^{\vartheta\xi} + c_7 \quad \text{for a.a. } t\in T, \text{ all } x\in\mathbb{R}^{2N}, \text{ with } c_6,c_7>0. \tag{5.269}$$

Recall that $x_n = \overline{x}_n + x_n^0 + \widehat{x}_n$ with $\overline{x}_n\in\overline{V}$, $x_n^0\in V_0$ ad $\widehat{x}_n\in\widehat{V}$. Exploiting the orthogonality of the component spaces, we have

$$\begin{aligned} \big(\varphi'(x_n),\widehat{x}_n\big)_V &= (A\widehat{x}_n,\widehat{x}_n)_V - \int_0^b \big(\nabla H(t,x_n),\widehat{x}_n\big)_{\mathbb{R}^{2N}}\mathrm{d}t \\ &\geq (A\widehat{x}_n,\widehat{x}_n)_V - \int_0^b \|\nabla H(t,x_n)\|\|\widehat{x}_n\|\mathrm{d}t \\ &\geq (A\widehat{x}_n,\widehat{x}_n)_V - c_8\Big(\int_0^b \|\nabla H(t,x_n)\|^\xi \mathrm{d}t\Big)^{1/\xi}\|\widehat{x}_n\|\mathrm{d}t \end{aligned} \tag{5.270}$$

for some $c_8>0$, all $n\geq 1$.

Using (5.269), we have

$$\begin{aligned}\Big(\int_0^b \|\nabla H(t,x_n)\|^{\xi}\mathrm{d}t\Big)^{1/\xi} &\leq \int_0^b (c_6\|x_n\|^{\vartheta\xi}+c_7)\mathrm{d}t\\ &\leq c_9\Big(\int_0^b \|x_n\|^r \mathrm{d}t\Big)^{1/r}\Big(\int_0^b \|x_n\|^{(\xi\vartheta-1)\frac{r}{r-1}}\mathrm{d}t\Big)^{1-1/r}\\ &\quad + c_{10},\ c_9, c_{10}>0\\ &\leq c_{11}\Big(\int_0^b \|x_n\|^r \mathrm{d}t\Big)^{1/r}\|x_n\|_V^{(\xi\vartheta-1)} + c_{10},\ c_{11}>0.\end{aligned} \tag{5.271}$$

If we combine (5.267), (5.268), and (5.271), we see that

$$\frac{1}{\|x_n\|_V}\Big(\int_0^b \|\nabla H(t,x_n)\|^{\xi}\mathrm{d}t\Big)^{1/\xi} \longrightarrow 0 \quad \text{as } n\to\infty.$$

Then from (5.264) and (5.270), we infer that

$$\begin{aligned}\frac{(A\widehat{x}_n,\widehat{x}_n)_V}{\|x_n\|_V\|\widehat{x}_n\|_V} &\leq \frac{\|\varphi'(x_n)\|_V\|\widehat{x}_n\|_V}{\|x_n\|_V\|\widehat{x}_n\|_V} + c_8\frac{1}{\|x_n\|_V}\Big(\int_0^b \|\nabla H(t,x_n)\|^{\xi}\mathrm{d}t\Big)^{1/\xi}\\ &\longrightarrow 0 \quad \text{as } n\to\infty,\end{aligned}$$

$$\Rightarrow \quad \frac{\|\widehat{x}_n\|_V}{\|x_n\|_V} \longrightarrow 0 \text{ as } n\to\infty. \tag{5.272}$$

In a similar fashion, we show that

$$\frac{\|\overline{x}_n\|}{\|x_n\|} \longrightarrow 0 \qquad \text{as } n\to\infty. \tag{5.273}$$

From hypothesis $\mathrm{H}_1(\mathrm{v})$, we have

$$c_{12}\|x\| - c_{13} \leq \big(\nabla H(t,x),x\big)_{\mathbb{R}^{2N}} - 2H(t,x) \quad \text{for a.a. } t\in T, \text{ all } x\in\mathbb{R}^{2N},$$
$$c_{12}, c_{13}>0.$$

So it follows that

$$\begin{aligned}2\varphi(x_n) - \big(\varphi'(x_n),x_n\big)_V &= \int_0^b \big[\big(\nabla H(t,x_n),x_n\big)_{\mathbb{R}^{2N}} - 2H(t,x_n)\big]\mathrm{d}t\\ &\geq \int_0^b (c_{12}\|x_n\| - c_{13})\mathrm{d}t\\ &\geq \int_0^b (c_{12}\|x_n^0\| - c_{12}\|\widehat{x}_n\| - c_{12}\|\overline{x}_n\| - c_{13})\mathrm{d}t\\ &\geq c_{13}\|x_n^0\|_V - c_{14}(\|\widehat{x}_n\|_V + \|\overline{x}_n\|_V) - c_{15},\\ &\quad c_{13}, c_{14}, c_{15}>0,\end{aligned} \tag{5.274}$$

because $V_0=\mathbb{R}^{2N}$. Therefore from (5.272) through (5.274), we deduce that

$$\frac{\|x_n^0\|_V}{\|x_n\|_V} \longrightarrow 0 \qquad \text{as } n\to\infty.$$

So finally

$$1=\frac{\|x_n\|_V}{\|x_n\|_V}\le\frac{\|\overline{x}_n\|_V+\|x_n^0\|_V+\|\widehat{x}_n\|_V}{\|x_n\|_V}\longrightarrow 0 \qquad \text{as } n\to\infty,$$

a contradiction. This proves that $\{x_n\}_{n\geq1}\subseteq V$ is bounded. So we may assume that

$$x_n\xrightarrow{w}x \quad \text{in } X \quad \text{and} \quad x_n\longrightarrow x \quad \text{in } L^2(T,\mathbb{R}^{2N})$$

(recall that $W^{(1/2),2}(S^1,\mathbb{R}^{2N})$ is embedded compactly in $L^2(T,\mathbb{R}^{2N})$). From (5.264), we have

$$\left|\frac{1}{2}(Ax_n,x_n-x)_V-\int_0^b\big(\nabla H(t,x_n),x_n-x\big)_{\mathbb{R}^{2N}}dt\right|\longrightarrow 0 \qquad \text{as } n\to\infty.$$

Evidently

$$\int_0^b\big(\nabla H(t,x_n),x_n-x\big)_{\mathbb{R}^{2N}}dt\longrightarrow 0.$$

So we have

$$\begin{aligned}&(Ax_n,x_n-x)_V\longrightarrow 0,\\ \Rightarrow\ &\|x_n\|_V\longrightarrow\|x\|_V \quad \text{as } n\to\infty.\end{aligned}$$

Because $x_n\xrightarrow{w}x$ in H and H is a Hilbert space, we conclude that $x_n\longrightarrow x$ in V, therefore φ satisfies the PS-condition. ☐

THEOREM 5.6.3 *If hypotheses* H_1 *hold, then problem (5.263) admits a nonconstant solution* $x\in C^1(T,\mathbb{R}^{2N})$.

PROOF: Hypotheses H_1 imply that for almost all $t\in T$ and all $x\in\mathbb{R}^N$, we have

$$|H(t,x)|\le c_1+c_2\|x\|^{\vartheta+1} \quad \text{with } c_1,c_2>0. \tag{5.275}$$

Also because of hypotheses H_1(iv), given $\varepsilon>0$, we can find $\delta=\delta(\varepsilon)>0$ such that

$$H(t,x)\le\varepsilon\|x\|^2 \quad \text{for a.a. } t\in T \text{ and all } \|x\|\le\delta. \tag{5.276}$$

From (5.275) and (5.276) it follows that we can find c_ε such that

$$H(t,x)\le\varepsilon\|x\|^2+c_\varepsilon\|x\|^{\vartheta+1} \quad \text{for a.a. } t\in T \text{ and all } x\in\mathbb{R}^{2N}. \tag{5.277}$$

Therefore by choosing $\varepsilon>0$ small, for every $x\in V^+$ we have

$$\begin{aligned}\varphi(x)&=\frac{1}{2}\int_0^b(-Jx',x)_{\mathbb{R}^{2N}}dt-\int_0^b H\big(t,x(t)\big)dt\\ &\ge c_3\|x\|_V^2-c_4\|x\|_V^{\vartheta+1}\end{aligned} \tag{5.278}$$

(see (5.277) and recall that $W^{(1/2),2}(S^1,\mathbb{R}^{2N})$ is embedded compactly in $L^{\vartheta+1}(S^1,\mathbb{R}^{2N})$).

Because of (5.278), we can find $\varrho>0$ small such that

$$\varphi(x)\ge\beta>0 \quad \text{for all } x\in V^+ \text{ with } \|x\|_V=\varrho. \tag{5.279}$$

Let $W=V^-\oplus V^0$ and consider $e\in V^+$ with $\|e\|_V=1$ and $E=W\oplus\mathbb{R}e$. We introduce

$$B_E=\{x\in E:\|x\|_V=1\},\quad \overline{\mu}=\inf\big[(A\overline{x},\overline{x})_V:\overline{x}\in V^-,\ \|\overline{x}\|_V=1\big]$$
$$\text{and}\quad \gamma=\Big(\frac{1}{\overline{\mu}}\|A\|\|\mathcal{L}\Big)^{1/2}. \tag{5.280}$$

For $x\in B_E$ we write $x=\overline{x}+x^0+\widehat{x}\in E$.

(a) If $\|\overline{x}\|_V>\gamma\|x^0+\widehat{x}\|_V$, then for any $\eta>0$ we have

$$\begin{aligned}\varphi(\eta x)&=\frac{1}{2}(A\eta\overline{x},\overline{x})_V+\frac{1}{2}(A\eta\widehat{x},\widehat{x})_V-\int_0^b H(t,\eta x)dt\\&\leq-\frac{\overline{\mu}}{2}\eta^2\|\overline{x}\|_V+\frac{1}{2}\|A\|_{\mathcal{L}}\eta^2\|\widehat{x}\|_V^2\quad\text{(see (5.280) and recall that } H\geq 0)\\&\leq 0\qquad(\text{because } \|\overline{x}\|_V>\gamma\|x^0+\widehat{x}\|_V).\end{aligned}$$

(b) If $\|\overline{x}\|_V\leq\gamma\|x^0+\widehat{x}\|_V$, then we have

$$\begin{aligned}1=\|x\|_V^2=\|\overline{x}\|_V^2+\|x^0+\widehat{x}\|_V^2&\leq(\gamma^2+1)\|x^0+\widehat{x}\|_V^2\\ \Rightarrow\ 0<\frac{1}{1+\gamma^2}&\leq\|x^0+\widehat{x}\|_V^2.\end{aligned}\tag{5.281}$$

Set $\widetilde{B}_E=\big\{x\in B_E:\|\overline{x}\|_V\leq\gamma\|x^0+\widehat{x}\|_V^2\big\}$.

Claim: We can find $\varepsilon_1>0$ such that for all $x\in\widetilde{B}_E$ we have

$$\big|\{t\in T:\|x(t)\|\geq\varepsilon_1\}\big|_1\geq\varepsilon_1. \tag{5.282}$$

Here by $|\cdot|_1$ we denote the Lebesgue measure on T.

Suppose that the claim is not true. Then for any $n\geq 1$, we can find $x_n\in\widetilde{B}_E$ such that

$$\big|\{t\in T:\|x_n(t)\|\geq\frac{1}{n}\}\big|_1<\frac{1}{n}.$$

We set $x_n=\overline{x}_n+x_n^0+\widehat{x}_n\in E$. Because $\dim(V_0\oplus\mathbb{R}e)<\infty$ and $\|x_n^0+\widehat{x}_n\|_V\leq 1$, by passing to a subsequence if necessary, we may assume that

$$\begin{aligned}&x_n^0+\widehat{x}_n\longrightarrow x^0+\widehat{x}\in V_0\oplus\mathbb{R}e\quad\text{as } n\to\infty,\\ \Rightarrow\ &0<\frac{1}{1+\gamma^2}\leq\|x^0+\widehat{x}\|_V^2\quad\text{(see (5.281))}.\end{aligned}\tag{5.283}$$

Also $\|\overline{x}_n\|_V\leq 1$ for all $n\geq 1$ and so because V is a Hilbert space, by the Eberlein–Smulian theorem, we may assume that $\overline{x}_n\xrightarrow{w}\overline{x}$ in V^- as $n\to\infty$. So finally

$$\begin{aligned}&x_n\xrightarrow{w}x=\overline{x}+x^0+\widehat{x}\quad\text{in } V \text{ as } n\to\infty,\\ \Rightarrow\ &x_n\longrightarrow x\quad\text{in } L^2(S^1,\mathbb{R}^{2N}),\end{aligned}$$

(because V is embedded compactly in $L^2(S^1,\mathbb{R}^{2N})$),

$$\begin{aligned}\Rightarrow\ &\|x\|_V>0\quad\text{(see (5.283))},\\ \Rightarrow\ &\|x\|_2>0.\end{aligned}$$

Therefore we can find $\delta_2\geq\delta_1>0$ such that

$$\big|\{t\in T:\|x(t)\|\geq\delta_1\}\big|_1\geq\delta_2. \tag{5.284}$$

Indeed, if (5.284) is not true, then for all $n \geq 1$, we must have

$$\begin{aligned} & \left|\left\{t \in T : \|x(t)\| \geq \frac{1}{n}\right\}\right|_1 = 0, \\ \Rightarrow\ & \left|\left\{t \in T : \|x(t)\| < \frac{1}{n}\right\}\right|_1 = 1, \\ \Rightarrow\ & 0 < \int_0^b \|x(t)\|^2 \mathrm{d}t < \frac{1}{n^2} \longrightarrow 0 \quad \text{as } n \to \infty, \end{aligned}$$

a contradiction. So we see that (5.284) is true.

We set $\widehat{T} = \{t \in T : \|x(t)\| \geq \delta_1\}$, $T_n = \{t \in T : \|x_n(t)\| < 1/n\}$ for all $n \geq 1$ and $T_n^c = T \setminus T_n$.

Because of (5.282) and (5.284), we have

$$\left|\widehat{T} \cap T_n\right|_1 = \left|\widehat{T} \setminus (\widehat{T} \cap T_n^c)\right|_1 \geq |\widehat{T}|_1 - |\widehat{T} \cap T_n^c|_1 \geq \delta_2 - \frac{1}{n}.$$

Let $n \geq 1$ be large enough so that $\delta_2 - 1/n \geq \delta_2/2$ and $\delta_1 - 1/n \geq \delta_1/2$. We have

$$\|x_n(t) - x(t)\|^2 \geq \left(\delta_1 - \frac{1}{n}\right)^2 \geq \left(\frac{\delta_1}{2}\right)^2 \qquad \text{for all } t \in \widehat{T} \cap T_n.$$

So it follows that for $n \geq 1$ large

$$\begin{aligned} \int_0^b \|x_n(t) - x(t)\|^2 \mathrm{d}t \geq \int_{\widehat{T} \cap T_n} \|x_n(t) - x(t)\|^2 \mathrm{d}t & \geq \left(\frac{\delta_1}{2}\right)^2 |\widehat{T} \cap T_n|_1 \\ & \geq \left(\frac{\delta_1}{2}\right)^2 \left(\delta_2 - \frac{1}{n}\right) \\ & \geq \left(\frac{\delta_1}{2}\right)^3 > 0. \end{aligned} \tag{5.285}$$

But recall that $x_n \longrightarrow x$ in $L^2(S^1, \mathbb{R}^{2N})$. Comparing this with (5.285), we reach a contradiction. This proves the claim (see (5.282)).

For $x = \overline{x} + x^0 + \widehat{x} \in \widetilde{B}_E$, let $T(x) = \{t \in T : \|x(t)\| \geq \varepsilon_1\}$. Hypothesis H_1(iii) implies that for $\beta_0 = \|A\|_{\mathcal{L}}/\varepsilon_1^3 > 0$, we can find $M_0 > 0$ such that

$$H(t, x) \geq \beta_0 \|x\|^2 \quad \text{for a.a. } t \in T \text{ and all } \|x\| \geq M_0.$$

Choose $\eta_0 \geq M_0/\varepsilon_1$. Then for $\eta \geq \eta_0$, we have

$$H\big(t, \eta x(t)\big) \geq \beta_0 \|\eta x(t)\|^2 \geq \beta_0 r^2 \varepsilon_1^2 \quad \text{for a.a. } t \in T(x).$$

Hence

$$\begin{aligned} \varphi(\eta x) & = \frac{1}{2}\eta^2 (A\overline{x}, \overline{x})_V + \frac{1}{2}\eta^2 (A\widehat{x}, \widehat{x})_V - \int_0^b H\big(t, \eta x(t)\big) \mathrm{d}t \\ & \leq \frac{1}{2}\eta^2 \|A\|_{\mathcal{L}} - \int_{T(x)} H\big(t, \eta x(t)\big) \mathrm{d}t \quad (\text{because } H \geq 0) \\ & \leq \frac{1}{2}\eta^2 \|A\|_{\mathcal{L}} - \beta_0 \eta^2 \varepsilon_1^2 |T(x)|_1 \\ & \leq \frac{1}{2}\eta^2 \|A\|_{\mathcal{L}} - \beta_0 \eta^2 \varepsilon_1^3 = -\frac{1}{2}\eta^2 \|A\|_{\mathcal{L}} < 0, \\ \Rightarrow \varphi(\eta x) & \leq 0 \qquad \text{for all } x \in B_E \text{ and all } \eta \geq \eta_0. \end{aligned} \tag{5.286}$$

Set $C = \{x \in W : \|x\|_V \leq 2\eta_0\} \oplus \{\eta e : 0 \leq \eta \leq 2\eta_0\}$ and $C_0 = \partial C$. Then from (5.286) we have

$$\varphi\big|_{C_0} \leq 0. \tag{5.287}$$

This together with (5.279) and Proposition 5.6.2, permit the use of the generalized mountain pass theorem (see Theorem 4.1.26), which gives $x \in V$ such that

$$\varphi(x) \geq \beta > 0 = \varphi(0) \quad \text{and} \quad \varphi'(x) = 0.$$

It follows that $x \neq 0$ and it is a solution of (5.263). Moreover, because $H \geq 0$ we see that x is nonconstant. Finally, it is clear from (5.263) that $x \in C(T, \mathbb{R}^{2N})$. □

Using the same technique we can also have an existence theorem for Hamiltonian systems for which the function $H(t,x)$ satisfies the (AR)-condition. In fact in this case the proofs are simplified.

So our hypotheses on the Hamiltonian $H(t,x)$ are the following.

H$_2$: $H\colon T \times \mathbb{R}^{2N} \longrightarrow \mathbb{R}$ is a function such that

(i) For all $x \in \mathbb{R}^{2N}$, $t \longrightarrow H(t,x)$ is measurable.

(ii) For almost all $t \in T$, $x \longrightarrow H(t,x)$ is a C^1-function.

(iii) $H(t,x) \geq 0$ for a.a. $t \in T$, all $x \in \mathbb{R}^{2N}$, and $\lim\limits_{x \to 0} \frac{H(t,x)}{\|x\|^2} = 0$ uniformly for a.a. $t \in T$.

(iv) Condition (AR) is satisfied (see (5.262)).

REMARK 5.6.4 Integrating the (AR)-condition (hypothesis H$_2$(iv)), we see that for a.a. $t \in T$ and all $x \in \mathbb{R}^{2N}$, we have $\widehat{c}_1\|x\|^{\mu} - \widehat{c}_2 \leq H(t,x)$ with $\widehat{c}_1, \widehat{c}_1 > 0$. Hence $H(t,\cdot)$ is superquadratic.

Following the reasoning of the proof of Theorem 5.6.3, via the generalized mountain pass theorem (see Theorem 4.1.26), we can have the following theorem. The reader can fill the details.

THEOREM 5.6.5 *If hypotheses H$_2$ hold, then problem (5.263) admits a nonconstant solution $x \in C^1(T, R^{2N})$.*

Next we consider a problem of a seemingly different nature. So we consider a Hamiltonian system but instead of fixing the period, the energy level is prescribed.

So we consider the following autonomous Hamiltonian system:

$$-Jx'(t) = \nabla H\big(x(t)\big) \qquad \text{for all } t \in T. \tag{5.288}$$

Let $x \in C^1(T, \mathbb{R}^{2N})$ be a solution of (5.288). Taking inner product of (5.288) with $x'(t)$, we obtain

$$\begin{aligned}
&\big(\nabla H\big(x(t)\big), x'(t)\big)_{\mathbb{R}^{2N}} \\
\Rightarrow\ & \frac{\mathrm{d}}{\mathrm{d}t} H\big(x(t)\big) = 0 \qquad \text{for all } t \in T, \\
\Rightarrow\ & H\big(x(t)\big) = \text{constant} \qquad \text{for all } t \in T.
\end{aligned}$$

This means that the energy is conserved. It is therefore natural to seek solutions of (5.288) and in particular periodic solutions with a prescribed energy level. The difficulty in dealing with this problem is that the period, and consequently the underlying solution space, is not a priori known. Nevertheless, with some assumptions on the Hamiltonian $H(x)$, we are able to reduce the fixed energy case to the fixed period case.

We start with a result, which says that under some conditions on the gradient map ∇H, the trajectories of (5.288) on the energy manifold S, are actually independent of the Hamiltonian H and depend only on the manifold S.

PROPOSITION 5.6.6 *If $H_k \in C^1(\mathbb{R}^{2N})$ and $c_k \in \mathbb{R}$, $k=1,2$, are such that*

$$\begin{aligned} & S = H_k^{-1}(c_k), \quad k=1,2 \\ \textit{and} \quad & \nabla H_k(x) \neq 0 \qquad \textit{for all } x \in S,\ k=1,2, \end{aligned}$$

then the trajectories of the Hamiltonian systems

$$-Jx'(t) = \nabla H_k\big(x(t)\big) \qquad \textit{for all } t \in T, \quad k=1,2,$$

are the same.

PROOF: Let $x_1 \in C^1(T,\mathbb{R}^{2N})$ be a solution of the Hamiltonian system corresponding to H_1. Note that the vectors $\nabla H_1\big(x_1(t)\big), \nabla H_2\big(x_1(t)\big)$, $t \in T$, are normal to S. So we can find $\mu : T \longrightarrow \mathbb{R}$ such that

$$\nabla H_2\big(x_1(t)\big) = \mu(t)\nabla H_1\big(x_1(t)\big) \qquad \text{for all } t \in T. \tag{5.289}$$

By hypothesis $\nabla H_1\big(x_1(t)\big), \nabla H_2\big(x_1(t)\big) \neq 0$ for all $t \in T$ and so $\mu(t) \neq 0$ for all $t \in T$. Moreover, because $t \longrightarrow \nabla H_k\big(x_1(t)\big)$ is continuous for $k=1,2$, it follows that $t \longrightarrow \mu(t)$ is continuous. Therefore $t \longrightarrow \mu(t)$ has constant sign on T. Let

$$\vartheta(t) = \int_0^t \frac{1}{\mu(s)} \mathrm{d}s.$$

Then ϑ is continuous and strictly monotone. We set $x_2 = x_1 \circ \vartheta^{-1}$ and we have

$$\begin{aligned} Jx_2'(t) &= J(x_1 \circ \vartheta^{-1})'(t) \\ &= J(x_1 \circ \vartheta^{-1})(t)\Big(\frac{1}{\vartheta'} \circ \vartheta^{-1}\Big)(t) \quad \text{(by the chain rule)} \\ &= -\nabla H_1\big(x_1\big(\vartheta^{-1}(t)\big)\big)\big(\mu\big(\vartheta^{-1}(t)\big)\big) \\ &= -\nabla H_2\big(x_1\big(\vartheta^{-1}(t)\big)\big) \qquad \text{for all } t \in \vartheta(T) \quad \text{(see 5.289))} \\ &= -\nabla H_2\big(x_2(t)\big) \qquad \text{for all } t \in \vartheta(T), \end{aligned}$$
$$\Rightarrow x_2 \in C^1\big(\vartheta(T),\mathbb{R}^{2N}\big) \quad \text{is a solution of} \quad -Jx'(t) = -\nabla H_2\big(x(t)\big),$$

for all $t \in \vartheta(T)$. □

REMARK 5.6.7 Evidently if $x_1 \in C^1(T,\mathbb{R}^{2N})$ is periodic, then so is $x_2 \in C^1(T,\mathbb{R}^{2N})$.

Now we use Proposition 5.6.6 and some convexity properties of H, to replace H by another Hamiltonian $\overline{H}$ that satisfies hypotheses H_1 and so Theorem 5.6.3 can be used to produce a periodic trajectory.

The hypotheses on the time-invariant (autonomous) Hamiltonian function $H(x)$, are the following.

H$_3$: $H \in C^1(\mathbb{R}^{2N})$, the energy surface $S = H^{-1}$ is the boundary of a relatively compact, star-shaped neighborhood of the origin and $\nabla H(x) \neq 0$ for all $x \in S$.

THEOREM 5.6.8 *If hypotheses H_3 hold, then system (5.288) admits a periodic trajectory.*

PROOF: Because $S = \partial U$ with U a relatively compact, star-shaped neighborhood of the origin in $\mathbb{R}^{2N}$, for every $x \in \mathbb{R}^{2N}$, we can find a unique $\xi(x) \in S$ and $\mu(x)>0$ such that

$$x = \mu(x)\xi(x)$$

(in fact $\mu(x) = x_n/\|\xi(x)\|$). Evidently μ is 1-homogeneous and C^1 on $\mathbb{R}^{2N} \setminus \{0\}$. We define

$$\overline{H}(x) = \begin{cases} 0 & \text{if } x = 0 \\ \mu(x)^4 & \text{if } x \neq 0 \end{cases}.$$

Then $\overline{H} \in C^1(\mathbb{R}^{2N}), \overline{H}^{-1}(1) = S, \overline{H} \geq 0, \lim\limits_{x\to 0}\big(H(x)\big)/\|x\|^2 = 0$ and since $\overline{H}$ is 4-homogeneous we have that $\big(\nabla\overline{H}(x), x\big)_{\mathbb{R}^{2N}} = 4\overline{H}(x)$ for all $x \in \mathbb{R}^{2N}$. Therefore in this case $\overline{H}$ satisfies the global (AR)-condition and so hypotheses H_3 are satisfied and we can apply Theorem 5.6.5. So for $b = 2\pi$ (for example), we can find a nonconstant 2π-periodic solution $u \in C^1(T, \mathbb{R}^{2N})$, $T = [0, 2\pi]$. We need not have $\overline{H}\big(u(t)\big) = 1$ for all $t \in T$. Nevertheless, we have that $\overline{H}\big(u(t)\big) = \lambda \neq 0$ (because $u \neq 0$). Set $\nu = \frac{1}{\lambda^{1/4}}$. Then $\overline{H}\big(\nu u(t)\big) = 1$ for all $t \in T$. Moreover,

$$-J\big(\nu u'(t)\big) = \nu\nabla\overline{H}\big(u(t)\big) = \frac{1}{\nu^2}\nabla\overline{H}\big(\nu u(t)\big) \qquad \text{for all } t \in T. \tag{5.290}$$

Set $\tau = \nu^{-2}t$ and $x(\tau) = \nu u(t)$. Then $x \in C^1\big(\frac{1}{\nu^2}T, \mathbb{R}^{2N}\big)$ is a $\frac{2\pi}{\nu^2}$-periodic map which by virtue of (5.290) satisfies

$$-Jx'(\tau) = \frac{1}{\nu^2}\nabla\overline{H}\big(\nu u(\nu^2\tau)\big)\frac{dt}{d\tau}(\tau) = \nabla\overline{H}\big(x(\tau)\big).$$

□

5.7 Remarks

5.1: Hypotheses $H(f)_1$ are such that when $p = 2$ (semilinear problem), they incorporate in our framework the so-called asymptotically linear problems at 0^+ and at $+\infty$. Since the appearance of the pioneering work of Amann–Zehnder [17], these problems have attracted a lot of interest and several papers dealing with them have appeared. Indicatively we mention the works of Stuart–Zhou [565], Tehrani [579], and Zhou [627] (for $p = 2$), and Costa–Magalhaes [159], Li–Zhou [379], Fan–Zhao–Huang [237], Huang–Zhou [318], and Hu–Papageorgiou [317] (for $p > 1$). In the last work the potential function $F(z,x)$ is nonsmooth, locally Lipschitz in $x \in \mathbb{R}$, and so the right-hand side nonlinearity in problem (5.1) is multivalued, namely the generalized subdifferential of the function $x \longrightarrow F(z,x)$. From the aforementioned nonlinear works with the exception of Huang–Zhou [318] and Hu–Papageorgiou [317], the rest deal with the situation considered here in hypotheses $H(f)_1$, namely they assume that the slope $\big(f(z,x)\big)/(x^{p-1})$ stays below $\lambda_1 > 0$ asymptotically as $x \longrightarrow 0^+$ and stays above $\lambda_1 > 0$ asymptotically as $\lambda \to +\infty$. However, their hypotheses are more restrictive than the ones used here. Huang–Zhou [318] and Hu–Papageorgiou [317] deal with the opposite situation.

Compared with the Dirichlet problem, the study of nonlinear Neumann problems involving the p-Laplacian differential operator is lagging behind. We mention the works of Binding–Drabek–Huang [76], Faraci [238], Filippakis–Gasiński–Papageorgiou [246], Motreanu–Papageorgiou [440], and Papalini [474].

When $A = 0$, problem (5.30) has been studied extensively and many solvability conditions are given, such as the coercivity condition (see Berger–Schechter [69]), the convexity condition (see Mawhin [414]), the subadditivity condition (see Tang [576]), and the sublinear condition (see Tang [577]). The case where $A(t) = k^2\omega^2 I_N$, with $k \in N$, $\omega = 2\pi/b$, and I_N the $N{\times}N$ identity matrix, was considered by Mawhin–Willem [415], under the assumption that the potential function $x \longrightarrow F(t,x)$ is a C^1 convex function (hence $x \longrightarrow \nabla F(t,x)$ is a continuous, monotone, hence maximal monotone operator from $\mathbb{R}^N$ into itself). The approach of Mawhin–Willem [415, p. 61], is based on a variant of the dual action principle. Also in the book of Mawhin–Willem [415, p. 88], we find the general problem, with the right-hand side nonlinearity $F(t,x)$ satisfying

$$|F(t,x)| \leq h(t) \quad \text{and} \quad \|\nabla F(t,x)\| \leq h(t)$$

for almost all $t \in T$, all $x \in \mathbb{R}^N$, and with $h \in L^1(T)_+$. The potential function $x \longrightarrow F(t,x)$ is not assumed to be convex. The work of Mawhin–Willem [415], was extended recently by Tang–Wu [578], who considered systems with a general Carathéodory, strictly sublinear nonlinearity. The opposite situation, of a strictly superlinear nonlinearity, was studied by Motreanu–Motreanu–Papageorgiou [438]. We should mention that by virtue of hypothesis $H(F)$(iv), problem (5.31) is classified as a strongly resonant (at the zero eigenvalue) problem. Such problems exhibit a certain partial lack of compactness, which is evident in Proposition 5.1.16. Finally we should mention that uses of the second deformation theorem (see Theorem 4.6.1) to obtain multiplicity results for boundary value problems, can be found in Papageorgiou–Papageorgiou [484] and Kyritsi–Motreanu–Papageorgiou [369].

Multiplicity results for resonant nonlinear elliptic problems with the p-Laplacian can be found in Costa–Magalhaes [159], Li–Zhou [379], and Liu–Su [390].

Finally Theorem 5.1.27, was first proved for $p = 2$ by Brezis–Nirenberg [106].

5.2: The method of upper–lower solutions, provides an effective tool to produce existence theorems for first- and second-order initial and boundary value problems and to generate monotone iterative techniques which provide constructive methods (amenable to numerical treatment) to obtain these solutions.

The question of existence of multiple solutions for problem (5.1) under condition of superlinear behavior of the nonlinearity $f(z,x)$ both at zero and $\pm\infty$, has been studied in the past only in the context of semilinear problems (i.e., $p = 2$). First Ambrosetti–Mancini [22] proved that if $\lambda > \lambda_1$, then the problem has two nontrivial solutions of constant sign (one positive and one negative). Soon thereafter Struwe [563] improved their result by showing that if $\lambda > \lambda_2$, then problem (5.1) (with $p=2$) has at least three nontrivial solutions. Ambrosetti–Lupo [23] slightly improved the result of Struwe and they also presented an approach based on Morse theory. This required that $f(z,\cdot) \in C^1(\mathbb{R})$. The most general result for the semilinear problem ($p = 2$), can be found in Struwe [564, p. 132], who succeeded in eliminating the differentiability condition on the nonlinearity f and simplified the argument of Ambrosetti–Lupo. Still though Struwe [564] requires that f is Lipschitz continuous in $x \in \mathbb{R}$. So even when $p = 2$, Theorem 5.2.10 is more general than the result of Struwe [564]. To our knowledge no other such multiplicity result for problem (5.1) with $p>1$, can be found in the literature.

Problem (5.33) has the interesting feature that it presents a compact formulation for the Dirichlet, Neumann, and Sturm–Liouville problems. Moreover, the method of proof for problem (5.33), can be repeated with only very minor changes in the case of the periodic problem (see problem (5.145)). Even more general nonlinear, multivalued boundary conditions can be found in the works of Halidias–Papageorgiou [282] and Gasiński–Papageorgiou [258]. The method of upper and lower solutions was used to study such problems by Bader–Papageorgiou [50] and Douka–Papageorgiou [204]. Further uses of the method to initial and boundary value problems can be found in the book of Heikkila–Lakshmikantham [287].

5.3: Degree theory has always been a powerful tool for the study of boundary value problems, especially for ordinary differential equations. The monograph of Mawhin [412] contains many such examples. The relatively recent generalizations of degree theory to nonlinear operators of monotone type (see Section 3.3) paved the way for the use of degree-theoretic methods to boundary value problems of nonlinear partial differential equations and to problems with unilateral constraints (such as variational and hemivariational inequalities). Our approach here in dealing with problem (5.146) is close to Amann [16] (where $p=2$) and Ambrosetti–Garcia Azorero–Peral Alonso [24]. However, instead of the Leray–Schauder degree, here we employ the degree for operators of type $(S)_+$ (see Section 5.3). This permits us to assume a nonsymmetric structure for our problem, in contrast to Ambrosetti–Garcia Azorero–Peral Alonso [24] (see hypothesis $\mathrm{H}(f)_1$). Other uses of degree-theoretic methods in the study of nonlinear elliptic problems with unilateral constraints can be found in Fillipakis–Papageorgiou [245] and Aizicovici–Papageorgiou–Staicu [6].

For the periodic problem (5.180), hypotheses $\mathrm{H}(f)_2$ assume a nonuniform nonresonance condition between two successive eigenvalues of the negative scalar p-Laplacian with periodic boundary conditions. Analogous situations for $p = 2$ (semilinear problems) were considered by Fonda–Mawhin [251], and Habets–Metzen [280], and for $p>1$ (nonlinear problems) and Dirichlet boundary value conditions by Zhang [623]. In hypotheses $\mathrm{H}(f)_3$ the asymptotic at 0 and $\pm\infty$ nonuniform nonresonance conditions (see hypotheses $\mathrm{H}(f)_3$(iv) and (v)), are replaced by Landesman–Lazer type conditions (see hypothesis $\mathrm{H}(f)_2$(iv)). In this direction results for $p=2$ were obtained by Cesari–Kannan [141], Iannacci–Nkashama [322]; also see the references therein. The results on problem (5.180) here are a particular case of those by Aizicovici–Papageorgiou–Staicu [5].

5.4: Problem (5.198) was studied for $p=2$ (semilinear case) by Rabinowitz [508]. Problem (5.198) with $p>1$ and $m=1$, was examined by Fan–Li [236]. They also used the Ambrosetti–Rabinowitz condition (see hypothesis $\mathrm{H}(f)_1$(iv)) and in addition they assumed that near the origin the right-hand side nonlinearity of the problem grows like $|x|^{r-1}$ with $r > p$. We should also mention the work of Garcia Melian–Sabina de Lis [257], who obtained bifurcation results but under radial symmetry and with a C^2-right-hand side nonlinearity.

Proposition 5.4.8 is due to Pohozaev [497]. In problem (5.231) the term causing problems is the last term $|x|^{2^*-2}x$. Recall that in the Sobolev embedding theorem the case $r = 2^*$ is the limiting case for the inclusion $H_0^1(Z) \subseteq L^r(Z)$ and it is not compact. Thus for $s = (N+2)/(N-2) - 2^* - 1$ the map $x \longrightarrow x^s$ is not compact from $H_0^1(Z)$ into $H^{-1}(Z)$. For this reason compactness arguments do not work for problem (5.231), the linear term λu may be replaced by other compact perturbations (see Brezis–Nirenberg [104]). Additional results for problems with critical nonlinearities can be found in the works of Ambrosetti–Struwe [21], Cerami–Fortunato–Struwe [137], Guedda–Veron [275], Miyagaki [437], and Willem [606].

5.5: Maximum principle–type results for the p-Laplacian can be found in Alegretto–Huang [12] and Garcia Melian–Sabina de Lis [257] who proved Theorem 5.1.8. Analogous results for some general nonlinear differential operators, can be found in Damascelli [172]. Also, we should mention the so-called *antimaximum principle.* So consider the following problem.

$$\left\{ \begin{array}{l} -\mathrm{div}\big(\|Dx(z)\|^{p-2}Dx(z)\big) = \lambda|x(z)|^{p-2}x(z) + h(z) \ \text{ a.e. on } Z, \\ Bx = 0 \end{array} \right\}. \tag{5.291}$$

Here B stands for the Dirichlet or Neumann boundary conditions, $1<p<\infty$ and $h \in L^\infty(Z)$.

THEOREM 5.7.1 *If $h \in L^\infty(Z)$, $h(z) \geq 0$ for almost all $z \in Z$, $h \neq 0$, then (5.291) there exists $\delta = \delta(h)$ such that for all $\lambda \in (\lambda_1, \lambda_1 + \delta)$ any solution $x \in C^1(\overline{Z})$ of (5.291) satisfies $x(z) < 0$ for all $z \in Z$ and $(\partial x/\partial n)(z) > 0$ for all $z \in \partial Z$.*

REMARK 5.7.2 It can be shown that $\delta > 0$ can be chosen independently of h for the scalar (i.e., $N = 1$) Neumann problem (uniform antimaximum principle).

For $p = 2$, the antimaximum principle was investigated by Clement–Peletier [155] and Godoy–Gossez–Paczka [267]. For $p \neq 2$ we refer to Godoy–Gossez–Paczka [268].

For comparison results involving the p-Laplacian, we refer to Garcia Melian–Sabina de Lis [257] (they obtained Theorem 5.5.9) and Guedda–Veron [275] (who proved Theorem 5.5.11). Further results in this direction can be found in Alegretto–Huang [12] and Damascelli [172].

5.6: Periodic solutions with prescribed minimal period were obtained by Clarke [150] (who introduced the dual action technique), Clarke–Ekeland [151], Girardi–Matzeu [265], Mawhin–Willem [413] and Rabinowitz [506] (who employed the (AR)-condition, see (5.262)). Periodic solutions on a given energy surface were first proved by Rabinowitz [506] and Weinstein [603]. Additional existence and multiplicity results in this direction can be found in Berestycki–Lasry–Mancini–Ruf [66], Ekeland–Hofer [225], and Rabinowitz [507]. Detailed studies of Hamiltonian systems can be found in the books of Chang [143], Ekeland [227], Mawhin–Willem [415], and Struwe [564].

6

Multivalued Analysis

Summary. *Multivalued analysis deals with the study of maps whose values are sets. Multivalued analysis is closely related to nonsmooth analysis and a symbiotic relationship exists between them which feeds both with new ideas and directions to grow. This chapter presents in detail the main aspects of nonsmooth analysis. We study the continuity and measurability properties of multifunctions (set-valued functions) and we present the main selection theorems, for both continuous and measurable selectors (Michael's theorem and the Kuratowski–Ryll Nardzewski and Yanlov–von Neumann–Aumann theorems). Then we deal with the set of integrable selectors of a multifunction and develop the main properties of the set-valued integral. We also prove fixed point theorems and study Carathéodory multifunctions. Finally we examine the different modes of convergence of sets and of multifunctions.

Introduction

Multivalued analysis deals with the study of the properties of the maps whose values are sets (elements in a hyperspace). The need for set-valued maps was recognized early in the twentieth century, but a systematic study started only in the mid-1960s. It is closely related to nonsmooth analysis (see Chapter 1). In fact the real explosion on multivalued analysis occurred at the exact same time that nonsmooth analysis made its appearance. Since then the two fields have moved in sychronization and provided each other with new tools, ideas, concepts, and results. This symbiotic relationship sustains their remarkable growth.

In this chapter we attempt a survey of some of the basic aspects of the theory of multivalued analysis. Needless to say our presentation omits certain parts of the theory. A more comprehensive treatment can be found in the two volumes of Hu–Papageorgiou [313, 316].

In Section 6.1, which is topological in nature, we introduce various continuity notions. We investigate their properties and also examine how they are related among them.

Section 6.2 is measure-theoretic in nature and deals with measurability properties of multifunctions.

N.S. Papageorgiou, S.Th. Kyritsi-Yiallourou, *Handbook of Applied Analysis*,
Advances in Mechanics and Mathematics 19, DOI 10.1007/b120946_6,

In Section 6.3 we address the fundamental problem of existence of measurable selectors for multifunctions. So we prove the Michael selection theorem (for continuous selectors) and the Kuratowski–Ryll Nardzewski and the Yankov–von Neumann–Aumann selection theorems.

In Section 6.4 we examine decomposable sets and the resulting set-valued integration theory. It turns out that in many situations decomposability is an effective substitute of convexity.

Section 6.5 contains some major fixed point theorems. Also we conduct a brief investigation of multifunctions of two variables (Carathéodory multifunctions).

Finally in Section 6.6 we introduce and study various modes of convergence of sets. We also obtain Fatou-type results for set-valued integrals and for sets of integrable selectors.

6.1 Continuity of Multifunctions

We start by fixing the notation. So let X be a Hausdorff topological space. We introduce the following hyperspaces,

$$\begin{aligned} P_f(X) &= \{A \subseteq X : A \text{ is nonempty and closed}\} \\ \text{and} \quad P_k(X) &= \{A \subseteq X : A \text{ is nonempty and compact}\}. \end{aligned}$$

If X is a normed space, we also consider the following hyperspaces

$$\begin{aligned} P_{fc}(X) &= \{A \in P_f(X) : A \text{ is convex}\} \\ P_{(w)kc}(X) &= \{A \subseteq X : A \text{ is nonempty, (weakly-) compact and convex}\} \\ P_{bf(c)}(X) &= \{A \in P_f(X) : A \text{ is bounded (and convex)}\}. \end{aligned}$$

In what follows for X a Hausdorff topological space and $x \in X$, by $\mathcal{N}(x)$ we denote the filter of all neighborhoods of x. If (X,d) is a metric space, $x \in X$, and $r>0$, then $B_r(x) = \{y \in X : d(x,y) < r\}$ (= the open r-ball with center at $x \in X$) and $\overline{B}_r(x) = \{y \in X : d(x,y) \leq r\}$ (= the closed r-ball with center at $x \in X$). When X is a normed space and $x = 0$, then as before we write $B_r = B_r(0)$ and $\overline{B}_r = \overline{B}_r(0)$.

For sets X, Y and a multifunction (set-valued function) $F : X \longrightarrow 2^Y$, given a set $C \subseteq Y$, we have two types of inverse images of C under the action of F, namely

$$\begin{aligned} F^+(C) &= \{x \in X : F(x) \subseteq C\} \quad \text{(the strong inverse image of } C) \\ \text{and} \quad F^-(C) &= \{x \in X : F(x) \cap C \neq \emptyset\} \quad \text{(the weak inverse image of } C). \end{aligned}$$

Evidently we have $F^+(C) \subseteq F^-(C) \subseteq X$.

It is straightforward to check that these inverse images obey the following calculus rules.

PROPOSITION 6.1.1 *Suppose that X, Y, Z are nonempty sets.*

(a) *If $F, G : X \longrightarrow 2^Y$ are two multifunctions and $(F \cup G)(x) = F(x) \cup G(x)$, $(F \cap G)(x) = F(x) \cap G(x)$ for all $x \in X$, then*

$$\begin{aligned} &(F \cup G)^+(C) = F^+(C) \cup G^+(C), \quad (F \cap G)^+(C) \supseteq F^+(C) \cap G^+(C) \\ &(F \cup G)^-(C) = F^-(C) \cup G^-(C), \quad (F \cap G)^-(C) \subseteq F^-(C) \cap G^-(C) \end{aligned}$$

for all $C \subseteq Y$.

(b) *If* $F: X \longrightarrow 2^Y$, $G: X \longrightarrow 2^Z$ *and* $x \longrightarrow (G \circ F)(x) = G\big(F(x)\big) = \bigcup_{y \in F(x)} G(y)$ *for all* $x \in X$*, then* $(G \circ F)^+(C) = F^+\big(G^+(C)\big)$*,* $(G \circ F)^-(C) = F^-\big(G^-(C)\big)$ *for all* $C \subseteq Y$*.*

(c) *If* $\{C_i, C\}_{i \in I}$ *are subsets of* Y *(I being an arbitrary index set), then*

$$\bigcup_{i\in I} F^+(C_i) \subseteq F^+\big(\bigcup_{i\in I} C_i\big), \quad \bigcap_{i\in I} F^+(C_i) \subseteq F^+\big(\bigcap_{i\in I} C_i\big)$$

$$\bigcup_{i\in I} F^-(C_i) = \bigcup_{i\in I} F^-(C_i), \quad F^-\big(\bigcap_{i\in I} C_i\big) \subseteq \bigcap_{i\in I} F^-(C_i).$$

(d) *If* $F: X \longrightarrow 2^Y$, $G: X \longrightarrow 2^Z$ *are two multifunctions and* $F \times G: X \longrightarrow 2^{Y\times Z}$ *is defined by* $(F \times G)(x) = F(x) \times G(x)$ *for all* $x \in X$*, then* $(F \times G)^+(C \times D) = F^+(C) \cap G^+(D)$*,* $(F \times G)^-(C \times D) = F^-(C) \cap G^-(D)$ *for all* $C \subseteq Y$*,* $D \subseteq Z$*; these equalities are still true for arbitrary products of multifunctions.*

Now we introduce the first continuity notions for multifunctions. In what follows X, Y are Hausdorff topological spaces. Additional hypotheses are introduced as needed.

DEFINITION 6.1.2 Let $F: X \longrightarrow 2^Y$ be a multifunction.

(a) We say that F is *upper semicontinuous* at $x_0 \in X$ (usc at x_0 for short), if for all $V \subseteq Y$ open such that $F(x_0) \subseteq V$, we can find $U \in \mathcal{N}(x_0)$ such that $F(U) \subseteq V$. If this true at every $x_0 \in X$, we say that F is *upper semicontinuous* (usc for short).
(b) We say that F is *lower semicontinuous* at $x_0 \in X$ (lsc at x_0 for short), if for all $V \subseteq Y$ open such that $F(x_0) \cap V \neq \emptyset$, we can find $U \in \mathcal{N}(x_0)$ such that $F(x) \cap V \neq \emptyset$ for all $x \in U$. If this is true at every $x_0 \in X$, we say that F is *lower semicontinuous* (lsc for short).
(c) We say that F is *continuous* or (*Vietoris continuous*) at $x_0 \in X$, if it is both usc and lsc at $x_0 \in X$. If this is true at every $x_0 \in X$, then we say that F is continuous or Vietoris continuous.

The following propositions are immediate consequences of the above definitions.

PROPOSITION 6.1.3 *Given a multifunction* $F: X \longrightarrow 2^Y$*, the following statements are equivalent.*

(a) F *is usc.*
(b) *For every* $C \subseteq Y$ *closed,* $F^-(C)$ *is closed in* X*.*
(c) *If* $x \in X$*,* $\{x_\alpha\}_{\alpha \in J} \subseteq X$ *is a net in* X*,* $x_\alpha \longrightarrow x$*,* $V \subseteq Y$ *is an open set with* $F(x) \subseteq V$*,*
then we can find $\alpha_0 \in J$ *such that* $F(x_\alpha) \subseteq V$ *for all* $\alpha \in J$ *with* $\alpha \geq \alpha_0$*.*

PROPOSITION 6.1.4 *Given a multifunction* $F: X \longrightarrow 2^Y$*, the following statements are equivalent.*

(a) F *is lsc.*
(b) *For every* $C \subseteq Y$ *closed,* $F^+(C)$ *is closed in* X*.*

(c) *If* $x \in X$, $\{x_\alpha\}_{\alpha \in J} \subseteq X$ *is a net in* X, $x_\alpha \longrightarrow x$, *and* $V \subseteq Y$ *is an open set with* $F(x) \cap V \neq \emptyset$, *then we can find* $\alpha_0 \in J$ *such that* $F(x_\alpha) \cap V \neq \emptyset$ *for all* $\alpha \in J$ *with* $\alpha \geq \alpha_0$.

(d) *If* $x \in X$, $\{x_\alpha\}_{\alpha \in J} \subseteq X$ *is a net in* X, $x_\alpha \longrightarrow x$, *and* $y \in F(x)$, *then for every* $\alpha \in J$ *we can find* $y_\alpha \in F(x_\alpha)$ *such that* $y_\alpha \longrightarrow y$ *in* X.

REMARK 6.1.5 It is clear from Definition 6.1.2, that when F is single-valued, the notions of upper and lower semicontinuity coincide with the usual notion of continuity of a map between two Hausdorff topological spaces. Also because of Proposition 6.1.1(c), we see that in the definition of lower semicontinuity (see Definition 6.1.2(b)), we can take V to be a basic open set in Y. In general the notions of upper and lower semicontinuity are distinct. Upper semicontinuity allows upward jumps (in the sense of inclusion), and lower semicontinuity allows downward jumps (in the sense of inclusion); see Propositions 6.1.3(c) and 6.1.4(c), respectively. For example

$$F_1(x)=\begin{cases} [0,1] & \text{if } x=0 \\ 1 & \text{if } x\neq 0 \end{cases}$$

is usc but not lsc, whereas

$$F_2(x)=\begin{cases} \{0\} & \text{if } x=0 \\ [0,1] & \text{if } x\neq 0 \end{cases}$$

is lsc but not usc. Note that if $F\colon \mathbb{R} \longrightarrow 2^{\mathbb{R}}$ is defined by $F(x) = [\psi(x), \varphi(x)] = \{y \in \mathbb{R} : \psi(x) \leq y \leq \varphi(x)\}$ and ψ is lower semicontinuous, φ is upper semicontinuous, then F is usc, whereas if ψ is upper semicontinuous, and φ is lower semicontinuous, then F is lsc.

Combining Propositions 6.1.3 and 6.1.4, we have the following.

PROPOSITION 6.1.6 *Given a multifunction* $F : X \longrightarrow 2^Y$, *the following statements are equivalent.*

(a) F *is continuous.*

(b) *For every* $C \subseteq Y$ *closed,* $F^+(C)$ *and* $F^-(C)$ *are both closed in* X.

(c) *If* $x \in X$, $\{x_\alpha\}_{\alpha \in J} \subseteq X$ *is a net in* X, $x_\alpha \longrightarrow x$, *and* $V \subseteq Y$ *is an open set with* $F(x) \subseteq V$ *or* $F(x) \cap V \neq \emptyset$, *then we can find* $\alpha_0 \in J$ *such that for all* $\alpha \in J$, $\alpha \geq \alpha_0$ *we have* $F(x_\alpha) \subseteq V$ *or* $F(x_\alpha) \cap V \neq \emptyset$.

DEFINITION 6.1.7 Given a multifunction $F : X \longrightarrow 2^Y$, its *graph* is the set

$$\mathrm{Gr}F = \{(x, y) \in X \times Y : y \in F(x)\}.$$

It is well known that a continuous map between two Hausdorff topological spaces, has a closed graph. The same is true for usc multifunctions.

PROPOSITION 6.1.8 *If* Y *is a regular topological space and* $F : X \longrightarrow P_f(Y)$ *is usc, then* $\mathrm{Gr}F$ *is closed in* $X \times Y$.

PROOF: Let $\{(x_\alpha, y_\alpha)\}_{\alpha \in J} \subseteq \text{Gr}F \subseteq X \times Y$ be a net and suppose that $(x_\alpha, y_\alpha) \longrightarrow (x, y)$ in $X \times Y$. Arguing by contradiction, suppose that $y \notin F(x)$. Then exploiting the regularity of the space Y, we can find $V_1 \in \mathcal{N}(y)$ and V_2 an open neighborhood of $F(x)$ such that $V_2 \supseteq F(x)$ and $V_1 \cap V_2 = \emptyset$. But because $y_\alpha \longrightarrow y$ and using Proposition 6.1.3, we can find $\alpha_0 \in J$ such that for all $\alpha \in J$, $\alpha \geq \alpha_0$ we have $y_\alpha \in V_1$ and $F(x_\alpha) \subseteq V_2$, hence $y_\alpha \notin F(x_\alpha)$, a contradiction. □

REMARK 6.1.9 It is clear from the above proof that if F has values in $P_k(Y)$, then in the above proposition we can drop the requirement that Y is regular. This is consistent with the fact that for single-valued maps continuity implies closedness of the graph without any additional conditions on the space Y.

Even for single-valued maps, the converse of Proposition 6.1.8 is not in general true.

PROPOSITION 6.1.10 *If $F : X \longrightarrow P_k(Y)$ has a closed graph and it is locally compact (i.e., for every $x \in X$ we can find $U \in \mathcal{N}(x)$ such that $\overline{F(U)}$ is compact in Y), then F is usc.*

PROOF: Given $C \subseteq Y$ we show that $F^-(C)$ is closed and this by virtue of Proposition 6.1.3 implies that F is usc. So let $\{x_\alpha\}_{\alpha \in J} \subseteq F^-(C)$ be a net and assume that $x_\alpha \longrightarrow x$. By hypothesis we can find $U \in \mathcal{N}(x)$ such that $\overline{F(U)}$ is compact in Y. We can find $\alpha_0 \in J$ such that if $\alpha \in J$, $\alpha \geq \alpha_0$, then $x_\alpha \in U$ and so if $y_\alpha \in F(x_\alpha)$, $\alpha \geq \alpha_0$, we can find a subnet $\{y_\beta\}_{\beta \in I}$ of $\{y_\alpha\}_{\alpha \in J}$ and $y \in \overline{F(U)}$ such that $y_\beta \longrightarrow y$. Evidently $(x, y) \in \text{Gr}F \cap (X \times C)$; that is, $y \in F(x)$ and $y \in C$. Therefore $x \in F^-(C)$ and so $F^-(C)$ is closed. □

DEFINITION 6.1.11 We say that a multifunction $F : X \longrightarrow 2^Y$ is *closed* (resp., *sequentially closed*), if $\text{Gr}F \subseteq X \times Y$ is closed (resp., sequentially closed).

REMARK 6.1.12 Evidently a closed multifunction $F : X \longrightarrow 2^Y \setminus \{\emptyset\}$ has values in $P_f(Y)$.

PROPOSITION 6.1.13 *If $F : X \longrightarrow P_k(Y)$ is usc and $K \subseteq X$ is compact, then $F(K) \subseteq Y$ is compact.*

PROOF: Let $\{y_\alpha\}_{\alpha \in J} \subseteq F(K)$ be a net. Then $y_\alpha \in F(x_\alpha)$, $x_\alpha \in K$ for all $\alpha \in J$. By virtue of the compactness of K we can find $\{x_\beta\}_{\beta \in I}$ a subnet of $\{x_\alpha\}_{\alpha \in J}$ such that $x_\beta \longrightarrow x$. We claim that $\{y_\beta\}_{\beta \in I}$ has a cluster point in $F(x)$. We argue indirectly. So suppose that for all $y \in F(x)$, we can find $\beta_0(y) \in J$ and $V(y) \subseteq \mathcal{N}(y)$ such that $y_\beta \notin V(y)$ for all $\beta \in I$, $\beta \geq \beta_0(y)$. Note that $\{V(y)\}_{y \in F(x)}$ is an open cover of the compact set $F(x)$. So we can find $\{V(y)\}_{k=1}^N$ an open subcover. Set $V = \bigcup_{k=1}^{N} V(y_k) \in \mathcal{N}(y)$. Then we can find $\beta \in I$ such that for all $\beta \in I$, $\beta \geq \beta_1$, we have $y_\beta \notin V = \bigcup_{k=1}^{N} V(y_k) \supseteq F(x)$, which contradicts Proposition 6.1.3(c). Therefore we can find a subnet $\{y_r\}_{r \in S}$ of $\{y_\beta\}_{\beta \in I}$ such that $y_r \longrightarrow y \in F(x) \subseteq F(K)$. This proves the compactness of $F(K)$. □

The following two functions associated with a set $A \subseteq X$ play a central role in what follows.

DEFINITION 6.1.14 (a) Let (X,d) be a metric space and $A \subseteq X$. For every $x \in X$, we define

$$d(x,A) = \inf[d(x,a) : a \in A].$$

As usual, we adopt the convention that $\inf_{\emptyset} = +\infty$. The distance function $x \longrightarrow d(x,A)$ is a contraction.

(b) Let X be a normed space. By X^* we denote the topological dual of X and by $\langle \cdot, \cdot \rangle$ the duality brackets for the pair (X, X^*). Given $A \subseteq X$, for every $x^* \in X^*$ we define

$$\sigma(x^*, A) = \sup[\langle x^*, a \rangle : a \in A].$$

As usual, we adopt the convention that $\sup_{\emptyset} = -\infty$.The function $x^* \longrightarrow \sigma(x^*, A)$ from X^* into $\mathbb{R} \cup \{\pm\infty\}$ is called the *support function* of the set A.

PROPOSITION 6.1.15 *Let $F : X \longrightarrow 2^Y \setminus \{\emptyset\}$ be a multifunction.*

(a) *If (Y,d) is a metric space, then $F(\cdot)$ is lsc if and only if for every $y \in Y$, $x \longrightarrow d\big(y, F(x)\big)$ is upper semicontinuous.*

(b) *If (Y,d) is a metric space and F is usc, then for every $y \in Y$ $x \longrightarrow d\big(y, F(x)\big)$ is lower semicontinuous; the converse is true if F is locally compact (see Proposition 6.1.10).*

(c) *If Y is a normed space furnished with the weak topology and F is usc, then for all $y^* \in Y^*$ $x \longrightarrow \sigma\big(y^*, F(x)\big)$ is an upper semicontinuous function from X into $\overline{\mathbb{R}} = \mathbb{R} \cup \{+\infty\}$.*

PROOF: (a) $\Rightarrow$: Suppose F is lsc. We need to show that for every $\lambda \in \mathbb{R}$ the upper level set $U_\lambda = \big\{x \in X : \vartheta_y(x) = d\big(y, F(x)\big) \geq \lambda\big\}$ is closed. So suppose that $\{x_\alpha\}_{\alpha \in J} \subseteq U_\lambda$ is a net and assume that $x_\alpha \longrightarrow x$. Given $\varepsilon > 0$ we can find $v \in F(x)$ such that $d(y,v) < \vartheta_y(x) + \varepsilon$. Because F is lsc, we can find $\alpha_0 \in J$ such that if $\alpha \in J$, $\alpha \geq \alpha_0$ we have $F(x_\alpha) \cap B_\varepsilon(v) \neq \emptyset$. So we can find $y_\alpha \in F(x_\alpha)$, $\alpha \geq \alpha_0$, such that $d(y, y_\alpha) \leq \vartheta_y(x) + 2\varepsilon$ and so $\vartheta_y(x_\alpha) \leq \vartheta_y(x) + 2\varepsilon$, hence $\lambda \leq \vartheta_y(x) + 2\varepsilon$. Because $\varepsilon > 0$ was arbitrary, we let $\varepsilon \downarrow 0$ to obtain $\lambda \leq \vartheta_y(x)$. Therefore U_λ is closed and so $x \longrightarrow d\big(y, F(x)\big)$ is upper semicontinuous.

$\Leftarrow$: Let $V \subseteq Y$ be open and let $x \in F^-(V)$. We choose $y \in F(x) \cap V$. Let $\varepsilon > 0$ such that $B_\varepsilon(y) \subseteq V$. Because $\vartheta_y(\cdot)$ is upper semicontinuous, we can find $U \in \mathcal{N}(x)$ such that

$$\begin{aligned} & \vartheta_y(x') < \vartheta_y(x) + \varepsilon = \varepsilon \qquad \text{for all } x' \in U, \\ \Rightarrow\ & F(x') \cap B_\varepsilon(y) \neq \emptyset \qquad \text{for all } x' \in U, \\ \Rightarrow\ & F(x') \cap V \neq \emptyset \qquad \text{for all } x' \in U. \end{aligned}$$

This means that F is lsc (see Definition 6.1.2(b)).

(b) We need to show that for every $\lambda \in \mathbb{R}$, the lower level set $L_\lambda = \{x \in X : \vartheta_y(x) = d\big(y, F(x)\big) \leq \lambda\}$ is closed. To this end let $\{x_\alpha\}_{\alpha \in J} \subseteq L_\lambda$ be a net such that $x_\alpha \longrightarrow x$. Because F is usc, given $\varepsilon > 0$, we can find $\alpha_0 \in J$ such that if $\alpha \in J$, $\alpha \geq \alpha_0$, then

$$\begin{aligned} & F(x_\alpha) \subseteq F(x)_\varepsilon = \{v \in Y : d\big(v, F(x)\big) < \varepsilon\}, \\ \Rightarrow\ & \vartheta_y(x) < \vartheta_y(x_\alpha) + \varepsilon, \\ \Rightarrow\ & \vartheta_y(x) < \lambda + \varepsilon. \end{aligned}$$

Because $\varepsilon > 0$ was arbitrary, we let $\varepsilon \downarrow 0$ to conclude that $\vartheta_y(x) \leq \lambda$ and so $x \in L_\lambda$. This proves the lower semicontinuity of the distance function $x \longrightarrow \vartheta_y(x) = d\big(y, F(x)\big)$.

Now suppose that $F : X \longrightarrow P_k(Y)$ is locally compact and the distance function $x \longrightarrow \vartheta_y(x)$ is lower semicontinuous for all $y \in Y$. We show that F is usc. By virtue of Proposition 6.1.10, it suffices to show that Gr F is closed in $X \times Y$. To this end let $\{(x_\alpha, y_\alpha)\}_{\alpha \in J} \subseteq$ Gr F be a net and assume that $(x_\alpha, y_\alpha) \longrightarrow (x, y)$ in $X \times Y$. We have $d\big(y, F(x)\big) \leq d(y, y_\alpha) \longrightarrow 0$. Because $\vartheta_y(\cdot)$ is lower semicontinuous we also have $d\big(y, F(x)\big) \leq \liminf_{\alpha \in J} d\big(y, F(x_\alpha)\big)$. Therefore it follows that $d\big(y, F(x)\big) = 0$ (i.e., $(x, y) \in$ Gr F).

(c) Fix $y^* \in Y^*$ and $\varepsilon > 0$ and define

$$W(y^*, \varepsilon) = \{y \in Y : \langle y^*, y\rangle < \varepsilon\}.$$

This is a weak neighborhood of the origin. By hypothesis F is usc from X into Y_w(= space Y endowed with the weak topology). So we can find $U \in \mathcal{N}(x)$ such that

$$\begin{aligned} & F(u) \subseteq F(x) + W(y^*, \varepsilon) \qquad \text{for all } u \in U, \\ \Rightarrow\ & \sigma\big(y^*, F(u)\big) \leq \sigma\big(y^*, F(x)\big) + \varepsilon \qquad \text{for all } y \in U, \\ \Rightarrow\ & x \longrightarrow \sigma\big(y^*, F(x)\big) \text{ is upper semicontinuous.} \end{aligned}$$

□

EXAMPLE 6.1.16 (a) In general lower semicontinuity of the distance function does not imply upper semicontinuity of the multifunction. To see this let $X = \mathbb{R}$, $Y = \mathbb{R}^2$, and $F : X \longrightarrow P_f(Y)$ is defined by $F(x) = \big\{(t, xt) : t \in \mathbb{R}\big\}$. Clearly $\vartheta_y(x) = d\big(y, F(x)\big)$ is continuous in $x \in \mathbb{R}$, but F is not usc.

(b) The converse of Proposition 6.1.15(c) is not in general true. To see this let $X = \mathbb{R}_+$, $Y = \mathbb{R}$ and consider $F : X \longrightarrow P_{kc}(Y)$ defined by

$$F(x) = \begin{cases} \{-1, 1\} & \text{if } x = 0 \\ [0, x] & \text{if } x \neq 0 \end{cases}.$$

For every $y^* \in Y^* = \mathbb{R}$ we have that $x \longrightarrow \sigma\big(y^*, F(x)\big)$ is upper semicontinuous, but $x \longrightarrow F(x)$ is not usc at $x = 0$.

However, if we strengthen the hypotheses on the multifunction F, we can have a converse to Proposition 6.1.15(c). For a proof of this result we refer to Hu–Papageorgiou [313, pp. 47–48].

PROPOSITION 6.1.17 *If Y is a normed space and $F : X \longrightarrow P_{wkc}(Y)$ is a multifunction such that for all $y^* \in Y^*$ the function $x \longrightarrow \sigma\big(y^*, F(x)\big)$ from X into $\overline{\mathbb{R}} = \mathbb{R} \cup \{+\infty\}$ is upper semicontinuous, then the multifunction F is usc from X into Y_w.*

The next result is useful in many applications. Consider a function $u : X \times Y \longrightarrow \mathbb{R} \cup \{\pm\infty\}$ and a multifunction $F : Y \longrightarrow 2^X \setminus \{\emptyset\}$. We consider the following parametric maximization problem,

$$v(y) = \sup[u(x,y) : x \in F(y)].$$

We call $v(\cdot)$ the value function depending on the parameter $y \in Y$. In addition to the value function, we also consider the solution multifunction $y \longrightarrow S(y)$ defined by

$$S(y) = \{x \in F(y) : u(x,y) = v(y)\}.$$

The next result, useful in a variety of applications, is often called the *Berge maximum theorem.*

THEOREM 6.1.18 *Let $X, Y, F, v(\cdot)$, and $S(\cdot)$ be as above and assume that for every $y \in Y$ we can find $x \in F(y)$ such that $u(x,y) \in \mathbb{R}$.*

(a) *If u is lower semicontinuous and F is lower semicontinuous, then the value function $v : Y \longrightarrow \overline{\mathbb{R}} = \mathbb{R} \cup \{+\infty\}$ is lower semicontinuous.*
(b) *If u is upper semicontinuous and F is upper semicontinuous with values in $P_k(X)$, then the value function $v : Y \longrightarrow \overline{\mathbb{R}} = \mathbb{R} \cup \{+\infty\}$ is upper semicontinuous.*
(c) *If $u : X \times Y \longrightarrow \mathbb{R}$ is continuous and F is continuous with values in $P_k(X)$, then the value function $v : Y \longrightarrow \mathbb{R}$ is continuous and the solution multifunction $S : Y \longrightarrow P_k(X)$ is upper semicontinuous.*

PROOF: (a) For every $\lambda \in \mathbb{R}$, we need to show the level set $L_\lambda = \{y \in Y : v(y) \leq \lambda\}$ is closed. So let $\{y_\alpha\}_{\alpha \in J} \subseteq L_\lambda$ be a net such that $y_\alpha \longrightarrow y$. By virtue of Proposition 6.1.4(d), for every $\alpha \in J$ we can find $x_\alpha \in F(y_\alpha)$ such that $x_\alpha \longrightarrow x$ in X. We have $u(x_\alpha, y_\alpha) \leq v(y_\alpha) \leq \lambda$ for all $\alpha \in J$ and because u is lower semicontinuous, we obtain $u(x,y) \leq \lambda$. Because $x \in F(y)$ was arbitrary, it follows that $v(y) \leq \lambda$, hence $y \in L_\lambda$. This proves the lower semicontinuity of $v(\cdot)$.

(b) For every $\lambda \in \mathbb{R}$, we need to show that the upper level set $U_\lambda = \{y \in Y : v(y) \geq \lambda\}$ is closed. So let $\{y_\alpha\}_{\alpha \in J}$ be a net such that $y_\alpha \longrightarrow y$. Since F has values in $P_k(X)$, for every $\alpha \in J$, we can find $x_\alpha \in F(y_\alpha)$ such that $v(y_\alpha) = u(x_\alpha, y_\alpha)$ (Weierstrass theorem). As in the proof of Proposition 6.1.13, we can show that $\{x_\alpha\}_{\alpha \in J}$ has a cluster point in $F(y)$. Therefore we can find $x \in F(y)$ and a subnet $\{x_\beta\}_{\beta \in I}$ of $\{x_\alpha\}_{\alpha \in J}$ such that $x_\beta \longrightarrow x$. We have

$$\lambda \leq v(y_\beta) = u(x_\beta, y_\beta) \qquad \text{for all } \beta \in I \text{ and } \limsup_{\beta \in I} u(x_\beta, y_\beta) \leq u(x,y),$$
$$\Rightarrow \lambda \leq u(x,y) \leq v(y) \qquad (\text{see } x \in F(y)).$$

This proves that U_λ is closed and so v is upper semicontinuous.

(c) The continuity of v follows from parts (a) and (b). So it remains to show that $S : Y \longrightarrow P_k(X)$ is usc. Because of Proposition 6.1.3, given $A \subseteq X$ closed, we need to show that $S^-(A) = \{y \in Y : S(y) \cap A \neq \emptyset\}$ is closed. So let $\{y_\alpha\}_{\alpha \in J} \subseteq S^-(A)$ be a net such that $y_\alpha \longrightarrow y$ in Y. For each $\alpha \in J$, we can find $x_\alpha \in F(y_\alpha) \cap A$ such that $u(x_\alpha, y_\alpha) = v(y_\alpha)$. As before $\{x_\alpha\}_{\alpha \in J}$ has a cluster point $x \in F(y)$. Thus we can find a subnet $\{x_\beta\}_{\beta \in I}$ of $\{x_\alpha\}_{\alpha \in J}$ such that $x_\beta \longrightarrow x$. Evidently $x \in F(y) \cap A$

and $u(x,y) = v(y)$ (because u and v are continuous). Therefore $y \in S^-(A)$ and we have that $S^-(A)$ is closed in Y, hence S is usc. $\square$

Given a multifunction $F : X \longrightarrow 2^Y$, note that for $V \subseteq Y$ open, $F(x) \cap V \neq \emptyset$ if and only if $\overline{F(x)} \cap V \neq \emptyset$. So we can state the following result.

PROPOSITION 6.1.19 *A multifunction $F : X \longrightarrow 2^Y \setminus \{\emptyset\}$ is lsc if and only if $x \longrightarrow \overline{F}(x) = \overline{F(x)}$ is lsc.*

The above result is not true for usc multifunctions.

EXAMPLE 6.1.20 Proposition 6.1.19 fails for usc multifunctions. To see this let $X = Y = \mathbb{R}$ and let $F : \mathbb{R} \longrightarrow 2^{\mathbb{R}} \setminus \{\emptyset\}$ be defined by $F(x) = (x-1, x+1)$. Note that $F^+\big((-1,1)\big) = \{0\}$ and so F is not usc. On the other hand $\overline{F}(x) = [x-1, x+1]$ and by virtue of Remark 6.1.5 $\overline{F}$ is continuous.

For usc multifunctions we have the following result.

PROPOSITION 6.1.21 *If Y is normal and $F : X \longrightarrow 2^Y \setminus \{\emptyset\}$ is usc, then so is $\overline{F}(\cdot)$.*

PROOF: Let $\{x_\alpha\}_{\alpha \in J}$ be a net such that $x_\alpha \longrightarrow x$ and let $V \subseteq Y$ be an open set such that $\overline{F}(x) \subseteq V$. Because of the normality of Y, we can find $V_0 \subseteq Y$ another open set such that

$$F(x) \subseteq \overline{F}(x) \subseteq V_0 \subseteq \overline{V}_0 \subseteq V. \tag{6.1}$$

Because F is usc, we can find $\alpha_0 \in J$ such that for $\alpha \in J$, $\alpha \geq \alpha_0$ we have

$$\begin{aligned} & F(x_\alpha) \subseteq V_0 \qquad \text{(see Proposition 6.1.3)}, \\ \Rightarrow\ & \overline{F}(x_\alpha) \subseteq V \qquad \text{for all } \alpha \geq \alpha_0 \qquad \text{(see (6.1))}, \\ \Rightarrow\ & \overline{F} \quad \text{is usc (see Proposition 6.1.3).} \end{aligned}$$

$\square$

PROPOSITION 6.1.22 *If Y is normal and $F_1, F_2 : X \longrightarrow P_f(Y)$ are both usc multifunctions such that for every $x \in X$, $(F_1 \cap F_2)(x) = F_1(x) \cap F_2(x) \neq \emptyset$, then $x \longrightarrow (F_1 \cap F_2)(x)$ is usc too.*

PROOF: We need to show that if $V \subseteq Y$ is open, then $(F_1 \cap F_2)^+(V)$ is open. By definition

$$(F_1 \cap F_2)^+(V) = \{x \in X : F_1(x) \cap F_2(x) \cap V^c = \emptyset\}. \tag{6.2}$$

Let $x \in (F_1 \cap F_2)^+(V)$. Then because of (6.2) the sets $F_1(x)$ and $F_2(x) \cap V^c$ are disjoint closed sets. Due to the normality of Y we can find $V_1, V_2 \subseteq Y$ disjoint open sets such that

$$F_1(x) \subseteq V_1 \quad \text{and} \quad F_2(x) \cap V^c \subseteq V_2.$$

Let $V_3 = V_2 \cup V$. Then $F_2(x) \subseteq V_3$. Because F_1, F_2 are usc, we can find $U \in \mathcal{N}(x)$ such that for all $x' \in U$, we have $F_1(x') \subseteq V_1$ and $F_2(x') \subseteq V_3$. Therefore

$$\begin{aligned} & F_1(x') \cap F_2(x') \subseteq V_1 \cap V_3 = V_1 \cap (V_2 \cup V) \subseteq V \qquad \text{for all } x' \in U, \\ \Rightarrow\ & (F_1 \cap F_2)^+(V) \text{ is open (i.e., } F_1 \cap F_2 \text{ is usc).} \end{aligned}$$

$\square$

Another result in this direction is given below.

PROPOSITION 6.1.23 *If $F_1{:}X\longrightarrow P_f(Y)$ is closed (see Definition 6.1.11), F_2: $X\longrightarrow P_k(Y)$ is usc and for all $x\in X$, $(F_1\cap F_2)(x)=F_1(x)\cap F_2(x)\neq\emptyset$, then $x\longrightarrow (F_1\cap F_2)(x)$ is usc.*

PROOF: We need to show that if $V\subseteq Y$ is open, then $(F_1\cap F_2)^+(V)$ is open. Let $x\in(F_1\cap F_2)^+(V)$. From (6.2) we have that the sets $F_1(x)$ and $F_2(x)\cap V^c$ are disjoint. Note that $F_2(x)\cap V^c$ is compact in Y. Let $y\in F_2(x)\cap V^c$. Then $(x,y)\notin$ Gr F_1 and because Gr F_1 is closed, we can find $U_y\in\mathcal{N}(x)$ and $W_y\in\mathcal{N}(y)$ such that $(U_y\times W_y)\cap \mathrm{Gr}\ F_1=\emptyset$. Hence for all $x'\in U_y$ we have $F_1(x')\cap W_y=\emptyset$. As we already observed, the set $F_2(x)\cap V^c$ is compact. So we can find $\{y_k\}_{k=1}^N\subseteq F_2(x)\cap V^c\subseteq \bigcup_{k=1}^N W_{y_k}=V_1\in\mathcal{N}(y)$. Set $U_1=\bigcap_{k=1}^N U_{y_k}\in\mathcal{N}(x)$. If $x'\in U_1$, then $F_1(x')\subseteq V_1^c$. Let $V_2=V\cup V_1$ and $U_2\in\mathcal{N}(x)$ be such that if $x'\in U_2$, then $F_2(x')\subseteq V_2$ (recall that F_2 is usc). Then if $x'\in U_1\cap U_2\in\mathcal{N}(x)$, we have $(F_1\cap F_2)(x')=F_1(x')\cap F_2(x')\subseteq V_1^c\cap V_2\subseteq V$, which proves that $(F_1\cap F_2)^+(V)$ is open, hence $x\longrightarrow(F_1\cap F_2)$ is usc. □

What about the intersection of two lsc multifunctions? In this case the situation is more delicate. First note that, if $F:X\longrightarrow 2^Y\backslash\{\emptyset\}$ is lsc and $V\subseteq Y$ a nonempty open set such that $F(x)\cap V\neq\emptyset$ for all $x\in X$, then $x\longrightarrow F(x)\cap V$ is lsc. Moreover, if $C\subseteq X$ is closed and

$$\widehat{F}(x)=\begin{cases}\overline{F(x)\cap V} & \text{for } x\in C\\ \overline{F}(x) & \text{for } x\in C^c\end{cases},$$

then it is straightforward to check that $\widehat{F}(\cdot)$ is lsc. More generally, we have the following result.

PROPOSITION 6.1.24 *If $F_1{:}X\longrightarrow 2^Y\backslash\{\emptyset\}$ is lsc, $F_2{:}X\longrightarrow 2^Y\backslash\{\emptyset\}$ has an open graph, and for every $x\in X$, $(F_1\cap F_2)(x)=F_1(x)\cap F_2(x)\neq\emptyset$, then $x\longrightarrow(F_1\cap F_2)(x)$ is lsc.*

PROOF: Let $V\subseteq Y$ be nonempty open, $x\in(F_1\cap F_2)^-(V)$, and $y\in F_1(x)\cap F_2(x)\cap V$. Then $(x,y)\in$ Gr $F_2\cap(X\times V)$. Since Gr F_2 is open we can find $U_1(x)\in\mathcal{N}(x)$ and $V_1(y)\in\mathcal{N}(y)$ such that $U_1(x)\times V_1(y)\subseteq$ Gr $F_2\cap(X\times V)$. Note that $F_1(x)\cap V_1(y)\neq\emptyset$ and because F_1 is lsc, we can find $U_2(x)\in\mathcal{N}(x)$ such that $F_1(x')\cap V_1(y)\neq\emptyset$ for all $x'\in U_2(x)$. Set $U(x)=U_1(x)\cap U_2(x)\in\mathcal{N}(x)$. Then for all $x'\in U(x)$ we have $F_1(x')\cap V_1(y)\neq\emptyset$ and $U(x)\times V_1(y)\subseteq$ Gr $F_2\cap(X\times V)$. Therefore for all $x'\in U(x)$, we have $F_1(x')\cap F_2(x')\neq\emptyset$ which means that $(F_1\cap F_2)^-(V)$ is open and so $x\longrightarrow(F_1\cap F_2)(x)$ is lsc. □

REMARK 6.1.25 Note that if Gr F is open, then F has open values. However, Proposition 6.1.24 fails if F_2 instead of an open graph has only open values.

If $Y=\mathbb{R}^N$, then the situation improves and we have the following two propositions whose proof can be found in Hu–Papageorgiou [313, pp. 55–56].

PROPOSITION 6.1.26 *If $F_1, F_2 : X \longrightarrow \mathbb{R}^N \backslash \{\emptyset\}$ are lsc multifunctions, F_2 has open convex values, and for all $x \in X$, $(F_1 \cap F_2)(x) = F_1(x) \cap F_2(x) \neq \emptyset$, then $x \longrightarrow (F_1 \cap F_2)(x)$ is lsc.*

PROPOSITION 6.1.27 *If $F_1, F_2 : X \longrightarrow \mathbb{R}^N \backslash \{\emptyset\}$ are lsc with convex values and for every $x \in X$, $(F_1 \cap \operatorname{int} F_2)(x) \neq \emptyset$, or $(\operatorname{int} F_1 \cap F_2)(x) \neq \emptyset$, then $x \longrightarrow (F_1 \cap F_2)(x)$ is lsc.*

REMARK 6.1.28 It is easy to check that if Y is a Banach space and $F : X \longrightarrow 2^Y \setminus \{\emptyset\}$ is lsc, then so are the multifunctions $x \longrightarrow \operatorname{conv} F(x)$, $x \longrightarrow \overline{\operatorname{conv}}\, F(x)$. Similarly if $F : X \longrightarrow P_k(Y)$ is usc, then so is the multifunction $x \longrightarrow \overline{\operatorname{conv}}\, F(x)$. Finally for X, Y Hausdorff topological spaces, then both upper and lower semicontinuity are preserved by the set-theoretic operation of union.

When X is a metric space, then we can exploit its metric structure, to introduce some additional continuity notions for multifunctions. For this purpose, we introduce the following notions.

DEFINITION 6.1.29 Let (X, d) be a metric space and $A, C \subseteq X$. We set

$$\begin{aligned} h^*(A, C) &= \sup\left[d(a, C) : a \in A\right] \\ &= \inf\left[\varepsilon > 0 : A \subseteq C_\varepsilon\right], \end{aligned}$$

where $C_\varepsilon = \{x \in X : d(x, C) < \varepsilon\}$ (the open ε-enlargement of C). We call $h^*(A, C)$ the excess of A over C. Then the *Hausdorff distance* between A and C is defined by

$$\begin{aligned} h(A, C) &= \max\left[h^*(A, C),\ h^*(C, A)\right] \\ &= \inf\left\{\varepsilon > 0 : A \subseteq C_\varepsilon \text{ and } C \subseteq A_\varepsilon\right\}. \end{aligned}$$

REMARK 6.1.30 It is easy to see that $h(A, C) = 0$ if and only if $\overline{A} = \overline{C}$ and so $\left(P_f(X) \cup \{\emptyset\}, h\right)$ is a (generalized) metric space. The empty set is an isolated point in this metric space and the metric $h(\cdot, \cdot)$ is called the *Hausdorff metric*.

The next proposition is a straightforward consequence of Definition 6.1.29.

PROPOSITION 6.1.31 *Let (X, d) be a metric space and $A, C \subseteq X$ nonempty.*

(a) $h^*(A, C) = \sup\left[d(x, C) - d(x, A) : x \in X\right]$
(b) $h(A, C) = \sup\left[|d(x, C) - d(x, A)| : x \in X\right]$
(c) *If X is a normed space and $A, C \in P_{bfc}(X)$, then* $h(A, C) = \sup\left[|\sigma(x^*; A) - \sigma(x^*; C)| : x^* \in X^*,\ \|x^*\| \leq 1\right]$

(Hörmander's formula)

The next proposition summarizes the basic facts about this metric. Detailed proofs can be found in Hu–Papageorgiou [313, p. 8].

PROPOSITION 6.1.32 (a) *If (X, d) is a complete metric space, then so is $(P_f(X), h)$.*

(b) *$P_{bf}(X)$ is a closed subset of $(P_f(X),h)$ and if X is separable, then so is $(P_k(X),h)$.*
(c) *If X is a normed space, then $P_{kc}(X) \subseteq P_{bfc}(X) \subseteq P_{fc}(X)$ and $P_k(X) \subseteq P_{bf}(X)$ are all closed subspaces of $\big(P_f(X),h\big)$.*

Now we are ready to introduce the new continuity concepts for multifunctions.

DEFINITION 6.1.33 Let X be a Hausdorff topological space, (Y,d) a metric space, and $F : X \longrightarrow 2^Y \setminus \{\emptyset\}$ a multifunction.

(a) We say that F is *h-upper semicontinuous* at $x_0 \in X$ (h-usc for short), if the function $x \longrightarrow h^*\big(F(x),F(x_0)\big)$ is continuous at $x_0 \in X$. If $F(\cdot)$ is h-usc at every $x_0 \in X$, then we say that F is h-upper semicontinuous (h-usc for short).
(b) We say that F is *h-lower semicontinuous* at $x_0 \in X$ (h-lsc for short), if the function $x \longrightarrow h^*\big(F(x_0),F(x)\big)$ is continuous at $x_0 \in X$. If $F(\cdot)$ is h-lsc at every $x_0 \in X$, then we say that F is h-lower semicontinuous (h-lsc for short).
(c) We say that F is *h-continuous* at $x_0 \in X$, if it is both h-usc and h- lsc at $x_0 \in X$. If $F(\cdot)$ is h-continuous at every $x_0 \in X$, then we say that F is h-continuous.

REMARK 6.1.34 Note that h-continuity of F, is continuity from X into the pseudometric space $\big(2^Y \setminus \{\emptyset\}, h\big)$.

It is natural to ask how these new continuity notions compare to those of Definition 6.1.2. In what follows X is a Hausdorff topological space and (Y,d) a metric space.

PROPOSITION 6.1.35 *If $F : X \longrightarrow 2^Y \setminus \{\emptyset\}$ is usc, then F is h-usc.*

PROOF: Because F is usc, for every $\varepsilon > 0$ and every $x \in X$ we have $F^+\big(F(x)_\varepsilon\big) = U_x \in \mathcal{N}(x)$ (recall that $F(x)_\varepsilon = \{y \in Y : d\big(y,F(x)\big) < \varepsilon\}$). For every $x' \in U_x$ we have $F(x') \subseteq F(x)_\varepsilon$, which means that $h\big(F(x'),F(x)\big) < \varepsilon$ and so we conclude that F is h-usc. □

The converse of the above proposition is not in general true as the following example illustrates.

EXAMPLE 6.1.36 h-usc $\not\Rightarrow$ usc: Let $X = [0,1]$, $Y = \mathbb{R}$ and consider the multifunction $F : X \longrightarrow 2^Y \setminus \{\emptyset\}$ defined by

$$F(x) = \begin{cases} [0,1] & \text{if } 0 \le x < 1 \\ [0,1) & \text{if } x = 1 \end{cases}.$$

Clearly F is h-usc, but it is not usc because $F^+\big((-1,1)\big) = \{1\}$.

PROPOSITION 6.1.37 *If $F : X \longrightarrow 2^Y \setminus \{\emptyset\}$ is h-lsc, then F is lsc.*

PROOF: We need to show that for every $C \subseteq Y$ closed, the set $F^+(C)$ is closed. To this end, let $\{x_\alpha\}_{\alpha \in J} \subseteq F^+(C)$ be a net and assume that $x_\alpha \longrightarrow x$. We have $F(x_\alpha) \subseteq C$ for all $\alpha \in J$. Because F is h-lsc, given $\varepsilon > 0$, we can find $\alpha_0 \in J$ such that for all $\alpha \in J$, $\alpha \geq \alpha_0$ we have $F(x) \subseteq F(x_\alpha) \subseteq C_\varepsilon$. Let $\varepsilon \downarrow 0$ to conclude that $F(x) \subseteq C$ (i.e., $x \in F^+(C)$). This proves that $F^+(C)$ is closed and so F is lsc. □

Again the converse of this proposition, in general fails.

EXAMPLE 6.1.38 lsc $\not\Rightarrow$ h-lsc: Let $X = [0,1]$, $Y = \mathbb{R}^2_+$, and let $F : X \longrightarrow 2^Y \setminus \{\emptyset\}$ be defined by $F(x) = \big\{[t, xt] : t > 0\big\}$. Then F is lsc but not h-lsc.

However, for $P_k(Y)$-valued multifunction the situation is better.

PROPOSITION 6.1.39 *If $F : X \longrightarrow P_k(Y)$, then*

(a) *F is usc if and only if F is h-usc.*
(b) *F is lsc if and only if F is h-lsc.*

PROOF: (a)$\Rightarrow$: This implication is Proposition 6.1.35.

$\Leftarrow$: We need to show that for every $C \subseteq Y$ closed, $F^-(C)$ is closed. To this end let $\{x_\alpha\}_{\alpha \in J} \subseteq F^-(C)$ be a net such that $x_\alpha \longrightarrow x$. Let $y_\alpha \in F(x_\alpha) \cap C$ for all $\alpha \in J$. We have

$$d\big(y_\alpha, F(x)\big) \leq h^*\big(F(x_\alpha), F(x)\big) \longrightarrow 0 \quad \text{(because } F \text{ is } h\text{-usc)}. \tag{6.3}$$

Because $F(x) \in P_k(Y)$, for every $\alpha \in J$ we can find $u_\alpha \in F(x)$ such that $d\big(y_\alpha, F(x)\big) = d(y_\alpha, u_\alpha)$. Due to the compactness of $F(x)$, we can find $\{u_\beta\}_{\beta \in I}$ a subnet of $\{u_\alpha\}_{\alpha \in J}$ such that $u_\beta \longrightarrow y \in F(x)$. Then because of (6.3), we see that $y_\beta \longrightarrow y \in F(x) \cap C$ (because C is closed). Therefore $x \in F^-(C)$ and we conclude that $F^-(C)$ is closed, hence F is usc.

(b)$\Rightarrow$: Suppose $\{x_\alpha\}_{\alpha \in J} \subseteq X$ is a net such that $x_\alpha \longrightarrow x$. Because $F(x) \in P_k(X)$, for every $\alpha \in J$ we can find $u_\alpha \in F(x)$ such that $d\big(u_\alpha, F(x_\alpha)\big) = h^*\big(F(x), F(x_\alpha)\big)$. Also we can find a subnet $\{u_\beta\}_{\beta \in I}$ of $\{u_\alpha\}_{\alpha \in J}$ such that $u_\beta \longrightarrow y$. Since F is lsc, given $\varepsilon > 0$, we can find $\beta_0 \in I$ such that for all $\beta \in I$, $\beta \geq \beta_0$, we have $F(x_\beta) \cap B_{\varepsilon/2}(y) \neq \emptyset$ and $u_\beta \in B_{\varepsilon/2}(y)$. So for $\beta \geq \beta_0$, we have

$$\begin{aligned} & h^*\big(F(x), F(x_\beta)\big) \leq d(u_\beta, y) + d\big(y, F(x_\beta)\big) < \varepsilon, \\ \Rightarrow \; & \lim_{\beta \in I} h^*\big(F(x), F(x_\beta)\big) = 0. \end{aligned} \tag{6.4}$$

Because every subnet of $\{x_\alpha\}_{\alpha \in J}$ has a further subnet so that (6.4) holds, we conclude that $h^*\big(F(x), F(x_\beta)\big) \longrightarrow 0$, hence F is h-lsc.

$\Leftarrow$: This implication is Proposition 6.1.37. □

COROLLARY 6.1.40 *If $F : X \longrightarrow P_k(Y)$, then F is continuous if and only if it is h-continuous.*

REMARK 6.1.41 If a multifunction $F : X \longrightarrow 2^Y \setminus \{\emptyset\}$ is h-usc (resp., h-lsc, h-continuous), then so are the multifunctions $x \longrightarrow \overline{F}(x)$, $x \longrightarrow \text{conv } F(x)$, $x \longrightarrow \overline{\text{conv}}\ F(x)$. Moreover, if $G : X \longrightarrow 2^Y \setminus \{\emptyset\}$ is another multifunction which too is h-usc (resp., h-lsc, h-continuous), then so is $x \longrightarrow (F \cup G)(x)$. The situation is more complicated with the operation of intersection.

To deal with the h-continuity properties of the intersection of two multifunctions, we need the following lemma, known as the *cancellation law lemma*.

LEMMA 6.1.42 *If V is a locally convex space, $D \subseteq V$ is nonempty bounded, and $C \subseteq V$ is convex, then $A + D \subseteq \overline{C + D}$ implies $A \subseteq \overline{C}$.*

PROOF: First we assume that $C \in P_{fc}(V)$ and $A + D \subseteq C + D$. Let $a \in A$ and $d_1 \in D$. By induction we can choose $c_k \in C$ and $d_{k+1} \in D$ such that $a + d_k = c_k + d_{k+1}$ for all $k \geq 1$. Then $c_k = a + d_k - d_{k+1}$. Summing up to n and then dividing by n, we obtain

$$a + \frac{d_1}{n} - \frac{d_{n+1}}{n} = \frac{1}{n} \sum_{k=1}^{n} c_k.$$

Because by hypothesis D is bounded, we have

$$a = \lim_{n \to \infty} \left(a + \frac{d_1}{n} - \frac{d_{n+1}}{n}\right) = \lim_{n \to \infty} \frac{1}{n} \sum_{k=1}^{n} c_k \in C,$$

because we have assumed that $C \in P_{fc}(V)$.

Now we consider the general case. Let $U \in \mathcal{N}(0)$ be convex and choose $U_1 \in \mathcal{N}(0)$ convex such that $U_1 + U_1 \subseteq U$. We have

$$A + D \subseteq \overline{C + D} \subseteq C + D + U_1 \subseteq \overline{C + U_1} + D.$$

Note that $\overline{C + U_1} \in P_{fc}(V)$. So we are in the first part of the proof, from which it follows that $A \subseteq \overline{C + U_1} \subseteq C + U_1 + U_1 \subseteq C + U$. Because $U \in \mathcal{N}(0)$ was arbitrary, we conclude that $A \subseteq \overline{C}$. $\square$

LEMMA 6.1.43 *If V is a normed space and $A \subseteq V$ is convex bounded with* $\text{int}\, A \neq \emptyset$*, then for every $\varepsilon > 0$ there exists $C \subseteq \text{int } A$ and $\delta > 0$ such that $C_\delta \subseteq A \subseteq C_\varepsilon$.*

PROOF: By translating things if necessary, we may assume that $0 \in \text{int } A$. Because A is bounded, given $\varepsilon > 0$ we can find $\lambda \in (0, 1)$ such that $\lambda \text{ int } A \subseteq B_{\varepsilon/2}$. Also because $0 \in \text{int } A$, we can find $\delta > 0$ such that $B_\delta \subseteq \lambda \text{ int } A$. Set $C = (1 - \lambda)\text{int } A$. Then $C + B_\delta = C_\delta \subseteq A$. On the other hand because A is convex, $\overline{A} = \overline{\text{int } A}$ and so $A \subseteq C + B_\varepsilon = C_\varepsilon$. $\square$

REMARK 6.1.44 This lemma fails if A is not bounded.

PROPOSITION 6.1.45 *If Y is a normed space and $F_1, F_2 : X \longrightarrow P_{bfc}(Y)$ are h-lsc and for all $x \in X$ we have*

$$\text{int } (F_1 \cap F_2)(x) \neq \emptyset,$$

then $x \longrightarrow F(x) = (F_1 \cap F_2)(x) = F_1(x) \cap F_2(x)$ is h-lsc.

PROOF: Fix $x \in X$ and let $\varepsilon > 0$ be given. Because of Lemma 6.1.43, we can find $C \subseteq \text{int}(F_1 \cap F_2)(x)$ and $\delta > 0$ such that $C_\delta \subseteq F(x) \subseteq C_\varepsilon$. Since F_1, F_2 are h-lsc, we can find $U \in \mathcal{N}(x)$ such that

$$\begin{aligned} & F_k(x) \subseteq F_k(x')_\delta \qquad \text{for all } x' \in U \text{ and for } k = 1, 2, \\ \Rightarrow\ & C_\delta \subseteq F_k(x')_\delta \qquad \text{for all } x' \in U. \end{aligned} \tag{6.5}$$

From (6.5) and Lemma 6.1.42, it follows that

$$\begin{aligned} & C \subseteq (F_1 \cap F_2)(x') = F(x') \qquad \text{for all } x' \in U, \\ \Rightarrow\ & F(x) \subseteq C_\varepsilon \subseteq F(x')_\varepsilon \qquad \text{for all } x' \in U, \\ \Rightarrow\ & F \text{ is } h\text{-lsc}. \end{aligned}$$

□

There is an analogous result for h-usc multifunctions.

PROPOSITION 6.1.46 *If $F_1 : X \longrightarrow P_f(Y)$ and $F_1 : X \longrightarrow P_k(Y)$ are h-usc and for all $x \in X$, $F_1(x) \cap F_2(x) \neq \emptyset$, then $x \longrightarrow (F_1 \cap F_2)(x) = F_1(x) \cap F_2(x)$ is h-usc.*

PROOF: It is easy to see that a $P_f(Y)$-valued, h-usc multifunction is closed. Then the proposition follows by combining Propositions 6.1.23 and 6.1.39(a). □

For h-continuity, if Y is an infinite-dimensional space, we cannot simply combine Propositions 6.1.45 and 6.1.46, because a normed compact set has empty interior. Nevertheless the result is still true for h-continuity (see Hu–Papageorgiou [313, p.66]).

PROPOSITION 6.1.47 *If Y is a normed space, $F_1, F_2 : X \longrightarrow P_{bfc}(Y)$ are h-continuous and for all $x \in X$*

$$\text{int}(F_1 \cap F_2)(x) \neq \emptyset,$$

then $x \longrightarrow F(x) = (F_1 \cap F_2)(x)$ is h-continuous.

We conclude this section with two weak versions of lower semicontinuity and of h-lower semicontinuity.

DEFINITION 6.1.48 Let X, Y be Hausdorff topological spaces and $F : X \longrightarrow 2^Y \setminus \{\emptyset\}$ a multifunction.

(a) We say that F is *almost lower semicontinuous* (a-lsc for short), if for every $x \in X$ and every $\varepsilon > 0$, we can find $U \in \mathcal{N}(x)$ such that $\bigcap_{x' \in U} F(x')_\varepsilon \neq \emptyset$.

(b) We say that F is *weakly h-lower semicontinuous* (h_w-lsc for short), if for every $x \in X$ and every $U \in \mathcal{N}(x)$, we can find $V \in \mathcal{N}(x)$, $V \subseteq U$ and a point $x' \in V$ such that $F(x') \subseteq F(u)_\varepsilon$ for all $u \in V$.

REMARK 6.1.49 If in Definition 6.1.48(b), $x = x'$, then we recover the definition of h-lower semicontinuity. Clearly h-lower semicontinuity implies h_w-lower semicontinuity and h_w-lower semicontinuity implies almost lower semicontinuity.

6.2 Measurability of Multifunctions

In the previous section we focused our attention to the topological properties of multifunctions. In this section we shift our investigation to the measure theoretic properties.

Throughout this section (Ω, Σ) is a measurable space and (X, d) a metric space. Additional hypotheses are introduced as needed.

DEFINITION 6.2.1 Let $F : \Omega \longrightarrow 2^X$ be a multifunction.

(a) We say that F is *measurable*, if for all $U \subseteq X$ open, we have $F^-(U) = \{\omega \in \Omega : F(\omega) \cap U \neq \emptyset\} \in \Sigma$.
(b) We say that F is *graph measurable*, if $\mathrm{Gr}\, F = \{(\omega, x) \in \Omega \times X : x \in F(\omega)\} \in \Sigma \times \mathbf{B}(X)$ with $\mathbf{B}(X)$ being the Borel σ-field of X.
(c) If X is a separable Banach space, we say that F is *scalarly measurable*, if for all $x^* \in X^*$, the function $\omega \longrightarrow \sigma\big(x^*, F(\omega)\big)$ is Σ-measurable.

REMARK 6.2.2 We define the domain of a measurable multifunction $F : \Omega \longrightarrow 2^X$ to be the set $\mathrm{dom}\, F = \{\omega \in \Omega : F(\omega) \neq \emptyset\}$. It is clear from Definition 6.2.1(a) that $\mathrm{dom}\, F \in \Sigma$. So in the sequel without any loss of generality, for measurable multifunctions, we always assume that $\mathrm{dom}\, F = \Omega$.

PROPOSITION 6.2.3 *If $F : \Omega \longrightarrow 2^X \setminus \{\emptyset\}$ and for every $C \subseteq X$ closed we have $F^-(C) \in \Sigma$, then F is measurable.*

PROOF: Recall that in a metric space, an open set U is F_σ. So $U = \bigcup_{n \geq 1} C_n$ with $C_n \subseteq X$ closed for all $n \geq 1$. Hence $F^-(U) = F^-\big(\bigcup_{n\geq 1} C_n\big) = \bigcup_{n \geq 1} F^-(C_n) \in \Sigma$ (see Proposition 6.1.1(c)). Therefore F is measurable. □

PROPOSITION 6.2.4 *A multifunction $F : \Omega \longrightarrow 2^X \setminus \{\emptyset\}$ is measurable if and only if for all $x \in X$, $\omega \longrightarrow d\big(x, F(\omega)\big)$ is Σ-measurable.*

PROOF: $\Rightarrow$: For every $\lambda > 0$, we need to show that $L_\lambda(x) = \{\omega \in \Omega : d\big(x, F(\omega)\big) < \lambda\} \in \Sigma$. Note that $L_\lambda(x) = F^-\big(B_\lambda(x)\big) \in \Sigma$.

$\Leftarrow$: For every $x \in X$ and every $\lambda > 0$, we have $F^-\big(B_\lambda(x)\big) \in \Sigma$. If $U \subseteq X$ is open, then $U = \bigcup_{n \geq 1} B_{\lambda_n}(x_n)$ (recall that X is separable, hence second countable). Then
$F^-(U) = F^-\big(\bigcup_{n\geq 1} B_{\lambda_n}(x_n)\big) = \bigcup_{n \geq 1} F^-\big(B_{\lambda_n}(x_n)\big) \in \Sigma$. □

In earlier sections we already used the notion of the Carathéodory function. Because it is central in our subsequent considerations, for easy reference we recall here its definition (in a general setting) and some fundamental properties of such functions.

DEFINITION 6.2.5 Let (Ω, Σ) be a measurable space and V, Y Hausdorff topological spaces. A function $\varphi : \Omega \times V \longrightarrow Y$ is said to be a *Carathéodory function*, if

(a) For every $v \in V$, $\omega \longrightarrow \varphi(\omega, v)$ is $(\Sigma, \mathbf{B}(Y))$-measurable and ($\mathbf{B}(Y)$ is the Borel σ-field of Y).
(b) For every $\omega \in \Omega$, $v \longrightarrow \varphi(\omega, v)$ is continuous.

The two theorems that follow reveal two basic and useful properties of Carathéodory functions. For their proof we refer to Denkowski–Migórski–Papageorgiou [194, pp. 188–189].

THEOREM 6.2.6 *If (Ω, Σ) is a measurable space, V is a separable metric space, Y a metric space, and $\varphi\colon \Omega \times V \longrightarrow Y$ a Carathéodory function, then φ is $\Sigma \times \mathbf{B}(V)$-measurable.*

REMARK 6.2.7 This theorem implies that φ is *superpositionally measurable* (*sup-measurable* for short), namely if $u : \Omega \longrightarrow V$ is Σ-measurable, then $\omega \longrightarrow \varphi(\omega, u(\omega))$ is Σ-measurable (i.e., the Nemitsky operator corresponding to φ maps measurable maps to measurable ones).

The second theorem is known in the literature as the *Scorza–Dragoni theorem* and is a parametric version of Lusin's theorem. First a definition.

DEFINITION 6.2.8 A Hausdorff topological space (V, τ) (τ being the Hausdorff topology of V), is said to be a *Polish space*, if it is separable and there exists a metric on V for which the topology τ is complete.

THEOREM 6.2.9 *If T, V are Polish spaces, Y is a separable metric space, μ a tight Borel measure on T, and $\varphi : T \times V \longrightarrow Y$ a Carathéodory function, then for every $\varepsilon > 0$, we can find $T_\varepsilon \subseteq T$ a compact subset with $\mu(T \setminus T_\varepsilon) < \varepsilon$ such that*

$$\varphi\big|_{T_\varepsilon \times V} \text{ is continuous.}$$

Returning to the measurability of multifunctions, we have the following result.

PROPOSITION 6.2.10 *If $F : \Omega \longrightarrow P_f(X)$ is measurable, then F is graph-measurable.*

PROOF: Note that $\mathrm{Gr}\, F = \{(\omega, x) \in \Omega \times X : d(x, F(\omega)) = 0\}$. From Proposition 6.2.4, we have that $(\omega, x) \longrightarrow \varphi(\omega, x) = d(x, F(\omega))$ is a Carathéodory function. Then by virtue of Theorem 6.2.9 $\varphi(\omega, x)$ is $\Sigma \times \mathbf{B}(X)$-measurable and so $\mathrm{Gr}\, F \in \Sigma \times \mathbf{B}(X)$ (i.e., F is graph-measurable). $\square$

The converse of the previous proposition is not in general true.

EXAMPLE 6.2.11 Graph-measurable $\not\Rightarrow$ measurable: Let $\Omega = [0, 1]$ with $\Sigma = \mathbf{B}([0, 1])$ (the Borel σ-field of $[0, 1]$) and $X = \mathbb{R} \setminus \mathbb{Q}$ = the set of irrationals. Recall that X is a Polish space. Consider C a closed subset of $\Omega \times X$ such that $\mathrm{proj}_\Omega C \notin \Sigma$ and $\mathrm{proj}_X C \neq X$. Choose $x_0 \in X \setminus \mathrm{proj}_X C$ and let $F : \Omega \longrightarrow P_f(X)$ be a multifunction defined by $F(\omega) = C(\omega) \cup \{x_0\}$. Then clearly F has closed graph, hence it is graph-measurable. On the other hand, if U is a neighborhood of $\mathrm{proj}_X C$ not containing x_0, then $F^-(U) = \mathrm{proj}_\Omega C \notin \Sigma$ and so F is not measurable.

For every open set $U \subseteq X$, we have $A \cap U \neq \emptyset$ if and only if $\overline{A} \cap U \neq \emptyset$. This leads at once to the following proposition.

PROPOSITION 6.2.12 *$F : \Omega \longrightarrow 2^X \setminus \{\emptyset\}$ is measurable if and only if $\overline{F} : \Omega \longrightarrow P_f(X)$ is measurable.*

In the next result we produce conditions under which the converse of Proposition 6.2.3 holds.

PROPOSITION 6.2.13 *If $F : \Omega \longrightarrow P_k(X)$, then F is measurable if and only if $F^-(C) \in \Sigma$ for all $C \subseteq X$ closed.*

PROOF: $\Rightarrow$: Let $C \subseteq X$ be closed and set $U = X \setminus C$. Then we know that $U = \bigcup\limits_{n \geq 1} \widehat{C}_n$ with $\widehat{C}_n = \{x \in X : d(x, C) \geq 1/n\}$ (i.e., U is an F_σ-set). Then $F^-(C) = \Omega \setminus F^+(U) = \Omega \setminus F^+\big(\bigcup\limits_{n \geq 1} \widehat{C}_n\big)$. From Proposition 6.1.1(c) we know that

$$F^+\big(\bigcup_{n \geq 1} \widehat{C}_n\big) \supseteq \bigcup_{n \geq 1} F^+(\widehat{C}_n). \tag{6.6}$$

We show that the opposite inclusion is also true. Indeed, if this is not true, we can find $\omega \in F^+\big(\bigcup\limits_{n \geq 1} \widehat{C}_n\big)$ such that $F(\omega) \cap \widehat{C}_n^c \neq \emptyset$ for all $n \geq 1$. Let $x_n \in F(\omega) \cap \widehat{C}_n^c$, $n \geq 1$. Because $F(\omega) \in P_k(X)$, by passing to a suitable subsequence if necessary, we may assume that $x_n \longrightarrow x$ and $x \in F(\omega) \subseteq U$. On the other hand $x_n \in \widehat{C}_n^c$ for all $n \geq 1$ and so $d(x_n, C) < 1/n$. Therefore in the limit as $n \to \infty$, we obtain $d(x, C) = 0$; that is, $x \in C = X \setminus U$, a contradiction. So the opposite inclusion in (6.6) holds and we have

$$\begin{aligned} F^+\big(\bigcup_{n \geq 1} \widehat{C}_n\big) &= \bigcup_{n \geq 1} F^+(\widehat{C}_n) \\ &= \bigcup_{n \geq 1} (\Omega \setminus F^-(\widehat{C}_n^c)) \in \Sigma \\ \Rightarrow\ F^-(C) \in \Sigma. & \end{aligned}$$

$\Leftarrow$: This implication is Proposition 6.2.3. ☐

PROPOSITION 6.2.14 *If $F : \Omega \longrightarrow P_f(X)$ is measurable, then $F^-(K) \in \Sigma$ for all $K \subseteq X$ compact.*

PROOF: Because X is a separable metric space, we can think of X as a dense subspace of a compact metric space Y (in fact it is homeomorphic to a subset of the Hilbert cube $[0,1]^N$ which is compact by Tychonov's theorem). Let $G : \Omega \longrightarrow P_k(X)$ be defined by $G(\omega) = \overline{F(\omega)}$. From Proposition 6.2.12 we have that G is measurable. Now let $K \subseteq X$ be compact. We have

$$\begin{aligned} F^-(K) &= \{\omega \in \Omega : F(\omega) \cap K \neq \emptyset\} \\ &= \{\omega \in \Omega : G(\omega) \cap X \cap K \neq \emptyset\} = G^-(K). \end{aligned} \tag{6.7}$$

But G is $P_k(Y)$-valued. So Proposition 6.2.13 implies that $G^-(K) \in \Sigma$, hence $F^-(K) \in \Sigma$ (see (6.7)). $\square$

If we enrich the structure of the space X, then we can strengthen our conclusions.

PROPOSITION 6.2.15 *If X is a σ-compact metric space and $F : \Omega \longrightarrow P_f(X)$, then the following statements are equivalent.*

(a) *$F^-(C) \in \Sigma$ for every $C \subseteq X$ closed.*
(b) *F is measurable.*
(c) *$F^-(K) \in \Sigma$ for every $K \subseteq X$ compact.*

PROOF: (a)$\Rightarrow$ (b): This implication is Proposition 6.2.3.

(b)$\Rightarrow$ (c): This implication is Proposition 6.2.14.

(c)$\Rightarrow$ (a): Because X is σ-compact, we have $X = \bigcup_{n\geq 1} K_n$ with $K_n \subseteq X$ is compact. Let $C \subseteq X$ be a closed set. Then we have $F^-(C) = F^-\big(\bigcup_{n\geq 1} K_n \cap C\big) = \bigcup_{n\geq 1} F^-(K_n \cap C) \in \Sigma$ (because $K_n \cap C$ is compact for every $n \geq 1$). $\square$

Next we introduce a class of spaces, broader than the class of Polish spaces, because they are not necessarily metrizable. This makes them useful in many concrete situations, for example, when we deal with certain Banach spaces furnished with their weak topology.

DEFINITION 6.2.16 A Hausdorff topological space X is said to be a *Souslin space*, if there exists a Polish space Y and a continuous surjection $u : Y \longrightarrow X$.

REMARK 6.2.17 Souslin subspaces of a Polish space, are known in the literature as analytic sets. A Souslin space is always separable but need not be metrizable. For example, consider an infinite-dimensional separable Banach space endowed with the weak topology. Evidently this is a nonmetrizable Souslin space. It can be shown that closed and open subsets of a Souslin space are Souslin, countable products of Souslin spaces are Souslin, and countable intersections and unions of Souslin subspaces of a Hausdorff topological space are Souslin. From these facts, it follows that if X is a separable Banach space and $X^*_{w^*}$ denotes the dual X^* of X endowed with the w^*–topology, then $X^*_{w^*}$ is Souslin (indeed note that $X^* = \bigcup_{n\geq 1} n\overline{B}^*_1$, where $\overline{B}^*_1 = \{x^* \in X^* : \|x^*\| \leq 1\}$ and recall that $(\overline{B}^*_1, w^*)$ is compact metrizable, hence Polish, in particular then Souslin).

It is well known that a Borel set in $\mathbb{R}^2$ does not in general project to a Borel set on $\mathbb{R}$. The next theorem determines more precisely the structure of this projection. The result is known as the *Yankov–von Neumann–Aumann projection theorem* and its proof can be found in Hu–Papageorgiou [313, p. 149]. Recall that if (Ω, Σ) is a measurable space, the universal σ-field corresponding to Σ is defined by $\widehat{\Sigma} = \bigcap_{\mu} \overline{\Sigma}_\mu$, where μ ranges over all finite measures on Σ and $\overline{\Sigma}_\mu$ denotes the μ-completion of Σ. If (Ω, Σ, μ) is a σ-finite complete measure space, then $\Sigma = \widehat{\Sigma}$.

THEOREM 6.2.18 *If (Ω, Σ) is a measurable space, X is a Souslin space, and $A \in \Sigma \times \mathbf{B}(X)$, then* $\operatorname{proj} A \in \widehat{\Sigma}$.

Using this theorem, we can have a partial converse to Proposition 6.2.10.

PROPOSITION 6.2.19 *If (Ω, Σ) is a complete measurable space (i.e., $\Sigma = \widehat{\Sigma}$), X is a Souslin space, and $F : \Omega \longrightarrow 2^X \setminus \{\emptyset\}$ is a multifunction such that* $\operatorname{Gr} F \in \Sigma \times \mathbf{B}(X)$, *then for every $D \in \mathbf{B}(X)$, we have $F^-(D) \in \Sigma$.*

PROOF: By definition $F^-(D) = \{\omega \in \Omega : F(\omega) \cap D \neq \emptyset\} = \operatorname{proj}[\operatorname{Gr} F \cap (\Omega \times D)] \in \widehat{\Sigma} = \Sigma$ (see Theorem 6.2.18). □

Now we can state a theorem summarizing the situation concerning the measurability of a multifunction with values in a separable metric space.

THEOREM 6.2.20 *Let (Ω, Σ) be a measurable space, X a separable metric space, and $F : \Omega \longrightarrow P_f(X)$. We consider the following properties.*

(a) *For every $D \in \mathbf{B}(X)$, $F^-(D) \in \Sigma$.*
(b) *For every $C \subseteq X$ closed, $F^-(C) \in \Sigma$.*
(c) *F is measurable.*
(d) *For every $x \in X$, $\omega \longrightarrow d(x, F(\omega))$ is Σ-measurable.*
(e) *F is graph-measurable.*

Then the following implications are true.

(1) (a)$\Rightarrow$(b)$\Rightarrow$(c)$\Leftrightarrow$ (d)$\Rightarrow$(e).
(2) *If X is σ-compact, then* (b)$\Leftrightarrow$ (c).
(3) *If $\Sigma = \widehat{\Sigma}$ and X is complete (i.e., X is a Polish space), then statements* (a) *to* (e) *are equivalent.*

It is easy to see that countable unions preserve measurability and graph measurability. This is no longer true with intersections. Additional conditions are needed to guarantee measurability of countable intersections.

PROPOSITION 6.2.21 *If $F_n : \Omega \longrightarrow P_f(X)$, $n \geq 1$, are measurable multifunctions and for some $n_0 \geq 1$ F_{n_0} has values in $P_k(X)$, then $\omega \longrightarrow F(\omega) = \bigcap\limits_{n\geq 1} F_n(\omega)$ is measurable.*

PROOF: First we assume that all multifunctions F_n have values in $P_k(X)$. We set $\mathcal{P}(\omega) = \prod\limits_{n\geq 1} F_n(\omega)$. It is easy to see that $\mathcal{P}$ is measurable. Because $\mathcal{P}$ is compact-valued, from Proposition 6.2.13 we have that $\mathcal{P}^-(C) \in \Sigma$ for every $C \subseteq X$ closed. Let $C \subseteq X$ be closed and $\Delta \subseteq X^{\mathbb{N}}$ be the diagonal set of the product space. Then $F^-(C) = \{\omega \in \Omega : \bigcap\limits_{n\geq 1} F_n(\omega) \cap C \neq \emptyset\} = \{\omega \in \Omega : \mathcal{P}(\omega) \cap \Delta \cap C^N \neq \emptyset\} \in \Sigma$ (because $\Delta \cap C^{\mathbb{N}}$ is closed in $X^{\mathbb{N}}$).

Next we assume that only one of the F_n's has compact values. Let Y be a metrizable compactification of X (recall that X is homeomorphic to a subset of the *Hilbert cube* $\mathcal{H} = [0,1]^{\mathbb{N}}$). Let $\widehat{F}_n : \Omega \longrightarrow P_k(Y)$ be defined by $\widehat{F}_n(\omega) = \overline{F}_n(\omega)$ (the closure in Y), for all $n \geq 1$. We know that $\widehat{F}_n$ is measurable (see Proposition

6.2.14). So from the first part of the proof, we know that if $\widehat{F}(\omega) = \bigcap_{n\geq 1} \widehat{F}_n(\omega)$; then $\widehat{F}^-(C) \in \Sigma$ for all $C \subseteq Y$ closed. Because $F_n = \widehat{F}_n$ for some $n \geq 1$, we have $F = \widehat{F}$ and so from Proposition 6.2.3 we conclude that F is measurable. $\square$

In order to study scalarly measurable multifunctions (see Definition 6.2.1(c)), we need to produce some results on the existence of measurable selectors for multifunctions. This is done in the next section, where we prove theorems on the existence of continuous and measurable selectors.

6.3 Continuous and Measurable Selectors

Given two sets X, Y and a multifunction $F : X \longrightarrow 2^Y \setminus \{\emptyset\}$, a selector of F is a single-valued map $f : X \longrightarrow Y$ such that $f(x) \in F(x)$ for all $x \in X$. When X, Y both have topological structure, it is natural to look for continuous selectors. If $X = \Omega$ has a measure-theoretic structure, then we seek to produce measurable selectors. In this section we study both cases. We start with continuous selectors. First a negative observation directs our efforts to the appropriate class of multifunctions.

EXAMPLE 6.3.1 An usc multifunction need not have a continuous selector. We consider the usc multifunction $F : \mathbb{R} \longrightarrow P_{fc}(\mathbb{R})$ defined by

$$F(x) = \begin{cases} -1 & \text{if } x < 0 \\ [0, 1] & \text{if } x = 0 \\ 1 & \text{if } x > 0 \end{cases}.$$

It is clear that F cannot have a continuous selector. Note that $F(x) = \partial\varphi(x)$ where $\varphi(x) = |x|$ (the subdifferential is in the sense of convex analysis, see Definition 1.2.28).

This example suggests that usc multifunctions are not the right class to consider. The next proposition indicates where we should look in order to produce continuous selectors. In what follows, for the study of continuous selectors of a multifunction we assume that X, Y are Hausdorff topological spaces. Additional hypotheses are introduced as needed.

PROPOSITION 6.3.2 *If $F : X \longrightarrow 2^Y \setminus \{\emptyset\}$ and for every $(x, y) \in \mathrm{Gr}\, F$ we can find $U \in \mathcal{N}(x)$ and f a continuous selector of $F\big|_U$ such that $f(x) = y$, then F is lsc.*

PROOF: Given $V \subseteq Y$ open set, we need to show that $F^-(V)$ is open. Let $(x, y) \in \mathrm{Gr}\, F \cap (X \times V)$. By hypothesis we know that there exist $U \in \mathcal{N}(x)$ and a continuous function $f : U \longrightarrow Y$ such that $f(u) \in F(u)$ for all $u \in U$ and $f(x) = y$. Set $U' = \{x' \in U : f(x') \in V\} \in \mathcal{N}(x)$. Then $U' \subseteq F^-(V)$ and so $F^-(V)$ is open. $\square$

REMARK 6.3.3 Multifunctions that satisfy the hypotheses of the above proposition are called *locally selectionable.*

PROPOSITION 6.3.4 *If X is paracompact, Y is a topological vector space, and $F : X \longrightarrow 2^Y \setminus \{\emptyset\}$ is a multifunction with convex values such that $F^-(y) = \{x \in X : y \in F(x)\}$ is open for all $y \in Y$, then F admits a continuous selector f (i.e., there exists a continuous map $f : X \longrightarrow Y$ such that $f(x) \in F(x)$ for all $x \in X$).*

PROOF: For every $(x, y) \in \mathrm{Gr}\, F$, we can take $U = F^-(y) \in \mathcal{N}(x)$ and $f : U \longrightarrow Y$ defined by $f(x') = y$ for all $x' \in U$. So F is locally selectionable (see Remark 6.3.3). Therefore for every $(x, y) \in \mathrm{Gr}\, F$ we can find $U_x \in \mathcal{N}(x)$ and $f_x : U_x \longrightarrow Y$ continuous. Because of paracompactness let $\{U_i\}_{i\in I}$ be a locally finite refinement of the cover $\{U_x\}_{x\in X}$. Let $\{g_i\}_{i\in I}$ be a continuous partition of unity subordinate to the cover $\{U_i\}_{i\in I}$. We define the map

$$f(x) = \sum_{i\in I} g_i(x) f_{x_i}(x) \qquad \text{for all } x \in X. \tag{6.8}$$

Here $x_i \in X$ is such that $U_i \subseteq U_{x_i}$ (recall that $\{U_i\}_{i\in I}$ is a refinement of $\{U_x\}_{x\in X}$). Note that because $\{U_i\}_{i\in I}$ is locally finite, at every $x \in X$, the summation in (6.8) is finite. Therefore f is continuous and because F is convex-valued, for every $x \in X$ we have $f(x) \in F(x)$. □

REMARK 6.3.5 If $\mathrm{Gr}\, F$ is open in $X \times Y$, then for every $y \in Y$ $F^-(y)$ is open.

Proposition 6.3.2 suggests that we should focus on lsc multifunctions. This leads to the celebrated *Michael's selection theorem.*

THEOREM 6.3.6 *If X is paracompact, Y is a Banach space, and $F : X \longrightarrow P_{fc}(Y)$ is lsc, then F admits a continuous selector.*

PROOF: In what follows $B_1 = \{y \in Y : \|y\| < 1\}$. First we produce a continuous function $f : X \longrightarrow Y$ such that

$$f(x) \in F(x) + \varepsilon B_1 \qquad \text{for all } x \in X, \text{ with } \varepsilon > 0. \tag{6.9}$$

For every $x \in X$, we choose $y_x \in F(x)$. Then due to the lower semicontinuity of F, the set $F^-(y_x + \varepsilon B_1)$ is open and so $\{F^-(y_x + \varepsilon B_1)\}_{x\in X}$ is an open cover of X. Due to the paracompactness of X, we can find $\{F^-(y_i + \varepsilon B_1)\}_{i\in I}$ a locally finite refinement of $\{F^-(y_i + \varepsilon B_i)\}_{i\in I}$. We set

$$f(x) = \sum_{i\in I} g_i(x) u_i \qquad \text{for all } x \in X,$$

where $u_i \in (y_i + \varepsilon B_i) \subseteq (y_{x_i} + \varepsilon B_1)$. As before, at every $x \in X$ the above sum is finite and so f is continuous and because of the convexity of the values of $x \longrightarrow F(x) + \varepsilon B_1$, we have

$$f(x) \in F(x) + \varepsilon B_1 \qquad \text{for all } x \in X.$$

Next, inductively we generate a sequence $\{f_n\}_{n\geq 1}$ of continuous functions $f_n : X \longrightarrow Y$ such that

$$f_n(x) \in F(x) + \frac{1}{2^n} B_1 \tag{6.10}$$

$$\text{and} \quad \|f_{n+1}(x) - f_n(x)\| < \frac{1}{2^{n-1}} \qquad \text{for all } x \in X \text{ and all } n \geq 1. \tag{6.11}$$

For $n = 1$, by the first part of the proof we have $f_1 : X \longrightarrow Y$ a continuous function such that $f_1(x) \in F(x) + \frac{1}{2}B_1$ for all $x \in X$. For the induction step, suppose we have produced functions $\{f_k\}_{k=1}^n$ that satisfy (6.10) and (6.11). We consider the multifunction

$$H(x) = F(x) \cap \big(f_n(x) + \frac{1}{2^n}B_1\big) \qquad \text{for all } x \in X.$$

Because of (6.10) we see that H has nonempty values, which are convex. Moreover, from Proposition 6.1.24 we have that H is lsc. Then from the first part of the proof we know that we can find $f_{n+1} : X \longrightarrow Y$ a continuous function such that

$$f_{n+1}(x) \in H(x) + \frac{1}{2^{n+1}}B_1 \qquad \text{for all } x \in X,$$

$$\Rightarrow \ f_{n+1}(x) \in f_n(x) + \frac{1}{2^n}B_1 + \frac{1}{2^{n+1}}B_1 \subseteq f_n(x) + \frac{1}{2^{n-1}}B_1 \qquad \text{for all } x \in X.$$

This completes the induction. Because $\sum\limits_{n\geq 1} 1/(2^{n-1}) < +\infty$, from (6.11) it follows that $\{f_n\}_{n\geq 1}$ is a Cauchy sequence in $C(X,Y)$. So we can find $f \in C(X,Y)$ such that $f_n(x) \longrightarrow f(x)$ uniformly in $x \in X$. Because F is closed-valued from (6.10) we deduce that $f(x) \in F(x)$ for all $x \in X$. Therefore f is a continuous selector of F. $\square$

REMARK 6.3.7 It can be shown that the existence of continuous selectors is in fact equivalent to the paracompactness of the space X. Namely, if X is a Hausdorff topological space, Y is a Banach space, and every lsc multifunction $F : X \longrightarrow P_{fc}(Y)$ admits a continuous selector, then X is necessarily paracompact.

We can use Theorem 6.3.6 to produce a continuous selector that passes from a prescribed point of Gr F.

COROLLARY 6.3.8 *If X is a paracompact space, Y is a Banach space, and $F : X \longrightarrow P_{fc}(Y)$ is lsc, then given $(x,y) \in \mathrm{Gr}F$ we can find a continuous selector f_0 of F such that $f_0(x) = y$.*

PROOF: Let $F_0 : X \longrightarrow P_{fc}(Y)$ be defined by

$$F_0(x') = \begin{cases} F(x') & \text{if } x' \neq x \\ \{y\} & \text{if } x' = x \end{cases}.$$

It is easy to see that F_o is lsc. So by virtue of Theorem 6.3.6 we can find $f_0 : X \longrightarrow Y$ a continuous function such that $f_0(x') \in F_0(x')$ for all $x' \in X$. Evidently f_0 is a continuous selector of F such that $f_0(x) = y$. $\square$

Another interesting consequence of Theorem 6.3.6, is the following corollary.

COROLLARY 6.3.9 *If X, Y are Banach spaces and $A \in \mathcal{L}(Y,X)$ is surjective, then there exists $f : X \longrightarrow Y$ a continuous (not necessarily linear) map such that*

$$A \circ f = \mathrm{Id}_X.$$

PROOF: Let $F(x)=A^{-1}(x)$, $x\in X$. Clearly F has nonempty, closed, and convex values. Moreover, from the open mapping theorem we know that A is an open map. Now, let $V\subseteq Y$ be open. We show that $F^-(V)$ is open in X, hence F is lsc. Indeed we have

$$F^-(V)=\{x\in X: F(x)\cap V\neq\emptyset\}:\{x\in X: A^{-1}(x)\cap V\neq\emptyset\}.$$

If $y\in A^{-1}(x)\cap V$, then $V\in\mathcal{N}(y)$ and so $U=A(V)\in\mathcal{N}(x)$ (recall A is an open map). Then for every $x'\in U$ we have $A^{-1}(x')\cap V\neq\emptyset$, hence $F^-(V)$ is open (i.e., F is lsc). We apply Theorem 6.3.6 and obtain $f: X\longrightarrow Y$ a continuous selector of F. Evidently $A\circ f=\mathrm{Id}_X$. □

REMARK 6.3.10 An interesting byproduct of the above proof is the fact that if X, Y are Hausdorff topological spaces and $\varphi: Y\longrightarrow X$ is an open map (i.e., it maps open sets to open sets), then $x\longrightarrow\varphi^{-1}(x):\{y\in Y:\varphi(y)=x\}$ is a multifunction which is lsc from X into $2^Y\setminus\{\emptyset\}$. In the same vein if $\varphi: Y\longrightarrow X$ is a closed map (i.e., it maps closed sets to closed sets), then $x\longrightarrow\varphi^{-1}(x)$ is usc from X into $2^Y\setminus\{\emptyset\}$.

We can improve the conclusion of Theorem 6.3.6 and produce a whole sequence $\{f_n\}_{n\geq1}$ of continuous selectors of F such that $\{f_n(x)\}_{n\geq1}$ is dense in $F(x)$ for all $x\in X$, provided we enrich the structure of the spaces X and Y.

THEOREM 6.3.11 *If X is metric space, Y is a separable Banach space, and $F: X\longrightarrow P_{fc}(Y)$ is lsc, then there exists $\{f_n\}_{n\geq1}$ a sequence of continuous selectors of F such that*

$$F(x)=\overline{\{f_n(x)\}}_{n\geq1}\qquad \text{for all } x\in X.$$

PROOF: Let $\{y_n\}_{n\geq1}$ be a dense set in Y and set $U_{nm}=F^-\Big(B_{1/2^m}(y_n)\Big)$, n, $m\geq1$. Due to the lower semicontinuity of F each set U_{nm} is open. Because X is a metric space, every open set is F_σ. So we have

$$U_{nm}=\bigcup_{k\geq1}C_{nmk}\quad\text{with } C_{nmk} \text{ is closed for all } k\geq1.$$

We define

$$F_{nmk}(x)=\begin{cases} F(x) & \text{if } x\notin C_{nmk}\\ \overline{F(x)\cap B_{1/2^m}(y_n)} & \text{if } x\in C_{nmk}\end{cases}.$$

Clearly F_{nmk} is lsc with values in $P_{fc}(Y)$. So by Theorem 6.3.6 we can find $f_{nmk}: X\longrightarrow Y$ a continuous selector of F_{nmk}. We claim that $\{f_{nmk}\}_{n,m,k\geq1}$ is the desired sequence. To see this let $y\in F(x)$ and $m\geq1$. We pick $y_n\in y+(1/2^m)B_1$. Then $x\in U_{n,m+2}$ and so $x\in C_{n,m+2,k}$ for some $k\geq1$. But then we have $f_{n,m+2,k}(x)\in y_n+\big(1/(2^{m+2})\big)\overline{B}_1\subseteq y_n+\big(1/(2^{m+1})\big)B_1\subseteq y+(1/2^m)B_1$. □

In general Theorem 6.3.6 (Michael's selection theorem) is optimal in the sense that all hypotheses are needed and can not be relaxed, namely the completeness of the space Y and the convexity and closedness of the values of the lsc multifunction

F. However, if the Banach space Y is separable, then we can drop the closedness of the values of $F(\cdot)$, provided we assume that they have a nonempty interior. More precisely we have the following result, whose proof can be found in Michael [426, Theorem 3.1$'''$, p. 368] (see also Hu–Papageorgiou [313, p. 97]).

THEOREM 6.3.12 *If X is a metric space, Y is a separable Banach space, and $F : X \longrightarrow 2^Y \setminus \{\emptyset\}$ is a lsc multifunction such that for all $x \in X$ $F(x)$ is convex and* int $F(x) \neq \emptyset$, *then F admits a continuous selector.*

Deutsch–Kenderov [196] determined the biggest class of convex-valued multifunctions that admit a continuous approximate selector. They proved the following result.

THEOREM 6.3.13 *If X is a Hausdorff topological space, Y is a normed space, and $F : X \longrightarrow 2^Y \setminus \{\emptyset\}$ is a multifunction with convex values, then F admits a continuous approximate selector (i.e., for every $\varepsilon > 0$ there exists $f_\varepsilon : X \longrightarrow Y$ a continuous map such that $f_\varepsilon(x) \in F(x) + \varepsilon B_1$ for every $x \in X$) if and only if $F(\cdot)$ is a-lsc (see Definition 6.1.48(a)).*

However, there exist multifunctions with nonempty, closed, and convex values values which are a-lsc and do not admit continuous selectors. Deutsch–Kenderov [196] and De Blasi–Myjal [179], produced examples in this direction. In addition De Blasi–Myjal [179] proved the following theorem.

THEOREM 6.3.14 *If X is a paracompact space, Y is a Banach space, and $F : X \longrightarrow P_{fc}(Y)$ is an h_w-lsc multifunction, then F admits a continuous selector.*

REMARK 6.3.15 Because a multifunction $F : X \longrightarrow 2^Y \setminus \{\emptyset\}$ that is lsc is not necessarily h_w-lsc and vice-versa, we see that Theorem 6.3.14 is distinct from Theorem 6.3.6.

Before passing to measurable selectors, let us have a look at what happens with an usc multifunction. Recall that we started this section with a simple example illustrating that in general we do not expect to have continuous selectors for an usc multifunction. However, we can have some kind of approximate continuous selector (verify this with the multifunction given in Example 6.3.1).

PROPOSITION 6.3.16 *If X is a metric space, Y is a Banach space, and $F : X \longrightarrow 2^Y \setminus \{\emptyset\}$ is usc with convex values, then given $\varepsilon > 0$, we can find a locally Lipschitz function $f_\varepsilon : X \longrightarrow Y$ such that*

$$f_\varepsilon(X) \subseteq \text{conv}\, F(X) \quad \textit{and} \quad h^*(\text{Gr}\, f_\varepsilon, \text{Gr}\, F) < \varepsilon.$$

PROOF: Fix $\varepsilon > 0$. Because F is usc, for every $x \in X$ we can find $0 < \delta < \delta(\varepsilon, x)$ such that $F(x') \subseteq F(x) + (\varepsilon/2) B_1$ for all $x' \in B_{\delta(x)}(x)$. The collection $\{B_{\delta/4}(x)\}_{x \in X}$ is an open cover of X. The space X being a metric space, it is paracompact and so we can find $\{U_\alpha\}_{\alpha \in J}$ a locally finite refinement and $\{g_\alpha\}_{\alpha \in J}$ a corresponding locally Lipschitz partition of unity subordinate to it. For each $\alpha \in J$ we choose

$(u_\alpha, y_\alpha) \in \text{Gr } F \cap (U_\alpha \times Y)$ and set $f_\varepsilon(x) = \sum_{\alpha \in J} g_\alpha(x) y_\alpha$. Clearly f_ε is locally Lipschitz and $f_\varepsilon(X) \subseteq \text{conv } F(X)$.

For $x \in X$ we can find $J(x) \subseteq J$ finite such that $g_\alpha(x) > 0$ if $\alpha \in J(x)$. For every $\alpha \in J(x)$, let $x_\alpha \in X$ be such that $U_\alpha \subseteq B_{\delta_\alpha/4}(x_\alpha)$ with $\delta_\alpha = \delta(x_\alpha)$. Let $\beta \in J(x_\alpha)$ and set $\delta_\beta = \max\{\delta_\alpha : \alpha \in J(x)\}$. We have $x_\alpha \in B_{\delta_\beta/2}(x_\beta)$ and so $U_\alpha \subseteq B_{\delta_\beta}(x_\beta)$. Hence for any $\alpha \in J(x)$, we have $y_\alpha \in F(U_\alpha) \subseteq F(x_\beta) + (\varepsilon/2)B_1$ and so $f_\varepsilon(x) \subseteq F(x_\beta) + (\varepsilon/2)B_1$. Because $x \in X$ was arbitrary we conclude that $h^*(\text{Gr } f_\varepsilon, \text{Gr } F) < \varepsilon$. □

Next we turn our attention to measurable selectors of measurable multifunctions. The first result in this direction is the so-called *Kuratowski–Ryll Nardzewski selection theorem.* The technique of its proof is similar to the proof of Michael's selection theorem (Theorem 6.3.6).

THEOREM 6.3.17 *If (Ω, Σ) is a measurable space, X is a Polish space, and $F : \Omega \longrightarrow P_f(X)$ is measurable, then F admits a measurable selector.*

PROOF: Let d be a compatible metric on X and let $\{x_n\}_{n \geq 1}$ be a dense subset of X. For every $\omega \in \Omega$ let $n \geq 1$ be the smallest integer such that $F(\omega) \cap B_1(x_n) \neq \emptyset$. We set $f_0(\omega) = x_n$. Because by hypothesis F is measurable, the function $f_0 : \Omega \longrightarrow X$ is Σ-measurable. Moreover, we have

$$d\big(f_0(\omega), F(\omega)\big) < 1 \qquad \text{for all } \omega \in \Omega.$$

Suppose that we have constructed measurable maps $f_k : \Omega \longrightarrow X$, $k = 0, 1, \ldots, m$ such that $f_k(\Omega) \subseteq \{x_n\}_{n \geq 1}$ and

$$d\big(f_k(\omega), F(\omega)\big) < \frac{1}{2^k} \qquad \text{for all } \omega \in \Omega \text{ and all } k \in \{0, 1, \ldots, m\} \tag{6.12}$$

$$d\big(f_k(\omega), f_{k+1}(\omega)\big) < \frac{1}{2^{k-1}} \qquad \text{for all } \omega \in \Omega \text{ and all } k \in \{0, 1, \ldots, m-1\}. \tag{6.13}$$

We consider the sets $C_n = f_m^{-1}(\{x_n\}) = \{\omega \in \Omega : f_m(\omega) = x_n\}$, $n \geq 1$. Evidently these sets are mutually disjoint and $\bigcup_{n \geq 1} C_n = \Omega$. Moreover, from (6.12) we have

$$F(\omega) \cap B_{1/2^m}(x_n) \neq \emptyset \qquad \text{for all } \omega \in C_n.$$

We fix $n \geq 1$ and $\omega \in C_n$. Consider the smallest integer $i \geq 1$ such that

$$F(\omega) \cap B_{1/2^m}(x_n) \cap B_{1/2^{m+1}}(x_i) \neq \emptyset.$$

We set $f_{m+1}(\omega) = x_i$. Then we have

$$d\big(f_m(\omega), f_{m+1}(\omega)\big) \leq \frac{1}{2^m} + \frac{1}{2^{m+1}} < \frac{1}{2^{m-1}}$$

$$\text{and} \quad d\big(f_{m+1}(\omega), F(\omega)\big) < \frac{1}{2^{m+1}}.$$

Moreover, the map $\omega \longrightarrow f_{m+1}(\omega)$ from Ω into $\{x_n\}_{n \geq 1}$ is Σ-measurable. So by induction we have produced a sequence of Σ-measurable functions $f_k : \Omega \longrightarrow$

$\{x_n\}_{n\geq 1}$ that satisfy (6.12) and (6.13) for all $k \geq 1$. From (6.13) we see that for all $\omega \in \Omega$ $\{f_k(\omega)\}_{k\geq 1}$ is a Cauchy sequence in X. Because X is complete $f_k(\omega) \longrightarrow f(\omega)$ as $k \longrightarrow \infty$, for all $\omega \in \Omega$ (see (6.12)). Then $f : \Omega \longrightarrow X$ is Σ-measurable and $f(\omega) \in F(\omega)$ for all $\omega \in \Omega$ (recall that F has closed values). $\square$

In analogy with Theorem 6.3.11, we can produce a whole sequence $f_n : \Omega \longrightarrow X$, $n \geq 1$, of measurable selectors of F, such that $\{f_n(\omega)\}_{n\geq 1}$ is dense in $F(\omega)$ for all $\omega \in \Omega$.

THEOREM 6.3.18 *If (Ω, Σ) is a measurable space, X is a Polish space and $F : \Omega \longrightarrow P_f(X)$ is measurable, then there exists a sequence $f_n : \Omega \longrightarrow X$ of measurable selectors of F such that*

$$F(\omega) = \overline{\{f_n(\omega)\}_{n\geq 1}} \qquad \textit{for all } \omega \in \Omega.$$

PROOF: Let $\{U_n\}_{n\geq 1}$ be a countable base for X (recall that X being separable it is second countable). We define $F_n : \Omega \longrightarrow 2^X \setminus \{\emptyset\}$ by

$$F_n(\omega) = \begin{cases} F(\omega) \cap U_n & \text{if } F(\omega) \cap U_n \neq \emptyset \\ F(\omega) & \text{otherwise} \end{cases}.$$

Let $\Omega_n = \big\{\omega \in \Omega : F(\omega) \cap U_n \neq \emptyset\big\}$. Because F is measurable, we have that $\Omega_n \in \Sigma$ for all $n \geq 1$.

Let $V \subseteq X$ be nonempty open. Then

$$F_n^-(V) = \big\{\omega \in \Omega_n : F(\omega) \cap U_n \cap V \neq \emptyset\big\} \cup \big\{\omega \in \Omega_n^c : F(\omega) \cap V \neq \emptyset\big\} \in \Sigma.$$

$\Rightarrow$ F_n is measurable for every $n \geq 1$.

$\Rightarrow$ $\overline{F}_n$ is measurable for every $n \geq 1$ (see Proposition 6.2.4).

Apply Theorem 6.3.17 to obtain $f_n : \Omega \longrightarrow X$ a Σ-measurable map such that $f_n(\omega) \in F_n(\omega)$ for all $\omega \in \Omega$ and all $n \geq 1$. Then clearly $F(\omega) = \overline{\{f_n(\omega)\}_{n\geq 1}}$ for all $\omega \in \Omega$. $\square$

Now we can complete Theorem 6.2.20 as follows.

THEOREM 6.3.19 *If (Ω, Σ) is a measurable space, X is a separable metric space, $F : \Omega \longrightarrow P_f(X)$, and we consider the following statements*

(a) *For every $D \in \mathbf{B}(X)$, $F^-(D) \in \Sigma$.*
(b) *For every $C \subseteq X$ closed, $F^-(C) \in \Sigma$.*
(c) *F is measurable.*
(d) *For every $x \in X$, $\omega \longrightarrow d\big(x, F(\omega)\big)$ is Σ-measurable.*
(e) *There exists a sequence $\{f_n\}_{n\geq 1}$ of Σ-measurable selectors of F such that $F(\omega) = \overline{\{f_n(\omega)\}_{n\geq 1}}$ for all $\omega \in \Omega$.*
(f) Gr $F \in \Sigma \times \mathbf{B}(X)$.

Then the following implications hold

(1) (a)$\Rightarrow$(b)$\Rightarrow$(c)$\Leftrightarrow$ (d)$\Rightarrow$(f).
(2) *If X is complete, then* (c)$\Leftrightarrow$(d) $\Leftrightarrow$(e).
(3) *If X is σ-compact, then* (b)$\Leftrightarrow$(c).

(4) *If* Σ *is complete* (*i.e.*, $\Sigma = \widehat{\Sigma}$) *and* X *is complete, then* (a)$\longrightarrow$(f) *are all equivalent.*

The other measurable selection theorem is graph-conditioned and is known in the literature as the *Yankov–von Neumann–Aumann selection theorem.* We state the result and for a proof of it, we refer to Hu–Papageorgiou [313, p.p. 158–159].

THEOREM 6.3.20 *If* (Ω, Σ) *is a complete measurable space,* X *is a Souslin space, and* $F : \Omega \longrightarrow 2^X \setminus \{\emptyset\}$ *is graph-measurable, then there exists a sequence* $\{f_n\}_{n\geq 1}$ *of* Σ*-measurable selectors of* F *such that*

$$F(\omega) \subseteq \overline{\{f_n(\omega)\}}_{n\geq 1} \qquad \text{for all } \omega \in \Omega.$$

REMARK 6.3.21 If Σ is not complete, then the selectors are $\widehat{\Sigma}$–measurable. So if (Ω, Σ, μ) is a σ-finite measure space, we can find a sequence $\{f_n\}_{n\geq 1}$ of Σ-measurable selectors of F such that $F(\omega) \subseteq \overline{\{f_n(\omega)\}}_{n\geq 1}$ μ-a.e. on Ω.

Now let us prove some useful consequences of these selection theorems. The first is about the existence of implicit measurable functions and is sometimes called *Filippov's implicit function theorem.*

THEOREM 6.3.22 *If* (Ω, Σ) *and* $(T, \mathcal{S})$ *are measurable spaces,* X *is a Souslin space,* $F:\Omega \longrightarrow 2^X \setminus \{\emptyset\}$ *and* $G:\Omega \longrightarrow 2^T \setminus \{\emptyset\}$ *are graph-measurable,* $h:\Omega \times X \longrightarrow T$ *is* $\big(\Sigma \times \mathbf{B}(X), \mathcal{S}\big)$*-measurable, and for all* $\omega \in \Omega$, $h\big(\omega, F(\omega)\big) \cap G(\omega) \neq \emptyset$, *then there exists a* $\widehat{\Sigma}$*-measurable selector* u *of* F *such that* $h\big(\omega, u(\omega)\big) \in G(\omega)$ *for all* $\omega \in \Omega$.

PROOF: Let $K(\omega) = \{x \in F(\omega) : h(\omega, x) \in G(\omega)\}$. By hypothesis $K(\omega) \neq \emptyset$ for all $\omega \in \Omega$. Also let $h_0 : \Omega \times X \longrightarrow \Omega \times T$ be defined by $h_0(\omega, x) = \big(\omega, h(\omega, x)\big)$. Evidently h_0 is $\big(\Sigma \times \mathbf{B}(X), \Sigma \times \mathcal{S}\big)$-measurable. We have

$$\mathrm{Gr}K = \mathrm{Gr}F \cap h_0^{-1}(\mathrm{Gr}G).$$

The measurability of h_0 and the graph-measurability of G imply that

$$\begin{aligned} & h_0^{-1}(\mathrm{Gr}G) \in \Sigma \times \mathbf{B}(X), \\ \Rightarrow \; & \mathrm{Gr}K \in \Sigma \times \mathbf{B}(X) \quad \text{(recall that } F \text{ is graph-measurable).} \end{aligned}$$

So we can apply Theorem 6.3.20 (see also Remark 6.3.21) and obtain $u : \Omega \longrightarrow X$, a $\widehat{\Sigma}$-measurable map such that $u(\omega) \in K(\omega)$ for all $\omega \in \Omega$. Then

$$u(\omega) \in F(\omega) \quad \text{and} \quad h\big(\omega, u(\omega)\big) \in G(\omega) \qquad \text{for all } \omega \in \Omega.$$

□

REMARK 6.3.23 Theorem 6.3.22 is useful in control theory, where $\Omega = S = [0, b]$ and $G(s) = \{v(s)\}$ with $v : S \longrightarrow T$ is a $(\Sigma, \mathcal{S})$-measurable map, usually corresponding to a velocity. If $v(s) \in h\big(s, F(s)\big)$ for all $s \in S$, then there exists a measurable control $u : S \longrightarrow X$ which is admissible; that is, $u(s) \in F(s)$ and $v(s) = h\big(s, u(s)\big)$ for all $s \in S$. In many cases the control constraint set is trivial, namely $F(s) = X$ for $s \in S$. If T and X are metrizable spaces and h is a Carathéodory function, then h is jointly measurable (see Theorem 6.2.6).

We can also use the measurable selection theorems to prove a measurable version of the *Berge maximum theorem* (see Theorem 6.1.18).

THEOREM 6.3.24 *If* (Ω, Σ) *is a measurable space,* X *is a Souslin space,* $\varphi : \Omega \times X \longrightarrow \mathbb{R}^* = \mathbb{R} \cup \{\pm\infty\}$ *is a* $\Sigma \times \mathbf{B}(X)$*-measurable function, and* $F : \Omega \longrightarrow 2^X \setminus \{\emptyset\}$ *is a graph-measurable multifunction, then*

(a) $\omega \longrightarrow m(\omega) = \sup\{\varphi(\omega, x) : x \in F(\omega)\}$ *is* $\widehat{\Sigma}$*-measurable.*
(b) *If for all* $\omega \in \Omega$*, the set* $S(\omega) = \{x \in F(\omega) : m(\omega) = \varphi(\omega, x)\}$ *is nonempty, then* $\mathrm{Gr}\, S \in \widehat{\Sigma} \times \mathbf{B}(X)$.

PROOF: (a) Let $\lambda \in \mathbb{R}$. Note that $m(\omega) > \lambda$ if and only if we can find $x \in F(\omega)$ such that $\varphi(\omega, x) > \lambda$. Hence

$$\{\omega \in \Omega : m(\omega) > \lambda\} = \mathrm{proj}_\Omega\{(\omega, x) \in \Omega \times X : \varphi(\omega, x) > \lambda\}.$$

Because φ is jointly measurable, we have

$$\{(\omega, x) \in \Omega \times X : \varphi(\omega, x) > \lambda\} \in \Sigma \times \mathbf{B}(X).$$

Invoking Theorem 6.2.18, we obtain

$$\begin{aligned} & \mathrm{proj}_\Omega\{(\omega, x) \in \Omega \times X : \varphi(\omega, x) > \lambda\} \in \widehat{\Sigma}, \\ \Rightarrow\ & \{\omega \in \Omega : m(\omega) > \lambda\} \in \widehat{\Sigma}. \end{aligned}$$

Because $\lambda \in \mathbb{R}$ was arbitrary, we conclude that $\omega \longrightarrow m(\omega)$ is $\widehat{\Sigma}$-measurable.

(b) Let $\psi(\omega, x) = m(\omega) - \varphi(\omega, x)$. Evidently ψ is $\widehat{\Sigma} \times \mathbf{B}(X)$-measurable (see part (a)). Then

$$\mathrm{Gr}\, S = \mathrm{Gr}\, F \cap \{(\omega, x) \in \Omega \times X : \psi(\omega, x) = 0\} \in \widehat{\Sigma} \times \mathbf{B}(X).$$

□

REMARK 6.3.25 Because of Theorem 6.3.24(b) and Theorem 6.3.20 we can have a $\widehat{\Sigma}$-measurable selector of S. If S has empty values, then we can look for ε-maximizers. More precisely, if m is $\mathbb{R}$-valued and $\varepsilon > 0$, then the multifunction $\omega \longrightarrow S_\varepsilon(\omega) = \{x \in F(\omega) : \varphi(\omega, x) \geq m(\omega) - \varepsilon\}$ has always nonempty values, its graph belongs in $\widehat{\Sigma} \times \mathbf{B}(X)$ and so admits a $\widehat{\Sigma}$-measurable selector.

PROPOSITION 6.3.26 *If* (Ω, Σ) *is a measurable space,* X *is a separable Banach space,* $u : \Omega \longrightarrow X$ *is a* Σ*-measurable map, and* $r : \Omega \longrightarrow \mathbb{R}_+$ *is a* Σ*-measurable function, then* $\omega \longrightarrow \overline{B}_{r(\omega)}\big(u(\omega)\big) = \big\{x \in X : \|x - u(\omega)\| \leq r(\omega)\big\}$ *is measurable.*

PROOF: Let $\{x_n\}_{n\geq 1}$ be dense in the closed unit ball of X. Set

$$v_n(\omega) = u(\omega) + r(\omega) x_n, \quad n \geq 1.$$

Evidently for every $n \geq 1$, v_n is Σ-measurable and

$$\begin{aligned} & \overline{B}_{r(\omega)}\big(u(\omega)\big) = \overline{\{v_n(\omega)\}}_{n\geq 1} \\ \Rightarrow\ & \omega \longrightarrow \overline{B}_{r(\omega)}\big(u(\omega)\big) \text{ is measurable (see Theorem 6.3.18(2)).} \end{aligned}$$

□

The measurable selection theorems also provide the necessary tools to deal with scalarly measurable multifunctions (see Definition 6.2.1(c)).

PROPOSITION 6.3.27 *If* (Ω, Σ) *is a complete measurable space,* X *is a separable Banach space, and* $F: \Omega \longrightarrow 2^X \setminus \{\emptyset\}$ *is graph-measurable, then* F *is scalarly measurable.*

PROOF: By Theorem 6.3.20 we can find a sequence $\{f_n\}_{n \geq 1}$ of Σ-measurable selectors of F such that $F(\omega) \subseteq \overline{\{f_n(\omega)\}}_{n \geq 1}$ for all $\omega \in \Omega$. Then for every $x^* \in X^*$, we have

$$\begin{aligned} & \sigma\big(x^*, F(\omega)\big) = \sup_{n \geq 1} \langle x^*, f_n(\omega) \rangle, \\ \Rightarrow \ & \omega \longrightarrow \sigma\big(x^*, F(\omega)\big) \text{ is } \Sigma\text{-measurable}, \\ \Rightarrow \ & F \text{ is scalarly measurable.} \end{aligned}$$

□

If F is P_{wkc}-valued, then measurability and scalar measurability are equivalent notions. To prove this, we need to do some preparatory work. We start with a lemma.

LEMMA 6.3.28 *If* V *is a normed space and* $C \subseteq V$ *is nonempty and convex, then for every* $v \in V$*, we have* $d(v, C) = \sup\big[\langle v^*, v \rangle - \sigma(v^*, C) : \|v^*\| \leq 1\big]$.

PROOF: Let $\varphi(v) = d(v, C)$. We know that φ is continuous convex. From Example 1.2.27(b), we know that $\varphi^* = \sigma(\cdot, C) + i_{\overline{B}_1^*}$, where $\overline{B}_1^* = \{v^* \in V^* : \|v^*\| \leq 1\}$ and

$$i_{\overline{B}_1^*}(v^*) = \begin{cases} 0 & \text{if } v^* \in \overline{B}_1^* \\ +\infty & \text{if } v^* \notin \overline{B}_1^* \end{cases} \quad \text{for all } v^* \in V^*.$$

Because φ is continuous convex, we have that $\varphi = \varphi^{**} = (\varphi^*)^*$ (see Theorem 1.2.21). Hence

$$\begin{aligned} \varphi(v) &= \sup\big[\langle v^*, v \rangle - \sigma(v^*, C) - i_{\overline{B}_1^*}(v^*) : v^* \in V^*\big] \\ &= \sup\big[\langle v^*, v \rangle - \sigma(v^*, C) : \|v^*\| \leq 1\big]. \end{aligned}$$

□

Also we use a particular topology on X^*, which for convenience of the reader we recall in the next definition.

DEFINITION 6.3.29 Let V be a locally convex space. The locally convex topology on V of uniform convergence on w^*-compact, convex balanced sets in V^* is called the *Mackey topology* on V. Recall that $C \subseteq X^*$ is balanced, if $\lambda C \subseteq C$ for all $|\lambda| \leq 1$.

REMARK 6.3.30 The strongest locally convex topology τ on V for which we have $(V,\tau)^* = V^*$, is the Mackey topology. So, if V is a Banach space and we consider the strongest topology τ on V^* such that $(V^*,\tau)^* = V$, this is the Mackey topology, whereas the weakest topology on V^* for which this is true is the weak*-topology. Of course in this case the Mackey topology is strictly weaker than the strong (norm) topology. However, on the Banach space V the Mackey and strong topologies coincide. Finally note that in a locally convex space V a set $C \subseteq V$ is $w(V,V^*)$–compact if and only if the function $v^* \longrightarrow \sigma(v^*, C)$ is $m(V^*,V)$–continuous ($m(V^*,V)$ is the Mackey topology on V^* for the pair (V^*,V)).

PROPOSITION 6.3.31 *If (Ω,Σ) is a measurable space, X is a separable Banach space, and $F : \Omega \longrightarrow P_{wkc}(X)$, then F is measurable if and only if it is scalarly measurable.*

PROOF: $\Rightarrow$: From Theorem 6.3.19(2), we know that we can find a sequence $\{f_n\}_{n\geq 1}$ of Σ-measurable selectors of F such that $F(\omega) = \overline{\{f_n(\omega)\}}_{n\geq 1}$ for all $\omega \in \Omega$. Then for all $x^* \in X^*$ we have $\sigma(x^*, F(\omega)) = \sup\limits_{n\geq 1} \langle x^*, f_n(\omega)\rangle$ and so we conclude that $\omega \longrightarrow \sigma(x^*, F(\omega))$ is Σ-measurable.

$\Leftarrow$: Since X is separable, the space $X^*_{w^*}$ (= the dual Banach space X^* endowed with the w^*-topology) is separable. It follows that for every other topology τ on X^* for which we have $(X^*_\tau)^* = X$, the space X^*_τ is separable. In particular this is true if $\tau = m(X^*,X)$. Because F is $P_{wkc}(X)$-valued, from Remark 6.3.30 we know that $x^* \longrightarrow \sigma(x^*, F(\omega))$ is $m(X^*,X)$-continuous on X^*. From Lemma 6.3.28 for every $x \in X$ and every $\omega \in \Omega$, we have

$$d(x, F(\omega)) = \sup_{n\geq 1} \big[\langle x_n^*, x\rangle - \sigma(x_n^*, F(\omega))\big], \tag{6.14}$$

where $\{x_n^*\}_{n\geq 1}$ is dense in $\overline{B}_1^* = \{x^* \in X^* : \|x^*\| \leq 1\}$ for the Mackey topology. From (6.14) and Theorem 6.3.19(2), we conclude that F is measurable. $\square$

Now we can state a version of Theorem 6.3.19 for nonmetrizable spaces.

THEOREM 6.3.32 *If (Ω,Σ) is a complete measurable space, X is a regular Souslin space, $F : \Omega \longrightarrow P_f(X)$ is a multifunction, and we consider the following statements,*

(a) *For every $D \in \mathbf{B}(X)$, $F^-(D) \in \Sigma$;*
(b) *For every $C \subseteq X$ closed, $F^-(C) \in \Sigma$;*
(c) *For every $U \subseteq X$ open, $F^-(U) \in \Sigma$;*
(d) *There exists a sequence $\{f_n\}_{n\geq 1}$ of Σ-selectors of F such that*

$$F(\omega) = \overline{\{f_n(\omega)\}}_{n\geq 1} \qquad \text{for all } \omega \in \Omega;$$

(e) Gr $F \in \Sigma \times \mathbf{B}(X)$;
(f) *For every continuous function $u : X \longrightarrow \mathbb{R}$, the function*

$$\omega \longrightarrow m(\omega) = \sup[u(x) : x \in F(\omega)] \text{ is } \Sigma\text{-measurable},$$

then

(1) (a)$\Leftrightarrow$(d)$\Leftrightarrow$(e)$\Leftrightarrow$(f)$\Rightarrow$(b) $\Rightarrow$(c).
(2) *If X is second countable, then* (a)$\longrightarrow$(f) *are all equivalent.*

REMARK 6.3.33 The proof of this theorem is based on the fact that there exists a metric d on X defining a Souslin metric topology finer that the original topology on X (see Saint-Pierre [536]). Then $\mathbf{B}(X) = \mathbf{B}(X_d)$ and we can apply Theorem 6.3.19 to the multifunction $F : \Omega \longrightarrow P_f(X_d)$ (X_d is the space X with the new Souslin metric topology).

Now, if X is a separable Banach space, then X_{w^*}(= the dual Banach space X^* with the w^*-topology) is a Souslin space (see Remark 6.2.17) which is regular and second countable. Then using Theorem 6.3.32 we can have the following "dual" version of Proposition 6.3.31.

PROPOSITION 6.3.34 *If (Ω, Σ) is a complete measurable space, X is a separable Banach space, and $F : \Omega \longrightarrow 2^X \setminus \{\emptyset\}$ is a multifunction with w^*-compact convex values, then for every $U \subseteq X^*$ open $F^-(U) \in \Sigma$ if and only if for every $x \in X$, $\omega \longrightarrow \sigma\big(x, F(\omega)\big)$ is Σ-measurable.*

Let (Ω, Σ) be a measurable space and X a separable Banach space. Given a multifunction $F : \Omega \longrightarrow 2^X \setminus \{\emptyset\}$, by ext F we denote the multifunction which to each $\omega \in \Omega$ assigns the set ext $F(\omega)$ of extreme points of the set $F(\omega)$. In the next proposition we establish the measurability properties of the multifunction $\omega \longrightarrow \text{ext } F(\omega)$.

PROPOSITION 6.3.35 *If (Ω, Σ) is a measurable space, X is a separable Banach space, and $F : \Omega \longrightarrow P_{wkc}(X)$ is a measurable multifunction, then $\omega \longrightarrow \text{ext } F(\omega)$ is graph-measurable.*

PROOF: From the Krein–Milman theorem we know that for all $\omega \in \Omega$, ext $F(\omega) \neq \emptyset$. Recall that X^* with the Mackey topology $m(X^*, X)$ is separable. Let $\{x_n^*\}_{n \geq 1} \subseteq \overline{B}_1^*$ be $m(X^*, X)$-dense and consider the function $\varphi_F : \Omega \times X \longrightarrow \overline{\mathbb{R}}_+$ defined by

$$\varphi_F(\omega, x) = \begin{cases} \sum\limits_{n \geq 1} \frac{\langle x_n^*, x \rangle^2}{2^n} & \text{if } x \in F(\omega) \\ +\infty & \text{otherwise} \end{cases}.$$

Evidently φ_F is $\Sigma \times \mathbf{B}(X)$-measurable and for every $\omega \in \Omega$, $\varphi_F(\omega, \cdot)\big|_{F(\omega)}$ is continuous. Let $\mathcal{A}$ be the set of all continuous affine function $\alpha : X \longrightarrow \mathbb{R}$. We consider

$$\widehat{\varphi}_F(\omega, x) = \inf\big[\alpha(x) : \alpha \in \mathcal{A},\ \alpha(v) > \varphi_F(\omega, v) \quad \text{for all } \ v \in F(\omega)\big].$$

Let $u_n : \Omega \longrightarrow X$, $n \geq 1$, be a sequence of Σ-measurable selectors of F such that

$$F(\omega) = \overline{\{u_n(\omega)\}}_{n \geq 1} \quad \text{(see Theorem 6.3.19).}$$

For every $(\omega, x^*) \in \Omega \times X^*$, let

$$r_{x^*}(\omega) = \sup[\ \varphi_F(\omega, v) - \langle x^*, v \rangle : v \in F(\omega)].$$

We have that $r_{x^*}(\omega) < +\infty$, for every $\omega \in \Omega$ $x^* \longrightarrow r_{x^*}(\omega)$ is continuous on X^* and

$$\begin{aligned} r_{x^*}(\omega) &= \sup \big[\varphi_F(\omega, x) - \langle x^*, x \rangle : x \in F(\omega)\big], \\ \Rightarrow\ r_{x^*}(\omega) &= \sup_{n \geq 1} \big[\varphi_F(\omega, u_n(\omega)) - \langle x^*, u_n(\omega) \rangle\big], \\ \Rightarrow (\omega, x^*) &\longrightarrow r_{x^*}(\omega) \text{ is } \Sigma \times \mathbf{B}(X^*)\text{-measurable.} \end{aligned}$$

If we set

$$\widehat{\varphi}_F(\omega, x) = \inf \big[\langle x^*, x \rangle + r_{x^*}(\omega) : x^* \in X^*\big],$$

then for $\{v_m^*\}_{m \geq 1} \subseteq X^*$ $m(X^*, X)$-dense set, we have

$$\begin{aligned} \widehat{\varphi}_F(\omega, x) &= \inf_{m \geq 1} \big[\langle v_m^*, x \rangle + r_{v_m^*}(\omega)\big], \\ \Rightarrow\ (\omega, x) &\longrightarrow \widehat{\varphi}_F(\omega, x) \text{ is } \Sigma \times \mathbf{B}(X)\text{-measurable.} \end{aligned}$$

From Choquet [144, Chapter 6], we know that

$$\begin{aligned} &\text{ext } F(\omega) = \{x \in X : \widehat{\varphi}_F(\omega, x) = \varphi_F(\omega, x)\}, \\ \Rightarrow\ &\text{Gr } F \in \Sigma \times \mathbf{B}(X). \end{aligned}$$

□

REMARK 6.3.36 If we set $\xi_F(\omega, x) = \widehat{\varphi}_F(\omega, x) - \varphi_F(\omega, x)$, then ξ_F is $\Sigma \times \mathbf{B}(X)$-measurable, for every $\omega \in \Omega$, $\xi_F(\omega, \cdot)$ is strictly concave on $F(\omega)$, concave on X, and upper semicontinuous. The function $\xi_F(\omega, \cdot)$ is known in the literature as the *Choquet function* of the set $F(\omega)$.

6.4 Decomposable Sets and Set-Valued Integration

Throughout this section the standing hypotheses are: (Ω, Σ, μ) is a σ-finite measure space and X is a separable Banach space. By $L^0(\Omega, X)$ we denote the space of all equivalence classes in the set of all Σ-measurable maps from Ω into X, for the equivalence relation of equality almost everywhere.

DEFINITION 6.4.1 A set $K \subseteq L^0(\Omega, X)$ is said to be *decomposable*, if for every triple $(A, f_1, f_2) \in \Sigma \times K \times K$ we have $\chi_A f_1 + \chi_{A^c} f_2 = \chi_A f_1 + (1 - \chi_A) f_2 \in K$.

EXAMPLE 6.4.2 If $F : \Omega \longrightarrow 2^X \setminus \{\emptyset\}$ is a multifunction and $S_F = \{f \in L^0(\Omega, X) : f(\omega) \in F(\omega)\ \mu\text{-a.e.}\}$, then S_F is decomposable (possibly empty). The same is true for the sets $S_F^p = S_F \cap L^p(\Omega, X)$, $1 \leq p \leq \infty$. We show in the sequel that within closure all decomposable sets are of this form.

The condition of decomposability looks similar to convexity but only formally, because χ_A is not a constant and does not assume values between zero and one. However, as we show in this section, decomposability has some implications analogous to those of convexity.

PROPOSITION 6.4.3 *If $F:\Omega\longrightarrow 2^X\setminus\{\emptyset\}$ is graph-measurable and $S_F^p\neq\emptyset$, $1\leq p\leq\infty$, then we can find a sequence $\{f_n\}_{n\geq 1}\subseteq S_F^p$ such that $F(\omega)\subseteq\overline{\{f_n(\omega)\}}_{n\geq 1}$ μ-a.e. on Ω.*

PROOF: From Theorem 6.3.19 (see also Remark 6.3.21), we can find a sequence $\{g_m\}_{m\geq 1}$ of Σ-measurable selectors of F such that $F(\omega)\subseteq\overline{\{g_m(\omega)\}}_{m\geq 1}$ μ-a.e. on Ω. Since μ is σ-finite on Ω, we can find $\{A_k\}_{k\geq 1}\subseteq\Sigma$ a countable partition of Ω with $\mu(A_k)<+\infty$ for all $k\geq 1$. Let $f\in S_F^p$ and define

$$\begin{aligned} C_{mki} &= \{\omega\in\Omega : i-1\leq \|g_m(\omega)\|<i\}\cap A_k \\ f_{mki} &= \chi_{C_{mki}}g_m+\chi_{C^c_{mki}}f \qquad \text{for all } \quad m,k,i\geq 1. \end{aligned}$$

Evidently $\{f_{mki}\}_{m,k,i\geq 1}\subseteq S_F^p$ and $F(\omega)\subseteq\overline{\{f_{mki}(\omega)\}}_{m,k,i\geq 1}$ μ-a.e. on Ω. ☐

COROLLARY 6.4.4 *If $F_1,F_2:\Omega\longrightarrow 2^X\setminus\{\emptyset\}$ are graph measurable and for some $1\leq p\leq\infty$ we have $S_{F_1}^p=S_{F_2}^p\neq\emptyset$, then $F_1(\omega)=F_2(\omega)$ μ-a.e. on Ω.*

LEMMA 6.4.5 *If $F:\Omega\longrightarrow 2^X\setminus\{\emptyset\}$ is graph-measurable, $1\leq p\leq\infty$, $\{f_n\}_{n\geq 1}\subseteq S_F^p$ satisfies $F(\omega)\subseteq\overline{\{f_n(\omega)\}}_{n\geq 1}$ μ-a.e. $f\subseteq S_F^p$, and $\varepsilon>0$, then we can find a finite Σ-partition $\{C_k\}_{k=1}^N$ of Ω such that*

$$\Big\|f-\sum_{k=1}^N\chi_{C_k}f_k\Big\|_p<\varepsilon.$$

PROOF: Without any loss of generality we assume that $f(\omega)\in F(\omega)$ for all $\omega\in\Omega$. Let $\vartheta\in L^1(\Omega)$, $\vartheta(\omega)>0$ for all $\omega\in\Omega$ and $\int_\Omega\vartheta d\mu<\varepsilon^p/3$. We can find $\{D_n\}_{n\geq 1}$ a Σ-partition of Ω such that

$$\|f(\omega)-f_n(\omega)\|^p<\vartheta(\omega) \qquad \text{for all } \omega\in D_n,\ n\geq 1.$$

We choose $N\geq 1$ large enough so that

$$\sum_{n\geq N+1}\int_{D_n}\|f(\omega)\|^p d\mu<\frac{\varepsilon^p}{3.2^p} \quad\text{and}\quad \sum_{n\geq N+1}\int_{D_n}\|f_1(\omega)\|^p d\mu<\frac{\varepsilon^p}{3.2^p}.$$

We define a finite Σ-partition $\{C_k\}_{k=1}^N$ of Ω by

$$C_1=D_1\cup\Big(\bigcup_{n\geq N+1}D_n\Big) \quad\text{and}\quad C_k=D_k \quad\text{for } k=2,\dots,N.$$

Then we have

$$\begin{aligned} \Big\|f-\sum_{k=1}^N\chi_{C_k}f_k\Big\|_p^p &= \sum_{k=1}^N\int_{D_k}\|f(\omega)-f_k(\omega)\|^p d\mu+\sum_{n\geq N+1}\int_{D_k}\|f(\omega)-f_1(\omega)\|^p d\mu \\ &\leq \int_\Omega\vartheta(\omega)d\mu+\sum_{n\geq N+1}2^{p-1}\int_{D_k}\|f(\omega)\|^p+\|f_1(\omega)\|^p d\mu<\varepsilon^p. \end{aligned}$$

☐

Using this lemma we can completely characterize closed decomposable sets in the Lebesgue–Bochner space $L^p(\Omega,X)$, $1\leq p<\infty$.

THEOREM 6.4.6 *If $K \subseteq L^p(\Omega, X)$, $1 \leq p<\infty$ is nonempty and closed, then K is decomposable if and only if $K=S^p_F$ for some $F:\Omega \longrightarrow P_f(X)$-measurable.*

PROOF: $\Rightarrow$: From Proposition 6.4.3 we know that we can find a sequence $\{f_n\}_{n\geq 1} \subseteq L^p(\Omega, X)$ such that

$$X = \overline{\{f_n(\omega)\}}_{n\geq 1} \quad \mu\text{-a.e. on } \Omega.$$

Let $\xi_n = \inf\left\{\|f_n - h\|_p : h \in K\right\}$ and let $\{h_{nm}\}_{m\geq 1} \subseteq K$ be such that

$$\|f_n - h_{nm}\|_p \downarrow \xi_n \quad \text{as } m \longrightarrow \infty.$$

We set $F(\omega)=\overline{\{h_{nm}(\omega)\}}_{n,m\geq 1}$ for all $\omega \in \Omega$. Evidently $\omega \longrightarrow F(\omega)$ is measurable from Ω into $P_f(X)$ (see Theorem 6.3.18). We claim that $K=S^p_F$. To this end let $f \in S^p_F$ and $\varepsilon>0$. From Lemma 6.4.5, we know that we can find a finite Σ-partition $\{C_k\}_{k=1}^N$ of Ω and $\{g_k\}_{k=1}^N \subseteq \{h_{nm}(\omega)\}_{n,m\geq 1}$ such that

$$\Big\|f - \sum_{k=1}^N \chi_{C_k} g_k\Big\|_p < \varepsilon.$$

Due to the decomposability of K we have

$$\begin{aligned} &\sum_{k=1}^N \chi_{C_k} g_k \in K, \\ \Rightarrow\ & f \in K \quad \text{(because } K \text{ is closed)}, \\ \Rightarrow\ & S^p_F \subseteq K. \end{aligned} \tag{6.15}$$

Suppose $K\neq S^p_F$. Then we can find $f \in K$, $A \in \Sigma$ with $\mu(A)>0$ and $\delta>0$ such that

$$\|f(\omega) - h_{nm}(\omega)\| \geq \delta \qquad \text{for all } \omega \in A \text{ and all } n,m \geq 1. \tag{6.16}$$

For what follows we fix $n\geq 1$ so that

$$E = A \cap \left\{\omega \in \Omega : \|f(\omega) - f_n(\omega)\| < \frac{\delta}{3}\right\} \tag{6.17}$$

has a positive measure. We define

$$g_m = \chi_E f + \chi_{E^c} h_{nm}, \quad m \geq 1.$$

We have $g_m \in K$ and

$$\begin{aligned} \|h_{nm}(\omega) - f_n(\omega)\| &\geq \|h_{nm}(\omega) - f(\omega)\| - \|f(\omega) - f_n(\omega)\| \\ &\geq \delta - \frac{\delta}{3} = \frac{2\delta}{3} \qquad \text{for all } \omega \in E \quad \text{(see (6.16) and (6.17))}. \end{aligned} \tag{6.18}$$

Therefore

$$\begin{aligned} &\|f_n - h_{nm}\|_p^p - \xi_n^p \\ \geq\ & \|f_n - h_{nm}\|_p^p - \|f_n - g_m\|_p^p \\ \geq\ & \int_E \left(\|f_n - h_{nm}\|^p - \|f_n - f\|^p\right) d\mu \\ \geq\ & \left[\left(\frac{2\delta}{3}\right)^p - \frac{\delta^p}{3^p}\right]\mu(E)>0, \quad m \geq 1 \quad \text{(see (6.18))}. \end{aligned} \tag{6.19}$$

If in (6.19) we let $m \longrightarrow \infty$, we have a contradiction. Hence

$$K \subseteq S_F^p. \tag{6.20}$$

From (6.15) and (6.20), we conclude that $K = S_F^p$. □

REMARK 6.4.7 The result is also true for $p=+\infty$, provided that K is boundedly closed; that is, if $\{f_n\}_{n\geq 1} \subseteq K$, $\sup_{n\geq 1} \|f_n\|_\infty < +\infty$ and $f_n(\omega) \longrightarrow f(\omega)$ μ-a.e. on Ω, then $f \in K$.

Bounded decomposable sets in $L^1(\Omega, X)$ are in fact uniformly integrable. To show this we need the following lemma, whose proof can be found in Hu–Papageorgiou [313, p. 178], or Neveu [458, p. 121].

LEMMA 6.4.8 *If (Ω, Σ, μ) is a finite space and $\mathcal{R}$ is a family of Σ-measurable, $\mathbb{R}_+ = \mathbb{R} \cup \{+\infty\}$-valued functions on Ω, then we can find a unique (modulo μ-a.e. equality) Σ-measurable function $h : \Omega \longrightarrow \mathbb{R}_+$ such that*

(a) *For every $f \in \mathcal{R}$, we have $f(\omega) \leq h(\omega)$ μ-a.e. on Ω.*
(b) *If $\widehat{h} : \Omega \longrightarrow \mathbb{R}_+$ is another Σ-measurable function such that $f(\omega) \leq \widehat{h}(\omega)$ μ-a.e. on Ω for all $f \in \mathcal{R}$, then $h(\omega) \leq \widehat{h}(\omega)$ μ-a.e. on Ω.*

Moreover, we can find a sequence $\{f_n\}_{n\geq 1} \subseteq \mathcal{R}$ such that

$$h(\omega) = \sup_{n\geq 1} f_n(\omega) \quad \mu\text{-a.e. on } \Omega.$$

Finally, if $\mathcal{R}$ is directed upwards (namely if for every $f_1, f_2 \in \mathcal{R}$, we can find $f_3 \in \mathcal{R}$ such that $f_1(\omega), f_2(\omega) \leq f_3(\omega)$ μ-a.e. on Ω), then $\{f_n\}_{n\geq 1}$ can be chosen to be increasing.

REMARK 6.4.9 The function h is denoted by ess sup $\mathcal{R}$ and is the least upper bound of $\mathcal{R}$ in the sense of inequality μ-a.e. The essential supremum coincides with the supremum modulo μ-null sets for countable families, but it is not the same for uncountable families. To see this let $A \subseteq [0,1]$ and $\mathcal{R} = \{\chi_{\{\alpha\}} : \alpha \in A\}$. Then ess sup $\mathcal{R} = 0$, but $\sup \chi_{\{\alpha\}} = \chi_A$ which need not be measurable if A is not or even if A is measurable but of positive Lebesgue measure, then $\chi_A \neq 0$. In a similar way we can define the essential infimum of $\mathcal{R}$, denoted by ess inf $\mathcal{R}$.

PROPOSITION 6.4.10 *If (Ω, Σ, μ) is a finite measure space, Y is a Banach space, and $K \subseteq L^1(\Omega, Y)$ is bounded and decomposable, then K is uniformly integrable.*

PROOF: We introduce the set $|K| = \{\|f(\cdot)\| : f \in K\} \subseteq L^1(\Omega)$ and let $h = \operatorname{ess\,sup} |K|$ (see Remark 6.4.9). From Lemma 6.4.8, we know that we can find $\{f_n\}_{n\geq 1} \subseteq K$ such that $h(\omega) = \sup_{n\geq 1} \|f_n(\omega)\|$ μ-a.e. on Ω. Moreover, the decomposability of K implies that $|K|$ is directed upwards and so $\{f_n\}_{n\geq 1}$ can be chosen such that $\|f_n(\omega)\| \uparrow h(\omega)$ μ-a.e. on Ω as $n \to \infty$. From the monotone convergence theorem and because by hypothesis K is bounded, it follows that $h \in L^1(\Omega)$. Because $\|f(\omega)\| \leq h(\omega)$ μ-a.e. on Ω for all $f \in K$, we conclude that K is uniformly integrable. □

An immediate consequence of this proposition and of the Dunford–Pettis theorem, is the following result.

COROLLARY 6.4.11 *If (Ω, Σ, μ) is a finite-measure space, Y is a reflexive Banach space, and $K \subseteq L^1(\Omega, Y)$ is a bounded decomposable set, then K is relatively weakly compact in $L^1(\Omega, Y)$.*

To prove the next proposition, we need to recall the following important result of measure theory, which has remarkable applications in control theory. The result is known as *Lyapunov's convexity theorem* and for a proof of it we refer to Diestel–Uhl [199, pp. 264 and 266].

THEOREM 6.4.12 *Let (Ω, Σ) be a measure space.*

(a) *If $m : \Sigma \longrightarrow \mathbb{R}^N$ is a nonatomic vector measure, then $m(\Sigma)$ is compact and convex in X.*
(b) *If X is a Banach space with the RNP* (Radon–Nikodym property) *and $m : \Sigma \longrightarrow X$ is a nonatomic measure of bounded variation, then $\overline{m(\Sigma)}$ is compact and convex.*

Using this theorem, we can prove the next proposition, which is another indication that decomposability and convexity are closely related notions.

PROPOSITION 6.4.13 *If (Ω, Σ, μ) is a finite nonatomic measure space, Y is a Banach space, $K \subseteq L^p(\Omega, Y)$ $1 \leq p < \infty$ is nonempty, and decomposable, V is a finite-dimensional Banach space, and $L \in \mathcal{L}\big(L^p(\Omega, Y), V\big)$, then $L(K)$ is convex.*

PROOF: Let $v_1, v_2 \in L(K)$. Then $v_1 = L(f_1)$ and $v_2 = L(f_2)$ with $f_1, f_2 \in K$. Consider the vector-valued set function $m : \Sigma \longrightarrow V$ defined by $m(A) = L\big(\chi_A(f_1 - f_2)\big)$ for all $A \in \Sigma$. We claim that m is a vector measure. To this end, let $\{A_n\}_{n\geq 1} \subseteq \Sigma$ be mutually disjoint sets and let $A = \bigcup_{n\geq 1} A_n$, $C_n = \bigcup_{k\geq n+1} A_k \in \Sigma$. We have

$$
\begin{aligned}
m(A) &= L\big(\chi_A(f_1 - f_2)\big) \\
&= \sum_{k=1}^{n} L\big(\chi_{A_k}(f_1 - f_2)\big) + L\big(\chi_{C_n}(f_1 - f_2)\big), \\
\Rightarrow \Big\|m(A) - \sum_{k=1}^{n} L\big(\chi_{A_k}(f_1 - f_2)\big)\Big\|_V &= \Big\|m(A) - \sum_{k=1}^{n} m(A_k)\Big\|_V \\
&= \big\|L\big(\chi_{C_n}(f_1 - f_2)\big)\big\|_V .
\end{aligned}
$$

Note that $C_n \downarrow \emptyset$ as $n \rightarrow \infty$ and so $\chi_{C_n}(f_1 - f_2) \longrightarrow 0$ in $L^p(\Omega, Y)$ and so $L\big(\chi_{C_n}(f_1 - f_2)\big) \longrightarrow 0$ in V. This proves that m is a vector measure.

Because μ is nonatomic and $m \ll \mu$, m is nonatomic too. Therefore we can apply Theorem 6.4.12(a) and deduce that $m(\Sigma)$ is convex. Then $D = m(\Sigma) + L(f_2)$ is convex in V. Note that $v_1, v_2 \in D$ and so for every $\lambda \in [0,1]$ we have $\lambda v_1 + (1-\lambda)v_2 \in D$. Because $\chi_A(f_1 - f_2) + f_2 = \chi_A f_1 + \chi_{A^c} f_2 \in K$ (due to the decomposability of K), we infer that $D \subseteq L(K)$. Therefore $\lambda v_1 + (1-\lambda)v_2 \in L(K)$ for all $\lambda \in [0,1]$. Since $v_1, v_2 \in L(K)$ were arbitrary, we conclude that $L(K)$ is convex. $\square$

PROPOSITION 6.4.14 *If (Ω, Σ, μ) is a finite measure space, Y is a Banach space, and $K \subseteq L^p(\Omega, Y)$ $1 \leq p < \infty$ is nonempty, decomposable and w-closed, then $L(K)$ is convex.*

PROOF: Let $g = \text{conv } K$. Then $g = \sum_{k=1}^{n} \lambda_k f_k$ with $\lambda_k \in [0,1]$, $\sum_{k=1}^{n} \lambda_k = 1$ and $f_k \in K$ for all $k \in \{1, \ldots, n\}$. Let $(\cdot, \cdot)_p$ denote the duality brackets for the pair $\big(L^p(\Omega, Y), L^{p'}(\Omega, Y^*_{w^*})\big)$, $\frac{1}{p} + \frac{1}{p'} = 1$ (see Ionescu–Tulcea [328, p. 99]). Consider the basic weak neighborhood of g defined by

$$V(g) = \big\{h \in L^p(\Omega, Y) : |(u_k, h - g)_p| < \varepsilon,\ k \in \{1, \ldots, N\}\big\},$$

where $N \geq 1$, $u_k \in L^{p'}(\Omega, Y^*_{w^*})$, $k \in \{1, \ldots, N\}$ and $\varepsilon > 0$. Let $L : L^p(\Omega, Y) \longrightarrow \mathbb{R}^N$ be defined by

$$L(h) = \big\{(u_k, h)\big\}_{k=1}^{N}.$$

From Proposition 6.4.13 we know that $L(K)$ is convex. Because $L(\text{conv } K) = \text{conv } L(K) = L(K)$, we can find $f \in K$ such that $L(g) = L(f)$, hence $V(g) \cap K \neq \emptyset$ and so $g \in \overline{K}^w = K$. Therefore $K = \text{conv } K$. $\square$

REMARK 6.4.15 Both Propositions 6.4.13 and 6.4.14 are also valid for σ-finite measure spaces (Ω, Σ, μ). In this case we work on the finite measure components of Ω.

The next theorem is a basic tool in what follows. It is also very useful in various optimization problems.

THEOREM 6.4.16 *If (Ω, Σ, μ) is a σ-finite measure space, X is a separable Banach space, $\varphi : \Omega \times X \longrightarrow \mathbb{R}^* = \mathbb{R} \cup \{\pm\infty\}$ is a $\Sigma \times \mathbf{B}(X)$-measurable function, $F : \Omega \longrightarrow 2^X \setminus \{\emptyset\}$ is graph-measurable, and the integral functional*

$$I_\varphi(u) = \int_\Omega \varphi\big(\omega, u(\omega)\big) d\mu(\omega), \quad u \in L^p(\Omega, X),\ 1 \leq p \leq \infty,$$

is defined for all $u \in S^p_F = \{u \in L^p(\Omega, X) : u(\omega) \in F(\omega)\ \mu\text{-a.e. on } \Omega\}$, and there exists $u_0 \in L^p(\Omega, X)$ such that $I_\varphi(u_0) > -\infty$, then $\sup\big[I_\varphi(u) : u \in S^p_F\big] = \int_\Omega \sup\big[\varphi(\omega, x) : x \in F(\omega)\big] d\mu$.

PROOF: Let $m(\omega) = \sup\big[\varphi(\omega, x) : x \in F(\omega)\big]$. From Theorem 6.3.24, we know that m is Σ_μ-measurable. Also for every $u \in S^p_F$ we have $\varphi\big(\omega, u(\omega)\big) \leq m(\omega)$ μ-a.e. on Ω. In particular then $\varphi\big(\omega, u_0(\omega)\big) \leq m(\omega)$ and so we see that $\int_\Omega m d\mu$ exists (possibly equals $+\infty$). We have

$$\sup\big[I_\varphi(u) : u \in S^p_F\big] \leq \int_\Omega m d\mu.$$

If $I_\varphi(u_0) = +\infty$, then we are done.

So assume that $I_\varphi(u_0) \in \mathbb{R}$. Then the function $\omega \longrightarrow \varphi\big(\omega, u_o(\omega)\big)$ belongs in $L^1(\Omega)$. Let $\xi < \int_\Omega m d\mu$ be given. We show that $\xi < I_\varphi(u)$ for some $u \in S^p_F$. Let $\{C_n\}_{n \geq 1} \subseteq \Sigma$ be such that $C_n \uparrow \Omega$ and $\mu(C_n) < \infty$ and consider a strictly positive function $\vartheta \in L^1(\Omega)$. We set

$$D_n = C_n \cap \{\omega \in \Omega : \varphi(\omega, u_o(\omega)) \leq n\}$$

and
$$m_n(\omega) = \begin{cases} m(\omega) - \frac{\vartheta(\omega)}{n} & \text{if } \omega \in D_n \text{ and } \vartheta(\omega) \leq n \\ n - \frac{\vartheta(\omega)}{n} & \text{if } \omega \in D_n \text{ and } \vartheta(\omega) > n \\ \varphi(\omega, u_o(\omega)) - \frac{\vartheta(\omega)}{n} & \text{if } \omega \in D_n^c \end{cases}.$$

Clearly $\{m_n\}_{n\geq 1} \subseteq L^1(\Omega)$ and $m_n \uparrow m$ as $n \to \infty$. Therefore by the monotone convergence theorem, we can find $n_0 \geq 1$ such that $\xi < \int_\Omega m_{n_0} d\mu$. Set $\eta = m_{n_0}$. Then $\xi < \int_\Omega \eta d\mu$ and $\eta(\omega) < m(\omega)$ μ-a.e. on Ω. Define

$$G(\omega) = F(\omega) \cap \{x \in X : \eta(\omega) \leq \varphi(\omega, x)\} \neq \emptyset \quad \text{for all } \omega \in \Omega. \tag{6.21}$$

Evidently $\operatorname{Gr} G \in \Sigma \times \mathbf{B}(X)$ and so by Theorem 6.3.20, we can find $u : \Omega \longrightarrow X$ a Σ–measurable function such that $u(\omega) \in G(\omega)$ for all $\omega \in \Omega$. We define

$$E_n = C_n \cap \{\omega \in \Omega : \|u(\omega)\| \leq n\}$$

and
$$v_n = \chi_{E_n} u + \chi_{E_n^c} u_0, \; n \geq 1.$$

Evidently $\{v_n\}_{n\geq 1} \subseteq S_F^p$ and we have

$$\begin{aligned} I_\varphi(v_n) &= \int_{E_n} \varphi(\omega, u(\omega)) d\mu + \int_{E_n^c} \varphi(\omega, u_0(\omega)) d\mu \\ &\geq \int_{E_n} \eta(\omega) d\mu + \int_{E_n^c} \varphi(\omega, u_0(\omega)) d\mu \quad \text{(see (6.21))} \\ &= \int_\Omega \eta(\omega) d\mu + \int_{E_n^c} \Big(\varphi(\omega, u_0(\omega)) - \eta(\omega)\Big) d\mu. \end{aligned} \tag{6.22}$$

Because $\int_\Omega \eta(\omega) d\mu > \xi$ and $E_n \uparrow \Omega$, from (6.22) it follows that for $n \geq 1$ large we have

$$I_\varphi(v_n) > \xi.$$

□

In what follows we study the set S_F^p $1 \leq p \leq \infty$ for a multifunction F. Throughout this study (Ω, Σ, μ) is a σ-finite measure space and X a separable Banach space. As always additional hypotheses are introduced as needed.

PROPOSITION 6.4.17 *If $F : \Omega \longrightarrow 2^X \setminus \{\emptyset\}$ is graph-measurable and $S_F^p \neq \emptyset$, $1 \leq p < \infty$, then $\overline{\operatorname{conv}}\, S_F^p = S_{\overline{\operatorname{conv}} F}^p$.*

PROOF: It is easy to see that $S_{\overline{\operatorname{conv}} F}^p \in P_{fc}(L^p(\Omega, X))$. So we have

$$\overline{\operatorname{conv}}\, S_F^p \subseteq S_{\overline{\operatorname{conv}} F}^p. \tag{6.23}$$

Suppose that the conclusion in (6.23) is strict. So we can find $f \in S_{\overline{\operatorname{conv}} F}^p$ such that $f \notin \overline{\operatorname{conv}} S_F^p$. Then from the strong separation theorem we can find $u^* \in L^{p'}(\Omega, X_{w^*}^*) = L^p(\Omega, X)^*$, $(1/p) + (1/p') = 1$ such that

$$\sigma(u^*, \overline{\operatorname{conv}}\, S_F^p) < \langle u^*, f \rangle_p. \tag{6.24}$$

From the definition of the support function (see Definition 6.1.14(b)), we have

$$\begin{aligned}
\sigma\big(u^*, \overline{\operatorname{conv}}\, S^p_F\big) &= \sigma\big(u^*, S^p_F\big) \\
&= \sup\big[\langle u^*, h\rangle_p : h \in S^p_F\big] \\
&= \sup\Big[\int_\Omega \langle u^*(\omega), h(\omega)\rangle\, \mathrm{d}\mu : h \in S^p_F\Big] \\
&= \int_\Omega \sup\big[\langle u^*(\omega), x\rangle : x \in F(\omega)\big] \mathrm{d}\mu \qquad \text{(see Theorem 6.4.16)} \\
&= \int_\Omega \sigma\big(u^*(\omega), F(\omega)\big) \mathrm{d}\mu \\
&= \int_\Omega \sigma\big(u^*(\omega), \overline{\operatorname{conv}}\, F(\omega)\big) \mathrm{d}\mu. \qquad (6.25)
\end{aligned}$$

From (6.24) and (6.25) and because $f \in S^p_{\overline{\operatorname{conv}} F}$, we have

$$\begin{aligned}
\int_\Omega \sigma\big(u^*(\omega), \overline{\operatorname{conv}}\, F(\omega)\big) \mathrm{d}\mu &< \int_\Omega \langle u^*(\omega), f(\omega)\rangle\, \mathrm{d}\mu \\
&\leq \int_\Omega \sigma\big(u^*(\omega), \overline{\operatorname{conv}}\, F(\omega)\big) \mathrm{d}\mu,
\end{aligned}$$

a contradiction. This proves that in (6.23) we have equality. □

A useful byproduct of the previous proof is the following result.

COROLLARY 6.4.18 *If $F : \Omega \longrightarrow 2^X \setminus \{\emptyset\}$ is graph-measurable and $S^p_F \neq \emptyset$, $1 \leq p < \infty$, then for every $u^* \in L^{p'}(\Omega, X^*_{w^*}) = L^p(\Omega, X)^*$, $\frac{1}{p} + \frac{1}{p'} = 1$, we have*

$$\sigma(u^*, S^p_F) = \int_\Omega \sigma\big(u^*(\omega), F(\omega)\big) \mathrm{d}\mu.$$

PROPOSITION 6.4.19 *If (Ω, Σ, μ) is a nonatomic, σ-finite measure space, $F : \Omega \longrightarrow 2^X \setminus \{\emptyset\}$ is graph-measurable, and $S^p_F \neq \emptyset$, $1 \leq p < \infty$, then $\overline{S^p_F}^w = S^p_{\overline{\operatorname{conv}} F}$ (here by $\overline{}^w$ we denote the closure in the weak topology).*

PROOF: Evidently $\overline{S^p_F}^w$ is nonempty, decomposable, and weakly closed. Therefore by Proposition 6.4.14 $\overline{S^p_F}^w$ is convex. It follows that $\overline{\operatorname{conv}}\, S^p_F = \overline{S^p_F}^w$. Invoking Proposition 6.4.17, we conclude that

$$S^p_{\overline{\operatorname{conv}} F} = \overline{S^p_F}^w.$$

□

PROPOSITION 6.4.20 *If $F : \Omega \longrightarrow 2^X \setminus \{\emptyset\}$ is graph-measurable, and $S^p_F \neq \emptyset$, $1 \leq p < \infty$, then $\overline{S^p_F} = S^p_{\overline{F}}$.*

PROOF: Because $S^p_{\overline{F}} \in P_f\big(L^p(\Omega, X)\big)$, we have $\overline{S^p_F} \subseteq S^p_{\overline{F}}$. Because μ is σ-finite, we can find $\{A_n\}_{n \geq 1} \subseteq \Sigma$ of finite μ-measure such that $A_n \subseteq A_{n+1}$ for all $n \geq 1$ and $\bigcup_{n \geq 1} A_n = \Omega$. Then we have $\mu(A_n) \uparrow +\infty$. Let $f \in S^p_{\overline{F}}$ and for every $n \geq 1$, we consider the multifunction

$$H_n(\omega) = \Big\{x \in F(\omega) : \|f(\omega) - x\| < \frac{1}{\mu(A_n)^2}\Big\}.$$

Clearly $H_n(\omega) \neq \emptyset$ for all $\omega \in \Omega$ and Gr $H_n \in \Sigma \times \mathbf{B}(X)$. Then by virtue of Theorem 6.3.20 (see also Remark 6.3.21), we can find $h_n : \Omega \longrightarrow X$ a Σ-measurable selector of F such that

$$\|f(\omega) - h_n(\omega)\| < \frac{1}{\mu(A_n)^2} \quad \mu\text{-a.e. on } \Omega. \tag{6.26}$$

Let $g \in S_F^p$ and set

$$u_n = \chi_{A_n} h_n + \chi_{A_n^c} g.$$

Then $u_n \in S_F^p$ and we have

$$\begin{aligned} \|f - u_n\|_p &= \int_{A_n} \|f(\omega) - h_n(\omega)\|^p d\mu + \int_{A_n^c} \|f(\omega) - g(\omega)\|^p d\mu \\ &< \frac{1}{\mu(A_n)^{2p}}\mu(A_n) + \int_{A_n^c} \|f(\omega) - g(\omega)\|^p d\mu. \end{aligned} \tag{6.27}$$

Note that $\|f(\cdot) - g(\cdot)\|^p \in L^1(\Omega)_+$ and $\mu(A_n^c) \downarrow 0$. So if we pass to the limit as $n \to \infty$ in (6.27), we conclude that

$$u_n \longrightarrow f \quad \text{in} \quad L^p(\Omega, X).$$

Because $u_n \in S_F^p$, it follows that $f \in \overline{S_F^p}$ and so $S_{\overline{F}}^p \subseteq \overline{S_F^p}$. Hence finally

$$\overline{S_F^p} = S_{\overline{F}}^p.$$

□

From the structure of the set S_F^p, we can extract information about the pointwise properties of the multifunction F.

PROPOSITION 6.4.21 *If $F:\Omega \longrightarrow 2^X \setminus \{\emptyset\}$ is graph-measurable and S_F^p, $1 \leq p < \infty$, is nonempty closed (resp., nonempty closed convex and μ is nonatomic), then for μ-a.a. $\omega \in \Omega$, $F(\omega) \in P_f(X)$ (resp., $F(\omega) \in F_{fc}(X)$).*

PROOF: First we assume that S_F^p is nonempty and closed. From Proposition 6.4.20 we have $S_F = S_{\overline{F}}$ and then Corollary 6.4.4 implies that $F(\omega) = \overline{F}(\omega)$ μ-a.e. on Ω.

Now assume that S_F^p is nonempty, closed, and convex and μ is nonatomic. Invoking Proposition 6.4.19, we have that $\overline{S_F^p}^w = S^p_{\overline{\text{conv}}\, F}$, hence $S_F^p = S^p_{\overline{\text{conv}}\, F}$ (recall that a convex set is closed if and only if it is w-closed). Once again Corollary 6.4.4 implies $F(\omega) = \overline{\text{conv}}\, F(\omega)$ μ-a.e. on Ω; that is, $F(\omega) \in P_{fc}(X)$ μ-a.e. on Ω. □

DEFINITION 6.4.22 A multifunction $F:\Omega \longrightarrow 2^X \setminus \{\emptyset\}$ is said to be $\mathbf{L}^p$-*integrably bounded* $(1 < p \leq \infty)$ and simply *integrably bounded* (for $p = 1$), if there exists $h \in L^p(\Omega)_+$ such that

$$|F(\omega)| = \sup\big\{\|x\| : x \in F(\omega)\big\} \leq h(\omega) \quad \mu\text{-a.e. on } \Omega.$$

For integrably bounded multifunctions we have a weak compactness result for the set S_F^1, which is a very valuable tool in many applications.

THEOREM 6.4.23 *If $F:\Omega\longrightarrow P_{wkc}(X)$ is a graph measurable and integrably bounded multifunction, then S_F^1 is nonempty, convex, and w-compact.*

PROOF: Clearly S_F^1 is nonempty and convex. Also let $u^*\in L^\infty(\Omega,X^*_{w^*})=L^1(\Omega,X)^*$. If by $\langle\cdot,\cdot\rangle_1$ we denote the duality brackets for the pair $\big(L^1(\Omega,X),L^\infty(\Omega,X^*_{w^*})\big)$, we have

$$\begin{aligned}\sup\big[\langle u^*,f\rangle_1 : f\in S_F^1\big] &= \sup\Big[\int_\Omega \langle u^*(\omega),f(\omega)\rangle\, d\mu : f\in S_F^1\Big]\\ &= \int_\Omega \sup\big[\langle u^*(\omega),x\rangle : x\in F(\omega)\big]d\mu,\end{aligned}$$

(see Theorem 6.4.16).

We introduce the multifunction $S:\Omega\longrightarrow 2^X$ defined by

$$S(\omega)=\big\{u\in F(\omega):\langle u^*(\omega),u\rangle=\sup\big[\langle u^*(\omega),x\rangle : x\in F(\omega)\big]=\sigma\big(u^*(\omega),F(\omega)\big)\big\}.$$

Because F has values in $P_{wkc}(X)$, we see that $S(\omega)\neq\emptyset$ for all $\omega\in\Omega$. Moreover, from Theorem 6.3.24 we know that $\omega\longrightarrow\sigma\big(u^*(\omega),F(\omega)\big)\big\}$ is Σ_μ-measurable (Σ_μ being the μ-completion of Σ). It follows that

$$\mathrm{Gr}S\in\Sigma_\mu\times\mathbf{B}(X).$$

Invoking Theorem 6.3.20 we can find $u:\Omega\longrightarrow X$ Σ-measurable such that

$$u(\omega)\in S(\omega)\quad \mu\text{-a.e.}$$

Evidently $u\in L^1(\Omega,X)$ and we have

$$\sup\big[\langle u^*,f\rangle_1 : f\in S_F^1\big]=\int_\Omega \langle u^*(\omega),u(\omega)\rangle\, d\mu=\langle u^*,u\rangle_1\,.$$

By virtue of the James theorem, because S_F^1 is closed, convex, and bounded, we conclude that S_F^1 is w-compact in $L^1(\Omega,X)$. □

The converse of this theorem is also true (see Klei [356, p. 313]).

THEOREM 6.4.24 *If $F:\Omega\longrightarrow P_f(X)$ is graph-measurable, integrably bounded, and $S_F^1\subseteq L^1(\Omega,X)$ is w-compact convex, then $F(\omega)\in P_{wkc}(X)$ μ-a.e. on Ω.*

PROPOSITION 6.4.25 *If $F:\Omega\longrightarrow P_{wkc}(X)$ is a scalarly measurable multifunction, then* $\operatorname{ext} S_F=S_{\operatorname{ext}F}\neq\emptyset$ *in $L^0(\Omega,X)$.*

PROOF: By virtue of Proposition 6.3.35 and Theorem 6.3.20, we have that $S_{\operatorname{ext}F}\neq\emptyset$. Moreover, we have $S_{\operatorname{ext}F}\subseteq\operatorname{ext} S_F$. Suppose that the inclusion is strict. So we can find $f\in\operatorname{ext} S_F$ and $A\in\Sigma$, $\mu(A)>0$ such that $f(\omega)\notin\operatorname{ext} F(\omega)$ for all $\omega\in A$. Consider the multifunction $S:A\longrightarrow 2^{X\times X}$ defined by

$$S(\omega)=\Big\{(x,y)\in F(\omega)\times F(\omega) : f(\omega)=\frac{1}{2}(x+y),\ x\neq y\Big\}.$$

Then $\mathrm{Gr}\, S \in (\mathrm{Gr}\, F \times \mathrm{Gr}\, F) \cap \{(\omega, x, y) \in A \times X \times X : f(\omega) = \frac{1}{2}(x+y)\} \cap (\Omega \times \Delta^c)$ where $\Delta = \{(x,y) \in X \times X : x = y\}$ (the diagonal in $X \times X$). Hence $\mathrm{Gr}\, S \in A \times \mathbf{B}(X)$ and we can apply Theorem 6.3.20 and obtain $\widehat{f}_1, \widehat{f}_2 : A \longrightarrow X$ $\Sigma_A = \Sigma \cap A$-measurable functions such that $(\widehat{f}_1(\omega), \widehat{f}_2(\omega)) \in S(\omega)$ μ-a.e. on A. Let $h \in S_F$ (it exists by virtue of Theorem 6.3.20) and set $f_1 = \chi_A \widehat{f}_1 + \chi_{A^c} h$, $f_2 = \chi_A \widehat{f}_2 + \chi_{A^c} h$. Evidently $f_1, f_2 \in S_F$ and $f = \frac{1}{2}(f_1 + f_2)$, $f_1 \neq f_2$, a contradiction to the fact that $f \in \mathrm{ext} S_F$. □

Using this proposition, we can prove the following theorem.

THEOREM 6.4.26 *If $F : \Omega \longrightarrow P_{wkc}(X)$ is a scalarly measurable multifunction and $S_F^p \neq \emptyset$, $1 \leq p \leq \infty$, then* $\mathrm{ext}\, S_F^p = S_{\mathrm{ext} F}^p$.

PROOF: Recall that $S_F^p = S_F \cap L^p(\Omega, X)$, $S_{\mathrm{ext} F}^p \cap L^p(\Omega, X)$ and from Proposition 6.4.25 we have $\mathrm{ext}\, S_F = S_{\mathrm{ext} F}$. So it suffices to show that $\mathrm{ext}\, S_F \cap L^p(\Omega, X) = \mathrm{ext}\big(S_F \cap L^p(\Omega, X)\big)$. It is easy to see that $\mathrm{ext}\, S_F \cap L^p(\Omega, X) \subseteq \mathrm{ext}\big(S_F \cap L^p(\Omega, X)\big)$. We show that the opposite inclusion is also true. To this end let $f \in \mathrm{ext}\big(S_F \cap L^p(\Omega, X)\big)$ and suppose that $f \notin \mathrm{ext}\, S_F$. So we can find $f_1, f_2 \in S_F$, $f_1 \neq f_2$ such that $f = \frac{1}{2}(f_1 + f_2)$. By translating things if necessary, we may assume that $f = 0$. Because μ is σ-finite, we can find $A \in \Sigma$ with $0 < \mu(A) < +\infty$ such that $f_1(\omega) \neq f_2(\omega)$ for all $\omega \in A$. We define

$$h_k(\omega) = \begin{cases} f_k(\omega) & \text{if } \|f_k(\omega)\| \leq 1,\ \omega \in A \\ \frac{f_k(\omega)}{\|f_k(\omega)\|} & \text{if } \|f_k(\omega)\| > 1,\ \omega \in A\ , \quad k = 1, 2. \\ 0 & \text{otherwise} \end{cases}$$

Clearly $h_k \in S_F \cap L^p(\Omega, X) = S_F^p$ and $0 = f = \frac{1}{2}(h_1 + h_2)$, which means that $0 = f \notin \mathrm{ext}\big(S_F \cap L^p(\Omega, X)\big)$. □

Combining Proposition 6.3.35 and the Krein–Milman theorem, we obtain the following result.

PROPOSITION 6.4.27 *If (Ω, Σ, μ) is a nonatomic, σ-finite measure space and $F : \Omega \longrightarrow P_{wkc}(X)$ is scalarly measurable with $S_F^p \neq \emptyset$, $1 \leq p \leq \infty$, then* $\overline{S_{\mathrm{ext} F}^p}^{w} = S_F^p$.

Before passing to set-valued integration, we state (without proofs which are rather long and technical) two continuous selection theorems for multifunctions that instead of convexity exhibit a decomposability property in their values. These theorems confirm the earlier observation that decomposability is a very effective substitute of convexity. For their proofs we refer to Hu–Papageorgiou [313, Section II.8].

The first theorem is a nonconvex-decomposable version of Michael's selection theorem (see Theorem 6.3.6).

THEOREM 6.4.28 *If Y is a metric space and $F : Y \longrightarrow P_f\big(L^1(\Omega, X)\big)$ is a lsc multifunction with decomposable values, then F admits a continuous selector.*

For the second continuous selection theorem, we need to do some preparatory work. So let $T = [0, b]$ furnished with the one-dimensional Lebesgue measure λ. Let Y be a Polish space and X a separable Banach space. For the Lebesgue–Bochner space $L^1(\Omega, X)$, we introduce a new norm, useful in the study of differential and evolution equations.

DEFINITION 6.4.29 The *weak norm* on the Lesesgue–Bochner space $L^1(T, X)$, is defined by

$$\|f\|_w = \sup\{\|\int_t^{t'} f(s)ds\| : 0 \leq t \leq t' \leq b\}, \quad f \in L^1(T, X).$$

REMARK 6.4.30 Clearly an equivalent definition of the weak norm, is the following.

$$\|f\|_w = \sup\{\|\int_0^{t} f(s)ds\| : 0 \leq t \leq b\}, \quad f \in L^1(T, X).$$

It is not immediately clear how the $\|\cdot\|_w$-norm topology and the weak topology on $L^1(T, X)$ are related. It turns out that the next concept is the link between the two topologies.

DEFINITION 6.4.31 A set $W \subseteq L^1(T, X)$ has *property* U, if

(a) W is uniformly integrable and
(b) For every $\varepsilon > 0$, we can find $K_\varepsilon \in P_k(X)$ such that for every $f \in W$ there exists $S_{\varepsilon,f} \subseteq T$-measurable subset with $\lambda(T \setminus S_{\varepsilon,f}) < \varepsilon$ such that $f(t) \in K_\varepsilon$ for all $t \in S_{\varepsilon,f}$.

REMARK 6.4.32 If $W \subseteq L^1(T, X)$ has property U, then W is relatively weakly compact in $L^1(T, X)$.

Using property U, we can compare the $\|\cdot\|_w$-norm and weak topologies on $L^1(T, X)$.

PROPOSITION 6.4.33 *If $W \subseteq L^1(T, X)$ has property U, then the $\|\cdot\|_w$-norm and weak topologies coincide on the set W.*

The second continuous selection theorem uses the $\|\cdot\|_w$-norm topology on $L^1(T, X)$. This theorem concerns multifunctions $F(t, x)$ of two variables $(t, x) \in T \times X$. The hypotheses on F are the following.

H(F): $F : T \times Y \longrightarrow P_{wkc}(X)$ is a multifunction such that

(i) For every $y \in Y$, $t \longrightarrow F(t, y)$ is measurable.

(ii) For almost all $t \in T$, $y \longrightarrow F(t, y)$ is h-continuous.

(iii) For every $C \in P_k(Y)$, we can find $\alpha_C \in L^1(T)_+$ such that for a.a. $t \in T$, all $y \in C$, and all $u \in F(t, y)$, we have $\|u\| \leq \alpha_C(t)$.

REMARK 6.4.34 In the next section we discuss in some detail multifunctions $F(t,x)$ of two variables. For the moment it suffices to mention that if F satisfies hypotheses H(F)(i), (ii), then F is jointly measurable, in particular then sup-measurable (compare with Theorem 6.2.6). This fact combined with hypothesis H(F)(iii), implies that if $y \in C(T,Y)$, then $S^1_{F(\cdot,y(\cdot))} \neq \emptyset$.

Now let $K \subseteq C(T,Y)$ be nonempty, and compact and consider the multifunction $\Gamma : K \longrightarrow P_{wkc}\big(L^1(T,X)\big)$ defined by $\Gamma(y) = S^1_{F(\cdot,y(\cdot))}$. Let CS^w_Γ (resp., $CS^w_{\text{ext}\Gamma}$) be the set of selectors of Γ (resp., of ext Γ) that are continuous from K into $L^1(T,X)$ endowed with the weak norm (denoted by $L^1_w(T,X)$, not to be confused with $L^1(T,X)_w$ which is the Lebesgue–Bochner space $L^1(T,X)$ with the weak topology).

THEOREM 6.4.35 *If $F(t,x)$ satisfies hypotheses $H(F)$, $K \subseteq C(T,Y)$ is compact, and $\Gamma(y) = S^1_{F(\cdot,y(\cdot))}$ for all $y \in K$, then $CS^w_\Gamma = \overline{CS^w_{\text{ext}\Gamma}}^{\|\cdot\|_w}$.*

REMARK 6.4.36 Theorems 6.4.28 and 6.4.35 are indispensable tools in the study of nonconvex differential inclusions (see Hu–Papageorgiou [316]).

Now we pass to set-valued integration. First of all the definition of the integral of a multifunction that we use is a straightforward continuous extension of the Minkowski sum of sets. Throughout this last part of this section (Ω, Σ, μ) is a σ-finite measure space, X is a separable Banach space, and $F : \Omega \longrightarrow 2^X \setminus \{\emptyset\}$ is a multifunction with $S^1_F \neq \emptyset$.

REMARK 6.4.37 For a graph-measurable multifunction $F : \Omega \longrightarrow 2^X \setminus \{\emptyset\}$, a straightforward measurable selection argument involving Theorem 6.3.20, reveals that $S^1_F \neq \emptyset$ if and only if there exists $h \in L^1(\Omega)_+$ such that $\inf\{\|u\| : u \in F(\omega)\} \leq h(\omega)$.

DEFINITION 6.4.38 The *set-valued integral* of a multifunction $F : \Omega \longrightarrow 2^X \setminus \{\emptyset\}$ with $S^1_F \neq \emptyset$, is defined by

$$\int_\Omega F \mathrm{d}\mu = \Big(\int_\Omega f \mathrm{d}\mu : f \in S^1_F\Big).$$

The following is an immediate consequence of Theorem 6.4.16.

PROPOSITION 6.4.39 *If $F : \Omega \longrightarrow 2^X \setminus \{\emptyset\}$ is graph measurable and $S^1_F \neq \emptyset$, then for every $x^* \in X^*$, we have $\sigma(x^*, \int_\Omega F \mathrm{d}\mu) = \int_\Omega \sigma(x^*, F) \mathrm{d}\mu$.*

Also from Proposition 6.4.17, we obtain the following.

PROPOSITION 6.4.40 *If $F : \Omega \longrightarrow 2^X \setminus \{\emptyset\}$ is graph-measurable and $S^1_F \neq \emptyset$, then* $\mathrm{cl} \int_\Omega \mathrm{conv}\, F \mathrm{d}\mu = \overline{\mathrm{conv}} \int_\Omega F \mathrm{d}\mu = \mathrm{cl} \int_\Omega \mathrm{conv}\, F \mathrm{d}\mu$.

The set-valued integral has some remarkable intrinsic convexity properties.

THEOREM 6.4.41 *If μ is nonatomic and $F:\Omega\longrightarrow 2^X\setminus\{\emptyset\}$ is graph-measurable with $S_F^1\neq\emptyset$, then* $\mathrm{cl}\int_\Omega F\mathrm{d}\mu$ *is convex.*

PROOF: Let $v_1,v_2\in\int_\Omega F\mathrm{d}\mu$. Then by Definition 6.4.38, we have

$$v_1=\int_\Omega f_1\mathrm{d}\mu \quad\text{and}\quad v_2=\int_\Omega f_2\mathrm{d}\mu,\quad \text{with } f_1,f_2\in S_F^1.$$

Consider the vector measure $m:\Sigma\longrightarrow X\times X$ defined by

$$m(A)=\Big(\int_A f_1\mathrm{d}\mu,\int_A f_2\mathrm{d}\mu\Big),\quad A\in\Sigma.$$

Invoking Theorem 6.4.12(b), we see that $\overline{m(\Sigma)}$ is convex. We have

$$m(\emptyset)=0\quad\text{and}\quad m(\Omega)=(v_1,v_2).$$

So given $\varepsilon>0$ and $t\in[0,1]$, we can find $A\in\Sigma$ such that

$$\Big\|t\int_\Omega f_k\mathrm{d}\mu-\int_A f_k\mathrm{d}\mu\Big\|<\frac{\varepsilon}{2},\quad k=1,2.$$

Set $f=\chi_A f_1+\chi_{A^c}f_2$. Then $f\in S_F^1$ and we have

$$\Big\|tv_1+(1-t)v_2-\int_\Omega f\mathrm{d}\mu\Big\|<\varepsilon,$$

$$\Rightarrow\ \mathrm{cl}\int_\Omega F\mathrm{d}\mu\ \text{ is convex.}$$

□

COROLLARY 6.4.42 *If μ is nonatomic and $F:\Omega\longrightarrow 2^X\setminus\{\emptyset\}$ is graph-measurable with $S_F^1\neq\emptyset$, then* $\mathrm{cl}\int_\Omega\overline{\mathrm{conv}}\,F\mathrm{d}\mu=\mathrm{cl}\int_\Omega F\mathrm{d}\mu$.

REMARK 6.4.43 So when μ is nonatomic, convexification of the multifunction F essentially does not add anything new to the set-valued integral.

If in Theorem 6.4.41, $\dim X<+\infty$, then we can exploit the fact that Lyapunov's convexity theorem is exact (i.e., no closure is needed; see Theorem 6.4.12(a)), in order to obtain the following result.

THEOREM 6.4.44 *If μ is nonatomic, X is finite-dimensional, and $F:\Omega\longrightarrow 2^X\setminus\{\emptyset\}$ is graph-measurable with $S_F^1\neq\emptyset$, then $\int_\Omega F\mathrm{d}\mu$ is convex.*

For $\mathbb{R}^N$-valued multifunctions, we have the following partial improvement of the above theorem. For a proof of this result see Hildenbrand [297, p. 48] and Hu–Papageorgiou [313, p. 202].

THEOREM 6.4.45 *If $F:\Omega\longrightarrow 2^{\mathbb{R}^N}\setminus\{\emptyset\}$ is graph-measurable, $S_F^1\neq\emptyset$, and $F(\omega)\subseteq\mathbb{R}_+^N$ for all $\omega\in\Omega$, then* $\int_\Omega\mathrm{conv}\,F\mathrm{d}\mu=\mathrm{conv}\int_\Omega F\mathrm{d}\mu$.

COROLLARY 6.4.46 *If μ is nonatomic, $F:\Omega\longrightarrow 2^{\mathbb{R}^N}\setminus\{\emptyset\}$ is graph-measurable, $S_F^1\neq\emptyset$, and $F(\omega)\subseteq\mathbb{R}_+^N$ for all $\omega\in\Omega$, then $\int_\Omega F d\mu = \int_\Omega \operatorname{conv} F d\mu$.*

REMARK 6.4.47 Corollary 6.4.46 fails if $\mathbb{R}^N$ is replaced by an infinite-dimensional Banach space. Similarly Theorem 6.4.45 fails if we remove the graph-measurability of F or the hypothesis that $F(\omega)\subseteq\mathbb{R}_+^N$ for all $\omega\in\Omega$.

As a straightforward consequence of Theorem 6.4.23, we have the following result.

PROPOSITION 6.4.48 *If $F:\Omega\longrightarrow P_{wkc}(X)$ is graph-measurable and integrably bounded, then $\int_\Omega F d\mu\in P_{wkc}(X)$.*

Continuing with the topological properties of the set-valued integral $\int_\Omega F d\mu$, we obtain conditions for $\int_\Omega F d\mu$ to be open.

PROPOSITION 6.4.49 *If $F:\Omega\longrightarrow 2^X\setminus\{\emptyset\}$ has open values, is graph-measurable, and $S_F^1\neq\emptyset$, then for every $A\in\Sigma$ the set $\int_A F d\mu$ is open in X.*

PROOF: Without any loss of generality, we may assume that $A=\Omega$. Let $f\in S_F^1$. By replacing F by $\omega\longrightarrow F(\omega)-f(\omega)$ if necessary, we may assume that $0\in F(\omega)$ for all $\omega\in\Omega$. So we need to show that there exists $\varepsilon>0$ such that $B_\varepsilon\subseteq\int_\Omega F d\mu$. For this purpose let $\xi_F(\omega)=d\big(0,F(\omega)^c\big)$. Let $\lambda>0$ and consider the set $L_\lambda=\{\omega\in\Omega:\xi_F(\omega)<\lambda\}$. If $G=\{(\omega,x)\in\Omega\times X: x\in B_\lambda\cap F(\omega)^c\}=(\Omega\times B_\lambda)\cap \operatorname{Gr}F^c\in\Sigma\times\mathbf{B}(X)$, then $L_\lambda=\operatorname{proj}_\Omega G\in\Sigma_\mu$ (see Theorem 6.2.18). This proves that the function ξ_F is Σ_μ-measurable. Because F has open values, we have $\xi_F(\omega)>0$ for all $\omega\in\Omega$ and so we can find $\varepsilon>0$ and $A\in\Sigma$, with $\mu(A)>0$ such that $\xi_F(\omega)\geq\varepsilon$ for all $\omega\in A$. This means that $B_\varepsilon\subseteq F(\omega)$ for all $\omega\in A$ and so $\mu(A)B_\varepsilon\subseteq\int_A F d\mu\subseteq\int_\Omega F d\mu$ (because $0\in F(\omega)$ for all $\omega\in\Omega$). $\square$

In fact under some reasonable conditions, we can establish that the interior and integral operators commute. To show this we need two auxiliary results.

LEMMA 6.4.50 *If V is a Banach space and $U_1\subseteq U_2\subseteq X$ are nonempty open sets with U_1 convex and dense in U_2, then $U_1=U_2$.*

PROOF: We have $U_2\subseteq\operatorname{int}\overline{U}_1$ and because U_1 is convex $U_1=\operatorname{int}\overline{U}_1$. Therefore $U_1=U_2$. $\square$

LEMMA 6.4.51 *If $F:\Omega\longrightarrow 2^X\setminus\{\emptyset\}$ is graph-measurable and $\operatorname{int}F(\omega)\neq\emptyset$ for all $\omega\in\Omega$, then $\operatorname{Gr}\operatorname{int}F\in\Sigma_\mu\times\mathbf{B}(X)$.*

PROOF: For every $\omega\in\Omega$ we have $\operatorname{int}F(\omega)=F(\omega)\setminus\partial F(\omega)$ with $\partial F(\omega)$ being the boundary of the set $F(\omega)$. Since $\partial F(\omega)=\overline{F(\omega)}\cap\overline{F(\omega)^c}$, we infer that $\operatorname{Gr}\partial F\in\Sigma_\mu\times\mathbf{B}(X)$ (see the proof of Proposition 6.4.49). Hence $\operatorname{Gr}\operatorname{int}F=\operatorname{Gr}F\cap(\operatorname{Gr}F)^c\in\Sigma_\mu\times\mathbf{B}(X)$. $\square$

Using these auxiliary results, we can prove a theorem on the commutation of the interior and integral operators.

THEOREM 6.4.52 *If μ is finite, $F:\Omega \longrightarrow 2^X\setminus\{\emptyset\}$ is graph-measurable, $S_F^1 \neq \emptyset$, $F(\omega)$ is convex for all $\omega \in \Omega$, and* int $F(\omega) \neq \emptyset$ *μ-a.e., then for all $A \in \Sigma$ we have* int $\int_A F d\mu = \int_A \text{int } F d\mu$.

PROOF: Clearly without any loss of generality, we may assume that int $F(\omega) \neq \emptyset$ for all $\omega \in \Omega$. We fix $A \in \Sigma$ with $\mu(A) > 0$. We show that $\int_A \text{int } F d\mu$. To this end let $f \in S_F^1$ and $\varepsilon > 0$. We introduce the multifunction $G_\varepsilon : \Omega \longrightarrow 2^X$ defined by

$$G_\varepsilon(\omega) = \left\{x \in \text{int} F(\omega) : \|x - f(\omega)\| < \frac{\varepsilon}{\mu(A)}\right\}.$$

Due to the convexity of $F(\omega)$ we have $\overline{F(\omega)} = \overline{\text{int } F(\omega)}$ for all $\omega \in \Omega$. This implies that $G_\varepsilon(\omega) \neq \emptyset$ for all $\omega \in \Omega$. Also due to Lemma 6.4.51 Gr $G_\varepsilon \in \Sigma_\mu \times \mathbf{B}(X)$. Applying Theorem 6.3.20 we can find $g : \Omega \longrightarrow X$ a Σ-measurable map such that $g(\omega) \in G_\varepsilon(\omega)$ μ-a.e. So we have

$$g(\omega) \in \text{int } F(\omega) \quad \text{and} \quad \|g(\omega) - f(\omega)\| < \frac{\varepsilon}{\mu(A)} \quad \mu\text{-a.e. on } \Omega,$$

$$\Rightarrow \ g \in S^1_{\text{int} F} \quad \text{and} \quad \left\| \int_A (g - f) d\mu \right\| < \varepsilon,$$

$$\Rightarrow \ \int_A \text{int } F d\mu \ \text{ is dense in } \int_A F d\mu.$$

But from Proposition 6.4.49, we have that $\int_A \text{int } F d\mu$ is open and so $\int_A \text{int } F d\mu \subseteq \text{int } \int_A F d\mu$. Invoking Lemma 6.4.50, we conclude that $\int_A \text{int } F d\mu = \text{int } \int_A F d\mu$. $\square$

Because of Theorem 6.4.44, when X is finite-dimensional, we can have the following stronger version of the above theorem.

THEOREM 6.4.53 *If μ is finite and nonatomic, X is-finite dimensional, $F : \Omega \longrightarrow 2^X\setminus\{\emptyset\}$ is graph-measurable, $S_F^1 \neq \emptyset$, and* int $F(\omega)$ *is dense in $F(\omega)$ for all $\omega \in \Omega$, then for every $A \in \Sigma$ we have $\int_A \text{int } F d\mu = \text{int } \int_A F d\mu$.*

EXAMPLE 6.4.54 In the above theorem we cannot drop the condition that int $F(\omega)$ is dense in $F(\omega)$ for all $\omega \in \Omega$. To see this let $\Omega = [0,1]$, $\mu = \lambda$ (λ is the Lebesgue measure on $\mathbb{R}$), and $X = \mathbb{R}$. Let $F : \Omega \longrightarrow P_f(X)$ be defined by $F(\omega) = [0, \frac{1}{2}] \cup \{1\}$ for all $\omega \in \Omega$. Then int $\int_\Omega F d\mu = (0,1)$, and $\int_\Omega \text{int } F d\mu = (0, \frac{1}{2})$.

In the next proposition, we use the set-valued integral to determine if $f \in L^1(\Omega, X)$ is an integrable selector of a multifunction F.

PROPOSITION 6.4.55 (a) *If $F:\Omega \longrightarrow P_{wkc}(X)$ is graph-measurable and integrably bounded, then $f \in S_F^1$ if and only if $\int_A f d\mu \in \text{cl} \int_A F d\mu = \int_A F d\mu$ for all $A \in \Sigma$.*

(b) *If X^* is separable and $F : \Omega \longrightarrow P_{fc}(X)$ is graph-measurable and integrably bounded, then $f \in S_F^1$ if and only if $\int_A f d\mu \in \text{cl} \int_A F d\mu$ for all $A \in \Sigma$.*

PROOF: (a) $\Rightarrow$: This implication follows from the definition of the set-valued integral (see Definition 6.4.38).

$\Leftarrow$: If by $\langle\cdot,\cdot\rangle$ we denote the duality brackets for the pair (X, X^*), for all $(x^*, A) \in X^* \times \Sigma$ we have

$$\left\langle x^*, \int_A f d\mu \right\rangle = \int_A \langle x^*, f\rangle \, d\mu \leq \sigma\Big(x^*, \int_A F d\mu\Big) = \int_A \sigma(x^*, F) d\mu.$$

So for all $\omega \in \Omega \setminus N(x^*)$, $\mu\big(N(x^*)\big) = 0$, we have

$$\langle x^*, f(\omega)\rangle \leq \sigma\big(x^*, F(\omega)\big).$$

Consider $\{x_n^*\}_{n\geq 1} \subseteq X^*$ m-dense. Because F is $P_{wkc}(X)$–valued $\sigma\big(\cdot, F(\omega)\big)$ is m-continuous. So

$$\langle x^*, f(\omega)\rangle \leq \sigma\big(x^*; F(\omega)\big)$$

for all $\omega \in \Omega \setminus N$, $\mu(N) = 0$ with $N = \bigcup_{n\geq 1} N(x_n^*)$ and all $x^* \in X^*$. Hence $f \in S_F^1$.

(b)$\Rightarrow$: Again this follows from Definition 6.4.38.

$\Leftarrow$: This follows similarly to the corresponding implication in part (a), using this time $\{x_n^*\}_{n\geq 1} \subseteq X^*$ strongly dense (because X^* is separable) and the fact that $\sigma\big(\cdot, F(\omega)\big)$ is strongly continuous on X^*. □

We conclude this section with two other approaches to set-valued integration. The first is the so-called *embedding method*, which concerns $P_{kc}(X)$-valued multifunctions and is based on the following embedding theorem, known as the *Rådström embedding theorem*.

THEOREM 6.4.56 *There exists a separable Banach space $\widehat{X}$ in which $P_{kc}(X)$ can be embedded as a closed convex cone. Moreover, the embedding $i_R : P_{kc}(X) \longrightarrow \widehat{X}$ is linear isometric.*

REMARK 6.4.57 Because X is separable, then $(\overline{B}_1^*, w^*)$ is compact metrizable. In what follows we consider $\overline{B}_1^*$ with the weak*-topology. We have that $C(\overline{B}_1^*)$ is a separable Banach space. Moreover, for any $C \in P_{bfc}(X)$ we have that $C \in P_{kc}(X)$ if and only if $\sigma(\cdot, C)$ is sequentially w^*-continuous. So in Theorem 6.4.56 $\widehat{X}$ is the closed subspace of $C(\overline{B}_1^*)$ generated by the set $\{\sigma(\cdot, C)\}_{C\in P_{kc}(X)}$.

Using Theorem 6.4.56 we can have the following alternative approach to set-valued integration.

THEOREM 6.4.58 *If $F : \Omega \longrightarrow P_{kc}(X)$ is graph-measurable and integrably bounded, then $\int_\Omega F d\mu = (B) - \int_\Omega i_R(F) d\mu$ with $(B) - \int_\Omega$ being the $\widehat{X}$-valued Bochner integral.*

The second alternative approach to set-valued integration requires topological structure on the set Ω and the multifunction F. So let $\Omega = Y$ with (Y, d) a metric space, μ a finite Borel measure on Y (recall that μ is regular), and $F : Y \longrightarrow P_{fc}(X)$ is lsc. We consider the set

$$CS_F = \{f \in C(Y,X) : f(y) \in F(y) \text{ for all } y \in Y\}.$$

From Theorem 6.3.6 (Michael's selection theorem), we know that $CS_F \neq \emptyset$. We can define a set-valued integral of F using the set CS_F.

DEFINITION 6.4.59 $\int_Y^c F d\mu = \{ \int_Y f(y) d\mu : f \in CS_F \}$.

REMARK 6.4.60 Clearly we always have $\int_Y^c F d\mu \subseteq \int_Y F d\mu$.

In the next theorem, we obtain equality for the closures of the two sets. The proof of this result can be found in Hu–Papageorgiou [313, p. 210].

THEOREM 6.4.61 *If $F : Y \longrightarrow P_{fc}(X)$ is lsc and integrably bounded, then for all $A \in \mathbf{B}(Y)$, $\mathrm{cl} \int_A^c F(y) d\mu = \mathrm{cl} \int_A F(y) d\mu$.*

REMARK 6.4.62 If F is $P_{wkc}(X)$-valued, then $\int_A^c F d\mu = \int_A F d\mu$ for all $A \in \mathbf{B}(Y)$. Moreover, if F is h-lsc, integrably bounded with open convex values, then $\int_A^c F d\mu = \int_A F d\mu$ for all $A \in \mathbf{B}(Y)$ (see Hu–Papageorgiou [313, p. 212]).

6.5 Fixed Points and Carathéodory Multifunctions

The purpose of this section is to extend some of the main fixed point theorems for single-valued maps to multifunctions and so determine some basic properties of multifunctions $F(\omega, x)$ that are measurable (in some sense; see Section 6.2) in $\omega \in \Omega$ and continuous (in some sense; see Section 6.1) in $x \in X$.

In Section 3.4 we proved metric fixed point theorems for single-valued maps. Undoubtedly the most important such result is the Banach contraction principle (see Theorem 3.4.3). The first theorem of this section generalizes the fixed point theorem of Banach to multifunctions.

DEFINITION 6.5.1 Let (X, d) be a metric space and h the corresponding Hausdorff metric on $P_f(X)$ (see Definition 6.1.29). A multifunction is an *h-contraction*, if

$$h(F(x), F(u)) \leq k\, d(x, u) \qquad \text{for all } x, u \in X \text{ with } 0 < k < 1.$$

THEOREM 6.5.2 *If (X, d) is a complete metric space and $F : X \longrightarrow P_f(X)$ is an h-contraction, then F has a fixed point; that is, there exists $x \in X$ such that $x \in F(x)$.*

PROOF: We choose $k < k_1 < 1$ and $x_0 \in X$. We pick $x_1 \in F(x_0)$ such that $d(x_0, x_1) > 0$. If no such $x_1 \in F(x_0)$ can be found, then $x_0 \in F(x_0)$ and we are done. Then

$$d(x_1, F(x_1)) \leq h(F(x_0), F(x_1)) \leq k d(x_0, x_1) < k_1 d(x_0, x_1).$$

Hence we can find $x_2 \in F(x_1)$ such that

$$d(x_1, x_2) < k_1 d(x_0, x_1).$$

By induction we can generate a sequence $\{x_n\}_{n \geq 1} \subseteq X$ such that

$$x_{n+1} \in F(x_n) \text{ and } d(x_n, x_{n+1}) < k_1^n d(x_0, x_1) \qquad \text{for all } n \geq 1. \tag{6.28}$$

From the inequality in (6.28) and since $k_1 < 1$, we deduce that $\{x_n\}_{n \geq 1} \subseteq X$ is Cauchy. Hence

$$x_n \longrightarrow x \in X \qquad \text{for some } x \in X.$$

From the inclusion in (6.28) and if we pass to the limit as $n \to \infty$, we obtain $x \in F(x)$. □

REMARK 6.5.3 In contrast to Theorem 3.4.3, in this case the fixed point is not unique. Indeed, if F is the trivial constant multifunction $F(x) = X$ for all $x \in X$, then every point in X is a fixed point of F. If $\text{Fix}(F) = \{x \in X : x \in F(x)\}$ (the set of fixed points of F), then it is easily seen that $\text{Fix}(F)$ is closed in X.

The next proposition proves a remarkable stability property of the set $\text{Fix}(F)$ with respect to the multifunction F.

PROPOSITION 6.5.4 *If (X,d) is a complete metric space, $F_1, F_2 : X \longrightarrow P_{bf}(X)$ are h-contractions with the same constant $0 < k < 1$, then $h\big(\text{Fix}(F_1), \text{Fix}(F_2)\big) \leq 1/(1-k) \sup_{x \in X} h\big(F_1(x), F_2(x)\big)$.*

PROOF: Let $\varepsilon > 0$ and also choose $\xi > 0$ such that $\xi \sum_{n \geq 1} nk^n < 1$. We set $\varepsilon_1 = \xi\varepsilon\big(1/(1-k)\big)$, pick $x_0 \in \text{Fix}(F_1)$, and then we choose $x_1 \in F_2(x_0)$ such that

$$d(x_0, x_1) \leq h\big(F_1(x_0), F_2(x_0)\big) + \varepsilon. \tag{6.29}$$

We can find $x_2 \in F_2(x_1)$ such that

$$d(x_2, x_1) \leq k\, d(x_0, x_1) + k\varepsilon_1$$

and then inductively we obtain $\{x_n\}_{n \geq 1} \subseteq X$ such that

$$x_{n+1} \in F_2(x_n) \quad \text{and} \quad d(x_{n+1}, x_n) \leq k^n d(x_0, x_1) + nk^n \varepsilon_1 \qquad \text{for all } n \geq 1, \tag{6.30}$$

$$\Rightarrow \sum_{n \geq m} d(x_{n+1}, x_n) \leq \frac{k^m}{1-k} d(x_0, x_1) + \varepsilon_1 \sum_{n \geq m} nk^n,$$

$\Rightarrow$ $\{x_n\}_{n \geq 1} \subseteq X$ is a Cauchy sequence and so $x_n \longrightarrow x \in X$.

Then from (6.30) in the limit as $n \to \infty$ we obtain $x \in \text{Fix}(F_2)$. Moreover,

$$\begin{aligned} d(x_0, x) \leq \sum_{n \geq 0} d(x_n, x_{n+1}) &\leq \frac{1}{1-k} d(x_0, x_1) + \varepsilon_1 \sum_{n \geq m} nk^n \\ &\leq \frac{1}{1-k}\Big(h\big(F_1(x_0), F_2(x_0)\big) + 2\varepsilon\Big) \quad \text{(see (6.29))}. \end{aligned}$$

Reversing the roles of F_1 and F_2 in the above argument, we obtain

$$h\big(\text{Fix}(F_1), \text{Fix}(F_2)\big) \leq \frac{1}{1-k} \sup_{x \in X} h\big(F_1(x), F_2(x)\big).$$

□

COROLLARY 6.5.5 *If (X,d) is a complete metric space, $F_n, F : X \longrightarrow P_{bf}(X)$ for $n \geq 1$ are h-contractions all with the same constant $0 < k < 1$, and* $\sup_{x\in X} h\big(F_n(x), F(x)\big) \longrightarrow 0$ *as $n \to \infty$, then $h\big(\mathrm{Fix}(F_n), \mathrm{Fix}(F)\big) \longrightarrow 0$ as $n \to \infty$.*

There is also a multivalued analogue of Theorem 3.4.23.

DEFINITION 6.5.6 Let (X,d) be a metric space and h the corresponding Hausdorff metric on $P_f(X)$. A multifunction $F : X \longrightarrow P_f(X)$ is said to be *nonexpansive*, if $h\big(F(x), F(y)\big) \leq d(x,y)$ for all $x, y \in X$.

The multivalued generalization of Theorem 3.4.23 is due to Lim [383].

THEOREM 6.5.7 *If X is a uniformly convex Banach space, $C \in P_{bfc}(X)$, and $F : C \longrightarrow P_f(C)$ is nonexpansive, then F has a fixed point.*

Now we turn our attention to topological fixed point theorems for multifunctions.

DEFINITION 6.5.8 Let X be a vector space.

(a) A subset $C \subseteq X$ is said to be *finitely closed*, if $C \cap Y$ is closed for every finite-dimensional affine subspace Y in X, when on Y we consider its Euclidean topology (recall that Y is an affine subspace of X, if $Y = y_0 + Y_0$ for some $y_0 \in X$ and some finite-dimensional subspace of X).
(b) A family $\{C_i\}_{i\in I}$ of nonempty sets in X is said to have the *finite intersection property*, if the intersection of every finite subfamily is nonempty.
(c) Let $C \subseteq X$ be nonempty. A multifunction $F : C \longrightarrow 2^X$ is a *Knaster–Kuratowski–Mazurkiewicz multifunction* (a KKM-multifunction for short), if for every finite set $\{x_k\}_{k=1}^m \subseteq C$, we have

$$\mathrm{conv}\{x_k\}_{k=1}^m \subseteq \bigcup_{k=1}^m F(x_k).$$

(see also Definition 2.3.6).

The next theorem gives the basic property of KKM-multifunctions. It extends Theorem 2.3.7.

THEOREM 6.5.9 *If X is a vector space, $C \subseteq X$ is nonempty, and $F : C \longrightarrow 2^X$ is a KKM-multifunction with values that are finitely closed, then the family $\{F(x)\}_{x\in C}$ has the finite intersection property (see Definition 6.5.8(b)).*

PROOF: We argue by contradiction. So suppose that $\bigcap_{k=1}^n F(x_k) = \emptyset$. Let $Y = \mathrm{span}\{x_k\}_{k=1}^n$, let d be the Euclidean metric on Y and $D = \mathrm{conv}\{x_k\}_{k=1}^n \subseteq Y$. We know that $Y \cap F(x_k)$ is closed for all $k \in \{1, \ldots, n\}$. We have $\bigcap_{k=1}^n \big(Y \cap F(x_k)\big) = \emptyset$ and so the function $\xi(x) = \sum_{k=1}^n d\big(x, Y \cap F(x_k)\big)$ satisfies $\xi(x) > 0$ for all $x \in C$. Then $\vartheta : D \longrightarrow D$ defined by

$$\vartheta(x) = \frac{1}{\xi(x)} \sum_{k=1}^{n} d\big(x, Y \cap F(x_k)\big) x_k$$

is continuous and so by Theorem 3.5.3 (Brouwer's fixed point theorem), we can find $x_0 \in D$ such that $\vartheta(x_0) = x_0$. Let $J = \big\{k \in \{1, \ldots, n\} : d\big(x_0, Y \cap F(x_k)\big) \neq 0\big\}$. Then $x_0 \notin \bigcup\limits_{k \in J} F(x_k)$. But

$$\vartheta(x_0) = x_0 \in \operatorname{conv}\{x_k\}_{k \in J} \subseteq \bigcup_{k \in J} F(x)$$

(because F is a KKM-multifunction), a contradiction. □

An immediate consequence of the previous theorem is the following result (see also Corollary 2.3.9).

THEOREM 6.5.10 *If X is a Hausdorff topological vector space, $C \subseteq X$ is nonempty, $F : C \longrightarrow 2^X$ is a KKM-multifunction with closed values, and for at least one $x_0 \in C$, $F(x_0)$ is compact, then $\bigcap\limits_{x \in C} F(x) \neq \emptyset$.*

The same conclusion can be reached in another way, which avoids any topological structure and any topological hypotheses and instead uses an auxiliary multifunction.

THEOREM 6.5.11 *If X is a vector space, $C \subseteq X$ is nonempty, $F : C \longrightarrow 2^X$ is a KKM-multifunction, $G : C \longrightarrow 2^X$ is another multifunction such that*

$$G(x) \subseteq F(x) \quad \text{for all } x \in C \quad \text{and} \quad \bigcap_{x \in C} G(x) = \bigcap_{x \in C} F(x),$$

and for some topology on X, G has compact values, then $\bigcap\limits_{x \in C} F(x) \neq \emptyset$.

PROPOSITION 6.5.12 *If X is a vector space, $C \subseteq X$ is nonempty, and $F : C \longrightarrow 2^X$ is a multifunction such that*

$$x \notin \operatorname{conv} F(x) \quad \text{for all } x \in C,$$

then $x \longrightarrow G(x) = X \setminus F^{-1}(x)$ is a KKM-multifunction.

PROOF: Let $\{x_k\}_{k=1}^{n} \subseteq X$. We have

$$X \setminus \bigcup_{k=1}^{n} G(x_k) = \bigcap_{k=1}^{n} F^{-1}(x_k).$$

Therefore we have

$$y \notin \bigcup_{k=1}^{n} G(x_k) \qquad \text{if and only if } y \in \bigcap_{k=1}^{n} F^{-1}(x_k)$$

and this is equivalent to

$$y \in C \quad \text{and} \quad x_k \in F(y) \quad \text{for all } k \in \{1, \ldots, n\}. \tag{6.31}$$

Next let $y \in \operatorname{conv}\{x_k\}_{k=1}^n$. We claim that $y \in \bigcup_{k=1}^n G(x_k)$. If this is not true, then from (6.31) we have

$$\begin{aligned} &\operatorname{conv}\, \{x_k\}_{k=1}^n \subseteq \operatorname{conv}\, F(y) \\ \Rightarrow\ & y \in \operatorname{conv}\, F(y), \quad \text{a contradiction.} \end{aligned}$$

Therefore, we have

$$\begin{aligned} &\operatorname{conv}\{x_k\}_{k=1}^n \subseteq \bigcup_{k=1}^n G(x_k), \\ \Rightarrow\ & G \text{ is a KKM-multifunction.} \end{aligned}$$

□

This proposition can be used to produce maximal elements with respect to certain irreflexive binary relations. The result is of importance in mathematical economics where the binary relation represents preference among pairs of goods.

THEOREM 6.5.13 *If X is a Hausdorff topological vector space, $C \subseteq X$ is a nonempty, compact, convex set, $\prec$ is an irreflexive binary relation on C, and*

(i) *For every $x \in C$, $x \notin \operatorname{conv}\{y \in C : x \prec y\}$.*
(ii) *For every $x \in C$, the lower section $\{y \in C : y \prec x\}$ is open in C,*

then the set of $\prec$-maximal elements in C, namely the set

$$\{x \in C : x \not\prec y \text{ for all } y \in C\}$$

is nonempty and compact.

PROOF: We consider the multifunction

$$x \longrightarrow F(x) = \{y \in C : x \prec y\} \qquad \text{for all } x \in C.$$

By virtue of hypothesis (i) and Proposition 6.5.12, we have that

$$x \longrightarrow G(x) = X \setminus F^{-1}(x) \ \text{ is a KKM-multifunction.}$$

Hence $H : C \longrightarrow 2^C$ defined by

$$x \longrightarrow H(x) = C \cap G(x) = C \setminus F^{-1}(x) \quad \text{is a KKM-multifunction too.}$$

Moreover, because of hypothesis (ii) the set $F^{-1}(x)$ is open and so H is compact-valued. So Theorem 6.5.10 implies that

$$\bigcap_{x \in C} H(x) = \bigcap_{x \in C} (C \setminus F^{-1}(x)) \neq \emptyset$$

and of course it is compact. But this is the set of $\prec$-maximal elements. □

This theorem leads to a fundamental existence theorem for variational inequalities.

THEOREM 6.5.14 *If X is a locally convex space, $C \subseteq X$ is nonempty, compact, and convex, and $u : C \longrightarrow X^*$ is a map such that $(x, y) \longrightarrow \langle u(x), y\rangle$ is jointly continuous on $C \times C$ (by $\langle \cdot, \cdot \rangle$ we denote the duality brackets for the dual pair (X, X^*)), then we can find $x \in C$ such that $\langle u(x), y - x\rangle \geq 0$ for all $y \in C$.*

PROOF: On C we define the irreflexive binary relation $\prec$ by

$$x \prec y \qquad \text{if and only if} \quad \langle u(x), x - y\rangle > 0.$$

Note that for every $x \in C$, we have

$$\begin{aligned} x \notin \operatorname{conv}\{y \in C : x \prec y\} &= \{y \in C : x \prec y\} \\ &= \big\{y \in C : \langle u(x), x - y\rangle > 0\big\}. \end{aligned}$$

Also note that because of the hypothesis on u, the set

$$\{y \in C : y \prec x\} = \big\{y \in C : \langle u(x), x - y\rangle < 0\big\}$$

is open. So we can apply Theorem 6.5.13 and obtain a $\prec$-maximal element $x \in C$. Hence

$$\langle u(x), x - y\rangle \geq 0 \qquad \text{for all } \ y \in C.$$

□

Next we use the results on KKM-multifunctions to deduce some basic topological fixed point theorems for set-valued maps.

DEFINITION 6.5.15 Let X be a vector space and $C \subseteq X$ a nonempty set.

(a) The *inward set* of $x \in C$ with respect to C is defined by

$$I_C(x) = \{x + \lambda(y - x) : \lambda \geq 0 \ \text{ and } \ y \in C\}.$$

(b) A multifunction $F : C \longrightarrow 2^X \setminus \{\emptyset\}$ is said to be *weakly inward* if

$$F(x) \cap \overline{I_C(x)} \neq \emptyset \qquad \text{for all } \ x \in C.$$

REMARK 6.5.16 If $F(C) \subseteq C$, then $F(x) \subseteq I_C(x)$ for all $x \in C$. Indeed, if $y \in F(x)$, then for $\lambda = 1$, $y = x + \lambda(y - x) = x + (y - x) = y$. Multifunctions satisfying $F(x) \subseteq \overline{I_C(x)}$ are often called *strongly inward*. For single-valued maps the geometric notion of weak inwardness has a metric equivalent, namely $f : C \longrightarrow X$ is weakly inward if and only if $\lim\limits_{\lambda \to 0^+} (1/\lambda) d\Big(x + \lambda\big(f(x) - x\big), C\Big) = 0$ for all $x \in C$.

THEOREM 6.5.17 *If X is a locally convex space, $C \subseteq X$ is nonempty, compact, and convex, and $F : C \longrightarrow P_{fc}(X)$ is usc and weakly inward, then F admits a fixed point; that is, there exists $x \in C$ such that $x \in F(x)$.*

PROOF: We argue indirectly. So suppose that F has no fixed point. Then for any given $x \in C$, we have $0 \notin x - F(x)$. Invoking the strong separation theorem for convex sets, we can find $x^* \in X^* \setminus \{0\}$ such that

$$\begin{aligned} &\langle x^*, x - y\rangle < 0 \qquad \text{for all } \ y \in F(x), \\ \Rightarrow \ &\sigma\big(x^*, x - F(x)\big) < 0. \end{aligned}$$

We set

$$U(x^*)=\{x \in C : \sigma(x^*, -F(x))<0\} \qquad \text{for all } x^* \in X^* \setminus \{0\}.$$

Because F is usc, we have that $U(x^*)$ is open (see Proposition 6.1.15(c)) and $\{U(x^*)\}_{x^* \in X^* \setminus \{0\}}$ is an open cover of C. Inasmuch as C is compact, we can find $\{U(x_k^*)\}_{k=1}^m$ a finite subcover and a corresponding continuous partition of unity $\{p_k\}_{k=1}^m$. We define

$$u(x) = \sum_{k=1}^{m} p_k(x) x_k^*.$$

Clearly the function $(x, y) \longrightarrow \langle u(x), y \rangle$ is continuous on $C \times C$ (as before by $\langle \cdot, \cdot \rangle$ we denote the duality brackets for the dual pair (X, X^*)). So we can apply Theorem 6.5.17 and obtain $x_0 \in C$ such that

$$\langle u(x_0), y - x_0 \rangle \geq 0 \qquad \text{for all } y \in C.$$

By hypothesis we can find $y \in F(x_0) \cap \overline{I_C(x_0)}$. Then

$$\begin{aligned} & y = x_0 + \lim_{n \to \infty} \lambda_n (y_n - x_0), \quad \text{with } \lambda_n \geq 0 \text{ and } y_n \in C, \\ \Rightarrow\ & y - x_0 = \lim_{n \to \infty} \lambda_n (y_n - x_0), \\ \Rightarrow\ & \langle u(x_0), y - x_0 \rangle = \lim_{n \to \infty} \lambda_n \langle u(x_0), y - x_0 \rangle \geq 0. \end{aligned}$$

But this contradicts the fact that

$$\sigma(u(x), x - F(x))<0 \qquad \text{for all } x \in C.$$

□

REMARK 6.5.18 It is easy to check that in the above theorem $\text{Fix}(F) = \{x \in C : x \in F(x)\}$ is compact.

If we combine Theorem 6.5.17 with Remarks 6.5.16 and 6.5.18 and if we recall that a locally compact multifunction is usc if and only if it is closed (i.e., its graph is closed; see Proposition 6.1.10), we can have the following multivalued generalization of the Schauder–Tychonov fixed point theorem (see Theorem 3.5.28). The result is known as the *Kakutani–Ky Fan fixed point theorem.*

THEOREM 6.5.19 *If X is locally convex space, $C \subseteq X$ is a nonempty, compact, and convex set, and $F : C \longrightarrow P_{fc}(C)$ is closed, then the set* Fix(F) *is nonempty and compact.*

A related topological fixed point theorem for multifunctions is the following.

THEOREM 6.5.20 *If X is a locally convex space, $C \subseteq X$ is a nonempty, compact, and convex set, and $F : C \longrightarrow 2^C \setminus \{\emptyset\}$ is a multifunction with convex values such that for each $y \in C$ the set $F^+(\{y\}) = \{x \in C : y \in F(x)\}$ is open, then F admits a fixed point.*

PROOF: The family $\{F^+(\{y\})\}_{y\in C}$ is an open cover of C. Due to the compactness of C, we can find a finite subcover $\{F^+(\{y_k\})\}_{k=1}^m$ and a corresponding continuous partition of unity $\{p_k\}_{k=1}^m$.

We set

$$u(x) = \sum_{k=1}^m p_k(x)y_k \qquad \text{for all } x \in C.$$

Clearly $u : C \longrightarrow C$ is continuous and $u(x) \in F(x)$ for all $x \in C$ (recall that F is convex-valued). Apply Theorem 3.5.28 to obtain $x \in C$ such that $u(x) = x \in F(x)$. ☐

Next let X, Y be Banach spaces and $C \subseteq X$, $D \subseteq Y$ are nonempty, closed, convex sets. In what follows by (D, w) we denote D furnished with the relative weak topology of Y. We consider multifunctions $G : C \longrightarrow 2^C \setminus \{\emptyset\}$ that admit the following decomposition

$$G = K \circ N, \tag{6.32}$$

where $N : C \longrightarrow 2^D \setminus \{\emptyset\}$ is usc from C with the relative strong (norm) topology into (D, w) and has weakly compact and convex values and $K : (D, w) \longrightarrow C$ is sequentially continuous, namely

$$y_n \xrightarrow{w} y \quad \text{in } D \Rightarrow K(y_n) \longrightarrow K(y) \quad \text{in } C.$$

Also we assume that G is compact; that is, it maps bounded sets in C into relatively compact sets in C.

We emphasize that with the above hypotheses G need not have convex values.

We have the following multivalued versions of the nonlinear alternative theorem (see Theorem 3.5.16) and of the Leray–Schauder alternative principle (see Corollary 3.5.18). Both theorems can be proved using degree-theoretic arguments. Details can be found in Bader [49].

THEOREM 6.5.21 *If X is a Banach space and $G : \overline{B}_R \longrightarrow 2^X \setminus \{\emptyset\}$ is a compact multifunction that admits the decomposition (6.32), then at least one of the following statements holds.*

(a) *There exists $x_0 \in \partial B_R$ and $\lambda \in (0,1)$ such that $x_0 \in \lambda G(x_0)$.*
(b) $\mathrm{Fix}(G) \neq \emptyset$.

THEOREM 6.5.22 *If X is a Banach space, $G : C \longrightarrow 2^C \setminus \{\emptyset\}$ is a compact multifunction that admits the decomposition (6.32), and $0 \in C$, then at least one of the following statements holds.*

(a) *G has a fixed point.*
(b) *The set $S = \{x \in C : x \in \lambda G(x) \text{ for some } 0<\lambda<1\}$ is bounded.*

Let us conclude our discussion of the fixed point theory of multifunctions, with two results on the topological structure of the set $\mathrm{Fix}(F)$ for h-contractions. First we have a definition.

DEFINITION 6.5.23 (a) Let A be a subset of the metric space X. Then A is called a *retract* if there exists a continuous map (retraction) $r : X \longrightarrow A$ such that $r\big|_A = \mathrm{id}_A$.

(b) A metric space E is said to be an *absolute retract* if for every metric space X and every nonempty closed set $A \subseteq X$, each continuous map $f : A \longrightarrow E$ admits a continuous extension on all of X.

REMARK 6.5.24 By virtue of Theorem 3.1.10 (Dugundji's extension theorem), every closed convex set of a normed space is a retract. Every homeomorphic image of an absolute retract is also an absolute retract. Moreover, if E is an absolute retract and A is a retract of E, then A is also an absolute retract.

The first structural theorem for the set $\mathrm{Fix}(F)$ is due to Ricceri [514].

THEOREM 6.5.25 *If X is a Banach space, $C \subseteq X$ is nonempty, closed, and convex and $F : C \longrightarrow P_{fc}(X)$ is an h-contraction, then* $\mathrm{Fix}(F)$ *is an absolute retract.*

The second structural theorem, is due to Bressan–Cellina–Fryszkowski [97].

THEOREM 6.5.26 *If (Ω, Σ, μ) is a finite nonatomic measure space, X is a separable Banach space, $Y = L^1(\Omega, X)$, and $F : Y \longrightarrow 2^Y \setminus \{\emptyset\}$ is an h-contraction with bounded, closed, and decomposable values, then* $\mathrm{Fix}(F)$ *is an absolute retract.*

We conclude this section with a few facts about Caratheodory multifunctions.

First suppose that T is a locally compact, σ-compact metric space equipped with a (regular) measure μ of bounded variation and defined on the σ-field $\mathcal{S}$ of a μ-measurable set (i.e., $\mathcal{S} = \mathbf{B}(T)_\mu$). Also X is a Polish space and Y is a separable metric space. Recall that for $P_k(Y)$–valued multifunctions continuity and h-continuity coincide (see Corollary 6.1.40) and that $\big(P_k(Y), h\big)$ is a separable metric space (see Proposition 6.1.32(b)). Therefore invoking Theorem 6.2.9, we obtain the following result.

PROPOSITION 6.5.27 *If $F : T \times X \longrightarrow P_k(Y)$ is a multifunction such that*

(i) *For every $x \in X$, $t \longrightarrow F(t,x)$ is measurable and*
(ii) *For every $t \in T$, $x \longrightarrow F(t,x)$ is continuous,*

then for every $\varepsilon > 0$ we can find $T_\varepsilon \subseteq T$ compact with $\mu(T \setminus T_\varepsilon) < \varepsilon$ such that $F\big|_{T_\varepsilon \times X}$ is continuous.

Another result in the same direction is the following.

PROPOSITION 6.5.28 *If $F : T \times X \longrightarrow 2^Y \setminus \{\emptyset\}$ is a multifunction such that*

(i) *$(t,x) \longrightarrow F(t,x)$ is measurable and*
(ii) *For every $t \in T$, $x \longrightarrow F(t,x)$ is lsc,*

then for every $\varepsilon > 0$ we can find $T_\varepsilon \subseteq T$ compact with $\mu(T \setminus T_\varepsilon) < \varepsilon$ such that $F\big|_{T_\varepsilon \times X}$ is lsc.

PROOF: Fix $y \in Y$ and consider the function $(t,x) \longrightarrow \varphi_y(t,x) = d\big(y, F(t,x)\big)$. Evidently φ_y is $\mathcal{L} \times \mathbf{B}(X)$-measurable (see Proposition 6.2.4) and for every $t \in T$ $\varphi_y(t,\cdot)$ is upper semicontinuous (see Proposition 6.1.15(a)). We can find a sequence $\{\varphi_n\}_{n \geq 1}$ of Carathéodory functions on $T \times X$ into Y such that $\varphi_n(t,x) \downarrow \varphi_y(t,x)$ for all

$(t,x) \in T \times X$. Invoking Theorem 6.2.9 given $\varepsilon > 0$, we can find $T_\varepsilon \subseteq T$ compact with $\mu(T \setminus T_\varepsilon) < \varepsilon$ such that $\varphi_n\big|_{T_\varepsilon \times X}$ is continuous. Hence $\varphi_y\big|_{T_\varepsilon \times X}$ is lower semicontinuous. A new application of Proposition 6.1.15(a) implies that $F\big|_{T_\varepsilon \times X}$ is lsc. □

REMARK 6.5.29 The result fails if for all $t \in T$, $F(t,\cdot)$ is usc.

Next we present two parametrized versions of Michael's selection (see Theorem 6.3.6).

PROPOSITION 6.5.30 *If Y is a separable reflexive Banach space and $F : T \times X \longrightarrow P_{fc}(Y)$ a multifunction such that*

(i) *$(t,x) \longrightarrow F(t,x)$ is measurable and*
(ii) *For every $t \in T$, $x \longrightarrow F(t,x)$ is lsc,*

then there exists a sequence $f_m : T \times X \longrightarrow Y$, $m \geq 1$, of Carathéodory functions such that for all $(t,x) \in T \times X$ we have $F(t,x) = \overline{\{f_m(t,x)\}}_{m \geq 1}$.

PROOF: By virtue of Proposition 6.5.28, for every $n \geq 1$ we can find $T_n \subseteq T$ compact such that $\mu(T \setminus T_n) < 1/n$ and $F\big|_{T_n \times X}$ is lsc. We apply Theorem 6.3.11 to obtain continuous functions $f_{n,m} : T_n \times X \longrightarrow Y$, $n, m \geq 1$, such that $F(t,x) = \overline{\{f_{nm}(t,x)\}}_{m \geq 1}$ for all $(t,x) \in T_n \times X$. Let $f_m : T \times X \longrightarrow Y$, $m \geq 1$, be defined by

$$f_m(t,x) = \begin{cases} f_{nm}(t,x) & \text{if } t \in T_n \text{ and } t \notin T_k \text{ for } k < n \\ 0 & \text{if } t \in T \setminus \bigcup\limits_{n \geq 1} T_n \end{cases}.$$

Clearly for every $m \geq 1$, f_m is a Carathéodory function and for all $(t,x) \in T \times X$, we have

$$F(t,x) = \overline{\{f_m(t,x)\}}_{m \geq 1}.$$

□

If $(T, \mathcal{S}, \mu)$ is replaced by (Ω, Σ) a measurable space (no topological structure on Ω), then the situation is more involved. Nevertheless by adapting the proof of Michael's selection theorem (see Theorem 6.3.6) to the present parametric situation, we can have the following theorem. For details we refer to Hu–Papageorgiou [313, p. 235]).

THEOREM 6.5.31 *If (Ω, Σ) is a complete measurable space, X is a Polish space, Y is a separable Banach space, and $F : \Omega \times X \longrightarrow P_{fc}(Y)$ is a multifunction such that*

(i) *$(\omega, x) \longrightarrow F(\omega, x)$ is measurable and*
(ii) *For all $\omega \in \Omega$, $x \longrightarrow F(\omega, x)$ is lsc,*

then we can find a sequence $f_n : \Omega \times X \longrightarrow Y$, $n \geq 1$, of Carathéodory functions such that

$$F(\omega, x) = \overline{\{f_n(\omega, x)\}}_{n \geq 1} \qquad \text{for all } (\omega, x) \in \Omega \times X.$$

REMARK 6.5.32 If $F(\omega, x)$ is measurable in $\omega \in \Omega$ and usc or lsc in $x \in X$, it is not in general jointly measurable.

6.6 Convergence of Sets

We start this section with the definitions of the different modes of set convergence, which we investigate.

DEFINITION 6.6.1 Let (X,d) be a metric space and $\{A_n, A\}_{n\geq 1} \subseteq P_f(X)$. We say that the A_ns converge to A in the *Hausdorff sense*, denoted by $A_n \xrightarrow{h} A$ or by $h\text{-}\lim_{n\to\infty} A_n = A$ if and only if $h(A_n, A) \longrightarrow 0$ as $n \to \infty$ (recall h is the Hausdorff metric on $P_f(X)$; see Definition 6.1.29).

REMARK 6.6.2 From Proposition 6.1.31(b), we see that $A_n \xrightarrow{h} A$ if and only if $d(\cdot, A_n) \longrightarrow d(\cdot, A)$ uniformly on X. Also if X is a normed space and $\{A_n, A\}_{n\geq 1} \subseteq P_{bfc}(X)$, then $A_n \xrightarrow{h} A$ if and only if $\sigma(\cdot; A_n) \longrightarrow \sigma(\cdot, A)$ uniformly on $\overline{B}_1^* = \{x^* \in X^* : \|x^*\| \leq 1\}$ (see Proposition 6.1.31(c)).

DEFINITION 6.6.3 Let (X,τ) be a Hausdorff topological space (τ denotes the Hausdorff topology) and let $\{A_n\}_{n\geq 1} \subseteq 2^X \setminus \{\emptyset\}$. We define

$$\begin{aligned} \tau\text{-}\liminf_{n\to\infty} A_n &= \{x \in X : x = \tau\text{-}\lim_{n\to\infty} x_n,\ x_n \in A_n,\ n \geq 1\} \quad \text{and} \\ \tau\text{-}\limsup_{n\to\infty} A_n &= \{x \in X : x = \tau\text{-}\lim_{n\to\infty} x_{n_k},\ x_{n_k} \in A_{n_k},\ n_k < n_{k+1},\ k \geq 1\}. \end{aligned}$$

The set $\tau\text{-}\liminf_{n\to\infty} A_n$ is the τ-*Kuratowski limit inferior* of the sequence $\{A_n\}_{n\geq 1}$ and $\tau\text{-}\limsup_{n\to\infty} A_n$ is the τ-*Kuratowski limit superior* of the sequence $\{A_n\}_{n\geq 1}$. If $A = \tau\text{-}\liminf_{n\to\infty} A_n = \tau\text{-}\limsup_{n\to\infty} A_n$, then A is the τ-*Kuratowski limit* of the sequence $\{A_n\}_{n\geq 1}$ and we write $A_n \xrightarrow{K_\tau} A$.

REMARK 6.6.4 If (X,d) is a metric space, then

$$\begin{aligned} d\text{-}\liminf_{n\to\infty} A_n &= \{x \in X : x = \lim_{n\to\infty} d(x, A_n) = 0\} \quad \text{and} \\ d\text{-}\limsup_{n\to\infty} A_n &= \{x \in X : x = \liminf_{n\to\infty} d(x, A_n) = 0\}. \end{aligned}$$

Note that we always have $\tau\text{-}\liminf_{n\to\infty} A_n \subseteq \tau\text{-}\limsup_{n\to\infty} A_n$ and if the space X is first countable, then $\tau\text{-}\limsup_{n\to\infty} A_n = \bigcap_{k\geq 1} \overline{\bigcup_{n\geq k} A_n}$ and so it is a closed set. If the topology τ is clearly understood, then we drop the letter τ.

The Kuratowki mode of set-convergence turns out to be suitable for locally compact spaces. In order to deal with sets defined in an infinite-dimensional Banach space (which is not locally compact), we need a new mode of set-convergence, which involves all the useful topologies defined on the Banach space.

DEFINITION 6.6.5 Let X be a Banach space. By w- (resp., s-) we denote the weak (resp. strong) topology on X. Let $\{A_n\}_{n\geq 1} \subseteq 2^X \setminus \{\emptyset\}$. We say that the A_ns converge to A in the *Mosco sense* if and only if $w\text{-}\liminf_{n\to\infty} A_n = s\text{-}\liminf_{n\to\infty} A_n = A$.

REMARK 6.6.6 Note that we always have $s\text{-}\liminf_{n\to\infty} A_n \subseteq w\text{-}\liminf_{n\to\infty} A_n$ and $s\text{-}\limsup_{n\to\infty} A_n \subseteq w\text{-}\limsup_{n\to\infty} A_n$. So comparing Definitions 6.6.3 and 6.6.5, we see that $A_n \xrightarrow{M} A$ if and only if $A_n \xrightarrow{K_s} A$ and $A_n \xrightarrow{K_w} A$.

DEFINITION 6.6.7 Let (X, d) be a metric space and $\{A_n, A\} \subseteq P_f(X)$. We say that the A_ns converge to A in the *Wijsman sense*, denoted by $A_n \xrightarrow{W} A$ if and only if for all $x \in X$ we have $d(x, A_n) \longrightarrow d(x, A)$.

REMARK 6.6.8 From Remark 6.6.2 we already know that $A_n \xrightarrow{h} A$ implies $A_n \xrightarrow{W} A$.

DEFINITION 6.6.9 Let X be a Banach space and $\{A_n, A\} \subseteq P_{fc}(X)$. We say that the A_ns converge to A *weakly* (or *scalarly*), denoted by $A_n \xrightarrow{w} A$, if and only if for all $x^* \in X^*$ we have $\sigma(x^*, A_n) \longrightarrow \sigma(x^*, A)$.

REMARK 6.6.10 From Remark 6.6.2, we know that if $\{A_n\}_{n\geq 1} \subseteq P_{bfc}(X)$, then $A_n \xrightarrow{h} A$ implies $A_n \xrightarrow{w} A$.

It is natural to ask what is the relation between these modes of set-convergence. Already in Remarks 6.6.8 and 6.6.10, we saw some first relations. In fact comparing the definitions and using Remark 6.6.6, we can have the following result which provides a first detailed general comparison between these notions. More specialized results are given later in the section.

PROPOSITION 6.6.11 *If X is a Banach space and $\{A_n\}_{n\geq 1} \subseteq 2^X \setminus \{\emptyset\}$, then*

(a) *$A_n \xrightarrow{h} A$ implies $A_n \xrightarrow{K_s} A$ and $A_n \xrightarrow{W} A$.*
(b) *If $\{A_n, A\}_{n\geq 1} \subseteq P_{bfc}(X)$, then $A_n \xrightarrow{h} A$ implies $A_n \xrightarrow{w} A$.*
(c) *$A_n \xrightarrow{M} A$ if and only if $A_n \xrightarrow{K_s} A$ and $A_n \xrightarrow{K_w} A$.*
(d) *$A_n \xrightarrow{W} A$ implies $A_n \xrightarrow{K_s} A$.*

Without additional hypotheses we cannot say more.

EXAMPLE 6.6.12 (a) *In general M-convergence does not imply h-convergence*: Let $X = l^2$ and let $\{e_n\}_{n\geq 1}$ be the standard orthonormal basis of the Hilbert space. We define $A_n = \{\lambda e_n : 0 \leq \lambda \leq 1\}$ and $A = \{0\}$. Evidently $A_n \xrightarrow{M} A$ but we do not have convergence in the Hausdorff metric because $h(A_n, A) = 1$ for all $n \geq 1$.

(b) *In general h-convergence does not imply M-convergence*: Let X be a reflexive Banach space and let $A_n = A = \partial B_1 = \{x \in X : \|x\| = 1\}$ for all k. The trivially $A_n \xrightarrow{h} A$. Recall that $\overline{\partial B_1}^{w} = \{x \in X : \|x\| \leq 1\}$. So we cannot have Mosco convergence.

(c) *In general K-convergence does not imply W-convergence*: Let $X = l^2$ and let $\{e_n\}_{n\geq 1}$ be the standard orthonormal basis of this Hilbert space. We define $A_n = \{x \in X : x = \lambda e_1 + (1-\lambda)e_n,\ 0 \leq \lambda \leq 1\}$, $n \geq 1$ and $A = \{e_1\}$. Evidently $A_n \xrightarrow{K} A$ but $d(0, A_n) = \frac{1}{2}\|e_1 + e_n\| = \sqrt{2}/2$ and so we cannot have W-convergence.

The next proposition explains the situation in Example 6.6.12(b) and also justifies the use of monotone techniques in variational analysis.

PROPOSITION 6.6.13 *If (X,d) is a metric space and $\{A_n\}_{n\geq 1} \subseteq 2^X \setminus \{\emptyset\}$, then*

(a) *The sets $\liminf\limits_{n\to\infty} A_n$, $\limsup\limits_{n\to\infty} A_n$ are both closed (possibly empty).*

(b) *If $\{A_n\}_{n\geq 1}$ is increasing, then $\liminf\limits_{n\to\infty} A_n = \limsup\limits_{n\to\infty} A_n = \overline{\bigcup\limits_{n\geq 1} A_n}$.*

(c) *If $\{A_n\}_{n\geq 1}$ is decreasing, then $\liminf\limits_{n\to\infty} A_n = \limsup\limits_{n\to\infty} A_n = \bigcap\limits_{n\geq 1} \overline{A_n}$.*

PROOF: (a): Note that $\limsup\limits_{n\to\infty} A_n = \bigcap\limits_{n\geq 1} \overline{\bigcup\limits_{n\geq 1} A_n}$ and so $\limsup\limits_{n\to\infty} A_n$ is closed. Also let $\{x_k\}_{k\geq 1} \subseteq \liminf\limits_{n\to\infty} A_n$ and assume that $x_k \longrightarrow x$ as $k \longrightarrow \infty$. We have $d(x, A_n) \leq \|x - x_k\| + d(x_k, A_n)$. Hence $\limsup\limits_{n\to\infty} d(x, A_n) \leq \|x - x_k\|$. Because this is true for all $k \geq 1$, we conclude that $d(x, A_n) \longrightarrow 0$ as $n \to \infty$, hence $x \in \liminf\limits_{n\to\infty} A_n$ (see Remark 6.6.4). Therefore $\liminf\limits_{n\to\infty} A_n$ is closed.

(a) and (c): Follow at once from Remark 6.6.4 and part (a). $\square$

REMARK 6.6.14 Note that in a metric space X, $\overline{\bigcup\limits_{k\geq 1} \bigcap\limits_{n\geq k} A_n} \subseteq \liminf\limits_{n\to\infty} A_n$.

PROPOSITION 6.6.15 *If X is a Banach space, $\{A_n\}_{n\geq 1} \subseteq P_{fc}(X)$, and $A_n \xrightarrow{h} A$, then $A_n \xrightarrow{M} A$.*

PROOF: From Proposition 6.6.11(a) we know that $A_n \xrightarrow{K_s} A$. So we need to show that $w\text{-}\limsup\limits_{n\to\infty} A_n \subseteq A$. To this end let $x \in w\text{-}\limsup\limits_{n\to\infty} A_n \subseteq A$. Then by virtue of Definition 6.6.3 we can find a subsequence $\{n_k\}$ on $\{n\}$ and $x_{n_k} \in A_{n_k}$ such that $x_{n_k} \xrightarrow{w} x$. Note that $A \in P_{fc}(X)$ and so $d(\cdot, A)$ is convex on X. So it is w-lower semicontinuous and we have $d(x, A) \leq \limsup\limits_{k\to\infty} d(x_{n_k}, A)$. On the other hand $d(x_{n_k}, A) \leq h(A_{n_k}, A) \longrightarrow 0$ as $k \longrightarrow \infty$. Therefore $d(x, A) = 0$ and so $x \in A$ (because $A \in P_{fc}(X)$). Hence we infer that $w\text{-}\limsup\limits_{n\to\infty} A_n \subseteq A$ and we conclude that $A_n \xrightarrow{M} A$. $\square$

PROPOSITION 6.6.16 *If (X,d) is a metric space, $\{A_n\}_{n\geq 1} \subseteq P_f(X)$, $A_n \xrightarrow{K} A$, and there exists a compact set $C \subseteq X$ such that $A_n \subseteq C$ for all $n \geq 1$, then $A_n \xrightarrow{h} A$ as $n \to \infty$.*

PROOF: First note that $A \in P_k(X)$. Let $x_n \in A$ be such that $d(x_n, A_n) = \sup\big[d(x, A_n) : x \in A\big] = h^*(A, A_n)$. Since $\{x_n\}_{n\geq 1} \subseteq C$, we can find a subsequence $\{x_{n_k}\}_{k\geq 1}$ of $\{x_n\}_{n\geq 1}$ such that $x_{n_k} \longrightarrow x \in A$. Also since $A_n \xrightarrow{K} A$, we can find $a_n \in A_n$, $n \geq 1$, such that $a_n \longrightarrow x$ in X. Then we have

$$h^*(A, A_{n_k}) = d(x_{n_k}, A_{n_k}) \leq d(x_{n_k}, a_{n_k}) \longrightarrow 0 \text{ as } k \longrightarrow \infty.$$

Next let $u_n \in A_n$, $n \geq 1$, such that $d(u_n, A) = \sup[d(x,a) : x \in A_n] = h^*(A_n, A)$. As before we can find a subsequence $\{u_{n_k}\}_{k\geq 1}$ such that $u_{n_k} \longrightarrow u$. Because $A_n \xrightarrow{K} A$, we have $u \in A$. Then

$$h^*(A_{n_k}, A) = d(u_{n_k}, A) \leq d(u_{n_k}, u) \longrightarrow 0 \quad \text{as } k \longrightarrow \infty.$$

From the above arguments, we have

$$h^*(A, A_n) \quad \text{and} \quad h^*(A_n, A) \longrightarrow 0 \quad \text{as } n \to \infty,$$
$$\Rightarrow A_n \xrightarrow{h} A.$$

□

Now suppose that X is a Banach space and $\{A_n\}_{n\geq 1} \subseteq 2^X \setminus \{\emptyset\}$. From Definition 6.6.3 it is clear that for every $x^* \in X^*$, we have

$$\sigma(x^*, w\text{-}\limsup_{n\to\infty} A_n) \leq \limsup_{n\to\infty} \sigma(x^*, A_n). \tag{6.33}$$

In the next proposition, we present a situation where equality holds in (6.33).

PROPOSITION 6.6.17 *If X is a Banach space, $\{A_n\}_{n\geq 1} \subseteq 2^X \setminus \{\emptyset\}$, and for all $n \geq 1$, $A_n \subseteq W \in P_{wk}(X)$, then w-$\limsup_{n\to\infty} A_n \neq \emptyset$ and for every $x^* \in X^*$ we have*

$$\limsup_{n\to\infty} \sigma(x^*, A_n) = \sigma(x^*, w\text{-}\limsup_{n\to\infty} A_n).$$

PROOF: Because $A_n \subseteq W \in P_{wk}(X)$, $n \geq 1$, from the Eberlein–Smulian theorem, we conclude that w-$\limsup_{n\to\infty} A_n \neq \emptyset$. Let $x^* \in X^*$ and choose $a_n \in A_n$ such that

$$\sigma(x^*, A_n) - \frac{1}{n} \leq \langle x^*, a_n \rangle. \tag{6.34}$$

By passing to a subsequence if necessary, we may assume that $a_n \longrightarrow a \in w$-$\limsup_{n\to\infty} A_n$. Then from (6.34), we have

$$\limsup_{n\to\infty} \sigma(x^*, A_n) \leq \langle x^*, a \rangle \leq \sigma(x^*, w\text{-}\limsup_{n\to\infty} A_n). \tag{6.35}$$

Comparing (6.33) and (6.35), we conclude that

$$\sigma(x^*, w\text{-}\limsup_{n\to\infty} A_n) = \limsup_{n\to\infty} \sigma(x^*, A_n).$$

□

A related result is the following.

PROPOSITION 6.6.18 *If X is a Banach space, $\{A_n, A\}_{n\geq 1} \subseteq 2^X \setminus \{\emptyset\}$ and for every $x^* \in X^*$ we have*

$$\limsup_{n\to\infty} \sigma(x^*, A_n) \leq \sigma(x^*, A),$$

then w-$\limsup_{n\to\infty} A_n \subseteq \overline{\text{conv}}\, A$.

PROOF: Let $a \in w\text{-}\limsup_{n\to\infty} A$. We can find a subsequence $\{n_k\}$ of $\{n\}$ and $a_{n_k} \in A_{n_k}$, $k \geq 1$, such that $a_{n_k} \xrightarrow{w} a$ in X. For every $x^* \in X^*$ we have $\langle x^*, a_{n_k}\rangle \longrightarrow \langle x^*, a\rangle$ and so $\langle x^*, a\rangle \leq \limsup_{k\to\infty} \sigma(x^*, A_{n_k}) \leq \sigma(x^*, A)$. Hence $a \in \overline{\text{conv}}\, A$ and we conclude that $w\text{-}\limsup_{n\to\infty} A_n \subseteq \overline{\text{conv}}\, A$. □

PROPOSITION 6.6.19 *If X is a Banach space, $\{A_n\}_{n\geq 1} \subseteq P_{fc}(X)$, for all $n \geq 1$ we have $A_n \in W \subseteq P_{wk}(X)$ and $A_n \xrightarrow{K_w} A$, then $A_n \xrightarrow{w} A$.*

PROOF: Evidently $A \in P_{fc}(X)$. We fix $x^* \in X^*$ and then choose $a_n \in A_n$ such that

$$\langle x^*, a_n\rangle = \sigma(x^*, A_n),\ n \geq 1. \tag{6.36}$$

By passing to a suitable subsequence if necessary, we may assume that $a_n \xrightarrow{w} a \in A$ (because $A_n \xrightarrow{K_w} A$). Hence

$$\begin{aligned} &\langle x^*, a\rangle = \lim \sigma(x^*, A_n) \leq \sigma(x^*, A) \quad \text{(see (6.36))} \\ \Rightarrow\ &\limsup_{n\to\infty} \sigma(x^*, A_n) \leq \sigma(x^*, A). \end{aligned} \tag{6.37}$$

From (6.33) and (6.37) and since $A = w\text{-}\limsup_{n\to\infty} A_n$, we conclude that

$$\begin{aligned} &\sigma(x^*, A_n) \longrightarrow \sigma(x^*, A) \qquad \text{for all } x^* \in X^*, \\ \Rightarrow\ &A_n \xrightarrow{w} A. \end{aligned}$$

□

REMARK 6.6.20 The previous proposition fails if we do not have the uniform boundedness of the sequence $\{A_n\}_{n\geq 1}$ by $W \in P_{wk}(X)$. Moreover, if in Proposition 6.6.19 X is separable, then $A_n \xrightarrow{K_w} A$ if and only if $A_n \xrightarrow{w} A$. For details we refer to Hu Papageorgiou [313, p. 676].

PROPOSITION 6.6.21 *If X is a Banach space, $\{A_n\}_{n\geq 1} \subseteq 2^X \setminus \{\emptyset\}$, then for every $x \in X$ we have $\limsup_{n\to\infty} d(x, A_n) \leq d(x, s\text{-}\liminf_{n\to\infty} A_n)$.*

PROOF: If $s\text{-}\liminf_{n\to\infty} A_n = \emptyset$, then $d(\cdot, s\text{-}\liminf_{n\to\infty} A_n) \equiv +\infty$ and so the proposition is trivially true. Hence we may assume that $s\text{-}\liminf_{n\to\infty} A_n \neq \emptyset$. Let $a \in s\text{-}\liminf_{n\to\infty} A_n$. We can find $a_n \in A_n$, $n \geq 1$, such that $a_n \longrightarrow a$ in X. Then $d(x, A_n) \leq \|x - a_n\|$ and so $\limsup_{n\to\infty} d(x, A_n) \leq \|x - a\|$. Because $a \in s\text{-}\liminf_{n\to\infty} A_n$ was arbitrary, we deduce that $\limsup_{n\to\infty} d(x, A_n) \leq d(x, s\text{-}\liminf_{n\to\infty} A_n)$. □

PROPOSITION 6.6.22 *If X is a Banach space, $\{A_n, A\}_{n\geq 1} \subseteq 2^X \setminus \{\emptyset\}$ and for every $x \in X$, we have*

$$\limsup_{n\to\infty} d(x, A_n) \leq d(x, A),$$

then $A \subseteq s\text{-}\liminf_{n\to\infty} A_n$.

PROOF: If $a \in A$, then $d(a, A_n) \longrightarrow 0$ and so $a \in s\text{-}\liminf_{n\to\infty} A_n$ (see Remark 6.6.4). Therefore we have

$$A \in s\text{-}\liminf_{n\to\infty} A_n.$$

□

If we combine Propositions 6.6.18 and 6.6.22, we obtain the following result.

PROPOSITION 6.6.23 *If X is a Banach space, $\{A_n, A\}_{n\geq 1} \subseteq P_{fc}(X)$ and $A_n \xrightarrow{W} A$, $A_n \xrightarrow{w} A$, then $A_n \xrightarrow{M} A$.*

In Propositions 6.6.21 and 6.6.22 we obtained results concerning $s\text{-}\liminf_{n\to\infty} A_n$. We want to have similar results for $w\text{-}\limsup_{n\to\infty} A_n$. For this purpose we introduce the following class of subsets of X.

DEFINITION 6.6.24 Let X be a Banach space. We define $\mathcal{A} = \{C \subseteq X :$ C is nonempty, weakly closed and for every $r > 0$, $C \cap \overline{B}_r \in P_{wk}(X)\}$, and $\mathcal{A}_c = \{C \in \mathcal{A} : C$ is also convex$\}$.

REMARK 6.6.25 Clearly the family $\mathcal{A}$ is closed under finite unions, arbitrary intersections, and of course contains the family of all weakly closed and locally weakly compact subsets of X. Moreover, if X is reflexive, then $\mathcal{A} = \{C \subseteq X :$ C is nonempty and weakly closed$\}$; that is, $\mathcal{A} = P_f(X_w)$.

PROPOSITION 6.6.26 *If X is a Banach space, $\{A_n\}_{n\geq 1} \subseteq 2^X \setminus \{\emptyset\}$ and $A_n \subseteq W \in \mathcal{A}$ for every $n \geq 1$, then for every $x \in X$, we have $d(x, w\text{-}\limsup_{n\to\infty} A_n) \leq \liminf_{n\to\infty} d(x, A_n)$.*

PROOF: Let $x \in X$ and $r = \liminf_{n\to\infty} d(x, A_n)$. If $r = +\infty$, then the proposition is trivially true. So let $r < +\infty$. Suppose that the result is not true. Then for some $x \in X$ we have

$$\liminf_{n\to\infty} d(x, A_n) < d(x, w\text{-}\limsup_{n\to\infty} A_n). \tag{6.38}$$

We can find a subsequence $\{n_k\}$ of $\{n\}$, such that

$$\lim_{k\to\infty} d(x, A_{n_k}) = \liminf_{n\to\infty} d(x, A_{n_k}) = r < \infty.$$

We pick $a_{n_k} \in A_{n_k}$ such that

$$\|x - a_{n_k}\| \leq d(x, A_{n_k}) + \frac{1}{k} \qquad \text{for all } k \geq 1. \tag{6.39}$$

For $k \geq 1$ large we have $a_{n_k} \in W \cap \overline{B}_{(r+\|x\|+1)} \in P_{wk}(X)$. So by passing to a further subsequence if necessary, we may assume that $a_{n_k} \xrightarrow{w} a \in w\text{-}\limsup_{n\to\infty} A_n \neq \emptyset$. Then

$$\begin{aligned}
\|x - a\| \leq \liminf_{k\to\infty} \|x - a_{n_k}\| &\leq \liminf_{n\to\infty} d(x, A_n) \quad \text{(see (6.39))} \\
&< d(x, w\text{-}\limsup_{n\to\infty} A_n) \quad \text{(see (6.38))} \\
&\leq \|x - a\|,
\end{aligned}$$

a contradiction. □

Hidden in the above proof is also the result that $w\text{-}\limsup\limits_{n\to\infty} A_n \neq \emptyset$. So we can state this separately.

PROPOSITION 6.6.27 *If X is a Banach space, $\{A_n\}_{n\geq 1} \subseteq 2^X \setminus \{\emptyset\}$, and $r = \liminf\limits_{n\to\infty} d(x, A_n)$, then*

(a) $w\text{-}\limsup\limits_{n\to\infty} A_n \neq \emptyset$ *implies* $r < +\infty$.

(b) *If* $A_n \subseteq W \in \mathcal{A}$ *for all* $n \geq 1$ *and* $r < +\infty$, *then* $w\text{-}\limsup\limits_{n\to\infty} A_n \neq \emptyset$.

REMARK 6.6.28 If X is reflexive, then in part (b), $W = X$. So in reflexive Banach spaces, for any sequence $\{A_n\}_{n\geq 1} \subseteq 2^X \setminus \{\emptyset\}$ we always have $d(x, w\text{-}\limsup\limits_{n\to\infty} A_n) \leq \liminf\limits_{n\to\infty} d(x, A_n)$ for all $x \in X$.

The next result is analogous to Proposition 6.6.22 but for w-lim sup this time. For a proof of this result, we refer to Hu–Papageorgiou [313, pp. 673–674].

PROPOSITION 6.6.29 *If X is a reflexive Banach space with X^* locally uniformly convex, $\{A_n, A\}_{n\geq 1} \subseteq P_{fc}(X)$ and for every $x \in X$ we have $d(x, A) \leq \liminf\limits_{n\to\infty} d(x, A_n)$, then $w\text{-}\limsup\limits_{n\to\infty} A_n \subseteq A$.*

Recalling that weakly convergent sequences in a Banach space are bounded, we see that

$$w\text{-}\limsup_{n\to\infty} A_n = \bigcup_{k\geq 1} w\text{-}\limsup_{n\to\infty} (A_n \cap kB_1).$$

This observation leads at once to the following result.

PROPOSITION 6.6.30 *If X is a Banach space, $\{A_n\}_{n\geq 1} \subseteq 2^X \setminus \{\emptyset\}$ and either (i) X^* is separable or (ii) X is separable and $A_n \subseteq W \in \mathcal{A}$ for all $n \geq 1$, then $w\text{-}\limsup\limits_{n\to\infty} A_n = \bigcup\limits_{k\geq 1} \bigcap\limits_{m\geq 1} \overline{\bigcup\limits_{n\geq m} A_n \cap kB_1}^{w}$.*

REMARK 6.6.31 Clearly in the above proposition $A_n \cap kB_1$ can be replaced by $A_n \cap k\overline{B}_1$ or by $\overline{A_n \cap kB_1}^{\tau}$ with $\tau = s$ or $\tau = w$. Note that in general $w\text{-}\limsup\limits_{n\to\infty} A_n$ is neither strongly nor weakly closed.

Finally in a finite-dimensional setting and under a compactness condition, the situation is very pleasant because we have the following.

PROPOSITION 6.6.32 *If X is a finite-dimensional Banach space and $\{A_n, A\}_{n\geq 1} \subseteq P_{fc}(X)$ with A compact, then $A_n \xrightarrow{K} A \Leftrightarrow A_n \xrightarrow{w} A \Leftrightarrow A_n \xrightarrow{h} A \Leftrightarrow A_n \xrightarrow{W} A$.*

In the last part of this section we deal with the convergence of the set-valued integrals and of the sets of L^p-selectors of sequences of measurable multifunctions. A major tool in our considerations is the following result on the pointwise behavior of a weakly convergent sequence in $L^p(\Omega, X)$, $(1 \leq p < \infty)$.

PROPOSITION 6.6.33 *If (Ω, Σ, μ) is a σ-finite measure space, X is a Banach space, $\{f_n, f\}_{n\geq 1} \subseteq L^p(\Omega, X,)$ $1 \leq p < \infty$, $f_n \xrightarrow{w} f$ in $L^p(\Omega, X)$, and for μ-almost all $\omega \in \Omega$ and all $n \geq 1$ $f_n(\omega) \in G(\omega) \in P_{wk}(X)$, then $f(\omega) \in \overline{\text{conv}}\ w\text{-}\limsup\limits_{n\to\infty}\{f_n(\omega)\}$ μ-a.e. on Ω.*

PROOF: By Mazur's lemma we have

$$f(\omega) \in \overline{\text{conv}} \bigcup_{n\geq k} \{f_n(\omega)\} \quad \mu\text{-a.e. on } \Omega.$$

So for every $k \geq 1$, $x^* \in X^*$ and $\omega \in \Omega \setminus N$, $\mu(N) = 0$, we have

$$\begin{aligned} &\langle x^*, f(\omega)\rangle \leq \sigma\Big(x^*, \bigcup_{n\geq k} \{f_n(\omega)\}\Big) = \sup_{n\geq k} \langle x^*, f_n(\omega)\rangle, \\ \Rightarrow\ &\langle x^*, f(\omega)\rangle \leq \limsup_{n\to\infty} \langle x^*, f_n(\omega)\rangle. \end{aligned}$$

Invoking Proposition 6.6.17, we obtain

$$\langle x^*, f(\omega)\rangle \leq \sigma\Big(x^*, w\text{-}\limsup_{n\to\infty}\{f_n(\omega)\}\Big),$$

for all $\omega \in \Omega \setminus N$ and all $x^* \in X^*$,

$$\Rightarrow\ f(\omega) \in \overline{\text{conv}}\ w\text{-}\limsup_{n\to\infty}\{f_n(\omega)\}.$$

$\square$

Our aim is to derive multivalued analogues of Fatou's lemma. To do this we need the following measurability result whose proof can be found in Hu–Papageorgiou [313, p. 692].

LEMMA 6.6.34 *If (Ω, Σ) is a measurable space, X is a separable Banach space, and $F_n : \Omega \longrightarrow 2^X \setminus \{\emptyset\}$, $n \geq 1$, are measurable multifunctions such that $F_n(\omega) \subseteq W(\omega) \in \mathcal{A}$ for all $\omega \in \Omega$ and all $n \geq 1$, then $\omega \longrightarrow \limsup\limits_{n\to\infty} F_n(\omega)$ is measurable.*

REMARK 6.6.35 If X is reflexive, then $W(\omega) = X$ for all $\omega \in \Omega$ and so $\omega \longrightarrow \limsup\limits_{n\to\infty} F_n(\omega)$ is always measurable.

PROPOSITION 6.6.36 *If (Ω, Σ, μ) is a nonatomic, σ-finite measure space, X is a separable Banach space and $F_n : \Omega \longrightarrow 2^X \setminus \{\emptyset\}$, $n \geq 1$, are graph-measurable multifunctions such that $F_n(\omega) \subseteq W(\omega)$ μ-a.e. on Ω, for all $n \geq 1$, with $W : \Omega \longrightarrow P_{wkc}(X)$ integrably bounded, then $w\text{-}\limsup\limits_{n\to\infty} \int_\Omega F_n d\mu \subseteq \overline{\int_\Omega w\text{-}\limsup\limits_{n\to\infty} F_n d\mu}$.*

PROOF: Let $x \in w\text{-}\limsup\limits_{n\to\infty} \int_\Omega F_n d\mu$. We can find a subsequence $\{n_k\}$ of $\{n\}$ and $x_{n_k} \in \int_\Omega F_{n_k} d\mu$ such that $x_{n_k} \xrightarrow{w} x$ in X. Then we have $x_{n_k} = \int_\Omega f_{n_k} d\mu$ with $f_{n_k} \in S^1_{F_{n_k}} \subseteq S^1_W$. From Theorem 6.4.23 we know that $S^1_W \subseteq L^1(\Omega, X)$ is w-compact. So we may assume that $f_{n_k} \xrightarrow{w} f$ in $L^1(\Omega, X)$. By virtue of Proposition 6.6.36 we have $f(\omega) \in \overline{\text{conv}}\ w\text{-}\limsup\limits_{n\to\infty} F_n(\omega) = \overline{\text{conv}}\ w\text{-}\limsup\limits_{n\to\infty} \overline{F_n(\omega)}$ and for every $n \geq 1$,$\omega \longrightarrow \overline{F_n(\omega)}$ is Σ_μ-measurable. So Lemma 6.6.34 implies that $\omega \longrightarrow w\text{-}\limsup\limits_{n\to\infty} F_n(\omega)$ is Σ_μ-measurable. Using Corollary 6.4.42, we have

$$
\begin{aligned}
&\text{cl} \int_\Omega \overline{\text{conv}}\ w\text{-}\limsup_{n\to\infty} F_n d\mu = \text{cl} \int_\Omega w\text{-}\limsup_{n\to\infty} F_n d\mu \\
\Rightarrow\ &\int_\Omega \overline{\text{conv}}\ w\text{-}\limsup_{n\to\infty} F_n d\mu = \text{cl} \int_\Omega w\text{-}\limsup_{n\to\infty} F_n d\mu \quad \text{(see Proposition 6.4.48)} \\
\Rightarrow\ &\int_\Omega f d\mu \in \overline{\int_\Omega w\text{-}\limsup_{n\to\infty} F_n d\mu}.
\end{aligned}
$$

□

REMARK 6.6.37 The closure on the set-valued integral $\int_\Omega w\text{-}\limsup\limits_{n\to\infty} F_n d\mu$ cannot be dropped generally. For a situation where this can be done, we refer to Hu–Papageorgiou [313, p. 697].

PROPOSITION 6.6.38 *If (Ω, Σ, μ) is a σ-finite measure space, X is a separable Banach space, and $F_n : \Omega \longrightarrow 2^X \setminus \{\emptyset\}$, $n \geq 1$, are graph measurable multifunctions such that $F_n(\omega) \subseteq W(\omega)$ μ-a.e. with $W(\omega) \in P_{wk}(X)$ and $\sup[\|u\| : u \in F(\omega)] \leq \varphi(\omega)$ μ-a.e. on Ω with $\varphi \in L^1(\Omega)_+$, then $w\text{-}\limsup\limits_{n\to\infty} S^1_{F_n} \subseteq \overline{\text{conv}}\ S^1_{w\text{-}\limsup\limits_{n\to\infty} F_n}$.*

PROOF: From Corollary 6.4.18, we know that for all $u \in L^\infty(\Omega, X^*_{w^*}) = L^1(\Omega, X)^*$ we have

$$
\begin{aligned}
&\sigma(u, S^1_{F_n}) = \int_\Omega \sigma(u, F_n) d\mu, \\
\Rightarrow\ &\limsup_{n\to\infty} \sigma(u, S^1_{F_n}) = \limsup_{n\to\infty} \int_\Omega \sigma(u, F_n) d\mu, \\
&\leq \int_\Omega \limsup_{n\to\infty} \sigma(u, F_n) d\mu \quad \text{(by Fatou's lemma)} \\
&= \int_\Omega \sigma(u, w\text{-}\limsup_{n\to\infty} F_n) d\mu \quad \text{(see Proposition 6.6.17)} \\
&= \sigma\big(u, S^1_{w\text{-}\limsup F_n}\big),
\end{aligned}
$$

(see Lemma 6.6.34 and Corollary 6.4.18)

$$
\Rightarrow\ w\text{-}\limsup_{n\to\infty} S^1_{F_n} \subseteq \overline{\text{conv}}\ S^1_{w\text{-}\limsup F_n} \quad \text{(see Proposition 6.6.18)}.
$$

□

REMARK 6.6.39 If μ is finite, then the pointwise boundedness condition on $|W(\omega)| = \sup\{\|u\| : u \in W(\omega)\}$ can be replaced by the condition that $\{|F_n(\cdot)|\}_{n\geq 1}$ is uniformly integrable.

We can have analogous Fatou-type results for $s\text{-}\liminf$.

PROPOSITION 6.6.40 *If (Ω, Σ, μ) is a σ-finite measure space, X is a separable Banach space, and $F_n : \Omega \longrightarrow 2^X \setminus \{\emptyset\}$, $n \geq 1$, are graph-measurable multifunctions such that $\sup\limits_{n\geq 1} d\big(0, F_n(\cdot)\big) \in L^1(\Omega)$, then $\int_\Omega s\text{-}\liminf\limits_{n\to\infty} F_n d\mu \subseteq s\text{-}\liminf\limits_{n\to\infty} \int_\Omega F_n d\mu$.*

PROOF: Let $x \in \int_\Omega s\text{-}\liminf\limits_{n\to\infty} F_n d\mu$. Then $x = \int_\Omega f d\mu$ with $f \in S^1_{s\text{-}\liminf F_n}$. Let $H_n(\omega) = \big\{x \in F_n(\omega) : \|f(\omega) - x\| \leq d\big(f(\omega), F_n(\omega)\big) + 1/n\big\}$. Clearly $\mathrm{Gr}\, H_n \in \Sigma_\mu \times \mathbf{B}(X)$ and so by Theorem 6.3.20, we can find $f_n : \Omega \longrightarrow X$ a Σ-measurable function such that $f_n(\omega) \in H_n(\omega)$ μ-a.e. on Ω. We have $\|f_n(\omega) - f(\omega)\| \longrightarrow 0$ μ-a.e. on Ω. Therefore by the dominated convergence theorem we have $f_n \longrightarrow f$ in $L^1(\Omega, X)$ and so $x_n = \int_\Omega f_n d\mu \longrightarrow x = \int_\Omega f d\mu$. Because $f_n \in S^1_{F_n}$, we conclude that

$$\int_\Omega s\text{-}\liminf_{n\to\infty} F_n d\mu \subseteq s\text{-}\liminf_{n\to\infty} \int_\Omega F_n d\mu.$$

□

PROPOSITION 6.6.41 *If (Ω, Σ, μ) is a σ-finite measure space, X is a separable Banach space, $F_n : \Omega \longrightarrow 2^X \setminus \{\emptyset\}$, $n \geq 1$, are graph-measurable multifunctions, and $\sup\limits_{n\geq 1} d\big(0, F_n(\cdot)\big) \in L^1(\Omega)$, then $S^1_{s\text{-}\liminf F_n} \subseteq s\text{-}\liminf\limits_{n\to\infty} S^1_{F_n}$.*

PROOF: Using Theorem 6.4.16, for every $u \in L^1(\Omega, X)$ and every $n \geq 1$, we have

$$\begin{aligned} d(u, S^1_{F_n}) &= \int_\Omega d\big(u(\omega), F_n(\omega)\big) d\mu, \\ \Rightarrow\ \limsup_{n\to\infty} d(u, S^1_{F_n}) &\leq \int_\Omega \limsup_{n\to\infty} d\big(u(\omega), F_n(\omega)\big) d\mu \quad \text{(by Fatou's lemma)} \\ &\leq \int_\Omega d\big(u(\omega), s\text{-}\liminf_{n\to\infty} F_n(\omega)\big) d\mu, \end{aligned} \tag{6.40}$$

(see Proposition 6.6.21). Note that

$$s\text{-}\liminf_{n\to\infty} F_n(\omega) = s\text{-}\liminf_{n\to\infty} \overline{F_n}(\omega) = \big\{x \in X : \lim_{n\to\infty} d\big(x, \overline{F_n}(\omega)\big) = 0\big\},$$

hence $\omega \longrightarrow s\text{-}\liminf\limits_{n\to\infty} F_n(\omega)$ is Σ_μ-measurable. Therefore by Theorem 6.4.16,

$$\begin{aligned} &\int_\Omega d\big(u(\omega), s\text{-}\liminf_{n\to\infty} F_n(\omega)\big) d\mu = d(u, S^1_{s\text{-}\liminf\limits_{n\to\infty} F_n}) \\ \Rightarrow\ &\limsup_{n\to\infty} d(u, S^1_{F_n}) \leq d(u, S^1_{s\text{-}\liminf F_n}) \quad \text{(see (6.40))} \\ \Rightarrow\ &S^1_{s\text{-}\liminf\limits_{n\to\infty} F_n} \subseteq s\text{-}\liminf_{n\to\infty} S^1_{F_n} \quad \text{(see Proposition 6.6.22).} \end{aligned}$$

□

REMARK 6.6.42 If μ is finite, then the hypothesis that $\sup\limits_{n\geq 1} d\big(0, F_n(\cdot)\big) \in L^1(\Omega)$, can be replaced by the weaker hypothesis that $\big\{d\big(0, F_n(\cdot)\big)\big\}_{n\geq 1}$ is uniformly integrable.

COROLLARY 6.6.43 *If (Ω, Σ, μ) is a σ-finite measure space, X is a separable Banach space, $F_n : \Omega \longrightarrow P_{wkc}(X)$, $n \geq 1$, are graph-measurable multifunctions with $F_n(\omega) \in W(\omega)$ μ-a.e. on Ω for all $n \geq 1$ where $W : \Omega \longrightarrow P_{wkc}(X)$ is integrably bounded and $F_n(\omega) \xrightarrow{M} F(\omega)$ μ-a.e. on Ω, then $\int_\Omega F_n d\mu \xrightarrow{M} \int_\Omega F d\mu$ and $S^1_{F_n} \xrightarrow{M} S^1_F$.*

COROLLARY 6.6.44 *If (Ω, Σ, μ) is a σ-finite measure space, X is a finite-dimensional Banach space, $F_n : \Omega \longrightarrow 2^X \setminus \{\emptyset\}$, $n \geq 1$, are graph-measurable multifunctions with $\sup_{n \geq 1} |F_n(\cdot)| \in L^1(\Omega)$ and $F_n(\omega) \xrightarrow{K} F(\omega)$ μ-a.e. on Ω, then $\int_\Omega F_n d\mu \xrightarrow{K} \int_\Omega F d\mu$.*

We conclude with two interesting facts concerning weakly convergent sequences in $L^1(\Omega, \mathbb{R}^N)$. A sequence $\{f_n\}_{n \geq 1} \subseteq L^1(\Omega)$ that converges weakly but not strongly in $L^1(\Omega)$, oscillates wildly around its weak limit. The next proposition is well known from the theory of Lebesgue spaces (see, e.g., Gasiński–Papageorgiou [259, p. 170]).

PROPOSITION 6.6.45 *If (Ω, Σ, μ) is a finite measure space, $\{f_n, f\}_{n \geq 1} \subseteq L^1(\Omega)$, $f_n \xrightarrow{w} f$ in $L^1(\Omega)$ and one of the following two conditions holds,*

(i) $f(\omega) \leq \liminf_{n \to \infty} f_n(\omega)$ *μ-a.e. or*

(ii) $\limsup f_n(\omega) \leq f(\omega)$ *μ-a.e., then $f_n \longrightarrow f$ in $L^1(\Omega)$.*

If $\mathbb{R}$ is replaced by $\mathbb{R}^N$, then conditions (i) and (ii) above are replaced by an extremality condition. The result is due to Visintin [596].

PROPOSITION 6.6.46 *If (Ω, Σ, μ) is a finite measure space and $\{f_n\}_{n \geq 1} \subseteq L^1(\Omega, \mathbb{R}^N)$ satisfy $f_n \xrightarrow{w} f$ in $L^1(\Omega, \mathbb{R}^N)$ with*

$$f(\omega) \in \text{ext}\big(\overline{\text{conv}} \limsup_{n \to \infty} \{f_n(\omega)\}\big) \quad \mu\text{-a.e. on } \Omega,$$

then $f_n \xrightarrow{w} f$ in $L^1(\Omega, \mathbb{R}^N)$.

Using this proposition we can have the following result for the extreme points of the set-valued integral in $\mathbb{R}^N$. For the proof of this result, we refer to Hu–Papageorgiou [313, p. 736].

PROPOSITION 6.6.47 *If (Ω, Σ, μ) is a finite nonatomic measure space, $F : \Omega \longrightarrow P_f(\mathbb{R}^N)$ is measurable, $e \in \text{ext} \int_\Omega F d\mu$, $\{f_n\}_{n \geq 1} \subseteq S^1_F$ is uniformly integrable, and $\int_\Omega f_n d\mu \longrightarrow e$ in $\mathbb{R}^N$, then we can find $f \in S^1_F$ such that $e = \int_\Omega f d\mu$ and $f_n \longrightarrow f$ in $L^1(\Omega, \mathbb{R}^N)$.*

6.7 Remarks

6.1: There are hyperspace topologies corresponding to the various kinds of continuity of multifunctions introduced in this section. For details in this direction we refer to the books of Beer [61] and Kuratowski [368] and to the paper of Michael

[425]. Continuity properties of multifunctions are studied at various levels of generality in the books of Aliprantis–Border [10], Aubin–Cellina [38], Aubin–Frankowska [40], Berge [67], Castaing–Valadier [134], Denkowski–Migórski–Papageorgiou [194], Hu–Papageorgiou [313], Kisielewicz [355], and Klein–Thompson [357].

6.2: The study of the measurability properties of multifunctions was initiated with the works of Castaing [133] and Jacobs [330]. The approach of Castaing [133] is topological because the multifunctions are defined on a locally-compact Hausdorff topological space equipped with a Radon measure and are compact valued, whereas Jacobs [330] drops this requirement. In Debreu [186] and Rockafellar [521, 529], the multifunctions are defined on a measure space (Ω, Σ, μ) (no topological structure is assumed) and have values in $\mathbb{R}^N$. In Debreu [186] the multifunctions are compact valued and in Rockafellar [521, 529] only closed valued. Further contributions to the theory were made by Valadier [591], Leese [373], Himmelberg [301], Himmelberg–Parthasarathy–Van Vleck [304]. When dealing with compact-valued multifunctions, the following result is helpful (see Debreu [186, p. 355]).

PROPOSITION 6.7.1 *If X is a separable metrizable space, then the Borel σ-field of the hyperspace $P_k(X)$, denoted by $\mathbf{B}\big(P_k(X)\big)$ is generated by the family $\{U^+ : U \subseteq X \text{ open}\}$ where $U^+ = \{C \in P_k(X) : C \subseteq U\}$; also it is generated by the family $\{U^- : U \subseteq X \text{ open}\}$, where $U^- = \{C \in P_k(X) : C \cap U \neq \emptyset\}$.*

6.3: Theorem 6.3.6 is due to Michael [426]. Michael [425, 426, 427, 428] made additional important contributions in this direction. Proposition 6.3.16 is due to Cellina [135], who was the first to conduct a systematic study of the existence of approximate continuous selectors and their use in differential inclusions. For $P_{kc}(\mathbb{R}^N)$-valued multifunction we can have a Lipschitz selector using the Steiner point map.

DEFINITION 6.7.2 The *Steiner point map* is a map $s : P_{kc}(\mathbb{R}^N) \longrightarrow \mathbb{R}^N$ that has the following properties.

(a) $s(C) \in C$ for all $C \in P_{kc}(\mathbb{R}^N)$.
(b) $s(C + D) = s(C) + s(D)$ for all $C, D \in P_{kc}(\mathbb{R}^N)$.
(c) $s(UC) = Us(C)$ for all $C \in P_{kc}(\mathbb{R}^N)$ and all $U \in \mathcal{O}(\mathbb{R}^N)$ where $\mathcal{O}(\mathbb{R}^N)$ is the group of orthogonal transformations on $\mathbb{R}^N$.
(d) $s(\lambda C) = \lambda s(C)$ for all $\lambda \in \mathbb{R}$ and $C \in P_{kc}(\mathbb{R}^N)$.

Using the Steiner point map, we have at once the following result on Lipschitz continuous selectors.

PROPOSITION 6.7.3 *If X is a metric space and $F : X \longrightarrow P_{kc}(\mathbb{R}^N)$ is h-Lipschitz, then F admits a Lipschitz continuous selector.*

This result cannot be generalized to infinite-dimensional Banach spaces. More precisely, we have the following result due to Yost [615].

PROPOSITION 6.7.4 *Let X be a metric space and Y a Banach space. Every h-Lipschitz multifunction $F : X \longrightarrow P_{bfc}(Y)$ admits a Lipschitz continuous selector if and only if* $\dim Y < +\infty$.

Theorem 6.3.17 is due to Kuratowski–Ryll–Nardzewski [367]. Theorem 6.3.20 in the form stated there is due to Saint-Beuve [535]. However, earlier versions with additional restrictions, primarily on X, were proved by Yankov [611], von Neumann [456], and Aumann [46]. Scalarly measurable multifunctions were first considered by Valadier [591]. For the graph measurability of the extremal multifunction $\omega \longrightarrow \text{ext}\, F(\omega)$ (see Proposition 6.3.35), we refer to Benamara [63] (who followed a different approach to prove the result), Castaing–Valadier [134, Chapter IV], and Hu–Papageorgiou [313, Section II.4]. A final useful observation concerning measurable selectors is included in the next proposition (see Hu–Papageorgiou [313, p. 173]).

PROPOSITION 6.7.5 *If (Ω, Σ, μ) is a separable finite measure space and $G : \Omega \longrightarrow P_f(X)$, then there exists $F : \Omega \longrightarrow P_f(X) \cup \{\emptyset\}$-measurable multifunction such that $F(\omega) \subseteq G(\omega)$ μ-a.e. and every measurable selector of G is also a selector of F.*

6.4: The importance of decomposability was evident from the beginning to people working in control theory and optimization; see Boltyanski–Gamkrelidze–Pontryagin [82] and Neustadt [457]. The name convex with respect to switching is also used in the control theory literature. Theorem 6.4.16 is a very valuable tool in applications and illustrates the power of the notion of decomposability. It was first proved by Rockafellar [529]. The form presented here is due to Hiai–Umegaki [293]. Theorem 6.4.23 is also useful in applications and was proved by Papageorgiou [475]. Systematic studies of decomposability can be found in Hiai–Umegaki [293], Hiai [294], Olech [468], and Hu–Papageorgiou [313, Section 2.3]. Decomposability is used to define conditional expectation of multifunctions and define multivalued martingales and their extensions. We refer to the papers of Hiai–Umegaki [293], Papageorgiou [475, 480], Wang–Xue [600], and Wang [599]. Property U (see Definition 6.4.31), was introduced by Bourgain [94]. Proposition 6.4.33 relating the weak norm (see Definition 6.4.29) and the weak topology is due to Gutman [279]. Theorem 6.4.35 was proved by Tolstonogov [584], using the Baire category method of De Blasi–Pianigiani [180, 181, 182].

The set-valued integral given in Definition 6.4.38 is due to Aumann [44]. The topological variant given in Definition 6.4.59 is due to Cornwall [157]. The approach based on the Rådström embedding theorem (see Theorem 6.4.56) is due to Debreu [186]. For more details about embedding theorems of hyperspaces of convex sets, we refer to Schmidt [544]. Studies of the set-valued integral can be found in Klein–Thompson [357], and Hu–Papageorgiou [313].

6.5: Theorem 6.5.2 was proved by Nadler [449] for multivalued contractions with nonempty, bounded, closed values. Soon thereafter Covitz–Nadler [165], extended the result to multivalued contractions whose values need not be bounded. The stability results presented in Proposition 6.5.4 and Corollary 6.5.5, are due to Lim [383]. Theorem 6.5.10 was first proved by Knaster–Kuratowski–Mazurkiewicz [358] in the special case where C is the set of vertices of a simplex in $\mathbb{R}^N$. Their approach is combinatorial based on Sprener's lemma. The infinite-dimensional version presented here is due to Fan [232]. The finite-dimensional forerunner of Theorem 6.5.13 is due to Hartman–Stampacchia [286], and the infinite-dimensional result presented here is due to Browder [121]. Weakly inward maps (see Definition 6.5.15(b)), were introduced by Halpern–Bergman [283]. Theorem 6.5.17 is due to Halpern [284]. Theorem 6.5.19 is due to Kakutani [337] when $X = \mathbb{R}^N$ and due to Fan [232] when X is a locally convex space. Theorem 6.5.20 is due to Browder and [118] and Theorems 6.5.21 and 6.5.22 are due to Bader [49]. Additional fixed point theorems for multifunctions can be found in the books of Andres–Gorniewicz [27], Border, [86], Gorniewicz [272], and Hu–Papageorgiou [313, 316].

6.6: The first systematic study of the limits in Definition 6.6.3 can be found in Kuratowski [366]. The Mosco convergence of sets was introduced by Mosco [445, 446]. From Section 1.5 we know that they two notions correspond to the Γ-convergence and Mosco-convergence of functions, respectively. The W-convergence (see Definition 6.6.7) and the weak or scalar convergence (see Definition 6.6.9), were both introduced by Wijsman [605] to deal with problems in mathematical statistics. A detailed study together with comparison results accompanied by examples and counterexamples, can be found in Hu–Papageorgiou [313, Chapters I and VII]. For the Hausdorff and Kuratowski convergences, we have the following two compactness results.

THEOREM 6.7.6 *If* (X,d) *is a metric space,* $\{A_n\}_{n\geq 1} \subseteq P_f(X)$*, and for every* $n \geq 1$, $A_n \subseteq K \in P_k(X)$*, then we can find a subsequence* $\{A_{n_k}\}_{k\geq 1} \subseteq \{A_n\}_{n\geq 1}$ *and* $A \in P_k(X)$*, such that* $A_{n_k} \xrightarrow{h} A$.

REMARK 6.7.7 This result is known as *Blaschke's theorem.*

THEOREM 6.7.8 *If* (X,d) *is a separable metric space and* $\{A_n\}_{n\geq 1} \subseteq 2^X \setminus \{\emptyset\}$*, then there exists a subsequence* $\{A_{n_k}\}_{k\geq 1}$ *of* $\{A_n\}_{n\geq 1}$ *and* $A \in 2^X$*, such that* $A_{n_k} \xrightarrow{K} A$.

For multivalued Fatou's lemmata, we refer to Papageorgiou [477] and Hu–Papageorgiou [313, Section VII.3]. Proposition 6.6.45 was extended to functions in the Lebesgue–Bochner space $L^1(\Omega, X)$, by Balder [53] and Rzezuchowski [533].

7

Economic Equilibrium and Optimal Economic Planning

Summary. *This chapter is devoted to mathematical economics. We start with the static model of an exchange economy with a measure space of agents (an abstract device to model perfect competition). We introduce the notions of core allocation and of Walras allocations. We show that in the context of perfect competition, the two sets coincide and they are nonempty (existence result). Then we pass to dynamic models and deal with discrete-time infinite horizon multisector growth models. We deal with both discounted and undiscounted model. For the latter, we introduce the optimality notion of "weak maximality". We prove existence theorems and we establish weak and strong "turnpike theorems". Turnpike programs are important because every optimal program eventually moves close to a turnpike one. Moreover, they are easier to compute and are relatively insensitive to the optimality criterion. Subsequently, we conduct an analogous study for models with uncertainty (nonstationary discounted and stationary undiscounted). Then we consider continuous-time models and finally we investigate the "expected utility hypothesis (EUH), the main hypothesis in the theory of decision making.

Introduction

Mathematical economics deals with the analytical and mathematical aspects of modern economic theory. The field has made remarkable progress in the last forty years and was also the source for significant developments in nonlinear analysis, such as nonsmooth analysis and multivalued analysis. Today the field has expanded very much and covers a variety of topics that cannot be surveyed in just one chapter. For this reason, here we focus on two particular topics that hold a central position in economic theory. We study the theory of competitive markets (in particular their equilibrium theory) and the theory of growth. At the end of the chapter we examine a notion, which is central in decision making and which brings us to the doorsteps of game theory, which is the topic of the next chapter.

In Section 7.1, we discuss a static model of an exchange economy. We assume that perfect competition prevails. This is modelled by a continuum of agents (a nonatomic measure space of agents). For such a model, we introduce two equilibrium concepts. The core allocation (which is price-independent) and the Walras

N.S. Papageorgiou, S.Th. Kyritsi-Yiallourou, *Handbook of Applied Analysis*,
Advances in Mechanics and Mathematics 19, DOI 10.1007/b120946_7,

allocation (which is price-competitive). We show that in the assumed environment of perfect competition, the sets of these allocations coincide (core–Walras equivalence theorem). We also prove the existence of a Walras allocation.

In Section 7.2 we turn our attention to growth theory and study infinite horizon discrete-time multisector growth models. First we deal with the discounted model. We show that it admits an optimal program (path) and then we characterize it via a system of supporting prices. Then we pass to the undiscounted model. In this case the intertemporal utility need not converge and so we need to have a new optimality concept. We introduce the notion of weak maximality and we prove that the model admits a weakly maximal program.

In Section 7.3, we prove weak and strong turnpike theorems. A turnpike program is a special stationary program which in general is easy to compute and optimal programs eventually move close to a turnpike. So in general terms, turnpike theory deals with the asymptotic properties of optimal programs.

In Section 7.4 we study growth under uncertainty. We investigate both nonstationary, discounted and stationary, undiscounted models. We obtain existence and characterization results that are analogous to the ones proved for the deterministic case (see Section 7.2).

In Section 7.5 we discuss a continuous-time, discounted growth model. We prove an existence theorem. Now the choice of the topology on the set of feasible programs is crucial in the analysis of the model.

Finally, in Section 7.6, we examine a basic hypothesis in the theory of decision making, namely the Expected Utility Hypothesis (EUH). Using a partial order in the space of measures, we characterize choice behavior that is consistent with EUH.

The models considered in this chapter are highly idealized versions of real-life situations and to some readers, may appear to be only of theoretical interest. However, we think this is incorrect. First, there are specific economic situations for which these mathematical models are reasonable and realistic approximations. Second, the postulated optimality of a particular allocation or path (mystical as it may be) can be used to find practical solutions. It can serve as the yardstick against which we measure all other real-life trajectories of the particular economic systems and decide how good our choices actually are.

7.1 Perfectly Competitive Economies: Core and Walras Equilibria

A perfectly competitive economy is one in which no individual agent can affect the social outcome of the economic activity by his or her individual decisions. To simplify our model we consider only pure exchange economies in which no production is possible. The prices at which the exchange takes place, are independent of the actions of the agents and are taken by the agents as given. With these prices, the agents can trade any amount of commodities, which of course implies that the supply and demand of each individual agent is negligible compared to the total volume of trade and so cannot influence the price. The mathematical device to express this economic situation is to use a continuum of agents (traders). In particular as we show in the sequel, we assume that the space of agents is a finite nonatomic measure space (e.g., the unit interval with the Lebesgue measure). This way we are able to model mathematically the situation in which an individual agent cannot influence the outcome of the collective activity.

There are two equilibrium concepts for a pure exchange economy. The first one is a core allocation, whose origins can be traced back to Edgeworth [219], under the name contract curve. To illustrate this notion, consider an economic agent who suggests a feasible allocation to the other agents. Suppose now a group of agents, using their initial endowments, can make each member better off than the proposed allocation. Then we say that this group can block the proposed allocation. A feasible allocation, is a core allocation if no group of agents can block it. The set of core allocations forms the core of the economy.

The second equilibrium concept goes back to Walras [598] and refers to the noncooperative allocation of resources via a price system. The idea behind the notion of Walras equilibrium is that when the agents are assumed to know only the price system and their own preferences and initial endowments and are allowed to trade freely among them in a decentralized market, then the result will be allocations that maximize the utilities of the agents (subject to their budget constraints) and equate supply and demand.

Comparing the two equilibrium concepts introduced above, we see that in contrast to the Walras (competitive) allocation, a core allocation allows for the possibility of cooperation among agents in the economy. Each Walras allocation belongs to the core and the core is in general larger than the set of Walras allocations. A classical conjecture, which dates back to Edgeworth [219] says that the core shrinks to the set of Walras allocations as the number of economic agents increases. The conjecture was made mathematically precise with the introduction of measure space agents. Then in such a setting Aumann [45] proved that the core coincides with the set of Walras allocations. In this section we prove this equivalence result for a perfectly competitive pure exchange (no production) economy.

We start with the description of the mathematical model for a perfectly competitive pure exchange economy. So let (Ω, Σ, μ) be a finite measure space, describing the agents of the economy. More precisely, Ω denotes the set of economic agents, the σ-field Σ represents the system of coalitions that can be formed among the agents, and the finite measure μ measures the size of all the feasible coalitions. The commodity space is $\mathbb{R}^N$. So there are N-commodities traded in the market. There is a consumption correspondence $F : \Omega \longrightarrow 2^{\mathbb{R}^N} \setminus \{\emptyset\}$. For each agent $\omega \in \Omega$, let $\prec_\omega$ denote an irreflexive binary relation defined as $F(\omega)$. This is the preference relation of ω and describes his consumption tastes. For two commodity vectors $v, \widehat{v} \in F(\omega)$ we write $v \prec_\omega \widehat{v}$ (resp., $v \not\prec_\omega \widehat{v}$), to express the fact that $\widehat{v}$ is preferred (resp., not preferred) to v by the agent $\omega \in \Omega$. The preference relation $\prec$ can be derived from an indifference relation $\preceq$, which is usually assumed to be a reflexive, transitive and complete binary relation. To complete the description of the economy, we are also given a function $e \in L^1(\Omega, \mathbb{R}^N)$ which assigns to each agent $\omega \in \Omega$, the agent's initial endowment vector $e(\omega) \in F(\omega)$. So we call e the initial endowment allocation. The economy under consideration is the quadruple $\big\{(\Omega, \Sigma, \mu), F, \prec, e\big\}$.

DEFINITION 7.1.1 Let $\mathcal{E} = \big\{(\Omega, \Sigma, \mu), F, \prec, e\big\}$ be an exchange economy as described above.

(a) An *allocation* for the economy $\mathcal{E}$ is a function $u \in L^1(\Omega, \mathbb{R}^N)$. We say that the allocation u is *feasible* if $\int_\Omega u d\mu = \int_\Omega e d\mu$.

(b) A coalition $A \in \Sigma$ can improve upon the allocation u if there is an allocation w for $\mathcal{E}$ such that
   (i) $u(\omega) \prec_\omega w(\omega)$ μ-a.e. on A.
   (ii) $\mu(A)>0$ and $\int_A w d\mu = \int_A e d\mu$.

REMARK 7.1.2 In Definition 7.1.1(b), the coalition $A \in \Sigma$ can block allocation u, because it can redistribute the initial endowment among its members in such a way so as to make each agent in the coalition better off (with respect to the allocation u).

DEFINITION 7.1.3 Given a pure exchange economy $\mathcal{E}=\{(\Omega,\Sigma,\mu),F,\prec,e\}$, the core $C(\mathcal{E})$ of the economy $\mathcal{E}$, is the set of all feasible allocations of $\mathcal{E}$, upon which no coalition in Σ can improve.

REMARK 7.1.4 The core of an economy is a fundamental concept in mathematical economics, because it provides a precise explanation of competitive behavior.

Now let $p \in \mathbb{R}^N_+ \setminus \{0\}$ be the price prevailing in the market. Given a commodity vector $v \in \mathbb{R}^N$, the real number $(p,v)_{\mathbb{R}^N}$ is the value of the vector v. Clearly two commodity vectors $v, \widehat{v} \in \mathbb{R}^N$ can be exchanged at price p, provided their values are equal; that is, $(p,v)_{\mathbb{R}^N} = (p,\widehat{v})_{\mathbb{R}^N}$.

DEFINITION 7.1.5 Let $\mathcal{E} = \{(\Omega,\Sigma,\mu),F,\prec,e\}$ be a pure exchange economy in which prevails the price system $p \in \mathbb{R}^N_+ \setminus \{0\}$.

(a) The *budget set* of agent $\omega \in \Omega$ is defined by

$$b(\omega,p) = \{v \in F(\omega) : (p,v)_{\mathbb{R}^N} \leq (p,e(\omega))_{\mathbb{R}^N}\}$$

and the *demand set* is defined by

$$d(\omega,p) = \{v \in b(\omega,p) : \text{there is no } y \in b(\omega,p) \text{ such that } v \prec_\omega y\}$$

(i.e., $d(\omega,p)$ consists of those elements in the budget set that are maximal in $b(\omega,p)$ with respect to $\prec_\omega$).

(b) The allocation-price pair $(u,p) \in L^1(\Omega,\mathbb{R}^N) \times (\mathbb{R}^N \setminus \{0\})$ is a *Walras* or *competitive equilibrium* if and only if
   (i) $u(\omega) \in d(\omega,p)$ μ-a.e. on Ω;
   (ii) $\int_\Omega u d\mu = \int_\Omega e d\mu$ (i.e., the allocation u is feasible).

We denote by $W(\mathcal{E})$ the set of all Walras equilibria for the economy $\mathcal{E}$. Next we make precise our hypotheses on the pure exchange $\mathcal{E}=\{(\Omega,\Sigma,\mu),F,\prec,e\}$.

H: (i) (Ω,Σ,μ) is a finite nonatomic measure space.

(ii) For all $\omega \in \Omega$, $F(\omega) = \mathbb{R}^N_+$.

(iii) $\int_\Omega e d\mu \in \text{int}\, \mathbb{R}^N_+$.

(iv) For every $\omega \in \Omega$, $\prec_\omega$ is irreflexive and transitive.

(v) For every $v \in \mathbb{R}_+^N$, the set $\{y \in \mathbb{R}^N : v \prec_\omega y\}$ is open.

(vi) For every $v \in \mathbb{R}_+^N$, the multifunction $\omega \longrightarrow G_v(\omega) = \{y \in \mathbb{R}^N : v \prec_\omega y\}$ is graph-measurable.

(vii) If $v \in \mathbb{R}_+^N$ and $x \in \mathbb{R}_+^N \setminus \{0\}$, then $v \prec_\omega v + x$ for all $\omega \in \Omega$.

REMARK 7.1.6 Hypothesis H(i) implies that we are in an economic situation of perfect competition. Hypothesis H(iii) implies that there are resources available to be traded among the agents. Hypotheses H(v), (vi) are continuity and measurability hypotheses on the preference relation and finally hypothesis H(vii) is a monotonicity relation.

THEOREM 7.1.7 *If $\mathcal{E} = \{(\Omega, \Sigma, \mu), F, \prec, e\}$ is a pure exchange economy satisfying hypotheses H, then $C(\mathcal{E}) = W(\mathcal{E})$.*

PROOF: First we show that $W(\mathcal{E}) \subseteq C(\mathcal{E})$.

To this end let $(u, p) \in L^1(\Omega, \mathbb{R}^N) \times (\mathbb{R}^N \setminus \{0\})$ be a Walras equilibrium for the economy $\mathcal{E}$. Assume that $u \notin C(\mathcal{E})$. Then according to Definition 7.1.3 we can find $A \in \Sigma$ and $w \in L^1(\Omega, \mathbb{R}^N)$ such that $\int_A w d\mu = \int_A e d\mu$ and $u(\omega) \prec_\omega w(\omega)$ μ-a.e. on A. Since $(u, p) \in W(\mathcal{E})$, from Definition 7.1.5(b) we have that $(p, e(\omega))_{\mathbb{R}^N} < (p, w(\omega))_{\mathbb{R}^N}$ μ-a.e. on A. Hence

$$\int_A (p, e(\omega))_{\mathbb{R}^N} d\mu < \int_A (p, w(\omega))_{\mathbb{R}^N} d\mu,$$

$$\Rightarrow \Big(p, \int_A e d\mu\Big)_{\mathbb{R}^N} < \Big(p, \int_A w d\mu\Big)_{\mathbb{R}^N},$$

which contradicts the fact that $\int_A e d\mu = \int_A w d\mu$. So we have proved that $W(\mathcal{E}) \subseteq C(\mathcal{E})$.

Next let $u \in C(\mathcal{E})$. We show that for some price system $p \in \mathbb{R}_+^N \setminus \{0\}$, $(u, p) \in W(\mathcal{E})$. For this purpose we introduce the multifunction.

$$\omega \longrightarrow G(\omega) = \{y \in \mathbb{R}^N : v \prec_\omega y\} \cup \{e(\omega)\}. \tag{7.1}$$

Evidently G is a graph-measurable multifunction. We claim that

$$\overline{\Big(\int_\Omega G d\mu - \int_\Omega e d\mu\Big)} \cap \operatorname{int}(-\mathbb{R}_+^N) = \emptyset. \tag{7.2}$$

Here $\int_\Omega G d\mu$ is the set-valued integral introduced in Definition 6.4.38. Note that (7.2) is equivalent

$$\Big(\int_\Omega G d\mu - \int_\Omega e d\mu\Big) \cap \operatorname{int}(-\mathbb{R}_+^N) = \emptyset. \tag{7.3}$$

We argue indirectly. So suppose that (7.3) is not true. Then we can find $y \in \operatorname{int} \mathbb{R}_+^N$ such that

$$\int_\Omega e d\mu - y \in \int_\Omega G d\mu,$$
$$\Rightarrow \int_\Omega e d\mu - y = \int_\Omega g d\mu \quad \text{with } g \in S^1_G \quad \text{(see Definition 6.4.38).} \tag{7.4}$$

We define

$$A = \{\omega \in \Omega : u(\omega) \prec_\omega g(\omega)\} \quad \text{and} \quad C = \{\omega \in \Omega : g(\omega) = e(\omega)\}. \tag{7.5}$$

Evidently $A, C \in \Sigma$ (see hypothesis H(vi)). Because $\int_\Omega e d\mu \neq \int_\Omega g d\mu$ and $g \in S^1_G$, from (7.1) we see that $\mu(A) > 0$. Let $\widehat{g} : A \longrightarrow \mathbb{R}^N_+$ be the Σ-measurable function defined by

$$\widehat{g}(\omega) = g(\omega) + \frac{1}{\mu(A)} y.$$

Because $y \in \operatorname{int} \mathbb{R}^N_+$, from hypothesis H(vi) we have

$$g(\omega) \prec_\omega \widehat{g}(\omega) \qquad \text{for all } \omega \in A. \tag{7.6}$$

Also from the definition of the multifunction G (see (7.1)), we have

$$u(\omega) \prec_\omega g(\omega) \qquad \text{for all } \omega \in A. \tag{7.7}$$

From (7.6), (7.7), and the transitivity of the binary relation $\prec_\omega$, $\omega \in A$, we have

$$u(\omega) \prec_\omega \widehat{g}(\omega) \qquad \text{for all } \omega \in A. \tag{7.8}$$

Moreover, note that

$$\begin{aligned}
\int_A \widehat{g} d\mu = \int_A g d\mu + y &= \int_\Omega g d\mu - \int_C g d\mu + y \\
&= \int_\Omega e d\mu - \int_C e d\mu \quad \text{(see (7.4))} \\
&= \int_A e d\mu \quad \text{(see (7.5))}
\end{aligned}$$
$$\Rightarrow \widehat{g} \in L^1(\Omega, \mathbb{R}^N) \text{ is a feasible allocation.} \tag{7.9}$$

Combining (7.8) and (7.9) we reach a contradiction to the fact that $u \in C(\mathcal{E})$. So (7.3) and equivalently (7.2) hold.

Because G is graph-measurable, $e \in S^1_G$, and the measure μ is nonatomic, we can apply Theorem 6.4.44 and infer that $\int_\Omega G d\mu - \int_\Omega e d\mu$ is convex. Because of (7.3), we can use the weak separation theorem and obtain $p \in \mathbb{R}^N_+ \setminus \{0\}$ such that

$$(p, \int_\Omega e d\mu)_{\mathbb{R}^N} \leq (p, v)_{\mathbb{R}^N} \qquad \text{for all } v \in \int_\Omega G d\mu,$$
$$\Rightarrow (p, \int_\Omega e d\mu)_{\mathbb{R}^N} \leq \inf \Big[\int_\Omega (p, g)_{\mathbb{R}^N} d\mu : g \in S^1_G \Big],$$
$$\Rightarrow \int_\Omega (p, e)_{\mathbb{R}^N} d\mu \leq \int_\Omega \inf \big[(p, x)_{\mathbb{R}^N} : x \in G(\omega) \big] d\mu \quad \text{(see Theorem 6.4.16).} \tag{7.10}$$

We claim that from (7.10) it follows that

$$\big(p, e(\omega)\big)_{\mathbb{R}^N} \leq \inf \big[(p,x)_{\mathbb{R}^N} : x \in G(\omega)\big] \quad \mu\text{-a.e. on } \Omega. \tag{7.11}$$

Suppose that (7.11) is not true. Then we can find $D \in \Sigma$ with $\mu(D) > 0$ such that

$$\inf \big[(p,x)_{\mathbb{R}^N} : x \in G(\omega)\big] < \big(p, e(\omega)\big)_{\mathbb{R}^N} \qquad \text{for all } \omega \in D.$$

From Theorem 6.3.24, we know that $\omega \longrightarrow m(\omega) = \inf \big[(p,x)_{\mathbb{R}^N} : x \in G(\omega)\big]$ is $\Sigma_\mu \cap D$-measurable on D. Let $\varepsilon : D \longrightarrow \mathbb{R}^N_+ \setminus \{0\}$ be a $\Sigma \cap D$-measurable function such that $m(\omega) + \varepsilon(\omega) < \big(p, e(\omega)\big)_{\mathbb{R}^N}$ and consider the multifunction $H : D \longrightarrow 2^{\mathbb{R}^N} \setminus \{\emptyset\}$ defined by

$$H(\omega) = \big\{x \in G(\omega) : (p,x)_{\mathbb{R}^N} \leq m(\omega) + \varepsilon(\omega)\big\}.$$

Because G is graph-measurable, it follows that

$$\mathrm{Gr} H \in (\Sigma_\mu \cap D) \times \mathbf{B}(\mathbb{R}^N).$$

So we can apply Theorem 6.3.20 (the Yankov–von Neumann–Aumann selection theorem) and obtain $h : D \longrightarrow \mathbb{R}^N_+$, $\Sigma \cap D$-measurable such that $h(\omega) \in H(\omega)$ μ-a.e. on D. We have

$$\big(p, h(\omega)\big) < \big(p, e(\omega)\big)_{\mathbb{R}^N} \quad \mu\text{-a.e. on } D \tag{7.12}$$

$$\text{and} \quad h(\omega) \in G(\omega) \quad \mu\text{-a.e. on } D. \tag{7.13}$$

Define $\widehat{h} : \Omega \longrightarrow \mathbb{R}^N_+$ by

$$\widehat{h}(\omega) = \begin{cases} h(\omega) & \text{if } \omega \in D \\ e(\omega) & \text{if } \omega \in D^c. \end{cases}$$

Evidently $\widehat{h} \in S^1_G$ (see (7.13)). Moreover, we have

$$\begin{aligned} \int_\Omega (p, \widehat{h})_{\mathbb{R}^N} \, d\mu &= \int_D (p,h)_{\mathbb{R}^N} \, d\mu + \int_{D^c} (p,e)_{\mathbb{R}^N} \, d\mu \\ &< \int_D (p,e)_{\mathbb{R}^N} \, d\mu + \int_{D^c} (p,e)_{\mathbb{R}^N} \, d\mu \quad \text{(see (7.12))} \\ &= \int_\Omega (p,e)_{\mathbb{R}^N} \, d\mu, \end{aligned}$$

which contradicts (7.10). This proves that (7.11) is true.

Next we show that

$$\big(p, u(\omega)\big)_{\mathbb{R}^N} = \big(p, e(\omega)\big)_{\mathbb{R}^N} \quad \mu\text{-a.e. on } \Omega. \tag{7.14}$$

Because of (7.11), we have that $\big(p, e(\omega)\big) \leq \big(p, u(\omega)\big)$ μ-a.e. on Ω. If this inequality is strict for all $\omega \in \Omega$, $\mu(E) > 0$, then

$$\begin{aligned} \Big(p, \int_\Omega u \, d\mu\Big)_{\mathbb{R}^N} &= \Big(p, \int_E u \, d\mu\Big)_{\mathbb{R}^N} + \Big(p, \int_{E^c} u \, d\mu\Big)_{\mathbb{R}^N} \\ &= \int_E (p,u)_{\mathbb{R}^N} \, d\mu + \int_{E^c} (p,u)_{\mathbb{R}^N} \, d\mu \\ &> \int_E (p,e)_{\mathbb{R}^N} \, d\mu + \int_{E^c} (p,e)_{\mathbb{R}^N} \, d\mu \\ &= \Big(p, \int_\Omega e \, d\mu\Big), \end{aligned}$$

a contradiction to the fact that $\int_\Omega u d\mu = \int_\Omega e d\mu$ (recall $u \in C(\mathcal{E})$). So we have proved that (7.14) holds and so $u(\omega) \in b(\omega, p)$ μ-a.e. on Ω. It remains to show that $u(\omega) \in d(\omega, p)$ μ-a.e. on Ω. By virtue of hypothesis H(iii), we have $\mu(S) > 0$ where $S = \{\omega \in \Omega : (p, e(\omega))_{\mathbb{R}^N} > 0\}$. Fix $\omega \in S$. We can find $v_0 \in \mathbb{R}^N_+ \setminus \{0\}$ such that $(p, v_0)_{\mathbb{R}^N} < (p, e(\omega))$. Let $y \in \mathbb{R}^N_+$ be such that $(p, y)_{\mathbb{R}^N} \leq (p, e(\omega))_{\mathbb{R}^N}$. We set $y_t = tv_0 + (1-t)y$ for $t \in (0,1)$. Note that $y_t \longrightarrow y$ as $t \longrightarrow 0$. But because of hypothesis H(v), $G(\omega)^c$ is closed and so $y \notin G(\omega)$. This proves that $u(\omega) \in d(\omega, p)$. From this and hypothesis H(vii) it follows that $(p, v)_{\mathbb{R}^N} > 0$ for all $v \in \mathbb{R}^N_+ \setminus \{0\}$, hence $p \in \operatorname{int} \mathbb{R}^N_+$. Therefore we have proved that $p \in \operatorname{int} \mathbb{R}^N_+$ and $u(\omega) \in d(\omega, p)$ for all $\omega \in S$. If $\omega \in S^c$, then $(p, e(\omega))_{\mathbb{R}^N} = 0$. But because $p \in \operatorname{int} \mathbb{R}^N_+$, we deduce that $b(\omega, p) = \{0\}$. Hence $u(\omega) = 0$ a.e. on S^c. Finally we can say that $u(\omega) \in d(\omega, p)$ μ-a.e. on Ω and $p \in \operatorname{int} \mathbb{R}^N_+$, which implies that $(u, p) \in W(\mathcal{E})$ (i.e., $C(\mathcal{E}) = W(\mathcal{E})$). □

Thus far we have introduced the notion of markets with a continuum of traders, we have demonstrated their significance as mathematical models for the intuitive concept of perfect competition, and we proved that under very general conditions, the core of such markets equals the set of all Walras (competitive) equilibria. However, we did not establish the existence of a Walras equilibrium allocation. It may well happen that both sets $C(\mathcal{E})$ and $W(\mathcal{E})$ are empty. In the second part of this section we fill-in this gap by showing that $W(\mathcal{E}) \neq \emptyset$ (hence $C(\mathcal{E}) \neq \emptyset$ because we always have $W(\mathcal{E}) \subseteq C(\mathcal{E})$). To do this we have to strengthen the hypotheses of our model.

We still consider markets with a continuum of agents. So (Ω, Σ, μ) is a nonatomic finite measure space. The nonatomicity of μ expresses the fact that we have perfect competition. We can think of atoms as being oligopolistic agents. The commodity space is $\mathbb{R}^N$; that is, there are N-different commodities traded in the market. Each trader can trade things in $\mathbb{R}^N_+$; that is, for all $\omega \in \Omega$, $F(\omega) \in \mathbb{R}^N_+$. For each trader (agent) $\omega \in \Omega$, there is defined on $\mathbb{R}^N_+$ (the set of feasible consumption plans of each trader $\omega \in \Omega$) a binary relation $\preceq_\omega$, known as the *preference-indifference relation*, which is assumed to be reflexive, transitive, and complete. From $\preceq_\omega$ we define the relations $\prec_\omega$ and $\sim_\omega$, called the *preference* and *indifference* relations, respectively, as follows.

$$x \prec_\omega y \quad \text{if and only if } x \preceq_\omega y \text{ but not } y \preceq_\omega x \tag{7.15}$$

$$x \sim_\omega y \quad \text{if and only if } x \preceq_\omega y \text{ and } y \preceq_\omega x. \tag{7.16}$$

DEFINITION 7.1.8 We say that the commodity vector $x \in \mathbb{R}^N_+$ *saturates* a trader's $\omega \in \Omega$ desire, if for all $y \in \mathbb{R}^N_+$, we have $y \preceq_\omega x$.

Also there is an initial endowment $e \in L^1(\Omega, \mathbb{R}^N_+)$. Recall that $e(\omega)$ presents the commodity vector with which trader $\omega \in \Omega$ comes to the market. So we have an economy $\mathcal{E} = \big((\Omega, \Sigma, \mu), \mathbb{R}^N_+, \preceq, e\big)$. The precise hypotheses on the characteristics of this economy, are the following.

H′: (i) (Ω, Σ, μ) is a finite nonatomic measure space.

(ii) For μ-a.a. $\omega \in \Omega$, the set $F(\omega)$ of all feasible consumption vectors is $\mathbb{R}^N_+$.

(iii) There is an initial endowment $e \in L^1(\Omega, \mathbb{R}^N_+)$ such that $e(\omega) \in \mathrm{int}\mathbb{R}^N_+$ μ-a.e. on Ω.

(iv) For μ-a.a. $\omega \in \Omega$, there is a reflexive, transitive, and complete binary relation $\preceq_\omega$ defined on $\mathbb{R}^N_+$; this is the preference–indifference relation for trader $\omega \in \Omega$ and from it we can define a preference relation $\prec_\omega$ and an indifference relation $\sim_\omega$ as indicated in (7.15) and (7.16), respectively.

(v) $\{(\omega, x, y) \in \Omega \times \mathbb{R}^N_+ \times \mathbb{R}^N_+ : x \prec_\omega y\} \in \Sigma \times \mathbf{B}(\mathbb{R}^N_+) \times \mathbf{B}(\mathbb{R}^N_+)$.

(vi) For μ-a.a. $\omega \in \Omega$ and all $y \in \mathbb{R}^N_+$, the sets $\{x \in \mathbb{R}^N_+ : y \prec_\omega x\}$ are open.

(vii) For μ-a.a. $\omega \in \Omega$, unless the commodity bundle $y \in \mathbb{R}^N_+$ saturates the agent's desire, we have that $x - y \in \mathrm{int}\,\mathbb{R}^N_+$ implies $y \preceq_\omega x$.

(viii) There is an allocation $v \in L^1(\Omega, \mathbb{R}^N_+)$ such that for μ-a.a. $\omega \in \Omega$, $v(\omega)$ saturates the desire of the trader

(ix) For μ-a.a. $\omega \in \Omega$, the commodity bundle $x \in \mathbb{R}^N_+$ can not saturate the desire of the agent unless $x - e(\omega) \in \mathrm{int}\,\mathbb{R}^N_+$.

REMARK 7.1.9 Hypothesis H$'$(vi) is a continuity in the commodity space for the preference relation $\prec_\omega$. In fact it can be shown that this hypothesis together with hypothesis H$'$(iv), implies that there is a continuous utility function $x \longrightarrow \widehat{u}(\omega, x)$ for each trader $\omega \in \Omega$ (see Debreu [185, p. 56]). Then hypothesis H$'$(v) implies that the utility function $(\omega, x) \longrightarrow \widehat{u}(\omega, x)$ is jointly measurable (a Carathéodory function). Hypothesis H$'$(vii) is a weak desirability hypothesis. The existence of the special allocation in hypothesis H$'$(viii) is intuitively very acceptable. It says that there is an upper bound on the amount of a commodity that can be profitably used by a trader, no matter what other commodities are or are not available. The fact that v is an allocation (i.e., $v \in L^1(\Omega, \mathbb{R}^N_+)$), simply means that the economy as a whole can be saturated, namely the commodity vector $\int_\Omega v d\mu$ can be distributed among the traders in such a way as to saturate each trader's desire.

Under these hypotheses, we can prove the following existence theorem.

THEOREM 7.1.10 *If the economy* $\mathcal{E} = \{(\Omega, \Sigma, \mu), \mathbb{R}^N_+, \preceq, e\}$ *satisfies hypotheses* H', *then* $W(\mathcal{E}) \neq \emptyset$.

PROOF: For each trader $\omega \in \Omega$ and each price vector $p \in \mathbb{R}^N_+$, we define the budget set

$$b(\omega, p) = \{x \in \mathbb{R}^N_+ : (p, x)_{\mathbb{R}^N} \leq (p, e(\omega))_{\mathbb{R}^N}\}$$

and the preferred set

$$\xi(\omega, p) = \{x \in \mathbb{R}^N_+ : y \preceq_\omega x \quad \text{for all } \ y \in b(\omega, p)\}.$$

We define $V(\omega) = \{x \in \mathbb{R}^N_+ : v(\omega) - x \in \mathbb{R}^N_+\}$ and then set

$$\vartheta(\omega, p) = V(\omega) \cap \xi(\omega, p).$$

Claim 1: For μ-a.a. $\omega \in \Omega$, $p \longrightarrow \vartheta(\omega, p)$ is continuous on $\mathbb{R}^N \setminus \{0\}$.

Fix $\omega \in \Omega \setminus N$, $\mu(N) = 0$. Then for every $p \in \mathbb{R}^N_+$, $\vartheta(\omega, p) \subseteq V(\omega) \in P_k(\mathbb{R}^N)$. So in order to prove that $p \longrightarrow \vartheta(\omega, p)$ is usc, it suffices to show that it has a closed graph (see Proposition 6.1.10). To this end let $\{(p_n, x_n)\}_{n \geq 1} \subseteq \text{Gr}\, \vartheta(\omega, \cdot)$ and assume that $p_n \longrightarrow p \neq 0$, $x_n \longrightarrow x$ in $\mathbb{R}^N_+$. Evidently $x \in V(\omega)$. Suppose that $x \notin \vartheta(\omega, p)$. Then $x \notin \xi(\omega, p)$ and this means that we can find $y \in b(\omega, p)$ such that $x \prec_\omega y$. By virtue of hypothesis H$'$(vi) we can find $z \in \mathbb{R}^N_+$ close to y such that

$$x \prec_\omega z \quad \text{and} \quad (p, z)_{\mathbb{R}^N} < \big(p, e(\omega)\big)_{\mathbb{R}^N}$$

(recall that because of hypothesis H$'$(iii) $\big(p, e(\omega)\big)_{\mathbb{R}^N} > 0$). Therefore we can find $n_0 \geq 1$ large such that

$$\begin{aligned} & x_n \prec_\omega z \quad \text{and} \quad (p_n, z)_{\mathbb{R}^N} \leq \big(p_n, e(\omega)\big)_{\mathbb{R}^N} \quad \text{for all } n \geq n_0, \\ \Rightarrow\ & x_n \notin \xi(\omega, p_n) \qquad \text{for all } n \geq n_0, \text{ a contradiction.} \end{aligned}$$

This proves that Gr $\vartheta(\omega, \cdot)$ is closed and so $p \longrightarrow \vartheta(\omega, p)$ is usc.

Next we show that $p \longrightarrow \vartheta(\omega, p)$ is lsc. Assume that $p_n \longrightarrow p \neq 0$ and let $x \in \vartheta(\omega, p)$. If x saturates the desire of trader $\omega \in \Omega$, then $x \in \xi(\omega, p_n)$ for all $n \geq 1$ and so if $x_n = x$ for all $n \geq 1$ we have $x_n \longrightarrow x$ with $x_n \in \xi(\omega, p_n)$ $n \geq 1$. So we may assume that x does not saturate the desire of trader $\omega \in \Omega$. Because $\vartheta(\omega, p_n) \in P_k(\mathbb{R}^N_+)$ we can find $x_n \in \vartheta(\omega, p_n)$ such that

$$d\big(x, \vartheta(\omega, p_n)\big) = \|x - x_n\|.$$

Let $\varepsilon > 0$ and set $s_\varepsilon = (\varepsilon, \ldots, \varepsilon) \in \text{int}\, \mathbb{R}^N_+$, $y_\varepsilon = x + s_\varepsilon$. Then by hypothesis H$'$(vii), we have

$$x \prec_\omega y_\varepsilon.$$

Then either for all $\varepsilon > 0$ and all $n \geq 1$ large, $y_\varepsilon \in \vartheta(\omega, p_n)$ or else for some $\varepsilon > 0$, we can find a subsequence $\{n_k\}_{k \geq 1}$ of $\{n\}$ such that $y_\varepsilon \notin \vartheta(\omega, p_n)$. In the first case, we have

$$\|x_n - x\| \leq \|y_\varepsilon - x\| \leq \varepsilon \sqrt{N}.$$

Because $\varepsilon > 0$ was arbitrary, it follows that $x_n \longrightarrow x$ and because $x_n \in \vartheta(\omega, p_n)$, $n \geq 1$, we are done.

In the second case, we can find $z_{n_k} \in \vartheta(\omega, p_n)$ such that $y_\varepsilon \prec_\omega z_{n_k}$. Because $\{z_{n_k}\}_{k \geq 1} \subseteq V(\omega) \in P_k(\mathbb{R}^N_+)$, we may assume that $z_{n_k} \longrightarrow z$ in $\mathbb{R}^N$. Since $(p_{n_k}, z_{n_k})_{\mathbb{R}^N} \leq \big(p_{n_k}, e(\omega)\big)_{\mathbb{R}^N}$, in the limit as $k \longrightarrow \infty$ we obtain $(p, z)_{\mathbb{R}^N} \leq \big(p, e(\omega)\big)_{\mathbb{R}^N}$ and so $z \in b(\omega, p)$. Also by virtue of hypothesis H$'$(vi), we have $y_\varepsilon \preceq_\omega z$ and because $x \prec_\omega y_\varepsilon$, hypothesis H$'$(iv) implies that $x \prec_\omega z$, a contradiction to the fact that $x \in \vartheta(\omega, p)$. This proves that $p \longrightarrow \vartheta(\omega, p)$ is lsc.

So we conclude that $p \longrightarrow \vartheta(\omega, p)$ is continuous (and h-continuous; see Corollary 6.1.40).

Claim 2: For all $p \in \mathbb{R}^N_+ \setminus \{0\}$, $\omega \longrightarrow \xi(\omega, p)$ and $\omega \longrightarrow \vartheta(\omega, p)$ are graph measurable multifunctions.

Clearly the budget multifunction $\omega \longrightarrow b(\omega, p)$ is $P_f(\mathbb{R}^N)$-valued and measurable. So by virtue of Theorem 6.3.18, we can find $h_n : \Omega \longrightarrow \mathbb{R}^N_+$, $n \geq 1$ Σ–measurable maps such that

$$b(\omega, p) = \overline{\{h_n(\omega)\}_{n \geq 1}} \qquad \text{for all } \omega \in \Omega.$$

Invoking hypothesis H$'$(vi) we see that

$$\xi(\omega,p)=\bigcap_{n\geq 1}\{x\in\mathbb{R}^N : h_n(\omega)\preceq_\omega x\}$$
$$\Rightarrow\ \omega\longrightarrow\xi(\omega,p)\ \text{ is graph-measurable (see hypothesis H}'\text{(v)),}$$
$$\Rightarrow\ \omega\longrightarrow\vartheta(\omega,p)=V(\omega)\cap\xi(\omega,p)\ \text{ is graph-measurable.}$$

Claim 3: $p\longrightarrow\int_\Omega\vartheta(\omega,p)\mathrm{d}\mu$ is $P_{kc}(\mathbb{R}^N)$ follows from Theorem 6.4.45 and 6.4.23.

Note that

$$\begin{aligned}&h\Big(\int_\Omega\vartheta(\omega,p)\mathrm{d}\mu,\int_\Omega\vartheta(\omega,p')\mathrm{d}\mu\Big)\\ &\leq\int_\Omega h\big(\vartheta(\omega,p),\vartheta(\omega,p')\big)\mathrm{d}\mu\qquad\text{for all } p,p'\in\mathbb{R}^N_+\setminus\{0\},\\ \Rightarrow\quad &p\longrightarrow\int_\Omega\vartheta(\omega,p)\mathrm{d}\mu\ \text{ is continuous by virtue of Claim 1.}\end{aligned}$$

Using Claim 3 we can consider the metric projection on the set $\int_\Omega\vartheta(\omega,p)\mathrm{d}\mu$ denoted by $\mathrm{proj}\big(\cdot;\int_\Omega\vartheta(\omega,p)\mathrm{d}\mu\big)$. Set $d(p)=\mathrm{proj}\big(\int_\Omega e\mathrm{d}\mu;\int_\Omega\vartheta(\omega,p)\mathrm{d}\mu\big)$.

Claim 4: $p\longrightarrow d(p)$ is continuous.

Suppose that $p_n\longrightarrow p$ and let $d(p_n)=\mathrm{proj}\big(\int_\Omega e\mathrm{d}\mu;\int_\Omega\vartheta(\omega,p_n)\mathrm{d}\mu\big)$, $n\geq 1$. By virtue of Proposition 6.1.13 and Claim 3, $\{d(p_n\}_{n\geq 1}$ is relatively compact. So we can find a subsequence $\big\{d(p_{n_k})\big\}_{k\geq 1}$ of $\big\{d(p_n)\big\}_{n\geq 1}$ such that $d(p_{n_k})\longrightarrow w$ in $\mathbb{R}^N_+$. We have

$$\Big\|\int_\Omega e\mathrm{d}\mu-d(p_{n_k})\Big\|=d\Big(\int_\Omega e\mathrm{d}\mu,\int_\Omega\vartheta(\omega,p_{n_k})\mathrm{d}\mu\Big).\tag{7.17}$$

From Claim 3 and Proposition 6.6.32, we have

$$d\Big(\int_\Omega e\mathrm{d}\mu,\int_\Omega\vartheta(\omega,p_{n_k})\mathrm{d}\mu\Big)\longrightarrow d\Big(\int_\Omega e\mathrm{d}\mu,\int_\Omega\vartheta(\omega,p)\mathrm{d}\mu\Big)\tag{7.18}$$

and
$$\Big\|\int_\Omega e\mathrm{d}\mu-d(p_{n_k})\Big\|\longrightarrow\Big\|\int_\Omega e\mathrm{d}\mu-w\Big\|.\tag{7.19}$$

Combining (7.17) through (7.19) we obtain

$$\begin{aligned}&\Big\|\int_\Omega e\mathrm{d}\mu-w\Big\|=d\Big(\int_\Omega e\mathrm{d}\mu,\int_\Omega\vartheta(\omega,p)\mathrm{d}\mu\Big),\\ \Rightarrow\ &w=\mathrm{proj}\Big(\int_\Omega e\mathrm{d}\mu;\int_\Omega\vartheta(\omega,p)\mathrm{d}\mu\Big)=d(p).\end{aligned}$$

Then by Urysohn's criterion for convergence of sequences we have $d(p_n)\longrightarrow w=d(p)$ and so we conclude that $p\longrightarrow d(p)$ is continuous.

Claim 5: For every $p\in\mathbb{R}^N_+\setminus\{0\}$, $d(p)-\int_\Omega e\mathrm{d}\mu\in\mathbb{R}^N_+$.

Suppose that the claim is not true. Then without any loss of generality we may assume that $d_1(p)<\int_\Omega e\mathrm{d}\mu$ (here $d(p)=\big(d_k(p)\big)_{k=1}^N\in\mathbb{R}^N$ and $e=(e_k)_{k=1}^N\in L^1(\Omega,\mathbb{R}^N_+)$). By definition $d(p)=\int_\Omega u\mathrm{d}\mu$ with $u\in S^1_{\vartheta(\cdot,p)}$. Let $y=(v_1,u_2,\ldots,u_N)$, where

$v = (v_k)_{k=1}^N$ (see hypothesis H′(viii)) and $u = (u_k)_{k=1}^N$. Then $y(\omega) - u(\omega) \in \mathbb{R}_+^N$ and $v(\omega) - y(\omega) \in \mathbb{R}_+^N$ for all $\omega \in \Omega$ and so it follows that $y(\omega) \in \vartheta(\omega, p)$ μ-a.e. on Ω. Hence

$$\Big(\int_\Omega y_1 \mathrm{d}\mu, d_2(p), \ldots, d_N(p)\Big) = \int_\Omega y \mathrm{d}\mu \in \int_\Omega \vartheta(\omega, p) \mathrm{d}\mu.$$

Recall that $d_1(p) < \int_\Omega e_1 \mathrm{d}\mu$ and by hypothesis H′(ix), we have $\int_\Omega e_1 \mathrm{d}\mu < \int_\Omega v_1 \mathrm{d}\mu$. So we can find $\lambda \in (0,1)$ such that $\int_\Omega e_1 \mathrm{d}\mu = \lambda \int_\Omega v_1 \mathrm{d}\mu + (1-\lambda) d_1(p)$. Set $w = \lambda y + (1-\lambda) u$ and $z = \int_\Omega w \mathrm{d}\mu$. Then $z \in \int_\Omega \vartheta(\omega, p) \mathrm{d}\mu$ (see Claim 3) and $z = \big(\int_\Omega e_1 \mathrm{d}\mu, d_2(p), \ldots, d_N(p)\big)$. We have

$$\begin{aligned}
\Big\|z - \int_\Omega e \mathrm{d}\mu\Big\|^2 &= \sum_{k=2}^N \Big(d_k(p) - \int_\Omega e_k \mathrm{d}\mu\Big)^2 \\
&< \sum_{k=1}^N \Big(d_k(p) - \int_\Omega e_k \mathrm{d}\mu\Big)^2 \quad \Big(\text{because } d_1(p) < \int_\Omega e_1 \mathrm{d}\mu\Big) \\
&= \Big\|d(p) - \int_\Omega e \mathrm{d}\mu\Big\|^2,
\end{aligned}$$

a contradiction to the definition of $d(p)$. This proves the claim.

Now let $\gamma(p) = d(p) - \int_\Omega e \mathrm{d}\mu \in \mathbb{R}_+^N$ for all p (see Claim 5) and $p \longrightarrow \gamma(p)$ is continuous. We define $c(p) = \mathrm{proj}\big(\int_\Omega e \mathrm{d}\mu; \int_\Omega \xi(\omega, p) \mathrm{d}\mu\big)$. This is well-defined because $\int_\Omega \xi(\omega, p) \mathrm{d}\mu \in P_{kc}(\mathbb{R}^N)$. We set $\eta(p) = c(p) - \int_\Omega e \mathrm{d}\mu$.

Claim 6: $\gamma(p) = \eta(p)$ for all p.

We fix p and write $\gamma = \gamma(p)$, $\eta = \eta(p)$, $c = c(p)$, and $d = d(p)$. If $\gamma = 0$, then we are done. So suppose that $\gamma \neq 0$. From the definition of γ, the hyperplane through d perpendicular to γ supports the compact convex set $\int_\Omega \vartheta(\omega, p) \mathrm{d}\mu$. This means that

$$(x, \gamma)_{\mathbb{R}^N} \geq \|\gamma\|^2 \qquad \text{for all } x \in \int_\Omega \vartheta(\omega, p) \mathrm{d}\mu - \int_\Omega e \mathrm{d}\mu. \tag{7.20}$$

Suppose that there is a point in $\int_\Omega \xi(\omega, p) \mathrm{d}\mu$ which is closer to $\int_\Omega e \mathrm{d}\mu$, than d is. Then we can find $y \in \int_\Omega \xi(\omega, p) \mathrm{d}\mu - \int_\Omega e \mathrm{d}\mu$ which is closer to the origin than γ is. We have

$$\|y\|^2 < \|\gamma\|^2. \tag{7.21}$$

Moreover, $\|y\|^2 - 2(y, \gamma)_{\mathbb{R}^N} + \|\gamma\|^2 = \|y - \gamma\|^2 > 0$. Hence $\|y\|^2 > (y, \gamma)_{\mathbb{R}^N} + \big[(y, \gamma)_{\mathbb{R}^N} - \|\gamma\|^2\big]$. If $(y, \gamma)_{\mathbb{R}^N} - \|\gamma\|^2 \geq 0$, then $\|y\|^2 > (y, \gamma)_{\mathbb{R}^N} \geq \|\gamma\|^2$, which contradicts (7.21). Therefore

$$(y, \gamma)_{\mathbb{R}^N} < \|\gamma\|^2. \tag{7.22}$$

We know that $y = \int_\Omega u \mathrm{d}\mu - \int_\Omega e \mathrm{d}\mu$ with $u \in S^1_{\xi(\cdot, p)}$.

If $\widehat{u}(\omega, x)$ is the utility function corresponding to $\omega \longrightarrow \preceq_\omega$ (see hypothesis H′(iv) and Remark 7.1.9). Then $\widehat{u}\big(\omega, u(\omega)\big) \in \vartheta(\omega, p)$ and $\widehat{u}\big(\omega, u(\omega)\big) \leq v(\omega)$ μ-a.e. on Ω. Set $w(\omega) = \widehat{u}\big(\omega, u(\omega)\big)$. We have $\int_\Omega w \mathrm{d}\mu \in \int_\Omega \vartheta(\omega, p) \mathrm{d}\mu$ and $\int_\Omega w \mathrm{d}\mu - \int_\Omega e \mathrm{d}\mu \leq y$ (i.e., $y - \int_\Omega w \mathrm{d}\mu + \int_\Omega e \mathrm{d}\mu \in \mathbb{R}_+^N$). Because $\gamma \in \mathbb{R}_+^N$ (see Claim 3), it follows that

$$\begin{aligned}
&\Big(\int_\Omega w \mathrm{d}\mu - \int_\Omega e \mathrm{d}\mu, \gamma\Big)_{\mathbb{R}^N} \leq (y, \gamma)_{\mathbb{R}^N}, \\
\Rightarrow\ &\Big(\int_\Omega w \mathrm{d}\mu - \int_\Omega e \mathrm{d}\mu, \gamma\Big)_{\mathbb{R}^N} < \|\gamma\|^2 \qquad (\text{see (7.22)}).
\end{aligned} \tag{7.23}$$

But because $\int_\Omega w\mathrm{d}\mu - \int_\Omega e\mathrm{d}\mu \in \int_\Omega \vartheta(\omega,p)\mathrm{d}\mu - \int_\Omega e\mathrm{d}\mu$, from (7.20) we have

$$\Big(\int_\Omega w\mathrm{d}\mu - \int_\Omega e\mathrm{d}\mu, \gamma\Big)_{\mathbb{R}^N} \geq \|\gamma\|^2. \tag{7.24}$$

Comparing (7.23) and (7.24), we reach a contradiction. This proves the claim.

Next let P be the standard price simplex; that is, $P=\big\{p=(p_k)_{k=1}^N \in \mathbb{R}_+^N : \sum\limits_{k=1}^N p_k = 1\big\}$. Let $\varphi : P \longrightarrow P$ be defined by

$$\varphi(p) = \frac{1}{1+\sum\limits_{k=1}^N \eta_k(p)}\,\big(p+\eta(p)\big).$$

By virtue of Claim 6, φ is continuous and by Brouwer's fixed point theorem (see Theorem 3.5.3), we can find $q \in P$ such that $\varphi(q) = q$. Then

$$\begin{aligned} &\Big(1+\sum_{k=1}^N \eta_k(q)\Big)\, q = q + \eta(q) \\ \Rightarrow\ \ &\eta(q) = \mu q \quad \text{with } \mu = \sum_{k=1}^N \eta(q) \geq 0 \quad \text{(see Claim 6)}. \end{aligned} \tag{7.25}$$

We show that $\eta(q)=0$. Suppose that this is not true. From the definition of η and the convexity of $\int_\Omega \xi(\omega,p)\mathrm{d}\mu$, we see that the hyperplane through $\eta(p)+\int_\Omega e\mathrm{d}\mu$ which is perpendicular to $\eta(p)$ supports the closed, convex set $\int_\Omega \xi(\omega,p)\mathrm{d}\mu$. So if $p=q$, we have

$$\Big(y - \int_\Omega e\mathrm{d}\mu, \eta(q)\Big)_{\mathbb{R}^N} \geq \|\eta(q)\|^2 \qquad \text{for all } y \in \int_\Omega \xi(\omega,p)\mathrm{d}\mu.$$

Because $\eta(q) \neq 0$, from (7.25) we have $\gamma > 0$ and

$$\Big(y - \int_\Omega e\mathrm{d}\mu, \eta(q)\Big)_{\mathbb{R}^N} \geq \mu^2\|q\|^2 > 0 \qquad \text{for all } y \in \int_\Omega \xi(\omega,p)\mathrm{d}\mu. \tag{7.26}$$

Using hypothesis H$'$(v) and the Yankov–von Neumann–Aumann selection theorem (see Theorem 6.3.20), we can find $u \in L^1(\Omega,\mathbb{R}_+^N)$ such that $u(\omega) \in b(\omega,q)$ μ-a.e. and is $\prec_\omega$-maximal for μ-a.e. $\omega \in \Omega$.

We have

$$\big(u(\omega)-e(\omega),q\big)_{\mathbb{R}^N} \leq 0 \quad \mu\text{-a.e. on } \Omega \tag{7.27}$$

$$\text{and} \quad u(\omega) \in \xi(\omega,p) \quad \mu\text{-a.e. on } \Omega. \tag{7.28}$$

Integrating (7.27) and (7.28) we obtain

$$\Big(\int_\Omega u\mathrm{d}\mu - \int_\Omega e\mathrm{d}\mu, q\Big)_{\mathbb{R}^N} \leq 0 \quad \text{with } \int_\Omega u\mathrm{d}\mu \in \int_\Omega \xi(\omega,p)\mathrm{d}\mu. \tag{7.29}$$

Comparing (7.26) and (7.29), we reach a contradiction. Therefore $\eta(q) = 0$. From this we infer that there exists $s \in S^1_{\xi(\cdot,q)}$ such that

$$\int_\Omega s d\mu = \int_\Omega e d\mu. \tag{7.30}$$

We claim that (s, q) is a Walras equilibrium (see Definition 7.1.5(b)). To establish this, it suffices to show $s \in S^1_{b(\cdot,q)}$. From hypotheses H$'$(viii) and (ix), we see that

$$\left(q, e(\omega)\right)_{\mathbb{R}^N} \leq \left(q, s(\omega)\right)_{\mathbb{R}^N} \quad \mu\text{-a.e. on } \Omega.$$

If this inequality is strict on a set of positive μ-measure, we contradict (7.30). Therefore

$$\begin{aligned} & \left(q, s(\omega)\right)_{\mathbb{R}^N} = \left(q, e(\omega)\right)_{\mathbb{R}^N} \quad \mu\text{-a.e. on } \Omega, \\ \Rightarrow \ & s(\omega) \in b(\omega, q) \quad \mu\text{-a.e. on } \Omega. \end{aligned}$$

□

REMARK 7.1.11 Walras equilibria are significant only in a perfect competition situation; that is, the market consists of a large number of individually insignificant traders. Theorem 7.1.10, illustrates that in a perfect competition situation, convex preferences are not needed to prove existence.

7.2 Infinite Horizon Multisector Growth Models

In this section we turn our attention to dynamic economic models and we examine optimal growth models in discrete time. An important feature of such models is the infinite (unbounded) planning horizon. This results from purely economic considerations. For example, one of the main problems in dynamic economic planning is that of determining what amounts of resources should be allocated to the production of consumption goods and what amounts should be allocated to the production of capital equipment that will be used in the production of future consumption goods. The only rational way to address this question is by considering an infinite time horizon. It is true that we usually hear about five or ten year plans, which set targets for the level of capital accumulation. But then it is natural to ask about the criteria that led to these target levels.

The only reasonable answer to this question can be some sort of trade-off between the sacrifices required to accumulate the capital during the planning period and the benefits that will result from the future use of the accumulated capital. In other words, a finite planning horizon requires some method of evaluating end-of-period capital stocks or assets and the only proper evaluation is their value in use in the subsequent future. But, if this is so, then the planning decisions have actually been based on considerations about times well beyond the period of the plan and this will be true independent of whether the planning period is short or long.

Therefore the proper way to model this mathematically is through an infinite horizon. So the open endedness of the future in the model is important, because it expresses the fundamental economic fact that the consequences of investment are long-lived. If the criterion to be maximized (the intertemporal utility) is discounted at every step, then an optimal program can be obtained easily using the direct method of calculus of variations. For such a situation the interesting question is to characterize the optimal programs (paths) using decentralizable conditions. If the discount factor δ equals 1 (i.e., we have an undiscounted instantaneous utility), then the intertemporal utility integral need not converge and we face some serious mathematical difficulties. For this reason we introduce some weaker notions of optimality, which we discuss in the second part of this section. In both the discounted and undiscounted cases, the growth model is a multisector model meaning that there is more than one commodity in the model.

We start the discussion with the discounted model. In this case the model is described by a triplet (G, u, δ), where $G \subseteq \mathbb{R}_+^N \times \mathbb{R}_+^N$ is the technology set, $u : \mathbb{R}_+^N \longrightarrow \mathbb{R}$ is the utility function, and $\delta \in (0,1)$ is the discount factor. So in our model the commodity space is $\mathbb{R}^N$, in other words there are N commodities. In the commodity space $\mathbb{R}^N$ we use the l^1-norm $\|x\| = \sum_{k=1}^{N} |x_k|$ for all $x = (x_k)_{k=1}^N$. Also for $x, y \in \mathbb{R}^N$ we use the following notation.

$$\begin{aligned} y \leq x &\quad \text{if and only if } \; y_k \leq x_k \; \text{ for all } \; k \in \{1, \ldots, N\}, \\ y < x &\quad \text{if and only if } \; y \leq x \; \text{ and } \; y \neq x, \\ y \ll x &\quad \text{if and only if } \; y_k < x_k \; \text{ for all } \; k \in \{1, \ldots, N\}, \end{aligned}$$

(i.e., $x - y \in \operatorname{int}\mathbb{R}_+^N$).

A pair $(x, y) \in G$ represents a technologically feasible production pair, where x stands for the initial stock of inputs and y stands for the final output, which can be produced using as inputs the vector x.

The hypotheses on the model (G, u, δ) are the following.

H₁: $G \subseteq \mathbb{R}_+^N \times \mathbb{R}_+^N$ is a nonempty closed set such that

(i) There exists $M > 0$ such that if $(x, y) \in G$ and $\|x\| \geq M$, then $\|y\| \leq \|x\|$.

(ii) If $(x, y) \in G$ and $x \leq x'$, $0 \leq y' \leq y$, then $(x', y') \in G$.

(iii) There exists $(\widehat{x}, \widehat{y}) \in G$ such that $\widehat{x} \ll \widehat{y}$.

REMARK 7.2.1 Note that we do not assume that the technology set is convex. Hypothesis H_1(i) in economic terms says that if the input exists in sufficiently large quantities, then we have losses due to depreciation. From a mathematical viewpoint, this hypothesis guarantees the boundedness of the set of feasible programs (paths). Hypothesis H_1(iv) is the free disposability hypothesis, very common in economic models. It says that we are always open to accept a supplement of goods, even if we do not have to do anything with them. Finally hypothesis H_1(iii) means that there is a feasible production pair in which the output is strictly bigger than the input.

H₂: $u : \mathbb{R}_+^N \longrightarrow \mathbb{R}$ is continuous.

REMARK 7.2.2 Note that no concavity assumption is made on u.

H₃: $0 < \delta < 1$.

DEFINITION 7.2.3 A *program* (*path*) starting at $\overline{y} \in \mathbb{R}^N_+$, is a sequence $\{(x_n, y_n)\}_{n\geq 0}$ such that $y_0 = \overline{y}$, $0 \leq x_n \leq y_n$, and $(x_n, y_{n+1}) \in G$ for all $n \geq 1$. Associated with such a program $\{(x_n, y_n)\}_{n\geq 0}$ is a *consumption sequence* $\{c_n\}_{n\geq 0}$ defined by $c_n = y_n - x_n \geq 0$ for all $n \geq 0$.

The utility function u at each stage is evaluated at the consumption level (i.e., we consider $u(c_n)$, $n \geq 0$) and it is discounted by the factor $\delta \in (0,1)$. So the intertemporal utility generated by a consumption sequence $\{c_n\}_{n\geq 0}$, is given by

$$\sum_{n\geq 0} \delta^n u(c_n). \tag{7.31}$$

LEMMA 7.2.4 *If hypotheses H_1 hold and $(x,y) \in G$, then* (a) $\|x\| \leq M$ *implies* $\|y\| \leq M$*;* (b) $\|y\| \leq \max\{\|x\|, M\}$.

PROOF: (a) We argue indirectly. So suppose that $\|x\| \leq M$ and $\|y\| > M$. Set $x' = x + (M - \|x\|)(e/N)$ with $e = (1, \ldots, 1) \in \text{int}\, \mathbb{R}^N_+$. Then $x' \geq x$ and so by hypothesis H_1(ii) (the free disposability hypothesis), we have $(x', y) \in G$. Note that $\|x'\| = M$ and so by virtue of hypothesis H_1(i) we have that $\|y\| \leq M$, a contradiction.

(b) If $\|x\| \leq M$, then from part (a) we have $\|y\| \leq M$. This combined with hypothesis H_1(iii) implies that $\|y\| \leq \max\{\|x\|, M\}$. □

LEMMA 7.2.5 *If hypotheses H_1 hold and $\{(x_n, y_n)\}_{n\geq 0}$ is a program starting at $\overline{y} \in \mathbb{R}^N_+$, then $\|x_n\|, \|y_n\|, \|c_n\| \leq M_1 = \max\{\|\overline{y}\|, M\}$ for all $n \geq 0$.*

PROOF: From Definition 7.2.3 we have that

$$0 \leq c_n = y_n - x_n \leq y_n \qquad \text{for all } n \geq 0.$$

Therefore it suffices to show that $\|y_n\| \leq M_1$ for all $n \geq 0$. Note that $y_0 = \overline{y}$ and so $\|y_0\| \leq M_1$. We proceed by contradiction. So suppose that $\|y_n\| \leq M_1$ for some $n \geq 1$. Then by virtue of Lemma 7.2.4(b) and because $(x_n, y_{n+1}) \in G$ (see Definition 7.2.3), we have

$$\begin{aligned} \|x_{n+1}\| \leq \|y_{n+1}\| &\leq \max\{\|x_n\|, M\} \\ &\leq M_1 \quad \text{(by the induction hypothesis).} \end{aligned}$$

□

This implies that for every program $\{(x_n, y_n)\}_{n\geq 0}$ emanating from $\overline{y} \in \mathbb{R}^N_+$, the intertemporal utility given in (7.31) is absolutely summable. So we can make the following definition.

DEFINITION 7.2.6 A program $\{(x^*_n, y^*_n)\}_{n\geq 0}$ starting at $\overline{y} \in \mathbb{R}^N_+$ is said to be *optimal* if

$$\sum_{n\geq 0} \delta^n u(c_n) \leq \sum_{n\geq 0} \delta^n u(c^*_n)$$

for every other program $\{(x_n, y_n)\}_{n\geq 0}$ starting from $\overline{y}$. Here $c_n = y_n - x_n$, $c^*_n = y^*_n - x^*_n \geq 0$ for all $n \geq 0$.

THEOREM 7.2.7 *If hypotheses H_1, H_2, and H_3 hold and $\overline{y} \in \mathbb{R}^N_+$, then there exists an optimal program starting at $\overline{y}$.*

PROOF: Let $F(\overline{y})$ be the set of all consumption sequences generated by programs that start at $\overline{y} \in \mathbb{R}^N_+$. Note that $F(\overline{y}) \subseteq l^\infty(\mathbb{R}^N)_+$ and from Lemma 7.2.4(b) we know that if $(c_n)_{n\geq 0} \subseteq F(\overline{y})$, then $\|c_n\| \leq M_1$ for all $n \geq 0$. Also because G is closed it is easy to see that $F(\overline{y})$ is w^*-closed in $l^\infty(\mathbb{R}^N)$. Therefore by Alaoglu's theorem $F(\overline{y})$ is w^*-compact in $l^\infty(\mathbb{R}^N)$. Similarly the intertemporal utility function $U: F(\overline{y}) \longrightarrow \mathbb{R}$ defined by $U(\widehat{c}) = \sum\limits_{n\geq 0} \delta^n u(c_n)$ for all $\widehat{c} = (c_n)_{n\geq 0} \in F(\overline{y})$ is w^*-continuous. Hence by the Weierstrass theorem we can find $\widehat{c}^* \in F(\overline{y})$ such that $U(\widehat{c}^*)$ realizes the maximum of U on $F(\overline{y})$. Then the program $\{(x_n^*, y_n^*)\}_{n\geq 0}$ which generates the consumption sequence $\{c_n^*\}_{n\geq 0}$ is an optimal program. $\square$

There is a close connection between price systems and optimal programs, which in some respects strongly resembles the relation existing between price systems and optimal programs in finite horizon problems. However, as we show there is a fundamental difference, which stems from the infinite horizon nature of the problem. In order to make all these facts precise, we need to introduce some additional concepts.

DEFINITION 7.2.8 (a) The *value function* $V: \mathbb{R}^N_+ \longrightarrow \mathbb{R}$ is defined by

$$V(y) = \sum_{n\geq 0} \delta^n u(c_n),$$

where $\{c_n\}_{n\geq 0}$ is the consumption sequence generated by an optimal program $\{(x_n, y_n)\}_{n\geq 0}$ starting at $y \in \mathbb{R}^N_+$ (see Definition 7.2.6).

(b) A sequence $\{(x_n, y_n, p_n)\}_{n\geq 0}\}$ is a *competitive program* starting at $\overline{y} \in \mathbb{R}^N_+$, if $\{(x_n, y_n)\}_{n\geq 0}\}$ is a program starting at $\overline{y} \in \mathbb{R}^N$, $p_n \in \mathbb{R}^N_+$ for all $n \geq 0$ and

$$\delta^n u(c) - (p_n, c)_{\mathbb{R}^N} \leq \delta^n u(c_n) - (p_n, c_n)_{\mathbb{R}^N} \qquad \text{for all } c \in \mathbb{R}^N_+, \text{ all } n \geq 0 \tag{7.32}$$

and

$$(p_{n+1}, y)_{\mathbb{R}^N} - (p_n, x)_{\mathbb{R}^N} \leq (p_{n+1}, y_{n+1})_{\mathbb{R}^N} - (p_n, x_n)_{\mathbb{R}^N} \tag{7.33}$$

for all $(x, y) \in G$ and all $n \geq 0$.

REMARK 7.2.9 Adding (7.32) and (7.33), we see that if $\{(x_n, y_n, p_n)\}_{n\geq 0}$ is a competitive program starting at $\overline{y} \in \mathbb{R}^N_+$; then for all $(x, y) \in G$, all $c \in \mathbb{R}^N_+$, and all $n \geq 0$ we have

$$\begin{aligned} &\delta^n u(c) + (p_{n+1}, y)_{\mathbb{R}^N} - (p_n, x + c)_{\mathbb{R}^N} \\ &\leq \delta^n u(c_n) + (p_{n+1}, y_{n+1})_{\mathbb{R}^N} - (p_n, y_n)_{\mathbb{R}^N}. \end{aligned} \tag{7.34}$$

In the definition of a competitive program, inequality (7.32) says that at every time period the consumption plan $\{c_n\}_{n\geq 0}$ maximizes the total utility among all other possible consumption vectors. The total utility is defined as $\delta^n u(c)$ minus the present cost of the consumption $(p_n, c)_{\mathbb{R}^N}$. In inequality (7.33), we interpret the quantity $(p_{n+1}, y)_{\mathbb{R}^N} - (p_n, x)_{\mathbb{R}^N}$ as the total profit from operating the technological process $(x, y) \in G$, when the price system $\{p_n\}_{n \geq 0}$ prevails in the market. Then inequality (7.33) says that a competitive program from $\overline{y}$ at every time period maximizes the total profit among all other technologically feasible input-output pairs. Finally inequality (7.34) says that a competitive program from $\overline{y}$ maximizes the sum of total utility and total profit at every time period in the set of all feasible input-output pairs.

DEFINITION 7.2.10 A competitive program $\{(x_n, y_n, p_n)\}_{n \geq 0}$ is said to satisfy the *transversality condition* if $\lim_{n \to \infty} (p_n, x_n)_{\mathbb{R}^N} = 0$.

REMARK 7.2.11 This condition means that the value of the input asymptotically equals zero. As we show in the sequel this is the condition that is characteristic of infinite horizon problems and which from an economic viewpoint is rather undesirable. But first let us state the theorem on the characterization of optimal programs. To do this we need to modify hypotheses H_1 and H_2 as follows.

H_1': $G \subseteq \mathbb{R}_+^N \times \mathbb{R}_+^N$ is a nonempty, closed, convex set that satisfies hypotheses H_1(i),(ii), and

(iii) $(0, 0) \in G$ and if $(0, y) \in G$, then $y = 0$.

(iv) There exists a pair $(\widehat{x}, y) \in G$ such that $y \in \operatorname{int} \mathbb{R}^N$.

REMARK 7.2.12 Note that now the feasibility set G is convex. This is common hypothesis in economic models and simply means that technological processes can be mixed in arbitrary proportions. The two conditions in hypothesis H_1'(iii) are very natural and simply mean that inaction is an option (first condition) and that a nonzero output cannot result from zero input (second condition). Finally hypothesis H_1'(iv) means that there is an input $\widehat{x} \in \mathbb{R}_+^N$ that produces a strictly positive quantity of each commodity. This condition is a weaker version of hypothesis H_1(iii). Moreover, the *input* $\widehat{x} \in \mathbb{R}_+^N$ is often called *sufficient*.

H_2': $u : \mathbb{R}_+^N \longrightarrow \mathbb{R}$ is continuous, concave, and strictly increasing in the sense that if $c' \ll c$ in $\mathbb{R}_+^N$, then $u(c') < u(c)$.

The next theorem characterizes optimal programs by means of competitive programs. Because the same result is proved in the more general context of stochastic models (see Theorem 7.4.10), here we simply state the theorem and postpone its proof until Section 7.4.

THEOREM 7.2.13 *If hypotheses H_1', H_2', H_3 hold and $\{(x_n, y_n)\}_{n \geq 0}$ is an optimal program starting at $\overline{y} \in \mathbb{R}_+^N$, then there exists a price sequence $\{p_n\}_{n \geq 0} \subseteq \mathbb{R}_+^N$ such that*

(a) *The sequence $\{(x_n, y_n, p_n)\}_{n \geq 0}$ is competitive (see Definition 7.2.8(b)).*

(b) $\delta^n V(y) - (p_n, y)_{\mathbb{R}^N} \leq \delta^n V(y_n) - (p_n, y_n)_{\mathbb{R}^N}$ *for any $y \in \mathbb{R}_+^N$ and $n \geq 0$.*

(c) $\lim_{n \to \infty} (p_n, x_n)_{\mathbb{R}^N} = 0$ *(i.e., the transversality condition holds).*

REMARK 7.2.14 According to (a) the optimal program $\{(x_n, y_n)\}_{n\geq 0}$ satisfies certain support properties for the technology and utility explained in Remark 7.2.9 (see (7.32) and (7.33)). Condition (b) implies that at every stage we have minimization of the cost among all programs producing no less future value. Finally condition (c) is the transversality condition, which distinguishes infinite horizon from finite horizon problems. Conditions (b) and (c) are not independent, as the next proposition shows.

PROPOSITION 7.2.15 *If hypotheses H_1', H_2, H_3 hold and $\{(x_n, y_n, p_n)\}_{n\geq 0}$ is a competitive program starting at $\overline{y} \in \mathbb{R}_+^N$, then (b) and (c) in Theorem 7.2.13 are equivalent.*

PROOF: (b)⇒(c): In statement (b) let $y = 0$. Then we have

$$0 \leq (p_n, y_n)_{\mathbb{R}^N} \leq \delta^n\big(V(y_n) - V(0)\big). \tag{7.35}$$

By hypothesis H_2, u is continuous. This combined with Lemma 7.2.5 implies that $\{V(y_n)\}_{n\geq 1} \subseteq \mathbb{R}$ is bounded. Moreover, from Theorem 7.2.7, we have that V is defined on all of $\mathbb{R}_+^N$. So from (7.35) we deduce that $\lim_{n\to\infty} (p_n, x_n)_{\mathbb{R}^N} = 0$ (recall that $0 < \delta < 1$; see hypothesis H_3).

(c)⇒(b): Choose any $k \geq 1$ and $\overline{y} \in \mathbb{R}_+^N$. Let $\{(x_n', y_n')\}_{n\geq 0}$ be a program starting at $\overline{y}$. Because by hypothesis $\{(x_n, y_n, p_n)\}_{n\geq 1}$ is a competitive program starting at $\overline{y}$, from (7.34) we have for all $n \geq k$

$$\begin{aligned}&\delta^n\big(u(c_{n-k}') - u(c_n)\big)\\ &\leq \big[(p_{n+1}, y_{n+1})_{\mathbb{R}^N} - (p_n, y_n)_{\mathbb{R}^N}\big] - \big[(p_{n+1}, y_{n-k+1}')_{\mathbb{R}^N} - (p_n, y_{n-k}')_{\mathbb{R}^N}\big].\end{aligned} \tag{7.36}$$

Summing up (7.36) from $n = k$ to $n = k + k_1$, $k_1 \geq 1$, we obtain

$$\begin{aligned}&\sum_{n=k}^{k+k_1} \delta^n\big(u(c_{n-k}') - u(c_n)\big)\\ &\leq (p_{k+k_1+1}, y_{k+k_1+1})_{\mathbb{R}^N} - (p_k, y_k)_{\mathbb{R}^N} + (p_k, \overline{y})_{\mathbb{R}^N}.\end{aligned}$$

From this we obtain

$$\begin{aligned}&\delta^k\Big(\sum_{n=0}^{k_1} \delta^n u(c_n') - \sum_{n=k}^{k+k_1} \delta^{n-k} u(c_n)\Big)\\ &\leq (p_{k+k_1+1}, y_{k+k_1+1})_{\mathbb{R}^N} - (p_k, y_k)_{\mathbb{R}^N} + (p_k, \overline{y})_{\mathbb{R}^N}.\end{aligned} \tag{7.37}$$

Note that as $k_1 \to +\infty$, the two sums in the left-hand side of (7.37) converge. So using the transversality condition (statement (c)), we obtain

$$\delta^k\Big(\sum_{n\geq 0} \delta^n u(c_n') - \sum_{n\geq k} \delta^{n-k} u(c_n)\Big) \leq (p_k, \overline{y})_{\mathbb{R}^N} - (p_k, y_k)_{\mathbb{R}^N}. \tag{7.38}$$

But from the principle of optimality, $\{(x_{n+k}, y_{n+k})\}_{n\geq 0}$ is an optimal program starting at $\overline{y} \in \mathbb{R}_+^N$ and

$$\sum_{n\geq k}\delta^{n-k}u(c_n)=\sum_{n\geq 0}\delta^n u(c_{n+k})=V(y_k). \tag{7.39}$$

Using (7.39) in (7.38) and because $\{(x_n', y_n')\}_{n\geq 0}$ is any program starting at $\overline{y}\in\mathbb{R}^N_+$, we have

$$\delta^k\big(V(\overline{y})-V(y_k)\big)\leq (p_k,\overline{y})_{\mathbb{R}^N}-(p_k,y_k)_{\mathbb{R}^N}.$$

□

Now we show that a competitive program that satisfies the transversality condition is in fact optimal.

THEOREM 7.2.16 *If hypotheses H_1, H_2, H_3 hold and $\{(x_n,y_n,p_n)\}_{n\geq 0}$ is a competitive program starting at $\overline{y}\in\mathbb{R}^N_+$ that satisfies the transversality condition, then $\{(x_n,y_n)\}_{n\geq 0}$ is an optimal program starting at $\overline{y}\in\mathbb{R}^N_+$.*

PROOF: From (7.34) we have

$$\begin{aligned}\sum_{n=0}^{k}\delta^n\big(u(c_n')-u(c_n)\big) &\leq \sum_{n=0}^{k}\big[(p_{n+1},y_{n+1}-y_{n+1}')_{\mathbb{R}^N}-(p_n,y_n-y_n')_{\mathbb{R}^N}\big]\\ &=\sum_{n=0}^{k}\big[(p_{n+1},x_{n+1}+c_{n+1}-x_{n+1}'-c_{n+1}')_{\mathbb{R}^N}\\ &\qquad -(p_n,x_n+c_n-x_n'-c_n')_{\mathbb{R}^N}\big]\\ &\leq (p_{k+1},x_{k+1})_{\mathbb{R}^N}+(p_{k+1},c_{k+1}-c_{k+1}')_{\mathbb{R}^N}.\end{aligned} \tag{7.40}$$

But from (7.32) we know that

$$(p_{k+1},c_{k+1}-c_{k+1}')_{\mathbb{R}^N}\leq\delta^k\big(u(c_{k+1})-u(c_{k+1}')\big). \tag{7.41}$$

Using (7.41) in (7.40), we obtain

$$\sum_{n=0}^{k+1}\delta^n\big(u(c_n')-u(c_n)\big)\leq (p_{k+1},x_{k+1})_{\mathbb{R}^N}. \tag{7.42}$$

Passing to the limit as $k\longrightarrow+\infty$ in (7.42) and using the transversality condition, we conclude that $\{(x_n,y_n)\}_{n\geq 0}$ is an optimal program starting at $\overline{y}\in\mathbb{R}^N_+$. □

As we already mentioned in Remark 7.2.11, the transversality condition is rather problematic from an economic viewpoint. Indeed, being asymptotic in nature, it is not a myopic behavior rule for an individual decision maker and for this reason it is unclear how this condition can be verified in a decentralized setting. So, we want to have a reformulation of Theorem 7.2.16 with the transversality hypothesis removed. For this purpose, we introduce the reachability condition.

DEFINITION 7.2.17 We say that the discounted multisector growth model (G,u,δ) satisfies the *reachability condition* (R), if given any $\overline{y}\in\mathbb{R}^N_+$ and any program $\{(x_n,y_n)\}_{n\geq 0}$ starting at $\overline{y}\in\mathbb{R}^N_+$, there is an integer $k_0\geq 1$ and a program $\{(x_n',y_n')\}_{n\geq 0}$ starting at $\widehat{y}$ such that $y_{k_0}'\geq y_{k_0}$.

REMARK 7.2.18 In essence the reachability condition (R) may be rephrased as follows. The technological production possibilities are such that, beginning with a capital stock from which expansion of stocks are feasible (hypothesis H_1(iii)), it is possible (if need be through pure accumulation of capital over a sufficiently long period), to attain the stocks along any feasible program, at some future date.

THEOREM 7.2.19 *If hypotheses H_1, H_2, H_3 hold, the reachability condition (R) is satisfied and $\{(x_n, y_n, p_n)\}_{n\geq 0}$ is a competitive program starting at any $y \in \mathbb{R}^N_+$, then $\{(x_n, y_n)\}_{n\geq 0}$ is an optimal program starting at $y \in \mathbb{R}^N_+$.*

PROOF: Let $c = 0$ and apply (7.34) to $(\widehat{x}, \widehat{y}) \in G,\ c = 0$. We obtain

$$\begin{aligned}
&\delta^n u(c) + (p_{n+1}, \widehat{y})_{\mathbb{R}^N} - (p_n, \widehat{x} + c)_{\mathbb{R}^N} \\
&= \delta^n u(0) + (p_{n+1}, \widehat{y})_{\mathbb{R}^N} - (p_n, \widehat{x})_{\mathbb{R}^N} \\
&= \delta^n u(0) + (p_{n+1}, \widehat{y} - \widehat{x})_{\mathbb{R}^N} + (p_{n+1} - p_n, \widehat{x})_{\mathbb{R}^N} \\
&\leq \delta^n u(c_n) + (p_{n+1}, y_{n+1})_{\mathbb{R}^N} - (p_n, y_n)_{\mathbb{R}^N}. \qquad (7.43)
\end{aligned}$$

For $k \geq 2$, we have

$$\sum_{n=0}^{k-1} \delta^n u(0) + \sum_{n=0}^{k-2} (p_{n+1}, \widehat{y} - \widehat{x})_{\mathbb{R}^N} + (p_k, \widehat{y})_{\mathbb{R}^N} - (p_0, \widehat{x})_{\mathbb{R}^N}. \qquad (7.44)$$

Consider the sequence $\{(x''_n, y''_n)\}_{n\geq 0}$ defined by

$$x''_n = x_{n+k}, \quad y''_n = y_{n+k} \qquad \text{for all } n \geq 0.$$

Evidently $\{(x''_n, y''_n)\}_{n\geq 0}$ is a program starting at y_k. Invoking the reachability condition (R) (see Definition 7.2.17), we can find a program $\{(x'_n, y'_n)\}_{n\geq 0}$ starting at $\widehat{y}$ and $k_0 \geq 0$ such that

$$y'_{k_0} \geq y''_{k_0} = y_{k_0+k}. \qquad (7.45)$$

Applying (7.34) to $(x'_n, y'_{n+1}) \in G\ c'_n = y'_n - x'_n$, we have for all $n \geq 0$.

$$\begin{aligned}
&\delta^{k+n} u(c'_n) + (p_{k+n+1}, y'_{n+1})_{\mathbb{R}^N} - (p_{k+n}, y'_n)_{\mathbb{R}^N} \\
&\leq \delta^{k+n} u(c_{k+n}) + (p_{k+n+1}, y_{n+k+1})_{\mathbb{R}^N} - (p_{k+n}, y_{n+k})_{\mathbb{R}^N}. \qquad (7.46)
\end{aligned}$$

We adopt the convention that if $k_0 = 0$, the summation from 0 to $k_0 - 1$, equals zero. Summing up (7.46) from $n = 0$ to $n = k_0 - 1$, we have

$$\begin{aligned}
&\sum_{n=0}^{k_0-1} \delta^{k+n} u(c'_n) + (p_{k+k_0}, y'_{k_0})_{\mathbb{R}^N} - (p_k, y'_0)_{\mathbb{R}^N} \\
&\leq \sum_{n=0}^{k_0-1} \delta^{k+n} u(c_{k+n}) + (p_{k+k_0}, y_{k+k_0})_{\mathbb{R}^N} - (p_k, y_k)_{\mathbb{R}^N}. \qquad (7.47)
\end{aligned}$$

Recall that $y'_0 = \widehat{y}$, $p_n \geq 0$ for all $n \geq 0$, and use (7.35). Then (7.47) yields

$$\begin{aligned}
&\sum_{n=0}^{k_0-1} \delta^{k+n} u(c'_n) - (p_k, \widehat{y})_{\mathbb{R}^N} \\
&\leq \sum_{n=0}^{k_0-1} \delta^{k+n} u(c_{k+n}) - (p_k, y_k)_{\mathbb{R}^N}. \qquad (7.48)
\end{aligned}$$

From (7.44) and (7.48), for every $k \geq 2$ we have

$$\sum_{n=0}^{k-1} \delta^n u(0) + \sum_{n=0}^{k_0-1} \delta^{k+n} u(c_n') + \sum_{n=0}^{k-2} (p_{n+1}, \widehat{y} - \widehat{x})_{\mathbb{R}^N} - (p_0, \widehat{x})_{\mathbb{R}^N}$$
$$\leq \sum_{n=0}^{k+k_0-1} \delta^n u(c_n) - (p_0, y_0)_{\mathbb{R}^N}. \tag{7.49}$$

From Lemma 7.2.5 we know that

$$\|c_n\| \leq M_1 = \max\{\|y\|, M\} \text{ and } \|c_n'\| \leq \widehat{M}_1 = \max\{\|\widehat{y}\|, M\} \qquad \text{for all } n \geq 0.$$

Define $M_2 = \max\{M_1, \widehat{M}_1\}$ and $M_3 = \max\left[|u(c)| : 0 \leq c \leq M_2 e\right]$ where $e = [1, 1, \ldots, 1] \in \mathbb{R}^N$. Then from (7.49), we have

$$\sum_{n=0}^{k-2} (p_{n+1}, \widehat{y} - \widehat{x})_{\mathbb{R}^N}$$
$$\leq \sum_{n=0}^{k+k_0-2} \delta^n u(c_n) - \sum_{n=0}^{k-1} \delta^n u(0) - \sum_{n=0}^{k_0-1} \delta^{k+n} u(c_n') + (p_0, \widehat{x} - y_0)_{\mathbb{R}^N}$$
$$\leq \frac{3M_3}{1-\delta} + (p_0, \widehat{x} - y_0)_{\mathbb{R}^N} < +\infty.$$

So for any $k \geq 2$

$$\sum_{n=0}^{k-2} (p_{n+1}, \widehat{y} - \widehat{x})_{\mathbb{R}^N} < +\infty. \tag{7.50}$$

Because $\widehat{y} - \widehat{x} \gg 0$ (see hypothesis H_1(iii)) and $p_n \geq 0$ for all $n \geq 0$, from (7.50) we infer that

$$\sum_{n \geq 0} p_n < +\infty,$$
$$\Rightarrow p_n \longrightarrow 0 \quad \text{as } n \to \infty.$$

Recall that $\|x_n\| \leq M_1 = \max\{\|y\|, M\}$ for all $n \geq 0$ (see Lemma 7.2.5). Therefore

$$(p_n, x_n)_{\mathbb{R}^N} \longrightarrow 0 \quad \text{as } n \to \infty.$$

Now we can apply Theorem 7.2.16 and conclude that $\{(x_n, y_n)\}_{n \geq 1}$ is an optimal program starting at $y \in \mathbb{R}^N$. □

Next we present an example of a linear model that satisfies the reachability condition (R).

EXAMPLE 7.2.20 *The dynamic Leontief model:* Let $y_n \in \mathbb{R}^N$ be a vector whose ith component y_n^i represents the output of the ith good in time period $n \geq 0$. Let $A = (a_{ij})_{i,j=1}^N$ be an $N \times N$-matrix, where a_{ij} denotes the current input of the ith good used to produce one unit of the jth good. Then $Ay = \sum_{j=1}^N a_{ij} y_j$ is the vector of inputs (raw materials). Also let $l = (l_k)_{k=1}^N \in \mathbb{R}^N$ with l_k representing the amount of labor required to produce one unit of the kth good. As before $u:\mathbb{R}^N \longrightarrow \mathbb{R}$ is a continuous function, representing the utility function and $0<\delta<1$ is the discount factor. Then the technology feasibility set $G \subseteq \mathbb{R}_+^N \times \mathbb{R}_+^N$ is defined by:

$$(x,y) \in G \text{ if and only if } x \geq Ay,\ x \geq 0,\ y \geq 0 \quad \text{and} \quad (l,y)_{\mathbb{R}^N} \leq 1. \tag{7.51}$$

We impose the following conditions on the input coefficient matrix.

(L1) $A \geq 0$ in the sense that $a_{ij} \geq 0$ for all $i,j \in \{1,\ldots,N\}$ and $l \gg 0$.
(L2) A is productive, meaning that there is $\widehat{y} \gg A\widehat{y}$ and $(l,\widehat{y})_{\mathbb{R}^N} \leq 1$.

From matrix theory we know that under the above hypothesis $A^n \longrightarrow 0$ as $n \to \infty$ and $(I-A)^{-1} = \sum_{n \geq 0} A^n$. Now we can verify that this model fits in the framework of the previous analysis.

Claim 1: If G is defined by (7.51) and A satisfies hypotheses (L1) and (L2), then G satisfies hypotheses H_1.

Proof: Clearly we have the free disposability property H_1(ii). Let $m = \min_{1 \leq k \leq N} l_k$. Because by (L1) $l \gg 0$, we have $m>0$. Set $M = 1/m$. If $(x,y) \in G$, then $(l,y)_{\mathbb{R}^N} \leq 1$ (see (7.51)). Hence $m\|y\| \leq \sum_{k=1}^{N} l_k y_k \leq 1$ and so $\|y\| \leq 1/m$. Therefore we have satisfied hypothesis H_1(i). Finally set $\widehat{x} = A\widehat{y}$. Then clearly $(\widehat{x},\widehat{y}) \in G$ and $\widehat{x} \ll \widehat{y}$ (see hypothesis (L2)). So we have satisfied hypothesis H_1(iii).

Claim 2: If $G \subseteq \mathbb{R}_+^N \times \mathbb{R}_+^N$ is defined by (7.51) and (L1) holds, then G satisfies hypotheses H_1 if and only if hypothesis (L2) holds.

Proof: Note that if G satisfies H_1(iii), then it satisfies (L2). This combined with Claim 1 produces the desired equivalence.

Claim 3: If $G \subseteq \mathbb{R}_+^N \times \mathbb{R}_+^N$ is defined by (7.51), hypotheses (L1), (L2) hold and $\{(x_n, y_n)\}_{n \geq 0}$ is a program starting at $y \in \mathbb{R}_+^N$, then there is a program $\{(x_n', y_n')\}_{n \geq 0}$ starting at $\widehat{y}$ and an integer $k_0 \geq 1$ such that $y_{k_0}' = y_{k_0}$.

Proof: From the proof of Claim 1, we know that $y_n \leq Me$ for all $n \geq 1$ with $e = (1,1,\ldots,1) \in \mathbb{R}_+^N$. Because $\widehat{y} \gg 0$ and $A^n \longrightarrow 0$ as $n \to \infty$, we can find $k_0 \geq 2$ such that

$$A^{k_0} Me \ll \widehat{y}. \tag{7.52}$$

We define a new program $\{(x_n', y_n')\}_{n \geq 0}$ as follows.

$$\begin{aligned} &x_0' = y_0' = \widehat{y}, \quad x_n' = A^{k_0-n} y_{k_0} = y_n' \qquad \text{for } 1 \leq n \leq k_0 - 1, \\ &x_n' = x_n, \quad y_n' = y_n' \quad \text{for } n \geq k_0. \end{aligned} \tag{7.53}$$

We also introduce the corresponding consumption sequence

$$c_n' = y_n' - x_n \quad \text{for } n \geq 0.$$

Clearly $x_n' \geq 0$, $y_n' \geq 0$ for $n \geq 0$, $c_n' = 0$ for $0 \leq n \leq k_0 - 1$, and $c_n' = c_n \geq 0$ for $n \geq k_0$. Also $y_0' = \widehat{y}$. So in order to have that $\{(x_n', y_n')\}_{n \geq 0}$ is a program starting at $\widehat{y}$, we need to verify that

$$Ay_{n+1}' \leq x_n' \qquad \text{for all } n \geq 0 \tag{7.54}$$

$$\text{and} \quad (l, y_n')_{\mathbb{R}^N} \leq 1 \qquad \text{for all } n \geq 1. \tag{7.55}$$

From (7.53), we see that (7.54) and (7.55) hold for all $n \geq k_0$. Because $A \geq 0$ (see hypothesis (L1)), we have

$$\begin{aligned} y_1' &= A^{k_0-1} y_{k_0} \leq A^{k_0-1} Me, \\ \Rightarrow\ Ay_1' &\leq A^{k_0} Me \ll \widehat{y} = x_0' \quad \text{(see (7.52))}. \end{aligned}$$

This proves (7.54) when $n = 0$.

Next if $1 \leq n \leq k_0 - 1$, then because of (7.53), we have

$$Ay_{n+1}' = AA^{k_0-(n+1)} y_{k_0} = x_n'.$$

This proves (7.54) when $1 \leq n < k_0 - 1$.

Finally, if $n = k_0 - 1$, then

$$Ay_{n+1}' = Ay_{k_0}' = Ay_{k_0} = A^{k_0-(k_0-1)} y_{k_0} = x_{k_0-1}' = x_n'.$$

This proves (7.54) when $n = k_0 - 1$ and so we have verified (7.54) for all $n \geq 0$.

It remains to verify (7.55) for $1 \leq n \leq k_0 - 1$. Because $\{(x_n, y_n)\}_{n \geq 0}$ is a program starting at $y \in \mathbb{R}_+^N$, we have

$$Ay_{n+1} \leq x_n \leq y_n \qquad \text{for all } n \geq 0.$$

So for any integer $i \geq 1$, we have

$$A^i y_{n+i} = A^{i-1} y_{n+i} \leq A^{i-1} y_{n+i-1} \quad \text{for all } n \geq 0.$$

Continuing this way until zero, we obtain

$$A^i y_{n+i} \leq y_n \quad \text{for all } n \geq 0 \text{ and all } i \geq 1. \tag{7.56}$$

So from (7.53) for $1 \leq n \leq k_0 - 1$, after setting $i = k_0 - n$ in (7.56), we obtain

$$y_n' = A^{k_0-n} y_{k_0} \leq y_n.$$

Because $l \gg 0$ (see hypothesis (L1)) and $(l, y_n)_{\mathbb{R}^N} \leq 1$ for all $n \geq 1$, we have

$$(l, y_n')_{\mathbb{R}^N} \leq (l, y_n)_{\mathbb{R}^N} \leq 1 \quad \text{for } 1 \leq n \leq k_0 - 1.$$

Therefore we have verified (7.55) and we have proved Claim 3.

Claim 3 implies that for the Leontief model the reachability condition (R) holds. So for the Leontief growth model Theorem 7.2.19 applies.

Next we turn our attention to growth models that are undiscounted (i.e., $\delta = 1$). In this case the sum describing the intertemporal utility along a program, need not converge. So in order to evaluate the performance of a program, we need to use a criterion different from the standard notion of optimality given in Definition 7.2.6. This leads us to the so-called weak maximality criterion. But first let us describe the model.

As before the planning horizon is infinite, namely $\mathbb{N}_0 = \{0, 1, 2, \ldots\}$ (i.e., we deal with a discrete time, infinite horizon growth model). Also the commodity space is $\mathbb{R}^N$ (there are N commodities in the market). The technological constraints are described by the following mathematical hypotheses.

H₄: $G : \mathbb{R}^N_+ \longrightarrow P_{fc}(\mathbb{R}^N_+)$ is a multifunction with a closed and convex graph such that

(i) There exists $M > 0$ such that $|G(x)| \leq \sup\{\|y\| : y \in G(x)\} \leq M$ for all $x \in \mathbb{R}^N_+$.

(ii) If $x \leq x'$, $y' \leq y$ and $y \in G(x)$, then $y' \in G(x')$.

(iii) There exists $(\widehat{x}, \widehat{y}) \in \operatorname{Gr} G$ such that $\widehat{x} \ll \widehat{y}$.

REMARK 7.2.21 Combining hypothesis H_4(i) with the fact that G has a closed graph, we infer that G is usc (see Proposition 6.1.10). Invoking the Kakutani–Ky Fan fixed point theorem (see Theorem 6.5.19), we have that $\Gamma = \{x \in \mathbb{R}^N_+ : x \in G(x)\} \neq \emptyset$. Hypothesis H_4(ii) is the free disposability hypothesis (see also hypothesis H_1(ii)). Hypothesis H_4(iii) is the same as hypothesis H_1(iii) and it says that the economy has an expansible commodity vector $\widehat{x} \in \mathbb{R}^N_+$.

Also we have a utility function $u(x, y)$ that satisfies the following hypotheses.

H₅: $u : \operatorname{Gr} G \longrightarrow \mathbb{R}$ is an upper semicontinuous concave function such that $x \longrightarrow u(x, y)$ is nondecreasing and $y \longrightarrow u(x, y)$ is nonincreasing.

DEFINITION 7.2.22 A *feasible program* is a sequence $\widehat{x} = \{x_n\}_{n \geq 0}$, $x_n \in \mathbb{R}^N_+$ for all $n \geq 0$ such that $x_{n+1} \in G(x_n)$ for all $n \geq 0$. In what follows by $F(x_0)$ we denote the set of feasible programs starting at $x_0 \in \mathbb{R}^N_+$.

REMARK 7.2.23 In contrast to the previous multisector growth model, in this one we have supressed the consumption sequence $\{c_n\}_{n \geq 0}$. In the new model a feasible input-output pair (x, y) includes consumption activities as well as production activities. One might object that this way we are obscuring the distinction between the consumption and production activities. Although this may have some minor implications, such as, for example, obscuring the role of consumer satiation, in general it does not create any conceptual problems and it is reasonable from an economic viewpoint. Moreover, it is mathematically convenient.

Due to the infinite planning horizon and the absence of a discount factor (in this model $\delta = 1$), we have difficulties in defining an optimal feasible program, because the sum of the intertemporal utility need not converge. As a first attempt to overcome this difficulty, we introduce the following notion.

DEFINITION 7.2.24 (a) Let $\{\alpha_n\}_{n \geq 0}, \{b_n\}_{n \geq 0}$ be two real sequences. We say that $\{b_n\}_{n \geq 0}$ *catches up* to $\{\alpha_n\}_{n \geq 0}$, if $\limsup\limits_{m \to \infty} \sum\limits_{n=0}^{m} (\alpha_n - b_n) \leq 0$.

(b) A feasible program $\widehat{x} = \{x_n\}_{n \geq 0}$ starting from $x_0 \in \mathbb{R}^N_+$ is said to be *strongly maximal* (or *optimal*) if and only if the sequence $\{u(x_n, x_{n+1})\}_{n \geq 0}$ catches up to the sequence $\{u(y_n, y_{n+1})\}_{n \geq 0}$ for any $\widehat{y} = (y_n)_{n \geq 0} \in F(x_0)$, in other words

$$\limsup_{m \to \infty} \Big(\sum_{n=0}^{m} \big(u(y_n, y_{n+1}) - u(x_n, x_{n+1}) \big) \Big) \leq 0$$

for every $\widehat{y} = (y_n)_{n \geq 0} \in F(x_0)$.

REMARK 7.2.25 The catching-up criterion introduces an ordering on the real sequences that is reflexive and transitive. In this ordering, two sequences $\{\alpha_n\}_{n\geq 0}, \{b_n\}_{n\geq 0}$ are equivalent (each of them catching up the other) if and only if $\lim_{m\to\infty} \sum_{n=0}^{m} (\alpha_n - b_n) = 0$. The ordering is *strict* if $\limsup_{m\to\infty} \sum_{n=0}^{m} (\alpha_n - b_n) < 0$ or $\limsup_{m\to\infty} \sum_{n=0}^{m} (\alpha_n - b_n) = 0$ and $\liminf_{m\to\infty} \sum_{n=0}^{m} (\alpha_n - b_n) < 0$. However, this ordering is not total. It may happen that both $\limsup_{m\to\infty} \sum_{n=0}^{m} (\alpha_n - b_n)$ and $\limsup_{m\to\infty} \sum_{n=0}^{m} (b_n - \alpha_n)$ may be strictly positive. To see this consider the sequences $\left\{\alpha_n = (-1)^n\right\}_{n\geq 0}$ and $\left\{b_n = 1/(2^{n+2})\right\}_{n\geq 0}$. In this case both lim sup are equal to $\frac{1}{2}$. For this reason instead of the strong maximality criterion, we prefer a weaker one, which nevertheless induces a total ordering.

DEFINITION 7.2.26 (a) Let $\{\alpha_n\}_{n\geq 0}, \{b_n\}_{n\geq 0}$ be two real sequences. We say that $\{b_n\}_{n\geq 0}$ *overtakes* $\{\alpha_n\}_{n\geq 0}$ if $\liminf_{m\to\infty} \sum_{n=0}^{m} (b_n - a_n) > 0$ (equivalently $\limsup_{m\to\infty} \sum_{n=0}^{m} (\alpha_n - b_n) < 0$).

(b) A feasible program $\widehat{x} = \{x_n\}_{n\geq 0}$ starting at $x_0 \in \mathbb{R}_+^N$ is said to be *weakly maximal* if and only if no element in $F(x_0)$ overtakes $\widehat{x}$, in other words

$$\liminf_{m\to\infty} \sum_{n=0}^{m} \big(u(y_n, y_{n+1}) - u(x_n, x_{n+1})\big) \leq 0$$

for every $\widehat{y} = (y_n)_{n\geq 0} \in F(x_0)$.

REMARK 7.2.27 Note that the overtaking criterion is total and another equivalent way to state it is to say that for every $\varepsilon > 0$ and every $m_0 \geq 1$, we can find $m = m(\varepsilon, m_0, \widehat{y}) \geq m_0$ such that

$$\sum_{n=0}^{m} u(y_n, y_{n+1}) - \varepsilon \leq \sum_{n=0}^{m} u(x_n, x_{n+1}).$$

In a multisector growth model, of special interest are stationary programs, because they permit a balanced growth of the economy. This justifies the next definition. First note that because $\Gamma = \text{Fix } G = \{x \in \mathbb{R}_+^N : x \in G(x)\} \in P_k(\mathbb{R}_+^N)$ and u is upper semicontinuous, the optimization problem

$$\widehat{u} = \sup[u(x, x) : x \in \Gamma] \tag{7.57}$$

has a solution. Moreover, due to hypothesis H_4(ii) (the free disposability) and hypothesis H_5, we can equivalently rewrite (7.57) as follows,

$$\widehat{u} = \max[u(x, y) : y \in G(x), x \leq y]. \tag{7.58}$$

Let S be the solution set of (7.57) (equivalently of (7.58)), (i.e., $S = \{x \in \Gamma : \widehat{u} = u(x, x)\} \neq \emptyset$).

DEFINITION 7.2.28 A *weakly maximal stationary program* is a vector $x^* \in S$ such that the constant (stationary) program $\widehat{x}^* = \{x_n^* = x^*\}_{n \geq 0}$ is weakly maximal within $F(x^*)$.

We are looking for weakly maximal programs. To this end we start by producing a stationary price system that supports the optimization problem (7.58).

PROPOSITION 7.2.29 *If hypotheses H_4 and H_5 hold, then there exists $p \in \mathbb{R}_+^N$ such that for all $(x, y) \in \mathrm{Gr}G$ we have*

$$u(x, y) + (p, y - x)_{\mathbb{R}^N} \leq \widehat{u}.$$

PROOF: Let $h : \mathbb{R}^N \times \mathbb{R}^N \longrightarrow \mathbb{R}^N$ be the continuous function defined by $h(x, y) = y - x$. Then we can write (7.58) as follows.

$$\widehat{u} = \max\{(x, y) : (x, y) \in \mathrm{Gr}\, G,\ h(x, y) \geq 0\}.$$

By hypothesis H_4(iii) $h(\widehat{x}, \widehat{y}) \gg 0$. So by the Kuhn–Tucker theorem, we can find $p \in \mathbb{R}_+^N$ such that

$$u(x, y) + (p, y - x)_{\mathbb{R}^N} \leq \widehat{u} \qquad \text{for all } (x, y) \in \mathrm{Gr}\, G.$$

□

Using this proposition, we can show that there is no feasible program starting at $x_0 \in \overline{B}_M^+ = \{v \in \mathbb{R}_+^N : \|v\| \leq M\}$ that is infinitely better than the value $\widehat{u}$ of problem (7.58).

PROPOSITION 7.2.30 *If hypotheses H_4 and H_5 hold, then there exists $M_1 > 0$ such that for any $\widehat{x} = (x_n)_{n \geq 0}$, $x_0 \in \overline{B}_M^+$, such that for all $m \geq 1$, we have*

$$\sum_{n=0}^{m} \big(u(x_n, x_{n+1}) - \widehat{u}\big) \leq M_1.$$

PROOF: By virtue of Proposition 7.2.29, we can find $p \in \mathbb{R}_+^N$ such that

$$u(x, y) + (p, y - x)_{\mathbb{R}^N} \leq \widehat{u} \qquad \text{for all } (x, y) \in \mathrm{Gr}\, G.$$

From this, we have

$$\begin{aligned} & u(x_n, x_{n+1}) - \widehat{u} = (p, x_n - x_{n+1})_{\mathbb{R}^N} \quad \text{for all } \widehat{x} = (x_n)_{n \geq 0} \in F(x_0) \\ \Rightarrow & \sum_{n=0}^{m} \big(u(x_n, x_{n+1}) - \widehat{u}\big) \leq (p, x_0)_{\mathbb{R}^N} - (p, x_m)_{\mathbb{R}^N} \leq \|p\| 2M = M_1, \end{aligned}$$

(see H_4(i)). □

So there is no feasible program that can produce an intertemporal utility infinitely better than the one generated by a stationary program $x^* \in S$. However, a program can be infinitely worse (e.g., any nonoptimal stationary program). To avoid this situation, we introduce the following definition.

DEFINITION 7.2.31 A feasible program $\widehat{x}=\{x_n\}_{n\geq 0}$ is said to be *good* if there exists $\gamma \in \mathbb{R}$ such that for all $m \geq 1$, we have $\sum_{n=0}^{m}\big(u(x_n, x_{n+1}) - \widehat{u}\big) \geq \gamma$; in other words

$$\liminf_{m\to\infty}\sum_{n=0}^{m}\big(u(x_n, x_{n+1}) - \widehat{u}\big) > -\infty.$$

REMARK 7.2.32 So a feasible program $\widehat{x} = (x_n)_{n\geq 0}$ is good, if it can sustain a utility level close to $\widehat{u}$. In the literature a good program is also called *eligible* (see Takayama [572]).

The next proposition links good programs with the set S of solutions of problem (7.58) (equivalently of (7.58)).

PROPOSITION 7.2.33 *If hypotheses* H_4 *and* H_5 *hold and* $\widehat{x} = (x_n)_{n\geq 0}$ *is a good program with* $x_0 \in \overline{B}_M^+$*, then* $L = \Big\{$*limit points of* $\{1/(n+1)\sum_{k=0}^{n} x_k\}_{n\geq 0}\Big\}$ *is nonempty and a subset of* S.

PROOF: Note that $1/(n+1)\sum_{k=0}^{n} x_k \in \overline{B}_M^+$ for all $n \geq 0$. So by the Heine–Borel theorem, we have that $L \neq \emptyset$. Let $x^* \in L$. Then we can find a sequence $m \longrightarrow n_m$ such that

$$\frac{1}{n_m+1}\sum_{k=0}^{n_m} x_k \longrightarrow x^* \quad \text{as } m \to \infty. \tag{7.59}$$

Exploiting the convexity of Gr G (see hypotheses H_4), we have

$$\Big(\frac{1}{n_m+1}\sum_{k=0}^{n_m} x_k, \ \frac{1}{n_m+1}\sum_{k=1}^{n_m+1} x_k\Big) \in \text{Gr } G. \tag{7.60}$$

Note that

$$\frac{1}{n_m+1}\sum_{k=1}^{n_m+1} x_k = \frac{1}{n_m+1}\sum_{k=0}^{n_m+1} x_k + \frac{1}{n_m+1}(x_{n_m+1} - x_0)$$

and $x_0, x_{n_m+1} \in \overline{B}_M^+$. Therefore

$$\frac{1}{n_m+1}\sum_{k=1}^{n_m+1} x_k \longrightarrow x^* \quad \text{as } m \to \infty. \tag{7.61}$$

Because Gr G is closed from (7.59), (7.61) and (7.60), we infer that

$$(x^*, x^*) \in \text{Gr } G \ \ (\text{i.e., } \ x^* \in \Gamma).$$

Moreover, because of H_5, Jensen's inequality, and the fact that by hypothesis $\widehat{x} = \{x_n\}_{n\geq 0}$ is a good program, we have (see Definition 7.2.31)

$$\frac{\gamma}{n_m+1} \le \frac{1}{n_m+1}\sum_{k=0}^{n_m}\big(u(x_k,x_{k+1})-\widehat{u}\big)$$
$$\le u\Big(\frac{1}{n_m+1}\sum_{k=0}^{n_m}x_k, \frac{1}{n_m+1}\sum_{k=1}^{n_m+1}x_k\Big)-\widehat{u}. \tag{7.62}$$

Passing to the limit as $m\longrightarrow\infty$ in (7.62), we obtain

$$\widehat{u} \le \limsup_{m\to\infty} u\Big(\frac{1}{n_m+1}\sum_{k=0}^{n_m}x_k, \frac{1}{n_m+1}\sum_{k=1}^{n_m+1}x_k\Big)$$
$$\le u(x^*,x^*) \quad \text{(because } u \text{ is upper semicontinuous; see } \mathrm{H}_4). \tag{7.63}$$

Because $x^*\in\Gamma$, from (7.63) it follows that

$$\widehat{u} = u(x^*,x^*)$$
$$\Rightarrow\ x^*\in S \ \text{ (i.e., } L\subseteq S).$$

□

This proposition suggests that we should look among the elements of S in order to produce a weakly maximal stationary program.

THEOREM 7.2.34 *If hypotheses H_4 and H_5 hold, then there exists a weakly maximal stationary program.*

PROOF: Let $p\in\mathbb{R}^N_+$ be as in Proposition 7.2.29 and let $x^*\in S$ such that $(p,x^*)_{\mathbb{R}^N}\le (p,y)_{\mathbb{R}^N}$ for all $y\in S$. It exists since S is compact. Consider the stationary program $\widehat{x}^*=\{x^*_n=x^*\}_{n\ge 0}$. We claim that this is a weakly stationary program in $F(x^*)$. We argue indirectly. So suppose that the claim is not true. Then there exist $\widehat{z}\in F(x^*)$, $\varepsilon>0$, and $m_0\ge 1$ such that for all $m\ge m_0$ we have

$$U_m(\widehat{x}^*)+\varepsilon \le U_m(\widehat{z}),$$

where $U_m(\widehat{x}^*)=\sum_{n=0}^{m} u(x^*_n,x^*_{n+1}) = \sum_{n=0}^{m} u(x^*,x^*)$ and $U_m(\widehat{z})=\sum_{n=0}^{m} u(z_n,z_{n+1})$ (see Remark 7.2.27). Then

$$\varepsilon \le U_m(\widehat{z})-U_m(\widehat{x}^*) \qquad \text{for all } m\ge m_0. \tag{7.64}$$

From (7.64) and Definition 7.2.31 it follows that $\widehat{z}$ is a good program. So by virtue of Proposition 7.2.33, we have (at least for a subsequence), that

$$\frac{1}{m+1}\sum_{n=0}^{m} z_n \longrightarrow y^*\in L\subseteq S \quad \text{as } m\longrightarrow\infty.$$

Hence

$$\lim_{m\to\infty}\frac{1}{m+1}\sum_{n=0}^{m}(p,z_n)_{\mathbb{R}^N} = \lim_{m\to\infty}\Big(p,\frac{1}{m+1}\sum_{n=0}^{m}z_n\Big)_{\mathbb{R}^N} \le \limsup_{m\to\infty}(p,z_m)_{\mathbb{R}^N},$$
$$\Rightarrow\ (p,y^*)_{\mathbb{R}^N}\le \limsup_{m\to\infty}(p,z_m)_{\mathbb{R}^N}. \tag{7.65}$$

Then from Proposition 7.2.29 and (7.64), we obtain

$$0<\varepsilon \leq (p,x^*)_{\mathbb{R}^N} - (p,z_{m+1})_{\mathbb{R}^N} \qquad \text{for all } m \geq m_0$$
$$\Rightarrow\ 0<\varepsilon \leq (p,x^*)_{\mathbb{R}^N} - \limsup_{m\to\infty}(p,z_{m+1})_{\mathbb{R}^N} \leq (p,x^*)_{\mathbb{R}^N} - (p,y^*)_{\mathbb{R}^N},$$

(see (7.64)),

$$\Rightarrow\ (p,y^*)_{\mathbb{R}^N} < (p,x^*)_{\mathbb{R}^N} \quad \text{and} \quad y^* \in S.$$

This contradicts the choice of $x^* \in S$. This proves that $\widehat{x}^* = \{x_n^* = x^*\}_{n\geq 0}$ is a weakly maximal stationary program. □

REMARK 7.2.35 The result fails if the weak maximality criterion is replaced by the strong maximality criterion (see Definition 7.2.24(b)).

7.3 Turnpike Theorems

In this section we continue the analysis of multisector growth models with an infinite planning horizon. As in the previous section, the model that we consider is stationary, namely the set of technological possibilities and the utility function are constant in time. The stationarity assumption is of course restrictive, but most realistic models such as the constant growth models, that is, $G_n = \lambda^n G$ with $\lambda > 0$ (the growth rate) and $u_n(x,y) = u\big((1/\lambda^n)x, (1/\lambda^n)y\big)$, can be transformed into stationary models. In general, any realistic growth model, will involve some kind of stationarity assumption because any probabilistic prediction of the behavior of the system over a long period is almost always based on some kind of stationarity hypothesis.

In the second part of Section 7.2, we saw that in stationary models the stationary programs are important (i.e., programs $\widehat{x} = \{x_n\}_{n\geq 0}$, where x_n is independent of time, namely $x_n = x^* \in \mathbb{R}_+^N$). In this section, we isolate among the stationary programs, a special subclass called turnpike programs (or simply turnpikes). A turnpike in a stationary model is a special stationary optimal program, distinguished by at least three properties.

(a) An optimal program eventually moves close to the turnpike.
(b) A turnpike is easier to compute than an arbitrary optimal program.
(c) A turnpike is relatively insensitive to the optimality criterion.

In this section we focus on property (a), which leads to the so-called turnpike theorems. These theorems for infinite programs maintain that these programs in a certain sense approach a turnpike. In fact this property motivated the term turnpike. In most books on the subject, the following analogy is given by way of further explanation. Suppose that we want to drive from town A to town B and the main highway (the turnpike) passes close to A and B. The time of an optimal journey is to drive from town A to the turnpike, follow the turnpike until the exit to town B, and then reach town B. Turnpike theorems for finite programs can have weak and strong forms. Weak turnpike theorems assert that for a sufficiently large planning horizon most of the time optimal programs remain close to the turnpike. However, these theorems say nothing about the whereabouts of the planning intervals where the optimal program deviates considerably from the turnpike. In contrast the strong turnpike theorems maintain that these distant points are concentrated at the beginning and at the end of the planning interval. So strong turnpike theorems, show that optimal programs of economic growth are analogous to the time of the optimal journey from town A to town B described above. Turnpike theorems are important because they establish the proximity of optimal programs to special programs with simple structure, the turnpikes (maximal stationary programs). Turnpikes are much easier to obtain compared to a general optimal program.

As in Section 7.2, the multisector growth model is described by a pair (G, u). The set $G \subseteq \mathbb{R}_+^N \times \mathbb{R}_+^N$ describes the technological possibilities of the economy. A pair $(x, y) \in G$ describes a technological process. So the expenditure of the vector $x \in \mathbb{R}_+^N$ (the input) at time $n-1$ $(n \geq 1)$, yields the production of the (not necessarily unique) output vector $y \in \mathbb{R}_+^N$ at time n $(n \geq 1)$. The function $u(x, y)$ measures the aggregate utility of the technological process (x, y). The hypotheses on the two items of the economy are the following.

H₁: $G \subseteq \mathbb{R}_+^N \times \mathbb{R}_+^N)$ is a nonempty, compact, and convex set such that

(i) $(0, 0) \in G$.

(ii) We can find $(\widehat{x}, \widehat{y}) \in G$ such that $\widehat{x} \ll \widehat{y}$.

REMARK 7.3.1 As before hypothesis H_1(i) means that inactivity is one option, and hypothesis H_1(ii) means that there is an expansible stock, namely we can find an input $\widehat{x} = (\widehat{x}_k)_{k=1}^N \in \mathbb{R}_+^N$ and a corresponding output $\widehat{y} = (\widehat{y}_k)_{k=1}^N \in \mathbb{R}^N$ such that $x_k < y_k$ for all $k \in \{1, \dots, N\}$. This is a productivity assumption and it is not that restrictive because the vectors $\widehat{x}, \widehat{y} \in \mathbb{R}_+^N$ may differ very little from each other and from the zero vector. Finally the convexity of the set G means that it is possible to mix technological processes in arbitrary proportions.

H₂: $u : G \longrightarrow \mathbb{R}$ is a continuous, strictly concave function.

REMARK 7.3.2 The concavity of the utility function reflects the nature of the preferences of the society for variety in general (half an apple and half an orange are preferred to either a whole apple or a whole orange).

From hypotheses H_1 and H_2 we see that the model is stationary, because both the technology set G and the utility function u are time-variant.

DEFINITION 7.3.3 A finite or infinite sequence $\{z_n\}_{n=1}^m, (m \leq +\infty)$, is said to be a *program* (*path*) of the economy if and only if

$$z_n = (x_{n-1}, y_n) \in G \quad \text{and} \quad x_n \leq y_n \quad \text{for all } n \geq 1. \tag{7.66}$$

If the vector $y_0 \in \mathbb{R}_+^N$ is fixed, then we have program originating from y_0.

REMARK 7.3.4 In (7.66), the first condition means that all pairs $z_n = (x_{n-1}, y_n)$ are technologically feasible. The second condition is a resource restriction and simply says that at each stage the input (or expenditures) can not exceed the output (available capital stock) from the previous stage. The vector $y_0 \in \mathbb{R}_+^N$ is called the initial vector for the program $\{z_n\}_{n=1}^m$, if $y \geq x_0$; that is, the input at stage zero does not exceed the initial resources.

The next definition describes how we measure the performance of a program $\{z_n\}_{n=1}^m$, $(m \leq \infty)$. We distinguish between finite programs $\{z_n\}_{n=1}^m$, $(m < \infty)$ and infinite programs $\{z_n\}_{n\geq 1}$ depending on whether we consider a finite or an infinite planning horizon. In the first case there is no issue of convergence of the intertemporal utility and so optimality is defined in the standard way. In the second case, we are faced with the possible nonconvergence of the intertemporal utility (because there is no discount; i.e., $\delta = 1$) and so to define optimality we use the notions introduced in the last part of Section 7.2.

DEFINITION 7.3.5 (a) A *finite program* $\{z_n\}_{n=1}^m$, $(m<\infty)$ starting at $y_0 \in \mathbb{R}_+^N$ is said to be *optimal* if and only if

$$\sum_{n=1}^m u(z_n') \leq \sum_{n=1}^m u(z_n)$$

for any finite program $\{z_n'\}_{n=1}^m$ starting at $y_0 \in \mathbb{R}_+^N$.

(b) An *infinite program* $\{z_n\}_{n\geq 1}$ starting at $y_0 \in \mathbb{R}_+^N$ is said to be *optimal* if and only if

$$\limsup_{m\to\infty} \sum_{n=1}^m \big(u(z_n') - u(z_n)\big) \leq 0$$

for any infinite program $\{z_n'\}_{n\geq 1}$ starting at $y_0 \in \mathbb{R}_+^N$.

REMARK 7.3.6 In part (b) the optimality of $\{z_n\}_{n\geq 1}$ is the same as saying that the program is weakly maximal in the sense of Definition 7.2.26(b). If for every program $\{z_n'\}_{n\geq 1}$ starting at $y_0 \in \mathbb{R}_+^N$, the intertemporal utility $\sum_{n\geq 1} u(z_n')$ converges, then an optimal program $\{z_n\}_{n\geq 1}$ (in the sense of Definition 7.3.5), maximizes the intertemporal utility (i.e., is optimal in the usual way of maximizing the performance criterion). Conversely, if exactly one program exists that maximizes the intertemporal utility, then it is optimal in the sense of Definition 7.3.5(b).

Because our model is stationary, the following definition is important.

DEFINITION 7.3.7 (a) A program $\{z_n\}_{n\geq 1}$ is said to be *stationary* if $z_n = z \in G$ for all $n \geq 1$.

(b) A stationary program $\{z_n = z\}_{n\geq 1}$ is a *turnpike* if it maximizes the utility function (i.e., $u(z) = \max[u(z') : z' \in G]$).

REMARK 7.3.8 Evidently a necessary and sufficient condition for a vector $z = (x, y) \in \mathbb{R}_+^N \times \mathbb{R}_+^N$ to define a stationary program is that $z = (x, y) \in G$ and $x \leq y$. So the stationary program z defines a turnpike if and only if it solves the maximization problem

$$\max\{u(z) : z = (x, y) \in G,\ x \leq y\}. \tag{7.67}$$

Because by hypothesis H_2, the utility function u is strictly concave, then a turnpike (which always exists due to the compactness of G), is unique.

From Proposition 7.2.29, we know that there is a stationary price system supporting the turnpike. Namely we have the following.

PROPOSITION 7.3.9 *If hypotheses H_1 and H_2 and $z^*=(x^*,y^*)\in\mathbb{R}_+^N\times\mathbb{R}_+^N$ is a turnpike, then there exists $p^*\in\mathbb{R}_+^N$ such that*

$$u(z)+(p^*,x-y)_{\mathbb{R}^N}\le u(z^*) \tag{7.68}$$

for all $z=(x,y)\in G$.

Let $\xi : G\longrightarrow\mathbb{R}$ be defined by

$$\xi(z)=u(z)+(p^*,x-y)_{\mathbb{R}^N}\qquad\text{for all } z\in(x,y)\in G.$$

Evidently $\xi(z)$ represents the total utility corresponding to the stationary prices $\{p_n^*=p^*\}_{n\ge1}$ (i.e., $\xi(z)$ is the utility $u(z)$ plus the profit resulting from realizing the pair $(x,y)\in G$, when the price system $p^*\ge 0$ prevails in the market).

We set

$$\varrho(z,z')=|\xi(z)-\xi(z')|\qquad\text{for all } z,z'\in G. \tag{7.69}$$

This is a pseudometric on G. In the formulation of the turnpike theorems, this pseudometric appears to be a more convenient measure of deviation from the turnpike z^* than the norm, because it facilitates the proofs.

Let us recall the following notion from Section 7.2 (see Definition 7.2.31). This notion is crucial in obtaining the weak turnpike theorem.

DEFINITION 7.3.10 A sequence of finite programs $\{z_n^m\}_{n=1}^m$ is said to be *good* if there exists $\gamma\in\mathbb{R}$ such that

$$\sum_{n=1}^m\big(u(z_n^m)-u(z^*)\big)\ge\gamma.$$

REMARK 7.3.11 The good programs can be worse than the turnpike, only by a maximum utility value equal to $-\gamma$.

In what follows $z^*=(x^*,y^*)$ is a turnpike, $p^*\in\mathbb{R}_+^N$ is the corresponding price system supporting z^* (see Proposition 7.3.9), and $\{z_n^m\}_{n=1}^m$, $m\ge1$ is a good sequence in the sense of Definition 7.3.10, with initial vector $y_0\in\mathbb{R}_+^N$.

LEMMA 7.3.12 *We can find $\beta>0$ independent of $m\ge1$ such that*

$$\sum_{n=1}^m\varrho(z_n^m,z^*)\le\beta\qquad\textit{for all } m\ge1.$$

PROOF: We have

$$\begin{aligned}\vartheta_m=\sum_{n=1}^m\varrho(z_n^m,z^*)&=\sum_{n=1}^m\big(\xi(z^*)-\xi(z_n^m)\big)\quad\text{(see (7.69))}\\&=\sum_{n=1}^m\big(u(z^*)-u(z_n^m)\big)+\sum_{n=1}^m(p^*,y^*-x^*)_{\mathbb{R}^N}\\&\quad-\sum_{n=1}^m(p^*,y_n^m-x_n^m)_{\mathbb{R}^N}+(p^*,x_0-y_m^m)_{\mathbb{R}^N}.\end{aligned}\tag{7.70}$$

Because $p^* \in \mathbb{R}_+^N$ is the stationary price system supporting the turnpike $z^* = (x^*, y^*)$, we have

$$(p^*, y^* - x^*)_{\mathbb{R}^N} = 0. \tag{7.71}$$

Also we have $x_n^m \leq y_n^m$ and because $p^* \in \mathbb{R}_+^N$, we obtain

$$(p^*, y_n^m - x_n^m)_{\mathbb{R}^N} \geq 0. \tag{7.72}$$

Moreover, because G is by hypothesis compact, we have

$$(p^*, x_0 - y_m^m)_{\mathbb{R}^N} \leq \eta \qquad \text{for some } \eta > 0, \text{ all } m \geq 1. \tag{7.73}$$

Returning to (7.70) and using (7.71) through (7.73), we infer that

$$\vartheta_m \leq \sum_{n=1}^{m} \big(u(z^*) - u(z_n^m)\big) + \eta. \tag{7.74}$$

From (7.74) and the fact that $\{z_n^m\}_{n=1}^m$, $m \geq 1$ is a good sequence (see Definition 7.3.10) we conclude that $\vartheta_m \leq \beta$ for some $\beta > 0$ (independent of $m \geq 1$) and all $m \geq 1$. □

LEMMA 7.3.13 *There exists a function* $\lambda : \mathbb{R}_+ \longrightarrow \mathbb{R}_+$ *nondecreasing such that* $\lambda(0) = 0$, $\lambda(\varepsilon) > 0$ *for all* $\varepsilon > 0$ *and* $p(z^*, z) \geq \lambda(\|z^* - z\|)$ *for all* $z \in G$.

PROOF: We set

$$\lambda(\varepsilon) = \inf\big[p(z^*, z) : z \in G,\ \|z^* - z\| \geq \varepsilon\big].$$

We only need to show that $\lambda(\varepsilon) > 0$ for $\varepsilon > 0$. If this is not true, then we can find $\{z_n\}_{n \geq 1} \subseteq G$ such that $\|z^* - z_n\| \geq \varepsilon$ for all $n \geq 1$, but $\varrho(z^*, z_n) \longrightarrow 0$ as $n \to \infty$. Because G is compact, we may assume that $z_n \longrightarrow \overline{z} \in G$. Clearly $\|z^* - \overline{z}\| \geq \varepsilon$.

For any $z \in G$, we have

$$\varrho(z^*, z) = |\xi(z^*) - \xi(z)| = \xi(z^*) - \xi(z) \geq 0. \tag{7.75}$$

On the other hand

$$\begin{aligned} 0 &= \lim_{n \to \infty} \varrho(z^*, z_n) = \xi(z^*) - \lim_{n \to \infty} \xi(z_n) = \xi(z^*) - \xi(\overline{z}) \quad \text{(see (7.75))} \\ \Rightarrow\ & \xi(z^*) = \xi(\overline{z}) = \max\{u(z) + (\mathbb{R}^*, y - x)_{\mathbb{R}^N} : z = (x, y) \in G\}. \end{aligned} \tag{7.76}$$

But ξ is strictly monotone because u is strictly concave (see hypothesis H_2). So from (7.76) it follows that

$$z^* = \overline{z}$$

which contradicts the fact that $\|z^* - \overline{z}\| \geq \varepsilon$. □

Now we can prove a weak turnpike theorem for good sequences. So we continue to have $z^* = (x^*, y^*)$ a turnpike and $\{z_n^m\}_{n=1}^m$, $m \geq 1$ a good sequence starting at $y_0 \in \mathbb{R}_+^N$. By $r_m(\varepsilon)$ we designate the number of the periods $k \in \{1, \ldots, m\}$ for which we have $\|z_k^m - z^*\| \geq \varepsilon$ (as in Section 7.2 on $\mathbb{R}^N \times \mathbb{R}^N$ we consider the l^1-norm, namely $\|z\| = \sum_{k=1}^{2N} |z_k|$).

DEFINITION 7.3.14 A vector $y_0 \in \mathbb{R}_+^N$ is said to be *sufficient* if a strictly positive vector can be produced in a finite number of steps starting from $y_0 \in \mathbb{R}_+^N$; that is, we can find a finite program $\{z_n = (x_{n-1}, y_n)\}_{n=1}^m$ starting from y_0 such that $y_m \gg 0$.

THEOREM 7.3.15 *If $z^* = (x^*, y^*) \in \mathbb{R}^N \times \mathbb{R}^N$ is a turnpike and $\{z_n\}_{n=1}^m$, $m \geq 1$, is a good sequence starting at a sufficient vector $y_0 \in \mathbb{R}^N$, then $\{r_m(\varepsilon)\}_{m\geq 1}$ is bounded.*

PROOF: Using Lemmata 7.3.12, and 7.3.13, we have

$$\begin{aligned} & r_m(\varepsilon)\lambda(\varepsilon) \leq \sum_{n=1}^m \varrho(z^*, z_n^m) \leq \beta, \qquad \text{for all } m \geq 1, \\ \Rightarrow\ & r_m(\varepsilon) \leq \frac{\beta}{\lambda(\varepsilon)} \qquad \text{for all } m \geq 1, \\ \Rightarrow\ & \{r_m\}_{n=1}^m \text{ is bounded.} \end{aligned}$$

□

However, our goal is to have this theorem for a sequence $\{z_n^m\}_{n=1}^m$, $m \geq 1$, where the finite program is optimal $\{z_n^m\}_{n=1}^m$, $m \geq 1$ for the finite horizon economy. So we need to check that the sequence $\{z_n^m\}_{n=1}^m$, $m \geq 1$, of optimal programs forms a good sequence. For this in turn it is enough to construct an infinite program $\{\overline{z}\}_{n\geq 1}$ starting at $y_0 \in \mathbb{R}_+^N$ for which we have

$$\sum_{n=1}^m \left[u(z^*) - u(\overline{z}_n)\right] \leq \beta_0 \qquad \text{for some } \beta_0 > 0, \text{ all } n \geq 1. \tag{7.77}$$

Indeed, because of the optimality of $\{z_n^m\}_{n=1}^m$, we have

$$\begin{aligned} & \sum_{n=1}^m \left[u(z^*) - u(z_n^m)\right] \leq \sum_{n=1}^m \left[u(z^*) - u(\overline{z}_n)\right] \qquad \text{for all } m \geq 1, \\ \Rightarrow\ & \sum_{n=1}^m \left[u(z^*) - u(z_n^m)\right] \leq \beta_0 \qquad \text{for all } m \geq 1, \\ \Rightarrow\ & \{z_n^m\}_{n=1}^m,\ m \geq 1, \text{ is a good sequence (see Definition 7.3.10).} \end{aligned}$$

An infinite program $\{\overline{z}_n\}_{n\geq 1}$ satisfying (7.77) is of course a good program (see Definition 7.2.31). In other words, it cannot be infinitely worse than the turnpike.

LEMMA 7.3.16 *If $y_0 \in \mathbb{R}_+^N$ is sufficient (see Definition 7.3.14), then a good program exists with initial vector $y_0 \in \mathbb{R}_+^N$.*

PROOF: Without any loss of generality, we may assume that $y_0 \gg 0$. Indeed, because y_0 is sufficient, according to Definition 7.3.14 with a finite program, we can reach a vector $y_m \gg 0$. Then we continue as described below.

So $y_0 \gg 0$. Let $\widehat{y} \in \mathbb{R}_+^N$ be the output of the expansible stock $\widehat{x} \in \mathbb{R}_+^N$ (see hypothesis H_3(ii)). We may assume that $\widehat{y} \ll y_0$. Because $(0,0) \in G$ and G is convex (see hypothesis H_1), we have that $(\mu\widehat{x}, \mu\widehat{y}) \in G$ and of course $\mu\widehat{x} \ll \mu\widehat{y}$ (i.e., $\mu\widehat{x}$ remains expansible).

Let $t \in (0,1)$ and define

$$\widetilde{z}_n = t^n \widehat{z} + (1-t^n) z^*, \quad \text{where } \widehat{z} = (\widehat{x}, \widehat{y}),\ n \geq 1.$$

We show that $\{\widetilde{z}_n\}_{n\geq 1}$ is a good sequence starting at y_0. Because G is convex, we see that $\widetilde{z}_n \in G$ for all $n \geq 1$. To establish that $\{\widetilde{z}_n\}_{n\geq 1}$ is a program we need to show that

$$\widetilde{x}_0 \leq y_0 \quad \text{and} \quad \widetilde{x}_n \leq \widetilde{y}_n \qquad \text{for all } n \geq 1, \tag{7.78}$$

(see Definition 7.3.3 and Remark 7.3.4). Note that $\widetilde{x}_0 = \widehat{x} \ll \widehat{y} \leq y_0$. So we have verified the first inequality in (7.78).

To check the second inequality in (7.78), by direct substitution we have

$$\widetilde{y}_n - \widetilde{x}_n = t^n \Big(\widehat{y} - \big(t\widehat{x} + (1-t)x^*\big)\Big) + (1-t^n)(y^* - x^*). \tag{7.79}$$

Recall that $x^* \leq y^*$. Also because $\widehat{x} \ll \widehat{y}$ (see hypothesis H_1(ii)), by choosing t close to 1, we see that $t\widehat{x} + (1-t)x^* \leq \widehat{y}$. Therefore from (7.79), we have $\widetilde{x}_n \leq \widetilde{y}_n$ for all $n \geq 1$. So we have also verified the second inequality in (7.78). This means that $\{\widetilde{z}_n\}_{n\geq 1}$ is a program starting at y_0.

Because u is concave (see hypothesis H_2), we have

$$\begin{aligned}
& t^n u(\widehat{z}) + (1-t^n) u(z^*) \leq u(\widetilde{z}_n) && \text{for all } n \geq 1, \\
\Rightarrow\ & u(z^*) - u(\widetilde{z}_n) \leq t^n \big(u(z^*) - u(\widehat{z})\big) && \text{for all } n \geq 1, \\
\Rightarrow\ & \sum_{n=1}^{m} \big[u(z^*) - u(\widetilde{z}_n)\big] \leq \beta_0 \quad \text{for some } \beta_0 > 0, \text{ all } n \geq 1, \\
\Rightarrow\ & \{\widetilde{z}_n\}_{n\geq 1} \text{ is a good program starting at } y_0 \in \mathbb{R}^N_+ .
\end{aligned}$$

□

Having this lemma, we can state the weak turnpike theorem for finite programs.

THEOREM 7.3.17 *If $\{z^m_n\}^m_{n=1}$, $m \geq 1$, is a sequence of optimal finite programs starting at $y_0 \in \mathbb{R}^N_+$, then for every $\varepsilon > 0$, the sequence $\{r_m(\varepsilon)\}_{m\geq 1}$ is bounded.*

REMARK 7.3.18 This result says that the number of time periods when the optimal program differs from the turnpike by at least $\varepsilon > 0$ is bounded by a constant $\widehat{\beta}(\varepsilon) > 0$ that is independent of $m \geq 1$. So the optimal program stays near the turnpike for all time except possibly for at most $\widehat{\beta}(\varepsilon)$ periods. In particular we have $\big((r_m(\varepsilon))/m\big) \longrightarrow 0$ as $m \longrightarrow \infty$. So the proportion of time when the optimal program deviates from the turnpike goes to zero as the planning horizon increases.

In fact we can extend the above weak turnpike theorem to infinite programs.

THEOREM 7.3.19 *If $\{z_n\}_{n\geq 1}$ is a good program, then $z_n \longrightarrow z^*$ as $n \to \infty$.*

PROOF: As in the proof of Lemma 7.3.12, we can show that

$$\sum_{n=1}^{m} \varrho(z^*, z_n) \leq \sum_{n=1}^{m} \big(u(z^*) - u(z_n)\big) + \eta, \quad \text{for some } \eta > 0, \text{ all } m \geq 1. \tag{7.80}$$

From (7.80) and because $\{z_n\}_{n\geq 1}$ is a good program, we infer that

$$\varrho(z^*, z_n) \longrightarrow 0 \quad \text{as } n \to \infty.$$

Suppose that for some subsequence $\{n_k\}$ of $\{n\}$, we have

$$\|z_{n_k} - z^*\| \geq \varepsilon \qquad \text{for all } k \geq 1.$$

Then from Lemma 7.3.13, we have

$$0 < \lambda(\varepsilon) \leq \varrho(z^*, z_{n_k}), \qquad \text{for all } k \geq 1,$$

a contradiction. Therefore we conclude that $z_n \longrightarrow z^*$ as $n \to \infty$. □

REMARK 7.3.20 So according to this theorem every good program, in particular every weakly maximal program, is asymptotic to the turnpike.

THEOREM 7.3.21 *If $y_0 \in \mathbb{R}_+^N$ is a sufficient vector, $\{z_n^m\}_{n \geq 1}$, $m \geq 1$, is a sequence of optimal finite programs starting at y_0, and the turnpike $z^* \in \operatorname{int} G$, then for any $\varepsilon > 0$, we can find $\vartheta = \vartheta(\varepsilon) > 0$ (independent of $m \geq 1$) such that for any $m > 2\vartheta$, we can find a finite program $\{\widetilde{z}_n^m\}_{n=1}^m$ starting at y_0 for which we have*

(a) $\widetilde{z}_n^m = z^*$ *for all* $n \in \{\vartheta, \ldots, m\text{-}\vartheta\}$.

(b) $\sum_{n=1}^{m} \big(u(z_n^m) - u(\widetilde{z}_n^m)\big) < \varepsilon$.

PROOF: Because $z^* \in \operatorname{int} G$, we can find $r > 0$ such that $B_r(z^*) \subseteq G$.

Let $\eta \in (0, r)$. By virtue of Theorem 7.3.17, we can find $\widehat{\beta} = \widehat{\beta}(\eta) > 0$ such that $m \geq 1$ there are at most $(\widehat{\beta} - 1)$ time periods $n \geq 1$ for which we have

$$\|z_n^m - z^*\| \geq \eta.$$

Fix $m > 2\widehat{\beta}$. Among the first $\widehat{\beta}$ periods $\{1, 2, \ldots, \widehat{\beta}\}$ and among the last $\widehat{\beta}$ periods $\{m - \widehat{\beta} + 1, \ldots, m\}$, there is at least one time period $k = k(m)$ and $i = i(m)$, respectively, such that

$$\|z_k^m - z^*\| < \eta \quad \text{and} \quad \|z_i^m - z^*\| < \eta.$$

We consider the following finite program:

$$\sigma_m = \big\{z_1^m, \ldots, z_{k-1}^m, (x_{k-1}^m, y^*), (x^*, y^*), \ldots, (x^*, y^*), (x^*, y_i^m), z_{i+1}^m, \ldots, z_m^m\big\}. \tag{7.81}$$

Note that in this finite sequence σ_m of length m, the initial and last parts (with lengths $k-1$ and $m-i-1$, resp.) coincide with the optimal program. The middle part coincides with the turnpike. The turnpike is linked with the other two parts using the pairs (x_{k-1}^m, y^*) and (x^*, y_i^m). It is straightforward to check that σ_m is indeed a finite program starting at y_0.

So for every $m > 2\widehat{\beta}$ we have a program $\sigma_m = \{\widetilde{z}_n^m\}_{n=1}^m$ as above. Then

$$\begin{aligned} \chi &= \sum_{n=1}^{m} \big(u(\widetilde{z}_n^m) - u(z_n^m)\big) \\ &= \sum_{n=k}^{m} \big(u(\widetilde{z}_n^m) - u(z_n^m)\big) \quad \text{(see (7.81))} \\ &= \sum_{n=k}^{i} \big(u(z^*) - u(z_n^m)\big) + \big(u(x_{k-1}^m, y^*) - u(z^*)\big) + \big(u(x^*, y_i^m) - u(z^*)\big), \end{aligned} \tag{7.82}$$

(see (7.81)).

Because $\{z_n^m\}_{n=1}^m$ is optimal, we have $\chi \leq 0$. Moreover, if $p^* \in \mathbb{R}_+^N$ is the price system supporting the turnpike (see Proposition 7.3.9), we have

$$\begin{aligned}\mu(\eta) \leq \big(u(x_{k-1}^m, y^*) - u(z^*)\big) + \big(u(x^*, y_i^m) - u(z^*)\big) \\ - (p^*, y^* - y_i^m)_{\mathbb{R}^N} + (p^*, x^* - x_i^m)_{\mathbb{R}^N} \leq \chi \leq 0 \end{aligned} \tag{7.83}$$

with $\mu(\eta)$ independent of $m \geq 1$ and $\mu(\eta) \longrightarrow 0$ as $\eta \longrightarrow 0$. Choose $\eta > 0$ small enough so that $|\mu(\eta)| < \varepsilon$. Then set $\vartheta = \vartheta(\varepsilon) = \widehat{\beta}(\eta) + 1$. We see that both statements (a) and (b) of the theorem are satisfied. □

REMARK 7.3.22 Statement (a) of the theorem says that the program $\{\widetilde{z}_n^m\}_{n=1}^m$ coincides with the turnpike except perhaps in the initial and final stages. The length of these parts is bounded by a constant independent of $m \geq 1$. Moreover, according to statement (b), this new program $\{\widetilde{z}_n^m\}_{n=1}^m$ is ε-optimal.

LEMMA 7.3.23 *If $\{z_n^m\}_{n=1}^m$ is an optimal program, $\{z_n\}_{n=1}^m$ is any finite program, and $1 \leq k \leq i \leq m$, then*

$$\begin{aligned}\sum_{n=k}^{i} \varrho(z^*, z_n^m) \leq \sum_{n=1}^{k-1}\big(u(z_n^m) - u(z_n)\big) + \sum_{n=i+1}^{n}\big(u(z_n^m) - u(z_n)\big) \\ + \sum_{n=k}^{i}\big(u(z^*) - u(z_n)\big) + (p^*, x^* - x_{k-1}^m)_{\mathbb{R}^N} + (p^*, y^* - y_i^m)_{\mathbb{R}^N}.\end{aligned} \tag{7.84}$$

PROOF: From (7.69) (definition of the pseudometric ϱ), we have

$$\begin{aligned}\sum_{n=k}^{i} \varrho(z^*, z_n^m) &= \sum_{n=k}^{i}\big(u(z^*) - u(z_n^m)\big) - \sum_{n=k}^{i-1}(p^*, y_n^m - x_n^m)_{\mathbb{R}^N} \\ &\quad + (p^*, y^* - y_i^m)_{\mathbb{R}^N} - (p^*, x^* - x_{k-1}^m)_{\mathbb{R}^N} \\ &\leq \sum_{n=k}^{i}\big(u(z^*) - u(z_n^m)\big) + (p^*, y^* - y_i^m)_{\mathbb{R}^N} - (p^*, x^* - x_{k-1}^m)_{\mathbb{R}^N}.\end{aligned} \tag{7.85}$$

Because $\{z_n^m\}_{n=1}^m$ is optimal, we have

$$\begin{aligned}&\sum_{n=1}^{m} u(z_n) \leq \sum_{n=1}^{m} u(z_n^m) \\ \Rightarrow &\sum_{n=1}^{k-1}\big(u(z_n) - u(z_n^m)\big) + \sum_{n=k}^{i} u(z_n) + \sum_{n=i+1}^{m}\big(u(z_n) - u(z_n^m)\big) \leq \sum_{n=k}^{i} u(z_n^m).\end{aligned} \tag{7.86}$$

Using (7.86) in (7.85), we obtain (7.84). □

Using this lemma we can prove the following result, which is crucial in the proof of the strong turnpike theorem.

LEMMA 7.3.24 *If $y_0 \in \mathbb{R}_+^N$ is a sufficient vector, $\{z_n^m\}_{n=1}^m$ is an optimal finite program starting at y_0, and $\vartheta > 0$, then we find $l = l(\vartheta) \geq 1$ such that $\sum\limits_{n=l}^{m-l} \varrho(z^*, z_n^m) < \vartheta$.*

PROOF: Using (7.84) (see Lemma 7.3.23) with $\{z_n^m\}_{n=1}^m = \sigma_m$, as in the proof of Theorem 7.3.21, we have

$$\begin{aligned}\sum_{n=\widehat{\beta}}^{m-\widehat{\beta}} \varrho(z^*, z_n^m) &\leq \sum_{n=k}^{i} \varrho(z^*, z_n^m) \\ &\leq \big(u(z^*) - u(z_k)\big) + \big(u(z^*) - u(z_i)\big) + 2\eta\|p^*\| \\ &\leq \varphi(\eta),\end{aligned}$$

where $\varphi(\eta) \longrightarrow 0$ as $\eta \longrightarrow 0$ because u is continuous on G and $\|z^* - z_k\|$, $\|z^* - z_i\| < \eta$. Choose $\eta > 0$ small enough so that $\varphi(\eta) < \vartheta$. Then $l = \widehat{\beta}(\eta)$ will do the job. $\square$

This lemma leads to the following *strong turnpike theorem.*

THEOREM 7.3.25 *If $y_0 \in \mathbb{R}_+^N$ is sufficient, $\{z_n^m\}_{n=1}^m$, $m \geq 1$, is a sequence of optimal finite programs, and the turnpike $z^* \in \operatorname{int} G$, then for any $\varepsilon > 0$, we can find $l = l(\varepsilon) > 0$ (independent of $m \geq 1$) such that for any $m \geq 2l$, we have $\|z^* - z_n^m\| < \varepsilon$ for all $h \in \{l, \ldots, m-l\}$.*

PROOF: Given $\varepsilon > 0$, let $\vartheta = \lambda(\varepsilon)$ where $\lambda(\cdot)$ is as in Lemma 7.3.13. Let $l = l(\vartheta) \geq 1$ be as in Lemma 7.3.24. If $\|z^* - z_n^m\| \geq \varepsilon$ for some $n \in \{l, \ldots, m - l\}$, then $\varrho(z^*, z_n^m) \geq \lambda(\varepsilon) = \vartheta$ (see Lemma 7.3.13) and this contradicts Lemma 7.3.24. So we have proved the strong turnpike theorem for the pseudometric ϱ. Then arguing as in the proof of Lemma 7.3.13, we can have it for the norm in $\mathbb{R}^N \times \mathbb{R}^N$. $\square$

REMARK 7.3.26 This theorem says that an optimal finite program is in the vicinity of the turnpike for all time periods not less than l periods distant from the endpoints. So compared to Theorem 7.3.17, now we identify those time periods when the optimal program deviates from the turnpike.

7.4 Stochastic Growth Models

In the previous sections we studied static and dynamic deterministic economic models. In this section, we examine discrete-time economic growth models with uncertainty. The uncertainty is present both in the utility and in the technological constraints. First we deal with the discounted nonstationary model and we prove the existence of an optimal program, which we characterize with a system of supporting prices. In the second half of the section, we deal with the undiscounted, stationary growth model for which we show that it has a weakly maximal program.

So first we describe and study the discounted, nonstationary growth growth model. For this purpose let (Ω, Σ, μ) be a complete probability space. Then $\omega \in \Omega$ represents a possible state of the environment, Σ is the collection of all possible events, and μ is the probability distribution of these events. We have a discrete-time, infinite planning horizon $\mathbb{N}_0=\{0,1,2,\dots\}$ (i.e., the model is a discrete-time, infinite horizon model). To describe mathematically the uncertainty of the system, we consider an increasing sequence $\{\Sigma_n\}_{n\geq 1}$ of complete sub-σ-fields of Σ and assume that $\Sigma = V_{n\geq 1}\Sigma_n$. The sub-$\sigma$-field Σ_n of Σ, represents the information about the state of the environment available up until time $n \geq 0$. There are N commodities in the economy. So the commodity space is $\mathbb{R}^N$ (multisector stochastic model). Finally there is a constant discount factor $0 < \delta < 1$ (discounted model).

At time period $n \geq 1$, the technological possibilities of the economy are described by a multifunction $G : \Omega \longrightarrow 2^{\mathbb{R}^N_+ \times \mathbb{R}^N_+} \setminus \{\emptyset\}$ which has a graph belonging in $\Sigma_n \times \mathbf{B}(\mathbb{R}^N_+) \times \mathbf{B}(\mathbb{R}^N_+)$ with $\mathbf{B}(\mathbb{R}^N_+)$ being the Borel σ-field. The set $G_n(\omega)$ describes all possible transformations of capital stock at time $n \geq 1$, when the state of the environment is $\omega \in \Omega$. More precisely, if $(k,y) \in G_n(\omega)$, when the state of the environment is $\omega \in \Omega$, we can transform a capital input k at time $n-1$ into a capital output y at time n. The uncertainty in this production process is manifested by the fact that the graph of G_n is Σ_n-measurable. Note that the technology feasibility set varies with time and so the model is nonstationary.

At every stage $n \geq 1$, the utility (gain) achieved by operating a particular technological process $(x,y) \in G_n(\omega)$, is measured by a utility function $u_n : \Omega \times \mathbb{R}^N_+ \times \mathbb{R}^N_+ \longrightarrow \mathbb{R}$ which is assumed to be $\Sigma_n \times \mathbf{B}(\mathbb{R}^N_+) \times \mathbf{B}(\mathbb{R}^N_+)$-measurable. Again the uncertainty is expressed by the Σ_n-measurability of the function. Finally there is a constant discount factor $\delta \in (0,1)$.

A sequence $\widehat{k} = \{k_n\}_{n\geq 0}$ with $k_n \in L^\infty(\Sigma_n, \mathbb{R}^N)$ is a program (or path). We say that program $\widehat{k} = \{k_n\}_{n\geq 0}$ is feasible if $\big(k_n(\omega), k_{n+1}(\omega)\big) \in G_{n+1}(\omega)$ μ-a.e. on Ω, for all $n \geq 0$. We say that $\widehat{k} = \{k_n\}_{n\geq 0}$ is a feasible program starting at $\overline{x}_0 \in L^\infty(\Sigma_0, \mathbb{R}^N)$, if $k_0 = \overline{x}_0$. The set of all feasible programs starting at $\overline{x}_0$, is denoted by $F(\overline{x}_0)$. Evidently $F(\overline{x}_0) \subseteq \prod\limits_{n\geq 0} L^\infty(\Sigma_n, \mathbb{R}^N)$.

Given an initial capital stock $\overline{x}_0 \in L^\infty(\Sigma_0, \mathbb{R}^N)_+$ (i.e. $\overline{x}_0(\omega) \geq 0$ μ-a.e. on Ω), we want to find an element in $F(\overline{x}_0)$ (i.e., a feasible program starting at $\overline{x}_0$) that maximizes the intertemporal utility

$$U(\widehat{k}) = \sum_{n\geq 0} \delta^{n+1} J_{n+1}(k_n, k_{n+1}), \tag{7.87}$$

where $\widehat{k} = \{k_n\}_{n\geq 0}$ and

$$J_{n+1}(v,w) = \int_\Omega u_{n+1}\big(\omega, v(\omega), w(\omega)\big)d\mu \tag{7.88}$$

for all $(v,w) \in L^\infty(\Sigma_n, \mathbb{R}^N)_+ \times L^\infty(\Sigma_{n+1}, \mathbb{R}^N)_+$, $n \geq 0$.

The mathematical hypotheses on the technological possibilities multifunction G_n are similar to those of the deterministic discounted multisector growth model.

H$_1$: $G_n : \Omega \longrightarrow P_{fc}(\mathbb{R}^N_+ \times \mathbb{R}^N_+)$ is a multifunction such that

(i) $\mathrm{Gr}\, G_n \in \Sigma_n \times \mathbf{B}(\mathbb{R}^N_+) \times \mathbf{B}(\mathbb{R}^N_+)$ (i.e., the multifunction G_n is Σ_n-graph measurable.

(ii) For every $\omega \in \Omega$, $(0,0) \in G_n(\omega)$ and $(0,y) \in G_n(\omega)$ implies $y = 0$.

(iii) For every $\omega \in \Omega$, if $(k, y) \in G_n(\omega)$, $k \leq k'$, and $y' \leq y$, then $(k', y') \in G_n(\omega)$ (free disposability).

(iv) There exists $M > 0$ such that if $\omega \in \Omega$ and $(k, y) \in G_n(\omega)$ with $\|k\| > M$, then $\|y\| \leq \|k\|$.

The hypotheses on the instantaneous utility function, are the following.

H₂: $u_n : \Omega \times \mathbb{R}_+^N \times \mathbb{R}_+^N \longrightarrow \mathbb{R}$, $n \geq 1$ is a function such that

(i) $(\omega, k, y) \longrightarrow u_n(\omega, k, y)$ is $\Sigma_n \times \mathbf{B}(\mathbb{R}_+^N) \times \mathbf{B}(\mathbb{R}_+^N)$-measurable.

(ii) For every $\omega \in \Omega$ $(k, y) \longrightarrow u_n(\omega, k, y)$ is concave and upper semicontinuous.

(iii) For all $(\omega, k, y) \in \Omega \times \mathbb{R}_+^N \times \mathbb{R}_+^N$, we have

$$|u_n(\omega, k, y)| \leq \varphi_n(\omega, \|k\|, \|y\|)$$

with $\varphi_n : \Omega \times \mathbb{R}_+ \times \mathbb{R}_+ \longrightarrow \mathbb{R}_+$ a $\Sigma_n \times \mathbf{B}(\mathbb{R}_+) \times \mathbf{B}(\mathbb{R}_+)$-measurable function, $\varphi_n(\omega, \cdot, \cdot)$ is nondecreasing, and $\sup\limits_{n\geq 1} \|\varphi_n(\cdot, v, v)\|_{L^1(\Sigma_n)} < +\infty$ for all $v \in \mathbb{R}_+$.

By a price system we mean a sequence $\widehat{p} = \{p_n\}_{n\geq 0} \subseteq L^1(\Sigma_n, \mathbb{R}^N)_+$. Normally prices belong to the dual of the space of production vectors. However, $L^\infty(\Sigma_n, \mathbb{R}^N)^*$ is too big and has no satisfactory economic interpretation. For this reason we limit ourselves to $L^1(\Sigma_n, \mathbb{R}^N) \subseteq L^\infty(\Sigma_n, \mathbb{R}^N)^*$. The precise description of $L^1(\Sigma_n, \mathbb{R}^N)$ as a subspace of $L^\infty(\Sigma_n, \mathbb{R}^N)^*$ is given by the Yosida–Hewitt theorem. For easy reference we state it here. First a definition.

DEFINITION 7.4.1 Let (Ω, Σ, μ) be a σ-finite measure space. A functional $u \in L^\infty(\Omega, \mathbb{R}^N)^*$ is said to be *absolutely continuous* if $u(g) = \int_\Omega \big(g(\omega), f(\omega)\big)_{\mathbb{R}^N} d\mu$ for some $f \in L^1(\Omega, \mathbb{R}^N)$ and all $g \in L^\infty(\Omega, \mathbb{R}^N)$. A functional $u \in L^\infty(\Omega, \mathbb{R}^N)^*$ is said to be *singular* if there exists a decreasing sequence $\{C_n\}_{n\geq 1} \subseteq \Sigma$ such that $\mu(C_n \cap C) \longrightarrow 0$ for every $C \in \Sigma$ with $\mu(C) < +\infty$ and u is supported by C_n; that is, $u(g) = 0$ for each $g \in L^\infty(\Omega, \mathbb{R}^N)$ which is identically zero on one of the C_ns (i.e., $u(g) = u(\chi_{C_n} g)$, $n \geq 1$, $g \in L^\infty(\Omega, \mathbb{R}^N)$).

REMARK 7.4.2 Absolutely continuous elements in $L^\infty(\Omega, \mathbb{R}^N)^*$ are identified with $L^1(\Omega, \mathbb{R}^N)$.

Using the notions of absolutely continuous and singular elements of $L^\infty(\Omega, \mathbb{R}^N)^*$, we can have a complete description of the dual space $L^\infty(\Omega, \mathbb{R}^N)^*$. The result is known as the *Yosida–Hewitt theorem* and for a proof of it, we refer to Levin [377].

THEOREM 7.4.3 *If (Ω, Σ, μ) is a σ-finite measure space and $u \in L^\infty(\Omega, \mathbb{R}^N)^*$, then u can be uniquely written as $u = u_a + u_s$, where u_a is absolutely continuous and u_s is singular. If $u \geq 0$, then $u_a, u_s \geq 0$. Moreover, we have*

$$\|u\| = \|u_a\| + \|u_s\|.$$

DEFINITION 7.4.4 A feasible program $\widehat{k}^* = \{k_n^*\}_{n\geq 0} \in F(\overline{x}_0)$ is said to be *optimal* if

$$U(\widehat{k}) \leq U(\widehat{k}^*) \qquad \text{for all } \widehat{k} \in F(\overline{x}_0).$$

To prove the existence of an optimal program for our model, we use the direct method of the calculus of variations. This requires that we prove the upper semicontinuity of the objective functional and the compactness of the constraint set in some useful topology and then apply the Weierstrass theorem.

THEOREM 7.4.5 *If hypotheses H_1, H_2 hold and $\overline{x}_0 \in L^\infty(\Sigma_0, \mathbb{R}^N)$, then there exists an optimal program in $F(\overline{x}_0)$.*

PROOF: Using hypotheses H_1 and arguing as in Lemma 7.2.5, we can have that

$$\|k_n\|_\infty \leq M_1 = \max\left\{\|\overline{x}_0\|_\infty, M\right\} \qquad \text{for all } \widehat{k} = \{k_n\}_{n\geq 0} \in F(\overline{x}_0).$$

Let $D_n = \left\{h \in L^1(\Sigma_n, \mathbb{R}^N) : \|h\|_\infty \leq M_1\right\}$, $n \geq 1$ (recall $L^\infty(\Sigma_n, \mathbb{R}^N) \subseteq L^1(\Sigma_n, \mathbb{R}^N)$). From the Eberlein–Smulian theorem, we infer that $D_n \subseteq L^1(\Sigma_n, \mathbb{R}^N)$ is weakly sequentially compact. Therefore $\prod\limits_{n\geq 1} D_n$ is weakly sequentially compact in $\prod\limits_{n\geq 1} L^1(\Sigma_n, \mathbb{R}^N)$. Recall that the weak topology on $\prod\limits_{n\geq 1} L^1(\Sigma_n, \mathbb{R}^N)$, is the product topology of the weak topologies on each factor space $L^1(\Sigma_n, \mathbb{R}^N)$, $n \geq 0$.

We define $\widehat{u}_n : \Omega \times \mathbb{R}_+^N \times \mathbb{R}_+^N \longrightarrow \overline{\mathbb{R}} = \mathbb{R} \cup \{-\infty\}$ by

$$\widehat{u}_n(\omega, k, y) = \begin{cases} u_n(\omega, k, y) & \text{if } (\omega, k, y) \in \mathrm{Gr}\, G_n \\ -\infty & \text{otherwise} \end{cases}.$$

It is easy to see that $\widehat{u}_n \in \Sigma_n \times \mathbf{B}(\mathbb{R}^N) \times \mathbf{B}(\mathbb{R}^N)$-measurable (see hypotheses H_1(i) and H_2(i)) and for every $\omega \in \Omega$, $\widehat{u}_n(\omega, \cdot, \cdot)$ is concave and upper semicontinuous. Moreover, we have

$$|\widehat{u}_n(\omega, k, y)| \leq \varphi_n(\omega, \|k\|, \|y\|) \qquad \text{for all } (\omega, k, y) \in \mathrm{Gr} G_n.$$

Let $\widehat{U} : \prod\limits_{n\geq 0} L^1(\Sigma_n, \mathbb{R}^N) \longrightarrow \overline{\mathbb{R}} = \mathbb{R} \cup \{-\infty\}$ be the expected intertemporal utility corresponding to $\widehat{u}_n$, $n \geq 1$; that is,

$$\widehat{U}(\widehat{k}) = \sum_{n\geq 0} \delta^{n+1} \widehat{J}_{n+1}(k_n, k_{n+1})$$

for all $\widehat{k} = \{k_n\}_{n\geq 0}$ and where

$$\widehat{J}_{n+1}(v, w) = \int_\Omega \widehat{u}_{n+1}\big(\omega, v(\omega), w(\omega)\big) d\mu$$

for all $(v, w) \in L^1(\Sigma_n, \mathbb{R}^N) \times L^1(\Sigma_{n+1}, \mathbb{R}^N)$.

We show that $\widehat{U}$ is weakly sequentially upper semicontinuous in $\prod\limits_{n\geq 0} L^1(\Sigma_n, \mathbb{R}^N)$. So suppose that $\widehat{k}^m \xrightarrow{w} \widehat{k}$ in $F(\overline{x}_0) \subseteq \prod\limits_{n\geq 0} L^1(\Sigma_n, \mathbb{R}^N)$, with $\widehat{k}^m = \{k_n^m\}_{n\geq 0}$ and

$\widehat{k} = \{k_n\}_{n\geq 0}$. We have $k_n^m \xrightarrow{w} k_n$ in $L^1(\Sigma_n, \mathbb{R}^N)$ as $m \longrightarrow \infty$, for all $n \geq 0$. Using Fatou's lemma, we have

$$\limsup_{m\to\infty} \widehat{J}_{n+1}(k_n^m, k_{n+1}^m) \leq \widehat{J}_{n+1}(k_n, k_{n+1}).$$

Given $\widehat{v} \in \prod_{n\geq 0} L^1(\Sigma_n, \mathbb{R}^N)$ and $K \geq 0$, we set

$$\widehat{U}_K(\widehat{v}) = \sum_{m=0}^{K} \delta^{n+1} \widehat{J}_{n+1}(v_n, v_{n+1}).$$

Using this notation we have

$$\begin{aligned}
\limsup_{m\to\infty} \widehat{U}_K(\widehat{k}^{\,m}) &= \limsup_{m\to\infty} \sum_{m=0}^{K} \delta^{n+1} \widehat{J}_{n+1}(k_n^m, k_{n+1}^m) \\
&\leq \sum_{m=0}^{K} \delta^{n+1} \limsup_{m\to\infty} \widehat{J}_{n+1}(k_n^m, k_{n+1}^m) \\
&\leq \sum_{m=0}^{K} \delta^{n+1} \widehat{J}_{n+1}(k_n, k_{n+1}) = \widehat{U}_K(\widehat{k}).
\end{aligned}$$

Because $\widehat{k} \in F(\overline{x}_0)$, we have

$$|\widehat{U}(\widehat{k}) - \widehat{U}_K(\widehat{k})| = \Big| \sum_{n=K+1}^{\infty} \delta^{n+1} \widehat{J}_{n+1}(k_n, k_{n+1}) \Big| \leq \sum_{n=K+1}^{\infty} \delta^{n+1}\beta,$$

where $\beta = \sup_{n\geq 1} \big\| \varphi_n(\cdot, M_1, M_1) \big\|_1$ (see hypothesis H_2(iii)). It follows that

$$\widehat{U}_K(\widehat{k}) \longrightarrow \widehat{U}(\widehat{k}) \quad \text{as } K \longrightarrow +\infty.$$

Then we can find a map $m \longrightarrow K(m)$ increasing to $+\infty$ such that

$$\begin{aligned}
\limsup_{m\to\infty} \widehat{U}_{K(m)}(\widehat{k}^{\,m}) &\leq \limsup_{K\to+\infty}\big(\limsup_{m\to\infty} \widehat{U}_K(\widehat{k}^{\,m})\big) \\
&\leq \limsup_{K\to+\infty} \widehat{U}_K(\widehat{k}) = \widehat{U}(\widehat{k}).
\end{aligned} \tag{7.89}$$

So we have

$$\begin{aligned}
\widehat{U}(\widehat{k}^{\,m}) - \widehat{U}(\widehat{k}) &= \big[\widehat{U}(\widehat{k}^{\,m}) - \widehat{U}_{K(m)}(\widehat{k}^{\,m})\big] + \big[\widehat{U}_{K(m)}(\widehat{k}^{\,m}) - \widehat{U}(\widehat{k})\big] \\
&\leq \sum_{n=K(m)+1}^{\infty} \delta^{n+1}\beta + \big[\widehat{U}_{K(m)}(\widehat{k}^{\,m}) - \widehat{U}(\widehat{k})\big]
\end{aligned}$$

$$\begin{aligned}
\Rightarrow \ \limsup_{m\to\infty}\big(\widehat{U}(\widehat{k}^{\,m}) - \widehat{U}(\widehat{k})\big) &\leq \limsup_{m\to\infty} \sum_{n=K(m)+1}^{\infty} \delta^{n+1}\beta \\
&\quad + \limsup_{m\to\infty} \big[\widehat{U}_{K(m)}(\widehat{k}^{\,m}) - \widehat{U}(\widehat{k})\big] \leq 0 \quad \text{(see (7.89))}
\end{aligned}$$

$$\Rightarrow \ \limsup_{m\to\infty} \widehat{U}(\widehat{k}^{\,m}) \leq \widehat{U}(\widehat{k}).$$

This proves the weak sequential upper semicontinuity of $\widehat{U}$ on $\prod_{n\geq 0} L^1(\Sigma_n, \mathbb{R}^N)$.

So by the Weierstrass theorem, we can find $\widehat{k}^* \in F(\overline{x}_0)$ such that

$$\widehat{U}(\widehat{k}^*) = U(\widehat{k}^*) = \sup\left[\widehat{U}(\widehat{k}) : \widehat{k} \in F(\overline{x}_0)\right].$$

□

Next we characterize an optimal program by generating a price system that supports it. More precisely, we prove that there exists a price system such that:

(a) At every time period, we have minimization of the cost among programs producing no less future value.

(b) At every time period, we have maximization of the total utility, defined as the sum of the instantaneous utility and the net profit resulting from operating a particular technological process at that period. The net profit is calculated as the value of the produced output minus the cost of the used input.

(c) The expected value of the optimal program goes to zero as the planning horizon expands to $+\infty$ (transversality condition).

Recall that the deterministic variant of such a result was given in Theorem 7.2.13.

In our effort to produce a price characterization of the optimal program, we need some quantities, which we introduce next.

So suppose that $f \in L^\infty(\Sigma_n, \mathbb{R}^N)$, $n \geq 0$. We define

$$\begin{aligned} D_n(f) = \big\{\widehat{h} = \{h_m\}_{m\geq n} : h_n = f \text{ and } \big(h_m(\omega), h_{m+1}(\omega)\big) \\ \in G_{m+1}(\omega)\ \mu\text{-a.e. on } \Omega\big\} \\ \text{and} \quad V_n(f) = \sup\Big[\sum_{m\geq n} \delta^{m+1} J_{m+1}(h_m, h_{m+1}) : \widehat{h} \in D_n(f)\Big]. \end{aligned}$$

Clearly V_n is the value (Bellman) function of the sequential optimization problem and as it is well-known it satisfies the dynamic programming functional equation, namely

$$\begin{aligned} V_n(f) = \sup\big[\delta^{m+1} J_{m+1}(f, g) + V_{n+1}(g) : g \in L^\infty(\Sigma_{n+1}, \mathbb{R}^N), \\ \big(f(\omega), g(\omega)\big) \in G_{n+1}(\omega) \quad \mu\text{-a.e. on } \Omega\big]. \end{aligned} \tag{7.90}$$

If $k^* = \{k_n^*\}_{n\geq 0} \in F(\overline{x}_0)$ is an optimal program, then

$$V_n(k_n^*) = \delta^{n+1} J_{n+1}(k_{n-1}^*, k_{n+1}^*) + V_{n+1}(k_{n+1}^*). \tag{7.91}$$

Also in what follows by S_{n+1} we denote the subset of $L^\infty(\Sigma_n, \mathbb{R}^N) \times L^\infty(\Sigma_{n+1}, \mathbb{R}^N)$ consisting of pairs of functions which pointwise correspond to a technological feasible production process, namely

$$\begin{aligned} S_{n+1} = \big\{(f, g) \in L^\infty(\Sigma_n, \mathbb{R}^N) \times L^\infty(\Sigma_{n+1}, \mathbb{R}^N) : \\ \big(f(\omega), g(\omega)\big) \in G_{n+1}(\omega)\ \mu\text{-a.e. on } \Omega\big\}. \end{aligned}$$

Note that hypotheses H_2 imply that the value function V_n is concave (i.e., $-V_n$ is convex). Then for every $z \in L^\infty(\Sigma_n, \mathbb{R}^N)$, we can define the subdifferential of V_n by

$$\partial V_n(z) = \big\{p \in L^\infty(\Sigma_n, \mathbb{R}^N)^* : V_n(y) - V_n(z) \le p_n(y - z) \\ \text{for all } \ y \in L^\infty(\Sigma_n, \mathbb{R}^N)\big\}$$

(see Definition 1.2.28).

To generate the desired supporting price system, we need to strengthen hypotheses H_2.

H_3: $u_n : \Omega \times \mathbb{R}^N_+ \times \mathbb{R}^N_+ \longrightarrow \mathbb{R}$ is a function such that

(i) For all $(k, y) \in \mathbb{R}^N_+ \times \mathbb{R}^N_+$, $\omega \longrightarrow u_n(\omega, k, y)$ is Σ_n-measurable.

(ii) For every $\omega \in \Omega$, $(k, y) \longrightarrow u_n(\omega, k, y)$ is continuous concave.

(iii) For every $(\omega, y) \in \Omega \times \mathbb{R}^N_+$, $k \longrightarrow u_n(\omega, k, y)$ is nondecreasing and for every $(\omega, k) \in \Omega \times \mathbb{R}^N_+$ $y \longrightarrow u_n(\omega, k, y)$ is nonincreasing.

(iv) For every $(\omega, k, y) \in \Omega \times \mathbb{R}^N_+ \times \mathbb{R}^N_+$, we have

$$|u_n(\omega, k, y)| \le \varphi_n\big(\omega, \|k\|, \|y\|\big),$$

with $\varphi_n : \Sigma_n \times \mathbf{B}(\mathbb{R}_+) \times \mathbf{B}(\mathbb{R}_+)$-measurable, for every $\omega \in \Omega$ $\varphi_n(\omega, \cdot, \cdot)$ is nondecreasing and $\sup\limits_{n \ge 1} \|\varphi_n(\cdot, v, v)\|_1 < \infty$ for all $v \ge 0$.

We also need a hypothesis concerning the value function V_n.

H_4: For every $n \ge 0$, V_n is continuous at some point in $L^\infty(\Sigma_n, \mathbb{R}^N)$ and $\partial V_0(\overline{x}_0) \ne \emptyset$.

REMARK 7.4.6 From Theorem 1.2.3, we know that the first part of hypothesis H_4 is equivalent to saying that V_n is bounded below in a neighborhood of a point. Also from Theorem 1.2.34, we know that if V_0 is continuous at $\overline{x}_0 \in L^\infty(\Sigma_0, \mathbb{R}^N)$, then $\partial V_0(\overline{x}_0) \ne \emptyset$. Note that if the technology is rich enough to admit a feasible program $\widehat{v} = \{v_n\}_{n \ge 0}$ such that $\big(v_n(\omega), v_{n+1}(\omega)\big) + B_{\varepsilon_n+1} \subseteq G_{n+1}(\omega)$ μ-a.e. on Ω (interior program), then hypothesis H_4 is satisfied.

To produce the supporting price system, we need to use the Mackey topology on the Lebesgue space $L^\infty(\Sigma_n, \mathbb{R}^N)$. We have already encountered the Mackey topology in Chapter 6 (see Definition 6.3.29). For easy reference we recall this definition here in the context of the space $L^\infty(\Sigma_n, \mathbb{R}^N)$, where it is used in the sequel.

DEFINITION 7.4.7 The *Mackey topology* m_∞ on $L^\infty(\Sigma_n, \mathbb{R}^N)$ induced by the pair $\big(L^\infty(\Sigma_n, \mathbb{R}^N), L^1(\Sigma_n, \mathbb{R}^N)\big)$, is the topology of uniform convergence on the weakly compact and convex subsets of $L^1(\Sigma_n, \mathbb{R}^N)$. So, if by $\langle \cdot, \cdot \rangle$ we denote the duality brackets for the pair $\big(L^\infty(\Sigma_n, \mathbb{R}^N), L^1(\Sigma_n, \mathbb{R}^N)\big)$ (recall $L^\infty(\Sigma_n, \mathbb{R}^N) = L^1(\Sigma_n, \mathbb{R}^N)^*$), then $k_a \xrightarrow{m_\infty} k$ if and only if for every $W \subseteq L^1(\Sigma_n, \mathbb{R}^N)$ weakly compact and convex, we have $\sup\big\{|\langle k_a - k, p\rangle| : p \in W\big\} \longrightarrow 0$.

REMARK 7.4.8 Clearly the w^*-topology on $L^\infty(\Sigma_n, \mathbb{R}^N) = L^1(\Sigma_n, \mathbb{R}^N)^*$ is weaker than the Mackey topology. We know that the Mackey topology is the strongest locally convex topology τ on $L^\infty(\Sigma_n, \mathbb{R}^N)$ such that $\big(L^\infty(\Sigma_n, \mathbb{R}^N)_\tau\big)^* = L^1(\Sigma_n, \mathbb{R}^N))$.

Convergence in the Mackey topology is related to the convergence in measure, denoted by $\xrightarrow{\mu}$.

PROPOSITION 7.4.9 *If* $\{h_k, h\}_{k\geq 1} \subseteq L^\infty(\Sigma_n, \mathbb{R}^N)$, $\sup\limits_{k\geq 1} \|h_k\|_\infty = \eta < +\infty$ *and* $h_k \xrightarrow{\mu} h$ *as* $k \longrightarrow \infty$, *then* $h_k \xrightarrow{m_\infty} h$ *in* $L^\infty(\Sigma_n, \mathbb{R}^N)$ *as* $k \longrightarrow \infty$.

PROOF: Let $W \subseteq L^\infty(\Sigma_n, \mathbb{R}^N)$ be a weakly compact and convex set. According to Definition 7.4.7, we need to show that

$$\sup\big[\,|\langle h_k - h, w\rangle| : w \in W\big] \longrightarrow 0 \quad \text{as } k \longrightarrow \infty.$$

Replacing h_k by $h_k - h$ if necessary, we may assume that $h = 0$. Because $W \subseteq L^1(\Sigma_n, \mathbb{R}^N)$ is weakly compact, from the Dunford–Pettis theorem, we have that W is uniformly integrable.

Hence the set $\{v = \|g\|_\infty \|w\|_1 : \|g\|_\infty \leq \eta,\ w \in W\}$ is uniformly integrable too. So given $\varepsilon > 0$, we can find $\varphi \in L^1(\Sigma_n)_+$, $\varphi > 0$ such that for all $g \in L^\infty(\Sigma_n, \mathbb{R}^N)_+$ with $\|g\|_\infty \leq \eta$ and all $w \in W$, we have

$$\int_{\{\|g\|\|w\|>\varphi\}} \|g(\omega)\|\ \|w(\omega)\| d\mu \leq \varepsilon.$$

So to prove the proposition, it suffices to show that

$$\lim_{m\to\infty} \sup_{w\in W} \int_{\{\|h_m\|\|w\|\leq\varphi\}} \|h_m(\omega)\|\ \|w(\omega)\| d\mu = 0.$$

We can always assume that $|W|_1 = \sup\big\{\|w\|_1 : w \in W\big\} \leq 1$.

Note that

$$\{\varphi = 0\} = \bigcap_{\lambda>0} \{\varphi < \lambda\}$$

and because $\varphi(\omega) > 0$ μ-a.e. on Ω, we see that we can find $\vartheta > 0$ small enough such that

$$\int_{\{\varphi<\vartheta\}} \varphi(\omega) d\mu \leq \varepsilon. \tag{7.92}$$

Also we can find $\delta = \delta(\varepsilon) > 0$ such that, if $A \in \Sigma_n$ and $\mu(A) \leq \delta$, then

$$\int_A \varphi(\omega) d\mu \leq \varepsilon. \tag{7.93}$$

Let $\gamma \in \big(0, \min\{\varepsilon, \delta\}\big)$. Since by hypothesis $h_k \xrightarrow{\mu} h$, we can find $k_0 \geq 1$ such that for all $k \geq k_0$, we have

$$\mu\big(\{\varphi \geq \vartheta,\ \|h_k\| \geq \gamma\}\big) \leq \gamma. \tag{7.94}$$

Then, for all $k \geq k_0$ and all $w \in W$, we have

$$\begin{aligned}&\int_{\{\|h_k\|\|w\|\leq\varphi\}\cap\{\varphi\geq\vartheta\}\cap\{\|h_k\|\geq\gamma\}}\|h_k(\omega)\|\,\|w(\omega)\|\mathrm{d}\mu\\&\leq\int_{\{\varphi\geq\vartheta\}\cap\{\|h_k\|\geq\gamma\}}\|h_k(\omega)\|\,\|w(\omega)\|\mathrm{d}\mu\leq\varepsilon\quad\text{(see (7.93) and (7.94)).}\end{aligned}\tag{7.95}$$

Also because we have assumed that $|W|_1 \leq 1$, for all $k \geq k_0$ and all $w \in W$, we have

$$\int_{\{\|h_k\|<\gamma\}}\|h_k(\omega)\|\,\|w(\omega)\|\mathrm{d}\mu<\gamma.\tag{7.96}$$

Hence, for all $k \geq k_0$ and all $w \in W$, we have

$$\begin{aligned}&\int_{\|h_k\|\|w\|\leq\varphi\}}\|h_k(\omega)\|\,\|w(\omega)\|\mathrm{d}\mu\\&=\int_{\{\|h_k\|\|w\|\leq\varphi\}\cap\{\varphi\geq\vartheta\}\cap\{\|h_k\|\geq\gamma\}}\|h_k(\omega)\|\,\|w(\omega)\|\mathrm{d}\mu+\int_{\{\|h_k\|\|w\|<\varphi\}\cap[\{\varphi<\vartheta\}\cup\{\|h_k\|<\gamma\}]}\|h_k(\omega)\|\,\|w(\omega)\|\mathrm{d}\mu\\&\leq\ 3\varepsilon\quad\text{(see (7.92) and 7.95)).}\end{aligned}$$

Therefore we conclude that

$$\lim_{k\to\infty}\sup_{w\in W}\int_{\{\|h_k\|\|w\|\leq\varphi\}}\|h_k(\omega)\|\,\|w(\omega)\|\mathrm{d}\mu=0$$

and as we already indicated in the beginning of the proof, this implies that $h_k \xrightarrow{m_\infty} h = 0$ in $L^\infty(\Sigma_n, \mathbb{R}^N)$ as $k \longrightarrow \infty$. $\square$

Using this proposition, we can now produce a price system supporting an optimal program and fulfilling the requirements described earlier.

THEOREM 7.4.10 *If hypotheses H_1, H_3, H_4 hold, $\overline{x}_0 \in L^\infty(\Sigma_0, \mathbb{R}^N)$, $\overline{x}_0 \geq 0$, and $\widehat{k}^* \in F(\overline{x}_0)$ is an optimal program, then we can find a price system $\widehat{p} = \{p_n\}_{n\geq 1}$, $p_n \in L^1(\Sigma_n, \mathbb{R}^N)_+$, $n \geq 1$, such that*

(a) $V_n(f) - V_n(k_n^*) \leq \langle p_n, f - k_n^*\rangle$ *for all* $f \in L^\infty(\Sigma_n, \mathbb{R}^N)$ *(hereafter by* $\langle\cdot,\cdot\rangle$ *we denote the duality brackets for the pair* $\big(L^1(\Sigma_n, \mathbb{R}^N), L^\infty(\Sigma_n, \mathbb{R}^N)\ = L^1(\Sigma_n, \mathbb{R}^N)^*\big)$, $n \geq 1$*),*

(b) *For every* $(f, g) \in S_{n+1}$, *we have*

$$\begin{aligned}&\delta^{n+1}J_{n+1}(f,g)-\langle p_n,f\rangle+\langle p_{n+1},g\rangle\\&\leq\ \delta^{n+1}J_{n+1}(k^*,k_{n+1}^*)-\langle p_n,k_n^*\rangle+\langle p_{n+1},k_{n+1}^*\rangle\,;\end{aligned}$$

(c) $\lim_{n\to\infty}\langle p_n, k_n^*\rangle = 0$ (transversality condition).

PROOF: Let $\eta_1^n : L^\infty(\Sigma_n, \mathbb{R}^N) \times L^\infty(\Sigma_{n+1}, \mathbb{R}^N) \longrightarrow L^\infty(\Sigma_n, \mathbb{R}^N)$ and $\eta_2^n : L^\infty(\Sigma_n, \mathbb{R}^N) \times L^\infty(\Sigma_{n+1}, \mathbb{R}^N) \longrightarrow L^\infty(\Sigma_{n+1}, \mathbb{R}^N)$ be defined by

$$\eta_1^n(f,g)=f\quad\text{and}\quad\eta_2^n(f,g)=g$$

(i.e., they are the projections on the factors of the Cartesian product $L^\infty(\Sigma_n, \mathbb{R}^N) \times L^\infty(\Sigma_{n+1}, \mathbb{R}^N)$). Then $(\eta_1^n)^* : L^\infty(\Sigma_n, \mathbb{R}^N)^* \longrightarrow L^\infty(\Sigma_n, \mathbb{R}^N)^* \times L^\infty(\Sigma_n, \mathbb{R}^N)^*$ and $(\eta_2^n)^* : L^\infty(\Sigma_{n+1}, \mathbb{R}^N)^* \longrightarrow L^\infty(\Sigma_n, \mathbb{R}^N)^* \times L^\infty(\Sigma_{n+1}, \mathbb{R}^N)^*$ are defined by

$$(\eta_1^n)^*(v) = (v, 0) \text{ and } (\eta_1^n)^*(w) = (0, w)$$

(i.e., they are the embedding operators of the factor spaces in the Cartesian product $L^\infty(\Sigma_n, \mathbb{R}^N)^* \times L^\infty(\Sigma_{n+1}, \mathbb{R}^N)^*$). Then we define

$$\xi_1^n = V_n \circ \eta_1^n \quad \text{and} \quad \xi_2^n = \delta^{n+1} \widehat{J}_{n+1} + V_{n+1} \circ \eta_2^n + i_{S_n+1},$$

where

$$i_{S_n+1} = \begin{cases} 0 & \text{if } (f, g) \in S_{n+1} \\ -\infty & \text{otherwise} \end{cases}$$

(the indicator function of the set S_{n+1}).

If $(f, g) \in L^\infty(\Sigma_n, \mathbb{R}^N) \times L^\infty(\Sigma_{n+1}, \mathbb{R}^N)$, then from (7.90), we have

$$\xi_2^n(f, g) \le \xi_1^n(f, g). \tag{7.97}$$

Because by hypothesis $\widehat{k} = \{k_n^*\}_{n \ge 0}$ is an optimal program, from (7.91) we have

$$\xi_2^n(k_n^*, k_{n+1}^*) = \xi_1^n(k_n^*, k_{n+1}^*). \tag{7.98}$$

From (7.97) and (7.98) and the definition of the concave subdifferential, we deduce that

$$\partial \xi_1^n(k_n^*, k_{n+1}^*) \subseteq \partial \xi_2^n(k_n^*, k_{n+1}^*). \tag{7.99}$$

Because of hypothesis H_4, we can use Theorem 1.2.40 (the nonsmooth chain rule) and obtain

$$\partial \xi_1^n(k_n^*, k_{n+1}^*) = (\eta_1^n)^* \partial V_2(k_n^*). \tag{7.100}$$

Recall that int $L^\infty(\Sigma_n, \mathbb{R}^N)_+ \neq \emptyset$. These facts combined with hypothesis H_4 permit the use of Theorem 1.2.38, which implies that

$$\begin{aligned} \partial \xi_2^n(k_n^*, k_{n+1}^*) &= \partial(\delta^{n+1} J_{n+1} + V_{n+1} \circ \eta_2^n + i_{S_{n+1}})(k_n^*, k_{n+1}^*) \\ &= \delta^{n+1} \partial J_{n+1}(k_n^*, k_{n+1}^*) + (\eta_2^n)^* \partial V_{n+1}(k_n^*) + \partial i_{S_{n+1}}(k_n^*, k_{n+1}^*). \end{aligned} \tag{7.101}$$

From (7.99) through (7.101), we see that if $p_n \in \partial V_n(k_n)$, then we can find $(z_n, z_{n+1}) \in \partial \widehat{J}_{n+1}(k_n^*, k_{n+1}^*)$, $p_{n+1} \in \partial V_{n+1}(k_{n+1}^*)$ and $(y_n, y_{n+1}) \in \partial i_{S_{n+1}}(k_n^*, k_{n+1}^*)$ such that

$$\begin{aligned} &(p_n, 0) = (\delta^{n+1} z_n, \delta^{n+1} z_{n+1}) + (0, p_{n+1}) + (y_n, y_{n+1}), \\ \Rightarrow\ & p_n = \delta^{n+1} z_n + y_n \quad \text{and} \quad -p_{n+1} = \delta^{n+1} z_{n+1} + y_{n+1}. \end{aligned}$$

Let $p_{n+1}^a \in L^1(\Sigma_{n+1}, \mathbb{R}^N)$ be the absolutely continuous part of $p_{n+1} \in L^\infty(\Sigma_{n+1}, \mathbb{R}^N)^*$ (see Theorem 7.4.3). We show that $p_{n+1}^a \in \partial V_{n+1}(k_{n+1}^*)$. Because $p_{n+1} \in \partial V_{n+1}(k_{n+1}^*)$, we have

$$V_{n+1}(w) - V_{n+1}(k_{n+1}^*) \le p_{n+1}(w - k_{n+1}^*) \qquad \text{for all } w \in L^\infty(\Sigma_{n+1}, \mathbb{R}^N)^*. \tag{7.102}$$

According to Definition 7.4.1, there is a sequence $\{C_m\}_{m \ge 1} \subseteq \Sigma_{n+1}$ which is decreasing, $\mu(C_m) \longrightarrow 0$ and p_{n+1}^s, the singular part of p_{n+1}, is supported on this sequence. We set

$$w_m = \chi_{C_m^c} w + \chi_{C_m} k_{n+1}^* \in L^\infty(\Sigma_{n+1}, \mathbb{R}^N) \qquad \text{for all } m \ge 1. \tag{7.103}$$

Returning to (7.102) and replacing w by w_m given by (7.103), we obtain

$$\begin{aligned} & V_{n+1}(w_m) - V_{n+1}(k^*_{n+1}) \\ \leq \; & \langle p^a_{n+1}, w_m - k^*_{n+1}\rangle + p^s_{n+1}(w_m - k^*_{n+1}) \\ = \; & \langle p^a_{n+1}, w_m - k^*_{n+1}\rangle + p^s_{n+1}\big(\chi_{C_m}(w_m - k^*_{n+1})\big), \end{aligned}$$

(see (7.103) and Definition 7.4.1)

$$= \langle p^a_{n+1}, w_m - k^*_{n+1}\rangle . \tag{7.104}$$

Evidently $w_m \xrightarrow{\mu} w$ as $m \longrightarrow \infty$. Invoking Proposition 7.4.9, we have

$$p^s_{n+1}(w_m - k^*_{n+1}) \longrightarrow 0 \quad \text{as } m \longrightarrow \infty. \tag{7.105}$$

We also claim that

$$V_{n+1}(w) \leq \liminf_{m\to\infty} V_{n+1}(w_m). \tag{7.106}$$

To this end, given $\varepsilon > 0$ we can find $y \in D_{n+1}(w)$ such that

$$V_{n+1}(w) - \frac{\varepsilon}{2} \leq \sum_{i=n+1}^{\infty} \delta^{i+1} J_{i+1}(y_i, y_{i+1}). \tag{7.107}$$

We set

$$\begin{aligned} y^m_i &= \chi_{C^c_m} y_i + \chi_{C_m} k^*_i \in L^\infty(\Sigma_i, \mathbb{R}^N) \\ \text{and} \quad y^m_{i+1} &= \chi_{C^c_m} y_{i+1} + \chi_{C_m} k^*_{i+1} \in L^\infty(\Sigma_{i+1}, \mathbb{R}^N). \end{aligned}$$

Clearly $y^m_i \xrightarrow{\mu} y_i$ and $y^m_{i+1} \xrightarrow{\mu} y_{i+1}$ as $m \longrightarrow \infty$. Because of hypotheses H_3, we can find $m_0 \geq 1$ such that for all $m \geq m_0$, we have

$$\Big| \sum_{i=n+1}^{\infty} \delta^{i+1} J_{i+1}(y^m_i, y^m_{i+1}) - \sum_{i=n+1}^{\infty} \delta^{i+1} J_{i+1}(y_i, y_{i+1}) \Big| < \frac{\varepsilon}{2}. \tag{7.108}$$

Using (7.108) in (7.107), we obtain

$$V_{n+1}(w) - \varepsilon \leq \sum_{i=n+1}^{\infty} \delta^{i+1} J_{i+1}(y^m_i, y^m_{i+1}) \leq V_{n+1}(w_m) \qquad \text{for all } m \geq m_0,$$

which implies (7.106).

We return to (7.104), pass to the limit as $m \longrightarrow \infty$, and use (7.105) and (7.106) to obtain

$$\begin{aligned} & V_{n+1}(w) - V_{n+1}(k^*_{n+1}) \leq \langle p^a_{n+1}, w - k^*_{n+1}\rangle \qquad \text{for all } w \in L^\infty(\Sigma_{n+1}, \mathbb{R}^N), \\ \Rightarrow \; & p^a_{n+1} \in \partial V_{n+1}(k^*_{n+1}). \end{aligned}$$

Arguing in a similar fashion, we can show that

$$(z_n, z^a_{n+1}) \in \partial J_{n+1}(k^*_n, k^*_{n+1}) \quad \text{and} \quad (y_n, y^a_{n+1}) \in \partial i_{S_{n+1}}(k^*_n, k^*_{n+1}).$$

Also we have $-p^a_{n+1} = \delta^{n+1} z^a_{n+1} + y^a_{n+1}$. Hence we have so far: $p_n \in \partial V_n(k^*_n)$ and

$$(z_n, z^a_{n+1}) \in \partial J_{n+1}(k^*_n, k^*_{n+1}), \tag{7.109}$$

$$p_{n+1}^a \in \partial V_{n+1}(k_{n+1}^*) \quad \text{and} \quad (y_n, y_{n+1}^a) \in \partial i_{S_{n+1}}(k_n^*, k_{n+1}^*) \tag{7.110}$$

and also $p_n = \delta^n z_n + y_n$, $-p_{n+1}^a = \delta^{n+1} z_{n+1}^a + y_{n+1}^a$. From (7.109), we obtain

$$\begin{aligned}
&\delta^{n+1} J_{n+1}(f,g) - \delta^{n+1} J_{n+1}(k_n^*, k_{n+1}^*) \\
&\leq \delta^{n+1} z_n(f - k_n^*) + \left\langle \delta^{n+1} z_{n+1}^a, g - k_{n+1}^* \right\rangle, \\
\text{for all } & (f,g) \in L^\infty(\Sigma_n, \mathbb{R}^N) \times L^\infty(\Sigma_{n+1}, \mathbb{R}^N) \\
&\Rightarrow \delta^{n+1} J_{n+1}(f,g) - \delta^{n+1} z_n(f) - \left\langle \delta^{n+1} z_{n+1}^a, g \right\rangle \\
&\leq \delta^{n+1} J_{n+1}(k_n^*, k_{n+1}^*) - \delta^{n+1} z_n(k_n^*) - \left\langle \delta^{n+1} z_{n+1}^a, k_{n+1}^* \right\rangle \\
\text{for all } & (f,g) \in L^\infty(\Sigma_n, \mathbb{R}^N) \times L^\infty(\Sigma_{n+1}, \mathbb{R}^N).
\end{aligned} \tag{7.111}$$

Moreover, from the second inclusion in (7.110), we have

$$0 \leq y_n(f - k_n^*) + \langle y_{n+1}^a, g - k_{n+1}^* \rangle \qquad \text{for all } (f,g) \in S_{n+1}. \tag{7.112}$$

Adding (7.111) and (7.112) and recalling that $p_n = \delta^{n+1} z_n + y_n$, $-p_{n+1}^a = \delta^{n+1} z_{n+1}^a + y_{n+1}^a$, we obtain

$$\begin{aligned}
&\delta^{n+1} J_{n+1}(f,g) - p_n(f) + \langle p_{n+1}^a, g \rangle \\
&\leq \delta^{n+1} J_{n+1}(k_n^*, k_{n+1}^*) - p_n(k_n^*) + \langle p_{n+1}^*, k_{n+1}^* \rangle \qquad \text{for all } (f,g) \in S_{n+1}.
\end{aligned} \tag{7.113}$$

In fact in (7.113), we can replace p_n by its absolutely continuous part p_n^a (see Definition 7.4.1). Indeed, let $\{C_m'\}_{m \geq 1} \subseteq \Sigma_n$ be a decreasing sequence such that $\mu(C_m') \longrightarrow 0$ as $m \longrightarrow \infty$ and which supports the singular functional p_n^s. We set

$$\begin{aligned}
f_m &= \chi_{C_m'^c} f + \chi_{C_m'} k_n^* \in L^\infty(\Sigma_n, \mathbb{R}^N) \\
\text{and} \quad g_m &= \chi_{C_m'^c} g + \chi_{C_m'} k_{n+1}^* \in L^\infty(\Sigma_{n+1}, \mathbb{R}^N), \quad m \geq 1.
\end{aligned}$$

Evidently, we have $f_m \xrightarrow{\mu} f$ and so by Proposition 7.4.9, $f_m \xrightarrow{m_\infty} f$. Therefore

$$p_n(f_m - k_n^*) = \langle p_n^\alpha, f_m - k_n^* \rangle + p_n^s(f_m - k_n^*) \longrightarrow \langle p_n^\alpha, f - k_n^* \rangle \quad \text{as } m \longrightarrow \infty.$$

Also because of hypotheses H_3, we have

$$J_{n+1}(f_m, g_m) \longrightarrow J_{n+1}(f,g) \quad \text{as } m \longrightarrow \infty.$$

So, if in (7.113), for (f,g) we use $(f_m, g_m) \in S_{n+1}$ and then pass to the limit as $m \longrightarrow \infty$, we obtain

$$\begin{aligned}
&\delta^{n+1} J_{n+1}(f,g) - \langle p_n^a, f \rangle + \langle p_{n+1}^a, g \rangle \\
&\leq \delta^{n+1} J_{n+1}(k_n^*, k_{n+1}^*) - \langle p_n^a, k_n^* \rangle + \langle p_{n+1}^a, k_{n+1}^* \rangle \quad \text{for all } (f,g) \in S_{n+1}.
\end{aligned} \tag{7.114}$$

Moreover, as we did for p_{n+1}, we can show that $p_n^a \in \partial V_n(k_n^*)$. So inductively we can generate a sequence $\{p_n\}_{n \geq 1}$, $p_n \in L^1(\Sigma_n, \mathbb{R}^N)$ that satisfies statements (a) and (b) of the theorem. We show that $p_n \geq 0$. To this end, let $z_n = k_n^* + e$ with $e \in L^\infty(\Sigma_n, \mathbb{R}^N)$, $e \geq 0$. Because of the free disposability hypothesis (see hypothesis H_1(iii)), we have $(z_n, k_{n+1}^*) \in S_{n+1}$. Hence from (7.113), we have

$$\begin{aligned}
&\delta^{n+1} J_{n+1}(z_n, k_{n+1}^*) - \langle p_n, z_n \rangle + \langle p_{n+1}, k_{n+1}^* \rangle \\
&\leq \delta^{n+1} J_{n+1}(k_n^*, k_{n+1}^*) - \langle p_n^a, k_n^* \rangle + \langle p_{n+1}^a, k_{n+1}^* \rangle, \\
&\Rightarrow \delta^{n+1}\big(J_{n+1}(z_n, k_{n+1}^*) - J_{n+1}(k_n^*, k_{n+1}^*)\big) \leq \langle p_n, e \rangle, \\
&\Rightarrow 0 \leq \langle p_n, e \rangle \quad \text{(see hypothesis } H_3\text{(iii))}.
\end{aligned} \tag{7.115}$$

Because $e \in L^\infty(\Sigma_n, \mathbb{R}^N)_+$ was arbitrary from (7.115) it follows that $p_n \geq 0$ for all $n \geq 1$.

Finally if in statement (a) we set $f = 0$ (see hypothesis H_3(ii)), we have

$$0 \leq \langle p_n, k_n^* \rangle \leq V_n(k_n^*) - V_n(0). \tag{7.116}$$

Because $V_n(k_n^*) - V_n(0) \longrightarrow 0$ as $n \to \infty$, from (7.116) we conclude that

$$\lim_{n \to \infty} \langle p_n, k_n^* \rangle = 0 \quad \text{(the transversality condition)}.$$

$\square$

REMARK 7.4.11 In Theorem 7.4.10 above, statements (a), (b), and (c) have significant economic interpretations. First of all note that if $(f, g) \in S_{n+1}$, then $\langle p_n, f \rangle$ is the cost of input f at time period $n \geq 0$ and $\langle p_{n+1}, g \rangle$ represents the value of the resulting output g at time period $n + 1$. The quantity $\delta^{n+1} J_{n+1}(f, g)$ is the direct utility resulting from operating the technological process $(f, g) \in S_{n+1}$ at time period $n + 1$. Therefore, we can view the quantity $\delta^{n+1} J_{n+1}(f, g) - \langle p_n, f \rangle + \langle p_{n+1}, g \rangle$ as the total expected utility generated by the process $(f, g) \in S_{n+1}$. Having this in mind, we can interpret the statements in Theorem 7.4.10 as follows.

(1) Statement (b) says that along the optimal program $\widehat{k}^* = \{k_n^*\}_{n \geq 0}$ we have at every time period maximization of the total utility, among all other feasible processes.

(2) Statement (a) says that the optimal program minimizes the cost of the input among all programs producing no less future value.

(3) Statement (c) (the transversality condition) says that the expected value of the input (and of the output) goes to zero as $n \to \infty$.

(4) Borrowing a definition from the deterministic models (see Definition 7.2.8(b)), we can say that if $\widehat{p} = \{p_n\}_{n \geq 1}$ is the price system produced in Theorem 7.4.10, then $(\widehat{k}^*, \widehat{p})$ is competitive.

In the next theorem, we prove the converse of Theorem 7.4.10. Namely we show that if a feasible program $\widehat{k}^* = \{k_n^*\}_{n \geq 0}$ admits a price system $\widehat{p} = \{p_n\}_{n \geq 1}$ such that $(\widehat{k}^*, \widehat{p})$ is competitive, then $\widehat{k}^*$ is an optimal program (compare with Theorem 7.2.19 for the deterministic model).

THEOREM 7.4.12 *If hypotheses H_1, H_3 hold and the program-price pair $\big(\widehat{k}^* = \{k_n^*\}_{n \geq 0}, \widehat{p} = \{p_n\}_{n \geq 1}\big)$ is competitive, then $\widehat{k}^* \in F(\overline{x}_0)$ is optimal.*

PROOF: From the hypothesis that $(\widehat{k}^*, \widehat{p})$ is competitive, we have

$$\sum_{m=0}^{K} \delta^{m+1}\big(J_{m+1}(y_m, y_{m+1}) - J_{m+1}(\widehat{k}_m^*, \widehat{k}_{m+1}^*)\big) \le \big\langle p_{K+1}, \widehat{k}_{K+1}^* \big\rangle$$

for all $\widehat{y} = \{y_n\}_{n\ge 0} \in F(\overline{x}_0)$ (recall that $p_{K+1} \ge 0$, $y_{K+1} \ge 0$). Let $K \longrightarrow +\infty$. Using the tranversality condition, we obtain

$$\sum_{m=0}^{\infty} \delta^{m+1} J_{m+1}(y_m, y_{m+1}) \le \sum_{m=0}^{\infty} \delta^{m+1} J_{m+1}(\widehat{k}_m^*, \widehat{k}_{m+1}^*) \quad \text{for all } \widehat{y} \in F(\overline{x}_0),$$

$\Rightarrow$ $\widehat{k}^* \in F(\overline{x}_0)$ is optimal. ∎

Next we turn our attention to the undiscounted stochastic economy, using as performance criterion the notion of weak maximality (see Definition 7.2.26(b)). To treat this case, we need to restrict ourselves to the stationary model. For this reason we assume that there exists a map $\tau : (\Omega, \Sigma_n) \longrightarrow (\Omega, \Sigma_{n-1})$ which is one-to-one onto and both τ and τ^{-1} are measurable. Also we assume that the transformation τ is measure preserving; namely for every $n \ge 1$ and every $A \in \Sigma_{n-1}$, we have $\mu(A) = \mu(\tau^{-1}(A))$. Under these hypotheses the probability spaces (Ω, Σ_n, μ) and $(\Omega, \Sigma_{n-1}, \mu)$ are isomorphic with isomorphism the transformation τ, $\tau(\Sigma_n) = \Sigma_{n-1}$ and for every $B \in \Sigma_n$ we have $\mu(\tau(B)) = \mu(B)$. So we see that the probability space $(\Omega, \tau(\Sigma_n), \mu)$ is in fact equivalent to the probability space $(\Omega, \Sigma_{n-1}), \mu)$. Therefore we can think of τ as a time shift operator. Suppose that $f : \Omega \longrightarrow \mathbb{R}^N$ is a Σ_n-measurable function, then we can find $g : \Omega \longrightarrow \mathbb{R}^N$ is a Σ_{n-1}-measurable function such that $f = g \circ \tau$, hence $g = f \circ \tau^{-1}$. Therefore any Σ_n-measurable function $f : \Omega \longrightarrow \mathbb{R}^N$ can be written as $f = g \circ \tau^n$ with $g : \Omega \longrightarrow \mathbb{R}^N$ Σ_0-measurable. Moreover, because τ is measure preserving, for every $C \in \mathbb{R}^{\mathbf{N}}$, we have

$$\begin{aligned}\mu(f^{-1}(C)) = \mu((g \circ \tau^n)^{-1}(C)) &= \mu(\tau^{-1} \circ (g \circ \tau^{n-1})^{-1}(C)) \\ &= \mu((g \circ \tau^{n-1})^{-1}(C)) = \cdots = \mu g^{-1}(C).\end{aligned}$$

So f and g are essentially the same function, only f is shifted in a later time period. For reasons of convenience in the presentation, we replace the feasibility set $G_n(\omega)$ by

$$L_n(\omega, k) = \{y \in \mathbb{R}_+^N : (k, y) \in G_n(\omega)\}, \quad n \ge 1.$$

We see that $\operatorname{Gr} L_n(\omega, \cdot) = G_n(\omega)$. The set $L_n(\omega, k)$ represents the set of all possible outputs when the state of the environment is $\omega \in \Omega$ and the input is $k \in \mathbb{R}_+^N$. This production multifunction and the instantaneous utility function, are stationary, namely

$$L_n(\omega, k) = L_1(\tau^{n-1}(\omega), k) \quad \text{and} \quad u_n(\omega, k, y) = u_1(\tau^{n-1}(\omega), k, y).$$

Therefore all the data of the model are described in terms of the corresponding quantity at time $n = 1$, by using the time shift operator τ. Hence L_1 and u_1 are the prototype production multifunction and utility function, respectively, and in the sequel we drop the subscript and simply denote them by L and u.

Let us give the precise mathematical hypotheses on L and u.

H$_5$: $L:\Omega\times\mathbb{R}_+^N \longrightarrow 2^{\mathbb{R}_+^N}\setminus\{\emptyset\}$ is a multifunction such that

(i) $\mathrm{Gr}\, L=\big\{(\omega,k,y)\in\Omega\times\mathbb{R}_+^N\times\mathbb{R}_+^N : y\in L(\omega,k)\big\}\in\Sigma_1\times\mathbf{B}(\mathbb{R}_+^N)\times\mathbf{B}(\mathbb{R}_+^N)$.

(ii) For every $\omega\in\Omega$, $\mathrm{Gr}L(\omega,\cdot)$ is closed and convex in $\mathbb{R}_+^N\times\mathbb{R}_+^N$.

(iii) There exists $M>0$ such that for all $(\omega,k)\in\Omega\times\mathbb{R}_+^N$, we have

$$|L(\omega,k)|=\sup\big\{\|y\|:y\in L(\omega,k)\big\}\leq M.$$

(iv) For every $\omega\in\Omega$, if $y\in L(\omega,k)$, $k'\geq k$, and $y'\leq y$, with $\|k\|\leq M$, then $y'\in L(\omega,k')$.

(v) For every $\omega\in\Omega$, there exists $(k,y)\in\mathrm{Gr}\ L(\omega,\cdot)$ such that $k\ll y$.

REMARK 7.4.13 Comparing this set of hypotheses with H$_1$, we see that they are essentially the same. Only we have included hypothesis H$_5$(v) on the existence of an expansible capital stock.

H$_6$: $u:\Omega\times\mathbb{R}_+^N\times\mathbb{R}_+^N \longrightarrow \overline{\mathbb{R}}=\mathbb{R}\cup\{-\infty\}$ is a function such that

(i) $(\omega,k,y)\longrightarrow u(\omega,k,y)$ is $\Sigma_1\times\mathbf{B}(\mathbb{R}_+^N)\times\mathbf{B}(\mathbb{R}_+^N)$-measurable.

(ii) For every $\omega\in\Omega$, $(k,y)\longrightarrow u(\omega,k,y)$ is upper semicontinuous and concave.

(iii) For all $k,y\in\mathbb{R}_+^N$ with $\|k\|,\|y\|\leq M$, we have $u(\omega,k,y)\leq\varphi(\omega)$ μ-a.e. on Ω with $\varphi\in L^1(\Omega)$.

(iv) For all $(\omega,y)\in\Omega\times\mathbb{R}_+^N$, $k\longrightarrow u(\omega,k,y)$ is nondecreasing and for all $(\omega,k)\in\Omega\times\mathbb{R}_+^N$, $y\longrightarrow u(\omega,k,y)$ is nonincreasing.

By a program, we understand a sequence $\widehat{k}=\{k_n\}_{n\geq 0}$, $k_n\in L^\infty(\Sigma_n,\mathbb{R}^N)$. The program is *feasible*, if we have

$$k_{n+1}(\omega)\in L\big(\omega,k_n(\omega)\big)\quad \mu\text{-a.e. on }\Omega,\ n\geq 0.$$

There is an initial capital stock $\overline{x}_0\in L^\infty(\Sigma_0,\mathbb{R}^N)$ and the set of all feasible programs starting at $\overline{x}_0$ is denoted by $F(\overline{x}_0)$.

DEFINITION 7.4.14 A program $\widehat{k}^{\,*}=\{k_n^*\}_{n\geq 0}\in F(\overline{x}_0)$ is said to be *weakly maximal* if for every $\overline{y}=\{y_n\}_{n\geq 0}\in F(\overline{x}_0)$, we have

$$\liminf_{m\to\infty}\sum_{n=0}^{m}\big(J(y_n\circ\tau^{-n},y_{n-1}\circ\tau^{-n-1})-J(k_n^*\circ\tau^{-n},k_{n+1}^*\circ\tau^{-n-1})\big)\leq 0,$$

where $J(f,g)=\int_\Omega u\big(\omega,f(\omega),g(\omega)\big)\mathrm{d}\mu$ for all $f,g\in L^\infty(\Omega,\mathbb{R}^N)$.

REMARK 7.4.15 Note that due to the stationarity of the model, we have

$$\begin{aligned} J_{n+1}(y_n,y_{n+1}) &= J(y_n\circ\tau^{-n},y_{n-1}\circ\tau^{-n-1})\\ \text{and}\quad J_{n+1}(k_n^*,k_{n+1}^*) &= J(k_n^*\circ\tau^{-n},k_{n+1}^*\circ\tau^{-n-1}),\quad n\geq 1. \end{aligned}$$

Therefore Definition 7.4.14 is Definition 7.2.26(b) adapted to the present stochastic setting.

We introduce the set

$$C=\{f\in L^1(\Sigma_0,\mathbb{R}^N):\|f(\omega)\|\leq M\ \mu\text{-a.e. on }\Omega\}.$$

In the light of hypothesis H_5(iii), C is the set where we expect to locate any possible weakly maximal programs. Clearly this set is convex and weakly compact in $L^1(\Sigma_0,\mathbb{R}^N)$.

We introduce the multifunction $R:C\longrightarrow 2^C\setminus\{\emptyset\}$ defined by

$$R(k)=\{y\in L^1(\Sigma_0,\mathbb{R}^N):y\big(\tau(\omega)\big)\in L\big(\omega,k(\omega)\big)\ \mu\text{-a.e. on }\Omega\}.$$

In what follows we use C endowed with the relative weak $L^1(\Sigma_0,\mathbb{R}^N)$-topology.

PROPOSITION 7.4.16 *If hypotheses H_5 hold, then R is usc from C into $P_{kc}(C)$.*

PROOF: Because $C\subseteq L^1(\Sigma_0,\mathbb{R}^N)$ is weakly compact and the weak topology on C is metrizable, to prove the desired upper semicontinuity of the multifunction R, it is enough to show that Gr R is sequentially closed in $C\times C$ (see Proposition 6.1.10). To this end let $(k_n,y_n)\in \mathrm{Gr}R$, $n\geq 1$ and assume that

$$k_n\xrightarrow{w}k\quad\text{and}\quad y_n\xrightarrow{w}y\quad\text{in }L^1(\Sigma_0,\mathbb{R}^N)\quad\text{as }n\to\infty.$$

In what follows by $\langle\cdot,\cdot\rangle$ we denote the duality brackets for the pair $\big(L^1(\Sigma_1,\mathbb{R}^N),L^\infty(\Sigma_1,\mathbb{R}^N)\big)$. For every $A\in\Sigma_1$ and every $x^*\in\mathbb{R}^N$, we have

$$\begin{aligned}&\langle\chi_A x^*,y_n\circ\tau\rangle\leq\sigma\big(\chi_A x^*,R(k_n)\circ\tau\big)=\sup\big[\langle\chi_A x^*,z\circ\tau\rangle:z\in R(k_n)\big]\\ \Rightarrow&\int_A\Big(x^*,y_n\big(\tau(\omega)\big)\Big)_{\mathbb{R}^N}\mathrm{d}\mu\leq\sup\Big[\int_A\big(x^*,(z\circ\tau)(\omega)\big)_{\mathbb{R}^N}\mathrm{d}\mu:z\in R(k_n)\Big]\\ &=\int_A\sup\big[(x^*,v)_{\mathbb{R}^N}:v\in L\big(\omega,k_n(\omega)\big)\big]\mathrm{d}\mu,\end{aligned}$$

(see Theorem 6.4.16)

$$=\int_A\sigma\Big(x^*,L\big(\omega,k_n(\omega)\big)\Big)\mathrm{d}\mu.$$

From hypotheses H_5(ii) and (iii), it follows that $z\longrightarrow\sigma\big(x^*,L(\omega,z)\big)$ is concave and upper semicontinuous (see Proposition 6.1.15(c)). So via Mazur's lemma and Fatou's lemma, we have

$$\begin{aligned}&\limsup_{n\to\infty}\int_A\sigma\Big(x^*,L\big(\omega,k_n(\omega)\big)\Big)\mathrm{d}\mu\\ \leq&\int_A\sigma\Big(x^*,L\big(\omega,k(\omega)\big)\Big)\mathrm{d}\mu=\sigma\big(\chi_A x^*,R(k)\circ\tau\big),\\ \Rightarrow&\int_A\Big(x^*,y\big(\tau(\omega)\big)\Big)_{\mathbb{R}^N}\mathrm{d}\mu\leq\int_A\sigma\Big(x^*,L\big(\omega,k(\omega)\big)\Big)\mathrm{d}\mu.\end{aligned}\tag{7.117}$$

Because $A\in\Sigma_1$ and $x^*\in\mathbb{R}^N$ were arbitrary, from (7.117) it follows that

$$\left(x^*, y\left(\tau(\omega)\right)\right)_{\mathbb{R}^N} \le \sigma\left(x^*, L\left(\omega, k(\omega)\right)\right) \qquad \text{for } \mu\text{-almost all } \omega \in \Omega$$

and all $x \in \mathbb{R}^N$,

$$\Rightarrow \; y\left(\tau(\omega)\right) \in L\left(\omega, k(\omega)\right) \quad \mu\text{-a.e. on } \Omega,$$
$$\Rightarrow \; y \in R(k).$$

□

Proposition 7.4.16 and Theorem 6.5.19 (the Kakutani–Ky Fan fixed point theorem), imply

$$S = \{k \in C : k \in R(k)\} \neq \emptyset.$$

Moreover, S is weakly compact and convex in $L^1(\Sigma_0, \mathbb{R}^N)$ (see hypothesis H_5(ii)).

We consider the following deterministic infinite-dimensional optimization problem,

$$\widehat{v} = \sup\left[J(k,k) : k \in R(k)\right] = \sup\left[J(k,k) : k \in S\right]. \tag{7.118}$$

Because of hypotheses H_6, $J(\cdot,\cdot)$ is upper semicontinuous on $C \times C$ and so problem (7.118) admits a solution $k^* \in S$. In addition, because of hypothesis H_6(iv), we see that

$$\widehat{v} = \sup\left[J(k,y) : y \in R(k),\; k \le y\right]. \tag{7.119}$$

Using the infinite-dimensional Kuhn–Tucker multiplier theorem we produce a stationary price system supporting the stationary program $k^* \in S$.

PROPOSITION 7.4.17 *If hypotheses H_5 and H_6 hold, then we can find $p \in L^1(\Sigma_0, \mathbb{R}^N)$, $p \ge 0$ such that for all $(k,y) \in \mathrm{Gr}\, R$, we have*

$$J(k,y) + \langle p, y - k\rangle \le J(k^*, k^*) = \widehat{v}.$$

PROOF: Let $G_1(\omega) = \left\{(k,y) \in \mathrm{Gr}L(\omega,\cdot) : k \ll y\right\}$. The existence of an expansible capital stock for every $\omega \in \Omega$ (see hypothesis H_5(v)), implies that $G_1(\omega) \neq \emptyset$ for all $\omega \in \Omega$. Let $\{x_n^*\}_{n\ge1}$ be dense in $\mathbb{R}_+^N$, $x_n^* \neq \emptyset$ for all $n \ge 1$. Then we have

$$G_1(\omega) = \bigcap_{n\ge1} \left\{(k,y) \in \mathrm{Gr}L(\omega,\cdot) : (x_n^*, y-k)_{\mathbb{R}^N} > 0\right\}$$
$$\Rightarrow \; \mathrm{Gr}G_1 \in \Sigma_1 \times \mathbf{B}(\mathbb{R}^N) \times \mathbf{B}(\mathbb{R}^N) \quad (\text{see hypothesis } \mathrm{H}_5(\mathrm{i})).$$

We can apply Theorem 6.3.20 (the Yankov–von Neumann–Aumann selection theorem) and obtain $k', y' : \Omega \longrightarrow \mathbb{R}^N$ Σ_1-measurable functions such that

$$\left(k'(\omega), y'(\omega)\right) \in G_1(\omega) \qquad \text{for all } \omega \in \Omega.$$

We set $k = k' \circ \tau^{-1}$ and $y = y' \circ \tau^{-1}$. Evidently $(k,y) \in L^\infty(\Sigma_0, \mathbb{R}^N) \times L^\infty(\Sigma_0, \mathbb{R}^N)$ and we have

$$k(\omega) \ll y(\omega) \qquad \text{for all } \omega \in \Omega,$$
$$\Rightarrow \; k \ll y \text{ in } L^\infty(\Sigma_0, \mathbb{R}^N) \quad (\text{recall int } L^\infty(\Sigma_0, \mathbb{R}^N)_+ \neq \emptyset).$$

So we can apply the infinite-dimensional Kuhn–Tucker multiplier rule and obtain $p \in L^\infty(\Sigma_0, \mathbb{R}^N)$, $p \ge 0$, such that

$$J(k,y)+p(y-k)\le J(k^*,k^*)=\widehat{v} \qquad \text{for all } (k,y)\in \text{Gr } R. \tag{7.120}$$

Arguing as in the proof of Theorem 7.4.10 and using Theorem 7.4.3, in (7.120), we can replace p by its absolutely continuous part $p^a\in L^1(\Sigma_0,\mathbb{R}^N)$, $p^a\ge 0$, which is the desired stationary price system. □

Next we give the definition of good program for this stochastic model (see also Definition 7.2.31).

DEFINITION 7.4.18 A *feasible program* $\widehat{k}=\{k_n\}_{n\ge 0}$ is a *good* program if

$$\limsup_{m\to\infty}\sum_{n=0}^{m}\left(\widehat{v}-J(k_n\circ\tau^{-n},k_{n+1}\circ\tau^{-n-1}\right)<+\infty.$$

Good programs exhibit a remarkable average asymptotic property.

PROPOSITION 7.4.19 *If hypotheses H_5, H_6 hold, and $\widehat{k}=\{k_n\}_{n\ge 0}$ is a good program, then the sequence of arithmetic averages*

$$\left\{\overline{k}_n=\frac{k_0+k_1\circ\tau^{-1}+\cdots+k_n\circ\tau^{-n}}{n+1}\right\}_{n\ge 0}$$

has weak limit points in $L^1(\Sigma_0,\mathbb{R}^N)$ and each such limit point solves the optimization problem (7.119) (hence it is an optimal stationary program).

PROOF: Because of hypothesis H_5(ii), we have

$$\left(\frac{1}{n+1}\sum_{m=0}^{n}(k_m\circ\tau^{-m})(\omega),\frac{1}{n+1}\sum_{m=1}^{n+1}(k_m\circ\tau^{-m})(\omega)\right)\in \text{Gr } L(\omega,\cdot)\ \ \mu\text{-a.e. on }\Omega. \tag{7.121}$$

We have

$$\frac{1}{n+1}\sum_{m=1}^{n+1}(k_m\circ\tau^{-m})(\omega)=\frac{1}{n+1}\sum_{m=0}^{n+1}(k_m\circ\tau^{-m})(\omega)-\frac{1}{n+1}k_0(\omega). \tag{7.122}$$

Also, because of hypothesis H_5(iii), we see that

$$\left\{\overline{k}_n=\frac{1}{n+1}\sum_{m=0}^{n}k_m\circ\tau^{-m}\right\}_{n\ge 0}\subseteq L^1(\Sigma_0,\mathbb{R}^N),$$

is relatively weakly sequentially compact. So we may assume that

$$\begin{aligned}&\overline{k}_n\xrightarrow{w}\widehat{k}^{\,*}\quad\text{in } L^1(\Sigma_0,\mathbb{R}^N),\\ \Rightarrow\ &\left(\widehat{k}^{\,*}(\omega),\widehat{k}^{\,*}(\omega)\right)\in \text{Gr}L(\omega,\cdot)\quad \mu\text{-a.e. on }\Omega,\quad (\text{see (7.121) and 7.121)}),\\ \Rightarrow\ &\widehat{k}^{\,*}\in S.\end{aligned}$$

Using Jensen's inequality and because by hypothesis $\widehat{k}$ is a good program,

$$\begin{aligned}
& J(k^*, k^*) - J\Big(\frac{1}{n+1}\sum_{m=0}^{n} k_m \circ \tau^{-m}, \frac{1}{n+1}\sum_{m=1}^{n+1} k_m \circ \tau^{-m}\Big) \\
\leq \; & \frac{1}{n+1}\sum_{m=0}^{n}\big(J(k^*, k^*) - J(k_m \circ \tau^{-m}, k_{m+1} \circ \tau^{-m-1})\big) \leq \frac{\gamma}{n+1}
\end{aligned} \tag{7.123}$$

for some $\gamma > 0$ and all $n \geq 1$. Recall that $\widehat{v} = J(k^*, k^*)$. Exploiting the weak upper semicontinuity of the expected utility function $J(\cdot, \cdot)$, we have

$$\limsup_{n \to \infty} J\Big(\frac{1}{n+1}\sum_{m=0}^{n} k_m \circ \tau^{-m}, \frac{1}{n+1}\sum_{m=1}^{n+1} k_m \circ \tau^{-m}\Big) \leq J(\widehat{k}^{\,*}, \widehat{k}^{\,*}). \tag{7.124}$$

So, if we pass to the limit as $n \to \infty$ in (7.123) and we use (7.124), we obtain

$$J(k^*, k^*) = \widehat{v} \leq J(\widehat{k}^{\,*}, \widehat{k}^{\,*}).$$

But recall that $\widehat{k}^{\,*} \in S$. So $J(k^*, k^*) = J(\widehat{k}^{\,*}, \widehat{k}^{\,*}) = \widehat{v}$. □

We introduce the value loss function $\xi(k, y)$ defined by

$$\xi(k, y) = \widehat{v} - J(k, y) + \langle p, y - k\rangle \qquad \text{for all } (k, y) \in \mathrm{Gr}\, R.$$

PROPOSITION 7.4.20 *If hypotheses* H_5, H_6 *hold and* $\{k_n\}_{n \geq 0}$, $\{z_n\}_{n \geq 0}$ *are two good programs that have sequences of arithmetic means* $\{\overline{k}_n\}_{n \geq 0}$, $\{\overline{z}_n\}_{n \geq 0}$ *which converge weakly in* $L^1(\Sigma_0, \mathbb{R}^N)$ *to the same limit* k^*, *then*

$$\begin{aligned}
& \liminf_{m \to \infty} \sum_{n=0}^{m}\big(J(k_n \circ \tau^{-n}, k_{n+1} \circ \tau^{-n-1}) - J(z_n \circ \tau^{-n}, z_{n+1} \circ \tau^{-n-1})\big) \\
\leq \; & \limsup_{m \to \infty} \sum_{n=0}^{m}\big(\xi(z_n \circ \tau^{-n}, z_{n+1} \circ \tau^{-n-1}) - \xi(k_n \circ \tau^{-n}, k_{n+1} \circ \tau^{-n-1})\big).
\end{aligned} \tag{7.125}$$

PROOF: We have

$$\begin{aligned}
& \sum_{n=0}^{m}\big(J(k_n \circ \tau^{-n}, k_{n+1} \circ \tau^{-n-1}) - J(z_n \circ \tau^{-n}, z_{n+1} \circ \tau^{-n-1})\big) \\
= \; & \sum_{n=0}^{m}\big(J(k_n \circ \tau^{-n}, k_{n+1} \circ \tau^{-n-1}) - \widehat{v} - \langle p, k_n \circ \tau^{-n} - k_{n+1} \circ \tau^{-n-1}\rangle \\
& \quad - J(z_n \circ \tau^{-n}, z_{n+1} \circ \tau^{-n-1}) + \widehat{v} + \langle p, z_n \circ \tau^{-n} - z_{n+1} \circ \tau^{-n-1}\rangle \\
& \quad + \langle p, k_n \circ \tau^{-n} - k_{n+1} \circ \tau^{-n-1}\rangle - \langle p, z_n \circ \tau^{-n} - z_{n+1} \circ \tau^{-n-1}\rangle\big) \\
\leq \; & \sum_{n=0}^{m}\big(\xi(z_n \circ \tau^{-n}, z_{n+1} \circ \tau^{-n-1}) - \xi(k_n \circ \tau^{-n}, k_{n+1} \circ \tau^{-n-1})\big) \\
& \quad + \langle p, z_{m+1} \circ \tau^{-m-1} - k_{m+1} \circ \tau^{-m-1}\rangle.
\end{aligned} \tag{7.126}$$

Note that because of Proposition 7.4.19, we have

$$\limsup_{m\to\infty} \langle p, z_{m+1} \circ \tau^{-n} - k_{m+1} \circ \tau^{-m-1} \rangle$$
$$\leq \lim_{m\to\infty} \frac{1}{m+2} \sum_{n=0}^{m+1} \langle p, z_n \circ \tau^{-n} - k_n \circ \tau^{-n} \rangle = 0. \qquad (7.127)$$

So, if we pass to the limit as $m \longrightarrow \infty$ in (7.126) and use (7.127), we obtain (7.125). □

REMARK 7.4.21 Note that the lim inf in the left-hand side of (7.125) is similar to the one involved in the definition of weak maximality (see Definition 7.4.18).

The above remark suggests that we should consider the following problem,

$$\sup[\vartheta(\widehat{k}) : \widehat{k} \in F(\overline{x}_0)], \qquad (7.128)$$

where $\vartheta(\widehat{k}) = \lim_{m\to\infty} \sum_{n=0}^{m} \big(-\xi(k_n \circ \tau^{-n}, k_{n+1} \circ \tau^{-n-1})\big)$, with $\widehat{k} = \{k_n\}_{n\geq 0}$. Because $-\xi \leq 0$, the limit in the definition of ϑ can be $-\infty$. However, this can not happen if $\widehat{k}$ is good. For this reason (in analogy with the deterministic case) we look at good programs to find a weakly maximal one. This requires that the solution set of (7.119) is a singleton.

H_7: There exists at least one good program $\widehat{k} = \{k_n\}_{n\geq 0} \in F(\overline{x}_0)$ and the solution set S_0 of (7.119) is a singleton.

REMARK 7.4.22 If for all $\omega \in \Omega$, $u(\omega, \cdot, \cdot)$ is strictly concave, the second part of hypothesis H_7 is automatically satisfied.

Now we can prove the existence of a weakly maximal program.

THEOREM 7.4.23 *If hypotheses H_5, H_6, and H_7 hold, then there exists a weakly maximal program $\widehat{k}^* = \{k_n^*\}_{n\geq 0} \in F(\overline{x}_0)$.*

PROOF: For every $m \geq 0$, we set

$$\vartheta_m(\widehat{k}) = \sum_{n=0}^{m} \big(-\xi(k_n \circ \tau^{-n}, k_{n+1} \circ \tau^{-n-1})\big)$$

for every $\widehat{k} = \{k_n\}_{n\geq 0} \in F(\overline{x}_0)$. Evidently ϑ_m is w^*-upper semicontinuous and concave. Moreover, we have $\vartheta_m \downarrow \vartheta$ as $m \longrightarrow \infty$. Hence ϑ is w^*-upper semicontinuous and concave. Also note that $F(\overline{x}_0) \subseteq \prod_{n=0}^{\infty} C \circ \tau^n$ and by Tychonov's theorem the product set is w^*-compact in $\prod_{n=0}^{\infty} L^\infty(\Sigma_n, \mathbb{R}^N)$. Therefore we can apply the theorem of Weierstrass and find $\widehat{k}^* = \{k_n^*\}_{n\geq 0} \in F(\overline{x}_0)$ such that

$$\vartheta(\widehat{k}^*) = \sup\big[\vartheta(\widehat{k}) : \widehat{k} \in F(\overline{x}_0)\big].$$

We claim that the maximizing feasible program $\widehat{k}^*$ is a good program. From hypothesis H_7 there is a good program $\widehat{k} = \{k_n\}_{n\geq 0} \in F(\overline{x}_0)$. Then given $\varepsilon > 0$, we can find $m_0 = m_0(\varepsilon) \geq 1$ such that for all $m \geq m_0$, we have

$$
\begin{aligned}
&\sum_{n=0}^{m}\big(J(k_n\circ\tau^{-n},k_{n+1}\circ\tau^{-n-1})\big)+\big\langle p,k_{n+1}\circ\tau^{-n-1}-k_n\circ\tau^{-n}\big\rangle\\
&\le \sum_{n=0}^{\infty}\big(J(k_n^*\circ\tau^{-n},k_{n+1}^*\circ\tau^{-n-1})\big)+\big\langle p,k_{n+1}^*\circ\tau^{-n-1}-k_n^*\circ\tau^{-n}\big\rangle+\varepsilon,\\
&\Rightarrow \sum_{n=0}^{m}\big(\widehat{v}-J(k_n^*\circ\tau^{-n},k_{n+1}^*\circ\tau^{-n-1})\big)\\
&\le \sum_{n=0}^{m}\big(\widehat{v}-J(k_n\circ\tau^{-n},k_{n+1}\circ\tau^{-n-1})\big)+\\
&\qquad\big\langle p,k_{m+1}^*\circ\tau^{-m-1}-k_{m+1}\circ\tau^{-m-1}\big\rangle+\varepsilon\\
&\le \gamma+4M\|p\|_1+\varepsilon\\
&\Rightarrow \widehat{k}^{\,*}=\{k_n^*\}_{n\ge 0}\in F(\overline{x}_0)\ \text{ is a good program (see Definition 7.4.18).}
\end{aligned}
$$

If $\widehat{k}^{\,*}$ is not weakly maximal, then we can find $\varepsilon>0$, $m_0\ge 1$, and $\widehat{z}\in F(\overline{x}_0)$ such that $V_m(\widehat{k}^{\,*})+\varepsilon\le V_m(\widehat{z})$, where

$$
\begin{aligned}
V_m(\widehat{k}^{\,*})&=\sum_{n=0}^{m}J(k_n^*\circ\tau^{-n},k_{n+1}^*\circ\tau^{-n-1})\\
\text{and}\quad V_m(\widehat{z})&=\sum_{n=0}^{m}J(z_n\circ\tau^{-n},z_{n+1}\circ\tau^{-n-1}).
\end{aligned}
$$

Then we have

$$
0<\varepsilon\le V_m(\widehat{z})-V_m(\widehat{k}^{\,*}). \tag{7.129}
$$

Recall that $\widehat{k}^{\,*}\in F(\overline{x}_0)$ too is a good program. In addition, because $\widehat{k}^{\,*}$ maximizes ϑ on $F(\overline{x}_0)$, given $\delta>0$, we can find $m_0=m_0(\delta)\ge 1$ such that for all $m\ge m_0$, we have

$$
\begin{aligned}
&\sum_{n=0}^{m}\big(\xi(k_n^*\circ\tau^{-n},k_{n+1}^*\circ\tau^{-n-1})-\xi(z_n\circ\tau^{-n},z_{n+1}\circ\tau^{-n-1})\big)\le\delta,\\
&\Rightarrow \sum_{n=0}^{m}\Big[\widehat{v}-J(k_n^*\circ\tau^{-n},k_{n+1}^*\circ\tau^{-n-1})+\big\langle p,k_{n+1}^*\circ\tau^{-n-1}-k_n^*\circ\tau^{-n}\big\rangle\\
&\qquad-\widehat{v}+J(z_n\circ\tau^{-n},z_{n+1}\circ\tau^{-n-1})-\big\langle p,z_{n+1}\circ\tau^{-n-1}-z_n\circ\tau^{-n}\big\rangle\Big]\le\delta,\\
&\Rightarrow \varepsilon\le\big\langle p,z_{m+1}\circ\tau^{-m-1}-k_{m+1}^*\circ\tau^{-m-1}\big\rangle+\delta.
\end{aligned}
$$

Passing to the limit as $m\longrightarrow\infty$ and using (7.127) (recall that $\{z_n\}_{n\ge0}$, $\{k_n^*\}_{n\ge0}$ are both good programs), we obtain

$$
0<\varepsilon\le\delta.
$$

But $\delta>0$ was arbitrary. So we let $\delta\downarrow 0$ to reach a contradiction. This proves that $\widehat{k}^*\in F(\overline{x}_0)$ is weakly maximal. ∎

A careful reading of the previous proof reveals that the hypothesis concerning the existence of a good program (see hypothesis H_7), was crucial. So we need to produce verifiable conditions that imply the existence of a good program in $F(\overline{x}_0)$. For this reason we introduce the following hypotheses.

H$_8$: $u:\Omega\times\mathbb{R}_+^N\times\mathbb{R}_+^N\longrightarrow\mathbb{R}$ is a function such that

(i) $(\omega,k,y)\longrightarrow u(\omega,k,y)$ is $\Sigma_1\times\mathbf{B}(\mathbb{R}_+^N)\times\mathbf{B}(\mathbb{R}_+^N)$-measurable.

(ii) For every $\omega\in\Omega$, $(k,y)\longrightarrow u(\omega,k,y)$ is continuous, concave.

(iii) For every $r>0$, there exists $\varphi_r\in L^1(\Omega)_+$ such that

$$|u(\omega,k,y)|\leq\varphi_r(\omega)$$

for μ-almost all $\omega\in\Omega$ and all $\|k\|,\|y\|\leq r$.

H$_9$: There exists $w\in L^\infty(\Sigma_0,\mathbb{R}^N)$ such that $(w\circ\tau)\in L\big(\omega,\overline{x}_0(\omega)\big)$ μ-a.e. on Ω and $\overline{x}_0(\omega)\ll w(\omega)$ μ-a.e. on Ω.

REMARK 7.4.24 This hypothesis means that the initial capital stock $\overline{x}$ is expansible.

THEOREM 7.4.25 *If hypotheses H_6, H_8, and H_9 hold, then there exists a good program $\widehat{k}=\{k_n\}_{n\geq 0}\in F(\overline{x}_0)$.*

PROOF: Hypothesis H$_9$ implies that there exists $0<\lambda<1$ such that

$$(1-\lambda)k^*+\lambda\overline{x}_0\leq w\quad\text{in the ordered Banach space } L^\infty(\Sigma_0,\mathbb{R}^N),$$

where $k^*\in S$ is a solution of problem (7.119). We set

$$k_n=(1-\lambda^n)k^*\circ\tau^n+\lambda^n\overline{x}_0\circ\tau^n\in L^\infty(\Sigma_n,\mathbb{R}^N),\quad n\geq 0.$$

We claim that $\widehat{k}=\{k_n\}_{n\geq 0}\in F(\overline{x}_0)$.

If $n=0$, then because of τ^0-identity, we have $k_0=\overline{x}_0$. For $n\geq 1$, because of the convexity of Gr $L(\omega,\cdot)$ (see hypothesis H$_5$(iii)), we have

$$\begin{aligned}&(1-\lambda^n)(k^*\circ\tau^n)(\omega)+\lambda^n(\overline{x}_0\circ\tau^n)(\omega)\\ \in\ &L\big(\tau^n(\omega),(1-\lambda^n)(k^*\circ\tau^n)(\omega)+\lambda(\overline{x}_0\circ\tau^n)(\omega)\big)\\ =\ &L\big(\tau^n(\omega),(1-\lambda^n)(k^*\circ\tau^n)(\omega)+\lambda(\overline{x}_0\circ\tau^n)(\omega)\big)\\ =\ &L_{n+1}\big(\omega,\overline{x}_0(\omega)\big)\quad\mu\text{-a.e. on }\Omega.\end{aligned}\tag{7.130}$$

Because $(1-\lambda)k^*+\lambda\overline{x}_0\leq w$ in $L^\infty(\Sigma_0,\mathbb{R}^N)$, we have

$$\begin{aligned}k_{n+1}&=(1-\lambda^{n+1})k^*\circ\tau^{n+1}+\lambda^{n+1}\overline{x}_0\circ\tau^{n+1}\\ &\leq\big((1-\lambda^{n+1})k^*\circ\tau^n+\lambda^n w\circ\tau^n\big)\circ\tau.\end{aligned}\tag{7.131}$$

Then from (7.130), (7.131), and the free disposability hypothesis (see hypothesis H$_5$(iv)), we have

$$k_{n+1}(\omega)\in L_{n+1}\big(\omega,k_n(\omega)\big)=L\big(\tau^n(\omega),k_n(\omega)\big)\quad\mu\text{-a.e. on }\Omega,$$
$$\Rightarrow\ \widehat{k}\in F(\overline{x}_0).$$

We claim that $\widehat{k}$ is good. From hypotheses H$_8$ we have that J is continuous, concave on $L^1(\Omega,\mathbb{R}^N)\times L^1(\Omega,\mathbb{R}^N)$. Therefore it is Lipschitz continuous on bounded sets, in particular on C. Therefore we have

$$\begin{aligned}\widehat{v}-J(k_n\circ\tau^{-n},k_{n+1}\circ\tau^{-n-1})&=J(k^*,k^*)-J(k_n\circ\tau^{-n},k_{n+1}\circ\tau^{-n-1})\\&\leq l_M(\lambda^n+\lambda^{n+1})4M,\end{aligned}$$

for some $l_M>0$, all $n\geq 1$.

Summing up, we obtain

$$\sum_{n\geq 0}\big(\widehat{v}-J(k_n\circ\tau^{-n},k_{n+1}\circ\tau^{-n-1})\big)<+\infty,$$

$$\Rightarrow\ \widehat{k}=\{k_n\}_{n\geq 0}\ \text{ is a good program (see Definition 7.4.18).}$$

□

7.5 Continuous-Time Growth Models

In this section, we turn our attention to continuous-time growth models. We consider a general deterministic model of capital accumulation, with a convex technology and a concave utility function. We establish the existence of an optimal growth path using the direct method of the calculus of variations. This requires a careful choice of topology on the space of feasible programs.

First let us describe the model. The planning horizon is $\mathbb{R}_+=[0,+\infty\}$ (continuous-time, infinite horizon model). There are N capital goods in the economy and so the commodity space is $\mathbb{R}^N$ (multisector growth model). The technological possibilities of the economy are described by a multifunction $G:\mathbb{R}_+\times\mathbb{R}_+^N\longrightarrow 2^{\mathbb{R}_+^N}\setminus\{\emptyset\}$. If $y\in G(t,x)$, then this means that if at time $t\geq 0$, we are given a capital stock x, then y can be accumulated as additional capital. Hence at every time instant $t\geq 0$, $G(t,\cdot)$ describes the technological capacity of the economy at that moment. For this reason, we call G the technology multifunction. Note that in our model this multifunction is time-varying and so the model (at least in its technology component) is nonstationary. The performance of a feasible technological process $(x,y)\in\operatorname{Gr}G(t,\cdot)$ is measured by an instantaneous utility function $u(t,x,y)$ and future utilities are discounted by a factor $\delta\in(0,1)$. So the intertemporal utility generated by a growth path x is given by

$$J(x)=\int_0^\infty e^{-\delta t}u\big(t,x(t),x'(t)\big)\mathrm{d}t.$$

Also, there is an initial capital stock $x_0\in\mathbb{R}_+^N$.

DEFINITION 7.5.1 A *capital accumulation program* x is an absolutely continuous map $x:\mathbb{R}_+\longrightarrow\mathbb{R}_+^N$. From Lebesgue's theorem we know that such a function is differentiable almost everywhere and for every $t\geq 0$ we have

$$x(t)=x(0)+\int_0^t x'(s)\mathrm{d}s.$$

A program x is *feasible* if $0\leq x(0)\leq x_0$ and $x'(t)\in G\big(t,x(t)\big)$ a.e. on $\mathbb{R}_+$. The set of all feasible programs is denoted by $F(x_0)$. We assume throughout this section that $F(x_0)\neq\emptyset$. Note that $F(x_0)\subseteq AC_{\mathrm{loc}}(\mathbb{R}_+,\mathbb{R}^N)$.

REMARK 7.5.2 The hypothesis that $F(x_0) \neq \emptyset$ is not at all restrictive. Indeed, we already know from the analysis of the discrete models, that it is reasonable to assume that inaction is an option (i.e., $0 \in G(t,0)$ a.e. on $\mathbb{R}_+$) and that free disposability holds. Hence for all $(t,x) \in \mathbb{R} \times \mathbb{R}_+^N$ we have $0 \in G(t,x)$ and so we see that $F(\overline{x}_0) \neq \emptyset$.

We introduce the measure μ on $\mathbb{R}_+$ defined by

$$d\mu = e^{-\delta t} dt.$$

Evidently μ is a finite measure on $\mathbb{R}_+$ which is absolutely continuous with respect to the Lebesgue measure. The Radon–Nikodym derivative of μ with respect to the Lebesgue measure is of course $e^{-\delta t}$. In what follows by $L^1(\mathbb{R}_+, \mu; \mathbb{R})$ we denote the space of functions from $\mathbb{R}_+$ into $\mathbb{R}$ whose absolute value is μ-integrable.

Now we can introduce our hypotheses on the technology G and the utility u:

$\mathbf{H_1}$: $G : \mathbb{R} \times \mathbb{R}_+^N \longrightarrow P_{kc}(\mathbb{R}^N)$ is a multifunction such that

(i) For all $x \in \mathbb{R}_+^N$, $t \longrightarrow G(t,x)$ is graph-measurable.

(ii) For almost $t \in T$, $x \longrightarrow G(t,x)$ has a closed graph.

(iii) There exists a function $\varphi \in L^1_{\text{loc}}(\mathbb{R}_+)$ such that for all $x \in F(x_0)$ we have

$$\|x'(t)\| \leq \varphi(t) \quad \text{a.e. on } \mathbb{R}_+.$$

REMARK 7.5.3 Hypothesis H_1(iii) is essentially a growth condition on the technology. For example, assume that for almost all $t \in T$ and all $x \in \mathbb{R}_+^N$ we have

$$|G(t,x)| = \sup \big[\|y\| : y \in G(t,x) \big] \leq \xi(t) \varphi(\|x\|),$$

with $\xi \in L^1_{\text{loc}}(\mathbb{R}_+)$ and φ a nonnegative continuous function on $\mathbb{R}_+$ such that $\int^\infty (dr/\varphi(r)) = +\infty$. Then hypothesis H_1(iii) is satisfied. To see this, fix $b > 0$ and then choose $M > \|x_0\| + 1$ such that

$$\int_{\|x_0\|+1}^{M} \frac{dr}{\varphi(r)} > \int_0^b \xi(t) dt. \tag{7.132}$$

We claim $\|x(t)\| \leq M$ for all $t \in T = [0,b]$. If this is not true, then we can find $t_1, t_2 \in T$ such that $0 \leq t_1 \leq t_2 \leq b$ and

$$x(t_1) = \|x_0\| + 1, \; x(t_2) = M \quad \text{and} \quad \|x_0\| + 1 \leq x(t) \leq M \quad \text{for all } t \in [t_1, t_2].$$

We have

$$\|x'(t)\| \leq \xi(t) \varphi(\|x(t)\|) \quad \text{a.e. on } T,$$

$$\Rightarrow \int_{\|x_0\|+1}^{M} \frac{dr}{\varphi(r)} \leq \int_{t_1}^{t_2} \xi(s) ds \leq \int_0^b \xi(s) ds$$

which contradicts (7.132). In fact we can assume a more general majorant; namely

$$|G(t,x)| \leq g(t, \|x\|) \quad \text{a.e. on } \mathbb{R}, \quad \text{for all } x \geq 0,$$

where $g : \mathbb{R}_+ \times \mathbb{R}_+ \longrightarrow \mathbb{R}_+$ is a function such that

- For all $r \geq 0$, $g(\cdot, r) \in L^1_{\text{loc}}(\mathbb{R}_+)$.
- For all $t \geq 0$, $g(t, \cdot)$ is nondecreasing.
- There exist $t_0 \geq 0$ and $u_0 > 0$ such that the scalar differential equation

$$u'(t) = g\big(t, u(t)\big), \quad u(t_0) = u_0$$

has a bounded solution on $[t_0, +\infty)$.

The hypotheses on the instantaneous utility function, are the following:

H₂: $u : \mathbb{R}_+ \times \mathbb{R}^N \times \mathbb{R}^N \longrightarrow \overline{\mathbb{R}} = \mathbb{R} \cup \{-\infty\}$ is a function such that

(i) $(t, x, y) \longrightarrow u(t, x, y)$ is Borel-measurable.

(ii) For almost all $t \geq 0$, $(x, y) \longrightarrow u(t, x, y)$ is upper semicontinuous.

(iii) For almost all $t \geq 0$ and all $x \in \mathbb{R}^N$, $y \longrightarrow u(t, x, y)$ is concave.

(iv) For almost all $t \geq 0$ and all $x, y \in \mathbb{R}^N$, we have

$$|u(t, x, y)| \leq \alpha(t) + c(\|x\| + \|y\|) \quad \text{with } \alpha \in L^1(\mathbb{R}_+, \mu; \mathbb{R})_+ \text{ and } c > 0.$$

REMARK 7.5.4 By allowing the utility function to take the value $-\infty$, we have incorporated in its definition constraints such as $u(t, x, y) = -\infty$ if $x \notin \mathbb{R}^N_+$. Another economically interesting situation accommodated by hypotheses H_2, is when $u(t, x, y) \longrightarrow -\infty$ as $\|x\|, \|y\| \longrightarrow 0$. Indeed, in many situations it is more natural to heavily penalize the inactivity option, more than simply setting $u(t, 0, 0) = 0$.

If $J(x) = \int_0^\infty e^{-\delta t} u\big(t, x(t), x'(t)\big) dt$, $x \in F(x_0)$, our goal is to find a solution of the following optimization problem,

$$V(x_0) = \sup[J(x) : x \in F(x_0)]. \tag{7.133}$$

As we already mentioned, we approach problem (7.133) using the direct method of the calculus of variations. This requires a careful choice of topology that will make the constraint set $F(x_0)$ compact and the objective functional (the discounted intertemporal utility) upper semicontinuous. The problem that we face is that even if we choose a good topology on the space of capital stocks, this translates into weak convergence of the corresponding investment flows. Weak convergence does not imply, in general, pointwise convergence (even for a subsequence). This lack of pointwise convergence poses several technical difficulties in the existence theory. For this reason, we have included convexity and concavity conditions in hypotheses H_1 and H_2.

First we focus on the constraint set and seek a topology that makes it compact. For this purpose, we introduce two topologies. The first one concerns the capital accumulation programs.

DEFINITION 7.5.5 The *c-topology* on $C(\mathbb{R}_+, \mathbb{R}^N)$ (compact-open topology) (hence on $AC_{\text{loc}}(\mathbb{R}_+, \mathbb{R}^N)$ too) is the topology of uniform convergence on compact sets.

REMARK 7.5.6 The above topology is defined by the family of seminorms

$$p_n(x) = \max\{\|x(t)\| : t \in [0,n]\},$$

where $\|\cdot\|$ is any of the equivalent l^p-norms on $\mathbb{R}^N$. It is well-known that $C(\mathbb{R}_+, \mathbb{R}^N)$ endowed with this topology becomes a Fréchet space. This topology makes the constraint set compact. But in order to be able to show this, we need to consider another topology related to the investment flows.

The investment flows are the derivatives of the capital accumulations. Because the latter by definition are absolutely continuous functions, by Lebesgue's theorem the investment flows belong in $L^1_{\text{loc}}(\mathbb{R}_+, \mathbb{R}^N)$. This too is a Fréchet space, for the topology defined by the sequence of seminorms

$$q_n(y) = \int_0^n \|y(t)\| dt, \quad n \geq 1.$$

Then using this fact we can define a second topology on the space of absolutely continuous functions from $\mathbb{R}_+$ into $\mathbb{R}^N$, namely the following.

DEFINITION 7.5.7 The *α-topology* on $AC_{\text{loc}}(\mathbb{R}_+, \mathbb{R}^N)$, is the Fréchet topology generated by the sequence of seminorms

$$r_n(x) = p_n(x) + q_n(x') \qquad \text{for all } n \geq 1.$$

Recall that $p_n(x) = \max\{\|x(t)\| : t \in [0,n]\}$ and $q_n(x') = \int_0^n \|x'(t)\| dt$.

REMARK 7.5.8 Evidently on $AC_{\text{loc}}(\mathbb{R}_+, \mathbb{R}^N)$, the a-topology is stronger than the c-topology. Note that the a-topology is equivalent to the one given by the natural metric on the direct sum $\mathbb{R}^N \oplus L^1_{\text{loc}}(\mathbb{R}_+, \mathbb{R}^N)$. So we can make the identification $AC_{\text{loc}}(\mathbb{R}_+, \mathbb{R}^N) = \mathbb{R}^N \oplus L^1_{\text{loc}}(\mathbb{R}_+, \mathbb{R}^N)$. Using this identification we can now determine the dual of $AC_{\text{loc}}(\mathbb{R}_+, \mathbb{R}^N)$ and therefore define the weak topology on $AC_{\text{loc}}(\mathbb{R}_+, \mathbb{R}^N)$. The dual of $\mathbb{R}^N$ with the l^p-norm, is of course $\mathbb{R}^N$ with the $l^{p'}$-norm, $(1/p) + (1/p') = 1$. The dual of $L^1_{\text{loc}}(\mathbb{R}_+, \mathbb{R}^N)$ is the space $\{y^* \in L^\infty(\mathbb{R}_+, \mathbb{R}^N) : \operatorname{supp} y^* \text{ is compact}\}$. So $x^* \in AC_{\text{loc}}(\mathbb{R}_+, \mathbb{R}^N)$ if and only if

$$\langle y^*, x \rangle = \big(v^*, x(0)\big)_{\mathbb{R}^N} + \int_0^b \big(w^*(t), x'(t)\big)_{\mathbb{R}^N} dt$$

for some $v^* \in \mathbb{R}^N$, some $b > 0$, and some $w^* \in L^\infty(T, \mathbb{R}^N)$, $T = [0,b]$. Therefore with this duality we can define a weak topology on $AC_{\text{loc}}(\mathbb{R}_+, \mathbb{R}^N)$. Using the Dunford–Pettis theorem, we can check that if $x_n \xrightarrow{w} x$ in $AC_{\text{loc}}(\mathbb{R}_+, \mathbb{R}^N)$, then $x_n \xrightarrow{c} x$. However, this is no longer true if sequences are replaced by nets. Recall that the generalized Lebesgue's dominated convergence theorem fails for nets.

Let $\xi \in L^1_{\text{loc}}(\mathbb{R}_+)$ and define

$$D(\xi) = \big\{x \in AC_{\text{loc}}(\mathbb{R}_+, \mathbb{R}^N) : \|x'(t)\| \leq \xi(t) \text{ a.e. on } T\big\}.$$

We show that when restricted to sets such as $D(\xi)$, the weak and c-topologies coincide. This permits us to show the c-compactness of the set $F(x_0)$. To this end, first we prove the following lemma.

LEMMA 7.5.9 *If $\{x_\alpha\}_{\alpha \in S}$ is a net in $D(\xi)$ and $x_\alpha \xrightarrow{c} x$ in $AC_{\mathrm{loc}}(\mathbb{R}_+, \mathbb{R}^N)$, then $x \xrightarrow{w} x$ in $AC_{\mathrm{loc}}(\mathbb{R}_+, \mathbb{R}^N)$ and $x \in D(\xi)$.*

PROOF: Fix $b > 0$. Given $\varepsilon > 0$, we can find $\delta > 0$ such that if $A \subseteq [0, b]$ is measurable and $|A| < \delta$ (by $|\cdot|$ we denote the Lebesgue measure on $\mathbb{R}$), then we have

$$\int_A \xi(t) \mathrm{d}t < \varepsilon. \tag{7.134}$$

We take $0 \leq s_1 \leq t_1 \leq \cdots \leq s_m \leq t_m \leq b$, with $\sum_{k=1}^m (t_k - s_k) < \delta$. Then

$$\begin{aligned}
&\sum_{k=1}^m \|x(t_k) - x(s_k)\| \\
\leq\ & \sum_{k=1}^m \Big(\|x(t_k) - x_a(t_k)\| + \|x_a(s_k) - x(s_k)\| + \int_{s_k}^{t_k} \|x_a'(\tau)\| \mathrm{d}\tau \Big) \\
\leq\ & 2m \|x - x_a\|_{C([0,b], \mathbb{R}^N)} + \sum_{k=1}^m \int_{s_k}^{t_k} \|x_a'(\tau)\| \mathrm{d}\tau \\
<\ & 2m \|x - x_a\|_{C([0,b], \mathbb{R}^N)} + \varepsilon < 2\varepsilon \quad \text{for } a \in S \text{ large (see (7.134))},
\end{aligned}$$

which implies that $x \in AC_{\mathrm{loc}}(\mathbb{R}_+, \mathbb{R}^N)$.

Next given $\varepsilon > 0$, we can find $\delta > 0$ such that

$$\int_C \xi \mathrm{d}t < \varepsilon \quad \text{and} \quad \Big\| \int_C k \mathrm{d}t \Big\| < \varepsilon$$

for all $C \subseteq [0, b]$ measurable with $0 < |C| < \delta$. Let $A \subseteq [0, b]$ be measurable with $0 < |A|$. We can find $U \subseteq [0, b]$ open such that $|C = A \triangle U| < \delta$. We know that $U = \bigcup_{k=1}^m (s_k, t_k)$. We have

$$\begin{aligned}
&\Big\| \int_A x_a' \mathrm{d}t - \int_A x' \mathrm{d}t \Big\| \\
\leq\ & \Big\| \int_U (x_a' - x') \mathrm{d}t \Big\| + \int_C \|x_a'\| \mathrm{d}t + \Big\| \int_C x' \mathrm{d}t \Big\| \\
\leq\ & \sum_{k=1}^m \big(\|x_a(t_k) - x(t_k)\| + \|x_a(s_k) - x_a(s_k)\| \big) + 2\varepsilon \\
\leq\ & 2m \|x_a - x\|_{C([0,b], \mathbb{R}^N)} + 2\varepsilon, \\
\Rightarrow\ & \int_A x_a' \mathrm{d}t \longrightarrow \int_A x' \mathrm{d}t.
\end{aligned}$$

Because $A \subseteq [0, b]$-measurable was arbitrary, we have

$$x_a' \xrightarrow{w} x' \ \text{ in } \ L^1([0, b], \mathbb{R}^N).$$

Also $x_a(0) \longrightarrow x(0)$. Because $b > 0$ was arbitrary, we conclude that

$$x_a \xrightarrow{w} x \quad \text{in } AC_{\mathrm{loc}}(\mathbb{R}_+, \mathbb{R}^N).$$

Moreover, from Mazur's lemma, we infer that

$$\|x'(t)\| \leq \xi(t) \quad \text{a.e. on } \mathbb{R}_+,$$
$$\Rightarrow x \in D(\xi).$$

□

LEMMA 7.5.10 *If* $\{x_a\}_{a\in S} \subseteq D(\xi)$ *and* $x_a \xrightarrow{w} x$ *in* $AC_{\mathrm{loc}}(\mathbb{R}_+,\mathbb{R}^N)$*, then* $x_a \xrightarrow{c} x$ *and* $x \in D(\xi)$.

PROOF: Let $\varepsilon > 0$ and $b > 0$ be given. We can find $\delta > 0$ such that

$$\Big|\int_s^t \xi(\tau)\mathrm{d}\tau\Big| < \varepsilon \quad \text{and} \quad \Big\|\int_s^t x'(\tau)\mathrm{d}\tau\Big\| < \varepsilon$$

for $0 \leq s \leq t \leq b$ and $(t-s) < \delta$. Because $x_a \xrightarrow{w} x$ in $AC_{\mathrm{loc}}(\mathbb{R}_+,\mathbb{R}^N)$ for $a \in S$ large, we have

$$\|x_a(\delta n) - x(\delta n)\| \leq \Big\|x_a(0) - x(0) + \int_0^{\delta n} (x_a' - x')\mathrm{d}\tau\Big\| < \varepsilon,$$

where n is an integer such that $0 \leq n \leq (1/\delta)b$. Let $t \in [0,b]$. We choose an integer $0 \leq n \leq (1/\delta)b$ with $|t - \delta n| < \delta$. Then for $a \in S$ large, we have

$$\|x_a(t) - x(t)\| \leq \|x_a(\delta n) - x(\delta n)\| + \Big\|\int_{\delta n}^t (x_a' - x')\mathrm{d}\tau < 3\varepsilon,$$
$$\Rightarrow x_a \xrightarrow{c} x \text{ and from Lemma 7.5.9, we have } x \in D(\xi).$$

□

From Lemmata 7.5.9 and 7.5.10, we infer the following.

PROPOSITION 7.5.11 *On the sets* $D(\xi)$*, the c-topology and the weak-α-topology coincide and of course are metrizable. Moreover,* $D(\xi)$ *is closed.*

PROPOSITION 7.5.12 *Let* $D_r(\xi) = D(\xi) \cap \{x \in AC_{\mathrm{loc}}(\mathbb{R}_+,\mathbb{R}^N) : \|x(0)\| \leq r\}$*. Then* $D_r(\xi)$ *is c-compact in* $AC_{\mathrm{loc}}(\mathbb{R}_+,\mathbb{R}^N)$.

PROOF: From Proposition 7.5.11 we know that $D_r(\xi)$ is c-closed. Also for every $b > 0$ and every $t \in [0,b]$ $D_r(\xi)(t) = \{x(t) : x \in D_r(\xi)\}$ is bounded and closed, hence compact. Moreover, if $t, s \in [0,b]$, $0 \leq s < t \leq b$, and $x \in D_r(\xi)$, we have

$$\|x(t) - x(s)\| \leq \int_s^t \|x'(\tau)\|\mathrm{d}\tau \leq \int_s^t \xi(\tau)\mathrm{d}\tau.$$

Since $\xi \in L^1_{\mathrm{loc}}(\mathbb{R}_+)$, given $\varepsilon > 0$, we can find $\delta > 0$ such that, if $t - s < \delta$, then $\int_s^t \xi(\tau)\mathrm{d}\tau < \varepsilon$. Hence, we have

$$\|x(t) - x(s)\| < \varepsilon \qquad \text{for all } x \in D_r(\xi) \text{ provided } t - s < \delta,$$
$$\Rightarrow D_r(\xi) \text{ is equicontinuous.}$$

Invoking the Arzela–Ascoli theorem, we conclude that $D_r(\xi)$ is c-compact. $\square$

In the past literature on the subject, a common mistake was to claim that weak convergence in $AC_{\text{loc}}(\mathbb{R}_+, \mathbb{R}^N)$ implies pointwise convergence of at least a subsequence of the derivatives. Of course this is not true in general.

EXAMPLE 7.5.13 For $1 \leq p < \infty$, consider the sequence $\{x_n\}_{n \geq 1} \subseteq L^p[0, 2\pi]$ defined by $x_n(t) = \cos(nt)$. The Riemann–Lebesgue lemma implies that $x_n \xrightarrow{w} 0$ in $L^p[0, 2\pi]$. However, because $\|x_n\|_2 = \pi$ for all $n \geq 0$, we cannot have $x_n \longrightarrow 0$ in the norm of $L^p[0, 2\pi]$, or pointwise, or in measure.

In order to get pointwise convergence (of at least a subsequence) out of a weakly convergent sequence, we need additional conditions (see Proposition 6.6.45 and Proposition 6.6.46). These conditions are suggested by the following observation, which is an immediate consequence of Proposition 6.6.33.

PROPOSITION 7.5.14 *If* $\{x_n\}_{n \geq 1} \subseteq L^1_{\text{loc}}(\mathbb{R}_+)$ *and* $x_n \xrightarrow{w} x$ *in* $L^1_{\text{loc}}(\mathbb{R}_+)$, *then* $x \in L^1_{\text{loc}}(\mathbb{R}_+)$ *and* $\liminf_{n \to \infty} x_n(t) \leq x(t) \leq \limsup_{n \to \infty} x_n(t)$ *a.e. on* $\mathbb{R}_+$.

Now we are ready to prove the compactness of the constraint set $F(x_0)$ in the c-topology (equivalently in the weak-α-topology).

THEOREM 7.5.15 *If hypotheses* H_1 *hold, then* $F(x_0)$ *is c-compact.*

PROOF: Note that $F(x_0) \subseteq D_{\|x_0\|}(\varphi)$ and from Proposition 7.5.12, we know that $D_{\|x_0\|}(\varphi)$ is c-compact. So it suffices to show that $F(x_0)$ is c-closed. To this end let $\{x_n\}_{n \geq 1} \subseteq F(x_0)$ and assume that $x_n \xrightarrow{c} x$. Then $x_n(0) \longrightarrow x(0)$ in $\mathbb{R}^N$, $0 \leq x_n(0) \leq x_0$ for all $n \geq 1$, hence $0 \leq x(0) \leq x_0$. Also because of Proposition 7.5.11 we have that $x_n' \xrightarrow{w} x'$ in $L^1_{\text{loc}}(\mathbb{R}_+, \mathbb{R}^N)$. For every $n \geq 1$, we have

$$x_n'(t) \in G\big(t, x_n(t)\big) \quad \text{a.e. on } T.$$

We have $x_n(t) \longrightarrow x(t)$ for every $t \geq 0$. Moreover, by virtue of hypotheses H_1(ii), we have

$$\limsup_{n \to \infty} G\big(t, x_n(t)\big) \subseteq G\big(t, x(t)\big) \qquad \text{for all } t \geq 0. \tag{7.135}$$

Invoking Proposition 6.6.33 and using (7.135) and the fact that G is convex-valued, we have

$$\begin{aligned} & x'(t) \in G\big(t, x(t)\big) \quad \text{a.e. on } T, \\ \Rightarrow \ & x \in F(x_0). \end{aligned}$$

Therefore $F(x_0)$ is c-closed, hence it is c-compact. $\square$

To establish the desired upper semicontinuity of $J(\cdot)$ in the c-topology, we will need the following general upper semicontinuity result, whose proof can be found in Hu–Papageorgiou [316, p. 31] (see also Theorem 2.1.28).

THEOREM 7.5.16 *If (Ω, Σ, μ) is a finite measure space, Y is a separable Banach space, V is a separable reflexive Banach space, and $\widehat{u} : \Omega \times Y \times V \longrightarrow \overline{\mathbb{R}} = \mathbb{R} \cup \{-\infty\}$ is a measurable integrand such that*

(i) *For μ-a.e. $\omega \in \Omega$, $(y,v) \longrightarrow \widehat{u}(\omega, y, v)$ is upper semicontinuous;*
(ii) *For μ-a.e. $\omega \in \Omega$ and every $y \in Y$, $v \longrightarrow \widehat{u}(\omega, y, v)$ is concave;*
(iii) *There exist $\alpha \in L^1(\Omega)_+$ and $c>0$ such that*

$$\widehat{u}(\omega, y, v) \leq a(\omega) + c(\|y\|_Y + \|v\|_V)$$

for almost all $\omega \in \Omega$ and all $(y,v) \in Y \times V$,

then $(y,v) \longrightarrow I_{\widehat{u}}(y,v) = \int_\Omega \widehat{u}(\omega, y(\omega), v(\omega))\mathrm{d}\mu$ is sequentially upper semicontinuous from $L^1(\Omega, Y) \times L^1(\Omega, V)_w$ into $\overline{\mathbb{R}} = \mathbb{R} \cup \{-\infty\}$ (by $L^1(\Omega, V)_w$ we denote the Lebesgue–Bochner space equipped with the weak topology).

Using this theorem, we can prove the upper semicontinuity of the objective functional.

THEOREM 7.5.17 *If hypotheses H_1 and H_2 hold, then the functional $J(\cdot)$ is c-upper semicontinuous on $F(x_0)$.*

PROOF: Note that $\big(\mathbb{R}_+, \mathbf{B}(\mathbb{R}_+), \mu\big)$ is a finite measure space. Suppose $x_n \xrightarrow{c} x$ with $x_n \in F(x_0)$. Then from Lemma 7.5.9, we have that $x_n \xrightarrow{w} x$ in $AC_{\mathrm{loc}}(\mathbb{R}_+, \mathbb{R}^N)$. Applying Theorem 7.5.16, we obtain

$$\begin{aligned} &\limsup_{n\to\infty} J(x_n) \leq J(x) \\ \Rightarrow\ &J(\cdot) \text{ is } c\text{-upper semicontinuous.} \end{aligned}$$

□

From Theorems 7.5.15 and 7.5.17, using the theorem of Weierstrass, we conclude that

THEOREM 7.5.18 *If hypotheses H_1 and H_2 hold, then there exists $x^* \in F(x_0)$ such that $J(x^*) = V(x_0)$ (see (7.133)).*

The open-endedness of the planning horizon, although conceptually and mathematically elegant, is impractical from a computational point of view. To actually solve such a problem, we must usually content ourselves with finite horizon approximates. So we need to know that the values of these finite horizon approximations converge to the value $V(x_0)$ of the original problem (see (7.133)).

Hence we consider the following finite horizon growth problem:

$$V_b(x_0) = \sup\Big\{ \int_0^b e^{-\delta t} u\big(t, x(t), x'(t)\big)\mathrm{d}t : x \in F_b(x_0)\Big\}. \tag{7.136}$$

Here the feasibility set $F_b(x_0)$, is defined by

$$\begin{aligned} F_b(x_0) = \big\{&x \in AC([0,b], \mathbb{R}^N_+) : x'(t) \in G\big(t, x(t)\big) \quad \text{a.e. on } [0,b], \\ &0 \leq x(0) \leq x_0\big\}. \end{aligned}$$

We want to determine conditions which guarantee that

$$V_b(x_0) \longrightarrow V(x_0) \quad \text{as } b \longrightarrow +\infty.$$

For this reason, we strengthen hypotheses H_2 as follows.

$\mathbf{H_3}$: $u: \mathbb{R}_+ \times \mathbb{R}^N \times \mathbb{R}^N \longrightarrow \overline{\mathbb{R}} = \mathbb{R} \cup \{-\infty\}$ is a function that satisfies hypotheses H_2 and also

(v) There exists $\psi \in L^1(\mathbb{R}_+, \mu)$ such that for almost all $t \geq 0$, all $x \geq 0$, and all $y \in \mathbb{R}^N$, we have
$$\psi(t) \leq u(t, x, y).$$

THEOREM 7.5.19 *If hypotheses H_1 and H_3 hold, then $b \longrightarrow V_b(x_0)$ is continuous on $\mathbb{R}_+$.*

PROOF: Note that by Theorem 7.5.18, we have $V(x_0) \in \mathbb{R}$. We can find $x \in F(x_0)$ such that

$$-\infty < V(x_0) - \varepsilon \leq \int_0^\infty e^{-\delta t} u\big(t, x(t), x'(t)\big) dt \leq V(x_0) < +\infty. \tag{7.137}$$

Then because of hypothesis $\mathrm{H}_3(\mathrm{v})$ we have

$$\lim_{b \to +\infty} \int_0^b e^{-\delta t} u\big(t, x(t), x'(t)\big) dt = \int_0^\infty e^{-\delta t} u\big(t, x(t), x'(t)\big) dt. \tag{7.138}$$

Let $b > 0$ and set $T_b = [0, b]$. We have $x\big|_{T_b} \in F_b(x_0)$ and so

$$\begin{aligned}
&\int_0^b e^{-\delta t} u\big(t, x(t), x'(t)\big) dt \leq V(x_0), \\
\Rightarrow \; &\lim_{b \to +\infty} \int_0^b e^{-\delta t} u\big(t, x(t), x'(t)\big) dt \leq \liminf_{b \to +\infty} V_b(x_0), \\
\Rightarrow \; &V(x_0) - \varepsilon \leq \int_0^\infty e^{-\delta t} u\big(t, x(t), x'(t)\big) dt \leq \liminf_{b \to +\infty} V_b(x_0)
\end{aligned}$$

(see (7.137)). Because $\varepsilon > 0$ was arbitrary, we let $\varepsilon \downarrow 0$ and obtain

$$V(x_0) \leq \liminf_{b \to +\infty} V_b(x_0). \tag{7.139}$$

Next let $\xi = \limsup_{b \to +\infty} V_b(x_0)$. We can find $b_n \uparrow +\infty$ such that

$$\xi - \frac{1}{n} \leq V_{b_n}(x_0) \qquad \text{for all } n \geq 1.$$

Similarly as for the infinite horizon problem, the finite horizon problem has a solution (see Theorem 7.5.18). So we can find $x_n \in F_{b_n}(x_0)$ such that

$$\xi - \frac{1}{n} \leq \int_0^b e^{-\delta t} u\big(t, x_n(t), x_n'(t)\big) dt. \tag{7.140}$$

Extend x_n on all of $\mathbb{R}_+$ by setting $x_n'(t) = 0$ for all $t > b_n$. Denote the extended function by $\widehat{x}_n$. Then $\widehat{x}_n \in AC_{\mathrm{loc}}(\mathbb{R}_+, \mathbb{R}^N)$ and in fact $\widehat{x}_n \in D_{\|x_0\|}(\varphi)$. From Proposition 7.5.12, we know that $D_{\|x_0\|}(\varphi)$ is c-compact. So we may assume that

$$\widehat{x}_n \xrightarrow{c} y \quad \text{with} \;\; y \in D_{\|x_0\|}(\varphi).$$

From this it follows that

$$\widehat{x}_n(t) \longrightarrow y(t) \quad \text{for all} \;\; t \geq 0, \text{ uniformly on compact sets in } \mathbb{R}_+$$

and $\qquad \widehat{x}_n' \xrightarrow{w} y'$ in $L^1_{\mathrm{loc}}(\mathbb{R}_+, \mathbb{R}^N)$ (see Lemma 7.5.9).

Fix $k \geq 1$. Then for every $n \geq k$, we have

$$\widehat{x}_n'(t) \in G\big(t, x_n(t)\big) \quad \text{a.e. on } T_{b_k}.$$

By virtue of Proposition 6.6.33, we obtain

$$y'(t) \in G\big(t, y(t)\big) \quad \text{a.e. on } \mathbb{R}_+. \tag{7.141}$$

Moreover, we have $0 \leq y(0) \leq x_0$ and so we conclude that $y \in F(x_0)$. In addition, using Theorem 7.5.17, we have

$$\limsup_{n \to \infty} \int_0^{\infty} e^{-\delta t} u\big(t, \widehat{x}_n(t), \widehat{x}_n'(t)\big) dt \leq \int_0^{\infty} e^{-\delta t} u\big(t, y(t), y'(t)\big) dt. \tag{7.142}$$

Using hypothesis H_3(v), we have

$$\begin{aligned} &\int_0^{b_n} e^{-\delta t} u\big(t, \widehat{x}_n(t), \widehat{x}_n'(t)\big) dt + \int_{b_n}^{\infty} e^{-\delta t} \psi(t) dt \\ &\leq \int_0^{\infty} e^{-\delta t} u\big(t, \widehat{x}_n(t), \widehat{x}_n'(t)\big) dt. \end{aligned} \tag{7.143}$$

Note that because $\psi \in L^1(\mathbb{R}_+, \mu)$, we have

$$\int_{b_n}^{\infty} \psi(t) \longrightarrow 0 \quad \text{as } n \to \infty. \tag{7.144}$$

Passing to the limit as $n \to \infty$ in (7.143) and using (7.144), we obtain

$$\begin{aligned} &\limsup_{n \to \infty} \int_0^{b_n} e^{-\delta t} u\big(t, \widehat{x}_n(t), \widehat{x}_n'(t)\big) dt \\ &\leq \limsup_{n \to \infty} \int_0^{\infty} e^{-\delta t} u\big(t, \widehat{x}_n(t), \widehat{x}_n'(t)\big) dt \\ &\leq \int_0^{\infty} e^{-\delta t} u\big(t, y(t), y'(t)\big) dt \quad \text{(see (7.142))}. \end{aligned} \tag{7.145}$$

From (7.140) and (7.145) it follows that

$$\begin{aligned} &\xi \leq \int_0^{\infty} e^{-\delta t} u\big(t, y(t), y'(t)\big) dt \leq V(x_0) < +\infty, \\ \Rightarrow \; &\limsup_{n \to \infty} V_{b_n}(x_0) \leq V(x_0). \end{aligned} \tag{7.146}$$

From (7.139) and (7.146), we conclude that

$V_{b_n}(x_0) \longrightarrow V(x_0)$. $\square$

Let $S_b(x_0) = \{x^* \in F(x_0) : V_b(x_0) = J_b(x^*)\}$ and $S(x_0) = \{x^* \in F(x_0) : V(x_0) = J(x^*)\}$. Here

$$J_b(x) = \int_0^b e^{-\delta t} u(t, x(t), x'(t)) \mathrm{d}t \qquad \text{for all } x \in AC_{\text{loc}}(\mathbb{R}_+, \mathbb{R}^N).$$

From the above proof, we deduce the following.

PROPOSITION 7.5.20 *If hypotheses H_1 and H_3 hold and $b_n \uparrow +\infty$, then $\limsup_{n\to+\infty} S_{b_n}(x_0) \subseteq S_b(x_0)$ in the c-topology.*

7.6 Expected Utility Hypothesis

In this section we deal with an issue that brings us closer to the subject of the next chapter, which is the theory of games. The Expected Utility Hypothesis (EUH for short), is the dominant theory of decision making under risk, employed in economics and finance. In simple terms it says that decision makers choose (or ought to choose) among lotteries, by computing the expected value of the utility of the lottery prizes and finally selecting the lottery with the largest expected utility. In this section we want to characterize the choice behavior, which is consistent with the EUH, in risky situations where there are objective probabilities and a ranking of prizes. This setting covers many situations of interest in economic theory.

The items chosen are not commodities, but lotteries. By a lottery we understand a probability distribution on a set of prizes. There is also a set of observations. To each observation corresponds a budget, which is a set of lotteries available to the decision maker and also a choice from each budget, which is a nonempty subset of the budget. If the choices are singletons, then we have a choice function. More generally, for at least some observations the choices may be more than one. This case we refer to as the choice multifunction. The choice function or multifunction determines the choice behavior of the decision maker. There is also a utility function. Since the prizes are monetary (remember we have assumed that there is an a priori ranking of prizes), it is reasonable to assume that the utility function is strictly increasing. Then a choice behavior is EU-rational if it maximizes the expected utility.

Let us make precise the mathematical setting for the items described above. There is a set observation T which is a compact Hausdorff topological space and a space of prizes X which is a compact subset of $\mathbb{R}$. A lottery is a probability measure μ on X (i.e., $\mu \in M_1^+(X)$). We topologize $M_1^+(X)$ as follows.

DEFINITION 7.6.1 The *weak topology* on $M_1^+(X)$ is the weakest topology on $M_1^+(X)$ that makes continuous all functions $\xi_h : M_1^+(X) \longrightarrow \mathbb{R}$, $h \in C(X)$ defined by

$$\xi_h(\mu) = \int_X h(x) \mathrm{d}\mu(x).$$

We denote the weak topology on $M_1^+(X)$ by $w(M_1^+(X), C(X))$ or simply by w.

REMARK 7.6.2 A subbasic element for the weak topology on $M_1^+(X)$ is given by

$$U_{h,\varepsilon}(\mu) = \left\{\mu' \in M_1^+(X) : \left| \int_X h \mathrm{d}\mu - \int_X h \mathrm{d}\mu' \right| < \varepsilon \right\},$$

where $h \in C(X)$, $\varepsilon > 0$. Evidently for a net $\{\mu_a\}_{a \in J}$ in $M_1^+(X)$, we have $\mu_a \xrightarrow{w} \mu$ if and only if for every $h \in C(X)$, we have

$$\int_X h \mathrm{d}\mu_a \longrightarrow \int_X h \mathrm{d}\mu.$$

Evidently the weak topology on $M_1^+(X)$ is the relative w^*-topology on $M_1^+(X)$, when the latter is viewed as a subspace of the dual Banach space $M(X) = C(X)^*$. Hence the weak topology on $M_1^+(X)$ is metrizable.

A *budget* is a set of lotteries. There is a *budget multifunction* $B : T \longrightarrow 2^{M_1^+(X)} \setminus \{\emptyset\}$. This multifunction provides to the decision maker a set of lotteries for each observation $t \in T$. A *choice function* is a measurable selector $t \longrightarrow \mu_t$ of the multifunction $t \longrightarrow B_t$. For each observation $t \in T$, μ_t represents the choice of the decision maker.

DEFINITION 7.6.3 A choice function $\mu : T \longrightarrow M_1^+(X)$ is said to be *EU-rational* if there is a strictly increasing utility function $u : X \longrightarrow \mathbb{R}$ (hence automatically Borel-measurable), such that for every observation $t \in T$, the choice μ_t maximizes the expected utility over the budget set

$$\int_X u(x) \mathrm{d}\nu(x) \leq \int_X u(x) \mathrm{d}\mu_t \quad \text{for every } \nu \in B_t \text{ and every } t \in T.$$

We say that u *rationalizes* μ.

Given a choice function μ, a compound lottery can be found by first choosing $t \in T$ at random according to a prior probability λ on T and then choosing a prize at random in B_t according to μ_t. This composition process generates a new lottery ν on X as follows. For every Borel set $A \subseteq X$, we have

$$\nu(A) - \int_T \mu_t(A) \mathrm{d}\lambda(t);$$

that is, $\nu = \int_T \mu_t \mathrm{d}\lambda(t)$, where the vector-valued integral is understood as a *Gelfand integral* (see Denkowski–Migòrski–Papageorgiou [194, p. 615]).

Recall that $M(X)$ is the Banach space of all finite Borel measures on X. We know that $M(X) = C(X)^*$.

DEFINITION 7.6.4 (a) If $\mu, \nu \in M(X)$, then we say that $\nu \prec \mu$ if and only if

$$\int_X u \mathrm{d}\nu \leq \int_X u \mathrm{d}\mu \qquad \text{for all } u \in C(X) \text{ nondecreasing}.$$

(b) A choice function $\mu : T \longrightarrow M_1^+(X)$ is *ex-ante dominated* if there exist another choice function $\nu : T \longrightarrow M_1^+(X)$ and a probability measure λ on T such that

$$\int_T \mu_t \mathrm{d}\lambda(t) \prec \int_T \nu_t \mathrm{d}\lambda(t) \qquad \text{(see (a))}.$$

In this case we say that μ is dominated by ν with respect to the prior λ.

REMARK 7.6.5 In what follows $K=\{\mu\in M(X):0\prec\mu\}$, the *positive cone* of $\prec$.

We characterize EU-rational choice functions. This requires a closer look at the order described by $\prec$. We start with a result of Nachbin [447, 448]. In what follows $M_+(X)$ denotes the nonnegative elements of $M(X)$ (i.e., $\mu(A)\geq 0$ for every Borel set $A\subseteq X$), $G=\{(x,y)\in X\times X : y\leq x\}$, and D is the diagonal of the Cartesian product $X\times X$.

THEOREM 7.6.6 *$\mu\in K$ (i.e., $0\prec\mu$) if and only if there exists $\xi\in M_+(G)$ such that*

$$\int_X g\mathrm{d}\mu = \int_G \big(g(x)-g(y)\big)\mathrm{d}\xi \qquad \textit{for all } g\in C(X).$$

This theorem has some remarkable consequences, which are useful in the sequel.

COROLLARY 7.6.7 *If $u\in C(X)$, it is strictly increasing, $\mu\in K$, and $\int_X u\mathrm{d}\mu=0$, then $\mu=0$.*

PROOF: According to Theorem 7.6.6, we can find $\xi\in M_+(G)$ such that

$$\int_X u\mathrm{d}\mu = \int_{G\setminus D} \big(u(x)-u(y)\big)\mathrm{d}\xi.$$

Because u is strictly increasing we have

$$u(x)-u(y)>0 \qquad \text{for all } (x,y)\in G\setminus D. \tag{7.147}$$

We have

$$\begin{aligned} 0 &= \int_X u\mathrm{d}\mu = \int_{G\setminus D}\big(u(x)-u(y)\big)\mathrm{d}\xi, \\ &\Rightarrow\ \xi(G\setminus D)=0 \quad \text{(see (7.147))}. \end{aligned}$$

Hence for any $g\in C(X)$, we have

$$\begin{aligned} &\int_X g\mathrm{d}\mu = \int_{G\setminus D}\big(g(x)-g(y)\big)\mathrm{d}\xi = 0 \\ &\Rightarrow\ \mu=0. \end{aligned}$$

□

From the above corollary, we get the following result at once.

COROLLARY 7.6.8 *The following statements are equivalent.*

(a) *For all $u\in C(X)$ strictly increasing, we have $\int_X u\mathrm{d}\nu<\int_X u\mathrm{d}\mu$.*
(b) *For all $u\in C(X)$ nondecreasing, we have $\int_X u\mathrm{d}\nu\leq\int_X u\mathrm{d}\mu$ and $\mu\neq\nu$.*

COROLLARY 7.6.9 *If $g\in C(X)$ and for every $\mu\in K\setminus\{0\}$ we have $\int_X g\mathrm{d}\mu>0$, then g is strictly increasing.*

PROOF: Suppose $y<x$ and let $\xi=\delta_{(x,y)} \in M_1^+(G)$ (by $\delta_{(x,y)}$ we denote the Dirac measure concentrated at (x,y)). We choose $\mu \in K \setminus \{0\}$ corresponding to ξ. We have

$$0<\int_X g\mathrm{d}\mu = \int_{G\setminus D} (g(x)-g(y))\mathrm{d}\xi = g(x)-g(y),$$

$\Rightarrow$ g is strictly increasing.

□

Finally from Corollaries 7.6.7 and 7.6.8, we obtain the following.

COROLLARY 7.6.10 *If $g_0 \in C(X)$ is strictly increasing, then the set $C=\{\mu \in K : \int_X g_0\mathrm{d}\mu=1\}$ is a w^*-closed and convex set in $M(X)=C(X)^*$ and*

$$K=\{\vartheta\mu : \vartheta \geq 0,\ \mu \in C\}.$$

Now we can give some alternative descriptions of $\prec$ restricted on $M_1^+(X)$.

PROPOSITION 7.6.11 *If $\mu,\nu \in M_1^+(X)$, then the following statements are equivalent.*

(a) $\nu \prec \mu$.

(b) *For every $u : X \longrightarrow \mathbb{R}$ nondecreasing, we have $\int_X u\mathrm{d}\nu \leq \int_X u\mathrm{d}\mu$.*

(c) *For every $u \in C(X)$ strictly increasing, we have $\int_X u\mathrm{d}\nu < \int_X u\mathrm{d}\mu$ or $\nu=\mu$.*

(d) *For every $u : X \longrightarrow \mathbb{R}$ strictly increasing, we have $\int_X u\mathrm{d}\nu < \int_X u\mathrm{d}\mu$ or $\nu=\mu$.*

(e) *For every $x \in X$, $\nu(\{y : x \leq y\}) \leq \mu(\{y : x \leq y\})$.*

PROOF: (b)$\Rightarrow$(a): Obvious (see Definition 7.6.4(a)).

(a)$\Leftrightarrow$(c): Follows from Corollary 7.6.8.

(d)$\Rightarrow$(c): Obvious.

(e)$\Rightarrow$(b): Let $\mathcal{S}(X)$ be the family of all increasing sets $A \subseteq X$; that is, χ_A (the indicator function of A) is increasing. Let $A \in \mathcal{S}(X)$. We need to show that $\nu(A) \leq \mu(A)$. For this purpose, given $\varepsilon > 0$ we can find $C_\varepsilon \subseteq A$ closed, such that $\mu(A)-\mu(C_\varepsilon)<\varepsilon$ and $\nu(A)-\nu(C_\varepsilon)<\varepsilon$ (regularity of the measures μ,ν). Let $E_\varepsilon=\{y \in X : x \leq y$ for some $x \in C_\varepsilon\}$. Evidently E_ε is closed and $E_\varepsilon \subseteq A$. Because of **(e)**, we have $\nu(E_\varepsilon) \leq \mu(E_\varepsilon)$ and because $C_\varepsilon \subseteq E_\varepsilon \subseteq A$, we have $\nu(A) \leq \nu(E_\varepsilon)+\varepsilon \leq \mu(E_\varepsilon)+\varepsilon$. But $\varepsilon>0$ was arbitrary. So we let $\varepsilon \downarrow 0$ and obtain $\nu(A) \leq \mu(A)$.

(b)$\Rightarrow$(e): This follows at once by taking u to be the indicator function of the set $\{y : x \leq y\}$.

(c)$\Rightarrow$(e): This implication holds because the indicator of $\{y : x \leq y\}$ is the poitwise limit of continuous nondecreasing functions.

(b)⇒(d): Obvious.

(e)⇒(d): From Kamae–Krengel–O'Brien [339, Theorem 1], we know that if (e) holds, then there are random variables S_μ and S_ν defined on $[0,1]$ equipped with the Lebesgue measure with values in X and distributions μ and ν, respectively, such that $S_\nu \leq S_\mu$ and the inequality is strict on a set of positive Lebesgue measure. Therefore if $u : X \longrightarrow \mathbb{R}$ is strictly increasing, then we have $u(S_\nu) \leq u(S_\mu)$ with strict inequality on a set of positive Lebesgue measure. From this we conclude that (d) holds. $\square$

REMARK 7.6.12 In the literature, the partial order induced on $M_1^+(X)$ by any of the equivalent statements of Proposition 7.6.11, is called *first-order stochastic dominance*.

PROPOSITION 7.6.13 *If V is a compact Hausdorff topological space, Y is a separable Banach space, $F:V\longrightarrow 2^{Y^*}\setminus\{\emptyset\}$ is a multifunction with w^*-compact and convex values which is Vietoris continuous (see Definition 6.1.2(c)) from V into Y^* endowed with the w^*-topology (denoted by $Y^*_{w^*}$), and*

$$R_c = \overline{\mathrm{conv}} \bigcup_{v\in V} F(v) \quad \text{and} \quad G = \bigcup_{\lambda\in M_1^+(V)} \int_V F(v)\mathrm{d}\lambda(v)$$

then $R_c = G$ and G is w^*-compact and convex.

PROOF: From Proposition 6.1.13, we know that $F(V)\subseteq Y^*$ is w^*-compact. Recall that $Y^*_{w^*}$ has the *convex compact property* (i.e., compact sets in $Y^*_{w^*}$ have a compact closed convex hull). Therefore we deduce that R_c is w^*-compact, convex.

First we show that

$$G \subseteq R_c. \tag{7.148}$$

We argue indirectly. So suppose that we can find $y^* \in G$, such that $y^* \notin R_c$. Then by the strong separation theorem, we can find $\widehat{y} \in Y$ such that

$$\sigma(\widehat{y}, R_c) < \langle y^*, \widehat{y}\rangle \tag{7.149}$$

(by $\langle\cdot,\cdot\rangle$ we denote the duality brackets for the pair (Y^*, Y)). By definition

$$y^* = \int_V u^*(v)\mathrm{d}\lambda(v),$$

where $\lambda\in M_1^+(V)$ and $u:V\longrightarrow X^*$ is w^*-measurable and $u^*(v)\in F(v)$ for all $v\in V$. Then

$$\langle y^*, \widehat{y}\rangle = \int_V \langle u^*(v), \widehat{y}\rangle\,\mathrm{d}\lambda(v) \leq \int_V \sigma\big(\widehat{y}, F(v)\big)\mathrm{d}\lambda(v) \leq \sigma(\widehat{y}, R_c). \tag{7.150}$$

Comparing (7.149) and (7.150), we reach a contradiction. Therefore (7.148) holds.

Next we show that

$$R_c \subseteq G. \tag{7.151}$$

If $y^* \in G$, then by definition we can find $\lambda \in M_1^+(V)$ such that

$$\langle y^*, \widehat{y}\rangle \leq \int_V \sigma\big(\widehat{y}, F(v)\big)\mathrm{d}\lambda(v) \qquad \text{for all } y \in Y, \tag{7.152}$$

if $y^* \in R_c$. Note that because Y is separable, the relative w^*-topology on R_c is compact and metrizable. So we can find $\{y_n^*\}_{n\geq 1} \subseteq \operatorname{conv} F(V)$ such that $y_n^* \xrightarrow{w^*} y^*$ in Y^*. There exist $\{\lambda_n\}_{n\geq 1} \subseteq M_1^+(V)$, each with finite support, such that

$$y_n^* = \int_V f(v)\mathrm{d}\lambda_n(v),$$

where $f : V \longrightarrow X^*$ is a w^*-measurable selector of F. We have

$$\langle y_n^*, \widehat{y}\rangle \leq \int_V \sigma\big(\widehat{y}, F(v)\big)\mathrm{d}\lambda_n(v). \tag{7.153}$$

We may assume that $\lambda_n \xrightarrow{w^*} \widehat{\lambda}$ in $M(V) = C(V)^*$ and $\widehat{\lambda} \in M_1^+(V)$. Because $v \longrightarrow \sigma\big(\widehat{y}, F(v)\big)$ is continuous (recall F is Vietoris continuous), we have

$$\int_V \sigma\big(\widehat{y}, F(v)\big)\mathrm{d}\lambda_n(v) \longrightarrow \int_V \sigma\big(\widehat{y}, F(v)\big)\mathrm{d}\widehat{\lambda}(v). \tag{7.154}$$

Therefore, if we pass to the limit as $n \to \infty$ in (7.153) and we use (7.154), we obtain

$$\begin{aligned} & \langle y^*, \widehat{y}\rangle \leq \int_V \sigma\big(\widehat{y}, F(v)\big)\mathrm{d}\widehat{\lambda}(v), \\ \Rightarrow\ & \widehat{y} \in G \qquad \text{(see (7.148))}, \\ \Rightarrow\ & R_c \subseteq G. \end{aligned} \tag{7.155}$$

Combining (7.151) and (7.155), we conclude that $R_c = G$. □

Now we can have the main theorem on the rationalization of the choice function.

THEOREM 7.6.14 *If*

(i) *The set of prizes X is a compact subset of $\mathbb{R}$;*

(ii) *The set of observations T is a compact Hausdorff topological space;*

(iii) *If on $M_1^+(X)$, we consider the weak topology (see Definition 7.6.1), the budget multifunction $B : T \longrightarrow 2^{M_1^+(X)} \setminus \{\emptyset\}$ has compact convex values and it is Vietoris continuous;*

and $\mu : T \longrightarrow M_1^+(X)$ is the choice function, then

(a) *If μ is ex-ante dominated, then it is not EU-rational.*
(b) *If μ is continuous, then one of the following alternatives holds.*
   (b₁) *μ is rationalized by a continuous strictly increasing utility function.*
   (b₂) *μ is ex-ante dominated.*

PROOF: (a) Clear from Definition 7.6.4.

(b) Suppose μ is continuous and it is ex-ante dominated. Then we can find a measurable selector ν of the budget multifunction such that for some $\lambda \in M_1^+(T)$, we have

$$\begin{aligned} &\int_T \mu_t \mathrm{d}\lambda(t) \prec \int_T \nu_t \mathrm{d}\lambda(t) \quad \text{(see Definition 7.6.4(b))}, \\ \Rightarrow \ &\int_T (\nu_t - \mu_t)\mathrm{d}\lambda(t) \succ 0. \end{aligned} \tag{7.156}$$

Set

$$G = \bigcup \Big\{ \int_T (B_t - \mu_t)\mathrm{d}\lambda(t) : \lambda \in M_1^+(T) \Big\}. \tag{7.157}$$

If μ is not ex-ante dominated, then $G \cap K \neq \emptyset$ (see (7.156)). Let $C \subseteq K$ be as in Corollary 7.6.10 (i.e., a w^*-closed convex base of the cone K). Then $G \cap C = \emptyset$.

Recall that the weak topology on $M_1^+(X)$, is the relative w^*-topology, that $M_1^+(X)$ inherits from the dual Banach space $M(X) = C(X)^*$. So $t \longrightarrow B_t - \mu_t$ is Vietoris continuous into $M(X)$ with the w^*-topology (denoted by $M(X)_{w^*}$) and has nonempty, compact, and convex values in $M(X)_{w^*}$. Because $G \cap C = \emptyset$, by the strong separation theorem we can find $u \in C(X) = \big(M(X)_{w^*}\big)^*$ such that

$$\langle \nu, u \rangle \leq 0 \qquad \text{for all } \nu \in G \tag{7.158}$$

$$\text{and} \quad \langle \xi, u \rangle > 0 \qquad \text{for all } \xi \in C. \tag{7.159}$$

Here by $\langle \cdot, \cdot \rangle$ we denote the duality brackets for the pair $\big(C(X), M(X)\big)$. From (7.159), we have

$$\langle \xi, u \rangle = \int_X u \mathrm{d}\xi > 0 \qquad \text{for all } \xi \in C.$$

This combined with Corollaries 7.6.10 and 7.6.9, implies that

$$u \in C(X) \quad \text{is strictly increasing.}$$

We claim that it rationalizes μ. To this end let $t \in T$ and let $\nu \in B_t$. Consider $\lambda \in M_1^+(T)$ defined by

$$\lambda = \delta_{\{t\}} \quad \text{(the Dirac measure concentrated on } t).$$

Note that $\nu - \mu_t \in B_t - \mu_t$ and so

$$\int_T (\nu - \mu_t)\mathrm{d}\lambda(t) = \nu - \mu_t \in G.$$

Then because of (7.158), we have

$$\begin{aligned} &\langle u, \nu - \mu_t \rangle \leq 0, \\ \Rightarrow \ &\int_X u \mathrm{d}\nu \leq \int_X u \mathrm{d}\mu_t, \\ \Rightarrow \ &u \quad \text{rationalizes } \mu \quad \text{(see Definition 7.6.3).} \end{aligned}$$

□

EXAMPLE 7.6.15 *The compactness of the space X of prizes cannot be dropped:* Indeed consider the situation where $T=\{t\}$, $X = \{0, 1, 2, \ldots\}$ and $B_t = \{\mu_k\}_{k \geq 0}$ with $\mu_0 = \delta_{\{t\}}$ and for $n \geq 1$, $\mu_n = (1/n)\delta_{\{0\}} + \big(1 - (1/n)\big)\delta_{\{2\}}$. The choice μ_0 is not $\prec$-dominated, but we can check that it is not EU-rational.

7.7 Remarks

7.1: The first to formulate and study of model of a pure exchange was Edgeworth [219]. He introduced a model with two economic agents and two commodities and using the recontracting adjustment process (i.e., no allocation occurs if it can be blocked – recontracted out – by a group of agents), he defined the well-known (Edgeworth) contract curve. This curve represents the locus of points where the marginal rate of substitution between the pair of goods, is the same for both agents. This curve can be illustrated, using a diagrammatic construction known as the Edgeworth box. The core, introduced in Definition 7.1.3 is a generalization of the contract curve. Note that the notion of core does not involve prices and it is the result of pure trading among the agents. In contrast, the other equilibrium notion, the Walras (or competitive) equilibrium, introduced in Definition 7.1.5(b), assumes that in the market prevail some prices for the commodities, which are accepted by all agents involved. But, intuitively speaking, we know that in reality, prices are a convenience and simply provide a common yardstick to measure all traded quantities. Therefore it is reasonable to claim that the two equilibrium notions should lead to the same allocations. A moment's reflection can lead to the conclusion that this should be the case in a perfect competition environment (see Theorem 7.1.7). This means that no individual agent or group of agents has enough power alone to influence the outcome of the economic process. The idea to model perfect competition using a continuum of agents (more precisely a nonatomic measure) is due to Aumann [43, 45]. In Aumann [43] we find the core-Walras equivalence theorem (see Theorem 7.1.7), and in Aumann [45], the author establishes the existence of a Walras (competitive) equilibrium (see Theorem 7.1.10). The results of Aumann were extended to economies with production by Hildenbrand [295]. There are some other equilibrium notions such as Pareto equilibrium, quasi-competitive equilibrium, and expenditure minimizimg equilibrium. A comparative study of all these different equilibrium notions can be found in Hildenbrand [296].

There is another approach in the core theory in which the primitive notion in the economic model is the coalition and not the individual agent. This changes the necessary mathematical tools and so multifunctions are replaced by set-valued measures (multimeasures). This approach was initiated by Vind [595].

There is still a third, more complicated pure exchange model, with "big" and "small"traders due to Dreze–Gabsewicz–Schmeidler–Vind [207].

In the last twenty years, a lot of effort was put into extending equilibrium theory to economies with an infinite-dimensional commodity space. In this direction we mention the works of Bewley [74], Aliprantis–Brown [8], Florenzano [250], Mas-Colell [411], Jones [333], Zame [618], and the book of Aliprantis–Brown–Burkinshaw [9], where a more detailed bibliography can be found.

The books of Debreu [185], Hildenbrand [297], Hildenbrand–Kirman [298], and Takayama [572] contain additional material on economic equilibrium theory. In particular, Hildenbrand [297] deals with economies described by a continuum of agents.

It does historical justice to say that the founding fathers of mathematical economics are Cournot, Walras, Pareto, Edgeworth, and Menger. Antoine Cournot (1801–1877) was French and is considered to have made the initial step in developing econometrics. Leon Walras (1834–1910) was French and the first holder in 1870 of the chair of political economy at the University of Lausanne in Switzerland. His two main contributions in the theory of mathematical economics are the development of the marginal utility approach to the theory of value (1873) and the development of the theory of general equilibrium (1874–1877). Vifredo Pareto (1848–1923) was Italian and the successor of Walras to the chair of political economy at the University of Lausanne (1892). He is better known for the Pareto optimum, which is described as a position from which it is impossible to improve anyone's welfare by altering production or exchange without impairing someone else's welfare. He is also the founder of positive economics, an economic science purged of all ethical elements. Pareto rejected socialism, justified the inequality of income, and viewed with sympathy the coming to power in Italy of the fascists (October 1922). Francis Ysidro Edgeworth (1845–1926) was British and professor of political economy at Oxford University from 1891 to 1922. He is credited with inventing indifference curves and the contract curve. Finally Carl Menger (1840–1921) was Austrian and one of the formulators (together with Walras) of the marginal utility theory of value. He is also the founder of the Austrian school of economic thought.

7.2: The question of price characterization of efficient capital accumulation and allocation of resources in infinite horizon economies ("efficient pricing") started with the work of Malinvaud [403, 404], and continued by Gale [254] in the context of optimal growth for multisector growth models ("competitive pricing"). Gale [254] introduced the notion of strong maximality (see Definition 7.2.24(b)) and assumed that the utility function is strictly concave. Brock [107] objected to this hypothesis, primarily because it excludes the von Neumann model. Brock [107] replaced the strict concavity of the utility function by the assumption that the optimal stationary program is unique and used the weak maximality criterion (see Definition 7.2.26(b)). Other contributions to the subject were made by Peleg [492], Peleg–Yaari [493], McKenzie [420, 423], Mitra [431, 432, 434], Mitra–Zilcha [433], Dechert–Nishimura [183], Khan–Mitra [350], and Joshi [334]. The books of Makarov–Rubinov [402] and Takayama [572] deal with discrete time growth models and contain additional references on the subject.

Starting with the work by Hurwicz [319] on informational decentralization, people raised the question of characterizing the optimality of competitive programs in terms of conditions that can be verified by agents in an informationally decentralized mechanism, which involves two basic restrictions; first that there is initial dispersion of information and each economic unit has only partial knowledge of the environment; second that there is limited communication and so it is impossible to completely centralize dispersed information about the environment. Significant contributions in this direction, were made by Brock–Majumdar [111], Hurwicz–Majumdar [320], Hurwicz–Weinberger [321], and Dasgupta–Mitra [177, 178] who introduced the reachability condition (R) (see Definition 7.2.17) and have additional particular economic models which satisfy this condition.

7.3: Turnpike theory originates with a simple remark in the book of Dorfman–Samuelson–Solow [203]. Radner [509] proved a weak turnpike theorem for the von Neumann–Gale model, and McKenzie [418] did this for a Leontief-type model. A strong turnpike theorem was first proved by Nikaido [461]; see also Tsukui [589] and Winter [607]. Additional results can be found in McKenzie [419, 420, 421, 422, 423], Araujo–Scheinkman [28], Benveniste–Scheinkman [65], Fershtman–Mullar [242], and Read [512], and in the books of Arkin–Evstingeev [29] and Takayama [572].

7.4: Optimal growth under uncertainty was first investigated in the early seventies. We mention the works of Brock–Mirman [108, 109], Dana [173], Dynkin [215], Evstingeev [230], Jeanjean [331], Radner [510], and Tacsar [570]. Brock–Mirman modelled the uncertainty in their model with a sequence of independent, identically distributed (iid) random variables. Dynkin–Evstingeev, Jeanjean, Radner, and Tascar considered a stationary Markov process modelling the uncertainty in the economy, considered the more general situation of a probability distribution on the set of sequences of states. This approach incorporates the other two via Kolmogorov's theorem (for the idd case) and the Ionescu–Tulcea disintegration theorem (for the Markovian case).

Soon thereafter Zilha [628, 629] considered the discounted model (see Zilha [628]) and the undiscounted model using the strong maximality criterion (see Zilha [629]) and established the existence of supporting prices for optimal programs.

More recent works on the subject are those by Evstingeev–Katyshev [231], Majumdar–Radner [400], Majumdar–Zilha [401], Nyarko [466], Zilha [630], and Pantelides–Papageorgiou [469]. In particular, Nyarko [466] extends to a stochastic economic growth model the work of Dasgupta–Mitra [177] on temporal and informational decentralization.

Various aspects of the theory of stochastic economic dynamics can be found in the book of Arkin–Evstingeev [29].

7.5: Continuous-time economic growth problems present more technical difficulties than the discrete ones. So the choice of the topology on the set of feasible programs is crucial. The investigation of the c-topology and the α-topology was conducted by Becker–Boyd–Sung [60], who this way were able to rectify some erroneous claims existing in the literature. Namely, Brock–Haurie [110] and Yano [612] claimed that stock convergence implies, at least for a subsequence, flow convergence almost everywhere and Balder [55] claimed that the c-topology and the weak α-topology coincide. We should mention that the problem studied by Becker–Boyd–Sung [60] involves a recursive utility. Undiscounted continuous-time growth models can be found in the works of Brock–Haurie [110], Papageorgiou [479], and in the book of Carlson–Haurie–Leizarowitz [130].

7.6: The model used in this section is due to Border [88]; see also Border [87]. Earlier works on the subject by Fishburn [248] and Ledyard [372] assumed the set of observations and the budget set are finite. Definition 7.6.4(a) is due to Nachbin [447]; see also Nachbin [448]. Proposition 7.6.11 is essentially due to Kamae–Krengel–O'Brien [339].

8

Game Theory

Summary. *Game-theoretic models provide a substantial amount of generalization of the basic notions of mathematical economics, such as core, equilibria, saddle points and intertempolar optimum. In this chapter, we deal with different models in game theory. We start with noncooperative games which lead to the notion of "Nash equilibria". Then we pass to cooperative games, for which we can define the notion of core. After that we consider random games with a continuum of players. For such games, we show the existence of Cournot–Nash equilibria. We also consider Bayesian games stochastic 2-player, and zero-sum games. Finally, using the notion of ε-subdifferential of convex functions, we prove the existence of approximate Nash equilibria for noncooperative games.

Introduction

Any discussion of economic models is incomplete if it is not accompanied by a presentation of game theory. This is because the models in game theory provide a substantial amount of generalization of the basic notions of mathematical economics such as core, equilibria, saddle point, and intertemporal optimum. A game is a decision problem for a group of players who may have cooperative or noncooperative (conflicting) objectives to optimize, with different or equal information structures and finite strategies from which to choose. In this chapter we present some basic issues from the theory of games. We deal with both static and dynamic (over an infinite discrete-time horizon) games and with deterministic and stochastic games.

In Section 8.1, we examine noncooperative games with n-players. The basic equilibrium concept for such games, is the so-called Nash equilibrium. We establish the existence of such equilibrium.

In Section 8.2, we deal with cooperative games with n-players. We consider the characteristic function form of such games and we study both side-payment games and no-side-payment games. For both the central stability concept is the core. We present theorems that guarantee the nonemptiness of the core.

In Section 8.3 we deal with random games, with a continuum of players (namely, atomless measure space of players) and, an infinite-dimensional strategy space. For

N.S. Papageorgiou, S.Th. Kyritsi-Yiallourou, *Handbook of Applied Analysis*,
Advances in Mechanics and Mathematics 19, DOI 10.1007/b120946_8,

such games, using notions and results from multivalued analysis, we prove the existence of Cournot–Nash equilibria.

Continuing with random games, in Section 8.4 we consider Bayesian games with an infinity of players and an infinite-dimensional strategy space. In such games the players have different information structures that are modelled as sub-σ-fields of the σ-field of all possible outcomes. Again we show that under reasonable hypotheses, Bayesian games admit Cournot–Nash equilibria.

In Section 8.5 we investigate stochastic 2-player zero-sum games, using the stochastic dynamic programming model, which we also present. We show the existence of a saddle point.

Finally in Section 8.6 we introduce and use the notion of ε-subdifferential of a convex function, in order to show that noncooperative n-player games with noncompact strategy sets admit approximate Nash equilibrium.

8.1 Noncooperative Games–Nash Equilibrium

In this section we study n-person games in normal form, in which the players exhibit a noncooperative (individualistic) behavior. The central concept for such games is the notion of Nash equilibrium.

We start with a description of the n-person game in normal (strategic) form. So let $N=\{1,2,\dots,n\}$, $n \geq 2$, be the set of players. Each player $k \in N$ has a strategy set X_k which describes all strategies available to him. We set

$$X = \prod_{k=1}^{n} X_k \quad \text{(the multistrategies set)}.$$

Also a set $S(N) \subseteq X$ is given describing the feasible multistrategies.

The preference relation of player $k \in N$ among the strategies available to him, is described by a utility function $u_k : S(N) \longrightarrow \mathbb{R}$, which to each feasible multistrategy $x \in S(N)$ assigns the number $u_k(x)$ measuring the utility produced by x. Of course he prefers multistrategies that yield a maximum utility. We can also think of u_k as a loss function, in which case $-u_k$ can be regarded as a utility function. The interpretation of u_k as a utility function, which we have adopted here, is motivated from the economic applications. From the viewpoint of player $k \in N$, the set X of multistrategies, is split as follows

$$X = \widehat{X} \times X_k \quad \text{with } \widehat{X} = \prod_{i \neq k} X_i.$$

We think of $k^c = N \setminus \{k\}$ as a coalition adverse or complementary to player $k \in N$. Also let p_k and $\widehat{p}$ be the projections from X onto X_k and $\widehat{X}$, respectively. The decomposition $(\widehat{x}, x_k) \in \widehat{X} \times X_k$ of a multistrategy $x \in X = \widehat{X} \times X_k$ is important for player $k \in N$, because this way he distinguishes the component x_k which he can control from the components $\widehat{x} = \{x_i\}_{i \neq k} \in \widehat{X}$ over which he has no control.

DEFINITION 8.1.1 A game in *normal form* is the collection $\{X_k, u_k\}_{k \in N}$.

Suppose that no player has any information about the choices made by the other players. Then a cautious way to proceed is for each player to maximize her minimum utility. More precisely, for every player $k \in N$, let

$$v_k(x_k) = \inf\big[u_k(x) : x \in S(N),\ p_k(x) = x_k\big]. \tag{8.1}$$

The quantity $v_k(x_k)$ in (8.1) represents the least utility for player k among all feasible multistrategies, when she employs strategy x_k. Then she will try to maximize $v_k(\cdot)$; that is,

$$\widehat{v}_k = \sup\big[v_k(x_k) : x_k \in p_k\big(S(N)\big)\big]. \tag{8.2}$$

Note that $\widehat{v}_k$ is the outcome of a sup inf process.

DEFINITION 8.1.2 The vector $\widehat{v} = \{\widehat{v}_k\}_{k=1}^n$ is the *conservative value* of the game and a multistrategy $x^* = \{x_k^*\}_{k=1}^n \in S(N)$ such that

$$\widehat{v}_k = v_k(x_k^*)$$

is a *conservative* or *maximin multistrategy*.

We have the following existence theorem for conservative multistrategies.

THEOREM 8.1.3 *If the strategy sets X_k, $k \in N$, are Hausdorff topological spaces, $S(N) \subseteq X$ is compact (for the product topology on X), and $u_k : S(N) \longrightarrow \mathbb{R}$, $k \in N$, is upper semicontinuous, then the game admits a conservative multistrategy $x^* \in S(N)$.*

PROOF: Note that

$$\begin{aligned} v_k(x_k) &= \inf\big[u_k(x) : x \in S(N),\ p_k(x) = x_k\big] \\ &= \inf\big[u_k(\widehat{x}, x_k) : (\widehat{x}, x_k) \in S(N)\big]. \end{aligned}$$

So from Theorem 6.1.18(a) we have that $v_k : p_k\big(S(N)\big) \longrightarrow \mathbb{R}$ is upper semicontinuous. Because $p_k\big(S(N)\big)$ is compact in X_k, we can find $x_k^* \in p_k\big(S(N)\big)$ such that

$$\widehat{v}_k = v_k(x_k^*),\ k \in N.$$

Evidently $x^* = \{x_k^*\}_{k \in N}$ is a conservative multistrategy for the game. □

REMARK 8.1.4 In general note that player $k \in N$ will reject any multistrategy x yielding a utility $u_k(x)$ smaller than $\widehat{v}_k$. So $\widehat{v}_k$ is the minimum that each player can tolerate.

In the above definition of conservative strategy, every player chooses his strategy independently of the rest of the players. We want to consider a different kind of behavior, in which every player varies his strategy based on the choice of the complementary coalition. Roughly speaking, what happens is the following: As soon as player k announces her intention to play, the complementary coalition $N \setminus \{k\}$ moves first and then player k responds. With such a rule, an equilibrium will be a multistrategy x^* in which no player k can increase her utility by moving away from strategy x_k^*, provided that the other players do not alter their choice. So we have

$$u_k(\widehat{x}^*, x_k^*) = \sup \left[u_k(\widehat{x}^*, y) : y = p_k(x),\ x \in S(N)\right]. \tag{8.3}$$

In other words, (8.3) says that given the multistrategy $\widehat{x}^*$ of the complementary coalition $N \setminus \{k\}$, the k-player responds by maximizing the utility

$$y \longrightarrow u_k(\widehat{x}^*, y) \quad y = p_k(x), \quad x \in S(N).$$

DEFINITION 8.1.5 Given a game $\{X_k, u_k\}_{k\in N}$ in normal form, a *Nash equilibrium* is a feasible multistrategy $x^* \in S(N)$, such that for every $k \in N$, we have

$$u_k(\widehat{x}^*, y) \leq u_k(\widehat{x}^*, x_k^*) \qquad \text{for all } y = p_k(x),\ x \in S(N).$$

REMARK 8.1.6 So in the Nash equilibrium there is no incentive for a player $k \in N$ to change her strategy x_k^* by herself. This is a noncooperative equilibrium concept.

To study the notion of Nash equilibrium, we introduce the function $\xi : S(N) \times S(N) \longrightarrow \mathbb{R}$ defined by

$$\xi(x, z) = \sum_{i=1}^{n} \left[u_i(x) - u_i(\widehat{x}, z_i)\right]. \tag{8.4}$$

PROPOSITION 8.1.7 *If $\{X_k, u_k\}_{k\in N}$ is a game in normal form and for $x \in S(N)$ we have*

$$\xi(x, z) \geq 0 \qquad \text{for all } z \in S(N), \tag{8.5}$$

then x is a Nash equilibrium for the game. The converse is true provided that $S(N)$ is a product space.

PROOF: Suppose that $x \in S(N)$ satisfies (8.5) and let $z \in S(N)$ such that $\widehat{p}(z) = \widehat{x} = \{x_i\}_{i\neq k}$. Using (8.5), we have

$$\begin{aligned} 0 \leq \xi(x, z) &= \sum_{i=1}^{n} \left[u_i(x) - u_i(\widehat{x}, z_i)\right] \\ &= \sum_{i\neq k} \left[u_i(x) - u_i(\widehat{x}, x_i)\right] + \left[u_k(x) - u_k(\widehat{x}, z_k)\right] \\ &= u_k(x) - u_k(\widehat{x}, z_k). \end{aligned}$$

There $x \in S(N)$ is a Nash equilibrium (see Definition 8.1.5).

Now suppose that $S(N) = \prod_{i=1}^{n} S_i \subseteq \prod_{i=1}^{n} X_i$ and let $x^* \in S(N)$ be a Nash equilibrium. Then by Definition 8.1.5, we have

$$u_i(x^*) - u_i(\widehat{x}^*, z_i) \geq 0 \qquad \text{for all } z_i \in S_i,\ i \in \{1, \ldots, n\} = N.$$

Adding these inequalities, we obtain $\xi(x^*, z) \geq 0$ for all $z \in S(N)$. □

To prove the existence of a Nash equilibrium, we need the following abstract theorem, known as *Ky Fan's inequality*. For a proof, we refer to Aubin–Ekeland [39, p. 327].

THEOREM 8.1.8 *If $K \subseteq \mathbb{R}^N$ is a nonempty, compact, convex set and $\xi : K \times K \longrightarrow \mathbb{R}$ is a function satisfying*

(i) *For every $z \in K$, $x \longrightarrow \xi(x,z)$ is upper semicontinuous;*
(ii) *For every $x \in K$, $z \longrightarrow \xi(x,z)$ is quasiconvex (see Definition 2.3.11);*
(iii) *For every $z \in K$, $\xi(z,z) \geq 0$,*

then we can find $x^* \in K$ such that

$$\inf_{z \in K} \xi(x^*, z) \geq 0.$$

Now we can state and prove the main existence theorem for Nash equilibria.

THEOREM 8.1.9 *If $\{X_k, u_k\}_{k \in N}$ is a game in normal form, $S(N) \subseteq \mathbb{R}^N$ is compact convex, for every $k \in N$ u_k is continuous, and for every $\widehat{x} \in \widehat{p}(S(N))$, $z \longrightarrow u_k(\widehat{x}, z)$ is quasiconcave, then there exists a Nash equilibrium.*

PROOF: The function

$$\xi(x,z) = \sum_{i=1}^{n} \left[u_i(x) - u_i(\widehat{x}, z_i)\right], \quad (x,z) \in S(N) \times S(N)$$

is continuous in x and quasiconvex in z. Moreover,

$$\xi(z,z) \geq 0 \qquad \text{for all } z \in S(N).$$

So we can apply Theorem 8.1.8 and obtain $x^* \in S(N)$ such that

$$\xi(x^*, z) \geq 0 \qquad \text{for all } z \in S(N).$$

This then by virtue of Proposition 8.1.7 implies that $x^* \in S(N)$ is a Nash equilibrium for the game. ☐

At this point it is instructive to briefly present a special case, known as a *bimatrix game*.

EXAMPLE 8.1.10 *Bimatrix Game*: This is a two-player game; that is, $N = \{1, 2\}$ and the two strategy sets X_1 and X_2 are finite sets; for example, $X_1 = \{1, \ldots, l_1\}$ and $X_2 = \{1, \ldots, l_2\}$. Also we have two utility functions u_1 and u_2. We set

$$a_{pq} = u_1(p,q) \quad \text{and} \quad b_{pq} = u_2(p,q) \qquad \text{for all } p \in X_1,\ q \in X_2.$$

So a_{pq} is the first player's payoff when the first player chooses a (pure) strategy $p \in X_1$ and the second chooses a (pure) strategy. Similarly for the payoff b_{pq} of the second player. Then the game $\{X_k, u_k\}_{k=1}^{2}$ reduces to the following bimatrix game:

$$\begin{bmatrix} (a_{11}, b_{11}) & \cdots & (a_{1l_2}, b_{1l_2}) \\ (a_{l_1 1}, b_{l_1 1}) & \cdots & (a_{l_1 l_2}, b_{l_1 l_2}) \end{bmatrix}.$$

Evidently a Nash equilibrium in general need not exist because the strategy sets X_1 and X_2 are not in general convex. The sets X_1 and X_2 are said to represent the pure strategies available to player 1 and player 2, respectively. When $a_{pq} + b_{pq} = 0$ for all $(p,q) \in \{1, \ldots, l_1\} \times \{1, \ldots, l_2\}$, we say that we have a *zero-sum game*. It represents a situation of direct conflict between the two players, because the gain of one is equal to the loss of the other. To be able to produce a Nash equilibrium, we need to consider mixed strategies. Namely, we are led to another game in normal form with strategy sets

$$\overline{X}_k = \left\{ x = (x_i)_{i=1}^{l_k} \in \mathbb{R}_+^{l_k} : \sum_{i=1}^{l_k} x_i = 1 \right\}, \quad k = 1, 2.$$

A point in the simplex $\overline{X}_k$ is a mixed strategy and each component x_i represents the probability attached to the pure strategy i. When the two players choose mixed strategies $(x, y) \in \overline{X}_1 \times X_2$, then the payoffs are

$$\sum_{p,q} a_{pq} x_p y_q \quad \text{for player 1} \tag{8.6}$$

$$\text{and} \quad \sum_{p,q} b_{pq} x_p y_q \quad \text{for player 2.} \tag{8.7}$$

In this case all hypotheses of Theorem 8.1.9 are satisfied and so a Nash equilibrium exists.

Now let us see a specific bimatrix game with a unique Nash equilibrium. This is the famous *prisoner's dilemma problem.* There are two players who are prisoners and have been charged with some joint crime. Each player has a strategy set with two elements (strategies). The first strategy is to plead innocent and the second is to plead guilty. If both prisoners plead guilty they will be punished. If both plead innocent, then there is no solid case against them and their punishment will be light. If one pleads guilty and other pleads innocent, then the first will be rewarded for his honesty and will be released, and the second will be punished severely. So let the bimatrix representing the game be the following one

$$\begin{bmatrix} (10, 10) & (0, 15) \\ (15, 0) & (5, 5) \end{bmatrix}. \tag{8.8}$$

The entries in this bimatrix are not the terms of imprisonment, but are the utility levels. Evidently the Nash equilibrium for this bimatrix game exists uniquely and is the pair of strategies that generates $(5, 5)$; that is, both prisoner's plead innocent. If we pass to the mixed strategies the situation does not change. The Nash equilibrium remains the same (a pure strategy) and is unique. Indeed let $(x^*, y^*) \in \overline{X}_1 \times \overline{X}_2$ be a Nash equilibrium. Then given y^*, x_1^* solves

$$\max_{0 \le x_1 \le 1} \left\{ 10.x_1 y_1^* + 0.x_1 y_2^* + 15.(1 - x_1) y_1^* + 5.(1 - x_1) y_2^* \right\} \text{ (see (8.6))}$$

$$\Rightarrow \; x_1^* = 0.$$

By using (8.7) we obtain $y_1^* = 0$. So indeed is the strategy producing utility $(5, 5)$ (see (8.8)) as before.

Note that in this game the incentive, for a rational player concerned only with his own survival, is to confess and let the other suffer the consequences. But, as both are motivated to act in this way, they end up with an outcome that is worse for both than if they had been able to make a binding agreement between themselves that no one confesses. This example illustrates how a rational behavior at a micro level leads to an apparently irrational macro outcome.

An interesting interpretation of the bimatrix game is in wage negotiations in a labor market. In this situation, player 1 is the labor union, player 2 is the management. Player 1 wants high wages, whereas management wishes to give as small a wage increase as possible. Hence the game is clearly a zero-sum game.

Next we go beyond games in normal form by introducing the notion of feasibility. This notion allows us to deal with situations in which the choices of players cannot be made independently (as was the case thus far). For example, consider several mining companies which extract coal from a field. Each company chooses to extract a certain amount x_k and to sell it. The price of the coal depends on the amount sold. So each company (producer) has a partial control of the price and consequently of its profits. On the other hand x_k can not be chosen independently, because $\sum_{k=1}^{n} x_k$ ($n =$ number of producers), cannot exceed the total amount of coal in the field. For this reason, for every player $k \in N$, we introduce a correspondence $F_k : X \longrightarrow X_k$, which describes those strategies available to player $k \in N$, when all players have chosen their strategy. The way we have defined F_k it also depends on the strategies of player k. This is done only for reasons of technical convenience. In concrete situations F_k is independent of players k choice. Also now, in contrast to the previous model, we do not assume that the preferences of the players are represented by utility functions. Instead we are more general and assume that each player $k \in N$, has a good reply multifunction U_k. So $U_k : X = \prod_{i=1}^{n} X_i \longrightarrow X_k$ and $y_k \in U_k(x)$, if y_k is a good reply for player $k \in N$ to multistrategy $x \in X$.

DEFINITION 8.1.11 A *generalized game* or *abstract economy* is a collection $\{X_k, F_k, U_k\}_{k \in N}$.

REMARK 8.1.12 This is not a game. No player can individually play this game, because he needs to know the strategies of the other players in order to determine his feasibility set. For this reason we call it a *generalized game.* A generalized game is a good mathematical tool to produce existence theorems in various applied contexts.

In this general setting, the notion of Nash equilibrium takes the following form:

DEFINITION 8.1.13
A *Nash equilibrium for an abstract economy* $\{X_k, F_k, U_k\}_{k \in N}$, is a multistrategy $x^* \in X$ such that $x^* \in F(x^*) = \prod_{k=1}^{n} F_k(x^*)$ and $F_k(x^*) \cap U_k(x^*) = \emptyset$ for all $k \in N$.

REMARK 8.1.14 So a Nash equilibrium for an abstract economy, is a multistrategy for which no player has a good reply.

PROPOSITION 8.1.15 *If $X_k \subseteq \mathbb{R}^{l_k}$ are nonempty, compact, convex sets and $U_k : \prod_{k=1}^{n} X_k \longrightarrow 2^{X_k}$ is a multifunction such that*

(i) *For every $x \in X$, $U_k(x)$ is convex and*

(ii) *For every* $\{x_k\}\in X$, $U_k(\{x_k\}) = \{x\in X : x_k \in U(x)\}$ *is open,*

then we can find $x = \{x_k\}_{k=1}^n \in X$ *such that*

$$x_k \in U_k(x) \quad \text{or} \quad U_k(x) \neq \emptyset \qquad \text{for all } k \in N = \{1,\dots,n\}.$$

PROOF: Let $\Gamma_k = \{x \in X : U_k(x) \neq \emptyset\}$. Because of hypothesis (ii), the set Γ_k is open in X. Let $\widehat{U}_k = U_k\big|_{\Gamma_k} : X \longrightarrow 2^{X_k} \setminus \{\emptyset\}$. Apply Proposition 6.3.4 to obtain a continuous function $u_k : \Gamma_k \longrightarrow X_k$ such that $u_k(x) \in U_k(x)$ for all $x \in \Gamma_k$. Consider the multifunction $G_k : X \longrightarrow 2^{X_k} \setminus \{\emptyset\}$ such that

$$G_k(x) = \begin{cases} \{u_k(x)\} & \text{if } x \in \Gamma_k \\ X_k & \text{otherwise} \end{cases}.$$

Evidently G_k is usc and has nonempty, compact, and convex values. Hence so is the multifunction $x \longrightarrow G(x) = \prod_{k=1}^n G_k(x)$ from X into the nonempty, compact, and convex subsets of X. So we can apply the Kakutani–Ky Fan fixed point theorem (see Theorem 6.5.19) and obtain $x^* \in X$ such that $x^* \in G(x^*)$. If $G_k(x^*) \neq X_k$, then $x_k^* \in G_k(x^*)$ ($x^* = \{x_k^*\}_{k=1}^n$) implies that $x_k^* = u_k(x^*) \in U_k(x^*)$. If $G_k(x^*) = X_k$, then $U_k(x^*) = \emptyset$. □

REMARK 8.1.16 A careful reading of the above proof reveals that if one of the U_ks is single-valued, then the result is automatically true (in this case the continuous selection γ_k is obtained trivially).

COROLLARY 8.1.17 *If* $X_k \subseteq \mathbb{R}^{l_k}$ *is nonempty, compact, and convex, and for every* $k \in N = \{1,\dots,n\}$ *the multifunction* $U_k : X \longrightarrow 2^{X_k}$ *has an open graph and satisfies* $x_k \notin \operatorname{conv} U_k(x)$ *for all* $x \in X$, *then we can find* $x \in X$ *such that* $U_k(x) = \emptyset$ *for all* $k \in N$.

PROOF: The multifunction $x \longrightarrow \operatorname{conv} U_k(x)$ satisfies hypotheses (i) and (ii) of Proposition 8.1.15. So by virtue of that result, we can find $x = \{x_k\}_{k=1}^n \in X$ such that $x_k \in \operatorname{conv} U_k(x)$ or $\operatorname{conv} U_k(x) = \emptyset$, $k \in N$. But by hypothesis $x_k \notin \operatorname{conv} U_k(x)$. Therefore $U_k(x) = \emptyset$ for all $k \in N$. □

THEOREM 8.1.18 *If* $\{X_k, F_k, U_k\}_{k\in N}$ *is an abstract economy and for every* $k \in N$,

(i) $X_k \subseteq \mathbb{R}^{l_k}$ *is nonempty, compact, and convex,*

(ii) $F_k : X \longrightarrow 2^{X_k}$ *has nonempty, closed, and convex values, is usc, for every* $x \in X$ *we have* $F_k(x) = \operatorname{cl}(\operatorname{int} F_k(x))$ *and* $x \longrightarrow \operatorname{int} F_k(x)$ *has an open graph,*

(iii) $\operatorname{Gr} U_k$ *is open in* $X \times X_k$,

(iv) *For every* $x \in X$, $x_k \notin \operatorname{conv} U_k(x)$,

then the abstract economy admits a Nash equilibrium $x^* \in X$ (*see Definition* 8.1.13).

PROOF: Let $V_0 = X$, $V_k = X_k$ for all $k \in N$ and $V = V_0 \times \prod_{k=1}^{n} V_k$. A generic element of V is denoted by $(x, y) \in V$ with $x \in V_0$ and $y \in \prod_{k=1}^{n} V_k$. For every $k \in N$, we introduce a multifunction $H_k : V \longrightarrow 2^{V_k}$ defined by

$$H_0(x, y) = \{y\} \tag{8.9}$$

$$\text{and for } k \in N, \quad H_k(x, y) = \begin{cases} \text{int } F_k(x) & \text{if } y_k \notin F_k(x) \\ \text{conv } U_k(y) \cap \text{int } F_k(x) & \text{if } y_k \in F_k(x) \end{cases}. \tag{8.10}$$

Clearly H_0 is single-valued continuous, whereas for $k \in N$, H_k has convex values (see hypothesis (ii)) and $y_k \notin H_k(x, y)$ (see hypothesis (iv)).

We claim that $\text{Gr}H_k \subseteq V \times V_k$ is open. To this end let

$$\begin{aligned} A_k &= \big\{(x, y, z_k) : z_k \in \text{int } F_k(x)\big\} \\ B_k &= \big\{(x, y, z_k) : y_k \notin F_k(x)\big\} \\ C_k &= \big\{(x, y, z_k) : z_k \in \text{conv } U_k(y)\big\}. \end{aligned}$$

Because by hypothesis (ii), $x \longrightarrow \text{int } F_k(x)$ has an open graph, we deduce that A_k is open. Suppose that $y_k \notin F_k(x)$. Then we can find a closed neighborhood D of y_k such that $F_k(x) \subseteq D^c$ and because F_k is usc, it follows that B_k is open. Moreover, C_k is open because of hypothesis (iii). Since

$$\text{Gr}H_k = (A_k \cap B_k) \cup (A_k \cap C_k),$$

we infer that indeed $\text{Gr } H_k \subseteq V \times V_k$ is open.

Apply Proposition 8.1.15 (see also Remark 8.1.16) to obtain $(x^*, y^*) \in V$ such that

$$x^* = y^* \quad \text{and} \quad H_k(x^*, y^*) = \emptyset \qquad \text{for all } k \in N \quad (\text{see } (8.9)). \tag{8.11}$$

From (8.10) and (8.11), we obtain

$$\begin{aligned} &\text{conv } U_k(x^*) \cap \text{int } F_k(x^*) = \emptyset \qquad \text{for all } k \in N, \\ \Rightarrow\ & U_k(x^*) \cap \text{int } F_k(x^*) = \emptyset \qquad \text{for all } k \in N. \end{aligned} \tag{8.12}$$

But by hypothesis (ii), $F_k(x^*) = \text{cl } [\text{int } F_k(x^*)]$. So from (8.12), we conclude that

$$\begin{aligned} &U_k(x^*) \cap F_k(x^*) = \emptyset \qquad \text{for all } k \in N, \\ \Rightarrow\ & x^* \in X \text{ is a Nash equilibrium (see Definition 8.1.13).} \end{aligned}$$

∎

8.2 Cooperative Games

In the previous section, we examined noncooperative games and we introduced the concept of Nash equilibrium according to which each player will consider only unilateral strategy changes in deciding whether or not she can be made better off. In this section we consider games in which players cooperate in order to improve their position. So coalitions of players may have more power to make their members better off than they would be by acting individually.

So let $N=\{1,\ldots,n\}$ be the set of players and $2^N\setminus\{\emptyset\}$ (all nonempty subsets of N) be the set of all possible coalitions of players. Throughout this section, we use the following notation. If $\{X_k\}_{k\in N}$ is a family of sets and $B\in 2^N\setminus\{\emptyset\}$, then $X^B=\prod_{k\in B} X_k$. Similarly we set $\mathbb{R}^B=\prod_{k\in B}\mathbb{R}$. By $p_B : X^N \longrightarrow X^B$ (resp., $p_B:\mathbb{R}^N \longrightarrow \mathbb{R}^B$), we denote the projection map. The simplest cooperative game is a game in characteristic function form with side payments (*a side-payment game*). This is defined as a function $v:2^N\setminus\{\emptyset\}\longrightarrow\mathbb{R}$, which for every coalition $B\in 2^N\setminus\{\emptyset\}$ assigns the maximal total utility of the coalition that its members can generate by their cooperation regardless of the actions of the players outside the coalition.

A side-payment game corresponds to a game in normal form (see Definition 8.1.1). Indeed, for each player $k\in N$, his strategy set is given by the mutually disjoint union

$$X_k=\bigcup\{X_{k,B} : B\in 2^N\setminus\{\emptyset\},\ k\in B\}.$$

If $x\in X_{k,B}$, then player $k\in N$ cooperates with the players in the coalition B. We set

$$u_k(\widehat{x},x_k)=-\infty \quad (\widehat{x}=p_{B^c}(x); \quad \text{see Section 8.1}),$$

if for the coalition B for which $x_k\in X_{k,B}$, there exists $m\in B$ such that $x_m\notin X_{m,B}$. Also let u_k be independent of $\{x_k\}_{k\in N\setminus B}$ where B is the coalition for which $x_k\in X_{k,B}$. Then the required game satisfies for every coalition $B\in 2^N\setminus\{\emptyset\}$

$$\Big\{(u_k)_{k\in B} : \sum_{k\in B} u_k\le v(B)\Big\}=\Big\{\big(u_k(\widehat{x},x_k)\big)_{k\in B} \text{ for all } k\in B,\ x_k\in X_{k,B}\Big\}.$$

But clearly the associated game in normal form is not easy to deal with and for this reason, we prefer to work directly with the value function ν.

The basic solution concept for cooperative games, is analogous to the notion of core of a pure competition economy (see Definition 7.1.3).

DEFINITION 8.2.1 The *core* of a side-payment game v is the set $C(v)$ of all $u=(u_k)_{k\in N}\in\mathbb{R}^N$ such that

(a) $\sum_{k\in N} u_k\le v(N)$.

(b) There is no coalition $B\in 2^N\setminus\{\emptyset\}$ such that $\sum_{k\in B} u_k<v(B)$.

REMARK 8.2.2 In the above definition, the requirement (a) corresponds to the feasibility of the utility vector $u=\{u_k\}_{k\in N}$, and the requirement (b) implies that no coalition can improve upon $u\in\mathbb{R}^N$. Note that the core $C(v)$, is the set of solutions of the 2^n-linear inequalities,

$$-(\chi_N,x)_{\mathbb{R}^n}\ge -v(N) \quad \text{and} \quad (\chi_B,x)_{\mathbb{R}^n}\ge v(B),$$

where for every $B\in 2^N\setminus\{\emptyset\}$, $\chi_B=\sum_{k\in B} e_k, e_k=(e_{k,l})_{l=1}^n$ with

$$e_{k,l}=\delta_{kl}=\begin{cases}1 & \text{if } k=l\\ 0 & \text{if } k\ne l\end{cases}.$$

The last observation leads us to consider systems of linear inequalities.

So let A be an $m \times n$-matrix and $b \in \mathbb{R}^m$. First we consider the following system of linear equalities with a nonnegativity constraint.

$$Ax = b \quad \text{with } x \geq 0. \tag{8.13}$$

DEFINITION 8.2.3 We say that system (8.13) is *consistent*, if there is $x \in \mathbb{R}^n_+$ (i.e., $x \in \mathbb{R}^n$, $x \geq 0$) that satisfies $Ax = b$. Such a vector is called a *solution* of (8.13).

PROPOSITION 8.2.4 *System (8.13) is consistent (i.e., it has a solution $x \in \mathbb{R}^n_+$) if and only if*

$$(y, b)_{\mathbb{R}^m} \geq 0 \qquad \textit{for all } y \in \mathbb{R}^m \textit{ such that } A^*y \geq 0.$$

PROOF: Let $\Gamma(A) = \{Ax : x \geq 0\}$. Clearly $\Gamma(A) \subseteq \mathbb{R}^m$ is a nonempty, closed, convex cone and $0 \in \Gamma(A)$. Note that the system (8.13) is consistent if and only if $b \in \Gamma(A)$, then $b = Ax$ for some $x \geq 0$ and so if $A^*y \geq 0$, then $(A^*y, x)_{\mathbb{R}^n} \geq 0$, hence $(y, Ax)_{\mathbb{R}^m} = (y, b)_{\mathbb{R}^m} \geq 0$.

Suppose $b \notin \Gamma(A)$. Then by the strong separation theorem we can find $y \in \mathbb{R}^m \setminus \{0\}$ and $\varepsilon > 0$ such that

$$\begin{aligned} &(y, b)_{\mathbb{R}^m} + \varepsilon \leq (y, u)_{\mathbb{R}^m} \qquad \text{for all } u \in \Gamma(A), \\ \Rightarrow\ &(y, b)_{\mathbb{R}^m} \leq -\varepsilon \quad (\text{because } 0 \in \Gamma(A)). \end{aligned}$$

On the other hand we have $(y, u)_{\mathbb{R}^m} \geq 0$ for all $u \in \Gamma(A)$. Indeed, if we can find $\widehat{u} \in \Gamma(A)$ such that $(y, \widehat{u})_{\mathbb{R}^m} < 0$, then because $\Gamma(A)$ is a convex cone, for $\vartheta > 0$ large enough, we have

$$\vartheta \widehat{u} \in \Gamma(A) \quad \text{and} \quad (y, \vartheta \widehat{u})_{\mathbb{R}^m} < (y, b)_{\mathbb{R}^m} + \varepsilon,$$

a contradiction. □

REMARK 8.2.5 This result is often called the *Minkowski–Farkas lemma.*

COROLLARY 8.2.6 *If A is an $m \times n$-matrix and $b \in \mathbb{R}^m$, then the system $Ax \geq b$ is consistent (i.e., there exists $x \in \mathbb{R}^n$ such that $Ax \geq b$) if and only if for every $y \in \mathbb{R}^m_+$ for which we have $A^*y = 0$, then $(y, b)_{\mathbb{R}^m} \leq 0$.*

PROOF: Let I_m be the $m \times m$-identity matrix. The system $Ax \geq b$ is consistent if and only if the system

$$\left[A \,\middle|\, -A \,\middle|\, -I_m \right] \begin{bmatrix} x_1 \\ x_2 \\ x_3 \end{bmatrix} = b \quad \text{with} \quad \begin{bmatrix} x_1 \\ x_2 \\ x_3 \end{bmatrix} \geq 0 \tag{8.14}$$

is consistent. Then by virtue of Proposition 8.2.4, (8.14) is consistent if and only if for every $y \in \mathbb{R}^m$ for which we have

$$(A^* | - A^* | - I_m) y \geq 0,$$

then $(y, b)_{\mathbb{R}^m} \geq 0$. □

Now we return to the side-payment game and introduce the following notion.

DEFINITION 8.2.7 A collection $\mathcal{S}$ of coalitions (i.e., $\mathcal{S} \subseteq 2^N \setminus \{\emptyset\}$) is said to be *balanced* if there exists a set $\{\lambda_B\}_{B\in\mathcal{S}} \subseteq \mathbb{R}_+$ such that

$$\sum_{B\in\mathcal{S}} \lambda_B \chi_B = \chi_N.$$

Recall that for every $B \in 2^N \setminus \{\emptyset\}$

$$\chi_B = \sum_{k\in B} e_k,$$

where $e_k = (e_{k,l})_{l=1}^n \in \mathbb{R}_+^n$ with

$$e_{k,l} = \delta_{kl} = \begin{cases} 1 & \text{if } k = l \\ 0 & \text{if } k \neq l \end{cases}.$$

Note that $\sum\limits_{B\in\mathcal{S}} \lambda_B \chi_B = \chi_N$ if and only for every $k \in N$, $\sum\limits_{\substack{B\in\mathcal{S} \\ k\in B}} \lambda_B = 1$. The set $\{\lambda_B\}_{B\in\mathcal{S}}$ is called the *associated balancing coefficients.* The side-payment game v is said to be *balanced* if for every balanced $\mathcal{S} \subseteq 2^N \setminus \{\emptyset\}$ with associated balancing coefficients $\{\lambda_B\}_{B\in\mathcal{S}}$, we have

$$\sum_{B\in\mathcal{S}} \lambda_B v(B) \leq v(B).$$

The notion of a balanced side-payment game is related to the nonemptiness of the core.

THEOREM 8.2.8 *If $v : 2^N \setminus \{\emptyset\} \longrightarrow \mathbb{R}$ is a side-payment game, then the core $C(v)$ of v is nonempty if and only if for every family $\{\lambda_B\}_{B\in 2^N\setminus\{\emptyset\}} \subseteq \mathbb{R}_+$ for which $\sum\limits_{B\in 2^N\setminus\{\emptyset\}} \lambda_B \chi_B = \chi_N$, we have*

$$\sum_{B\in 2^N\setminus\{\emptyset\}} \lambda_B v(B) < v(N).$$

PROOF: By virtue of Corollary 8.2.6, we have that $C(v) \neq 0$ if and only for every family $\{\lambda'_B\}_{B\in 2^N\setminus\{\emptyset\}} \cup \xi \subseteq \mathbb{R}_+$ for which

$$\sum_{B\in 2^N\setminus\{\emptyset\}} \lambda'_B \chi_B + \xi(-\chi_N) = 0,$$

we have

$$\sum_{B\in 2^N\setminus\{\emptyset\}} \lambda'_B v(B) + \xi(-v(N)) \leq 0. \tag{8.15}$$

If $\xi = 0$, then $\lambda'_B = 0$ for all $B \in 2^N \setminus \{\emptyset\}$, satisfies (8.15).

If $\xi > 0$, then we set $\lambda_B = \lambda'_B/t$ and we obtain

$$\sum_{B\in 2^N\setminus\{\emptyset\}} \lambda_B v(B) \leq v(N).$$

□

A shortcoming of the model of a side-payment game is that the notion of total utility $v(B)$ for a coalition $B \in 2^N \setminus \{\emptyset\}$ is hard to interpret, unless the players have utility functions. If the players have preferences over outcomes that are not representable by utility functions, then the game must specify the physical outcomes that a coalition can guarantee for its members. The preferences can then be described as binary relations on vectors of physical outcomes and it is not necessary to rely on a utility function.

To accommodate this situation, we introduce a more general game. So for every coalition $B \in 2^N \setminus \{\emptyset\}$, we set

$$\mathbb{R}^B = \{x = (x_k)_{k=1}^n \in \mathbb{R}^n : x_k = 0 \quad \text{for every} \quad k \in B\}.$$

A game in characteristic function form without side-payments (or simply a non-side-payment game) is a multifunction $\widehat{V} : 2^N \setminus \{\emptyset\} \longrightarrow 2^{\mathbb{R}^n} \setminus \{\emptyset\}$ such that

$$\widehat{V}(B) \subseteq \mathbb{R}^B \quad \text{for every} \quad B \in 2^N \setminus \{\emptyset\}.$$

We have that $u \in \widehat{V}(B)$ if and only if cooperation among the members of coalition B, can bring about the utility allocation $(u_k)_{k \in B}$ to the members of B.

An equivalent definition of a non-side-payment game is a multifunction $V : 2^N \setminus \{\emptyset\} \longrightarrow 2^{\mathbb{R}^n} \setminus \{\emptyset\}$ such that

$$\begin{array}{llll} u, u' \in \mathbb{R}^n & \text{and} \quad u_k = u'_k & \text{for all } k \in B & \text{implies} \\ u \in V(B) & \text{if and only if} \quad u' \in V(B). & & \end{array}$$

Evidently the set $V(B)$ is a cylinder with basis $\widehat{V}(B)$; that is,

$$V(B) = \{u \in \mathbb{R}^n : (u_k, \delta_{kB})_{k \in B} \in \widehat{V}(B)\}$$

where

$$\delta_{kB} = \begin{cases} 1 & \text{if } k \in B \\ 0 & \text{if } k \in N \setminus B \end{cases}.$$

We adopt this definition of a non-side-payment game. Again we can associate with a game in normal form. However, its formulation is very cumbersome and so it is much easier to work with the value multifunction $B \longrightarrow V(B)$.

For such a game, the notion of core is defined as follows.

DEFINITION 8.2.9 The *core* of a non-side-payment game V, is the set $C(V)$ of all $u = (u_k)_{k=1}^n \in \mathbb{R}^n$ such that

(a) $u \in V(N)$.

(b) We can find coalition $B \in 2^N \setminus \{\emptyset\}$ and $u' \in V(B)$ such that

$$u_k < u'_k \qquad \text{for all } k \in B.$$

REMARK 8.2.10 In the above definition, requirement (a) is a feasibility condition on the core vector and requirement (b) simply says that no coalition can improve upon u.

DEFINITION 8.2.11 A non-side-payment game V is said to be *balanced* if for every balanced collection of coalitions $\mathcal{S}$ (see Definition 8.2.7) we have

$$\bigcap_{B\in\mathcal{S}} V(S) \subseteq V(N).$$

In what follows, we show that the notion of a *balanced game*, is sufficient for the nonemptiness of the core. However, in contrast to side-payment games (see Theorem 8.2.8) it is not necessary. But first we need to state a version of the KKM-theorem (see Theorem 6.5.9) due to Shapley [551], which we need in the proof of the theorem for the nonemptiness of the core. For every coalition B, $\Delta_B = \operatorname{conv}\{e_k : k \in B\}$.

PROPOSITION 8.2.12 *If $\{C_B : B \in 2^N \setminus \{\emptyset\}\}$ is a family of nonempty closed subsets of Δ_N such that for each $B \in 2^N \setminus \{\emptyset\}$, $\Delta_B \subseteq \bigcup_{D\subseteq B} C_D$, then there is a balanced collection $\mathcal{S}$ such that*

$$\bigcap_{B\in\mathcal{S}} C_B \neq \emptyset.$$

Using this proposition we can now establish the nonemptiness of the core $C(V)$.

THEOREM 8.2.13 *If V is a non-side-payment game, $b=(b_k)_{k=1}^n \in \mathbb{R}^n$ is defined by*

$$b_k = \sup\left\{u_k : u = (u_i)_{i=1}^n \in V(\{k\})\right\}$$

and we have

(i) *For every $B \in 2^N \setminus \{\emptyset\}$, $V(B) - \mathbb{R}^n_+ = V(B)$;*

(ii) *There exists $M>0$ such that for every $B \in 2^N \setminus \{\emptyset\}$*

$$u \in V(B) \cap \left[\{b\} + \mathbb{R}^n_+\right] \quad \textit{implies } u_k < M \qquad \textit{for all } k \in B;$$

(iii) *For every $B \in 2^N \setminus \{\emptyset\}$, $V(B)$ is closed in $\mathbb{R}^n$;*

(iv) *V is balanced,*

then $C(V)$ is nonempty.

PROOF: We may assume without any loss of generality that $b = 0$.

Let $M>0$ be as in hypothesis (ii) and for every coalition B define

$$D_B = \operatorname{conv}\{-Mne_k : k \in B\}.$$

We consider the function $\xi : D_N \longrightarrow \mathbb{R}$ defined by

$$\xi(y) = \max\left\{\lambda \in \mathbb{R} : y + \lambda\chi_N \in \bigcup_{B\in 2^N\setminus\{\emptyset\}} V(B)\right\}.$$

The function ξ is continuous (see Theorem 6.1.18(c)). Hence the function

$$y \longrightarrow g(y) = y + \xi(y)\chi_N$$

is continuous too. We set

$$\begin{aligned} C_B &= \{y \in D_N : g(y) \in V(B)\} = g^{-1}(V(B)), \\ \Rightarrow\ C_B &\subseteq D_N \quad \text{is closed.} \end{aligned} \tag{8.16}$$

Claim: If B, B' are coalitions and $D_{B'} \cap C_B \neq \emptyset$, then $B \subseteq B'$.

Clearly the claim is true if $B' = N$. So suppose that $|B'| < n$ and let $y \in D_{B'} \cap C_B$. We have

$$\sum_{k \in B'} y_k = -Mn,$$

hence, we can find $k_0 \in B'$ such that

$$y_{k_0} \leq -\frac{Mn}{|B'|} < -M, \tag{8.17}$$

$$\Rightarrow\ y_{k_0} + \xi(y) \geq 0 \quad (\text{because } g(y) \in \mathbb{R}^n_+),$$

$$\Rightarrow\ \xi(y) > M \quad (\text{see (8.17)}). \tag{8.18}$$

On the other hand, we have

$$\begin{aligned} & g(y) \in V(B) \quad (\text{see (8.16)}), \\ \Rightarrow\ & y_k + \xi(y) < M \qquad \text{for all } k \in B \quad (\text{see hypothesis (ii)}), \\ \Rightarrow\ & y_k < 0 \qquad \text{for all } k \in B \quad (\text{see (8.18)}), \\ \Rightarrow\ & B \subseteq B'. \end{aligned}$$

Thus we have satisfied all the hypotheses of Proposition 8.2.12. According to that result we can find a balanced collection $\mathcal{S}_0$ such that

$$\bigcap_{B \in \mathcal{S}_0} C_B \neq \emptyset.$$

Let $y_0 \in \bigcap_{B \in \mathcal{S}_0} C_B$. Then we have

$$\begin{aligned} g(y_0) \in g\Big(\bigcap_{B \in \mathcal{S}_0} C_B\Big) &\subseteq \bigcap_{B \in \mathcal{S}_0} g(C_B) \\ &\subseteq \bigcap_{B \in \mathcal{S}_0} V(B) \quad (\text{see (8.16)}) \\ &\subseteq V(N) \quad (\text{because } \mathcal{S}_0 \text{ is balanced}). \end{aligned} \tag{8.19}$$

From (8.19) and the definition of ξ, we see that $g(y_0)$ is on the boundary of $\bigcup\big(V(B) : B \in 2^N \setminus \{\emptyset\}\big)$. This together with hypothesis (i) imply that $g(y_0)$ can not be improved upon by any coalition. Therefore $g(y_0) \in C(V)$; that is, $C(V) \neq \emptyset$. □

REMARK 8.2.14 Hypothesis (i) is usually called the *comprehensiveness condition* (free disposability in economic models). Also hypothesis (iii) can be replaced by the weaker condition: (iii)′ $V(N) \subseteq \mathbb{R}^n$ is closed.

8.3 Cournot–Nash Equilibria for Random Games

In Section 8.1 we considered noncooperative games and introduced an equilibrium notion for them, due to Nash (see Definition 8.1.5). The game had a finite number of players each of whom is described by a strategy set and a utility (payoff) function defined on the Cartesian product of the strategy sets. It was assumed that the strategy sets were subsets of $\mathbb{R}$ and in fact without any additional effort we can also assume that each strategy set is a subset of a finite-dimensional Banach space.

In this section we present a twofold generalization of this model. We allow the strategy set of each player to be a subset of an infinite dimensional Banach space. On the other hand, we consider a continuum of players, in analogy to the model of perfect competition investigated in Section 7.1. So the space of players is a finite measure space. Each player has a preference multifunction, which describes his or her preference relation on the set of feasible strategies. This relation need not be transitive and complete and so may not be representable by a utility function.

So let $(T, \mathcal{T}, \mu)$ be a complete finite nonatomic measure space. Here T is the set of players, $\mathcal{T}$ is the set of all possible coalitions among the players, and μ is a set function measuring the size of each coalition. The space of all possible strategies (decisions) of the players is a separable Banach space X. More precisely, there is a multifunction $F: T \longrightarrow 2^X \setminus \{\emptyset\}$ which for each player $t \in T$ describes his strategy set $F(t)$ (i.e., $F(t) \subseteq X$ is the set of all strategies (actions, decisions) available to player $t \in T$). The randomness of the game is described by a complete probability space (Ω, Σ, μ), with Ω corresponding to the set of all possible states of nature, and Σ to the collection of all possible outcomes. Also given is a multifunction $P: \Omega \times T \times S_F^1 \longrightarrow 2^X$ which describes the random preferences of the players. More precisely, for every $(\omega, t) \in \Omega \times T$, $P(\omega, t, u) \subseteq F(t)$, and for μ-almost every player $t \in T$, the set $P(\omega, t, u)$ consists of all the strategies (actions), that he/she strictly prefers to her own strategy $u(t)$, given that $\omega \in \Omega$ is the state of the environment and given that the decisions of all players (modulo null-coalitions), are represented by $u \in S_F^1$. So we see that each player's preference pattern is influenced by the strategies of the other players (modulo null coalitions) and by the realized state of nature. However, each player must choose his strategy independently, after having observed the realized state $\omega \in \Omega$ of nature. So the players' actions can be modelled by a map $g : \Omega \longrightarrow S_F^1$, which prescribes for each possible state of nature $\omega \in \Omega$, the strategies of all players modulo null coalitions. Note that the players act independently and noncooperatively.

So the random game under consideration is described by the quadruple

$$\mathcal{R} = \big\{(T, \mathcal{T}, \mu), (\Omega, \Sigma, \nu), F, P\big\}.$$

DEFINITION 8.3.1 A *Cournot–Nash equilibrium* for the random game $\mathcal{R}$, is a map $\widehat{g} : \Omega \longrightarrow S_F^1$ which is $\Big(\Sigma, \mathbf{B}\big(L^1(T, X)\big)\Big)$-measurable such that for ν-a.a. $\omega \in \Omega$, we have

$$P\big(\omega, t, \widehat{g}(\omega)\big) = \emptyset \quad \text{for } \mu\text{-a.a. } t \in T.$$

REMARK 8.3.2 Every $\Big(\Sigma, \mathbf{B}\big(L^1(T, X)\big)\Big)$-measurable map $g : \Omega \longrightarrow S_F^1$, is called a *strategy rule*. According to the above definition, no player $t \in T$ outside a powerless (due to the nonatomicity of μ) null coalition, can produce a strategy, which he/she strictly prefers to the equilibrium strategy $\widehat{g}(\omega)(t)$.

Now we are ready to introduce the precise mathematical hypotheses on the data of the random game. In what follows (S_F^1, w) is the set S_F^1 with the relative weak $L^1(T, X)$-topology.

H₁: $(T, \mathcal{T}, \mu)$ is a complete, finite, nonatomic measure space.

REMARK 8.3.3 As was the case with the economic model of pure competition considered in Section 7.1, the nonatomicity of the measure space of players expresses the fact that there is no coalition of players which has more influence on the game than the others.

H₂: $F : T \longrightarrow P_{wkc}(X)$ is graph-measurable and integrably bounded (see Definition 6.4.22).

H₃: dom $P = \{(\omega, t, u) \in \Omega \times T \times S_F^1 : P(\omega, t, u) \neq \emptyset\} \in \Sigma \times \mathcal{T} \times \mathbf{B}(L^1(T, X))$ and $\mathrm{dom}(\mathrm{conv}\, P(\omega, t, \cdot))$ is w-open.

H₄: There is no $\big(\Sigma, \mathbf{B}(L^1(T, X))\big)$-measurable map $f : \Omega \longrightarrow S_F^1$ (a decision rule), such that for ν-a.a. $\omega \in \Omega$, we have $f(\omega)(t) \in \mathrm{conv}\, P(\omega, t, f(\omega))$ μ-a.e. on T.

H₅: There exists a multifunction $H : \mathrm{dom}P \longrightarrow P_{fc}(X)$ that is graph-measurable and

(i) For every $(\omega, t, u) \in \Omega \times T \times S_F^1$, $H(\omega, t, u) \subseteq \mathrm{conv}\, P(\omega, t, u) \subseteq F(t)$.

(ii) For every $(\omega, t) \in \Omega \times T$, $u \longrightarrow H(\omega, t, u)$ is usc from (S_F^1, w) into X_w.

REMARK 8.3.4 If $(\omega, t, u) \longrightarrow \mathrm{conv}\, P(\omega, t, u)$ is graph measurable and for all $(\omega, t) \in \Omega \times T$, $u \longrightarrow P(\omega, t, u)$ is usc with closed values, then we can take $H = \mathrm{conv}P$.

Let (Y, d) be a separable metric space and consider a multifunction $G : \Omega \times T \times Y \longrightarrow P_{fc}(X)$ such that for every $(\omega, t, y) \in \Omega \times T \times Y$ we have

$$G(\omega, t, y) \subseteq F(t).$$

Then we define $\Gamma : \Omega \times Y \longrightarrow 2^{S_F^1}$ by

$$\Gamma(\omega, y) = \{u \in S_F^1 : u(t) \in G(\omega, t, y) \ \mu\text{-a.e. on } T\}.$$

PROPOSITION 8.3.5 *If hypotheses H_1, H_2 hold and for every $(\omega, t) \in \Omega \times T$, $G(\omega, t, \cdot)$ is usc from Y into X furnished with the weak topology (denoted by X_w), then*

(a) *For every $\omega \in \Omega$, $\Gamma(\omega, \cdot)$ is usc from Y into (S_F^1, w).*

(b) *If* Gr $G \in \Sigma \times T \times \mathcal{T} \times \mathbf{B}(Y) \times \mathbf{B}(X)$, *then we can find a multifunction $\Gamma' : \Omega \times Y \longrightarrow 2^{S_F^1}$ such that* $\mathrm{Gr}\Gamma' \in \Sigma \times \mathbf{B}(S_F^1)$ *and for ν-a.a. $\omega \in \Omega$ we have*

$$\Gamma(\omega, \cdot) = \Gamma'(\omega, \cdot).$$

PROOF: (a) By virtue of hypothesis H_1 and Theorem 6.4.23, we have that S_F^1 is nonempty, convex, and w-compact. Moreover, (S_F^1, w) is metrizable. So in order to prove the desired upper semicontinnuity of $\Gamma(\omega,\cdot)$, it suffices to show that $\mathrm{Gr}\,\Gamma(\omega,\cdot)$ is sequentially closed in $Y \times (S_F^1, w)$ (see Proposition 6.1.10). To this end let $\{(y_n, u_n)\} \subseteq Y \times S_F^1$ and suppose that

$$y_n \longrightarrow y \quad \text{in } (Y,d) \quad \text{and} \quad u_n \xrightarrow{w} u \quad \text{in } L^1(T,X).$$

We have

$$u_n(t) \in G(\omega, t, y_n) \quad \mu\text{-a.e. on } T.$$

Invoking Proposition 6.6.33, we obtain

$$u(t) \in \overline{\mathrm{conv}}\; w\text{-}\limsup G(\omega,t,y_n) \quad \mu\text{-a.e. on } T.$$

But since $G(\omega,t,\cdot)$ is w-usc and has closed convex, values, we obtain

$$\begin{aligned} & u(t) \in G(\omega,t,y) \quad \mu\text{-a.e. on } T, \\ \Rightarrow\ & u \in \Gamma(\omega,y). \end{aligned}$$

This proves the upper semicontinuity of $y \longrightarrow \Gamma(\omega,y)$ from Y into (S_F^1, w).

(b) Let

$$\vartheta(\omega,t,y,x) = i_{G(\omega,t,y)}(x) = \begin{cases} 0 & \text{if } x \in G(\omega,t,y) \\ +\infty & \text{otherwise} \end{cases},$$

the indicator function of the set $G(\omega,t,y)$. By virtue of the graph measurability of G, we have that ϑ is measurable. Moreover, because $\mathrm{Gr}\, G(\omega,t,\cdot)$ is closed in $Y \times X_w$ (due to the upper semicontinuity of $y \longrightarrow G(\omega,t,y)$ from Y into X_w), we have that $\vartheta(\omega,t,\cdot)$ is lower semicontinuous on $Y \times X_w$ and of course it is convex in $x \in X$. Therefore we can find an increasing sequence $\vartheta_n \colon \Omega \times T \times Y \times X \longrightarrow \mathbb{R}$, $n \geq 1$, of measurable functions such that for every $n \geq 1$ and every $(\omega,t) \in \Omega \times T$, $\vartheta_n(\omega,t,\cdot,\cdot)$ is Lipschitz continuous on $Y \times X$ with Lipschitz constant $n \geq 1$ and $\vartheta_n \uparrow \vartheta$ as $n \to \infty$. Let

$$I_{\vartheta_n}(\omega,y,u) = \int_T \vartheta_n(\omega,t,y,u(t))\,d\mu \quad \text{and} \quad I_\vartheta(\omega,y,u) = \int_T \vartheta(\omega,t,y,u(t))\,d\mu.$$

Using Theorem 2.1.28, we can check that for every $\omega \in \Omega$, $I_{\vartheta_n}(\omega,\cdot,\cdot)$ is continuous on $Y \times L^1(T,X)$. Also for every $n \geq 1$ and every $(y,u) \in Y \times S_F^1$, $\omega \longrightarrow I_{\vartheta_n}(\omega,y,u)$ is Σ-measurable. Thus by virtue of Theorem 6.2.6 for every $n \geq 1$, $(\omega,y,u) \longrightarrow I_{\vartheta_n}(\omega,y,u)$ is $\Sigma \times \mathbf{B}(Y) \times \mathbf{B}(S_F^1)$-measurable. Note that from the monotone convergence theorem, we have

$$\begin{aligned} & I_{\vartheta_n} \uparrow I_\vartheta \quad \text{as } n \to \infty, \\ \Rightarrow\ & I_\vartheta \quad \text{is } \Sigma \times \mathbf{B}(Y) \times \mathbf{B}(S_F^1)\text{-measurable} \end{aligned}$$

(recall that $\mathbf{B}(S_F^1) = \mathbf{B}\big((S_F^1, w)\big)$). Finally let Γ' be defined by

$$\mathrm{Gr}\Gamma' = \big\{(\omega,y,u) \in \Omega \times Y \times S_F^1 : \lim_{n\to\infty} I_{\vartheta_n}(\omega,y,u) \leq 0\big\}.$$

Then $\Gamma'(\omega,y,u)$ has all the desired properties. $\square$

Now we are ready to establish the existence of a Cournot–Nash equilibrium (see Definition 8.3.1 for the random game under consideration.

THEOREM 8.3.6 *If hypotheses H_1, H_2, H_3, H_4, and H_5 hold, then the random game $\mathcal{R}$ admits a Cournot–Nash equilibrium.*

PROOF: As we already pointed out (S_F^1, w) is compact, and metrizable (see the proof of Proposition 8.3.5). Let $G:\Omega\times T\times S_F^1 \longrightarrow 2^X$ be defined by

$$G(\omega,t,y)=\begin{cases} H(\omega,t,y) & \text{if } y\in \operatorname{dom} P(\omega,t,\cdot) \\ F(t) & \text{otherwise} \end{cases}. \tag{8.20}$$

Evidently $G(\omega,t,y)\in P_{wkc}(X)$ and $G(\omega,t,y)\subseteq F(t)$ for all $(\omega,t,y)\in\Omega\times T\times S_F^1$. Moreover, because of hypothesis H_5(ii), we have that for all $(\omega,t)\in\Omega\times T$, $G(\omega,t,\cdot)$ is usc from (S_F^1,w) into X_w. So, if we define

$$\Gamma(\omega,y)=\big\{u\in S_F^1 : u(t)\in G(\omega,t,y)\ \ \mu\text{-a.e. on } T\big\},$$

then we can apply Proposition 8.3.5 with

$$Y=(S_F^1,w)$$
$$\text{and}\qquad G,\Gamma \text{ as above,}$$

to deduce that for every $\omega\in\Omega$, the multifunction

$$y\longrightarrow \Gamma(\omega,y)$$

is usc from (S_F^1,w) into (S_F^1,w). Because $\Gamma(\omega,\cdot)$ has nonempty, compact, convex values in (S_F^1,w), we can apply Theorem 6.5.19 (the Kakutani–Ky Fan fixed point theorem) and deduce that for every $\omega\in\Omega$, we can find $u\in S_F^1$ (depending on $\omega\in\Omega$) such that

$$u\in\Gamma(\omega,u). \tag{8.21}$$

From Proposition 8.3.5(b), there is a multifunction $\Gamma':\Omega\times S_F^1\longrightarrow 2^{S_F^1}$ such that Γ' is graph-measurable and for ν-a.a. $\omega\in\Omega$, we have

$$\Gamma(\omega,\cdot)=\Gamma'(\omega,\cdot). \tag{8.22}$$

So from (8.21) and (8.22) and the μ-completeness of Σ, we have

$$D=\big\{(\omega,u)\in\Omega\times S_F^1 : u\in\Gamma(\omega,u)\big\}\in\Sigma\times\mathbf{B}(S_F^1).$$

Invoking Theorem 6.3.20 (the Yankov–von Neumann–Aumann selection theorem), we obtain a $\big(\Sigma\times\mathbf{B}(S_F^1)\big)$-measurable map $u:\Omega\longrightarrow S_F^1$ such that

$$u(\omega)\in\Gamma\big(\omega,u(\omega)\big)\quad \nu\text{-a.e. on } \Omega.$$

This implies that for ν-a.a. $\omega\in\Omega$, we have

$$u(\omega)(t)\in G\big(\omega,t,u(\omega)\big)\quad \mu\text{-a.e. on } T.$$

Suppose that we can find $B\in\Sigma$ with $\nu(B)>0$ such that for every $\omega\in B$

$$\operatorname{dom}P\big(\omega,t,u(\omega)\big)\neq\emptyset\quad \mu\text{-a.e. on } T.$$

Then for every $\omega\in B$, we have

$$u(\omega)(t)\in H\big(\omega,t,u(\omega)\big)\subseteq \operatorname{conv} P\big(\omega,t,u(\omega)\big)\quad \mu\text{-a.e. on } T\quad \text{(see (8.20))},$$

which contradicts hypothesis H_4. Therefore we conclude that for ν-a.a. $\omega\in\Omega$,

$$P\big(\omega,t,u(\omega)\big)=\emptyset\quad\text{for } \mu\text{-a.a. } t\in T,$$
$$\Rightarrow\ u\in S_F^1 \text{ is a Cournot-Nash equilibrium for } \mathcal{R}.$$

∎

8.4 Bayesian Games

In the previous section, we studied the existence of equilibria for random games in which all players shared the same information about the state of the environment. In this section, we consider the situation in which the players have private information (imperfect information in the terminology of Harsanyi [285] or differential information). Motivated by the analogous situation in statistics, we call such games Bayesian games. We consider the situation with a countable number of players and an infinite-dimensional strategy space. The precise mathematical model of the game is the following.

Let (Ω, Σ, μ) be a complete probability space. Here Ω denotes the set of all possible states of the environment, the σ-field Σ corresponds to the collection of all possible events and the measure μ describes the distribution of these events. Also T is a countable set and its elements represent the players of the game. The players choose their strategies (actions) from a separable Banach space X. Each player $t \in T$, has a strategy (action) multifunction

$$F_t : \Omega \longrightarrow 2^X \setminus \{\emptyset\}.$$

If the state of the environment is $\omega \in \Omega$, the set $F_t(\omega) \subseteq X$ represents the set of strategies (actions) available to player $t \in T$ when the state of the environment is $\omega \in \Omega$. Also each player $t \in T$ has a utility function $u_t(\omega, \cdot) : \prod\limits_{s \in T} F_s(\omega) \longrightarrow \mathbb{R}$ that depends on the state of the environment $\omega \in \Omega$. This utility function describes the preference of the player among the various feasible strategies, when the state is $\omega \in \Omega$. Moreover, to every player $t \in T$ corresponds a complete sub-σ-field Σ_t of Σ, which represents the private information of that player and a prior $p_t : \Omega \longrightarrow \mathbb{R}_+$, which is a Radon–Nikodym derivative with respect to the probability μ such that $\int_\Omega p_t(\omega) d\mu = 1$.

Then the Bayesian game (or game with private information or game with differential information), is the quadruple.

$$\mathcal{B} = \{(F_t, u_t, \Sigma_t, p_t)_{t \in T}\}.$$

In what follows we use the following notation.

$$S^1_{F_t}(\Sigma_t) = \{v_t \in L^1(\Omega, \Sigma_t; X); v_t(\omega) \in F_t(\omega) \ \mu\text{-a.e. on } \Omega\}.$$

We set

$$S^1_F = \prod_{t \in T} S^1_{F_t}(\Sigma_t)$$

$$\text{and} \quad \widehat{S}^1_{F_t} = \prod_{\substack{s \in T \\ s \neq t}} S^1_{F_s}(\Sigma_s).$$

An element of $S^1_{F_t}(\Sigma_t)$ is a strategy for player $t \in T$.

For each player $t \in T$, we assume that there exists a finite or countable partition $\mathcal{P}_t$ of Ω and $\Sigma_t = \sigma(\mathcal{P}_t)$ (i.e., $\mathcal{P}_t$ generates the sub-σ-field Σ_t containing $\omega \in \Omega$. We assume that

$$\int_{A_t(\omega)} p_t(\omega') d\mu > 0.$$

We set

$$p_t\big(\omega'\big|A_t(\omega)\big) = \begin{cases} 0 & \text{if } \omega' \notin A_t(\omega) \\ \frac{p_t(\omega')}{\int_{A_t(\omega)} p_t(\omega')\mathrm{d}\mu} & \text{if } \omega' \in A_t(\omega) \end{cases}. \tag{8.23}$$

DEFINITION 8.4.1 The *conditional expected utility* of player $t \in T$, $U_t(\omega, \cdot, \cdot) : \widehat{S}^1_{F_t} \times F_t(\omega) \longrightarrow \mathbb{R}$, is defined by

$$U_t(\omega, \widehat{y}, x) = \int_{A_t(\omega)} u_t\big(\omega', \widehat{y}(\omega'), x\big) p_t\big(\omega'\big|A_t(\omega)\big)\mathrm{d}\mu \quad \text{(see (8.23)).} \tag{8.24}$$

REMARK 8.4.2 In the above definition we understand $U_t(\omega, \widehat{y}, x)$ as the conditional expected utility of player $t \in T$; when the state of the environment is $\omega \in \Omega$, the player chooses the strategy $x \in F_t(\omega)$ and the other players have chosen the strategy profile $\widehat{y} \in \widehat{S}_{F_t}$.

DEFINITION 8.4.3 A *Bayesian Nash equilibrium* for the Bayesian game $\mathcal{B}$ is a strategy profile $y^* \in S^1_F$ such that for all $t \in T$, we have

$$U_t\big(\omega, \widehat{y}^*, y_t(\omega)\big) = \max\big[U_t(\omega, \widehat{y}^*, v) : v \in F_t(\omega)\big] \quad \mu\text{-a.e. on } \Omega.$$

We introduce some hypotheses on the data of the Bayesian game $\mathcal{B}$.

H₁: $F : \Omega \times T \longrightarrow P_{wkc}(X)$ is a multifunction such that for every $t \in T$, $\mathrm{Gr}\, F_t \in \Sigma_t \times \mathbf{B}(X)$, and F_t is integrably bounded.

Let $\widehat{F}_t(\omega) = \prod_{s \neq t} F_s(\omega)$ and $X_t = X$ for all $t \in T$.

H₂: $u : \Omega \times \prod_{t \in T} X_t,\ \omega \longrightarrow \overline{\mathbb{R}} = \mathbb{R} \cup \{-\infty\}$ is a function such that

(i) For every $(t, x) \in T \times \prod_{t \in T} X_t$, $\omega \longrightarrow u_t(\omega, x)$ is Σ-measurable.

(ii) For every $(\omega, t) \in \Omega \times T$, $u_t(\omega, \cdot, \cdot) : \widehat{F}_t(\omega) \times F_t(\omega) \longrightarrow \mathbb{R}$ and it is continuous when $\widehat{F}_t(\omega)$ is endowed with the relative product weak topology of $\prod_{s \neq t} X_s$ and $F_t(\omega)$ is endowed with the norm topology.

(iii) For every $(\omega, t, \widehat{x}) \in \Omega \times T \times \prod_{s \neq t} F_s(\omega)$, the function $x \longrightarrow u_t(\omega, \widehat{x}, x)$ is concave on $F_t(\omega)$.

(iv) There exists $h \in L^1(\Omega, \Sigma)$ such that for μ-a.a. $\omega \in \Omega$ and all $(t, x) \in T \times \widehat{F}_t(\omega) \times F_t(\omega)$ we have $|u_t(\omega, x)| \leq h(\omega)$.

We start with an auxiliary result that establishes the continuity properties of the conditional expected utility (see (8.24)). Let $\mu'_t(A) = \int_A p_t(\omega)\mathrm{d}\mu$ for every $t \in T$ and $A \in \Sigma$. Evidently this is a probability measure on (Ω, Σ).

PROPOSITION 8.4.4 *If hypotheses H_1, H_2 hold and for every fixed $\omega \in \Omega$, $t \in T$ and $A \in \Sigma$, we set*

$$\overline{U}_t^A(\widehat{y}, x) = \int_A u_t(\omega', \widehat{y}(\omega'), x) d\mu_t'(\omega') \quad \text{with } \widehat{y} \in \widehat{S}^1_{F_t},\ x \in F_t(\omega),$$

then $\overline{U}_t^A$ is continuous on $\widehat{S}^1_{F_t} \times F_t(\omega)$, when $\widehat{S}^1_{F_t}$ is equipped with the relative product weak topology and $F_t(\omega)$ is equipped with the norm topology.

PROOF: Suppose that $\widehat{y}^{\,n} \xrightarrow{w} \widehat{y}$ in $\widehat{S}^1_{F_t}$ and $x_n \longrightarrow x$ in $F_t(\omega)$. We know that $\widehat{S}^1_{F_t}$ is weakly compact (see hypothesis H_1 and Theorem 6.4.23). We have $\widehat{y}^{\,n} = (y_s^n)_{\substack{s \in T \\ s \neq t}}$ and $y_s^n \xrightarrow{w} y_s$ in $S^1_{F_s}$ for all $s \in T, s \neq t$.

Claim: For every $\omega \in \Omega$ and every $s \in T, s \neq t$, we have $y_s^n(\omega) \xrightarrow{w} y_s(\omega)$ in $F_s(\omega)$.

We fix $\omega \in \Omega$. Let $\mathcal{P}_s$ be the countable Σ_s-partition of Ω such that $\sigma(\mathcal{P}_s) = \Sigma_s$. We set

$$\mathcal{P}_s = \{D_s^k\}_{k \geq 1}.$$

Hence we have

$$y_s^n = \sum_{k \geq 1} z_s^{n,k} \chi_{D_s^k} \quad \text{and} \quad y_s = \sum_{k \geq 1} z_s^k \chi_{D_s^k},$$

where $z_s^{n,k}, z_s^k \in F_s(\omega)$. For every $s \in T$, we can find a unique $D_s^{k(\omega)} \in \mathcal{P}_s$ such that

$$\omega \in D_s^{k(\omega)}.$$

Then for every $x^* \in X^*$, we have

$$\begin{aligned} \langle x^*, y_s^n(\omega)\rangle &= \int_\Omega \frac{\langle x^*, y_s^n(\omega)\rangle}{\mu(D_s^{k(\omega)})} \chi_{D_s^{k(\omega)}} d\mu(\omega') \\ &= \int_\Omega \frac{\langle x^*, y_s^n(\omega')\rangle}{\mu(D_s^{k(\omega)})} \chi_{D_s^{k(\omega)}}(\omega') d\mu(\omega') \end{aligned} \tag{8.25}$$

(because $y_s^n(\omega') = y_s^n(\omega)$ for all $\omega' \in D_s^{k(\omega)}$).

Evidently $\big(x^*/\mu(D_s^{k(\omega)})\big) \in L^\infty(\Omega, X^*)$ and $y_s^n \chi_{D_s^{k(\omega)}} \in L^1(\Omega, X)$. Therefore from (8.25) and because $y_s^n \xrightarrow{w} y_s$ in $L^1(\Omega, X)$, we obtain

$$\begin{aligned} &\langle x^*, y_s^n(\omega)\rangle \longrightarrow \langle x^*, y_s(\omega)\rangle, \\ \Rightarrow\ & y_s^n(\omega) \xrightarrow{w} y_s(\omega) \quad \text{in } F_s(\omega) \subseteq X. \end{aligned}$$

This proves the claim. Then using the claim and hypothesis H_2(ii), we have

$$u_t\big(\omega, \widehat{y}^{\,n}(\omega), x_n\big) \longrightarrow u_t\big(\omega, \widehat{y}(\omega), x\big).$$

From this and the dominated convergence theorem (it can be used here because of hypothesis H_2(iv)), we have

$$\overline{U}_t^A(\widehat{y}^{\,n}, x_n) \longrightarrow \overline{U}_t^A(\widehat{y}, x).$$

□

Now we can have the equilibrium result for game $\mathcal{B}$.

THEOREM 8.4.5 *If hypotheses* H_1 *and* H_2 *hold, then the Bayesian game* $\mathcal{B}$ *admits a Bayesian Nash equilibrium.*

PROOF: By virtue of Proposition 8.4.4, for every $\omega \in \Omega$

$$U_t(\omega, \cdot, \cdot) : \widehat{S}^1_{F_t} \times F_t(\omega) \longrightarrow \mathbb{R}$$

is continuous when $\widehat{S}^1_{F_t}$ is furnished with the relative product weak topology and $F_t(\omega)$ is furnished with the norm topology. Moreover, because of hypothesis H_2(iii) for every $(\omega, \widehat{y}) \in \Omega \times \widehat{S}^1_{F_t}$, $U_t(\omega, \widehat{y}, \cdot)$ is concave.

For every $t \in T$, we consider the multifunction $E_t : \Omega \times \widehat{S}^1_{F_t} \longrightarrow 2^X$ defined by

$$E_t(\omega, \widehat{y}) = \{x \in F_t(\omega) : U_t(\omega, \widehat{y}, x) = \max\{U_t(\omega, \widehat{y}, v) : v \in F_t(\omega)\}\}.$$

Due to the concavity and continuity of $U_t(\omega, \widehat{y}, \cdot)$, we have that it is weakly upper semicontinuous. Hence because the set $F_t(\omega) \in P_{wkc}(X)$, we deduce that

$$E_t(\omega, \widehat{y}) \neq \emptyset.$$

Moreover, the concavity of $U_t(\omega, \widehat{y}, \cdot)$ implies that

$$E_t(\omega, \widehat{y}) \subseteq F_t(\omega) \text{ is convex.}$$

Theorem 6.1.18(b) implies that for every $\omega \in \Omega$, $y \longrightarrow E_t(\omega, y)$ is usc from $\widehat{S}^1_{F_t}$ with the relative product weak topology into the nonempty, closed, and convex subsets of $F_t(\omega)$, the latter equipped with the norm topology.

From hypothesis H_1, we know that we can find $v_n : \Omega \longrightarrow X$, $n \geq 1$, Σ_t-measurable selectors of $F_t(\cdot)$ such that

$$F_t(\omega) = \overline{\{v_n(\omega)\}}_{n \geq 1} \quad \text{for all } \omega \in \Omega \quad \text{(see Theorem 6.3.20)}.$$

Then exploiting the norm continuity of $U_t(\omega, \widehat{y}, \cdot)$, we have

$$\begin{aligned} &\max\left[U_t(\omega, \widehat{y}, v) : v \in F_t(\omega)\right] = \max_{n \geq 1} U_t(\omega, \widehat{y}, v_n(\omega)), \\ \Rightarrow\ &\omega \longrightarrow \max\left[U_t(\omega, \widehat{y}, v) : v \in F_t(\omega)\right] = \xi_t(\omega, \widehat{y}) \text{ is } \Sigma_t\text{-measurable}, \\ \Rightarrow\ &\mathrm{Gr} E_t(\cdot, \widehat{y}) = \{(\omega, x) \in \mathrm{Gr} F_t : U_t(\omega, \widehat{y}, x) = \xi_t(\omega, \widehat{y})\} \in \Sigma_t \times \mathbf{B}(X) \end{aligned}$$

(hypothesis H_1).

So applying Theorem 6.3.20 (the Yankov–von Neumann–Aumann selection theorem), we obtain $f_t : \Omega \longrightarrow X$ a Σ_t-measurable map such that

$$f_t(\omega) \in E_t(\omega, \widehat{y}) \qquad \text{for all } \omega \in \Omega.$$

Since $E_t(\omega, \widehat{y}) \subseteq F_t(\omega)$ and by hypothesis H_1, F_t is integrably bounded, it follows that $f_t \in S^1_{F_t}(\Sigma_t)$. So, if we consider the multifunction $\Gamma_t : \widehat{S}^1_{F_t} \longrightarrow 2^{S^1_{F_t}(\Sigma_t)}$ defined by

$$\Gamma_t(\widehat{y}) = \{x \in S^1_{F_t}(\Sigma_t) : x(\omega) \in E_t(\omega, \widehat{y}) \ \mu\text{-a.e. on } \Omega\};$$

then we see that Γ_t has nonempty values. We claim that it is usc from $\widehat{S}^1_{F_t}$ with the relative product weak topology into $S^1_{F_t}(\Sigma_t)$ with the weak topology. Because $S^1_{F_t}(\Sigma_t)$ is weakly compact in $L^1(\Omega, \Sigma_t; X)$ (see Theorem 6.4.23), it suffices to show that Gr Γ_t is closed in $\widehat{S}^1_{F_t} \times S^1_{F_t}(\Sigma_t)$ (see Proposition 6.1.10). Because the weak topologies on $\widehat{S}^1_{F_t}$ and $S^1_{F_t}(\Sigma_t)$ are metrizable, we can use sequences. So suppose that $\widehat{y}_n \xrightarrow{w} \widehat{y}$ in $\widehat{S}^1_{F_t}$, $x_n \xrightarrow{w} x$ in $S^1_{F_t}(\Sigma_t)$, and $x_n \in \Gamma_t(\widehat{y}_n)$. Then

$$x_n(\omega) \in E_t(\omega, \widehat{y}_n) \quad \mu\text{-a.e. on } \Omega \qquad \text{for all } n \geq 1.$$

Invoking Proposition 6.6.33, we obtain

$$\begin{aligned} x(\omega) &\in \overline{\text{conv}}\, w\text{-}\limsup_{n\to\infty} E_t(\omega, \widehat{y}_n) \\ &\subseteq E_t(\omega, \widehat{y}), \end{aligned} \tag{8.26}$$

where the inclusion in (8.26) follows from the fact that $E_t(\omega, \cdot)$ is usc on $\widehat{S}^1_{F_t}$ with the relative product weak topology and it has closed and convex values. Therefore it follows that

$$\begin{aligned} &(\widehat{y}, x) \in \text{Gr}\, \Gamma_t, \\ \Rightarrow\ &\Gamma_t \text{ is usc as claimed.} \end{aligned}$$

Let $\Gamma : S^1_F \longrightarrow P_{wkc}(S^1_F)$ be defined by

$$\Gamma(y) = \prod_{t\in T} \Gamma_t(\widehat{y}_t)$$

where $y = (y_t)_{t\in T}$ with $y_t \in S^1_{F_t}(\Sigma_t)$ and $\widehat{y}_t = (y_s)_{\substack{s\in T \\ s\neq t}} \in \widehat{S}^1_{F_t}$. Evidently Γ is usc when S^1_F is furnished with the relative product weak topology. We apply Theorem 6.5.19 (the Kakutani–Ky Fan fixed point theorem), to obtain $y^* \in S^1_F$ such that

$$y^* \in \Gamma(y^*).$$

Clearly then

$$\begin{aligned} &U_t\big(\omega, \widehat{y}^*, y_t^*(\omega)\big) = \max\big[U_t(\omega, \widehat{y}^*, v) : v \in F_t(\omega)\big] \quad \mu\text{-a.e. on } \Omega, \\ \Rightarrow\ &y^* \text{ is a Bayesian Nash equilibrium for } \mathcal{B}. \end{aligned}$$

□

8.5 Stochastic Games

In this section, we consider discounted stochastic games in which the state space is a Borel space (i.e., a Borel subset of a Polish space) and the action spaces of the players are compact metric spaces. Under reasonable continuity hypotheses on the reward function and the transition probability describing the law of motion of the system, we show that the discounted stochastic game has a value and both players have optimal stationary policies. To do this we use the so-called Bellman's principle of optimality for the discounted dynamic programming problem. For this reason our discussion begins with a brief presentation of the dynamic programming model and of the associated Bellman's principle of optimality.

The discounted dynamic programming model is determined by the following objects.

- A state space S, which is assumed to be Borel space (i.e., it is the Borel subset of a Polish space).

- A space of actions X, which is a Borel space too.

- A constraint multifunction $F : S \longrightarrow 2^X \setminus \{\emptyset\}$; this multifunction assigns to each state $s \in S$ a nonempty feasible (permissible) set of actions $F(s)$.

- A law of motion q, which to every pair $(s, x) \in \text{Gr } F$ assigns a probability measure $q(\cdot|s, x)$ on the Borel sets of S. This probability measure describes the distribution of the state next visited by the system if the system is in state $s \in S$ and action $x \in X$ is taken.

- A bounded reward function $r : \text{Gr } F \longrightarrow \mathbb{R}$.

- A discount factor $\delta \in (0, 1)$.

  We make the following mathematical hypotheses on the above items:.

H$_1$: $\text{Gr } F \in \mathbf{B}(S \times X) = \mathbf{B}(S) \times \mathbf{B}(X)$ and contains the graph of a Borel map from S to X.

H$_2$: $q(\cdot|s, x)$ is a Borel-measurable transition probability on the Borel σ-field $\mathbf{B}(S)$; that is, for every $\mathcal{B} \in \mathbf{B}(S)$, the function $(s, x) \longrightarrow q(\mathcal{B}|s, x)$ is Borel-measurable from $S \times X$ into $[0, 1]$.

H$_3$: The reward function $r : \text{Gr } F \longrightarrow \mathbb{R}$ is bounded and Borel-measurable.

Let $H_1 = S$ and $H_n = \text{Gr } F \times H_{n-1}$ for $n \geq 2$. These are the sets of all possible histories of the system up to the stage $n \geq 1$. Then a policy π is a sequence $\{\pi_n\}_{n \geq 1}$, where for each $n \geq 1$, π_n is a conditional probability on $\mathbf{B}(X)$ given the past history H_n of the system. We assume that π_n satisfies the constraint

$$\pi_n\big(F(s_n)\big|h_n\big) = 1$$

for all histories $h_n = (s_1, x_1, \ldots, s_n)$.

DEFINITION 8.5.1 A *history* π is said to be *stationary* if there exists a Borel measurable selector of f of F such that $\pi_n = f$; that is, $\pi_n\big(f(s_n)\big|h_n\big) = 1$ for all histories $h_n = (s_1, x_1, \ldots, s_n)$. The stationary policies are identified with S_F = set of Borel selectors of F.

Any policy π together with the law of motion q, defines a conditional probability p_n on the set $X \times S \times X \times S \times \cdots$ of futures of the system, given the initial state $s \in S$; that is,

$$p_\pi(\cdot|s) = \pi_1 q \pi_2 q \ldots \tag{8.27}$$

The expected total discounted reward is defined by

$$L(\pi)(s) = E_\pi\Big(\sum_{n=1}^{\infty} \delta^{n-1} r(s_n, x_n)\big|s\Big), \quad s \in S. \tag{8.28}$$

Here $E_\pi(\cdot|s)$ denotes the conditional expectation with respect to $p_n(\cdot|s)$ (see (8.27)).

DEFINITION 8.5.2 A policy π^* is *optimal* if $L(\pi)(s) \le L(\pi^*)(s)$ for all policies π and all states $s \in S$ (see (8.28)).

The next theorem, known as *Bellman's optimality principle*, gives a necessary and sufficient condition for optimality of a policy π^*. We assume that hypotheses H_1, H_2, and H_3 are in effect.

THEOREM 8.5.3 *A policy π^* is optimal if and only if its reward $L(\pi^*)$ satisfies the optimality equation*

$$u(s) = \sup\left[r(s,x) + \delta \int_S u(s')\mathrm{d}q(s'|s,x) : x \in F(s)\right], \quad s \in S. \tag{8.29}$$

PROOF: *Necessity:* Suppose that π^* is optimal. Fix a state $s \in S$. For any $x \in F(s)$, we define

$$G(x) = \{s \in S : (s,x) \in \mathrm{Gr}\ F\}. \tag{8.30}$$

Evidently Gr $G \in \mathbf{B}(X) \times \mathbf{B}(S)$ (see hypothesis H_1). Let f be a Borel-measurable selector of F. It exists by virtue of hypothesis H_1. For $x \in F(s)$, we consider the map $f_x : S \longrightarrow X$ defined by

$$f_x(s) = \begin{cases} x & \text{if } s \in G(x) \\ f(s) & \text{otherwise} \end{cases}.$$

Clearly f is Borel-measurable and a selector of F (see (8.30)). If, by (f_x, π^*) we denote the policy $\{f_x, \pi_1^*, \pi_2^*, \ldots\}$, then by virtue of the optimality of π^*, we have

$$L(f_x, \pi^*)(s) = r(s,x) + \delta \int_S L(\pi^*)(s')\mathrm{d}q(s'|s,x) \le L(\pi^*)(s). \tag{8.31}$$

Because $x \in F(s)$ was arbitrary, from (8.31) it follows that

$$\sup\left[r(s,x) + \delta \int_S L(\pi^*)(s')\mathrm{d}q(s'|s,x) : x \in F(s)\right] \le L(\pi^*)(s), \quad s \in S. \tag{8.32}$$

Given any policy $\pi = \{\pi_n\}_{n \ge 1}$, we define $\pi'_n(\cdot|h_n) = \pi_{n+1}(\cdot|s,x,h_n)$, $n \ge 1$ and $\pi_{s,x} = \{\pi'_n\}_{n \ge 1}$. We set

$$v(s,x,s') = L(\pi_{s,x})(s') \tag{8.33}$$

and we have

$$L(\pi)(s) = \int_{F(s)} \left[r(s,x) + \delta \int_S v(s,x,s')\mathrm{d}q(s'|s,x)\right]\mathrm{d}\pi_1(x|s), \quad s \in S. \tag{8.34}$$

From (8.34) it follows that

$$L(\pi^*)(s) \le \sup\left[r(s,x) + \delta \int_S L(\pi^*)(s')\mathrm{d}q(s'|s,x) : x \in F(s)\right]. \tag{8.35}$$

Comparing (8.33) and (8.35), we conclude that

$$L(\pi^*)(s) = \sup\left[r(s,x) + \delta \int_S L(\pi^*)(s')\mathrm{d}q(s'|s,x) : x \in F(s)\right], \quad s \in S.$$

Sufficiency: Suppose that $L(\pi^*)$ satisfies the functional equation (8.29). Let $h_n^1 = (s, x_1, \dots, s_n, x_n)$ and $u(s) = L(\pi^*)(s),\ s \in S$. Then, for any policy π and any state $s \in S$, we have

$$E_\pi\Big[\sum_{n=1}^{N}\Big(\delta^n u(s_{n+1}) - E_\pi\big(\delta^n u(s_{n+1})|h_n'\big)\Big)\Big|s\Big] = 0. \tag{8.36}$$

Here $E_\pi(\cdot|h_n')$ denotes the conditional expectation given the history h_n'. Then for $p_\pi(\cdot|s)$-almost all futures, we have

$$\begin{aligned} & E_\pi\big(\delta^n u(s_{n+1})|h_n'\big) \\ &= \delta^n \int_S u(s')\mathrm{d}q(s'|s_n, x_n) \\ &= \delta^{n-1}\Big[r(s_n, x_n) + \delta\int_S u(s')\mathrm{d}q(s'|s_n, x_n)\Big] - \delta^{n-1} r(s_n, x_n) \\ &\le \delta^{n-1}u(s_n) - \delta^{n-1}r(s_n, x_n). \end{aligned} \tag{8.37}$$

Using (8.37) in (8.36), we obtain

$$\begin{aligned} & E_\pi\Big[\sum_{n=1}^{N}(\delta^n u(s_{n+1}) - \delta^{n-1}u(s_n) + \delta^{n-1}r(s_n, x_n))\big|s\Big] \\ =\ & E_\pi\Big[\delta^N u(s_{N+1}) - u(s) + \sum_{n=1}^{N}\delta^{n-1}r(s_n, x_n)\big|s\Big] \le 0. \end{aligned} \tag{8.38}$$

We pass to the limit as $N \longrightarrow \infty$ in (8.38). Using the dominated convergence theorem, we have

$$\begin{aligned} & E_\pi\Big[\sum_{n\ge 1}\delta^{n-1}r(s_n, x_n)\big|s\Big] = L(\pi)(s) \le u(s) = L(\pi^*)(s), \quad s \in S, \\ \Rightarrow\ & \pi^* \text{ is an optimal policy.} \end{aligned}$$

□

Bellman's optimality principle leads to the existence of optimal stationary policies.

H₁′: $F : S \longrightarrow P_k(X)$ is a measurable multifunction.

REMARK 8.5.4 By Theorem 6.3.17 (the Kuratowski–Ryll Nardzewski selection theorem), if F satisfies hypothesis H_1', it admits a Borel-measurable selector.

H₂′: For every $s \in S$ and every bounded Borel measurable function $v : S \longrightarrow \mathbb{R}$, the function

$$x \longrightarrow \int_S v(s')\mathrm{d}q(s'|s, x)$$

is continuous.

H₃′ : The reward function $r : \mathrm{Gr}\, F \longrightarrow \mathbb{R}$ is bounded (i.e., $|r(s, x)| \le M$ for some $M > 0$ and all $(s, x) \in S \times X$), Borel-measurable and for every $s \in S,\ x \longrightarrow r(s, x)$ is continuous.

First let us make a straightforward but useful observation concerning the solution of the optimality functional equation. In what follows by $\widehat{B}_0(S)$ we denote the space of all bounded Borel-measurable functions on S. This is a Banach space for the supremum norm $\|u\|_\infty = \sup\limits_{s\in S}|u(s)|$.

PROPOSITION 8.5.5 *If hypotheses H_1', H_2', and H_3' hold, then the optimality functional equation (8.29) has a unique solution $u^* \in \widehat{B}_0(S)$.*

PROOF: Let $u \in \widehat{B}_0(S)$. We consider the optimization problem

$$\sup\bigl[r(s,x) + \delta \int_S u(s')\mathrm{d}q(s'|s,x) : x \in F(s)\bigr] = \xi(s). \tag{8.39}$$

By virtue of hypothesis H_1' we can find a sequence $f_n : S \longrightarrow X$, $n \geq 1$, of Borel-measurable functions such that

$$F(s) = \overline{\{f_n(s)\}_{n\geq 1}} \qquad \text{for all } s \in S.$$

Because $x \longrightarrow \int_S u(s')\mathrm{d}q(s'|s,x)$ is continuous (hypothesis H_2') and $x \longrightarrow r(s,x)$ is continuous (see hypothesis H_3'), from (8.39) we have

$$\xi(s) = \sup_{n\geq 1}\bigl[r\bigl(s, f_n(s)\bigr) + \delta \int_S u(s')\mathrm{d}q\bigl(s'|s, f_n(s)\bigr)\bigr],$$

$\Rightarrow$ ξ is Borel-measurable.

In addition it is clear that ξ is bounded (see hypothesis H_3'). So we can define the operator $P : \widehat{B}_0(S) \longrightarrow \widehat{B}_0(S)$ by

$$P(u)(s) = \sup\,\bigl[r(s,x) + \delta \int_S u(s')\mathrm{d}q(s'|s,x) : x \in F(s)\bigr], \quad s \in S.$$

For every $u, v \in \widehat{B}_0(S)$ and every $s \in S$, we have

$$\begin{aligned} &|P(u)(s) - P(v)(s)| \\ \leq\ & \sup\ \delta\bigl|\int_S \bigl(u(s') - v(s')\bigr)\mathrm{d}q(s'|s,x)\bigr| \\ \leq\ & \delta\|u - v\|_\infty. \end{aligned}$$

Therefore P is a contraction map and by the Banach fixed point theorem (see Theorem 3.4.3), P has a unique fixed point, which is the unique solution of the Bellman's optimality equation (8.29). □

REMARK 8.5.6 In the above proof, the operator P is known as the *dynamic programming operator*.

Now we can use Theorem 8.5.3 to establish the existence of an optimal stationary policy.

THEOREM 8.5.7 *If hypotheses H_1', H_2', and H_3' hold, then there exists an optimal stationary policy.*

PROOF: Let $u^* \in \mathbf{B}(S)$ be the unique fixed point of the dynamic programming operator. We have

$$u^*(s) = \sup\left[r(s,x) + \delta \int_S u^*(s')\mathrm{d}q(s'|s,x) : x \in F(s)\right] = \xi(s).$$

Let $\Gamma(s) = \{x^* \in F(s) : \xi(s) = r(s,x^*) + \delta \int_S u(s')\mathrm{d}q(s'|s,x^*)\}$. Because $F(s) \in P_k(X)$ (see hypothesis H_1') and the function $x \longrightarrow r(s,x) + \delta \int_S u^*(s')\mathrm{d}q(s'|s,x)$ is continuous (see hypotheses H_1' and H_3'), from the Weierstrass theorem, we have that $\Gamma(s) \neq \emptyset$ for all $s \in S$ and in fact $\Gamma(s) \in P_k(X)$. Moreover, recalling that a Carathéodory function is jointly measurable (see Theorem 6.2.6) and because $s \longrightarrow \xi(s)$ is Borel-measurable (see the proof of Proposition 8.5.5), we conclude that

$$\mathrm{Gr}F \in \mathbf{B}(S) \times \mathbf{B}(X).$$

Because Γ is compact valued, it follows that Γ is measurable. Therefore by the Kuratowski–Ryll–Nardzewski selection theorem (see Theorem 6.3.17), we can find $f^* \in S_F$ such that

$$\begin{aligned} f^*(s) &\in \Gamma(s) \qquad \text{for all } s \in S, \\ u^*(s) &= r(s, f^*(s)) + \delta \int_S u^*(s')\mathrm{d}q(s'|s, f^*(s)) \\ &= \sup\left[r(s,x) + \delta \int_S u^*(s')\mathrm{d}q(s'|s,x) : x \in F(s)\right] \end{aligned}$$

and so by Theorem 8.5.3 we infer that $f^* \in S_F$ is optimal and $L(f^*) = u^*$. □

REMARK 8.5.8 In fact we can strengthen the above theorem and say that $f^* \in S_F$ is an optimal policy if and only if $u^*(s) = r(s, f^*(s)) + \delta \int_S u^*(s')\mathrm{d}q(s'|s, f^*(s))$ for all $s \in S$. Here again $u^* \in \mathbf{B}(S)$ is the unique fixed point of the dynamic programming operator P (see Remark 8.5.6). For details we refer to Hernandez–Lerma [290].

Now we pass to discounted stochastic games. We consider a two-player stochastic game. The items determining such a stochastic game are the following.

- S is a nonempty Borel subset of a Polish space (a Borel space) and corresponds to the set of all possible states of the system.
- X and Y are compact metric spaces and correspond to the set of actions available to player I and player II, respectively.
- $F_1 : S \longrightarrow P_f(X)$ and $F_2 : S \longrightarrow P_f(Y)$ are two measurable multifunctions that restrict the actions of the two players; so if the state of the system is $s \in S$, then player I can choose an action from the set $F_1(s)$ and the player II from the set $F_2(s)$.
- q is the law of motion (transition law) of the system and with every triple $(s,x,y) \in S \times X \times Y$ associates probability measure $q(\cdot|s,x,y)$ on the Borel subsets $\mathbf{B}(S)$ of S; so if the system is in state $s \in S$ and the two players have chosen actions $x \in X$ and $y \in Y$, respectively, the system moves to a new state according to the probability distribution $q(\cdot|s,x,y)$.

- $r: S \times X \times Y \longrightarrow \mathbb{R}$ is a bounded reward function.
- $\delta \in (0,1)$ is the discount factor; so the unit income today is worth δ^n at the nth period in the future.

The game is played as follows. Periodically (say, once a day), players I and II observe the current state $s \in S$ of the system and choose actions $x \in F_1(s)$ and $y \in F_2(s)$, respectively. This choice is made with full knowledge of the history of the system as it has evolved up to the present time. Then as a consequence of the actions chosen by the two players, player II pays player I an amount equal to $r(s,x,y)$. Subsequently the system moves to a new state s' according to the distribution $q(\cdot|s,x,y)$. Then the whole process is repeated from the new state $s' \in S$. The goal of player I is to maximize the expected discounted reward (income) as the game proceeds over the infinite future and the goal of player II is to minimize the expected discounted loss.

For each $n \geq 0$, we define the space H_n of all admissible histories of the systems up to time n (the n-histories), by

$$H_0 = S \quad \text{and} \quad H_n = \mathrm{Gr}(F_1 \times F_2)^n \times S = \mathrm{Gr}(F_1 \times F_2) \times H_{n-1} \quad \text{for} \quad n \geq 1.$$

The generic element $h_n \in H_n$, is a vector

$$h_n = (s_0, x_0, y_0, \ldots, s_{n-1}, x_{n-1}, y_{n-1}, s_n),$$

where $(s_k, x_k, y_k) \in \mathrm{Gr}(F_1 \times F_2)$ for all $k = 0, 1, \ldots, n-1$ and $s_n \in S$.

DEFINITION 8.5.9 (a) A *randomized, admissible policy* π for player I, is a sequence $\pi = \{\pi_n\}_{n \geq 0}$ of stochastic kernels (transition probabilities) π_n on the Borel σ-field $\mathbf{B}(X)$ given H_n and satisfying the constraint

$$\pi_n\big(F_1(s)|h_n\big) = 1 \qquad \text{for all } h_n \in H_n \quad \text{and all } n \geq 0.$$

Similarly a (randomized admissible) policy β for player II is a sequence $\beta = \{\beta_n\}_{n \geq 0}$ of stochastic kernels (transition probabilities) β_n on the Borel σ-field $\mathbf{B}(Y)$ given H_n and satisfying the constraint

$$\beta_n\big(F_2(s)|h_n\big) = 1 \qquad \text{for all } h_n \in H_n \quad \text{and all } n \geq 0.$$

We denote the set of all policies of player I (resp., of player II) by Δ_1 (resp. Δ_2).

(b) A *stationary policy* for player I (resp., for player II) is a Borel-measurable map $f : S \longrightarrow M_+^1(X)$ (resp., a Borel-measurable map $g : S \longrightarrow M_+^1(Y)$).

REMARK 8.5.10 A stationary policy is a particular case of a randomized admissible policy. Indeed, if $p_n : H_n \longrightarrow S$ is the projection map defined by $p_n(h_n) = s_n$ ($h_n = (s_0, x_0, y_0, \ldots, s_{n-1}, x_{n-1}, y_{n_1}, s_n)$), then $\pi = \{\pi_n = f \circ p_n\}_{n \geq 0} \in \Delta_1$. So π_n is the Dirac measure concentrated at $f(s_n)$. Similarly for player II. A stationary policy is a special case of a Markov policy. Clearly the stationary policies of player I (resp., of player II), can be identified with S_{F_1} (resp., with S_{F_1}).

A pair (π, β) of policies for players I and II associates with every initial state $s \in S$, an $n^{\underline{\text{th}}}$-day expected reward (gain) $r_n(\pi, \beta)(s)$ for player I (see the dynamic programming model). The total expected discounted reward for player I, is given by

$$L(\pi,\beta)(s)=\sum_{n\geq 0}\delta^n r_n(\pi,\beta)(s). \tag{8.40}$$

The rewards $r_n(\pi,\beta)(\cdot)$ are Borel-measurable and so from (8.40) it follows that $L(\pi,\beta)(\cdot)$ is Borel-measurable.

DEFINITION 8.5.11 A policy π^* is *optimal for player I* if

$$\inf_{\beta\in\Delta_2}\sup_{\pi\in\Delta_1} L(\pi,\beta)(s)\leq L(\pi^*,\beta)(s) \qquad \text{for all } \beta\in\Delta_2,\ s\in S.$$

A policy β^* is *optimal for player II* if $\sup\limits_{\pi\in\Delta_1}\inf\limits_{\beta\in\Delta_2} L(\pi,\beta)(s)\geq L(\pi,\beta^*)(s)$ for all $\pi\in\Delta_1,\ s\in S$.

The stochastic game is said to have a value if

$$\inf_{\beta\in\Delta_2}\sup_{\pi\in\Delta_1} L(\pi,\beta)(s)=\sup_{\pi\in\Delta_1}\inf_{\beta\in\Delta_2} L(\pi,\beta)(s) \quad \text{for every } s\in S.$$

In this case the function

$$v(s)=\inf_{\beta\in\Delta_2}\sup_{\pi\in\Delta_1} L(\pi,\beta)(s)=\sup_{\pi\in\Delta_1}\inf_{\beta\in\Delta_2} L(\pi,\beta)(s), \quad s\in S,$$

is called the *value function* of the stochastic game.

To be able to establish the existence of optimal policies for the two players, we need some mathematical hypotheses on the data $\{F_1,F_2,q,r\}$ of the stochastic game.

$\mathbf{H_4}$: $F_1:S\longrightarrow P_f(X)$ and $F_2:S\longrightarrow P_f(Y)$ are measurable multifunctions.

REMARK 8.5.12 Because X and Y are compact metric spaces, the measurability of F_1,F_2 is equivalent to Borel-measurability into the separable metric spaces $\big(P_k(X),h\big)$ and $\big(P_k(Y),h\big)$, respectively.

Let $C_b(S)$ denote the space of bounded continuous functions on S. This is a Banach space for the supremum norm $\|u\|_\infty=\sup\limits_{s\in S}|u(s)|$, $u\in C_b(S)$. Also by $M_+^1(S)$ we denote the space of probability measures on S.

$\mathbf{H_5}$: $q:S\times X\times Y\longrightarrow M_+^1(S)$ is a map such that

(i) For every $(x,y)\in X\times Y$ and every $B\in\mathbf{B}(S)$, $s\longrightarrow q(B|s,x,y)$ is Borel-measurable.

(ii) For every $s\in S$, $(x,y)\longrightarrow q(\cdot|s,x,y)$ is continuous in the sense that if $(x_n,y_n)\longrightarrow(x,y)$ in $X\times Y$, then for every $u\in\widehat{B}_0(S)$ we have

$$\int_S u(s')dq(s'|s,x_n,y_n)\longrightarrow\int_S u(s')dq(s'|s,x,y) \quad \text{as } n\to\infty.$$

REMARK 8.5.13 By approximating a function in $C_b(S)$ uniformly by simple functions, we can check that hypothesis H_5(i) is in fact equivalent to saying that for every $(x,y) \in X \times Y$, $s \longrightarrow q(\cdot|s,x,y)$ is Borel-measurable from S into $M_+^1(S)$ furnished with the weak topology $w\big(M_+^1(S), C_b(S)\big)$. Recall that $M_+^1(S)$ topologized this way is a Borel space (hence a separable metrizable space).

H_6: $r : S \times X \times Y \longrightarrow \mathbb{R}$ is a function such that

(i) For every $(x,y) \in X \times Y$, $s \longrightarrow r(s,x,y)$ is measurable.

(ii) For every $s \in S$, $(x,y) \longrightarrow r(s,x,y)$ is continuous.

REMARK 8.5.14 The reward function r is a Carathéodory function, thus Borel-measurable (see Theorem 6.2.6).

For $u \in \mathbf{B}(S)$ and $(s,\lambda,\mu) \in S \times M_+^1(X) \times M_+^1(Y)$, we define

$$K_u(s,\lambda,\mu) = r(s,\lambda,\mu) + \delta \int_S u(s')\mathrm{d}q(s'|s,\lambda,\mu), \tag{8.41}$$

where

$$r(s,\lambda,\mu) = \int_Y \int_X r(s,x,y)\mathrm{d}\lambda(x)\mathrm{d}\mu(y)$$
$$\text{and} \quad q(B|s,\lambda,\mu) = \int_Y \int_X q(B|s,x,y)\mathrm{d}\lambda(x)\mathrm{d}\mu(y)$$

for all $B \in \mathbf{B}(S)$. Because of hypotheses H_5 and H_6, the function $(s,\lambda,\mu) \longrightarrow K_u(s,\lambda,\mu)$ is a Carathéodory function on $S \times M_+^1(X) \times M_+^1(Y)$ into $\mathbb{R}$; that is, for every $(\lambda,\mu) \in M_+^1(X) \times M_+^1(Y)$, $s \longrightarrow K_u(s,\lambda,\mu)$ is Borel-measurable and for every $s \in S$, $(\lambda,\mu) \longrightarrow K_u(s,\lambda,\mu)$ is continuous on $M_+^1(X) \times M_+^1(Y)$, when $M_+^1(X)$ and $M_+^1(Y)$ are endowed with the weak topologies $w\big(M_+^1(X), C(X)\big)$ and $w\big(M_+^1(Y), C(Y)\big)$, respectively. In what follows $M_+^1(X)$ and $M_+^1(Y)$ are topologized this way.

Let $P_1 : S \longrightarrow P_k\big(M_+^1(X)\big)$ and $P_2 : S \longrightarrow P_k\big(M_+^1(Y)\big)$ be defined by

$$P_1(s) = \big\{\lambda \in M_+^1(X) : \lambda\big(F_1(s)\big) = 1\big\}$$
$$\text{and} \quad P_2(s) = \big\{\mu \in M_+^1(Y) : \mu\big(F_2(s)\big) = 1\big\}.$$

Note that the function $K_u(s,\cdot,\cdot)$ defines a two-person, zero-sum game $\eta(s,u)$ and $P_1(s)$ and $P_2(s)$ are the spaces of pure strategies in $\eta(s,u)$ for players I and II, respectively, and $K_u(s,\cdot,\cdot)$ is the reward (payoff, utility) function of the game.

The next theorem, known as the *portmanteau theorem*, helps us establish the existence of a pair of optimal stationary strategies for the two players. For a proof of the portmanteau theorem, we refer to Denkowski–Migòrski–Papageorgiou [194, p. 195] or Parthasarathy [487, p. 40]. Recall that if Z is a separable metric space, then $M_+^1(Z)$ endowed with the weak topology $w\big(M_+^1(Z), C_b(Z)\big)$ is separable, and metrizable.

THEOREM 8.5.15 *If Z is a separable metric space and $\{\mu_n\}_{n\geq 1} \subseteq M_+^1(Z)$, then the following statements are equivalent.*

(a) $\mu_n \xrightarrow{w} \mu$ *in* $M_+^1(Z)$ ($\xrightarrow{w}$ *denotes convergence in the weak topology* $w\big(M_+^1(Z), C_b(Z)\big)$).

(b) $\int_Z u d\mu_n \longrightarrow \int_Z u d\mu$ *for all* $u \in U_b(Z)$ = *space of bounded,* $\mathbb{R}$*-valued uniformly continuous functions.*

(c) $\limsup_{n\to\infty} \mu_n(C) \leq \mu(C)$ *for every closed set* $C \subseteq Z$.

(d) $\liminf_{n\to\infty} \mu_n(U) \geq \mu(U)$ *for every open set* $U \subseteq Z$.

(e) $\mu_n(A) \longrightarrow \mu(A)$ *for every Borel set* $A \subseteq Z$ *such that* $\mu(\partial A) = 0$.

PROPOSITION 8.5.16 *If hypothesis* H_4 *holds, then the multifunctions* $P_1 : S \longrightarrow 2^{M_+^1(X)} \setminus \{\emptyset\}$ *and* $P_2 : S \longrightarrow 2^{M_+^1(Y)} \setminus \{\emptyset\}$ *have compact values and are Borel-measurable.*

PROOF: We do the proof for P_1. The proof for P_2 is similar.

Because X is a compact metric space, then so is $M_+^1(X)$ furnished with the weak topology. Hence using Theorem 8.5.15 (in particular the equivalence of (a) and (c)), we deduce at once that P_1 has compact values. Also the multifunction F_1 is Borel-measurable from S into $\big(P_k(X), h\big)$ and the latter is a Polish space. Moreover, for every $C \in P_k(X)$, the map $\lambda \longrightarrow \lambda(C)$ is continuous on $M_+^1(X)$ (see Theorem 8.5.15). Therefore if we consider the function $\xi_1 : S \times M_+^1(X) \longrightarrow \mathbb{R}$ defined by

$$\xi_1(s, \lambda) = \lambda\big(F_1(s)\big)$$

we see that ξ_1 is a Carathéodory function, hence jointly Borel-measurable. But

$$\mathrm{Gr}P_1 = \big\{(s, \lambda) \in S \times M_+^1(X) : \xi_1(s, \lambda) = 1\big\} \in \mathbf{B}(S) \times \mathbf{B}\big(M_+^1(X)\big).$$

$\square$

Now we can prove the main existence theorem for the stochastic game.

THEOREM 8.5.17 *If hypotheses* H_4, H_5, H_6 *hold, then the discounted stochastic game has a value, the value function is Borel-measurable, and the two players have optimal stationary policies.*

PROOF: Recall that for every $u \in \widehat{B}_0(S)$, the function

$$(s, \lambda, \mu) \longrightarrow K_u(s, \lambda, \mu) \quad \text{(see (8.41))}$$

is a Carathéodory function from $S \times M_+^1(X) \times M_+^1(Y)$ into $\mathbb{R}$. Hence it is jointly Borel-measurable. Also because of Proposition 8.5.16, we can find $\lambda_n : S \longrightarrow M_+^1(X)$, $n \geq 1$, Borel-measuarble maps such that

$$P_1(s) = \overline{\{\lambda_n(s)\}}_{n\geq 1} \qquad \text{for all } s \in S.$$

So for fixed $\mu \in M_+^1(Y)$, we have

$$\vartheta_1^u(s,\mu) = \max\left[K_u(s,\lambda,\mu) : \lambda \in P_1(s)\right]$$
$$= \sup_{n\geq 1} K_u\big(s,\lambda_n(s),\mu\big),$$
$$\Rightarrow s \longrightarrow \vartheta_1^u(s,\mu) \text{ is Borel-measurable.}$$

Moreover, Theorem 6.1.18(c) implies that

$$\mu \longrightarrow \vartheta_1^u(s,\mu) \quad \text{is continuous on } M_+^1(Y).$$

Because of Proposition 8.5.16, we can find $\mu_n : S \longrightarrow M_+^1(Y)$, $n \geq 1$, Borel-measurable maps such that $P_2(s) = \overline{\{\mu_n(s)\}}_{n\geq 1}$ for all $s \in S$. Therefore

$$v_1(s) = \min\left[\vartheta_1^u(s,\mu) : \mu \in M_+^1(Y)\right]$$
$$= \inf_{n\geq 1} \vartheta_1^u\big(s,\mu_n(s)\big),$$
$$\Rightarrow s \longrightarrow v_1^u(s) = \min_{\mu\in P_2(s)} \max_{\lambda\in P_1(s)} K_u(s,\lambda,\mu) \text{ is Borel-measurable.}$$

Similarly we show that

$$s \longrightarrow v_2^u(s) = \max_{\lambda\in P_1(s)} \min_{\mu\in P_2(s)} K_u(s,\lambda,\mu) \text{ is Borel-measurable.}$$

Moreover, Theorem 2.3.13, implies that

$$v_1^u(s) = v_2^u(s) = v^u(s) \qquad \text{for all } s \in S. \tag{8.42}$$

Consider the operator $V : \widehat{B}_0(S) \longrightarrow \widehat{B}_0(S)$ defined by

$$V(u)(s) = v^u(s).$$

We have that V is a contraction map (see the proof of Proposition 8.5.5). So by the Banach fixed point theorem (see Theorem 3.4.3), there exists unique $u^* \in \widehat{B}_0(S)$ such that

$$V(u^*) = u^*.$$

Then as in the proof of Theorem 8.5.7, we can find Borel maps

$$f^* : S \longrightarrow M_+^1(X) \quad \text{and} \quad g^* : S \longrightarrow M_+^1(Y),$$

such that

$$f^*(s) \in P_1(s) \quad \text{and} \quad g^*(s) \in P_2(s) \qquad \text{for all } s \in S.$$

$$u^*(s) = \min\left[r\big(s,f^*(s),\mu\big) + \delta\int_S u^*(s')dq\big(s'|s,f^*(s),\mu\big) : \mu \in P_2(s)\right] \tag{8.43}$$

$$u^*(s) = \max\left[r\big(s,\lambda,g^*(s)\big) + \delta\int_S u^*(s')dq\big(s'|s,\lambda,g^*(s)\big) : \lambda \in P_1(s)\right] \tag{8.44}$$

$$u^*(s) = r\big(s,f^*(s),g^*(s)\big) + \delta\int_S u^*(s')dq\big(s'|s,f^*(s),g^*(s)\big). \tag{8.45}$$

The system of equations (8.43) through (8.45) may be interpreted as follows. For the game, $u^*(s)$ is the value of the game and $f^*(s), g^*(s)$ are, respectively, the optimal strategies for players I and II.

We prove that $u^*(\cdot)$ is the value function of the stochastic game and f^*, g^* are optimal stationary policies for players I and II, respectively. To this end for every $(f', g') \in S_{P_1} \times S_{P_2}$ (recall S_{P_1} (resp., S_{P_2}) is the set of Borel-measurable selectors of P_1 (resp., of P_2)), we set

$$W(f', g')(u)(s) = r\big(s, f'(s), g'(s)\big) + \delta \int_S u(s') dq\big(s'|s, f'(s), g'(s)\big), \quad s \in S.$$

We interpret $W(f', g')(u)(s)$ as the expected amount player II pays player I. When the initial state of the system is $s \in S$, player I uses strategy $f'(s)$, player II uses strategy $g'(s)$, and the game is terminated at the beginning of the second day with player II paying player I the amount $u(s')$ with $s' \in S$ being the state of the system on the second day.

It is easy to check that $W(f', g') : \widehat{B}_0(S) \longrightarrow \widehat{B}_0(S)$ is a contraction and so it admits a unique fixed point, which is $L(f^*, g^*)$. Therefore we have

$$\begin{aligned} L(f^*, g^*)(s) &= \max \Big[r\big(s, \lambda, g^*(s)\big) + \delta \int_S L(f^*, g^*)(s') dq\big(s'|s, \lambda, g^*(s)\big) : \\ &\qquad \lambda \in P_1(s)\Big] \\ &= \min \Big[r\big(s, f^*(s), \mu\big) + \delta \int_S L(f^*, g^*)(s') dq\big(s'|s, f^*(s), \mu\big) : \\ &\qquad \mu \in P_2(s)\Big]. \end{aligned}$$

Note that if we fix the stationary policy g^* of player II in the stochastic game (i.e., player II is allowed to use only this policy), then the stochastic game reduces to a dynamic programming model such as the one considered in the first part of this section. More precisely, the state space is S, the action space $M_+^1(X)$, the constraint multifunction is P_1, the law of motion q' is given by $q'\big(\cdot|s, \lambda, g^*(s)\big)$, and the reward function r' defined by $r'(s, \lambda) = r\big(s, \lambda, g^*(s)\big)$. Hypotheses H_4, H_5, H_6 imply that these quantities are well-defined and we can use Theorem 8.5.7 to conclude that f^* is an optimal stationary policy for the dynamic programming problem. From this it follows easily that

$$L(f^*, g^*)(s) = \sup \big[L(\pi, g^*)(s) : \pi \in \Delta_1\big], \quad s \in S. \tag{8.46}$$

In a similar fashion, we show that

$$L(f^*, g^*)(s) = \inf \big[L(f^*, \beta)(s) : \beta \in \Delta_2\big], \quad s \in S. \tag{8.47}$$

Therefore from (8.46) and (8.47) we deduce that

$$L(f^*, g^*)(s) = \inf_{\Delta_2} \sup_{\Delta_1} L(\pi, \beta)(s) = \sup_{\Delta_1} \inf_{\Delta_2} L(\pi, \beta)(s), \quad s \in S. \tag{8.48}$$

From (8.48) it follows that the discounted stochastic game has a value, the value function $s \longrightarrow L(f^*, g^*)(s) = u^*(s)$ is bounded Borel-measurable (i.e., it belongs in $\widehat{B}_0(S)$), and f^*, g^* are optimal stationary policies for players I and II, respectively.□

8.6 Approximate Equilibria

In this section, we return to the first topic of this chapter, namely noncooperative games. The individual stability in such games was described by the fundamental notion of Nash equilibrium (see Definition 8.1.5). In Theorem 8.1.9 we established the existence of such an equilibrium for cooperative games. However, in order to do that, we had to assume that the strategy set of each player is compact. We want to weaken this compactness condition. Excluding the compactness condition on the strategy set of each player, leads to the notion of ε-equilibrium (approximate equilibrium), because an equilibrium need not exist anymore.

We consider a noncooperative n-player game in normal form as in Section 8.1. So let $N = \{1, \ldots, n\}$ be the set of players. Each player has a strategy set X_k which is assumed to be a subset of a Banach space Z_k, $k \in N$. In contrast to Section 8.1, in order to make use of the approximate convex subdifferential, we assume that each player $k \in N$, has a loss function $u_k : X \longrightarrow \mathbb{R}$. Then $u = (u_k)_{k=1}^n : X \longrightarrow \mathbb{R}^n$ is the multiloss map. We emphasize that this is done only for reasons of convenience and in what follows if $\varepsilon = 0$ and u_k is replaced by $-u_k$, we recover the setting of Section 8.1. As before, we consider the following splitting of the multistrategy set $X = \prod\limits_{k \in N} X_k$,

$$X = \widehat{X} \times X_k \quad \text{with } \widehat{X} = \prod_{i \neq k} X_i.$$

As before, we think of $k^c = N \setminus \{k\}$ as the coalition adverse (complementary) to player k. By $\widehat{p}_k : X \longrightarrow \widehat{X}$ and $p_k : X \longrightarrow X_k$ we denote the projection maps of X onto $\widehat{X}$ and X_k, respectively. Every multistrategy $x \in X$ can be written as $x = (\widehat{x}, x_k)$ with $\widehat{x} = \widehat{p}_k(x) \in \widehat{X}$ and $x_k = p_k(x) \in X_k$.

For each $k \in N$, we set

$$\beta_k = \inf\big[u_k(x) : x \in X\big] \tag{8.49}$$

and throughout this section we assume that $\beta_k > -\infty$ for all $k \in N$. Then the vector $\beta = (\beta_k)_{k=1}^n$ is called the *shadow minimum* of the game. Note that

$$u(X) \subseteq \beta + \mathbb{R}_+^n.$$

If $\beta \in u(X)$, then $\beta = u(x^*)$ with $x^* \in X$ and so x^* realizes the infimum in (8.49) for every player $k \in N$. However, this situation is rare and for this reason, we need to introduce approximate equilibria.

Recall that in Section 8.1, we introduced the following quantity

$$v_k(x_k) = \inf\big[u_k(x) : x \in X,\ p_k(x) = x_k\big]. \tag{8.50}$$

This quantity represents the least utility for player $k \in N$ among all feasible multistrategies, when she employs strategy x_k.

DEFINITION 8.6.1 A multistrategy $\overline{x} = (\overline{x}_k)_{k=1}^n \in X$ is said to be an ε-*equilibrium* if for some $\varepsilon \geq 0$ and for all $k \in N$, we have

$$u_k(\overline{x}) \leq v_k(\overline{x}_k) + \varepsilon.$$

REMARK 8.6.2 If $\varepsilon = 0$, then we recover the notion of Nash equilibrium (see Definition 8.1.5 with u_k there replaced by $-u_k$). In this case, given the strategy $\widehat{\overline{x}} \in \widehat{X}$ of the adverse coalition k^c, the player $k \in N$ responds by choosing the strategy $\overline{x}_k \in X_k$ that minimizes v_k on X_k; that is,

$$u_k(\widehat{\overline{x}}, \overline{x}_k) = \inf \left[v_k(x_k) : x \in X,\ x_k \in X_k \right]. \tag{8.51}$$

Moreover, if $N = \{1, 2\}$ and $u_1(x) + u_2(x) = 0$ (zero-sum game), then the ε-equilibrium is an ε-saddle point (a saddle point if $\varepsilon = 0$).

As in Section 8.1, we introduce the function $\xi : X \times X \longrightarrow \mathbb{R}$ defined by

$$\xi(x, z) = \sum_{i=1}^{n} \left[u_i(x) - u_i(\widehat{x}, z_i) \right].$$

PROPOSITION 8.6.3 *If for some $\varepsilon \geq 0$ and $\overline{x} \in X$ we have*

$$\sup \left[\xi(\overline{x}, z) : z \in X \right] \leq \varepsilon, \tag{8.52}$$

then $\overline{x}$ is an ε-equilibrium multistrategy.

PROOF: Let $z = (\widehat{\overline{x}}, z_k)$. Then we have

$$\begin{aligned} & \xi(\overline{x}, z) = u_k(\overline{x}) - u_k(z) \leq \varepsilon \quad \text{(see (8.52),} \\ \Rightarrow\ & u_k(\overline{x}) \leq \inf \left[u_k(z) : z \in X,\ p_k(z) = \overline{x}_k \right] + \varepsilon, \qquad \text{for all } k \in N, \\ \Rightarrow\ & \overline{x} \in X \text{ is an } \varepsilon\text{-equilibrium.} \end{aligned}$$

□

PROPOSITION 8.6.4 *If $\overline{x} \in X$ is an ε-equilibrium, then for every multistrategy $z \in X$, we have $\xi(\overline{x}, z) \leq n\varepsilon$.*

PROOF: From Definition 8.3.1, we have

$$u_k(\overline{x}) - u_k(\widehat{\overline{x}}, z_k) \leq \varepsilon \qquad \text{for all } k \in N \text{ and all } z_k \in X_k \quad \text{(see (8.51).}$$

Adding these inequalities over $k \in N$, we obtain

$$\xi(\overline{x}, z) \leq n\varepsilon \qquad \text{for all } z \in X.$$

□

We introduce the conjugate function of $z_k \longrightarrow u_k(\widehat{x}, z_k)$ (see Definition 1.2.15). So

$$u_k^*(\widehat{x})(z_k^*) = \sup \left[\langle z_k^*, z_k \rangle_{Z_k} - u_k(\widehat{x}, z_k) : z \in Z = \prod_{k \in N} Z_k,\ \widehat{p}_k(z) = \widehat{x} \right],\quad z_k^* \in Z_k^*. \tag{8.53}$$

Here by Z_k^* we denote the topological dual of the Banach space Z_k and by $\langle \cdot, \cdot \rangle_{Z_k}$ the duality brackets for the pair (Z_k^*, Z_k). Recall that $u_k^*(\widehat{x})(\cdot)$ is convex and lower semicontinuous.

In addition to the subdifferential of a convex function introduced in Definition 1.2.28, we can define the approximate subdifferential (ε-subdifferential), which is also a useful tool in convex analysis.

DEFINITION 8.6.5 Let Y be a Banach space and $\varphi : Y \longrightarrow \overline{\mathbb{R}} = \mathbb{R} \cup \{+\infty\}$ be a proper convex function. For each $\varepsilon \geq 0$ the *ε-subdifferential* of φ at $y \in \operatorname{dom} \varphi$, is defined to be the set

$$\partial_\varepsilon \varphi(y) = \big\{ y^* \in Y^* : \langle y^*, v - y \rangle - \varepsilon \leq \varphi(v) - \varphi(y) \qquad \text{for all } \; v \in \operatorname{dom}\varphi \big\}.$$

Here $\langle \cdot, \cdot \rangle$ denotes the duality brackets for the dual pair (Y^*, Y).

REMARK 8.6.6 If $\varepsilon = 0$, then the above definition coincides with Definition 1.2.28 (i.e., $\partial_0 \varphi = \partial \varphi$). However there is a basic difference between the subdifferential ($\varepsilon = 0$) and the ε-subdiffrential ($\varepsilon > 0$). Although $\partial \varphi = \partial_0 \varphi$ is a local notion, for $\varepsilon > 0$ $\partial_\varepsilon \varphi$ is a global one, namely the behavior of φ on all of X may be relevant for the construction of $\partial_\varepsilon \varphi$. This explains why $\partial \varphi$ and $\partial_\varepsilon \varphi$ have in general different properties, with the most basic difference being that if $\varphi \in \Gamma_0(Y)$ and $y \in \operatorname{dom} \varphi$, for every $\varepsilon > 0$ the set $\partial_\varepsilon \varphi(y)$ is nonempty, w^*-closed, and convex. For the function $u_k^*(\widehat{x}) \in \Gamma_0(Z_k^*)$, the ε-subdifferential $\partial_\varepsilon u_k^*(\widehat{x})(z_k^*)$ is defined by

$$\partial_\varepsilon u_k^*(\widehat{x})(z_k^*) = \big\{ z_k^{**} \in Z_k^{**} : \langle z_k^{**}, v^* - z_k^* \rangle - \varepsilon \leq u_k^*(\widehat{x})(v_k^*) - u_k^*(\widehat{x})(z_k^*) \\ \text{for all } \; v_k^* \in Z_k^* \big\}.$$

Using the notion of ε-subdifferential, we can produce necessary and, under some additional hypotheses, also sufficient conditions for a multistrategy to be an ε-equilibrium.

THEOREM 8.6.7 *If $\overline{x} = (\overline{x}_k)_{k=1}^n$ is an ε-equilibrium when $X_k = Z_k$ for all $k \in N$ (i.e., $X = Z$), then $\overline{x}_k \in \partial_\varepsilon u_k^*(\widehat{\overline{x}})(0)$.*

PROOF: From (8.53), we see that

$$\begin{aligned} u_k^*(\widehat{\overline{x}})(0) &= -\inf \big[u_k(\widehat{\overline{x}}, z_k) : z \in Z, \; \widehat{p}_k(z) = \widehat{\overline{x}} \big] \\ &\leq -u_k(\overline{x}) + \varepsilon \quad (\text{because } \; \overline{x} \; \text{ is an } \varepsilon\text{-equilibrium}) \\ &\leq \langle -v_k^*, \overline{x}_k \rangle_{Z_k} + u_k^*(\widehat{\overline{x}})(v_k^*) + \varepsilon \\ \Rightarrow \overline{x}_k &\in \partial_\varepsilon u_k^*(\widehat{\overline{x}})(0) \quad (\text{see Definition 8.6.5}). \end{aligned}$$

□

THEOREM 8.6.8 *If for every $k \in N$ and for some multistrategy $\overline{x} \in X$, $u_k(\widehat{\overline{x}}, \cdot)$ is convex and lower semicontinuous on Z_k and*

$$\overline{x}_k \in \partial_\varepsilon u_k^*(\widehat{\overline{x}})(0),$$

then $\overline{x} \in X$ is an ε-equilibrium strategy.

PROOF: Because by hypothesis $\overline{x}_k \in \partial u_k^*(\widehat{\overline{x}})(0)$, for all $k \in N$ and all $v_k^* \in Z_k^*$, we have

$$\begin{aligned} & \langle v_k^*, \overline{x}_k \rangle_{Z_k} - \varepsilon \leq u_k^*(\widehat{\overline{x}})(v_k^*) - u_k^*(\widehat{\overline{x}})(0), \\ \Rightarrow \; & \langle v_k^*, \overline{x}_k \rangle_{Z_k} - u_k^*(\widehat{\overline{x}})(v_k^*) \leq \varepsilon - u_k^*(\widehat{\overline{x}})(0) \\ \Rightarrow \; & u_k^{**}(\widehat{\overline{x}})(\overline{x}_k) \leq \varepsilon + \inf \big[u_k(\widehat{\overline{x}}, z_k) : z \in Z, \; \widehat{p}_k(z) = \widehat{\overline{x}} \big] \\ \Rightarrow \; & u_k(\overline{x}) \leq \varepsilon + \inf \big[u_k(\widehat{\overline{x}}, z_k) : z \in Z, \; \widehat{p}_k(z) = \widehat{\overline{x}} \big] \end{aligned}$$

(because $u_k(\widehat{\overline{x}}, \cdot) \in \Gamma_0(Z_k)$).

This implies that $\overline{x} \in X$ is an ε-equilibrium. $\square$

Now we assume that the loss function u_k of each player k, is Gâteaux differentiable and we examine how the Gâteaux derivative and the ε-equilibria multistrategies are related.

THEOREM 8.6.9 *If the strategy set $X_k \subseteq Z_k$ of each player $k \in N$ is closed, with nonempty interior, $\overline{x} = (\overline{x}_k)_{k=1}^n \in \operatorname{int} X = \prod_{k=1}^n \operatorname{int} X_k \neq \emptyset$ is an ε-equilibrium with $\varepsilon > 0$ and for every $k \in N$, $u_k(\widehat{\overline{x}}, \cdot)$ is lower semicontinuous and Gâteaux differentiable on Z_k, then there exists a multistrategy $y = (y_k)_{k=1}^n \in X$ such that*

$$\|y - \overline{x}\| \leq n\sqrt{\varepsilon}$$

$$\text{and} \quad \Big\|\frac{\partial u_k}{\partial z_k}(\widehat{\overline{x}}, y_k)\Big\|_{Z_k^*} \leq \sqrt{\varepsilon} \quad \text{for every } k \in N.$$

PROOF: Recall that in the beginning of the section, we have assumed that the game is bounded below. So we can apply Theorem 2.4.1 (the Ekeland variational principle) and obtain $y_k \in X_k$ such that

$$\|\overline{x}_k - y_k\| \leq \sqrt{\varepsilon} \tag{8.54}$$

$$\text{and} \quad u_k(\widehat{\overline{x}}, y_k) \leq u_k(\widehat{\overline{x}}, z_k) + \sqrt{\varepsilon}\|z_k - y_k\| \quad \text{for all } z_k \in X_k. \tag{8.55}$$

Let $h_k \in Z_k$ and set $z_k = y_k + th_k$ for $t > 0$ small so that $z_k \in X_k$ (recall that $\overline{x}_k \in \operatorname{int} X_k$). Then

$$-\sqrt{\varepsilon}\|h_k\| \leq \frac{1}{t}\big[u_k(\widehat{\overline{x}}, y_k + th_k) - u_k(\widehat{\overline{x}}, y_k)\big].$$

We let $t \longrightarrow 0$ to obtain

$$-\sqrt{\varepsilon}\|h_k\| \leq \Big\langle \frac{\partial u_k}{\partial z_k}(\widehat{\overline{x}}, y_k), h_k \Big\rangle_{Z_k}. \tag{8.56}$$

In (8.56), we take infimum of both sides with respect to all $h_k \in Z_k$ with $\|h_k\|_{Z_k} = 1$. It follows that

$$\Big\|\frac{\partial u_k}{\partial z_k}(\widehat{\overline{x}})\Big\|_{Z_k} \leq \sqrt{\varepsilon}, \quad k \in N.$$

Also, if we add inequalities (8.56), we conclude that

$$\|\overline{x} - y\|_X \leq n\sqrt{\varepsilon}.$$

$\square$

THEOREM 8.6.10 *If each player $k \in N$ has a reflexive strategy space Z_k, $X_k = Z_k$, $\overline{x} = (\overline{x}_k)_{k=1}^n \in X = Z$ is an ε-equilibrium multistrategy with $\varepsilon > 0$ and $u_k(\widehat{\overline{x}}, \cdot) \in \Gamma_0(Z_k)$, then we can find $y_\varepsilon = (y_{\varepsilon,k})_{k=1}^n \in X = Z$ and $y_\varepsilon^* = (y_{\varepsilon,k}^*)_{k=1}^n \in Z^* = \prod_{k=1}^n Z_k^*$ such that for every player $k \in N$, we have*

$$\|\overline{x}_k - y_{\varepsilon,k}\|_{Z_k} \leq \sqrt{\varepsilon} \tag{8.57}$$

$$\|y_{\varepsilon,k}^*\|_{Z_k^*} \leq \sqrt{\varepsilon}; \tag{8.58}$$

moreover, $y_\varepsilon = (y_{\varepsilon,k})_{k=1}^n \in X = Z$ *is a Nash equilibrium of the game* $\{\widetilde{u}_k, X_k\}_{k\in N}$*, where*

$$\widetilde{u}_k(x) = u_k(\widehat{\overline{x}}, x_k) - \langle y^*_{\varepsilon,k}, x_k \rangle_{Z_k} \quad \textit{for every } x = (x_k)_{k=1}^n \in X = \prod_{k=1}^n X_k.$$

PROOF: From Theorem 8.6.7, we have

$$\overline{x}_k \in \partial_\varepsilon u^*_k(\widehat{\overline{x}})(0) \qquad \text{for all } k \in N. \tag{8.59}$$

Let $\vartheta_k(\widehat{\overline{x}})(v^*_k) = u^*_k(\widehat{\overline{x}}) - \langle v^*_k, \overline{x}_k \rangle_{Z_k}$, $k \in N$. Then $\vartheta_k(\widehat{\overline{x}}) \in \Gamma_0(Z^*_k)$ and from (8.59) we have

$$\vartheta_k(\widehat{\overline{x}})(0) \leq \inf\left[\vartheta_k(\widehat{\overline{x}})(v^*_k) : v^*_k \in Z^*_k\right] + \varepsilon. \tag{8.60}$$

From (8.60) it follows that $\vartheta_k(\widehat{\overline{x}})(\cdot)$ is bounded below on Z^*_k. So we can apply Theorem 2.4.1 (the Ekeland variational principle) and obtain $y^*_{\varepsilon,k} \in Z^*_k$ such that for all $v^*_k \neq y^*_{\varepsilon,k}$

$$\vartheta_k(\widehat{\overline{x}})(y^*_{\varepsilon,k}) - \sqrt{\varepsilon}\|v^*_k - y^*_{\varepsilon,k}\|_{Z^*_k} < \vartheta_k(\widehat{\overline{x}})(v^*_k) \tag{8.61}$$

$$\text{and} \quad \vartheta_k(\widehat{\overline{x}})(y^*_{\varepsilon,k}) \leq \vartheta_k(\widehat{\overline{x}})(0) - \sqrt{\varepsilon}\|y^*_{\varepsilon,k}\|_{Z^*_k}. \tag{8.62}$$

From (8.61) it follows that $y^*_{\varepsilon,k}$ minimizes the function

$$f(\widehat{\overline{x}})(\cdot) = \vartheta_k(\widehat{\overline{x}})(\cdot) + \sqrt{\varepsilon}\| \cdot - y^*_{\varepsilon,k}\|_{Z^*_k}$$

defined on Z^*_k. So we have

$$0 \in \partial\big(\vartheta_k(\widehat{\overline{x}})(\cdot) + \sqrt{\varepsilon}\| \cdot - y^*_{\varepsilon,k}\|_{Z^*_k}\big)(y^*_{\varepsilon,k}).$$

Invoking Theorem 1.2.38, we have

$$0 \in \partial\vartheta_k(\widehat{\overline{x}})(y^*_{\varepsilon,k}) + \partial j(y^*_{\varepsilon,k}), \tag{8.63}$$

where $j(v^*_k) = \sqrt{\varepsilon}\|v^*_k - y^*_{\varepsilon,k}\|_{Z^*_k}$. From Example 1.2.41(c), we know that

$$\partial j(y^*_{\varepsilon,k}) = \sqrt{\varepsilon}\overline{B}^{**}_{1,k} \tag{8.64}$$

with $\overline{B}^{**}_{1,k}$ being the closed unit ball in Z^{**}_k. The space Z_k is reflexive, therefore we have $Z^{**}_k = Z_k$ and so from (8.63), (8.64), and the definition of $\vartheta_k(\widehat{\overline{x}})(\cdot)$, we deduce that there exists $y_{\varepsilon,k} \in Z_k$ such that

$$y_{\varepsilon,k} \in \partial u^*_k(\widehat{\overline{x}})(y^*_{\varepsilon,k}) \tag{8.65}$$

$$\text{and} \quad y_{\varepsilon,k} = \overline{x}_k - \sqrt{\varepsilon}\|v_k\|_{Z_k},\ \|v_k\|_{Z_k} \leq 1. \tag{8.66}$$

From (8.66), we obtain

$$\|\overline{x}_k - y_{\varepsilon,k}\| \leq \sqrt{\varepsilon}$$

which proves inequality (8.57) of the theorem.

From (8.60) and (8.62), we have

$$\|y^*_{\varepsilon,k}\|_{Z^*_k} \leq \sqrt{\varepsilon},$$

which proves inequality (8.58) of the theorem.

Next from (8.65), we have

$$
\begin{aligned}
&\left\langle v_k^* - y_{\varepsilon,k}^*, y_{\varepsilon,k}\right\rangle_{Z_k} \le u_k^*(\widehat{\overline{x}})(v_k^*) - u_k^*(\widehat{\overline{x}})(y_{\varepsilon,k}^*) \qquad \text{for all } v_k^* \in Z_k^*,\\
\Rightarrow\ &\left\langle v_k^*, y_{\varepsilon,k}\right\rangle_{Z_k} - u_k^*(\widehat{\overline{x}})(v_k^*) - \left\langle y_{\varepsilon,k}^*, y_{\varepsilon,k}\right\rangle_{Z_k} \le -u_k^*(\widehat{\overline{x}})(y_{\varepsilon,k}^*),\\
\Rightarrow\ &u_k^{**}(\widehat{\overline{x}})(y_{\varepsilon,k}) - \left\langle y_{\varepsilon,k}^*, y_{\varepsilon,k}\right\rangle_{Z_k} \le -\sup\left[\left\langle y_{\varepsilon,k}^*, z_k\right\rangle_{Z_k} - u_k(\widehat{\overline{x}}, z_k) : z_k \in Z_k\right]\\
&\qquad\qquad\qquad\qquad\qquad\qquad = \inf\left[u_k(\widehat{\overline{x}}, z_k) - \left\langle y_{\varepsilon,k}^*, z_k\right\rangle_{Z_k} : z_k \in Z_k\right]\\
\Rightarrow\ &u_k(\widehat{\overline{x}}, y_{\varepsilon,k}) - \left\langle y_{\varepsilon,k}^*, y_{\varepsilon,k}\right\rangle_{Z_k} \le \inf\left[u_k(\widehat{\overline{x}}, z_k) - \left\langle y_{\varepsilon,k}^*, z_k\right\rangle_{Z_k} : z_k \in Z_k\right]
\end{aligned}
\tag{8.67}
$$

Because $u_k(\widehat{\overline{x}}, \cdot) \in \Gamma_0(Z_k)$. From (8.67) it follows that

$$
\begin{aligned}
&\widetilde{u}_k(y_\varepsilon) = \min\left[\widetilde{u}_k(z) : z \in X,\ \widehat{p}_k(z) = \widehat{y}_\varepsilon\right],\\
\Rightarrow\ &y_\varepsilon = (y_{\varepsilon,k})_{k=1}^n \in Z \text{ is a Nash equilibrium for the game } \{X_k, \widetilde{u}_k\}_{k\in N}.
\end{aligned}
$$

□

8.7 Remarks

8.1: Noncooperative games and the associated equilibrium notion (see Definition 8.1.5) were introduced by Nash [451, 452]. Nash proved Theorem 8.1.9 for games where the preferences of the players are described by continuous quasiconcave utility functions and the strategy sets are simplexes. The work of Nash generalized the pioneering work of von Neumann [454], who proved the first minimax theorem (using Brouwer's fixed point theory) and initiated game theory. The work of Nash achieved the shift of attention from the minimax concept (saddle point) of von Neumann to the new concept of Nash equilibrium (noncooperative equilibrium). It is only with this new notion that the theory of noncooperative games could move with full generality to the case of n-players with $n > 2$. Theorem 8.1.9 assumes that the utility functions $u_k(\widehat{x}, y_k)$ are continuous in both variables. A partial generalization in this direction, can be found in Nikaido–Isoda [460]. Abstract economies (or generalized games; see Definition 8.1.11), were first introduced by Debreu [184], who was influenced by the work of Nash. He introduced and studied a model in which the agents' (players') preference relations are described by a utility function (hence they are transitive). Later Mas-Colell [410] and Shafer–Sonnenschein [549] dropped the transitivity requirement and described preferences using a preference multifunction (see Theorem 8.1.18). For further discussion of noncooperative games and abstract economies, with a rich bibliography, we refer to the books of Aubin [36], Border [86], and Ichiishi [324].

8.2: The first side-payment game was introduced by von Neumann–Morgenstern [455], but had a specific interpretation of the characteristic function v of the game. The notion of core of a side-payment game (see Definition 8.2.1) was introduced by Shapley [550]. The concept of a balanced side-payment game was introduced by Bondavera [84, 85] and for non-side-payment games (see Definition 8.2.9) by Scarf [537]. Theorem 8.2.8 is independently due to Bondavera [84, 85], and Shapley [551]. Side-payment games with infinitely many players have been studied. We mention the works of Schmeidler [541, 542], Kannai [340], and Delbaen [193]. The first to formulate non-side-payment games, were Aumann–Peleg [41]. Aumann [42] defined the core for such games (see Definition 8.2.9). Theorem 8.2.13, was first proved by Scarf [537]. However, the proof presented here, which is based on the special version of the KKM-Theorem (see Proposition 8.2.12), is due to Shapley [552]. Non-side-payment games with a continuum of players, can be found in Aumann–Shapley [47], Ichiishi [324], and in the paper of Ichiishi–Weber [323].

8.3: The first to study the existence of Cournot–Nash equilibria for games with an atomless measure space of players (a continuum game model), was Schmeidler [543]. In Schmeidler's model the strategy spaces are finite-dimensional. Khan [349] extended this to the infinite-dimensional case. His work and the subsequent extensions by Khan–Papageorgiou [351] and Balder–Yannelis [54] considered models in which each player has a preference multifunction instead of a utility function. So her preference relation need not be transitive or complete. Cournot–Nash equilibrium distributions for games that are viewed as a probability measure on the space of payoff (utility) function were established by Mas Collel [411], and Khan [352].

8.4: For games with a finite number of players and actions and with private information (imperfect information in the terminology of Harsanyi [285]), equilibrium results were proved by Radner–Rosenthal [511], Milgram–Weber [429], and Mamer–Schilling [405]. Bayesian games with an infinite number of players, were first introduced by Palfrey–Srivastava [473], and Postlewaite–Schmeidler [501], who introduced the information partition approach used here. Bayesian games with an infinite-dimensional strategy space can be found in Balder–Yannelis [54], and Kim–Yannelis [353].

8.5: The dynamic programming model has its origins in the pioneering work of Bellman [62]. A rigorous foundation of stochastic dynamic programming, was given by Blackwell [77, 78], and Strauch [561]. Advanced treatment of stochastic dynamic programming can be found in the books of Hinderer [305], Striebel [562], Bertsekas–Shreve [72], Dynkin–Yushkevich [216], Arkin–Evstingeev [29], and Maitra–Sudderth [399]. Theorem 8.5.3 is basic in the theory of stochastic dynamic programming and it is due to Blackwell [77]. Discounted stochastic games were investigated by many authors. Indicatively we mention the works of Maitra–Parthasarathy [397, 398], Parthasarathy [488, 489], Himmelberg–Parthasarathy–Raghavan–Van Vleck [302], Himmelberg–Parthasarathy–Van Vleck [303], Nowak [462, 463], and Whitt [604]. Adaptive control problems can be found in the book of Hermandez Lerma [290], and gambling games in the book of Maitra–Sudderth [399].

8.6: Approximate equilibria for noncooperative games were studied by Tanaka–Yokoyama [575]. The ε-subdifferential of convex functions (see Definition 8.6.5) was studied in a systematic way by Hiriart–Urruty [306, 307].

9

Uncertainty, Information, Decision Making

Summary. *In this chapter, we study how information can be incorporated as a variable in various decision models. In a decision model with uncertainty, it is natural to model information by sub-σ-fields of a probability space which represents the set of all possible states of the "world". For this reason we topologize the set of sub-σ-fields with two topologies, using tools from probability theory and functional analysis. Then we examine the "ex-post view" and the "ex-ante view". In both cases, we prove continuous dependence of the model on the information variable. Then we introduce a third mode of convergence of the information variable and we study prediction sequences. Finally we study games with incomplete information or unbounded cost and general state space.

Introduction

The goal of this chapter is to study how information can be incorporated as a variable in various decision models. Such a formulation allows us to treat situations with asymmetric information.

In Section 9.1 we present a mathematical framework that permits the analytical treatment of the notion of information. We produce two such comparable metric topologies using tools from probability theory and functional analysis.

In Section 9.2, we examine the ex-post view, in modelling systems with uncertainty. According to this view, every agent chooses an action after observing his or her information and updating his or her belief. Using the topologies of Section 9.1, we prove the continuity of this model on the information variable.

In Section 9.3, we examine an alternative view of uncertain decision systems, known as the ex-ante view. In this alternative approach, an agent formulates a plan of what action to choose at each state before observing her information, subject to the constraint that the plan must be measurable with respect to her information.

In Section 9.4, we introduce a third mode of convergence of information, distinct from the ones introduced and studied in Section 9.1. We show that this new mode of convergence is suitable in the analysis of prediction sequences.

Section 9.5 formulates and studies a general two-person, zero-sum game with incomplete information. The hypotheses on the data of the model are minimal.

N.S. Papageorgiou, S.Th. Kyritsi-Yiallourou, *Handbook of Applied Analysis*,
Advances in Mechanics and Mathematics 19, DOI 10.1007/b120946_9,

Finally in Section 9.6, we treat games with a general state space and unbounded cost. We prove the existence of equilibria.

9.1 Mathematical Space of Information

In many economic models with uncertainty, the information structure is an endogenous variable or an exogenous parameter. A natural and general way to model information is to represent it by sub-σ-fields of a probability space (Ω, Σ, μ), which describes the set of all possible states of the world (Ω), the family (σ-field) of all possible events (Σ), and the distribution of those events (μ). In order to make formal statements on the relation of information to the other variables of the model, we need to define precisely the mathematical space of information and endow it with a suitable topology. This topology should not be too weak, or otherwise various decision variables that depend on information (such as consumer demand) may fail to be continuous. On the other hand the topology should not be too strong (too rich), or otherwise information structures (fields) that are known to lead to similar behavior, may fail to be topologically close to each other. So we need to achieve a rather delicate balance with the topology on the space of information. In this section, we outline a topology that achieves this balance.

Let (Ω, Σ, μ) be a probability space. It represents the uncertainty present in the microeconomic model. By $\mathcal{S}^*$ we denote the set of all sub-σ-fields of Σ. On $\mathcal{S}^*$ we introduce the relation $\sim$, defined by

> for $\Sigma_1, \Sigma_2 \in \mathcal{S}^*$, we have $\Sigma_1 \sim \Sigma_2$ if and only if they differ only by μ-null sets.

So for every $A \in \Sigma_1$, there is a $C \in \Sigma_2$ such that $\mu(A \triangle C) = 0$ and vice versa. The space of information $\mathcal{S}_0$ is defined to be the set of $\sim$-equivalence classes of $\mathcal{S}^*$; that is,

$$\mathcal{S}_0 = \mathcal{S}^* / \sim .$$

In what follows, given $\Sigma' \in \mathcal{S}^*$ and $f \in L^1(\Omega, \Sigma)$, by $E(f|\Sigma')$ we denote the conditional expectation of f with respect to the sub-σ-field Σ'. Using the standard approximation of $L^1(\Omega, \Sigma)$ functions by simple functions, we obtain the following result.

PROPOSITION 9.1.1 *$\Sigma_1 \sim \Sigma_2$ if and only if $E(f|\Sigma_1) = E(f|\Sigma_2)$ for all $f \in L^1(\Omega, \Sigma)$.*

REMARK 9.1.2 Note that for every $\Sigma' \in \mathcal{S}^*$ and every $f \in L^1(\Omega, \Sigma)$, $E(f|\Sigma')$ belongs in $L^1(\Omega, \Sigma') \subseteq L^1(\Omega, \Sigma)$.

Now let $\mathcal{L}\big(L^1(\Omega, \Sigma)\big)$ be the Banach space of all bounded linear operators $L : L^1(\Omega, \Sigma) \longrightarrow L^1(\Omega, \Sigma)$. Then for $\Sigma' \in \mathcal{S}_0$, $E(\cdot|\Sigma') \in L^1(\Omega, \Sigma') \subseteq L^1(\Omega, \Sigma)$ (see Remark 9.1.2) and the map $\Sigma' \longrightarrow E(f|\Sigma')$ from $\mathcal{S}_0$ into $L^1(\Omega, \Sigma)$ is injective (see Proposition 9.1.1). So we can embed $\mathcal{S}_0$ into $\mathcal{L}\big(L^1(\Omega, \Sigma)\big)$ and topologize it with the subspace topology of any topology we consider on the space $\mathcal{L}\big(L^1(\Omega, \Sigma)\big)$. The space $\mathcal{L}\big(L^1(\Omega, \Sigma)\big)$ is a Banach space with the norm

$$\|L\|_{\mathcal{L}} = \sup\left[\frac{\|L(f)\|_1}{\|f\|_1} : f \in L^1(\Omega,\Sigma),\ f \neq 0\right].$$

The induced norm topology is called the *uniform operator topology*. In this topology the map $\{L, K\} \longrightarrow L \circ K \in \mathcal{L}\big(L^1(\Omega,\Sigma)\big)$ is jointly continuous.

We now introduce another topology on $\mathcal{L}\big(L^1(\Omega,\Sigma)\big)$ that is weaker than the uniform operator topology and more suitable for our purposes.

DEFINITION 9.1.3 The *strong operator topology* or *topology of pointwise convergence* is the weakest topology on $\mathcal{L}\big(L^1(\Omega,\Sigma)\big)$ such that the maps

$$E_f : \mathcal{L}\big(L^1(\Omega,\Sigma)\big) \longrightarrow L^1(\Omega,\Sigma)$$

defined by $E_f(L) = L(f)$ are continuous for every $f \in L^1(\Omega,\Sigma)$.

REMARK 9.1.4 A neighborhood basis of the origin is given by the sets

$$\big\{L \in \mathcal{L}\big(L^1(\Omega,\Sigma)\big) : \|L(f_k)\|_1 < \varepsilon,\ k = 1,\dots,n\big\},$$

where $\{f_k\}_{k=1}^n$ is a finite collection of elements in $L^1(\Omega,\Sigma)$ and $\varepsilon > 0$. In this topology a net $\{L_a\}_{a\in J}$ of operators converges to an operator $L \in \mathcal{L}\big(L^1(\Omega,\Sigma)\big)$ denoted by $L_a \xrightarrow{s} L$ if and only if $\|L_a(f) - L(f)\|_1 \longrightarrow 0$ for all $f \in L^1(\Omega,\Sigma)$. The map $\{L, K\} \longrightarrow L \circ K$ is separately but not jointly continuous.

In general the strong operator topology on $\mathcal{L}\big(L^1(\Omega,\Sigma)\big)$ is not well-behaved and in particular it is not metrizable. However, the relative topology on $\mathcal{S}_0 \subseteq \mathcal{L}\big(L^1(\Omega,\Sigma)\big)$ is much better behaved because $\mathcal{S}_0$ is a uniformly equicontinuous subset. In what follows we assume that the Lebesgue space $L^1(\Omega,\Sigma)$ is separable. This is true when Σ is countably generated and this in turn holds if Σ is the Borel σ-field of a second-countable Hausdorff topological space (e.g., of a separable metric space). In the sequel by τ_s we denote the strong operator topology on $\mathcal{L}\big(L^1(\Omega,\Sigma)\big)$.

THEOREM 9.1.5 *If $L^1(\Omega,\Sigma)$ is separable, then $(\mathcal{S}_0, \tau_s)$ is a Polish space and the metric is given by*

$$d_s(\Sigma_1,\Sigma_2) = \sum_{k=1}^{\infty} \frac{1}{2^k} \min\big\{\|E(f_k|\Sigma_1) - E(f_k|\Sigma_2)\|_1, 1\big\}, \tag{9.1}$$

where $\{f_k\}_{k\geq 1}$ is a countable dense subset of $L^1(\Omega,\Sigma)$. Any two metrics using different countable dense subsets of $L^1(\Omega,\Sigma)$ are uniformly equivalent.

PROOF: First we show that d_s is a metric.

Clearly d_s satisfies the triangle inequality. Also from Proposition 9.1.1 we see that

$$\Sigma_1 = \Sigma_2 \Rightarrow d_s(\Sigma_1,\Sigma_2) = 0.$$

On the other hand, if $d_s(\Sigma_1,\Sigma_2) = 0$, then from (9.1) we see that

$$\|E(f_k|\Sigma_1) - E(f_k|\Sigma_2)\|_1 = 0 \qquad \text{for all } k \geq 1. \tag{9.2}$$

Given any $h \in L^1(\Omega,\Sigma)$ and $\varepsilon > 0$, we can find $k \geq 1$ such that $\|h - f_k\| < \varepsilon$. Then we have

$$\|E(h|\Sigma_1) - E(h|\Sigma_2)\|_1 \leq \|E(h|\Sigma_1) - E(f_k|\Sigma_1)\|_1 + \|E(f_k|\Sigma_2) - E(h|\Sigma_2)\|_1$$

(see (9.2))

$$\leq \|h - f_k\|_1 + \|f_k - h\|_1 < 2\varepsilon.$$

Because $\varepsilon > 0$ was arbitrary, we let $\varepsilon \downarrow 0$ to obtain

$$\|E(h|\Sigma_1) - E(h|\Sigma_2)\|_1 = 0 \quad \text{for all } h \in L^1(\Omega, \Sigma)$$
$$\Rightarrow \Sigma_1 = \Sigma_2 \quad \text{(see Proposition 9.1.1).}$$

Finally it is obvious that $d_s(\Sigma_1, \Sigma_2) = d_s(\Sigma_2, \Sigma_1)$ for all $\Sigma_1, \Sigma_2 \in \mathcal{S}_0$. Therefore we have shown that d_s is a metric on $\mathcal{S}_0$.

Note that for all $f \in L^1(\Omega, \Sigma)$ and all $\Sigma' \in \mathcal{S}_0$, we have

$$\|E(f|\Sigma')\|_1 \leq \|f\|_1,$$
$$\Rightarrow \mathcal{S}_0 \text{ is uniformly equicontinuous.}$$

So from a result of point-set topology (see, e.g., Kelley [344, p. 238]), we have that the topologies d_s and τ_s on $\mathcal{S}_0$ coincide and in fact d_s is independent of the particular dense sequence $\{f_k\}_{k\geq 1}$ we use. Moreover, the topology is separable.

It remains to show that the metric topology is complete. To this end let $\{\Sigma_n\}_{n\geq 1}$ be a d_s-Cauchy sequence. Then given $\varepsilon \in (0,1)$ and $k \geq 1$, there exists N such that for $m, n \geq N$, we have

$$d_s(\Sigma_n, \Sigma_m) < \frac{\varepsilon}{2^k},$$
$$\Rightarrow \|E(f_k|\Sigma_n) - E(f_k|\Sigma_m)\|_1 < \varepsilon \quad \text{(see (9.1)),}$$
$$\Rightarrow \{E(f_k|\Sigma_n)\}_{n\geq 1} \subseteq L^1(\Omega, \Sigma) \quad \text{is a Cauchy sequence.}$$

Therefore we can find $L \in \mathcal{L}(L^1(\Omega, \Sigma))$ such that

$$L(h) = \lim_{n\to\infty} E(h|\Sigma_n) \quad \text{for every } h \in L^1(\Omega, \Sigma). \tag{9.3}$$

We show that L is also a conditional expectation with respect to some $\widehat{\Sigma} \in \mathcal{S}_0$. To this end we need to show that

$$L(gL(h)) = L(g)L(h) \qquad \text{for all } g \in L^\infty(\Omega, \Sigma), \text{ all } h \in L^1(\Omega, \Sigma) \tag{9.4}$$

$$\text{and} \qquad \int_\Omega L(h)d\mu = \int_\Omega h d\mu \qquad \text{for all } h \in L^1(\Omega, \Sigma) \tag{9.5}$$

(see Neveu [459, p. 12]). So we have

$$\begin{aligned} L(gL(h)) &= L(g \lim_{n\to\infty} E(h|\Sigma_n)) \qquad \text{(see (9.3))} \\ &= \lim_{n\to\infty} L(gE(h|\Sigma_n)) \qquad \text{(because } L \text{ is continuous)} \\ &= \lim_{n\to\infty} E(gE(h|\Sigma_n)|\Sigma_n) \qquad \text{(see (9.3))} \\ &= \lim_{n\to\infty} E(g|\Sigma_n)E(h|\Sigma_n) = L(g)L(h) \end{aligned}$$

which proves (9.4). Also

$$\int_{\Omega} L(h)\mathrm{d}\mu = \int_{\Omega} \lim_{n\to\infty} E(h|\Sigma_n)\mathrm{d}\mu = \lim_{n\to\infty} \int_{\Omega} E(h|\Sigma_n)\mathrm{d}\mu = \int_{\Omega} h\mathrm{d}\mu,$$

which proves (9.5). Therefore we can find $\widehat{\Sigma} \in \mathcal{S}_0$ such that $L(h) = E(h|\widehat{\Sigma})$ for all $h \in L^1(\Omega, \Sigma)$ and so $(\mathcal{S}_0, d_s)$ is a Polish space. $\square$

In the propositions that follow, we more closely examine the convergence in the Polish space $(\mathcal{S}_0, d_s)$.

PROPOSITION 9.1.6 $\Sigma_n \xrightarrow{d_s} \widehat{\Sigma}$ *if and only if* $E(\chi_A|\Sigma_n) \xrightarrow{\mu} E(\chi_A|\widehat{\Sigma})$ *for all* $A \in \Sigma$.

PROOF: $\Rightarrow$: This implication follows from the definition of the topology $\tau_s = d_s$ (see Definition 9.1.3).

$\Leftarrow$: We know that for every $f \in L^1(\Omega, \Sigma)$, $\{E(f|\Sigma') : \Sigma' \in \mathcal{S}_0\} \subseteq L^1(\Omega, \Sigma)$ is uniformly integrable. So from the hypothesis and the extended dominated convergence theorem (Vitali's theorem) we have that $E(s|\Sigma_n) \longrightarrow E(s|\widehat{\Sigma})$ for every simple function $s \in L^1(\Omega, \Sigma)$. But simple functions are dense in $L^1(\Omega, \Sigma)$. So given $f \in L^1(\Omega, \Sigma)$, we can find simple functions $\{s_m\}_{m\geq 1} \subseteq L^1(\Omega, \Sigma)$ such that $s_m \longrightarrow f$ in $L^1(\Omega, \Sigma)$. We have

$$\begin{aligned}
&\|E(f|\Sigma_n) - E(f|\widehat{\Sigma})\|_1 \leq \\
&\|E(f|\Sigma_n) - E(s_m|\Sigma_n)\|_1 + \|E(s_m|\Sigma_n) - E(s_m|\widehat{\Sigma})\|_1 + \|E(s_m|\widehat{\Sigma}) - E(f|\widehat{\Sigma})\|_1 \\
&\qquad\leq 2\|f - s_m\|_1 + \|E(s_m|\Sigma_n) - E(s_m|\widehat{\Sigma})\|_1 \longrightarrow 0, \\
&\qquad\qquad \text{as} \quad m \longrightarrow \infty,
\end{aligned}$$

$\Rightarrow \Sigma_n \xrightarrow{d_s} \widehat{\Sigma}$. $\square$

In a similar fashion we also show the following result.

PROPOSITION 9.1.7 $\Sigma_n \xrightarrow{d_s} \widehat{\Sigma}$ *if and only if for every* $D \subseteq L^1(\Omega, \Sigma)$ *dense set* $E(f|\Sigma_n) \xrightarrow{\mu} E(f|\widehat{\Sigma})$ *for all* $f \in D$.

DEFINITION 9.1.8 (a) For $\Sigma_1, \Sigma_2 \in \mathcal{S}_0$, $\Sigma_1 \vee \Sigma_2$ is the smallest σ-field containing both Σ_1 and Σ_2.
(b) For $\{\Sigma_n\}_{n\geq 1} \subseteq \mathcal{S}_0$, we define

$$\liminf_{n\to\infty} \Sigma_n = \bigvee_{m\geq 1} \bigcap_{n\geq m} \Sigma_n \quad \text{and} \quad \limsup_{n\to\infty} \Sigma_n = \bigcap_{m\geq 1} \bigvee_{n\geq m} \Sigma_n.$$

PROPOSITION 9.1.9 $\widehat{\Sigma} = \liminf_{n\to\infty} \Sigma_n = \limsup_{n\to\infty} \Sigma_n$ *implies* $\Sigma_n \xrightarrow{d_s} \widehat{\Sigma}$.

PROOF: Let $\Gamma_m = \bigcap_{n\geq m} \Sigma_n$ and $\Lambda_m = \bigvee_{n\geq m} \Sigma_n$. Evidently $\{\Gamma_m\}_{m\geq 1}$ is an increasing sequence and $\{\Lambda_m\}_{m\geq 1}$ is a decreasing sequence of sub-σ-fields of Σ. From the martingale and reverse martingale convergence theorems, for every $f\in L^1(\Omega,\Sigma)$ we have

$$E(f|\Gamma_m)\longrightarrow E(f|\widehat{\Sigma}) \quad \text{and} \quad E(f|\Lambda_m)\longrightarrow E(f|\widehat{\Sigma}) \quad \text{in } L^1(\Omega,\Sigma) \quad \text{as } m\longrightarrow\infty.$$

Given $\varepsilon>0$, we can find $m_0=m_0(\varepsilon)\geq 1$ such that

$$\|E(f|\Gamma_m)-E(f|\Lambda_m)\|_1<\frac{\varepsilon^2}{2} \qquad \text{for all } m\geq m_0. \tag{9.6}$$

Let $m\geq m_0$ and $A\in\Sigma_m$. Because $\Sigma_m\subseteq\Lambda_m$, we have

$$\begin{aligned}
&\Big|\int_A \big(E(f|\Gamma_m)-E(f|\Sigma_m)\big)d\mu\Big| \\
&=\Big|\int_A \big(E(f|\Gamma_m)-E(f|\Lambda_m)\big)d\mu\Big| \\
&\leq \int_A \big|E(f|\Gamma_m)-E(f|\Lambda_m)\big|d\mu \\
&\leq \|E(f|\Gamma_m)-E(f|\Lambda_m)\|_1<\frac{\varepsilon^2}{2} \quad \text{(see (9.6))}.
\end{aligned} \tag{9.7}$$

Let $A_m=\big\{\omega\in\Omega: E(f|\Gamma_m)(\omega)-E(f|\Sigma_m)(\omega)>\varepsilon\big\}\in\Sigma_m$ (because $\Gamma_m\subseteq\Sigma_m$). Using this set in (9.7) we obtain

$$\begin{aligned}
&\varepsilon\mu(A_m)\leq\Big|\int_A \big(E(f|\Gamma_m)-E(f|\Sigma_m)\big)d\mu\Big|<\frac{\varepsilon^2}{2}, \\
&\Rightarrow\ \mu(A_m)<\frac{\varepsilon}{2} \qquad \text{for all } m\geq m_0.
\end{aligned} \tag{9.8}$$

Similarly, if $C_m=\big\{\omega\in\Omega: E(f|\Gamma_m)(\omega)-E(f|\Sigma_m)(\omega)<-\varepsilon\big\}$, we can say that

$$\mu(C_m)<\frac{\varepsilon}{2} \qquad \text{for all } m\geq m_0. \tag{9.9}$$

Combining (9.8) and (9.9), we obtain

$$\begin{aligned}
&\mu\big\{\omega\in\Omega: \big|E(f|\Gamma_m)(\omega)-E(f|\Sigma_m)(\omega)\big|>\varepsilon\big\}<\varepsilon \qquad \text{for all } m\geq m_0, \\
&\Rightarrow\ E(f|\Gamma_m)-E(f|\Sigma_m)\xrightarrow{\mu} 0 \quad \text{as } m\longrightarrow\infty.
\end{aligned} \tag{9.10}$$

Also from the martingale convergence theorem, we have

$$E(f|\Gamma_m)(\omega)\longrightarrow E(f|\widehat{\Sigma})(\omega) \quad \mu\text{-a.e. on } \Omega \quad \text{as } m\longrightarrow\infty. \tag{9.11}$$

From (9.10) and (9.11), we infer that

$$E(f|\Sigma_m)\xrightarrow{\mu} E(f|\widehat{\Sigma}) \quad \text{as } m\longrightarrow\infty. \tag{9.12}$$

Also, again from the martingale convergence theorem, we have

$$\|E(f|\Gamma_m)\|_1,\ \|E(f|\Lambda_m)\|_1\longrightarrow\|E(f|\widehat{\Sigma})\|_1 \quad \text{as } m\longrightarrow\infty. \tag{9.13}$$

Because $\Gamma_m\subseteq\Sigma_m\subseteq\Lambda_m$, we have

$$\|E(f|\Gamma_m)\|_1\leq\|E(f|\Sigma_m)\|_1\leq\|E(f|\Lambda_m)\|_1 \quad \text{for all } m\geq 1. \tag{9.14}$$

From (9.13) and (9.14), we infer that

$$\|E(f|\Sigma_m)\|_1 \longrightarrow \|E(f|\widehat{\Sigma})\|_1 \quad \text{as } m \longrightarrow \infty. \tag{9.15}$$

From (9.12) and (9.15), we conclude that

$$E(f|\Sigma_m) \longrightarrow E(f|\widehat{\Sigma}) \quad \text{in } L^1(\Omega, \Sigma) \quad \text{as } m \longrightarrow \infty.$$

Because $f \in L^1(\Omega, \Sigma)$ was arbitrary, we have

$$\Sigma_m \xrightarrow{d_s} \widehat{\Sigma} \quad \text{as } m \longrightarrow \infty.$$

□

The strong operator topology (topology of pointwise convergence) on $\mathcal{S}_0$ has the nice property that the subset of information consisting of all finite partitions of the state space is dense. This is nice because in models in which the information variable appears in a continuous fashion, finite-dimensional approximations can be used.

PROPOSITION 9.1.10 *If $\widehat{\Sigma} \in \mathcal{S}_0$ and $\varepsilon > 0$ are given, then we can find a finite Σ-partition $\Sigma_f \in \mathcal{S}_0$ of Ω with $\Sigma_f \subseteq \widehat{\Sigma}$ such that*

$$d_s(\widehat{\Sigma}, \Sigma_f) < \varepsilon.$$

PROOF: Let $\{f_k\}_{k=1}^N \subseteq L^\infty(\Omega, \Sigma)$ and let $c > \max\limits_{1 \le k \le N} \|f_k\|_\infty$. Choose $r > 0$ such that $c < r\varepsilon/4$. For every $k = \{1, \dots, N\}$ we set

$$A_{ki} = \Big\{\omega \in \Omega : \frac{\varepsilon i}{2} \le E(f_k|\widehat{\Sigma})(\omega) \le \frac{\varepsilon(i+1)}{2}\Big\}$$

and set $\quad \Sigma_k = \sigma\big\{A_{k1}, \dots, A_{kr}, A_{k,(-1)}, \dots, A_{k,(-r)}\big\}$ and $\Sigma_f = \sigma\{\Sigma_k\}_{k=1}^N$. Then,

$$\big|E(f_k|\widehat{\Sigma})(\omega) - E(f_k|\Sigma_f)(\omega)\big| < \varepsilon \quad \mu\text{-a.e. on } \Omega.$$

□

Next we show how the information topology depends on the probability measure μ. This way, we are able to handle certain asymmetries of the agents' prior beliefs about uncertainty.

Recall that the collection of null-sets affects the information space. For this reason, to be able to proceed further in our discussion we need the following definition.

DEFINITION 9.1.11 Let μ, ν be two probability measures on the measurable space (Ω, Σ). We say that μ and ν are *equivalent*, if they produce the same null sets.

REMARK 9.1.12 For equivalent measures there is no ambiguity in the expression *almost everywhere*. So the space of information $\mathcal{S}^*$ is the same. Also $L^\infty(\Omega, \Sigma, \mu) = L^\infty(\Omega, \Sigma, \nu)$ for μ, ν equivalent. However, it is not true that $L^1(\Omega, \Sigma, \mu) = L^1(\Omega, \Sigma, \nu)$ (although the common L^∞-space is dense in both). Moreover, the conditional expectations $E_\mu(\cdot|\Sigma'), E_\nu(\cdot|\Sigma'), \Sigma \in \mathcal{S}_0$, with respect to μ and ν, are different. The equivalent measures μ, ν are mutually absolutely continuous and so we can define the Radon–Nikodym derivatives $(\mathrm{d}\mu)/(\mathrm{d}\nu), (\mathrm{d}\nu)/(\mathrm{d}\mu)$.

LEMMA 9.1.13 *If μ, ν are equivalent probability measures on (Ω, Σ), $\Sigma' \in \mathcal{S}_0$, and $f \in L^\infty(\Omega, \Sigma)$, then $E_\mu(f|\Sigma') = E_\mu\big(\frac{\mathrm{d}\nu}{\mathrm{d}\mu}|\Sigma'\big)E_\nu\big(f\frac{\mathrm{d}\mu}{\mathrm{d}\nu}|\Sigma'\big)$ a.e. on Ω.*

PROOF: For $A \in \Sigma'$, we have

$$\begin{aligned}
&\int_A E_\mu(\frac{\mathrm{d}\nu}{\mathrm{d}\mu}|\Sigma')E_\nu(f\frac{\mathrm{d}\mu}{\mathrm{d}\nu}|\Sigma')\mathrm{d}\mu \\
&= \int_A E_\mu\Big(\frac{\mathrm{d}\nu}{\mathrm{d}\mu}E_\nu(f\frac{\mathrm{d}\mu}{\mathrm{d}\nu}|\Sigma')\Big)\mathrm{d}\mu \\
&= \int_A E_\nu(f\frac{\mathrm{d}\mu}{\mathrm{d}\nu}|\Sigma')\mathrm{d}\nu = \int_A f\mathrm{d}\mu, \\
\Rightarrow \quad & E_\mu(f|\Sigma') = E_\mu\big(\frac{\mathrm{d}\nu}{\mathrm{d}\mu}|\Sigma'\big)E_\nu\big(f\frac{\mathrm{d}\mu}{\mathrm{d}\nu}|\Sigma'\big) \quad \text{a.e. on } \Omega.
\end{aligned}$$

□

Taking $f = 1 \in L^\infty(\Omega, \Sigma)$, we obtain the following corollary.

COROLLARY 9.1.14 *If μ, ν are equivalent probability measures on (Ω, Σ) and $\Sigma' \in \mathcal{S}_0$, then $E_\nu\big(\frac{\mathrm{d}\mu}{\mathrm{d}\nu}|\Sigma'\big) = \big[E_\mu\big(\frac{\mathrm{d}\nu}{\mathrm{d}\mu}|\Sigma'\big)\big]^{-1}$ a.e. on Ω.*

Recall that throughout this section, we assume that $L^1(\Omega, \Sigma)$ is separable.

LEMMA 9.1.15 *If μ, ν are equivalent probability measures on (Ω, Σ), and $\{f_k\}_{k\geq 1} \subseteq L^\infty(\Omega, \Sigma)$ is dense in $L^1(\Omega, \Sigma, \mu)$, then $\big\{f_k\frac{\mathrm{d}\mu}{\mathrm{d}\nu}\big\}_{k\geq 1} \subseteq L^\infty(\Omega, \Sigma)$ is dense in $L^1(\Omega, \Sigma, \nu)$.*

PROOF: Let $h \in L^1(\Omega, \Sigma, \nu)$. We have

$$\begin{aligned}
\|h - f_k\frac{\mathrm{d}\mu}{\mathrm{d}\nu}\|_{L^1(\nu)} &= \int_\Omega \Big|h - f_k\frac{\mathrm{d}\mu}{\mathrm{d}\nu}\Big|\mathrm{d}\nu = \int_\Omega \Big|h\frac{\mathrm{d}\nu}{\mathrm{d}\mu} - f_k\Big|\frac{\mathrm{d}\mu}{\mathrm{d}\nu}\mathrm{d}\nu \\
&= \|h\frac{\mathrm{d}\nu}{\mathrm{d}\mu} - f_k\|_{L^1(\mu)}
\end{aligned}$$

$$\Rightarrow \Big\{f_k\frac{\mathrm{d}\mu}{\mathrm{d}\nu}\Big\}_{k\geq 1} \subseteq L^\infty(\Omega, \Sigma) \quad \text{is dense in} \quad L^1(\Omega, \Sigma, \nu).$$

□

Using these auxiliary results, we can now compare the metrics $d_s(\mu)$ and $d_s(\nu)$ generating the strong operator topologies (topologies of pointwise convergence) for the two probability measures μ and ν.

THEOREM 9.1.16 *If μ, ν are equivalent probability measures on (Ω, Σ) and $d_s(\mu)$, $d_s(\nu)$ are the metrics generating the corresponding strong operator topologies on $\mathcal{S}_0$, then $d_s(\mu)$ and $d_s(\nu)$ are uniformly equivalent.*

PROOF: Let $\{f_k\}_{k\geq 1} \subseteq L^\infty(\Omega, \Sigma)$ be dense in $L^1(\Omega, \Sigma, \mu)$ with $f_1 = 1$. From Theorem 9.1.5, we know that

$$d_s(\nu)(\Sigma_1, \Sigma_2) = \sum_{k=1}^\infty \frac{1}{2^k} \min\big\{\|E(f_k|\Sigma_1) - E(f_k|\Sigma_2)\|_{L^1(\nu)}, 1\big\}$$

is a metric generating the strong operator topology (topology of pointwise convergence) on $\mathcal{S}_0$ for the prior ν. Given $\varepsilon>0$, we choose $k_0 \geq 1$ and $c>1$ such that

$$\frac{1}{2^{k_0-3}} < \varepsilon \quad \text{and} \quad \|f_k\|_\infty < c \qquad \text{for all } k<k_0.$$

Due to the fact that $\nu \ll \mu$, we can find $\delta = \delta(\varepsilon)>0$ such that for all $A \in \Sigma$

$$\nu(A)<\delta \quad \text{implies } \mu(A)<\frac{\varepsilon}{4c}.$$

We have

$$\int_\Omega E_\nu\Big(\frac{d\nu}{d\mu}|\Sigma_1\Big)\frac{d\mu}{d\nu}d\nu = \int_\Omega E_\mu\Big(\frac{d\nu}{d\mu}|\Sigma_1\Big)d\mu = 1.$$

So, if we set

$$A = \Big\{\omega \in \Omega : E_\mu\Big(\frac{d\nu}{d\mu}|\Sigma_1\Big)(\omega)\frac{d\mu}{d\nu}(\omega) \leq \frac{1}{\delta}\Big\},$$

we have

$$\mu(A) \geq 1 - \frac{\varepsilon}{4c} \quad \text{(by Chebyshev's inequality).}$$

If we set $\vartheta = \delta\varepsilon\big(1/(2^{k_0+2}c)\big)$, then $d_s(\nu)(\Sigma_1,\Sigma_2)<\vartheta$ implies

$$\Big\|E_\nu\Big(f_k\frac{d\mu}{d\nu}|\Sigma_1\Big) - E_\nu\Big(f_k\frac{d\mu}{d\nu}|\Sigma_2\Big)\Big\|_{L^1(\nu)} < \frac{\delta\varepsilon}{8c} \qquad \text{for all } k<k_0$$

and so using Lemma 9.1.15, we obtain

$$\|E_\mu(f_k|\Sigma_1) - E_\mu(f_k|\Sigma_2)\|_{L^1(\mu)} < \frac{\varepsilon}{2} \qquad \text{for all } k \geq 1,$$

$$\Rightarrow \; d_s(\mu)(\Sigma_1,\Sigma_2)<\varepsilon.$$

Reversing the roles of μ, ν in the above argument, we conclude that the metrics $d_s(\mu)$ and $d_s(\nu)$ are uniformly equivalent. □

There is another topology on $\mathcal{S}_0$, which is actually the first topology in the literature introduced for the purpose of topologizing the space of information.

DEFINITION 9.1.17 For $\Sigma_1, \Sigma_2 \in \mathcal{S}_0$, we define the quantity

$$d^*(\Sigma_1,\Sigma_2) = \sup_{A\in\Sigma_1} \inf_{C\in\Sigma_2} \mu(A \triangle C)$$

and then set

$$\widehat{d}(\Sigma_1,\Sigma_2) = \max\big\{d^*(\Sigma_1,\Sigma_2), d^*(\Sigma_2,\Sigma_1)\big\}.$$

It can be shown that $\widehat{d}$ is a metric on $\mathcal{S}_0$, known in the literature as the *Boylan metric*.

REMARK 9.1.18 On Σ we can define the relation $\sim'$, by $A \sim' C$ if and only if $\mu(A \triangle C) = 0$. This is an equivalence relation. We set $\dot{\Sigma} = \Sigma/\sim'$. Note that if $A \sim' C$, then $\mu(A) = \mu(C)$. Thus a function $\dot{\mu}$ is unambiguously defined on $\dot{\Sigma}$ by setting $\dot{\mu}([A]) = \mu(A)$ where $[A]$ is the $\sim'$-equivalence class of $A \in \Sigma$. Then $\dot{d}([A],[C]) = \dot{\mu}([A \triangle C])$ is a metric on $\dot{\Sigma}$. A probability space (Ω,Σ,μ) is said to be *separable*, if the associated metric space $(\dot{\Sigma},\dot{d})$ is separable. The metric space $(\dot{\Sigma},\dot{d})$ is complete. We emphasize that completeness of the probability space (Ω,Σ,μ) has nothing to do with the completeness of the associated metric space $(\dot{\Sigma},\dot{d})$.

Consider the Banach space $\mathcal{L}\big(L^\infty(\Omega,\Sigma),L^1(\Omega,\Sigma)\big)$ of bounded linear operators from $L^\infty(\Omega,\Sigma)$ into $L^1(\Omega,\Sigma)$. For every $f\in L^\infty(\Omega,\Sigma)$ and every $\Sigma'\in\mathcal{S}_0$, we have $E(f|\Sigma')\in L^\infty(\Omega,\Sigma')\subseteq L^1(\Omega,\Sigma)$. So we can consider that $\mathcal{S}_0$ is embedded in $\mathcal{L}\big(L^\infty(\Omega,\Sigma),L^1(\Omega,\Sigma)\big)$. We have the following result due to Allen [13, Fact 9.3].

PROPOSITION 9.1.19 *The $\widehat{d}$-metric topology on $\mathcal{S}_0$ coincides with the relative uniform operator topology of $\mathcal{L}\big(L^\infty(\Omega,\Sigma),L^1(\Omega,\Sigma)\big)$ on $\mathcal{S}_0$. So $\widehat{d}(\Sigma_n,\widehat{\Sigma})\longrightarrow 0$ if and only if*

$$\sup\big[\|E(f|\Sigma_n)-E(f|\widehat{\Sigma})\|_1 : \|f\|_\infty\le 1\big]\longrightarrow 0 \quad \textit{as } n\to\infty.$$

This proposition together with the results of Allen [13, Proposition 13.1], leads to the following comparison of the d_s and $\widehat{d}$-metric topologies.

PROPOSITION 9.1.20 *The $\widehat{d}$-metric topology on $\mathcal{S}_0$ is finer (stronger) than the d_s-topology. The two coincide when μ is purely atomic and in this case $\mathcal{S}_0$ is a compact metric space.*

In general the $\widehat{d}$-metric topology has some serious drawbacks. One such drawback is illustrated in the next example.

EXAMPLE 9.1.21 *An increasing sequence $\{\Sigma_n\}_{n\ge1}\subseteq\mathcal{S}_0$ need not converge in the metric $\widehat{d}$:* To see this, for every $n\ge1$ let $\Omega_n=[0,1)$, $\mathcal{B}_n=\mathbf{B}\big([0,1)\big)$ and $\mu_n=\lambda=$the Lebesgue measure on $[0,1)$. We set

$$\Omega=\prod_{n\ge1}\Omega_n,\ \Sigma=\prod_{n\ge1}\mathcal{B}_n \quad\text{and}\quad \mu=\prod_{n\ge1}\mu_n.$$

Let $\mathcal{T}_n=\{\emptyset,\Omega_n\}$, $n\ge1$, be the trivial σ-field on Ω_n and set

$$\Sigma_n=\prod_{k=1}^{n}\mathcal{B}_n\times\prod_{k>n+1}\mathcal{T}_k \qquad \text{for all } n\ge1.$$

Evidently $\Sigma_n\uparrow\Sigma$. However, we claim that

$$\frac{1}{2}\le\widehat{d}(\Sigma_n,\Sigma_{n+1}) \quad \text{for every } n\ge1. \tag{9.16}$$

Let A,C be independent subsets of Ω. We have

$$\begin{aligned}\mu(A\triangle C)&=\mu(A\cap C^c)+\mu(C\cap A^c)\\&=\mu(A)-\mu(A\cap C)+\mu(C)-\mu(A\cap C)\\&=\mu(A)+\mu(C)-2\mu(A)\mu(C),\end{aligned}$$

(due to the independence of A and C),

$$\Rightarrow \mu(A\triangle C)=\mu(A)\,\big[1-2\mu(C)\big]+\mu(C).$$

If $\mu(C)=\frac{1}{2}$, then we have $\mu(A\triangle C)=\mu(C)$. Choose $C_n\in\Sigma_{n+1}$ independent of all the sets in Σ_n with $\mu(C_n)=\frac{1}{2}$. We have

$$\frac{1}{2}=\inf\big[\mu(A\triangle C_n):A\in\Sigma_n\big]\le\widehat{d}(\Sigma_n,\Sigma_{n+1}) \qquad \text{for all } n\ge1,$$

(see Definition 9.1.17), which proves (9.16). Hence $\{\Sigma_n\}_{n\geq 1}$ is not a $\widehat{d}$-Cauchy sequence, although $\Sigma_n \uparrow \Sigma$ as $n \to \infty$. Note also that in this example $(\mathcal{S}_0, \widehat{d})$ is not compact.

In Proposition 9.1.10 we showed that the family of all finite partitions is dense in $(\mathcal{S}_0, d_s)$. This is important because it permits the use of finite-dimensional approximations in models which are d_s-continuous with respect to the information variable. This is no longer true for the $\widehat{d}$-metric topology as the next example illustrates.

EXAMPLE 9.1.22 *The family of all finite partitions of Ω is not $\widehat{d}$-dense in $\mathcal{S}_0$:* Let $\Omega = [0,1]$, $\Sigma = \mathbf{B}([0,1])$, and $\mu = \lambda =$ the Lebesgue measure on $[0,1]$. Let $\{A_k\}_{k=1}^N$ be a Borel partition of Ω. We know that we can find $C \in \Sigma$ such that $\mu(A_k \cap C) = \frac{1}{2}\mu(A_k)$ for all $k \in \{1, \ldots, N\}$. Therefore, if $\Sigma_f = \sigma(\{A_k\}_{k=1}^N)$, we have

$$\frac{1}{2} \leq \widehat{d}(\Sigma, \Sigma_f).$$

REMARK 9.1.23 In fact the $\widehat{d}$-closure of all finite Σ-partitions of Ω is $\{\Sigma' \in \mathcal{S}_0 : \mu|_{\Sigma'}$, is purely atomic$\}$.

Example 9.1.22 shows that the $\widehat{d}$-metric topology is not suitable to consider finite-dimensional approximations of the model. However, the $\widehat{d}$-metric topology being a relative uniform operator topology, has the property that the map $\xi : \mathcal{S}_0 \times \mathcal{S}_0 \longrightarrow \mathcal{S}_0$ defined by

$$\xi(\Sigma_1, \Sigma_2) = \Sigma_1 \vee \Sigma_2 = \sigma(\Sigma_1 \cup \Sigma_2) \qquad \text{(see Definition 9.1.8(a))}$$

is jointly $\widehat{d}$-continuous. This is no longer true for the d_s-metric topology. We can only say that if Σ_f is a finite Σ-partition of Ω, then $\Sigma' \longrightarrow \Sigma_f \vee \Sigma'$ is d_s-continuous. In general the $\vee$-operation is not even separately continuous.

We conclude with one more observation emphasizing the differences between the two metrics d_s and $\widehat{d}$. We start with a definition.

DEFINITION 9.1.24 The elements $\{\Sigma_k\}_{k=1}^N \subseteq \mathcal{S}_0$ are said to be *independent* if for every $A_k \in \Sigma_k$, $k = 1, \ldots, N$, we have $\mu(A_1 \cap \ldots \cap A_N) = \mu(A_1) \ldots \mu(A_N)$. A subset of $\mathcal{S}_0$ is said to be *independent* if every finite collection of elements in it is independent.

PROPOSITION 9.1.25 *If $\{\Sigma_n\}_{n\geq 1} \subseteq \mathcal{S}_0$ is independent, then*

(a) $\Sigma_n \xrightarrow{d_s} \{\emptyset, \Omega\} = \mathcal{T}$.

(b) *If* $\sup\limits_{n\geq 1} \inf\limits_{A\in\Sigma_n} \left|\mu(A) - \frac{1}{2}\right| < \frac{1}{2}$, *then* $\Sigma_n \nrightarrow \mathcal{T}$ *in the $\widehat{d}$-metric.*

PROOF: (a) From Kolmogorov's $(0,1)$ law, we have

$$\limsup_{n\to\infty} \Sigma_n = \mathcal{T}.$$

Because $\liminf\limits_{n\to\infty} \Sigma_n \subseteq \limsup\limits_{n\to\infty} \Sigma_n$, from Proposition 9.1.9 we conclude that $\Sigma_n \xrightarrow{d_s} \mathcal{T}$.

(b) For every $n \geq 1$, we have

$$\widehat{d}(\Sigma_n, \mathcal{T}) = \sup_{A \in \Sigma_n} \min\{\mu(A), 1-\mu(A)\} = \frac{1}{2} - \sup_{A \in \Sigma_n} \left|\mu(A) - \frac{1}{2}\right|,$$

for all $n \geq 1$,

$$\Rightarrow \Sigma_n \nrightarrow \mathcal{T} \quad \text{in the } \widehat{d}\text{-metric.}$$

□

EXAMPLE 9.1.26 Let $\Omega = [0,1]$, $\Sigma = \mathbf{B}([0,1])$, and $\mu = \lambda =$ the Lebesgue measure on $[0,1]$. Set

$$A_n = \{x \in [0,1] : \text{the } n\text{th entry in the binary expansion of } x \text{ is } 0\}$$

and

$$\Sigma_n = \{\emptyset, \Omega, A_n^c, A_n\}, \quad n \geq 1.$$

So Σ_n reveals the nth binary entry of the true state. It is easy to see that $\{\Sigma_n\}_{n \geq 1}$ is independent. Hence by Proposition 9.1.25(a), we have

$$\Sigma_n \xrightarrow{d_s} \mathcal{T}.$$

On the other hand $\mu(A_n) = \frac{1}{2}$ for all $n \geq 1$. Therefore by Proposition 9.1.25(b), we have

$$\Sigma_n \nrightarrow \mathcal{T} \quad \text{in the metric } \widehat{d}.$$

9.2 The ex-Post View

There are two ways we can view information and how it affects the behavior of the model. In this section, we examine the first way, which is known as the ex-post view or Bayesian view. According to it an agent chooses an action after observing his information and updating his belief. Then information affects state-dependent actions through posteriors (transition measures). When the state-dependent posteriors are not well-defined, then the idea is to show that the mapping from information to conditional expected utility is continuous and then use the fact that the mapping from utility to actions is continuous.

We establish the continuity of consumer demand under uncertainty with respect to his private information. So consider the commodity space $\mathbb{R}^N$. We set

$$\Delta = \left\{p = (p_k)_{k=1}^N \in \mathbb{R}_+^N : \sum_{k=1}^N p_k = 1\right\} \quad \text{(the usual price simplex).}$$

There is a consumption set $C \subseteq \mathbb{R}^N$ which is assumed to be closed, convex, and bounded from below. Let

$$\mathcal{U} = \{u : C \longrightarrow \mathbb{R} : u \text{ is continuous, strictly increasing and strictly concave}\}.$$

We furnish $\mathcal{U}$ with the topology of uniform convergence on compacta (c-topology or compact-open topology). Recall that $C(C, \mathbb{R})$ with the c-topology is a separable Fréchet space. Therefore $\mathcal{U}$ with the c-topology is a separable metric space. There is uncertainty in the model described by the probability space (Ω, Σ, μ). The dependence of the utilities on the state of the world is described by a $\big(\Sigma, \mathbf{B}(\mathcal{U})\big)$-measurable function $U : \Omega \longrightarrow \mathcal{U}$ (by $\mathbf{B}(\mathcal{U})$ we denote the Borel σ-field of the separable metric space $\mathcal{U}$). We interpret $U(\omega) \in \mathcal{U}$ as the consumer's ex-post utility function when $\omega \in \Omega$ is the true state of the world. We write

$$U(\omega)(x) = u(\omega, x) \qquad \text{for all } (\omega, x) \in \Omega \times C. \tag{9.17}$$

Recall that the evaluation maps

$$e : \mathcal{U} \times C \longrightarrow \mathbb{R} \quad \text{and} \quad e_x : \mathcal{U} \longrightarrow \mathbb{R} \quad \text{for } x \in C$$

defined by

$$e(u, x) = u(x) \quad \text{and} \quad e_x(u) = u(x)$$

are both continuous. Therefore from (9.17) and the measurability of $U(\cdot)$, we see that $(\omega, x) \longrightarrow u(\omega, x)$ is a Carathéodory function; that is, for every $x \in C$, $\omega \longrightarrow u(\omega, x)$ is Σ-measurable and for every $\omega \in \Omega$, $x \longrightarrow u(\omega, x)$ is continuous. Therefore by virtue of Theorem 6.2.6 $(\omega, x) \longrightarrow u(\omega, x)$ is $\Sigma \times \mathbf{B}(C)$-measurable.

The consumer faces uncertainty only about her preferences for commodities. So only the utility function is state dependent. Her initial endowment $e \in \operatorname{int} C$ and the price vector $p \in \Delta$ are both known and state independent (i.e., non-stochastic). The information of the consumer about the uncertainty she faces is described by $\Sigma_0 \in \mathcal{S}_0$.

The budget constraint of the consumer is the set

$$B(e, p) = \{x \in C : (p, x)_{\mathbb{R}^N} \leq (p, e)_{\mathbb{R}^N}\}.$$

The consumer first observes Σ_0. This means that if the true state of the world is $\omega \in \Omega$, she observes the utility function

$$x \longrightarrow E\big(u(\cdot, x)|\Sigma_0\big)(\omega)$$

and then she performs the maximization

$$\sup\big[E\big(u(\cdot, x)|\Sigma_0\big)(\omega) : x \in B(e, p)\big].$$

To make sure that expected utilities are well defined, we must assume that as states of the world vary with probability one, utilities remain in some compact subset of $\mathcal{U}$. So we introduce the following hypothesis.

(H): The image measure (distribution) $\mu \circ U^{-1}$ has compact support in $\mathcal{U}$.

REMARK 9.2.1 The support of a measure ν is the smallest closed set whose complement is ν-null.

Based on hypothesis (H), we introduce the space

$$\mathcal{D} = \big\{U : \Omega \longrightarrow \mathcal{U} : U \text{ is } \big(\Sigma, \mathbf{B}(\mathcal{U})\big)\text{-measurable, } \mu \circ U^{-1} \text{ has compact support}\big\}.$$

If ϱ is the metric on $\mathcal{U}$, we endow $\mathcal{D}$ with the metric

$$\varrho_0(U, U') = E\big(\varrho(U, U')\big) = \int_\Omega \varrho(U, U') d\mu.$$

Now the excess demand map $d_e : \Delta \times \mathcal{S}_0 \times \operatorname{int} C \times \mathcal{D} \longrightarrow L^1(\Omega, \mathbb{R}^N)$ is defined by

$$d_e(\varrho, \Sigma_0, e, U)(\omega) = \arg\max\left[E\big(u(\cdot, x)|\Sigma_0\big)(\omega) : x \in B(e,p)\right] - e. \tag{9.18}$$

We want to determine the continuity properties of the excess demand function d_e when $\mathcal{S}_0$ is furnished with the d_s-metric topology (topology of pointwise convergence).

Before continuing our analysis to this end, we need to clarify the conditional expected utility involved in (9.18). First note that for each $x \in C$ $E\big(u(\cdot, x)|\Sigma_0\big) \in L^1(\Omega)$. Because the Lebesgue dominated convergence theorem holds for conditional expectations, we see that

$$(\omega, x) \longrightarrow w(\omega, x) = E\big(u(\cdot, x)|\Sigma_0\big)(\omega)$$

is a Σ_0-Carathéodory function (i.e., $w(\cdot, x)$ is Σ_0-measurable). Also from the general theory of vector-valued integration, we have

$$E(U|\Sigma_0)(\omega) \in \mathcal{U} \quad \text{for } \mu\text{-a.a. } \omega \in \Omega.$$

So immediately we deduce the following.

PROPOSITION 9.2.2 *For any $U \in \mathcal{D}$ and $\Sigma_0 \in \mathcal{S}_0$, we have $E(U|\Sigma_0) \in \mathcal{D}$ which is Σ_0-measurable and for all $h \in L^1(\Omega, \Sigma_0, \mathbb{R}^N)$, we have*

$$E(U|\Sigma_0)(\omega)\big(h(\omega)\big) = E\Big(u\big(\omega, h(\omega)\big)|\Sigma_0\Big) \quad \mu\text{-a.e. on } \Omega.$$

PROPOSITION 9.2.3 *If $U \in \mathcal{D}$ and $\varepsilon > 0$ are given, then there exists $\Omega_0 \in \Sigma$ with $\mu(\Omega_0) = 1$ such that*

$$\{E(U|\Sigma_0)(\omega) : \omega \in \Omega_0,\ \Sigma_0 \in \mathcal{S}_0\}$$

is equicontinuous on C and uniformly equicontinuous on every compact $K \subset C$.

PROOF: Let $x \in X$ and $\delta > 0$ be given. Because for all $\omega \in \Omega_0$, $\mu(\Omega_0) = 1$, $u(\omega, \cdot)$ is continuous, we can find $\eta = \eta(x, \delta) > 0$ such that

$$x' \in X \quad \text{and} \quad \|x' - x\| < \eta \ \text{ imply } \ |u(\omega, x') - u(\omega, x)| < \delta.$$

Then for all $\omega \in \Omega_0$ we have

$$\begin{aligned} &\left|E\big(u(\cdot, x)|\Sigma_0\big)(\omega) - E\big(u(\cdot, x')|\Sigma_0\big)(\omega)\right| \\ \leq\ & E\big(|u(\cdot, x) - u(\cdot, x')|\,|\Sigma_0\big)(\omega) < \delta. \end{aligned}$$

This proves the first assertion of the proposition.

The second assertion follows from the fact that on compact sets equicontinuity is in fact uniform equicontinuity. □

Now we can have the first result on the continuous dependence on the information. Here $L^1(\Omega, \mathcal{U}) = \{U : \Omega \longrightarrow \mathcal{U} : U$ is Borel-measurable and $E\big(\varrho(U, 0)\big) < \infty\}$.

THEOREM 9.2.4 *If* $\{U_n, U\}_{n\geq 1} \subseteq \mathcal{D}$, $U_n(\omega) \longrightarrow U(\omega)$ *μ-a.e. in* $\mathcal{U}$, *for every* $K \subseteq C$ *compact there exists* $M_K > 0$ *such that for all* $x \in K$ *and all* $n \geq 1$ *we have*

$$|u_n(\omega, x)| \leq M_K \quad \textit{and} \quad |u(\omega, x)| \leq M_K \quad \mu\textit{-a.e. on } \Omega,$$

and $\Sigma_n \xrightarrow{d_s} \widehat{\Sigma}$ *as* $n \to \infty$, *then* $E(U_n|\Sigma_n) \longrightarrow E(U|\widehat{\Sigma})$ *as* $n \to \infty$ *in* $L^1(\Omega, \mathcal{U})$ *and in probability.*

PROOF: Let $K \subseteq X$ compact and $\varepsilon > 0$ be given. By hypothesis $U_n(\omega) \longrightarrow U(\omega)$ μ-a.e. in $\mathcal{U}$, therefore we have

$$\lim_{n\to\infty} \left(\sup_{x\in K} |u_n(\omega, x) - u(\omega, x)| \right) = 0 \quad \mu\text{-a.e. on } \Omega.$$

So we can find $n_0(\varepsilon) \geq 1$ such that

$$\sup_{x\in K} \|u_n(\cdot, x) - u(\cdot, x)\|_1 < \frac{\varepsilon}{2} \qquad \text{for all } n \geq n_0. \tag{9.19}$$

Also we can find $n_1(\varepsilon) \geq 1$ such that

$$\sup_{x\in K} \left\| E\big(u(\cdot, x)|\Sigma_n\big) - E\big(u(\cdot, x)|\widehat{\Sigma}\big) \right\|_1 < \frac{\varepsilon}{3}. \tag{9.20}$$

Then for every $n \geq \max\{n_0(\varepsilon), n_1(\varepsilon)\}$, we have

$$\begin{aligned}
&\sup_{x\in K} \left\| E\big(u_n(\cdot, x)|\Sigma_n\big) - E\big(u_n(\cdot, x)|\widehat{\Sigma}\big) \right\|_1 \\
\leq\ &\sup_{x\in K} \left\| E\big(u_n(\cdot, x)|\Sigma_n\big) - E\big(u(\cdot, x)|\Sigma_n\big) \right\|_1 \\
&+ \sup_{x\in K} \left\| E\big(u(\cdot, x)|\Sigma_n\big) - E\big(u(\cdot, x)|\widehat{\Sigma}\big) \right\|_1 \\
&+ \sup_{x\in K} \left\| E\big(u(\cdot, x)|\widehat{\Sigma}\big) - E\big(u_n(\cdot, x)|\widehat{\Sigma}\big) \right\|_1 \\
<\ &\frac{\varepsilon}{3} + 2 \sup_{x\in K} \|u_n(\cdot, x) - u(\cdot, x)\|_1 \quad \text{(see (9.20))} \\
<\ &\varepsilon \quad \text{(see (9.19))}.
\end{aligned} \tag{9.21}$$

Because of Proposition 9.2.3, there exist $\Omega_0 \in \Sigma$, $\mu(\Omega_0) = 1$, and for $n \geq 1$ a $\delta_n > 0$ such that, $\omega \in \Omega_0$, $x, x' \in K$ with $\|x - x'\| < \delta_n$ imply

$$|E\big(u_n(\cdot, x)|\Sigma_n\big)(\omega) - E\big(u_n(\cdot, x')|\Sigma_n\big)(\omega)| < \varepsilon \tag{9.22}$$

$$\text{and} \quad |E\big(u_n(\cdot, x)|\widehat{\Sigma}\big)(\omega) - E\big(u_n(\cdot, x')|\widehat{\Sigma}\big)(\omega)| < \varepsilon. \tag{9.23}$$

Because K is compact it is totally bounded and so we can find $\{x_k^n\}_{k=1}^m \subseteq K$ such that $K \subseteq \bigcup_{k=1}^m B_{\delta_n}(x_k^n)$. We have

$$\begin{aligned}
&\int_\Omega \sup_{x\in K} |E\big(u_n(\cdot,x)|\Sigma_n\big)(\omega) - E\big(u_n(\cdot,x)|\widehat{\Sigma}\big)(\omega)|\mathrm{d}\mu \\
&\le \int_\Omega \sup_{x\in K} |E\big(u_n(\cdot,x)|\Sigma_n\big)(\omega) - E\big(u_n(\cdot,x_k^n)|\Sigma_n\big)(\omega)|\mathrm{d}\mu \\
&\quad + \int_\Omega |E\big(u_n(\cdot,x_k^n)|\Sigma_n\big)(\omega) - E\big(u_n(\cdot,x_k^n)|\widehat{\Sigma}\big)(\omega)|\mathrm{d}\mu \\
&\quad + \int_\Omega \sup_{x\in K} |E\big(u_n(\cdot,x_k^n)|\widehat{\Sigma}\big)(\omega) - E\big(u_n(\cdot,x)|\widehat{\Sigma}\big)(\omega)|\mathrm{d}\mu \\
&\le 3\varepsilon \quad \text{(see (9.21), (9.22), and (9.23))}.
\end{aligned} \tag{9.24}$$

Recall that $U_n(\omega) \longrightarrow U(\omega)$ μ-a.e. on Ω in $\mathcal{U}$. So we can find $n_2(\varepsilon) \ge 1$ such that for all $n \ge n_2(\varepsilon)$, we have

$$\begin{aligned}
&\int_\Omega \sup_{x\in K} |E\big(u_n(\cdot,x)|\widehat{\Sigma}\big)(\omega) - E\big(u(\cdot,x)|\widehat{\Sigma}\big)(\omega)|\mathrm{d}\mu \\
&\le \int_\Omega \sup_{x\in K} E\big(|u_n(\cdot,x)| - u(\cdot,x)|\widehat{\Sigma}\big)(\omega)|\mathrm{d}\mu < \varepsilon.
\end{aligned} \tag{9.25}$$

Finally letting $\widehat{n}(\varepsilon) = \max\big\{n_0(\varepsilon), n_1(\varepsilon), n_2(\varepsilon)\big\}$, for $n \ge \widehat{n}(\varepsilon)$ we have

$$\begin{aligned}
&\int_\Omega \sup_{x\in K} |E\big(u_n(\cdot,x)|\Sigma_n\big)(\omega) - E\big(u(\cdot,x)|\widehat{\Sigma}\big)(\omega)|\mathrm{d}\mu < 3\varepsilon \\
&\Rightarrow \; E(U_n|\Sigma_n) \longrightarrow E(U|\widehat{\Sigma}) \quad \text{in } L^1(\Omega,\mathcal{U}) \text{ and in probability.}
\end{aligned}$$

∎

COROLLARY 9.2.5 *If* $\{U_n, U\}_{n\ge 1} \subseteq \mathcal{D}$, $U_n \longrightarrow U$ *in* $L^1(\Omega,\mathcal{U})$ *and* $\Sigma_n \xrightarrow{d_s} \widehat{\Sigma}$, *then* $E(U_n|\Sigma_n) \longrightarrow E(U|\widehat{\Sigma})$ *in* $L^1(\Omega,\mathcal{U})$ *and in probability.*

Next we establish the continuity of the excess demand function on the information variable.

THEOREM 9.2.6 *The excess demand function* $d_e : \Delta \times \mathcal{S}_0 \times \operatorname{int} C \times \mathcal{D} \longrightarrow L^1(\Omega,\mathbb{R}^N)$ *is continuous when* $\mathcal{S}_0$ *is endowed with the topology of pointwise convergence.*

PROOF: Suppose that $(p_n, \Sigma_n, e_n, U_n) \longrightarrow (p, \widehat{\Sigma}, e, U)$ in $\Delta \times \mathcal{S}_0 \times \operatorname{int} C \times \mathcal{D}$ as $n \to \infty$, with $e \in \operatorname{int} C$. Then by virtue of Corollary 9.2.5, we have that

$$E(U_n|\Sigma_n)(\omega) \longrightarrow E(U|\widehat{\Sigma})(\omega) \quad \text{in } \mathcal{U} \text{ for } \mu\text{-a.a. } \omega \in \Omega. \tag{9.26}$$

Let $\widehat{x}_n \in B(e_n, p_n)$ such that

$$E\big(u_n(\cdot,\widehat{x}_n)|\Sigma_n\big)(\omega) = \max\big[E\big(u_n(\cdot,x)|\Sigma_n\big)(\omega) : x \in B(e_n,p_n)\big].$$

Note that $\overline{\bigcup_{n\ge 1} B(e_n,p_n)} \in P_k(\mathbb{R}^N)$ and so we may assume that $\widehat{x}_n \longrightarrow \widehat{x}$ in $\mathbb{R}^N$. From Proposition 9.2.2 and (9.26) we have

$$E\big(u_n(\cdot,\widehat{x}_n)|\Sigma_n\big)(\omega) \longrightarrow E\big(u(\cdot,\widehat{x})|\widehat{\Sigma}\big)(\omega) \quad \text{for } \mu\text{-a.a. } \omega \in \Omega.$$

Because $B(e_n, p_n) \longrightarrow B(e,p)$, is given any $x \in B(e,p)$ we can find $x_n \in B(e_n, p_n)$, $n \geq 1$, such that $x_n \longrightarrow x$ in C and so

$$E\big(u_n(\cdot, x_n)|\Sigma_n\big)(\omega) \longrightarrow E\big(u(\cdot, x)|\widehat{\Sigma}\big)(\omega) \quad \text{for } \mu\text{-a.a. } \omega \in \Omega.$$

We have

$$\begin{aligned} & E\big(u_n(\cdot, x_n)|\Sigma_n\big)(\omega) \leq E\big(u_n(\cdot, \widehat{x}_n)|\Sigma_n\big)(\omega) \quad \text{for } \mu\text{-a.a. } \omega \in \Omega, \\ \Rightarrow \ & E\big(u(\cdot, x)|\widehat{\Sigma}\big)(\omega) \leq E\big(u(\cdot, \widehat{x})|\widehat{\Sigma}\big)(\omega) \quad \text{for } \mu\text{-a.a. } \omega \in \Omega. \end{aligned}$$

Because $x \in B(e,p)$ was arbitrary, it follows that

$$\begin{aligned} & \widehat{x} = \arg\max \big\{E\big(u(\cdot, x)|\widehat{\Sigma}\big)(\omega) : x \in B(e,p)\big\}, \\ \Rightarrow \ & d_e \quad \text{is continuous as claimed by the theorem.} \end{aligned}$$

□

DEFINITION 9.2.7 The *value of information* of the choice problem is the function $V : \Delta \times \mathcal{S}_0 \times \operatorname{int} C \times \mathcal{D} \longrightarrow \mathbb{R}$ defined by

$$V(p, \Sigma_0, e, U) = \int_\Omega U(\omega)\big(d_e(p, \Sigma_0, e, U)(\omega) + e\big) d\mu.$$

Concerning the function V, we have the following continuity result.

THEOREM 9.2.8 *The value of information map $V : \Delta \times \mathcal{S}_0 \times \operatorname{int} C \times \mathcal{D} \longrightarrow \mathbb{R}$ is continuous when $\mathcal{S}_0$ is endowed with the topology of pointwise convergence.*

PROOF: Suppose $(p_n, \Sigma_n, e_n, U_n) \longrightarrow (p, \widehat{\Sigma}, e, U)$ in $\Delta \times \mathcal{S}_0 \times \operatorname{int} C \times \mathcal{D}$ with $e \in \operatorname{int} C$. Then from Theorem 9.2.6, we have that

$$d_e(p_n, \Sigma_n, e_n, U_n) \longrightarrow d_e(p, \widehat{\Sigma}, e, U) \quad \text{in } L^1(\Omega, \mathcal{U}) \quad \text{as } n \to \infty. \tag{9.27}$$

Evidently the integral functional

$$h \longrightarrow \int_\Omega U(\omega)\big(h(\omega) + e\big) d\mu$$

is continuous on $L^1(\Omega, \mathcal{U})$. Therefore because of (9.27) we have

$$\begin{aligned} & V(p_n, \Sigma_n, e_n, U_n) \longrightarrow V(p, \widehat{\Sigma}, e, U), \\ \Rightarrow \ & V \quad \text{is continuous.} \end{aligned}$$

□

REMARK 9.2.9 Of course Theorems 9.2.6 and 9.2.8 are also valid if on $\mathcal{S}_0$ we consider the $\widehat{d}$-metric topology.

9.3 The ex-ante View

In this section we investigate the second way of viewing information in a decision problem and how it affects behavior. This is the ex-ante view or measurability view. According to this approach, the agent formulates a plan of what action to take at each state before observing his information, subject to the constraint that the plan be measurable with respect to his information. So information affects the actions through this measurability constraint.

The decision problem has the following structure. There is uncertainty represented by a probability space (Ω, Σ, μ) with Σ countably generated so that $L^1(\Omega)$ is separable. The information observed by the agent is represented by a sub-σ-field Σ_0 of Σ. We emphasize that this is an ex-ante representation of information and so it does not denote what the agent has observed, but rather what an agent would observe in each state. Before observing his information, the agent formulates a state-contingent plan $f:\Omega \longrightarrow X$ where $X \subseteq \mathbb{R}^N$ is a nonempty, compact, and convex set, representing the set of all feasible actions. We must have that f is Σ_0-measurable in order for the plan to be informationally feasible. The agent also faces for the plan f an additional constraint $B(p)$ depending on a variable $p \in \mathcal{P}$ (p can be, e.g., the price system prevailing in the market). The set $B(p)$ denotes all those plans that are economically feasible when the measurability constraint is not taken into account. The agent has some preferences among the feasible plans, which is described by the complete preorder $\preceq$. Then the decision problem facing the agent, is the following.

$$\max\big\{ \preceq: f:\Omega \longrightarrow X,\ f \text{ is } \Sigma_0\text{-measurable},\ f \in B(p)\big\}. \tag{9.28}$$

In problem (9.28), we view $p \longrightarrow B(p)$ as the economic constraint multifunction. On the set $\mathcal{E}$ of all Σ-measurable functions $f:\Omega \longrightarrow X$, we introduce the equivalence relation $\sim$ of equality μ-a.e. We set $\mathcal{E}_0 = \mathcal{E}/\sim$.

DEFINITION 9.3.1 A *plan* is an element of $\mathcal{E}_0$.

The measurability constraint multifunction $M : \mathcal{S}_0 \longrightarrow 2^{\mathcal{E}_0}$ is defined by

$$M(\Sigma_0) = \big\{f \in \mathcal{E}_0 : f \text{ is } \Sigma_0\text{-measurable}\big\}. \tag{9.29}$$

Then the overall constraint multifunction $\xi : \mathcal{P} \times \mathcal{S}_0 \longrightarrow 2^{\mathcal{E}_0}$ is defined by

$$\xi(p, \Sigma_0) = B(p) \cap M(\Sigma_0). \tag{9.30}$$

The solution multifunction $S : \mathcal{P} \times \mathcal{S}_0 \longrightarrow 2^{\mathcal{E}_0}$ for decision problem (9.28) is defined by

$$S(p, \Sigma_0) = \big\{f \in \xi(p, \Sigma_0) : g \preceq f \text{ for all } g \in \xi(p, \Sigma_0)\big\}. \tag{9.31}$$

We want to characterize the continuity of S.

DEFINITION 9.3.2 Let Z be a Hausdorff topological space and $\preceq$ a complete preorder on Z. Then

(a) $\preceq$ is *lower semicontinuous* (lsc for short) if for every $z \in Z$, the set

$$L(z) = \{z' \in Z : z' \preceq z\}$$

is closed.

(b) $\preceq$ is *upper semicontinuous* (usc for short) if for every $z \in Z$, the set

$$U(z) = \{z' \in Z : z \preceq z'\}$$

is closed.

(c) $\preceq$ is *continuous* if it is both lsc and usc.

REMARK 9.3.3 Recall that a general binary relation R on a set X is a preorder (or quasiorder), if it is reflexive (i.e., xRx) and transitive (i.e., xRy, yRz imply xRz). If in addition xRy and yRx imply $x = y$, then R is called a *partial order* (or simply an *ordering*). We should mention that some authors do not require a partial order to be reflexive. If for a preorder R we necessarily have xRy or yRx or both, then we say that R is a complete or total preorder. Evidently a complete preorder is reflexive. A total partial order is called a *linear order*. Given a preorder $\preceq$ on X, we can define the associated strict preorder $\prec$ by

$$x \prec y \quad \text{if and only if } x \preceq y \quad \text{and} \quad y \npreceq x.$$

Evidently $\prec$ is irreflexive and transitive. If the set X is a Hausdorff topological space, then in analogy to Definition 9.3.2 we can say that $\prec$ is lsc (resp., usc), if for every $x \in X$, the set $L_s(x) = \{x' \in X : x \prec x'\}$ (resp., the set $U_s(x) = \{x' \in X : x' \prec x\}$) is open. Note that for a total preorder $\preceq$, lower semicontinuity of the corresponding $\prec$ is equivalent to the lower semicontinuity of $\preceq$ (see Definition 9.3.2(a)). Similarly for the upper semicontinuity.

The parameter space $\mathcal{P}$ is assumed to be Hausdorff topological space. Now we want to topologize $\mathcal{E}_0$. We consider two topologies on $\mathcal{E}_0$. The first is the norm topology of $L^1(\Omega, \mathbb{R}^N)$, denoted by s-. The second is the weak topology of $L^1(\Omega, \mathbb{R}^N)$, denoted by w-.

PROPOSITION 9.3.4 *If* $u : \Omega \times X \longrightarrow \mathbb{R}$ *is a function such that*

(i) *For all* $x \in X$, $\omega \longrightarrow u(\omega, x)$ *is* Σ*-measurable,*

(ii) *For* μ*-a.a.* $\omega \in \Omega$, $x \longrightarrow u(\omega, x)$ *is continuous,*

(iii) $\omega \longrightarrow \max_{x \in X} |u(\omega, x)|$ *belongs in* $L^1(\Omega)$,

then the integral operator $U : \mathcal{E}_0 \longrightarrow \mathbb{R}$ *defined by*

$$U(x) = \int_\Omega u(\omega, x(\omega)) d\mu$$

is s*-continuous.*

PROOF: This is an immediate consequence of a dominated convergence theorem and from the fact that from a strongly convergent sequence in $L^1(\Omega, \mathbb{R}^N)$, we can extract a μ-a.e. convergent subsequence. $\square$

REMARK 9.3.5 Note, however, that U is not in general w-continuous. Let $\Omega = [0,1]$, $\Sigma = \mathbf{B}([0,1])$, and $\mu = \lambda =$ the Lebesgue measure. Also suppose $N = 1$ and $X = [0,1] \subseteq \mathbb{R}$. Consider the utility function $u(x) = x^2$. Then if $\{x_n\}_{n \geq 1}$ is the sequence of Rademacher functions; that is,

$$x_n(t) = \begin{cases} 1 & \text{if } t \in \left[\frac{k}{2^n}, \frac{k+1}{2^n}\right] \; k = \text{even} \\ -1 & \text{otherwise} \end{cases},$$

then $x_n \xrightarrow{w} 0$ in $L^1([0,1])$ but $U(x_n) = 1 \not\to U(0) = 0$.

Our stated goal is to determine the continuity properties of the solution multifunction $S(p, \Sigma_0)$ (see (9.31)). A reasonable approach is to use the Berge maximum theorem (see Theorem 6.1.18). Then according to that result if ξ is lower semicontinuous and has a closed graph and the preference relation $\preceq$ is continuous (in the sense of Definition 9.3.2(c)), then S has closed graph. This requires that on $\mathcal{E}_0$ we employ the w-topology, which makes $\mathcal{E}_0$ w-compact and assume that $\preceq$ is w-continuous, a rather restrictive requirement as was observed in Remark 9.3.5. So we see that although the weak topology is a good choice for $\mathcal{E}_0$ because it gives compactness of that space (see Theorem 6.4.23), however, this choice is too restrictive for the preference relation $\preceq$, where the natural choice is the strong topology. Hence we have conflicting requirements and eventually what we need is a version of Theorem 6.1.18 when two different topologies are involved. Such a result was proved by Horsley–Van Zandt–Wrobel [312, Corollary 8].

PROPOSITION 9.3.6 *If $(\mathcal{P}, \tau)$ is a Hausdorff topological space, V is a linear space on which we have two linear topologies s, w, whose restrictions on any straight line in V are identical to the usual topology of $\mathbb{R}$, X is a convex and w-compact subset of V, $B : \mathcal{P} \longrightarrow 2^X \setminus \{\emptyset\}$ has w-compact values and it is τ-to-w usc and τ-to-s lsc, $\preceq$ is a total preorder on X which is s-lsc and w-usc, then the optimal action correspondence*

$$S(p) = \{x \in B(p) : \textit{for all } z \in B(p) \;\; z \preceq x\}, \quad p \in \mathcal{P},$$

from $\mathcal{P}$ into $2^X \setminus \{\emptyset\}$ is τ-to-w usc with w-compact values.

Using this proposition in the case of our decision problem, we obtain the following result.

PROPOSITION 9.3.7 *If $(\mathcal{P}, \tau)$ is a Hausdorff topological space, on $\mathcal{S}_0$ we have a Hausdorff topology τ_0, ξ is $\tau \times \tau_0$-to-s lower semicontinuous, and $\operatorname{Gr} \xi$ is closed for the product topology $\tau \times \tau_0 \times w$, $\preceq$ is s-lsc and w-usc, then $\operatorname{Gr} S$ is closed for the product topology $\tau \times \tau_0 \times w$ and has nonempty and w-compact values.*

The above result is missing the concrete topology τ_0 on $\mathcal{S}_0$ that will make the measurability constraint multifunction M (see (9.29)) s-lsc and w-closed. In Section 9.1, we defined and discussed the d_s-metric topology on $\mathcal{S}_0$ (the topology of pointwise convergence).

PROPOSITION 9.3.8 *If on $\mathcal{S}_0$ we consider the d_s-metric topology, then the map $(f, \Sigma_0) \longrightarrow E(f|\Sigma_0)$ is $w \times d_s$-to-w continuous.*

PROOF: Recall that the weak topology on $\mathcal{E}_0$ is metrizable. So we can work with sequences. Suppose $(f_n, \Sigma_n) \xrightarrow{w \times d_s} (f, \widehat{\Sigma})$. For every $h \in L^\infty(\Omega)$, we have

$$\int_\Omega E(f_n|\Sigma_n)h d\mu = \int_\Omega f_n E(h|\Sigma_n) d\mu.$$

Because $L^\infty(\Omega) \subseteq L^1(\Omega)$, we have $E(h|\Sigma_n) \longrightarrow E(h|\widehat{\Sigma})$ in $L^1(\Omega)$. Also note that because X is compact and $f_n \xrightarrow{w} f$ in $L^1(\Omega)$, we also have that $f_n \xrightarrow{w^*} f$ in $L^\infty(\Omega)$. Then

$$\int_\Omega f_n E(h|\Sigma_n) d\mu \longrightarrow \int_\Omega f E(h|\widehat{\Sigma}) d\mu = \int_\Omega E(f|\widehat{\Sigma}) h d\mu,$$
$$\Rightarrow \int_\Omega E(f_n|\Sigma_n) h d\mu \longrightarrow \int_\Omega E(f|\widehat{\Sigma}) h d\mu.$$

Because $h \in L^\infty(\Omega)$ was arbitrary, we conclude that

$$E(f_n|\Sigma_n) \xrightarrow{w} E(f|\widehat{\Sigma}) \quad \text{in } L^1(\Omega).$$

□

PROPOSITION 9.3.9 *If on $\mathcal{S}_0$ we consider the d_s-metric topology, then $M : \mathcal{S}_0 \longrightarrow 2^{\mathcal{E}_0}$ has nonempty, w-compact, convex values and it is d_s-to-s lower semicontinuous and* Gr M *is $d_s \times w$-closed.*

PROOF: First we show the d_s-to-s lower semicontinuity. To this end suppose $\Sigma_n \xrightarrow{d_s} \widehat{\Sigma}$. It suffices to show that

$$M(\widehat{\Sigma}) \subseteq s\text{-}\liminf M(\Sigma_n). \tag{9.32}$$

So let $g \in M(\widehat{\Sigma})$. Then $g_n = E(g|\Sigma_n) \in M(\Sigma_n)$ for all $n \geq 1$ and $g_n \longrightarrow g$ in $L^1(\Omega)$. This proves (9.32) and so we have the s-lower semicontinuity of M.

Next we show that Gr M is $d_s \times w$-closed. To this end let $\{(\Sigma_n, g_n)\}_{n\geq 1} \subseteq$ Gr M and $\Sigma_n \xrightarrow{d_s} \widehat{\Sigma}$, $g_n \xrightarrow{w} g$ in $L^1(\Omega)$. From Proposition 9.3.8 we have

$$g_n = E(g_n|\Sigma_n) \xrightarrow{w} E(g|\widehat{\Sigma}) \quad \text{in } L^1(\Omega),$$
$$\Rightarrow g = E(g|\widehat{\Sigma}) \ \ (\text{i.e., } g \in M(\widehat{\Sigma})).$$

Therefore Gr M is $d_s \times w$-closed. □

PROPOSITION 9.3.10 *If $(\mathcal{P}, \tau)$ is a Hausdorff topological space, $\mathcal{S}_0$ is endowed with the d_s-metric topology, and $B : \mathcal{P} \longrightarrow 2^{\mathcal{E}_0} \setminus \{\emptyset\}$ has a graph that is $\tau \times w$-closed, then $\xi : \mathcal{P} \times \mathcal{S}_0 \longrightarrow 2^{\mathcal{E}_0} \setminus \{\emptyset\}$ has a graph that is $\tau \times d_s \times w$-closed.*

DEFINITION 9.3.11 We say that the constraint multifunction $B : \mathcal{P} \longrightarrow 2^{\mathcal{E}_0} \setminus \{\emptyset\}$ is *state independent* if, for all $p \in \mathcal{P}$, all $g \in B(p)$, and all $\Sigma_0 \in \mathcal{S}_0$, we have $E(g|\Sigma_0) \in B(p)$.

REMARK 9.3.12 One way to interpret the above condition is to say that the economic constraint multifunction B does not reveal information. Another interpretation is to say that information is not necessary in order to satisfy the constraint. There are economic models in which this condition is not satisfied (e.g., the microeconomic rational expectations general equilibrium model).

PROPOSITION 9.3.13 *If $(\mathcal{P},\tau)$ is a Hausdorff topological space, $\mathcal{S}_0$ is endowed with the d_s-metric topology, and $B : \mathcal{P} \longrightarrow 2^{\mathcal{E}_0}\setminus\{\emptyset\}$ is τ-to-s lower semicontinuous, then $\xi : \mathcal{P}\times\mathcal{S}_0 \longrightarrow 2^{\mathcal{E}_0}\setminus\{\emptyset\}$ is $\tau\times d_s$-to-s lower semicontinuous.*

PROOF: First we show that

$$\xi(p,\Sigma_0) = E\big(B(p)|\Sigma_0\big) \qquad \text{for all } (p,\Sigma_0)\in\mathcal{P}\times\mathcal{S}_0. \tag{9.33}$$

Let $g \in \xi(p,\Sigma_0)$. Then from (9.30) we see that $g \in B(p)$ and because g is Σ_0-measurable, we have $g = E(g|\Sigma_0)$. Therefore $g \in E\big(B(p)|\Sigma_0\big)$ and we have proved

$$\xi(p,\Sigma_0) \subseteq E\big(B(p)|\Sigma_0\big). \tag{9.34}$$

On the other hand, if $g \in E\big(B(p)|\Sigma_0\big)$, then we can find $h \in B(p)$ such that

$$g = E(h|\Sigma_0).$$

Because B is by hypothesis state-independent, we have $g \in B(p)$. Because g is Σ_0-measurable we have $g \in M(\Sigma_0)$, hence $g \in B(p)\cap M(\Sigma_0) = \xi(p,\Sigma_0)$. Therefore we have proved that

$$E\big(B(p)|\Sigma_0\big) \subseteq \xi(p,\Sigma_0). \tag{9.35}$$

From (9.34) and (9.35), we conclude that (9.33) holds.

Let $\eta_1 : \mathcal{P}\times\mathcal{S}_0 \longrightarrow \mathcal{E}_0\times\mathcal{S}_0$ be defined by

$$\eta_1(p,\Sigma_0) = B(p)\times\{\Sigma_0\}.$$

Evidently this is $\tau\times d_s$-to-$s\times d_s$ lower semicontinuous. Also let $\eta_2 : \mathcal{E}_0\times\mathcal{S}_0 \longrightarrow \mathcal{E}_0$ be defined by

$$\eta_2(g,\Sigma_0) = E(g|\Sigma_0).$$

We know that η_2 is $s\times d_s$-to-s continuous (see Proposition 9.3.8).

From (9.33) we see that

$$\begin{aligned} &\xi = \eta_2\circ\eta_1, \\ \Rightarrow\ &\xi \quad \text{is } \tau\times d_s\text{-to-}s \text{ lower semicontinuous.} \end{aligned}$$

∎

REMARK 9.3.14 The result remains true if instead we assume that the economic constraint multifunction $B : \mathcal{P} \longrightarrow 2^{\mathcal{E}_0}\setminus\{\emptyset\}$ is τ-to-w lower semicontinuous (see Proposition 9.3.8).

The above results yield the following theorem on the continuity of the solution multifunction $\xi(p,\Sigma_0)$.

THEOREM 9.3.15 *If $(\mathcal{P},\tau)$ is a Hausdorff topological space, on $\mathcal{S}_0$ we consider the d_s-metric topology, and*

(i) $B : \mathcal{P} \longrightarrow 2^{\mathcal{E}_0}\setminus\{\emptyset\}$ *is τ-to-s lower semicontinuous and* Gr B *is $\tau\times w$-closed,*

(ii) *B is state-independent (see Definition 9.3.11),*

(iii) $\preceq$ *is s-lower semicontinuous and w-upper semicontinuous,*

then the solution multifunction $S : \mathcal{P} \times \mathcal{S}_0 \longrightarrow 2^{\mathcal{E}_0}$ *has nonempty, w-compact values, and* Gr S *is* $\tau \times d_s \times w$*-closed.*

PROOF: From Propositions 9.3.10 and 9.3.13 we have that $(p, \Sigma_0) \longrightarrow \xi(p, \Sigma_0)$ is s-lsc and w-closed and clearly has nonempty values. These facts in conjunction with hypothesis (iii) permit the use of Proposition 9.3.6, from which we get the conclusion of the theorem. □

9.4 Convergence of σ-Fields and Prediction Sequences

In this section motivated from the work in Section 9.1 and in particular from Definition 9.1.8(b), we define still another mode of convergence of sub-σ-fields, which turns out to be suitable for the convergence of prediction sequences.

So let (Ω, Σ, μ) be a complete probability space. To simplify our considerations throughout this section we assume that all sub-σ-fields considered are μ-complete.

DEFINITION 9.4.1 Let $\{\Sigma_n\}_{n\geq 1}$ be a sequence of sub-σ-fields of Σ

(a) $\mu\text{-}\liminf\limits_{n\to\infty} \Sigma_n$ is the sub-σ-field Σ_0 such that for every $f \in L^\infty(\Omega)$, we have

$$\|E(f|\Sigma_0)\|_1 \leq \liminf_{n\to\infty} \|E(f|\Sigma_n)\|_1. \tag{9.36}$$

and Σ_0 is maximal among all sub-σ-fields satisfying (9.36), namely if Σ_0' is a sub-σ-field for which (9.36) is true (with Σ_0 replaced by Σ_0'), then $\Sigma_0' \subseteq \Sigma_0$.

(b) $\mu\text{-}\limsup\limits_{n\to\infty} \Sigma_n$ is the sub-σ-field Σ_0 such that for every $f \in L^\infty(\Omega)$, we have

$$\limsup_{n\to\infty} \|E(f|\Sigma_n)\|_1 \leq \|E(f|\Sigma_0)\|_1. \tag{9.37}$$

and Σ_0 is minimal among all sub-σ-fields satisfying (9.37), namely if Σ_0' is a sub-σ-field for which (9.37) is true (with Σ_0 replaced by Σ_0'), then $\Sigma_0 \subseteq \Sigma_0'$.

Immediately from this definition, we have the following.

PROPOSITION 9.4.2 *If* $\{\Sigma_n\}_{n\geq 1} \in \mathcal{S}_0$*, then we always have* $\mu\text{-}\liminf\limits_{n\to\infty} \Sigma_n \subseteq \mu\text{-}\limsup\limits_{n\to\infty} \Sigma_n$.

REMARK 9.4.3 The inclusion in the above result can be strict.

PROPOSITION 9.4.4 *If* $\{\Sigma_n\}_{n\geq 1} \in \mathcal{S}_0$*, then* $\mu\text{-}\liminf\limits_{n\to\infty} \Sigma_n = \Big\{A \in \Sigma : \lim\limits_{n\to\infty} \inf\big[\mu(A \triangle C) : C \in \Sigma_n\big] = 0\Big\}$.

PROOF: In what follows $\Sigma_l = \{A \in \Sigma : \lim_{n\to\infty} \inf [\mu(A \triangle C) : C \in \Sigma_n] = 0\}$.

For any sets $A_1, A_2, C_1, C_2 \in \Sigma$ we have

$$\mu((A_1 \cap A_2) \triangle (C_1 \cap C_2)) \leq \mu(A_1 \triangle C_1) + \mu(A_2 \triangle C_2) \tag{9.38}$$

and $$\mu((A_1 \cup A_2) \triangle (C_1 \cap C_2)) \leq \mu(A_1 \triangle C_1) + \mu(A_2 \triangle C_2). \tag{9.39}$$

From (9.38) and (9.39), we see that

$$A_1, A_2 \in \Sigma_l \Rightarrow A_1 \cup A_2 \in \Sigma_l \quad \text{and} \quad A_1 \cap A_2 \in \Sigma_l. \tag{9.40}$$

Suppose that $\{C_n\}_{n\geq 1} \subseteq \Sigma_l$ and set

$$\widehat{C} = \bigcup_{n\geq 1} C_n, \quad \widehat{C}_m = \bigcup_{n=1}^{m} C_n.$$

Given $\varepsilon > 0$, we can find $m = m(\varepsilon) \geq 1$ such that

$$\mu(\widehat{C} \triangle \widehat{C}_m) < \frac{\varepsilon}{2}.$$

Because of (9.40), we have that $\widehat{C}_m \in \Sigma_l$. So from the definition of Σ_l, we can find $n_0 \geq 1$ and $D_n \in \Sigma_n$ such that for $n \geq n_0$ we have

$$\mu(D_n \triangle \widehat{C}_m) < \frac{\varepsilon}{2}$$

$$\Rightarrow \inf [\mu(\widehat{C} \triangle C_n) : C_n \in \Sigma_n] \leq \mu(\widehat{C} \triangle C_n) \leq (\widehat{C} \triangle \widehat{C}_n) + \mu(\widehat{C}_m \triangle C_n) < \varepsilon$$

for $n \geq n_0$

$\Rightarrow \Sigma_l$ is closed under countable unions.

Clearly Σ_l is closed under complementation and so Σ_l is a σ-field.

It is easy to see that any set $A \in \Sigma_l$ satisfies

$$\lim_{n\to\infty} \int_\Omega \left|\frac{1}{2} - \mu(A|\Sigma_n)\right| d\mu = \int_\Omega \left|\frac{1}{2} - \mu(A|\Sigma_l)\right| d\mu \tag{9.41}$$

and so from (9.41), it follows that for every $f \in L^\infty(\Omega)$, we have

$$\|E(f|\Sigma_l)\|_1 \leq \liminf_{n\to\infty} \|E(f|\Sigma_n)\|_1. \tag{9.42}$$

Next we show that if $\Sigma_l' \in \mathcal{S}_0$ also satisfies (9.42), then $\Sigma_l' \subseteq \Sigma_l$. Let $C \in \Sigma_l'$ and take $f = \frac{1}{2} - \chi_C$ in (9.42). We have

$$\begin{aligned}\frac{1}{2} = \int_\Omega \left|\frac{1}{2} - \chi_C\right| d\mu = \int_\Omega \left|\frac{1}{2} - \mu(C|\Sigma_l')\right| d\mu \\ &\leq \liminf_{n\to\infty} \int_\Omega \left|\frac{1}{2} - \mu(C|\Sigma_n)\right| d\mu \\ &\leq \limsup_{n\to\infty} \int_\Omega \left|\frac{1}{2} - \mu(C|\Sigma_n)\right| d\mu \leq \frac{1}{2}\end{aligned}$$

$$\Rightarrow \lim_{n\to\infty} \int_\Omega \left|\frac{1}{2} - \mu(C|\Sigma_n)\right| d\mu = \frac{1}{2},$$

$$\Rightarrow \lim_{n\to\infty} \inf [\mu(C \triangle C_n) : C_n \in \Sigma_n] = 0,$$

$$\Rightarrow \Sigma_l' \subseteq \Sigma_l.$$

Therefore we conclude that $\Sigma_l = \mu\text{-}\liminf_{n\to\infty} \Sigma_n$. $\square$

Note that $\{L^2(\Omega, \Sigma_n)\}_{n\geq 1}$ is a sequence of closed subspaces of $L^2(\Omega)$. So we can define $s\text{-}\liminf_{n\to\infty} L^2(\Omega, \Sigma_n)$ and $w\text{-}\liminf_{n\to\infty} L^2(\Omega, \Sigma_n)$ in the sense of Definition 6.6.3.

THEOREM 9.4.5 *If $\{\Sigma_n\}_{n\geq 1} \subseteq \mathcal{S}_0$, then $s\text{-}\liminf_{n\to\infty} L^2(\Omega, \Sigma_n) = L^2(\Omega, \mu\text{-}\liminf_{n\to\infty} \Sigma_n)$.*

PROOF: First we show that we can find $\widehat{\Sigma} \in \mathcal{S}_0$ such that

$$s\text{-}\liminf_{n\to\infty} L^2(\Omega, \Sigma_n) = L^2(\Omega, \widehat{\Sigma}).$$

To do this, we need to show that it is a lattice and it contains the constant functions (see e.g., Schaefer [539, p. 210]). So let $f, g \in s\text{-}\liminf_{n\to\infty} L^2(\Omega, \Sigma_n)$. By virtue of Definition 6.6.3 we can find sequences $\{f_n\}_{n\geq 1}, \{g_n\}_{n\geq 1} \subseteq L^2(\Omega, \Sigma_n)$ such that

$$f_n \longrightarrow f \quad \text{and} \quad g_n \longrightarrow g \quad \text{in } L^2(\Omega) \quad \text{as } n\to\infty.$$

Let $\vee$ denote the maximum operation and recall that for $a, b, c, d \in \mathbb{R}$, we have

$$(a \vee b - c \vee d)^2 \leq (a-c)^2 + (b-d)^2. \tag{9.43}$$

Using (9.43), we obtain

$$\begin{aligned} &\|f_n \vee g_n - f \vee g\|_2^2 \leq \|f_n - f\|_2^2 + \|g_n - g\|_2^2 \longrightarrow 0 \quad \text{as } n\to\infty, \\ \Rightarrow\ & f \vee g \in s\text{-}\liminf_{n\to\infty} L^2(\Omega, \Sigma_n). \end{aligned} \tag{9.44}$$

Because $f_n \wedge g_n = f_n \vee g_n - |f_n - g_n|$, we see that

$$f \wedge g \in s\text{-}\liminf_{n\to\infty} L^2(\Omega, \Sigma_n). \tag{9.45}$$

From (9.44) and (9.45), it follows that $s\text{-}\liminf_{n\to\infty} L^2(\Omega, \Sigma_n)$ is indeed a lattice. Clearly it contains the constant functions. So as we already said, we can find $\widehat{\Sigma} \in \mathcal{S}_0$ such that

$$s\text{-}\liminf_{n\to\infty} L^2(\Omega, \Sigma_n) = L^2(\Omega, \widehat{\Sigma}).$$

Let $A \in \widehat{\Sigma}$. Then we can find $\{f_n\}_{n\geq 1} \subseteq L^2(\Omega, \Sigma_n)$ such that $f_n \longrightarrow \chi_A$ in $L^2(\Omega)$ and of course in probability. Set

$$A_n = \{\omega \in \Omega : f_n(\omega) \geq \frac{1}{2}\}.$$

We have

$$\begin{aligned} A_n \triangle A = \{\omega \in \Omega : &\big(\chi_A(\omega) = 0 \quad \text{and} \quad f_n(\omega) \geq \tfrac{1}{2}\big) \\ &\text{or} \quad \big(\chi_A(\omega) = 1 \quad \text{and} \quad f_n(\omega) < \tfrac{1}{2}\big)\} \\ \subseteq \{\omega \in \Omega : &|\chi_A(\omega) - f_n(\omega)| \geq \tfrac{1}{2}\} \qquad \text{for all } n \geq 1, \end{aligned}$$

$$\begin{aligned} &\Rightarrow \mu(A_n \triangle A) \longrightarrow 0 \quad \text{as } n\to\infty, \\ &\Rightarrow A \in \mu\text{-}\liminf_{n\to\infty} \Sigma_n \end{aligned}$$

(because $A_n \in \Sigma_n$ for every $n \geq 1$, see Proposition 9.4.4).

Now suppose that $A \in \mu\text{-}\liminf_{n\to\infty} \Sigma_n$. Then by Proposition 9.4.4 we can find $A_n \in \Sigma_n$, $n \geq 1$, such that $\mu(A_n \triangle A) \longrightarrow 0$. Then we have

$$\begin{aligned}
&\|\chi_{A_n} - \chi_A\|_2^2 = \int_\Omega |\chi_{A_n} - \chi_A|^2 \mathrm{d}\mu = \int_\Omega \chi_{A_n \triangle A} \mathrm{d}\mu = \mu(A_n \triangle A) \longrightarrow 0, \\
&\Rightarrow \chi_A \in s\text{-}\liminf_{n\to\infty} L^2(\Omega, \Sigma_n), \\
&\Rightarrow A \in \widehat{\Sigma}.
\end{aligned}$$

□

To further characterize $\mu\text{-}\liminf_{n\to\infty} \Sigma_n$ we need the following general result about the Mosco convergence of closed subspaces of a Hilbert space H.

PROPOSITION 9.4.6 *If H is a Hilbert space, and $\{H_n\}_{n\geq 1}$ is a sequence of closed subspaces of H, then*

(a) *$s\text{-}\liminf_{n\to\infty} H_n$ is the maximum among all closed subspaces $\widehat{H}'$ of H such that*

$$\|p_{\widehat{H}'}(x)\| \leq \liminf_{n\to\infty} \|p_{H_n}(x)\| \qquad \text{for all } x \in H. \tag{9.46}$$

(by $p_{H'}$ we denote the orthogonal projection on a closed subspace H' of H).

(b) *$w\text{-}\limsup_{n\to\infty} H_n$ is the minimum subspace among subspaces H' of H such that*

$$\limsup_{n\to\infty} \|p_{H_n}(x)\| \leq \|p_{H'}(x)\| \qquad \text{for all } x \in H. \tag{9.47}$$

PROOF: (a) Let $\widehat{H}' = s\text{-}\liminf_{n\to\infty} H_n$. We have

$$\begin{aligned}
\|p_{H_n}(x)\|^2 &= \|p_{H_n} p_{\widehat{H}'}(x) + p_{H_n}(1 - p_{\widehat{H}'})(x)\|^2 \\
&= \big(p_{H_n} p_{\widehat{H}'}(x), p_{\widehat{H}'}(x)\big) + \big(p_{H_n} p_{\widehat{H}'}(x), (1 - p_{\widehat{H}'})(x)\big) \\
&\quad + \big((1 - p_{\widehat{H}'})(x), p_{H_n} p_{\widehat{H}'}(x)\big) + \|p_{H_n}(1 - p_{\widehat{H}'})(x)\|^2 \\
&\geq \big(p_{H_n} p_{\widehat{H}'}(x), p_{\widehat{H}}(x)\big) + \big(p_{H_n} p_{\widehat{H}'}(x), (1 - p_{\widehat{H}'})(x)\big) \\
&\quad + \big((1 - p_{\widehat{H}'})(x), p_{H_n} p_{\widehat{H}'}(x)\big) \longrightarrow \|p_{\widehat{H}'}(x)\|^2
\end{aligned}$$

(recall $p_{H_n}(x) \longrightarrow x$ for all $x \in H$). Therefore $s\text{-}\liminf H_n$ satisfies (9.36).

On the other hand, let $\widehat{H}'$ be any closed subspace of H satisfying (9.36). We have

$$\|p_{\widehat{H}'}(x)\|^2 \leq \liminf_{n\to\infty} \|p_{H_n} p_{\widehat{H}'}(x)\|^2 \leq \limsup_{n\to\infty} \|p_{H_n} p_{\widehat{H}'}(x)\|^2 \leq \|p_{\widehat{H}'}(x)\|$$
$$\text{and} \quad \|p_{H_n} p_{\widehat{H}'}(x)\| \longrightarrow \|p_{\widehat{H}'}(x)\| \quad \text{as } n \to \infty.$$

Hence we have

$$\begin{aligned}
\|p_{\widehat{H}'}(x) - p_{H_n} p_{\widehat{H}'}(x)\|^2 &= \|p_{\widehat{H}'}(x)\|^2 - \big(p_{\widehat{H}'}(x), p_{H_n} p_{\widehat{H}'}(x)\big) \\
&\quad - \big(p_{H_n} p_{\widehat{H}'}(x), p_{\widehat{H}'}(x)\big) + \big(p_{H_n} p_{\widehat{H}'}(x), p_{\widehat{H}'}(x)\big) \\
&= \|p_{\widehat{H}'}(x)\|^2 - \|p_{H_n} p_{\widehat{H}'}(x)\|^2 \longrightarrow 0 \quad \text{as } n \to \infty.
\end{aligned}$$

(b) Let $H' = w\text{-}\limsup_{n\to\infty} H_n$. Then we have

$$(H')^\perp = s\text{-}\liminf_{n\to\infty} H_n^\perp$$

and from (a) it follows that

$$\begin{aligned}
&\|(1 - p_{\widehat{H}'})(x)\|^2 \le \liminf_{n\to\infty} \|(1 - p_{H_n})(x)\|^2 \quad \text{for any } x \in H, \\
\Rightarrow\ & \|x\|^2 - \|p_{\widehat{H}'}(x)\|^2 \le \|x\|^2 - \limsup_{n\to\infty} \|p_{H_n}(x)\|^2, \\
\Rightarrow\ & \limsup_{n\to\infty} \|p_{H_n}(x)\|^2 \le \|p_{\widehat{H}'}(x)\|^2 \qquad \text{for all } x \in H.
\end{aligned}$$

On the other hand, let H' be a closed subspace of H satisfying (9.47). By following the previous argument backwards, we obtain

$$w\text{-}\limsup_{n\to\infty} H_n \subseteq H'.$$

□

Using this proposition we can further characterize $\mu\text{-}\liminf_{n\to\infty} \Sigma_n$.

THEOREM 9.4.7 *If $\{\Sigma_n\}_{n\geq 1} \subseteq \mathcal{S}_0$, then $\mu\text{-}\liminf_{n\to\infty} \Sigma_n$ is the maximum among all the sub-σ-fields $\widehat{\Sigma}'$ of Σ such that*

$$\|E(f|\widehat{\Sigma}')\|_2 \le \liminf_{n\to\infty} \|E(f|\Sigma_n)\|_2 \qquad \textit{for all } f \in L^2(\Omega). \tag{9.48}$$

PROOF: Recall that the conditional expectation $E(\cdot|\Sigma_n)$ is the orthogonal projection of $L^2(\Omega)$ onto $L^2(\Omega, \Sigma_n)$. From Theorem 9.4.5, we see that $E(\cdot|\mu\text{-}\liminf_{n\to\infty} \Sigma_n)$ is the orthogonal projection of $L^2(\Omega)$ onto $s\text{-}\liminf_{n\to\infty} L^2(\Omega, \Sigma_n)$. So by virtue of Proposition 9.4.6, we have

$$\|E(f|s\text{-}\liminf_{n\to\infty} \Sigma_n)\|_2 \le \liminf_{n\to\infty} \|E(f|\Sigma_n)\|_2.$$

On the other hand if $\widehat{\Sigma}' \in \mathcal{S}_0$ satisfies (9.48), then from Proposition 9.4.6 and Theorem 9.4.5, we have

$$\begin{aligned}
& L^2(\Omega, \widehat{\Sigma}') \subseteq s\text{-}\liminf_{n\to\infty} L^2(\Omega, \mu\text{-}\liminf_{n\to\infty} \Sigma_n), \\
\Rightarrow\ & \widehat{\Sigma}' \subseteq s\text{-}\liminf_{n\to\infty} \Sigma_n.
\end{aligned}$$

□

THEOREM 9.4.8 *If $\{\Sigma_n\}_{n\geq 1} \subseteq \mathcal{S}_0$, then for any $f \in L^1(\Omega)$ the following statements are equivalent.*

(a) *f is $\mu\text{-}\liminf_{n\to\infty} \Sigma_n$-measurable.*

(b) $\|E(f|\Sigma_n) - f\|_1 \longrightarrow 0$.

PROOF: (a)$\Rightarrow$(b): Given $\varepsilon>0$ we can find $g \in L^2(\Omega, \mu\text{-}\liminf_{n\to\infty} \Sigma_n)$ such that

$$\|f-g\|_1 < \frac{\varepsilon}{3}.$$

We have

$$\|E(g|\Sigma_n) - g\|_2 \longrightarrow 0 \quad \text{as } n\to\infty \text{ (see Theorem 9.4.5).}$$

So we can find $n_0 = n_0(\varepsilon) \geq 1$ such that

$$\|E(g|\Sigma_n) - g\|_2 < \frac{\varepsilon}{3} \qquad \text{for all } n \geq n_0.$$

Hence we have

$$\begin{aligned}
&\|E(f|\Sigma_n) - f\|_1 \\
\leq\ &\|E(f|\Sigma_n) - E(g|\Sigma_n)\|_1 + \|E(g|\Sigma_n) - g\|_1 + \|g-f\|_1 \\
\leq\ &2\|f-g\|_1 + \|E(g|\Sigma_n) - g\|_2 \ \text{(because } \|\cdot\|_1 \leq \|\cdot\|_2) \\
<\ &\frac{2\varepsilon}{3} + \frac{\varepsilon}{3} = \varepsilon \qquad \text{for all } n \geq n_0.
\end{aligned}$$

Therefore we conclude that

$$\|E(g|\Sigma_n) - f\|_1 \longrightarrow 0 \quad \text{as } n\to\infty.$$

(b)$\Rightarrow$(a): Let $V = \big\{ f \in L^1(\Omega) : \|E(g|\Sigma_n) - f\|_1 \longrightarrow 0 \ \text{ as } n\to\infty \big\}$. Let $f,g \in V$. Then as above (see the proof of Theorem 9.4.5), we have

$$\|E(f|\Sigma_n) \vee E(g|\Sigma_n) - f \vee g\|_1 \longrightarrow 0 \quad \text{as } n\to\infty.$$

So we have

$$\begin{aligned}
\|E(f\vee g|\Sigma_n) - f\vee g\|_1 &\leq \|E(f\vee g|\Sigma_n) - E(f|\Sigma_n)\vee E(g|\Sigma_n)\|_1 \\
&\quad + \|E(f|\Sigma_n)\vee E(g|\Sigma_n) - f\vee g\|_1 \\
&= \|E\big(f\vee g - E(f|\Sigma_n)\vee E(g|\Sigma_n)|\Sigma_n\big)\|_1 \\
&\quad + \|E(f|\Sigma_n)\vee E(g|\Sigma_n) - f\vee g\|_1 \\
&\leq 2\|E(f|\Sigma_n)\vee E(g|\Sigma_n) - f\vee g\|_1 \longrightarrow 0 \quad \text{as } n\to\infty,
\end{aligned}$$

$\Rightarrow f \vee g \in V$.

Similarly we show that $f \wedge g \in V$. Also note that V is a linear subspace, which is closed and contains the constant functions. Therefore we can find $\widehat{\Sigma}' \in \mathcal{S}_0$ such that

$$V = L^1(\Omega, \widehat{\Sigma}').$$

For every $f \in L^\infty(\Omega, \widehat{\Sigma}')$, we have

$$\begin{aligned}
\|E(f|\Sigma_n) - f\|_2^2 &\leq \|E(f|\Sigma_n) - f\|_\infty \|E(f|\Sigma_n) - f\|_1 \\
&\leq 2\|f\|_\infty \|E(f|\Sigma_n) - f\|_1 \quad \text{as } n\to\infty
\end{aligned}$$

$$\Rightarrow\ f \in L^2(\Omega, \mu\text{-}\liminf_{n\to\infty} \Sigma_n),$$

$$\Rightarrow\ L^\infty(\Omega, \widehat{\Sigma}') \subseteq L^2(\Omega, \mu\text{-}\liminf_{n\to\infty} \Sigma_n).$$

Taking the L^1-closure of both sides, we conclude that

$$V \subseteq L^1(\Omega, \mu\text{-} \liminf_{n\to\infty} \Sigma_n).$$

□

Let $\{\Sigma_n\}_{n\geq 1}$ be a sequence in $\mathcal{S}_0$ and define

$$\mathcal{S}_0^p = \Big\{\Sigma' \in \mathcal{S}_0 : \limsup_{n\to\infty} \|E(f|\Sigma_n)\|_p \leq \|E(f|\Sigma')\|_p \\ \text{for all } \; f \in L^\infty(\Omega)\Big\}, \; p = 1, 2.$$

LEMMA 9.4.9 *If* $\Sigma' \in \mathcal{S}_0^1 \cup \mathcal{S}_0^2$ *then* $\limsup_{n\to\infty} \|E\big(E(f|\Sigma')|\Sigma_n\big)\|_p = \limsup_{n\to\infty} \|E(f|\Sigma_n)\|_p$ *for every* $f \in L^\infty(\Omega)$.

PROOF: Let $\Sigma' \in \mathcal{S}_0^1$ and $f \in L^\infty(\Omega)$. We have

$$\begin{aligned} &\limsup_{n\to\infty} \|E\big(E(f|\Sigma') - f|\Sigma_n\big)\|_1 \leq \|E\big(E(f|\Sigma') - f|\Sigma'\big)\|_1 = 0, \\ \Rightarrow\; &\Big|\limsup_{n\to\infty} \|E\big(E(f|\Sigma')|\Sigma_n\big)\|_1 - \limsup_{n\to\infty} \|E(f|\Sigma_n)\|_1\Big| \\ \leq\; &\limsup_{n\to\infty} \|E\big(E(f|\Sigma')|\Sigma_n\big)\|_1 = 0. \end{aligned}$$

Also we have

$$\begin{aligned} &\limsup_{n\to\infty} \|E\big(E(f|\Sigma') - f|\Sigma_n\big)\|_2^2 \\ \leq\; &\limsup_{n\to\infty} \|E\big(E(f|\Sigma') - f|\Sigma_n\big)\|_\infty \|E\big(E(f|\Sigma') - f|\Sigma_n\big)\|_1 \\ \leq\; &\|E(f|\Sigma') - f\|_\infty \limsup_{n\to\infty} \|E\big(E(f|\Sigma') - f|\Sigma_n\big)\|_1 = 0. \end{aligned}$$

Therefore

$$\begin{aligned} &\Big|\limsup_{n\to\infty} \|E\big(E(f|\Sigma')|\Sigma_n\big)\|_2^2 - \limsup_{n\to\infty} \|E(f|\Sigma_n)\|_2\Big| \\ \leq\; &\limsup_{n\to\infty} \|E\big(E(f|\Sigma')|\Sigma_n\big) - E(f|\Sigma_n)\|_2 = 0. \end{aligned}$$

The proof for $\Sigma' \in \mathcal{S}_0^2$ is done similarly. □

LEMMA 9.4.10 *If* $\Sigma' \in \mathcal{S}_0^1$ *and* $\Sigma'' \in \mathcal{S}_0^2$, *then* $\Sigma' \cap \Sigma'' \in \mathcal{S}_0^1 \cap \mathcal{S}_0^2$.

PROOF: To simplify the notation, we set $E' = E(\cdot|\Sigma')$ and $E'' = E(\cdot|\Sigma'')$. We claim that

$$\limsup_{n\to\infty} \|E(f|\Sigma_n)\|_1 \leq \|(E'E'')^m f\|_1 \quad \text{for every } \; f \in L^\infty(\Omega), \; m \geq 1. \tag{9.49}$$

Note that (9.49) is true for $m = 1$. Suppose it is true for $m = k$. Then using Lemma 9.4.9, we obtain

$$\begin{aligned} \limsup_{n\to\infty} \|E(f|\Sigma_n)\|_1 &= \limsup_{n\to\infty} \|E(E'E''f|\Sigma_n)\|_1 \\ &\leq \|(E'E'')^k E'E''f\|_1 \\ &= \|(E'E'')^{k+1} f\|_1. \end{aligned}$$

So (9.49) is also true for $m=k+1$ and by induction we have proved it for all $m \geq 1$. Passing to the limit as $m \longrightarrow \infty$ in (9.49), we obtain

$$\limsup_{n\to\infty} \|E(f|\Sigma_n)\|_1 \leq \|E(f|\Sigma' \cap \Sigma'')\|_1.$$

Similarly we prove (9.49) for $\|\cdot\|_2$. □

Using the two lemmata we can characterize μ-$\limsup_{n\to\infty} \Sigma_n$.

THEOREM 9.4.11 *If $\{\Sigma_n\}_{n\geq 1} \subseteq \mathcal{S}_0$, then μ-$\limsup_{n\to\infty} \Sigma_n$ is the minimum sub σ-field Σ' such that*

$$\limsup_{n\to\infty} \|E(f|\Sigma_n)\|_2 \leq \|E(f|\Sigma')\|_2 \qquad \text{for all } f \in L^2(\Omega). \tag{9.50}$$

PROOF: By a density argument, we see that (9.50) is satisfied for every $f \in L^2(\Omega)$ if and only if it is satisfied for every $f \in L^\infty(\Omega)$. Also by Lemma 9.4.10, we have $\mathcal{S}_0' = \mathcal{S}_0^2$. This proves the theorem. □

Using this theorem together with Proposition 9.4.6(b), we obtain the following.

THEOREM 9.4.12 *If $\{\Sigma_n\}_{n\geq 1} \subseteq \mathcal{S}_0$, then the closed linear lattice generated by w-$\limsup_{n\to\infty} L^2(\Omega, \Sigma_n)$ is equal to $L^2(\Omega, \mu\text{-}\limsup_{n\to\infty} \Sigma_n)$.*

REMARK 9.4.13 The above results can be extended to general L^p-spaces ($1 \leq p < \infty$).

9.5 Games with Incomplete Information

The specification of a game with incomplete information provides a very general framework for the analysis of noncooperative decision making under uncertainty. One can describe such an n-player game with general hypotheses. So let (Ω, Σ, μ) be a probability space, with Ω a Polish space, $\Sigma = \mathbf{B}(\Omega) =$ the Borel σ-field of Ω, and μ the probability measure. This probability space models the uncertainty in the system. There are n-players to whom correspond n-action spaces X_k, $k=1,\ldots,n$. We assume that each X_k is a compact metric space. Also there are n-random variables f_k on Ω, $k=1,\ldots,n$, which we assume are bounded and measurable. Each player $k \in \{1,\ldots,n\}$ before choosing his action, consults his private information random variable f_k, which leads him to an assessment of the state of nature $\omega \in \Omega$. This then dictates his choice of action from the action space X_k. As soon as all these are finalized, he receives a reward $u_k(\omega, x_1, \ldots, x_n)$.

In this section we investigate the existence of Nash equilibrium for a two-person, zero-sum game with incomplete information. Our hypotheses on the data are minimal and in particular for the utility (payoff) function $u(\omega, x_1, x_2)$, we only assume that it is bounded, measurable, and continuous in $x_k, k=1,2$, separately for each fixed $\omega \in \Omega$. The hypotheses on the model do not include a continuity of information condition, which typically requires the joint distribution of the information available to the players be absolutely continuous with respect to a product measure on the product of the spaces of types. In general such a condition is difficult to verify.

Now let us give the precise mathematical description of the game denoted by Γ. The game Γ is determined by the following mathematical items.

(1) The compact metric spaces X and Y representing the action spaces of the two players.
(2) Two Polish spaces Ω_1 and Ω_2 called the types of the two players.
(3) Another Polish space Ω_0 called the type of nature. We set $\Omega = \Omega_0 \times \Omega_1 \times \Omega_2$.
(4) There is a probability measure μ on $\big(\Omega, \mathbf{B}(\Omega)\big)$; the probability space $\big(\Omega, \mathbf{B}(\Omega), \mu\big)$ represents the information structure of the game.
(5) There is a utility (payoff) function $u:\Omega \times X \times Y \longrightarrow \mathbb{R}$ which is assumed to be jointly Borel-measurable.

The game begins with nature choosing a vector $\omega = (\omega_k)_{k=0}^2 \in \Omega$ according to the probability μ. Then player I observes $\omega_1 \in \Omega_1$ and based on that, chooses her action $x \in X$, whereas player II observes ω_2 and chooses action $y \in Y$. At the end player II pays player I the amount $u(\omega, x, y)$ (zero-sum game).

DEFINITION 9.5.1 (a) A *distributional strategy* σ for player I is a probability measure on $\Omega_1 \times X$ such that the marginal probability of σ on Ω_1 is equal to the marginal probability of μ on Ω_1 (recall that for a measure ν on a product space $S \times T$, the *marginals* ν_S and ν_T on S and T, respectively, are defined by $\nu_S(A) = \nu(A \times T)$ for all $A \in \Sigma(S)$ and by $\nu_T(C) = \nu(S \times C)$ for all $C \in \Sigma(T)$.
(b) A *distributional strategy* τ for player II, is a probability measure on $\Omega_2 \times Y$ such that the marginal probability of τ on Ω_2 is equal to the marginal probability of μ on Ω_2.

REMARK 9.5.2 Informally, the conditional probability $\sigma(x|\omega_1)$ represents the probability that player I will choose action $x \in X$, provided that she observes the type $\omega_1 \in \Omega_1$. Similarly for the conditional $\tau(y|\omega_2)$.

DEFINITION 9.5.3 The *expected payoff* of the game Γ is given by

$$U(\sigma, \tau) = \int_{\Omega \times X \times Y} u(\omega, x, y)\sigma(\mathrm{d}x|\omega_1)\tau(\mathrm{d}y|\omega_2)\mathrm{d}\mu$$

when player I uses the distributional strategy σ and player II uses the distributional strategy τ.

An equilibrium point of game Γ is defined in the usual way.

DEFINITION 9.5.4 An *equilibrium* of game Γ is a saddle point of the expected payoff, namely a pair $(\widehat{\sigma}, \widehat{\tau})$ of distributional strategies for the two players such that

$$U(\widehat{\sigma}, \widehat{\tau}) = \sup_{\sigma} U(\sigma, \widehat{\tau}) = \inf_{\tau} U(\widehat{\sigma}, \tau).$$

Our goal is to establish the existence of an equilibrium for game Γ. To do this we need some auxiliary results.

PROPOSITION 9.5.5 *If (Ω, Σ, μ) is a probability space, with Ω a Polish space, $\Sigma = \mathbf{B}(\Omega) =$ the Borel σ-field of Ω, and μ the probability measure, Z is a compact metric space, and $h : \Omega \times Z \longrightarrow \mathbb{R}$ is a Carathéodory function (i.e., for all $z \in Z$, $\omega \longrightarrow h(\omega, z)$ is Borel-measurable and for all $\omega \in \Omega$, $z \longrightarrow h(\omega, z)$ is continuous) which is bounded, then given $\varepsilon > 0$ we can find a set $D_\varepsilon \in \Sigma$ with $\mu(\Omega \setminus D_\varepsilon) < \varepsilon$ such that*

$$\{h(\omega, \cdot) : \omega \in D_\varepsilon\}$$

is equicontinuous.

PROOF: Recall that μ is tight. Therefore we can apply Theorem 6.2.9 (the Scorza-Dragoni theorem) and obtain $D_\varepsilon \subseteq \Omega$ compact subset such that $\mu(\Omega \setminus D_\varepsilon) < \varepsilon$ and $h\big|_{D_\varepsilon \times Z}$ is continuous. Therefore it follows at once that $\{h(\omega, \cdot) : \omega \in D_\varepsilon\}$ is equicontinuous. $\square$

Keeping the setting of the previous proposition, we define

$$M^1_{+,\mu}(\Omega \times Z) = \{\vartheta \in M^1_+(\Omega \times Z) : \vartheta_\Omega = \mu\};$$

that is, $M^1_{+,\mu}(\Omega \times Z)$ are all probability measures on $\Omega \times Z$, such that their Ω-marginal ϑ_Ω equals μ. Clearly $M^1_{+,\mu}(\Omega \times Z)$ inherits the weak topology $w\big(M^1_{+,\mu}(\Omega \times Z), C_b(\Omega \times Z)\big)$ from the space $M^1_+(\Omega \times Z)$. As in Proposition 9.5.5, we consider a Carathéodory function $h : \Omega \times Z \longrightarrow \mathbb{R}$ and then define $H : M^1_{+,\mu}(\Omega \times Z) \longrightarrow \mathbb{R}$ by

$$H(\vartheta) = \int_{\Omega \times Z} h(\omega, z) d\vartheta.$$

PROPOSITION 9.5.6 *The function $H : M^1_{+,\mu}(\Omega \times Z) \longrightarrow \mathbb{R}$ is weakly continuous.*

PROOF: Let $\vartheta_n \xrightarrow{w} \vartheta$ in $M^1_{+,\mu}(\Omega \times Z)$. By virtue of Theorem 6.2.9 we can find $D_\varepsilon \subseteq \Omega$ compact with $\mu(\Omega \setminus D_\varepsilon) < \varepsilon$ such that $h\big|_{D_\varepsilon \times Z}$ is continuous and so bounded. We have

$$\int_{\Omega \times Z} h d\vartheta_n = \int_{D_\varepsilon \times Z} h d\vartheta_n + \int_{(\Omega \setminus D_\varepsilon) \times Z} h d\vartheta_n \tag{9.51}$$

and

$$\int_{D_\varepsilon \times Z} h d\vartheta_n \longrightarrow \int_{D_\varepsilon \times Z} h d\vartheta \quad \text{as } n \to \infty. \tag{9.52}$$

Also, if $\xi > 0$ is the bound of h (recall that by hypothesis h is bounded), then

$$\Big|\int_{(\Omega \setminus D_\varepsilon) \times Z} h d\vartheta_n\Big| \le \int_{(\Omega \setminus D_\varepsilon) \times Z} |h| d\vartheta_n \le \xi\varepsilon \qquad \text{for all } n \ge 1. \tag{9.53}$$

So, if we pass to the limit as $n \to \infty$ in (9.51) and we use (9.52) and (9.53), we obtain

$$\int_{D_\varepsilon \times Z} h d\vartheta - \xi\varepsilon \le \liminf_{n \to \infty} \int_{\Omega \times Z} h d\vartheta_n \le \limsup_{n \to \infty} \int_{\Omega \times Z} h d\vartheta_n \le \int_{D_\varepsilon \times Z} h d\vartheta + \xi\varepsilon. \tag{9.54}$$

Because $\varepsilon > 0$ was arbitrary we let $\varepsilon \downarrow 0$. From (9.54) we infer that

$$\int_{\Omega\times Z} h d\vartheta_n \longrightarrow \int_{\Omega\times Z} h d\vartheta,$$
$$\Rightarrow \ H \text{ is weakly continuous.}$$

□

Now we are ready to prove an equilibrium theorem for game Γ.

THEOREM 9.5.7 *If the game Γ is as above and in addition the payoff function $u:\Omega\times X\times Y\longrightarrow\mathbb{R}$ is bounded and*

(i) *For all $(\omega,y)\in\Omega\times Y$, $x\longrightarrow u(\omega,x,y)$ is continuous;*

(ii) *For all $(\omega,x)\in\Omega\times X$, $y\longrightarrow u(\omega,x,y)$ is continuous,*

then the game Γ has an equilibrium.

PROOF: First we show that the expected payoff function $U(\varepsilon,\tau)$ is weakly continuous in each variable separately.

For fixed $\widehat{\omega}_1\in\Omega_1$ and distributional strategy τ^* for player II, we define the Borel measure $\eta_{\widehat{\omega}_1}$ on $\Omega_0\times\Omega_2\times Y$ by

$$\eta_{\widehat{\omega}_1}(A)=\int_A 1\tau^*(\mathrm{d}y|\omega_2)\mu\big(d(\omega,\omega_2)|\widehat{\omega}_1\big).$$

We define $H:\Omega_1\times X\longrightarrow\mathbb{R}$ by

$$H(\widehat{\omega}_1,x)=\int_{\Omega_0\times\Omega_2\times Y} u(\omega,x,y)\mathrm{d}\eta_{\widehat{\omega}_1}.$$

Claim: For every fixed $\widehat{\omega}_1\in\Omega_1$, $x\longrightarrow H(\widehat{\omega}_1,x)$ is continuous on X.

By virtue of Proposition 9.5.5, for every $n\geq 1$, we can find $E_n\subseteq\Omega_0\times\Omega_2\times Y$ a Borel set with $\eta_{\widehat{\omega}_1}(E_n)>1-(1/n)$ such that

$$\big\{u(\omega,\widehat{\omega}_1,\omega_2,\cdot,y):(\omega,\omega_2,y)\in E_n\big\}$$

is equicontinuous. Define $H_n:X\longrightarrow\mathbb{R}$ by

$$H_n(x)=\int_{E_n} u(\omega,\widehat{\omega}_1,\omega_2,x,y)\mathrm{d}\eta_{\widehat{\omega}_1}.$$

Evidently H_n is continuous. Because u is bounded and $\eta_{\widehat{\omega}_1}(E_n)>1-(1/n)$, it follows that

$$H_n\longrightarrow H(\widehat{\omega}_1,\cdot) \ \text{uniformly},$$
$$\Rightarrow \ H(\widehat{\omega}_1,\cdot) \ \text{is continuous.}$$

Next let

$$J(\sigma)=\int_{\Omega_1\times X} H(\omega_1,x)\mathrm{d}\sigma.$$

From Proposition 9.5.6 we have that J is weakly continuous. We have

$$\begin{aligned} J(\sigma) &= \int_{\Omega_1 \times X} \Big[\int_{\Omega_0 \times \Omega_2 \times X} u(\omega, x, y) d\eta_{\omega_1} \Big] d\sigma \\ &= \int_{\Omega \times X \times Y} u(\omega, x, y) \sigma(dx|\omega_1) \tau^*(dy|\omega_2) d\mu \\ &= U(\sigma, \tau^*). \end{aligned}$$

Therefore $U(\sigma, \tau)$ is weakly continuous separately on each variable.

Now observe that the sets of distributional strategies are convex and weakly compact. So we can apply Theorem 2.3.13 and produce a saddle point $(\widehat{\sigma}, \widehat{\tau})$ of U, namely

$$U(\widehat{\sigma}, \widehat{\tau}) = \inf_{\tau} \sup_{\sigma} U(\sigma, \tau) = \sup_{\sigma} \inf_{\tau} U(\sigma, \tau).$$

This pair $(\widehat{\sigma}, \widehat{\tau})$ is the desired equilibrium of game Γ. □

REMARK 9.5.8 It is interesting to extend this result to the case of nonzero-sum games.

9.6 Markov Decision Chains with Unbounded Costs

In this section we consider a general state space Markov decision chain (MDC), with finite action sets and possibly unbounded cost. We prove the existence of optimal stationary discounted policies.

So let (Ω, Σ) be a measurable space for which singletons belong to Σ. Also X denotes the action set that is assumed to be countable. Given a state $\omega \in \Omega$ and an action $x \in X$, the next state is chosen according to a transition probability (stochastic kernel), $p(\cdot|\omega, x)$ defined on Σ. So for every $(\omega, x) \in \Omega \times X$, we have:

(a) $p(\cdot|\omega, x)$ is a probability measure on Σ.
(b) For every $A \in \Sigma$, $p(A|\cdot, \cdot) : \Omega \times X \longrightarrow [0,1]$ is $(\Sigma \times 2^X)$-measurable.

Let μ be a probability measure on Ω, known as the initial measure. Usually μ is the Dirac probability measure concentrated at the known initial state. A policy π is a rule for choosing actions, given the history of the process, and it may be randomized. So, if the policy is π, the process develops as follows. The initial state ω_0 is chosen according to the distribution μ. Subsequently the initial action x_0 is chosen according to the rule π. Then the next state ω_1 is chosen according to the distribution $p(\cdot|\omega_0, x_0)$ and the process continues indefinitely.

Of course the above description of a policy (strategy) π is informal. To rigorously define it, we proceed as follows. Let $\Omega_0, \Omega_1, \Omega_2, \Omega_3, \dots$ be copies of the state space Ω and $X_0, X_1, X_2, X_3, \dots$ be copies of the action set X. The history of the process up until time $n \geq 0$, is defined inductively as follows.

(a) $H_0 = \Omega_0$.
(b) Given Ω_n, we set $H_{n+1} = H_n X_n \Omega_{n+1}$.

Clearly H_n represents the information available to the decision maker at stage $n \geq 0$. Let $H = \bigcup_{n \geq 0} H_n$ and let $M_+^1(X)$ be the space of all probability measures on X.

DEFINITION 9.6.1 A *policy* (*strategy*) is a map $\pi : H \longrightarrow M_+(X)$ such that for each $n \geq 0$ and $x \in X$, $h_n \longrightarrow \pi\big|_{H_n}(h_n)(x)$ is measurable on H_n with respect to the product σ-field $\Sigma \times 2^X \times \Sigma \times \cdots \times \Sigma$.

REMARK 9.6.2 By the disintegration theorem (see, e.g., Ash [30, p. 109]), there is a probability measure on the set of sample paths $S = \prod_{k \geq 0} \Omega_k \times X_k$ under the policy π. Equip S with the σ-field $\Sigma \times 2^X \times \Sigma \times 2^X \times \ldots$ with initial measure μ on $(\Omega_0 = \Omega, \Sigma)$. Given $(\omega_0, x_0, \ldots, \omega_n, x_n)$ we need a measure on Σ. This is the law of motion $p(\cdot|\omega_n, x_n)$. Also given $(\omega_0, x_0, \ldots, \omega_n)$ we need a measure on 2^X. This measure is induced by the probability distribution $\pi(\omega_0, x_0, \ldots, \omega_n)$. Note that we do not assume that the state space is a Borel space (see Section 8.5).

DEFINITION 9.6.3 A *stationary policy* is a measurable map $g : \Omega \longrightarrow X$.

REMARK 9.6.4 So whenever $\omega_n = \omega$, the stationary policy π_s chooses the action $f(\omega)$. For this reason the stationary policy π_s is identified with f.

We assume that a cost function is given, namely a function $u : \Omega \times X \longrightarrow \overline{\mathbb{R}}_+ = \mathbb{R}_+ \cup \{+\infty\}$. We make the following hypothesis concerning u.

H$_1$: $u : \Omega \times X \longrightarrow \overline{\mathbb{R}}_+ = \mathbb{R}_+ \cup \{+\infty\}$ is a measurable map such that for every $\omega \in \Omega$, the set

$$\{x \in X : u(\omega, x) < +\infty\} \tag{9.55}$$

is finite.

REMARK 9.6.5 If an action x is infeasible, then $u(\omega, x) = +\infty$. With a slight abuse of notation, when a stationary policy g is employed, the cost function evaluated at the point ω is denoted by $u(\omega, g)$.

There is a discount factor $0 < \beta < 1$. Using it we can now define the total discounted cost under a policy π. For $n \geq 0$, we define $u_n : S \longrightarrow \overline{\mathbb{R}} = \mathbb{R}_+ \cup \{+\infty\}$ by

$$u_n(\omega_0, x_0, \ldots) = u(\omega_n, x_n).$$

Evidently these functions are measurable. We consider

$$U(\omega_0, x_0, \ldots) = \sum_{n \geq 0} \beta^n u(\omega_n, x_n).$$

Then U is measurable on S. Conditioned on $\omega_0 = \omega$, we denote the conditional expectation by

$$V_\beta(\pi, \omega).$$

This quantity may be $+\infty$ for some policies. Let

$$V_\beta(\omega) = \inf\left[V_\beta(\pi, \omega) : \pi = \text{policy}\right].$$

We make the following hypothesis.

H$_2$: $V_\beta(\omega) < +\infty$ for every $(\beta, \omega) \in (0, 1) \times \Omega$.

REMARK 9.6.6 This hypothesis implies that given an initial state $\omega \in \Omega$, there must be at least one action $x \in X$ such that $u(\omega, x) < +\infty$. Hence in hypothesis H_1, the set given by (9.55) is nonempty.

As we already saw in Section 8.5, the proof of the existence of an optimal stationary policy depends on a measurable selection argument, which permits the realization of a stationary policy that achieves the minimum in the discounted optimality equation. The next proposition paves the way to obtain such a measurable selector.

PROPOSITION 9.6.7 *If $h_x : \Omega \longrightarrow \overline{\mathbb{R}}_+ = \mathbb{R}_+ \cup \{+\infty\}$, $x \in X$, are measurable functions such that for each $\omega \in \Omega$, the set $\{x \in X : h_x(\omega) < +\infty\}$ is finite, $h = \min_{x \in X} h_x$, the elements of the countable set X are ordered as $x_0 < x_1 < x_2 < \cdots$ and we define $g : \Omega \longrightarrow X$ by $g(\omega) =$ smallest $x \in X$ such that $h_x(\omega) = h(\omega)$, then g is measurable.*

PROOF: Evidently h is measurable. We need to show that for every $k \geq 0$, $g^{-1}(\{x_k\}) \in \Sigma$. Note that

$$g^{-1}(\{x_k\}) = \bigcap_{i=0}^{k-1} \{\omega \in \Omega : h_{x_k}(\omega) < h_{x_i}(\omega)\} \cap \bigcap_{i \geq k+1} \{\omega \in \Omega : h_{x_k}(\omega) \leq h_{x_i}(\omega)\},$$

$$\Rightarrow g^{-1}(\{x_k\}) \in \Sigma.$$

If for some $\omega \in \Omega$, $h(\omega) = +\infty$, then the result also holds. □

PROPOSITION 9.6.8 (a) *If $h : \Omega \times \Omega \longrightarrow \overline{\mathbb{R}}_+$ is jointly measurable, then for every $x \in X$, $\omega \longrightarrow \int_\Omega h(\omega, s) p(ds|\omega, x)$ is measurable.*
(b) *If $h : \Omega \longrightarrow \mathbb{R}_+$ is Σ-measurable and $g : \Omega \longrightarrow X$ is a stationary policy, then $\omega \longrightarrow \int_\Omega h(s) p(ds|\omega, g)$ is Σ-measurable.*

PROOF: (a) Let $h(\omega, s) = \sum_{k=1}^{N} \lambda_k \chi_{C_k}(\omega) \chi_{D_k}(s)$ with $C_k, D_k \in \Sigma$, $k \in \{1, \ldots, N\}$. We have

$$\int_\Omega h(\omega, s) p(ds|\omega, x) = \sum_{k=1}^{N} \lambda_k \chi_{C_k}(\omega) p(D_k|\omega, x),$$

$$\Rightarrow \omega \longrightarrow \int_\Omega h(\omega, s) p(ds|\omega, x) \quad \text{is } \Sigma\text{-measurable.}$$

Because every measurable function $h : \Omega \times \Omega \longrightarrow \mathbb{R}_+$ can be approximated by an increasing sequence of simple functions, we conclude that $\omega \longrightarrow \int_\Omega h(\omega, s) p(ds|\omega, x)$ is Σ-measurable.

(b) Exploiting as above the approximation of a measurable function $h : \Omega \times \Omega \longrightarrow \mathbb{R}_+$ by an increasing sequence of simple functions, it suffices to show that for each $A \in \Sigma$, $\omega \longrightarrow p(A|\omega, g(\omega))$ is Σ-measurable. To this end note that $\xi : \Omega \longrightarrow \Omega \times X$ defined by

$$\xi_1(\omega) = (\omega, g(\omega))$$

is $(\Sigma, \Sigma \times \mathbf{B}(X))$-measurable. Recall that $\xi_2 : \Omega \times X \longrightarrow [0, 1]$ defined by

$$\xi_2(\omega, x) = p(A|\omega, x)$$

is $\Sigma \times \mathbf{B}(X)$-measurable. Therefore $\xi_2 \circ \xi_1$ is Σ-measurable. But note that

$$\xi_2 \circ \xi_1(\omega) = p\big(A|\omega, f(\omega)\big).$$

□

DEFINITION 9.6.9 If π is a policy and $(\omega, x) \in \Omega \times X$, then the *shifted policy* $\pi(\omega, x)$ is defined by

$$\pi(\omega, x)(\omega_0, x_0, \dots, x_n) = \pi(\omega, x, \omega_0, x_0, \dots, x_n).$$

REMARK 9.6.10 According to this definition the decision made by the shifted policy $\pi(\omega, x)$ at the nth stage is the same as the decision made by π at stage $n+1$, given that the initial state and the initial decision are $(\omega, x) \in \Omega \times X$.

PROPOSITION 9.6.11 *For every policy π, the value function $\omega \longrightarrow V_\beta(\pi, \omega)$ is Σ-measurable. Moreover, for every fixed $x \in X$, $(\omega, s) \longrightarrow V_\beta\big(\pi(\omega, x), s\big)$ is $\Sigma \times \Sigma$-measurable on $\Omega \times \Omega$.*

PROOF: To prove the first part of the proposition, it suffices to show that for all $k \geq 0$, the function $\omega \longrightarrow E_\pi(u_k|\omega_0 = \omega)$ is Σ-measurable. We have

$$\begin{aligned} E_\pi(u_k|\omega_0 = \omega) = \sum_{x \in X} \pi(\omega)(x) \int_\Omega \sum_{y_1 \in X} \pi(\omega, x, \omega_1)(y_1) p(\mathrm{d}\omega_1|\omega, x) \\ \cdots \int_\Omega \sum_{y_{k-1} \in X} \pi(\omega, x, \omega_1, \dots, \omega_{k-1})(x_{k-1}) p(\mathrm{d}\omega_{k-1}|\omega_{k-2}, x_{k-2}) \\ \int_\Omega \sum_{y_k \in X} u(\omega_k, x_k) \pi(\omega, x, \omega_1, \dots, \omega_k)(x_k) p(\mathrm{d}\omega_k|\omega_{k-1}, x_{k-1}). \end{aligned}$$

From Definition 9.6.1 and Proposition 9.6.8, we see that the first part of the proposition holds.

Similarly we deduce the second part of the proposition, by establishing in a similar fashion the joint measurability of $(\omega, s) \longrightarrow E_{\pi((\omega),x)}(u_k|s_0 = s)$. □

Now we can prove a theorem on the existence of a β-discounted optimal stationary policy.

THEOREM 9.6.12 *If hypotheses H_1 and H_2 hold, then the finite value function $\omega \longrightarrow V_\beta(\omega)$ is the minimal nonnegative measurable solution of the optimality functional equation*

$$V_\beta(\omega) = \min_{x \in X} \Big[u(\omega, x) + \beta \int_\Omega V_\beta(s) p(\mathrm{d}s|\omega, x)\Big], \quad \omega \in \Omega. \tag{9.56}$$

If the minimum in (9.56) is realized at $g(\omega)$, then $\omega \longrightarrow g(\omega)$ is a stationary optimal policy.

PROOF: For each $n \geq 0$, we define inductively a sequence of functions $V_n : \Omega \longrightarrow \overline{\mathbb{R}}_+ = \mathbb{R}_+ \cup \{+\infty\}$ as follows:

$$V_0 = 0 \quad \text{and} \quad V_n(\omega) = \min_{x \in X} \big[u(\omega, x) + \beta \int_\Omega V_{n-1}(s) p(\mathrm{d}s|\omega, x)\big], \quad \omega \in \Omega,\ n \geq 1. \tag{9.57}$$

We show the following.

- $\omega \longrightarrow V_n(\omega)$ is Σ-measurable. (9.58)
- $\{V_n\}_{n\geq 0}$ is increasing. (9.59)
- $V_n \leq V_\beta(\pi, \cdot)$ for all $n \geq 0$ and all policies π. (9.60)

Clearly (9.58) follows from (9.57), Proposition 9.6.8(a), and the fact that X is countable.

Also (9.59) follows immediately from the inductive definition in (9.57).

To prove (9.60), let $\widehat{V}_n(\pi, \omega)$ be the extended discounted cost if we follow policy π for n steps and then the process is terminated with terminal cost zero. We claim

$$V_n \leq \widehat{V}_n(\pi, \cdot) \qquad \text{for all } n \geq 0. \tag{9.61}$$

Evidently (9.61) implies (9.60). So we need to prove (9.61). We have

$$V_1(\omega) \leq \sum_{x \in X} u(\omega, x) \pi(\omega)(x) = \widehat{V}_1(\pi, \omega).$$

Suppose (9.61) is true for $n - 1$. We can easily check that

$$\widehat{V}_n(\pi, \omega) = \sum_{x \in X} \pi(\omega)(x) \big[u(\omega, x) + \beta \int_\Omega \widehat{V}_{n-1}\big(\pi(\omega, x), s\big) p(\mathrm{d}s|\omega, x)\big],$$

with $\pi(\omega, x)$ being the shifted policy (see Definition 9.6.9). From Proposition 9.6.11 we know that $(\omega, s) \longrightarrow V_{n-1}\big(\pi(\omega, x), s\big)$ is $\Sigma \times \Sigma$-measurable on $\Omega \times \Omega$ and so Proposition 9.6.8(a) implies that $V_n(\pi, \cdot)$ is Σ-measurable. Then by the induction hypothesis

$$\begin{aligned} & V_{n-1}(\omega) \leq \widehat{V}_{n-1}\big(\pi(\omega, x), s\big) \\ & \Rightarrow V_n(\omega) \leq u(\omega, x) + \beta \int_\Omega \widehat{V}_{n-1}\big(\pi(\omega, x), s\big) p(\mathrm{d}s|\omega, x), \end{aligned}$$

and this proves (9.61) (hence (9.60) too).

Then $\lim_{n\to\infty} V_n = v$ exists and is Σ-measurable (see (9.58) and (9.59)). Moreover, from the monotone convergence theorem, we see that v satisfies

$$v(\omega) = \min_{x \in X} \big[u(\omega, x) + \beta \int_\Omega v(s) p(\mathrm{d}s|\omega, x)\big], \quad \omega \in \Omega. \tag{9.62}$$

By virtue of Proposition (9.61), there exists a stationary policy $g : \Omega \longrightarrow X$ realizing the minimum in (9.62). So

$$v(\omega) = u(\omega, g) + \beta \int_\Omega v(s) p(\mathrm{d}s|\omega, x). \tag{9.63}$$

From Proposition 9.6.8(b), the integral in the right-hand side of (9.63) produces a Σ-measurable function. Iterating (9.63), we obtain

$$\begin{aligned} & V_n(g,\omega) \leq v(\omega), \\ \Rightarrow\ & V_\beta(g,\omega) \leq v(\omega). \end{aligned}$$

This combined with (9.60) implies that

$$v = V_\beta \quad \text{and} \quad g = \text{stationary optimal policy}.$$

Finally let w be a nonnegative Σ-measurable solution of (9.56) and let g_w be the stationary policy realizing the minimum in (9.56) for w (instead of u). Then following the above argument, we obtain

$$\begin{aligned} & V_\beta \leq V_\beta(g_w,\cdot) \leq w, \\ \Rightarrow\ & V_\beta \ \text{ is indeed the minimal solution of (9.56).} \end{aligned}$$

□

9.7 Remarks

9.1: The first to apply topologies on sub-σ-fields to study the dependence of economic models on information was Allen [13], who employed the $\widehat{d}$-metric topology (see Definition 9.1.17). The $\widehat{d}$-metric was first defined by Boylan [96], who used it to study the convergence of certain martingale like sequences. The weaker topology of pointwise convergence (see Definition 9.1.3), was introduced by Cotter [160]. Proposition 9.1.9 is due to Fetter [243]. Other topologies considered in probability theory, can be found in Allen [14]. Stinchcombe [559, 560] introduced some Bayesian topologies, in which two information structures are similar if there is only small probability that the posteriors are far apart. Stinchcombe [559, 560], explored the Bayesian topologies on sub-σ-fields that are generated by different vector space topologies on the space of distributions.

9.2: The ex-post (or Bayesian) view was adopted by Allen [13], Cotter [160, 161] and Stinchcombe [559]. In Allen [13] and Cotter [160, 161] the state-dependent posteriors are not well defined and so instead they show that the map from information to expected utility is continuous and then use the fact that the map from utility to actions is continuous. In Stinchcombe [559], the author imposes assumptions on the probability space so that the state-dependent posteriors (regular conditional probabilities) given a sub-σ-field, are well defined. Then he showed that the mapping from information to state-dependent posteriors is continuous under suitable topologies on the space of information. The mapping from posteriors to actions is also continuous. Hence the composition of the two maps is continuous. In this section we present the first approach and outline the continuous dependence results due to Cotter [160] and Allen [13].

9.3: The ex-ante view was adopted by Van Zandt [592]. Proposition 9.3.6 is due to Horsley–Van Zandt–Wrobel [312]. This approach can be found also in Cotter

[162] who proved that the set of type-correlated equilibria (an extension of correlated equilibria to Bayesian games) is upper semicontinuous with respect to the information variable.

9.4: Definition 9.4.1 was first introduced by Kudo [365], who also proved Proposition 9.4.4. If $\widehat{\Sigma} = \mu\text{-}\liminf_{n\to\infty} \Sigma_n = \mu\text{-}\limsup_{n\to\infty} \Sigma_n$, then we say that the sequence of Σ_ns strongly converges to $\widehat{\Sigma}$. The use of these limits of sub-σ-fields, to study the convergence of prediction sequences, is due to Tsukada [588].

9.5: The first general equilibrium result for games with incomplete information was proved by Milgram–Weber [429]. They made two key hypotheses. The first (called the continuity of payoff), is that for fixed $\omega \in \Omega$, $\{u_k(\omega, \cdot)\}_{k=1}^n$ are uniformly equicontinuous. The second (called continuity of information), is that the joint distribution of the information available to the players, is absolutely continuous with respect to a product measure on the product of the spaces of types. This last hypothesis is in general difficult to verify, unless the information random variables are either independent or discrete and finite-valued. Here we present the result of Mamer–Schilling [405], who were able to weaken the continuity of payoffs hypothesis to continuity of $u(\omega, \cdot)$ on each action variable separately and they eliminated completely the continuity of information hypothesis.

9.6: The first work of Markov decision chains with an arbitrary state space and unbounded costs, is due to Ritt–Sennott [515]. Previous works had a general state space but a bounded cost, see for example Bertsekas–Shreve [72] and Hernandez Lerma [290].

10

Evolution Equations

Summary. *Evolution equations are the abstract formulation of dynamic partial differential equations. In this chapter, we examine both semilinear and nonlinear evolution equations. We start by developing the mathematical tools which are necessary in this study. Among them, the notion of evolution triple plays central role and always allows us to use different spaces within the analysis of a single evolution equation. First we consider semilinear evolutions, which we analyze using the semigroup method. Subsequently, we pass to nonlinear evolutions. We consider two such classes. Evolutions of the subdifferential type (which incorporate variational inequalities) and evolution formulated in the framework of evolution triples with operators of monotone type. The Galerkin method is crucial here. We conclude with the study of second-order evolutions.

Introduction

Evolution equations are the abstract formulation of dynamic partial differential equations (i.e., partial differential equations with time t as one of the independent variables). Such equations arise in many branches of science and engineering and describe mathematically important physical processes and phenomena. The basic theoretical question related to such problems is whether for given initial data the equation has a solution at least locally in time and whether this solution is unique. It is also important to know how this solution depends on the initial data. We investigate these issues in the context of certain broad classes of semilinear and nonlinear evolution equations.

In Section 10.1 we present the mathematical tools needed to study evolution equations. Central in this section is the notion of evolution triple of spaces. Evolution triples provide a natural and flexible framework for the study of evolution equations. Evolution triples realize the modern strategy to study pdes, which is to use many different spaces in the same problem. In Section 10.1, we examine some basic function spaces associated with evolution triples.

Section 10.2 deals with semilinear evolution equations. We employ the semigroup method and prove existence, uniqueness, and blow-up of solutions for such evolutions. We also apply the abstract results to some parabolic initial-boundary value problems.

N.S. Papageorgiou, S.Th. Kyritsi-Yiallourou, *Handbook of Applied Analysis*,
Advances in Mechanics and Mathematics 19, DOI 10.1007/b120946_10,

In Section 10.3 we study nonlinear evolution equations. We consider two classes of such equations. The first class involves time-invariant subdifferential operators and so it incorporates certain variational inequalities. The second class involves problems formulated in the framework of evolution triples. The Galerkin method is the basic tool here.

Finally in Section 10.4 we deal with second-order evolutions. The same two classes of equations are studied in this section too.

10.1 Lebesgue–Bochner Spaces

In this section we introduce some spaces of Banach space–valued functions that are needed in the study of evolution equations. By a Banach space–valued function, we mean any map that takes values in a Banach space X, for example, a map $f\colon T=[0,b] \longrightarrow X$.

Throughout this section (Ω, Σ, μ) is a finite measure space and X a Banach space. Additional hypotheses are introduced as needed. By $\|\cdot\|$ we denote the norm of X and by X^* its topological dual.

DEFINITION 10.1.1 (a) A function $f\colon\Omega \longrightarrow X$ is a *simple function* if there exist finitely many sets $\{C_k\}_{k=1}^n \subseteq \Sigma$ that are mutually disjoint and $\{x_k\}_{k=1}^n \subseteq X$ such that

$$f(\omega) = \sum_{k=1}^{n} x_k \chi_{C_k}(\omega) \qquad \text{for all } \omega \in \Omega.$$

(b) A function $f\colon\Omega \longrightarrow X$ is said to be *strongly measurable* if there exists a sequence of simple function $s_n : \Omega \longrightarrow X$, $n \geq 1$ such that

$$\lim_{n\to\infty} \|f(\omega) - s_n(\omega)\| = 0 \quad \mu\text{-a.e. on } \Omega.$$

(c) A function $f\colon\Omega \longrightarrow X$ is said to be *weakly measurable* if for any $x^* \in X^*$ the $\mathbb{R}$-valued function $\omega \longrightarrow \langle x^*, f(\omega)\rangle$ is Lebesgue measurable. Hereafter, by $\langle\cdot,\cdot\rangle$ we denote the duality brackets for the pair (X^*, X).

The next theorem, usually known as the *Pettis measurability theorem* relates the notions of strong and weak measurability (see Definition 10.1.1(b),(c)) and its proof can be found, for example, in Gasiński–Papageorgiou [259, p. 109].

THEOREM 10.1.2 *A function $f\colon\Omega \longrightarrow X$ is strongly measurable if and only if*

(a) *f is almost separably valued (i.e., there exists a μ-null set $C \subseteq \Omega$ such that $f(\Omega \setminus C)$ is separable in X) and*
(b) *f is weakly measurable.*

In particular, if X is separable, then the notions of strong and weak measurability are equivalent.

DEFINITION 10.1.3 Let $f:\Omega\longrightarrow X$ be strongly measurable. We say that f is *Bochner integrable* if there exists a sequence $\{s_n\}_{n\geq 1}$ of simple functions such that

$$\lim_{n\to\infty}\int_\Omega \|f(\omega)-s_n(\omega)\|\mathrm{d}\mu=0.$$

In this case $\int_\Omega f(\omega)\mathrm{d}\mu=\lim\limits_{n\to\infty}\int_\Omega s_n(\omega)\mathrm{d}\mu$ and it is called the *Bochner integral* of f.

REMARK 10.1.4 It is easy to see that the above definition of the Bochner integral is independent of the choice of the sequence $\{s_n\}_{n\geq 1}$ of simple functions. For any $C\in\Sigma$ we set $\int_C f(\omega)\mathrm{d}\mu=\int_\Omega \chi_C(\omega)f(\omega)\mathrm{d}\mu$.

An easy consequence of Definition 10.1.3 is the following convenient criterion for Bochner integrability.

PROPOSITION 10.1.5 *If $f:\Omega\longrightarrow X$ is strongly measurable, then f is Bochner integrable if and only if $\|f(\cdot)\|\in L^1(\Omega)$. Moreover, we have*

$$\|\int_\Omega f(\omega)\mathrm{d}\mu\|\leq\int_\Omega\|f(\omega)\|\mathrm{d}\mu.$$

The Bochner integral possesses almost the same properties as the Lebesgue integral. In the next theorem we have gathered the basics of those properties. We omit the proofs which can be found in Gasiński–Papageorgiou [259, Section 2.1].

THEOREM 10.1.6 (a) *If $f_n:\Omega\longrightarrow X$, $n\geq 1$, is a sequence of Bochner integrable functions and $f_n(\omega)\xrightarrow{w} f(\omega)$ μ-a.e. in X, then f is Bochner integrable and $\int_\Omega\|f(\omega)\|\mathrm{d}\mu\leq\liminf\limits_{n\to\infty}\int_\Omega\|f_n(\omega)\|\mathrm{d}\mu$.*

(b) *If $f_n:\Omega\longrightarrow X$, $n\geq 1$, is a sequence of Bochner integrable functions, $f_n\xrightarrow{\mu} f$ (i.e., for every $\varepsilon>0$, $\mu(\{\omega\in\Omega:\|f_n(\omega)-f(\omega)\|\geq\varepsilon\})\longrightarrow 0$ as $n\to\infty$) and there exists $h\in L^1(\Omega)_+$ such that $\|f_n(\omega)\|\leq h(\omega)$ μ-a.e. on Ω, then f is Bochner integrable and $\int_C f_n\mathrm{d}\mu\xrightarrow{s}\int_C f\mathrm{d}\mu$ for all $C\in\Sigma$. In fact, $\lim\limits_{n\to\infty}\int_\Omega\|f_n-f\|\mathrm{d}\mu=0$.*

(c) *If $m(C)=\int_C f\mathrm{d}\mu$, $C\in\Sigma$, then m is a vector measure (i.e., $m(\emptyset)=0$) and if $\{C_n\}_{n\geq 1}\subseteq\Sigma$ is a sequence of pairwise disjoint sets and $C=\bigcup\limits_{n\geq 1}C_n$, then $m(C)=\sum\limits_{n\geq 1}m(C_n)$ (the sum is actually absolutely convergent). Moreover, $m\ll\mu$ (i.e., $\lim\limits_{\mu(C)\to 0}\int_C f\mathrm{d}\mu=0$) and $|m|(C)=\int_C\|f\|\mathrm{d}\mu$ for all $C\in\Sigma$, with $|m|$ being the total variation of the vector measure m.*

COROLLARY 10.1.7 *If $f,g:\Omega\longrightarrow X$ are Bochner integrable and $\int_C f\mathrm{d}\mu=\int_C g\mathrm{d}\mu$ for all $C\in\Sigma$, then $f(\omega)=g(\omega)$ μ-a.e. on Ω.*

PROOF: Let $m(C)=\int_C(f-g)\mathrm{d}\mu$, $C\in\Sigma$. Then by hypothesis $m(C)=0$ for every $C\in\Sigma$, hence $|m|(C)=0$. But $|m|(C)=\int_C\|f-g\|\mathrm{d}\mu$ for all $C\in\Sigma$. So $\int_C\|f-g\|\mathrm{d}\mu=0$ for all $C\in\Sigma$, hence $\|f(\omega)-g(\omega)\|=0$ μ-a.e. on Ω. $\square$

The next theorem states a useful property of the Bochner integral, which has no counterpart in the Lebesgue integration. For a proof of this result we refer to Gasiński–Papageorgiou [259, p. 116].

THEOREM 10.1.8 *If Y is another Banach space, $A : \mathcal{D}(A) \subseteq X \longrightarrow Y$ is a closed linear operator, and $f : \Omega \longrightarrow \mathcal{D}(A) \subseteq X$ and $Af : \Omega \longrightarrow Y$ are both Bochner integrable, then $A\big(\int_\Omega f d\mu\big) = \int_\Omega Af d\mu$.*

Given $f, g : \Omega \longrightarrow X$ strongly measurable, we say that $f \sim g$ if and only if $f(\omega) = g(\omega) \mu$-a.e. Evidently this is an equivalence relation. Now we can define the so-called *Lebesgue–Bochner spaces*.

DEFINITION 10.1.9 (a) Let $1 \leq p < \infty$. By $L^p(\Omega, X)$ we denote the space of all the equivalence classes for the relation $\sim$ of the strongly measurable functions $f : \Omega \longrightarrow X$ such that $\|f(\cdot)\| \in L^p(\Omega)_+$. We set

$$\|f\|_{L^p(\Omega,X)} = \Big(\int_\Omega \|f(\omega)\|^p d\mu\Big)^{1/p}.$$

(b) The space $L^\infty(\Omega, X)$ is defined to be the space of all equivalence classes for the relation $\sim$ of the strongly measurable functions $f : \Omega \longrightarrow X$ such that $\|f(\cdot)\| \in L^\infty(\Omega)_+$. We set

$$\|f\|_{L^\infty(\Omega,X)} = \text{ess sup}\|f(\cdot)\|.$$

REMARK 10.1.10 For every $1 \leq p \leq \infty$, $L^p(\Omega, X)$ is a Banach space. Moreover, if Σ is countably generated and X is a separable Banach space, then $L^p(\Omega, X)$, $1 \leq p < \infty$ is separable. Recall that, if Ω is an open or a closed subset of $\mathbb{R}^N$, then the Borel σ-field $\mathbf{B}(\Omega)$ is countably generated. For $1 < p < \infty$ $L^p(\Omega, X)$ is uniformly convex if and only if X is uniformly convex. Simple functions are dense in $L^p(\Omega, X)$ for $1 \leq p < \infty$ and countably-valued functions of $L^\infty(\Omega, X)$ are dense in $L^\infty(\Omega, X)$. Finally if $\Omega \subseteq \mathbb{R}^N$ is bounded open, then $C^\infty(\overline{\Omega}, X)$ is dense in $L^p(\Omega, X)$, $1 \leq p < \infty$.

The characterization of the dual of these spaces is not easy. The reason for this is the fact that the Radon–Nikodym theorem is not in general true for vector measures. This leads to the following classification of Banach spaces.

DEFINITION 10.1.11 A Banach space X has the *Radon–Nikodym property* (RNP for short) if for every finite measure space (Ω, Σ, μ) and every vector measure $m : \Sigma \longrightarrow X$ of bounded variation such that $m \ll \mu$, we can find $f \in L^1(\Omega, X)$ such that $m(C) = \int_C f(\omega) d\mu$ for all $C \in \Sigma$.

Large classes of Banach spaces have the RNP (see Diestel–Uhl [199, pp. 79–82]).

THEOREM 10.1.12 (a) *Reflexive Banach spaces have the RNP.*
(b) *Separable dual Banach spaces have the RNP.*

Using this notion we can have a Riesz representation theorem for the Lebesgue–Bochner spaces. For a proof see Diestel–Uhl [199, pp. 89–100].

THEOREM 10.1.13 *If X is a Banach space such that X^* has the RNP and $1 \le p < \infty$, then $L^p(\Omega, X)^* = L^{p'}(\Omega, X^*)$, $(1/p) + (1/p') = 1$.*

There is a version of this theorem when X^* does not satisfy the RNP. We state here the case $p = 1$ which appears often in applications.

DEFINITION 10.1.14 Given two functions $g, h : \Omega \longrightarrow X^*$ that are w^*-measurable (i.e., for every $x \in X$, $\omega \longrightarrow \langle g(\omega), x\rangle$, $\omega \longrightarrow \langle h(\omega), x\rangle$ are Σ-measurable), we have

$$g \sim h \quad \text{if and only if } \langle g(\omega), x\rangle = \langle h(\omega), x\rangle \quad \mu\text{-a.e. on } \Omega \quad \text{for every } x \in X. \tag{10.1}$$

Note that in (10.1) the μ-null set in general depends on $x \in X$. Also $\sim$ is an equivalence relation. By $L^\infty(\Omega, X^*_{w^*})$, we denote the linear space of the $\sim$-equivalence classes of functions $g : \Omega \longrightarrow X^*$ that are w^*-measurable and there exists $c \ge 0$ such that

$$|\langle g(\omega), x\rangle| \le c\|x\| \quad \mu\text{-a.e. on } \Omega \text{ for every } x \in X. \tag{10.2}$$

Again the μ-null set in (10.2) may depend on $x \in X$. The infimum of all $c \ge 0$ such that (10.2) holds is denoted by $\|g\|_{L^\infty(\Omega, X^*_{w^*})}$ and it is easily seen to be a norm for the space $L^\infty(\Omega, X^*_{w^*})$.

REMARK 10.1.15 If X is separable and $g \in L^\infty(\Omega, X^*_{w^*})$, then $\omega \longrightarrow \|g(\omega)\|_{X^*}$ belongs in $L^\infty(\Omega)_+$ and

$$\|g\|_{L^\infty(\Omega, X^*_{w^*})} = \operatorname*{ess\,sup}_{\Omega} \|g(\cdot)\|_{X^*}.$$

The next theorem is usually called the *Dinculeanu–Foias theorem* (Ionescu–Tulcea [328, p. 95]).

THEOREM 10.1.16 *If X is an arbitrary Banach space, then $L^1(\Omega, X)^* = L^\infty(\Omega, X^*_{w^*})$ and the duality pairing between the two spaces is given by*

$$\langle g, f\rangle_0 = \int_\Omega \langle g(\omega), f(\omega)\rangle \, d\mu.$$

Using the notion of RNP, we can also extend the fundamental theorem of the Lebesgue calculus to vector-valued functions.

DEFINITION 10.1.17 A function $f : T = [0, b] \longrightarrow X$ is said to be *absolutely continuous* if for every $\varepsilon > 0$, we can find $\delta > 0$ such that for every sequence of disjoint intervals $(\alpha_n, c_n) \subseteq T$ verifying $\sum_{n \ge 1} (c_n - \alpha_n) < \delta$, we have $\sum_{n \ge 1} \|f(c_n) - f(\alpha_n)\| < \varepsilon$.

THEOREM 10.1.18 *If X has the RNP and $f : T \longrightarrow X$ is absolutely continuous, then f is strongly differentiable a.e. on T, $f' \in L^1(T, X)$, and*

$$f(t) = f(0) + \int_0^t f'(s) ds \qquad \text{for all } t \in T.$$

In applications this theorem is used in the context of reflexive Banach spaces (see Theorem 10.1.12(a)). For this reason in the next definition we assume that X is reflexive.

DEFINITION 10.1.19 Suppose X is a reflexive Banach space.

(a) By $AC^{1,p}(T,X)$, $1 \leq p \leq \infty$, we denote the space of all absolutely continuous functions $f: T \longrightarrow X$ such that $f' \in L^p(T,X)$ (see Theorem 10.1.18).

(b) By $\mathcal{D}'\big((0,b),X\big)$ we denote the space of all continuous linear operators from $\mathcal{D}(0,b) = C_c^\infty(0,b)$ into X. An element $u \in \mathcal{D}'\big((0,b),X\big)$ is an X-valued distribution on $(0,b)$. If $u \in \mathcal{D}'\big((0,b),X\big)$, then for a positive integer $k \geq 1$

$$D^k u(\vartheta) = (-1)^k u\Big(\frac{d^k\vartheta}{dt^k}\Big) \qquad \text{for all } \vartheta \in \mathcal{D}(0,b) = C_c^\infty(0,b),$$

defines another vectorial distribution $D^k u \in \mathcal{D}'\big((0,b),X\big)$, known as the k-vectorial distributional derivative of u. If $k=1$, then we write $D^1 = D$.

(c) Let $1 \leq p \leq \infty$. The vectorial Sobolev space $W^{1,p}\big((0,b),X\big)$ is defined to be the space of all functions $u \in L^p(T,X)$ such that the vectorial distributional derivative Du also belongs in $L^p(T,X)$.

As in the case for $\mathbb{R}$-valued functions, the spaces $AC^{1,p}(T,X)$ and $W^{1,p}\big((0,b),X\big)$ can be identified (see Barbu [58, p. 19] and Gasiński–Papageorgiou [259, p. 138]).

THEOREM 10.1.20 *If X is a reflexive Banach space and $u \in L^p(T,X)$, $1 \leq p \leq \infty$, then the following statements are equivalent.*

(a) $u \in W^{1,p}\big((0,b),X\big)$.

(b) *There exists $\widehat{u} \in AC^{1,p}(T,X)$ such that $\widehat{u}(t) = u(t)$ a.e. on $(0,b)$.*

Suppose X and Y are Banach spaces and X is embedded continuously and densely into Y. Then we can easily verify that

- Y^* is embedded continuously into X^*. (10.3)
- If X is reflexive, then the above embedding is also dense. (10.4)

With this in mind we make the following definition, which is basic in the study of evolution equations.

DEFINITION 10.1.21 By an *evolution triple* (or *Gelfand triple*), we understand a triple of spaces

$$X \subseteq H \subseteq X^*$$

such that

(a) X is a separable, reflexive Banach space.

(b) H is a separable Hilbert space that is identified with its dual (pivot space).

(c) X is embedded continuously and densely into H.

REMARK 10.1.22 By virtue of (10.3) and (10.4), for an evolution triple (X, H, X^*) we have $X \subseteq H$ and $H^* = H \subseteq X^*$ with the embedding being continuous and dense. In what follows by $\langle \cdot, \cdot \rangle$ we denote the duality brackets for the pair (X^*, X) and by $(\cdot, \cdot)$ the inner product of H. Also by $\|\cdot\|$ (resp., $|\cdot|, \|\cdot\|_*$), we denote the norm of X (resp., of H, X^*). We have

$$\langle \cdot, \cdot \rangle \big|_{H \times X} = (\cdot, \cdot) \quad \text{and} \quad \|x\|_* \leq c|x| \quad \text{for some } c>0 \text{ and all } x \in H. \tag{10.5}$$

Using evolution triples we can define the following Banach space, which is a useful tool in the study of evolution equations.

DEFINITION 10.1.23 $W_p(0,b) = \left\{x \in L^p(T,X) : x' \in L^{p'}(T,X^*)\right\}$, $1 < p < \infty$, $(1/p) + (1/p') = 1$. The space $W_p(0,b)$ is equipped with the norm

$$\|x\|_{W_p(0,b)} = \|x\|_{L^p(T,X)} + \|x'\|_{L^{p'}(T,X^*)}$$

which clearly makes $W_p(0,b)$ a Banach space.

REMARK 10.1.24 If in the above definition, we interpret the derivative of x as a vectorial distributional derivative into X^* (weak derivative), by virtue of Theorem 10.1.20, we see that it is a strong derivative if we view x as an X^*-valued function. Evidently $W_p(0,b) \subseteq AC^{1,p'}(T,X^*) = W^{1,p'}\big((0,b),X^*\big)$.

THEOREM 10.1.25 *If (X, H, X^*) is an an evolution triple and $1<p<\infty$, then $W_p(0,b)$ is embedded continuously and densely in $C(T,H)$.*

PROOF: Let $c<0<b<d$ and $x \in W_p(0,b)$. Because $x \in AC^{1,p}(T,X^*)$, we can extend x on $(c,0)$ and on (b,d) by setting $x(t)=x(0)$ for $t \in (c,0)$ and $x(t) = x(b)$ for $t \in (b,d)$. We choose $\varphi \in C_c^\infty(c,d)$ such that $\varphi\big|_T=1$. We set $\widehat{x}(t) = x\big(\varphi(t)\big)$ for all $t \in (c,d)$. Then clearly $\widehat{x} \in W_p(c,d)$ and $\widehat{x}\big|_T=x$. Also we have

$$\|x\|_{W_p(0,b)} \leq \|\widehat{x}\|_{W_p(c,d)} \leq c(\varphi)\|x\|_{W_p(0,b)} \quad \text{with } c(\varphi)>0$$

and we can find $\varepsilon>0$ small such that

$$\widehat{x}\big|_{(c,c+\varepsilon)}=0 \quad \text{and} \quad \widehat{x}\big|_{(d-\varepsilon,d)}=0.$$

Let $\vartheta \in C_c^\infty(-1,1)$ be a mollifier (i.e., $\vartheta \geq 0$, $\int_{-1}^1 \vartheta(t)dt$). We set $\vartheta_m(s) = m\vartheta(ms)$ and define the corresponding mollification of $\widehat{x}$; that is,

$$x_m(t) = \int_c^d \widehat{x}(s)\vartheta_m(t-s)ds.$$

We know that $x_m \in C_c^\infty\big((c,d),X\big)$ for $m \geq 1$ large and $x_m \longrightarrow \widehat{x}$ in $W_p(c,d)$ with

$$\|x_m\|_{W_p(c,d)} \leq \|\widehat{x}\|_{W_p(c,d)}.$$

We have

$$\begin{aligned}\frac{1}{2}\frac{\mathrm{d}}{\mathrm{d}t}|x_m(t)|^2 &= \big(x_m'(t), x_m(t)\big) = \big\langle x_m'(t), x_m(t)\big\rangle \quad \text{(see 10.5)},\\ \Rightarrow\ \frac{1}{2}|x_m(t)|^2 &= \int_c^t \big\langle x_m'(s), x_m(s)\big\rangle\, \mathrm{d}s\\ &\leq \int_c^t \|x_m'(s)\|_*\|x_m(s)\|\mathrm{d}s\\ &\leq \|x_m\|^2_{W_p(c,d)} \quad \text{(by the Cauchy–Schwartz inequality).}\end{aligned} \tag{10.6}$$

From (10.6), we infer that for $m,n\geq 1$, we have

$$\begin{aligned}&\frac{1}{2}|x_m(t)-x_n(t)|^2 \leq \|x_m - x_n\|^2_{W_p(c,d)} \qquad \text{for all } t\in[c,d]\\ \Rightarrow\ &\{x_m\}_{m\geq 1}\subseteq C(T,H) \quad \text{is Cauchy.}\end{aligned}$$

Therefore we have $x_m \longrightarrow \widehat{x}$ in $C(T,H)$, hence $\widehat{x} = x\in C(T,H)$.

So we have proved that $W_p(0,b)$ is embedded continuously in $C(T,H)$. The embedding is also dense because the set of all X-valued polynomials is dense in both $W_p(0,b)$ and $C(T,H)$. □

From the above proof, we deduce the following corollary.

COROLLARY 10.1.26 *If $x,y\in W_p(0,b)$, then $t\longrightarrow \big(x(t),y(t)\big)$ is absolutely continuous and*

$$\frac{\mathrm{d}}{\mathrm{d}t}\big(x(t),y(t)\big) = \big\langle x'(t), y(t)\big\rangle + \big\langle y'(t), x(t)\big\rangle \ \ \textit{a.e. on } T. \tag{10.7}$$

REMARK 10.1.27 We call (10.7) the integration by parts formula for the elements in $W_p(0,b)$. The embedding of $W_p(0,b)$ in $C(T,H)$ is not in general compact. We can have compact embedding of $W_p(0,b)$ into $L^p(T,H)$. This is a particular case of Theorem 10.1.29 below.

First we need the following interpolation lemma, sometimes called in the literature *Ehrling's inequality*.

LEMMA 10.1.28 *If X,Y,Z are Banach spaces such that $X\subseteq Y\subseteq Z$, the embedding of X into Y is compact, and the embedding of Y into Z is continuous, then given $\varepsilon>0$, we can find $c(\varepsilon)>0$ such that for all $x\in X$ we have*

$$\|x\|_Y \leq \varepsilon\|x\|_X + c(\varepsilon)\|x\|_Z. \tag{10.8}$$

PROOF: Suppose that (10.8) is false. Then we can find $\varepsilon>0$ and sequence $\{x_n\}_{n\geq 1}\subseteq X$ such that

$$\|x_n\|_Y > \varepsilon\|x_n\|_X + n\|x_n\|_Z \qquad \text{for all } n\geq 1.$$

We set $y_n = x_n/\|x_n\|_X$, $n\geq 1$. Evidently $\|y_n\|_X = 1$ for all $n\geq 1$ and we have

$$\|y_n\|_Y \geq \varepsilon + n\|y_n\|_Z \qquad \text{for all } n\geq 1. \tag{10.9}$$

Because $\{y_n\}_{n\geq 1} \subseteq X$ is bounded and by hypothesis X is embedded compactly in Y and Y is embedded continuously in Z, by passing to a suitable subsequence if necessary, we may assume that

$$y_n \longrightarrow y \quad \text{in } Y \quad \text{and in} \quad Z \quad \text{as } n \to \infty. \tag{10.10}$$

From (10.9) and (10.10), we obtain

$$\|y\|_Z = 0 \quad \text{and} \quad \|y\|_Y \geq \varepsilon > 0,$$

a contradiction. □

THEOREM 10.1.29 *If X, Y, Z are Banach spaces with X, Z reflexive such that $X \subseteq Y \subseteq Z$, the embedding of X into Y is compact, the embedding of Y into Z is continuous, and for $1 < p,\ q < \infty$, we define the Banach space*

$$W_{pq}(0,b) = \big\{x \in L^p(T,X) : x' \in L^q(T,Z)\big\},$$

then $W_{pq}(0,b)$ is embedded compactly in $L^p(T,Y)$.

PROOF: Clearly $W_{pq}(0,b)$ is a reflexive Banach space. Let $\{x_n\}_{n\geq 1} \subseteq W_{pq}(0,b)$ be a bounded sequence. We may assume that $x_n \xrightarrow{w} x$ in $W_{pq}(0,b)$ as $n \to \infty$. We have

$$x_n \xrightarrow{w} x \quad \text{in } L^p(T,X) \quad \text{and} \quad x_n' \xrightarrow{w} x' \quad \text{in } L^q(T,Z) \quad \text{as } n \to \infty.$$

Without any loss of generality we may assume that $x = 0$.

First we show that $x_n(t) \longrightarrow 0$ in Z for every $t \in T$. We show this convergence for $t = 0$ and for the other points in T the proof is similar. Because $x_n \in AC^{1,r}(T,X)$, $r = \min\{p,q\}$, we have

$$x_n(0) = x_n(t) - \int_0^t x_n'(s)\mathrm{d}s, \quad t \in T.$$

Integrating this inequality we obtain

$$\begin{aligned} x_n(0) &= \frac{1}{s}\int_0^s x_n(t)\mathrm{d}t - \int_0^s \int_0^t x_n'(\tau)\mathrm{d}\tau\mathrm{d}t \\ &= a_n + c_n, \end{aligned} \tag{10.11}$$

with

$$a_n = \frac{1}{s}\int_0^s x_n(t)\mathrm{d}t \quad \text{and} \quad c_n = -\frac{1}{s}\int_0^s (s-\tau)x_n'(\tau)\mathrm{d}\tau, \ n \geq 1.$$

Given $\varepsilon > 0$, we choose $s > 0$ small so that

$$\|c_n\|_Z \leq \int_0^s \|x_n'(\tau)\|_Z \mathrm{d}\tau \leq \frac{\varepsilon}{2}. \tag{10.12}$$

For this fixed $s > 0$ and because $x_n \xrightarrow{w} 0$ in $L^p(T,X)$ (recall $x = 0$), we see that

$$\begin{aligned} & a_n \xrightarrow{w} 0 \quad \text{in } X \quad \text{as } n \to \infty, \\ \Rightarrow\ & a_n \longrightarrow 0 \quad \text{in } Z \quad \text{as } n \to \infty. \end{aligned}$$

So we can find $n_0 = n_0(\varepsilon) \geq 1$ such that for all $n \geq n_0$, we have

$$\|a_n\|_Z \leq \frac{\varepsilon}{2}. \tag{10.13}$$

Returning to (10.11) and using (10.12) and (10.13), we have

$$\begin{aligned} &\|x_n(0)\|_Z \leq \varepsilon \qquad \text{for all } n \geq n_0, \\ \Rightarrow\ & x_n(0) \longrightarrow 0 \ \text{ in } Z \ \text{ as } n \to \infty. \end{aligned} \tag{10.14}$$

Note that $W_{pq}(0,b)$ is embedded continuously in $C(T,Z)$ (see Theorem 10.1.20). So it follows that $\{x_n\}_{n\geq 1} \subseteq C(T,Z)$ is bounded. Therefore by virtue of the dominated convergence theorem (see (10.14)), we have

$$x_n \longrightarrow 0 \ \text{ in } L^p(T,Z) \ \text{ as } n \to \infty. \tag{10.15}$$

Invoking Lemma 10.1.28, for any $\delta > 0$ we have

$$\begin{aligned} \|x_n\|_{L^p(T,Y)} &\leq \delta \|x_n\|_{L^p(T,X)} + c(\delta)\|x_n\|_{L^p(T,Z)} \\ &\leq \delta M + c(\delta)\|x_n\|_{L^p(T,Z)} \quad (\text{with } M = \sup_{n\geq 1} \|x_n\|_{L^p(T,X)} < \infty) \end{aligned}$$

$\Rightarrow \|x_n\|_{L^p(T,Y)} \longrightarrow 0$ as $n \to \infty$ (see (10.15).

This proves that $W_{pq}(0,b)$ is embedded compactly in $L^p(T,Y)$. □

Another useful result in this direction is included in the next proposition.

PROPOSITION 10.1.30 *If X, Y are Banach spaces, X is reflexive and is embedded continuously in Y, and $x \in L^\infty(T,X) \cap C(T,Y_w)$, then $x \in C(T,X_w)$ (here X_w, resp., Y_w, denotes the space X, resp., Y, endowed with the weak topology).*

PROOF: By replacing Y with $\overline{X}^{\|\cdot\|_Y}$ if necessary, we may assume that the embedding of X into Y is also dense. Then Y^* is embedded continuously and densely in X^* (see (10.3) and (10.4)).

If $t_n \longrightarrow t$ in T, then because by hypothesis $x \in C(T,Y_w)$, we have

$$\langle y^*, x(t_n)\rangle_Y \longrightarrow \langle y^*, x(t)\rangle_Y \quad \text{as } n \to \infty, \qquad \text{for all } y^* \in Y^*. \tag{10.16}$$

Here by $\langle \cdot,\cdot \rangle_Y$ we denote the duality brackets for the pair (Y^*, Y).

First we show that $x(t) \in X$ for all $t \in T$ and

$$\|x(t)\|_X \leq \|x\|_{L^\infty(T,X)} \qquad \text{for all } t \in T. \tag{10.17}$$

To see this, extend x by zero outside T, denote this extension by $\overline{x}$ and then regularize $\overline{x}$. So we can find a sequence $\{x_n\}_{n\geq 1} \subseteq C^1(T,X)$ such that

$$\begin{aligned} &\|x_n(t)\|_X \leq \|x\|_{L^\infty(T,X)} \qquad \text{for all } t \in T, \text{ all } n \geq 1 \\ \text{and} \quad &\langle y^*, x_n(t)\rangle_Y \longrightarrow \langle y^*, x(t)\rangle_Y \quad \text{as } n \to \infty, \qquad \text{for all } y^* \in Y. \end{aligned}$$

We have

$$\begin{aligned} |\langle y^*, x_n(t)\rangle_Y| &= |\langle y^*, x_n(t)\rangle_X| \\ &\leq \|y^*\|_{X^*} \|x\|_{L^\infty(T,X)}. \end{aligned} \tag{10.18}$$

Here by $\langle \cdot,\cdot \rangle_X$ we denote the duality brackets for the pair (X^*, X). Passing to the limit as $n \to \infty$ in (10.18), we obtain

$$\begin{aligned} |\langle y^*, x(t)\rangle_Y| &= |\langle y^*, x(t)\rangle_X| \\ &\leq \|y^*\|_{X^*}\|x\|_{L^\infty(T,X)} \qquad \text{for all } t\in T \text{ and all } y^* \in Y^*. \end{aligned} \tag{10.19}$$

Because Y^* is dense in X^*, from (10.19) it follows that

$$x(t) \in X \qquad \text{for all } t\in T \quad \text{and} \quad (10.17) \text{ holds.} \tag{10.20}$$

Let $x^* \in X^*$. Because Y^* is dense in X^*, we can find $\{y_m^*\}_{m\geq 1} \subseteq Y^*$, $y_m^* \longrightarrow x^*$ in X^* as $m \longrightarrow \infty$. We have

$$\langle y_m^*, x(t_n)\rangle_X \longrightarrow \langle y_m^*, x(t)\rangle_X \quad \text{as } n\to\infty \qquad \text{for all } m \geq 1 \quad (\text{see } (10.20)). \tag{10.21}$$

Also

$$\langle y_m^*, x(t)\rangle_X \longrightarrow \langle x^*, x(t)\rangle_X \quad \text{as } m \longrightarrow \infty. \tag{10.22}$$

Because of (10.21) and (10.22) we can find a sequence $n \longrightarrow m(n)$ not necessarily strictly increasing, such that $m(n) \longrightarrow \infty$ as $n\to\infty$ and

$$\big\langle y_{m(n)}^*, x(t_n)\big\rangle_X \longrightarrow \langle x^*, x(t)\rangle_X \quad \text{as } n\to\infty. \tag{10.23}$$

Then we have

$$\begin{aligned} & |\langle x^*, x(t_n)\rangle_X - \langle x^*, x(t)\rangle_X| \\ \leq\ & \big|\langle x^*, x(t_n)\rangle_X - \big\langle y_{m(n)}^*, x(t_n)\big\rangle_X\big| + \big|\big\langle y_{m(n)}^*, x(t_n)\big\rangle_X - \langle x^*, x(t)\rangle_X\big| \\ \Rightarrow\ & \langle x^*, x(t_n)\rangle_X \longrightarrow \langle x^*, x(t)\rangle_X \quad \text{as } n\to\infty \quad (\text{see } (10.23)) \end{aligned} \tag{10.24}$$

Because in (10.24) $x^* \in X^*$ was arbitrary, we conclude that $x \in C(T, X_w)$. □

10.2 Semilinear Evolution Equations

In this section we examine semilinear evolution equations using semigroups of linear operators (see Section 3.2).

So let X be a Banach space, $A : \mathcal{D}(A) \subseteq X \longrightarrow X$ be the infinitesimal generator of a C_0-semigroup $\{S(t)\}_{t\geq 0}$ and $f : [0,b] \times X \longrightarrow X$ be a function (the nonlinear perturbation). We consider the following semilinear Cauchy problem:

$$\left\{ \begin{array}{l} x'(t) = Ax(t) + f\big(t, x(t)\big),\ t\in T = [0,b], \\ x(0) = x_0 \in X \end{array} \right\}. \tag{10.25}$$

We introduce the following hypotheses on the nonlinearity $f(t,x)$.

H(f): $f : T\times X \longrightarrow X$ is a function such that

(i) For every $x\in X$, $t \longrightarrow f(t,x)$ is strongly measurable.

(ii) There exists $k \in L^1(T)_+$ such that for almost all $t \in T$ and all $x, u \in X$, we have

$$\begin{aligned} &\|f(t,x) - f(t,u)\| \leq k(t)\|x-u\| \\ \text{and} \quad &\|f(t,0)\| \leq k(t). \end{aligned}$$

DEFINITION 10.2.1 A function $x \in C(T,X)$ is said to be a *mild solution* of (10.25) if it is a solution of the following integral equation

$$x(t) = S(t)x_0 + \int_0^t S(t-\tau)f\big(\tau, x(\tau)\big)\mathrm{d}\tau, \quad t \in T = [0,b]. \tag{10.26}$$

REMARK 10.2.2 Hereafter we do not distinguish between the mild solutions of (10.25) and the solutions of (10.26).

PROPOSITION 10.2.3 *If A is the infinitesimal generator of a contraction semigroup $\{S(t)\}_{t\geq 0}$; that is,*

$$\|S(t)\|_{\mathcal{L}} \leq 1 \qquad \textit{for all } t \geq 0,$$

$f(t,x)$ satisfies hypotheses $H(f)$, and $x_0 \in X$, then problem (10.25) admits a unique mild solution $x(\cdot; x_0) \in C(T,X)$ satisfying

$$\|x(t;x_0)\| \leq \exp\Big(\int_0^t k(\tau)\mathrm{d}\tau\Big)\|x_0\| \tag{10.27}$$

$$\textit{and} \quad \|x(t;x_0) - x(t;\overline{x}_0)\| \leq \exp\Big(\int_0^t k(\tau)\mathrm{d}\tau\Big)\|x_0 - \overline{x}_0\|. \tag{10.28}$$

PROOF: On $C(T,X)$ we introduce the following equivalent norm

$$|u| = \max\Big\{\exp\Big(-L\int_0^t k(\tau)\mathrm{d}\tau\Big)\|x(t)\| : t \in T\Big\} \tag{10.29}$$

with $L > 1$. We consider the nonlinear operator $\xi : C(T,X) \longrightarrow C(T,X)$ defined by

$$\xi(x)(t) = S(t)x_0 + \int_0^t S(t-\tau)f\big(\tau, x(\tau)\big)\mathrm{d}\tau.$$

We show that ξ is a $|\cdot|$-contraction. To this end, for $x, u \in C(T,X)$, we have

$$\begin{aligned} &\|\xi(x)(t) - \xi(u)(t)\| \\ = &\ \Big\|\int_0^t S(t-\tau)\Big(f\big(\tau, x(\tau)\big) - f\big(\tau, u(\tau)\big)\Big)\mathrm{d}\tau\Big\| \\ \leq &\int_0^t k(\tau)\|x(\tau) - u(\tau)\|\mathrm{d}\tau. \end{aligned} \tag{10.30}$$

Multiplying (10.30) with $\exp\big(-L\int_0^t k(\tau)\mathrm{d}\tau\big)$, we obtain

$$\exp\big(-L\int_0^t k(\tau)\mathrm{d}\tau\big)\|\xi(x)(t)-\xi(u)(t)\|$$
$$\leq \int_0^t \exp\big(-L\int_\tau^t k(s)\mathrm{d}s\big)\exp\big(-L\int_0^\tau k(s)\mathrm{d}s\big)k(\tau)\|x(\tau)-u(\tau)\|\mathrm{d}\tau$$
$$\leq |x-u|\int_0^t \exp\big(-L\int_\tau^t k(s)\mathrm{d}s\big)k(\tau)\mathrm{d}\tau \quad \text{(see (10.29))}$$
$$\leq \frac{1}{L}|x-u|$$
$$\Rightarrow |\xi(x)-\xi(u)| \leq \frac{1}{L}|x-u|.$$

Because $L>1$, it follows that ξ is a $|\cdot|$-contraction and so by Banach's contraction principle, we deduce that there exists a unique $x(\cdot;x_0)\in C(T,X)$ such that

$$x(\cdot;x_0)=\xi\big(x(\cdot;x_0)\big).$$

Clearly this is the unique mild solution of (10.25). We have

$$\begin{aligned}\|x(t;x_0)\| &\leq \|x_0\|+\int_0^t \|f\big(\tau,x(\tau)\big)\|\mathrm{d}\tau\\ &\leq \|x_0\|+\int_0^t \big(\|f(\tau,0)\|+k(\tau)\|x(\tau)\|\big)\mathrm{d}\tau\\ &\leq \|x_0\|+\int_0^t k(\tau)\big(1+\|x(\tau)\|\big)\mathrm{d}\tau.\end{aligned}$$

Invoking Gronwall's inequality, we infer that

$$\|x(t;x_0)\|\leq \exp\big(\int_0^t k(\tau)\mathrm{d}\tau\big)(1+\|x_0\|).$$

Moreover, we have

$$\begin{aligned}\|x(t;x_0)-x(t;\overline{x}_0)\| &\leq \|S(t)(x_0-\overline{x}_0)\|+\|\int_0^t S(t-\tau)\big(x(\tau;x_0)-x(\tau;\overline{x}_0)\big)\mathrm{d}\tau\|\\ &\leq \|x_0-\overline{x}_0\|+\int_0^t k(\tau)\|x(t;x_0)-x(t;\overline{x}_0)\|\mathrm{d}\tau.\end{aligned}$$

Once again Gronwall's inequality implies that

$$\|x(t;x_0)-x(t;\overline{x}_0)\|\leq \exp\big(\int_0^t k(\tau)\mathrm{d}\tau\big)\|x_0-\overline{x}_0\|.$$

□

REMARK 10.2.4 Of course the result is still true if A is the infinitesimal generator of a C_0-semigroup $\{S(t)\}_{t\geq 0}$ satisfying

$$\|S(t)\|_{\mathcal{L}}\leq Me^{\omega t} \quad \text{for all } t\geq 0 \quad \text{with } M\geq 1,\ \omega>0.$$

In this case the estimates (10.27) and (10.28) have the following form,

$$\|x(t;x_0)\|\leq M\exp\big(\omega t+M\int_0^t k(\tau)\mathrm{d}\tau\big)(1+\|x_0\|)$$

$$\text{and}\quad \|x(t;x_0)-x(t;\overline{x}_0)\|\leq M\exp\big(\omega t+M\int_0^t k(\tau)\mathrm{d}\tau\big)\|x_0-\overline{x}_0\| \quad \text{for all } t\in T.$$

In Proposition 10.2.3 we assumed that $x_0 \in X$. For any $x_0 \in X$ that does not belong in $\mathcal{D}(A)$, in general $t \longrightarrow S(t)x_0$ is not necessarily differentiable and neither belongs in $\mathcal{D}(A)$ for $t > 0$. Therefore, if $x_0 \in \mathcal{D}(A)$ (smooth initial condition), we expect to deduce some regularity for the mild solution.

PROPOSITION 10.2.5 *If A is the infinitesimal generator of a C_0-semigroup and $f(t,x)$ satisfies hypotheses $H(f)$, with $k \in L^\infty(T)_+$ and $x_0 \in \mathcal{D}(A)$, then the unique mild solution $x(\cdot; x_0)$ is Lipschitz continuous.*

PROOF: Without any loss of generality we assume that $\{S(t)\}_{t\geq 0}$ is a contraction semigroup. For any $r > 0$, we set $\widehat{x}(t) = x(t+r)$. Evidently $\widehat{x}$ is a mild solution of (10.25) with initial condition $x(r)$. So from (10.28) it follows that

$$\|x(t+r) - x(t)\| \leq \exp\Big(\int_0^t k(\tau)\mathrm{d}\tau\Big)\|x(r) - x_0\|. \tag{10.31}$$

Also note that

$$\begin{aligned}\|x(r) - x_0\| &= \|S(r)x_0 - x_0 + \int_0^r S(t-\tau)f\big(\tau, x(\tau)\big)\mathrm{d}\tau\| \\ &\leq \|\int_0^r S(\tau)Ax_0\mathrm{d}\tau\| + \int_0^r \|f(\tau, x_0)\|\mathrm{d}\tau + \int_0^r k(\tau)\|x(\tau) - x_0\|\mathrm{d}\tau\end{aligned}$$

(because $x_0 \in \mathcal{D}(A)$)

$$\begin{aligned}&\leq \|Ax_0\|r + \int_0^r k(\tau)(1 + \|x_0\|)\mathrm{d}\tau + \int_0^r k(\tau)\|x(\tau) - x_0\|\mathrm{d}\tau \\ &\leq \big(\|Ax_0\| + (1 + \|x_0\|)\|k\|_\infty\big)r + \int_0^t k(\tau)\|x(\tau) - x_0\|\mathrm{d}\tau.\end{aligned} \tag{10.32}$$

From (10.32) and Gronwall's inequality, we obtain

$$\|x(r) - x_0\| \leq cr \quad \text{for some } c > 0 \text{ (independent of } r \in T). \tag{10.33}$$

Using (10.33) in (10.31), we conclude that $x(\cdot; x_0)$ is Lipschitz continuous. □

COROLLARY 10.2.6 *If X is a reflexive Banach space, A is the infinitesimal generator of a C_0-semigroup, f satisfies hypotheses $H(f)$ with $k \in L^\infty(T)_+$, and $x_0 \in \mathcal{D}(A)$, then the mild solution $x(\cdot; x_0)$ is a strong solution; that is, it satisfies (10.25) pointwise almost everywhere. Moreover, if f is t-independent, then the mild solution is a classical solution; that is, $x(\cdot; x_0) \in C(T, X) \cap C^1\big((0,b), X\big)$.*

Proposition 10.2.3 is still valid if $T = [0,b]$ is replaced by $\mathbb{R}_+$. So we have the following.

PROPOSITION 10.2.7 *If A is the infinitesimal generator of C_0-semigroup, $f(t,x)$ satisfies hypotheses $H(f)$, and $x_0 \in X$, then problem (10.25) admits a unique global mild solution $x(\cdot; x_0) \in C(\mathbb{R}_+, X)$ and*

$$\|x(t; x_0) - x(t; \overline{x}_0)\| \leq \exp\Big(\int_0^t k(\tau)\mathrm{d}\tau\Big)\|x_0 - \overline{x}_0\|.$$

PROOF: It is similar to the proof of Proposition 10.2.3, so we only sketch it here. Again we introduce the map $\xi : C(\mathbb{R}_+, X) \longrightarrow C(\mathbb{R}_+, X)$, defined by

$$\xi(x)(t) = S(t)x_0 + \int_0^t S(t-\tau)f\big(\tau, x(\tau)\big)\mathrm{d}\tau, \quad t \geq 0.$$

Let $V = \Big\{x \in C(\mathbb{R}_+, X) : \sup_{t\geq 0} \exp\big(-L\int_0^t k(\tau)\mathrm{d}\tau\big)\|x(t)\| < \infty\Big\}$, with $L > 1$. In V we introduce the norm

$$|x| = \sup_{t\geq 0} \exp\Big(-L\int_0^t k(\tau)\mathrm{d}\tau\Big)\|x(t)\|.$$

Clearly V equipped with this norm is a Banach space. Moreover, $\xi(V) \subseteq V$. Indeed, for $x \in V$, we have

$$\begin{aligned}
\|\xi(x)(t)\| &\leq \|S(t)x_0\| + \Big\|\int_0^t S(t-\tau)f\big(\tau, x(\tau)\big)\mathrm{d}\tau\Big\| \\
&\leq \|x_0\| + \int_0^t \big\|f\big(\tau, x(\tau)\big)\big\|\mathrm{d}\tau \\
&\leq \|x_0\| + \int_0^t k(\tau)(1 + \|x(\tau)\|)\mathrm{d}\tau
\end{aligned}$$

$$\Rightarrow \ \exp\Big(-L\int_0^t k(\tau)\mathrm{d}\tau\Big)\|\xi(x)(t)\| \leq c_1 + \int_0^t \exp\Big(-L\int_0^t k(\tau)\mathrm{d}\tau\Big)k(\tau)\|x(\tau\|\mathrm{d}\tau,$$
$$c_1 > 0,$$
$$\Rightarrow \ |\xi(x)| < \infty \ \text{ (i.e., } \xi(x) \in V\text{)}.$$

As in the proof of Proposition 10.2.3, we can check that ξ is a contraction. So by the Banach contraction principle it has a unique fixed point in V, which is a mild solution of (10.25). Moreover, the uniqueness also holds in $C(\mathbb{R}_+, X)$. Indeed if $x, u \in C(\mathbb{R}_+, X)$ are mild solutions of (10.25) and $y = x - u$, then

$$\begin{aligned}
y(t) &= \int_0^t S(t-\tau)\Big(f\big(\tau, x(\tau)\big) - f\big(\tau, u(\tau)\big)\Big)\mathrm{d}\tau, \\
\Rightarrow \ \|y(t)\| &\leq \int_0^t k(\tau)\|y(\tau)\|\mathrm{d}\tau, \ t \geq 0, \\
\Rightarrow \ y(t) &= 0 \qquad \text{for all } t \geq 0 \ \text{ (by Gronwall's inequality)}, \\
\Rightarrow \ x &= u.
\end{aligned}$$

The Lipschitz dependence on the initial condition follows via Gronwall's inequality, as in the proof of Proposition 10.2.3. □

As before, for simplicity we assume that A is the infinitesimal generator of a contraction semigroup. Then from the Hille–Yosida theorem (see Theorem 3.2.87), we know that for every $\lambda > 0$, the operator $I - \lambda A$ is one-to-one from $\mathcal{D}(A) \subseteq X$ onto X and so by Banach's theorem we know that

$$J_\lambda = (I - \lambda A)^{-1} \in \mathcal{L}(X). \tag{10.34}$$

This operator is called the *resolvent* of A. Using J_λ we also define

$$A_\lambda = \frac{1}{\lambda}(I - J_\lambda) \in \mathcal{L}(X) \tag{10.35}$$

which is the so-called Yosida approximation of A (see also Definition 3.2.42, where these notions were defined in the context of nonlinear monotone operators defined on a pivot Hilbert space). As in Proposition 3.2.44, we can show that the following are true for the operators J_λ and A_λ defined in (10.34) and (10.35), respectively.

PROPOSITION 10.2.8 (a) *For any $\lambda>0$ and $x\in X$, we have $A_\lambda(x)=A(J_\lambda x)$.*

(b) *For any $\lambda > 0$ and $x \in \mathcal{D}(A)$, we have $A_\lambda x = J_\lambda(Ax)$.*

(c) *For any $\lambda>0$ and $x \in \mathcal{D}(A)$, we have $\|A_\lambda x\| \leq \|Ax\|$.*

(d) *For any $x\in X$, we have $\lim\limits_{\lambda\to 0} J_\lambda x = x$.*

(e) *For any $x \in \mathcal{D}(A)$, we have $\lim\limits_{\lambda\to 0} A_\lambda x = Ax$.*

A is unbounded, therefore a mild solution need not be strongly differentiable. This may be an inconvenience in many applications. The following approximation result helps us to overcome such an inconvenience.

PROPOSITION 10.2.9 *If A is the infinitesimal generator of a contraction semigroup, $f(t,x)$ satisfies hypotheses $H(f)$, $x_0 \in X$, $x \in C(T,X)$ is a mild solution of problem (10.25), and $x_\lambda \in C(T,X)$ is the solution of*

$$x_\lambda(t) = S_\lambda(t)x_0 + \int_0^t S_\lambda(t-\tau)f\big(\tau, x_\lambda(\tau)\big)\mathrm{d}\tau, \quad t\in T,$$

where $S_\lambda(t) = e^{A_\lambda t} = \sum\limits_{k\geq 0} \frac{(A_\lambda t)^k}{k!}$ (the semigroup generated by $A_\lambda \in \mathcal{L}(X)$) then $\lim\limits_{\lambda\to 0} \max\limits_{t\in T} \|x_\lambda(t) - x(t)\| = 0$.

PROOF: As before using hypothesis II(f)(ii) and Gronwall's inequality, we obtain

$$\|x_\lambda(t) - x(t)\| \leq \int_0^b \sup_{r\in T} \|\big(S_\lambda(r) - S(r)\big)f\big(\tau, x(\tau)\big)\|\mathrm{d}\tau. \tag{10.36}$$

But we know that for any $x\in X$.

$$S_\lambda(t)x \longrightarrow S(t)x \quad \text{as } \lambda \downarrow 0 \quad \text{uniformly in } t\in T.$$

So from (10.36) and the dominated convergence theorem, we conclude that

$$\lim_{\lambda\to 0} \max_{t\in T} \|x_\lambda(t) - x(t)\| = 0.$$

□

In Section 3.2, we saw the Hille–Yosida characterization of the infinitesimal generator of a C_0-semigroup. In the next theorem we produce a different characterization for the infinitesimal generators of contraction semigroups. The result is known as the *Lumer–Phillips theorem* and its proof can be found in Yosida [614, p. 250].

THEOREM 10.2.10 *If A is a closed and densely defined linear operator in a Banach space X, then A is the infinitesimal generator of a contraction semigroup if and only if the operator $-A$ is m-accretive.*

Recall that if A is a closed operator in a Banach space X with domain $\mathcal{D}(A)$, then we can consider $\mathcal{D}(A)$ with the graph norm

$$\|x\|_{\mathcal{D}(A)} = \|x\| + \|Ax\|. \tag{10.37}$$

Due to the closedness of A, $(\mathcal{D}(A), \|\cdot\|_{\mathcal{D}(A)})$ is a Banach space that is embedded continuously in X.

We consider the following Cauchy problem:

$$\left\{ \begin{array}{l} x'(t) = Ax(t) + g(t),\ t \geq 0, \\ x(0) = x_0 \end{array} \right\}. \tag{10.38}$$

We assume that $-A$ is closed, densely defined, and m-accretive, hence it is the infinitesimal generator of a contraction semigroup $\{S(t)\}_{t\geq 0}$.

We have the following existence result for problem (10.38).

PROPOSITION 10.2.11 *If A is closed, densely defined, $-A$ is m-accretive, $g \in C^1(\mathbb{R}_+, X)$, and $x_0 \in \mathcal{D}(A)$, then problem (10.38) admits a unique classical solution x such that*

$$x \in C^1(\mathbb{R}_+, X) \cap C\big(\mathbb{R}_+, \mathcal{D}(A)\big)$$

which can be expressed as

$$x(t) = S(t)x_0 + \int_0^t S(t-\tau)g(\tau)\mathrm{d}\tau.$$

PROOF: We know that $t \longrightarrow S(t)x_0$ (the solution of the homogeneous equation) belongs in $C^1(\mathbb{R}_+, X) \cap C\big(\mathbb{R}_+, \mathcal{D}(A)\big)$. So we need to check that

$$u(t) = \int_0^t S(t-\tau)g(\tau)\mathrm{d}\tau$$

belongs in $C^1(\mathbb{R}_+, X) \cap C\big(\mathbb{R}_+, \mathcal{D}(A)\big)$ and satisfies (10.38). To this end we consider

$$\begin{aligned} \frac{u(t+r) - u(t)}{r} &= \frac{1}{r}\Big[\int_0^{t+r} S(t+r-\tau)g(\tau)\mathrm{d}\tau - \int_0^t S(t-\tau)g(\tau)\mathrm{d}\tau\Big] \\ &= \frac{1}{r}\Big[\int_t^{t+r} S(t+r-\tau)g(\tau)\mathrm{d}\tau + \int_0^t \big(S(t+r-\tau) - S(t-\tau)\big)g(\tau)\mathrm{d}\tau\Big] \end{aligned} \tag{10.39}$$

$$= \frac{1}{r}\int_t^{t+r} S(z)g(t+h-z)\mathrm{d}z + \frac{1}{r}\int_0^t S(z)\big(g(t+h-z) - g(t-z)\big)\mathrm{d}z. \tag{10.40}$$

We let $r \longrightarrow 0$ and in the right side of (10.40), we obtain

$$S(t)g(0) + \int_0^t S(z)g'(t-z)\mathrm{d}z \in C(\mathbb{R}_+, X).$$

So it follows that $u \in C^1(\mathbb{R}_+, X)$. Moreover, the right-hand side in (10.39) gives

$$S(0)g(t) - Au(t) = g(t) - Au(t).$$

Hence $u \in C(\mathbb{R}_+, \mathcal{D}(A))$ and we have proved the proposition. □

COROLLARY 10.2.12 *If A is closed, densely defined, $-A$ is m-accretive, $g \in C(\mathbb{R}_+, \mathcal{D}(A))$, and $x_0 \in \mathcal{D}(A)$, then problem (10.38) has a classical solution x given by*

$$x(t) = S(t)x_0 + \int_0^t S(t-\tau)g(\tau)\mathrm{d}\tau, \quad t \geq 0.$$

COROLLARY 10.2.13 *If A is closed, densely defined, $-A$ is m-accretive, $g \in C(\mathbb{R}_+, X)$, g is strongly differentiable with $g' \in L^1_{\mathrm{loc}}(\mathbb{R}_+, X)$, and $x_0 \in \mathcal{D}(A)$, then problem (10.38) has a classical solution x given by*

$$x(t) = S(t)x_0 + \int_0^t S(t-\tau)g(\tau)\mathrm{d}\tau, \quad t \geq 0.$$

PROOF: Let $b>0$, $T = [0, b]$. By hypothesis $g \in L^1(T, X)$. We set

$$u(t) = \int_0^t S(t-\tau)g(\tau)\mathrm{d}\tau, \quad t \in T.$$

We can easily check that $u \in C(T, X)$. Also from (10.40) we see that u is strongly differentiable almost everywhere on T and at points of strong differentiability we have

$$\begin{aligned} u'(t) &= S(t)g(0) + \int_0^t S(\tau)g'(t-\tau)\mathrm{d}\tau \\ &= S(t)g(0) + \int_0^t S(t-\tau)g'(\tau)\mathrm{d}\tau \in C(T, X). \end{aligned}$$

Therefore for almost all $t \in \mathbb{R}_+$, we have

$$\begin{aligned} & u'(t) = Au(t) + f(t), \\ \Rightarrow \quad & u \in C^1(T, X) \cap C(T, \mathcal{D}(A)). \end{aligned}$$

□

REMARK 10.2.14 The assumption $g \in C(\mathbb{R}_+, X)$ alone is not enough to guarantee the existence of a classical solution.

COROLLARY 10.2.15 *If X is a reflexive Banach space, A is closed, and densely defined, $-A$ is m-accretive, $g : \mathbb{R}_+ \longrightarrow X$ is Lipschitz continuous, and $x_0 \in \mathcal{D}(A)$, then problem (10.38) has a classical solution x given by*

$$x(t) = S(t)x_0 + \int_0^t S(t-\tau)g(\tau)\mathrm{d}\tau, \quad t \geq 0.$$

Now we use Proposition 10.2.11 to obtain an existence result for classical solutions for problem (10.25).

PROPOSITION 10.2.16 *If A is closed, densely defined, $-A$ is m-accretive, $f \in C^1(X,X)$, and there exists $k>0$ such that*

$$\|f(x)-f(u)\| \leq k\|x-u\| \qquad \text{for all } x,u \in X,$$

then for every $x_0 \in \mathcal{D}(A)$ problem (10.25) admits a unique global classical solution.

PROOF: By virtue of Proposition 10.2.7, problem (10.25) has a unique global mild solution $x \in C(\mathbb{R}_+,X)$. Because of the hypotheses on f and Proposition 10.2.11, it suffices to show that $x \in C^1(\mathbb{R}_+,X)$. To this end, we consider the following auxiliary problem,

$$\left\{ \begin{array}{l} u'(t) = Au(t) + f'\big(x(t)\big)u(t),\ t \geq 0, \\ u(0) = f(x_0) + Ax_0 \in X \end{array} \right\}. \tag{10.41}$$

Problem (10.41) has a unique global mild solution $u \in C(\mathbb{R}_+,X)$ given by

$$u(t) = S(t)\big(f(x_0) - Ax_0)\big) + \int_0^t S(t-\tau)f'\big(x(\tau)\big)u(\tau)\mathrm{d}\tau, \quad t \geq 0.$$

We have

$$\begin{aligned}
&\Big\|\frac{x(t+r)-x(t)}{r} - u(t)\Big\| \\
=\ & \Big\|\frac{1}{r}\Big(S(t)\big(S(r)-I\big)x_0 + \int_0^{t+r} S(t+r-\tau)f\big(x(\tau)\big)\mathrm{d}\tau \\
&- \int_0^t S(t-\tau)f\big(x(\tau)\big)\mathrm{d}\tau - S(t)\big(f(x_0)+Ax_0\big) - \int_0^t S(t-\tau)f'\big(x(\tau)\big)u(\tau)\mathrm{d}\tau\Big)\Big\| \\
\leq\ & \Big\|\frac{S(r)x_0-x_0}{r} - Ax_0\Big\| + \int_0^t \Big\|\frac{f\big(x(\tau+r)\big)-f\big(x(\tau)\big)}{r} - f'\big(x(\tau)\big)u(\tau)\Big\|\mathrm{d}\tau \\
&+ \Big\|\frac{1}{r}\int_0^r S(t+r-\tau)f\big(x(\tau)\big)\mathrm{d}\tau - S(t)f(x_0)\Big\| \\
=\ & \xi_1 + \xi_2 + \xi_3.
\end{aligned} \tag{10.42}$$

Because by hypothesis $x_0 \in \mathcal{D}(A)$, we have

$$\xi_1 \longrightarrow 0 \quad \text{as } r \downarrow 0. \tag{10.43}$$

Also due to the continuity of x and f, we have

$$\xi_3 \longrightarrow 0 \quad \text{as } r \downarrow 0. \tag{10.44}$$

Next note that because f is strongly differentiable, we have

$$\begin{aligned}
&\Big\|\frac{f\big(x(\tau+r)\big)-f\big(x(\tau)\big)}{r} - f'\big(x(\tau)\big)\Big(\frac{x(\tau+r)-x(\tau)}{r}\Big)\Big\| \\
\leq\ & \Big\|\frac{x(\tau+r)-x(\tau)}{r}\Big\|\vartheta\big(\|x(\tau+r)-x(\tau)\|\big) \\
&\text{(with } \vartheta(s) \longrightarrow 0 \quad \text{as } s \longrightarrow 0).
\end{aligned} \tag{10.45}$$

So we have

$$\begin{aligned}
&\Big\|\frac{f\big(x(\tau+r)\big)-f\big(x(\tau)\big)}{r}-f'\big(x(\tau)\big)u(\tau)\Big\| \\
\le\ &\Big\|\frac{f\big(x(\tau+r)\big)-f\big(x(\tau)\big)}{r}-f'\big(x(\tau)\big)\Big(\frac{x(\tau+r)-x(\tau)}{r}\Big)\Big\| \\
&+\Big\|f'\big(x(\tau)\big)\Big(\frac{x(\tau+r)-x(\tau)}{r}-u(\tau)\Big)\Big\| \\
\le\ &\Big\|\frac{x(\tau+r)-x(\tau)}{r}\Big\|\vartheta\big(\|x(\tau+r)-x(\tau)\|\big)+k\Big\|\frac{x(\tau+r)-x(\tau)}{r}-u(t)\Big\|
\end{aligned} \tag{10.46}$$

using (10.46) and recalling that $\|f'\big(x(\tau)\big)\|_{\mathcal{L}}\le k$ for all $\tau\ge 0$.

Because of Proposition 10.2.5, we see that

$$\Big\|\frac{x(\tau+r)-x(\tau)}{r}\Big\|\vartheta\big(\|x(\tau+r)-x(\tau)\|\big)\longrightarrow 0 \quad \text{as } r\longrightarrow 0. \tag{10.47}$$

Using (10.43), (10.44), (10.46), and (10.47), we see that given $\varepsilon>0$, we can find $r>0$ small such that

$$\Big\|\frac{x(t+r)-x(t)}{r}-u(t)\Big\|\le\varepsilon+k\int_0^t\Big\|\frac{x(\tau+r)-x(\tau)}{r}-u(\tau)\Big\|d\tau.$$

Invoking Gronwall's inequality and because $\varepsilon>0$ was arbitrary, we conclude that

$$\begin{aligned}
&\lim_{r\to 0}\Big\|\frac{x(t+r)-x(t)}{r}-u(t)\Big\|=0, \\
\Rightarrow\ & x'(t)=u(t) \qquad \text{for all } t\ge 0, \\
\Rightarrow\ & x\in C^1(\mathbb{R}_+,X).
\end{aligned}$$

$\square$

Thus far all the existence results that we proved required that the nonlinearity $f(x)$ be globally Lipschitz. This is a rather restrictive hypothesis. In particular it implies that

$$\|f(x)\|\le\|f(0)\|+k\|x\|.$$

This means that f has at most linear growth and so we rule out several interesting nonlinearities (such as quadratic). For this reason we would like to be able to relax this global Lipschitzness hypothesis. In this direction we have the following existence result.

PROPOSITION 10.2.17 *If A is a closed, densely defined operator, $-A$ is m-accretive, $f: X\longrightarrow X$ is Lipschitz continuous on bounded sets, and $x_0\in X$, then we can find $b>0$ depending on $\|x_0\|$ such that problem (10.25) admits a unique local solution $x\in C(T,X)$. Moreover, if $x_0\in\mathcal{D}(A)$, then x is Lipschitz continuous on T. Finally if X is reflexive and $x_0\in\mathcal{D}(A)$, then the mild solution x is in fact classical.*

PROOF: Let $r=1+\|x_0\|$ and let $k_r>0$ be the Lipschitz constant of f on $\overline{B}_r=\{x\in X:\|x\|\le r\}$. Let $0<b<1/k_r$ and define

$$C_b=\big\{x\in C(T,X):\|x(t)\|\le r,\ t\in T=[0,b]\big\}.$$

Clearly C_b is a closed and convex subset of $C(T,X)$. We consider the nonlinear map $\xi: C_b \longrightarrow C(T,X)$ defined by

$$\xi(x)(t) = S(t)x_0 + \int_0^t S(t-\tau) f\big(x(\tau)\big) \mathrm{d}\tau \qquad \text{for all } t \in T.$$

We have

$$\|\xi(x)(t)\| \leq \|x_0\| + b\big(\|f(0)\| + k_r r\big) \qquad \text{for all } t \in T.$$

If we choose $b > 0$ such that

$$b < \min\left\{\frac{1}{k_r}, \frac{1}{\|f(0)\| + k_r r}\right\}, \tag{10.48}$$

then we deduce that

$$\|\xi(x)(t)\| \leq \|x_0\| + 1 = r.$$

Therefore, $\xi: C_b \longrightarrow C_b$. Moreover, for every $x, u \in C_b$, we have

$$\begin{aligned} \|\xi(x) - \xi(u)\|_{C(T,X)} &= \sup_{t \in T} \left\| \int_0^t S(t-\tau)\Big(f\big(x(\tau)\big) - f\big(u(\tau)\big)\Big) \mathrm{d}\tau \right\| \\ &\leq b\, k_r \|x - u\|_{C(T,X)}. \end{aligned}$$

Because $bk_r < 1$ (see (10.48)), we see that ξ is a contraction on C_b. So by Banach's fixed point theorem, we can find a unique fixed point of ξ. It is easily seen that this is the unique mild solution of (10.25).

If $x_0 \in \mathcal{D}(A)$, Proposition 10.2.5 implies that x is Lipschitz continuous.

Finally when X is reflexive and $x_0 \in \mathcal{D}(A)$, Corollary 10.2.6 implies that the unique mild solution is in fact a classical one. □

As is the case with ordinary differential equations in $\mathbb{R}^N$, we can always produce a maximal solution; that is, a solution defined on a maximal time interval.

PROPOSITION 10.2.18 *If A is a closed, densely defined operator, $-A$ is m-accretive, $f : X \longrightarrow X$ is Lipschitz continuous on bounded sets, and $x_0 \in X$, then problem (10.25) has a unique mild solution on a maximal time interval $[0, b_{\max})$ such that either*

(a) $b_{\max} = +\infty$; *that is, the unique mild solution is global, or*

(b) $b_{\max} < +\infty$ *and* $\lim\limits_{t \to b_{\max}^-} \|x(t)\| = +\infty$ *(i.e., we have a blow-up of the solution in finite time).*

PROOF: Suppose x_1, x_2 are two mild solutions defined on $T_1 = [0, b_1]$ and $T_2 = [0, b_2]$, respectively, and assume that $b_1 < b_2$. Then $T_1 \subseteq T_2$ and due to the uniqueness of the mild solution on T_1 we have

$$x_2\big|_{T_1} = x_1.$$

Invoking Zorn's lemma, we can find a maximal interval $T_{\max} = [0, b_{\max})$ on which the unique mild solution exists.

We need to show that if $b_{\max} < +\infty$, then

$$\lim_{t \to b_{\max}^-} \|x(t)\| = +\infty. \tag{10.49}$$

We proceed indirectly. So suppose that (10.49) is not true. We show that we can continue x beyond $b_{\max}$, a contradiction to the maximality of the interval $T_{\max} = [0, b_{\max})$.

So suppose we can find $\{t_n\}_{n \geq 1} \subseteq T_{\max}$ such that

$$t_n \longrightarrow b_{\max}^- \quad \text{as } n \to \infty \quad \text{and} \quad \|x(t_n)\| \leq c \quad \text{for some } c > 0, \text{ all } n \geq 1.$$

We consider the following Cauchy problem

$$\left\{ \begin{array}{l} u'(t) = Au(t) + f\big(u(t)\big), \ t \geq 0, \\ u(0) = x(t_n), \ n \geq 1 \end{array} \right\}. \tag{10.50}$$

By virtue of Proposition 10.2.17, problem (10.50) has a unique mild solution on some interval $[0, b]$, with $b > 0$ depending on $c > 0$. Let $n \geq 1$ be large so that $b_{\max} < b + t_n$. We set

$$y(t) = \begin{cases} x(t) & \text{if } t \in [0, t_n] \\ u(t - t_n) & \text{if } t \in [t_n, t_n + b] \end{cases}. \tag{10.51}$$

We show that y is a mild solution of (10.25) on the interval $[0, b + t_n]$. To this end note that

$$y(t) = S(t)x_0 + \int_0^t S(t-\tau)f\big(y(\tau)\big)d\tau \quad \text{for } t \in [0, t_n] \ \text{(see (10.51))} \tag{10.52}$$

$$\text{and} \quad u(t) = S(t)x(t_n) + \int_0^t S(t-\tau)f\big(u(\tau)\big)d\tau \quad \text{for } t \in [0, b]. \tag{10.53}$$

From (10.52) we see that $y \in C([0, b+t_n], X)$ is a mild solution of (10.25) on the time interval $[0, t_n]$. Now, if $t \in [0, b]$, we have

$$\begin{aligned} y(t+t_n) = u(t) &= S(t)x(t_n) + \int_0^t S(t-\tau)f\big(u(\tau)\big)d\tau \quad \text{(see (10.51) and (10.53))} \\ &= S(t)\Big(S(t_n)x_0 + \int_0^{t_n} S(t-\tau)f\big(x(\tau)\big)d\tau\Big) \\ &\quad + \int_0^t S(t-\tau)f\big(u(\tau)\big)d\tau \\ &= S(t+t_n)x_0 + \int_0^{t_n} S(t+t_n-\tau)f\big(x(\tau)\big)d\tau \\ &\quad + \int_{t_n}^{t_n+t} S(t+t_n-\tau)f\big(u(\tau-t_n)\big)d\tau \\ &= S(t+t_n)x_0 + \int_0^{t+t_n} S(t+t_n-\tau)f\big(y(\tau)\big)d\tau \quad \text{(see (10.51))}. \end{aligned} \tag{10.54}$$

From (10.54) we conclude that y is a mild solution of (10.25) on $[0, b+t_n]$ which strictly contains $T_{\max}$, a contradiction to the maximality of $T_{\max}$. $\square$

Using Propositions 10.2.17 and 10.2.18, we can have an existence and uniqueness result for a global mild solution, under conditions that are slightly weaker than the global Lipschitz condition considered in the first part of this section.

COROLLARY 10.2.19 *If A is a closed, densely defined operator, $-A$ is m-accretive, and $f : X \longrightarrow X$ is Lipschitz continuous on bounded sets, there exist $c_1, c_2 > 0$ such that*

$$\|f(x)\| \leq c_1 + c_2\|x\| \qquad \text{for all } x \in X \tag{10.55}$$

and $x_0 \in X$, then problem (10.25) admits a unique maximal mild solution $x \in C(\mathbb{R}_+, X)$.

PROOF: From Proposition 10.2.18 we know that problem (10.25) admits a unique maximal mild solution $x \in C(T_{\max}, X)$, $T_{\max} = [0, b_{\max})$. We have

$$\begin{aligned} x(t) &= S(t)x_0 + \int_0^t S(t-\tau)f\big(x(\tau)\big)\mathrm{d}\tau \quad \text{for } t \in T_{\max} = [0, b_{\max}), \\ \Rightarrow \ \|x(t)\| &\leq \|x_0\| + \int_0^t \|f\big(x(\tau)\big)\|\mathrm{d}\tau \\ &\leq \|x_0\| + \int_0^t \big(c_1 + c_2\|x(\tau)\|\big)\mathrm{d}\tau \quad \text{(see (10.55))}. \end{aligned}$$

If $b_{\max} < +\infty$, via Gronwall's inequality, we obtain

$$\|x(t)\| \leq M \quad \text{for some } M > 0 \quad \text{and all } t \in T_{\max}. \tag{10.56}$$

But from Proposition 10.2.18, we know that we must have

$$\lim_{t \to b_{\max}^-} \|x(t)\| = +\infty,$$

a contradiction to (10.56). ∎

COROLLARY 10.2.20 *If $X = H =$ Hilbert space, $-A$ is maximal monotone, $f : H \longrightarrow H$ is Lipschitz continuous on bounded sets, and there exist $c_1, c_2 > 0$ such that*

$$(f(x), x)_H \leq c_1 + c_2\|x\|^2 \qquad \text{for all } x \in H, \tag{10.57}$$

then for every $x_0 \in X$ problem (10.25) admits a unique global mild solution $x \in C(\mathbb{R}_+, X)$. Moreover, if $x_0 \in \mathcal{D}(A)$, then this unique global mild solution is in fact a classical one.

PROOF: First suppose that $x_0 \in \mathcal{D}(A)$. Then from Propositions 10.2.17 and 10.2.18, problem (10.25) admits a maximal classical solution $x \in C(T_{\max}, H)$ with $T_{\max} = [0, b_{\max})$. We have

$$\begin{aligned}
& x'(t) = Ax(t) + f\big(x(t)\big) \qquad \text{for all } t \in T_{\max}, \\
\Rightarrow\ & \big(x'(t), x(t)\big)_H = \big(Ax(t), x(t)\big)_H + \Big(f\big(x(t)\big), x(t)\Big)_H \\
\Rightarrow\ & \frac{1}{2}\frac{\mathrm{d}}{\mathrm{d}t}\|x(t)\|^2 \leq c_1 + c_2\|x(t)\|^2
\end{aligned}$$

(because $-A$ is monotone and using (10.57))

$$\Rightarrow \|x(t)\|^2 \leq c_3 + c_2\int_0^t \|x(\tau)\|^2 \mathrm{d}\tau \qquad \text{for all } t \in T_{\max}. \tag{10.58}$$

As in the proof of Corollary 10.2.19, (10.58) implies that $b_{\max} = +\infty$.

Now suppose that $x_0 \in X$.Then we can find $\{x_0^n\}_{n\geq 1} \subseteq \mathcal{D}(A)$ such that

$$x_0^n \longrightarrow x_0 \quad \text{in } X \quad \text{as } n \to \infty.$$

Let $x_n \in C(\mathbb{R}_+, X)$ be the global classical solution emanating from x_0^n (see the first part of the proof). As above, using (10.57), we show that for every $b>0$, we have

$$\|x_n(t)\| \leq M \qquad \text{for all } n \geq 1, \text{ all } t \in T = [0, b],$$

with $M>0$ independent of $n\geq 1$.

Also if $n, m \geq 1$ are fixed and we set

$$u(t) = x_n(t) - x_m(t), \quad t \in T,$$

then we have

$$\left\{\begin{array}{l} u'(t) = Au(t) + f\big(x_n(t)\big) - f\big(x_m(t)\big),\ t \geq 0, \\ u(0) = x_0^n - x_0^m,\ n \geq 1 \end{array}\right\}. \tag{10.59}$$

From (10.59) as above, using (10.57) and the monotonicity of $-A$, we obtain

$$\frac{1}{2}\frac{\mathrm{d}}{\mathrm{d}t}\|u(t)\|^2 \leq k_M\|u(t)\|^2 \tag{10.60}$$

with $k_M>0$ being the Lipschitz constant of f on $\overline{B}_M$. From (10.60), via Gronwall's inequality, we obtain

$$\|u(t)\|^2 = \|x_n(t) - x_m(t)\|^2 \leq \widehat{c}\|x_0^n - x_0^m\|^2,$$

for some $\widehat{c}>0$ independent of $n, m \geq 1$. Therefore $\{x_n\}_{n\geq 1} \subseteq C(T, X)$ is Cauchy and so

$$x_n \longrightarrow x \quad \text{in } C(T, X).$$

Evidently x is mild solution of (10.25) on $T = [0, b]$. Because $b>0$ is arbitrary, it follows that x is a global mild solution. As before we can easily check that x is unique. □

We conclude this section by applying the abstract results to some concrete parabolic problems.

So let $Z \subseteq \mathbb{R}^N$ be a bounded domain with a C^∞-boundary. First we consider the following initial-boundary value problem:

$$\left\{ \begin{array}{ll} \frac{\partial x}{\partial t} + \triangle x(t,z) = g(t,z) & \text{on } \mathbb{R}_+ \times Z, \\ \\ x\big|_{\mathbb{R}_+ \times Z} = 0, \ x(0,z) = x_0(z), \ z \in Z \end{array} \right\}. \tag{10.61}$$

We make the following hypotheses on g.

H(g): $g: T \times Z \longrightarrow \mathbb{R}$ is a function such that $g(0,\cdot) \in L^2(Z)$ and

(i) For all $t \in T$, $z \longrightarrow g(t,z)$ is measurable.

(ii) For almost all $z \in Z$ and all $t, t' \in \mathbb{R}_+$ we have

$$|g(t,z) - g(t',z)| \leq k(z)|t - t'| \quad \text{with } k \in L^2(Z).$$

Then we have the following result concerning problem (10.61).

PROPOSITION 10.2.21 *If $g(t,z)$ satisfies hypotheses H(g) and $x_0 \in H_0^1(Z) \cap H^2(Z)$, then problem (10.61) admits a unique global classical solution*

$$x \in C\big(\mathbb{R}_+, L^2(Z)\big) \cap C^1\big((0,\infty), H_0^1(Z) \cap H^2(Z)\big).$$

PROOF: Let $X = H = L^2(Z)$ and $A: \mathcal{D}(A) \subseteq H \longrightarrow H$ defined by

$$Ax = -\triangle x$$

with $x \in \mathcal{D}(A) = H_0^1(Z) \cap H^2(Z)$. Then we can rewrite (10.61) as the following equivalent semilinear evolution equation

$$\left\{ \begin{array}{l} x'(t) = Ax(t) + \widehat{g}(t) \\ x(0) = x_0 \end{array} \right\}, \tag{10.62}$$

where $\widehat{g}(t) = g(t,\cdot) \in H$ for all $t \geq 0$.

We claim that $-A$ is maximal monotone. To this end let $h \in H = L^2(Z)$ and consider the following stationary problem

$$\left\{ \begin{array}{ll} u(z) - \triangle u(z) = h(z) & \text{a.e. on } Z, \\ u\big|_{\partial Z} = 0 \end{array} \right\}. \tag{10.63}$$

Let $V \in \mathcal{L}\big(H_0^1(Z), H^{-1}(Z)\big)$ be defined by

$$\big(V(x), y\big)_H = \int_Z (Dx, Dy)_{\mathbb{R}^N} dz.$$

Evidently V is monotone (hence maximal monotone) and coercive. Therefore it is surjective (see Corollary 3.2.28) and we can find $x \in H_0^1(Z)$ such that

$$V(x) = h \in H.$$

Then x is a solution of (10.63) and from the linear regularity theory, we have

$$\begin{aligned} & x \in H^2(Z), \\ \Rightarrow \ & x \in \mathcal{D}(A), \\ \Rightarrow \ & \mathcal{R}(I - A) = L^2(Z) = H. \end{aligned}$$

Because $-A$ is clearly monotone, it follows that $-A$ is maximal monotone (see Theorem 3.2.30). Note that by virtue of hypothesis H(g)(ii), $\widehat{g} : \mathbb{R}_+ \longrightarrow H$ is Lipschitz continuous. So we can apply Corollary 10.2.15 and obtain a classical solution x such that

$$x \in C\big(\mathbb{R}_+, L^2(Z)\big) \cap C^1\big((0,\infty), H_0^1(Z) \cap H^2(Z)\big).$$

□

Next we consider the semilinear heat equation:

$$\left\{\begin{array}{ll} \frac{\partial x}{\partial t} + \triangle x(t,z) = f_0\big(x(t,z)\big) & \text{on } \mathbb{R}_+ \times Z, \\ x\big|_{\mathbb{R}_+ \times Z} = 0, \; x(0,z) = x_0(z), \; z \in Z & \end{array}\right\}. \tag{10.64}$$

PROPOSITION 10.2.22 *If $f_0 \in C^1(\mathbb{R})$ and f_0' is uniformly bounded, then for every $x_0 \in L^2(Z)$, problem (10.64) admits a unique global mild solution*

$$x \in C\big(\mathbb{R}, L^2(Z)\big).$$

Moreover, if $x_0 \in H_0^1(Z) \cap H^2(Z)$, then the above solution is classical and

$$x \in C^1\big(\mathbb{R}_+, L^2(Z)\big) \cap C\big(\mathbb{R}_+, H_0^1(Z) \cap H^2(Z)\big).$$

PROOF: As before, let $X = H = L^2(Z)$, $A : \mathcal{D}(A) \subseteq H \longrightarrow H$ is defined by

$$Ax = -\triangle x$$

for all $x \in \mathcal{D}(A) = H_0^1(Z) \cap H^2(Z)$, and $f : L^2(Z) \longrightarrow L^2(Z)$ is defined by

$$f(x)(\cdot) = f_0\big(x(\cdot)\big).$$

Evidently $f \in C^1\big(L^2(Z), L^2(Z)\big)$ and also it is Lipschitz continuous. Then problem (10.64) can be equivalently rewritten as the following abstract evolution equation

$$\left\{\begin{array}{l} x'(t) = Ax(t) + f\big(x(t)\big) \\ x(0) = x_0 \end{array}\right\}. \tag{10.65}$$

Invoking Proposition 10.2.16 and Corollary 10.2.15, we obtain the conclusions of the proposition. □

PROPOSITION 10.2.23 *If $N \leq 3$, $f_0(x) = -x^3 + x$, and $x_0 \in H_0^1(Z) \cap H^2(Z)$, then problem (10.64) admits a unique classical solution*

$$x \in C^1\big(\mathbb{R}_+, L^2(Z)\big) \cap C\big(\mathbb{R}_+, H_0^1(Z) \cap H^2(Z)\big).$$

PROOF: Again $X = H = L^2(Z)$, $A : \mathcal{D}(A) \subseteq H \longrightarrow H$ is defined by $Ax = -\triangle x$ for all $x \in \mathcal{D}(A) = H_0^1(Z) \cap H^2(Z)$ and $f : L^2(Z) \longrightarrow L^2(Z)$ is defined by $f(x)(\cdot) = f_0\big(x(\cdot)\big)$. Then problem (10.64) is equivalent to the semilinear evolution equation (10.65).We claim that $f : \mathcal{D}(A) \longrightarrow \mathcal{D}(A)$. Indeed, because $N \leq 3$, by the Sobolev embedding theorem, we have that $H_0^1(Z)$ is embedded continuously in $L^6(Z)$ and $H^2(Z)$ is embedded continuously into $C(\overline{Z})$. Because $f_0 \in C^\infty(\mathbb{R})$, we see that for all $u \in \mathcal{D}(A)$, $f(u) \in H_0^1(Z)$ and from the chain rule we have

$$D f_0\big(u(z)\big) = f_0'\big(u(z)\big) Du(z)$$

and $$D^2 f_0\big(u(z)\big) = f_0'\big(u(z)\big) D^2 u(z) + f_0''\big(u(z)\big)\big(Du(z), Du(z)\big)_{\mathbb{R}^N}.$$

Therefore we see that $f(u) \in H^2(Z)$ and so we have $f : \mathcal{D}(A) \longrightarrow \mathcal{D}(A)$.

Now let $H_1 = \mathcal{D}(A)$ (with the graph norm) and $A_1 : \mathcal{D}(A_1) \subseteq H_1 \longrightarrow H_1$ is defined by

$$A_1 x = Ax$$

with $x \in \mathcal{D}(A_1) = \mathcal{D}(A^2)$. Then applying Proposition 10.2.18 on H_1 with operator A_1, we obtain a unique maximal classical solution

$$x \in C^1\big(T_{\max}, L^2(Z)\big) \cap C\big(T_{\max}, H_0^1(Z) \cap H^2(Z)\big)$$

with $T_{\max} = [0, b_{\max})$. We show that $b_{\max} = +\infty$. To this end, it suffices to show that $\|x(t)\|_{H^2(Z)}$ is bounded uniformly in $t \geq 0$. So we multiply (10.64) with x_t and then integrate over Z. We obtain

$$\frac{\mathrm{d}}{\mathrm{d}t} \int_Z \Big(\frac{1}{2}\|Dx\|^2 + \frac{1}{4}x^4 - \frac{1}{2}x^2\Big)\mathrm{d}z + \Big\|\frac{\partial x}{\partial t}\Big\|_{L^2(Z)}^2 = 0. \tag{10.66}$$

Note that

$$x(z)^2 \leq \frac{1}{4}x(z)^4 + 1.$$

Using this in (10.66) we obtain

$$\frac{\mathrm{d}}{\mathrm{d}t} \int_Z \frac{1}{2}(\|Dx\|^2 + \|x\|^2)\mathrm{d}z + \Big\|\frac{\partial x}{\partial t}\Big\|_{L^2(Z)}^2 = 0,$$

$$\Rightarrow \|x\|_{H_0^1(Z)} \leq M \quad \text{and} \quad \int_0^t \Big\|\frac{\partial x}{\partial t}\Big\|_{L^2(Z)}\mathrm{d}t \leq M \quad \text{for some } M > 0.$$

For $r > 0$, let

$$u(t) = x(t + r).$$

Evidently we have

$$u \in C^1\big([0, b_{\max} - r), L^2(Z)\big) \cap C\big([0, b_{\max} - r), H_0^1(Z) \cap H^2(Z)\big)$$

and $$\left\{\begin{array}{l} \frac{\partial u}{\partial t} + \Delta u(t,z) = -u(t,z)^3 + u(t,z) \\ u(t,\cdot)\big|_{\partial Z} = 0,\ u(0,z) = x(r,z) \end{array}\right\}. \tag{10.67}$$

We set $y(t,z) = u(t,z) - x(t,z) = x(t+r,z) - x(t,z)$. From (10.64) and (10.67), we have

$$\left\{\begin{array}{l} \frac{\partial y}{\partial t} + \Delta y(t,z) + y(t,z)\big(u(t,z)^2 + u(t,z)x(t,z) + x(t,z)^2\big) - y(t,z) = 0 \\ y(t,\cdot)\big|_{\partial Z} = 0,\ y(0,z) = x(r,z) - x_0(z) \end{array}\right\}. \tag{10.68}$$

We multiply (10.68) with $y(t,z)$ and then integrate over Z. We obtain

$$\frac{1}{2}\frac{\mathrm{d}}{\mathrm{d}t}\|y(t,\cdot)\|_{L^2(Z)}^2 + \|Dy(t,\cdot)\|_{L^2(Z)}^2 + \int_Z y^2(x^2 + xu + u^2)\mathrm{d}z = \|y(t,\cdot)\|_{L^2(Z)}^2. \tag{10.69}$$

Note that

$$x^2 + xu + u^2 \geq 0,$$
$$\Rightarrow \int_Z y^2(x^2 + xu + u^2)\mathrm{d}z \geq 0. \tag{10.70}$$

Using (10.70) in (10.69), we have

$$\frac{1}{2}\frac{\mathrm{d}}{\mathrm{d}t}\|y(t,\cdot)\|^2_{L^2(Z)} \leq \|y(t,\cdot)\|^2_{L^2(Z)}$$
$$\Rightarrow \|y(t,\cdot)\|^2_{L^2(Z)} = \|x(t+r,\cdot) - x(t,\cdot)\|^2_{L^2(Z)} \leq M_1\|x(r,\cdot) - x_0\|^2_{L^2(Z)} \tag{10.71}$$

for some $M_1 > 0$ (by Gronwall's inequality). The constant $M_1 > 0$ depends only on $b_{\max}$. From (10.71) it follows that

$$\|\frac{\partial x}{\partial t}(t,\cdot)\|^2_{L^2(Z)} \leq M_1\| - \Delta u_0 - u_0^3 + u_0\|^2_{L^2(Z)} < +\infty \tag{10.72}$$

(recall $u_0 \in H^1_0(Z) \cap H^2(Z)$). We have

$$\begin{aligned}\|u(t,\cdot)\|_{H^1_0(Z)} &\leq M\|\Delta u(t,\cdot)\|_{L^2(Z)} \\ &\leq M_1\Big(\|\frac{\partial x}{\partial t}(t,\cdot)\|^2_{L^2(Z)} + \|x(t,\cdot)^3\|_{L^2(Z)} + \|x(t,\cdot)\|_{L^2(Z)}\Big) \\ &\leq M_2 \quad \text{(see (10.72))}\end{aligned} \tag{10.73}$$

with $M_2 > 0$ depending only on $b_{\max} > 0$. So, if $b_{\max} < +\infty$, from (10.73) it follows that

$$\lim_{t \to b_{\max}^-} \|x(t,\cdot)\|_{L^2(Z)} < +\infty,$$

a contradiction. Therefore $b_{\max} = +\infty$. $\square$

REMARK 10.2.24 A careful reading of the above proof reveals that the result is more generally true if $N \leq 3$ and $f_0 \in C^3(\mathbb{R})$ with $f_0(0) = 0$.

10.3 Nonlinear Evolution Equations

In this section we deal with nonlinear evolution equations. We focus on two broad classes of nonlinear evolution equations that arise often in applications. The first class concerns dynamic problems driven by subdifferential operators. It contains gradient flows and certain variational inequalities and it is related to the nonlinear semigroup theory briefly mentioned in Section 3.2. The second class considers evolutions monitored by general nonlinear operators of monotone type that are defined within the framework of an evolution triple of spaces (see Section 10.1).

We start with the evolution inclusions of the subdifferential type. To prove the main existence and regularity results for such systems, we need some preparatory work.We start with a Gronwall-type lemma, which is a basic tool in deriving a priori estimates.

LEMMA 10.3.1 *If $h \in L^1(T)_+$, $T=[0,b]$, $c \geq 0$ and $x \in C(T)$ satisfies*

$$\frac{1}{2}x^2(t) \leq \frac{1}{2}c^2 + \int_0^t h(\tau)x(\tau)\mathrm{d}\tau \qquad \text{for all } t \in T,$$

then $|x(t)| \leq c + \int_0^t h(\tau)\mathrm{d}\tau$ for all $t \in T$.

PROOF: For any $\varepsilon \geq 0$, let

$$\vartheta_\varepsilon(t) = \frac{1}{2}(c+\varepsilon)^2 + \int_0^t h(\tau)x(\tau)\mathrm{d}\tau.$$

We have

$$\vartheta_\varepsilon'(t) = h(t)x(t) \quad \text{for a.a. } t \in T \tag{10.74}$$

$$\text{and} \quad \frac{1}{2}x^2(t) \leq \vartheta_0(t) \leq \vartheta_\varepsilon(t) \qquad \text{for all } t \in T. \tag{10.75}$$

From (10.74) and (10.75) it follows that

$$\vartheta_\varepsilon'(t) \leq h(t)\sqrt{2}\sqrt{\vartheta_\varepsilon(t)} \quad \text{for a.a. } t \in T. \tag{10.76}$$

The function $t \longrightarrow \vartheta_\varepsilon(t)$ is absolutely continuous and the function $r \longrightarrow \sqrt{r}$ is locally Lipschitz. So from the *Serrin–Vallée Poussin chain rule*, we have

$$\frac{\mathrm{d}}{\mathrm{d}t}\sqrt{\vartheta_\varepsilon(t)} = \frac{1}{2\sqrt{\vartheta_\varepsilon(t)}}\vartheta_\varepsilon'(t) \quad \text{for a.a. } t \in T,$$

$$\Rightarrow \frac{\mathrm{d}}{\mathrm{d}t}\sqrt{\vartheta_\varepsilon(t)} \leq \frac{1}{\sqrt{2}}h(t) \quad \text{a.e. on } t \in T \quad \text{(see (10.75))},$$

$$\Rightarrow \sqrt{\vartheta_\varepsilon(t)} \leq \sqrt{\vartheta_\varepsilon(0)} + \frac{1}{\sqrt{2}}\int_0^t h(\tau)\mathrm{d}\tau \qquad \text{for all } t \in T. \tag{10.77}$$

Combining (10.75) and (10.77) we obtain

$$\begin{aligned}|x(t)| \leq \sqrt{2}\sqrt{\vartheta_\varepsilon(t)} &\leq \sqrt{2\vartheta_\varepsilon(0)} + \int_0^t h(\tau)\mathrm{d}\tau \\ &= c + \varepsilon + \int_0^t h(\tau)\mathrm{d}\tau \qquad \text{for all } t \in T.\end{aligned}$$

Because $\varepsilon > 0$ was arbitrary, we let $\varepsilon \downarrow 0$, to obtain

$$|x(t)| \leq c + \int_0^t h(\tau)\mathrm{d}\tau \qquad \text{for all } t \in T.$$

□

Let H be a Hilbert space with inner product denoted by $(\cdot,\cdot)_H$. The next lemma extends the Serrin–Vallée Poussin chain rule.

LEMMA 10.3.2 *If $\varphi \in \Gamma_0(H)$, $x \in W^{1,2}\big((0,b),H\big)$, $x(t) \in \mathcal{D}(\partial\varphi)$ a.e. on $(0,b)$, and we can find $h \in L^2(T,H)$ such that $h(t) \in \partial\varphi\big(x(t)\big)$ a.e. on T, then the function $t \longrightarrow \varphi\big(x(t)\big)$ is absolutely continuous on T and*

$$\frac{\mathrm{d}}{\mathrm{d}t}\varphi\big(x(t)\big) = \big(u, x'(t)\big)_H \quad \text{a.e. on } T \text{ for all } u \in \partial\varphi\big(x(t)\big).$$

PROOF: From Corollary 3.2.51(a), we know that for every $\lambda > 0$ the function $t \longrightarrow \varphi_\lambda(x(t))$ is differentiable a.e. on T and we have

$$\begin{aligned}\frac{\mathrm{d}}{\mathrm{d}t}\varphi_\lambda(x(t)) &= \big(\varphi'_\lambda(x(t)), x'(t)\big)_H \\ &= \big((\partial\varphi)_\lambda(x(t)), x'(t)\big)_H \quad \text{a.e. on } (0,b),\end{aligned}$$

$$\Rightarrow \varphi_\lambda(x(t)) - \varphi_\lambda(x(s)) = \int_s^t \big((\partial\varphi)_\lambda(x(\tau)), x'(\tau)\big)_H \mathrm{d}\tau \qquad \text{for all } s,t \in T. \tag{10.78}$$

From Proposition 3.2.44(d)(e) we know that

$$\|(\partial\varphi)_\lambda(x(t))\| \le \|(\partial\varphi)^0(x(t)) \le \|h(t)\| \quad \text{a.e. on } T$$

$$\text{and} \quad (\partial\varphi)_\lambda(x(t)) \longrightarrow (\partial\varphi)^0(x(t)) \quad \text{a.e. on } T \text{ as } \lambda \downarrow 0.$$

So, from the dominated convergence theorem, we have

$$(\partial\varphi)_\lambda(x(\cdot)) = \partial\varphi_\lambda(x(\cdot)) \longrightarrow (\partial\varphi)^0(x(\cdot)) \quad \text{in } L^2(T,H) \text{ as } \lambda \downarrow 0.$$

Then passing to the limit as $\lambda \downarrow 0$ in (10.78), we obtain

$$\varphi(x(t)) - \varphi(x(s)) = \int_s^t \big((\partial\varphi)^0(x(\tau)), x'(\tau)\big)_H \mathrm{d}\tau \qquad \text{for all } s,t \in T$$

(see Corollary 3.2.51(c)). Therefore the function $t \longrightarrow \varphi(x(t))$ is absolutely continuous on T. Let $t_0 \in T$ be a differentiability point for both $x(\cdot)$ and $\varphi(x(\cdot))$ and $x(t_0) \in \mathcal{D}(\partial\varphi)$. Then for every $u \in \partial\varphi(x(t_0))$, we have

$$\varphi(x(t_0)) - \varphi(y) \le (u, x(t_0) - y)_H \qquad \text{for all } y \in H.$$

Let $y = x(t_0 \pm \varepsilon)$, $\varepsilon > 0$ and divide by ε. We have

$$\frac{1}{\varepsilon}\big(\varphi(x(t_0)) - \varphi(x(t_0 \pm \varepsilon))\big)_H \le \frac{1}{\varepsilon}(u, x(t_0) - x(t_0 \pm \varepsilon))_H.$$

Passing to the limit as $\varepsilon \longrightarrow 0$, we obtain

$$\varphi'(x(t_0)) = (u, x'(t_0))_H.$$

$\square$

Now let $A : \mathcal{D}(A) \subseteq H \longrightarrow 2^H \setminus \{\emptyset\}$ be a possibly multivalued nonlinear operator, $x_0 \in H$, and $g \in L^1(T,H)$. We consider the following Cauchy problem,

$$\begin{Bmatrix} -x'(t) \in A(x(t)) + g(t),\ t \in T, \\ x(0) = x_0 \end{Bmatrix}. \tag{10.79}$$

DEFINITION 10.3.3 (a) A function $x \in C(T,H)$ is said to be a *strong solution* (or *solution*) for the Cauchy problem (10.79), if $t \longrightarrow x(t)$ is absolutely continuous on every compact subinterval of $(0,b)$, $x(t) \in \mathcal{D}(A)$ a.e. on T, $x(0) = x_0$ and $x(\cdot)$ satisfies the differential equation in (10.79) for almost all $t \in T$.

(b) A function $x \in C(T,H)$ is said to be a *weak solution* for the Cauchy problem (10.79), if there exist sequences $\{x_n\}_{n\geq 1} \subseteq W^{1,1}\big((0,b),H\big)$ and $\{g_n\}_{n\geq 1} \subseteq L^1(T,H)$ such that for every $n\geq 1$, $-x_n(t) \in A\big(x_n(t)\big)+g_n(t)$ a.e. on T, $g_n \longrightarrow g$ in $L^1(T,H)$, and $x_n \longrightarrow x$ in $C(T,H)$ as $n\to\infty$ with $x(0)=x_0$.

The following existence theorem for problem (10.79) is proved in Barbu [58, p. 124] and in Brezis [102, p. 54].

THEOREM 10.3.4 *If $A:\mathcal{D}(A)\subseteq H \longrightarrow 2^H$ is maximal monotone, $x_0 \in \mathcal{D}(A)$, and $g\in W^{1,1}\big((0,b),H\big)$, then problem (10.79) has a unique (strong) solution $x\in C(T,H)$, which is Lipschitz continuous (i.e., $x\in W^{1,\infty}\big((0,b),H\big)$), it is right differentiable at every point $t\in[0,b)$, and*

$$-\frac{\mathrm{d}^+x}{\mathrm{d}t}(t) = \Big(A\big(x(t)\big) + g(t)\Big)^0 \qquad \textit{for all } t \in [0,b) \tag{10.80}$$

$$\|\frac{\mathrm{d}^+x}{\mathrm{d}t}(t)\| \leq \|\Big(A\big(x(t)\big) + g(t)\Big)^0\| + \int_0^t \|g'(\tau)\|\mathrm{d}\tau \qquad \textit{for all } t \in [0,b). \tag{10.81}$$

Moreover, if $x,y\in W^{1,\infty}\big((0,b),H\big)$ are solutions for the data (x_0,g_1), $(y_0,g_2)\in \mathcal{D}(A)\times W^{1,1}\big((0,b),H\big)$, respectively, we have

$$\|x(t)-y(t)\| \leq \|x_0-y_0\| + \int_0^t \|g_1(\tau)-g_2(\tau)\|\mathrm{d}\tau \qquad \textit{for all } t\in T. \tag{10.82}$$

REMARK 10.3.5 The result is still true if $T=[0,b]$ is replaced by $\mathbb{R}_+$. Note that (10.80) means that for the multivalued system, among all choices the system tends to minimize its velocity. We can also slightly generalize Theorem 10.3.4 and assume that for some $\vartheta>0$, $x \longrightarrow A(x)+\vartheta x$ is maximal monotone. Then the existence theorem is still valid, only the estimates in (10.81) and (10.82) need to be modified accordingly.

The next theorem concerns weak solutions.

THEOREM 10.3.6 *If $A:\mathcal{D}(A)\subseteq H \longrightarrow 2^H$ is a maximal monotone operator, $x_0 \in \overline{\mathcal{D}(A)}$, and $g\in L^1(T,H)$, then there exists a unique weak solution $x\in C(T,H)$ of (10.79) such that*

$$\frac{1}{2}\|x(t)-v\|^2 \leq \frac{1}{2}\|x(s)-v\|^2 + \int_s^t \big(g(\tau)-w, x(\tau)-v\big)_H \mathrm{d}\tau \tag{10.83}$$

for all $0\leq s\leq t\leq b$ and all $(v,w)\in \mathrm{Gr}\, A$. Moreover, if $x,y\in C(T,X)$ are weak solutions corresponding to the data (x_0,g_1),$(y_0,g_2)\in \mathcal{D}(A)\times L^1(T,H)$, respectively, then we have

$$\frac{1}{2}\|x(t)-y(t)\|^2 \leq \frac{1}{2}\|x(s)-y(s)\|^2 + \int_s^t \big(g_1(\tau)-g_2(\tau), x(\tau)-y(\tau)\big)_H \mathrm{d}\tau, \tag{10.84}$$

for all $0\leq s\leq t\leq b$.

PROOF: Fix $(x_0, g_0) \in \overline{\mathcal{D}(A)} \times L^1(T,H)$. We can find $\{x_0^n\}_{n\geq 1} \subseteq \mathcal{D}(A)$ and $\{g_n\}_{n\geq 1} \subseteq W^{1,1}\big((0,b),H\big)$ such that

$$x_0^n \longrightarrow x_0 \quad \text{in } H \quad \text{and} \quad g_n \longrightarrow g \quad \text{in } L^1(T,H) \quad \text{as } n \to \infty.$$

For every $n \geq 1$, we consider the following Cauchy problem

$$\left\{ \begin{array}{l} -x_n'(t) \in A\big(x_n(t)\big) + g_n(t) \text{ a.e. on } T, \\ x_n(0) = x_0^n \end{array} \right\}. \tag{10.85}$$

By virtue of Theorem 10.3.4 problem (10.85) has a unique strong solution $x_n \in W^{1,\infty}\big((0,b),H\big)$. Moreover, (10.82) implies that for all $n, m \geq 1$, we have

$$\|x_n(t) - x_m(t)\| \leq \|x_0^n - x_0^m\| + \int_0^b \|g_n(\tau) - g_m(\tau)\| d\tau,$$
$$\Rightarrow \{x_n\}_{n\geq 1} \subseteq C(T,H) \text{ is Cauchy.}$$

So there exists $x \in C(T,H)$ such that

$$x_n \longrightarrow x \quad \text{in } C(T,H).$$

Evidently $x \in C(T,H)$ is a weak solution for problem (10.79). Moreover, because for every $n \geq 1$ $x_n(\cdot)$ satisfies (10.83) and (10.84), passing to the limit as $n \to \infty$, we deduce that $x(\cdot)$ satisfies them too.

Finally the uniqueness of the weak solution follows at once from (10.84). □

From Theorems 10.3.4 and 10.3.6 we see that in the general case, in order to have a strong solution we need to restrict the data of the problem, namely we need to assume that $x_0 \in \mathcal{D}(A)$ and $g \in W^{1,1}\big((0,b),H\big)$; that is, we need a "smooth" initial condition and "smooth" forcing term. We show that in the case of subdifferential evolution inclusions these restrictive requirements can be removed.

THEOREM 10.3.7 *If $A = \partial\varphi : \mathcal{D}(\partial\varphi) \subseteq H \longrightarrow 2^H$ with $\varphi \in \Gamma_0(H)$, $x_0 \in \overline{\mathcal{D}(\partial\varphi)}$, and $g \in L^2(T,H)$, then problem (10.79) admits a unique strong solution $x \in C(T,H)$ such that*

$$t \longrightarrow \sqrt{t}\, x'(t) \quad \textit{belongs in } L^2(T),$$

$t \longrightarrow \varphi\big(x(t)\big)$ belongs in $L^1(T,H)$, and it is absolutely continuous on $[\delta, b]$, $\forall \delta > 0$.

Moreover, if $x_0 \in \operatorname{dom} \varphi$, then $x' \in L^2(T,H)$ and $t \longrightarrow \varphi\big(x(t)\big)$ is absolutely continuous on T.

PROOF: First note that if $(v_0, w_0) \in \operatorname{Gr} \partial\varphi$ and we introduce the function

$$\varphi_0(x) = \varphi(x) - \varphi(v_0) - (w_0, x - v_0)_H \geq 0, \quad x \in H,$$

then the subdifferential inclusion $-x'(t) \in \partial\varphi\big(x(t)\big) + g(t)$ is equivalent to

$$-x'(t) \in \partial\varphi_0\big(x(t)\big) + g(t) - w_0.$$

So, without any loss of generality, we may assume that

$$\varphi(v_0) = \inf_H \varphi = 0. \tag{10.86}$$

Now suppose that $x_0 \in \mathcal{D}(\partial\varphi)$ and $g \in W^{1,2}\big((0,b),H\big)$. Then Theorem 10.3.4 implies that there exists a unique strong solution $x \in W^{1,\infty}\big((0,b),H\big)$. We multiply the equation with $tx'(t)$ and use Lemma 10.3.2. We obtain

$$t\|x'(t)\|^2 + t\frac{\mathrm{d}}{\mathrm{d}t}\varphi\big(x(t)\big) = t\big(g(t),x'(t)\big)_H \quad \text{a.e. on } T,$$

$$\Rightarrow \int_0^b t\|x'(t)\|^2\mathrm{d}t + b\,\varphi\big(x(b)\big) = \int_0^b t\big(g(t),x'(t)\big)_H + \int_0^b \varphi\big(x(t)\big)\mathrm{d}t. \tag{10.87}$$

Note that

$$t\big(g(t),x'(t)\big)_H \le t\|g(t)\|\,\|x'(t)\| \le \frac{1}{2}t\,\|g(t)\|^2 + \frac{1}{2}t\,\|x'(t)\|^2 \qquad \text{for all } t\in T. \tag{10.88}$$

Using (10.88) and the fact that $\varphi \ge 0$ (see (10.86)) in (10.87), we obtain

$$\int_0^b t\|x'(t)\|^2\mathrm{d}t \le \int_0^b t\|g(t)\|^2\mathrm{d}t + 2\int_0^b \varphi\big(x(t)\big)\mathrm{d}t. \tag{10.89}$$

Also from the definition of the convex subdifferential, we have

$$\varphi\big(x(t)\big) \le \big(-x'(t) - g(t), x(t) - v_0\big)_H$$

(recall that $\phi(v_0) = 0$, see (10.86))

$$\begin{aligned}\Rightarrow \int_0^b \varphi\big(x(t)\big)\mathrm{d}t &\le \int_0^b \big(-x'(t), x(t)-v_0\big)_H\mathrm{d}t + \int_0^b \|g(t)\|\,\|x(t)-v_0\|\mathrm{d}t \\ &= -\int_0^b \frac{1}{2}\frac{\mathrm{d}}{\mathrm{d}t}\|x(t)-v_0\|^2\mathrm{d}t + \int_0^b \|g(t)\|\,\|x(t)-v_0\|\mathrm{d}t \\ &\le \frac{1}{2}\|x_0 - v_0\|^2 + \int_0^b \|g(t)\|\,\|x(t)-v_0\|\mathrm{d}t. \end{aligned} \tag{10.90}$$

Next take the inner product of the differential equation with $x(t) - v_0$ and then integrate over $[0,s]$. We obtain

$$\begin{aligned}\int_0^s \big(-x'(t), x(t)-v_0\big)_H\mathrm{d}t \in{}& \int_0^s \Big(\partial\varphi\big(x(t)\big), x(t)-v_0\Big)_H\mathrm{d}t \\ &+ \int_0^s \big(g(t), x(t)-v_0\big)_H\mathrm{d}t. \end{aligned} \tag{10.91}$$

We have

$$\int_0^s \big(-x'(t), x(t)-v_0\big)_H\mathrm{d}t = \frac{1}{2}\|x_0 - v_0\|^2 - \frac{1}{2}\|x(s) - v_0\|^2 \tag{10.92}$$

$$\int_0^s \Big(\partial\varphi\big(x(t)\big), x(t)-v_0\Big)_H\mathrm{d}t \ge 0 \ \text{(because } 0 \in \partial\varphi(v_0) \text{ and } \partial\varphi \text{ is monotone)} \tag{10.93}$$

$$\text{and} \quad \int_0^s \big(g(t), x(t)-v_0\big)_H\mathrm{d}t \ge -\int_0^s \|g(t)\|\,\|x(t)-v_0\|\mathrm{d}t. \tag{10.94}$$

Using (10.92) through (10.94) in (10.91), we obtain

$$\frac{1}{2}\|x(s)-v_0\|^2 \le \frac{1}{2}\|x_0-v_0\|^2 + \int_0^s \|g(t)\|\,\|x(t)-v_0\|\mathrm{d}t,$$

$$\Rightarrow \|x(s)-v_0\| \le \|x_0-v_0\| + \int_0^b \|g(t)\|\mathrm{d}t \qquad \text{for all } s\in T. \tag{10.95}$$

Using (10.95) in (10.90), we obtain

$$\int_0^b \varphi\big(x(t)\big)\mathrm{d}t \le \Big(\|x_0-v_0\| + \int_0^b \|g(t)\|\mathrm{d}t\Big)^2. \tag{10.96}$$

We use (10.96) in (10.89) and have

$$\int_0^b t\|x'(t)\|^2\mathrm{d}t \le \int_0^b t\|g(t)\|^2\mathrm{d}t + 2\Big(\|x_0-v_0\| + \int_0^b \|g(t)\|\mathrm{d}t\Big)^2. \tag{10.97}$$

Now suppose that $x_0\in\overline{\mathcal{D}(\partial\varphi)}=\overline{\mathrm{dom}\,\varphi}$ and $g\in L^2(T,H)$. Then we can find

$$\{x_0^n\}_{n\geq 1}\subseteq\mathcal{D}(\partial\varphi) \quad\text{and}\quad \{g_n\}_{n\geq 1}\subseteq L^2(T,H)$$

such that

$$x_0^n\longrightarrow x_0 \quad\text{in } H \quad\text{and}\quad g_n\longrightarrow g \quad\text{in } L^2(T,H) \quad\text{as } n\to\infty.$$

Let $x_n\in W^{1,\infty}\big((0,b),H\big)$, $n\geq 1$, be the unique strong solution of the Cauchy problem with data (x_0^n,g_n). From (10.82), we see that $\{x_n\}_{n\geq 1}\subseteq C(T,H)$ is Cauchy. So we can find $x\in C(T,H)$ such that

$$x_n\longrightarrow x \quad\text{in } C(T,H) \text{ as } n\to\infty.$$

So we have that $x_n'\longrightarrow x'$ in the sense of H-valued distributions on $(0,b)$. Because (10.97) is true for $\{x_n,x_0^n,g_n\}_{n\geq 1}$, we deduce that $t\longrightarrow tx'(t)$ belongs in $L^2(T,H)$ and

$$\sqrt{t}\,x_n' \xrightarrow{w} \sqrt{t}\,x' \quad\text{in } L^2(T,H) \text{ as } n\to\infty.$$

In particular then, for any $0<\delta<b$, we have

$$x_n'\longrightarrow x' \quad\text{in } L^2([\delta,b],H) \text{ as } n\to\infty.$$

If $\widehat{A}$ is the realization of A in $L^2([\delta,b],H)$, then $\widehat{A}$ is maximal monotone and

$$(x_n,-x_n'-g_n)\in\mathrm{Gr}\,\widehat{A} \qquad\text{for all } n\geq 1.$$

From Proposition 3.2.7 it follows that

$$(x,-x'-g)\in\mathrm{Gr}\,\widehat{A},$$

$$\left\{\begin{array}{ll} -x'(t)\in\partial\varphi\big(x(t)\big)+g(t) & \text{a.e. on } T \text{ (because } \delta\geq 0 \text{ is arbitrary),} \\ x(0)=x_0 & \end{array}\right\}.$$

Next, let $J_\varphi: L^1(T,H)\longrightarrow\overline{\mathbb{R}}=\mathbb{R}\cup\{+\infty\}$ be the integral functional defined by

$$J_\varphi(u)=\begin{cases}\int_0^b\varphi\big(u(t)\big)\mathrm{d}t & \text{if } \varphi\big(u(\cdot)\big)\in L^1(T,H)\\ +\infty & \text{otherwise}\end{cases}.$$

We have that $J_\varphi \in \Gamma_0\big(L^1(T,H)\big)$. Also because of (10.96) we see that we can find $c_0>0$ such that for $n\geq 1$

$$\begin{aligned}
&\Rightarrow\ J_\varphi(x_n) \leq c_0,\\
&\Rightarrow\ J_\varphi(x) \leq \liminf_{n\to\infty} J_\varphi(x_n) \leq c_0,\\
&\Rightarrow\ x \in \text{dom } J_\varphi.
\end{aligned}$$

So using Lemma 10.3.2, we have that $t \longrightarrow \varphi\big(x(t)\big)$ belongs in $L^1(T,H)$ and it is absolutely continuous on $[\delta,b]$ for all $0<\delta<b$.

Finally suppose that $x_0 \in \text{dom } \varphi$. We have

$$\|x'(t)\|^2 + \frac{\mathrm{d}}{\mathrm{d}t}\varphi\big(x(t)\big) \leq \|g(t)\|\,\|x'(t)\| \quad \text{a.e. on } T,$$

$$\Rightarrow\ \frac{1}{2}\|x'(t)\|^2 + \frac{\mathrm{d}}{\mathrm{d}t}\varphi\big(x(t)\big) \leq \frac{1}{2}\|g(t)\|^2 \quad \text{a.e. on } T, \tag{10.98}$$

$$\Rightarrow\ t \longrightarrow \varphi\big(x(t)\big) - \frac{1}{2}\int_0^t \|g(\tau)\|^2 \mathrm{d}\tau \quad \text{is decreasing on } T. \tag{10.99}$$

Because $x_0 \in \text{dom } \varphi$, we obtain

$$\varphi\big(x(t)\big) \leq \varphi(x_0) + \frac{1}{2}\int_0^t \|g(\tau)\|^2 \mathrm{d}\tau \qquad \text{for all } t \in [0,b). \tag{10.100}$$

For $\delta>0$, from (10.98), we obtain

$$\begin{aligned}
\frac{1}{2}\int_\delta^b \|x'(t)\|^2 \mathrm{d}t &\leq \varphi\big(x(\delta)\big) + \frac{1}{2}\int_\delta^b \|g(t)\|^2 \mathrm{d}t\\
&\leq \varphi(x_0) + \frac{1}{2}\int_0^b \|g(t)\|^2 \mathrm{d}t \quad \text{(see (10.98))},
\end{aligned}$$

$$\Rightarrow\ x' \in L^2(T).$$

Finally from (10.100), we have $\varphi\big(x(\cdot)\big) \in L^\infty(T)$ and from Lemma 10.3.2 we conclude that $t \longrightarrow \varphi\big(x(t)\big)$ is absolutely continuous. □

REMARK 10.3.8 If $g \in W^{1,1}\big((0,b),H\big)$, then x is right differentiable at every point in $(0,b)$, $x(t)\in\mathcal{D}(A)$ for every $t\in(0,b]$ and $-\frac{\mathrm{d}^+x}{\mathrm{d}t}(t) \in \Big(\partial\varphi\big(x(t)\big)+g(t)\Big)^0$ for all $t \in (0,b]$.

Subdifferential evolution inclusions exhibit a remarkable smoothing effect. In what follows by $\{S(t)\}_{t\geq 0}$ we denote the semigroup of nonlinear contractions $S(t): \overline{\mathcal{D}(A)} \longrightarrow \overline{\mathcal{D}(A)}$, $t \geq 0$, generated by A (see Theorem 3.2.93).

THEOREM 10.3.9 *If $A = \partial\varphi : \mathcal{D}(\partial\varphi) \subseteq H \longrightarrow 2^H$, $\varphi \in \Gamma_0(H)$, and $x_0 \in \overline{\mathcal{D}(A)}$, then*

(a) *$Sx_0 \in \mathcal{D}(A)$ for all $t>0$ (smoothing effect).*

(b) $\left\|\frac{\mathrm{d}^+}{\mathrm{d}t}S(t)x_0\right\| = \|A^0\big(S(t)x_0\big)\| \leq \|A^0(x_0)\| + \frac{1}{t}\|x_0 - v_0\|$ *for all $t>0$ and all $v_0 \in \mathcal{D}(\partial\varphi)$.*

PROOF: (a) As in the proof of Theorem 10.3.7, let $(v_0, w_0) \in \operatorname{Gr} \partial\varphi$ and define the function

$$\varphi_0(x) = \varphi(x) - \varphi(v_0) - (w_0, x - v_0)_H, \quad x \in H.$$

We have $\varphi_0 \geq 0$ (recall the definition of the subdifferential) and

$$0 = \varphi_0(v_0) = \min_H \varphi_0.$$

We set $x(t) = S(t)x_0,\ t \geq 0$. Then $x \in C(T, H)$ is a strong solution of the Cauchy problem

$$\left\{ \begin{array}{l} -x'(t) \in \partial\varphi_0\big(x(t)\big) + w_0, \\ x(0) = x_0 \end{array} \right\}. \tag{10.101}$$

Then by virtue of Remark 10.3.8, we have $x(t) = S(t)x_0 \in \mathcal{D}(A)$ for all $t > 0$.

(b) Also, again from Remark 10.3.8, we have

$$\frac{\mathrm{d}^+ x}{\mathrm{d}t}(t) \in \Big(-\partial\varphi\big(x(t)\big) - w_0\Big)^0 \qquad \text{for all } t>0 \tag{10.102}$$

$$\begin{aligned} \Rightarrow \int_0^t s\Big\|\frac{\mathrm{d}^+ x}{\mathrm{d}t}(s)\Big\|^2 \mathrm{d}s &\leq \int_0^t \varphi_0\big(x(s)\big)\mathrm{d}s - \int_0^t s\Big(w_0, \frac{\mathrm{d}^+ x}{\mathrm{d}t}(s)\Big)_H \mathrm{d}s \\ &= \int_0^t \varphi_0\big(x(s)\big)\mathrm{d}s - t\big(w_0, x(t)\big)_H + \int_0^t \big(w_0, x(s)\big)\mathrm{d}s \end{aligned} \tag{10.103}$$

(integration by parts).

Because $-w_0 - x'(t) \in \partial\varphi_0\big(x(t)\big)$ a.e on $\mathbb{R}_+$, we have

$$\begin{aligned} &\varphi_0\big(x(t)\big) \leq (-w_0 - x'(t), x(t) - v_0)_H \quad \text{a.e. on } T, \\ \Rightarrow &\int_0^t \varphi_0\big(x(s)\big)\mathrm{d}s \leq \int_0^t (-w_0 - x'(s), x(s) - v_0)_H \mathrm{d}s \\ &\qquad = \frac{1}{2}\|x_0 - v_0\|^2 - \frac{1}{2}\|x(t) - v_0\|^2 + t(v_0, w_0)_H \\ &\qquad\quad - \int_0^t \big(w_0, x(s)\big)_H \mathrm{d}s, \ t \geq 0. \end{aligned} \tag{10.104}$$

Using (10.104) in (10.103), we obtain

$$\begin{aligned} \int_0^t s\Big\|\frac{\mathrm{d}^+ x}{\mathrm{d}t}(s)\Big\|^2 \mathrm{d}s &\leq \frac{1}{2}\|x_0 - v_0\|^2 + t(w_0, v_0 - x(t))_H - \frac{1}{2}\|v_0 - x(t)\|^2 \\ &\leq \frac{1}{2}\big(\|x_0 - v_0\|^2 + t^2\|w_0\|^2\big), \quad t \geq 0. \end{aligned} \tag{10.105}$$

Note that

$$\begin{aligned} &\frac{\mathrm{d}^+}{\mathrm{d}t}\|x(t+h) - x(t)\|^2 \leq 0 \qquad \text{for all } t>0 \ \text{ and all } h>0, \\ \Rightarrow\ &\frac{\|x(t+h) - x(t)\|}{h} \leq \frac{\|x(s+h) - x(s)\|}{h} \qquad \text{for all } 0<s \leq t,\ h>0, \\ \Rightarrow\ &t \longrightarrow \Big\|\frac{\mathrm{d}^+ x}{\mathrm{d}t}(t)\Big\| \ \text{ is decreasing on } (0, +\infty). \end{aligned}$$

From this fact and (10.105), we obtain

$$\begin{aligned}
&\Big\|\frac{\mathrm{d}^+x}{\mathrm{d}t}(t)\Big\|^2 \leq \frac{1}{2}\Big(\frac{1}{t^2}\|x_0-v_0\|^2+\|w_0\|^2\Big), \quad t>0,\\
\Rightarrow\ &\Big\|\frac{\mathrm{d}^+}{\mathrm{d}t}S(t)x_0\Big\| \leq \frac{1}{t}\|x_0-v_0\|+\|w_0\|, \quad t>0.
\end{aligned}$$

□

Let us present an application of Theorem 10.3.7. So let $\beta:\mathcal{D}(\beta)\subseteq\mathbb{R}\longrightarrow 2^{\mathbb{R}}\setminus\{\emptyset\}$ be a maximal monotone function. We know that $\beta=\partial j$ with $j\in\Gamma_0(\mathbb{R})$. Also let $Z\subseteq\mathbb{R}^N$ be a bounded domain with a C^2-boundary ∂Z. We consider the following heat equation with Neumann boundary condition.

$$\left\{\begin{array}{l} -\frac{\partial x}{\partial t}-\Delta x(t,z)=\overline{g}(t,z) \ \text{ a.e. on } (0,b)\times Z,\\ -\frac{\partial x}{\partial n}\big|_{(0,b)\times\partial Z}\in\beta\big(x(t,z)\big),\ x(0,z)=x_0(z)\end{array}\right\}. \tag{10.106}$$

PROPOSITION 10.3.10 *If $\overline{g}\in L^2(T\times Z)$ and $x_0\in L^2(Z)$, then problem (10.106) has a unique solution $x\in C\big(T,L^2(Z)\big)$ such that*

$$\sqrt{t}x\in L^2\big(T,H^2(Z)\big), \quad \sqrt{t}\,\frac{\partial x}{\partial t}\in L^2(T\times Z).$$

Moreover, if $x_0\in H^1(Z)$ and $j\big(x_0(\cdot)\big)\in L^1(Z)$, then

$$x\in L^2\big(T,H^2(Z)\big), \quad \frac{\partial x}{\partial t}\in L^2(T\times Z).$$

PROOF: Let $H=L^2(Z)$ and let $\varphi:H=L^2(Z)\longrightarrow\overline{\mathbb{R}}=\mathbb{R}\cup\{+\infty\}$ be defined by

$$\varphi(x)=\left\{\begin{array}{ll}\frac{1}{2}\int\limits_Z\|Dx(z)\|^2\mathrm{d}z+\int\limits_{\partial Z}j\big(x(z)\big)\mathrm{d}\sigma & \text{if } x\in H^1(Z) \text{ and } j\big(x(\cdot)\big)\in L^1(\partial Z)\\ +\infty & \text{otherwise.}\end{array}\right.$$

Let $A:L^2(Z)\longrightarrow L^2(Z)$ be the operator defined by

$$Ax=-\Delta x \qquad \text{for all } x\in\mathcal{D}(A)=\Big\{x\in H^2(Z):-\frac{\partial x}{\partial n}(z)\in\beta\big(x(z)\big) \text{ a.e. on } \partial Z\Big\}.$$

Evidently this is a nonlinear operator and the definition of $\mathcal{D}(A)$ makes sense because $x\in H^2(Z)$ implies that $x\big|_{\partial Z}\in H^{3/2}(\partial Z)$ and $(\partial x/\partial n)\in H^{1/2}(\partial Z)\subseteq L^2(\partial Z)$. Using Green's identity, we obtain

$$\begin{aligned}
\int_Z -\Delta x(y-x)\mathrm{d}z &= \int_Z (Dx,Dy-Dx)_{\mathbb{R}^N}\mathrm{d}z-\int_{\partial Z}\frac{\partial x}{\partial n}(y-x)\mathrm{d}\sigma\\
&\leq \varphi(y)-\varphi(x) \qquad \text{for all } x\in\mathcal{D}(A),\ y\in L^2(Z),
\end{aligned}$$

$$\Rightarrow\ A\subseteq\partial\varphi.$$

To show that $A=\partial\varphi$, it suffices to show that A is maximal monotone in $L^2(Z)$ and this follows if we show that $\mathcal{R}(I+A)=L^2(Z)$. To this end, we consider the following elliptic (stationary) problem.

$$\left\{\begin{array}{ll} -\Delta x_\lambda(z) + x_\lambda(z) = h(z) & \text{a.e. on } Z, \\ -\frac{\partial x_\lambda}{\partial n}(z) = \beta_\lambda(x_\lambda(z)) & \text{a.e. on } \partial Z \end{array}\right\}, \tag{10.107}$$

with $h \in L^2(Z)$ and $\beta_\lambda = (1/\lambda)\big(1 - (1+\lambda\beta)^{-1}\big)$ (the Yosida approximation of β). Consider the operator $K : L^2(\partial Z) \longrightarrow L^2(\partial Z)$, where $K(x) = y\big|_{\partial Z}$ with y being the unique solution of

$$\left\{\begin{array}{ll} -\Delta y(z) + y(z) = h(z) & \text{a.e. on } Z, \\ y(z) + \lambda\frac{\partial y}{\partial n}(z) = (1+\lambda\beta)^{-1}x(z) & \text{a.e. on } \partial Z \end{array}\right\}. \tag{10.108}$$

Problem (10.108) has a unique solution $y \in H^2(Z)$ and using Green's identity we can have

$$\|K(x) - K(u)\|_{L^2(\partial Z)} \le \vartheta\|x-u\|_{L^2(\partial Z)} \quad 0<\vartheta<1, \qquad \text{for all } x,u \in L^2(Z).$$

So by Banach's fixed point theorem we can find $y \in L^2(\partial Z)$ such that

$$K(y) = y.$$

Then the corresponding solution $x_\lambda \in H^2(Z)$ of (10.108) also solves (10.107). Moreover, from Brezis [100], we know that

$$\|x_\lambda\|_{H^2(Z)} \le c\big(1 + \|h\|_{L^2(Z)}\big) \quad \text{for some } c>0 \quad \text{and all } \lambda>0.$$

Recalling that the trace map $x \longrightarrow \big(x\big|_{\partial Z}, \frac{\partial x}{\partial n}\big|_{\partial Z}\big)$ is continuous from $H^2((Z))$ into $H^{3/2}(\partial Z) \times H^{1/2}(\partial Z)$, we conclude that $\big\{x_\lambda, \frac{\partial x_\lambda}{\partial n}\big\}_{\lambda>0} \subseteq L^2(\partial Z) \times L^2(\partial Z)$ is bounded. Also we know that $H^2(Z)$ is embedded compactly in $L^2(Z)$. Hence we may assume that

$$x_\lambda \xrightarrow{w} x \ \text{ in } H^2(Z),\ x_\lambda \longrightarrow x \ \text{ in } L^2(Z),\ \frac{\partial x_\lambda}{\partial n} \xrightarrow{w} \frac{\partial x}{\partial n} \ \text{ in } L^2(\partial Z) \ \text{ as } \lambda \downarrow 0.$$

Moreover, because $\big\{\beta_\lambda(x_\lambda(\cdot))\big\}_{\lambda>0} \subseteq L^2(Z)$ is bounded, we have

$$(1+\lambda\beta)^{-1}x_\lambda \longrightarrow x \quad \text{in } L^2(Z) \ \text{ as } \lambda \downarrow 0. \tag{10.109}$$

Therefore, in the limit as $\lambda \downarrow 0$, we obtain

$$-\Delta x(z) + x(z) = h(z) \quad \text{a.e. on } Z.$$

Also, if $\widehat{\beta} : \mathcal{D}(\widehat{\beta}) \subseteq L^2(\partial Z) \longrightarrow 2^{L^2(\partial Z)}$ is defined by

$$\widehat{\beta}(x) = \big\{y \in L^2(\partial Z) : y(z) \in \beta(x(z)) \ \text{ a.e. on } \partial Z\big\}, \quad x \in L^2(\partial Z)$$

(the realization (lifting) of β on $L^2(\partial Z)$), then $\widehat{\beta}$ is maximal monotone and

$$\widehat{\beta}_\lambda(x)(z) = \beta_\lambda(x(z)) \quad \text{a.e. on } \partial Z.$$

Because $\widehat{\beta}_\lambda(x) \in \widehat{\beta}\big((I+\lambda\beta)^{-1}x\big)$, we obtain

$$\widehat{\beta}_\lambda(x_\lambda) \xrightarrow{w} \frac{\partial x}{\partial n} \quad \text{in } L^2(\partial Z) \quad \text{as } \lambda \downarrow 0 \quad \text{(see (10.109))}.$$

Because $\widehat{\beta}$ is maximal monotone, we deduce that

$$\frac{\partial x}{\partial n} \in \widehat{\beta}(x).$$

Hence we conclude that A is maximal monotone and so $A = \partial\varphi$.

Now we equivalently rewrite (10.106) as the following nonlinear evolution inclusion

$$\left\{ \begin{array}{ll} -x'(t) \in \partial\varphi\big(x(t)\big) + g(t) & \text{a.e. on } T, \\ x(0) = x_0 & \end{array} \right\}. \tag{10.110}$$

with $g(t) = \overline{g}(t,\cdot) \in L^2(Z)$ for all $t \in T$. Then the conclusion of Proposition 10.3.10 follows from Theorem 10.3.7 applied to the Cauchy problem (10.110). $\square$

Now we turn our attention to evolution inclusions driven by operators of monotone type defined in the framework of an evolution triple.

So let $T = [0,b]$ and let (X, H, X^*) be an evolution triple of spaces (see Definition 10.1.21). We consider the following Cauchy problem,

$$\left\{ \begin{array}{ll} -x'(t) + A\big(t, x(t)\big) = f\big(t, x(t)\big) & \text{a.e. on } T, \\ x(0) = x_0 & \end{array} \right\}. \tag{10.111}$$

The hypotheses on the data of (10.111) are the following.

H(A): $A : T \times X \longrightarrow X^*$ is a map such that

(i) For every $x \in X$, $t \longrightarrow A(t,x)$ is measurable.

(ii) For almost all $t \in T$, $x \longrightarrow A(t,x)$ is hemicontinuous monotone.

(iii) For almost all $t \in T$ and all $x \in X$, we have

$$\|A(t,x)\|_* \leq \alpha(t) + c\|x\|^{p-1}$$

with $2 \leq p < \infty$, $\alpha \in L^{p'}(T)$, $\frac{1}{p} + \frac{1}{p'} = 1$, $c > 0$.

(iv) For almost all $t \in T$ and all $x \in X$, we have

$$\langle A(t,x), x \rangle \geq c_1 \|x\|^p \quad \text{with } c_1 > 0.$$

REMARK 10.3.11 Hypothesis H(A)(ii) implies that for almost all $t \in T$, $A(t,\cdot)$ is maximal monotone.

H(f): $f : T \times H \longrightarrow H$ is a function such that

(i) For all $x \in H$, $t \longrightarrow f(t,x)$ is measurable.

(ii) For almost all $t \in T$, $x \longrightarrow f(t,x)$ is sequentially continuous from H into H_w (by H_w we denote the pivot Hilbert space H furnished with the weak topology).

(iii) For almost all $t \in T$ and all $x \in H$, we have

$$\|f(t,x)\| \leq \alpha_2(t) + c_2|x|^{2/p'}$$

with $\alpha_2, c_2 \in L^{p'}(T)$, $\left(\frac{1}{p} + \frac{1}{p'} = 1\right)$.

By a solution of (10.111), we understand a function $x \in W_p(0,b)$ (see Definition 10.1.23) that satisfies (10.111). In what follows by $S(x_0)$ we denote the solution set of (10.111). Then

$$S(x_0) \subseteq W_p(0,b) \subseteq C(T,H)$$

(see Theorem 10.1.25).

THEOREM 10.3.12 *If hypotheses H(A), H(f) hold, $x_0 \in H$, and X is embedded compactly into H, then $S(x_0)$ is nonempty, weakly compact in $W_p(0,b)$ and compact in $C(T,H)$.*

PROOF: The proof of the existence is based on the Galerkin method.

Because X is separable reflexive, we can find a basis $\{v_k\}_{k\geq 1}$ for X (hence for H and X^* too). We set

$$H_n = \operatorname{span}\{v_k\}_{k=1}^n, \quad n \geq 1$$

and endow H_n with the inner product inherited from H. By $p_n : H \longrightarrow H_n$ we denote the orthogonal projection on H_n. Also let X_n be the finite dimensional space H_n equipped with the X-norm (i.e., we view H_n as a subspace of X rather than of H). Then the dual X_n^* of X_n is the space H_n furnished with the X^*-norm.

For every $n \geq 1$, we consider $A_n : T \times X_n \longrightarrow X_n^*$ to be restriction (Galerkin approximation) of $A(t,\cdot)$ on X_n. So we have

$$A_n(t,x) = y \quad \text{for } x \in X_n \quad \text{with } y \in X_n^* \text{ satisfying } \langle A(t,x), u\rangle = \langle y, u\rangle\,, \quad \text{for all } u \in X_n;$$

that is, $\langle A_n(t,x), u\rangle_{X_n} = \langle A(t,x), u\rangle$ for all $x, u \in X_n$.

Here by $\langle\cdot,\cdot\rangle_{X_n}$, we denote the duality brackets for the pair (X_n^*, X_n). Clearly for every $x \in X_n$, $t \longrightarrow A_n(t,x)$ is measurable and for almost all $t \in T$, $x \longrightarrow A_n(t,x)$ is continuous. Also we set $f_n(t,x) = p_n f(t,x)$ for $(t,x) \in T \times H_n$. We see that f_n is a Carathéodory function and

$$\begin{aligned} |f_n(t,x)| = |p_n f(t,x)| &\leq \|p_n\|_{\mathcal{L}} |f(t,x)| \\ &\leq \alpha_2(t) + c_2(t)|x|^{2/q} \quad \text{for a.a. } t \in T, \text{ all } x \in H. \end{aligned}$$

We consider the following finite-dimensional approximation (Galerkin approximation), of problem (10.111).

$$\left\{ \begin{array}{l} x_n'(t) + A_n\big(t, x_n(t)\big) = f\big(t, x_n(t)\big) \quad \text{a.e. on } T, \\ x_n(0) = p_n x_0 = x_0^n \in H_n \end{array} \right\}. \tag{10.112}$$

From Carathéodory's existence theorem, we know that problem (10.112) has at least one solution $x_n \in W_p(0,b)$. Next we derive some a priori bounds for the sequence $\{x_n\}_{n\geq 1}$. For this purpose we take the duality brackets with $x_n(t)$ and obtain

$$\langle x_n'(t), x_n(t)\rangle + \langle A_n(t, x_n(t)), x_n(t)\rangle = \langle p_n f(t, x_n(t)), x_n(t)\rangle$$

a.e. on $T=[0,b]$

$$\Rightarrow \frac{\mathrm{d}}{\mathrm{d}t}|x_n(t)|^2 + 2c_1\|x_n(t)\|^2 \leq 2|p_n f(t, x_n(t))||x_n(t)|,$$

(see Corollary 10.1.26 and H(A)(iv))

$$\leq 2\xi\Big(\frac{\varepsilon^{p'}}{p'}|p_n f(t, x_n(t))|^{p'} + \frac{1}{\varepsilon^p p}\|x_n(t)\|^p\Big)$$

(by Young's inequality) with $\varepsilon>0$ and $\xi>0$ such that $|\cdot| \leq \xi\|\cdot\|$. We choose $\varepsilon = \left(\frac{\varepsilon}{c_1 p}\right)^{1/p} > 0$ and obtain

$$\begin{aligned}\frac{\mathrm{d}}{\mathrm{d}t}|x_n(t)|^2 &\leq c_3\big|p_n f(t, x_n(t))\big|^{p'} \quad \text{a.e. on } T \text{ with } c_3>0\\ &\leq c_3\Big(\alpha_2(t) + c_2|x_n(t)|^{2/p'}\Big)^{p'}\\ &\leq 2^{p'-1}c_3\alpha_2(t)^{p'} + 2^{p'-1}c_3c_2^{p'}|x_n(t)|^2 \quad \text{a.e. on } T,\end{aligned}$$

(see H(f)(iii)).

We integrate this inequality over $[0,t]$ and because $|x_0^n| = |p_n x_0| \leq \|p_n\|_{\mathcal{L}}|x_0| = |x_0|$, we obtain

$$|x_n(t)|^2 \leq |x_0|^2 + 2^{p'-1}c_3^{p'}\|\alpha_2\|_{p'}^{p'} + 2^{p'-1}c_3^{p'}\int_0^t c_2(s)^{p'}|x_n(s)|^2\mathrm{d}s. \tag{10.113}$$

From (10.113) using Gronwall's inequality, we obtain

$$|x_n(t)| \leq M_1 \qquad \text{for all } n\geq 1 \quad \text{and all } t\in T=[0,b]. \tag{10.114}$$

Recall that

$$\begin{aligned}\frac{\mathrm{d}}{\mathrm{d}t}|x_n(t)|^2 + 2c_1\|x_n(t)\|^2 &\leq 2\big|p_n f(t, x_n(t))\big|\;|x_n(t)| \quad \text{a.e. on } T,\\ &\leq 2M_1\big|p_n f(t, x_n(t))\big| \quad \text{a.e. on } T,\end{aligned}$$

(see (10.114)),

$$\begin{aligned}\Rightarrow\; 2c_1\int_0^b \|x_n(t)\|\mathrm{d}t &\leq |x_0|^2 + 2M_1\int_0^b |f(t, x_n(t))|\mathrm{d}t\\ &\leq |x_0|^2 + 2M_1\int_0^b \big(\alpha_2(t) + c_2(t)M_1^{2/p'}\big)\mathrm{d}t\end{aligned}$$

$$\Rightarrow\; \|x_n\|_{L^2(T,X)} \leq M_2 \quad \text{for some } M_2>0 \quad \text{and all } n\geq 1. \tag{10.115}$$

Also we know that

$$x_n'(t) = -A_n(t, x_n(t)) + p_n f(t, x_n(t)) \quad \text{a.e. on } T \quad \text{for all } n\geq 1. \tag{10.116}$$

From (10.116) and hypotheses H(A)(iii) and H(f)(iii), we obtain that

$$\|x_n'\|_{L^2(T,X^*)} \leq M_3 \quad \text{for some } M_3>0 \text{ and all } n\geq 1. \tag{10.117}$$

Motivated from (10.115) and (10.117), we introduce the set

$$C = \{x \in W_p(0,b) : \|x\|_{L^p(T,X)} \leq M_2 \quad \text{and} \quad \|x'\|_{L^{p'}(T,X^*)} \leq M_3\}.$$

Evidently C is bounded, closed, and convex (hence w-compact, convex) in $W_p(0,b)$. Invoking Theorem 10.1.29, we deduce that C is compact in $L^p(T,H)$. Let S_n be the solution set of the Galerkin approximation problem (10.113). Note that $S_n \subseteq C$ for all $n \geq 1$. By passing to a subsequence if necessary, we may assume that

$$\begin{aligned} x_n \xrightarrow{w} x \quad \text{in } W_p(0,b),\ x_n \longrightarrow x \quad \text{in } L^p(T,H) \\ \text{and } x_n(t) \longrightarrow x(t) \quad \text{a.e. on } T \text{ in } H. \end{aligned}$$

Also, if $N_n(x_n)(\cdot) = f_n(\cdot, x_n(t)) = p_n f_n(\cdot, x_n(t))$ (the Nemitsky operator), then

$$N_n(x_n) \xrightarrow{w} g \quad \text{in } L^{p'}(T,H) \quad \text{as } n \to \infty.$$

For every $D \subseteq T$ measurable and every $h \in H$, we have

$$\begin{aligned} \int_D (N_n(x_n)(t), h)\mathrm{d}t &= \int_D \big(p_n f(t, x_n(t)), h\big)\mathrm{d}t \\ &= \int_D \big(f(t, x_n(t)), p_n h\big)\mathrm{d}t, \\ \Rightarrow \int_D (g(t), h)\mathrm{d}t &= \int_D \big(f(t, x(t)), h\big)\mathrm{d}t \quad \text{(because } p_n h \longrightarrow h \text{ in } H). \end{aligned}$$

Because $D \subseteq T$ and $h \in H$, were arbitrary, we infer that

$$g(t) = f(t, x(t)) \quad \text{a.e. on } T. \tag{10.118}$$

Let $\widehat{A}_n : L^p(T, X_n) \longrightarrow L^{p'}(T, X_n)$, $n \geq 1$, be the Nemitsky operator corresponding to $A_n(t,x)$; that is, $\widehat{A}_n(x)(\cdot) = A_n(\cdot, x(\cdot))$ for all $x \in L^p(T, X_n)$. Also let $((\cdot,\cdot))$ denote the duality brackets for the pair $(L^{p'}(T,X^*), L^p(T,X))$; that is,

$$((u^*, x)) = \int_0^b \langle u^*(t), x(t)\rangle\, \mathrm{d}t \qquad \text{for all } x \in L^p(T,X),\ u^* \in L^{p'}(T,X^*).$$

Then for every $y \in L^p(T,X)$, we have

$$((x_n', y)) + ((\widehat{A}_n(x_n), y)) = ((N_n(x_n), y)). \tag{10.119}$$

Because of hypothesis H(A)(iii) and (10.115), we may assume that

$$\widehat{A}_n(x_n) \xrightarrow{w} v \quad \text{in } L^{p'}(T,X^*) \quad \text{as } n \to \infty.$$

So, if we pass to the limit as $n \to \infty$ in (10.119), we obtain

$$\begin{aligned} &((x', y)) + ((v, y)) = ((N(x), y)) \qquad \text{for all } y \in L^p(T,X), \\ &\quad \text{with } N(x)(\cdot) = f(\cdot, x(\cdot)), \quad \text{(see (10.118))}, \\ &\Rightarrow x'(t) + v(t) = f(t, x(t)) \quad \text{a.e. on } T. \end{aligned} \tag{10.120}$$

From Theorem 10.1.25, we know that $W_p(0,b)$ is embedded continuously in $C(T,H)$. So $x_n(0) = x_0^n \xrightarrow{w} x(0) = x_0$ and $x_n(b) \xrightarrow{w} x(b)$ in H. Also

$$\begin{aligned}
&\big((N_n,(x_n),x_n)\big) - \big((x_n',x_n)\big) = \big((\widehat{A}_n(x_n),x_n)\big)\\
\Rightarrow\ &\big((N_n(x_n),x_n)\big) - \frac{1}{2}|x_n(b)|^2 + \frac{1}{2}|x_0^n|^2 = \big((\widehat{A}_n(x_n),x_n)\big)
\end{aligned}$$

(see Corollary 10.1.26),

$$\Rightarrow\ \big((N(x),x)\big) - \frac{1}{2}|x_n(b)|^2 + \frac{1}{2}|x_0|^2 \geq \limsup_{n\to\infty}\big((\widehat{A}_n(x_n),x_n)\big). \tag{10.121}$$

From (10.120), with $y = x$ and (10.121), we have

$$\begin{aligned}
&\limsup_{n\to\infty}\big((\widehat{A}_n(x_n),x_n)\big) \leq \big((v,x)\big),\\
\Rightarrow\ &\limsup_{n\to\infty}\big((\widehat{A}_n(x_n),x_n - x)\big) \leq 0.
\end{aligned}\tag{10.122}$$

But it is easy to see that $\widehat{A}$ is hemicontinuous and monotone, hence maximal monotone. So it is generalized pseudomonotone and from (10.122) we have $\widehat{A}(x) = v$; that is, $\widehat{A}_n(x_n) \xrightarrow{w} \widehat{A}(x)$ in $L^{p'}(T,X^*)$. Thus finally

$$\begin{aligned}
&x' + \widehat{A}(x) = N(x), \quad x(0) = x_0,\\
\Rightarrow\ &x'(t) + A\big(t,x(t)\big) = f\big(t,x(t)\big) \quad \text{a.e. on } T,\ x(0) = x_0,\\
\Rightarrow\ &x \in S(x_0) \neq \emptyset.
\end{aligned}$$

Next we show the topological properties of $S(x_0) \subseteq W_p(0,b)$. From (10.114), (10.115), and (10.117), we know that we can find $M = \max\{M_1, M_2, M_3\} > 0$ such that

$$\|x\|_{L^p(T,X)},\ \|x'\|_{L^{p'}(T,X^*)},\ \|x\|_{C(T,X)} \leq M \qquad \text{for all } x \in S(x_0).$$

If we set $\vartheta(t) = \alpha_2(t) + c_2(t)M^{2/p'}$, then $\vartheta \in L^{p'}(T)$ and we may assume that

$$|f(t,x)| \leq \vartheta(t) \quad \text{for a.a. } t\in T, \text{ all } x \in H.$$

Otherwise we replace $f(t,x)$ by $f\big(t,r_M(x)\big)$, with $r_M : H \longrightarrow H$ being the M-radical retraction; that is,

$$r_M(x) = \begin{cases} x & \text{if } |x| \leq M \\ \frac{Mx}{|x|} & \text{if } |x| > M \end{cases} \quad \text{for all } x \in H.$$

Then we introduce the set

$$K = \big\{h \in L^{p'}(T,H) : |h(t)| \leq \vartheta(t) \quad \text{a.e. on } T\big\}.$$

In what follows we consider K equipped with the relatively weak $L^{p'}(T,H)$-topology. Then K becomes a compact metrizable space. For every $h \in K$ we consider the auxiliary Cauchy problem

$$\left\{\begin{array}{ll} -x'(t) + A\big(t,x(t)\big) = h(t) & \text{a.e. on } T, \\ x(0) = x_0 & \end{array}\right\}.$$

This problem has a unique solution $x = \eta(h) \in W_p(0,b)$. We consider the solution map $\eta : L^{p'}(T,H) \longrightarrow C(T,H)$.

Claim: $\eta(K)$ is compact in $C(T,H)$.

Let $\{x_n\}_{n\geq 1} \subseteq \eta(K)$. Then $x_n = \eta(h_n)$ with $h_n \in K$, $n\geq 1$. From the a priori estimation of the first part of the proof, we know that $\{x_n\}_{n\geq 1} \subseteq W_p(0,b)$ is bounded. So we may assume that

$$\begin{aligned} & x_n \xrightarrow{w} x \quad \text{in } W_p(0,b), \quad x_n \longrightarrow x \quad \text{in } L^p(T,H) \quad \text{(see Theorem 10.1.29)} \\ & x_n(t) \longrightarrow x(t) \qquad \text{for all } t\in T\setminus N_0,\ |N_0|=0 \quad \text{in } H \\ \text{and} \quad & h_n \xrightarrow{w} h \quad \text{in } L^{p'}(T,H) \quad \text{as } n\to\infty. \end{aligned}$$

The sequence $\{\langle x_n'(\cdot), x_n(\cdot) - x(\cdot)\rangle\}_{n\geq 1}$ is uniformly integrable. Therefore given $\varepsilon>0$, we can find $t\in T\setminus N_0$ such that

$$\int_t^b |\langle x_n'(s), x_n(s) - x(s)\rangle| \mathrm{d}s < \varepsilon \qquad \text{for all } n\geq 1. \tag{10.123}$$

For $t \in T$, let $((\cdot,\cdot))_t$ denote the duality brackets for the pair $\big(L^{p'}([0,t],X^*), L^p([0,t],X)\big)$. Using Corollary 10.1.26, we have

$$((x_n', x_n - x))_t = \frac{1}{2}|x_n(t) - x(t)|^2 + ((x', x_n - x))_t. \tag{10.124}$$

If $t\in T\setminus N_0$, then $|x_n(t)-x(t)|^2 \longrightarrow 0$. Also because $x_n \xrightarrow{w} x$ in $W_p(0,b)$, we have $((x', x_n - x))_t \longrightarrow 0$ as $n\to\infty$. So from (10.124), we infer that

$$((x_n', x_n - x))_t \longrightarrow 0 \quad \text{as } n\to\infty.$$

We have

$$\begin{aligned} ((x_n', x_n - x)) &= ((x_n', x_n - x))_t + \int_t^b \langle x_n'(s), x_n(s) - x(s)\rangle \,\mathrm{d}s \\ &\geq ((x_n', x_n - x))_t - \varepsilon \quad \text{(see (10.123))}, \\ \Rightarrow \ \liminf_{n\to\infty} ((x_n', x_n - x)) &\geq 0 \quad (\text{because } \varepsilon>0 \text{ was arbitrary}). \end{aligned} \tag{10.125}$$

In a similar fashion, we show that

$$\limsup_{n\to\infty} ((x_n', x_n - x)) \leq 0. \tag{10.126}$$

Therefore, from (10.125) and (10.126) we conclude that

$$((x_n', x_n - x)) \longrightarrow 0 \quad \text{as } n\to\infty. \tag{10.127}$$

Note that

$$\begin{aligned} & ((x_n', x_n - x)) + ((\widehat{A}(x_n), x_n - x)) = \int_0^b (h_n(t), x_n(t) - x(t))\mathrm{d}t, \\ \Rightarrow \ & \lim_{n\to\infty} ((\widehat{A}(x_n), x_n - x)) = 0. \end{aligned} \tag{10.128}$$

As before from (10.128), it follows that

$$\widehat{A}(x_n) \xrightarrow{w} \widehat{A}(x) \quad \text{in } L^{p'}(T, X^*).$$

So, in the limit as $n \to \infty$, we have

$$\begin{aligned} & x' + \widehat{A}(x) = h, \ h \in K, \ x(0) = x_0, \\ \Rightarrow \ & x \in \eta(K). \end{aligned}$$

We show that $x_n \longrightarrow x$ in $C(T, H)$. We have

$$\begin{aligned} & \frac{1}{2}|x_n(t) - x(t)|^2 = \big((h_n - h, x_n - x)\big) - \big((\widehat{A}(x_n) - \widehat{A}(x), x_n - x)\big)_t, \\ \Rightarrow \ & \frac{1}{2}|x_n(t) - x(t)|^2 \leq \int_0^b |(h_n(s) - h(s), x_n(s) - x(s))|\mathrm{d}s \\ & \qquad + \int_0^b |\langle A(s), x_n(s) - x(s)\rangle|\mathrm{d}t + \big((\widehat{A}(x), x_n - x)\big)_t. \end{aligned} \tag{10.129}$$

Note that

$$\int_0^b |(h_n(s) - h(s), x_n(s) - x(s))|\mathrm{d}s \longrightarrow 0 \quad \text{as } n \to \infty. \tag{10.130}$$

Let $\beta_n(t) = \big\langle A\big(t, x_n(t), x_n(t) - x(t)\big)\big\rangle$. If $t \in T \setminus \overline{N}_0$, $N_0 \subseteq \overline{N}_0$, $|\overline{N}_0| = 0$, then

$$\beta_n(t) \geq c_1\|x_n(t)\|^p \alpha(t) + c\|x_n(t)\|^{p-1}\|x(t)\|. \tag{10.131}$$

Let $D = \big\{t \in T : \liminf\limits_{n\to\infty} \beta_n(t) < 0\big\}$. This is a Lebesgue measurable subset of T. Suppose $|D| > 0$. For all $t \in C \cap (T \setminus \overline{N}_0) \neq \emptyset$ we have that $\{x_n(t)\}_{n\geq 1} \subseteq X$ is bounded (see (10.131)). Because $x_n \xrightarrow{w} x$ in $W_p(0, b)$ and $W_p(0, b)$ is embedded continuously in $C(T, H)$, we have $x_n \xrightarrow{w} x$ in $C(T, H)$ and so $x_n(t) \xrightarrow{w} x(t)$ in H for all $t \in T$. Therefore, exploiting the reflexivity and separability of X, we have $x_n(t) \xrightarrow{w} x(t)$ in X as $n \to \infty$ for all $t \in T$. Then because of the monotonicity of $A(t, \cdot)$, we have

$$\liminf_{n\to\infty} \big\langle A\big(t, x_n(t)\big), x_n(t) - x(t)\big\rangle \geq \lim_{n\to 0} \big\langle A\big(t, x(t)\big), x_n(t) - x(t)\big\rangle = 0$$

for all $t \in D \cap (T \setminus N_0)$, a contradiction to the definition of D. Therefore $|D| = 0$ and we have

$$0 \leq \liminf_{n\to\infty} \beta_n(t) \quad \text{a.e. on } T.$$

So from Fatou's lemma, we obtain

$$0 \leq \int_0^b \liminf_{n\to\infty} \beta_n(t)\mathrm{d}t \leq \liminf_{n\to\infty} \int_0^b \beta_n(t)\mathrm{d}t = \lim\big((\widehat{A}(x_n), x_n - x)\big) = 0,$$

(see (10.128)),

$$\Rightarrow \int_0^b \beta_n(t)\mathrm{d}t \longrightarrow 0.$$

From (10.131), we see that there exists $\{\gamma_n\}_{n\geq 1}$ uniformly integrable sequence such that

$$\begin{aligned}
&\gamma_n(t) \leq \beta_n(t) \quad \text{a.e. on } T, \qquad \text{for all } n \geq 1,\\
\Rightarrow\ & 0 \leq \beta_n^-(t) \leq \gamma_n^-(t) \quad \text{a.e. on } T, \qquad \text{for all } n \geq 1\\
\Rightarrow\ & \int_0^b \beta_n^-(t)dt \longrightarrow 0 \quad \text{as } n \to \infty.
\end{aligned}$$

Therefore, we have

$$\begin{aligned}
&\int_0^b |\beta_n(t)|dt = \int_0^b \beta_n(t)dt + 2\int_0^b \beta_n^-(t)dt \longrightarrow 0,\\
\Rightarrow\ & \beta_n \longrightarrow 0 \quad \text{in } L^1(T),\\
\Rightarrow\ & \int_0^b \left| \left\langle A\big(t, x_n(t), x_n(t)\big), x_n(t) - x(t) \right\rangle \right| \longrightarrow 0. \qquad (10.132)
\end{aligned}$$

Finally let $\xi_n(t) = \big((A(x), x_n - x)\big)_t$. Then $\xi_n \in C(T)$ and let $t_n \in T$ such that

$$\begin{aligned}
\xi_n(t_n) = \max_T \xi_n &= \int_0^{t_n} \left\langle A\big(s, x(s)\big), x_n(s) - x(s) \right\rangle ds\\
&= \int_0^b \chi_{[0,t_n]} \left\langle A\big(s, x(s)\big), x_n(s) - x(s) \right\rangle ds\\
&= \big((\chi_{[0,t_n]} \widehat{A}(x), x_n - x)\big) \longrightarrow 0,\\
\Rightarrow\ \sup_T \big((A(x), x_n - x)\big)_t &\longrightarrow 0. \qquad (10.133)
\end{aligned}$$

Returning to (10.129) and using (10.130), (10.132), and (10.133), we obtain

$$\begin{aligned}
&x_n \longrightarrow x \quad \text{in } C(T,H) \quad \text{as } n \to \infty \quad \text{and} \quad x \in \eta(K)\\
\Rightarrow\ & \eta(K) \quad \text{is compact in } C(T,H).
\end{aligned}$$

This proves the claim.

Note that $S(x_0) \subseteq \eta(K)$. Also from the previous part of the proof, we have that $S(x_0) \subseteq W_p(0,b)$ is weakly closed. So $S(x_0) \subseteq W_p(0,b)$ is weakly compact in $W_p(0,b)$ and compact in $C(T,H)$. □

Let $\widehat{\eta} : L^{p'}(T,H) \times H \longrightarrow C(T,H)$ be the map that for each pair $(h, x_0) \in L^{p'}(T,H) \times H$ assigns the unique solution of

$$\left\{ \begin{array}{l} -x'(t) + A\big(t, x(t)\big) = h(t) \quad \text{a.e. on } T,\\ x(0) = x_0 \end{array} \right\}. \qquad (10.134)$$

From the proof of Theorem 10.3.12, we have as a byproduct the following continuous dependence result.

PROPOSITION 10.3.13 *If hypotheses $H(A)$ hold and X is embedded compactly into H, then $\widehat{\eta}: L^{p'}(T,H)_w \times H \longrightarrow C(T,H)$ is sequentially continuous.*

Recall that $S_n \subseteq W_p(0,b)$ is the solution set of the Galerkin approximation (10.109). Another byproduct of the proof of Theorem 10.3.12, is the following result.

PROPOSITION 10.3.14 *If hypotheses $H(A)$, $H(f)$ hold, $x_0 \in H$, and X is embedded compactly in H, then* $\limsup_{n\to\infty} S_n \subseteq S(x_0)$ *and* $h^*\big(S_n, S(x_0)\big) \longrightarrow 0$ *in* $C(T,H)$.

In particular, if $S(x_0)$ is a singleton (this is the case if $f(t,\cdot)$ is Lipschitz continuous) and $x_n \in W_p(0,b)$ is the unique solution of the Galerkin approximation (10.109), then $x_n \longrightarrow x$ in $C(T,H)$ with $x \in W_p(0,b)$ the unique solution of (10.108).

We can generalize hypotheses H(A) and still have an existence theorem for problem (10.74). In concrete parabolic problems, the new hypotheses on $A(t,x)$ permit the differential operator to have lower-order terms of nonmonotone nature. In this extended existence theorem, we need the following notion.

DEFINITION 10.3.15 Let X be a reflexive Banach space, $L : \mathcal{D}(L) \subseteq X \longrightarrow X^*$ is a linear maximal monotone operator, and $A : X \longrightarrow 2^{X^*}$. We say that A is *L-pseudomonotone* if the following conditions hold.

(a) For every $x \in X$, $A(x) \subseteq X^*$ is nonempty, weakly compact, and convex.
(b) A is usc from every finite-dimensional subspace of X into X^* furnished with the weak topology.
(c) If $\{x_n\}_{n\geq 1} \subseteq \mathcal{D}(L)$, $x_n \xrightarrow{w} x \in \mathcal{D}(L)$ in X, $Lx_n \xrightarrow{w} Lx$ in X^*, $x_n^* \in A(x_n)$ for all $n \geq 1$, $x_n^* \xrightarrow{w} x^*$ in X^* and $\limsup_{n\to\infty} \langle x_n^*, x_n - x\rangle \leq 0$, then $x^* \in A(x)$ and $\langle x_n^*, x_n\rangle \longrightarrow \langle x^*, x\rangle$.

The next result is the analogue of Theorem 3.2.60 for L-pseudomonotone operators. Its proof can be found in Papageorgiou–Papalini–Renzacci [482].

THEOREM 10.3.16 *If X is a reflexive Banach space that is strictly convex, $L : \mathcal{D}(L) \subseteq X \longrightarrow X^*$ is a linear maximal monotone operator, and $A : X \longrightarrow 2^{X^*}$ is bounded, L-pseudomonotone and coercive, then $L + A$ is surjective; that is, $\mathcal{R}(L + A) = X^*$.*

For the extended version of Theorem 10.3.12, we employ the following conditions on the map $A(t,x)$. Here X is a separable reflexive Banach space and X^* its topological dual.

H(A)′: $A : T \times X \longrightarrow X^*$ is a map such that

(i) For all $x \in X$, $t \longrightarrow A(t,x)$ is measurable.

(ii) For almost all $t \in T$, $x \longrightarrow A(t,x)$ is pseudomonotone.

(iii) For almost all $t \in T$ and all $x \in X$, we have

$$\|A(t,x)\|_* \leq \alpha(t) + c\|x\|^{p-1},$$

with $2 \leq p < \infty$, $\alpha \in L^{p'}(T)_+$, $\left(\frac{1}{p} + \frac{1}{p'} = 1\right)$ and $c > 0$.

(iv) For almost all $t \in T$ and all $x \in X$, we have

$$\langle A(t,x), x\rangle \geq c_1\|x\|^p - \vartheta(t),$$

with $c_1 > 0$, $\vartheta \in L^1(T)_+$.

REMARK 10.3.17 Note that the pseudomonotonicity of $A(t,\cdot)$ for almost all $t\in T$ (see H(A)(ii)) and the p-growth condition (see H(A)(iii)), imply that for almost all $t\in T$, $A(t,\cdot)$ is demicontinuous.

Using an argument similar to that in the last part of the proof of Theorem 10.3.12 with the functions $\{\beta_n\}_{n\geq 1}$ and the set D, we can show the following "lifting" theorem (see also Hu–Papageorgiou [316, p. 41]).

THEOREM 10.3.18 *If hypotheses H$(A)'$ hold and $\widehat{A}: L^p(T,X)\longrightarrow L^{p'}(T,X^*)$ is defined by*

$$\widehat{A}(x)(\cdot)=A(\cdot,x(\cdot))$$

(the Nemitsky operator corresponding to A), then $\widehat{A}$ is demicontinuous and if $x_n\xrightarrow{w} x$ in $L^p(T,X)$ and

$$\limsup_{n\to\infty}\big((\widehat{A}(x_n),x_n-x)\big)\leq 0,$$

we have $\widehat{A}(x_n)\xrightarrow{w}\widehat{A}(x)$ in $L^{p'}(T,X^*)$ and $\big((\widehat{A}(x_n),x_n-x)\big)\longrightarrow\big((\widehat{A}(x),x)\big)$ as $n\to\infty$.

H(f)$'$: $f: T\times H\longrightarrow H$ is a function such that

(i) For all $x\in H$, $t\longrightarrow f(t,x)$ is measurable.

(ii) For almost all $t\in T$, $x\longrightarrow f(t,x)$ is sequentially continuous from H into H_w.

(iii) For almost all $t\in T$ and all $x\in H$, we have

$$|f(t,x)|\leq \alpha_2(t)+c_2|x|^{2/q}$$

with $\alpha_2\in L^{p'}(T)_+$, $c_2>0$ if $p>2$ and $\alpha_2\in L^{p'}(T)_+$, $c_2>0$, $c_2\beta^2<c_1/2$, where $\beta>0$ is such that $|\cdot|\leq\beta\|\cdot\|$ when $p=2$.

Using Theorems 10.3.16 and 10.3.18, we can have the following extension of Theorem 10.3.12. For the proof see Hu–Papageorgiou [316, p. 42].

THEOREM 10.3.19 *If in the evolution triple (X,H,X^*) the embedding of X into H is also compact, hypotheses H$(A)'$ and H$(f)'$ hold, and $x_0\in H$, then the solution set $S(x_0)$ of problem (10.74) is nonempty, weakly compact in $W_p(0,b)$, and compact in $C(T,H)$.*

Let us see an application to a nonlinear parabolic problem.

So let $T=[0,b]$ and let $Z\subseteq\mathbb{R}^N$ be a bounded domain with a C^2-boundary ∂Z. We consider the following nonlinear parabolic initial boundary value problem.

$$\left\{\begin{array}{l}\frac{\partial x}{\partial t}-\operatorname{div}\big(a(t,z)\|Dx(t,z)\|^{p-2}Dx(t,x)\big)=f_0\big(t,z,x(t,z)\big)\quad\text{in } T\times Z,\\ x(0,z)=x_0(z)\quad\text{a.e. on } Z,\\ x\big|_{T\times\partial Z}=0,\ 2\leq p<\infty\end{array}\right\}.\tag{10.135}$$

We impose the following conditions on the data of (10.135):

H(α): $\alpha : T \times Z \longrightarrow \mathbb{R}$ is a measurable function such that

$$0 < \gamma_1 \leq \alpha(t,z) \leq \gamma_2 \quad \text{for almost all } (t,z) \in T \times Z.$$

H(f₀): $f_0 : T \times Z \times \mathbb{R} \longrightarrow \mathbb{R}$ is a function such that

(i) For all $x \in \mathbb{R}$, $(t,z) \longrightarrow f_0(t,z,x)$ is measurable.

(ii) For almost all $(t,z) \in T \times Z$, $x \longrightarrow f_0(t,z,x)$ is continuous.

(iii) $|f_0(t,z,x)| \leq \eta_1(t,z) + \eta_2 |x|^{2/p'}$ for a.a. $(t,z) \in T \times Z$, with $\eta_1 \in L^{p'}\big(T, L^2(Z)\big)$ and $\eta_2 \in L^{p'}\big(T, L^\infty(Z)\big)$.

PROPOSITION 10.3.20 *If hypotheses $H(\alpha)$, $H(f_0)$ hold and $x_0 \in L^2(Z)$, then problem (10.135) has a solution $x \in L^2\big(T, W_0^{1,p}(Z)\big) \cap C\big(T, L^2(Z)\big)$ such that $\partial x / \partial t \in L^2\big(T, W^{-1,p'}(Z)\big)$ and the solution set of the problem is compact in $C\big(T, L^2(Z)\big)$.*

PROOF: We consider the evolution triple (X, H, X^*), consisting of the following spaces.

$$X = W_0^{1,p}(Z),\ H = L^2(Z),\ X^* = W^{-1,p'}(Z)\ \Big(\frac{1}{p} + \frac{1}{p'} = 1\Big).$$

From the Sobolev embedding theorem, we know that X is embedded compactly into H (recall $p \geq 2$).

Consider the time-dependent Dirichlet form $\widehat{\alpha} : T \times X \times X \longrightarrow \mathbb{R}$ defined by

$$\widehat{\alpha}(t,x,y) = \int_Z \alpha(t,z) \|Dx\|^{p-2} (Dx, Dy)_{\mathbb{R}^N} dz.$$

Using hypotheses $\mathrm{H}(\alpha)$, we can easily check that

$$|\widehat{\alpha}(t,x,y)| \leq c_3 \|x\|^{p-1} \|y\| \quad \text{for some } c_3 > 0.$$

So we can introduce a map $A : T \times X \longrightarrow X^*$ defined by

$$\langle A(t,x), y \rangle = \widehat{\alpha}(t,x,y).$$

Then clearly we have.

- For all $x \in X$, $t \longrightarrow A(t,x)$ is measurable (see Theorem 10.1.2).
- For almost all $t \in T$, $x \longrightarrow A(t,x)$ is demicontinuous, monotone.
- For almost all $t \in T$ and all $x \in X$, we have

$$\|A(t,x)\|_* \leq c \|x\|^{p-1}.$$

- For almost all $t \in T$ and all $x \in X$, we have

$$\langle A(t,x), x \rangle \geq c_1 \|x\|^p.$$

Also let $f : T\times H \longrightarrow H$ be the Nemitsky operator corresponding to the function $f_0(t,\cdot,\cdot)$; that is,

$$f(t,x)(\cdot) = f_0\big(t,\cdot,x(\cdot)\big).$$

It is straightforward to check that f satisfies hypotheses H(f). We equivalently rewrite (10.135) as the following nonlinear evolution equation.

$$\left\{\begin{array}{ll} x'(t) + A\big(t,x(t)\big) = f\big(t,x(t)\big) & \text{a.e. on } T, \\ x(0) = x_0 \in H & \end{array}\right\}.$$

Invoking Theorem 10.3.12, we obtain all the conclusions of the proposition. □

10.4 Second-Order Nonlinear Evolution Equations

In this section we deal with second-order nonlinear evolution equations. We consider two classes of problems, both defined in the framework of evolution triples. In the first class, the evolution triple consists of Hilbert spaces, the evolution equation involves multivalued terms, and the analysis is conducted using the material from the first half of Section 10.3. As a particular case of this class, we obtain hyperbolic variational inequalities. The second class of problems is formulated on a general evolution triple, involves maximal monotone coercive operators, and the method of proof is based on the results of the second half of Section 10.3.

So let $T = [0,b]$ and let (X,H,X^*) be an evolution triple consisting of Hilbert spaces. We are interested in the following second-order nonlinear evolution equation.

$$\left\{\begin{array}{ll} x''(t) + Ax(t) + B\big(x'(t)\big) \ni g(t) & \text{a.e. on } T, \\ x(0) = x_0,\ x'(0) = y_0 & \end{array}\right\}. \tag{10.136}$$

The hypotheses on the data of (10.136) are the following.

H(A): $A\in\mathcal{L}(X,X^*)$ and $\langle Ax,x\rangle + c_1|x|^2 \geq c_2\|x\|^2$ for all $x\in X$ with $c_1\in\mathbb{R}$, $c_2>0$.

H(B): $B : \mathcal{D}(B)\subseteq X \longrightarrow 2^{X^*}$ is maximal monotone.

THEOREM 10.4.1 *If hypotheses H(A), H(B) hold, $g\in W^{1,1}\big((0,b),H\big)$, $x_0\in X$, $y_0\in\mathcal{D}(B)$, and*

$$\big(Ax_0 + B(y_0)\big)\cap H \neq \emptyset,$$

then problem (10.136) admits a unique strong solution $x(\cdot)$ such that

$$x\in W^{1,\infty}\big((0,b),X\big)\cap W^{2,\infty}\big((0,b),H\big)$$

$\frac{d^+}{dt}x(t)$ exists in X, $\frac{d^+}{dt}x'(t)$ exists in H, both for all $t\in[0,b)$ and

$$\frac{d^+}{dt}x'(t) + Ax(t) + B\big(x'(t)\big) \ni g(t) \qquad \textit{for all } t\in[0,b).$$

PROOF: Let $H_0 = X\times H$. This is a Hilbert space with inner product

$$(u_1,u_2)_0 = \langle Ax_1,x_2\rangle + c_1(x_1,x_2) + (y_1,y_2)$$

for all $u_1 = [x_1,y_1]$, $u_2 = [x_2,y_2]\in X\times H$.

Let $E:\mathcal{D}(E)\subseteq H_0 \longrightarrow 2^{H_0}$ be the nonlinear operator defined by

$$E(u) = E([x,y]) = \big[-y, \big(Ax + B(y)\big)\cap H\big] + \xi[x,y]$$

for all $[x,y] \in \mathcal{D}(E) = \big\{[x,y] \in X \times H : \big(Ax + B(y)\big)\cap H \neq \emptyset\big\}$ and with

$$\xi = \sup\Big\{\frac{c_1(x,y)}{\langle Ax, y\rangle + c_1|x|^2 + |y|^2} : x\in X,\ y \in H,\ \|x\| + |y| \neq 0\Big\}. \tag{10.137}$$

Then we can equivalently rewrite problem (10.136), as the following first-order evolution equation in the Hilbert space $H_0 = X\times H$:

$$\left\{\begin{array}{ll} u'(t) \in -E\big(u(t)\big) + \xi u(t) + \widehat{g}(t) & \text{on } T, \\ u(0) = u_0 & \end{array}\right\}, \tag{10.138}$$

where $u(t) = [x(t), x'(t)]$, $\widehat{g}(t) = \big[0, g(t)\big]$ and $u_0 = [x_0, y_0]$.

It is easy to check that E is monotone in H_0. We claim that in fact E is maximal monotone. It suffices to show that $\mathcal{R}(I+E) = H_0$. Let $(v,h) \in X \times H$ be arbitrary. Then the conclusion

$$u + E(u) \ni (v,h)$$

is actually equivalent to the following system

$$\left\{\begin{array}{l} x + (-y + \xi x) = v, \\ y + \big(Ax + B(y) + \xi y\big) \ni h \end{array}\right\},$$

$$\Rightarrow \left\{\begin{array}{l} (1+\xi)x - y = v, \\ (1+\xi)y + Ax + B(y) \ni h \end{array}\right\}. \tag{10.139}$$

From the first equation in (10.139) we have

$$x = \frac{1}{1+\xi}(v+y).$$

Using this in the second inclusion of system (10.139), we obtain

$$(1+\xi)y + \frac{1}{1+\xi}Ay + B(y) \ni -\frac{1}{1+\xi}Av. \tag{10.140}$$

Note that $y \longrightarrow (1+\xi)y + 1/(1+\xi)Ay$ is continuous, monotone, and coercive from X into X^*. Then $y(1+\xi)y + 1/(1+\xi)Ay + B(y)$ is maximal monotone, and coercive (see Theorem 3.2.33). Therefore by virtue of Corollary 3.2.28 it is also surjective. So the inclusion (10.140) has a unique solution $y \in \mathcal{D}(E)$. Hence $[v,h] \in \mathcal{R}(I+E)$. Therefore, E is maximal monotone and we can apply Theorem 10.3.6 to finish the proof of the theorem. □

If $B = \partial\varphi$ with $\varphi \in \Gamma_0(X)$, then we have an existence result for nonlinear hyperbolic variational inequalities (see Theorem 10.3.7).

COROLLARY 10.4.2 *If hypothesis H(A) holds, $B = \partial\varphi$ with $\varphi \in \Gamma_0(X)$, $g \in W^{1,1}\big((0,b), H\big)$, $x_0 \in X$, $y_0 \in \mathcal{D}(\partial\varphi)$, and*

$$\big(Ax_0 + B(y_0)\big)\cap H \neq \emptyset,$$

then there exists a unique function $x \in C(T,X)$ *such that*

$$x' \in C(T,H) \cap L^\infty(T,X),\ x'(t) \in \operatorname{dom}\varphi \quad \textit{for all } t \in (0,b)$$
$$x'' \in L^\infty(T,H)$$
$$\big\langle -x''(t) - Ax(t) + g(t), y - x'(t)\big\rangle \leq \varphi(y) - \varphi\big(x(t)\big),$$

for a.a. $t \in T$, *all* $y \in X$

$$x(0) = x_0,\ x'(0) = y_0.$$

As an application consider the following hyperbolic initial-boundary value problem. Let $T = [0,b]$ and $Z \subseteq \mathbb{R}^N$ is a bounded domain with a C^2-boundary ∂Z.

$$\left\{ \begin{array}{ll} \frac{\partial^2 x}{\partial t^2} - \Delta x(t,z) + \beta\Big(\frac{\partial x}{\partial t}\Big) \ni g(t,z) & \text{a.e. on } T \times Z, \\ x(z,0) = x_0(z),\ \frac{\partial x}{\partial t}(z,0) = y_0(z) & \text{a.e. on } Z, \\ x\big|_{T\times\partial Z} = 0 & \end{array} \right\}. \tag{10.141}$$

PROPOSITION 10.4.3 *If* $\beta : \mathbb{R} \longrightarrow 2^{\mathbb{R}}$ *is a maximal monotone function (hence* $\beta = \partial j$ *with* $j \in \Gamma_0(\mathbb{R})$*) with* $0 \in \mathcal{D}(\beta)$, $g \in W^{1,2}\big((0,b), L^2(Z)\big)$, $x_0 \in H_0^1(Z) \cap H^2(Z)$, *and* $y_0 \in H_0^1(Z)$ *with* $y_0(z) \in \overline{\mathcal{D}(\beta)}$ *a.e. on* Z, *then problem (10.141) has a unique solution* $x(t,z)$ *that satisfies*

$$x \in C\big(T, H_0^1(Z)\big) \cap L^\infty\big(T, H^2(Z)\big)$$
$$\frac{\partial x}{\partial t} \in C\big(T, L^2(Z)\big) \cap L^\infty\big(T, H_0^1(Z)\big)$$
$$\sqrt{t}\frac{\partial^2 x}{\partial t^2} \in L^2(T \times Z),\ t\frac{\partial^2 x}{\partial t^2} \in L^\infty\big(T, L^2(Z)\big).$$

Moreover, if $j\big(y_0(\cdot)\big) \in L^1(Z)$, *then* $\partial^2 x/\partial t^2 \in L^2(T \times Z)$.

PROOF: Let $X = H_0^1(Z)$, $H = L^2(Z)$ and $X^* = H^{-1}(Z)$. Let $A \in \mathcal{L}\big(H_0^1(Z), H^{-1}(Z)\big)$ be defined by

$$\langle A(x), y\rangle = \int_Z (Dx, Dy)_{\mathbb{R}^N} dz \qquad \text{for all } x, y \in H_0^1(Z).$$

Hence $Ax = -\Delta x$ (by integration by parts). Also let $\varphi \in \Gamma_0\big(H_0^1(Z)\big)$ be defined by

$$\varphi(x) = \int_Z j\big(x(z)\big)dz \qquad \text{for all } x \in H_0^1(Z) \ \text{(recall } \beta = \partial j \text{ with } j \in \Gamma_0(\mathbb{R})\text{)}.$$

Note that $\{x \in L^2(Z) : A \in L^2(Z)\} = H_0^1(Z) \cap H^2(Z)$. Therefore, we have

$$\big(Ax_0 + \partial\varphi(y_0)\big) \cap H \neq \emptyset.$$

Thus we can apply Corollary 10.4.2 and establish the conclusions of the proposition. □

Next we consider a second-order evolution equation defined in the framework of a general evolution triple.

So let $T = [0,b]$ and (X, H, X^*) a general evolution triple of spaces (see Definition 10.1.21). We examine the following second-order Cauchy problem:

$$\left\{\begin{array}{l} x''(t) + A(t, x'(t)) + Bx(t) = f(t, x(t), x'(t)) \quad \text{a.e. on } T, \\ x(0) = x_0, \ x'(0) = y_0 \end{array}\right\}. \tag{10.142}$$

In problem (10.142), $A : T \times X \longrightarrow X^*$ is a nonlinear in general operator, $B \in \mathcal{L}(X, X^*)$, $f : T \times H \times H \longrightarrow H$, and $(x_0, y_0) \in X \times H$.

DEFINITION 10.4.4 By a *solution* of problem (10.142), we mean a function $x \in C(T, X)$ such that $x' \in W_p(0, b)$, $(1 < p < \infty)$, it satisfies the equation in problem (10.142) for almost all $t \in T$ and $x(0) = x_0$, $x'(0) = y_0$. We denote the set of solutions of problem (10.142) by $\mathcal{S}(x_0, y_0)$.

REMARK 10.4.5 Because $W_p(0, b) \subseteq C(T, H)$ (see Theorem 10.1.25), it follows that $\mathcal{S}(x_0, y_0) \subseteq C^1(T, H)$.

We impose the following hypotheses on the maps $A(t, x)$, B, and $f(t, x, y)$.

H(A)$'$: $A : T \times X \longrightarrow X^*$ is a map such that

(i) For all $x \in X$, $t \longrightarrow A(t, x)$ is measurable.

(ii) For almost all $t \in T$, $x \longrightarrow A(t, x)$ is demicontinuous monotone.

(iii) For almost all $t \in T$ and all $x \in X$, we have

$$\|A(t, x)\|_* \leq \alpha(t) + c\|x\|^{p-1},$$

with $2 \leq p < \infty$, $\alpha \in L^{p'}(T)_+$, $(1/p) + (1/p') = 1$ and $c > 0$.

(iv) For almost all $t \in T$ and all $x \in X$, we have

$$\langle A(t, x), x \rangle \geq c_1 \|x\|^p - \alpha_1(t),$$

with $c_1 > 0$ and $\alpha_1 \in L^1(T)_+$.

H(B)$'$: $B \in \mathcal{L}(X, X^*)$, $\langle Bx, x \rangle \geq 0$ for all $x \in X$ and $\langle Bx, u \rangle = \langle x, Bu \rangle$ for all $x, u \in X$.

H(f): $f : T \times H \times H \longrightarrow H$ is a function such that

(i) For all $x, y \in H$, $t \longrightarrow f(t, x, y)$ is measurable.

(ii) For almost all $t \in T$, $(x, y) \longrightarrow f(t, x, y)$ is sequentially continuous from $H \times H$ into H_w (here H_w denotes the Hilbert space H furnished with the weak topology).

(iii) For almost all $t \in T$ and all $x, y \in H$, we have

$$|f(t, x, y)| \leq \alpha_2(t) + c_2\left(|x|^{2/p'} + |y|^{2/p'}\right),$$

with $\alpha_2 \in L^{p'}(T)_+$, $c_2 > 0$.

THEOREM 10.4.6 *If hypotheses H(A)′, H(B)′, H(f) hold and $(x_0, y_0) \in X \times H$, then the solution set $S(x_0, y_0)$ is nonempty, weakly compact in $W^{1,p}\big((0,b), X\big)$, and compact in $C^1(T, H)$.*

PROOF: First we derive some a priori bounds for the elements of $S(x_0, y_0)$.
So let $x \in S(x_0, y_0)$. We know that $x' \in W_p(0,b) \subseteq C(T, H)$. We have

$$\begin{aligned}&\big\langle x''(t), x'(t)\big\rangle + \big\langle A\big(t, x'(t)\big), x'(t)\big\rangle + \big\langle B\big(x(t)\big), x'(t)\big\rangle \\ = \;& \big(f\big(t, x(t), x'(t)\big), x'(t)\big) \quad \text{a.e. on } T.\end{aligned} \tag{10.143}$$

Using Corollary 10.1.26, we have

$$\big\langle x''(t), x'(t)\big\rangle = \frac{1}{2}\frac{\mathrm{d}}{\mathrm{d}t}|x'(t)|^2 \quad \text{a.e. on } T. \tag{10.144}$$

Also because of hypotheses H$(A)'$(iv) and H$(B)'$, we have

$$\big\langle A\big(t, x'(t)\big), x'(t)\big\rangle \geq c_1\|x'(t)\|^p - \alpha_1(t) \quad \text{a.e. on } T \tag{10.145}$$

$$\text{and } \big\langle B\big(x(t)\big), x'(t)\big\rangle = \frac{1}{2}\frac{\mathrm{d}}{\mathrm{d}t}\big\langle B\big(x(t)\big), x(t)\big\rangle \quad \text{a.e. on } T. \tag{10.146}$$

So, if we return to (10.143) and use (10.144) through (10.146), we obtain

$$\begin{aligned}&\frac{1}{2}\frac{\mathrm{d}}{\mathrm{d}t}|x'(t)|^2 + c_1\|x'(t)\|^p + \frac{1}{2}\frac{\mathrm{d}}{\mathrm{d}t}\big\langle B\big(x(t)\big), x(t)\big\rangle \\ \leq \;& \big(f\big(t, x(t), x'(t)\big), x'(t)\big) + \alpha_1(t) \quad \text{a.e. on } T.\end{aligned}$$

Integrating, we have

$$\begin{aligned}&\frac{1}{2}|x'(t)|^2 + c_1\int_0^t \|x'(s)\|^p \mathrm{d}s + \frac{1}{2}\big\langle B\big(x(t)\big), x(t)\big\rangle \\ \leq \;& \int_0^t \big(f\big(s, x(s), x'(s)\big), x'(s)\big)\mathrm{d}s + c_3 \quad \text{for some } c_3 > 0, \\ \Rightarrow \;& \frac{1}{2}|x'(t)|^2 + c_1\int_0^t \|x'(s)\|^p \mathrm{d}s \\ \leq \;& \int_0^t \Big[\alpha_2(s) + c_2\big(|x(s)|^{2/p'} + |x'(s)|^{2/p'}\big)\Big]|x'(s)|\mathrm{d}s + c_3\end{aligned} \tag{10.147}$$

(see hypotheses H$(B)'$ and H(f)(iii)). Because $x' \in W_p(0,b)$, we have

$$\begin{aligned}&x(t) = x_0 + \int_0^t x'(s)\mathrm{d}s \quad \text{in } X \qquad \text{for all } t \in T, \\ \Rightarrow \;& |x(t)|^2 \leq 2|x_0|^2 + 2\int_0^t |x'(s)|^2 \mathrm{d}s.\end{aligned} \tag{10.148}$$

So, if on the integral in the right-hand side of (10.147) we use Young's inequality with $\varepsilon > 0$ and (10.148), after some calculations, we obtain

$$\begin{aligned}&\frac{1}{2}|x'(t)|^2 + c_1\int_0^t \|x'(s)\|^p \mathrm{d}s \\ \leq \;& c_4(\varepsilon) + c_5(\varepsilon)\int_0^t |x'(s)|^2 \mathrm{d}s + c_6(\varepsilon)\int_0^t\int_0^s |x'(\tau)|^2 \mathrm{d}\tau\mathrm{d}s + \frac{\varepsilon^p\beta^p}{p}\int_0^t \|x'(s)\|^p \mathrm{d}s\end{aligned} \tag{10.149}$$

with $c_4(\varepsilon), c_5(\varepsilon), c_6(\varepsilon) > 0$ and $\beta > 0$ is such that $|\cdot| \leq \beta\|\cdot\|$. We choose $\varepsilon > 0$ such that $(\varepsilon^p \beta^p)/p < c_1$. We have

$$\frac{1}{2}|x'(t)|^2 \leq c_4(\varepsilon) + c_5(\varepsilon)\int_0^t |x'(s)|^2 \mathrm{d}s + c_6(\varepsilon)\int_0^t\int_0^s |x'(\tau)|^2 \mathrm{d}\tau \mathrm{d}s. \tag{10.150}$$

From (10.150) and using the generalized Gronwall inequality (see Denkowski–Migórski–Papageorgiou [195, p. 128]), we obtain $M_1>0$ such that

$$|x'(t)| \leq M_1 \qquad \text{for all } t \in T \quad \text{and all } x \in \mathcal{S}(x_0, y_0). \tag{10.151}$$

From (10.148) and (10.151), it follows that we can find $M_2 > 0$ such that

$$|x(t)| \leq M_2 \qquad \text{for all } t \in T \quad \text{and all } x \in \mathcal{S}(x_0, y_0). \tag{10.152}$$

So, if in (10.149), we use (10.151) and (10.152) and recalling the choice of $\varepsilon > 0$, we see that there exists $M_3 > 0$ such that

$$\|x'\|_{L^p(T,X)} \leq M_3 \qquad \text{for all } x \in \mathcal{S}(x_0, y_0). \tag{10.153}$$

Recall that

$$x(t) = x_0 + \int_0^t x'(s)\mathrm{d}s \quad \text{in } X, \text{ for all } t \in T.$$

Hence there exists $M_4 > 0$ such that

$$\|x(t)\| \leq M_4 \qquad \text{for all } t \in T \text{ and all } x \in \mathcal{S}(x_0, y_0). \tag{10.154}$$

Using (10.152) through (10.154), directly from the equation in (10.142), we can find $M_5 > 0$ such that

$$\|x''\|_{L^{p'}(T,X^*)} \leq M_5 \qquad \text{for all } x \in \mathcal{S}(x_0, y_0). \tag{10.155}$$

As in the proof of Theorem 10.3.12, because of (10.151) and (10.152), we can replace $f(t,x,y)$ by $f_1(t,x,y)$ defined by

$$f_1(t,x,y) = f\big(t, r_{M_1}(x), r_{M_2}(y)\big)$$

(r_{M_k} is the M_k-radial retraction in H). Note that for almost all $t \in T$ and all $x, y \in H$ we have

$$|f_1(t,x,y)| \leq \psi(t) \quad \text{with } \psi \in L^{p'}(T)_+.$$

Let $K : L^p(T,X) \longrightarrow C(T,X)$ be the integral operator defined by

$$(Ky)(t) = x_0 + \int_0^t y(s)\mathrm{d}s \qquad \text{for all } t \in T.$$

With the use of K, we can equivalently recast (10.142) as the following first order evolution equation.

$$\left\{\begin{array}{l} y'(t) + A\big(t, y(t)\big) + B\big((Ky)(t)\big) = f_1\big(t, (Ky)(t), y(t)\big) \quad \text{a.e. on } T, \\ y(0) = y_0 \end{array}\right\}. \tag{10.156}$$

If y solves (10.156), then $x(t) = x_0 + \int_0^t y(s)\mathrm{d}s$ solves (10.142). To solve (10.156) we use the material of Section 10.3.

First assume that $y_0 \in X$ and set $A_1(t,x) = A(t, x + y_0)$. We can easily verify that A_1 still satisfies hypotheses such as those in $H(A)'$.

Also let $\widehat{B}_1 \in \mathcal{L}\big(L^p(T,X), L^{p'}(T,X^*)\big)$ be defined by

$$\widehat{B}_1(y)(\cdot) = B\big(K(y + y_0)(\cdot)\big)$$

and let $F_1 : L^p(T,X) \longrightarrow L^{p'}(T,X^*)$ be defined by

$$F_1(y)(\cdot) = f_1\big(\cdot, K(y + y_0)(\cdot), y(\cdot) + y_0\big).$$

We consider the following Cauchy problem

$$\left\{\begin{array}{ll} y'(t) + A_1\big(t, y(t)\big) + \widehat{B}_1(y)(t) = F_1(y)(t) & \text{a.e. on } T = [0,b], \\ y(0) = y_0 & \end{array}\right\}. \tag{10.157}$$

Note that $y \in W_p(0,b)$ solves problem (10.156) if and only if $y(\cdot) - y_0$ solves problem (10.157).

Let $L : \mathcal{D}(L) \subseteq L^p(T,X) \longrightarrow L^{p'}(T,X^*)$ be the linear operator defined by

$$Ly = y'$$

for all $y \in \mathcal{D}(L) = \big\{y \in W_p(0,b) : y(0) = 0\big\}$. It is easy to see that L is densely defined, maximal monotone. Also let $\widehat{A}_1 : L^p(T,X) \longrightarrow L^{p'}(T,X^*)$ be defined by

$$\widehat{A}_1(y)(\cdot) = A_1\big(\cdot, y(\cdot)\big).$$

Clearly $\widehat{A}_1$ is demicontinuous, monotone, hence it is maximal monotone. We set $V(y) = \widehat{A}_1(y) + \widehat{B}_1(y) - F_1(y)$, $V : L^p(T,X) \longrightarrow L^{p'}(T,X^*)$.

Claim 1: V is L-pseudomonotone (see Definition 10.3.15).

Clearly V is bounded and demicontinuous. Suppose $y_n \xrightarrow{w} y$ in $W_p(0,b)$ and assume that

$$\limsup_{n\to\infty} \big((V(y_n), y_n - y)\big) \leq 0. \tag{10.158}$$

From the definition of V, we have

$$\limsup_{n\to\infty} \big((\widehat{A}_1(y_n), y_n - y)\big) = \limsup_{n\to\infty} \big((V(y_n) - \widehat{B}_1(y_n), y_n - y)\big). \tag{10.159}$$

Also, if by $(\cdot,\cdot)_{p'p}$ we denote the duality brackets for the pair $\big(L^{p'}(T,H), L^p(T,H)\big)$, then we have

$$\big((F_1(y_n), y_n - y)\big) = \big(F_1(y_n), y_n - y\big)_{p'p} \longrightarrow 0 \quad \text{as } n \to \infty. \tag{10.160}$$

Moreover, because of $\mathrm{H}(B)'$, we have

$$0 = \lim_{n\to\infty} \big((\widehat{B}_1(y), y_n - y)\big) \leq \liminf_{n\to\infty} \big((\widehat{B}_1(y_n), y_n - y)\big). \tag{10.161}$$

If in (10.159), we use (10.158), (10.160), and (10.161), we obtain

$$\limsup_{n\to\infty} \big((\widehat{A}_1(y_n), y_n - y)\big) \leq 0. \tag{10.162}$$

By virtue of the maximal monotonicity of $\widehat{A}_1$, from (10.162), we have

$$\begin{aligned}&\big((\widehat{A}_1(y_n),y_n)\big)\longrightarrow\big((\widehat{A}_1(y),y)\big)\quad\text{and}\quad \widehat{A}_1(y_n)\xrightarrow{w}\widehat{A}_1(y)\ \text{ in } L^{p'}(T,X^*),\\ \Rightarrow\ & V(y_n)\xrightarrow{w}V(y)\quad\text{in } L^{p'}(T,X^*).\end{aligned}\tag{10.163}$$

Also because of (10.158), we have

$$\limsup_{n\to\infty}\big((\widehat{B}_1(y_n),y_n-y)\big)\leq 0.\tag{10.164}$$

As before from (10.164) and because $\widehat{B}_1$ is maximal monotone, we have

$$\begin{aligned}&\big((\widehat{B}_1(y_n),y_n)\big)\longrightarrow\big((\widehat{B}_1(y),y)\big)\\ \Rightarrow\ &\big((V(y_n),y_n)\big)\xrightarrow{w}\big((V(y),y)\big).\end{aligned}\tag{10.165}$$

From (10.163) and (10.165), we conclude that V is L-pseudomonotone.

Claim 2: V is coercive.

Because of hypothesis H$(A)'$(iv), we have

$$\big((\widehat{A}_1(y),y)\big)\geq c_7\|y\|^p_{L^p(T,X)}-c_8\quad\text{for some } c_7,c_8>0.\tag{10.166}$$

Also hypothesis H$(B)'$ implies that

$$\big((\widehat{B}_1(y),y)\big)\geq 0,\tag{10.167}$$

whereas for F_1, we have

$$\big|\big((F_1(y),y)\big)\big|=\big|\big(F_1(y),y\big)_{p'p}\big|\geq-\|\psi\|_{p'}\|y\|_{L^p(T,H)}.\tag{10.168}$$

Combining (10.166) through (10.168) and because $2\leq p<\infty$, we conclude that V is coercive.

Because of Claims 1 and 2 and using Theorem 10.3.16, we can find $y\in\mathcal{D}(L)\subseteq W_p(0,b)$ such that

$$L(y)+V(y)=0$$

when $y_0\in X$. Then $\overline{y}=y+y_0$ solves (10.157).

Now we remove the restriction that $y_0\in X$. So suppose $y_0\in H$. We can find $\{y_0^n\}_{n\geq 1}\subseteq X$ such that $y_0^n\longrightarrow y_0$ in H as $n\to\infty$. Let $y_n\in W_p(0,b)$ be the solution of (10.157) for y_0^n obtained above. An a priori estimation as in the proof of Theorem 10.3.12 (note that $\{y_0^n\}_{n\geq1}\subseteq H$ is bounded), implies that $\{y_n\}_{n\geq1}\subseteq W_p(0,b)$ is bounded. Therefore we may assume that $y_n\xrightarrow{w}y$ in $W_p(0,b)$. Then as in the proof of Theorem 10.3.12 (with the Galerkin solutions), we check that y is a solution of (10.157) with $y_0\in H$. This proves the solvability of (10.157), hence the solvability of (10.142) too.

Next we prove the compactness properties of the solution set $\mathcal{S}(x_0,y_0)$. The proof follows the steps of the corresponding part of the proof of Theorem 10.3.12. So let

$$\Sigma=\big\{h\in L^p(T,H):|h(t)|\leq\psi(t)\quad\text{a.e. on } T\big\}.$$

Endowed with the relative weak $L^{p'}(T,H)$-topology, Σ becomes a compact metrizable space. Let $\Gamma:\Sigma\longrightarrow 2^{C(T,H)}$ be the multifunction which to each $h\in L^{p'}(T,H)$ assigns the set of solutions of

$$\left\{ \begin{array}{ll} y'(t) + A\big(t, y(t)\big) + B\big((Ky)(t)\big) = h(t) & \text{a.e. on } T, \\ y(0) = y_0 \end{array} \right\}.$$

We saw that $\Gamma(h) \neq \emptyset$ and in fact $\Gamma(h)$ is a singleton.

Claim 3: $\Gamma(\Sigma)$ is weakly in $W_p(0,b)$ and compact in $C(T,H)$.

To show the weak compactness in $W_p(0,b)$, it suffices to show that $\Gamma(\Sigma)$ is weakly sequentially closed. So suppose $\{y_n\}_{n\geq 1} \subseteq \Gamma(\Sigma)$ and assume that $y_n \xrightarrow{w} y$ in $W_p(0,b)$. Then as in proof of Theorem 10.3.12, we can show that

$$\big((x_n', x_n - x)\big) \longrightarrow 0.$$

Also we have

$$0 = \lim_{n\to\infty} \big((\widehat{B}(Ky), y_n - y)\big) \leq \liminf_{n\to\infty} \big((\widehat{B}(Ky_n), y_n - y)\big)$$

and $$\lim_{n\to\infty} \big((h_n, y_n - y)\big) = \lim_{n\to\infty} (h_n, y_n - y)_{p'p} = 0.$$

Therefore finally we have

$$\limsup_{n\to\infty} \big((\widehat{A}(y_n), y_n - y)\big) \leq 0.$$

Because of the maximal monotonicity of $\widehat{A}$, we infer that

$$\widehat{A}(y_n) \xrightarrow{w} \widehat{A}(y) \quad \text{in } L^{p'}(T, X^*).$$

Because $y_n' + \widehat{A}(y_n) + \widehat{B}(Ky_n) = h_n$ for all $n \geq 1$, passing to the limit as $n \to \infty$, we have

$$\begin{aligned} & y' + \widehat{A}(y) + \widehat{B}(Ky) = h, \ y(0) = y_0, \quad (\text{where } h_n \xrightarrow{w} H \text{ in } L^{p'}(T,H)) \\ \Rightarrow \ & y \in \Gamma(\Sigma) \ \text{(i.e., } \Gamma(\Sigma) \text{ is weakly compact in } W_p(0,b)). \end{aligned}$$

Next we show that $\Gamma(\Sigma)$ is compact in $C(T,H)$. So suppose $\{y_n\}_{n\geq 1} \subseteq \Gamma(\Sigma)$. For every $n \geq 1$, $y_n = \Gamma(h_n)$, $h_n \in \Sigma$ and we may assume that $h_n \xrightarrow{w} h$ in $L^{p'}(T,H)$. For every $t \in T$ and if $y = \Gamma(h)$, we have

$$\begin{aligned} \frac{1}{2}|y_n(t) - y(t)|^2 = & \big((h_n - h, y_n - y)\big)_t - \big((\widehat{A}(y_n) - \widehat{A}(y), y_n - y)\big)_t \\ & - \big((\widehat{B}(Ky_n) - \widehat{B}(Ky), y_n - y)\big)_t. \end{aligned} \tag{10.169}$$

Note that

$$\big((\widehat{B}(Ky_n) - \widehat{B}(Ky), y_n - y)\big)_t \geq 0 \ \text{(see hypothesis H}(B)'). \tag{10.170}$$

Using (10.170) in (10.169), we obtain

$$\begin{aligned} \frac{1}{2}|y_n(t) - y(t)|^2 \leq & \int_0^b |(h_n(t) - h(t), y_n(t) - y(t))| \\ & + \int_0^b |\langle A(t, y_n(t)), y_n(t) - y(t)\rangle| dt + \big((\widehat{A}(y), y_n - y)\big)_t. \end{aligned}$$

Because $y_n \xrightarrow{w} y$ in $W_p(0,b)$, we have $y_n \longrightarrow y$ in $L^p(T,H)$ (see Theorem 10.1.29). Hence

$$\int_0^b |\big(h_n(t)-h(t), y_n(t)-y(t)\big)| \longrightarrow 0.$$

Also from the proof of Theorem 10.3.12 we have

$$\int_0^b |\big\langle A\big(t,y_n(t)\big), y_n(t)-y(t)\big\rangle|dt \longrightarrow 0$$

$$\text{and} \quad \sup_{t\in T}\big((\widehat{A}(y), y_n-y)\big)_t \longrightarrow 0.$$

So finally

$$\sup_{t\in T}|y_n(t)-y(t)|^2 \longrightarrow 0 \ \text{(i.e.,} \ y_n \longrightarrow y \ \text{in} \ C(T,H)),$$
$$\Rightarrow \ \Gamma(\Sigma) \ \text{is compact in} \ C(T,H).$$

This proves Claim 3.

Let $\mathcal{S}'(x_0,y_0)=\{x' : x\in\mathcal{S}(x_0,y_0)\}$. Evidently $\mathcal{S}'(x_0,y_0)\subseteq\Gamma(\Sigma)$ and it is weakly closed in $W_p(0,b)$ and strongly closed in $C(T,H)$. Because $\mathcal{S}(x_0,y_0) = K\big(\mathcal{S}'(x_0,y_0)\big)$, we conclude that $\mathcal{S}(x_0,y_0)$ is weakly compact in $W_p(0,b)$ and strongly compact in $C^1(T,H)$. □

REMARK 10.4.7 The above theorem remains valid if instead of H(A)(ii), we assume the following more general condition

(ii)$'$ for almost all $t\in T$, $x\longrightarrow A(t,x)$ is pseudomonotone.

Moreover, if for almost all $t\in T$, $f(t,\cdot,\cdot)$ is Lipschitz continuous, then the solution set $\mathcal{S}(x_0,y_0)$ is a singleton.

We conclude this section with an example of a nonlinear hyperbolic initial boundary value problem. So let $T=[0,b]$ and $Z\subseteq\mathbb{R}^N$ be a bounded domain with a C^2-boundary ∂Z. We consider the following hyperbolic problem.

$$\left\{\begin{array}{l} \frac{\partial^2 x}{\partial t^2} - \Delta x(t,z) - \operatorname{div}(\|Dx_t\|^{p-2}Dx_t) = f_0\big(t,z,x(t,z),x_t(t,z)\big) \quad \text{in } T\times Z, \\ x(z,0)=x_0(z), \ x_t(0,z)=y_0(z) \quad \text{a.e. on } Z, \\ x\big|_{T\times\partial Z}=0 \end{array}\right\}. \tag{10.171}$$

Here $x_t=\partial x/\partial t$. The hypotheses on the function $f_0(t,z,x,y)$ are the following.

H(f₀): $f_0 : T\times Z\times\mathbb{R}\times\mathbb{R}\longrightarrow\mathbb{R}$ is a function such that

(i) For all $x,y\in\mathbb{R}$, $(t,z)\longrightarrow f_0(t,z,x,y)$ is measurable.

(ii) For almost all $(t,z)\in T\times Z$, $(x,y)\longrightarrow f_0(t,z,x,y)$ continuous .

(iii) For almost all $(t,z)\in T\times Z$ and all $x,y\in\mathbb{R}$, we have

$$|f_0(t,z,x,y)| \le \widehat{\alpha}_2(t,z)+c_2\big(|x|+|y|\big),$$

with $\widehat{\alpha}_2\in L^{p'}\big(T,L^2(Z)\big)$, $c_2>0$, $2\le p<\infty$, $\frac{1}{p}+\frac{1}{p'}=1$.

DEFINITION 10.4.8 By a *solution* of problem (10.171), we mean a function $x \in C\big(T, W_0^{1,p}(Z)\big)$ such that

$$\frac{\partial x}{\partial t} \in L^p\big(T, W_0^{1,p}(Z)\big), \quad \frac{\partial^2 x}{\partial t^2} \in L^{p'}\big(T, W^{-1,p'}(Z)\big)$$

and
$$\frac{\mathrm{d}^2}{\mathrm{d}t^2}\int_Z x(t,z)u(z)\mathrm{d}z + \int_Z \|Dx_t\|^{p-2}(Dx_t, Du)_{\mathbb{R}^N}\mathrm{d}z + \int_Z (Dx, Du)_{\mathbb{R}^N}\mathrm{d}z = \int_Z f(t,z,x,x_t)u(z)\mathrm{d}z$$

for all $u \in W_0^{1,p}(Z)$ (weak solution).

PROPOSITION 10.4.9 *If hypotheses* $H(f_0)$ *hold,* $x_0 \in W_0^{1,p}(Z)$*, and* $y_0 \in L^2(Z)$*, then the set of (weak) solutions of problem (10.171) is nonempty and compact in* $C^1\big(T, L^2(Z)\big)$.

PROOF: The evolution triple is $X = W_0^{1,p}(Z)$, $H = L^2(Z)$, $X^* = W^{-1,p'}(Z)$.
Also $A: X \longrightarrow X^*$ is defined by

$$\langle A(x), y\rangle = \int_Z \|Dx\|^{p-2}(Dx, Dy)_{\mathbb{R}^N}\mathrm{d}z \qquad \text{for all } x, y \in W_0^{1,p}(Z).$$

Clearly it satisfies hypotheses H(A). Also $B \in \mathcal{L}(X, X^*)$ is given by

$$\langle Bx, y\rangle = \int_Z (Dx, Dy)_{\mathbb{R}^N}\mathrm{d}z \qquad \text{for all } x, y \in W_0^{1,p}(Z).$$

It satisfies hypothesis H(B). Finally we set

$$f(t,x,y)(\cdot) = f_0\big(t, \cdot, x(\cdot), y(\cdot)\big).$$

Then $f: T \times H \times H \longrightarrow H$ satisfies hypotheses H(f). Using A, B, and f, we can equivalently rewrite (10.171) in the form of (10.136). We apply Theorem 10.4.6 to conclude the proof. ☐

10.5 Remarks

10.1: The origins of the Bochner integral can be traced in the works of Bochner [81] and Dunford [213]. Some people call it Dunford's second integral. A more detailed study of the Bochner integral can be found in Diestel–Uhl [199], Denkowski–Migórski–Papageorgiou [194], and Gasiński–Papageorgiou [259]. There is also a weak vector-valued integral known as the Pettis integral. A detailed investigation of it can be found in Talagrand [573]. The RNP (see Definition 10.1.11), is investigated in a systematic way in the books of Diestel–Uhl [199] and Bourgin [95]. Theorem 10.1.16 is due to Dinculeanu–Foias [201]. For $X = H =$ a Hilbert space, Theorem 10.1.18 was proved by Komura [360]. Evolution triples (see Definition 10.1.21) are also known as Gelfand triples , because it was Gelfand who first made systematic use of them (see Gelfand–Shilov [261]). Lemma 10.1.28 appears in the literature under the names Lions lemma or Ehrling's inequality (see Ehrling [220]). Theorem 10.1.29

is due to Aubin [35]. More on evolution triples and the related function spaces, can be found in the paper of Simon [555] and in the books of Gasiński–Papageorgiou [259], Showalter [554], Wloka [608], and Zeidler [621, 622].

10.2: Semilinear evolution equations using the semigroup method can be found in the books of Amann [19], Henry [291], Pazy [491], Showalter [553, 554], Tanabe [574], and Zheng [625]. The case of a time-dependent operator A, in which case we are dealing with the so-called evolution operators, can be found in Amann [19] and Tanabe [574]. For the applications to parabolic problems, we follow Zheng [625].

10.3: Evolution equations associated with maximal monotone operators (Hilbert space case) or accretive operators (Banach space case), were the starting point for the introduction of nonlinear semigroups: see Komura [360] (Hilbert spaces) and Crandall–Liggett [166]) (Banach spaces) (see also Segal [546]). Subdifferential evolution inclusion, with time-invariant φ, was first investigated by Brezis [99, 100]. Brezis [99] proved Theorem 10.3.7 and also established the regularizing effect on the initial condition. A comprehensive treatment of such evolution equations can be found in the books of Barbu [58] and Brezis [102]. The case of time-dependent subdifferentials, is studied by Attouch–Damlamian [32], Hu–Papageorgiou [314, 315], Kenmochi [347, 348], Yamada [609], and Yotsutani [616]. For the periodic problem, we refer to Bader–Papageorgiou [51]. Further results can be found in Chapter 2 of the book of Hu–Papageorgiou [316]. The Galerkin method together with the monotonicity method was used extensively in the framework of evolution triples by Lions [387]. Extensions can be found in Papageorgiou–Papalini–Yannakakis [483]. We also refer to the books of Showalter [554] and Zeidler [622].

10.4: Theorem 10.4.1 was originally by Lions–Strauss [386], however the proof given here is due to Brezis (see also Barbu [58, pp. 268–269]). Similar results can be found in Brezis [100]. Second-order evolution equations defined in the framework of a general evolution triple can be found in Zeidler [622, Chapter 33]. The work of Zeidler was generalized by Papageorgiou–Yannakakis [485, 486]. Our presentation here is based on these papers.

References

[1] Acerbi, E.–Fusco, N., *Semicontinuity problems in the calculus of variations*, Arch. Rat. Mech. Anal. **86** (1984), 125–145

[2] Agarwal, R.–Filippakis, M.–O'Regan D.–Papageorgiou N. S., *Solutions for nonlinear Neumann problems via degree theory for multivalued perturbations of $(S)_+$-maps*, Adv. Diff. Eqns. **II** (2006), 961–980

[3] Ahmed, N. U.–Teo, K. L. H., *Optimal Control of Disturbed Parameter Systems*, North–Holland, New York (1981)

[4] Ahmed, N. U., *Optimization and Identification of Systems Governed by Evolution Equations in Banach spaces*, Longman, Essex, UK (1988)

[5] Aizicovici, S.–Papageorgiou, N. S.–Staicu, V. *Periodic solutions for second order differential inclusions with the scalar p-Laplacian*, J. Math. Anal. Appl. **322** (2006), 913–929

[6] Aizicovici, S.–Papageorgiou, N. S.–Staicu, V. *Degree theory for operators of monotone type and nonlinear elliptic equations with inequality constraints*, Memoirs AMS–December 2008

[7] Alexeev, V.–Tichomirov, V.–Fomin, S., *Commande Optimale*, Mir, Moscow (1982)

[8] Aliprantis, C.–Brown, D., *Equilibria in markets with a Riesz space of commodities*, J. Math. Economics **11** (1983), 189–207

[9] Aliprantis, C.–Brown, D.–Burkinshaw, O., *Existence and Optimality of Competitive Equilibria*, Springer-Verlag, New York (1989)

[10] Aliprantis, C.–Border, K., *Infinite Dimensional Analysis*, Springer-Verlag, Berlin (1994)

[11] Allegretto, W.–Huang, Y–X., *Principal eigenvalues and Sturm comparison via Picone's identity*, Nonlin. Anal. **32** (1998), 819–830

[12] Allegretto, W.–Huang, Y–X., *A Picone's identity for the p-Laplacian and applications*, J. Diff. Eqns. **156** (1999), 427–438

[13] Allen, B., *Neighboring information and distributions of agents characteristics under uncertainty*, J. Math. Economics **12** (1983), 63–101

[14] Allen, B., *Convergence of σ-fields and application to mathematical economics*, Eds. K. Hammer–D. Pallaschke, Springer, Berlin, 161–174

[15] Amann, H.–Weiss, S., *On the uniqueness of the topological degree*, Math. Zeits **130** (1973), 39–54

[16] Amann, H., *Fixed point equations and nonlinear eigenvalue problems in ordered Banach spaces*, SIAM Rev. **18** (1976), 620–709

[17] Amann, H.–Zehnder, E., *Nontrivial solutions for a class of nonresonance problems and applications to nonlinear differential equations*, Ann. Scuola Norm. Sup. Pisa **7** (1980), 539–603

[18] Amann, H., *A note on degree theory for gradient mappings*, Proc. AMS **85** (1982), 591–595

[19] Amann, H., *Linear and Quasilinear Parabolic Problems*, Vol. I:Abstract Linear Theory, Birkhäuser, Basel (1995)

[20] Ambrosetti, A.–Rabinowitz, P., *Dual variational methods in critical point theory and applications*, J. Funct. Anal. **14** (1973), 349–381

[21] Ambrosetti, A.–Struwe, M., *A note on the problem* $-\triangle u = \lambda u + |u|^{2^*-2}u$, Manuscr Math. **54** (1986), 373–379

[22] Ambrosetti, A.–Mancini, G., *Sharp nonuniqueness results for some nonlinear problems*, Nonlin. Anal. **73**, (1979), 635–648

[23] Ambrosetti, A.–Lupo, D., *On a class of nonlinear Dirichlet problems with multiple solutions*, Nonlin. Anal. **8** (1984), 1145–1150

[24] Ambrosetti, A.–Garcia Azorero, J.–Peral Alonso, I., *Multiplicity results for some nonlinear elliptic equations*, J. Funct. Anal. **137** (1996), 219–242

[25] Anane, A., *Simplicité et isolation de la première valeur propre du p-Laplacian avec poids*, CRAS Paris, t.**305** (1987), 725–728

[26] Anane, A.–Tsouli, N., *On the second eigenvalue of the p-Laplacian* in *Nonlinear Partial Differential Equations*, Eds. A. Benikrane–J.-P. Gossez, Pitman Research Notes in Math, Vol. **343** (1996), Longman, Harlow, 1–9

[27] Andres, J.–Gorniewicz, L., *Topological Fixed Point Principles for Boundary Value Problems*, Kluwer, (2003)

[28] Araujo, A.–Scheinkman, J., *Smoothness, comparative dynamics and the turnpike property*, Econometrica **45** (1979), 601–620

[29] Arkin, V.–Evstigneev, I., *Stochastic Models of Control and Economic Dynamics*, Academic Press, New York (1987)

[30] Ash, R. B., *Real analysis and probability*, Probability and Mathematical Statistics, No. **11**, Academic Press, New York–London, (1972)

[31] Asplund, E., *Fréchet differentiability of convex functions*, Acta Math. **121** (1968), 31–47

[32] Attouch, H.–Damlamian, A., *Problémes d'evolution dans les hilberts et applications*, J. Math. Pures Appl. **54** (1975)

[33] Attouch, H., *On the maximality of the sum of two maximal monotone operators*, Nonlin. Anal. **5** (1981), 143–147

[34] Attouch, H., *Variational Convergence for Functions and Operators*, Pitman London (1984)

[35] Aubin, J.–P., *Un théorème de compacité*, CRAS Paris, t. **256** (1963), 5042–5044

[36] Aubin, J.–P., *Mathematical Methods of Control and Economic Dynamics*, Academic Press, New York (1979)

[37] Aubin, J.–P., *Further properties of Lagrange multipliers in nonsmooth optimization*, Appl. Math. Optim. **6** (1980), 79–90

[38] Aubin, J.–P.–Cellina, A., *Differential Inclusions*, Springer-Verlag, Berlin (1984)

[39] Aubin, J.–P.–Ekeland, I., *Applied Nonlinear Analysis*, Wiley, New York (1984)

[40] Aubin, J.-P.-Frankowska, H., *Set-Valued Analysis*, Birkhäuser, Boston (1990)

[41] Aumann, R.-Peleg, B., *von Neumann-Morgenstein solutions to cooperative games without side payments*, Bull. AMS, **66** (1960), 173–179

[42] Aumann, R., *The core of a cooperative game without side payments*, Trans. AMS **98** (1961), 539–552

[43] Aumann, R., *Markets with a continuum of traders*, Econometrica **32** (1964), 39–50

[44] Aumann, R., *Integral of set-valued functions*, J. Math. Anal. Appl. **12** (1965), 1–12

[45] Aumann, R., *Existence of competitive equilibrium in markets with a continuum of traders*, Econometrica, **34** (1966), 1–17

[46] Aumann, R., *Measurable utility and the measurable choice theorem*, Actes Colloq. Internat. du CNRS Aix-en-Provence, (1969), 15–26

[47] Aumann, R.-Shapley, L., *Values of Non-Atomic Games*, Princeton Univ. Press, Princeton, NJ (1974)

[48] Averbukh, V.-Smolyanov, O., *The theory of differentiation in linear topological spaces*, Russian Math Surveys **22** (1967), 201–258

[49] Bader, R., *A topological fixed-point index theory for evolution inclusions*, Z. Anal. Anwend., **20** (2001), 3–15

[50] Bader, R.-Papageorgiou, N. S., *Nonlinear boundary value problems for differential inclusions*, Math. Nachr., **244** (2002), 5–25

[51] Bader, R.-Papageorgiou, N. S., *On a problem of periodic evolution inclusions of the subdifferential type*, Z. Anal. Anwendungen, **21** (2002), 963–984

[52] Balder, E., *A general denseness result for relaxed control theory*, Bull. Austr. Math. Soc., **30** (1984), 463–475

[53] Balder, E., *Necessary and sufficient conditions for L_1-strong-weak lower semicontinuity of integral functionals*, Nonlin. Anal. **11** (1987), 1399–1404

[54] Balder, E.-Yannelis, N., *Equilibria in random and bayesian games with a continuum of players* in *Equilibrium Theory in Infinite Dimensional Games*, Eds. M. A. Khan-N. Yannelis, Springer-Verlag, Berlin (1991), 333–350

[55] Balder, E., *Existence of solutions for control and variational problems with recursive objectives*, J. Math. Anal., **178** (1993), 418–437

[56] Ball, J.-Zhang, K., *Lower semicontinuity of multiple integrals and the biting lemma*, Proc. Royal. Soc. Edinburgh, **114A** (1990), 367–379

[57] Banach, S., *Sur les opérations dans les ensembles abstraits et leur application aux équations intégrales*, Fund. Math., **3** (1922), 133–181

[58] Barbu, V., *Nonlinear Semigroups and Differential Equations in Banach Spaces*, Noordhoff International, Leyden, The Netherlands (1976)

[59] Barbu, V.-Precupanu, T., *Convexity and Optimization in Banach Spaces*, Reidel, Dordecht, The Netherlands (1986)

[60] Becker, R.-Boyd, J.-Sung, B., *Recursive utility and optimal capital accumulation I: Existence*, J. Economic Theory **47** (1989), 76–100

[61] Beer, G., *Topologies on Closed and Closed Convex Sets*, Kluwer, Dordrecht (1993)

[62] Bellman, R., *Dynamic Programming*, Princeton Univ. Press, Princeton, NJ (1957)

[63] Benamara, M., *Points Extremaux, Multi-Applications et Fonctionalles Integrales*, Thése du 3eme Cycle, Univ. de Grenoble, France (1975)

[64] Benci, V.–Rabinowitz, P., *Critical point theorems for indefinite functionals*, Invent. Math., **52** (1979), 241–273

[65] Benveniste, L.–Scheinkman, J., *On the differentiability of the value function in dynamic models of economics*, Econometrica, **47** (1979), 728–732

[66] Berestycki, H.–Lasry, M.–Mancini, G.–Ruf, B., *Existence of multiple periodic orbits on star-shaped Hamiltonian surfaces*, Comm. Pure Appl. Math., **38** (1985), 253–289

[67] Berge, C., *Espaces Topologiques, Fonctions Multivoques* (2–nd edition), Dunod, Paris (1966)

[68] Berger, M., *Nonlinearity and Functional Analysis*, Academic Press, New York (1977)

[69] Berger, M.–Schechter, M., *On the solvability of semilinear gradient operator equations*, Adv. Math., **25** (1977), 97–132

[70] Berkovitz, L., *Optimal Control Theory*, Springer-Verlag, New York (1974)

[71] Berkovitz, L., *Lower semicontinuity of integral functionals*, Trans AMS, **192** (1974), 51–57

[72] Bertsekas, D.–Shreve, S., *Stochastic Optimal Control: The Discrete Time Case*, Academic Press, New York (1978)

[73] Beurling, M.–Livingston, A., *A theorem on duality mapping in Banach spaces*, Ark. för Math., **4** (1962), 405–411

[74] Bewley, T., *Existence of equilibria in economics with infinitely many commodities*, J. Economic Theory, **4** (1972), 514–540

[75] Bielecki, A., *Une remarque sur la méthode de Banach–Caccioppoli–Tichomirov*, Bull. Acad. Polon. Sci., **4** (1956), 261–268

[76] Binding, P.–Drabek, P.–Huang, Y.–X., *On Neumann boundary value problems for some quasilinear elliptic equations*, Electr. J. Diff. Eqns., Vol. **1997**, No. **5** (1997), 1–11

[77] Blackwell, D., *Discounted dynamic programming*, Ann. Math. Stat., **36** (1966), 226–235

[78] Blackwell, D., *Positive dynamic programming* in *Proceedings of the Fifth Berkeley Symp. on Math. Stat. Prob.*, Ann. Math. Stat., Vol. **2**, Univ. of California, Berkeley, (1967), 415–418

[79] Blanchard, P.–Brüning, E., *Variational Methods in Mathematical Physics*, Springer-Verlag (1992)

[80] Bliss, G., *Lectures on the Calculus of Variations*, Univ. of Chicago Press, Chicago (1946)

[81] Bochner, S., *Integration von Funktionen deren Werte die Elemente eines Vectorraumes sind*, Fund. Math., **20** (1933), 262–272

[82] Boltyanski, V. G.–Gamkrelidze, R. V.–Pontryagin, L. S., *Theory of optimal processes I. Maximum principle*, (Russian) Izv. Akad. Nauk. SSSR, Ser. Mat. **24** (1960), 3–42

[83] Bolza, O., *Lectures on the Calculus of Variations*, Dover, New York (1961)

[84] Bondavera, O. N., *Theory of the core in the n-person game*, Vestnik Leningrand, **13** (1962), 141–142 (in Russian)

[85] Bondareva, O. N., *Some applications of linear programming methods to the theory of cooperative games*, Problemy Kibernetik, **10** (1963), 119–139 (in Russian)

[86] Border, K., *Fixed Point Theorems with Applications to Economics and Game Theory*, Cambridge Univ. Press, Cambridge, UK (1985)

[87] Border, K., *Functional analytic tools for expected utility theory* in *Positive Operartors, Riesz Spaces and Economics*, Eds. C. Aliprantis–D. Brown–W. Luxemburg, Springer-Verlag (1991)

[88] Border, K., *Revealed preference, stochastic dominance and the expected utility hypothesis*, J. Economic Theory, **56** (1992), 20–42

[89] Borel, E., *Les "paradoxes" de la théorie des ensembles*, Ann. Sci. Ecole Norm. Sup., **25** (1908), 443–448

[90] Borsuk, K., *Sur les retractes*, Fund. Math., **17** (1931), 152–170

[91] Borsuk, K., *Drei Sätze über die n–dimensionale euklidische Sphäre*, Fund. Math., **20** (1933), 177–190

[92] Borwein, J.–Preiss, D., *A smooth variational principle with applications to subdifferentiability and to differentiability of convex functions*, Trans. AMS, **303** (1987), 517–527

[93] Bouligand, G., *Introduction à la Geometrie Infinitésimale Directe*, Gauthier Villars, Paris (1932)

[94] Bourgain, J., *An averaging result for l^1-sequences and applications to weakly conditionally compact sets in $L^1(X)$*, Israel J. Math., **32** (1979), 289–298

[95] Bourgin, R., *Geometric Aspects of Convex Sets with the Radon–Nikodym Property*, Lecture Notes in Math., Vol. **993**, Springer-Verlag, Berlin (1983)

[96] Boylan, E., *Equiconvergence of martingales*, Ann. Math. Stat., **42** (1971), 552–559

[97] Bressan, A.–Cellina, A.–Fryszkowski, A., *A class of absolute retracts in spaces of integrable functions*, Proc. Amer. Math. Soc., **112** (1991), 413–418

[98] Brezis, H., *Equations et inequations non lineaires dans les espaces vectoriels en dualité*, Ann. Inst. Fourier, **18** (1968), 115–175

[99] Brezis, H., *Propriétés régularisantes de certains semi–groupes nonlinéaires*, Israel J. Math., **9**, (1971), 513–534

[100] Brezis, H., *Problèmes unilatéraux*, J. Math. Pures Appl., **51** (1972), 1–164

[101] Brezis, H.–Nirenberg, L.–Stampacchia, G., *A remark of Ky Fan's minimax principle*, Bolletino UMI, **6** (1972), 293–300

[102] Brezis, H., *Operateurs Maximaux Monotones et Semi-Groupes de Contractions dans les Espaces de Hilbert*, North Holland, Amsterdam (1973)

[103] Brezis, H.–Browder, F., *A general principle on ordered sets in nonlinear functional analysis*, Adv. Math., **21** No. **3** (1976), 355–364

[104] Brezis, H.–Nirenberg, L., *Positive solutions of nonlinear elliptic equations involving critical Sobolev exponents*, Comm. Pure Appl. Math., **36** (1983), 437–477

[105] Brezis, H.–Nirenberg, L., *Remarks on finding critical points*, Comm. Pure Appl. Math., **44** (1991), 939–963

[106] Brezis, H.–Nirenberg, L., *H^1 versus C^1-local minimizers*, CRAS Paris, t. **317** (1993), 465–472

[107] Brock, W., *On the existence of weakly maximal programmes in a multisector economy*, Review of Economic Studies, **37** (1970), 275–280

[108] Brock, W.–Mirman, L., *Optimal economic growth and uncertainty: The discounted case*, J. Economic Theory, **4** (1972), 479–513

[109] Brock, W.–Mirman, L., *Optimal economic growth and uncertainty: The non-discounting case*, Intern. Economic Review, **14** (1973), 560–573

[110] Brock, W.–Haurie, A., *On the existence of overtaking optimal trajectories over an infinite time horizon*, Math. Operat. Res., **1** (1976), 337–346

[111] Brock, W.–Majumdar, M., *On characterizing optimal competitive programs in terms of decentralizable conditions*, J. Economic Theory, **45**, (1988), 262–273

[112] Brondsted, A., *Conjugate convex functions in topological vector spaces*, Mat–Fys. Medd. Danske Vid. Selsk, **34** (1964), 1–26

[113] Brodskii, M.–Milman, D., *On the center of a convex set*, Dokl. Akad. Nauk., USSR, **59** (1948), 837–840 (Russian)

[114] Brouwer, L. E. J., *On continuous one-to-one transformations of surfaces into themselves*, Proc. Kon. Ned. Ak. V. Wet., Series A **11** (1909), 788–798

[115] Brouwer, L. E. J., *Über abblildung von mannigfaligkeiten*, Math. Ann., **71**, (1912), 97–115

[116] Browder, F., *Nonexpansive nonlinear operators in a Banach space*, Proc. Nat. Acad. Sci., USA, **54** (1965), 1041–1044

[117] Browder, F., *Infinite dimensional manifolds and nonlinear eigenvalue problems*, Ann. of Math., **82** (1965), 459–477

[118] Browder, F., *The fixed point theory of multivalued mappings in topological vector spaces*, Math. Ann., **177** (1968), 283–301

[119] Browder, F., *Nonlinear maximal monotone mappings in Banach spaces*, Math. Ann., **175** (1968), 81–113

[120] Browder, F.–Hess, P., *Nonlinear mappings of monotone type in Banach Spaces*, J. Funct. Anal., **11** (1972), 251–294

[121] Browder, F., *Nonlinear Operators and Nonlinear Equations of Evolution in Banach Spaces*, Proc. of Symposia in Pure Math., Vol. **XVII**, Part 2, AMS, Providence (1976)

[122] Browder, F., *Fixed point theory and nonlinear problems*, Bull. AMS (N.S.), **9** (1983), 1–39

[123] Browder, F., *Degree Theory for Nonlinear Mappings* in *Nonlinear Functional Analysis and its Applications*, Ed. F. Browder, Proc. of Symposia in Pure Math., Vol. **45**, Part I, AMS, Providence (1986)

[124] Brown, R., *A Topological Introduction to Nonlinear Analysis*, Birkhäuser, Boston (1993)

[125] Buttazzo, G.–Dal Maso, G., *Γ-convergence and optimal control problems*, J. Optim. Th. Appl., **38** (1982), 385–407

[126] Buttazzo, G., *Semicontinuity, Relaxation and Integral Representation in the Calculus of Variations*, Pitman Research Notes in Math., vol. **207**, Longman Scientific and Technical, Essex, UK. (1989)

[127] Caklovic, L.–Li, S.–Willem, M., *A note on Palais–Smale condition and coercivity*, Diff. Integral Eqns., **3** (1990), 799–800

[128] Caristi, J.–Kirk, W., *Geometric fixed point theory and inwardness conditions*, Lecture Notes in Math., Vol. **490**, Springer-Verlag (1975)

[129] Caristi, J., *Fixed point theorems for mappings satisfying inwardness conditions*, Trans. AMS, **215** (1976), 241–251

[130] Carlson, D.–Haurie, A.–Leizarowitz, A., *Infinite Horizon Optimal Control: Deterministic and Stochastic Systems*, Springer-Verlag, New York (1991)

[131] Cartan, H., *Calcul Différentiel*, Herman, Paris (1967)

[132] Casas, E.–Fernadez, L., *A Green's formula for quasilinear elliptic operators*, J. Math. Anal. Appl., **142** (1989), 62–73

[133] Castaing, C., *Sur les multiapplications mesurables*, Revue Francaise Inform. Rech. Operat., **1** (1967), 91–126

[134] Castaing, C.–Valadier, M., *Convex Analysis and Measurable Multifunctions*, Lecture Notes in Math., Vol. **580**, Springer-Verlag, Berlin (1977)

[135] Cellina, A., *The role of approximation in the theory of multivalued mappings* in *Differential Games and Related Topics*, Eds. H. W. Kuhn–G. P. Szego, North–Holland, Amsterdam, (1971), 209–220

[136] Cerami, G., *Un criterio di existenza per i punti critici su varieta illimitate*, 1st Lombardo Accad. Sci. Lett. Rend. A, **112** (1978), 332–336

[137] Cerami, G.–Fortunato, D.–Struwe, M., *Bifurcation and multiplicity results for nonlinear elliptic problems involving critical Sobolev exponents*, Ann. Inst. H. Poincaré, Analyse Non Linéaire, **1** (1984), 341–350

[138] Cesari, L., *Closure theorems for orietor fields and weak convergence*, Arch. Rational Mech. Anal., **55** (1974), 332–356

[139] Cesari, L., *Lower semicontinuity and lower closure theorems without seminormality conditions*, Annali di Mat. Pura Appl., **98** (1974), 381–397

[140] Cesari, L., *Optimization Theory and Applications*, Springer-Verlag, New York (1983)

[141] Cesari, L.–Kannan, R., *Existence of solutions of a nonlinear differential equation*, Proc. AMS, **88** (1983), 605–613

[142] Chang, K.–C., *Solutions of asymptotically linear operator equations via Morse theory*, Comm. Pure Math., **34** (1981), 693–712

[143] Chang, K.–C., *Infinite Dimensional Morse Theory and Multiple Solution Problems*, Birkhäuser, Boston (1993)

[144] Choquet, G., *Lectures on Analysis I,II,III*, Benjamin, Reading, MA (1969)

[145] Christensen, J., *Topology and Borel Structure*, North Holland, Amsterdam (1974)

[146] Cioranescu, I., *Geometry of Banach Spaces, Duality Mappings and Nonlinear Problems*, Kluwer, Dordrecht, The Netherlands (1990)

[147] Clark, D., *A variant of the Lusternik–Schnirelmann theory*, Indiana Univ. Math. J., **22** (1972), 65–74

[148] Clarke, F., *Generalized gradients and applications*, Trans. AMS, **205** (1975), 247–262

[149] Clarke, F., *A new approach to Lagrange multipliers*, Math. Oper. Res., **1** (1976), 165–174

[150] Clarke, F., *A classical variational principle for periodic Hamiltonian trajectories*, Proc. AMS, **76** (1979), 186–188

[151] Clarke, F.–Ekeland, I., *Hamiltonian trajectories with prescribed minimal period*, Comm. Pure Appl. Math., **33** (1980), 103–116

[152] Clarke, F., *Generalized gradients of Lipschitz functionals*, Adv. Math., **40** (1981), 52–67

[153] Clarke, F., *Optimization and Nonsmooth Analysis*, Wiley, New York (1983)

[154] Clarke, F., *Methods of Dynamic and Nonsmooth Optimization*, CBMS–NSF, Regional Conf. in Appl. Math., Vol. **57**, SIAM, Philadelphia (1989)

[155] Clement, P.–Peletier, L., *An antimaximum principle for second order elliptic operators*, J. Diff. Eqns., **34** (1979), 218–229

[156] Coffman, C. V., *A minimum–maximum principle for a class of nonlinear integral equations*, J. Anal. Math., **22** (1969), 391–419

[157] Cornwall, R., *Conditions on the graph of the integral of a correspondence to be open*, J. Math. Anal. Appl., **39** (1972), 771–792

[158] Costa, D.–Silva, E., *The Palais–Smale condition versus coercivity*, Nonlin. Anal., **16** (1991), 371–381

[159] Costa, D.–Magalhaes, C., *Existence results for perturbations of the p-Laplacian*, Nonlin. Anal., **24** (1995), 409–418

[160] Cotter, K., *Similarity of information and behavior with a pointwise convergence topology*, J. Math. Economics, **15** (1986), 25–38

[161] Cotter, K., *Convergence of information, random variables and noise*, J. Math. Economics, **16** (1987), 39–51

[162] Cotter, K., *Type correlated equilibria for games with payoff uncertainty*, (English summary), Economic Theory, **4** (1994), 617–627

[163] Courant, R., *Dirichlet's Principle, Conformal Mapping and Minimal Surfaces*, Appendix by M. Schiffer, Interscience, New York (1950)

[164] Courant, R.–Hilbert, D., *Methods of Mathematical Physics I*, Interscience, New York (1953)

[165] Covitz, H.–Nadler, S., *Multivalued contraction mappings in generalized metric spaces*, Israel J. Math., **8** (1970), 5–11

[166] Crandall, M.–Liggett, T., *Generation of semigroups of nonlinear transformations on general Banach spaces*, Amer. J. Math., **93** (1971), 265–298

[167] Cuesta, M.–De Figueiredo, D.–Gossez, J.–P., *The beginning of the Fučik spectrum for the p-Laplacian*, J. Diff. Eqns., **159**, No. **1** (1999), 212–238

[168] Cuesta, M., *Eigenvalue problems for the p-Laplacian with indefinite weights*, Electronic J. Diff. Eqns., Vol. **2001**, No. **33** (2001), 1–9

[169] Cuesta, M., *Minimax theorems on C^1-manifolds via Ekeland variational principle*, Abstr. Appl. Anal., **13** (2003), 757–768

[170] Dacorogna, B., *Direct Methods in the Calculus of Variations*, Springer-Verlag, New York (1989)

[171] Dal Maso, G., *An Introduction to Γ-Convergence*, Birkhäuser, Boston (1993)

[172] Damascelli, L., *Comparison theorems for some quasilinear degenerate elliptic operators and applications to symmetry and monotonicity results*, Ann. Inst. H. Poincaré, Analyse Non Linéaire, **15** (1998), 493–516

[173] Dana, R. A., *Evaluation of development programs in a stationary stochastic economy with bounded primary resources* in *Mathematical Models in Economics*, Eds. J. Los–M. Los, North Holland, Amsterdam, (1974), 179–205

[174] Dancer, E., *Bifurcation theory in real Banach spaces*, Proc. London Math. Soc., **23** (1971), 699–734

[175] Danes, J., *A geometric theorem useful in nonlinear functional analysis*, Bollettino UMI, **6** (1972), 369–375

[176] Darbo, G., *Punti uniti in transformazioni a condominio non compatto*, Rend. Sem. Mat. Univ. Padova, **24** (1955), 84–92

[177] Dasgupta, S.–Mitra, T., *Characterization of intertemporal optimality in terms of decentralization conditions: The discounted case*, J. Economic Theory, **45** (1988), 274–287

[178] Dasgupta, S.–Mitra, T., *Optimal and competitive programs in reachable multisector models*, Economic Theory, **14** (1999), 565–582

[179] De Blasi, F. S.–Myjal, J., *The Baire category method in existence problems for a class of multivalued differential equations with nonconvex right hand side*, Funkc. Ekvac, **28** (1985), 139–156

[180] De Blasi, F. S.–Pianigiani, G., *The Baire category method in existence problems for a class of multivalued differential equations with nonconvex right hand side*, Funkc. Ekvac, **28** (1985), 139–156

[181] De Blasi, F. S.–Pianigiani, G., *Nonconvex valued differential inclusions in Banach spaces*, J. Math. Anal. Appl., **157** (1991), 469–494

[182] De Blasi, F. S.–Pianigiani, G., *Topological properties of nonconvex differential inclusions*, Nonlin. Anal., **20**, (1993), 871–894

[183] Dechert, W. D.–Nishimura, K., *A complete characterization of optimal growth paths in an aggregated model with a non-concave production function*, J. Economic Theory, **31** (1983), 332–354

[184] Debreu, G., *A social equilibrium existence theorem*, Proc. Nat. Acad. Sci. USA, **38** (1952), 886–893

[185] Debreu, G., *Theory of Value*, Cowles Foundation Monograph, Yale Univ. Press, New Haven, CO (1959)

[186] Debreu, G., *Integration of correspondences*, Proc. of the fifth Berkley Symp. on Math. statistics and Probability (1965/66), Vol. **II**, Univ. of California Press, Berkley, 351–372

[187] Deimling, K., *Fixed points of condensing maps* in *Volterra equations*, Lecture Notes in Math., Vol. **737**, (1979) Springer-Verlag, New York, 67–82

[188] Deimling, K., *Nonlinear Functional Analysis*, Springer-Verlag, Berlin (1985)

[189] De Giorgi, E., *Teoremi di Semicontinuita nel Calcolo delle Variazioni*, Notes of a course held in INDAM, Rome (1968)

[190] De Giorgi, E.–Franzoni, T., *Su un tipo di convergenza variazionale*, Atti. Accad. Naz. Lincei. Rend Cl. Sci. Fis. Mat. Nat., **58** (1975), 842–850

[191] De Giorgi, E.–Franzoni, T., *Su un tipo di convergenza variazionale*, Rend Sem. Mat. Brescia, **3** (1979), 63–101

[192] Degiovanni, M.–Marzocchi, M., *A critical point theory for nonsmooth functionals*, Ann. Mat. Pura ed Appl., **167** (1994), 73–100

[193] Delbaen, F., *Convex games and extreme points*, J. Math. Anal. Appl., **45** (1974), 210–233

[194] Denkowski, Z.–Migórski, S.–Papageorgiou N. S., *An Introduction to Nonlinear Analysis: Theory*, Kluwer Academic, Boston (2003a)

[195] Denkowski, Z.–Migórski, S.–Papageorgiou N. S., *An Introduction to Nonlinear Analysis: Applications*, Kluwer Academic, Boston (2003b)

[196] Deutsch, F.–Kenderov, P., *Continuous selections and approximate selection for set–valued mappings and applications to metric projections*, SIAM J. Math. Anal., **14** (1983), 185–194

[197] Deville, R.–Godefroy, R.–Zizler, V., *Smoothness and Renorming in B–spaces*, Longman, Essex, UK (1993)

[198] E. Di Benedetto, E., *Degenerate Parabolic Equations*, Springer-Verlag, New York(1993)

[199] Diestel, J.–Uhl, J., *Vector Measures*, Math. Surveys, Vol. **15** (1977), AMS, Providence, RI

[200] Dieudonné, J., *Foundations of Modern Analysis*, Pure and Applied Mathematics, Vol. **10–I**, Academic Press, New York (1969)

[201] Dinculeanu, N.–Foias, C., *Sur la representation intégrale des certaines opérations linéaires IV*, Canad. J. Math., **13** (1961), 529–556

[202] Dontchev, A.–Zolezzi, T., *Well-Posed Optimization Problems*, Lecture Notes in Math., Vol. **1404**, Springer-Verlag, Berlin (1994)

[203] Dorfman, R.–Samuelson, P.–Solow, R., *Linear Programming and Economic Analysis*, McGraw-Hill, New York (1958)

[204] Douka, P.–Papageorgiou, N. S., *Extremal solutions for nonlinear second order differential inclusions*, Math. Nachr., **247** (2004), 30–45

[205] Drabek, P.–Manasevich, R., *On the closed solution to some nonhomogeneous eigenvalue problems with p-Laplacian*, Diff. Integral Eqns., **12**, (1999), 773–788

[206] Drabek, P.–Robinson, S., *Resonance problems for the p-Laplacian*, J. Funct. Anal., **169** (1999), 189–200

[207] Dreze, G.–Gabszewicz, J.–Schmeidler, D.–Vind, K., *Cores and prices in an exchange economy with an atomless sector*, Econometrica, **40** (1972), 1091–1108

[208] Dubovitskii, A.–Milyutin, A., *Extermal problems with side conditions*, USSR Comput. Math. Phys., **5** (1965), 1–80

[209] Dugundji, J., *An extension of Tietze's theorem*, Pacific J. Math., **1** (1951), 353–367

[210] Dugundji, J., *Topology*, Allyn and Bacon, Boston (1966)

[211] Dugundji, J.–Granas, A., *KKM–maps and variational inequalities*, Ann. Scuola Norm. Sup. Pisa, **5** (1978), 679–682

[212] Dugundji, J.–Granas, A., *Fixed Point Theory*, Polish Scientific, Warsaw (1982)

[213] Dunford, N., *Integration in general analysis*, Trans. AMS, **37** (1935), 441–453

[214] Dunford, N.–Schwartz, *Linear Operators I*, Wiley, New York (1958)

[215] Dynkin, E., *Optimal programs and stimulating prices in stochastic models of economic growth* in *Mathematical Models in Economics*, Eds. J. Los–M. Los, North Holland, Amsterdam, (1974), 207–218

[216] Dynkin, E.–Yushkevich, A., *Controlled Markov Processes*, Springer-Verlag, Berlin (1979)

[217] Edelstein, M., *On fixed and periodic points under contractive mappings*, J. London Math. Soc., **37** (1962), 74–79

[218] Edelstein, M., *The construction of an asymptotic center with a fixed point property*, Bull. AMS, **78** (1972), 206–208

[219] Edgeworth, F. Y., *Mathematical Physics*, Paul Kegan, London (1881)

[220] Ehrling, G., *On a type of eigenvalue problems for certain differential operators*, Math. Scand., **2** (1954), 267–287

[221] Ekeland, I., *On the variational principle*, J. Math. Anal. Appl., **47** (1974), 324–353

[222] Ekeland, I.–Temam R., *Convex Analysis and Variational Problems*, North–Holland, Amsterdam (1976)

[223] Ekeland, I., *Nonconvex minimization problems*, Bull. AMS (N. S.), **1** (1979), 443–474

[224] Ekeland, I.–Lasry, M., *On the number of periodic trajectories for a Hamiltonian flow on a convex energy surface*, Ann. Math., **112**, (1980), 283–319

[225] Ekeland, I.–Hofer, H., *Convex Hamiltonian energy surfaces and their periodic trajectories*, Comm. Math. Phys., **113** (1987), 419–469

[226] Ekeland, I., *The ε–variational principle revisited* in *Methods of Nonconvex Analysis*, ed. A. Cellina, Lecture Notes in Math., Vol. **1446**, Springer-Verlag, Berlin, (1989), 1–15

[227] Ekeland, I., *Convexity Methods in Hamiltonian Mechanics*, Springer-Verlag, Berlin (1990)

[228] Evans, I.–Gariepy, R., *Measure Theory and Fine Properties of Functions*, CRC Press, Boca Raton, FL (1992)

[229] Evans, G., *Partial Differential Equations*, Graduate Studies in Mathematics, Vol. **19**, AMS, Providence, (1998)

[230] Evstigneev, I., *Optimal stochastic programs and their stimulating prices* in *Mathematical Models in Economics*, Eds. J. Los–M. Los, North Holland, Amsterdam, (1974), 219–252

[231] Evstigneev, I.–Katyshev, P., *Equilibrium points in stochastic models of economic dynamics*, Theory Prob. Appl., **27**, (1982), 127–135

[232] Fan, K., *Fixed points and minimax theorems in locally convex linear spaces*, Proc. Nat. Acad. Sci. USA, **38** (1952), 121–126

[233] Fan, K., *A generalization of Tychonoff's fixed point theorem*, Math. Ann., **142** (1961), 305–310

[234] Fan, K., *Sur un théorème minimax*, CRAS Paris, t. **259** (1964), 3925–3928

[235] Fan, K., *Applications of a theorem concerning sets with convex sections*, Math. Ann., **163** (1966), 189–206

[236] Fan, X.–Li, Z., *Linking and existence results for perturbations of the p-Lapalcian*, Nonlin. Anal., **42** (2000), 1413–1420

[237] Fan, X.–L.–Zhao, Y.–Z.–Huang, G.–F., *Existence of solutions for the p-Laplacian with crossing nonlinearity*, Disc. Cont. Dynam. Systems, **8** (2002), 1019–1024

[238] Faraci, F., *Multiplicity results for a Neumann problem involving the p-Laplacian*, J. Math. Anal. Appl., **277** (2003), 180–189

[239] Fattorini, H., *Infinite Dimensional Optimization and Control Theory*, Cambridge Univ. Press, Cambridge, UK (1999)

[240] Feller, W., *On the generalization of unbounded semigroups of bounded linear operators*, Ann. of Math., **58** (1953), 166–174

[241] Fenchel, W., *Convex Cones, Sets and Functions*, Princeton University Press, Princeton, NJ (1951)

[242] Fershtman, C.–Mullar, E., *Turnpike properties of capital accumulation games*, J. Economic Theory, **38** (1986), 167–177

[243] Fetter, H., *On the continuity of conditional expectations*, J. Math. Anal. Appl., **61** (1977), 227–231

[244] de Figueiredo, D., *The Ekeland Variational Principle with Applications and Detours*, Tata Institute of Fundamental Research, Bombay (1989)

[245] Filippakis, M.–Papageorgiou, N. S., *Solvability of nonlinear variational–hemivariational inequalities*, J. Math. Anal. Appl., **311** (2005), 162–181

[246] Filippakis, M.–Gasiński, L.–Papageorgiou, N. S., *Multiplicity results for nonlinear Neumann problems*, Canad. J. Math., **58** (2006), 64–92

[247] Filippov, A. F., *On certain questions in the theory of optimal control*, SIAM J. Control, **1** (1962), 76–84

[248] Fishburn, P., *Separation theorems and expected utility*, J. Economic Theory, **II** (1975), 16–34

[249] Fleming, W.–Richel, R., *Deterministic and Stochastic Optimal Control*, Springer-Verlag, New York (1975)

[250] Florenzano, M., *On the existence of equilibria in economies with an infinite dimensional commodity space*, J. Math. Economics, **12** (1983), 207–219

[251] Fonda, A.–Mawhin, J., *Quadratic forms, weighted eigenfunctions and boundary value problems for nonlinear second order differential equations*, Proc. Royal Soc. Edinburgh, **112A** (1989), 145–163

[252] Fonseca, I.–Gangbo, W., *Degree Theory in Analysis and Applications*, Clarendon Press, Oxford, UK (1995)

[253] Führer, L., *Ein elementarer analytischer Beweis zu Eindeutigkeit des Abbildungsgrades in* $\mathbb{R}^N$, Math. Nach., **52** (1972), 259–267

[254] Gale, D., *An optimal development in a multisector economy*, Review of Econ. Studies, **34** (1967), 1–18

[255] Gamkreilidze, R. V., *Principles of Optimal Control Theory*, Plenum Press, New York (1978)

[256] Garcia Azorero, J.–Manfredi, J.–Peral Alonso, I., *Sobolev versus Hölder local minimizers and global multiplicity for some quasilinear elliptic equations*, Comm. Contemp. Math., **2** (2000), 385–404

[257] Garcia Melian, J.–Sabina de Lis, J., *Maximum and comparison principles for operators involving the p-Laplacian*, J. Math. Anal. Appl., **218** (1998), 49–65

[258] Gasiński, L.–Papageorgiou, N. S., *Nonsmooth Critical Point Theory and Nonlinear Boundary Value Problems*, Chapman & Hall/CRC Press, Boca Raton, Fl (2005)

[259] Gasiński, L.–Papageorgiou, N. S., *Nonlinear Analysis*, Chapman & Hall/CRC Press, Boca Raton, Fl (2006)

[260] Gelfand, I. M.–Fomin, S. V., *Calculus of Variations*, Prentice Hall, Englewood Cliffs, NJ (1963)

[261] Gelfand, I.–Shilov, G., *Generalized Functions I: Properties and Operations*, Academic Press, New York (1977)

[262] Ghoussoub, N., *Duality and Perturbation Methods in Critical Point Theory*, Cambridge Univ. Press, Cambridge, UK (1993)

[263] Giaquinta, M., *Multiple Integrals in the Calculus of Variations and Nonlinear Elliptic Systems*, Princeton Univ. Press, Princeton (1983)

[264] Giles, J., *Convex Analysis with Applications in Differentation of Convex Functions*, Pitman, Boston (1982)

[265] Girardi, M.–Matzeu, M., *Solutions of prescribed maximal period to convex and nonconvex Hamiltonian systems*, Bollettino UMI, **4** (1985), 951–967

[266] Girsanov, I. V., *Lectures on Mathematical Theory of Extremum Problems*, Springer-Verlag, New York (1972)

[267] Godoy, T.–Gossez, J.–P.–Paczka, S., *Antimaximum principle for elliptic problems with weight*, Electr. J. Diff. Eqns., Vol. **1999**, No. **22**, (1999), 1–15

[268] Godoy, T.–Gossez, J.–P.–Paczka, S., *On the antimaximum principle for the p-Laplacian with indefinite weight*, Nonlin. Anal., **51** (2002), 449–467

[269] Goebel, K.–Kirk, W., *Topics in Metric Fixed Point Theory*, Cambridge Univ. Press, Cambridge, UK (1990)

[270] Goeleven, D., *A note on Palais–Smale condition in the sense of Szulkin*, Diff. Integral Eqns., **6** (1993), 1041–1043

[271] Göhde, D., *Zum Prinzip der kontraktiven Abbildung*, Math. Nachr., **30** (1965), 251–258

[272] Gorniewicz, L., *Topological Fixed Point Theory of Multivalued Mappings*, Kluwer, Dordrecht (1999)

[273] Granas, A., *The theory of compact vector fields and some of its applications to the topology of functional spaces*, Rozprawy Mathematyczne, No. **30**, Warsaw, (1962), 93

[274] Gromes, W., *Ein einfacher Beweis des Satzes von Borsuk*, Math. Zeits., **178** (1981), 399–400

[275] Guedda, M.–Veron, L., *Quasilinear elliprtic equations involving critical Sobelev exponents*, Nonlin. Anal., **13** (1989), 879–902

[276] Guo, D., *Some fixed point theorems of expansion and compression type with applications* in *Nonlinear Analysis and Applications*, Ed. V. Lakshmikantham, Lecture Notes in Pure and Applied Mathematics, Vol. **109**, Marcel Dekker, New York, (1987), 213–221

[277] Guo, D.–Lakshmikantham, *Nonlinear Problems in Abstract Cones*, Academic Press, Boston (1988)

[278] Guo, D.–Sun J., *Some global generalizations of the Birkhoff–Kellog theorem and applications*, J. Math. Anal. Appl., **129** (1988), 231–242

[279] Gutman, S., *Topological equivalence in the space of integrable vector–valued functions*, Proc. AMS, **93** (1985), 40–42

[280] Habets, R.–Metzen, G., *Existence of periodic solutions of Duffing equations*, J. Diff. Eqns., **78** (1989), 1–32

[281] Hadamard, J., *Sur quelques applications de l'indice de Kronecker*, Hermann Paris, (1910), 437–477

[282] Halidias, N.–Papageorgiou, N. S., *Existence and relaxation results for nonlinear second–order multivalued boundary value problems in $\mathbb{R}^N$*, J. Diff. Eqns., **147** (1998), 123–154

[283] Halpern, B. R.–Bergman, G. M., *A fixed point theorem for inward and outward maps*, Trans. AMS, **130** (1968), 353–358

[284] Halpern, B. R., *Fixed point theorems for set–valued maps in infinite dimensional spaces*, Math. Anal., **189** (1970), 87–89

[285] Harsanyi, J. C., *Games with randomly disturbed payoffs*, Intern. J. Game Theory, **2** (1973), 1–23

[286] Hartman, P.–Stampacchia, G., *On some nonlinear elliptic differential equations*, Acta Math., **115** (1966), 271–310

[287] Heikkila, S.–Lakshmikantham, V., *Monotone Iterative Techniques for Discontinuous Nonlinear Differential Equations*, Marcel Dekker, New York (1994)

[288] Heinz, E., *An elementary theory of the degree of a mapping in n–dimensional space*, J. Math. Mech., **8** (1959), 231–247

[289] Hermes, H.–La Salle, J., *Functional Analysis and Time Optimal Control*, Academic Press, New York (1969)

[290] Hernandez Lerma, O., *Adaptive Markov Control Processes*, Springer-Verlag, New York (1989)

[291] Henry, D, *Geometric Theory of Semilinear Parabolic Equations*, Lecture Notes in Math., Vol. **840**, Springer-Verlag, New York (1981)

[292] Hestenes, M., *Calculus of Variations and the Optimal Control Theory*, Wiley, New York (1966)

[293] Hiai, F.–Umegaki, H., *Integrals, conditional expectations and martingales of multivalued functions*, J. Multiv. Anal., **7** (1977), 149–182

[294] Hiai, F., *Representation of additive functionals on vector–valued normed Köthe spaces*, Kodai Math.–J., **2** (1979), 303–313

[295] Hildenbrand, W., *On the core of an economy with a measure space of agents*, Review of Econ. Studies, **35** (1968), 443–452

[296] Hildenbrand, W., *Existence of equilibria for economies with production and a measure space of consumers*, Econometrica, **38** (1970), 608–623

[297] Hildenbrand, W., *Core and Equilibria of a Large Economy*, Princeton Univ. Press, Princeton, NJ (1974)

[298] Hildenbrand, W.–Kirman, A., *Introduction to Equilibrium Analysis*, North–Holland, Amsterdam (1976)

[299] Hille, E., *Representation of one–parameter semigroup of linear transformations*, Proc. Nat. Acad. Sci., USA, **28** (1942), 175–178

[300] Hille, E.–Phillips, R., *Functional Analysis and Semigroups*, AMS Colloquium Publications, Vol. **31**, AMS, Providence (1957)

[301] Himmelberg, C., *Measurable relations*, Fund. Math., **87** (1975), 53–72

[302] Himmelberg, C.–Parthasarathy, T.–Raghavan, T.–Van Vleck, F., *Existence of p-equilibrium and optimal stationary srategies in stochastic games*, Proc. AMS, **60** (1976), 245–251

[303] Himmelberg, C.–Parthasarathy, T.–Van Vleck, F., *Optimal plans for dynamic programming problems*, Math. Oper. Res., **1** (1976), 390–394

[304] Himmelberg, C.–Parthasarathy, T.–Van Vleck, F., *On measurable relations*, Fund. Math., **111**, (1981), 161–167

[305] Hinderer, K., *Foundations of Nonstationary Dynamic Programming with Discrete–Time Parameter*, Lecture Notes in Oper. Res. and Math. Systems, Vol. **33**, Springer-Verlag, New York (1970)

[306] Hiriart-Urruty, J.–B., *Lipschitz r-continuity of the approximate subdifferential of a convex function*, Math. Scand., **47** (1980), 123–134

[307] Hiriart-Urruty, J.–B., *ε-subdifferential calculus* in *Convex Analysis and Optimization*, Eds. J.–P. Aubin–I. Ekeland, Research Notes in Math., Vol. **57**, (1982), Pitman, London, 43–92

[308] Hiriart-Urruty, J.–B.–Lemarechal, C., *Convex Analysis and Minimization Algorithms I*, Die Grundlehren der Mathematischen Wissenschaften, Vol. **306**, Springer-Verlag, New York (1993)

[309] Hiriart-Urruty, J.–B.–Lemarechal, C., *Convex Analysis and Minimization Algorithms II*, Die Grundlehren der Mathematischen Wissenschaften, Vol. **306**, Springer-Verlag, New York (1993)

[310] Holmes, R., *Geometric Functional Analysis and Applications*, Graduate Texts in Mathematics, Vol. **24** Springer-Verlag, New York (1975)

[311] Hörmander, L., *Sur la fonction d'appui des ensembles convexes dans un espace localement convexe*, Arckiv für Math., **3**, (1955), 180–186

[312] Horsley, A.–Van Zandt, T.–Wrobel, A., *Berge's maximum theorem with two topologies on actions*, Econ. Letters, **61** (1998), 285–291

[313] Hu, S.–Papageorgiou, N. S., *Handbook of Multivalued Analysis. Volume I: Theory*, Kluwer, Dordrecht, The Netherlands, (1997)

[314] Hu, S.–Papageorgiou, N. S., *On nonlinear, nonconvex evolution inclusions*, Kodai Math. J., **18** (1995), 169–186

[315] Hu, S.–Papageorgiou, N. S., *Time-Dependent Subdifferential Evolution Inclusions and Optimal Control*, Memoirs AMS, Vol. **632**, (1998)

[316] Hu, S.–Papageorgiou, N. S., *Handbook of Multivalued Analysis Volume II: Applications*, Kluwer, Dordrecht, The Netherlands, (2000)

[317] Hu, S.–Papageorgiou, N. S., *Solutions and multiple solutions for problems with the p-Laplacian*, Monatsch. Math., **150** (2007), 309–326

[318] Huang, Y.–Zhou, H.–S., *Positive solution for* $-\triangle_p u = f(x,u)$ *with* $f(x,u)$ *growing as* u^{p-1} *at infinity*, Appl. Math. Letters, **17** (2004), 881–887

[319] Hurwicz, L., *On informationally decentralized systems. Decision and organization*, (a volume in honor of Jacob Marschak), pp. 297–336. Studies in Mathematical and Managerial Economics, Vol. **12**, North–Holland, Amsterdam (1972)

[320] Hurwicz, L.–Majumdar, M., *Optimal intertemporal allocation mechanisms and decentralization of decisions*, J. Economic Theory, **45** (1988), 228–261

[321] Hurwicz, L.–Weinberger, H., *A necessary condition for decentralization and an application to intertemporal allocation*, J. Economic Theory, **51** (1990), 313–345

[322] Iannacci, R.–Nkashama, M. N., *Unbounded perturbations of forced second order ordinary differential equations at resonance*, J. Diff. Eqns., **69** (1987), 289–309

[323] Ichiishi, T.–Weber, S., *Some theorems on the core of a non side payment game with a measure space of players*, Intern. J. Game Theory, **7** (1978), 95–112

[324] Ichiishi, T., *Game Theory for Economic Analysis*, Academic Press, New York (1983)

[325] Ize, J., *Bifurcation theory for Fredholm operators*, Memoirs AMS, Vol. **174**, AMS, Providence, RI (1976)

[326] Ioffe, A., *On lower semicontinuity of integral functionals I,II*, SIAM J. Control Optim., **15** (1977), 169–186 & 991–1000

[327] Ioffe, A.–Tichomirov, V., *Theory of Extremal Problems*, North–Holland, Amsterdam (1979)

[328] Ionescu–Tulcea, A. and C., *Topics in the Theory of Lifting*, Springer-Verlag, Berlin (1969)

[329] Istratescu, V., *Fixed Point Theory*, Kluwer, Dordrecht, The Netherlands (1981)

[330] Jacobs, M., *Measurable multivalued maps and Lusin's theorem*, Trans. AMS, **134** (1968), 471–481

[331] Jeanjean, P., *Optimal development programs under uncertainty: The undiscounted case*, J. Economic Theory, **7** (1974), 66–92

[332] Jiu, Q.–Su, J., *Existence and multiplicity results for Dirichlet problems with p-Laplacian*, J. Math. Anal. Appl., **281** (2003), 587–601

[333] Jones, S., *Existence of equilibria with infintely many commodities: Banach lattices reconsidered*, J. Math. Economics, **16** (1987), 89–104

[334] Joshi, S., *Existence in undiscounted non-stationary non-convex multisector environments*, J. Math. Economics, **28** (1997), 111–126

[335] Jost, J., *Partial Differential Equations*, Graduate Texts in Mathematics, Vol. **214**, Springer, New York (2002)

[336] Kachurovski, R., *On monotone operators and convex functionals*, Uspekhi Mat. Nauk., **15** (1960), 213–215 (in Russian)

[337] Kakutani, S., *A generalization of Brouwer's fixed point theorem*, Duke Math. J., **8** (1941), 457–459

[338] Kakutani, S., *Topological properties of the unit sphere of a Hilbert space*, Proc. Imp. Acad. Tokyo, **19** (1943), 269–271

[339] Kamae, T.–Krengel, U.–O'Brien, G. L., *Stochastic inequalities on partially ordered spaces*, Ann. Prob., **5** (1977), 899–912

[340] Kannai, Y., *Countably additive measures in cores of games*, J. Math. Anal. Appl., **27** (1967), 227–240

[341] Kato, T., *Nonlinear semigroups and evolution equations*, J. Math. Soc. Japan, **19** (1967), 508–520

[342] Kato, T., *Accretive operators and nonlinear evolution equations in Banach spaces* in *Nonlinear Functional Analysis*, Ed. F. Browder, Proc. of Symposia in Pure Math., Vol. **18**, AMS Providence (1968)

[343] Kato, T., *Perturbation Theory for Linear Operators*, (2nd edition) Springer-Verlag, Berlin (1976)

[344] Kelley, J., *General Topology*, Springer-Verlag, New York (1955)

[345] Kenmochi, N., *Nonlinear operators of monotone type in reflexive Banach spaces and nonlinear perturbations*, Hiroshima Math. J., **4** (1974), 229–263

[346] Kenmochi, N., *Pseudomonotone operators and nonlinear elliptic boundary value problems*, J. Math. Soc. Japan, **27** (1975), 121–149

[347] Kenmochi, N., *Some nonlinear parabolic variational inequalities*, Israel J. Math., **22** (1975), 304–331

[348] Kenmochi, N., *Solvability of nonlinear evolution equations with time-dependent constraints and applications*, Bull. Fac. Edu., Chiba Univ., **30** (1980), 1–87

[349] Khan, M. A., *Equilibrium points of nonatomic games over a Banach space*, Trans. AMS, **293** (1986), 737–749

[350] Khan, M. A.–Mitra, T., *On the existence of a stationary optimal stock of a multisector economy: A primal approach*, J. Economic Theory, **40** (1986), 319–328

[351] Khan, M. A.–Papageorgiou, N. S., *On Cournot–Nash equilibria in generalized quantitive games with an atomless measure space of agents*, Proc. AMS, **100** (1987), 505–510

[352] Khan, M. A., *On Cournot–Nash equilibrium for games with a nonmetrizable action space and upper semicontinuous payoffs*, Trans. AMS, **315** (1989), 127–146

[353] Kim, T.–Yannelis, N., *Existence of equilibrium in Bayesian games with infinitely many players*, J. Economic Theory, **77** (1997), 330–353

[354] Kirk, W., *A fixed point theorem for mappings which do not increase distances*, Amer. Math. Monthly, **72** (1965), 1004–1006

[355] Kisielewicz, M., *Differential Inclusions and Optimal Control*, Kluwer, Dordrecht (1991)

[356] Klei, H. A., *A compactness criterion in $L^1(E)$ and Radon–Nikodym theorems for multimeasures*, Bull. Sci. Math., **112** (1988), 305–324

[357] Klein, E.–Thompson, A., *Theory of Correspondences*, Wiley, New York (1984)

[358] Knaster, B.–Kuratowski, K.–Mazurkiewicz, S., *Ein Beweis des Fixpunktsatze für n–dimensionale Simplexe*, Fund. Math., **14** (1929), 132–137

[359] Kobayashi, J.–Otani, M., *An index formula for the degree of $(S)_+$–mappings associated with one–dimensional p-Laplacian*, Abstr. Appl. Anal., Vol. **2004**, No. **11** (2004), 981–995

[360] Komura, Y., *Nonlinear semigroups in Hilbert spaces*, J. Math. Soc. Japan, **19** (1967), 493–507

[361] Krasnoselskii, M. A., *Positive Solutions of Operator Equations*, Noordhoff, Groningen, The Netherlands (1964)

[362] Krasnoselskii, M. A., *Topological Methods in the Theory of Nonlinear Integral Equations*, Pergamon Press, Oxford, UK (1964)

[363] Krasnoselskii, M. A.–Zabreiko, P., *Geometrical Methods of Nonlinear Analysis*, Springer-Verlag, New York (1984)

[364] Krawcewicz, W.–Wu, J., *Theory of Degrees with Applications to Bifurcations and Differential Equations*, Wiley, New York (1997)

[365] Kudo, H., *A note on the strong convergence of σ–algebras*, Ann. Prob., **2** (1974), 76–83

[366] Kuratowski, K., *Les fonctions semi–continues dans l'espace des ensembles fermés*, Fund. Math., **18** (1931), 148–159

[367] Kuratowski, K.–Ryll–Nardzewski, C., *A general theorem on selectors*, Bull. Acad. Polon. Sci. Ser. Sci. Math., Astron., Phys. **13** (1965), 397–403

[368] Kuratowski, K., *Topology, Vol. I*, Academic Press, New York (1966)

[369] Kyritsi, S. T.–Motreanu D.–Papageorgiou, N. S., *Two nontrivial solutions for strongly resonant nonlinear elliptic equations*, Arch. Math., **83** (2004), 60–69

[370] Laurent, P., *Approximation et Optimisation*, Hermann, Paris (1972)

[371] Lebourg, G., *Valeur moyenne pour gradient généralisé*, CRAS Paris, t. **281** (1975), 795–797

[372] Ledyard, J., *The scope of the hypothesis of Bayesian equilibrium*, J. Economic Theory, **39** (1986), 59–82

[373] Leese, S., *Multifunctions of Souslin type*, Bull. Austr. Math. Soc., **11** (1974), 395–411

[374] Leggett, L.–Williams, L., *Multiple positive fixed points of nonlinear operators on ordered Banach spaces*, Indiana Univ. Math. J., **28** (1979), 673–688

[375] Leray, J.–Schauder, J., *Topologie et equations fonctionelles*, Ann. Sci. Ecole Norm. Super., **51**, (1934), 45–78

[376] Levin, V., *Subdifferentials of convex integral functionals and liftings that are the identity on subspaces in L^∞*, Soviet Math. Dokl., **14** (1973), 1163–1166

[377] Levin, V., *The Lebesgue decomposition for functions on the vector space L^∞_X*, Funct. Anal. Appl., **8** (1974), 314–317

[378] Levin, V., *Convex integral functionals and lifting theory*, Russian Math. Surveys, **30** (1975), 119–184

[379] Li, G.–Zhou, H.–S., *Asymptotically linear Dirichlet problem for the p-Laplacian*, Nonlin. Anal., **43** (2001), 1043–1055

[380] Li, S.–Willem, M., *Applications of local linking to critical point theory*, J. Math. Anal. Appl., **189** (1995), 6–32

[381] Li, X.–Yong, J., *Optimal Control Theory for Infinite Dimensional Systems*, Birkhäuser, Boston (1995)

[382] Lieberman, G., *Boundary regularity for solutions of degenerate elliptic equations*, Nonlin. Anal., **12** (1988), 1203–1219

[383] Lim, T.–C., *On fixed point stability for set–valued contractive mappings with applications to generalized differential equations*, J. Math. Anal. Appl., **110** (1985), 436–441

[384] Lindqvist, P., *On the equation div($\|\nabla x\|^{p-2}\nabla x+\lambda|x|^{p-2}x = 0$*, Proc. AMS, **109** (1990), 157–164

[385] Lindqvist, P., *Addendum to: On the equation div($\|\nabla x\|^{p-2}\nabla x+\lambda|x|^{p-2}x = 0$*, Proc. AMS, **116** (1992), 583–584

[386] Lions, J.–L.–Strauss, W., *Some nonlinear evolution equations*, Bull. Soc. Math. France, **93** (1965), 43–96

[387] Lions, J.–L., *Quelques Méthodes de Résolution des Problèmes aux Limites Non–Linéaires*, Dunod, Paris (1969)

[388] Lions, J.–L., *Optimal Control of Systems Governed by Partial Differential Equations*, Springer-Verlag, New York (1971)

[389] Liu, J.–Li, S., *An existence theorem of multiple critical points and its applications*, Kexue Tongbao, **29** (1984), 1025–1027 (in Chinese)

[390] Liu, J.–Q.–Su, J., *Remarks on multiple nontrivial solutions for quasilinear resonant problems*, J. Math. Anal. Appl., **258** (2001), 209–222

[391] Ljusternik, L., *Topologishe Grundlagen der allgemeinen Eigenwertheorie*, Monatsh. Math. Phys., **37** (1930), 125–130

[392] Ljusternik, L., *On constrained extrema of functionals*, Mat. Sb., **41** (1934), 390–401

[393] Ljusternik, L.–Schnirelmann, L., *Methodes Topologiques dans les Problémes Variationelles*, Hermann, Paris (1934)

[394] Ljusternik, L.–Schnirelmann, L., *Topological methods in variational problems and their application to the differential geometry of surfaces*, Uspekhi Mat. Nauk., **2** (1947), 166–217 (in Russian)

[395] Ljusternik, L.–Sobolev, V., *Elements of Functional Analysis*, Gordon and Breach Science, New York (1968)

[396] Lloyd, N. G., *Degree Theory*, Cambridge Univ. Press, Cambridge, UK (1978)

[397] Maitra, A.–Parthasarathy, T., *On stochastic games*, J. Optim. Theory Appl., **5** (1970), 289–300

[398] Maitra, A.–Parthasarathy, T., *On stochastic games II*, J. Optim. Theory Appl., **8** (1971), 154–160

[399] Maitra, A.–Sudderth, W., *Discrete Gambling and Stochastic Games*, Springer, New York (1996)

[400] Majumdar, M.–Radner, R., *Stationary optimal policies with discounting in a stochastic activity analysis model*, Econometrica, **51** (1983), 1821–1837

[401] Majumdar, M.–Zilha, I., *Optimal growth in a stochastic environment: Some sensitivity and turnpike results*, J. Economic Theory, **43** (1987), 116–133

[402] Makarov, V.–Rubinov, A., *Mathematical Theory of Economic Dynamics and Equilibria*, Springer-Verlag, New York (1977)

[403] Malinvaud, E., *Capital accumulation and efficient allocation of resources*, Econometrica, **21** (1953), 233–268

[404] Malinvaud, E., *Efficient capital accumulation: A corringendum*, Econometrica, **30** (1962), 570–573

[405] Mamer, J.–Schilling, K., *A zero–sum game with incomplete information and compact action spaces*, Math. Oper. Res., **11** (1986), 627–631

[406] Manasevich, R.–Mawhin, J., *Periodic solutions for nonlinear systems with p-Laplacian–like operators*, J. Diff. Eqns., **145** (1998), 367–393

[407] Manasevich, R.–Mawhin, J., *Boundary value problems for nonlinear perturbations of vector p-Laplacian–like operators*, J. Korean Math. Soc., **37** (2000), 665–685

[408] Marino, A.–Prodi, G., *Metodi perturbativi nella teoria di Morse*, Bollettino UMI, **3** (1975), 1–32

[409] Martio, O., *Counterexamples on the unique continuation*, Manuscripta Math., **60** (1988), 21–47

[410] Mas Collel, A., *An equilibrium existence theorem without complete or transitive preferences*, J. Math. Economics, **1** (1974), 237–246

[411] Mas Collel, A., *On a theorem of Schmeidler*, J. Math. Economics, **13** (1984), 201–206

[412] Mawhin, J., *Topological Degree Methods in Nonlinear Boundary Value Problems*, CBMS, Regional Conference Series in Math., Vol. **40** (1979), AMS, Providence, RI

[413] Mawhin, J.–Willem, M., *Critical points of convex perturbations of some indefinite forms and semilinear boundary value problems at resonance*, Ann. Inst. H. Poincaré, Analyse Non–Linéaire **3** (1986), 431–453

[414] Mawhin, J., *Semicoercive monotone variational problems*, Acad. Royal Belg., Bull. Cl. Sci., **73** (1987), 118–130

[415] Mawhin, J.–Willem, M., *Critical Point Theory and Hamiltonian Systems*, Springer-Verlag, New York (1989)

[416] Mawhin, J., *Periodic solutions of systems with p-Laplacian–like operators* in *Nonlinear Analysis and its Applications to Differential Equations*, Progress in Nonlinear Differential Equations, Vol. **45** (2001), Birkhäuser, Boston, 37–63

[417] Mazur, S., *Über konvexe Mengen in linearen normierten Räumen*, Studia Math., **4** (1933), 70–84

[418] McKenzie, L., *Turnpike theorems for a generalized Leontief model*, Econometrica, **31** (1963), 165–180

[419] McKenzie, L., *Turnpike theorems with technology and welfare function variable* in *Mathematical Models in Economics*, Eds. J. Los–M. Los, North–Holland, Amsterdam (1974), 120–145

[420] McKenzie, L., *Turnpike theory*, Econometrica, **44** (1976), 841–856

[421] McKenzie, L., *A primal route to turnpike and Liapunov stability*, J. Economic Theory, **27** (1982), 194–209

[422] McKenzie, L., *Turnpike theory, discounted utility and the von Neumann Facet*, J. Economic Theory, **30** (1983), 330–352

[423] McKenzie, L., *Optimal economic growth, turnpike theorems and comparative dynamics* in *Handbook of Mathematical Economics*, Vol. **III**, Eds. K. Arrow–M. Intriligator, North–Holland, Amsterdam (1986), 1281–1355

[424] Megginson, R., *An Introduction to Banach Space Theory*, Graduate Text in Math., Vol. **183**, Springer, New York (1998)

[425] Michael, E., *Topologies on spaces of subsets*, Trans. AMS, **71** (1951), 152–182

[426] Michael, E., *Continuous selections I*, Ann. Math., **63** (1956), 361–382

[427] Michael, E., *Continuous selections II*, Ann. Math., **64** (1956), 562–580

[428] Michael, E., *A survey of continuous selections* in *Set-valued Mappings, Selections and Topological Properties of* 2^X, Ed. W. M. Fleischman, Lecture Notes in Math., Vol. **171**, Springer-Verlag, Berlin (1970), 54–57

[429] Milgram, P.–Weber, R., *Distributional strategies for games with incomplete information*, Math. Oper. Res., **10** (1985), 619–632

[430] Minty, G., *Monotone (nonlinear) operators in a Hilbert space*, Duke Math. J., **29** (1962), 341–346

[431] Mitra, T., *On the value maximizing property of infinite horizon efficient programs*, Inter. Economic Review, **20** (1979), 635–642

[432] Mitra, T., *Efficiency, weak value maximality and weak value optimality in a multisector model*, Review of Economic Studies, **48** (1981), 643–647

[433] Mitra, T.–Zilcha, I., *An optimal economic growth with changing technology and tastes: Characterization and stability results*, Inter. Economic Review, **22** (1981), 221–238

[434] Mitra, T., *Sensitivity of optimal programs with respect to changes in target stocks: The case of irreversible investment*, J. Economic Theory, **29** (1983), 172–184

[435] Miyadera, I., *Nonlinear Semigroups*, Translation of Math. Monographs, Vol. **109** (1992), AMS, Providence, RI

[436] Miyadera, I., *Generation of a strongly continuous semigroup of operators*, Tôhoku Math. J., **4** (1952), 109–114

[437] Miyagaki, O., *On a class of semilinear elliptic problems in* $\mathbb{R}^N$ *with critical growth*, Nonlin. Anal., **29** (1997), 773–781

[438] Motreanu, D.–Montreanu, V.–Papageorgiou, N. S., *Periodic solutions for nonautonomous systems with nonsmooth quadratic or superquadratic potential*, Topol. Meth. Nonlin. Anal., **24** (2004), 269–296

[439] Motreanu, D.–Radulescu, V., *Variational and Nonvariational Methods in Nonlinear Analysis and Boundary Value Problems*, Kluwer, Dordrecht (2003)

[440] Motreanu, D.–Papageorgiou, N. S., *Existence and multiplicity of solutions for Neumann problems*, J. Diff. Eqns., **232** (2007), 1–35

[441] Moreau, J.–J., *Proximitè et dualité dans un espace hilbertien*, Bull. Soc. Math. France, **93** (1965), 273–299

[442] Moreau, J.–J., *Fonctionelles Convexes*, Seminaire sur les equations aux derivèes partielles, **II**, College de France, Paris (1966–67)

[443] Morrey, C. B., *Multiple Integrals in the Calculus of Variations*, Springer-Verlag, Berlin (1966)

[444] Morse, M., *The Calculus of Variations in the Large*, AMS, Providence, RI (1934)

[445] Mosco, U., *Convergence of convex sets and of solutions of variational inequalities*, Adv. Math., **3** (1969), 510–585

[446] Mosco, U., *On the continuity of the Young–Fenchel transform*, J. Math. Anal. Appl., **35** (1971), 518–535

[447] Nachbin, L., *Linear continuous funtionals positive on the increasing continuous functions*, Summa Brasil Math., **2** (1951), 135–150

[448] Nachbin, L., *Topology and Order*, Van Nostrand, Princeton, NJ (1965)

[449] Nadler, S., *Multivalued contraction mappings*, Pacific J. Math., **30** (1969), 475–488

[450] Nagumo, M., *Degree of mapping in convex linear topological spaces*, Amer. J. Math., **73** (1951), 497–511

[451] Nash, J., *Equilibrium points in n–person game*, Proc. Nat. Acad. Sci. USA, **36** (1950), 48–49

[452] Nash, J., *Non–cooperative games*, Ann. Math., **54** (1951), 286–295

[453] Nashed, Z., *Differentiability and related properties of nonlinear operators: Some aspects of the role of differentials in nonlinear functional analysis* in *Nonlinear Functional Analysis and Applications*, Academic Press, New York (1971)

[454] Neumann von, J., *Zur Theorie der Gesellschaftsspiele*, Math. Ann., **100** (1928), 295–320 (English Translation: *On the theory of games of strategy* in *Contributions to the Theory of Games*, Vol. **IV**, Eds. A. W. Tucker–R. D. Luce, Princeton Univ. Press, Princeton, NJ 13–42

[455] Neumann von, J.–Morgenstern, O., *Theory of Games and Economic Behavior*, Princeton Univ. Press, Princeton, NJ (1947)

[456] Neumann von, J., *On rings of operators, reduction theory*, Ann. Math., **50** (1949), 401–485

[457] Neustadt, L., *Optimization. A Theory of Necessary Conditions*, Princeton Univ. Press, Princeton, NJ (1976)

[458] Neveu, J., *Calcul des Probabilités*, Masson, Paris (1970)

[459] Neveu, J., *Discrete Parameter Martingales*, North Holland, Amsterdam (1975)

[460] Nikaido, H.–Isoda, K., *Note on noncooperative convex games*, Pacific J. Math., **5** (1955), 807–815

[461] Nikaido, H., *Persistence of continual growth near the von Neumann ray: a strong version of the Radner turnpike theorem*, Econometrica, **32** (1964), 151–163

[462] Nowak, A., *Universally measurable strategies in zero–sum stochastic games*, Ann. Prob., **13** (1985), 269–287

[463] Nowak, A., *On the weak topology on a space of probability measures induced by policies*, Bull. Polish Acad. Sci., **36** (1988), 181–188

[464] Nussbaum, R. D., *Degree theory for local condensing maps*, J. Math Anal. Appl., **37** (1972), 741–766

[465] Nussbaum, R. D., *A global bifurcation theorem with application to functional differential equations*, J. Funct. Anal., **19** (1975), 319–338

[466] Nyarko, Y., *On characterizing optimality of stochastic competitive processes*, J. Economic Theory, **45** (1988), 316–329

[467] Olech, C., *A characterization of L_1–weak lower semicontinuity of integral functionals*, Bull. Acad. Polon. Sci., **25** (1977), 135–142

[468] Olech, C., *Decomposability as a substitute of convexity* in *Proceedings of the Conference on Multifunctions and Integrands*, ed. G. Salinetti, Lecture Notes in Math., Vol. **1091**, Springer-Verlag, Berlin (1984)

[469] Pantelides, G.–Papageorgiou, N. S., *Weakly maximal stationary programs for a descrete time stochastic growth model*, Japan Jour. Indust. Appl. Math., **10** (1993), 471–486

[470] Palais, R.–Smale, S., *A generalized Morse theory*, Bull. AMS, **70** (1964), 165–172

[471] Palais, R., *Lusternik–Schnirelmann theory on Banach Manifolds*, Topology, **5** (1966), 115–132

[472] Palais, R., *The principle of symmetric criticality*, Comm. Math. Phys., **69** (1979), 19–30

[473] Palfrey, T.–Srivastava, S., *Private information in large economies*, J. Economic Theory, **39** (1986), 34–58

[474] Papalini, F., *A quasilinear Neumann problem with discontinuous nonlinearity*, Math. Nachr., **250** (2003), 82–97

[475] Papageorgiou, N. S., *On the theory of Banach space valued multifunctions, part 1: Integration and conditional expectation*, J. Multiv. Anal., **17** (1985), 185–206

[476] Papageorgiou, N. S., *On the theory of Banach space valued multifunctions, part 2: Set valued martingales and set valued measures*, J. Multiv. Anal., **17** (1985), 207–227

[477] Papageorgiou, N. S., *On Fatou's lemma and parametric integrals of set-valued functions*, J. Math. Anal. Appl., **187** (1994), 809–825

[478] Papageorgiou, N. S., *Optimal programs and their price characterizations in a multisector growth model with uncertainty*, Proc. AMS, **122** (1994), 227–240

[479] Papageorgiou, N. S., *Weakly maximal programs for a multisector growth model*, J. Math. Anal. Appl., **183** (1994), 624–630

[480] Papageorgiou, N. S., *On the conditional expectation and convergence properties of random sets*, Trans. AMS, **347** (1995), 2495–2515

[481] Papageorgiou, N. S., *On parametric evolution inclusions of the subdifferential type with applications to optimal control problems*, Trans. AMS, **347** (1995), 203–231

[482] Papageorgiou, N. S.–Papalini, F.–Renzacci, F., *Existence of solutions and periodic solutions for nonlinear evolution inclusions*, Rend. Circ. Mat. Palermo (2), **48** (1999), 341–364

[483] Papageorgiou, N. S.–Papalini, F.–Yannakakis, N., *Nonmonotone, nonlinear evolution inclusions*, Math. Comp. Modelling, **32** (Special Issue on *Nonlinear Operator Theory*, Eds. R. Agarwal–D. O'Regan), (2000), 1345–1366

[484] Papageorgiou, E.–Papageorgiou, N. S., *Two nontrivial solutions for quasilinear periodic problems*, Proc. AMS, **132** (2004), 429–434

[485] Papageorgiou, N. S.–Yannakakis, N., *Second order nonlinear evolution inclusions I: Existence and relaxation results*, Acta Math. Sinica, **21** (2005), 977–996

[486] Papageorgiou, N. S.–Yannakakis, N., *Second order nonlinear evolution inclusions II: Structure of the solution set*, Acta Math. Sinica, **22** (2006), 195–206

[487] Parthasarathy, K. R., *Probability Measures on Metric Spaces*, Academic Press, New York (1967)

[488] Parthasarathy, T., *Discounted, positive and non-cooperative stochastic games*, Intern. J. Game Theory, **2** (1973), 25–37

[489] Parthasarathy, T., *Existence of equilibrium stationary strategies in discounted stochastic games*, Sankhya, **44** (1982), 114–127

[490] Pascali, D.–Sburlan, S., *Nonlinear Mappings of Monotone Type*, Sijthoff and Noordhoff, The Netherlands (1978)

[491] Pazy, A., *Semigroups of Linear Operators and Partial Differential Equations*, Springer-Verlag, New York (1983)

[492] Peleg, B., *Efficiency prices for optimal consumption IV*, SIAM J. Control, **10** (1972), 414–433

[493] Peleg, B.–Yaari, M., *Price properties of optimal consumption programs* in *Models of Economic Growth*, Eds. I. Mirelees–N. Stern, Macmillan, London (1973), 306–317

[494] Penot, J.–P., *The drop theorem, the petal theorem and Ekeland's variational principle*, Nonlin. Anal., **10** (1986), 813–822

[495] Phelps, R., *Convex Functions, Monotone Operators and Differentiablity*, Lecture Notes in Math., Vol. **1364**, Springer-Verlag, Berlin (1993)

[496] Phillips, R., *Perturbation theory for semigroups of linear operators*, Trans. AMS, **74** (1953), 199–221

[497] Pohozaev, S., *Eigenfunctions of the equation* $\triangle u + \lambda f(u) = 0$, Soviet Math. Dokl., **6** (1965), 1408–1411

[498] Poincaré, H., *Les figures equilibrium*, Acta Math., **7** (1885), 259–302

[499] Pontryagin, L. S., *Optimal regulation processes*, Amer. Math. Soc. Transl. (2), **18** (1961), 321–339

[500] Pontryagin, L. P.–Boltyanski, V.–Gamkrelidze, R.–Mischenko, E., *Mathematical Theory of Optimal Processes*, Wiley, New York (1962)

[501] Postlewaite, A.–Schmeidler, D., *Implementation in differential information economies*, J. Economic Theory, **39** (1986), 14–33

[502] Rabinowitz, P., *Some global results for nonlinear eigenvalue problems*, J. Funct. Anal., **7** (1971), 487–513

[503] Rabinowitz, P., *Some aspects of nonlinear eigenvalue problems*, Rocky Mountain Consortium, Symposium on Nonlinear Eigenvalue Problems (Santa Fe, N. M., 1971), Rocky Mountain J. Math., **3** (1973), 161–202

[504] Rabinowitz, P., *Some minimax theorems and applications to partial differential equations* in *Nonlinear Analysis–Selection of Papers in Honor of E. Röthe*, Eds. L. Cesari–R. Kannan–H. Weinberg, Academic Press, New York (1978), 161–177

[505] Rabinowitz, P., *Some critical point theorems and applications to semilinear elliptic partial differential equations*, Ann. Scuola Norm. Sup. Pisa Cl. Sci., **5(4)** (1978), 215–223

[506] Rabinowitz, P., *Periodic solutions of Hamiltonian systems*, Comm. Pure Appl. Math., **31** (1978), 157–184

[507] Rabinowitz, P., *Periodic solutions of a Hamiltonian system on a prescribed energy surface*, J. Diff. Eqns., **33** (1979), 336–352

[508] Rabinowitz, P., *Minimax Methods in Critical Point Theory with Applications to Differential Equations*, CBMS, Regional Conference Series in Math., No. **65**, AMS, Providence, RI (1986)

[509] Radner, R., *Paths of economic growth that are optimal with regard only to final states: A turnpike theorem*, Review of Economic Studies, **28** (1961), 98–104

[510] Radner, R., *Optimal stationary consumption with stochastic production and resources*, J. Economic Theory, **6** (1973), 68–90

[511] Radner, R.–Rosenthal, R. W., *Private information and pure-strategy equilibria*, Math. Oper. Res., **7** (1982), 401–409

[512] Read, T., *Balanced growth without constant returns to scale*, J. Math. Economics, **15** (1986), 171–178

[513] Reed, M.–Simon, B., *Functional Analysis*, Academic Press, New York (1972)

[514] Ricceri, B., *Une propriete toplogique de l'ensemble des points fixes d'une contraction multivoque a valeurs convexes*, Atti Accad. Naz. Lincei., **81** (1987), 283–286

[515] Ritt, R.–Sennott, L., *Optimal stationary policies in general state space, Markov decision chains with finite action sets*, Math. Oper. Res., **17** (1992), 901–909

[516] Roberts, A.–Varberg, D., *Convex Functions*, Pure and Applied Mathematics, Vol. **57**, Academic Press, New York (1973)

[517] Rockafellar, R. T., *An extension of Fenchel's duality theorem for convex functions*, Duke Math. J., **33** (1966), 81–90

[518] Rockafellar, R. T., *Duality and stability in extremum problems involving convex functions*, Pacific J. Math., **21** (1967), 167–187

[519] Rockafellar, R. T., *Integrals which are convex functionals*, Pacific J. Math., **24** (1968), 525–539

[520] Rockafellar, R. T., *Local boundedness of nonlinear monotone operators*, Michigan Math. J., **16** (1969), 397–407

[521] Rockafellar, R. T., *Measurable dependence of convex sets and functions on parameters*, J. Math. Anal. Appl., **28** (1969), 4–25

[522] Rockafellar, R. T., *Convex Analysis*, Princeton Math. Series, Vol. **28**, Princeton Univ. Press, Princeton, NJ (1970)

[523] Rockafellar, R. T., *On the maximal monotonicity of subdifferential mappings*, Pacific J. Math, **24** (1970), 525–539

[524] Rockafellar, R. T., *On the maximal monotonicity of sums of nonlinear monotone operators*, Trans. AMS, **149** (1970), 75–88

[525] Rockafellar, R. T., *Convex integral functionals and duality* in *Contributions to Nonlinear Analysis*, Ed. E. Zarantonello, Academic Press, New York (1971)

[526] Rockafellar, R. T., *Integrals which are convex functionals II*, Pacific J. Math., **39** (1971), 439–469

[527] Rockafellar, R. T., *Weak compactness of level sets of integral functionals* in *Troisème Colloque sur l'Analyse Fonctionelle*, Ed. H. Garnir, Vander, Louvain (1971)

[528] Rockafellar, R. T., *Conjugate Duality and Optimization*, CBMS, Vol. **16**, Regional Conference Series in Math., SIAM, Philadelphia (1974)

[529] Rockafellar, R. T., *Integral functionals, normal integrands and measurable selections* in *Nonlinear Operators and the Calculus of Variations*, ed. I. Waelhroek, Lecture Notes in Math., Vol. **543**, Springer-Verlag, New York (1976)

[530] Rockafellar, R. T.–Wets, R., *Variational Analysis*, Die Grundlehren der Mathematischen Wissenschaften, Vol. **317**, Springer-Verlag, Berlin (1998)

[531] Röthe, E., *Morse Theory in Hilbert space*, Rocky Mountain J. Math., **3** (1975), 251–274

[532] Roubicek, T., *Relaxation in Optimization Theory and Variational Calculus*, W. De Gruyter, Berlin (1997)

[533] Rzezuchowski, T., *Strong convergence of selectors implied by weak*, Bull. Austr. Math. Soc., **39** (1989), 201–214

[534] Sadovskii, B. N., *Limit compact and condensing operators*, Russian Math. Surveys, **27** (1972), 85–155

[535] Saint-Beuve, M.–F., *On the extension of von Neumann–Aumann's theorem*, J. Funct. Anal., **17** (1974), 112–129

[536] Saint-Pierre, J., *Une remarque sur les espaces sousliniens reguliers*, CRAS Paris, t. **282** (1976), 1425–1427

[537] Scarf, H., *The core os an N-person game*, Econometrica, **35**, (1967), 50–69

[538] Schaefer, H., *Über die Methode der a priori Schranken*, Math. Ann., **129** (1955), 415–416

[539] Schaefer, H., *Banach Lattices and Positive Operators*, Springer-Verlag, New York (1974)

[540] Schauder, J., *Der Fixpunftsatz in Funktionalräumen*, Studia Math, **2** (1930), 171–180

[541] Schmeidler, D., *On balanced games with infintely many players*, Research Program in Game Theory and Mathematical Economics, Research Memorandum No. **28**, Dept. of Mathematics, The Hebrew Univ. of Jerusalem (1967)

[542] Schmeidler, D., *Cores of exact games I*, J. Math. Anal. Appl., **40**, 214–225

[543] Schmeidler, D., *Equilibrium points of non-atomic games*, J. Stat. Phys., **7**, 295–300

[544] Schmidt, K. D., *Embedding theorems for classes of convex sets*, Acta Appl. Math., **5** (1986), 209–237

[545] Schwartz, J., *Nonlinear Functional Analysis*, Gordon & Breach, New York (1969)

[546] Segal, I., *Nonlinear Semigroups*, Ann. Math., **78** (1963), 339–364

[547] Seierstadt, A.–Sydsaeter, K., *Optimal Control Theory with Economic Applications*, North Holland, Amsterdam (1987)

[548] Serrin, J., *A Harnack inequality for nonlinear equations*, Bull. AMS, **55** (1963), 481–486

[549] Shafer, W.–Sonnenschein, H., *Equilibrium in abstract economies without ordered preferences*, J. Math. Economics, **2** (1975), 345–348

[550] Shapley, L., *A value for n-person games* in *Contributions to the Theory of Games*, Vol. **II**, Eds. H. W. Kuhn–A. W. Tucker, Princeton Univ. Press, Princeton (1953), 307–317

[551] Shapley, L., *On balanced sets and cores*, Naval Res. Logist. Quart., **14** (1967), 453–460

[552] Shapley, L., *On balanced games without side payments* in *Mathematical Programming*, Eds. T. C. Hu–S. Robinson, Academic Press, New York (1973), 261–290

[553] Showalter, R., *Hilbert Space Methods for Partial Differential Equations*, Pitman, London (1977)

[554] Showalter, R., *Monotone Operators in Banach Space and Nonlinear Partial Differential Equations*, Math. Surveys and Monographs, Vol. **49**, AMS, Providence, RI (1997)

[555] Simon, J., *Compact sets in the space* $L^p(O,T;B)$, Ann. Mat. Pura ed Appl., **146** (1987), 65–96

[556] Simons, S., *An upward-downward minimax theorem*, Arch. der Math., **55** (1990), 275–279

[557] Sion, M., *On general min–max theorems*, Pacific J. Math., **8** (1958), 171–176

[558] Smart, D. R., *Fixed Point Theorems* (2nd Edition), Cambridge Univ,. Press, Cambridge, UK (1980)

[559] Stinchcombe, M., *Bayesian information topologies*, J. Math. Economics, **19** (1990), 233–253

[560] Stinchcombe, M., *A further note on Bayesian information topologies*, J. Math. Economics, **22** (1993), 189–193

[561] Strauch, R., *Negative dynamic programming*, Ann. Math. Stat., **37** (1966), 871–890

[562] Striebel, C., *Optimal Control of Discrete Time Stochastic Systems*, Lecture Notes in Econ. Math. Systems, Vol. **110**, Springer-Verlag, Berlin (1975)

[563] Struwe, M., *A note on a result of Ambrosetti–Mancini*, Ann. Mat. Pura ed Appl., **81** (1982), 107–115

[564] Struwe, M., *Variational Methods*, Springer-Verlag, Berlin (1990)

[565] Stuart, C. A.–Zhou, H. S., *Applying the mountain pass theorem to an asymptotically linear elliptic equation in* $\mathbb{R}^N$, Comm. PDEs, **24** (1999), 1731–1758

[566] Sun, J.–Sun, Y., *Some fixed point theorems for increasing operators*, Appl. Anal., **23** (1986), 23–27

[567] Sun, Y., *A fixed point theorem for mixed monotone operators with applications*, J. Math. Anal. Appl., **156** (1991), 240–252

[568] Szulkin, A., *Minimax principles for lower semicontinuous functions and applications to nonlinear boundary value problems*, Ann. Inst. H. Poincaré, Analyse Non Linéaire, **3** (1986), 77–109

[569] Szulkin, A., *Ljusternik–Schnirelmann theory on* C^1*-manifolds*, Ann. Inst. H. Poincaré, Analyse Non Linéaire, **5** (1988), 119–139

[570] Tacsar, M. L., *Optimal planning over infinite time interval under random factors* in *Mathematical Models in Economics*, Eds. J. Los–M. Los, North Holland, Amsterdam (1974), 289–298

[571] Takahashi, W., *Existence theorems generalizing fixed point theorems of multivlaued mappings* in *Fixed Point Theory and Applications*, Eds. J. Baillon–M. Thera, Pitman Res. Notes in Math., Vol. **252**, Longman Scientific and Technical, Harlow (1991)

[572] Takayama, A., *Mathematical Econonics* (2nd edition), Cambridge Univ. Press, Cambridge, UK (1985)

[573] Talagrand, M., *Pettis Integral and Measure Theory*, Memoirs AMS, Vol. **51** (1984)

[574] Tanabe, H., *Equations of Evolution*, Pitman, London (1979)

[575] Tanaka, K.–Yokoyama, K., *On* ε*-equilibrium point in a noncooperative n-person game*, J. Math. Anal. Appl., **160** (1991), 413–423

[576] Tang, C.–L., *Periodic solutions of nonautonomous second order systems with* γ*-quasisubadditive potential*, J. Math. Anal. Appl., **189** (1995), 671–675

[577] Tang, C.–L., *Periodic solutions for nonautonomous second order systems with sublinear nonlinearity*, Proc. Amer. Math. Soc., **126**, No. **11** (1998), 3263–3270

[578] Tang, C.–L.–Wu, X.–P. *Periodic solutions for a class of nonautonomous subquadratic second order Hamiltonian systems*, J. Math. Anal. Appl., **275** (2002), 870–882

[579] Tehrani, H.–T., *A note on asymptotically linear elliptic problems in* $\mathbb{R}^N$, J. Math. Anal. Appl., **271** (2002), 546–554

[580] Terkelsen, F., *Some minimax theorems*, Math. Scad., **31** (1972), 405–413

[581] Tiba, D., *Optimal Control of Nonsmooth Distributed Parameter Systems*, Lecture Notes in Math., Vol. **1459**, Springer-Verlag, New York (1997)

[582] Tichomirov, V., *Fundamental Principles of the Theory of Extremal Problems*, Wiley, New York (1986)

[583] Tolksdorf, P., *Regularity for a more general class of quasilinear elliptic equations*, J. Diff. Eqns., **51** (1984), 126–150

[584] Tolstonogov, A., *Extremal selections of multivalued mappings and the bang–bang principle for evolution inclusions*, Soviet Math. Doklady, **43** (1991), 481–485

[585] Tonelli, L., *Fondamenti di Calcolo delle Variazioni*, Zanichelli, Bologna (1921)

[586] Troutman, J., *Variational Calculus with Elementary Convexity*, Springer-Verlag, New York (1983)

[587] Troyanski, S. L., *On equivalent norms and minimal systems in nonseparable Banach spaces*, Studia Math., **43** (1972), 125–138

[588] Tsukada, M., *Convergence of closed convex sets and σ-fields*, Z. Wahr. verw. Gab., **62** (1983), 137–146

[589] Tsukui, J., *Turnpike theorem in a generalized dynamic input–output system*, Econometrica, **34** (1966), 396–407

[590] Vainberg, M., *Variational and Method of Monotone Operators in the Theory of Nonlinear Equations*, Halsted Press, New York (1973)

[591] Valadier, M., *Multiapplications measurables a valeurs convexes compactes*, J. Math. Pures Appl., **50** (1971), 265–297

[592] Van Zandt, T., *Information, measurability and continuous behavior*, J. Math. Economics, **38** (2002), 293–309

[593] Vazquez, J. L., *A strong maximum principle for some quasilinear elliptic equations*, Appl. Math. Optim. **12** (1984), 191–208

[594] Vidossich, G., *On the topological properties of the set of fixed points of nonlinear operators*, Confer. Sem. Mat. Univ. Bari, **126** (1971), 3–63

[595] Vind, K., *Edgeworth allocations in an exchange economy with many traders*, Inter. Economic Review, **5** (1964), 165–177

[596] Visintin, A., *Strong convergence results related to strict convexity*, Comm. PDEs, **9** (1987), 439–466

[597] Vrabie, I., *Compactness Methods for Nonlinear Evolutions*, Longman Scientific and Technical, Essex, UK (1987)

[598] Walras, L., *Eléments d'Economie Pure*, Corbaz, Lausanne (1874)

[599] Wang, R., *Essential (convex) closure of a family of random sets and its applications*, J. Math. Anal. Appl., **262** (2001), 667–687

[600] Wang, Z.–Xue, X., *On the convergence and closedness of multivalued martingales*, Trans. AMS, **341** (1994), 807–827

[601] Warga, J., *Optimal Control of Differential and Functional Equations*, Academic Press, New York (1972)

[602] Webster, R., *Convexity*, Oxford Science, Oxford University Press, New York (1994)

[603] Weinstein, A., *Periodic orbits for convex Hamiltonian systems*, Annals Math., **108** (1978), 507–518

[604] Whitt, W., *Representation and approximation of noncooperative sequential games*, SIAM J. Control Optim., **28** (1990), 1148–1161

[605] Wijsman, R., *Convergence of sequences of convex sets, cones and functions II*, Trans. AMS, **123** (1966), 32–45

[606] Willem, M., *Minimax Theorems*, Birkhäuser, Boston (1996)

[607] Winter, S. G., *The norm of a closed technology and the straight–down–the turnpike theorem*, Review Economic Studies, **34** (1967), 67–84

[608] Wloka, J., *Partial Differential Equations*, Cambridge Univ. Press, Cambridge, UK (1987)

[609] Yamada, Y., *On evolution equations generated by subdifferential operators*, J. Fac. Sci. Univ. Tokyo, **23** (1976), 491–515

[610] Yang, C.–T., *On theorems of Borsuk–Ulam, Kakutani–Yamabe–Yujobô and Dyson II*, Anal. Math., **62** (1955), 271–283

[611] Yankov, V., *On the uniformization of A-sets*, Doklady Akad. Nauk. USSR, **30** (1941), 591–592

[612] Yano, M., *A note on the existenxe of an optimal capital accumulation in the continuous–time horizon*, J. Economic Theory, **27** (1982), 421–429

[613] Yosida, K., *On the differentiability and representation of one–parameter semigroups of linear operators*, J. Math. Soc. Japan, **1** (1948), 15–21

[614] Yosida, K., *Functional Analysis* (Fifth Edition), Springer-Verlag (1978)

[615] Yost, D., *There can be no Lipschitz version of Michael's selection theorem* in *Proceedings of the Analysis Conference (Singapore, 1986)*, Eds. S. Choy–J. Jesudason–P. Lee, North–Holland, Amsterdam (1988), 295–299

[616] Yotsutani, S., *Evolution equations associated with subdifferentials*, J. Math. Soc. Japan, **31** (1978), 623–646

[617] Young, L. C., *Lectures on the Calculus of Variations and Optimal Control Theory*, Saunders, Philadelphia (1969)

[618] Zame, W., *Competitive equilibria in production economies with an infinite dimensional commodity space*, Econometrica, **55** (1987), 1075–110

[619] Zeidler, E., *Nonlinear Functional Analysis and Its Applications I: Fixed Point Theorems*, Springer-Verlag, New York (1985)

[620] Zeidler, E., *Nonlinear Functional Analysis and Its Applications III: Variational Methods and Optimization*, Springer-Verlag, New York (1985)

[621] Zeidler, E., *Nonlinear Functional Analysis and Its Applications II/A: Linear Monotone Operators*, Springer-Verlag, New York (1990)

[622] Zeidler, E., *Nonlinear Functional Analysis and Its Applications II/B: Nonlinear Monotone Operators*, Springer-Verlag, New York (1990)

[623] Zhang, M., *Nonuniform nonresonance of semilinear differential equations*, J. Diff. Eqns., **166** (2000), 35–50

[624] Zhang, M., *The rotation number approach to eigenvalues of the one–dimensional p-Laplacian with periodic potentials*, J. London Math. Soc. (2), **64** (2001), 125–143

[625] Zheng, S., *Nonlinear Evolution Equations*, Chapman & Hall/CRC Press, Boca Raton, FL (2004)

[626] Zhong, C. K., *On Ekeland's variational principle and a minimax theorem*, Nonlin. Anal., **44** (2001), 909–918

[627] Zhou, H.–S., *Existence of asymptotically linear Dirichlet problem*, J. Math. Anal. Appl., **205** (1997), 239–250

[628] Zilha, I., *Characterization by prices of optimal programs under uncertainty*, J. Math. Economics, **3** (1976), 173–183

[629] Zilha, I., *On competitive prices in a multisector economy with stochastic production and resources*, Review of Economic Studies, **43** (1976), 431–438

[630] Zilha, I., *Efficiency in economic growth models under uncertainty*, J. Economic Dynam. Control, **16** (1992), 27–38

List of Symbols

Symbol	Meaning
$\mathcal{L}(X,Y)$	space of linear operators from X to Y
$\Gamma_0(X)$	proper, convex, lower semicontinuous
φ^*	conjugate function
$\partial\varphi$	convex subdifferential
φ^0	generalized directional derivative
T_C	tangent cone of C
N_C	normal cone of C
T'_C	Clarke tangent cone of C
N'_C	Clarke normal cone of C
$\Gamma\text{-}\lim$	Γ-limit
$\mathbf{B}(X)$	Borel σ-field of X
$\Gamma_{\mathrm{seq}}(X_1^{\beta_1}, X_2^{\beta_2})$	multiple Γ-operator
$\mathcal{K}(C,Y)$	compact maps from C into Y
$\mathcal{K}_f(C,Y)$	linear space of finite rank mappings from C to Y
$\sigma(A)$	spectrum of A
$\mathcal{L}_c(X,Y)$	compact linear operators from X to Y
$\mathcal{L}_f(X,Y)$	finite rank operators from X to Y
$\Phi(X,Y)$	Fredholm operators from X to Y
$\Phi_+(X,Y)$	semi-Fredholm operators from X to Y
$\widehat{d}$	Leray–Schauder degree map

α	Kuratowski measure of noncompactness
β	ball measure of noncompactness
$\widehat{d}_0$	Nussbaum–Sadovskii degree
$d_{(S)_+} = \deg_{(S)_+}$	degree for $(S)_+$ maps
$\mathrm{cat}_Y A$	Ljusternik–Schnirelmann category of A in Y
γ	genus
$C_0^1(\overline{Z})$	space of C^1-functions on $\overline{Z}$ s.t. on ∂Z are zero
$W^{1,2}_{\mathrm{per}}\big((0,b),\mathbb{R}^N\big)$	$\big\{x \in W^{1,2}\big((0,b),\mathbb{R}^N\big) : x(0) = x(b)\big\}$
$P_f(X), P_k(X)$	spaces of sets
$P_{fc}(X), P_{wkc}(X)$	spaces of sets
$P_{kc}(X), P_{bfc}(X)$	spaces of sets
$m(X^*, X)$	Mackey topology on X^* for (X, X^*)
$F^+(C), F^-(C)$	inverse images of multifunctions
S_F^p	L^p-selectors of F
J_λ	resolvent operator
A_λ	Yosida approximation of A

Index

Lightning Source UK Ltd.
Milton Keynes UK
UKOW031802081111

181640UK00008B/14/P

9 780387 789064